Florence Heri-Tech 2018

The Future of Heritage Science and Technologies

IOP Conference Series: Materials Science and Engineering
Volume 364

Florence, Italy
16 – 18 May 2018

ISBN: 978-1-5108-6445-0
ISSN: 1757-8981

Printed from e-media with permission by:

Curran Associates, Inc.
57 Morehouse Lane
Red Hook, NY 12571

CURRAN ASSOCIATES INC. proceedings.com

Some format issues inherent in the e-media version may also appear in this print version.

Printed by Curran Associates, Inc. (2018)

Institute of Physics
Dirac House, Temple Back
Bristol BS1 6BE UK

Phone: 44 1 17 929 7481
Fax: 44 1 17 920 0979

techtracking@iop.org

Additional copies of this publication are available from:

Curran Associates, Inc.
57 Morehouse Lane
Red Hook, NY 12571 USA
Phone: 845-758-0400
Fax: 845-758-2633
Email: curran@proceedings.com
Web: www.proceedings.com

Table of contents

Volume 364

Florence Heri-Tech - The Future of Heritage Science and Technologies

16–18 May 2018, Florence, Italy

Accepted papers received: 10 May 2018
Published online: 18 June 2018

Preface

Papers

Engineering

Materials Science

GEOMATICS AND STRUCTURAL MODELLING FOR CH

Plenary European Session

Poster Session

IOP Conf. Series: Materials Science and Engineering **364** (2018) 011001 doi:10.1088/1757-899X/364/1/011001

Preface

Under the patronage of the University of Florence and organized with the support of the Department of Industrial Engineering, Florence Heri-Tech gathered researchers and experts in the field of "heritage science and related technologies" to disseminate their recent research at an international level as well as to draw new inspiration.

The Conference was part of the Florence International Biennal for Art and Restoration, an international event attracting prestigious institutions and companies and creating a unique opportunity to bring together the academic world with industry.

Florence Heri-Tech ultimate goal was to foster European mobility and co-operation between students and staff, to enhance Europe's development as a multi-cultural society and to encourage the concept that scientific-cultural research must be an integral part of society. Moreover, the Conference aimed to promote international networks between universities, training institutes and companies to create long-term collaboration opportunities.

The topics cover experiences in the use of innovative technologies and methods for documenting, managing, preserving and communicating cultural heritage.

Eighty three experts, coordinated by the General Chairs and supported by a high-level Scientific Committee, were involved in the review process which led to the selection of 112 papers (out of 173 submissions) focusing on multi-disciplinary and inter-disciplinary research concerning future trends in material science, engineering-related methods for Cultural Heritage, diagnostic and monitoring as well as digital heritage.

The Honorary Chairs, Prof. Piero Baglioni, Prof. Giorgio Bonsanti, Prof. Vito Cappellini, Prof. Sharon Cather, Prof. Marco Ciatti, Prof. Maria Perla Colombini, Prof. Luigi Dei, Prof. Rodrigues Jose Delgado, Prof. Mario Primicerio, Prof. Pedro Santos, Prof. Antonio Sgamellotti, which are renowned experts in their research fields, further enhanced the Conference prestige.

General Chairs extend their thanks to all authors, speakers, and those persons whose labour, financial support and encouragement made the Florence Heri-Tech event possible.

Florence Heri-Tech Chairs

Rocco Furferi

Rodorico Giorgi

Lapo Governi

Mauro Matteini

Yary Volpe

IOP Conf. Series: Materials Science and Engineering **364** (2018) 011002 doi:10.1088/1757-899X/364/1/011002

Organizing Committee

Conference General Chairs

Prof. Rocco Furferi, Department of industrial Engineering, Università di Firenze

Prof. Rodorico Giorgi, Department of Chemistry "Ugo Schiff", Università di Firenze

Prof. Lapo Governi, Department of industrial Engineering, Università di Firenze

Prof. Mauro Matteini, formerly – CNR Institute for the Conservation and Promotion of Cultural Heritage, Firenze

Prof. Yary Volpe, Department of industrial Engineering, Università di Firenze

Honorary Chairs

Prof. Baglioni, Piero – University of Florence

Prof. Bonsanti, Giorgio – University of Florence

Prof. Emeritus Cappellini, Vito – University of Florence

Prof. Cather, Sharon – The Courtauld Institute of Art, London

Prof. Ciatti, Marco – Opificio delle Pietre Dure, Florence

Prof. Colombini Maria Perla – CNR-ICVBC. National Research Council of Italy – Institute for the Conservation and Valorization of Cultural Heritage

Prof. Dei, Luigi – Rector of the University of Florence

Prof. Delgado Rodrigues, Jose – Laboratorio Nacional De Engenharia Civil, Lisbon

Prof. Primicerio, Mario – University of Florence

Prof. Santos, Pedro – Fraunhofer Institute for Computer Graphics Research IGD, Darmstadt

Prof. Sgamellotti, Antonio – University of Perugia

Technical Program Committee

Prof. Balocco, Carla – Industrial Engineering Department, University of Florence

Dr. Bernardi, Adriana – University of Padova

Prof. Bianconi, Francesco – University of Perugia

Dr. Camilli, Andrea – Soprintendenza per i Beni Archeologici della Toscana, Department Member

Prof. Cassar, JoAnn – University of Malta

Prof. Cavalieri, Duccio – Department of Biology, University of Florence

Prof. Cheung, Sidney – The Chinese University of Hong Kong, Hong Kong

Dr. Del Bianco, Corinna – Director of the International Institute Life Beyond Tourism and adviser of the Romualdo Del Bianco Foundation

IOP Conf. Series: Materials Science and Engineering **364** (2018) 011002 doi:10.1088/1757-899X/364/1/011002

Peer review statement

All papers published in this volume of *IOP Conference Series: Materials Science and Engineering* have been peer reviewed through processes administered by the proceedings Editors. Reviews were conducted by expert referees to the professional and scientific standards expected of a proceedings journal published by IOP Publishing.

IOP Conf. Series: Materials Science and Engineering **364** (2018) 012001 doi:10.1088/1757-899X/364/1/012001

Motion magnification analysis applied to the dynamic identification of historic constructions

Vincenzo Fioriti, Ivan Roselli, Angelo Tatì, Roberto Romano and Gerardo De Canio

ENEA Italian National Agency for New Technologies, Energy and Sustainable Economic Development, Italy

Vincenzo.fioriti@enea.it

Abstract. We applied a new methodology, namely the motion magnification (MM), to the dynamic identification of historic constructions. MM acts like a microscope amplifying small motions in video sequences, therefore tiny motion patterns are made visible with the naked eye. This technique provides advantages of particular interest to the dynamic identification: no wires, no physical contact, simplicity and low costs. In this paper, we investigated the ambient vibration monitoring of historic structures in urban environments, which is a relevant issue for the health surveying and early damage detection, particularly for ancient buildings. We give an introduction to the MM methodology and describe its practical application to the frequency domain through two case-studies: the so-called Temple of Minerva Medica of Rome and the Ponte delle Torri of Spoleto. Since in the outdoor environment the MM is much more prone to noise because of the wind, shadows, atmospheric refraction, light reflection, distance from the object, weak vibration sources, the case study of the Ponte delle Torri represents a hard test-bed for MM. However, the MM estimates of the first modes showed a good agreement with the experimental data of contact velocimeters.

1. Introduction

Vibration monitoring is a key issue for the structural health analysis of the cultural heritage. To this end, today a new image processing tool, the motion magnification (MM) is available. Motion magnification acts like a microscope for micro-motions in videos, but saving the topology of the image sequence and unveiling visual patterns hardly visible with the naked eye. The motion magnification use the spatial resolution of the video-camera to extract physical properties from images to make inferences about the dynamical behavior of the object. Researchers are very interested in assessing the MM analysis (MMA) feasibility, since conventional vibrometers are surely more precise, but expensive and much less practical. Moreover, evaluating the health of large structures, such as historical monuments in a short time span and possibly by contactless devices, may produce a major breakthrough for civil engineering. The usage of video sequences in the field of civil engineering is not new. For many years attempts to produce qualitative and even quantitative analysis by means of videos of large structures have been conducted, but with poor results. This because of the resolution in terms of pixels, of the camera frame rate, computer time and finally because of the lack of proper algorithms, able to deal with the extremely small motions related to building displacements. These and others limitations have restricted in the past the application of digital video methodologies, till the

IOP Conf. Series: Materials Science and Engineering **364** (2018) 012001 doi:10.1088/1757-899X/364/1/012001

important advances obtained during the last years at the Massachusetts Institute of Technology (MIT) by Freeman and coworkers [3]. Their algorithm, named motion magnification, works within a reasonably short elaboration time. The point is crucial, as it is well known that video processing takes a lot of time and resources, preventing a viable application to the real world.

Recently, a number of indoor experiments conducted on simple geometries like rods and other small objects at MIT, as well as on masonries tested on the shaking table at the ENEA Casaccia facilities, showed the reliability of this methodology compared to lasers vibrometers or contact sensors. However, in the outdoor environment the analysis is much more difficult.

Moreover, although revealing the smallest movements of the structure is per se an interesting feature, we need a quantitative analysis to assess the health of the monument by the classical frequency domain methods. To this end, we have chosen the environment of the so-called temple of Minerva Medica and the Ponte delle Torri as a demanding outdoor test-bed for MMA, using as a reference the modal analysis provided by [1].

In [1] experimental velocimetry data and the finite element model (FEM) are in strong accordance, representing an appropriate benchmark for MMA. Therefore, in this paper we use extensively the results of [1] to evaluate the performance of the motion magnification in the frequency domain.

2. The Ponte delle Torri of Spoleto

The Ponte delle Torri is a large historical construction that connects Colle Sant'Elia with Mount Monteluco in Spoleto, Italy (Figure 1). It was probably built in the 13th century, possibly on Etruscan or Roman ruins [1]. The bridge superstructure is made up of a pedestrian deck, provided with a water canal on one side, supported by lancet arcades and stone piers known as "towers".

The bridge has a stone structure with shoulders and piers made up of rubble walls in a square matrix assembled by mortar and lime. The structure is only apparently regular. In fact, piers and arches have all different shapes and sizes, as well as the walls towards Colle Sant'Elia and Monteluco are also diverse in the masonry texture, due to their different construction timing and subsequent rebuilding interventions. The bridge has an overall length of about 230 m, while the highest tower is over 70 m. The bridge has a state of damage typical of historic masonry bridges. At the top of the arches, especially those on the slope of Monteluco, there are heavy water infiltrations resulting in losses of mortar binder and wall apparatus skiving. Moreover, a widespread damaged layer, especially in the lower part of the piers, was detected. This has caused the expulsion of some cornerstones of the piers, the disarticulation of the masonry texture and the collapse of portions of the outer layers [2].

2.1. The OMA identification

In order to identify the dynamic properties of the bridge through experimental data, three seismographs were used to acquire the ambient vibration on site [1].

Figure 1. The Ponte delle Torri of Spoleto (Northeast view).

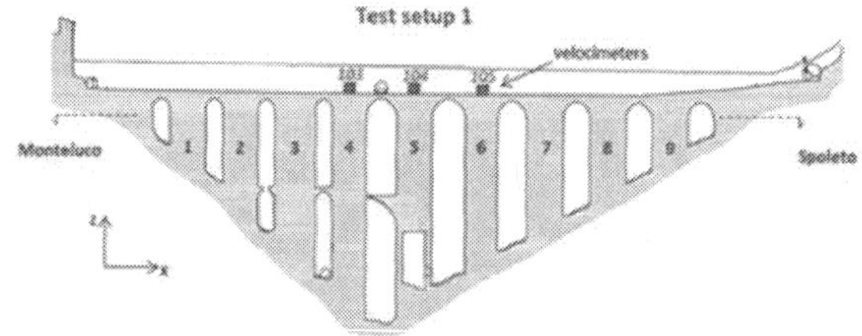

Figure 2. Positions of velocimeters (in red) in the first experimental setup [1].

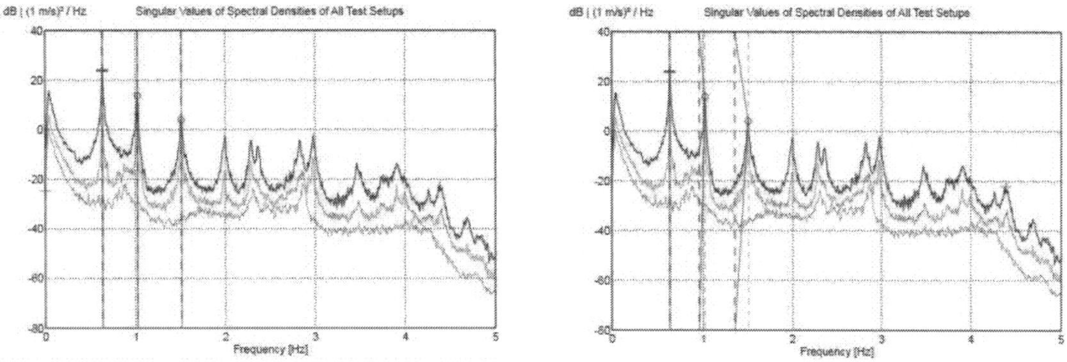

Figure 3. Modal estimation graph by the Frequency Domain Decomposition and the Enhanced Frequency Domain Decomposition methods. Although these two modal calculation methods are slightly different, results are the same.

The data presented here were acquired on 29 May 2017. The used seismographs, SL06 recorders (SARA Instruments) equipped with triaxial electrodynamic velocimeters, were set to 200 Hz sampling frequency. Multiple Test Setups Measurement Procedures (MTSP) were performed with one common measurement point on top of the central pier of the bridge, as reference sensor (Figure 2).

As the reference sensor was staying in the same positions during all setups, it basically measured the mode shapes in this position over and over, while other sensors were moved to different positions on the bridge. Such reference position was determined as the measurement point where the modes of interest were supposed to have the highest response level according to FEM modal shapes. Data were acquired in eight configurations, each representing a test setup in the MTSP. All configurations were acquired for at least 20 minutes.

The Operational Modal Analysis (OMA) techniques implemented in the ARTeMIS Modal Pro software were used to elaborate the experimental data of the velocimeters (Figure 3 and Table 1). Among the several OMA techniques available, the Frequency Domain Decomposition (FDD), the Enhanced Frequency Domain Decomposition (EFDD) and the Stochastic Subspace Identification (SSI) were utilized. EFDD provides also an estimation of modal damping, SSI is a more sophisticated and automatic procedure based on time-domain approach [1, 2].

The results of the identification of the modal frequencies averaged over the OMA techniques are resumed in Table 1. Also a finite element analysis (FEM) has been carried out along with the experimental campaign, furtherly confirming results of Table 1 [1, 2].

Table 1. Averaged modal frequencies by OMA techniques (data acquired on 29 May 2017).

Mode	Averaged modal frequency (Hz)
1	0.632
2	1.011
3	1.496
4	1.975

3. The magnified motion algorithm

Since our intention is only to give a general idea of the potentiality of the motion magnification we will not enter into the full formal description of the algorithms, see [3, 4, 5] for an extensive discussion. Videos are made of a temporal sequence of 2D images, whose pixel intensity is $I(x, t)$. The

2D array of color intensity is the spatial domain, while the time domain corresponds to the temporal sequence. Here, in order to describe the Eulerian version of the magnification algorithm, we consider only a 1D translating image with displacement $\delta(t)$.

At the image-position x and video-time $t = 0$, it is: $I(x, 0) = f(x)$ (for the treatment of the general problem, see [3]). Translating for the quantity $\delta(t)$, we have:

$$I(x, t) = f(x - \delta(t) \tag{1}$$

The final expression of the magnified motion by the constant α is defined as:

$$\Delta I = f(x - (1 + \alpha)\, \delta(t)) \tag{2}$$

Now, if the displacement δ(t) is small enough, it is possible to expand the relation (1) as the Taylor's first order series around x at the time t:

$$I(x, t) = f(x) - \delta(t)\, (\partial f / \partial x) + \varepsilon \tag{3}$$

where ε is the error due to the Taylor's approximation. The intensity change at each pixel can be expressed as:

$$\Delta(x, t) = I(x, t) - I(x, 0) \tag{4}$$

that taking into account (3), becomes:

$$\Delta(x, t) = f(x) - \delta(t)\, (\partial f / \partial x) + \varepsilon - f(x) \tag{5}$$

and finally disregarding the error ε:

$$\Delta(x, t) \approx - \delta(t)\, (\partial f / \partial x) \tag{6}$$

meaning that the absolute pixel intensity variation Δ is proportional to the displacement and to the spatial gradient. Therefore, the pixel intensity can be written as follows:

$$I(x, t) \approx I(x, 0) + \Delta(x, t) \tag{7}$$

Magnifying the motion by a given constant α using equations (3) and (4), simply means that pixel intensity $I(x, t)$ is replaced by the magnified pixel intensity $I_{magn}(x, t)$ according to the following:

$$I_{magn}(x, t) \approx I(x, 0) + \alpha\, \Delta(x, t) \approx f(x) - \delta(t)\, (\partial f / \partial x) - \alpha\, \delta(t)\, (\partial f / \partial x) + O(\varepsilon, \delta) \tag{8}$$

where $O(\varepsilon, \delta)$ is the remainder of the Taylor series. Then, the magnified intensity can be calculated as:

$$I_{magn}(x, t) \approx f(x) - (1 + \alpha)\, \delta(t)\, (\partial f / \partial x) \tag{9}$$

But (9) is immediately derived from the first order Taylor's expansion of the exact magnified motion:

$$\Delta I = f(x - (1 + \alpha)\, \delta(t)) \tag{10}$$

and therefore:

$$f(x - (1 + \alpha)\,\delta(t)) \approx f(x) - (1 + \alpha)\,\delta(t)\,(\partial f/\partial x) \tag{11}$$

It is important to observe that (6) is obtained by a band-pass derivation. Therefore, we can say that to magnify the motion displacement it suffices to add $\alpha\Delta(x, t)$ to $I(x, t)$, as long as the Taylor's expansion (9) is valid, that is until its remainder $O(\varepsilon, \alpha)$ is small. This limitation depends on the linear approach entailed in the Taylor's expansion, either if the initial expansion (3) or the amplification α are too large.

In practice, to remain into the linearity bound, we need slowly changing the images and small amplifications. Moreover, here we do not consider the noise of variance σ^2 to be added to the intensity, that is amplified too, resulting in an amplified noise variance $2\sigma^2\alpha^2$, thus the error to be evaluated should be $O(\varepsilon, \alpha, 2\sigma^2\alpha^2)$ actually.

Therefore, if the video is a long-lasting one, the computational time may be a major problem. On the other hand, the Nyquist sampling theorem requires the frame rate of the video f_{fps} to be at least double than highest frequency of interest (f_{max}). Thus, to reproduce correctly a signal it is necessary that:

$$f_{sampling} \geq 2f_{max} \tag{12}$$

where f_{max} is the maximum frequency of the signal in the temporal domain, $f_{sampling}$ is the sampling frequency. Then (11) becomes:

$$f_{fps} \geq 2f_{max} \tag{13}$$

with f_{fps} acting as a sampling frequency. Using a 50 fps video camera, the maximum frequency allowed is 25 Hz: above this threshold spurious frequencies are introduced because of the aliasing. Furthermore, since in our elaborations the video duration is about 10 seconds, the *theoretical* frequency resolution is equal to 0.1 Hz. It must also be said that the short time-span of the video of course has a negative impact on the PSD calculations.

Other physical limitations, such as illumination, shadows, camera unwanted vibrations, poor pixel resolution, low frame rate, presence of large motion, distance from the object, decrease severely the quality of the motion magnification and should be taken into account in order to achieve high-quality results. All these difficulties are present in the case-study of the Ponte delle Torri.

4. MMA modal analysis

Firstly, we recorded a video of the monument during the campaign of the 29 May 2017, taking care to avoid large motions such as people passing by in front of the camera and swinging objects. The presence of large motions is one of the most important sources of noise for the MM, requiring the experimenter to isolate part of the image, although usually this is not a feasible option. We recorded by means of a commercial camera, (pixel resolution 360 x 445, frame rate 50 fps), and then videos were processed by the MM algorithm [3]. The basic methodology is to take advantage of the large number of pixels contained in an image. Theoretically, we could have160200 "virtual sensors", meaning that each pixel provides a time history of intensity variation (colour or grey scale), from the first frame to the last one. These time series contain the information about the displacements of the physical points related to the pixels (although they are not *real* displacements). Of course, not all the surface of the structure generates useful information, therefore we identify a small area possibly with high signal-to-noise ratio (SNR). The identification of this region of interest (ROI) is a crucial point of the procedure. The angle of the camera is much less than 90° (Figure 4). This circumstance is unfavorable, since the magnification will be affected by a distortion. Also the air refraction and the wind provide a substantial source of noise, affecting the algorithm. No attempt to compensate for these distortions has been made.

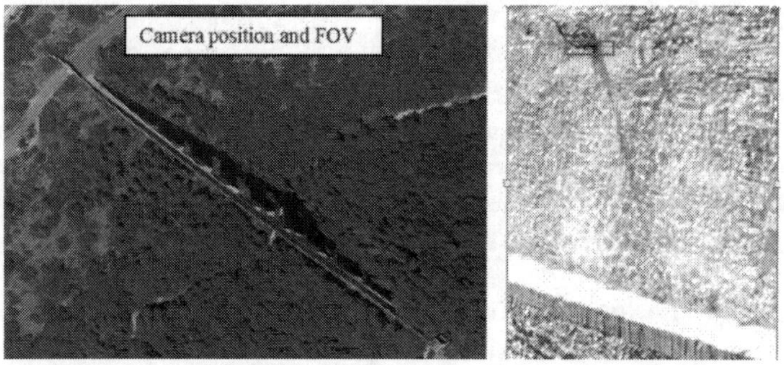

Figure 4. Camera position and field of view (left). The selected region of interest (red box) used to extract the MM intensity time-series (right).

The main source of vibration is the wind, although a road flanks the monuments on the right side. In any case, man-made vibrations may be considered of low intensity. The camera is positioned on the road near the monument as in Figure 5, hence wind and traffic affect its stability too.

To run the MIT algorithm, we need to set some parameters. In particular, the magnification band and the amplification factor are the more important. We chose a band of 0.5 – 2.5 Hz and an amplification factor of 140. Similar values would be also acceptable, nevertheless, since the MM acts as a pass-band filter, choosing a band close to the actual modes gives better results. In our case we have been supported by the previous analysis [1], but generally one must rely on experience to decide a suitable band.

After the MIT phase based magnification step [3, 4, 5], we have to select the ROI of the virtual sensors, meaning to pick some pixels whose time history of color variation will be translated into the frequency domain. We choose the area as in the red box of Figure 5, because it is morphologically different from the wall and provides some edges to the algorithm. The presence of marked edges or texture would be very helpful for the MMA, but unfortunately the surface of the monument is rather homogeneous. These circumstances produce a large amount of noise, to be added to other disturbances.

The ROI is part of a lamppost, in-built to the wall about 40 m from the camera; however only the red box pixels are able to supply low-noise information, because are close to the structure. In fact, the upper part of the lamppost spreads spurious frequencies, due to its own resonances. Other choices of the ROI have not improved the measurements. Signals provided by MM contain displacement information, but they cannot be used immediately as real displacements. Anyway, acceleration may be calculated just like it can be done with the velocimeter displacement measurements, and the PSD obtained as well. Every pixel provides a signal, therefore an average over all the PSDs is calculated.

4.1. Results

The ROI area of Figure 4 is identified qualitatively by trial and error considering the number of peaks in the PSD and their inter-distance; a dense presence of peaks suggests a high level of noise and of course is misleading, therefore only PSD spectra with identifiable frequencies are acceptable. In practice, to avoid a biased choice, one selects manually the area several times and repeats the calculation procedure. In our case, we find that ten iterations of the MM-PSD procedure is a good trade-off between speed and precision (intended as closeness to the mean value).

Table 2 contains the MM modal frequencies calculated over ten runs, compared to the experimental results by OMA. Table 2 shows a good agreement with the experimental measurements obtained both from the FEM method and data collect by standard velocimeters [1].

In Figure 5 a typical MM-PSD is shown with passband filter in the range 0.5 – 2.5 Hz. The smooth appearance of the MM-PSD depends on the short time span of the video that decreases the PSD frequency resolution.

IOP Conf. Series: Materials Science and Engineering **364** (2018) 012001 doi:10.1088/1757-899X/364/1/012001

Table 2. Numerical values of first 4 modal frequencies by OMA and by magnified motion analysis. The last column is the percentage error.

Mode	Averaged OMA frequency (Hz)	Averaged MM frequency (Hz)	error %
1	0.632	0.685	-8.39 %
2	1.011	1.014	-0.30 %
3	1.496	1.311	12.37 %
4	1.975	2.068	-4.71 %

Figure 6 illustrates all the values obtained over the ten runs of the MM-PSD procedure, along with their averages and the OMA values. Note that some outliers are evidently present, therefore removing them by statistical methods would improve these results.

An analogous experiment was carried out on 19 April 2017 at the temple of Minerva Medica in Rome [6]. In Figure 7 we can see the south side of the temple, a velocimeter in the blue circle, and the ROI area of the virtual sensors in the red box. A tramway flanks the monument at only 2 m, producing very strong vibrations [7]. In this preliminary analysis only the first mode was analyzed. The estimated value of the modal frequency by MMA was 1.950 Hz, which is roughly close to the value obtained through OMA of velocimeters data (2.057 Hz) that leads to an error of 5.2 %.

To explain why the MMA appears less precise for the Ponte delle Torri, it should be taken into account the adverse acquisition conditions of the Ponte delle Torri, that prevented a correct recording angle for the camera, the distance from the target, the absence of suitable edges and of course the weak vibration sources. This last point seems the decisive factor. The MM is able to amplify sufficiently even the micro-displacements caused by the wind, but the signal-to-noise ratio (SNR) remains low if the vibration sources are weak, since further increasing the amplification factor would produce artifacts in the magnified video. In other words, the displacements can be recovered, but amid a *relatively* high noise level, that affects the MM-PSD elaborations.

In the case of the temple of Minerva Medica case-study, the displacements were two orders of magnitude larger, producing a much more favourable SNR for the ROI. Moreover, the camera angle was about 90° and the distance reduced to 8 m.

Yet, having tested the outdoor MM over four modes with an error less than 12.5% in the worst case, results can be regarded as a very promising step toward an outdoor digital video modal analysis.

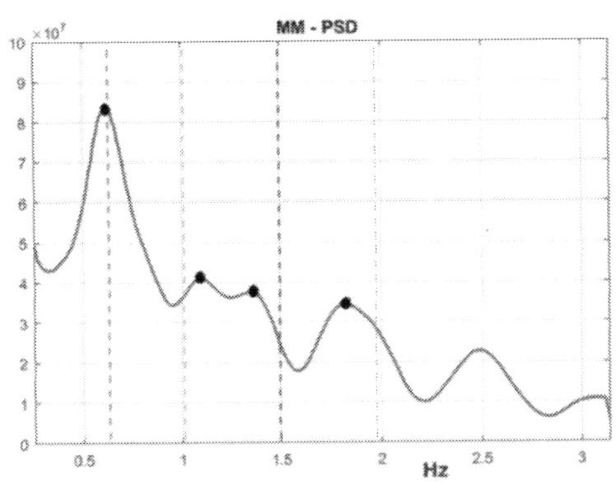

Figure 5. A typical MM-PSD. Bold black dots are the first four peaks on the MMA curve. The coloured dashed lines indicate the average OMA frequencies.

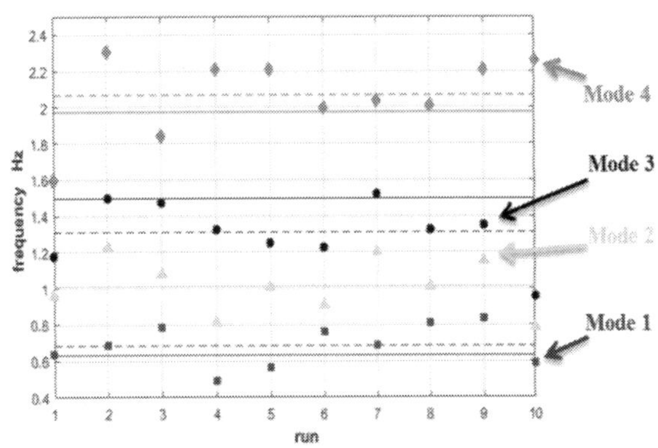

Figure 6. Modal frequencies obtained by all MM-PSD runs (dots) and average values (dashed lines) compared to OMA values (continuous lines). Each colour indicates a different mode.

Figure 7. South side of the temple of Minerva Medica with the position of the velocimeter used as refrence (blue circle) and the area of the frame used as "virtual sensors" (red box).

5. Conclusions

We have investigated the application of the magnified motion, a novel video processing technique, to the Ponte delle Torri of Spoleto and to the temple of Minerva Medica, in order to develop a video driven modal analysis. We have discussed the role of noise and of the virtual sensors, that require careful boundary conditions to the video recording. Given the adverse scenario produced by a low SNR, the estimates of the first four modes for the Ponte delle Torri are more than satisfying, while for the temple of Minerva Medica case a better SNR allowed an almost exact identification. When a dense sensor monitoring is needed or it is impossible to deploy laser vibrometers and velocimeters, the magnified motion technique offers a viable, contactless, low cost solution. Many issues still deserve to be deepened: noise reduction, operative guidelines for outdoor analysis and finally the design of a hardware customization for real-time applications. However, it is probable that all these problems will be fixed over the next few years.

Acknowledgments

The authors wish to thank dr. Valerio Cirulli, for useful discussions and support, and the CO.B.RA. project funded by Lazio Region, within which this work was financed.

References

[1] Roselli I, Malena M, Mongelli M, Cavalagli N, Gioffrè M, De Canio G and De Felice G 2018 Health assessment and ambient vibration testing of the "Ponte delle Torri" of Spoleto during the 2016-17 Central Italy seismic sequence *Int. J. Civil Structural Health Monitoring* **8**(2)

[2] De Canio G, Mongelli M, Roselli I, Tatì A, Addessi D, Nocera M and Liberatore D 2016 Numerical and operational modal analyses of the "Ponte delle Torri", Spoleto, Italy *Proc. Int. Conf. fn Structural Analysis of Historical Constructions SAHC16 (Leuven)* pp 752–7

[3] Wadhwa N, Wu H, Davis A, Rubinstein, M, Shih E, Mysore G, Chen J, Buyukozturk O, Guttag J, Freeman W and, Durand F 2017 Eulerian Video Magnification and Analysis *Communications of the ACM* **60** pp 87- 95

[4] Yu-Wu H et al. 2017 Eulerian Video Magnification for Revealing Subtle Changes in the World, https://people.csail.mit.edu/mrub/papers/vidmag.pdf. Accessed on 2 March 2017

[5] Wadhwa N et al. 2017 Motion microscopy for visualizing and quantifying small motions *PNAS* **114** 44 pp 11639–44

[6] Barbera M, Magnani Cianetti M and Barrano S 2014 Da Massenzio a Costantino: le indagini in corso nel c. d. tempio di Minerva Medica *Proc. Int. Conf. of CISEM - La villa restaurata e i nuovi studi sull'edilizia residenziale tardoantica (in Italian)*

[7] Roselli I, Fioriti V, Bellagamba I, Mongelli M, Tatì A, Barbera M, Magnani Cianetti M and De Canio G 2017 Urban transport vibrations and cultural heritage sites in Rome: the cases of the temple of Minerva Medica and of the Catacomb of Priscilla *WIT Transactions on Ecology and the Environment* **223** pp 335–43

IOP Conf. Series: Materials Science and Engineering **364** (2018) 012002 doi:10.1088/1757-899X/364/1/012002

Results of diagnostic campaign promoted by AIAr in the deposits of the Archaeological Museum of Paestum

M. E Oddo[1]*, P. Ricci[11], D. Angelici[2], F. Fantino[2], E. Sibilia[3], M. F Alberghina[4], S. Schiavone[4], C. Grifa[5], M. Mercurio[5], C. Germinario[5], F. Izzo[5], A. Langella[5], E. Massa[6], S. Bracci[7], D. Magrini[7], R. Costa[8], A. Pelagotti[9], G. Zuchtriegel[10], C. Lubritto[11]

[1] IMT Lucca, Lucca, Italy, maria.oddo@imtlucca.it
[2]TecnArt S.r.l. Italy, Network AIAr-Aziende, tecnart@unito.it
[3]Centro Universitario Datazioni e Archeometria, Università di Milano Bicocca e INFN, sezione di Milano Bicocca, emanuela.sibilia@unimib.it
[4]S.T.Art-Test s.a.s, Network AIAr-Aziende, Italy, info@start-test.it
[5]Dipartimento di Scienze e Tecnologie - Università degli Studi del Sannio, Italy, celestino.grifa@unisannio.it, mariano.mercurio@unisannio.it, chiara.germinario@unisannio.it, francesco.izzo@unisannio.it, langella@unisannio.it
[6] Art-Test Firenze, Network AIAr-Aziende, Italy, info@art-test.com
[7]ICVBC-CNR, Istituto per la Conservazione e la Valorizzazione dei Beni Culturali, Firenze, Italy, bracci@icvbc.cnr.it, magrini@icvbc.cnr.it
[8] Dipartimento di Fisica, Università degli Studi di Salerno, Italy
[9] National Institute of Optics, CNR Firenze, Italy
[10] Parco Archeologico di Paestum, MIBACT, gabriel.zuchtriegel@beniculturali.it
[11] Dartimento di Scienze e Tecnologie Ambientali, Biologiche e Farmaceutiche, Università della Campania "Luigi Vanvitelli", Caserta, Italy, carmine.lubritto@unicampania.it

Abstract. Thirty artefacts from the Archaeological Park of Paestum were investigated by means of scientific techniques on the occasion of the 2016 exhibition 'Possessione. Trafugamenti e falsi di antichità a Paestum'. The multi-analytic diagnostic campaign was aimed at identifying forgeries. Results provided a deeper understanding of both ancient technology and contemporary forgery techniques.

1. The diagnostic campaign

During the last 25 years the "Associazione Italiana di Archeometria" (AIAr) has been committed to promoting and disseminating research, education and professional development in the field of Heritage Science. As the main Association of Archaeometry in Italy, it aims at fostering cooperation among researchers in the Humanities and in the Natural Sciences in order to promote interdisciplinary studies on Cultural Heritage. Indeed, AIAr coordinates numerous events and activities whose goal is to disseminate conservation science and archaeometry.

In the occasion of the exhibition "Possessione. Trafugamenti e falsi di antichità a Paestum" (Possession. Robberies and forgeries of Antiquities in Paestum) held at the Archaeological Park of Paestum in 2016, a wide set of chemical analyses and imaging techniques were used to characterize the artefacts on display [1]. The results of this diagnostic campaign were subsequently presented at the second edition of the workshop "Arte e Scienza. Difendi il bello, combatti il falso" (Art and science. Protect beauty and fight forgery) dedicated to young audiences.

The antiquities market is constantly flooded with illegally excavated artefacts and forgeries. When illicit collections are seized by law enforcements, they are delivered to museums. Therefore, several inauthentic objects happen to enter the collections of public institutions. Some of the forgeries are sloppy copies of one specific original work or they are eclectic (and impossible) combinations of different traits taken from various ancient artefacts. In this case they can be detected by the expert's eye [2, 3, 4]. Most often, however, forgeries are reproductions of a common artefact's type and their authenticity can only be tested through scientific analyses [5]. Forgeries' detection is a fast-evolving branch of both archaeology and archaeological science, as the refinement of forgers' methods has required scholars to find always new solutions, constantly improving on existing techniques [a recent and complete handbook on forgeries' detection is 6].

A large number of artefacts (pottery vases, bronze statuettes, coins and painted stone slabs) were selected to be displayed in the exhibition. The storage of the Archaeological Park of Paestum hosts several thousands of illegally excavated artefacts, mostly grave goods from the rich necropolises outside the ancient walls. The analytic campaign regarding the artefacts on display aimed at integrating archaeological methods and hard sciences. The selected artefacts were first analysed from a typological and stylistic point of view, allowing archaeologists to express their ideas and doubts about their authenticity. The most doubtful artefacts were then selected to be the object of a multi-analytic campaign aimed at dating, authenticating and characterising the artworks. The results were used to validate archaeological assumptions or, in some cases, to disprove them. The latter were the most interesting findings, as they induced archaeologists to investigate further, resulting in an advancement in knowledge.

2. Thermoluminescence dating

Thermoluminescence dating is a trapped charge dating technique that measures signals from negative charges trapped in the crystalline structure of samples due to exposure to ionizing radiation, estimating the time elapsed since the 'traps' were last voided, which in the case of pottery usually corresponds to the last firing of the pot [for a more detailed account of TL dating see 7, for artificial radiation and other techniques used by forgers to deceive TL analyses cf.8]

Eight vases of various shape, seized between the 80s and the 90s, were analyzed by means of thermoluminescence (TL). Three modern reproductions of red-figured vases were identified. The samples present very low natural TL signal, high sensitivity to irradiation, and linear TL dose response. In all three cases, the absorbed dose is lesser than 0.5 Gy, a value that is not compatible with ancient artefacts (Figg.1a&b – 2a&b – 3a&b – 4a&b).

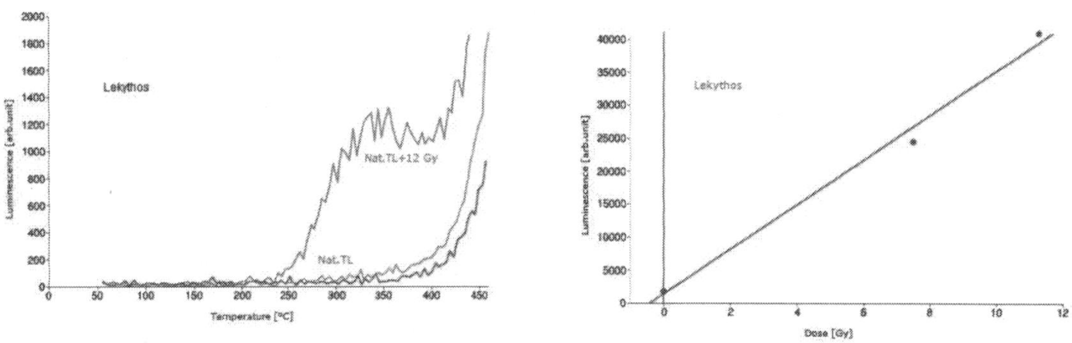

Fig 1 – a) TL curves of the squat *lekythos* PAE 000063; b) absorption rate of the squat lekythos PAE 000063. (CUDAM, Centro Universitario Datazioni e Archeometria Milano Bicocca)

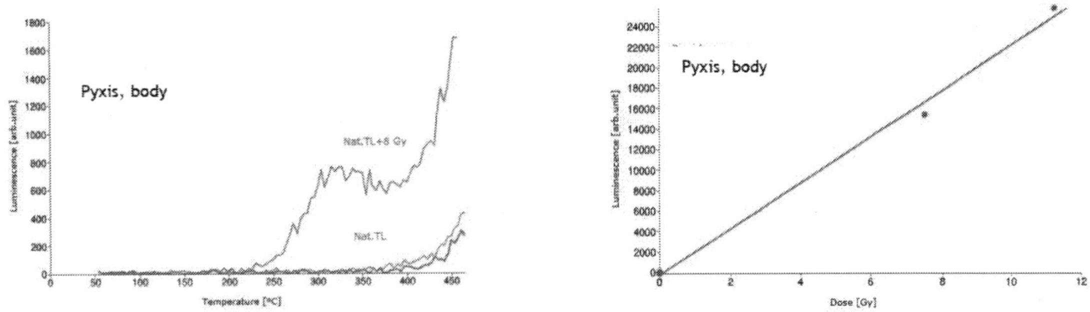

Fig. 2 – a) TL curves of the sample taken from the body of the globular pyxis PAE 000061; b) absorption rate of the sample taken from the body of the globular pyxis PAE 000061. (CUDAM, Centro Universitario Datazioni e Archeometria Milano Bicocca).

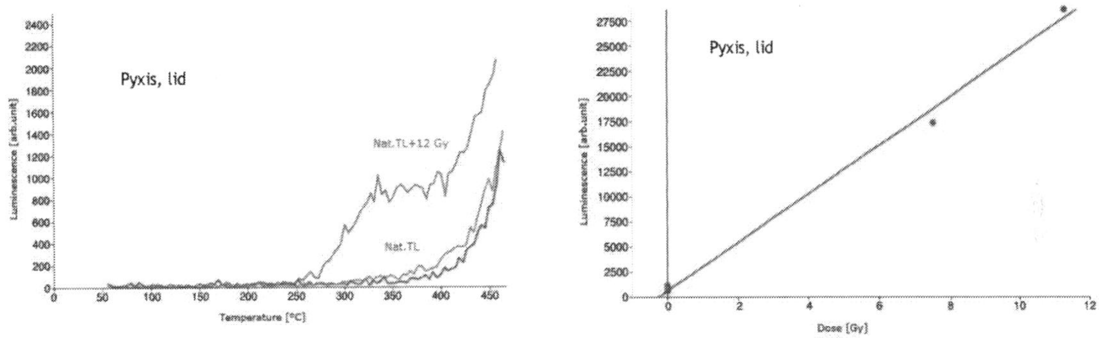

Fig. 3 – a) TL curves of the sample taken from the lid of the globular pyxis PAE 000061; b) absorption rate of the sample taken from the lid of the globular pyxis PAE 000061. (CUDAM, Centro Universitario Datazioni e Archeometria Milano Bicocca).

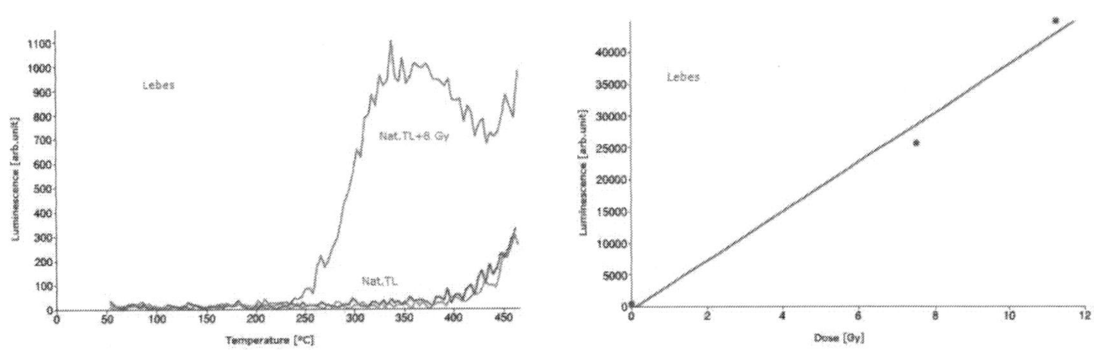

Fig. 4 – a) TL curves of the sample taken from the lebes PAE 000062; b) absorption rate of the sample taken from the lebes PAE 000062. (CUDAM, Centro Universitario Datazioni e Archeometria Milano Bicocca).

A surprising result came from the thermoluminescence dating of a red figured Lucanian panathenaic amphora, whose authenticity was largely questioned by scholars. The vase had been in storage for over fifty years and it had never been displayed to the public because of these concerns regarding its authenticity. The analysis shows moderate natural TL signal, moderate sensitivity to irradiation, and linear TL dose response. The absorbed dose is 9.9+0.9 Gy, a value that is compatible with ancient productions [1, p. 136]. This result induced archaeologists to investigate further. Indeed, the reason why some scholars had doubted its authenticity was the disproportion and coarseness of the figures and the oddity of the composition. On further inspection this vase was understood to be the work of a painter close to the Choephoroi Painter [9, p.26-27] and the Primato Painter [9, p.33-37], from the third quarter of the IV century B.C. This amphora can be considered a very accentuated instance of a general tendency of late Lucanian vase painters, who disregard natural proportions and balanced composition while enhancing symbolic details [10].

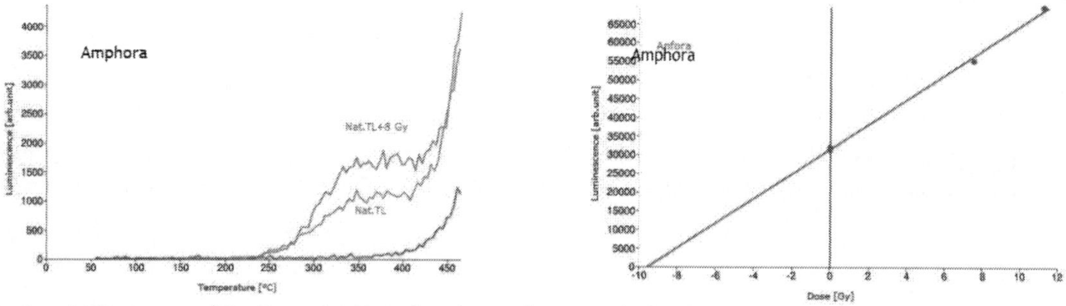

Fig 5 – a) TL curves of the Lucanian Panathenaic amphora, Zuchtriegel 2016 n°13.1; b) absorption rate of the Lucanian Panathenaic amphora, Zuchtriegel 2016 n°13.1 (CUDAM, Centro Universitario Datazioni e Archeometria Milano Bicocca)

A second vase, whose TL curve is compatible with ancient dating, posed a similar problem. A first sample (sample A) was taken from its base: it was treated with HCl (10% wt) and analysed through fine-grains method. Effective dose value is 9.5+1.1 Gy, compatible with a date between the IV and the II century B.C. However, even after further inspection, the shape of the vessel could not be related to any known typological group and it remains an isolated occurrence.

Since the sample for TL dating was taken from the bottom of the pot, a hypothesis has been advanced that the upper part of the vessel had been added in recent times to 'fix' a fragmentary ancient vase. Indeed, the body of the vessel corresponds to the normal shape of a lebes, whereas its neck resembles those found in amphoras [11].

To test this hypothesis, a second sample (sample B) was taken from the neck of the vase and analysed with matching protocol: effective dose value is 10.0±1.3 Gy, which confirms the authenticity of the vessel. A new hybrid ceramic shape was thus discovered in virtue of these investigations. This artefact constitutes one more piece of evidence confirming the tendency of late Paestan potters to deconstruct and combine traditional vessels' types.

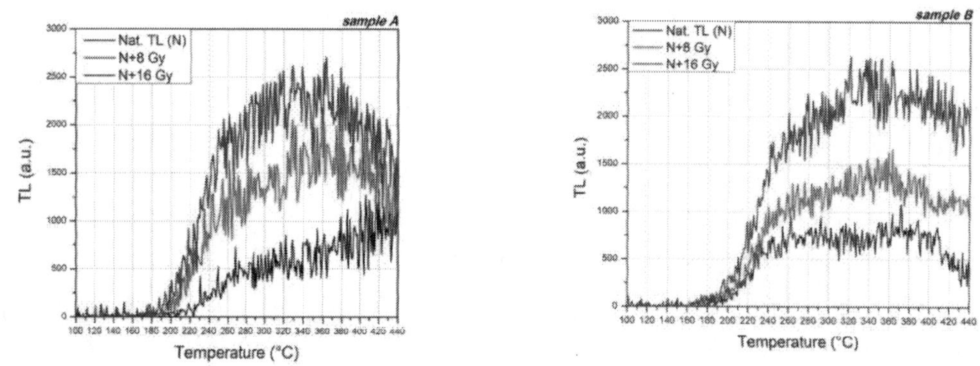

Fig. 6 – a) TL curve of the so called 'amphora-lebes' vase, sample A; a) TL curve of the so called 'amphora-lebes' vase, sample B, inv. PAE 000132. (TecnArt S.r.l.).

IOP Conf. Series: Materials Science and Engineering **364** (2018) 012002 doi:10.1088/1757-899X/364/1/012002

3. An 'embellished tomb'?

The so-called 'T2 Sequestro Finanza' is a Paestan painted tomb that was seized in 1980 by the Guardia di Finanza. The stone slab had been 'slimmed' with mechanical means to be transportable. The antiquities' dealer who was found in possession of the painting, was known for his habit of consolidating and creatively restoring ancient artworks. Therefore, a multi-analytic campaign was required to discriminate possible 'embellishments' to the original painting. UV fluorescence photography was used because, when exposed to ultraviolet light, different materials (pigments, binders, consolidants, etc.) respond with different colours, identifying intrusions and areas suitable for further chemical analyses [more on UV fluorescence in 12]. Fiber Optics Reflectance Spectrometry (FORS) and X-ray fluorescence (XRF) were used to identify the pigments [further information on chemical analyses on painted surfaces can be found in 13 and 14, and bibliography therein]. UV fluorescence images allowed identifying both organic and inorganic non-original materials (Figg. 7a&b – 8a&b). These intrusions were probably applied in more than one phase. They sometimes overlap and are made of different materials (some of which synthetic), as it was established by FORS and XRF analyses. A certain number of interventions should be attributed to the 2005-2006 restoration campaign promoted by the museum after the restitution of the slabs. However, it is impossible to decide whether earlier intrusions are to be attributed to the dealer's action or to an earlier phase of the same restoration campaign [1, p.108].

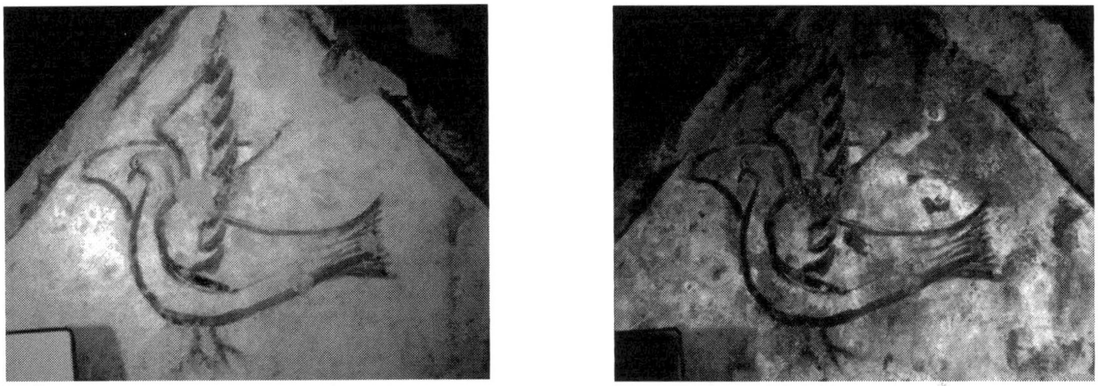

Fig. 7 – a) UV fluorescence image RGB of a detail from the West slab of the T2 Sequestro Finanaza, PAE 000057; b) visible light image RGB of a detail from the West slab of the T2 Sequestro Finanza, PAE 000057.

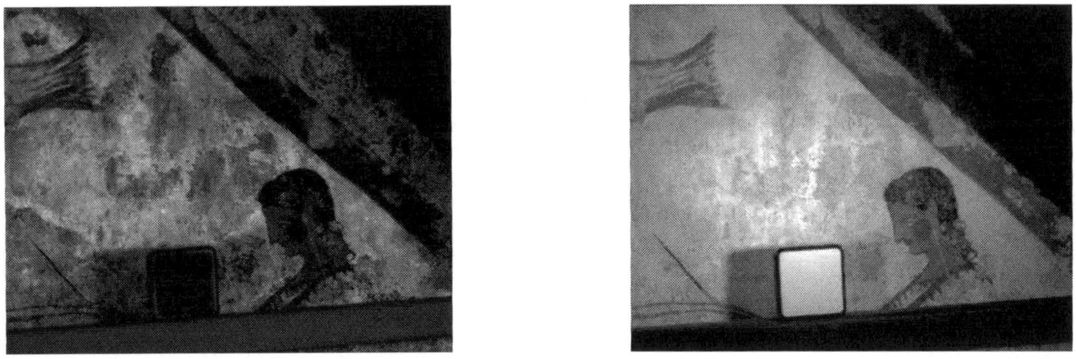

Fig. 8 – a) UV fluorescence image RGB of a detail from the West slab of the T2 Sequestro Finanaza, PAE 000057; b) visible light image RGB of a detail from the West slab of the T2 Sequestro Finanza, PAE 000057.

IOP Conf. Series: Materials Science and Engineering **364** (2018) 012002 doi:10.1088/1757-899X/364/1/012002

4. Chemical analyses of bronze statuettes and coins

Five bronze statuettes resembling Etruscan and Roman exemplars, six metallic surgical instruments and three coins were selected for analysis because of their stylistic anomalies. Non-destructive methods were used to identify the alloy and the occurrence of superficial patinas, in order to compare them with published data on the composition of known original artefacts. X-ray Fluorescence (XRF) was used to acquire a qualitative (and in some cases quantitative) spectrum of chemical elements in the alloys and patinas [on the use of XRF for forgeries' detection see 6 and bibliography therein].

Sample	Cu	Sn	Pb	Fe	Ag	Co	Zn	Ni	Re	Mn	Bi	Cr	Cd	TOT
	% (±2)	% (±0.02)	% (±0.3)	% (±0.3)	% (±0.01)	% (±0.01)	% (±1)	% (±0.3)	% (±0.01)	% (±0.01)	% (±0.01)	% (±0.01)	% (±0.01)	%
SI 1	73.60		0.47	0.41	0.02		25.39	0.07	0.03					99.99
SI 2	69.62	0.24	0.55	0.60	0.03	0.04	28.57	0.28	0.04			0.03		100.00
SI 3	72.07	0.24	1.17	1.38	0.04		24.85	0.15	0.04	0.02			0.04	100.00
SI 4	69.51	0.20	0.90	0.68	0.13		28.21	0.27	0.02	0.04	0.03			99.99
SI 5	68.05	0.02	1.31	0.52		0.05	29.79	0.17	0.05	0.04				100.00
SI 6	67.29	0.17	2.10	2.40			27.78	0.11	0.03	0.10			0.03	100.01

Table 1 – Compositional analyses on modern imitations of Roman surgical instruments (PAE 000069)

In particular, a curious set of Roman surgical instruments (Fig. 9) was investigated by means of micro-EDXRF Bruker Mistral M1 (spot 0.3*0.1 mm, HV 50 kV, current 800 µA, time 150s). Results are reported in Table 1. The surgical instruments were found to be made of a Cu-Zn alloy. However, Zn concentration was higher than expected for Roman melting technology, thus validating the stylistic judgment of inauthenticity [15, 16].

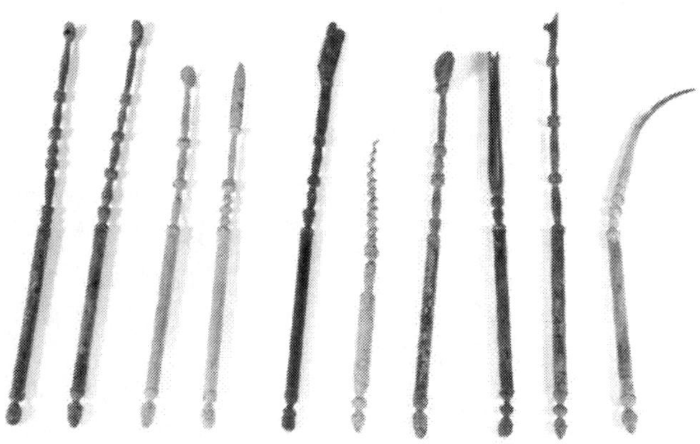

Fig. 9 – Modern imitations of Roman surgical instruments (PAE 000069)

XRF analyses were also performed on a statuette representing a kouros discophoros. The patina has been found to be a very thick layer of basic carbonates and copper chlorides/hydrochlorides, without any trace of zinc. However, measurements taken in areas with no or thinner patina layer revealed Zn in the bulk alloy. Archaeologists have interpreted this result as a possible indicator of an artificial patina used by forgers to give an 'older' look to their artefacts [17,18].

A fragmentary statuette of Venus pudica was analysed by means of a portable XRF spectrometer (AmpteK Mini-X; X-123SDDetector; parameters: 0.5 mm Al filter, spot 0.25 mm2, HV 35 kV, current 80 µA, time 100 sec). The spectrum (Fig. 10b) revealed the typical composition of brass alloy and allowed identifying traces of titanium in the outer surface. Titanium was discovered by William Gregor in England in 1789, but was first used in pure form in 1825 by the Swedish chemist Jacob Berzelius. Therefore, its presence can be attributed to a chemical treatment used by the forger to make the artefact look 'older' [17].

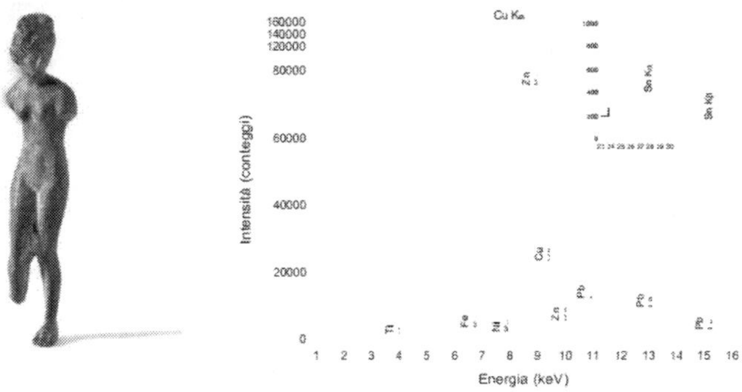

Fig. 10 – a) Venus pudica statuette (PAE 000067); b) Typical spectrum acquired from alloy analysis of a Venus pudica statuette (PAE 000067)

A coin resembling the mint produced in 193 C.E. by Didia Clara (a famous Paestan citizen) was seized in 1989 together with hundreds of other artefacts: this coin along with several other objects from this collection were found to be forgeries. Compositional analyses were performed by means of a portable XRF spectrometer (AmpteK Mini-X; X-123SDDetector, parameters: 0.5 mm Al filter, spot 0.25 mm2, HV 35 kV, current 80 µA, time 100 sec). The spectrum (Fig. 11b) revealed the presence of zinc, which is not compatible with historical data on the chemical composition of denarii in the II century C.E. [19, 20].

Fig. 11 – a) Modern imitation of a Denarius of Didia Clara (PAE 000072); b) Typical spectrum acquired from alloy analysis of Denarius of Didia Clara (PAE 000072).

The same collection that hosted the aforementioned metallic artefacts also included two coins resembling Nero's Port of Ostia sestertii, coined between 66 and 68 C.E. Because of its high market value, this coinage is often reproduced by modern forgers. Therefore XRF and micro-EDXRF analyses were performed both on the patina and on the alloy in order to compare their composition

IOP Conf. Series: Materials Science and Engineering **364** (2018) 012002 doi:10.1088/1757-899X/364/1/012002

with historical data. The two coins have different chemical composition. PAE 000073.1 is made of copper (70.90% ±1), tin (6.23% ± 0.5), lead (22.5% ±1), iron (0.09% ± 0.02), silver (0.06% ± 0.02), cobalt (0.17% ±0.05). PAE 000073.2 is made of copper (91.52% ±1), zinc (7.5% ±0.5), tin (0.08 ±0.02), lead (0.14% ±0.5), iron (0.39% ±0.02) and traces of nickel, rhenium, silver and bismuth.

Studies on the composition of Nero's coinage showed that sestertii were made of orichalcum, an alloy of copper and zinc, with the latter in a 15-20% percentage [21]. This information allows identifying PAE 000073.1 sestertius as a modern forgery. The second coin could be a late exemplar of the Port of Ostia coinage, re-melted over and over from old coins: this process would explain zinc depletion, as Zn is a very volatile element. However, they are more likely accurate modern forgeries, due to the small timespan over which this coinage was produced.

5. Conclusion

The diagnostic campaign promoted by AIAr and the Archaeological Park of Paestum on the occasion of the 2016 exhibition 'Possessione. Trafugamenti e falsi di antichità a Paestum' was a successful example of integration of stylistic and scientific methods, as it allowed for a deeper understanding of both ancient technology and contemporary forgery techniques. Archaeometric results were included both in the exhibition catalogue and in the explanatory apparatus, in order to foster awareness on the contribution of science to Cultural Heritage studies.

References

[1] Zuchtriegel G (ed.) 2016, *Possessione. Trafugamenti e falsi di antichità a Paestum,* Napoli.
[2] Bottini A 2016, Antico ad ogni costo..., in Zuchtriegel G (ed.) 2016, *Possessione. Trafugamenti e falsi di antichità a Paestum,* Napoli, 47-49.
[3] Borrelli V 2002, I falsi in archeologia, in *Il mondo dell'Archeologia,* Enciclopedia Treccani on-line.
[4] Fontannaz D 1999, Falsare humanum est. Un atelier de faussers en Italie méridionale, in *Ostraka* 8.1 35-98.
[5] *Vero o falso nelle opere d'arte e nei materiali storici: il ruolo dell'archeometria* (Atti della giornata di studio Roma 8 novembre 2006) [Quaderni AIAr Vol.118] Roma 2008
[6] Craddock P T 2009 *Scientific Investigation of copies, fakes and forgeries,* Oxford.
[7] Wintle A G 2008, Fifty Years of Luminescence Dating, *Archaeometry* 50.2, 276-312.
[8] Rogers F E 1973, Chemistry and Art. Thermoluminescence and forgery, *Journal of Chemical Education* 50.6, 388
[9] Todisco L 2012 (ed.) *La ceramica a figure rosse della Magna Grecia e della Sicilia,* I vol, Roma
[10] Roscino C 2016, Anfora di tipo panatenaico lucana a figure rosse, in Zuchtriegel G (ed.) 2016, *Possessione. Trafugamenti e falsi di antichità a Paestum,* Napoli,128-129
[11] Scafuro M 2016, Vasi a figure rosse, in Zuchtriegel G (ed.) 2016, *Possessione. Trafugamenti e falsi di antichità a Paestum,* Napoli, 132-133.
[12] Pelagotti et al. 2006, Multispectral UV fluorescence analysis of painted surfaces, *Proceedings of the XIV Signal Processing European Conference*
[13] Bacci M. et al. 2005 Fiber Optics Reflectance Spectroscopy in the Entire VIS-IR Range: a Powerful Tool for the Non-invasive Characterization of Paintings, *Proceedings of the Mathematical Research Society Symposium* 852 (2005) OO.2.4.1 - OO.2.4.6.
[14] Shackley M S 2011, *An Introduction to X-Ray Fluorescence (XRF) Analysis in Archaeology,* in Shackley M S *X-Ray Fluorescence Spectrometry (XRF) in Geoarchaeology,* New York, 7-44
[15] Bayley J 1998, The production of brass in Antiquity with particular reference to Roman Britain, in Craddock P T (eds.) *2000 Years of Zinc and Brass* [British Museum Occasional Papers 50], London, 7-26.

IOP Conf. Series: Materials Science and Engineering **364** (2018) 012002 doi:10.1088/1757-899X/364/1/012002

[16] Craddock P T and Eckstein K 2003, Production of brass in Antiquity by direct reduction, in Craddock P T and Lang J (eds.), *Mining and Metal Production Through the Ages,* London, 216-230.
[17] Foresta S 2016, Statuette bronzee. Imitazioni moderne, in Zuchtriegel G (ed.) 2016, *Possessione. Trafugamenti e falsi di antichità a Paestum,* Napoli, 110 – 111.
[18] Robbiola L, Blengino J M, Fiaud C 1998, Morphology and mechanisms of formation of natural patinas on archaeological Cu-Sn alloys, in *Corrosion Science* 40, 2083-2111.
[19] Beck L 2004, Silver surface enrichment of silver-copper alloys, in *Nuclear Instruments and Methods in Physics Research Section B* 226, 153-362.
[20] Sellars J 2013, *The monetary system of the Romans,* Melbourne.
[21] MacDowall D W 1963-1967, The quality of Nero's orichalcum, in *Gazette numismatique Suisse,* 13-17, 101-106.

IOP Conf. Series: Materials Science and Engineering **364** (2018) 012003 doi:10.1088/1757-899X/364/1/012003

Methodology for tensile testing historic tapestries

P Máximo Rocha[1], D D'Ayala[1] and C Vlachou-Mogire[2]

[1] CEGE, UCL, London, UK

[2] HRP, Hampton Court, London, UK

pedro.rocha.16@ucl.ac.uk

Abstract. Historic tapestries are very important and complex artworks. Their heterogeneous structure composed with different materials is weaved in warp and weft threads with the objective of render an image to be displayed. Often on open display these objects are exposed to forces caused by their own weight as well as by the strain experience due to moisture adsorption and desorption cycles when temperature and relative humidity of the surrounding environment fluctuate. Since physical damage is usually supported by conservation stitching techniques there is a need to understand how certain decisions would influence future mechanical behaviour of a tapestry. This requires the need of establishing a methodology that helps to understand how different sections of these heterogeneous objects respond when forces are applied to them.

Past research considered tensile testing of primarily new woven aged sections which are not representative of the historic tapestries [1]. Digital image correlation (DIC) was also successfully used in order to characterize the strain states of woven tapestries [2]. However, information on how wool and silk behaves as well as the influence of damage and stitching in some areas is still scarce and there is the lack of a methodology for a full strain analysis.

In this research, tensile mechanical behaviour is being studied in order to inform on how different fragments of historic tapestries behave when a certain tensile force is applied on them. For this, a new methodology of analysis is proposed. Force strain curves collected from tensile tests are analysed together with complementary DIC data to understand the strain states that were present in different stages of the mechanical tests. Results proved that a holistic methodology, using both techniques required for successful interpretation of data. Furthermore, this experimental work demonstrated that tapestry areas behave differently when a tensile force is applied.

1. Introduction

Historic tapestries are one of the finest artistic works that exists in European art. Their manufacture process in a loom usually required the work of several high skilled weavers that transposed into textile what was their own interpretation of a cartoon previously designed by an artist [3]. After mounting the warp threads in the loom, the weaver will then interweave the dyed weft threads against the undyed warp structure, beating down each one of the weft threads until the warp structure is completely hidden in the final result [4]. In the process of weaving a tapestry slits are formed when there is a break in colour on the line of one warp. When a tapestry was finished, the slits were usually stitched together so the tapestry could be displayed by hanging in its weft direction. This meticulous work not only required a big workload effort from all the people involved in the tapestry manufacture but most of the times it also required the use of expensive materials including silk, silver and gold.

Tapestry production which had its high peak of production from the 15th to the 18th centuries were a symbol of status meant to be hanged only on special occasions [5]. However with the change of trends in the end of the 18th century, tapestry manufacture fell on decay and some of the best sets of tapestries were being used in permanent display as wallpapers to hung paintings or even burned to extract its precious metal threads [6, 5].

Notwithstanding the story of neglect that many tapestries have experienced, during the last decades there was an increasing interest towards their conservation and preservation. Nowadays tapestries are regarded as precious historic objects, however to preserve them to future generations there is the need of understanding conservation techniques implemented as well as how displaying conditions affect their mechanical behavior and therefore their deterioration processes.

Tapestries are usually displayed in a long-term basis without a complete understanding on how environmental factors cause mechanical damage to their structure. When on display, in their weft direction, these objects are always supporting their own weight. However, cycles of moisture adsorption and desorption makes weight change through time making the research of tensile loads in tapestry structures an area with major interest.

Past research studied how DIC can be used to monitor strain in historic tapestries [2] as well as how tensile tests are useful to extract tensile properties from tapestry samples [7].

This work aims to test a methodology of using together tensile testing with digital image correlation in historic samples of tapestries. Since the purpose of a tapestry is to render an image with interweaving threads, every section in an historic tapestry has always a high level of heterogeneity in its structure. This heterogeneity in the weaving is even more enhanced when there are slits, damaged areas or stitching repairs that most of the times use different materials than those used in historic weaving. Therefore its mechanical behavior cannot be evaluated according to only tensile tests performed in some weaved samples since this can lead to incorrect interpretations of what is really being quantified.

Previous research on tensile testing of tapestries considered mainly new woven aged samples [1], or performed tests on single yarns [8] which are not representative of the complex structure of the historic tapestries. This study explores the results of a methodology obtained for testing historic tapestry samples selected from Historic Royal Palaces (HRP) collection of tapestry fragments made available for the purpose of this study. This paper discusses the development and testing of a proposed methodology as well as informs of some unique mechanical characteristics that can be found in historic tapestry structures.

2. Materials and Methods

A tapestry by its nature, has a heterogeneous structure composed of warp and weft threads. Considering this heterogeneity, 5 different fragments were selected to be tested in this study. In each one of these fragments, 5 different zones measuring 76 mm per 80 mm were tested using the grab method for tensile testing. Since the average results of the tensile tests were used to characterize the tensile response of each fragment, sample zones were selected with the objective to be representative of the whole fragment. Test samples were chosen accordingly with characteristics that were most common across all the fragments. However, in some cases these 5 samples shared common areas between each other due to the reduced size in some fragments as well as the difficulty in the selection of 5 representative test zones. Notwithstanding, the selection process of samples was a crucial step in this study once that the 5 samples needed to be characteristic of the whole fragment, otherwise it would not be suitable to work with the average results from the 5 tensile load-strain curves.

Physical properties of thickness, fineness, percentage of silk and a ratio for slits were characterized for each of the selected sample zones.

The methodology used in this study results from a combination of two common methods applied in textile testing. These two methods are the tensile testing using a Universal Tensile Machine (UTM) and Digital Image Correlation (DIC). Each method has been proved to be useful in the understanding of the mechanical behaviour of tapestry structures [7, 9]. However, what is now explored is the simultaneous use of these two methods in a more holistic way by characterizing the heterogeneous physical properties of historic samples by clarifying the understanding on how these complex tapestry structures behave. This methodology is summarized in Figure 1.

Samples used were selected and named according to their file number in the collection of tapestry fragments at HRP. For this selection, 2 homogeneous fragments were chosen, fragment 44 with weft threads woven entirely from wool and fragment 143 from silk. Fragment 131 is also entirely made from wool, yet it has a heterogeneous structure typical of historic tapestries. Fragment 154 was chosen since it has similar quantities of silk and wool in its weft direction. Finally fragment 68 had a large number of silts in its structure. Destructive sampling on these woven textiles was not considered by the fact that these fragments are historic. Instead, grab test method for tensile testing selected sample areas on each fragment was carried out.

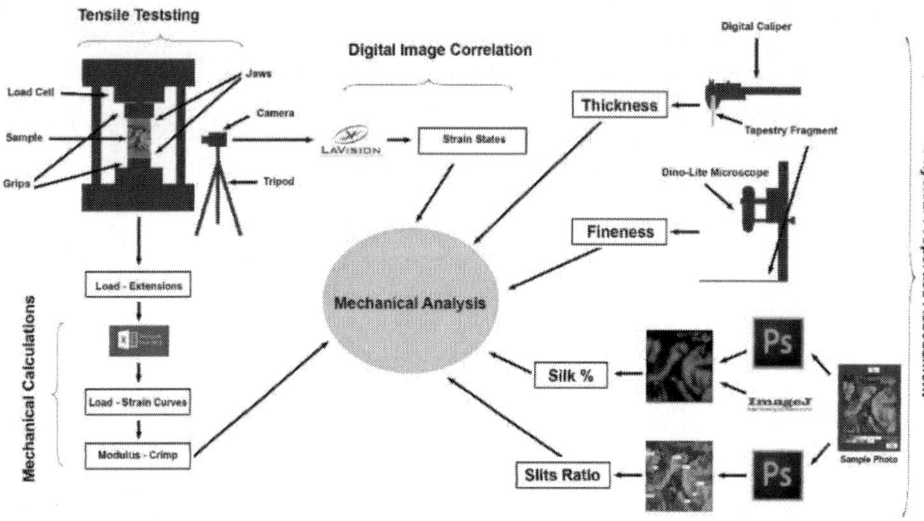

Figure 1. Methodology used to characterize and test the samples.

2.1. Physical Properties assessment

Tapestries are usually weaved with multiple materials being wool and silk the most common ones. To differentiate and quantify the different materials present on each sample, a material assessment was carried out. This assessment was done considering photographs taken on the perpendicular direction of the selected samples. After distinguishing silk from wool, silk weaved areas were manually highlighted using the software Adobe Photoshop CS6 and had latter their percentages measured under ImageJ. To characterize the number of slits that were present in each sample, photographs of the samples were used to measure the length of horizontal projection of each slit to create ratio. This ratio was calculated as $S = \frac{\sum_{1}^{n}(W)}{76}$ where S is the slit ratio, w the width of slits in mm, n the number of slits measured and 76 is the width of each sample in mm.

In each fragment, the thickness was characterized using a digital calliper HITEC 101-45 having 0.01mm precision. Initially with the jaws joint together in the closed position the calliper was set to zero. After, the jaws were open and then closed sequentially on each measurement point dispersed around the fragment. A total of 10 measurement points were considered in each fragment being 5 in the weft and 5 in the warp direction. A period of time of 30±5 seconds was given after closing the jaws and before reading each measurement in accordance with the current standard [10]. Each one of these points were measured 3 times and the results were then averaged.

IOP Conf. Series: Materials Science and Engineering **364** (2018) 012003 doi:10.1088/1757-899X/364/1/012003

2.2. Tensile Testing

Since tapestries are displayed by hanged in the weft direction, supporting their own weight, tensile tests only considered this orientation.

To characterize the tensile properties of the selected samples a universal testing machine (UTM) Instron 3365 connected to a pressurised pump was used. To ensure the homogeneity of every experiment, every specification used during the tests was the same on every sample. Tensile tests used serrated jaws in the holding grips with the purpose to avoid slippage of samples during testing. However, to prevent the damage that can be caused by weakening of the samples in the holding areas of the serrated jaws, an adhesive tape was used on each jaw. Each sample was mounted in its relaxation state with the weft threads perpendicular to the edges of the Jaws. Just before closing the grip's jaws, the sample was carefully adjusted to ensure that sample areas of each fragment placed on the UTM matched the previously recorded areas. After sample adjustment, the pressurised air pump was activated to hold samples at a pressure of 3.5 bar.

In these tests, the rate of extension was 2 mm/min and a load cell of 5KN was used. The accuracy of the measurements was 0.25N for the force and 0.12mm for the extensions.

These tensile tests were done under controlled environmental conditions of 25,0±2,0 °C for temperature and 45,0±4,0% for relative humidity. All samples were conditioned in this environment during 24h before testing.

The stopping criteria was defined as the break of any thread except for samples 44-1 and 143-1 that were tested to a level where there was a defined linear zone.

Results are presented in load-strain curves to be more comparable with other values on literature, however since stress is defined by $\sigma = \frac{F}{A}$ where F is the force applied and A the area, it means that stress is dependent on the thickness of material and some considerations regarding interpretation of results were taken.

To measure stiffness in each curve, Young modulus was considered as the slope of the linear zone in a stress-strain curve [11]. However due to the fact that textiles are mechanically different from other materials [7], in this study the slope of the linear region was called the apparent linear modulus and was calculated by the slope of the linear zone in the load-strain curve as suggested in the literature [1]. Units of KN/m were used to compare results with others available in literature [1,7]. To confirm the linearity of the linear zone in each curve a linear regression line was calculated on excel in steps of 1% strain, starting from 0% until the end of the curve. When the coefficient of determination R^2 of the regression line was higher than 0.9985 the slope was considered linear and the value of the apparent modulus was recorded. Strain where this linearity started was taken as a measure of the end of the crimp zone.

2.2.1. DIC

A DSLR camera Canon EOS 700D with an ISO 100 was used to capture a sequence of photographs along the test. Just before the test started, two photographs were taken in order to inform about the noise values that were taking place under DIC.

Each photograph was captured manually using the remote control of EOS Utility software. This was made using a levelled tripod which hold the camera lenses without vibrations in a parallel direction of the tensile tested sample. For light conditions, flash was not considered, however good and constant illumination of the samples was satisfied. To ensure the best possible quality, these photographs were shot using RAW format, which was later converted into a .tiff format and grey level image using both dcraw [12] and Adobe Photoshop CS6 software respectively.

All photographs were processed with the software Davis 8 from La Vision using 2D DIC. The strain states on each one of the samples were measured by comparing the distorted images with the reference photographs of the samples before the test. Strain distribution on both vertical and horizontal directions was calculated and extracted using the "sum of differential" as correlation mode in Davis software.

3. Results and Discussion

Measured physical properties and calculated apparent linear modulus and end of crimp strain are displayed in Table 1. Load-strain curves of the averaged 5 tensile tests on 5 historic fragments are presented on Figure 2. Figure 2 also shows the tangent at the last stage of each load-strain curve. As a first observation on these curves it can be noticed that they follow the standard tensile behaviour that characterizes other textile woven samples that are found on the literature [7]. This behaviour is composed of three different stages. Initially there is a zone where the samples are adjusting to the pulling grips and that is commonly known as "slack" [7]. In tested samples of Figure 2, this is a very subtle change that happens before 1 % strain. After this primary stage, the weft threads are aligned with the pulling grip and they start to lose their crimp by becoming more and more stretched as the strain increases. This region is characterised of being mainly curvilinear and corresponds to the transition from a relaxed into a stretched state. This curvilinear region is placed in the load-strain curves between the end of the slack region until a stage where the behaviour is linear. This linear stage is named in the literature as the linear region [7]. In the linear zone, the woven textile, having its weft threads under tension and without crimp, behaves as an elastic material until its weft thread structure starts to fail. Tangents on Figure 1 show that for most of historic fragments tested, except for 44, there is not a clearly defined transition from crimp to the elastic region. Since fragment 44 is very homogeneous, this suggests that in historic tapestries the progression from crimp region to the elastic phase doesn't seem to have a clearly defined transition point.

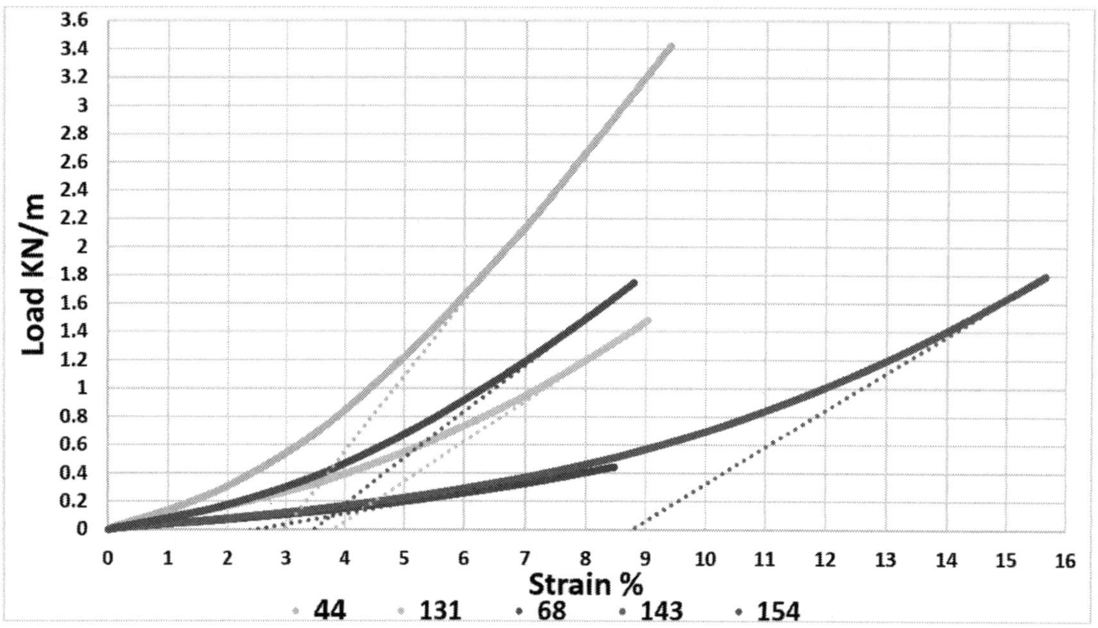

Figure 2. Averaged force-strain curves for 5 samples tested in each fragment.

Despite having some minor damaged zones, sample 44 and sample 143 are both very homogeneous. However, by observing Table 1 it can be noticed that while sample 44 has its weft structure composed entirely of wool, sample 143 is composed of silk. It can be observed that the sample made from wool is stiffer and has its elastic zone starting at around 6% strain, while on the one made from silk the elastic area only starts at around 14% strain. With this, it seems that samples made from wool are stiffer and had less crimp in the weft threads when compared to those made from silk. However, when looking at the other samples this could not to explain why wool sample 68 is behaving as the silk sample 143 and 154 and 131 although with different material compositions have stiffness between values of wool sample 44 and silk sample 143.

IOP Conf. Series: Materials Science and Engineering **364** (2018) 012003 doi:10.1088/1757-899X/364/1/012003

Table 1. Physical and mechanical properties of tested samples.

Tapestry Fragment	Direction	Fineness (Threads/cm)	Thickness (mm)	Testing Zones (Samples)	Silk %	Slits Ratio	Apparent Linear Modulus (KN/m)	End of Crimp (Strain %)
44	Weft	10.1±6.2	1.7±0.1	1	0	0.00	0.55	4
				2	0	0.00	0.55	6
	Warp	6.2±0.5		3	0	0.00	0.41	5
				4	0	0.00	0.60	5
				5	0	0.00	0.47	6
68	Weft	9.3±3.9	2.5±0.2	1	0	1.26	0.10	8
				2	0	1.45	0.07	5
	Warp	3.9±0.3		3	0	1.17	0.11	8
				4	0	1.47	0.12	8
				5	0	0.91	0.13	8
131	Weft	6.9±4.4	2.9±0.2	1	0	0.00	0.24	7
				2	0	0.00	0.27	8
	Warp	4.4±0.4		3	0	0.00	0.31	7
				4	0	0.00	0.29	7
				5	0	0.00	0.3	8
143	Weft	10.8±7.6	1.3±0.1	1	100	0.00	0.25	15
				2	100	0.00	0.28	12
	Warp	7.6±0.4		3	100	0.00	0.22	14
				4	100	0.00	0.27	14
				5	100	0.00	0.37	14
154	Weft	15.0±5.0	2.1±0.2	1	59	0.00	0.21	4
				2	56	0.36	0.30	7
	Warp	5.0±0.4		3	61	0.00	0.30	7
				4	58	0.00	0.25	5
				5	63	0.00	0.39	6

Images of the DIC results for strain both on vertical and horizontal directions are presented on Figure 3. Here one single sample was selected to be representative of the 5 tested samples. This sample is identified by its fragment number followed by its sample number.

With results taken from DIC, interpretation of the tensile behaviour in each sample starts to emerge. First of all it is noticeable that homogenous samples 44 and 143 have fairly homogeneous strain distribution across its surface when a tensile force is applied on them. Moreover only with the exception of the vertical direction in sample 143 that is very homogeneous, damaged and creased areas became visible under DIC both on horizontal and vertical directions.

Sample 154 has 59% of silk, however its load-strain curve in terms of stiffness is more similar with the sample 44 made from wool. An observation of strain distribution on DIC results suggests that this behaviour happens since it's mainly the brown wool threads in sample 154 that are absorbing most of the tensile load applied in the vertical direction. On the horizontal direction, these areas are also the ones that have more compression which agrees with the Poisson ratio effect of shrinkage in the horizontal direction when the vertical is being stretched.

Regarding wool sample 131, although is thicker than the rest of the samples, it seems to have a very similar behaviour to sample 154. It is interesting however that this happens and this can mean two things. Once that in sample 154 it is the wool areas that seems to be absorbing most of the load applied, this could mean that both samples are in the range of the usual behaviour of heterogeneous wool samples or it can be that due to the small damage of sample 131, shown by the white warp threads in Figure 3, this wool sample has more crimp that what is usual in wool samples. However, to clarify this point and completely characterize the behaviour of wool samples more tests with different samples should be done. By observing the load-strain curves in Figure 2, a curious tensile behaviour of a wool sample is the one of sample 68 which its curve is very identical to silk sample 143. Yet an observation to the properties and strain states of this sample suggests the reason why this happens. Properties on Table 1 show that sample 68 is the only that has a considerable amount of slits in its structure. Moreover, a closer look at this sample on Figure 3 shows that this slits have some stiches closing them. This means that the crimp experienced by this sample is higher than the other ones since slits enable this sample to have an increased freedom of movement. Furthermore, DIC results show that areas with stiches are those that have more strain justifying the level of crimp present on this sample. Stiffness can also find an explanation by these observations since stiches are made with yarns of a different material than the one used to weave wool and silk areas.

As a general observation on DIC results it is noticeable that horizontal strain showed more compression in areas of damage and slits like in samples 44 and 68 and on areas of interrupted threads in samples 131 and 154.

It is also noticeable that the average load-strain curves showed correlation in values of strain with the individual DIC results, being consistent with them.

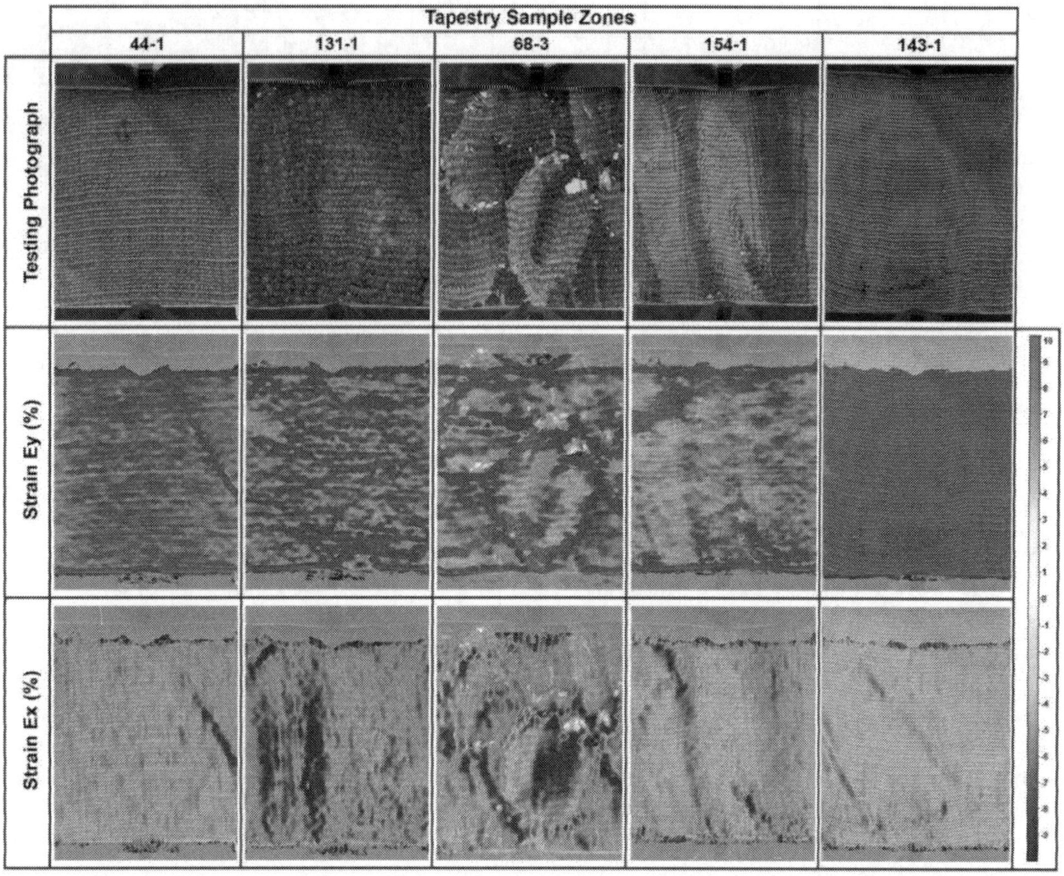

Figure 3. DIC results for vertical and horizontal strain distribution in 5 tested samples.

4. Conclusions

Performed tests seems to suggest that under hanging, different areas of an historic tapestry will behave differently according to number of factors. These areas will experience different levels of strain distribution and load concentration making some zones more susceptible to breaking when exposed to hanging loads. It is also suggested that the areas with more strain applied correspond to threads that have originally less crimp applied on them. Notwithstanding, to have solid conclusions about the behaviour of historic tapestries its required to test more examples of historic fabric.

Digital image correlation enabled not only to increase the understanding of the strain states that are present on a sample section but also the establishment of a relationship with the load-strain curves obtained on the tensile tests.

This methodology, implies that both techniques are being used at the same time. Furthermore, it has been showed that there is a consistency between results from tensile testing and from DIC. This testing procedure proved to be beneficial for interpreting the results of different heterogeneous historic tapestry samples with good repeatability.

References

[1] Duffus P. Manufacture, Analysis and Conservation Strategies for Historic Tapestries. University of Manchester; 2013.
[2] Khennouf D, Dulieu-Barton JM, Chambers AR, Lennard FJ, Eastop DD. Assessing the feasibility of monitoring strain in historical tapestries using digital image correlation. Strain. 2010;46(1):19–32.
[3] Tabard F. The Weaver's Art. In: The Book of Tapestry. New York: Vendome Press; 1978. p. 188–225.
[4] Bosworth D, Caroline C. Development of a couching technique for the treatment of historic tapestries. In: Lennard F, Hayward M, editors. Tapestry Conservation: Principles and Practice. Elsevier; 2006. p. 91–6.
[5] Band J. The survival of Henry VIII's History of Abraham tapestries: an account of how they were perceived, used and treated over the centuries. In: Lennard, F., Hayward M, editor. Tapestry Conservation – Principles and Practice. 2006. p. 20–7.
[6] Campbell T. Tapestry in the Renaissance Art and Magnificence. The Metropolitan Museum of Art; 2002. p. 606.
[7] Ł. Bratasz, M. Łukomski, A. Klisińska-Kopacz, W. Zawadzki, K. Dzierżęga, M. Bartosik, J. Sobczyk FJL and RK. Risk of Climate-Induced Damage in Historic Textiles. Strain. 2014;51:78–88.
[8] McCullough L. A non-destructive method for assessing the degradation of wool yarns within historic tapestries while in situ. UCL; 2014.
[9] Lennard F, Eastop D, Dulieu-Barton J, Chambers A, Khennouf D, Ye C-C, et al. Strain monitoring of tapestries: results of a three-year research project. In: ICOM-CC 16th triennial conference Lisbon 19-23 September 2011: preprints. 2011. p. 1–8.
[10] British standards. BS EN ISO 5084:1997 Textiles — Determination of thickness of textiles and textile products. 1997.
[11] ASTM. ASTM E111-04(2010), Standard Test Method for Young's Modulus, Tangent Modulus, and Chord Modulus. 2010.
[12] Coffin D. dcraw [Internet]. 1997. Available from: https://www.cybercom.net/~dcoffin/dcraw/

Mutual validation between different modal analysis techniques for dynamic identification of the so-called Temple of Minerva Medica, Rome

Ivan Roselli, Vincenzo Fioriti, Marialuisa Mongelli, Irene Bellagamba and Gerardo De Canio

ENEA Italian National Agency for New Technologies, Energy and Sustainable Economic Development, Italy

ivan.roselli@enea.it

Abstract. The dynamic identification by ambient vibration data is widely used to supply information on the global health of structures through the investigation of changes in their modal parameters. It can be used even for verification of the state of damage of structures after hazardous threats, for example seismic activity. Therefore, it can play a crucial role to integrate and support conservation strategies for historic architectural assets. Sometimes, in historic constructions only a limited number of positions are accessible or usable to install sensors, and so modal analysis must be based on data from few measurement points. Moreover, they might not be the optimal positions for the studied structure, so the obtained results would need further verification. In such circumstances, the mutual validation between different modal analysis techniques can be useful to assess the reliability of results. In the present paper a case study of application to the so-called Temple of Minerva Medica, Rome, is described. Ambient vibration data were acquired in four rowing acquisition sessions carried out from July 2016 to July 2017, which is a timespan usable to assess the impact of the recent Central Italy seismic sequence. For problems related to the installation of the scaffolding only few points were available for instruments positioning. A variety of techniques were applied, including FRF, FDD, EFDD, SSI, HVSR and complex modal models. The variance of the modal parameters obtained by each different technique was utilized to provide indications on the reliability of the average values.

1. Introduction

The diagnosis and monitoring of the structural state of damage are crucial issues to a proper planning of restoration, preventive conservation and maintenance interventions of cultural heritage constructions. In particular, in archaeological sites located in areas with high natural hazard, such as in seismic zones, relevant damages may be caused in case of extreme events, such as earthquakes. Unfortunately, this is a quite typical situation in most Italian historic sites, as recent events in Central Italy demonstrated [1]. On the other hand, many Italian archaeological sites are located in historic city centers where also other relevant sources of strong vibration, such as the ones due to urban traffic vehicles, excavation works or other anthropic activities, add to the risk of structural damage [2].

Within this context, methods based on ambient vibration measurements constitute valuable tools to understand the dynamic behavior of a historic building, which is related to its state of damage [3].

A variety of techniques are available to analyze the vibration data, but they are all aimed at the modal identification of the structure, which essentially means at the extraction of the modal parameters that characterize the dynamic behavior of the building. As well known, such parameters can be monitored to give indications on the degradation of the state of health of the structure [4, 5].

In order to have the most accurate and comprehensive understanding of the overall behavior of the studied building, a high number of sensors are usually located on the structure and their positioning is studied so as to maximize the measurable dynamic response [6, 7], e.g. with the help of preliminary numerical estimates of the modal shapes by means of finite element models (FEMs) [8]. Sometimes, though, historic constructions are not completely reachable and only a limited number of positions are accessible or usable to install sensors. In such cases, ambient vibration data from only very few measurement points, possibly not the optimal positions for the studied structure, can be obtained, so results of modal analysis may be very rough and their reliability needs to be verified.

A simple way to provide indications on the reliability of research findings is to validate them through a statistical comparison with other results obtained with more consolidated and accurate data, when such data are available. In other circumstances, a variety of consolidated methods based on different theoretical assumptions can be applied to the same dataset. This idea of mutual validation of data through the application of different methods was fruitfully applied in the present paper to assess the reliability of identified modal frequencies of the so-called Temple of Minerva Medica, an ancient masonry construction located in Rome [9], studied within the framework of the CO.B.RA. project, which focused on the development of advanced technologies and methods for the conservation of cultural heritage assets. For problems related to the positioning of the scaffolding during data acquisitions only limited portions of the monument were accessible and a limited number of measurement points were available for instruments installation. After data acquisition, a variety of modal analysis techniques were applied, both Experimental Modal Analysis (EMA) [10] and Operational Modal Analysis (OMA) [11] techniques, including time-domain [12] and frequency-domain techniques [13], as well as unconventional methods derived from other disciplines that have been proved capable of approximate estimates of the modal frequencies of structures.

2. Modal analysis methods

Modal analysis methods are methods aimed at providing the modal parameters, which are characteristic of the dynamic behavior of the studied building. The main modal parameters are the resonance frequency, the damping ratio and the modal shape associated to each mode of vibration of the studied object. The most used modal parameter for damage detection and monitoring is the natural frequency, which is substantially related to the structural stiffness resulting from the building's shape and from the mechanical properties of the materials it is made of. The basic rationale behind the monitoring of the modal frequencies is that the arising of fractures in the material provokes a loss of stiffness of the entire structure. The more the fracturing processes proceed with time, the more the structure stiffness decays and modal frequencies shift to lower values. Consequently, the modal frequencies can provide indications on the effect of the evolution of the crack pattern of the construction.

The extraction of the modal parameters from a dataset of vibration data acquired by sensors located at measurement points of the structure can be achieved through a variety of techniques. In the following subsections the techniques applied for the estimate of modal frequencies in the present study are illustrated and their characteristics are discussed.

2.1. EMA techniques

The vibration-based analysis techniques aimed at finding the dynamic response of a structure subjected to a known or measurable excitation input are defined as Experimental Modal Analysis (EMA) techniques, because this is generally the case of data acquired in the context of laboratory or field experiments, where vibration input is controlled by test operators. Any external vibration source exciting the structure can be used in these methods, even if some input characteristics (waveform, duration, intensity etc.) may affect accuracy in the dynamic response calculation. The most used EMA techniques are based on frequency domain approaches. In particular, they are commonly based on the implementation of a frequency response function (FRF), i.e. an input-output relationship between two points on a structure as a function of frequency based on the Transmissibility Function H [14]. Since both force and motion have directions associated with them (vector quantities), therefore, an FRF is actually defined in each direction. Moreover, FRFs can be calculated for each couple of signals at input and output points (single-input/single-output or SISO approach). When the input is measured in only one point (single-input), while the response can be measured at several points of the structure (multi-output) FRF calculation can take into account the global behaviour of the structure (SIMO approach), which was the one used in the present work.

2.2. OMA techniques

In contrast to EMA methods, OMA approaches are based on output-only data, i.e. measuring only the response signals of a structure and using the ambient and natural operating forces as unmeasured input. OMA techniques are used instead of classical mobility-based modal analysis for accurate modal identification under actual operating conditions, and in situations where it is practically difficult, inconvenient or even dangerous to artificially excite the structure, such as in the case of many historic buildings.

OMA algorithms generally assume that the input forces are stochastic in nature, so that they can be considered as white-noise signals. This is widely accepted in the case of ambient vibrations of civil engineering structures, which are mainly excited by ambient forces like wind, traffic, urban activities or seismic micro-tremors. One of the most consolidated OMA techniques is the Frequency Domain Decomposition (FDD) [15]. This technique is based on the computation of the Singular Value Decomposition (SVD) of the estimated spectral densities of the measured dynamic response of the structure. After calculating the SVD plots, modes are identified by identifying the peaks in such plots (peak-picking) in the frequency domain. As the FDD technique is based on using a single frequency line from the Fast Fourier Transform (FFT) analysis, the accuracy of the estimated modal frequencies depends on the FFT spectral resolution. In FDD no modal damping is calculated, but in the present paper only the modal frequency will be taken into account.

The Enhanced Frequency Domain Decomposition (EFDD) is a more recent evolution of the FDD with extended potentialities and accuracy [16]. In particular, compared to FDD, the EFDD gives an improved estimate of both the modal frequencies and shapes.

In EFDD, for each mode a power spectral density function, identified around a peak in the SVD, is calculated. After taking it back to the time domain by the Inverse Discrete Fourier Transform (IDFT), the corresponding modal frequency is obtained by computing the number of zero-crossing as a function of time. Besides, the EFDD provides also estimates of the modal damping by the logarithmic decrement of the normalized auto correlation function of the Single Degree Of Freedom (SDOF) corresponding to the identified mode. The SDOF function is estimated using the modal shape determined through the peak-picking method similarly to the FDD. The Subspace Stochastic Identification (SSI) is a time-domain estimator of modal parameters that is based on a difference equation obtained by the raw time-history data. In particular, the SSI method takes advantage a State Space Model that is used to convert higher order problems into 1st order problems [17]. In fact, while estimating at different over-specified model orders, it can be observed that physical modes remain quite constant. Contrarily, spurious modes vary in such conditions so that they can be distinguished using a stabilization diagrams [17].

The use of the stabilization diagram for the extraction of the identified modes implies a high order needed for the identified system. Thus, a very long sequence of nodes is required. An evolution of this method, called Cristal Clear (CCI-SSI), implements the covariance subspace identification [18], through which it is the possible to handle non-stationary data and consequently avoid the choice of too long sequences of measurements. In the following, the CCI-SSI was used in order to obtain clear stabilization diagrams to identify with high accuracy the frequencies of the physical modes.

2.3. HVSR technique

The Horizontal-to-Vertical Spectral Ratio (HVSR) method is applied by calculating the ratio between the amplitude of the Fourier spectra of horizontal and vertical components recorded on a structure. Modes are identified by picking the peaks of the obtained HVSR graphs in the frequency domain.

This method was initially introduced by Nakamura to determine the predominant period of ground motion signals to evaluate site amplification characteristics during earthquakes [19]. Subsequently, it was successfully explored to estimate also the modal frequencies of buildings, on the basis of the simple observation that resonance at lower structural modes (generally the most relevant ones for the overall dynamic behavior of the building) causes vibration amplification in horizontal directions, while the system remains much stiffer and substantially unamplified in the vertical direction. As this is the most common situation of the main bending and torsional modes identified in civil structures, the findings of several studies confirmed the good agreement between the frequencies identified by HVSR and by other most consolidated dynamic identification techniques [20, 21].

3. Experimental application to the so-called temple of Minerva Medica

The above mutual validation approach was applied to the case study of the ambient vibration data acquired on 4th July 2016 at the so-called temple of Minerva Medica. This monument is an ancient ruined building located in the city-center of Rome. Traditionally interpreted as a temple dedicated to Minerva Medica ("Minerva the Doctor") since 16th century, it was more recently hypothesized to be an ancient nymphaeum part of the Horti Liciniani, but attribution is still discussed [22]. The initial construction could be traced back to the Imperial Rome (early 4th century A.D.), but later several restoration and reconstruction interventions were carried out until 2013 [22].

The main building is a majestic hall with decagonal polylobate plan (diameter of 25 m and overall height originally of 32 m, but currently reduced to only 24 m due to past partial collapses of the dome. Initially, it was entirely built in *opus latericium*, a typical roman technique of that time. After early structural problems arose, it was restored and reinforced through works in *opus mixtum* of tuff bricks and Roman bricks, as well as through the addition of buttresses in the southeast side of the monument, which testify a serious structural weakness in this part of the structure, confirmed by the major historically documented damages concentrated at the south side [22]. In consequence of several historic collapses and reconstructions, the current masonry presents quite heterogeneous materials and irregular shapes.

3.1. Ambient vibration acquisitions and analysis

The vibration data were collected by the use of 5 digital recorders positioned at the measurement points depicted in figures 1 and 2.

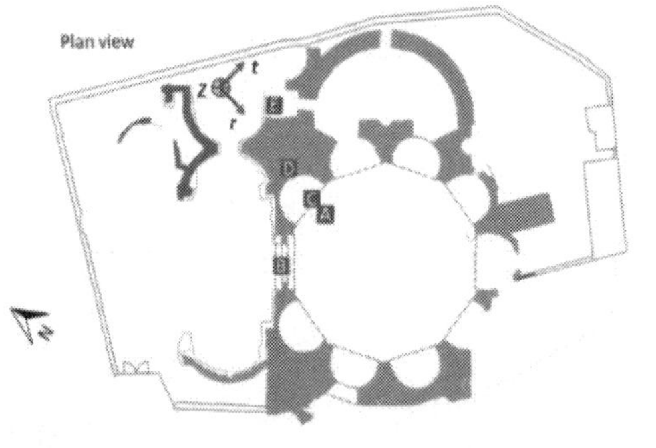

Figure 1. Plan view of the monument with positions of instruments (in red squares).

Figure 2. Measurement points on the main façade (view from North).

Figure 3. Installation of measurement point A at the dome curb on the north side.

Figure 4. Instrumentation setup: position C.

Each digital recorder was equipped with a triaxial velocimeters, whose time synchronization is guaranteed by a GPS antenna. Data were acquired for 20 minutes at a sampling frequency of 200 Hz, as in previous similar studies [23].

According to the instruments setup procedure, the digital recorders were simply laid on the horizontal surfaces available and reachable by proper scaffoldings (figures 3 and 4). Consequently, they could be located only at the two reachable windows (positions B and C), at the dome curb on the north side (A), on the base wall (D) and on the ground foundation (E). Measurement at E was adopted as input signal for FRF calculations, as the ground vibration due to tram and train passages very close to the monument was assumed as the main responsible of dynamic excitation to the structure. This assumption was taken after verifying that vibration intensity at E revealed strong peaks in correspondence of tram and train passages. All other positions (A, B, C and D) were taken as output signal (SIMO approach). The HVSR was calculated as the average response of A, C and B, as E and D substantially gave the ground response. Both FRF and HVSR were calculated in north (N) and east (E) directions.

4. Results

The modal frequencies for the first five modes covering a frequency range of 0-5 Hz were considered in the analyses. Figure 5 illustrates the results obtained by the application of FRF with SIMO approach between 0 and 5 Hz.

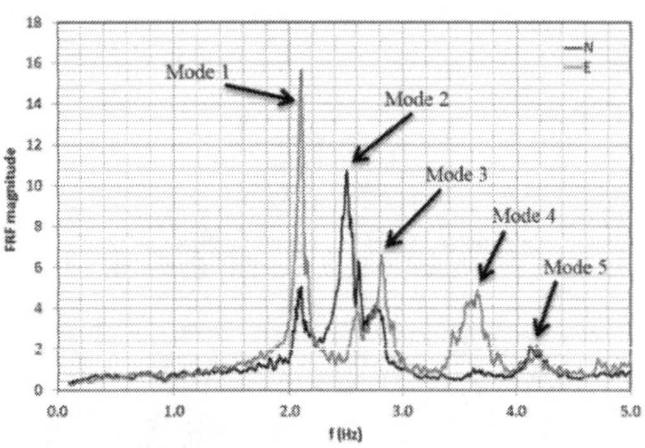

Figure 5. Modal frequencies identified by FRF magnitude peaks in North (N) and East (E) directions.

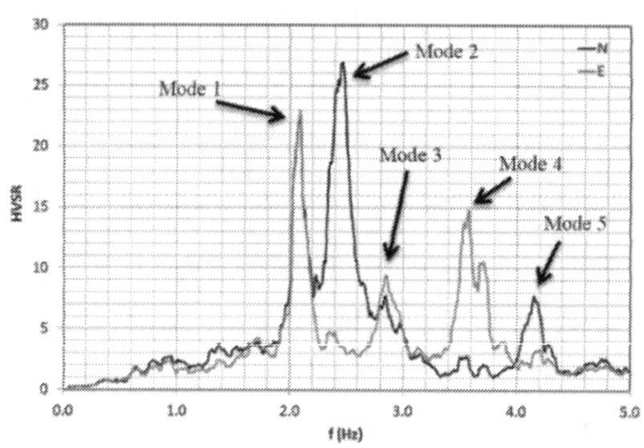

Figure 6. Modal frequencies identified by peaks in the Horizontal-to-Vertical Spectral Ratio (HVSR) in North (N) and East (E) directions.

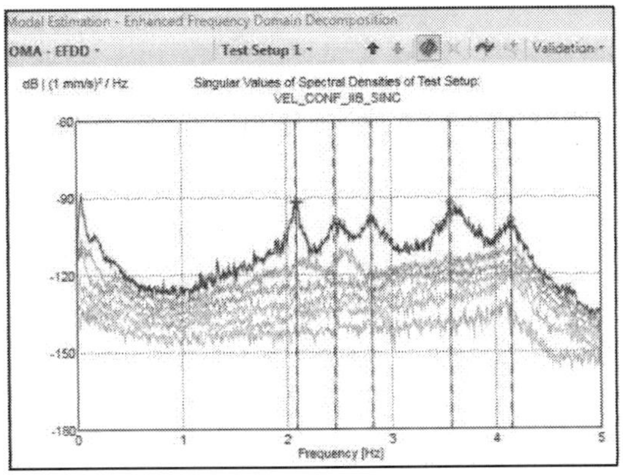

Figure 7. Modal frequencies by Enhanced Frequency Domain Decomposition (EFDD)

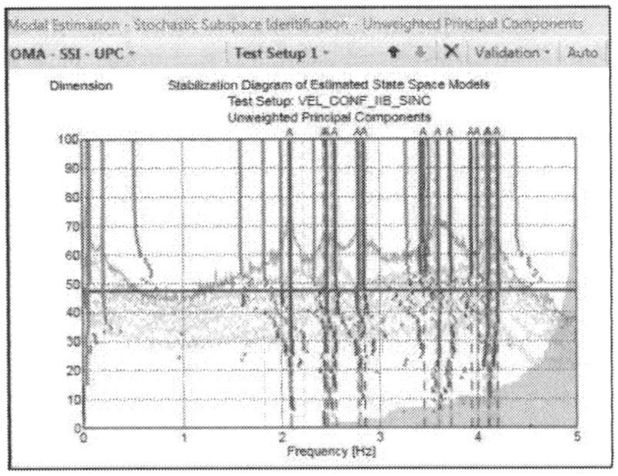

Figure 8. Modal frequencies by Cristal Clear Stochastic Subspace Identification (CC-SSI).

It shows evident peaks in North (N) and East (E) directions, which clearly identify the first f modes. Similarly, figure 6 illustrates the peaks of the HVSR computed by averaging the curves at A, B and C positions. As for the OMA techniques, all the applied methods provided very similar modal frequencies. In particular, the identification of modes by EFDD through the analysis of the SVD is depicted in the figure 7, while the modes found by CC-SSI through the analysis of the stabilization diagram is shown in figure 8. In order to quantify the results convergence some statistical parameters were used. In particular, in the following, the modal frequencies calculated from each technique were compared through their standard deviation (SD) and their coefficient of variation (CV), which is defined as the percentage ratio of SD to the corresponding average frequency value. Moreover, In order to take into account that OMA techniques are not strictly independent methods, as they are based on similar assumptions, especially about the unknown input, the frequencies obtained by OMA methods were averaged and correlated separately to FRF and HVSR results, respectively.

On the basis of the linear regression obtained in these correlations a mutual evaluation of the validity of the results was carried out. In particular, the higher the correlation of the linear regression and the closer the correlation equation to the identity line, the more the results can be considered validated. All the obtained results are summarized in table 1, which shows the modal frequencies extracted with all the applied methods, along with some basic statistics (average, standard deviation SD, coefficient of variation CV).

Table 1. Modal frequencies calculated with all applied modal analysis technique and basic statistics (average, standard deviation SD, coefficient of variation CV) for the first five modes.

	Modal frequencies (Hz) by analysis technique							Statistics		
Mode	FDD	EFDD	CC-SSI	HVSR (N)	HVSR (E)	FRF (N)	FRF (E)	average	SD	CV
1	2.100	2.092	2.094	2.090	2.090	2.097	2.090	2.093	0.004	0.19%
2	2.476	2.454	2.473	2.470	--	2.506	--	2.476	0.019	0.76%
3	2.803	2.816	2.789	2.814	2.850	2.777	2.812	2.809	0.023	0.83%
4	3.555	3.568	3.591	--	3.570	--	3.662	3.589	0.043	1.19%
5	4.150	4.147	4.120	4.160	--	4.120	4.178	4.146	0.023	0.55%

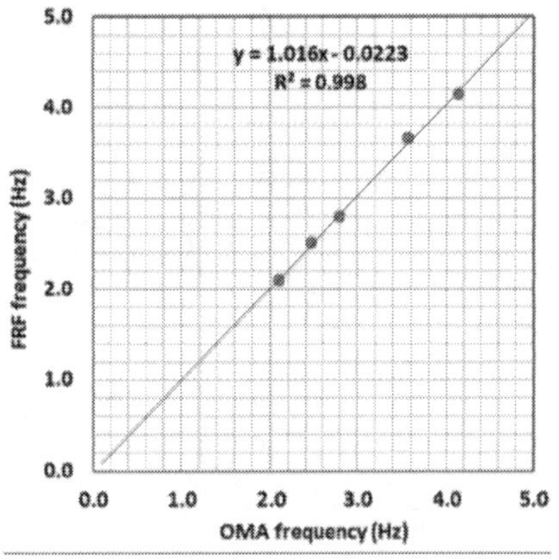

Figure 9. Correlation between OMA and FRF results.

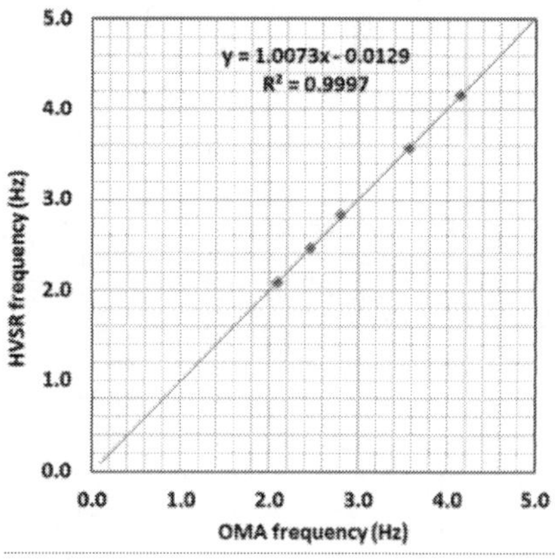

Figure 10. Correlation between OMA and HVSR results.

The figures 9 and 10 illustrate the correlation found of the average OMA to the average FRF and HVSR frequencies, respectively. Both correlations showed very high coefficients of determination R^2 with linear regressions very close to the identity line expressed by the relation $y = x$. These statistical comparisons substantially indicated that OMA, FRF and HVSR provided almost identical outcomes, which offers a solid basis for mutual validation of results.

5. Conclusions

The present work was conducted with the aim to assess the validity of the modal frequencies obtained by ambient vibration data acquired at very few measurement points of the so-called temple of Minerva Medica. The relevance, accuracy and validity of the provided modal frequencies of a monument are provided by the high reliability of the modal parameters for diagnostic and monitoring of the structure's state of damage. To such purposed a variety of modal analysis techniques were applied to the recorded dataset, namely OMA, FRF and HVSR methods, which are based on different theoretical assumptions. The results obtained through the applied methods were compared by basic statistical analysis for mutual validation. According to such statistical comparison, a very good agreement was found for the frequencies computed through all above methods, which substantially provided almost identical outcomes. Consequently, a solid basis for mutual validation of the results could be demonstrated, corroborating the reliability of the obtained findings.

Acknowledgments

The authors conducted the present study within the framework of the CO.B.RA. project funded by Lazio Region.

References

[1] Fasan M, Magrin A, Romanelli CF and Vaccari F 2016 A seismological and engineering perspective on the 2016 Central Italy earthquakes *Int. J. Earthquake and Impact Engineering* **1**(4)

[2] Roselli I, Fioriti V, Bellagamba I, Mongelli M, Tatì A, Barbera M, Magnani Cianetti M and De Canio G 2018 Impact of traffic vibration on the temple of Minerva Medica, Rome: preliminary study within the CO.B.RA. project *International Journal of Heritage Architecture* **2**(1) pp 102-14

[3] Gentile C and Saisi A 2007 Ambient vibration testing of historic masonry towers for structural identification and damage assessment *Construction and Building Materials* **21** pp 1311-21

[4] Garescì F, Catalano L and Petrone F 2006 Experimental Results of a Damage Detection Methodology using Variations in Modal Parameters *Exp. Mech.* **46**(4) pp 441-51

[5] Altunışık AC, Okur FY and Kahya V 2017 Modal parameter identification and vibration based damage detection of a multiple cracked cantilever beam, *Engineering Failure Analysis* **79** p 154-170

[6] Li B and Der Kiureghian A 2016 Robust optimal sensor placement for operational modal analysis based on maximum expected utility *Mechanical Systems and Signal Processing* **75** pp 155-175.

[7] Li DS and Li HN 2006 The state of the art of sensor placement methods in structural health monitoring *Smart Structures and Materials 2006: Sensors and Smart Structures Technologies for Civil, Mechanical, and Aerospace Systems* **6174** pp 1217–27

[8] Mongelli M, Roselli I, De Canio G and Ambrosino F 2016 Quasi real-time FEM calibration by 3D displacement measurements of large shaking table tests using HPC resources *Advances in Engineering Software* (in press)

[9] Roselli I, Fioriti V, Bellagamba I, Mongelli M, Tatì A, Barbera M, Magnani Cianetti M and De Canio G 2017 Urban transport vibrations and cultural heritage sites in Rome: the cases of the temple of Minerva Medica and of the Catacomb of Priscilla *WIT Transactions on Ecology and the Environment* **223** pp 335-43

[10] Capecchi D and D'Ambrogio W 1993 Experimental modal analysis and damage detection on an ancient masonry building *Meccanica* **28** pp 13-25

[11] Brincker R and Ventura CE 2015 *Introduction to Operational Modal Analysis* (Wiley Online Library)

[12] Yuen KV and Katafygiotis LS 2001 Bayesian time-domain approach for modal updating using ambient data *Probabilistic Engineering Mechanics* **16**(3) p 219–31

[13] Yuen KV and Katafygiotis LS 2001 Bayesian spectral density approach for modal updating using ambient data *Earthq. Eng. Struct. D.* **30**(8) pp 1103-23

[14] Rocklin GT, Crowley J and Vold H 1985 A Comparison of H1, H2, and HV Frequency Response Functions *Proc. 3rd Int. Modal Analysis Conf. (Orlando, Florida)*

[15] Gade S, Møller N, Herlufsen H and Konstantin-Hansen H 2005 Frequency Domain Techniques for Operational Modal Analysis *Proc. 1st Int. Operational Modal Analysis Conf. (Copenhagen, Denmark)*

[16] Jacobsen NJ, Andersen P and Brincker R 2006 Using Enhanced Frequency Domain Decomposition as a Robust Technique to Harmonic Excitation in Operational Modal Analysis *Proc. Int. Conf. on Noise and Vibration Engineering ISMA2006 (Heverlee, Belgium)*

[17] Møller N, Herlufsen H and Gade S 2005 Stochastic Subspace Identification Techniques in Operational Modal Analysis *Proc. 1st Int. Operational Modal Analysis Conf. (Copenhagen, Denmark)*

[18] Goursat M and Mevel L 2009 Algorithms for Covariance Subspace Identification: a Choice of Effective Implementations., *Proc. 27th Int. Modal Analysis Conf. (Orlando, Florida)*

[19] Nakamura Y 1989 A method for dynamic characteristics estimation of subsurface using microtremor on the ground surface *Quarterly Report of RTRI* **30** pp 25-33

[20] Gallipoli MR, Mucciarelli M and Vona M 2009 Empirical estimate of fundamental frequencies and damping for Italian buildings *Earthq. Eng. Struct. D.* **38**(8) pp 973–88

[21] De Canio G, de Felice G, De Santis S, Giocoli A, Mongelli M, Paolacci F and Roselli I 2016 Passive 3D motion optical data in shaking table tests of a SRG-reinforced masonry wall *Earthquakes and Structures* **40**(1) pp 53-71

[22] Barbera M, Magnani Cianetti M and Barrano S 2014 Da Massenzio a Costantino: le indagini in corso nel c. d. tempio di Minerva Medica *Proc. Int. Conf. of CISEM - La villa restaurata e i nuovi studi sull'edilizia residenziale tardoantica (in Italian)*

[23] Roselli I, Malena M, Mongelli M, Cavalagli N, Gioffrè M, De Canio G and De Felice G 2018 Health assessment and ambient vibration testing of the "Ponte delle Torri" of Spoleto during the 2016-17 Central Italy seismic sequence *Int. J. Civil Structural Health Monitoring* **8**(2)

IOP Conf. Series: Materials Science and Engineering **364** (2018) 012005 doi:10.1088/1757-899X/364/1/012005

A New Methodology for the Structural Analysis of 3D Digitized Cultural Heritage through FEA

Sara Gonizzi Barsanti, Gabriele Guidi
Department of Mechanics, Politecnico di Milano, Via La Masa, 1, 20156 – Italy
sara.gonizzi@polimi.it

Abstract. Finite Elements Analysis (FEA) is widely used for modelling stress behaviour in any mechanical system. The processing workflow starts from CAD 3D models representing the ideal shape of the object to be simulated. Such models are typically made of mathematical elements defining its geometrical components. Those are pre-processed before the simulation for creating a volumetric mesh out of the CAD model. Recently the use of FEA has also been extended to the simulation of ancient structures and artefacts, revealing significant potentialities for the conservation of Cultural Heritage. Unlike modern mechanical systems, heritage objects are usually altered by the time passed since their original creation, and the representation with a schematic CAD model may introduce an excessive level of approximation leading to wrong simulation results. In the last two decades, 3D documentation of CH has been developed through reality-based approaches. However, the related mesh models of the exterior surfaces are not proper for direct use in FEA. Such high-resolution surface meshes has to be converted to volumetric meshes made of tetrahedral or hexahedral elementary volumes and a limited number of external and internal nodes. The focus of this paper is on a new method aiming at generating the best possible 3D solid representation of a real artefact from its accurate reality-based surface model by reducing its number of nodes of several orders of magnitude while maintaining a geometrical coherence in the order of the measurement uncertainty of the 3D capturing technique used. The approach proposed is based on wise use of retopology procedures and a transformation of this retopologized model to a mathematical one made by NURBS surfaces, suitable for being processed by a volumetric mesh generator typically embedded in any standard FEM package. The resulting volumetric mesh allows obtaining FEA of ancient structures, providing a far better accurate simulation than those attainable by a rough CAD redrawing of the heritage asset of interest.

1. Introduction

For research, documentation, preservation and conservation issues in archaeology and Cultural Heritage, it is very important to appropriately record, document and survey artefacts and sites as an accurate and complete 3D digital documentation is a prerequisite for further analysis and interpretations.

In modern times the atmospheric agents, the urban growing, the density of constructions, the bombing during the two world wars, the negligence during the centuries and the present situation of political instability and danger in certain areas, have all contributed to damage and strongly influence stability of our Heritage. In this framework, it is fundamental to define a proper methodology to preserve our Heritage. In order to achieve diagnostic studies aiming at understanding the level of decay of Cultural Heritage for selecting the appropriate preservation methods and materials, a scientific base for allowing correct interventions would be fundamental. However, it's always difficult to predict how a historical object or structure, built up to a few thousand years ago, will suffer for environmental agents such as earthquakes, pollution, wind and rain, or human factors like constructions in the surroundings, vehicular traffic, or heavy tourism. Conservation of Cultural Heritage is a key topic and structural changes and damages can affect the structural behaviour of Cultural Heritage artefacts and buildings. Considering that the application of technologies can help in preserving, conserving and restoring ancient structures, it is mandatory to find the best pipeline to produce the correct analysis.

The analysis of the structural behaviour of a Cultural object can be provided with the use of Finite Element Analysis, a well-known engineering technique used in modelling stress behaviour of objects and structures. The typical workflow of such analysis involves the use of CAD 3D models made by Non-Uniform Rational B-splines (NURBS) surfaces, representing the ideal shape of the object to be simulated [1]. The major FEA packages has meshing modules capable to transform a NURBS model, made only of its exterior surfaces, to a volumetric mesh, which differently from a surface mesh typically generated with 3D capture methods, has nodes distributed both on the exterior surface and into the interior volume, connected each other by elementary volumes such as tetrahedron, pyramids, prisms or hexahedrons. This workflow is appropriate in the mechanical field, where a physical element to be simulated is very close to its ideal drawing within strict tolerances. Conversely, when applied to 3D models of Cultural Heritage (CH) objects or structures, often altered by the time passed since their original creation, the representation with a schematic CAD model may introduce an excessive level of approximation leading to wrong simulation results. Nowadays, 3D documentation of CH has been widely developed through active sensors or passive approaches like photogrammetry, but the models, formed by the exterior surfaces of the objects captured at high resolution and therefore represented by millions of polygons, are not suitable for a direct use in FEA. A few preliminary experiments have been done on real CH structures digitized with active or passive methods, whose models have been processed for simulating stress behaviour and predicting damage to artefacts considered critical within the field of conservation. The results are very promising but some issues have been made evident: a) the way for obtaining a volumetric mesh suitable for FEA from the raw 3D data is not yet clearly defined and may greatly influence the final result, b) the balance between geometric resolution and confidence level of the simulated results is often not compliant with the shape of a 3D model originated by a 3D acquisition process.

2. State of the Art

Preliminary experiments have been done on real CH structures digitized with active or passive methods, whose models have been processed for simulating stress behaviour and predicting possible damages to improve their conservation. By analysing the previous works, different approaches have been followed: a) redrawing with a CAD modeller a new surface model following the superficial mesh originated by the acquired 3D cloud; b) using directly the triangular mesh as starting point for volumetric models; c) generating a volumetric mesh from the point cloud with no preliminary surface meshing.

The first approach has been used for simulating the behaviour of the Trajan's Markets [2], and in many other applications [3;4]. In some cases, the reconstructed mathematical model was refined with the insertion of limited patches of reality-based superficial meshes [5]. An option of using a 3D CAD model extrapolated from a reality-based model is to extract or draw cross-sections or profiles from the 3D model to produce the CAD one as shown in [6], referred to the walls surrounding Montagnana, a small ancient city near Padua. An alternative option is passing through the procedures for producing BIM/HBIM models starting from the acquired 3D data, to be used for FEA. In [10] for example the use of BIM model as a support for sites monitoring and management is explored, extrapolating thematic information for structural analysis from the model, even if the authors didn't provide a Finite Element Analysis on the structure. The Masegra Castle, located near the city of Sondrio in Italy, was also used as a test object for a novel procedure called Cloud-to-BIM-to-FEM [11]. The BIM was turned into a finite element model with a geometric rationalization taking into consideration the preservation of irregularities and anomalies, such as verticality deviation and variable thickness. If this procedure, with all its accuracy limitations due to the re-draw of the model in a CAD software, can be applied to CH buildings because the geometry of the structure can be replicated through a CAD drawing using profiles, cannot be instead used for statues, whose geometry is more complex and that cannot be simplified through elements as bean, truss or shell, used for the modelling in FEA.

The second approach has a range of slightly different methods, such as bare simplification of the triangular surface mesh before converting it in a volumetric one, that may give significant deviations between the actual shape and the simulated one [7]. The simplified description of the shape in this case are discretized profiles giving a low-resolution representation of the interior and the exterior of the structure, from which produce a volumetric model or the fitting of the acquired model with a parametric model suitable to be converted in volumetric mesh. A recent project used the triangular mesh to perform the overall stability analysis of a marble statue when placed on its new pedestal and the mechanical stresses generated on the statue and the pedestal materials [8]. This paper presented a new way of creating the geometrical model of the statue, by subtracting the mesh from a prismatic block shape, according to a Boolean operation performed with an open source CAD software, to preserve the spatial positioning of the statue. Then, another open source software was used to create the tetrahedral mesh.

The third strategy does not even consider the mesh, generating a volumetric approximation of the shape from the raw 3D cloud of points [9]. A further evolution of this approach [12] is called FEA *in situ*, meaning that the algorithm performs the analysis directly on the 3D data without passing neither through the surface nor the volumetric mesh model. Some test on mechanical objects has been performed to validate the method.

3. Test objects

The tower of the carceres of the roman circus in Milan (fig. 1a) and two statue from the Uffizi Gallery (Figure 1a-b) have been used as test objects. The two statues were surveyed through photogrammetry using a SONY ILCE 6000 camera coupled with a 16mm lens while the tower with both photogrammetry for the interior with a Panasonic DMC-GH4 coupled with a 14mm lens (Table 1) and with laser scanning for the exterior with a FARO Focus 3D scanner.

(a) (b) (c)

Figure 1. The tower of the carceres in Milan (a) and the two statues held in the Uffizi Gallery: (b) the Gladiator, (c) the Buontalenti horse.

	ISO	Focal Length	GSD
Tower of the carceres	800	2.8	1mm
Horse	1000	5.6	0.5mm
Gladiator	1000	5.6	1mm

Table 1. The settings of the photogrammetric surveys of the three test objects.

4. Methodology

The starting point of the methodology proposed is the consideration of the complexity of doing structural analysis on Cultural Heritage. The difficulty lies on the geometry of the objects, compromised by the time passed, on the construction techniques and on the different materials used for buildings. Starting from the previous analysis and considering the lack of information that the methodology used since now carries on in performing structural analysis, the intrinsic characteristics of the different element involved in the pipeline and the non-linear and non-symmetric geometry of the structures, the methodology proposed took into account the use of retopology for the decimation of the reality-based 3D models. The pipeline starts from the post-processing of the acquired reality based models considering the correction of the topological errors and of the orientation of the model on a suitable reference system, usually with the XY plane corresponding to the base of the model, and the Z-axis passing from the centre of gravity of the artefact. This was considered a proper positioning for the following finite element analysis. In this

way, the high-resolution models used were closed meshes properly reoriented in order to post process them to suitable volumetric models for FEA. The second step regarded the use of retopology for generating a brand-new quad-mesh following the shape of the high-resolution models , but with an optimized arrangement of the polygons following the characteristic lines of the original model, and a strong reduction of the mesh density. The resulting quad mesh is a 3D objects lighter than the initial one and suitable for the transformation to NURBS. The conversion of a polygonal model in a NURBS model is available on some CAD packages as automatic process, that in general tends to produce a higher number of small patches when the original mesh is topologically unorganized. As a result, by rearranging the initial topology of the mesh, a preliminary condition for minimizing the number of NURBS patches of the converted model is set. In addition, the quadrangular structure of the polygons matches the intrinsic quadrangular structure of a NURBS patch, making easier the following mesh-to-NURBS transformation. Therefore, the resulting NURBS model represents a better starting point for the volumetric mesher embedded in any standard FEM packages.
After all these passages, the 3D models are finally ready for the finite element analysis.

4.1. Retopology

The retopology process is available both in open source packages such InstantMeshes [13] or Blender, or in commercial software packages such as ZBrush by Pixologic. Here, the software ZBrush was used. The main reason for the choice was the fact that this software is built specifically for rebuilding the topology of the models. Blender showed some problems in managing big files while InstantMeshes, even if powerful, gave as results sometimes inadequate meshes with holes and missing parts, especially when strongly decreasing the number of nodes. Moreover, ZBrush has the option of projecting the retopologized model on the high resolution one, increasing the adherence of the two models. The tool used for retopology was the ZRemesher that permits to strongly lower the number of polygons and reorder them on the surface, transforming also the typical triangular structure of an acquired mesh in a quad based one (Figure 2). The tool is optimized to work on all kind of structures and shapes but will by default produce better results with organic shapes. In the ZRemesher palette, there is the possibility to select the number of polygons desired for the retopology and the choice of increasing the coherence of the two models. This can be done selecting the "adapt" button and increasing the value of the "adaptive size" slider. This is an important parameter for ZRemesher, defines the polygon distribution on the model and can drastically increase the quality of the topology by giving more flexibility to the algorithm.

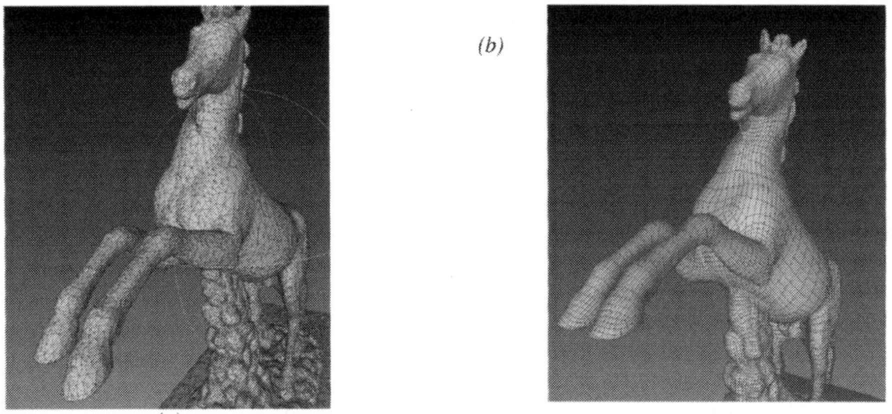

(a) *(b)*

Figure 2. Comparison among triangular decimation (a) and retopologized model (b) of the statue of the Buontalenti Horse in the Uffizi gallery. The horse was simplified from more than 1.1 mil triangular polygons to 35K quads.

This function defines a vertex ratio based on the curvature of the mesh. A low setting will provide polygons that are as much square as possible and almost the same size, an amount of final polygons closer to the number set in the selection tool but can introduce topology irregularities where the geometry is much complex. A high adaptive size means obtaining polygons that are rectangular in shape to best fit the mesh's curvature and which density can vary along the mesh surface even if the program creates smaller polygons where the geometry requires. So, increasing the value of the adapt size, the quality of retopology will increase but the program is more elastic about the target number of final polygons. This happens because when the desired number of polygons is set, the software distributes them equally on the surface and then analyses the curvature deforming the shape of the polygons or changing their density to be more adherent to the initial mesh.

4.2. NURBS conversion

A mesh represents 3D surfaces with a series of discreet faces, more likely as pixel form an image. NURBS, on the contrary, are mathematical surfaces, able to represent complex shapes with no granularity as in the mesh. The conversion from a mesh to a NURBS in implemented in CAD software or similar (e.g. 3DMax, Blender, Rhinoceros, Maya, Grasshopper, etc.) and it transforms a mesh composed by polygons or faces to a faceted NURBS surface. In details, it creates one NURBS surface for each face of the mesh and then merge everything into a single polysurface. A NURBS surface (patch) is obtained by a series of NURBS curves in two directions (called "U" and "V") interpolated to create a surface as the tensor product of two NURBS curves, originating a quadrangular patch. Depending on the mesh, the conversion works in different ways. If the starting point is a triangular mesh, and while, by definition, triangles are plane, the conversion creates trimmed or untrimmed planar patches. The degree of the patches is 1x1 surface trimmed in the middle to form a triangle. If the starting point is a quadrangular mesh, the conversion creates a 4-sided untrimmed degree1 NURBS patches, meaning that the edges of the mesh are the same as the outer boundaries of the patches. Considering the theory, a quadrangular mesh is more suitable to be converted in NURBS and by rearranging the initial topology of the mesh, a preliminary condition for minimizing the number of NURBS patches of the converted model is set, and this represents a better starting point for the volumetric mesher embedded in any standard FEM package.

The use of NURBS for creating volumetric models for FEA depends on the ability of NURBS to represent exactly a given geometry at low order with a high refinement of the mesh, that allow to have a higher accuracy in the Finite Element Analysis without changing the geometry [14].

4.3. FEA analysis

The two statues were retopologiezed, compared to the high resolution one, and then converted in NURBS and imported in the FEA software for two different analysis: a static one imposing the gravitational load and a modal one, fixing 10 frequencies to determine the natural mode shapes and frequencies of the object during free vibration. The comparison on the horse between the high resolution model of 2.2Mvertices gave a 2mm mean and a standard deviation of 2mm (Figure 3) on the 26K retopologized model. The conditions set for the Finite element analysis were (Figure 4):

- Density: 2500 kg/m^3,
- Young's Modulus for marble 78000 MPa,
- Poisson Ratio 0.3, Gravity on -Z axis,
- Fixed Support under the basement of the statue as boundary condition,
- Meshing element: 10-nodes Tetrahedrons
- Element size: 50 mm;

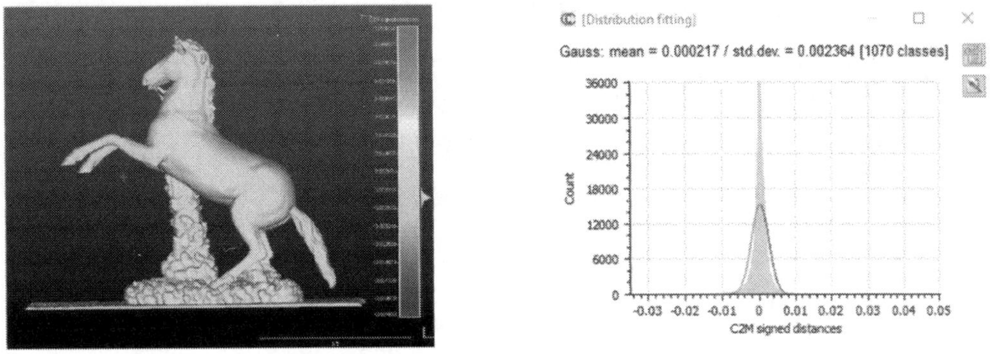

Figure 3. A comparison between the high resolution and the retopologized model of the horse.

(a) (b) (c)

Figure 4. The analysis on the horse statue: static structural imposing the gravity on –Z axis, maximum principal stress (a), minimum principal stress (b) and the modal analysis (c).

The high-resolution model of statue of the gladiator had 1,1M faces and was simplified to 30K nodes. The comparison gave a mean of 0.1mm and a standard deviation of 0.4mm (Figure 5). The same condition was imposed in the structural analysis with a dimension of the volumetric element of 16mm (Figure 6).

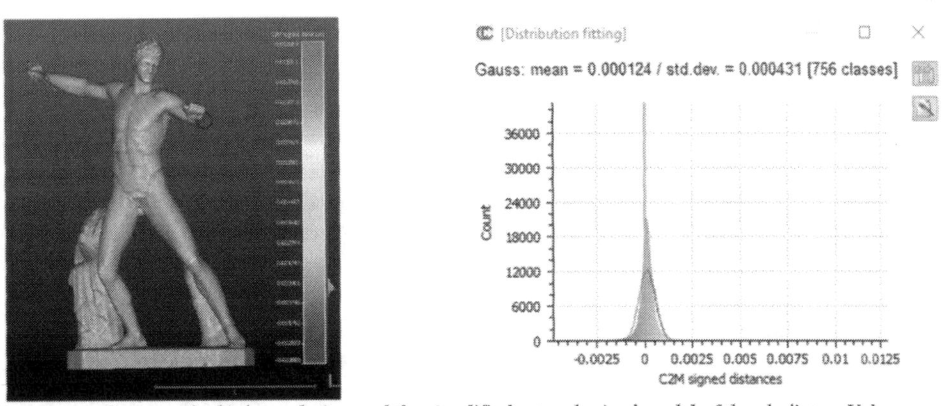

Figure 5. Comparison between the high resolution and the simplified retopologized model of the gladiator. Values expressed in meters.

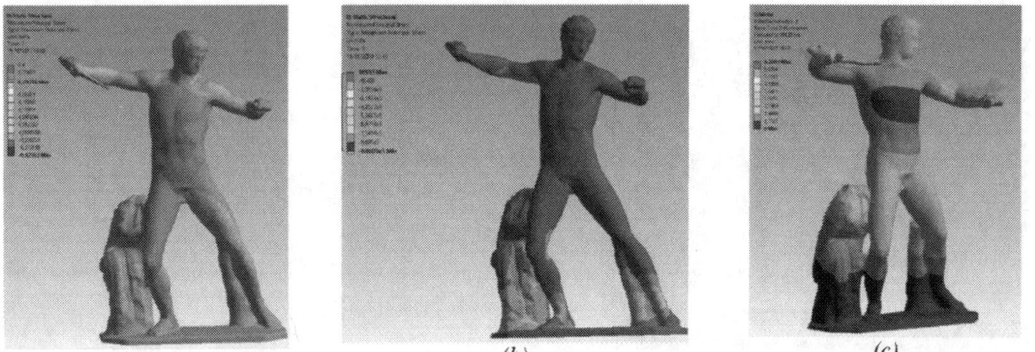

(a) (b) (c)

Figure 6. The static structural analysis imposing the gravity on the –Z axis (a) showing the maximum principal stress, (b) the minimum principal stress and (c) the modal analysis for the statue of the gladiator.

The high-resolution model of the tower of the carceres counted more than 7M vertices, while the retopologized one after all the topological correction was composed by 56K vertices. The comparison between the high-resolution model and the retopologized one gave a mean of 0.03mm and a standard deviation of 0.2mm, both acceptable for a structure 30 m high and considering the fact that the final model is the ensemble of two techniques (Figure 7). A static structural and a modal analysis were done with the following parameters: Young modulus for masonry 8000 MPa, Density 2000 kg/m^3 and Poisson Ratio 0.15. The dimension of the 10-nodes tetrahedral element in the meshing part of the FEA was 10mm, coherent with the dimension of the superficial mesh element and the static structural analysis was performed imposing the gravity on – Z axis (Figure 8).

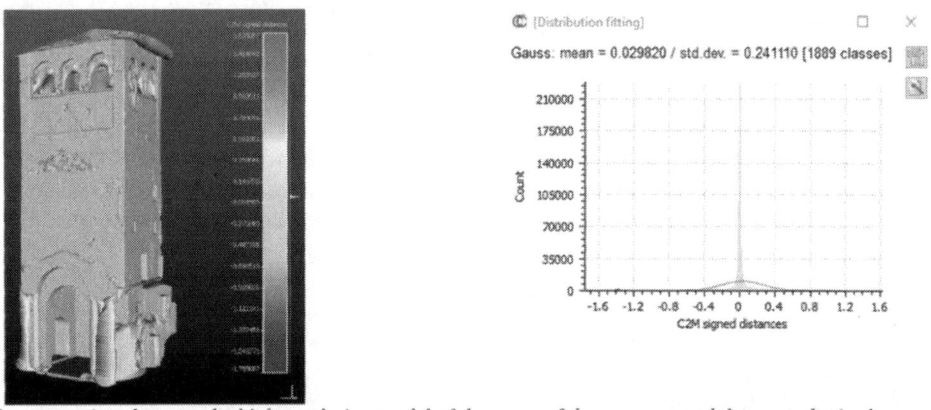

Figure 7. The comparison between the high-resolution model of the tower of the carceres and the retopologized one.

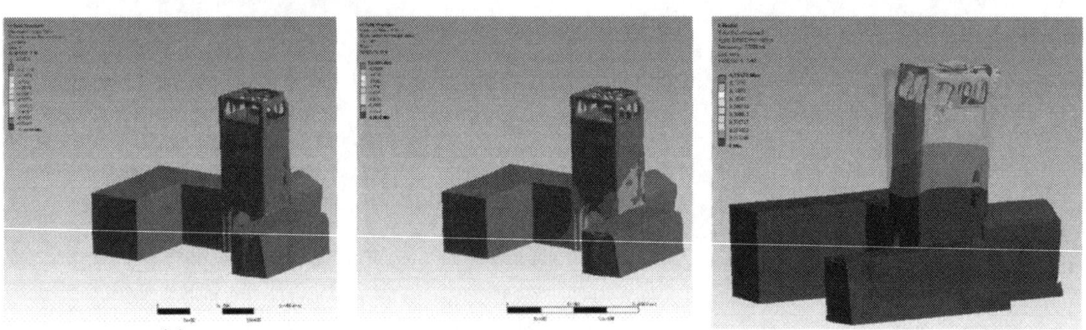

(a) (b) (c)

Figure 8. The FEA on the tower: (a) static structural imposing gravity showing the Maximum Principal Stress; (b) the same analysis showing the Minimum Principal Stress; (c) the modal analysis.

IOP Conf. Series: Materials Science and Engineering **364** (2018) 012005 doi:10.1088/1757-899X/364/1/012005

5. Conclusions

The use of retopology and the method proposed showed great performances and presented interesting properties in the process as a consistent pre-processing step to directly use a reality-based model for Finite Element Analysis. The use of the 3D measurement uncertainty as simplification criterion, allowed to dramatically reduce the mesh size maintaining a high accuracy of the simplified model compared to the high resolution one. This was the first important point to be achieved in the research. Starting from the idea to take advantage of the high resolution and of all the geometrical details of the reality-based models for the structural analysis of complex artefacts as Cultural Heritage objects and buildings, the possibility to transform all these information into a model that can be processed in FEA software is the most important result. When dealing with complex objects, a proper topological analysis is needed and this can be easily done with specific software as for example Meshlab. The important thing is to obtain a clean, closed mesh to be converted in a closed NURBS, in order to have a volumetric model for FEA.

References

[1] Höllig, K. (2003). Finite Element Methods with B-Splines. Society for Industrial and Applied Mathematics, Stuttgart, Germany.

[2] Brune, P., Perucchio, R. (2012). Roman Concrete Vaulting in the Great Hall of Trajan's Markets: A Structural Evaluation. J. Archit. Eng. 18, 332–340.

[3] Erkal, A., Ozhan, H.O. (2014). Value and vulnerability assessment of a historic tomb for conservation. Sci. World J. 2014, Article ID 357679.

[4] Milani, G., S. Casolo, A. Naliato, and A. Tralli. 2012. Seismic assessment of a medieval masonry tower in northern Italy by limit, non-linear static and full dynamic analyses. International Journal of Architectural Heritage 6(5):489–524.

[5] Zvietcovich, F., Castaneda, B., Perucchio, R. (2014). 3D solid model updating of complex ancient monumental structures based on local geometrical meshes. Digit. Appl. Archaeol. Cult. Herit. 2, 12–27.

[6] Guarnieri, A., Pirotti, F., Pontin, M. & Vettore, A. 2005. Combined 3D surveying techniques for structural analysis applications. International archives of photogrammetry, remote sensing and spatial information sciences, 36, 6.

[7] Riccardelli, C., Morris, M., Wheeler, G., Soultanian, J., Becker, L., Street, R., 2014. The Treatment of Tullio Lombardo's Adam: A New Approach to the Conservation of Monumental Marble

[8] Bagnéris, M., Cherblanc, F., Bromblet, P., Gattet, E., Gügi, L., Nony, N., Mercurio, V., Pamart, A., A complete methodology for the mechanical diagnosis of statue provided by innovative uses of 3D model. Application to the imperial marble statue of Alba-la-Romaine (France), In Journal of Cultural Heritage, 2017 , ISSN 1296-2074,

[9] Shapiro, V., Tsukanov, I. (1999). Mesh free simulation of deforming domains. CAD Comput. Aided Des. 31, 459–471.

[10] Oreni, D., Brumana, R.,Cuca, B., "Towards a methodology for 3D content models: The reconstruction of ancient vaults for maintenance and structural behaviour in the logic of BIM management," 2012 18th International Conference on Virtual Systems and Multimedia, Milan, 2012, pp. 475-482.

[11] Barazzetti, L., Banfi, F., Brumana, R., Gusmeroli, G., Previtali, M., Schiantarelli, G., Cloud-to-BIM-to-FEM: Structural simulation with accurate historic BIM from laser scans, In Simulation Modelling Practice and Theory, Vol. 57, 2015, pp 71-87, ISSN 1569-190X.

[12] Freytag, M., Shapiro, V., Tsukanov, I., Finite element analysis in situ, In Finite Elements in Analysis and Design, Volume 47, Issue 9, 2011, Pages 957-972, ISSN 0168-874X.

[13] Jakob, W., Tarini, M., 2015. Instant Field-Aligned Meshes. Siggraph Asia 34, 15

[14] Hughes, T., Cottrell, J, A., Bazilevs, Y., Isogeometric analysis: CAD, finite elements, NURBS, exact geometry and mesh refinement, Comput. Methods Appl. Mech. Engrg. 194 (2005) 4135–4195, 2005

IOP Conf. Series: Materials Science and Engineering **364** (2018) 012006 doi:10.1088/1757-899X/364/1/012006

Numerical evaluation of seismic-induced damages in masonry elements of historical arch viaduct

I Drygala[1, a], J Dulinska[1], Ł Bednarz[2] and J Jasienko[2]

[1]Cracow University of Technology, Warszawska 24, PL 31-155

[2]Wroclaw University of Science and Technology, Plac Grunwaldzki 11, Pl 50-377

[a] Corresponding author's e-mail address: idrygala@pk.edu.pl

Abstract. The main purpose of this research was to estimate the seismic response of the historical masonry arch viaduct located in Krakow (Southern Poland) subjected to the representative strong seismic event. For the numerical evaluation a 3D finite element model (FEM) of the masonry viaduct was assembling with the ABAQUS/Standard software program. To represent nonlinear behaviour of the masonry structural elements of the viaduct under the strong seismic event, a Barcelona Model (BM) was assumed as a constitutive material model. Three components of the registered strong seismic shock were applied as kinematic excitation acting in three directions. The results of the numerical simulation showed that evident nonlinear behaviour of the masonry structural elements of the viaduct was observed under the earthquake. The plastic strains as well as the tensile damage (cracking) were obtained in some zones of the walls of the structure after seismic shock. On the basis of the obtained results authors also discussed the methodology for structural health monitoring (SHM) as well as seismic retrofitting which could be dedicated for historical masonry arch viaducts and bridges.

1. Introduction

Historical arch masonry viaducts as well as bridges erected over the centuries are usually monuments of bridge engineering of significant historic value. In the fast majority of cases, these structures are still used in public road infrastructures. Because of their historical value as well as serviceableness the protection of these type of masonry bridges which are erected in the seismic activity areas is significant task for modern civil engineers [1,2].

The studies of destructions of masonry arch bridges located in areas of strong seismic activity have disclosed that the safety of these objects turned out to be a substantial technical problem [3]. On the basis of the observations it is clearly visible that the numerical studies of the masonry bridges exposed to seismic shocks are reasonable and recommended. Additionally, the detection of damages in masonry elements of structures which is supported by numerical tools is effective as well as sufficient, i.e. this technique makes possible to create project of the reinforcement of the structures elements which are threatened [4].

The main objective of this investigation was the evaluation of the dynamic response of an arch masonry viaduct located in Krakow (Southern Poland) subjected to the strong representative seismic event registered in the region of Abruzzo (central Italy) [4]. In order to evaluation the impact of the shock on the structure, the Barcelona Model of material, which describes multi-hardening plasticity and progressive damage (cracking), was implemented [5]. On the basis of the obtained results, the scenario of reinforcement and health structure monitoring (SHM) solutions for similar structures was proposed.

2. Dimensions of a historical arch bridge and its finite element model (FEM)

The masonry bridge which was chosen as a case study is located in Krakow (Southern Poland). The structure was erected in 1863 and was designed by the Cracow architect Leon Mikucki. Primary purpose of the bridge was conduct the railway above the historical riverbed of Vistula. Nowadays, since the old riverbed of Vistula was buried, the bridge works as a viaduct above the road. The studied viaduct is a one of the oldest bridges in Krakow.

The structure consists of five masonry arch spans of approximately 11.4 m span. In the Fig. 1 the structural layout of the viaduct is presented. The structure of the viaduct consists with two types of parts, i.e. masonry as well as stone elements.

For the research purposes the 3D finite element model (FEM) of the structure was prepared with the ABAQUS/Standard software program. In the 3D FEM only parts that that constitute the primary structural system were taken into consideration. For all structural parts the quadratic hexahedral elements of type C3D20R were applied. The total number of elements is equal to 71820, the total number of nodes is equal to 350498 respectively. The FEM of the viaduct is shown in Fig. 2.

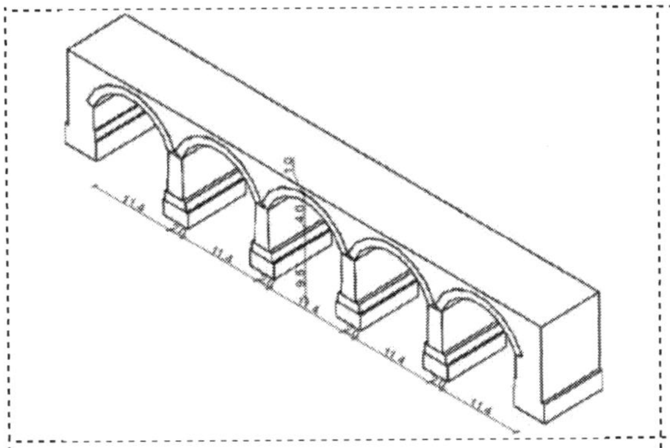

Figure 1. Structural layout of the arch bridge.

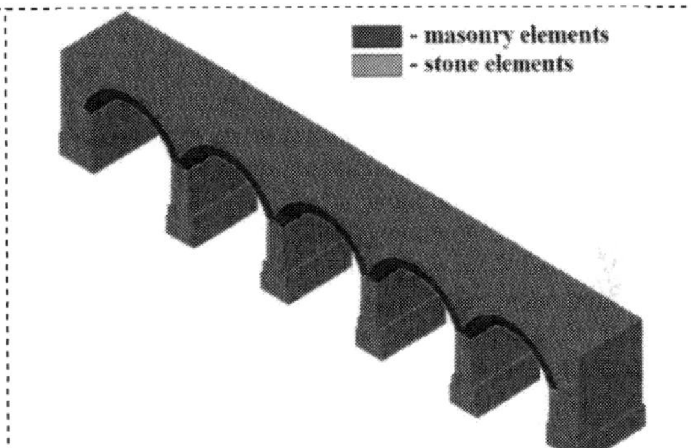

Figure 2. FEM of the arch bridge.

The considered viaduct is assembled with two types of materials, i.e. masonry and stone elements. The main intention of the presented research was investigation of seismic-induced damages in masonry elements of the arch viaduct. Because of this only for masonry arches the non-linear material behaviour was examined. In the numerical simulations, which were conducted in ABAQUS/Standard software program, the concrete damage plasticity model (CDP), also known as the Barcelona Model (BM), was applied [6,7]. This non-linear material model is dedicated to evaluation the response of structures subjected to cyclic loading. It must be pointed out that the CDP is very useful tool in evaluation of seismic-induced damages in structures, especially bridges. The CDP (BM) is dedicated to reinforced concrete structural elements and it represents the isotropic character of the concrete material. The most important assumption of the CDP (BM) is based on the mechanism of failure, i.e. tensile cracking and compressive crushing of the analysed structure. Although the nature of masonry elements is different than reinforcement concrete elements (masonry is anisotropic material), the BM has found the useful application in seismic numerical simulations of masonry structures. However, using simply rescaling of its yield stress is not appropriate solution. The suggestions and conditions for using BM for masonry elements were presented in [5]. The constitutive parameters of the BM were adopted on the basis of literature [5,8,9] and were also compared with results obtained from own research [10]. The parameters which were used in the numerical simulations are summarized in Table 1. For the stone parts of the viaduct the material parameters were assumed as follows: E=10.00 GPa, ϱ=2300 kg/m^3 and ν=0.18 [11].

Table 1. The constitutive parameters of the BM.

Compressive strength f_c *[MPa]*	Tensile strength f_t *[MPa]*	Elasticity modulus *E [GPa]*	Poisson's ratio ν *[-]*	Dilation angle φ *[°]*	Mass density ρ *[kg/m^3]*

12.3	2.0	9.48	0.16	38°	2000

3. Modal parameters of the bridge and characteristic of the seismic event

As products of the first stage of the numerical simulations, the four natural frequencies and mode shapes of the viaduct are presented. The obtained modes were compared with the results provided by different authors [11]. The values of the natural frequencies and modes of vibrations are summarised in Fig. 3.

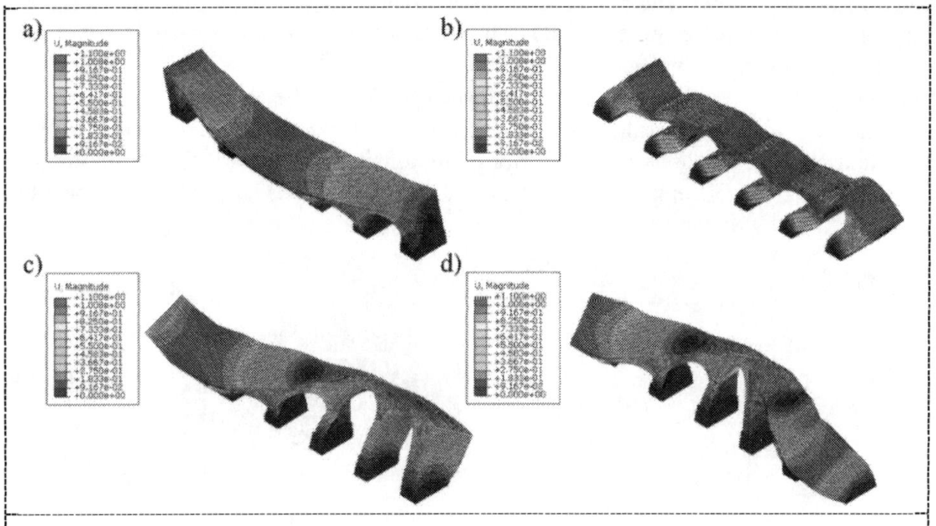

Figure 3. The mode shapes of vibrations of the viaduct:
a) f_1=7.84 *[Hz]*; b) f_2=7.94 *[Hz]*; c) f_3=9.83 *[Hz]*; d) f_4=12.40 *[Hz]*.

In the presented study, the representative earthquake data registered in the region of Abruzzo (central Italy) were applied as kinematic excitations of the viaduct (ITACA 2017) [12]. The magnitude of the applied seismic event was 5.3. The phase of the strong ground motion approximately lasted 10 s. The frequency characteristics of the seismic shock are presented in Fig. 7. It is clearly visible, that dominant frequencies of the studied earthquake (1-5 Hz, see Fig. 5) are lower than first natural frequency of the viaduct (f_1=7.84 Hz). Time histories of accelerations of the in three directions are presented in Fig. 4. In the Tab. 2 the peak ground accelerations (PGA) of the seismic shock are summarized.

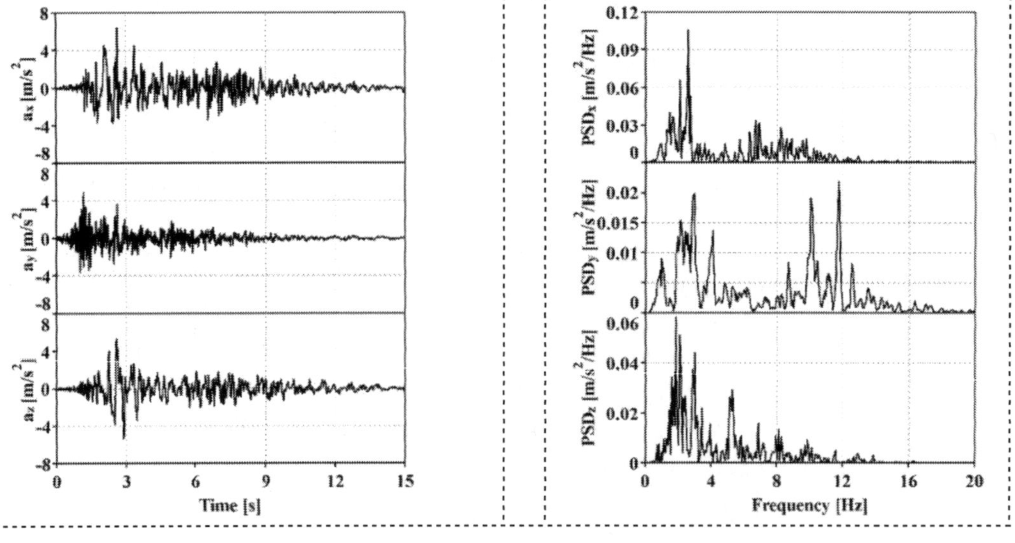

Figure 4. Time histories of the seismic shock accelerations [m/s²] in three directions: WE direction (a_x), vertical direction (a_y) and NS direction (a_z).

Figure 5. The frequency characteristics of the seismic shock PSD [m/s²/Hz].

Table 2. The peak ground accelerations (PGA) of the seismic excitation [12].

The peak ground accelerations (PGA) [m/s²]		
WE direction	vertical direction	NS direction
6.44	4.87	5.35

4. The evaluation of seismic-induced damages in masonry elements of the viaduct

For the numerical simulations the Rayleigh model of stiffness and mass proportional damping was applied. In this damping model the coefficients α and β refer to mass and stiffness matrices are used. The damping coefficients α = 2.7391 and β = 0.0009 were estimated for damping ratios of 5.0 % for the first (7.84 Hz) and the third (9.83 Hz) natural frequencies. Damping ratio values was adopted for masonry bridges [11]. Fallowing equation (1) was used to calculate the damping coefficients α and β:

$$\frac{1}{2}\begin{bmatrix} \frac{1}{2\pi f_i} & 2\pi f_i \\ \frac{1}{2\pi f_j} & 2\pi f_j \end{bmatrix}\begin{Bmatrix}\alpha\\ \beta\end{Bmatrix}=\begin{Bmatrix}\varsigma_i\\ \varsigma_j\end{Bmatrix} \tag{1}$$

The non-linear dynamic response of the viaduct to the seismic shock was evaluated with the ABAQUS/Standard software program. The Hilber-Hughes-Taylor time integration method dedicated for a direct step-by-step solution was used for full time history analysis. As the BM of the masonry material implements strong nonlinearity a step of integration in numerical procedure was not fixed. The step varied from 10^{-8} to 10^{-2} s, according to convergence requirements. The time of the dynamic analysis was 10 s. In the numerical investigation, the geometric nonlinearity was also taken into account.

As a measurements of damages in masonry elements of the viaduct the time histories of BM variables were obtained, as follows:

- LEP (principal logarithmic strain).
- PEEQT (tensile equivalent plastic strain).
- DAMAGET (tensile damage variable).
- PEEQ (compressive equivalent plastic strain).
- DAMAGEC (compressive damage variable).
- SDEG (stiffness degradation variable).

For strongly geometry as well as material non-linearity one of the most useful variable, which takes into account the continuous variation of length, is principal logarithmic strain (LEP). In the case of the BM implemented in ABAQUS/Standard software, equivalent plastic strains in tension (PEEQT) and in compression (PEEQ) are separate measures of plastic action under tension and compression loading. The tensile damage variable (DAMAGET) indicates damage occurring in the elements due to cracking. The compressive damage variable (DAMADEC) shows damages in areas of studied elements due to crushing. The total loss of strength is detect by value of the stiffness degradation variable (SDEG).

As a final products of the numerical simulations, the distributions of BM variables of the studied parts related to the masonry arches of the viaduct are presented (see Fig. 6). It was observed, that the most damages occurred on the outer arches of the structure (see Fig. 6e). It is also clearly visible in Fig. 6 that the crack appeared on typical areas for masonry arch type structures [1,2,3].

In the presented research, plasticisation areas on the arches of the viaduct caused by the discussed strong seismic event appeared due to tensile as well as compressive loading (see Fig. 6a and 6b). Taken into consideration the distribution of damage variables, i.e. DAMAGET and DAMADEC, the tensile harm areas are dominant (see Fig. 6c). However, the damages caused by compression effects also appeared (see Fig.6d). To show the plastic behaviour and the damages of the selected areas of the arch elements

IOP Conf. Series: Materials Science and Engineering **364** (2018) 012006 doi:10.1088/1757-899X/364/1/012006

the time histories of BM variables are presented. As a representative elements of the investigated viaduct, the areas located around the points A, B, C and D were chosen to discuss (see Fig. 6).

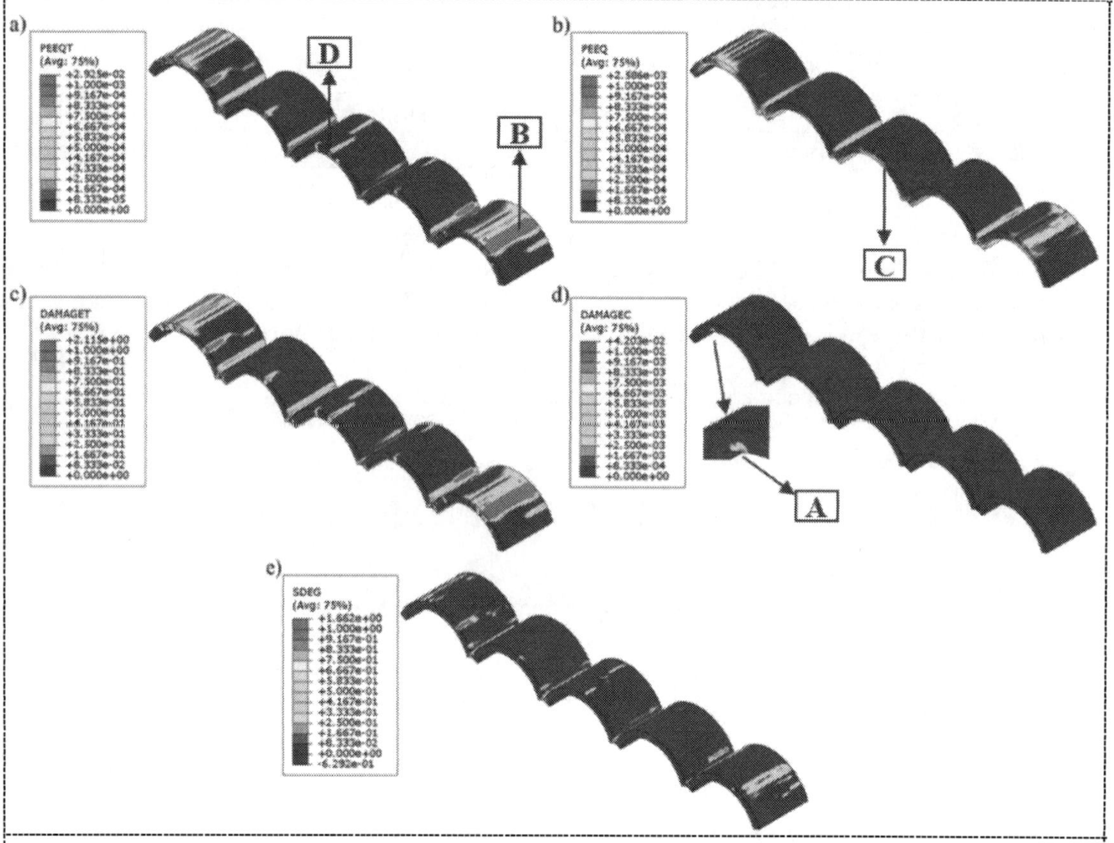

Figure 6. Distribution of BM variables on arches: a) PEEQT – tensile equivalent plastic strain; b) PEEQ – compressive equivalent plastic strain; c) DAMAGET – tensile damage variable, d) DAMAGEC – compression damage variable; e) SDEG - stiffness degradation variable.

For all selected points, i.e. A, B, C and D (see Fig. 6), the time histories of BM variables as follows are delivered: SDEG, PEEQT, DAMAGET, PEEQ, DAMAGEC and LEP. In the Fig. 7 the scenario for areas around the point A (see Fig. 6d) is showed. For this part of the masonry arch, the destructions and the loss of stiffness caused by tensile as well as compressive loading were observed. However, the maximum value of the DAMAGEC variable is much more lower than maximum value of the DAMAGET variable. For the point A the SDEG, PEEQT, DAMAGET and PEEQ variables increased with the biggest progress between 2nd and 4th second of the seismic shock (see Fig. 7). In this range of time the seismic event had the highest intensity (see Fig. 4). For the DAMAGEC variable the growth of its value was observed in 8th second of the seismic shock (see Fig. 7). The intensity of the shock in this space of time was much lower than in the first range (see Fig. 4). For zones around B, C and D points (see Fig. 6) damages caused by the compression loading were not observed. For this parts of the structure the damages described by the SDEG, PEEQT, DAMAGET and PEEQ variables occurred between 2nd and 4th second of the seismic shock (see Fig. 8-10).

Taken into consideration the numerical results, especially the distribution of damages on the structure (see Fig. 6), the preparation of the proposal of strengthening is possible. The type of the studied structure as well as its historical value caused that the techniques which would allow the least possible intervention in the historical tissue of the building are possible. This was the reason for opting for the C-FRCM system, which consists of a carbon mesh embedded in mineral matrix [13]. This solution could be used to strengthen the masonry elements, especially arches. In this method, the mesh is applied to properly prepared structure surface and requires also adequate access and work space [14,15]. The SHM methodology should include an initial survey and a visual assessment, especially of the zones with the highest number of damages.

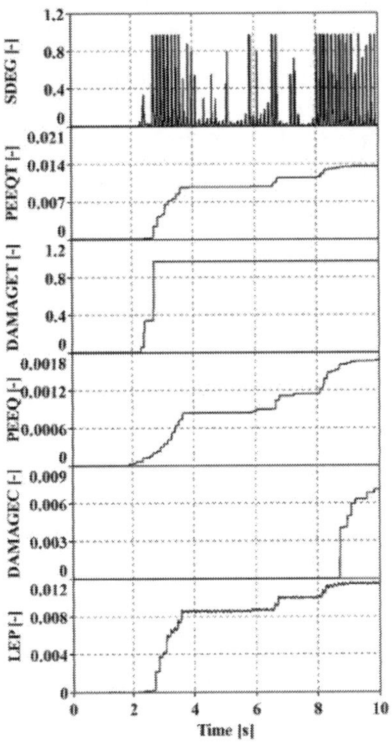

Figure 7. Time history of BM variables at point A due to the seismic shock.

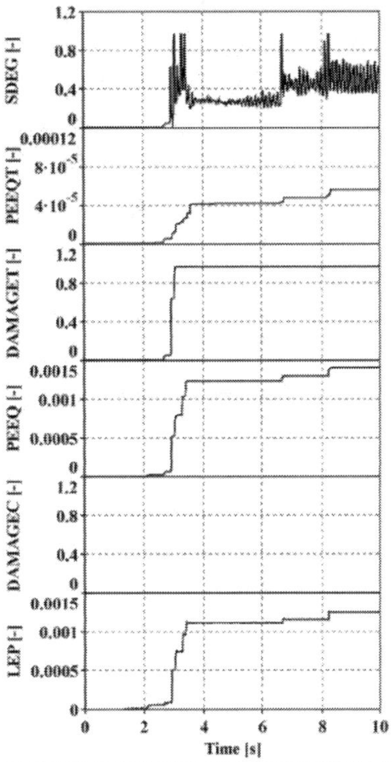

Figure 8. Time history of BM variables at point B due to the seismic shock.

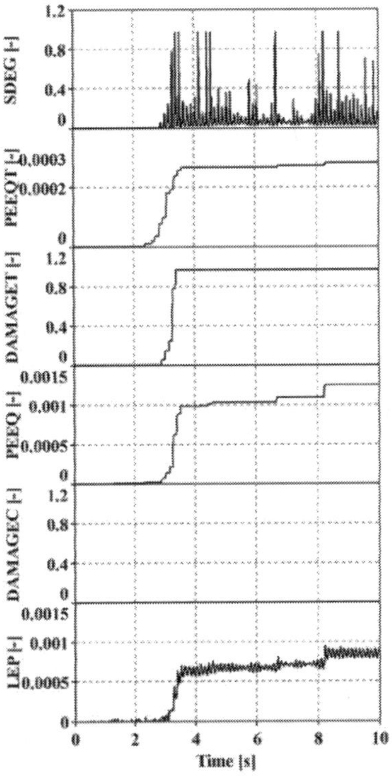

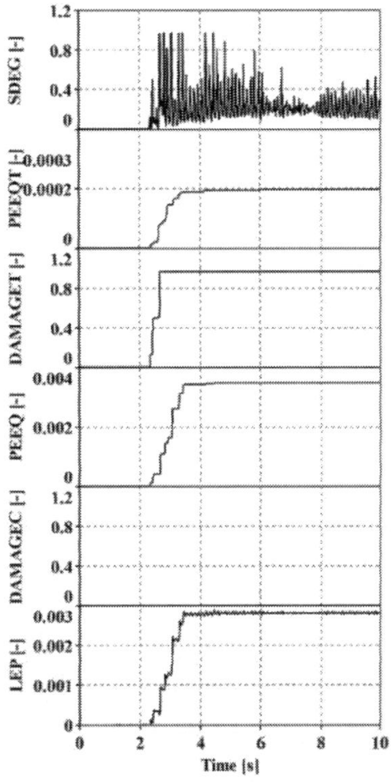

IOP Conf. Series: Materials Science and Engineering **364** (2018) 012006 doi:10.1088/1757-899X/364/1/012006

Figure 9. Time history of BM variables at point C due to the seismic shock.

Figure 10. Time history of BM variables at point D due to the seismic shock.

5. Conclusions

The main objective of the presented research was the evaluation of seismic-induced damages in masonry elements of historical arch viaduct. The existing masonry viaduct, erected in Krakow (Poland), was chosen as a case study. The numerical simulations with BM of material were conducted with ABAQUS/Standard software. The delivered analysis proved that a typical strong seismic event can caused significant damages in structural elements of this type of bridges. The following conclusions concerning seismic behaviour of masonry arch type viaduct under the strong seismic event can be formulated:

- The numerical simulations with the non-linear material models are efficient and practical tools. Taken into consideration the results (see Fig. 6-10), it is possible to prepare the proposal of strengthening of the structure as well as the SHM methodology.
- Implementation of the Barcelona Model (BM) allowed to analyze strongly nonlinear seismic performance of masonry elements of the investigated viaduct in detail. Different scenarios of the dynamic performance of the masonry material occurred in arches of the viaduct.
- The propagation of plastic and damage measures allowed to assess the impact of the seismic shock on masonry arches of the viaduct. It turned out that the investigated seismic event which PGA is 0.65 g was strong enough to cause cracking in the masonry elements of the structure.

6. References

[1] Karaton M Aksoy H S Sayın E and Calayır Y 2017 *Nonlinear seismic performance of a 12th century historical masonry bridge under different earthquake levels* Engineering Failure Analysis (vol 79) pp 408-421

[2] Santis S and Felice G 2012 *Seismic analysis of masonry arches* Proc. 15th World Conference on Earthquake Engineering

[3] Kiyono J and Furukawa A 2012 *Seismic Assessment of Stone Arched Bridges* 15th World Conference on Earthquake Engineering

[4] Drygala I Dulinska J Bednarz Ł and Jasienko J 2017 *Seismic performance of a historical apartment building using a Barcelona Model (BM) adapted for masonry walls* Key Engineering Materials (vol 747) pp 646-652

[5] Wawrzynek A and Cińcio A 2006 *Numerical verification of the Barcelona Model adapted for brick walls* Proc. of the 7th International Masonry Conference

[6] Lee J and Fenves GL *Plastic-damage model for cyclic loading of concrete structures* Journal of Engineering Mechanics (vol 124) pp 892-900

[7] Jankowiak T and Łodygowski T *2005 Identification of parameters of concrete damage plasticity constitutive model* Foundations of Civil and Environmental Engineering (vol 6) pp 53-69

[8] Cińcio A 2004 *Numerical analysis of dynamic resistance on semi-seismic tremors of low buildings with application of spatial object models* PhD Thesis Library of SUoT

[9] Passowicz A Norberciak D and Sielicki P 2005 *Analiza numeryczna konstrukcji Kościoła Najświętszej Marii Panny na Ostrowie Tumskim w Poznaniu (in Polish)* Master Thesis Library of Poznań University of Technology

[10] Bednarz Ł 2008 *Praca statyczna zabytkowych, zakrzywionych konstrukcji ceglanych poddanych zabiegom naprawy i wzmacniania (in Polish)* PhD Thesis Library of Wrocław University of Science and Technology

[11] Dulińska J 2006 *Odpowiedź dynamiczna budowli wielopodporowych na nierównomierne wymuszenie parasejmiczne pochodzenia górniczego* (in Polish) Wydawnictwo PK

[12] ITACA - The Italian Accelerometric Archive, version 2.1. [Internet]. c2016. ITACA working group; [cited 2017 February 20]. Available from: http://itaca.mi.ingv.it/.

[13] Bednarz Ł Jasieńko J Rutkowski M and Nowak 2014 *Strengthening and long-term monitoring of the structure of an historical church presbytery* Engineering Structures (vol 81) pp 62-75

[14] Barbieri A Borri A Corradi M and Di Tommaso A 2002 *Dynamic behaviour of masonry vaults repaired with FRP: experimental analysis* Proc. 6th International Masonry Conference
[15] Mrozek M 2012 *Numeryczna symulacja wzmacniania matami CFRP konstrukcji murowych z cegły (in Polish)* PhD Thesis Library of SUoT

Light, Information and Perception inside Historical Buildings. A Case Study.

Carla Balocco[1], Alessandro Farini[2], Elisabetta Baldanzi[2] and Giulia Volante[1]

1 Industrial Engineering Department, University of Florence, Firenze, Italy

2 CNR-INO of Florence, Italy

Corresponding author: carla.balocco@unifi.it

Abstract. Many literature studies demonstrated that technologies belonging to cognitive neuroscience can be used in museums. Physics, optics, thermodynamics and neurosciences applications can provide an important support to applied and practical research for lighting our Cultural Heritage. In our research we provided a multidisciplinary integrated approach for the study of the luminous climate inside a historical building: Villa La Quiete in Florence is the case study. The quantitative measurements of the relationship between observer and artworks were performed, with eye-tracking technique application. The Information Theory, read on a thermodynamic basis, the ergonomy of the multi-perceptive learning and optical physics, were the fundamental tools for the assessment of the correct light sources in terms of spectral emission and colour light temperature. The eye-tracking technique combined with the results of lighting parameter quantification, allowed checking how the colour of light changes the observer's perception, and from the information theory point of view, the communication and interpretation process of the signals due to different lighting. At the same time, it was also possible to measure by assessing the perceptive data of the visual path, and thus the neg-entropy, the informative content of the interaction of light with works of art.

Keywords: Lighting, Eye-tracking, Perception, Information, thermodynamics, LED

1. Introduction

The Sciences of the Cultural Heritage have oriented many studies to the relationship between observer and work of art, the latter meant in the broad sense (space, building, object of historical artistic value, etc). The measure of this relationship can be made with the eye-tracking technique and therefore aimed at a project based on forms of fruition and enhancement that will be convincing, effective and sustainable [1,2,3]. It is well known that the interaction between observer and object is a complex system of *top-down* cognitive processes (guided by knowledge, empirical, tactile, visual and previous kinesthetic experiences, semantic/cognitive representations and syntactic representations of the subject), and *bottom up* (guided by the optical, photometric, colorimetric properties of the object, but above all by light, luminance, spectral distribution and colour that invest it in the "perceptive neutrality" of the subject; [3,4]). Most of the literature concerns eye-tracking applications for ophthalmology, psychology, commercial research, medicine and psychiatry [5,6]. It is also used for people suffering from serious diseases or neurodegenerative diseases, because it allows communication only with ocular movements.

IOP Conf. Series: Materials Science and Engineering **364** (2018) 012007 doi:10.1088/1757-899X/364/1/012007

Lately the eye tracker is also used for tool usability: e.g. to evaluate how and how much a website is visited [7,8,9].

The aim of our present research is to investigate the physical connection between light, perception and information/communication by an experimental methodological approach. The method here proposed can be a useful tool for lighting design oriented to light control and quality rather than quantity.

Our study starts from the strong belief that any efficient and effective lighting design, especially within historic buildings and museums, requires knowledge of light and its connected physical, optical, colorimetric, photometric and radiometric aspects. As a matter of fact, some recent studies concern the assessment of visual appreciation of light sources and LED in museum environment. In particular, the colour appearance and colour quality have been studied under different light sources and LED lighting settings [10,11,12]. Different approaches, based on Shannon entropy, have been proposed to study light and visual perception [13,14,15]. A "visual entropy" method combined with eye-tracker and spectral emission of light sources measurements has been used to assess the order of fixation transitions of driving-related eye-movements [13]. Others authors have provided a method for choosing informative lighting directions for physical objects through Shannon entropy [14]. Krejtz et al. [15] have suggested a method for quantifying eye movement transitions between different AOIs of works of art, using Shannon entropy for the calculation of stationary and transitional distribution of fixations.

Our present research aims at the application of eye-tracking technique for eye movement assessment (saccades and fixations), of an observers sample during the observation of frescoes in order to evaluate the influence of the different types of light on vision and perception. It was possible to verify how the interaction between light, information and perception, that constitutes the visual/perceptual experience of each subject can be modified by different light sources. Information Theory, on a thermodynamic basis (neg-entropy), the ergonomy of multi-perceptive learning and optical physics, were basilar for our proposed integrated approach. Eye-tracking measurements were the first step in testing visual material: fixations are necessary for gathering visual information and without them perception is impossible.

Following the approach suggested in [14,15] we could analyse results on perception and preferences of observers, patterns of their visual scanning, locally oriented viewing works of art under different LED sources, through entropy and then the information content of images. Combining these data analysis with photometric and colorimetric measurements, it was possible to choice the correct LED sources allowing the transmission of information content of the images (i.e. lower entropy production due to visual patterns and eye movements in different AOIs that is higher neg-entropy and for the observers higher attractiveness).

2. Material and Methods

2.1. Experimental set-up, measurement protocol and methodology

One of the two frescoed rooms, called "Boscherecce" and known as the "Rooms of the Lady" of Villa La Quiete, is our case study. In order to define the luminous climate of the room and consequently investigate the interaction between light and perception of works of art, the geometric relief and measurements of the architectural features were carried out. A laser distance meter DISTO D2- Leica Geosystem was used. This allowed us to derive the room-index necessary to obtain the total numbers of measurement points and the experimental grid dimensions referring to the method suggested by the UNI EN 12464-1 [16]: this index value is 0,77. The room is 6,80 meters wide and 8,90 meters long, with a height at the base of the vaulted ceiling of 4 m, 5,60 m at the highest point and 5,42 m at the intermediate point corresponding to the change in the vault curvature. The room has only one window which is always closed. During the experimental measuring campaign, the wooden entrance door was always kept closed to eliminate interference effects due to the light coming from the adjacent corridor. Only the presence of artificial light due to the antique chandelier in the centre of the ceiling, provided with 9 LED sources, was considered. In particular, these light sources are ZafiroLED with the following nominal data: 4W power, E14 connection, 470 lm of luminous flux emitted and 2700 K colour temperature. Information on thermo-physical properties of the existing materials were collected: this was possible by means of a thorough research combined with direct observations and a specific photographic documentation whose results were compared with literature evidences. This was necessary to distinguish organic materials according to photosensitivity classes in relation to their degree of aging and photochemical degradation, in the perspective of their protection and preventive conservation as suggested by current standards.

2.2. *Photometric and colorimetric measurements*

In the present section the experimental photometric and colorimetric measurements are presented and discussed. The illuminance levels were measured at each point of the mesh by the standard technique with digital luxmeter. The measurements of the horizontal illuminance were carried out on grids of different dimensions, on the floor and all the perimeter walls, to which a number of different measuring points corresponds. An operative and procedural methodological scheme to reduce the executive time and improve the sequencing was defined. Three series of experimental measurements of illuminance levels, were carried out on each point of the grid, waiting for the stabilization and calibration of the instrument. In Figure 1 experimental meshes used and the corresponding measurement points are shown.

The measurement technique with spectrophotometer, applied on surface portion with dominant colour of the frescoes, was the method for measuring the fraction of the visible radiation intensity that the same surface reflects by diffusion at different wavelengths. The change of spectral distribution of the reflected radiation, with respect to that due to incident radiation, provides the surface colour. In particular, to measure the characteristics of a painting/object illuminated by a specific source, it is necessary to choose a repeatable and standardized coded measurement method. This is crucial for measurement of a diffusing surface/object. We used an experimental set-up for which the light source was placed at 45° with respect to the object. Measurements were performed at the front (configuration 45/0), so that the specular light component was not present in the visual field of the instrument, that is the part of the reflective specular radiation of a particularly shiny object. Figure 2 shows the experimental apparatus used for the measurements. The instrument, equipped with a camera lens, "looks" at a given area on the pictorial surface and analyses diffused light coming from that area. Quality and quantity of diffused radiation depend on the measuring point and light, with which the surface is illuminated, and the reflective features of the pictorial pigments (e.g. blue, red, gold, etc.). If the illuminated surface is not a picture or an object but a white diffusing surface (the white reference tile, Spectralon® Labsphere), the reflected light from it, is identical to that emitted from the light source: spectral characteristics of each tested light source were obtained. The spectrophotometer was connected, through a RS-232 serial cable, to a notebook, for continuous data acquisition. The numerical values that describe the spectral curve, luminance (LV) and chromatic coordinates (x, y), that are a measure of the light colour, were deduced. It was used together with its own white reference tile, to measure the spectral distribution, luminance, chromaticity and colour temperature of the tested light sources. Accurate measurements of electromagnetic radiation in the visible spectrum were also carried out.

Some tests were performed at the CNR-INO laboratory to verify method reproducibility, before achieving experimental measurements in situ. Five consecutive measurements were conducted on the same white stimulus: the luminance value obtained was $L = 79.11 \pm 0.04$ cd/m^2. Then, five different measurements were made: after each one there was a five minute pause and after reproducing the white stimulus at the PC monitor, the measurement was performed again. In this last case the luminance of the white stimulus was $L = 79.00 \pm 0.08$ cd/m^2. These two measurements are comparable: this guarantees measurement repeatability and the method validity. Standard deviation of all the experimental data, provided an error that is much lower than 1% corresponding to instrument accuracy. The experimental error is not shown because it is smaller than the point size in the corresponding graph. Technical data of instruments used are: Eye Tracker-Tobii-Glasses 2, with dark pupil Corneal reflection, 30 Hz sampling rate and four sensors; Spectrophotometer/radiometer CS 1000 Minolta, with 380-780 nm wavelength range, 5 nm spectral bandwidth, 0,9 nm wavelength resolution, ± 0,3 nm spectral accuracy and 0,01-80000 cd/m^2 Luminance range; Luxmeter CL-200 A Konica Minolta, with a silicon photocell of spectral responsivity within 6%, 0,01-299 900 lx measuring range and ± 2% ± 1 accuracy.

2.3. Perception experimental measurements

The observation (vision and perception) of works of art is a complex process conditioned by different factors. There are two approaches for understanding the perception process: the *top-down* processing concerns the pattern recognition development by use of contextual information; in the *bottom-up* processing perception starts at the sensory input (stimulus) and can be described as data-driven from the retina to the visual cortex in the brain. The salient features of an image (contrast, luminance, colour, dynamism) influence ocular movements: in particular, colour improves the perception of details and the number of observed elements, increasing their complexity. Starting from fundamental studies concerning psychology, Gestalt-psychology and neuro-(brain)-sciences, eye-tracking technique application was used to understand how and how much a specific light source affects vision and perception processes. The Tobii Pro Glasses2 [17], equipped with a functional unit for data collection Tobii Pro Glasses Controller and the software Tobii Pro Analyzer for the output data processing, was used. The eye tracking measurements were performed on a sample of 15 participants (women and men), aged between 20 and 25 with normal vision. The eye-tracker instrument records all the eye movements and behaviours related to perceptive and cognitive mental activity. Tobii eye-tracker wearable glasses recording data at a frequency of 30 Hz were used: they are easy to wear and do not require that the participant be accompanied or supported. Using Tobii Studio software the visual scan path with the corresponding stimulus can be overlapped. Before proceeding with measurements the Tobii Pro Glasses2 was calibrated on each observer for recognizing the convergence coordinates of gaze. The following sentence at the beginning of each test was read: "*During the experiment you should concentrate on the wall painting, trying to exclude, as far as possible, the boundary and expressing your own preference for the fresco, as the selected light scenography changes*". The LED sources were positioned at a distance of 3 meters from the wall paintings; the distance between fresco and participants was 1,5 meter. The experiment was divided into three tests corresponding to the three tested light sources. During each test the fresco was lit in sequence with each different LED for 15 seconds. At the end, each observer expressed her/his preference for the first, second or third light source. The method was validated comparing experimental setting, measurement protocol, boundary conditions and results with basilar literature researches on eye-tracking technology applications [2, 3, 18, 19].

3. Results and analysis

3.1. Photometric and colorimetric experimental measurement results

Photometric measurement results showed darkness of the room: the chandelier was kept on, for all the time of the measurements, in order to maintain the real conditions, the mean illuminance value is 12,5 lx, in the presence of 24,2 lx maximum and 3,4 lx minimum value. The mean illuminance values, obtained as the average over the three series of measurements at each grid point (refer to Figure 1 for experimental grids and measuring points) are: for the wall that divides the two rooms 15,9 lx; for the wall of two inaccessible rooms 11,6 lx; for the floor 10,5 lx. Balance and uniformity of illuminance is completely absent because its limit ratios are less than those suggested by the standard: the calculated ratio between the minimum and average illuminance value, and between maximum and average illuminance values are respectively 0,27 and 1,94. The experimental error data analysis, using Pearson standard deviation calculation, provided an average error of 4,5% over all the measures of illuminance: therefore, measurements have good reliability and validity. Three lighting sources were used for optical-perceptive measurements: a LED Vivid Warm Plus (VWP), a LED Vivid White (VWh) and a LED Standard (STD). Their main component is a COB (Chip On Board) light source with a combination of specific LEDs with specific phosphor mix and with nominal light beam opening up to 120°. Their secondary optics are provided by a combined diffusing/shielding, diffractive/shielding system that works with a front closure in extra-clear tempered glass, up to 60 ° nominal opening of the light beam. Nominal colour temperature of the three LED sources is: 3200 K for VWP; 3000 K for STD and 4100 K for VWh LED. Results obtained for the three LED sources illuminating the white reference tile, and their spectral characteristics are provided in Figures 3-5.

Figure 1. The meshes used for measurements.

Their spectrum is the typical LED spectrum, i.e. with a first peak marked around the 450 nm wavelength and a second peak around 600-650 nm wider, due to the fluorophores (yellow) deposition whose emission is excited by the same blue light at 450 nm wavelength. Comparing the results obtained for the VWP and STD, differences between their spectra are due to the greater presence, especially in VWP, of radiation in the wavelengths between 500 nm and 550 nm, and also in the displacement of the highest peak from 600 nm to 640 nm. This fact, although it does not produce appreciable colour temperature differences (i.e. from 3062 K to 3091 K), however moves the colour point on the chromaticity diagram CIE1931, with respect to the black body curve, the x coordinate from 0.43 to 0.42 and the y coordinate from 0.40 to 0.38. Since x and y lower values correspond to spectra with more green-blue light. From measurement results it can also be noted that the tested LEDs have a very similar colour temperature but different from the nominal ones.

From comparison between the emission spectra normalized on a 450 nm wavelength for the three LEDs and their correspondent colour point of light, it can be deduced that: VWP has a maximum emission peak between 620-650 nm, corresponding to red colour; STD has a maximum emission peak between 580-610 nm in correspondence of the yellow-orange colour; VWh has a maximum emission peak between 440-460 nm in correspondence of light blue/indigo colour. Comparing the colour point of the three light sources on the CIE chromaticity diagram, it is possible to deduce their colour temperature: for VWP is 3047 K, for VWh 4049 K and for STD 3782 K. It can be noted that VWP has a strong yellow/red component compared with blue. In VWh there is a strong blue component compared with red and in STD the two components are quite well balanced.

Figure 2. Experimental set-up with apparatus used (left) and the enlargement of spectrophotometer (right).

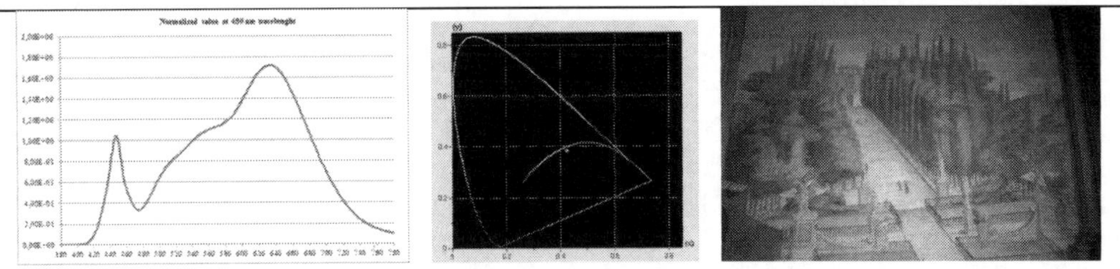

Figure 3. Spectral emission normalized at 450 nm wavelenght (left); Chromaticity diagram xyY CIE (center); real wall fresco lighting (right) - VWh 4100.

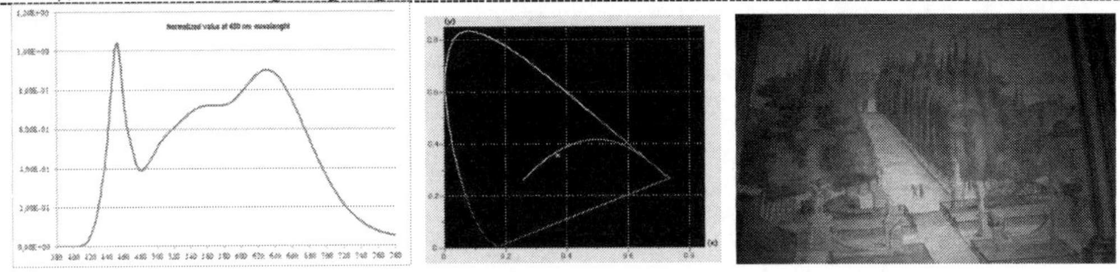

Figure 4. Spectral emission normalized at 450 nm wavelength (left); Chromaticity diagram xyY CIE (center); real wall fresco lighting (right) – VWP

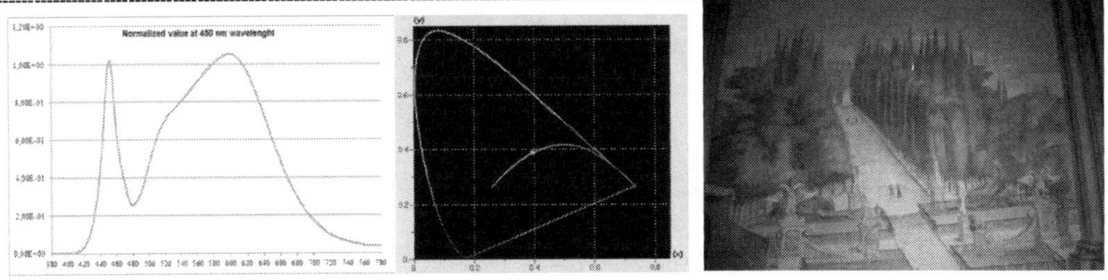

Figure 5. Spectral emission normalized at 450 nm wavelength (left); Chromaticity diagram xyY CIE (center); real wall fresco lighting (right) - STD 4000

3.2. Light experience and perceptive experimental measure results

From the Tobii-Studio program, the data post-processing was made and results were mapped. The following indexes were obtained: visit number, visit duration, gaze number (number of times that the eye stops on different areas), fixation number (number of micro-movements of fovea occurring during visual path), time from the first gaze and time from the first fixation. The assessment of the above indexes allowed the knowledge of the exact instant in which the eye has focused a specific zone of the fresco connected to the visit numbers and then the activation of the neuro-cognitive processes (i.e. interpretation and representation of the visual stimulus). Gaze data comparison allowed the definition of the areas of interest (AOI, targeted by Fixation Count and visualized by the HeatMap) that is the fresco zones with the highest number of visits. There are three identified AOIs: the landscape-park; the courtyard with the statues; the avenue with the building at the back (Figure 6). Referring to the time of the first fixation, the visit path of the observers was evaluated: for all the subjects the visual path moves from the centre of the fresco, towards the sides, to go down to the courtyard with the statues. Gaze plots show the location, order, and time spent looking at locations: they provide time sequence of looking expressed as fixation duration, quantified by the diameter of the fixation circles (i.e. the longer the look, the larger the circle; Figure 7). An initial general analysis suggests that data collected from 15 visitors follow a consistent pattern when observing the fresco. The AOIs identified on the fresco, were shared by most visitors and then the order (visual pathway) was similar for each of them. In particular, 54% of observers preferred VWh. The ratio between higher fixations number and less visits number provided indication about the lower entropy (i.e. higher neg-entropy or information content) because fixations tend to be concentrated on a certain AOI that appears to be more interesting.

As a matter of fact, the value of this ratio for the three different LED lighting, is: for VWh 5,26 AOI-courtyard, 2,6 AOI-threes-right, 2,2 AOI three-left, 3,6 AOI-avenue, 2,4 AOI-sky; for VWP 3,5 AOI-courtyard, 2,3 AO- threes-right, 1,8 AOI-three-left , 2,5 AOI-avenue, 2,1 AOI-sky; for STD 3,9 AOI-courtyard, 2,4 AOI threes-right, 2 AOI three-left, 2,8 AOI-avenue, 2 AOI-sky. The AOI-courtyard shows the highest gazes (Figure 7). The percentage of time, during which an AOI was fixed and observed under the three LED lighting, compared to the total time of analysis is: for AOI- courtyard 9,79% VWh, 9,14% STD, 8,71% VWP; for AOI-threes-right 2,95% VWh, 2,17% STD, 1,83% VWP; for AOI three-left 4,85% VWh, 4,01% STD, 3,78% VWP; for AOI-avenue 8,94% VWh, 7,18% STD, 6,78% VWP; for AOI-sky 1,18% VWh, 0,89% STD, 0,78% VWP. The greatest variations in time and fixation numbers of each area depend on the type of light source: higher concentration, processing of visual stimulus, visual scanning patterns with many fixations oriented to an AOI before moving to another, perception and then lower transition entropy (i.e. higher neg-entropy or information content transmission; [14,15]) are obtained with colder light, i.e. the VWh (Figure 8).

Figure 6. Areas of interest AOI

Figure 7. Gaze Plot: STD-LED, top left; bottom centered, WP-LED; top right Vivid-White-LED

Figure 8. Heat Maps: STD-LED, top left; bottom centered, WP-LED; top right Vivid-White-LED

4. Conclusions

The post-processing of experimental data showed that the highest amount of visual explorations and fixations are detected in those AOIs where contrast, duration and number of fixations are higher, and then where the information content (neg-entropy) is greater, consisting of simple but complex visual stimuli, because they are immediately recognizable when they recall to the memory of the subject his/her past experience and activate his/her embodied simulation (i.e. the emotional visual experience due to the activity of mirror neurons that allow mapping on the same nervous substrate images/actions performed and observed, feelings/emotions experienced and observed on ourselves and on others).

Comparison between the eye-tracking results and LED lighting parameters allowed checking of how the colour of light changes the observer's perception, and from the information theory point of view, the communication and interpretation process of different signals due to different light.

This study allowed us to evaluate how light colour affects perception: In particular, if the AOI were common to most visitors, the order and observation priority (visual pathway), was different: this constitutes that "common sense" that light, as a visual-perceptive experience can return in the transmission-communication of neg-entropy. Method and results of our research belong to a pilot project that we would like to develop in order to provide guidelines for effective lighting solutions (not only for museums) because they are imagined, structured and designed for the quality of vision and perception and, in other words, information transmission or visual light communication.

References

[1] Gallese V, Gilli G and Marchetti A 2012 When Art Moves the Eyes: A Behavioral and Eye Tracking Study *PLOS ONE-Open Acces* Volume 7, issue 5, pp 1-11.
[2] Wade N J 2010 Pioneers of Eye Movement Research *I-Perception* Volume 1, issue 2, pp 33-68.
[3] Villani D, Morganti F, Cipresso P, Ruggi S, Riva G and Gilli G 2015 Visual exploration patterns of human figures in action: an eye tracker study with art paintings *Frontiers in Pshicology* Volume 6, pp 1-10.
[4] Balocco C, Baldanzi E, Farnini A, Tito A and Raffaelli M 2016 Light and Art: perceptual stimuli and eye-tracking techniques *LUCE*, Volume 320, pp 101-106.
[5] Harezlak K, Kasprowski P 2017 Application of eye tracking in medicine: A survey, research issues and challenges *Comp Med Imaging Graph* pp 1-15.
[6] Cerrolaza J J, Villanueva A and Cabeza R 2012 Study of polynomial mapping functions in video-oculography eye trackers *ACM Trans. Comput.-Hum. Interact.* Volume 19, issue 2, pp 1-25.
[7] Cutrell E, Guan Z 2007 What are you looking for? An eye-tracking study of information usage *in Web Search. Proceedings of CHI 2007, Human Factors in Computing Systems*, pp. 407–416.
[8] Ewing K 2005 Studying web pages using Eye Tracking, Tobii Technology, USA.
[9] Oya Çınar M. 2009 Eye Tracking Method to Compare the Usability of University Web Sites: A Case Study *Int. Conf. on Human Centered Design HCD 2009*, pp 671-678.
[10] Magnusso Y 2014 Museum Lighting and LED Technology *E-Conservation Journal*, pp 34-42
[11] Smet K, Ryckaert WR, Pointer MR, Deconinck G, Hanselaer P 2012 A memory colour quality metric for white light sources *En. Build*. Vol. 49, pp 216-225.
[12] Feltrin F, Leccese F, Hanselaer P, Smet K 2017 Analysis of painted artworks'color appearance under various lighting settings, IEEE 2017, 17th Int. Conf. on Environment and Electrical Engineering, Milan, 6-9 June 2017.
[13] Gilland J Driving, eye-tracking and visual entropy: exploration of age and task effects, PhD Thesis, Dep. of Psychology University of South Dakota May, 2008.
[14] Malzbender T, Ordentlich E 2005 Maximum Entropy Lighting for Physical Objects, HP Laboratories Palo Alto external publication
[15] Krejtz K, Szmidt T, Duchowski AT, Krejtz I. 2014 Entropy-based statistical analysis of eye movement transitions Proceeding ETRA '14, Symposium on Eye Tracking Research and Applications pp 159-166, Safety Harbor, Florida, March 26-28, 2014
[16] UNI EN 12464-1 2010 Light and lighting - Lighting of work places - Part 1: Indoor work places.
[17] Tobii srl, Tobii Studio. User Manual, Darderyd, Sweden:Tobii Technology AB, 2014.
[18] Filippini Fantoni S, Jaebker K, Bauer D and Stofer K 2013 Capturing Visitors' Gazes: Three Eye Tracking Studies in Museums *Museums and the Web 2013, Silver Spring, MD: Museums and the Web.*
[19] Eghbal-Azar K., Widlok T 2012 Potentials and limitations of mobile eye tracking in visitor studies: Evidence from field research at two museum exhibitions in Germany *Social Science Computer Review*, Vol.31 (1), pp 103-118.

Acknowledgements

The authors thank the Natural History Museum of the University of Florence and the Institute for Research and Studies in Optics and Optometry in Vinci, for supporting their research.

IOP Conf. Series: Materials Science and Engineering **364** (2018) 012008 doi:10.1088/1757-899X/364/1/012008

Robust microclimate control for artwork preservation in response to extreme climatic conditions: simulation of museum halls for temporary exhibitions with a validated dynamic tool

E Schito, P Conti and D Testi

BETTER (Building Energy Technique and Technology Research Group), DESTEC (Department of Energy, Systems, Territory and Constructions Engineering), University of Pisa, Largo Lucio Lazzarino, 56122 Pisa, Italy

E-mail: eva.schito@for.unipi.it

Abstract. In museums, artworks preservation is hampered by several factors, in particular temperature and relative humidity as inappropriate values of these parameters can cause biological, mechanical, and chemical degradation. Thus, HVAC systems must be able to maintain specific suitable indoor microclimate also in critical conditions, such as variation of external climate and huge presence of visitors. In this paper, we first present the results of a monitoring campaign conducted in some rooms of an Italian museum in the 2015-2016 winter exhibition. The current HVAC system was able to maintain the required microclimate. A dynamic model (using TRNSYS for the envelope modeling and an in-house developed tool for the HVAC system) has been validated through the measured data. The dynamic model allows performing additional simulations characterized by critical boundary conditions (e.g., low external temperature and/or high relative humidity, and high number of visitors). In current conditions, these conditions would have caused unacceptable microclimate in the exhibition rooms, meaning higher risks for artworks. Through a rational rearrangement of the current HVAC system (both generators and terminal units), the microclimate was found to be suitable also in critical situations. This outcome is particularly important in exhibition rooms with temporarily borrowed artworks, since climatic conditions are basically unpredictable for the season of exhibition and so a robust HVAC system is needed.

1. Introduction

There are several factors that can hamper artworks preservation, such as temperature, relative humidity, light, and pollutants [1–3]. In particular, temperature (T) and relative humidity (RH) unsuitable values are responsible for biological, mechanical, and chemical degradation. In the last decades, conservation experts and heritage scientists have focused an increased attention to the identification of correct values of T and RH to prevent artwork deterioration. This attention led to the development of guidelines and technical standards for the correct preservation of artworks [4–8]; in some other cases, scientists have developed risk assessment methods for specific artwork typologies [9, 10]. In Italy, the technical standard UNI 10829 [4] includes, for different artwork typologies, suggested microclimate values for a correct preservation. Two ranges are identified: the "optimal range" represents the values of T and RH related to an optimal preservation of artworks; the "tolerant range" represents a still suitable T and RH range to be applied if it is not possible to install an optimal HVAC system for the microclimate control (for example when the museum is a protected historic building).

In the light of these studies, the importance of HVAC systems in museum exhibition rooms (but also in all the rooms where artworks are stored, not exposed) is clear. The objective of HVAC system in these environments is to maintain correct values of T and RH: usually, water terminal units (e.g., fan coils, radiant panels) control temperature in the rooms, while air handling units (AHUs) control relative humidity. It is important that HVAC systems fast react to changes in the external climate or indoor systems: for example, in case of peaks of visitors, sensible and latent gains due to individuals could cause an increase in internal temperature and relative humidity. HVAC systems should be then robust, in the sense of being able to cope with abrupt and potentially dangerous changes. This feature is particularly important in the case of HVAC systems used in temporary exhibitions museums: the borrowed artworks should be maintained in optimal conditions to avoid to pay penalty fees and so HVAC system should cope with unpredictable external climate and visitors turn-out.

In this paper, we present the results of a monitoring campaign carried out in a museum in Pisa, Italy, highlighting that current HVAC system performed well in the maintenance of the correct microclimate values (Section 2). We then used a dynamic simulation to verify the response of the system to changes in the external climate and in the visitors' presence (Section 3), showing that a rational reorganization of the current HVAC system is suitable to maintain correct indoor climate also in critical boundary conditions (Section 4). Finally, Section 5 presents the conclusions and future prospects of the work.

2. Description of the case study

2.1. Monitoring campaign

A temperature and relative humidity monitoring campaign was carried out in a museum in Pisa, Italy, for three months during a temporary exhibition of prints, posters and drawings on paper. For this artwork typology, Italian technical standard UNI 10829 suggests a temperature between 18 and 22 °C and a relative humidity between 40 and 55%. As there were several exhibition rooms, we here focus only on the most visited room, which was the one with the most famous artworks of the exhibition. This room, to which we refer as Reference Room (RR), has an area of 100 m^2 and a volume of 400 m^3. The envelope material is rubble masonry (average total thickness: 70 cm, thermal transmittance: $0.9\frac{W}{m^2K}$) and the roof has a typical Tuscan structure, with wood beams, bricks, and an insulating layer (thermal transmittance: $0.7\frac{W}{m^2K}$). There is no glazed area. The RR is heated in winter through fan coils, whereas fresh air is provided by an AHU. The AHU includes a preheat coil, a cooling coil, a vaporizer and a reheat coil. A dedicated heat pump provides hot water at the AHU heating coils and fan coils. As the heat pump can operate only in heating mode in winter, the AHU cooling coil is not used. It is worth noting that the T and RH sensors that control fan coils and AHU switching on/off are not positioned in the RR, but in an adjacent room, which is less crowded. We refer to this room as the Control Room (CR). The HVAC system operates from 8 a.m. to 9 p.m. every day, as the high envelope capacitance allows to switch off the system during the night without significant variation of temperature and relative humidity.

In accordance to technical standard UNI 10829 [4], EN 15758 [11], and EN 16242 [12], T and RH sensors were positioned in points representative of an area where the thermal environment is substantially homogeneous, avoiding AHU dampers and fan coils. Table 1 shows the position and the characteristics of the sensors. Sensors gathered T and RH at a 1.5 m height in proximity of the artworks every 6 minutes. Additional sensors gathered data of the external climate and of the HVAC system (e.g., AHU dampers, heat pump supply and return water temperature). Finally, for two reference days, we performed 15-minute observation of visitors' presence in RR.

Table 1: Specifics of the sensors used in the monitoring campaign.

Model	Position	Monitored parameter	Measuring range	Accuracy	Acquisition period
Siemens QAA24	RR and CR	T	-50...+80 °C	±0.6 K	15 min.
SEMITEC NTC	RR and CR	T	-50...+110 °C	±0.2 K	6 min.
4-noks THL-M	RR and CR	T RH	-40 ...+120 °C 0 ... 100 %	±0.2 K ±3 %	6 min.
Siemens QFM21160	AHU dampers	T RH	0 ... 50 °C 0 ... 100 %	±0.7 K ±3 %	15 min.
DICKSON TM325	AHU entrance	T RH	-20 ...+70 °C 0 ... 95 %	±0.5 K ±3 %	15 min.

Further details on the case study can be found in [13].

The suitability of the indoor thermal environment has been assessed through the evaluation of specific indexes, discussed in [13] and [14]. In particular, we used PI and PI* indexes, where:

- PI is the percentage of time in which T and RH values are within the optimal range;
- PI* identifies the percentage of time in which T and RH occurs within the tolerant range in the monitored environment.

These indexes were evaluated considering December month, which was a central month of the exhibition. The results of the monitoring campaign showed that the current HVAC system (heat pump, AHU, and fan coils) was able to maintain the correct microclimate inside the RR in December. For this exhibition, PI and PI* are, respectively, 87% and 100%, highlighting that the indoor climate is suitable for artwork preservation.

2.2. Dynamic simulation of the RR

We developed a detailed dynamic model of the RR in TRNSYS 17 [15], using the characteristics of the envelope and of the HVAC system. In particular, we simulated the entire zone of the museum that includes the RR and the CR, to more realistically represent the control of the HVAC system. The outputs of the simulation were the profiles of T and RH in the RR during December.

The whole model was validated using a one-to-one approach: single models (e.g., envelope, visitors' presence, AHU, heat pump) were validated through a comparison of the simulation results with monitoring data. In particular:

- the TRNSYS envelope model was validated comparing internal T and specific humidity in the RR with monitored data, using as inputs monitored external climate, visitors' profile, AHU T and RH of the supply air, supply and return water T of the heat pump;
- the visitors' profile model in the RR, based on the sold tickets at the box office, was validated comparing specific humidity from simulation and from monitoring, using as inputs monitored external climate, AHU T and RH of the supply air, supply and return water T of the heat pump [16];
- the AHU model, based on the ϵ-NTU method [17] of the heat exchangers, was validated comparing the inlet specific humidity and the return water T at the heat pump from

IOP Conf. Series: Materials Science and Engineering **364** (2018) 012008 doi:10.1088/1757-899X/364/1/012008

simulation and monitoring, using as inputs monitored external climate and supply water T of the heat pump [18] .

3. Analysis of the system in critical conditions

3.1. Sensitivity analysis: discussion

Changes in external climate or visitors' number in RR can cause a variation of risk assessment. Through the validated model of the building-HVAC system, variation of PI and PI* can be estimated when varying boundary conditions. The main criticality in the analyzed museum is the absence of dehumidification in winter, so we verified the current system performance varying the parameters that are responsible for an internal relative humidity increase. Three parameters are identified: a lower external T, a higher external RH, and a higher number of visitors. Also synergistic effects have been considered. External temperature, relative humidity and visitors' number were modified using three parameters:

- a_T, varying in the range [0.8; 1.0], in the expression:

$$T_{ext,new} = T_{ext} - abs[(T_{ext} - T_{ext,ref})(1 - a_T)]$$

  where $T_{ext,new}$ is the new value of external temperature used in the simulation, T_{ext} the actual monitored value of external temperature, and $T_{ext,ref}$ is a reference temperature, namely 0 °C;

- b_{RH}, varying in the range [1.0; 1.1] in the expression:

$$RH_{ext,new} = RH_{ext} + [(RH_{ext,ref} - RH_{ext})(b_{RH} - 1)]$$

  where $RH_{ext,new}$ is the new value of external RH used in the simulation, RH_{ext} the actual monitored value of external RH, and $RH_{ext,ref}$ is a reference RH, namely 100%.

- c_{vis}, varying in the range [1;2], in the expression:

$$n_{vis,new} = n_{vis} c_{vis}$$

  where $n_{vis,new}$ is the new number of visitors in the RR, used in the simulation, while n_{vis} is the number of visitors in the RR as evaluated through the dynamic model in [16].

Accordingly to the paper aim, the chosen ranges of a_T, b_{RH}, and c_{vis} include the most critical conditions, as expected for the museum location and analysis period. The performed simulations considered also synergistic effects, as in some cases a_T, b_{RH}, and c_{vis} concurrently vary. Simulations were performed for the December month, using the 2015 monitored external climate data as reference.

3.2. Results and discussion

The simulations results show that the variation of external T and RH and visitors' number are responsible for a worse microclimate in the RR.

Figure 1 shows the simulation results, in terms of PI, varying the three parameters a_T, b_{RH}, and c_{vis}. The figure shows that PI rapidly decreases when more critical conditions occur. In particular, the most critical parameter is the number of visitors in the room, which causes a significant reduction of PI: when $c_{vis} = 2$, $a_T = 1$ and $b_{RH} = 1$, PI value is about 66%, and decreases up to 44% in the worst condition ($a_T = 0.8$, $b_{RH} = 1.1$, and $c_{vis} = 2$). Visitors are in fact responsible for high internal gains, particularly latent ones: this causes a high increase of relative humidity in the RR. Also the external temperature decrease causes a reduction of suitability parameters: a lower external temperature increases heat exchange through the

IOP Conf. Series: Materials Science and Engineering **364** (2018) 012008 doi:10.1088/1757-899X/364/1/012008

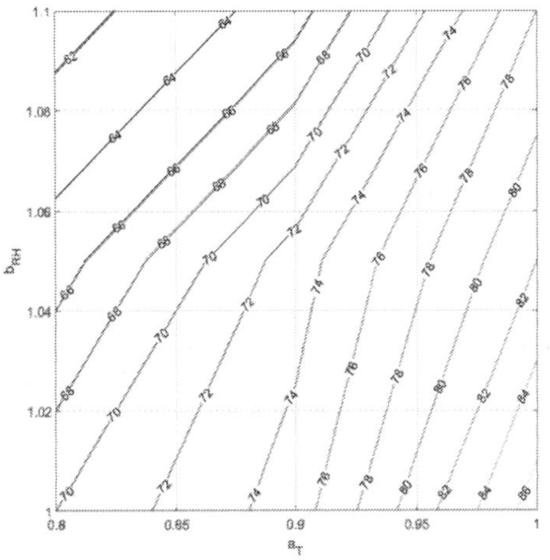

(a) Case with $c_{vis} = 1$.

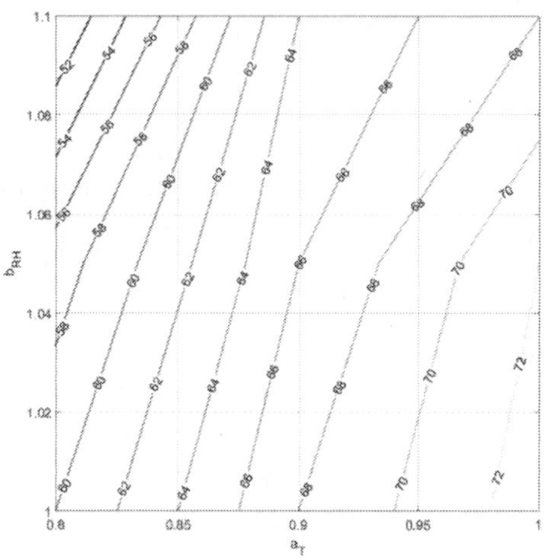

(b) Case with $c_{vis} = 1.5$.

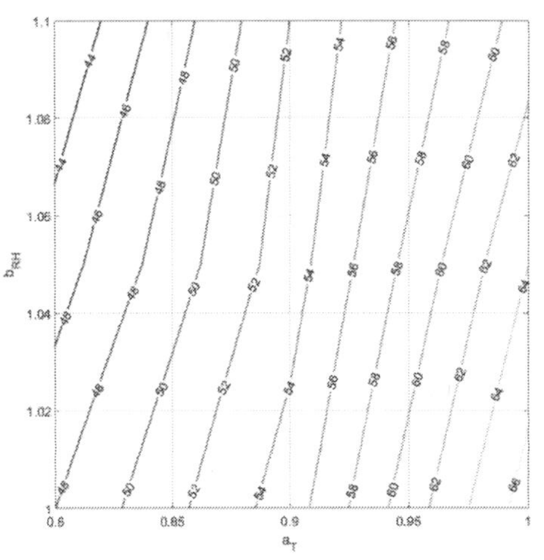

(c) Case with $c_{vis} = 2$.

Figure 1: PI indexes as results of the sensitivity analysis in the current HVAC configuration.

building envelope, so the reduction of internal microclimate suitability is due to the reduction of internal temperature during nights, when HVAC system is switched off. RH is high in the RR due to visitors' presence and lack of dehumidification and it even increases during night, as T decreases: thus, it exceeds the threshold limit. Also PI* decreases when worse conditions occur: the 100%-value of PI* is found when the three sensitivity parameters are equal to 1, but lower values (around 70-75%) are reached when visitors' number increase or external temperature decreases. These results clearly highlights that current HVAC system is potentially unable to cope with critical boundary conditions, hampering artworks preservation: so a variation of the HVAC system is needed.

4. Reorganization of the HVAC system and application of the sensitivity analysis parameters

4.1. Implementation of the new HVAC system: discussion

Usually, the refurbishment of a museum hosted in a historic building is difficult, as several classical solutions for optimal HVAC system design cannot be used because of the constraints for the protection of the envelope. The most recommended solutions are identified after an in-depth analysis of the whole building-HVAC system. In the analyzed case study, we considered the whole building-HVAC system, which includes also other rooms and generators. We found that another temporary exhibition area is heated through a condensing boiler, which is largely oversized (nominal power: 500 kW). In fact, the analysis of this adjacent zone showed that it needs the 50% of the condensing boiler power for heating purposes. This condensing boiler is positioned near to the RR generation system. Then, we considered the possibility of a rearrangement of the HVAC system: the condensing boiler can be used for the generation of heated water for fan coils and heating coils of the AHU, while the current heat pump can be used only for cooling water of AHU cooling coils in winter and for fan coils and AHU cooling coils in summer. In the decision of the HVAC system rearrangement, we considered the following aspects:

- the solution respects the constraints due to the museum protection by the local agency for the historic building protection (i.e. Soprintendenza), as it does not affect the envelope;
- it allows the modification of the microclimate in the RR without modifying it in the other rooms of the temporary exhibition area;
- as the condensing boiler, heat pump and AHU are in adjacent equipment rooms, it should be easy to connect the generators, the terminal units and the AHU.

The current dynamic model of the HVAC system in RR was modified, implementing the use of AHU cooling coils: the cooled water is provided by the currently-used heat pump, while the heated water (at both AHU and fan coils) is provided by the condensing boiler. The HVAC system is considered switched off during night, as in the current case. The sensitivity analysis is then applied, to verify if the new system is more suitable in maintaining correct microclimate in the RR. The sensitivity parameters a_T, b_{RH} and c_{vis} are supposed to vary in the same ranges as discussed in previous section. The simulation is performed for the same reference month as above.

4.2. Results of the sensitivity analysis with the new HVAC system

The results of the new simulations show that the retrofit solution allow a good control of internal microclimate also when both external climate and visitors profile significantly change. Figure 2 shows the simulation results when the three sensitivity parameters vary.

The results show a small PI decrease when a_T decreases and when c_{vis} increases (the T and RH control sensors are positioned in a less crowded room, not in the RR):

- when a_T decreases, the external temperature is lower also during night, when the HVAC system is turned off; the night heat losses through the envelope are higher and the indoor temperature decreases, thus resulting in a lower PI than in $a_T = 1$ case;
- PI decreases when c_{vis} increases because the values of AHU supply specific humidity are chosen to meet the hygrothermal setpoints on the basis of microclimate in the CR, which is less crowded than the RR; then, higher values of RH are found in the RR, resulting in a worse microclimate when the visitors' number is high.

In any case, high values of PI are reached in almost all the analyzed cases: when $a_T = 1$, $b_{RH} = 1$, and $c_{vis} = 1$, PI value is about 90%, while in the current HVAC configuration PI is about 84%. The improvement of PI value is clearly visible in the results of the simulations where $c_{vis} = 2$, which were the most critical cases. In fact, in the current HVAC configuration, PI value

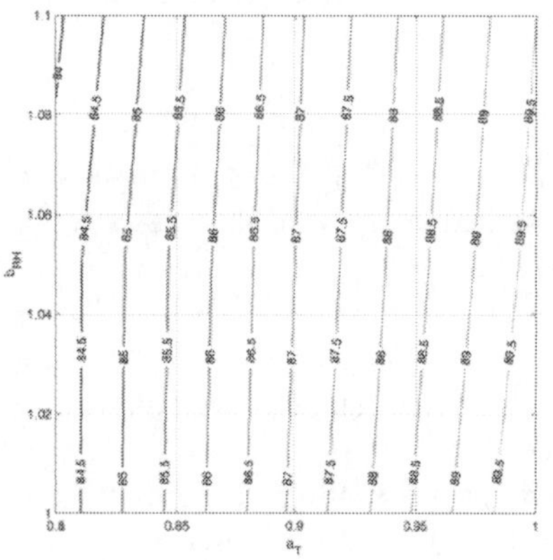

(a) Case with $c_{vis} = 1$.

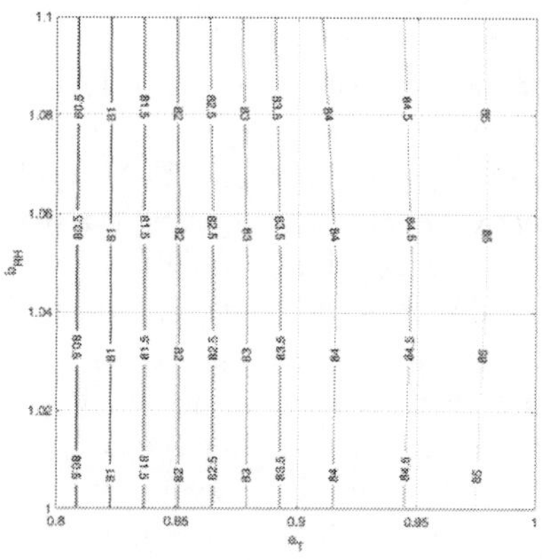

(b) Case with $c_{vis} = 1.5$.

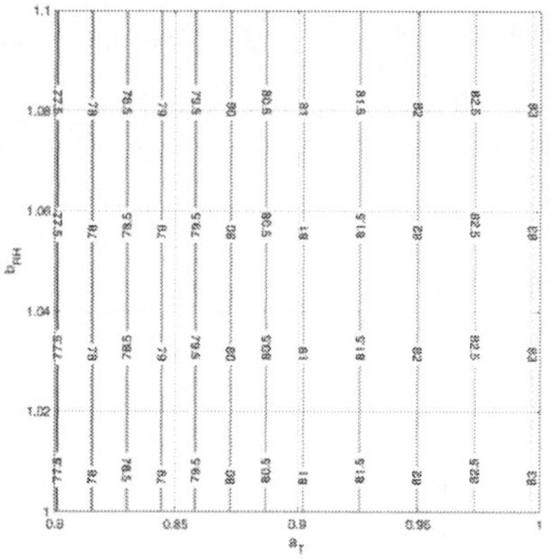

(c) Case with $c_{vis} = 2$.

Figure 2: PI indexes as results of the sensitivity analysis in the new HVAC configuration.

ranges between 44 and 66%, while, with the new HVAC configuration, PI ranges between 77 and 83%. Also PI* value has improved, being equal to 100% in all the simulations.

The higher values of microclimate suitability indexes are due to the improved RH control. During the opening hours, in fact, AHU maintains indoor RH around the 45%-setpoint; only in the cases with high visitors' crowd, the indoor RH increases, with values around 55%. In these cases, the HVAC turning off during nights causes the microclimate to go beyond the optimal limits, thus decreasing PI. In any case, the variation of microclimate between opening and closing hours is limited, with a maximum T daily span of 2 °C and a maximum RH daily span of 5%. According to [19], for artwork preservation, suitable daily variation is 2 °C for T and 8% for RH. Hence, the indoor microclimate variation due to the nocturnal HVAC turning off does not hamper artwork integrity.

It is worth noting that these good outcomes are reached without additional costs except the ones for the new connections between generators and distribution system.

5. Conclusions

In this paper, we verified that current HVAC system of our case study maintained correct microclimate during a temporary exhibition in 2015-2016, but, in presence of more critical conditions (lower external T, higher external RH and higher visitors' number), it would be unable to maintain suitable microclimate due to the current working technical equipment. Through an analysis of the whole museum, we have identified a new configuration of the HVAC system using only the currently-present components, so without additional installation costs. The new configuration has been tested through dynamic simulations, verifying the improvement of all the suitability indexes, which indicates a better control of internal microclimate also in critical conditions. Further research will investigate also the microclimate improvements due to change in the HVAC control system (e.g., setpoints, operating periods, setback temperature).

This study highlights the importance of a flexible HVAC system to maintain T and RH in the correct ranges for artwork preservation. In museums hosted in historic buildings, it is not always possible to have an optimal envelope-HVAC system, because of the constraints due to the building protection. Furthermore, the HVAC should be also "resilient", that is being effective when boundary conditions become critical. This is particularly important for HVAC systems in temporary exhibition museums, where HVAC systems should be able to maintain different setpoints depending on the exposed artworks, regardless of unpredictable external conditions and internal gains. In this perspective, an accurate analysis of the building-HVAC system and dynamic simulation support stakeholders and experts for correct design and management.

Indoor conditions are also important to ensure visitors' comfort, though different T and RH values are generally required. The efficiency of energy production has to be considered as well. The same implemented models of the building and generators can be used to investigate this three-multi-objective problem: we will deal with this additional analysis in future works.

References

[1] Camuffo D 1998 *Microclimate for Cultural Heritage* (Amsterdam: Elsevier).
[2] Bernardi A 2003 *Conservare opere d'arte. Il microclima negli ambienti museali* (Saonara, IT: Il Prato).
[3] Pavlogeorgatos G 2003 *Build. Environ.* **38** 1457–62.
[4] UNI 1999 UNI 10829
[5] CEN 2010 EN 15757
[6] Italian Cultural Heritage Ministry 2001 D.M. 10/05/2001
[7] National Museum Directors' Conference 2009
[8] Publicly Available Specifications 2012 PAS 198
[9] Schito E and Testi D 2017 *Build. Environ.* **123** 585–600.
[10] Silva H E, Henriques F M A, Henriques T A S and Coelho G 2016 *Build. Environ.* **104** 21–34.
[11] CEN 2010 EN 15758
[12] CEN 2012 EN 16242
[13] Schito E, Testi D and Grassi W 2016 *Buildings* **6** 41.
[14] Corgnati S P, Fabi V and Filippi M 2009 *Build. Environ.* **44** 1253–60.
[15] Klein S A et al 2017 *TRNSYS 17: A Transient System Simulation Program* (Madison, USA).
[16] Schito E and Testi D 2017 *Build. Simul* **10** 977–87.
[17] Incropera F P, DeWitt D P, Bergman T L and Lavine A s 2007 *Fundamentals of Heat and Mass Transfer* (Hoboken, USA: John Wiley & Sons).
[18] Schito E 2017 *Methods and tools for a rational and efficient use of energy in museum environments* (PhD Thesis, University of Pisa, Pisa, IT).
[19] Thomson G 1978 *The museum environment* (Amsterdam: Elsevier).

Computational simulation of hygrothermal processes in historical building envelopes provided with interior thermal insulation

Jan Kočí, Jiří Maděra, Miloš Jerman, Magdaléna Doleželová, Robert Černý

Department of Materials Engineering and Chemistry, Faculty of Civil Engineering, Czech Technical University in Prague, Thákurova 7, 166 29 Prague 6, Czech Republic

jan.koci@fsv.cvut.cz

Abstract. Interior thermal insulation systems are an integral part of retrofitting works, especially when historical buildings are taken into account. In this paper, three different insulation materials (mineral wool, calcium silicate, and sheep wool) were investigated in order to assess their suitability for the application in historical buildings. First, the basic physical, hygric and thermal properties of those materials were measured in laboratory conditions and then a series of computational simulations was carried out. Finally, based on the simulation results, the hygrothermal response was analyzed and several conclusions were drawn. The presented results proved all materials suitable for application in interior thermal insulation systems but some differences in the hygrothermal response of individual materials were observed.

1. Introduction

Historical buildings are considered in most nations as valuable heritages from past generations and cultures. In some cases, it can be talked about structures of incalculable value. However, in the modern world, those buildings often fail to meet current standards and therefore require some retrofits to be applied.

A proper decision on renovation measures is very important for the overall effectiveness of retrofitting works. When thermal insulation is considered as a renovation measure for an old historical building, the engineers are usually restricted to the applications of additional insulation layers on the interior side of the building enclosures. However, the interior thermal insulations may increase a potential water vapor condensation hazard in the buildings' interior and thus to act counterproductively, as the quality of indoor environment may be significantly worsened. In order to prevent the negative actions, a serious hygrothermal assessment must be done. Due to high complexity of the assessment procedure, the advanced technique based on perfect knowledge on thermal and hygric properties of insulation materials together with the application of computational modelling seems to be most effective for this kind of task.

In this paper a series of computational simulations is carried out and the hygrothermal response of interior thermal insulation layers in reconstructions of historical buildings on the Czech territory is assessed. For that reason, three different insulating materials were selected (calcium silicate, hydrophilic mineral wool, and sheep wool) and their thermal and hygric properties were analyzed in the laboratories, in order to obtain accurate input data for the computational model.

IOP Conf. Series: Materials Science and Engineering **364** (2018) 012009 doi:10.1088/1757-899X/364/1/012009

2. Studied insulation systems

The cross-section of a historical building wall is shown in figure 1. In that cross-section following materials were assumed: sandstone from Libnava quarry (SL), interior thermal insulation (calcium silicate, hydrophilic mineral wool, or sheep wool), interior and exterior plaster, and connecting layer between load-bearing structure and thermal insulation. The interior plaster and exterior plaster were made from the same material that was developed for renovation of historical masonry.

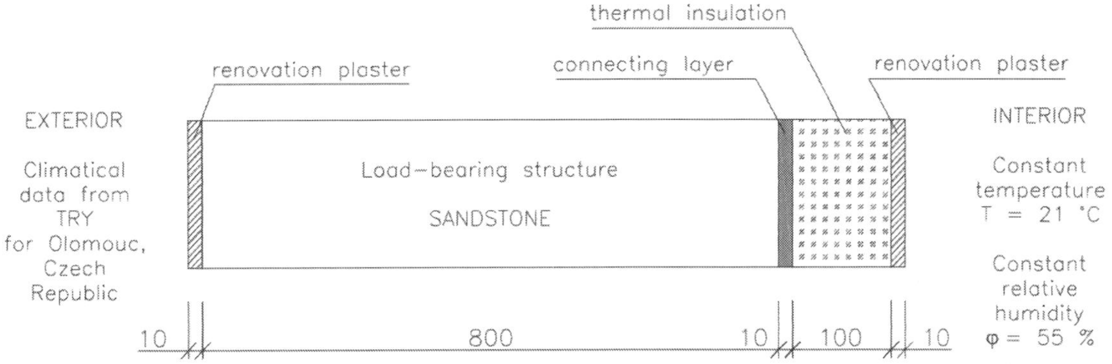

Figure 1. Horizontal cross-section of investigated historical wall

3. Experimental

The involved materials were subjected to laboratory analysis in order to obtain basic material parameters that are necessary for further investigation of their hygrothermal response. The material properties of calcium silicate (CS) and sheep wool (SW) were determined in the laboratories of the Department of Materials Engineering and Chemistry, Faculty of Civil Engineering, Czech Technical University in Prague within the frame of this project and were stored in the material database. The other materials namely sandstone (SL) [1], mineral wool (MW) [2], connecting layer (M2) [3], interior (IP) and exterior (EP) plasters [4] were analyzed in the same laboratories, but have been published earlier. All material parameters were measured according to the methodology described in [5]. The basic physical, heat and moisture transport and storage properties of sandstones are given in table 1, where ϱ is the bulk density, ψ the porosity, c_{dry} the specific heat capacity in dry conditions, c_{sat} the specific heat capacity in saturated conditions, $\mu_{dry\ cup}$ the water vapor diffusion resistance factor in dry conditions, $\mu_{wet\ cup}$ the water vapor diffusion resistance factor in wet conditions, λ_{dry} the thermal conductivity in dry conditions, λ_{sat} the thermal conductivity in water saturated conditions.

In addition to data presented in table 1, the sorption isotherms and moisture diffusivities of all involved materials were measured as well.

Table 1. Basic physical properties involved materials

Parameter	SL	MW	CS	SW	M2	IP/EP
ϱ (kg m^{-3})	2191	270	231	15	1430	1690
ψ (%)	17.9	96.5	90.7	98.7	42.6	34.1
c_{dry} (J kg^{-1} K^{-1})	721	810	1248	972	1020	877
c_{sat} (J kg^{-1} K^{-1})	1127	3850	3523	972	1020	877

$\mu_{dry\ cup}$ (-)	11.60	2.62	2.11	2.83	6.60	29.00
$\mu_{wet\ cup}$ (-)	7.40	1.34	1.65	2.36	6.20	11.60
λ_{dry} (W m^{-1} K^{-1})	2.71	0.036	0.0682	0.045	0.481	0.886
λ_{sat} (W m^{-1} K^{-1})	4.64	0.705	0.5026	0.045	2.022	2.190

4. Computational

4.1. Mathematical model

For the investigation of hygrothermal performance of studied constructions a modified Künzel's mathematical model was used [6]. The modified mathematical model is able to distinguish more precisely between liquid and gaseous phase of moisture transport and thus brings higher accuracy to the calculations [7]. Balance equations are expressed as

$$\frac{dH}{dT}\frac{\partial T}{\partial t} = div(\lambda grad T) + L_v div(\delta_p grad\ p_v), \tag{1}$$

$$\left[\rho_w \frac{dw}{dp_v} + (n-w)\frac{M}{RT}\right]\frac{\partial p_v}{\partial t} = div\left[D_g \ \mathrm{grad}\, p_v\right], \tag{2}$$

where H (J·m^{-3}) is the enthalpy density, T (K) the absolute temperature, λ (W·m^{-1}·K^{-1}) the thermal conductivity, L_v (J·kg^{-1}) the latent heat of evaporation of water, δ_p (s) the water vapor diffusion permeability, p_v (Pa) the partial pressure of water vapor in the porous space, ϱ_w (kg·m^{-3}) the density of water, w (m^3·m^{-3}) the moisture content by volume, n (-) the porosity, M (kg·mol^{-1}) the molar mass of water vapor, and R (J·K^{-1}·mol^{-1}) is the universal gas constant. D_g (s) is the global moisture transport function defined as

$$D_g = B \cdot D_w \rho_w \frac{dw}{dp_v} + A \cdot \delta_p, \tag{3}$$

where A and B in Eq. (3) are the membership functions defining the transition between particular phases of water, which can be formulated as

$$B = \begin{cases} 0 & \varphi \in \langle 0; 0.9) \\ 32\left[\left(\frac{1}{p_{v2} - p_{v1}}\right)(p_v - p_{v1})\right]^6 & \varphi \in \langle 0.9; 0.938) \\ 1 - 32\left[\left(\frac{1}{p_{v2} - p_{v1}}\right)(p_{v2} - p_v)\right]^6 & \varphi \in \langle 0.938; 0.976) \\ 1 & \varphi \in \langle 0.976; 1 \rangle \end{cases}, \tag{4}$$

$$A = 1 - B, \tag{5}$$

where the partial pressures of water vapor p_{v1} and p_{v2} (Pa) define the transition region. In this paper the values of p_{v1} and p_{v2} corresponded to the values of relative humidity of 90 % and 97.6 %, respectively.

IOP Conf. Series: Materials Science and Engineering **364** (2018) 012009 doi:10.1088/1757-899X/364/1/012009

The simulations were carried out on the one-dimensional cross section depicted in figure 1 using the TempMoist2D simulation tool. The tool is based on the general finite element package SIFEL (SImple Finite Elements) [8].

4.2. Boundary conditions
Boundary conditions in the interior were set constant as 21 °C and 55 % relative humidity. On the exterior side, the climatic load in a form of Test Reference Year (TRY) for Olomouc, Czech Republic was applied. The TRY was obtained from the Czech Hydrometeorological Institute (CHMI) which is the official authority for climatology, hydrology and atmosphere quality in the Czech Republic. The weather data obtained from CHMI as well as other sources are stored in weather database [9]. The data contained hourly values of temperature, relative humidity, wind speed, wind direction, precipitation and solar radiation. The summary of temperature, relative humidity and amount of precipitation is provided in table 2.

Table 2. Summary of applied weather data

Month	Temperature (° C)	Relative humidity (%)	Rainfall (mm)
Jan	-2.02	85.52	30.70
Feb	2.40	78.76	8.10
Mar	4.54	69.08	71.50
Apr	10.47	68.16	42.20
May	15.09	68.15	38.09
Jun	18.49	69.23	67.21
Jul	20.16	69.26	107.80
Aug	20.73	69.09	63.00
Sep	14.52	77.80	20.50
Oct	10.30	81.17	18.00
Nov	4.44	90.23	68.50
Dec	-1.64	86.50	35.30

5. Results and discussion
The results of computational simulations performed by HEMOT code are presented below in figures 2 and 3. All the computer simulations were accomplished for 10 years in order to limit the effect of initial conditions in the studied walls. The hygrothermal performance was captured in four different days representing all periods of the year. The data are presented for 1 February (winter), 1 May (spring), 1 August (summer) and 1 November (autumn).

IOP Conf. Series: Materials Science and Engineering **364** (2018) 012009 doi:10.1088/1757-899X/364/1/012009

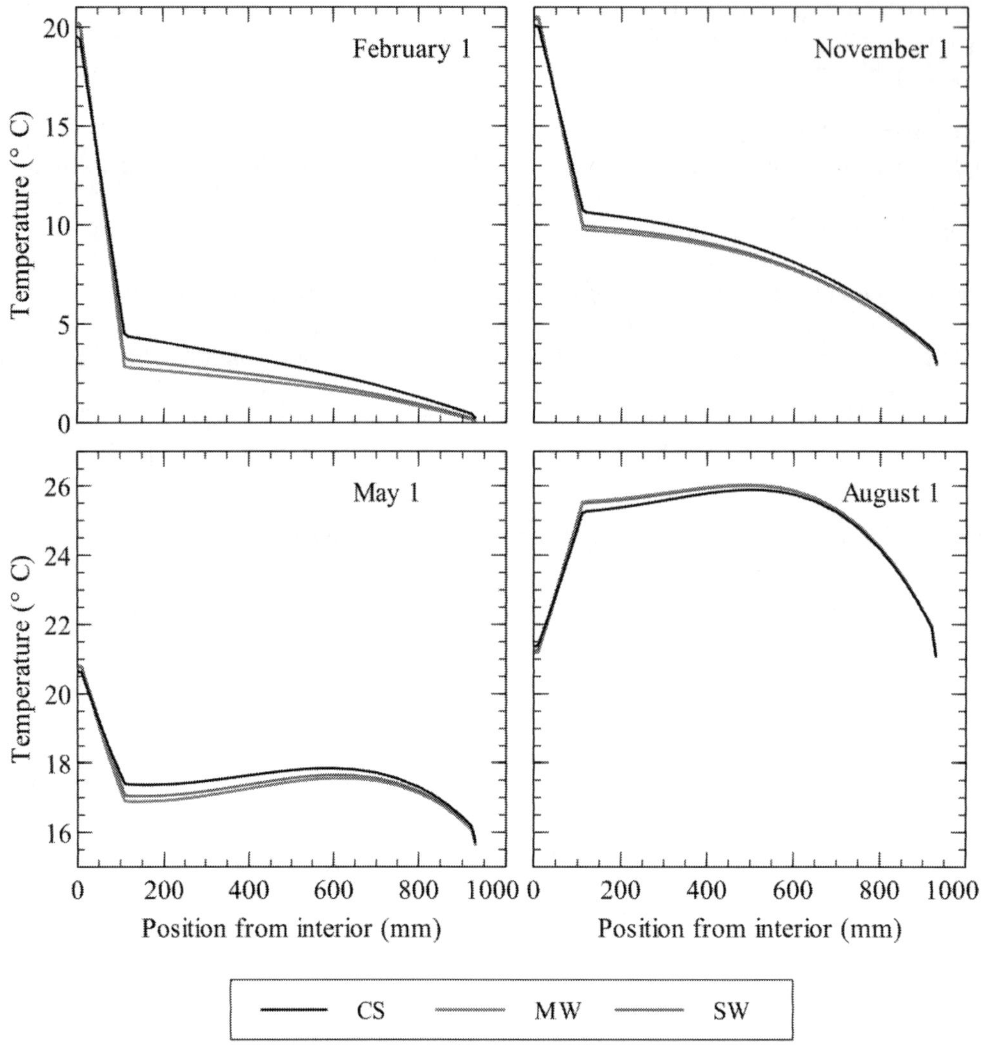

Figure 2. Temperature distribution in the studied envelopes

As it is obvious from figure 2 the highest temperatures in the wall can be observed mostly for calcium silicate and lowest for mineral wool. This might be crucial for the possible condensation of water vapor. However, the differences in thermal performance are not significant as they vary about 1 – 3 °C. Furthermore, the closer to the exterior surface the more the values approach each other.

When the interior surface temperature is considered, the highest values can be observed for mineral wool (20.24 °C in February and 20.54 ° C in November) and lowest for calcium silicate (19.53 °C in February and 20.09 °C in November). The sheep wool is very close to the mineral wool in this respect.

IOP Conf. Series: Materials Science and Engineering **364** (2018) 012009 doi:10.1088/1757-899X/364/1/012009

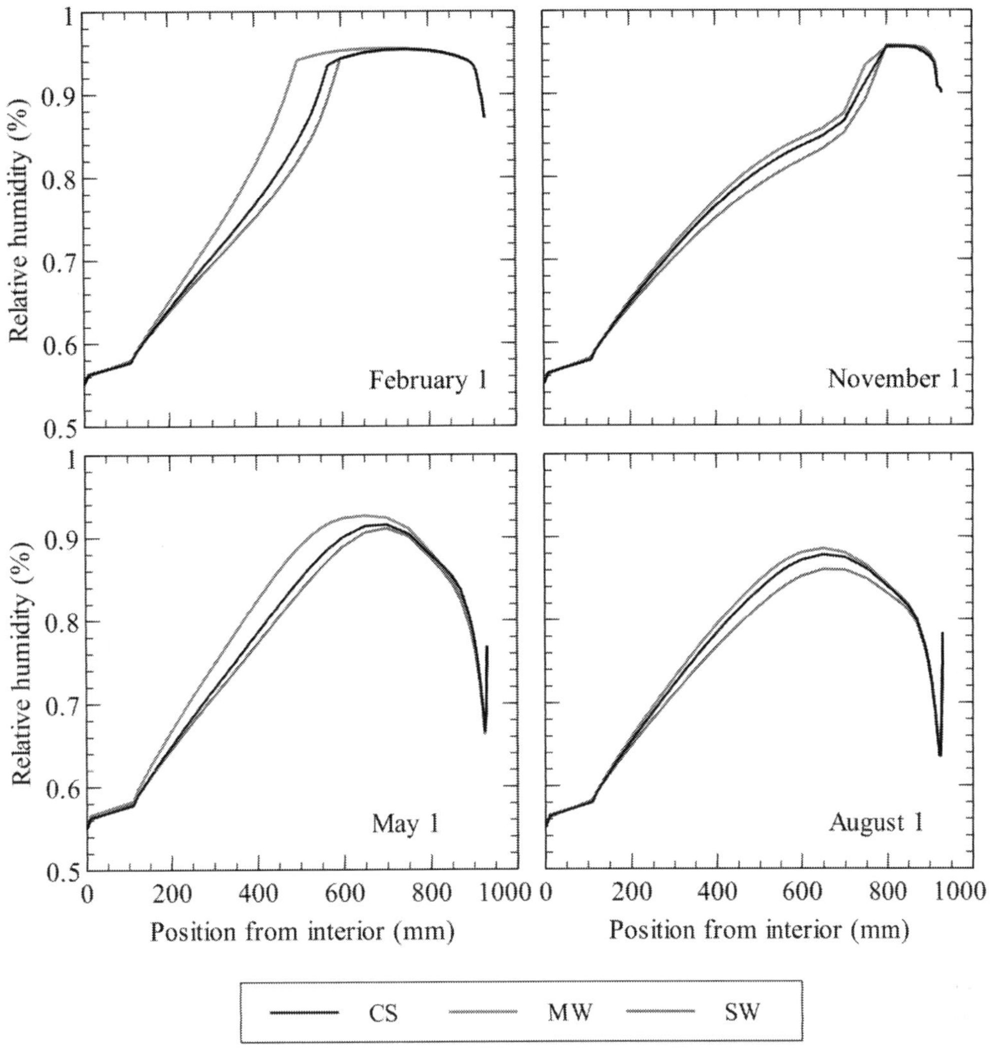

Figure 3. Relative humidity distribution in the studied envelopes

The hygric performance is probably more important than thermal as all studied thermal insulation materials proved very similar thermal parameters. However, the hygric properties of those materials can significantly influence the response of whole structure. From that point of view the best results can be observed for sheep wool which exhibits the lowest values of relative humidity in all investigated seasons. On the other hand, the highest values were observed for mineral wool which implies the highest risk of water vapor condensation.

6. Conclusions

Three different thermal insulation materials were investigated in this paper as possible candidates for interior thermal insulation systems of historical buildings. Based on the presented results it can be concluded that all of them were suitable for the application within retrofitting works but some small differences in the hygrothermal response were observed:

- Mineral wool proved best thermal insulation capabilities among the investigated materials
- The highest interior surface temperature was observed for mineral wool and sheep wool

- The best hygric performance was observed for sheep wool, whereas mineral wool showed the worst performance, especially in the winter period.

Acknowledgements
This research has been supported by the Ministry of Culture of the Czech Republic, under project No DG16P02H046.

References

[1] Kočí V, Maděra J, Fořt J, Žumár J, Pavlíková M, Pavlík Z and Černý, R 2014 Sci World J Article ID 802509, 12 p
[2] Jiřičková M and Černý R 2006 Constr. Build. Mater. **20** 425-34
[3] Kočí V, Jerman M, Fiala L and Černý R 2017 AIP Conference Proceedings **1906**, 140005
[4] Vejmelková E, Máca P, Konvalinka P and Černý R 2011 Advanced Materials Research **324** 372-75
[5] Černý R 2013 *Complex System of Methods for Directed Design and Assessment of Functional Properties of Building Materials and its Applications* (Prague: CTU Prague) p 509
[6] Künzel HM 1995 *Simultaneous Heat and Moisture Transport in Building Components* (Stuttgart: IRB Verlag) PhD thesis
[7] Maděra J, Kočí J, Kočí V and Kruis J 2017 Adv. Eng. Softw. **113** 47-53
[8] Kruis J, Koudelka T and Krejčí T 2010 Math and Comput Simulat **80** 1578-88
[9] Ďurana K, Maděra J and Černý R 2013 WIT Transactions on the Built Environment **131** 245-56

IOP Conf. Series: Materials Science and Engineering **364** (2018) 012010 doi:10.1088/1757-899X/364/1/012010

The 3D virtual reconstruction of an engineering work of the past

Di Angelo L, Di Stefano P, Pane C

Department of Industrial and Information Engineering, and of Economics, University of L'Aquila, via G. Gronchi 18 L'Aquila, Italy,

{luca.diangelo,paolo.distefano,caterina.pane}@univaq.it

Abstract. The construction of artificial emissary of Lake Fucino's represents one of the most significant engineering challenges that took place in antiquity. Unfortunately, the imposing structure of the Roman emissary was almost completely erased from the building of the late nineteenth century, constructed for the final drying of the lake. This article presents its virtual reconstruction, to understand the technological issues met by Romans and to visually offer reconstruction theories starting from interpreting of the few Roman remains. So that, engineering resources are used to try to understand unresolved issues in archaeology.

1. Introduction

The emissary of Lake Fucino in the Abruzzo region of Italy, build between 41 and 52 B.C. can be considered one of the most important hydraulic construction in antiquity. It operated until Middle Age when the tunnel collapsed. Its length, about 5650 m, is a record unmatched until the 19th century. The tunnel construction required to adopt new technologies in order to deal with unusual problems never faced before. This emissary was replaced in the 19th century, when the old Roman emissary was obliterated by the construction of a modern tunnel. The modern emissary was made using the old one as a track, but its section presents different shape and size, more suitable to cope with the different pressures of the soil, and capable to sustain of greater water flow. The result is a tunnel much larger than the Roman one that completely erased the Roman emissary.

The first detailed records about the Roman construction are written by Raffaele Fabretti [1] and Carlo Afan De Rivera [2]. The first reports some illustrations of the tunnel and its accessory structures. The second one proposes a complete work describing in detail the conditions of the Roman emissary at the beginning of 18th century. The most complete record is written by Alexandre Brisse and Léon De Rotrou [4]. This consists of two books and an Atlas, which report the Roman emissary's path and its sections. The analysis of these sources, taken individually, does not permit the complete comprehension of this hydraulic construction. Furthermore, a joint study of these sources, without adding engineering resources does not solve some open questions.

The aim of this paper is to understand the specific technologies used to plan and realize the emissary. It is useful, to fully comprehend this work and the technologies applied to its construction, to collect in a single representation all the documentary sources. These concern the territorial geography and the archaeological remains, such as wells and other auxiliary structures. The main goal is to evidence the complex functional connections among the wells, the tortuosity of the main tunnel's path, the secondary tunnels network and the territory's orography, and the subsoil. Together with these purposes, the

IOP Conf. Series: Materials Science and Engineering **364** (2018) 012010 doi:10.1088/1757-899X/364/1/012010

exteriority of the construction is shown. The 3D geometric relations that all these sources show, hold the technical reasons for such a complex work. Since a lot of the technologies used to realize the emissary apply geometry's principles, a 3D virtual reconstruction of the Roman emissary gives the opportunity to achieve information about the methods Romans used to realize such a difficult task. Advanced techniques for 3D reconstruction can be used for specifically understanding several issues Romans met in realizing the emissary: surveying methods, hydraulic technologies, the constructive phases together.

3D reality-based technologies are typically used to reconstruct ancient monumental buildings, focusing on their aesthetic and historical value. The Fucino's emissary can be an opportunity to evaluate how virtual reality technologies are able to support the archaeological research of scientific and technological aspects, through the analysis of measured data, combined with documentary and iconographic sources.

2. Historical events: a brief summary

Lake Fucino was an endorheic lake without any natural emissary, whose waters were affected by great variations, both in surface and depth. Therefore, the lake overflowed the lands surrounding its shores, causing several problems to the nearby towns. This was the main reason that brought local people asking to somehow regulate the Lake level.

The first idea of controlling Fucino's waters and exploiting its territories was expressed by Roman administration. The drying project was conceived during the first century BC by Caesar. He wanted to make new arable land available, through the draining of the Pontine Marshes and the Fucino. Caesar's death prevented the realization of this project, that was taken over a century later by Claudius. The emperor decided to build an artificial emissary to regulate the water's level. Even if this wasn't the first time that Romans build artificial emissaries and hydraulic construction, the Lake Fucino's emissary represented a more difficult challenge. This work was complex and hard to achieve, demanded the best methods in tunnels building and innovative technologies. With no written records, it's very difficult to retrace the several phases of the survey and the construction. The emissary was completed in AD 52 and its use continued with Claudius' successors. Trajan and Hadrian took care of the restoration and maintenance of the emissary, as documented by historical and archaeological sources. As a result of indifference, and natural events such as earthquakes, the Roman tunnel lost, in the following centuries, its function as an emissary.

A renewed interest in Lake Fucino emerged in Medieval times. The first one, who took into consideration the idea of restoring the emissary, was Frederick II, in 1240 ca. However, his death cancelled any plan concerning Fucino and there are no reliable reports about practical intervention in this area. In the meantime, issues related to the endorheic nature of Fucino and the lack of any emissary occurred again. The level of water increased without interruption from 1787 to 1816 and the lake reached 35 m in depth.

The Kingdom of Naples, which in those days exercised its government in the Fucino area, considered to carry out such a project with Carlo Afan De Rivera. The military engineer was appointed Director of Public Works of the Kingdom of the Two Sicilies in 1824, but the lack of funding made more difficult the work continuation and then, the death of Afan De Rivera in 1845, temporarily put an end to the Bourbon efforts to figure out Fucino's problem. The definitive solution came with the involvement of Prince Alessandro Torlonia, who thought of radical modifying the project by substituting the Roman emissary with a modern hydraulic construction. The new tunnel differentiates itself from the ancient one, being much larger and, due to its construction, the Roman emissary was almost completely erased.

3. Documentary and iconographic sources

Torlonia's work helped to increase knowledge about Claudio's emissary, together with all its pertaining structures. Nevertheless, this knowledge comes from its definitive and complete destruction. Moreover, even if this is one of the most challenging hydraulic work realized in antiquity, involving the application of new technologies, the Fucino's emissary remains quite unknown throughout the centuries.

IOP Conf. Series: Materials Science and Engineering **364** (2018) 012010 doi:10.1088/1757-899X/364/1/012010

To research what is left about this construction and its functions, it is necessary to collect and pull together in a single representation, data that can be directly measured in site and the ones coming from sources and documentation.

The first documentation available comes from classical sources like Plinius, Tacitus and Cassius Dio, who mainly described only the emissary's inauguration and the historical events involving Claudius and his court. These sources do not contain engineering information about the hydraulic constructions and used technologies. This kind of information can be found in the descriptions of the Roman emissary made by Raffaele Fabretti [1] and Carlo Afan De Rivera [2]. The first one briefly described the Roman construction, including plate and drawing in his work. The last one reported his works to recover the ancient construction in 1836, with the aim to use again the original Roman tunnel and outlet. A. De Rivera describes the methods he used for the underground excavation and the machines needed for it, together with the solutions adopted. These methods and techniques probably didn't diverge too much from the ones used in antiquity. It gives also a very precise description about working environment and condition at beginning of 18th century.

The books published by A. Brisse and L. De Rotrou [4] are the most important and detailed document nowadays left. They supervised the building of the modern tunnel and can be considered as the last witnesses of the Roman emissary, before its final demolition. The ancient construction is described in two volumes and an illustrated Atlas, which also contains the technical description of the modern works to drain Lake Fucino (figure 1). Detailed plates describe the tunnel's itinerary and its longitudinal section, together with wells and secondary tunnels (cunicola). Moreover, the Atlas presents a series of detailed charts with sections of the Roman gallery every ten meters. As regards the *cunicola*, sections and plans are reported in [7].

In the preparation of this work, other sources were used. One is the G. Agricola's treaty, De re metallica [3], which is not a specific work about Fucino's emissary, but it can be used as a frame of reference for wells and tunnels function, as well for working techniques and machinery. A further special evidence for the equipment used for the Roman works is a bas-relief found during the nineteenth-century Torlonia construction. Other elements were pointed out by studying the few remains of the tunnels and the other structures.

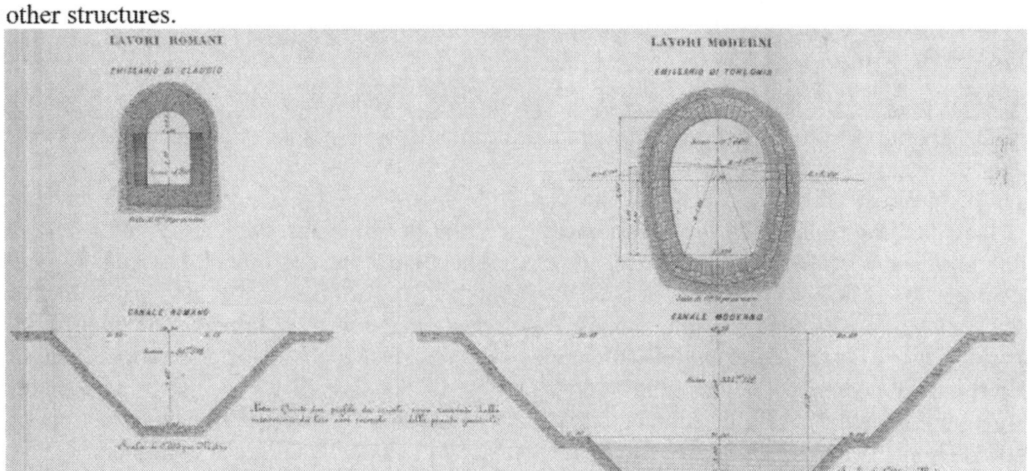

Figure 1. Drawings sections of the Roman emissary compared to the modern one as illustrated in [4].

The documentation above illustrated gives the possibility to offer a series of reconstruction hypothesis about the original method used to project such a difficult challenge, and about the use of suitable types of machinery for lifting and transporting construction materials.

4. Virtual Reconstruction

In this virtual reconstruction, it is necessary to manage a considerable number and complex amount of information. Moreover, the available data here are mostly provided by written sources, especially the

Torlonia's Atlas. Different tools have been used to provide a reliable virtual reconstruction of the ancient hydraulic construction.

The information about the territory's topography are given by IGM cartography (1:25000) and aerial photographs. These were use as basis for the digital terrain reconstruction. The drawing of the tunnel described in the Torlonia Atlas [4] has been digitalized and added to the modern 2D representation, together with the acquisition of topographic points on site. An image processing software erased maps and digital image distortions. It was also necessary to use 2D drafting software, for providing dimensional relation among the different maps used as basis for this work (figure 2).

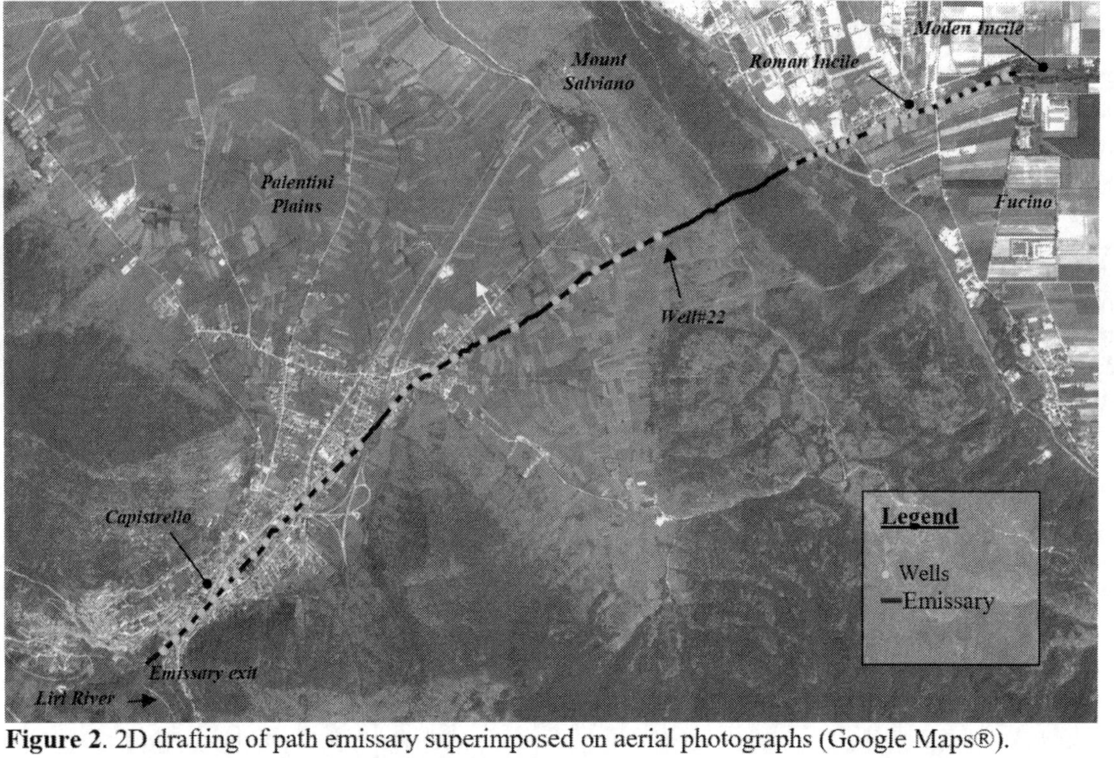

Figure 2. 2D drafting of path emissary superimposed on aerial photographs (Google Maps®).

The 3D tunnel's itinerary too has been modeled following the Torlonia's engineers documentation [4]. In the Atlas, the emissary is well represented and described by plans, together with the heights of its bottom and 600 sections disposed every ten meters. Then, the secondary tunnels *cunicola* [7] and the wells were modeled. *Cunicola* entrances are the only parts of the original roman cladding still existing today. Resulting from detailed researches, textures and rendering tools were applied to 3D models, using a specific software for rendering and 3D animation. Textures related to wells and material construction has been determined based on the study and analyses of cladding in use at that time for similar construction. The textures of the geological strata are symbolic and representative, and they are taken from the Torlonia Atlas (figure 3). They help to understand the tunnel's failure and problems.

4.1. The results

Claudio's work consists in several constructions aimed to drain the Fucino's Lake and functional to the tunnel construction:

- The incile, the artificial outlet;
- The emissary, the main tunnel;
- The cunicola, the secondary tunnels and the wells.

The first one is an opencast collecting canal for conveying Fucino's waters towards a system of tanks: the so-called Incile (figure 4).

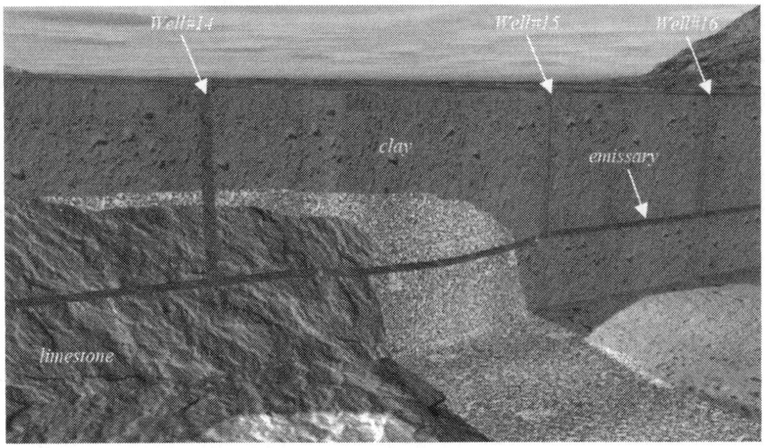

Figure 3. Emissary section in correspondence of Palentini Plains.

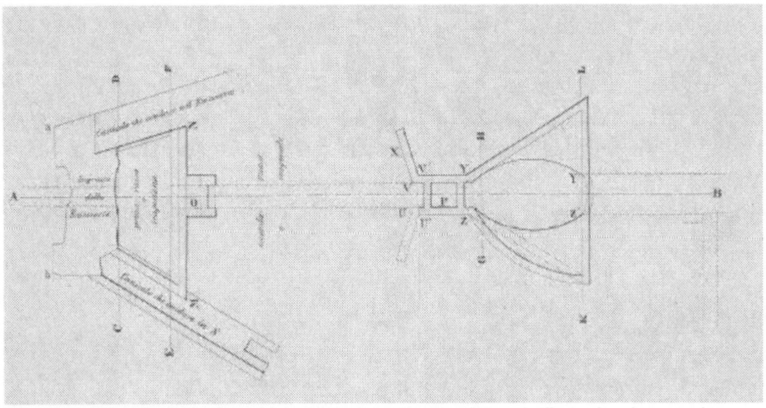

Figure 4. The "incile" as documented in Torlonia's Atlas before definitely destroying and erasing it.

Then, there was the excavation of an underground tunnel, with a route that crossed Mount Salviano and Palentini Plain, ending in the Liri River. Probably original Roman plan didn't consider the draining of all the lake's waters, as happened centuries later with Torlonia's works. Emperor Claudius wanted to regulate the water level of the lake maintaining, therefore, a part of the basin. The activity needed to be planned with a careful levelling, considering the level of Liri river lower than Fucino Lake. Once that the direction was decided, and the length measured, the outlets quotes were fixed, setting out the inclination necessary for the tunnel to be functional.

The main tunnel had to be built following a path far from straight, by setting both the starting and final points. This is illustrated by the reconstruction in figure 5, showing the topography of the area and the itinerary of the emissary (as a red line). Reconstructing and representing this itinerary in 3D allows to understand the technologies used to plan and execute the works, especially as regard the used topography technique [5].

The construction of the main emissary tunnel was difficult to pursue, especially where the path crossed Mount Salviano. To overcome several difficulties, a system of wells and secondary tunnels (*cunicola*) were realized. The system made of wells and secondary tunnels increased the spots where the horizontal tunnel excavation started. Once the well was dug to an appropriate depth, the tunnel excavation started, proceeding on the two opposite fronts (Figure 6).

Different functions can be assigned to these structures:

- They were used as start points for digging the main tunnel, sub-divided into several segments, simultaneously working as direction's reference;
- Facilitated tunnel's ventilation and air flow during excavation's work;
- They were the fastest system to introduce building materials and to dispose of water and resulting material.

The excavation of the wells is located in areas of different soil consistency, so some could suffer unsustainable side pressures. It was therefore necessary to arm them with wooden or wall structures, and both systems were found and illustrated by the engineers involved in Torlonia's work (figure 6b).

At least 40 wells were built in a square section together with the tunnels. 29 of these wells are positioned on the side where the town of Capistrello is located, and 11 beyond the mountain Salviano, on the eastern slope. The depth of the wells is variable from 19 m to the deepest well (number 22), the last on the side of Capistrello, which reaches 122 m in depth. The wells on the western side of Mount Salviano have a depth generally greater than 80m. The wells had wooden scaffolds like the ones represented in the virtual reconstruction in the figure 6a.

The *cunicola* were tunnels directed towards the main gallery, with the same function of the vertical wells, but placed on the two slopes of Mount Salviano. These passages found below Salviano's slopes are seven and constitute the expedient used to deal with one of the most difficult segment of the project. So, they have a function similar to the wells, plus they were used to transport excavation materials and as an entrance for workers and means.

The analysis of 3D textured emissary model helps to understand the tunnel's failure and problems. The section under Mount Salviano, where the tunnel is dug into the compact rock, it crosses a section of rocky debris mixed with pebbles. Here the Romans ran into great difficulties and needed to make a deviation from the preset path, probably because of landslides. The emissary then crosses a long stretch of clay and sand, where the Roman gallery suffered, in the centuries following its realization, the collapse of many tunnel's parts.

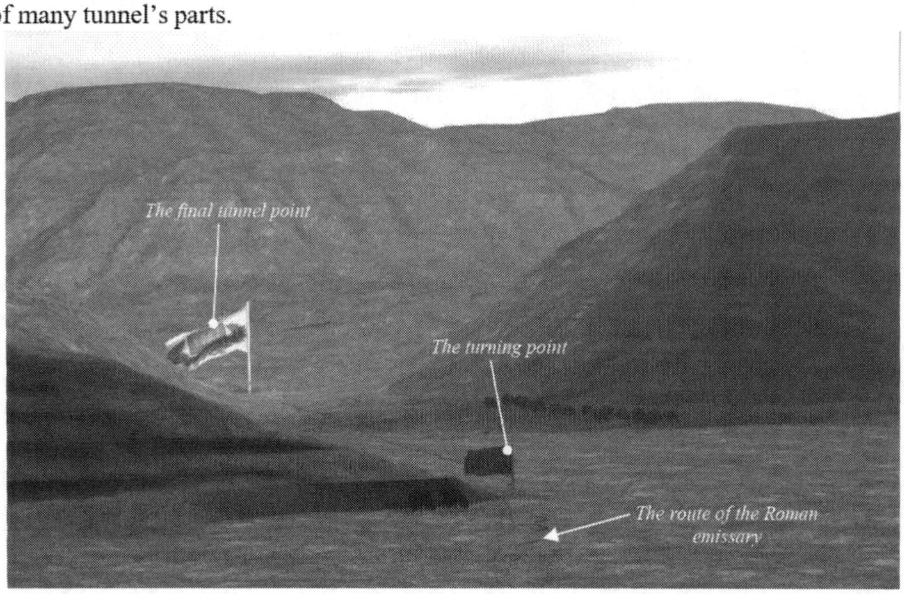

Figure 5. Virtual reproduction of the Palentini Plains territory overlapping the route of the Roman emissary.

IOP Conf. Series: Materials Science and Engineering **364** (2018) 012010 doi:10.1088/1757-899X/364/1/012010

a

c

b

Figure 6a. Virtual reproduction of the centering used in the Roman works.

Figure 6b. Virtual reconstruction of the Roman emissary under construction.

Figure 6c. Reproduction of the Roman wells scaffolding, found intact during the nineteenth century works.

The tunnels located on the side of the lake are Cunicolo Imperiale, Cunicolo del Ferraro, and Cunicolo Maggiore. The latter is the most monumental, with an entrance made on three arches in vertical succession. It is the last descender on the east side of Salviano and is connected, transversely, to the Ferraro's tunnel (figure 7). After the Mount Salviano, there are three descents, sequentially excavated and intercepting well 22, called Salviano 1, Salviano 2 and Salviano 3, or Calderaro tunnel. Not far away there is another tunnel, the Salviano 4, connected to the well no. 21 through a short deviation.

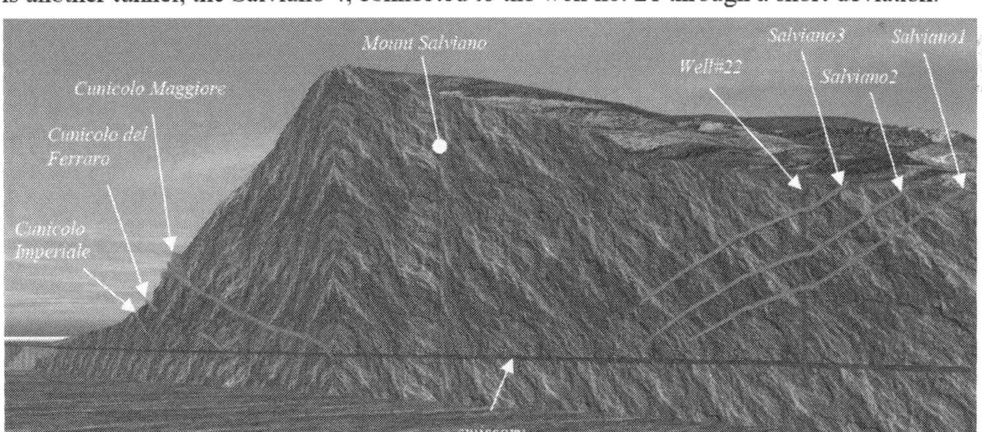

Figure 7. Reproduction of M. Salviano and the tunnels system. The virtual sections of the emissary are achieved following the tunnel route.

The final tunnel will result in an impressive work by the length of approximately 5,650 meters.

The exit of the emissary is on a bend of the Liri River, in a narrow gorge in Capistrello town. It is a monumental exit surmounted by an arch of great height, made of masonry in opus incertum and in opus reticulatum.

5. Conclusion

The virtual reconstruction of the Fucino's Roman emissary here presented aims to investigate and to study the Roman hydraulic work and the technologies applied in its realization. Virtual reality contributes to further understand the different steps in the emissary's construction: how the surveying techniques were applied in such a complex construction project, the knowledge of digging technology, the realization of the main tunnel and the *cunicola*. The 3D modelled tunnels system shows a complex functionality, which the modern virtual reality systems enhanced, stressing for example the tunnels orientation in space. This allowed to revise the technical assessment made later by the Torlonia engineer, which believed the roman project and emissary construction affected by errors that cause several issues through the years.

The virtual reality in cultural heritage is usually applied architectural or landscape reconstruction contexts. In this case, the application is closer to the original use of 3D techniques (Computer Aided Design) used as a tool for functional analysis of artifacts rather than for their historical or artistic representation. The here presented work can be compared to a reverse engineering process, which extracts knowledge and design information from ancient sources, documentation, and drawings of the Roman emissary's archaeological remains. In this work, virtual reality completes the missing data and it is used to reconstruct the model of an ancient hydraulic construction, illustrating its original functionality and giving hypothesis about the Roman organization of work and the building site.

The 3D geometric model could be used to investigate the hydraulic construction considering a different technical point of view. By using, for example, numerical fluid dynamic models, with regard to water flows, but also to study the *cunicola* and wells function analyzing the air flows necessary for the survival of the workers who built the construction. In this way, the engineering resource becomes the means for a deeper understanding of archaeological questions, combining the tools of engineering and archeology and thus promoting the development of the knowledge of ancient technology.

The result of this work was alongside with the production of a film documentary, which describes and summarizes the history of this construction, from the origin to present day. It has an educational purpose, and it is nowadays used to introduce visitors to site visits, but it's still useful to fully understand the technologies behind this hydraulic construction.

Aknowledgments: The city of Capistrello, Italy, contributed to realizing this work by partally funding it.

References

[1] Fabretti R.G. *De columna Traiani syntagma Accesserunt Explicatio veteris tabellae anaglyphae Homeri Iliadem atque ex Stesichoro Arctino et Lesche Ilii excidium continentis et Emissarii lacus Fucini descriptio*. (Roma 1960) pp. 385-420.

[2] C. Afan de Rivera 1836. *Progetto della restaurazione dello emissario di Claudio e dello scolo del Fucino*. (Dalla Stamperia e cartiera del Fibreno Napoli).

[3] Agricola G. 1556 *De Re Metallica* (Basilea)

[4] A. Brisse, L. De Rotrou 1876 *Desséchement du lac Fucino exécuté par S. E. Alexandre Torlonia: précis historique et technique* (Roma, Impr. De la Propagande)

[5] Grewe K. 1998 *Licht am Ende des tunnels. Planung und trassierung im antiken tunnelbau.* (Mainz am Rehin, Von Zabern)

[6] Pietrantoni A. 2007 *Il Fucino e il suo emissario, cenni storici-critici*. (Girardi Editore)

[7] AA. VV. 1994 Rilievi dei cunicoli romani. In Burri E. Campanelli A. *Il lago Fucino e il suo emissario: sulle rive della memoria* (Pescara)

[8] Giuliani C.F. 2007 Sfidando gli inferi: problemi di cantiere dell'emissario del Fucino. In Camporeale, S.–Dessales, H.–Pizzo, A. *Arqueología de la construcción, I, Los procesos constructivos en el mundo romano: Italia y provincias occident*ales. (Mérida, Instituto de Arqueología, 25-26 de octubre de 2007) pp. 33-47.

ADAPTIVE EXHIBITION TOPOLOGIES FOR PERSONALIZED VIRTUAL MUSEUMS

Vasileios Komianos
Dept. of Informatics, Ionian University, Corfu, Greece
E-mail: vkomianos@ionio.gr

Konstantinos Oikonomou
Dept. of Informatics, Ionian University, Corfu, Greece
E-mail: okon@ionio.gr

Abstract. The paper addresses the issue of virtual museums' space personalization. Personalization techniques are poorly integrated in virtual museums implementations and when applied little attention is given in the abilities, requirements and needs of their visitors. In this paper, the concept of adaptive exhibition topologies is introduced. Adaptive exhibition topologies are patterns of space structure that are specifically designed in order to provide unlimited exhibition space for personalized exhibitions based on users' interests and visiting preferences. Adaptive exhibition topologies focus on the time that visitors are willing to spend for their visit. In contrast to conventional implementations, adaptive exhibition topologies instances are created according to their visitors' interests and their available time while taking into consideration certain curatorial guidelines. A number of topologies is introduced as well as appropriate metrics for their study and comparison reasons. The effect of different exhibition topologies on users' navigation and interaction with the exhibits is studied through the conducted usability and evaluation tests.

Keywords: adaptive exhibition topologies, virtual museums, personalization, navigation

1. Introduction

A significant number of works aiming at virtual cultural heritage corresponds to the development of *virtual museums* and *virtual exhibitions* represented in 3D virtual environments [1]. Virtual museums and virtual exhibitions have been proposed as effective media for promotion of cultural heritage as they provide solutions for many issues including the lack of available exhibition spaces, their high economic cost and their limited accessibility imposed by geographical constraints [2]. In addition, traditional exhibitions are mostly static and any curatorial interventions are difficult to be performed due to the time, effort and cost needed [3, 4]. Due to their static nature, traditional exhibitions, provide limited personalization abilities which are mostly focused on the exhibition route that is followed and on the narration of the observed artifacts [5-7].

Virtual exhibitions on the other hand, provide theoretically unlimited space, are comparatively less costly, they are world-wide accessible, curatorial changes can be performed immediately [8], provide ease of access especially for the disabled, all the artifacts can be exhibited and users are able to interact with [2]. In addition, virtual exhibitions can be personalized promoting artifacts that will be of interest to their visitors while their design can be adapted to their needs or further preferences. Personalization provides mitigation of information overload [9] and the ability to control the amount of time dedicating for a information related task [10, 11]. Information overload is considered to be a factor of museum fatigue [12] and museum fatigue is also considered to be increased in time [13].

In this paper, the concept of *adaptive exhibition topologies* [14] which aim to provide automated construction of virtual exhibitions is presented. Adaptive exhibition topologies provide exhibition spaces that are able to host a variable number of artifacts where the size of the constructed virtual exhibitions and the number of hosted artifacts are determined according to visitors' available time and visiting preferences. The topologies are designed in order to facilitate visitors' navigation and interaction with the exhibited artifacts. Focusing on navigation, special attention is payed on the ability of the presented topologies to provide alternative exhibition routes of reduced distance given the fact that distance in both real and virtual environments is closely linked with visiting time and fatigue [15].

In Section 2, relevant literature and recent related work is reviewed. In Section 3, the proposed adaptive exhibition topologies are presented, their adaptive size properties and their ability to enhance visitors navigation by providing alternative routes of reduced distance are discussed. In Section 4, conclusions on the adaptive exhibition topologies are discussed and guidelines for future work are drawn.

2. Literature Review

Museums' and exhibitions' design has been thoroughly studied as it affects visitors behavior and thus is considered an important factor of their success [16]. Psarra [17] study the effect of spatial structure on visitors' experience while Wineman and Peponis [18] argues that spatial structure shapes the ways in which visitors explore, engage, and understand museums and museum exhibitions. In [19] two, almost opposite, exhibition layouts are studied: (i) the *single sequence layout*; and (ii) the *grid layout*. In the single sequence layout, visitors have to traverse a specific sequence of space components. In the grid layout visitors for each position they are placed they are able to choose between the available neighboring space components resulting maximized randomness in the pattern of movement and exploration. In [20] it is stated that museum layouts can be separated in two models: (i) the deterministic; and (ii) the probabilistic, the deterministic model corresponds to the single sequence layout and the grid layout falls into the probabilistic model.

Exhibition design, apart from its spatial characteristics, is closely linked with the arrangement of the exhibited artifacts. Choi [20] notes that the objects on display are usually grouped according to classificatory principles which may be as direct as chronological or geographical origin, or as elaborate as a refined stylistic discrimination and that museum space provides a physical realization of classificatory principles which are supposed to make the collections accessible to understanding.

Regarding virtual museums and virtual exhibitions there are numerous related works but despite their number the literature review reveals that these are not focused on the spatial structure of exhibitions and the user navigation constraints. Kadobayashi et al. [21] propose personalizing museum exhibitions to fit the needs of each visitor by having agents mediate between curators and museum visitors and having the exhibitions be reconstructed for the visitors based on their interests. Wojciechowski et al. [3] present a system that allows museums to build and manage virtual and augmented reality exhibitions based on pre-designed

IOP Conf. Series: Materials Science and Engineering **364** (2018) 012011 doi:10.1088/1757-899X/364/1/012011

visualization templates that allows content designers to create virtual exhibitions very efficiently. White et al. [8] present a tool that allows virtual exhibitions to be designed by the use of a presentation manager where the structure of the exhibitions is defined by the structure of exhibition spaces. Bonis et al. [22] presents a content personalization platform for virtual museums which is based on a semantic description of content and on information implicitly collected about the users through their interactions with the museum. Kiourt et al. [23] present a web-based virtual museum framework that provides a complete authoring interface, in which users can create customized virtual exhibitions. Hayashi et al. [24] present a system for automatic generation of personalized virtual museums capable of exhibiting arbitrary planar artifacts, e.g. paintings, which have been specified by the user. The system based on the input automatically creates a 3D environment with flexible walls changing its space in accordance with the specified paintings displayed.

3. Adaptive Exhibition Topologies

In this paper, adaptive exhibition topologies are proposed for the construction of virtual exhibitions' spaces. The proposed adaptive exhibition topologies provide exhibition designs of variable size in which a certain number of objects are allocated according to curatorial [20] or recommendation instructions based on user's interests [25]. It is stressed that an exhibition's size as well as the number of recommended arranged objects can be determined based on user's available time for the specific visiting session.

Graphs are used to represent the proposed topologies. The topology graphs consist of vertices representing exhibition objects O and edges representing the paths P enabling the transition among them. An object could be a distinct entity of interest or even a room of a museum containing multiple entities of interest belonging to a particular category or context.

The following exhibition topologies are presented in the next paragraphs: (i) sequential; (ii) looped sequential; and (iii) grid.

3.1. Sequential Topology

In sequential topologies the objects are organized in sequence according to specific instructions, i.e., recommendation ranking, chronological arrangement, etc., having only one available route which users are guided to follow, Figure 1. The benefits of the specific layout is that the objects are arranged according to the specified instructions, the visitor needs no further guiding instructions in order to navigate in it and the complete sequential distance can be kept under a certain limit.

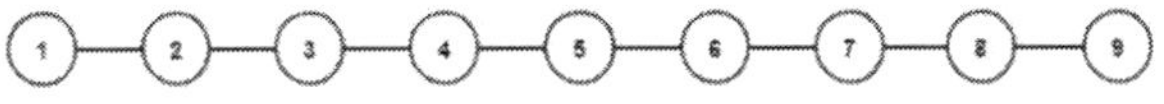

Figure 1. Sequential layout.

3.2. Looped Sequential Topology

Looped sequential topology is proposed as an advancement of the sequential topology that provides alternative routes of reduced distance. In addition to the sequential topology the sequence forms a loop by adding a path (edge), $p_n = (o_n, o_1)$, connecting the n_{th} object, o_n, with the first object of the sequence, o_1 Figure 2.

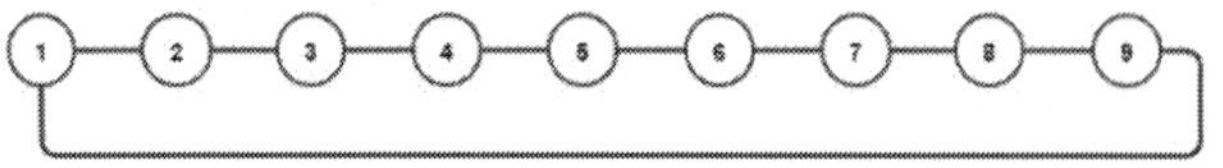

Figure 2. Looped Sequential layout.

3.3. Grid Topology

In order to further reduce the overall distance of a setting additional paths have to be introduced. The sequential topology is enhanced by separating the sequence in segments of equal distance, the segments are then connected forming a two-dimensional grid, Figure 3. A grid's size and its capacity is measured in number of objects O and is defined by each dimension's size given by $O = m \times n$. The objects are allocated in the available positions in each of the m rows and n columns.

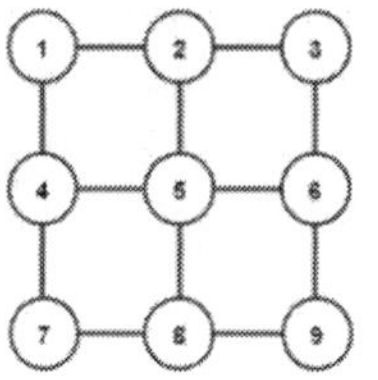

Figure 3. Grid topology instance of dimensional size 3 × 3 containing 9 objects.

3.4. Topology metrics and properties

In order to study the properties of the presented topologies and to be able to compare them a set of metrics related to the traversed distance for a movement from a starting object o_s to a target object o_t is introduced. All the introduced distance metrics are measured in edges (paths) which are considered to be of the same length $p_{1...n}(l) = 1$.

The *sequential pair distance*,

$$D_S(o_i, o_j) \tag{1}$$

is used to calculate the distance that a visitor has to traverse in order to move from and object o_i to an object o_j given that he/she follows the exhibition sequence.

When visitors visit an exhibition there are cases that they do not follow the recommended exhibition sequence. In such cases there should be alternative exhibition routes of reduced distance. The distances of these alternative routes are given by the *minimum pair distance* metric,

$$D_M(o_i, o_j) \tag{2}$$

In order to measure the ability of a topology to provide alternative routes of reduced distance, the *overall minimum distance*, D_{OM}, is introduced. The overall minimum distance D_{OM}, is calculated as the sum of the distances of the shortest routes connecting every pair of objects.

$$D_{OM} = \sum_{i=1}^{i=n-1} \sum_{j=i+1}^{j=n} D_M(o_i, o_j) \tag{3}$$

3.5. Comparison

The presented topologies are compared for their ability to provide alternative routes of reduced distance. For the comparison the overall minimum distance (Equation 3.4) for topologies instances of varying size is calculated. Figure 4 shows that sequential topology provides the maximum overall distance as it does not provide alternative routes. Looped sequential topology provides reduction of the overall minimum distance (average reduction of 24%) and further reduction is provided by spiral topology (average reduction of 58%).

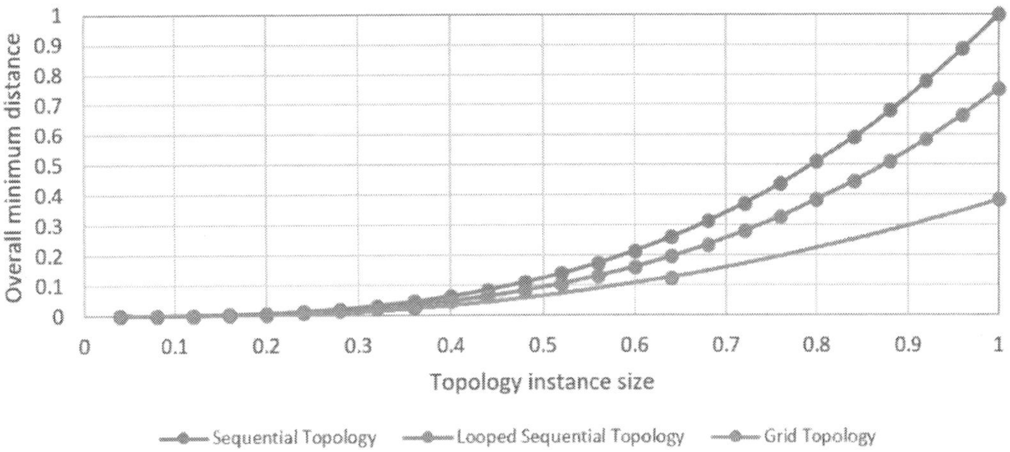

Figure 4. Overall minimum distance (Equation 3.4) of (i) a sequential layout; (ii) a looped sequential layout; and (iii) a grid layout. The X axis depicts the size of the layout and the axis Y the overall minimum distance. Sequential layout dose not provide alternative routes and does not reduce the overall minimum distance, looped sequential provides average reduction by 24% and grid topology provides average reduction by 58%.

3.6. Navigation Scenarios Evaluation

A testing application (Figure 5) incorporating the described exhibition topologies is developed and used in order to evaluate the efficiency of each exhibition topology in real navigation scenarios. A series of tests including 10 random navigation tasks in: (i) sequential topology; (ii) looped sequential topology; and (iii) grid topology of 25 objects is given to test users. Each random navigation task is tested on all the described exhibition topologies and the time needed in order to complete the task is automatically recorded by the application. For each topology, the average transition time resulting by the random navigation tasks is calculated. The maximum of the average values (the average time of sequential topology) is used as the base value for normalizing the resulted average values for comparison reasons. The results (Figure 6) confirmed that looped sequential topology and grid topology offer alternative routes of reduced distance which can be used by user in real navigation scenarios. More specifically, looped sequential topology provided reduced navigation time by 39% and grid topology provided reduced navigation time by 81%. It is noted that tests resulted greater reduction values than those calculated in Paragraph 3.5. This is attributed to the fact that the tested navigation scenarios include pairs of object of relatively large distance. In case of more navigation scenarios including pairs of objects of shorter distance the average navigation time is expected to be increased accordingly to the results listed in Paragraph 3.5.

Figure 5. Test application. Users are requested to visit specific exhibits assisted by a target indication method. The amount of time needed for their navigation is automatically recorded by the test application.

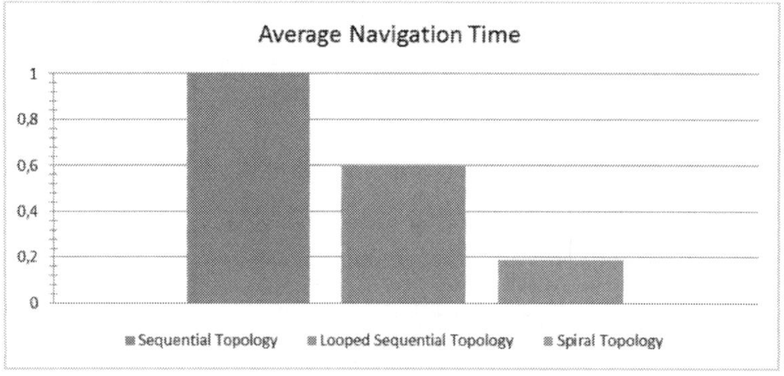

Figure 6. Visiting scenarios average navigation time comparison. In the conducted tests: (i) looped sequential topology provided reduced navigation time by 39%; and (ii) grid topology provided reduced navigation time by 81%.

4. Conclusions

The present paper's purpose is to propose adaptive exhibition topologies as a method for automatic construction of personalized virtual museum exhibitions. Adaptive exhibition topologies aim to provide flexible exhibition spaces that enhance visiting experience by focusing on facilitation of visitors navigation. Sequential topology appears to be appropriate solution in cases that visitors follow a sequence route and interaction with exhibits is mostly focused on consecutive exhibits. Looped sequential topology is proposed for cases of sequence routes that for a visitor to return to start point no further distance should be added to the route. Grid topology enhances visiting experience by providing alternative routes of reduced distance in cases that visitors navigate through the exhibition without following a strict sequence route. Finally, this work indicates that alternative adaptive exhibition topologies can be used depending on the exhibition purpose or certain curatorial approaches.

5. Bibliography

[1] Komianos, V., Kavvadia, E. and Oikonomou, K. (2014). Efficient and realistic cultural heritage presentation in large scale virtual environments, *Information, Intelligence, Systems and Applications*, IISA 2014, The 5th International Conference on, IEEE, pp. 1-6.
[2] Tsichritzis, D., Gibbs, S. J. et al. (1991). Virtual museums and virtual realities., *ICHIM*, pp. 17–25.
[3] Wojciechowski, R., Walczak, K., White, M. and Cellary, W. (2004). Building virtual and augmented reality museum exhibitions, *Proceedings of the ninth international conference on 3D Web technology*, ACM, pp. 135–144.
[4] Styliani, S., Fotis, L., Kostas, K. and Petros, P. (2009). Virtual museums, a survey and some issues for consideration, *Journal of cultural Heritage* 10(4): 520–528.
[5] Terrenghi, L. and Zimmermann, A. (2004). Tailored audio augmented environments for museums, *Proceedings of the 9th international conference on Intelligent user interfaces*, ACM, pp. 334–336.
[6] Wang, Y., Stash, N., Sambeek, R., Schuurmans, Y., Aroyo, L., Schreiber, G. and Gorgels, P. (2009). Cultivating personalized museum tours online and on-site, *Interdisciplinary science reviews* **34**(2-3): 139–153.
[7] Sparacino, F. (2002). The museum wearable: Real-time sensor-driven understanding of visitors' interests for personalized visually-augmented museum experiences.
[8] White, M., Mourkoussis, N., Darcy, J., Petridis, P., Liarokapis, F., Lister, P., Walczak, K., Wojciechowski, K., Cellary, W., Chmielewski, J. et al. (2004). Arco-an architecture for digitization, management and presentation of virtual exhibitions, *Computer Graphics International*, 2004. Proceedings, IEEE, pp. 622–625.
[9] Bawden, D. and Robinson, L. (2009). The dark side of information: overload, anxiety and other paradoxes and pathologies, *Journal of information science* **35**(2): 180–191.
[10] Radev, D. R., Fan, W. and Zhang, Z. (2001). Webinessence: A personalized web-based multi-document summarization and recommendation system, *Ann Arbor* 1001: 48103.
[11] Liang, T.-P., Lai, H.-J. and Ku, Y.-C. (2006). Personalized content recommendation and user satisfaction: Theoretical synthesis and empirical findings, *Journal of Management Information Systems* **23**(3): 45–70.
[12] Bitgood, S. (2009). Museum fatigue: A critical review, *Visitor Studies* **12**(2): 93–111.

[13] Davey, G. (2005). What is museum fatigue, *Visitor Studies Today* 8(3): 17–21.
[14] Komianos, V. (2017). Content personalization approaches in cultural heritage (real, virtual and mixed) environments, Doctoral Thesis, Department of Informatics, Ionian University .
[15] Komianos, V. and Oikonomou, K. (2015). Constrained interest-based tour recommendations in large scale cultural heritage virtual environments, *Information, Intelligence, Systems and Applications (IISA), 2015 6th International Conference on*, IEEE, pp. 1-6.
[16] Bitgood, S., Patterson, D. and Benefield, A. (1988). Exhibit design and visitor behavior: Empirical relationships, *Environment and behavior* **20**(4): 474–491.
[17] Psarra, S. (2005). Spatial culture, way-finding and the educational message, The Impact of Layout on the Spatial, Social and Educational Experiences of Visitors to Museums and Galleries In: McLeod, S ed. Reshaping Museum Space: Architecture, Design, Exhibitions. London and New York: Routledge pp. 78–94.
[18] Wineman, J. D. and Peponis, J. (2010). Constructing spatial meaning: Spatial affordances in museum design, *Environment and Behavior* 42(1): 86–109.
[19] Tzortzi, K. (2007). Museum building design and exhibition layout, Proceedings of the 6th International Space Syntax Symposium, Istanbul, Turkey, Vol. **1215**, p. 072.
[20] Choi, Y. K. (1999). The morphology of exploration and encounter in museum layouts, *Environment and Planning B: Planning and Design* 26(2): 241–250.
[21] Kadobayashi, R., Nishimoto, K., Sumi, Y. and Mase, K. (1998). Personalizing museum exhibition by mediating agents, Tasks and methods in applied artificial intelligence pp. 648–657.
[22] Bonis, B., Stamos, J., Vosinakis, S., Andreou, I. and Panayiotopoulos, T. (2009). A platform for virtual museums with personalized content, *Multimedia tools and applications* **42**(2): 139–159.
[23] Kiourt, C., Koutsoudis, A. and Pavlidis, G. (2016). Dynamus: a fully dynamic 3d virtual museum framework, *Journal of Cultural Heritage* **22**: 984–991.
[24] Hayashi, M., Bachelder, S. and Nakajima, M. (2016). *Automatic generation of personal virtual museum, Cyberworlds (CW), 2016 International Conference on*, IEEE, pp. 219–222.
[25] Komianos, V., Kavvadia, E. and Oikonomou, K. (2015). Cultural heritage recommendations and user navigation in large scale virtual environments, *International Journal of Computational Intelligence Studies* **4**(2): 151–172.

IOP Conf. Series: Materials Science and Engineering **364** (2018) 012012 doi:10.1088/1757-899X/364/1/012012

The Issues of Humanoid Robots with Artificial Creativity in terms of Cultural Heritage and Art

Melik Ziya Yakut
Mechatronics Engineering, Faculty of Technology, Suleyman Demirel University
Isparta, Turkey
ziyayakut@sdu.edu.tr

Abstract. Modern life is changing ever more complex and technologically. The technology is utilized in many ways in the life of humans, and even in many of them, human is used as supportive devices inside and outside of the human body. The greatest change in our lives is that human beings can produce more humanoid robots that can learn, plan, reasoning, problem solve, comment, feel, have intelligence. Because of this situation, many things will change in this life which is designed according to human beings. Cultural heritage and art will have an important place in these subjects. Nowadays, there is still discuss on the issues and limitations of cultural heritage and art subjects. Humanoid robots will add problems that await many solutions. When humanoid robots become so productive and become acceptable in society, there is no doubt that many complicated ethical issues will arise. These issues should be determined, discussed and solutions should be given as detailed as possible. The biggest problem of human being is that the solutions will not only be observed by human rights, but also the humanoid robots' rights to be taken into consideration at this time.

1. Introduction

Today, the influence of technology on human life is increasing day by day in modern life and many different complex technological products are being developed. Hardware and software products started to build strong links with the Internet. They are designed to help human works that are using internet and artificial intelligence technology. As a result of the advancement of these technology humanoid robots are doing works for humanity.

Recently humanoid robots are discussed with the development of their emotional intelligence that they can now do jobs based on creative intelligence. They work in many different areas that are difficult to do, especially artistic work, and this causes some new problems to arise and discuss.

How the art can be defined has been discussed for years, and is still debated. The content of art is expanding historically. The art evolved as it challenged the age, the values, taboos, rules and limitations of the era.

For the first time in 1956, the expression of "artificial intelligence", used by a group of researchers, has become one of the most common concepts in recent years. Artificial intelligence, as is known, is a computer or computer-controlled robot that analyses human thinking methods and develops similar artificial directives. Artificial intelligence is increasingly being used in more areas of daily life. Elon Musk, Stephen Hawking, Mark Zuckerberg, Bill Gates, and Ray Kurzweil have various ideas about artificial intelligence in the shaping of today's world. Many companies, automobile brands and social media networks have begun to benefit from the potential of artificial intelligence [1].

Artificial creativity is a multi-disciplinary software that combines the areas of artificial intelligence, cognitive psychology, philosophy and art.

Art is the study of ethics and aesthetics at the core, using knowledge as well. This research is easier to understand than scientific research at the same time. Art is influenced by science and technology. Today's technology offers possibilities that begin to use too. This is why a "Photoshop art" and "digital art" are mentioned.

The new generation humanoid robots can produce art-related products, thanks to their high-level experience, to comment on events. Nowadays, there is still a humanoid robot with a local culture. Because they interpret the events with the culture of the creators who created them.

There will be no humanoid robots produced by humans in a very short time, because there are other robots that produce and program humanoid robots. These trials have already begun, but they need more time to develop. In the near future, humanoid robots that can perform artificial creativity will be seen more often. In this study, it was evaluated what humanoid robots can produce for art and what kind of problems they face.

2. The relationship of technology to art

With the development of technology in the late 1980s and early 1990s, cybernetic organisms (cyborg), a combination of human and machine beings, were thought to be things that would be possible with the technology of the future. But many people around the world have gradually begun to turn themselves into a cyborg [2].

The Spanish avant-garde artist Moon Ribas is able to connect the chip on the inside of her elbow to an application on his phone and can feel the earthquakes happening in the world at that time. Figure 1 also appears to present her performance on the stage. The more severe the quake is, the more the device on Ribas's arm is vibrating. Turning these vibrations into a dance on stage, the artist is dancing with a kind of world by developing improvised choreography based on the severity of earthquakes [3].

Figure 1. Presenting performance on stage [4].

Performance artist Stearc implanted an artificial ear in the inner part of his left arm to transform his body into a listening device. The artist, who wants to put a microphone in this ear and connect the microphone to the Internet, aims to ensure that anyone who wants the voices in the environment can listen at any time [5].

Nowadays, many Internet-connected devices and smart applications emerged, so many branches of art have been very impressed with these technological products. Some of these branches of art are still debated whether they are artistic or not. In particular, the new generation of high-quality camera systems, which are also used in cinema, photography, and the drones used in some shots, have high-capacity

IOP Conf. Series: Materials Science and Engineering **364** (2018) 012012 doi:10.1088/1757-899X/364/1/012012

computers that work. In addition, virtual reality glasses, smart mobile phones, virtual reality applications, wearable technological equipment have been developed to provide users with a high sense of expertise.

Technology, which gives new ideas to the artist with its supporting technological products connected to many Internet sites. However, today, software-supported technologies are more prominent.

3. Artificial intelligence in art

In the recent past, now and in the future, artificial intelligence continues to help the artist and critics in many areas of art. It is becoming more and more beneficial in every stage. Many films and novels can predict the continuation of the series.

Artificial intelligence was used in novel writing. Harry Potter, a legendary fantasy novel series, was read by an artificial intelligence, and then using this software to write a storyline as a continuation of the series. The story was about Harry's close friend, Harmonie and his family [6].

After thinking, it is capable of transforming photos and pictures in the use of imagination. Researchers have designed an artificial intelligence program that can switch images from a picture to a computer that can change the photos or paintings they upload in a variety of ways. In this way, they can make many famous paintings a real photograph, while also placing the different type or colour of the objects found in the photos.

Figure 2. Image to image translation samples [7].

Their algorithm learns to automatically "translate" an image from one into the other and vice versa their method applied to several translation problems shown in Figure 2 [7].

In another article, a newly developed software is used to distinguish the origins of famous tableware of historical value from fake ones. Fraud in artwork is difficult to understand and requires a number of professional processes that require high cost. Art historians benefitted from technology for many years already to distinguish fake tables. Infrared scanning systems and many electronic inspection techniques were not able to fully reveal fraud. With artificial intelligence, fake tables can be distinguished without the need for human interpretation. The software for this, looking at the brush strokes on the charts, gives the result about the original condition of the table in a short time. Researchers at Rutgers University in Netherlands and the government-affiliated art organizations; Around 300 paintings by Picasso, Matisse,

Modigliani and other famous artists examined the original parts of the artist, which allocated to 80000. Then these analyses were taught to artificial intelligence software. The software has reached its capacity to identify each original brush stroke. At the same time, the software is receiving support from machine learning to distinguish the shapes of brush marks based on specific characteristics. So as the painting is introduced to the system, artificial intelligence is starting to give better results. Because the machine learning algorithm is trained on certain features, it is said to be the most ambitious artificial intelligence project ever made in this area.

In all the brush strokes that are unique to the artists themselves, there are different strengths. So it's even harder for the software to distinguish the original ones. Researchers proved how successful the system was by obtaining an 80% accuracy in the first stage. In the later stages, artificial intelligence has been made to recognize a single brush stroke to distinguish the original paintings from the fake ones. As it is known, it takes years to specialize in paintings as a human art history expert and to the same level as the software. In fact, the research team says the software exceeds the level that a human can reach [8].

In fact, now it's thinking and able to distinguish and physically apply. A robot can write with very difficult and complex kinematic control algorithms and image processing application on nowadays. The first created font database and human movements propose a robotic calligraphic system that uses Chinese characters and English to write words. The three-dimensional motion detection is working an input device to create the font database and capture the human arm orbits. Thus, this study created a humanoid robot that can write handwriting like humans [9].

Google continues to work to speed up the learning process. One of them is the website named AI Experiments (artificial intelligence experiments). In this study, people can play with artificial intelligence programs from the website and as they are played, these virtual machines develop themselves. Thus, with the data collected from humans, it is used to produce more humanoid outcomes. All these technological products, demands and the curiosity of human beings, caused the emergence of humanoid robots and their artificial creativity.

4. Artificial creativity of humanoid robots

Thanks to technology, humanoid robots ' senses and their ability to interpret will begin to increase. Furthermore, researchers at MIT's Computer Science and Artificial Intelligence Laboratory (CSAIL) are able to learn how human feel emotionally through wireless signals [10].

In this way, a humanoid robot will be able to measure how humans react to events at that time. It will be able to feel human's instant feelings from news in the media, from their statements on the street, and reflect their feelings into productive work as they interact. This will have a positive effect on artificial intelligence, giving artificial creativity a spirit. These sensors and databases can create artificial conscience. It can create instant feelings through instant events, various parametric data, and social interaction. Based on these feelings, it can make a picture, it can write a poem, it can prepare a music, or a different kind of products. It could be an emotional expression, a physical expression. It will even improve itself by watching the effects of these outputs on humans. It will soon be possible to measure and change the effectiveness of the sensory music, then it is preparing and to get a better result. This will ensure that there are as many alternatives as possible.

Creativity is the phenomenon of creating something new and valuable in a way. Creativity is the ability to solve problems outside of existing options. In psychology and cognitive science, this research area is called creative problem solving. Artificial creativity should be a program that can model and continually improve the creative intelligence that fulfills the role of the potential of the creative minds of human as a target.

Based on its historical development, the subject of creativity has been tried in many artistic fields and every study show that they can be more successful in getting humanoid results. First of all, computational creativity in the generation of visual art has had some notable successes in the creation of both abstract art and representational art. The most famous program in this domain is Harold Cohen's AARON, which has been continuously developed and augmented since 1973. Other software artists of note include the NEvAr system (for "Neuro-Evolutionary Art") of Penousal Machado. NEvAr uses a genetic algorithm to derive a mathematical function that is then used to generate a coloured three-dimensional surface. The artist Krasimira Dimtchevska and the software developer Svillen Ranev have created a computational system combining a rule-based generator of English sentences and a visual composition builder that converts sentences generated by the system into abstract art. The Painting Fool, developed by Simon Colton originated as a system for overpainting digital images of a given scene in a choice of different painting styles, colour palettes and brush types. Given its dependence on an input source image to work with, the earliest iterations of the Painting Fool raised questions about the extent of, or lack of, creativity in a computational art system. In August 2015 researchers from Tübingen, Germany created a convolutional neural network that uses neural representations to separate and recombine content and style of arbitrary images which is able to turn images into stylistic imitations of works of art by artists such as a Picasso or Van Gogh in about an hour. Their algorithm is put into use in the website DeepArt that allows users to create unique artistic images by their algorithm. In early 2016, a global team of researchers explained how a new computational creativity approach known as the Digital Synaptic Neural Substrate (DSNS) could be used to generate original chess puzzles that were not derived from endgame databases. The DSNS is able to combine features of different objects (e.g. chess problems, paintings, music) using stochastic methods in order to derive new feature specifications which can be used to generate objects in any of the original domains [11].

With the addition of many technology companies, these developments have become more widespread. A new study by Google's artificial intelligence company DeepMind revealed that "imagination" was added to artificial intelligence. This innovation; it means that artificial intelligence systems are able to think about the movements they are doing and gain the ability of reasoning. This system contains the "imaginary coder", which helps you to decide what estimates are useful or useless about what is happening around artificial intelligence. Google's artificial Intelligence laboratory, AI Lab and DeepMind are working to make artificial intelligence, plan for the future, and even dream. Given the fact that artificial intelligence is based on algorithms, it is not possible for now to replace human intelligence. Artificial intelligence does not gain human skills, such as reasoning or critical thinking, but develops a profound understanding of effects and outcomes. The Google's artificial intelligence is that it uses the basic skills of human used. This system uses a combination of reinforced learning that involves learning by means of trial and error, so deep learning can process large amounts of data in a very similar way to the human brain [12].

In one example, the company's project is from Magenta. Magenta, which Google has developed with art and music focus, is trying to develop artificial intelligence's music ear. Neural networks and artificial intelligence, which learned four music notes, formed a piano melody with these notes. After learning the artificial intelligence by listening to the notes, it tries to play the music instrument himself. Developing an application that helps the artist in a creative way, from an intelligence that is targeted with the magenta, to a self-made art. These algorithms, which are capable of painting and music for the moment, seem to force mathematics in the future [13].

The program, Google Deep Dream, is already drawing beautifully from many of us. The Program scans millions of photos by pixel pixels, defining entities, identifying the colours and all the tones, and then scanning the boundaries between objects. It's making a catalogue of all the artifacts and photographs it has scanned over time. It organizes and categorizes objects with similar characteristics and then uses them to create compositions on landscape templates shown in Figure3.

Figure 3. A sample of Google Deep Dream algorithm [14].

An algorithm developed with a joint study of Rutgers University, Charleston College and Facebook Artificial Intelligence labs, works with a new technique 2-step. First, the section called "distinctive" is loaded with tens of thousands of works by thousands of artists. Then the second part called "creative" begins to create images randomly. These two parts are trying to trick each other. The "distinctive" eliminates the works produced by the "creative" by making comparisons with the works that are uploaded to it. As a result, artificial intelligence produced so convincing works of art that even art experts were surprised. The experts couldn't figure out which paintings were produced by human, which were produced by artificial intelligence. A machine that works with artificial intelligence managed to produce unusual-style pictures and the products that were introduced won the admiration of many people. When artificial intelligence began to produce paintings, people were invited to investigate online to evaluate these products. These people didn't know that the painting they were going to take were made by a machine. The participants were asked how complex and original each picture was. They were also asked if the paintings inspired them and if they enjoyed looking at these works. A surprising result came out, and the paintings made by the machines received higher scores in most cases than those made by human [15].

These studies show that more soulful and insightful robots will always be created. These robots will reveal the product with more humanoid emotions. In addition, these humanoid robots will evolve into various levels according to their level of sophistication and interpreter coding. This will lead to the creation of a digital culture of humanoid robots. This will evolve by changing the level of technology in the age and the culture of the creators ' perspectives.

5. Conclusions:

At present, it is not possible to determine what boundaries and rules of art should be. Even cinema and photography are still being discussed. Even the 'Dada' movement, which ignores the essence, form and language conception of art theories, and which is close to the futurism movement, has laid the groundwork for the surrealism movement, should be discussed according to many critics. From this point of view, if the humanoid robots provide the art world, they will find themselves more difficult on the verge of such discussions. In the near future, the works of humanoid robots can only be regarded as technological products that can help artists.

Humanoid robots will be able to give better results if they live in this world like humans with their high sensors and a well-coded evolutionary software. Perhaps it can be experienced in an age when humanoid robots can produce and evaluate their own works in a world that is more active in the far future.

Furthermore, these products may also become a digital cultural heritage for the future. In this regard, however, another problem is faced. It's ethical. The copyright of the productions is the humanoid robot itself? Or programmer? Or are they a common product of the owners that the uploaded data? At this point, it may seem too simple, but what happens if artificial intelligence starts writing its own software? If the programmer or the software company of artificial intelligence becomes the owner of the emerging products, an artificial intelligence that can think about things that a human can't even imagine, can become the owner of everything over time.

Patents are often given to inventions that are the product of the human mind. But what if the most important question of the recent time is to invent a non-human resource like artificial intelligence? Should computers be able to obtain patents as inventors? Ryan Abbott, professor of law at the University of Surrey [16], thinks computers should be able to get patents. He is also a member of the US Patent and Trademark Office (USPTO), who is requesting a fundamental change in this field. He also emphasized that computers should be defined as inventors, thus contributing to the development of more creative computers and the development of more innovative in society. It turns out that the existing patent laws do not accept that non-human inventors are producing products that can be challenged, that creative computers, the foundations of the inventor's concept, and perhaps the entire patent system need to be re-examined. It is not difficult to get a patent if there is something that can be protected by a patent. A patent may be granted if the criteria are met. According to the current rules, there are three basic criteria for getting a patent: the invention must be innovative, it must be very different from the existing technology and be useful. If these criteria were produced entirely by a computer or artificial intelligence, the patent was awarded. As a simple solution to this situation, the patent can be given to people who develop that artificial intelligence. But if more than one artificial intelligence comes together to produce a common product, then who will be given the patent? These issues are uncertain for now. However, it is clear that a solution should be produced soon.

With advances in artificial intelligence development, it is necessary to be treated with foresight. Artificial intelligence will radically change the labour market and social interactions. People need to think more about this issue and find ways to make everyone benefit from these developments as a society. All this work means we are at the crossroads. A new limit of artificial intelligence will become soon. These recent developments seem to make artificial intelligence more powerful, more human-like and we are not ready this technological progress. We need to more discuss and change something about this new life standard.

IOP Conf. Series: Materials Science and Engineering **364** (2018) 012012 doi:10.1088/1757-899X/364/1/012012

6. References

[1] https://en.wikipedia.org/wiki/History_of_artificial_intelligence

[2] https://en.wikipedia.org/wiki/Cybernetics

[3] https://qz.com/677218/this-woman-a-self-described-cyborg-can-sense-every-earthquake-in-real-time

[4] https://www.montag.wtf/content/images/2017/06/earthquakes.jpeg

[5] http://stelarc.org/?catID=20242

[6] https://www.digitaltrends.com/cool-tech/harry-potter-ai-generation/

[7] Jun-Yan Zhu, Taesung Park, Phillip Isola, Alexei A. Efros, 24 Nov 2017, arxiv 1703.10593, 2017, Unpaired Image-to-Image Translation using Cycle-Consistent Adversarial Networks.

[8] Ahmed Elgammal, Yan Kang, Milko Den Leeuw, arXiv:1711.03536v1 [eess.IV] 8 Nov 2017, Picasso, Matisse, or a Fake? Automated Analysis of Drawings at the Stroke Level for Attribution and Authentication, November 13, 2017.

[9] Fei Chaoa, Yuxuan Huanga, Xin Zhanga, Changjing Shangb, Longzhi Yangc, Changle Zhoua, Huosheng Hua, Chih-Min Lin, Engineering Applications of Artificial Intelligence 59 (2017) 1–14, A robot calligraphy system: From simple to complex writing by human gestures.

[10] http://news.mit.edu/2016/detecting-emotions-with-wireless-signals-0920

[11] https://en.wikipedia.org/wiki/Computational_creativity

[12] http://www.ibtimes.com/creative-robots-googles-deepmind-artificial-intelligence-getting-imagination-2571174

[13] https://magenta.tensorflow.org/2016/07/15/lookback-rnn-attention-rnn

[14] https://deepdreamgenerator.com/#gallery

[15] Ahmed Elgammal, Bingchen Liu, Mohamed Elhoseiny, Marian Mazzone, arXiv:1706.07068v1 [cs.AI] 21 Jun 2017, CAN: Creative Adversarial Networks Generating "Art" by Learning About Styles and Deviating from Style Norms.

[16] Ryan Abbott, 09-28-2016, Boston College Law Review, I Think, Therefore I Invent: Creative Computers and the Future of Patent Law, Volume 57, Issue 4, Article 2, p.1079-1126.

IOP Conf. Series: Materials Science and Engineering **364** (2018) 012013 doi:10.1088/1757-899X/364/1/012013

A digital reconstruction of the sunken "Villa con ingresso a protiro" in the underwater archaeological site of Baiae

B Davidde Petriaggi[1], R Petriaggi[1], F Bruno[2], A Lagudi[3], R Peluso[3] and S Passaro[4]

[1] Istituto Superiore per la Conservazione ed il Restauro, San Michele 23, 00153 Roma, Italy
[2] Department of Mechanical, Energy and Management Engineering, University of Calabria, P. Bucci Cubo 46 C, 87036, Arcavacata di Rende (CS), Italy
[3] 3D Research S.R.L., P. Bucci Cubo 45 C, 87036, Arcavacata di Rende (CS), Italy
[4] Institute for the Coastal and Marine Environment of the Italian National Research Council, Calata Porta di Massa, 80133 Napoli, Italy

E-mail: barbara.davidde@beniculturali.it

Abstract. The Underwater Cultural Heritage represents a key aspect of our historical memory still little known due to a number of limitations imposed by the underwater environment. The aim of this paper is to explore the use of digital three-dimensional reconstructions to support the research about this immeasurable archaeological and historical resource. The whole virtual reconstruction process is described step by step, focusing on the iterative feedback allowing for reaching the best virtual reconstruction solutions, helping the archaeologists to better focus their reasoning through a detailed visual representation, and the technical experts to avoid misleading details in the final virtual reconstruction.

1. Introduction

The Underwater Cultural Heritage (UCH) is an immeasurable archaeological and historical resource, with extensive and varied assets (sunken cities, ancient shipwrecks, prehistoric submerged landscapes), but it cannot be easily accessible by humans due to a number of limitations imposed by the underwater environment. As a result, it is quite evident how Computers graphics techniques like 3D Reconstruction, Virtual (VR) and Augmented (AR) Reality are a still underrated way to leverage the enormous potential of the UCH. In fact, these technologies can be efficiently applied in this field in order to improve, as much as possible, the accessibility of the UCH to the public without any constraint given by distance or time.

Today, optical and acoustic surveys are the main acquisition techniques to obtain a 3D representation of underwater archaeological sites and the related remains. Such information is traditionally used in order to have an accurate documentation of the status of the site. In particular, these high-resolution data represent a valuable source of information for interpreting architectural structures with a great level of detail.

An accurate 3D representation can be also used as the starting point for creating a scientifically sound virtual reconstruction of the site, embedding historical information coming from different sources. It can be useful for a careful interpretation of the existing archaeological remains but sometimes they might also be capable to suggest new archeological discoveries.

In this paper, we present the results obtained from the virtual reconstruction of the complex of the "Villa con ingresso a protiro - Villa with Vestibule" dated to the first half of the II century AD, Hadrian Emperor era, located in the Marine Protected Area - Underwater Park of Baiae (Naples). The applied method exploits the high-resolution data acquired by using the latest techniques for the construction of acoustic microbathymetric maps together with drawings and other historical and archaeological information to build a suggestive 3D digital reconstruction of underwater architectures not anymore existing. It is based on an iterative feedback that allows for reaching the best reconstruction results, helping the archaeologists to better focus their reasoning through a detailed visual representation, and the technical experts to avoid misleading details in the final virtual model. This workflow involves several professional figures, such as 3D graphic designers, archaeologists, and art historians.

2. Methodology

The reconstructive workflow (from fieldwork to the virtual reconstruction) was carried out taking into account the results of the theoretical and methodological discussion built around the definition of Virtual archaeology [1]. In fact, in the last years, a lot of efforts have been spent in defining guidelines for the implementation of methods to construct virtual interpretations, in specific documents such as the London Charter [2] and the Principle of Seville [3] with the aim of supporting and promoting interpretation and simulation based on a theoretical and multidisciplinary scientific approach.

The main steps of the proposed methodology are shown in Figure 1:

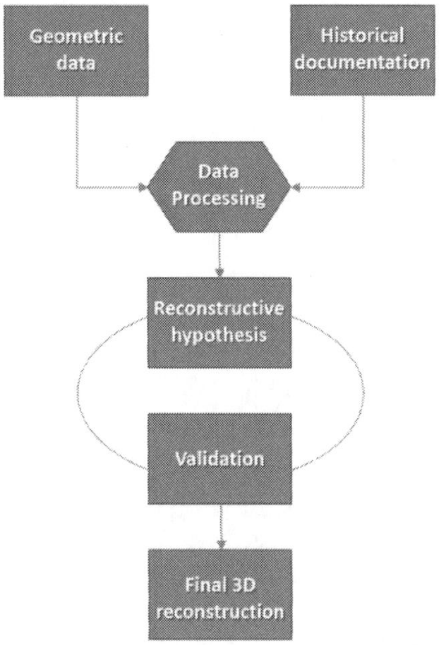

Figure 1. Iterative feedback reconstruction process based on the integration of 3D data and historical analysis.

The first stage of the process is the data collection. It is focused on the acquisition of historical documentation, scientific literature and geometric data gathered during the survey campaign. The aim is to collect all the information needed to create a hypothetical reconstruction of the archaeological remains with a high level of consistency.

Subsequently, all the acquired data (archaeological maps, illustrations, photos, Digital Terrain Model, etc.) are analyzed and put in relation. Several experts have to be involved in order to discuss different interpretation hypotheses. The discussion is supported by the employment of a 3D platform capable of merging the 3D reconstruction and the acquired model in the same 3D space.

The last steps of the process consist of modeling the architectural remains and validating it in line with the interpretation process. The approach is based on interleaving a phase of technical reconstruction with a strong critical revision in order to generate a feedback process, iterating the construction/correction loop as much as needed. Finally, in order to map the evolution of the virtual interpretation, several 3D layers are saved together with the final model, examined and approved by the scientific experts.

3. Historical documentation

3.1. The underwater archaeological park of Baiae

Baiae was a famous coastal settlement in the volcanic Phlegraean Fields, much appreciates by the Roman aristocracy since the Late Republic (II-I century BC). The thermal waters and the climate favored the development of a monumental and residential architecture which was improved by the Emperors during the following centuries (I-IV AD). At the end of the fourth century, beginning of the fifth century AD, several ground movements (bradyseism) caused the sinking of a portion of the city below the sea level. Many luxurious villas built in Baiae and a large part of the town, became Emperor's property under Augustus and his successors (such as Julius Caesar, Augustus, Pompeius the Great, Marcus Antonius, Caligula, Nero, Hadrian, *Septimius Severus* and the list goes on).

Even after the decline of the Roman Empire, the baths of Baiae continued to be used by the Visigoths, Vandals and Gots. The historian Cassiodorus in the sixth century affirm that Baiae is still a pleasant resort. However, the settlement was devastated by Muslim raiders in the VIII century AD.

Baiae overcame the crisis and in the Middle Ages, it was still famous for its thermal waters. We know for example that the Swabian King Friedric II 7 frequented its thermal waters in 1227 and restored some facilities. The importance of the port of Baiae in the fifteenth century is well highlighted by the construction of the castle and of the coastal defense systems by the Aragonese.

In the following centuries Baiae continued to be a frequented center and, with alternating vicissitudes also due to the seismic nature of the territory, remained a center of attraction and the charm of its ruins made it one of the favorite destination by travelers from all over Europe up to our days.

Nowadays, most of the buildings of the city lie on the bottom of the sea covering an area of about 177 hectares that since 2002 has been declared a Marine Protected Area. The scuba divers can visit several underwater paths : 1) the Nymphaeum of Claudius lying off Punta Epitaffio with the copies of the original statues, now exposed in the Phlegraean Fields Museum; 2) the remains of the Villa dei Pisoni, with the spectacular architectural changes due to emperor Hadrian; 3) the remains of houses and storehouses of Portus Iulius; 4) the "Villa con ingresso a protiro – (Villa with Vestibule)", that is the subject of this paper, siding a road along which there were thermae, tabernae, and other buildings.

3.2. The "Villa con ingresso a protiro"

The "Villa con ingresso a protiro" (Figure 2) has two clearly defined areas, the residential quarters with the garden (the squared area in the centre of Figure 1) and the baths on the left side of the garden.

Two red-plastered masonry benches - framed by stucco pilasters - marked out the threshold along the road. The name '*prothyrum*' comes from the presence of two stuccoed columns – no longer existing – that were edging two short parting walls built on the sides of the threshold.

The vestibule, with doorways to the '*ostiarius* quarters' (gatekeeper's lodge), led to the *atrium* with *impluvium* in the center, whose walls were decorated in the lower part with marble slabs (Tenaro's red marble fragments were recovered during underwayter surveys). Also the walls of the adjacent rooms were decorated with marble, many of which had mosaic flooring. A mosaic floor in black and white *tesserae* with geometrical motifs (hexagons and pseudo-emblemas with circles and *peltae*) is positioned in a room at the north-eastern corner of the *atrium*.

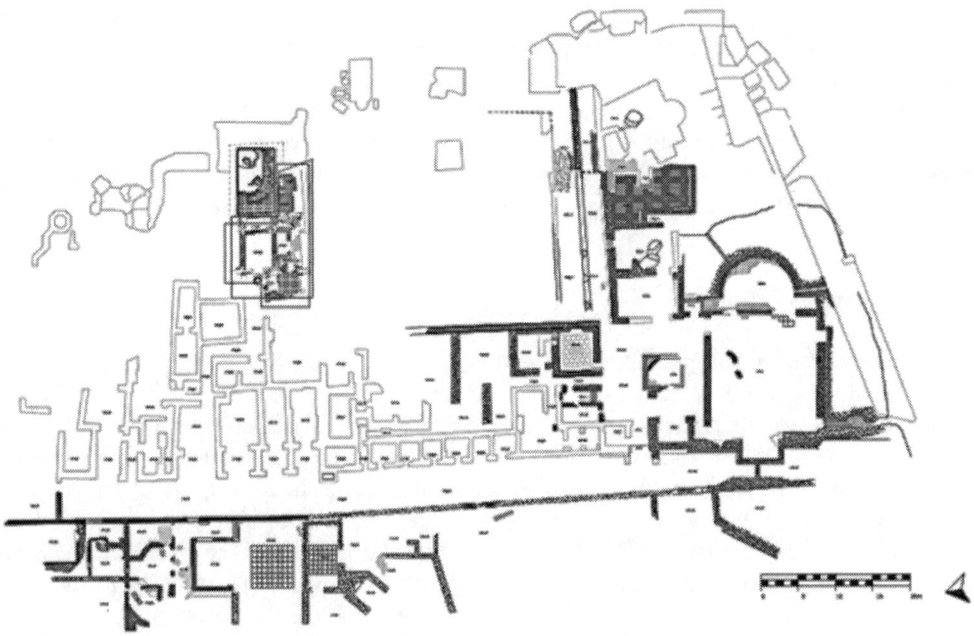

Figure 2. Archaeological map of "Villa con ingresso a protiro".

This mosaic, classifies by ISCR as n. VIP 19, lies inside a rectangular area measuring 4.77 metres by 5.55 metres in the northwestern part of the Villa. It consists of a background of white tesserae, with the ornamental motifs bordered by a double line of black tesserae. The principal geometrical design is a series of hexagons covering the whole floor, with a small square of four black tesserae at the centre of each hexagon. Black tesserae also make up the double border at the edges. The central section is a pseudo-emblema consisting of four circles inside a large square. Each circle contains four peltae and a central quincunx (the five-spot motif on six-sided dice). There are three equidistant solid black equilateral triangles outside the circles on each of the inside edges of the square. The point of each pelta terminates in a small flower with three heart-shaped petals separated by a central dot. The central space between the four circles contains a small square enclosing a cruciform floral motif with a central dot. The floral motif has lobed petals with arrowhead tips. The mosaic dates back to the second century AD [4, 5].

A large hall with a 10.37-metre wide apse lies to the South of the atrium (a nymphaeum?). This was probably not part of the original design but at the moment it is impossible to establish the date of construction and its relation with it. Its general appearance, and in particular the sumptuous large-scale marble panelling, is reminiscent of the grand halls of the Late-Imperial domus of Ostia.

In 1992, in the garden of the Villa, a statue similar in style to the Aphrodite of the Gardens by the Greek artist Alcamenes, was recovered and is now housed in the Archaeological Museum of Campi Flegrei.

The first archaeological documentation of the Villa was done in the second half of the 1980s, thanks to a group of advocationals that produced the extended survey and mapping of the archaeological remains of Submerged Baiae [6-11].

Then, several rooms of the Villa were documented and restored in the frame of the ISCR "Restoring Underwater" Project, funded by the Ministry of Cultural Heritage and Activities and Tourism. This project, designed by the archaeologist Roberto Petriaggi, started in 2001 with the conservation of three fishponds of the Roman Villa of Torre Astura (Nettuno-Rome), then moved to Baiae in 2003 and it is on course. It is dedicated to testing methodologies and tools for documenting, conserving and valorizing *in situ* the underwater cultural heritage [12-14].

The "Restoring Underwater" Project took care of the Villa con ingresso a protiro for the first time in 2003, when Roberto Petriaggi directed the restoration of a room with a white mosaic floor measuring around 4.5 metres long and 2.5 metres wide (classified as n. 14 by ISCR) [15]. Then, on 2010 a thermal room (caldarium) were restored under the direction of Roberto Petriaggi [16].

In 2012 started another campaign of 3D documentation [17-19] and conservation under the direction of Barbara Davidde Petriaggi, with the restoration of some sectors of the archaeological complex and in particular the room paved with the black and white mosaic floor (Figure 3).

Figure 3. ISCR restoration activities in the "Villa con ingresso a protiro".

The DTM of the underwater archaeological site has been obtained through a morpho-bathymetric survey carried out by the Institute for Coastal Marine Environment (IAMC) of the Italian National Research Council (CNR) [20]. The DTM of the seafloor allowed to draw the main archaeological features of the area, that also include the military complex of Portus Iulius and the ancient thermal and villa complex of Baianus Lacus. The extraction of archaeological features from DTM has achieved thanks to the use of a GIS-derived tool based on the profile curvature algorithm.

4. Source-based 3D modeling

The sources for the virtual reconstruction of the "Villa con ingresso a protiro" were the manual and 3D graphics documentation, digital photo made by ISCR during the conservation campaigns and the acoustic bathymetry provided by IAMC.

To improve the understanding of the underwater architectural complex by the underwater and non-underwater visitors, given the poor conservation of ancient submerged structures, it was necessary to hypothesize much more than the structures relieved *in situ*. A three-dimensional hypothetical reconstruction based on archaeological evidence has been realized and, for what has not been preserved, reconstructions have been proposed on the coeval examples known from the archaeological research and literature.

4.1. Volumetric model

Once all the needed information about the villa has been acquired and processed, it is possible to start the modelling task. At the beginning, a volumetric model has been created showing the walls of the main structure. It has been developed using as reference the acoustic DTM and the archaeological map. In particular, the archaeological map, has been used as constraint to verify the correct scale of the model. Contrary to other known roman villas, the "Villa con ingresso a protiro" has not a canonical shape. In fact, its shape was conditioned on the front by the presence of the road, while in the back by the line of the coast. The main entrance is opened on the west side of the atrium and it is not perfectly in axe with the impluvium basin. Usually, in the Roman Villas, on the opposite side of the entrance there was the tablinium, a richly decorated room where the *pater familias* used to receive his *clientes* and from where it was possible to enter the garden. Instead, in this Villa, the entrance to the garden is an open passage in the north side of the hall's wall.

Moreover, the most part of Roman villas has a garden, surrounded by a 4 or 3-sided column portico; in this villa, until now, we have found traces of columns only on the South side. Nevertheless, in the reconstruction hypothesis, we suggest to build the portico on the whole perimeter of the garden, but it is a hypothesis that we hope will be confirmed by future archaeological excavations.

On the right side of the hall, there is the large hall with a wide apse that we could hypothetically identify as a nymphaeum. As at the moment, the relation with the original design of the villa and this room is unknown, we did not include it in the 3D hypothesis. We will consider adding this room in some future update as its function and dating will be more clear and sure.

4.2. *Architectural model*

In order to build the architectural model, we have added on the conceptual model structure details like roofs, columns, doors and windows, referring to the images and documents collected during the data acquisition task, and taking as an example the Roman villas dated at the Second Century AD [21, 22] (Figure 4). The archaeological reconstruction of the roof covering, of the wall paintings and of the furnishings as of the garden, are purely hypothetical because no traces of these archaeological data have been documented underwater. In fact, the remains of the walls of the Villa are almost completely disappeared. For example, the room paved with the mosaic floor n. 19 has the perimeter walls measuring about 5-10 cm high, along the seaward sides to the south and the east while the walls in the northern and western side measure about 50 cm in height (Figure 5).

Figure 4. Architectural model of the "Villa con ingresso a protiro"

5. Conclusion

This paper describes a process of acquisition and modeling applied to create a virtual reconstruction of the complex of the "Villa con ingresso a protiro - Villa with Vestibule" dated around the II century AD, located in the Marine Protected Area - Underwater Park of Baiae (Naples).

A commonly used 3D acquisition and modeling approach along with some different topics are discussed, suggesting both optimized procedures for data processing and communication. In particular, as synthesized in Figure 1, an iterative methodology is suggested to support the interpretative modeling step, simplifying the feedback on virtual models by the archeologists. This solution gives a significant improvement on the complex procedure for reconstructing in 3D archaeological structures whose remains are often very few with respect to its original shape. In this way both an instrument of communication and valorization of the cultural heritage site is provided, helping the interpretation of archeological and architectural ruins.

Figure 5. Images representing different portions of the Villa: a) room with pelte mosaic; b) *atrium* with *impluvium*; c) *peristilium*; d) garden.

Acknowledgments

This work was partially supported by i-MARECULTURE project. It has received funding from the European Union's Horizon 2020 research and innovation programme under grant agreement No 727153

References

[1] Munster S and Koehler T 2016 3D Reconstruction of Cultural Heritage Artifacts. A Literature Based Survey of Recent Projects and Workflows, In book: Virtual Palaces, Part II. Lost Palaces and their Afterlife. Virtual Reconstruction between Science and Media, Publisher: Paladium, Editors: Stephan Hoppe, Stefan Breitling pp. 87-102.

[2] Charter L 2009 London charter for the computer-based visualisation of cultural heritage. Online at: www. londoncharter. org.

[3] Lopez-Menchero V M and Grande A 2011 The principles of the Seville Charter. In CIPA symposium proceedings Vol. 2011 pp. 2-6.

[4] Rinaldi F 2007 Mosaici e pavimenti del Veneto. Province di Padova, Rovigo, Verona e Vicenza (I sec. a.C. - VI sec. d.C.) pp. 167-169.

[5] BUENO M. 2011 Mosaici e pavimenti della Toscana (II sec. a.C.-V sec. d.C.), Roma AntenorQuaderni, 22.

[6] Di Fraia G 1993 Baia sommersa. Nuove evidenze topografiche e monumentali, in Gianfrotta P. A., Pelagatti P. (ed.), Archeologia subacquea. Studi, ricerche e documenti I, Istituto Poligrafico e Zecca dello Stato pp. 21 - 48.

[7] Lombardo N 1993 Le terme di Punta dell'Epitaffio a Baia, in Gianfrotta P. A., Pelagatti P. (ed.), Archeologia subacquea. Studi, ricerche e documenti I, Istituto Poligrafico e Zecca dello Stato pp. 55 - 63.

[8] Lombardo N 1993 Un documento epigrafico dalla «Villa dei Pisoni» a Baia, in Gianfrotta P. A., Pelagatti P. (ed.), Archeologia subacquea. Studi, ricerche e documenti I, Istituto Poligrafico e Zecca dello Stato pp. 49 - 53.

[9] Scognamiglio E 1993 Il rilievo di Baia sommersa: note tecniche e osservazioni, in Gianfrotta P. A., Pelagatti P. (ed.), Archeologia subacquea. Studi, ricerche e documenti I, Istituto Poligrafico e Zecca dello Stato pp. 65-70.

[10] Scognamiglio E 2002 Nuovi dati su Baia sommersa, in Gianfrotta P. A., Pelagatti P. (ed.), Archeologia Subacquea Studi, ricerche e documenti III, Istituto Poligrafico e Zecca dello Stato pp. 47-55.

[11] Scognamiglio E 1997 Aggiornamenti per la topografia di Baia sommersa, in Gianfrotta P. A., Pelagatti P. (ed.), Archeologia subacquea. Studi, ricerche e documenti II, Istituto Poligrafico e Zecca dello Stato pp. 35-45.

[12] Petriaggi R and Davidde B 2012 The ISCR Project "Restoring Underwater": an evaluation of the results after ten years from the beginning 4th Conference on Preserving Archaeological Remains in Situ PARIS4, in 'Conservation and Management of Archaeological Sites', vol. 14 pp. 192-199.

[13] Petriaggi R and Davidde B 2007 Restaurare sott'acqua: cinque anni di sperimentazione del NIAS-ICR, in 'Bollettino dell'Istituto Centrale per il Restauro', nuova serie, n. 14, pp. 127-141.

[14] Petriaggi R 2008 Restaurer sous l'eau: le cas de Baïes (Naples), Actes du colloque international Les Grands chantiers de restauration en Europe, Paris, Institut National du Patrimoine, 28-29 juin 2008 pp. 80-85.

[15] Petriaggi R and Mancinelli R 2004 An experimental conservation treatment on the mosaic floor and perimeter walls of room n. 1 of the so-called «Villa con ingresso a protiro» in the underwater archaeological park of Baia (Naples), in 'Archaeologia Maritima Mediterranea' vol 1 pp. 109-126.

[16] Petriaggi R 2005 Nuove esperienza di restauro conservativo nel Parco Sommerso di Baia, in 'Archaeologia Maritima Mediterranea' vol 2 pp. 135-147.

[17] Davidde Petriaggi B, Petriaggi R and Gomez de Ayala G 2014 3D Documentation for the Assessment of Underwater Archaeological Remains, in Graeme E., Sly T., Chrysanthi A., Murrieta-Flores P., Papadopoulos C., Romanowska I., and Wheatley D. (eds.), Archaeology in the Digital Era, Volume II, e-Papers from the 40th Conference in Computer Applications and Quantitative Methods in Archaeology, Southampton, United Kingdom, 26-30 March 2012, Amsterdam University Press, 2014, pp. 174-180.

[18] Davidde B. 2017, Documentazione 3D per la conservazione del patrimonio archeologico sommerso: l'esperienza dell'ISCR, in A. Colombo, V. Perzolla (a cura di), Le tecnologie digitali al servizio della conservazione, Atti del Convegno (Vicenza, 19 marzo 2016), Padova pp. 27-31.

[19] Davidde Petriaggi B and Gomez de Ayala G 2017 Laser Scanner Reliefs of selected archaeological structures in the submerged Baiae (Naples), in 'The International Archives of the Photogrammetry, Remote Sensing and Spatial Information Sciences' pp. 79-83.

[20] Passaro S, Barra M, Saggiomo R, Di Giacomo S, Leocca A, Uhlen H and Mazzola S 2013 Multi-resolucion morphobachymecric survey resulcs at che Pozzuoli-Baia underwater archaeological site (Naples, Italy). J Archaeol Sci. vol40(2) pp. 1268-78.

[21] Baldassarre I, Pontrandolfo A, Rouveret A and Salvadori M 2006 Pittura Romana, Milano.

[22] Bragantini I and Sampaolo V 2009 (a cura di), La pittura pompeiana, Verona.

IOP Conf. Series: Materials Science and Engineering **364** (2018) 012014 doi:10.1088/1757-899X/364/1/012014

CAD-CAE methods to support restoration and museum exhibition of bronze statues: the "Principe Ellenistico"

M Bici[1], F Campana[1 4], O Colacicchi[2] and G D'Ercoli[3]
[1] Dipartimento di Ingegneria Meccanica e Aerospaziale, Sapienza Università di Roma. Via Eudossiana 18, 00184 Roma, Italy
[2] Museo Nazionale Romano. Roma. Italy
[3] Istituto Superiore per la Conservazione e il Restauro, Roma, Italy

Abstract. Ancient bronze statues mainly require material integrity assessment and restoration. Restoration may include also the update of the museum exhibition, defining new structural frames and fragment re-composition to preserve the statue and improve the interpretation of the original aspect. This paper proves how engineering methods (such as Finite Element Analysis, Computer Aided Design modelling, Reverse Engineering) may assist cultural heritage experts and restorers in these tasks. It presents the activities made together with the Museo Nazionale Romano and the Istituto Superiore per la Conservazione e il Restauro, on the so-called "Principe Ellenistico" (Hellenistic Prince). This bronze was found in pieces (body, left arm and right leg), at the end of 19th century during an excavation made in Rome. No visual or reference sources can say its origin and its final posture was defined by restorers at the end of the 19th century according to their hypothesis and studies. In the 20th century, a further restoration was made on the critical areas of the surface, together with some structural improvement of the inner frame. Nowadays, after a review of its position inside the Museum, new experimental and numerical analyses have been carried out to better understand surface weakness and correct left arm positioning.

1. Introduction

The conservation of archaeological bronze sculpture in Italy has evolved in several phases. During the 19th century, the main focus was the reconstruction of the formal integrity of the sculpture by reassembling its fragments and fractured elements and the integration of missing parts. Examples of this approach are the bronze statues found at Pompeii and Herculaneum now at the Archaeological Museum in Naples, and the "Pugilatore"(Boxer) and the "Principe Ellenistico" (Hellenistic Prince) both in the Museo Nazionale Romano (MNR), and the statue of Treboniano Gallo, Metropolitan Museum of Art, New York [1,2]. Starting from the second half of the 20th century, attention shifted towards preserving the integrity of the surface layers, by investigating the compositional elements of the original surface and studying preventive methods for the stabilization of corrosion processes and the preservation of ancient patinas that would at the same time respect and enhance the appearance of the sculpture [2]. Thanks to the establishment in the 1940's of an Institute for Conservation (Istituto Centrale del Restauro, ICR), strongly inspired by the ideas of Giulio Carlo Argan, and founded by Cesare Brandi, a new concept of 'preventive conservation' was defined and described in the publication of Brandi's 'Teoria del Restauro' (1963), where methods of conservation were defined in a systematic treaty, [3,4].

In the 1970's, important discoveries of ancient bronze statues such as the Riace bronzes, and conservation interventions on other bronze sculptures present in Museum collections (the Cartoceto bronzes and the Dionysus of the MNR) or originally exposed outdoors (The San Marco horses and the

[4] Author to whom any correspondence should be addressed: francesca.campana@uniroma1.it

Colossus of Barletta), generated the opportunity of applying innovative non-destructive diagnostic methods along with traditional chemical analysis, initiating new strategies capable of identifying compositional elements and, at the same time, defining new methods of preventive conservation and monitoring with non-destructive procedures.

During the same period the artisanal techniques and reconstruction methods of the 19th century were replaced by new reassembling methods involving reversible materials and new technological approaches. Innovative solutions were developed at the end of the1970's by researching and designing internal support systems. These new support techniques were used for the Selinunte Ephebus and the Eros of the MNR. These solutions are still widely adopted today to connect the fractured parts of large sculptures and for the creation of static base elements for sculptures with a limited ground support area, as in the case of statues who stand solely on feet soles or horses' hooves. A specific example of this application is the Satiro Danzante of Mazara del Vallo [5], where the absence of the inferior part of the body required the re-composition of the fragments over a reconfigurable frame able to sustain and secure the statue. Anti-seismic solutions have also been investigated and defined, like in the case of Bronzi di Riace and the statue of "Germanico da Amelia".

Concerning non-destructive tests and diagnostic, new techniques were applied like for example eddy current, holographic interferometry and acoustic emission. They are used to improve the accuracy of the radiographic inspection with a more detailed view of the solid parts against the cracks, and new qualitative and quantitative concerns about stress and strain distribution according to the environment agents. The quantification of the stress and the strain field has also been approached through Finite Element Analysis (FEA). One of the first studies was made on the Marco Aurelio statue to evaluate the stress field due to the gravity and the critical areas of the horse according to its preservation state. These results were correlated to experimental measurements by means of gauges, moving the knight over the horse [6, 7]. Generally speaking FEA for cultural heritage involves studies about static behaviour or seismic response of buildings, statues and their support systems [8, 9]. More recently, thanks to the enhancement of virtual modeling and shape digitalization, new applications of Computer-Aided technologies have been made. They are focused on fragment reconstruction, digital modeling for data archive and communication and numerical analysis via virtual prototyping. In particular a detailed reproduction of the restoration chronology via virtual model, together with virtual prototyping results may help the experts to assess and evaluate critical parts of a work, improving its interpretation and exhibition.

This paper concerns with this field of applications, applying a Computer Aided Engineering (CAE) approach to the ancient bronze statue, known as the "Principe Ellenistico". More in details, it reviews the chronological steps of its restoration and presents CAE analysis with the aim of assessing criticality in terms of mechanical response and evaluating how it can aid its preservation and exhibition.

2. The statue

The Hellenistic Prince, is currently placed in Rome at Palazzo Massimo, part of the Museo Nazionale Romano. It was found in 1885 on the hillside of Quirinale, Rome, probably in the ancient area of the Costantin's thermae. It is 2.040 m tall and depicts a naked hero as usually made in case of sovereigns or warlords (figure 1.a). Although its original representation is still investigated, several experts recognise it as an hellenistic prince of the III-II century b.C., carried in Rome as spoils of war.

The first restoration was made at the end of the 19th century to recompose fragments and to guarantee structural support with an ad-hoc frame. The fragment reconstruction involved front part of the right foot, right leg at the thigh (fragment found separated from the statue), left leg at the knee, left arm (fragment found separated, with lacuna). Conjunctions were made through plates of brass that were jointed to the statue transversally to cracks and edges, by brass screws (Ø 6mm). Missed parts and holes were also rebuilt and filled with bronze. The spear in the left arm is not original but added according to the final posture and iconographic studies.

IOP Conf. Series: Materials Science and Engineering **364** (2018) 012014 doi:10.1088/1757-899X/364/1/012014

(a)

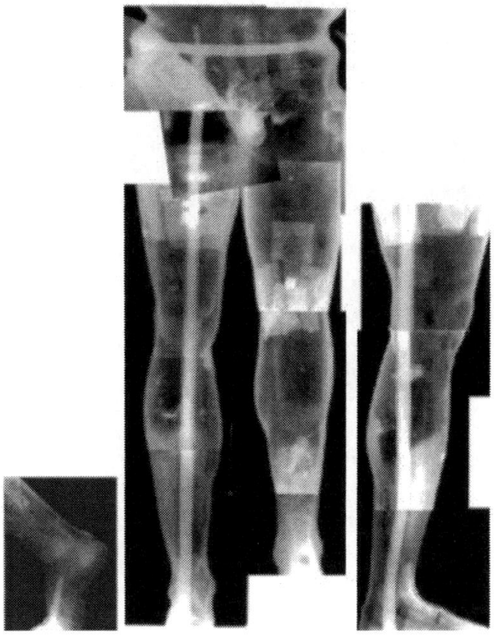
(b)

Figure 1 (a) the "Principe Ellenistico"; (b) RX for support frame and left foot

At the end of the last century, the left arm was taken apart and reviewed due to cracks along the stucco used to joint it. At the same time, a new inner support for the left arm was specifically designed; and a testing campaign was defined according to ENEA and ICR. It was oriented to evaluate chemical composition and to acquire by radiography and endoscopy bronze integrity, frame and junctions status. In 2003 active corrosion started again, mainly in form of pitting. This was approached by applying corrosion inhibitors and protective coatings. This work was carried out in 2004, in the framework of a maintenance intervention. At the present, new surface checks have been made together with the renewal of the exhibition set-up that involves a new basement and air-monitoring equipment for the room, so that climate environment changes may not have effect. In addition, CAE investigations have been planned starting from a 3D model achieved through a digital acquisition.

3. CAE investigations

Scope of CAE activities concerns with an evaluation of critical areas and a check of the posture, considering the changes made to the frame anchorage and the left arm, during the restorations.

3.1 Modeling the support frame

No technical drawings of the support frame are present, so that CAE modelling has been made according to the evaluation based on a pictorial drawing, RX shown in figure 1.b and visualization of the inlays for the conjunction. The inner frame to anchor the statue to the ground basement, is made with two bars with square section (30 x 30 mm). One is inserted in the right leg, it is rather vertical although it was bent to follow the shape of the leg. The other bar is at the hips. In addition, a pin was inserted in the left foot to connect it to the basement.

In the 19th century, the vertical bar and the pin were constrained at the bottom of the statue, filling the empty space by different kind of materials (stucco, wood,…). Filling empty space was a common practice of that period.

IOP Conf. Series: Materials Science and Engineering **364** (2018) 012014 doi:10.1088/1757-899X/364/1/012014

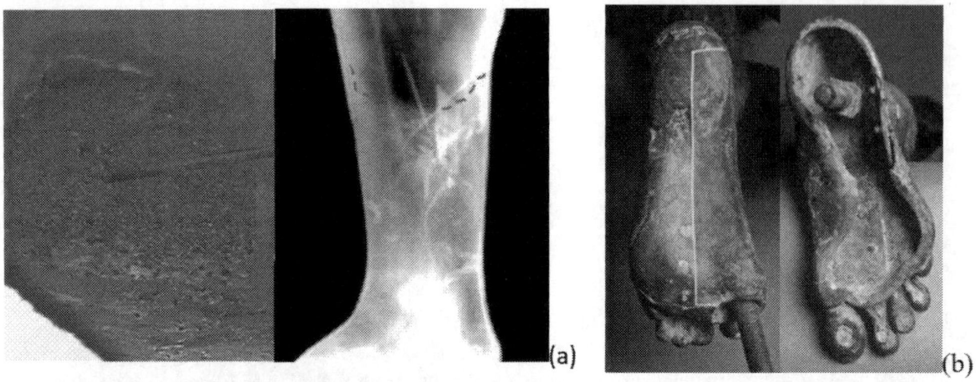

Figure 2 (a) RX and endoscopy of the filler in the left foot; (b) Details of the feet and their anchorages

At the end of the last century, the filler was removed from the right leg, whereas it was impossible for the left foot. As result, the vertical bar in the right leg may be compliant in the respect of right leg anchorage. Figure 2.b shows the details of the feet to give evidence of: (i) the small areas that are involved in the contact with the ground; (ii) the filler still present in the left foot.

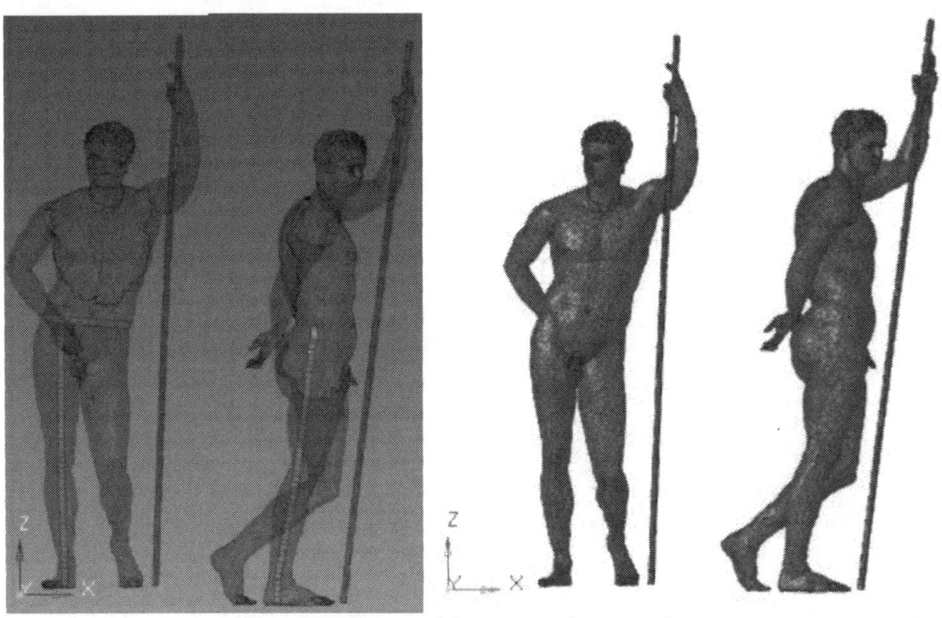

Figure 3 FEA model: support frame and statue

3.2 Modeling the statue

The surface modeling has been made starting from the digital acquisition of the bronze, made by a 3D Scanner Artec Eva. It has more than $2x10^6$ points that have been filtered, according to curvatures, so that, a triangular tessellation of about $26x10^3$ 2D elements has been achieved through Delaunay's algorithms. A restriction on the element distortion (connected to minimum length and angle) has been imposed to avoid elongated or distorted elements. Doing so, a regular mesh has been obtained, consistent with FEA mesh requirements and, finally, tessellated elements have been converted into shell elements. Figure 3, on the right, shows the results of these procedures, showing a higher density of elements in zones with high variety of features and curvatures.

The final FEA model is made by three parts: the statue, the inner frame assuming the left arm part of the shoulder, the spear in the left arm in contact, but not connected, with the forearm and the hand.

Thickness distribution of the statue has several uncertainties. Investigations have been made through ultrasound technique by ICR. In the statue, 5 major castings can be found: trunk and left leg, right leg, right arm, left arm, head. The results gives a thickness in the range of 5.0-6.5 mm. The thickest areas are that related to major details of the shape, muscles of the back (6.3±1 mm), face (6.75±0.07 mm). Considering that the overall weight without the basement is about 180 kg, we assume to investigate two configurations: one with uniform thickness equal to 5.9 mm and another with three set of thickness 8.0 mm and 6.5 mm, for the head and the back respectively, and 5.7 mm elsewhere (non-uniform thickness distribution). Major thickness has been given to the head, considering that the hair represent a part with thickness larger than the rest of the face, but extremely difficult to be measured. In this case, the weight in the upper part of the statue increases of about 24%, moving the centre of mass along the inclination of shoulders and head.

Simulations have been made using Optistruct, inside Hyperworks2017, finding both linear elastic and modal solutions. As first hypothesis, the statue has been assumed without discontinuities among the edges of the fragments, as well as the material between shoulder and left arm. Concerning the Boundary Constraints (BC) we have investigated different variants:

- BC#1: right foot in contact with ground but free in the respect of the inner wireframe; anterior part of the left foot rigid with the pin inside; no added support frame between left arm and shoulder.
- BC#2: right foot rigid with the inner support frame, up to the filler present before the 1970s restoration.

In both cases, the spear was considered in contact but not integral, so that it can react only if contact with the arm is forced.

4. Results and discussion

4.1 Static Analysis

In the case of BC#1, final displacement of the static analysis confirms the tendency towards a rotation around the axis defined by the feet. The maximum displacement is 3.56 mm in case of non-uniform thickness distribution, 3.34 mm in case of uniform thickness, that means about 6.5% of difference. Figure 4 shows the displacement field for the statue and the inner frame, in the most severe condition for the displacement (BC#1 non-uniform thickness distribution).

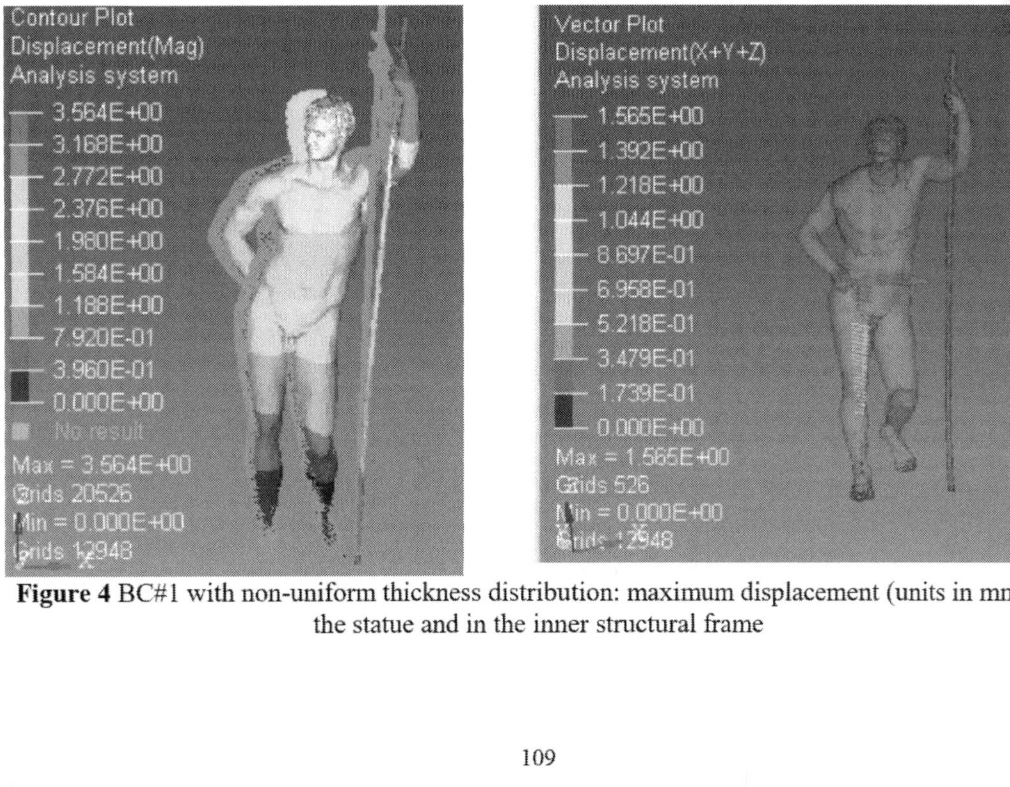

Figure 4 BC#1 with non-uniform thickness distribution: maximum displacement (units in mm) in the statue and in the inner structural frame

IOP Conf. Series: Materials Science and Engineering **364** (2018) 012014 doi:10.1088/1757-899X/364/1/012014

The spear does not react axially to the statue weight, but it is bent due to the arm rotation under load. Loads are distributed between left foot and right foot plus the frame in the ratio 2:3. Changes in thickness distribution do not change statue response, significantly.

Concerning the stress distribution, higher values are always located at the feet. A quantification of the stress level is shown in Figure 5 in terms of equivalent stress, computed according to the Von Mises formula:

$$\sigma_0 = \sqrt{\frac{(\sigma_1-\sigma_2)^2+(\sigma_2-\sigma_3)^2+(\sigma_3-\sigma_1)^2}{2}} \quad (1)$$

where σ_i stands for the ith principal stress tensor component.

In this case, an average value has been taken between inner and outer layer of the statue. Filtering the maximum value above 10 MPa other gradients may be highlighted in the left arm area and from the thighs up to the hips.

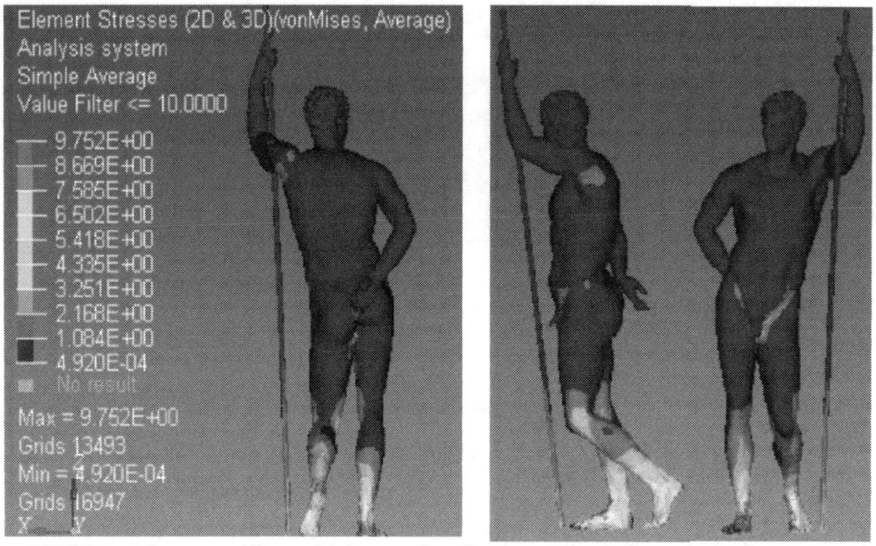

Figure 5 BC#1 with non-uniform thickness distribution: Von Mises equivalent stress (averaged on the layers, units are in MPa)

Although maximum values are far from failure stresses (qualitative value for yield 100-130 MPa), static analysis confirms critical areas where original cracked fragments were located (figure 6.a). In particular, Figure 6.b shows a plot of the principal stress tensor at the legs, to give evidence of the changes in direction and values. At the knees up to the thighs, a transition occurs with both compression and tension. It becomes axial stress at the ankles, in traction in the back and compression in the front according to the BC on the ground and to the displacement that occurs under the load condition.

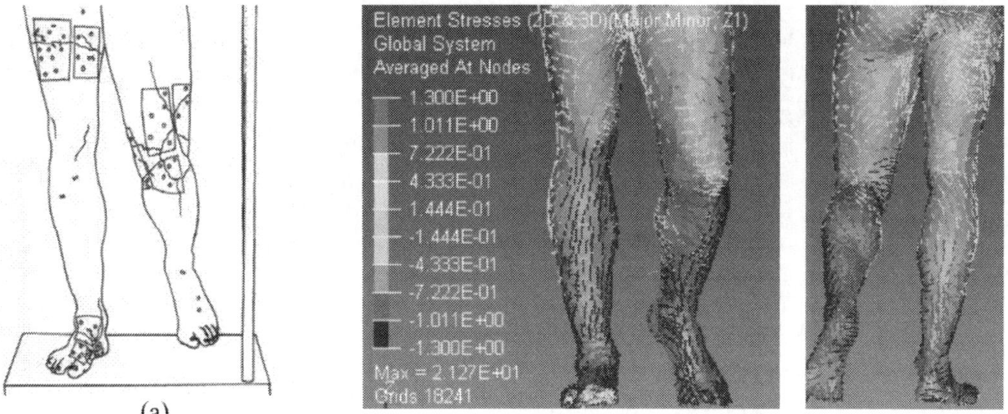

(a) (b)

Figure 6 (a) qualitative representation of cracked areas and fragments after the first restoration; (b) BC#1 with non-uniform thickness distribution: principal stress tensor with values (units are in MPa)

Assuming the boundary conditions of BC#2, maximum displacements decrease of about 25%. Stress distribution becomes more severe in terms of magnitude and amplitude of the gradients, although they are still lower than failure stress.

4.2 Modal Analysis and hypothesis of fragment re-alignment

CAE investigations may also help considerations about statue exhibition. In particular, via modal analysis, natural frequencies may be found for preliminary anti-seismic evaluations. Moreover, modal analysis may also give information about deformations under occasional loads like those related to carriage and handling.

In this case, modal analysis gives similar results for each BC and thickness distribution: 5 Hz and 10 Hz for bending forward, like in static deformation, and in its orthogonal direction. Over 20 Hz for transversal bending (ZX plane) and over 25 Hz up to 32 Hz torsion modes.

As shown in Figure 7, the arm was found as a separated fragment with lacuna. Figure 7.a also shows a view from the top of the principal direction of the stress tensor around the left arm. Back part of the shoulder is in tension due to bending, effect that may increase the risk of fracture.

Its reconstruction was made two times: the first one adding a bronze detail (19th century), the second one via reconfigurable inner-support plus resin addition. Photographic evidences show different postures among these interventions, so that, major investigations about possible reconstructions seem necessary. A preliminary result is shown in Figure 7.b. It is obtained using the 3D model to re-align the arm with a translation in the point A of the RX of figure 7, and a rotation made along the axis defined by the feet. This direction has been chosen imagining that the crack has been propagated due to an exasperation of the stress during bending. Figure 7.b shows 2 rotations of 3.5° and 7 ° respectively (the red and the green).

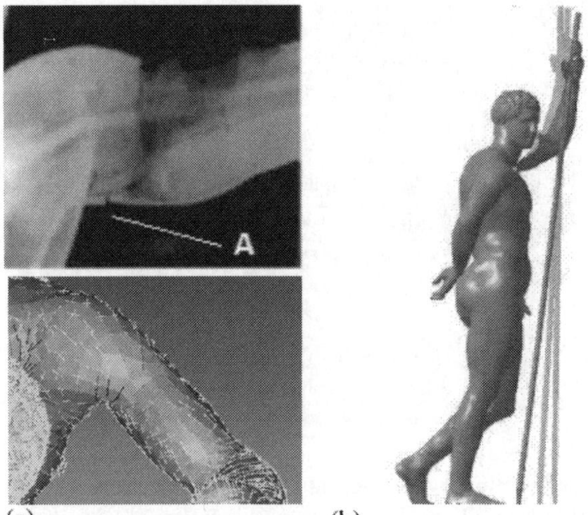

(a) (b)

Figure 7 (a) RX of the left arm and qualitative principal stress distribution; (b) arm rotation, preliminary hypothesis

5. Conclusions

This paper shows some results of CAE investigations on an ancient bronze statue found in Rome. Many restorations have been done to improve material conservation and guarantee structural stability. Although some simplified hypotheses are necessary, the static analysis confirms that fragments are associated to areas with higher stress concentration. They are ankles, thighs and left arm nearby the shoulder. The left arm is also critical for the bending natural mode of vibration, which is related to the first frequency (about 5 Hz). Major simplified hypotheses are related to the thickness distribution and the definition of the constraints. A sensitivity analysis has been made according to: compliance of the constraints at the ground; uniform versus non-uniform thickness distribution. Changing the type of constraints, static deformation increases with the compliance between inner frame and right leg (up to 35% concerning the deformations); lower effects are due to the non-uniform thickness distribution.

Other effects not investigated are: actual values of the material characteristics; role of the discontinuities along the fragments (presence of plates to make the conjunctions).

The low effect due to changes in the thickness distribution is probably related to the adoption of a 3D model of the bronze statue obtained via digitalization. This allows high accuracy in shape, better estimating change of surface curvature and, thus, bronze final area.

Robustness of the results, changing the constraint conditions, and preliminary correlations with the original fractures may encourage the adoption of CAE as analysis method of critical conditions. Moreover, its combination with CAD modeling and digital acquisition may assist fragment reconstruction and posture optimization.

References

[1] Hemingway S, McGregor S and Smith D 2013 The bronze statue of Trebonianus Gallus in the Metropolitan Museum of Art. Restoration, technique, and interpretation The Restoration of Ancient Bronzes: Naples and Beyond E Risser and D Saunders (curated by), (Los Angeles) 113-36 URL consulted on 16th of December 2017

[2] Colacicchi O, Ferretti M and Formigli M 2006 Il Pugile del Museo Nazionale Romano: risultanze conoscitive dell'intervento di manutenzione, Materiali e Strutture (nuova serie year II), 3-4/2004, 48-61

[3] Brandi C 1963 Teoria del restauro, first ed (Rome)

[4] Vlad Borrelli L 2012 Attualità della "Teoria del restauro" di Cesare Brandi Etica della Conservazione e tutela del passato curated by Basile G, Lauro G and Migliosi Tantillo A

(Rome) 77-90

[5] Petriaggi R (curated by) 2004 Il satiro danzante da Mazara del Vallo. Diario di un restauro (Rome - Istituto Centrale per il Restauro)

[6] Accardo G, Amodio D, Bennici A, Cappa P, Santucci G and Torre M 1990 Structural analysis of the equestrian monument to Marcus Aurelius in Rome Software for Engineering Workstation 6 (2) 58-61

[7] Accardo G, Amodio D, Bennici A, Cappa P, Santucci G and Torre M 1990 Strain fields on the statue of Marcus Aurelius Experimental Mechanics 30 (4) 372-6. DOI: 10.1007/BF02321507.

[8] Borri A and Grazini A 2006 Diagnostic analysis of the lesions and stability of Michelangelo's David Journal of Cultural Heritage 7 (4) 273-85.

[9] Wittich C E and Hutchinson T C 2016 Experimental modal analysis and seismic mitigation of statue-pedestal systems Journal of Cultural Heritage 20 641-8. ISSN 1296-2074

[10] Fontana R, Greco M, Materazzi M, Pampaloni E, Pezzati L, Rocchini C and Scopigno R 2002 Three-dimensional modelling of statues: the Minerva of Arezzo Journal of Cultural Heritage 3 (4) 325-31. ISSN 1296-2074, https://doi.org/10.1016/S1296-2074(02)01242-6.

IOP Conf. Series: Materials Science and Engineering **364** (2018) 012015 doi:10.1088/1757-899X/364/1/012015

Massive 3D digitization of sculptures: Methodological approaches for improving efficiency

Umair Shafqat Malik and Gabriele Guidi

Department of Mechanics, Politecnico di Milano, Italy
(umairshafqat.malik; gabriele.guidi)@polimi.it

Abstract. This paper describes a methodology for efficient massive 3D digitization of ancient sculptures under the scope of IU-Uffizi project. The project is sponsored by Indiana University (USA) with the technical support of Politecnico di Milano and the advice of a scientific committee of experts in the fields of art history and archaeology. The project aims at digitizing 1250 ancient sculptures in the "Uffizi museum", Pitti Palace", "Boboli Gardens" and a storage facility inside Villa Corsini in Castello near Florence. The authors have experienced a variety of different issues in the museum, including working with different lighting conditions, the placements of sculptures at difficult places for shooting images, inaccessible height and different ways for posing camera on the scene in an environment where the use of drones is prohibited. To solve such issues, specific technical choices were needed to reduce the digitization time and costs while maintaining a high coherence between the physical artefact and its digital counterpart. These technical solutions are being discussed, in a context where the purpose is to massively digitize complex objects in their original setting by minimizing the impact on the museum. Furthermore, several methodologies are being discussed to improve the efficiency of digitization regarding image capturing, 3D model creation, scaling and mesh editing.

1. Introduction

The digital era has led us to witness a *revolution of relevancy*, based on understanding of its effect and the boundaries of its use [1]. Nowadays, digital applications are used to promote the museum's heritage in several fields such as education, promotion, advertising and research. The IU-Uffizi Project was developed by the museum in an effort to highlight an important collection of ancient Greek and Roman sculpture, which is often not given the attention to its merits by the visitors of the Uffizi, who understandably, are mostly interested in seeing the masterpieces of painters such as Giotto, Leonardo and Botticelli. Therefore, the project confronts two challenges: (i) amplifying the visibility and comprehensibility of the sculptural heritage for museum visitors and (ii) using digital publication on the Internet to make the sculptures better known to students, scholars, and the general public wherever they may live. To deal with these challenges in operational terms, one must select appropriate digital solution and methodology, so that 3D modeling not only becomes attraction but also a tool for analyses.

This paper describes a complete pipeline and several technical choices to achieve the final goal for which the Indiana University-Uffizi Project was developed: to respect the museum's needs, the contemporary visitor's desire for engagement, and to increase knowledge and spread cultural awareness beyond the confines of the museum itself by utilizing the Internet. The results of two years activities are already showing that people are more eager to visit the archaeological collection and are paying more attention to its masterpieces. Visitors have also had the chance to talk directly to the scientists engaged in the project thanks to several public lectures and meetings held at the museum and open to the general public, in which various aspects of the project were discussed. The project has also prepared a website on which all the sculptures modeled can be viewed. It is expected that the website will be open to the public shortly. Meanwhile, the Uffizi is in the process of embedding the project's newly created 3D models onto its own publicly available website.

2. Developments of digital technologies in cultural heritage

The advancements in digital photography [2]–[5] have led to a whole new role of image-based 3D imaging in several areas, including cultural heritage documentation. Based on the international experiences starting from the late 1990's to the early 2000's [6]–[9], several projects are being developed, especially at the European level, to improve technologies and methodologies for systematic digitization of the museum contents.

Starting from the EPOCH network of excellence (2004-2008) [10][11], several areas have been identified for concentrating research efforts to achieve an efficient and sustainable application of digital technologies to archaeological research and the presentation of cultural information in museums and historical sites. In 2008, the 3DCOFORM project was started to develop some of the main tools to create reality-based digital models of sculptures and museum objects in general. The process uses a series of well-established processing steps for each museum object of interest i.e. the 3D acquisition, the processing of the texturized model, its visualization and the creation of metadata with which to describe the different descriptive and technical aspects of the model [12]. This method of the digitization process was supported by several tools developed during that project, such as: (i) the InHand Scanner, a manual device to quickly acquire the 3D shape of a museum object, (ii) the MultiView Dome to capture the visual appearance of an object such as color and reflectivity, (iii) the well-known software in the public domain Meshlab, which is now being widely used in the production and post-production of 3D models originating from 3D acquisition [13] and (v) the Ingestion tool to load the metadata of a 3D model on the EUROPEANA platform.

During the past five years, some initiatives for massive 3D digitization of complete collections of archeological artifacts have been started worldwide. The 3DICONS (2012-2015) has provided EUROPEANA with over 3000 objects, environments, buildings and specially digitized archaeological sites, which include for example the entire contents of the Civic Archaeological Museum of Milan [14].The Cultlab3D was developed in Germany by the Fraunhofer Institute for Computer Graphics Research IGD, which is an automated digitization system for small museum artefacts [15]. It is based on a conveyor belt and a series of devices positioned along the path, which capture both the shape and the visual appearance of the object under examination, significantly reducing the digitization time of a museum piece. The 3D Petrie Museum project, developed by University College London (UCL), dealt with the systematic digitization of the Egyptian Petrie art in London and networking of the relevant 3D models [16]. It was similar to the Smithsonian 3D project developed by Smithsonian Institute [17].

3. Materials and Methods

3.1. Principal technique

Even though in some cases laser scanning was used (scaling of models for example, see section 4.1.), but the main 3D capturing technique used in the IU Uffizi project is Structure from Motion (SfM) / Image Matching (IM) photogrammetry, which allows to acquire both the geometrical and surface texture of the captured objects. Even though the photogrammetric process is highly automatic, the results can still be affected by the photographic quality of the images, the proper distribution of camera locations in space around the artifact being surveyed and skills of the post-processing operator.

The 3D acquisition pipeline based on active devices [18], requires a considerable amount of manual work in some of these steps, with particular emphasis on alignment, mesh and texture editing. Contrarily, the SFM based approach, is fully automatic after acquiring the images in correct sequence. The output of this process is a textured 3D mesh for which a moderate amount of editing might still be needed, but the extent of the editing work is smaller to that for scanned models. Even though high resolution can be achieved by using triangulation based laser scanning, the texture still has to be re-created from images on a 3D scanned model. This process requires to orient all of the texturing pictures by the 3D model orientation and projecting them on the portions of mesh that are visible at the corresponding point of view of images. Finally, the overlapped images are blended in terms of exposure and color temperature. As a result, generating a texturized model from a mesh obtained from an active device is much more time consuming than the automatically generated texturized model from SFM [19]. Furthermore, compared to scanning, much less equipment is required for photogrammetry, which can easily be carried to reach the sculptures irrespective of their size and placement. Triangulation based 3D scanning devices are not be suitable for capturing the whole sculpture geometry because of their different sizes and cannot be moved because the museum is open for public display.

3.2. Photographic Devices

A careful selection of the camera and mounted lenses is fundamental for achieving optical image quality for photogrammetry. The camera used for most of the photogrammetric survey in this project was a mirrorless Sony α6000 with a 24.3 megapixel APS-C CMOS sensor and E-mount lenses. Depending on the camera object distance (the placement of sculptures in the museum) and the surface quality of the surveyed objects, two different lenses were used: a Zeiss Sonnar 24mm f/1.8 and a Zeiss Touit 12mm f/2.8 Lens. Thanks to the combination of one of the widest fields of view in APS-C format, the latter lens was especially useful for digitizing sculptures near the wall which will be explained further in section 4.2.

All of the images for the outdoor sculptures were captured with natural light. Same approach was adopted for the sculptures placed inside the gallery or museum rooms, wherever possible, the use of artificial lighting was avoided. In situations, especially inside the rooms with no windows, where enough natural light was not present, standard photographic illuminators with controlled color temperature were used. In addition, for acquiring the back sides of the sculptures, close to the walls, a HVL-RL1 LED ring light was used. It is a device designed for video recording, therefore it is a continuous illuminator and not a flash. This particular feature is useful for guaranteeing the proper functionality of the autofocus function.

3.3. 3D capturing procedures

The illumination in the rooms and galleries of the museum was a mixture of artificial spotlights and the natural lights from the windows. Other sculptures placed outside in "Loggia dei lanzi" and "Boboli Garden" were illuminated by the uniform natural lighting. Based on different lighting conditions, four different approaches were used to adjust the white balance: (i) an Automatic White Balance (AWB) mode was used when the sculpture was illuminated by a uniform mixture of natural and artificial light (in the rooms with a lot of windows) or uniformly illuminated by the natural light (sculptures placed outdoors); (ii) a White Balance set for artificial lighting was used for the sculptures that were predominantly illuminated by the artificial lighting; (iii) for the sculptures illuminated mainly by the exterior light coming from the windows, the White Balance was set for the exterior light temperature and (iv) for the sculptures that cannot be moved and are placed in a position where the light temperature is extremely different on the back side than the front side, a color-checking panel was placed in the scene to adjust the color temperature on raw images before processing.

As no zoom lenses were used in this project, therefore the focal length remained the same for all sets of images in order to avoid a calibration parameter for focal length. Operating mode was set to be "Automatic" with Aperture priority, because distortions are very much influenced by the aperture size. The f-value was chosen based on the desired depth of field. In any situation even with very low lighting conditions, the f-value was not set to be more than 5.6.

The sensitivity level (ISO) was set to be as low as possible based on the lighting conditions in order to avoid noise grains on the images, which in turn can reduce the accuracy of 3D measurement. However, for less illuminated objects, the ISO value was set to be higher than minimum to avoid any movement blurring due to long exposure time while capturing images without a tripod. For all of the shootings in this project, the "ISO" value was not set to be higher than 1000 for any lighting condition, instead a tripod was used wherever necessary to avoid increasing the "ISO" value. For outdoor shooting in normal daylight where generally enough light is available, a minimum ISO value was used to obtain minimal grains in the photos.

As, the software used for aligning images can process only "jpeg" format of images (which allow a maximum dynamic range of 8 bits per channel), therefore to achieve a higher dynamic range, "raw" format of images was selected for shooting which allowed us to work on the color or exposure corrections in post processing phase without losing useful information.

3.4. Data processing and post processing

"Agisoft Photoscan", which is a semiautomatic software package, was used for orienting the images to create dense cloud and meshes. The camera orientation and the internal calibration are made automatically in this software, allowing a little interaction to the user. The software implements image orientation and mesh generation through SFM and dense multi-view stereo-matching algorithms.There are several kind of objects of different shape, measures and materials in the museum, therefore several potentially critical situations may occur. Since the aim of the project is to produce a high number of models in a short period of time, one of the main factor was to identify the best pipeline with a reasonable tradeoff between processing time and accuracy. Several test on different types of objects were performed for deciding a reasonable setting of the meshing parameters.

For post-processing, i.e. mesh/texture editing, model orientation, possible fine-scaling and model analysis, different open source and commercial software packages were used. These include: (i) Autodesk meshmixer; Pixologic ZBrush; InnovMetric Polyworks for Mesh editing; (ii) Pixologic ZBrush and Adobe Photoshop 3D for texture editing and (iii) MeshLab and CloudCompare for the final model analysis and orientation. In Figure 1 some examples of the final results are shown. The project is still in progress with more than 300 3D models already produced, with an average rate higher than 20 new 3D models per month, which is compliant with the massive 3D digitization operation needed in the framework of this project.

(a) (b) (c)

Figure 1. Some examples of final models with high resolution texture (16 mega pixel); (a) Laocoön and his sons In the Uffizi gallery; (b) Horse Buontalenti in the basement of Uffizi gallery; (c) Ancient Medici lion in loggia dei Lanzi.

4. Technical choices

4.1. Scaling

In the beginning, for first year of the project, scaling has been done by following the standard procedure of photogrammetry, i.e. by placing several targets in space around the object to be digitized. Even though accurate scaling can be achieved by utilizing this method, but it is very time consuming in terms of placing the targets in the scene and manually measuring precise distances between them. Therefore, this scaling approach was changed later on in the project. To speed up the scaling procedure, we have used FARO Focus 3D laser scanner, working on the principle of range sensing based on phase shift detection. As the points cloud generated by a laser scanner has an absolute scale: each point represents the actual position of corresponding point in space with some measurement uncertainty, the model generated by the photogrammetric process (which has a relative scale) can be scaled by aligning it with the points cloud from laser scanning.

To determine the accuracy of this method of scaling, a 2m tall sculpture of Demeter was digitized by photogrammetry and scaled by using physical targets, which resulted in a scaling error of less than 0.5mm (Figure 2a). Correspondingly, a single scan was made for a part of the same sculpture (front part in this case) by the laser scanner mentioned above. The polygonal mesh generated and scaled by the standard procedure was then imported into CloudCompare (an open source 3D point cloud processing software) along with the point cloud of front part of the sculpture generated by the laser scanner (Figure 2b). The mesh was then aligned and scaled with reference to the point cloud by using an iterative closest point (ICP) algorithm, which resulted in a scaling of less than 1% of the polygonal mesh, which represents the closeness of sizes after scaling by traditional procedure and scaling by the method described here (Figure 2c).

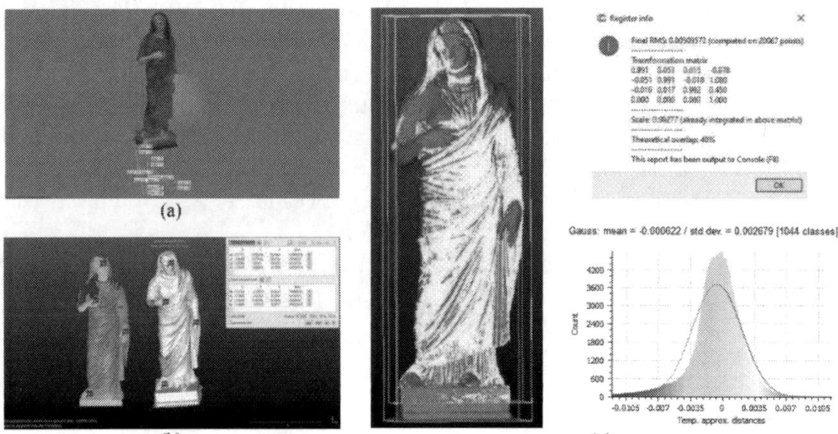

Figure 2. Comparison of different scaling procedures: (a) 3D mesh of a sculpture of Demeter generated and scaled by conventional method in Agisoft Photoscan. The scaling error is 5mm in this case; (b) sculpture of Demeter: imported points cloud from laser scanning (on left) and mesh generated and scaled in Agisoft (on right) (c) mesh aligned and scaled with reference to the point cloud from laser scanner, demonstrating final scaling results and error histogram.

A standard deviation of error of 2.68mm was observed when the distance between the aligned mesh and the points cloud was computed. A part of this error is due to the error associated to the photogrammetric model (i.e. 0.5 mm in this case). Remaining error is directly related to the measurement uncertainty of the laser scanner whose theoretical ranging uncertainty is 2 mm on calibrated targets. In this case the material characteristics of the sculpture (marble) tend to increase this uncertainty [20], [21], making the actual observed standard deviation of error reasonable. However, this local uncertainty does not affect the overall scaling, which is done on the whole shape of the sculpture.

More tests were performed with same procedure on different sculptures of varying sizes and surface qualities: (i) sculpture of Demeter (2m tall with dark rough marble surface); (ii) sculpture of shewolf (1m long with maroon shiny marble surface); bust of Vespasian (0.9 m high with partially shiny white marble surface) and (iii) bust of Domitia (0.8 m high with rough white marble surface). The scaling and alignment errors are compiled in Table 1.

Table 1. Scaling errors and standard deviation of error after distance computation between the scaled models and the points cloud from laser scanner for different sculptures.

	Scaling error %	**Standard deviation mm**
Demeter	0.72	2.68
She Wolf	0.50	4.61
Bust of Vespasian	0.22	3.34
Bust of Domitia	0.18	3.30

It can be noted that the standard deviation of error between the scaled model and points cloud from laser scanner, is solely dependent on the surface quality and measurement uncertainty, as it is highest in the case of the sculpture with most shiny surface (She Wolf) and least in the case of the sculpture with roughest surface (Demeter). On the other hand, the scaling error is maximum for the tallest sculpture and minimum for the smallest sculpture. Hence, it can be concluded that the scaling error is independent of the quality of the laser scan but depends on the whole shape and size of the sculpture.

The scaling of models by the traditional method and the method described here, resulted in almost same output (less than 1% of difference) in all of the examples demonstrated here, therefore this method can be used for scaling photogrammetric models. Even though, this method requires more equipment availability on site, but the data required for the scaling step are represented by a single laser scan image taken at a few meters from the sculpture, which in terms of time takes much less than preparing targets network around the statue and measuring all the relative distances needed for the traditional scaling phase. Therefore, to save time in placing physical targets and measuring their relative position during the photogrammetric survey, we are using the approach of scaling of photogrammetric models with laser scanning.

4.2. Sculptures close to walls

Working in a museum without closing it to the public means adopting the technological choices, as much as possible, according to the situation. The position of the sculpture in the exhibition environment, in many cases, might be very unsuitable for shooting images all around it. This is particularly true for a museum like the Uffizi Gallery, where the number pieces being displayed are enormous, and the space available for exhibition is relatively limited. As a result, many important statues are located close to the walls like the example shown in figure 4a.

In these cases, a possible solution could be to physically move the artifact in the middle of the room to allow a proper photographic shooting all around. But relocating such objects is not an easy task, both for their weight, ranging from hundreds to thousands of kilograms and for their intrinsic value, which makes it mandatory to use the extraordinarily costly and time-consuming processes and precautions.

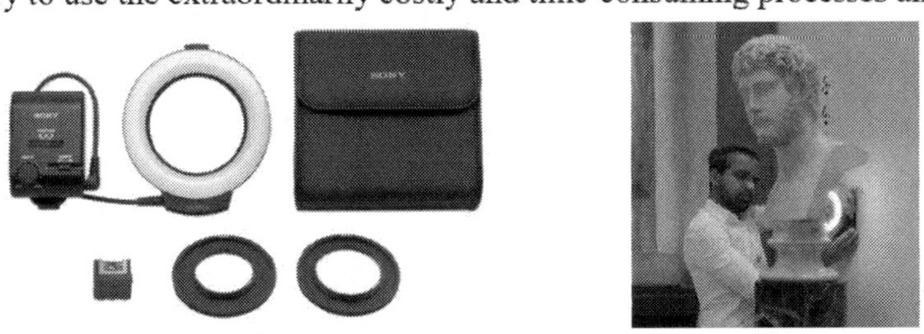

(a) (b)

Figure 3. (a) HVL-RL1 LED ring light with attachments for different size of lenses; (b) shooting the backside of a sculpture in the museum with ring light attachment.

By analyzing the current situation of the sculptures in the museum located close to the wall, it was found that the 90% were at about 20cm from the wall. Such space allows entering behind the sculpture with a small camera like the one being used in this project, even if with the lens at a very short distance from the sculpted surface. This implies a limited field of view and consequently the need of covering the surface with thousands of images. Also, the zone behind a statue is usually very dark and operating without any additional lighting would have required increasing the ISO setting and opening the diaphragm. But this would have grown the image grain and reduced the depth of field, which is exactly the opposite of what a shooting from a short distance would need.

The solution adopted was to use a wide-angle lens suitable to focus at a very short distance from the surface (18 cm), coupled with a ring illuminator with continuous light and specially designed for video recording (figure 3). This configuration was first tested in the lab on a test object in order to find the optimal balance between light intensity, shooting time and aperture, suitable to capture at that distance with the most limited blurring. By using this combination of wide angle lens and illumination, we were able to capture the back surfaces of the sculptures near to the wall also in very dark areas, which otherwise was not possible.

IOP Conf. Series: Materials Science and Engineering **364** (2018) 012015 doi:10.1088/1757-899X/364/1/012015

(a) (b) (c)

Figure 4. Sculptures close to the walls: a) example of sculpture having only 20 cm between the sculpted surface and the wall; b) model obtained with a 24mm lens, where a huge lacking part is present; c) model obtained with the 12 mm lens and the ring illuminator, completed in all areas close to the wall.

4.3. Camera positioning in high places

Carrying out a photogrammetric project on large artifacts, like those owned by Uffizi gallery (up to 3m high), requires the movement of the camera all around the object and therefore also in positions not easy to reach. Two different approaches are usually used to reach higher places for a photogrammetric survey: climbing up the scaffoldings all around the artifact to be digitized or using a drone.

The first approach, generally used for restorations of the statues, involves a lot of additional work for mounting the structure of the scaffolding and for moving it around the object resulting in a solution too slow and costly especially for a systematic digitization process involving so many pieces. The drone was instead problematic regarding the related flight authorizations. The strict regulations about unmanned flying vehicles in the Italian art cities, where the areas hosting the sculptures are visited by hundreds of tourists every day, prevents the authorities to allow any flight unless the area is closed to the public. And this condition was not practical to be reached unless a huge amount of time would have been spent in dealing with the associated bureaucratic issues. Therefore, for the IU-Uffizi project, none of the two approaches were followed.

Therefore, the simplest solution was to use a tall monopod on top of which the camera was mounted. The camera used for this purpose, has a feature to be remotely controlled with a portable device like a smartphone or a tablet. Both the viewfinder and all the photographic controls can be transferred to a handheld device, and an operator can remotely decide to expose, focus and shoot. This specific monopod is 8m aluminum pole made by 4 sections of 2m each, which can be used modularly for reaching 2, 4, 6 or 8m of height (Figure 5). It was remarkably suitable for the gigantic sculptures in the "Loggia dei Lanzi," in front of the Uffizi gallery and for a part of the Uffizi collection. Such statues are from 2 to 3 meters high, and lie on a 1.5 to 2m basement, reaching easily 4 to 5 meters of absolute height from the floor. Imaging these artifacts from 8 m allows to capture all the details of the top surfaces that would be otherwise lost. In addition, reducing the height at steps of 2m allows to capture images all around the sculpture from different heights. Even though this method requires extra personnel, at least one person for holding the pole and one for capturing images remotely, but the overall time of survey is much more less than using other methods.

Figure 5. Monopod for shooting of tall sculptures with four section structure.

5. Conclusions

An ongoing massive digitization project has been explained in terms of the methodologies and technical choices adopted to digitize maximum number of museum artifacts in minimum time and to create models containing maximum information in terms of geometry and surface characteristics. The project has already produced more than 300 models of the most complex sculptures till now. The website of this project (www.digitalscupture-uffizi.org) offers interactive 3D models, published using the web service Sketchfab, embedded on a page which also includes the traditional information typically provided for a work of art (name, artist, date, material, dimensions, principal restorations, bibliography) as well as technical metadata about the author of the model, the method of 3D data capture, and the software used for 3D modeling ad 3D editing. It is anticipated that at the end of each academic year, the database will be updated to include the newly modeled sculptures. When it will be launched, this year, the preliminary website should present ca. 300 sculptures.

References

[1] CENSIS, *L'economia della disintermediazione digitale. 12° Rapporto Censis/Ucsi sulla comunicazione*. Rome: CENSIS, 2015.
[2] D. Lowe, "Object Recognition fromLocal Scale-Invariant Features", *IEEE Int. Conf. Comput. Vis.*, bll 1150–1157, 1999.
[3] S. Keypoints en D. G. Lowe, "Distinctive Image Features from", *Int. J. Comput. Vis.*, vol 60, no 2, bll 91–110, 2004.
[4] Yan Ke en R. Sukthankar, "PCA-SIFT: a more distinctive representation for local image descriptors", *Proc. 2004 IEEE Comput. Soc. Conf. Comput. Vis. Pattern Recognition, 2004. CVPR 2004.*, vol 2, bll 506–513, 2004.
[5] H. Bay, T. Tuytelaars, en L. Van Gool, "SURF: Speeded Up Robust Features", in *Computer Vision -- ECCV 2006: 9th European Conference on Computer Vision, Graz, Austria, May 7-13, 2006. Proceedings, Part I*, A. Leonardis, H. Bischof, en A. Pinz, Reds Berlin, Heidelberg: Springer Berlin Heidelberg, 2006, bll 404–417.
[6] M. Levoy, K. Pulli, B. Curless, S. Rusinkiewicz, D. Koller, L. Pereira, M. Ginzton, S. Anderson, J. Davis, J. Ginsberg, J. Shade, en D. Fulk, "The Digital Michelangelo Project: 3D Scanning of Large Statues", in *Proceedings of the 27th Annual Conference on Computer Graphics and Interactive Techniques*, 2000, bll 131–144.
[7] G. Godin, J.-A. Beraldin, J. Taylor, L. Cournoyer, M. Rioux, S. F. El-Hakim, R. Baribeau, F. Blais, P. Boulanger, J. Domey, en M. Picard, "Active optical 3D imaging for heritage applications", *IEEE Comput. Graph. Appl.*, vol 22, no 5, bll 24–35, Sep 2002.
[8] M. Pieraccini, G. Guidi, en C. Atzeni, "3D digitizing of cultural heritage", *J. Cult. Herit.*, vol 2, no 1, bll 63–70, Mrt 2001.
[9] G. Guidi, J.-A. Beraldin, en C. Atzeni, "High-accuracy 3-D modeling of cultural heritage: the digitizing of Donatello's 'Maddalena'.", *IEEE Trans. Image Process.*, vol 13, no 3, bll 370–380, Mrt 2004.
[10] G. Müller, G. H. Bendels, en R. Klein, "Rapid Synchronous Acquisition of Geometry and Appearance of Cultural Heritage Artefacts", in *Archaeology and Cultural Heritage VAST*, 2005.
[11] D. Arnold, F. Niccolucci, D. Pletinckx, L. Van Gool, S. Havemann, V. Settgast, D. Fellner, G. Willems, G. Müller, M. Schneider, en R. Klein, "The Presentation of Cultural Heritage Models in Epoch", in *EPOCH Conference on Open Digital Cultural Heritage Systems*, 2008.
[12] David Arnold, "3D COFORM - Project Final Report", 2013.
[13] R. Scopigno, M. Callieri, P. Cignoni, M. Corsini, M. Dellepiane, F. Ponchio, en G. Ranzuglia, "3D Models for Cultural Heritage: Beyond Plain Visualization", *Computer*, vol 44, no 7. bll 48–55, 2011.
[14] G. Guidi, P. R. Navarro, L. L. Micoli, S. G. Barsanti, en M. Russo, "3D Digitizing a whole museum: a metadata centered workflow", in *2013 Digital Heritage International Congress (DigitalHeritage)*, 2013, bll 307–310.
[15] G. Singh, "CultLab3D - Digitizing Cultural Heritage", *IEEE Comput. Graph. Appl.*, vol 34, no 3, bll 4–5, 2014.
[16] S. Robson, S. MacDonald, G. Were, en M. Hess, "3D recording and museums", in *Digital Humanities in Practice*, C. Warwick, M. Terras, en J. Hyhan, Reds London: Facet Publishing, 2012, bll 91–115.
[17] "Digitization Program Office, 'Smithsonian X 3D,' Smithsonian Institution, 2013". [Online]. Available at: https://3d.si.edu/. [Toegang verkry: 23-Jan-2018].
[18] F. Bernardini en H. Rushmeier, "The 3D Model Acquisition Pipeline", *Comput. Graph. Forum*, vol 21, no 2, bll 149–172, Jun 2002.
[19] F. Fassi, L. Fregonese, S. Ackerman, en V. De Troia, "Comparison Between Laser Scanning and Automated 3D Modelling Techniques To Reconstruct Complex and Extensive Cultural Heritage Areas", in *3D-ARCH 2013 - 3D Virtual Reconstruction and Visualization of Complex Architectures*, 2013, bll 73–80.
[20] S. El-Hakim, J. A. Beraldin, F. Remondino, M. Picard, L. Cournoyer, en E. Baltsavias, "Using terrestrial laser scanning and digital images for the 3D modelling of the Erechteion, Acropolis of Athens", *DMACH Conf. Proceedings, Amman, Jordan*, no November, bll 3–16, 2008.
[21] G. Guidi, F. Remondino, M. Russo, en A. Spinetti, "Range sensors on marble surfaces: quantitative evaluation of artifacts", in *Proceedings of SPIE on Videometrics, Range Imaging, and Applications X*, 2009, vol 7447, bll 744703–744712.

IOP Conf. Series: Materials Science and Engineering **364** (2018) 012016 doi:10.1088/1757-899X/364/1/012016

An engineering study of a Bronze Age war chariot

A Mazzù[1,3], F M Gambari[2], S Uberti[1], I Bodini[1], S Pasinetti[1] and G Sansoni[1]

[1] University of Brescia, Department of Mechanical and Industrial Engineering, via Branze, 38, 20123 Brescia (Italy)
[2] Museum of Civilisation, P.za Guglielmo Marconi, 14 – 00144 Rome EUR

[3] Corresponding author: angelo.mazzu@unibs.it

Abstract. The Bronze Age crossbar wheel found in the XIX century in Mercurago (Italy) is an amazing example of the technical innovations stimulated by the diffusion of horse draught war chariots in Europe across the third and second millennium b.C. In this paper the tools of modern engineering were used to study the structural issues concerning the wheel and the chariot it should be attached to. In particular, the laser-scanner technology and finite element analysis were used for investigating the dimensional, shape and assembly issues of the wheel, as well as for assessing its structural integrity under operating conditions. The role of the inserted nave, which could be similar to modern bushings, was particularly emphasised, as it is a very innovative solution for that time. The performance of a war chariot equipped by this wheel was studied by means of a vehicle dynamics simulation software, hypothesizing two different chariot structures on the basis of Armenian and Egyptian evidences respectively. The former is probably more similar to the chariot to whom the Mercurago wheel was attached; the latter is technically much more advanced. The results of the analysis allowed obtaining important information about the chariot stability, reliability and structural integrity.

1. Introduction

The invention of the wheel was a key factor in the development of prehistoric cultures in the Old World. The archeological evidences suggest that important social, economic and technological changes occurred across the fourth and the third millennium B.C., providing the preconditions for the development of the earliest wheeled vehicles [1]. In particular, cereal cultivation and animal domestication spread out all over the continent, stimulating the employment of the ox as draught animal. The availability of this animal allowed the development of sledges and wheeled vehicles for transport. The earliest wheels were massive wooden single pieces, with integral nave, obtained from planks cut from tree trunks through wedges. A natural evolution of this technology was the development of tripartite wheels, obtained by assembling three planks cut out from a narrower log. The three parts of the wheel could be assembled by internal dowels and/or external battens. In some cases, the wheels were turning around the fixed axles, and could have integral or inserted nave. In other cases the wheels were fixed to the turning axle through square mortises. The evidences of vehicles equipped with these wheels are usually four wheeled wagons and, more rarely, two wheeled carts, suitable for heavy and slow transport by means of couples of yoked oxen.

The scenario of the wheeled vehicles changed by the end of the third millennium B.C. with the diffusion of domesticated horses throughout Europe. The new motive force provided by this animal allowed dramatically increase the ground transport speed, stimulating the development of lighter and faster vehicles. The new invention was the chariot, a light, resistant two-wheeled vehicle, aimed at rapidly displacing one or two persons. This innovation likely started from the Near East, mainly for war and ceremonial purposes, subsequently spreading all over Egypt and Europe and becoming a key device up to the ancient empires [2,3]. In particular, the six complete chariots found in 1922 in the tomb of the Egyptian Pharaoh Tutankhamun, and two other similar ones (the "Florence chariot" and the "Yuia-Tuiu" chariot) attracted the attention of scientists of ancient technologies, because of their high efficiency and the surprising level of technological knowledge required for realizing them [4,5]. The chariots of Near East and Egypt communities in the second millennium B.C. had spoked wheels, particularly efficient in terms of strength/weight ratio.

At an intermediate technological level between the earlier bulky wheels and the advanced spoked ones stands the crossbar wheel, whose a well-known evidence is the one found in the XIX century in Mercurago, in northern Italy. Nowadays, a plaster clast of this wheel remains, conserved in the Antiquity Museum of Turin (Italy), as the technologies available at the time of the finding did not allow the conservation of wooden evidences. This wheel, dated between the XVIII and the XIII century B.C., is characterized by a diametric bar, thick enough in the central region to accommodate an inserted nave, with four thinner rods departing eccentrically on each side of the nave, and terminating with their ends mortised into the felloe. The overall shape and assembly of this wheel, as well as the constructive complexity and optimized structure, makes the hypothesis that it was destined to a war chariot the most probable.

In this paper, the results of a technical study conducted on this evidence are presented. In particular, laser-scanner technology and finite element structural analyses were used for investigating some open questions about the wheel structural and functional concept, the assembly techniques and the materials. Subsequently, preliminary analyses of the performances of a chariot supported by this wheel were conducted by means of a vehicle dynamics simulation software, considering two chariot configurations based on Armenian and Egyptian evidences.

2. Analysis of the wheel of Mercurago

The original plaster clast of the wheel of Mercurago was analysed by means of optical laser scanner acquisition, aimed at evaluating and measuring its overall geometry and particulars. Subsequently, a solid virtual model of the wheel, as it should be when it was in use, was built and analysed by finite element simulations, in order to investigate constructive and structural issues. The details of such analysis are published in [6]; here the main data about the methodology and results are summarized.

2.1. Laser scanner acquisition of the original plaster clast

Optical acquisition of the wheel surfaces was carried out by the system illustrated in Figure 1. It consists of a laser beam, expanded by a cylindrical lens, scanning the whole object; a video camera for the deformed laser beam acquisition; a software for the elaboration of the acquired point cloud.

Figure 2 shows the 3D model of the wheel, with the nave virtually inserted. The crossbar thickness increases as far as approaching the wheel center, therefore optimizing its resistance to the bending moment due to a lateral force. The crossbar ends are spade shaped, in effects integrating the rim structure. The ends of the felloes are connected to the crossbar ends by internal dowels, whose slots are visible in the free crossbar end. The rods end with a tapered wedge shape; signs of somewhat like putty or glue are visible at the entrance of the mortice: this reveals the willing of realizing a very tight joint. The rest of a broken rod is even visible: in the mid of the cave a prominence of circular section with two wings on the sides appears: maybe it was a metallic insert, aimed at enforcing the joint. Other relevant information can be obtained by the observation of the coupling between the nave and the wheel hub. The features of the coupled surfaces were quantitatively examined on the virtual model. The external radius of the nave was is about 48–49 mm in the central zone; the internal radius of the wheel hub is larger near the ends, but it lowers down to less than 49 mm in the central region. These measurements show that the two members should be coupled by interference, even necessary to prevent axial pulling out of the wheel. Even some longitudinal grooves on both of the coupled surfaces, shown in Figure 3 seem to confirm this hypothesis, as they could be the marks of the nave insertion and extraction.

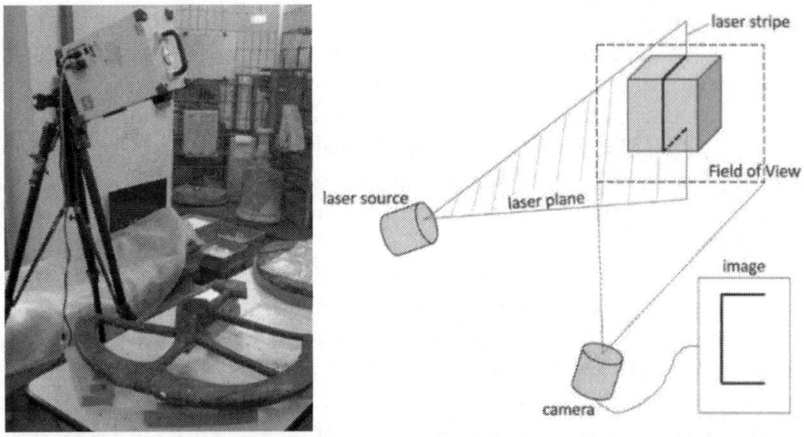

Figure 1. Optical acquisition set-up

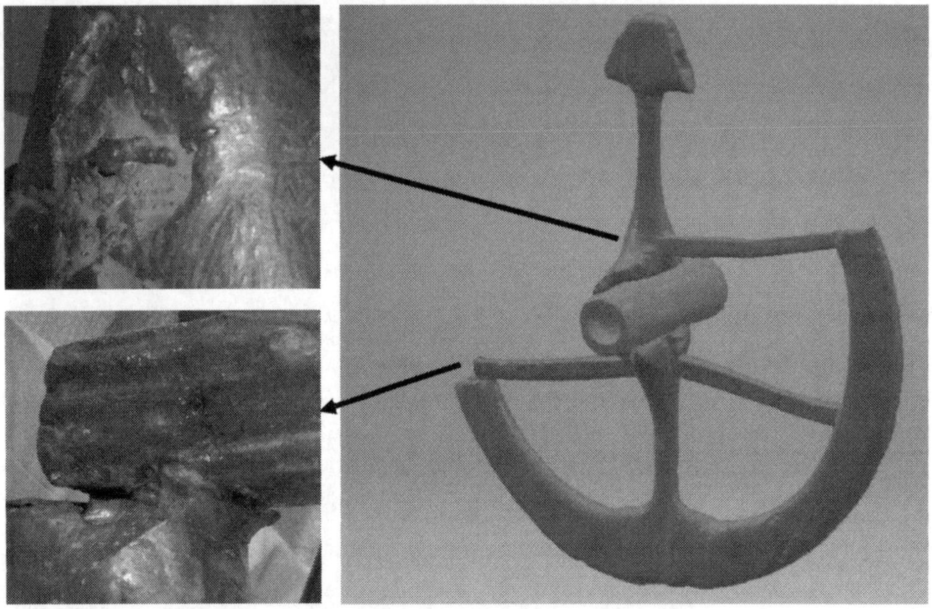

Figure 2. Virtual 3D model of the Mercurago wheel, with particulars

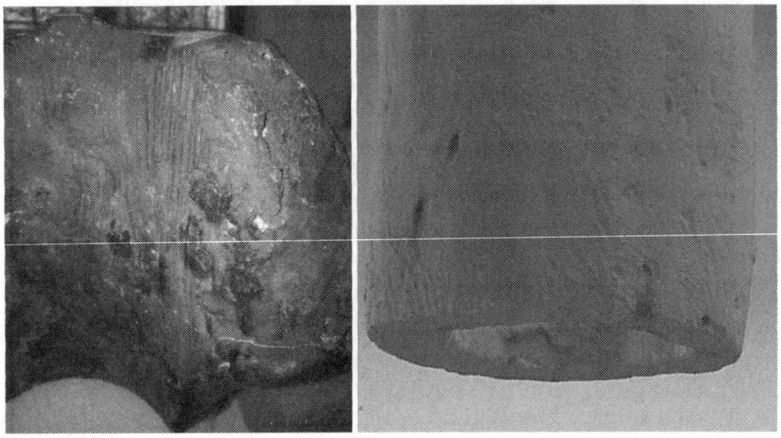

Figure 3. Longitudinal marks on the hub and the nave, likely due to insertion with interference.

IOP Conf. Series: Materials Science and Engineering **364** (2018) 012016 doi:10.1088/1757-899X/364/1/012016

2.2. Finite element analysis

Several finite elements analyses were carried out with the ABAQUS 6.14 software in order to investigate the structural issues. Figure 4 shows the solid model of the integral wheel reconstructed from the digitalized model by means of the SolidWorks 2013 CAD software. The internal surface of the hub is tapered; the external diameter of the nave was set so that an interference ranging from 1 to 2 mm in the central zone was established.

It is thought that both the felloes and the crossbar were in some hardwood, as they were called to give structural solidity to the wheel and to resist penetration by external bodies in the contact with soil. The same hypothesis can be assumed for the dowels, given their crucial role in the wheel structural integrity. Verisimilarly, a particular walnut species, nowadays extinct, could be employed. More uncertain is the material used for the nave and the rods: the implications of using a softwood or a hardwood were investigated. Nowadays, it's impossible to exactly know the mechanical properties of the materials employed for the wheel construction; furthermore, the wood properties are highly changing depending on the wood age, treatment, atmospheric conditions, etc. In the present paper an estimation was done on the basis of the properties of some woods presently in use, but it has to be taken as an approximation able to give a trend about the wheel behaviour. In this paper, the possible properties of a hardwood and a softwood were approximated with the average properties of the black walnut and the northern white cedar respectively, with 12% moisture content, according to the data published in [7]. Two loading conditions were set, hypothesizing two occupants on the chariot: one wheel bumping on a soil asperity at 40 km/h in rectilinear run, and running a curve of 5 m radius at 20 km/h. Figure 5 shows the wheel deformation in curve, amplified by a factor of 5. Other analyses considering a leather tyre wrapping the felloes were carried out, in order to investigate the effect of such technical device which was proven to be in use in the antiquity [1].

Figure 4. From the acquired 3D model to the reconstructed full wheel model.

Figure 5. Deformed wheel in curve (amplified 5 times).

The main findings of the finite element analyses were the following:

- The wheel structure was very efficient in terms of stiffness/weight ratio (its weight was only 7-8 kg) when loaded in radial direction (i.e. in straight running). In curve, the lower stiffness of the rods with respect to the crossbar was a weak point: these had to be realised in hardwood for supporting the lateral forces, but this could not be enough in the simulated tight cornering.
- The leather tires had the effect of significantly reduce the stresses on the felloes, therefore it is very probable that these were used.
- The inserted nave had probably a more functional rather than structural role: if made in softwood, being subjected to higher wear, it could more easily adapt itself to the rolling coupling

with the axle, preventing the risk of wheel blocking; furthermore, it would prevent wear of the axle. At the end of the battle it could be easily replaced by a new one. All these considerations allow supposing that it was a sort of spare part, with a role similar to that of the modern bushings. This would be a very innovative technology for that time, unfound even in more advanced chariots such as the Tutankhamun class ones.

3. Dynamical analysis of the chariot

A dynamical analysis of the behavior of a war chariot equipped with the Mercurago wheel was carried out with the aim of verifying the performance in battle. As the original chariot is not known, two configurations, based on archaeological evidences, were analyzed, in order to compare different design solutions.

The first solution is based on the Tutankhamun chariots [3-5], which are the most advanced ancient devices: these are made by very complex components, whose construction required highly specialized workforce. For the aim of the present study, its main feature is that the axle was beyond the occupants: this way, the elasticity of the pole, coupled with the damping of the floor which was made by a texture of leather stripes, should provide a very good stability. Furthermore, longitudinal oscillations, due to the horses gallop, were damped by the connection between the pole and the axle.

The second solution is based on the evidence of the Armenian chariots found in Lchashen [1]: in this chariot the occupants stood above the axle, and the damping effect relied on the floor. No longitudinal oscillation damping was present as well. This solution, much simpler than the former, was likely more similar to the chariot to whom the Mercurago wheel was attached, being more suitable for small societies as the northern Italian ones were. For the aim of this paper, this solution differs from the previous for the position of the axle. The schemes of the two chariots are shown in Figure 6.

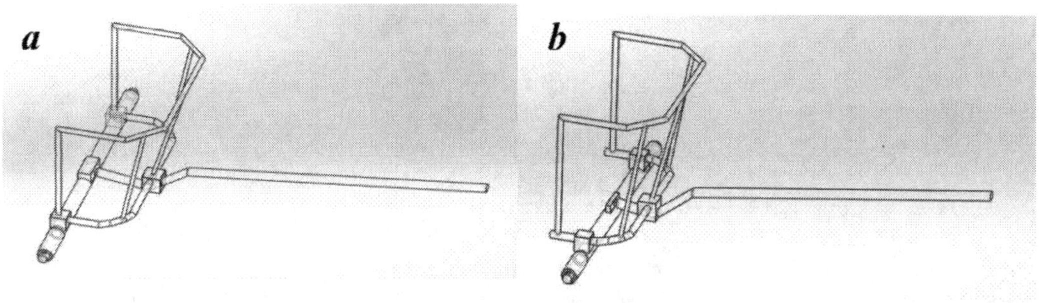

Figure 6. Schemes of the analysed chariots: a) Egyptian; b) Armenian

All joints between the axle, the pole and the cockpit are realized by deformable cubes, aimed at simulating the behavior of bindings realized by leather laces. The axle end diameter is known through the internal diameter of the nave (60 mm); however, it is not known whether such diameter was constant along the whole axle, or if it was reduced at the ends for housing the wheels and blocking them axially. For each chariot, therefore, two different solutions were tested (see Figure 7): the first one with a constant diameter, equal to the internal diameter of the nave, along the whole axle length; the second one, with the diameter equal to the external diameter of the nave (100 mm) in the central zone, reduced to 60 mm at both of the ends at the coupling with the naves.

The dynamical analysis was performed by the CARSIM software, normally used for simulations of vehicle dynamics, which requires a schematization of the vehicle as an assembly of rigid bodies connected by springs and dampers (concentrated parameters model). Therefore, the elastic parameters of the two vehicles had to be preliminary determined by means of finite element analyses: they were subjected to vertical, longitudinal, lateral and torsional loads, and the corresponding displacements were measured in order to determine the equivalent spring constants by the load/displacement ratio. The materials considered were a stiffer wood for the axle (longitudinal modulus of elasticity $E_L = 14190$

N/mm^2) and a slightly less stiff wood for the framework and the pole (E_L = 12320 N/mm^2). Both materials were considered orthotropic elastic, but as all component are subjected mainly to bending stresses, the influence of the elasticity constants in the other directions is negligible. Again, these material elasticity constants have to be taken as indicative.

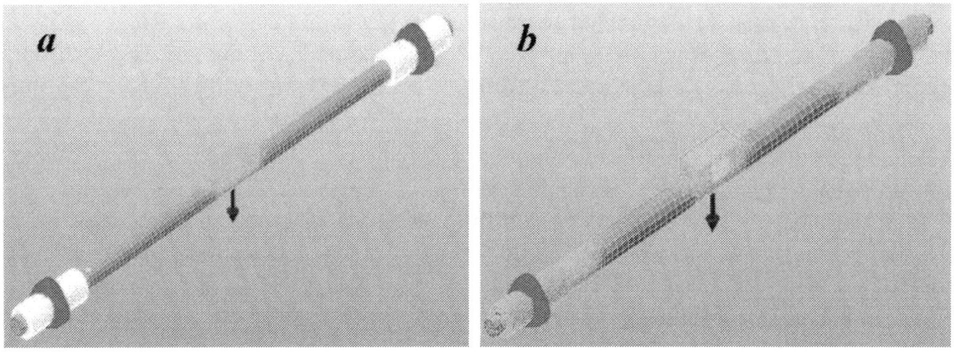

Figure 7. Schemes of the analysed axles: a) thin axle; b) reinforced axle

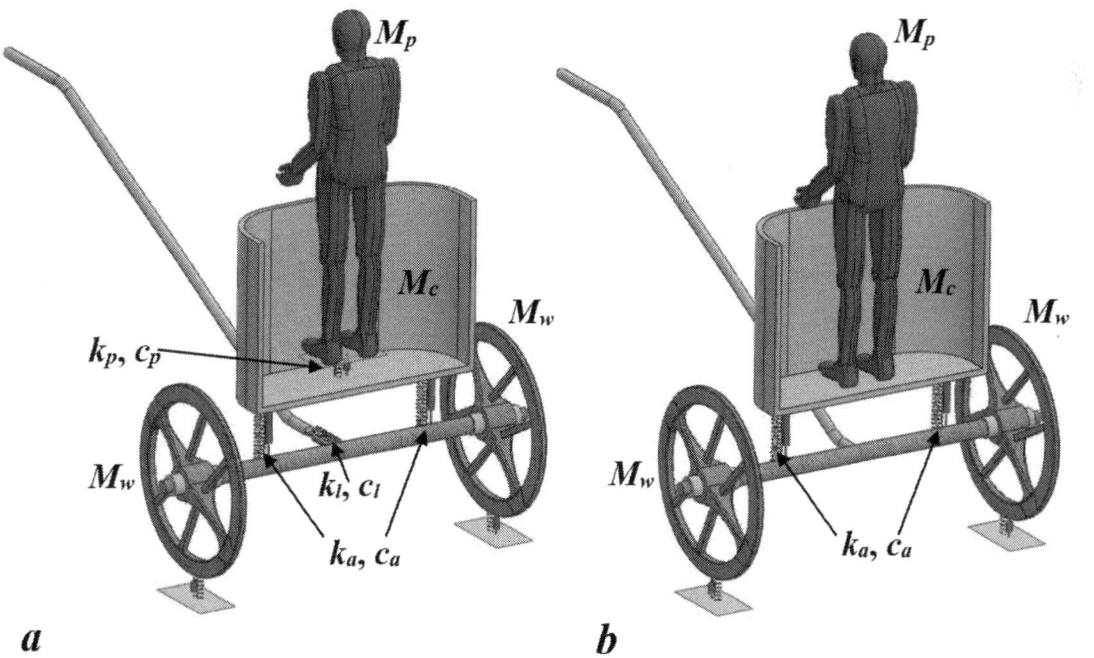

Figure 8. Concentrated parameters schemes of the simulated chariots: *a*) Egyptian; *b*) Armenian.

Figure 8*a* shows the schematization of the Egyptian chariot. In this scheme, the transported mass M_p is connected to the chariot by a spring of constant k_p and a damper of constant c_p, which represents the vertical elasticity and damping of the framework and the pole. The chariot mass M_c is connected to the wheels by means of two spring/damper couples of constant k_a and c_a respectively, which represent the properties of the two half-axles. Furthermore, a longitudinal spring/damper couple of constants k_l and c_l connects the whole draught mass to the horses: such constants are given by the properties of the chariot frame and the pole in longitudinal direction, as well as by the pole-axle joint. The wheels, of mass M_w, should be connected to the soil by means of a springs/damper couple representing their own properties; however, these effects were considered negligible with respect to the action of the other components.

Figure 8*b* shows the scheme of the Armenian chariot. There is not a suspension system between the chariot and the transported mass because this stands directly on the axle. The system for longitudinal oscillation damping is absent as well. The values of the masses, spring constants and damping coefficients for the two chariots in both of the axle diameters are shown in Table 1. M_p was estimated with a reasonable value for an armed man; M_c and M_w were estimated by means of the finite element pre-processors; k_p, k_a and k_l were obtained by the preliminary finite element stiffness analyses cited above; the damping coefficients c_i (where $i = p, a, l$) were reasonably estimated as 1/6 of the critical damping coefficient, which is a function of the sprung mass M_i and the spring constant k_i according to the formula $c_i = 2\sqrt{k_i M_i}$.

Two running conditions were simulated: bumping with both wheels over an obstacle of 50 mm height at 40 km/h and bumping over the same obstacle at the same speed with a single wheel. Many data were extracted from the analyses, but here a focus on the vertical acceleration of the transported mass is reported, as this parameter determines the stability of the occupant and the risk to be thrown out from the chariot.

Table 1. Masses, elastic constants and damping coefficients introduced in the dynamical simulation.

	Egyptian		Armenian	
	Thin axle	Reinforced axle	Thin axle	Reinforced axle
M_p (kg)	80	80	80	80
M_c (kg)	27	33.2	27	33.2
M_w (kg)	7.4	7.4	7.4	7.4
k_p (N mm^{-1})	100	100	-	-
k_a (N mm^{-1}	250	675	250	675
k_l (N mm^{-1})	86	86	-	-
c_p (N s mm^{-1})	0.771	0.771	-	-
c_a (N s mm^{-1})	0.862	1.457	1.140	1.903
c_l (N s mm^{-1})	1.046	1.073	-	-

Figure 9 shows the vertical acceleration of the transported mass M_p (normalized by the gravity acceleration) in all of the considered cases. In general, the a_z is lower with the Egyptian chariot than with the Armenian one. The advantage is felt mainly in double wheel bumping, where the maximum acceleration with the Armenian chariot is nearly twice as with the Egyptian one. With the reinforced axle the vertical acceleration increases, but not so dramatically. With both of the chariots, however, the double wheel bumping could be a very critical condition, as the vertical inertia significantly exceeded gravity. This suggests that the floor of interwoven leather stripes, which was not considered in this analysis, could have a crucial effect in absorbing the dynamic loads; furthermore, the passage over an elongated obstacle, such as a surface root, could be fatal. The analysis allowed calculating even the forces at the wheels. Their maxima were about 2200 N in vertical direction and 400 in longitudinal direction, obtained in both of the conditions. With these forces, it was calculated that the thin axle could not resist, thus likely the reinforced one was used.

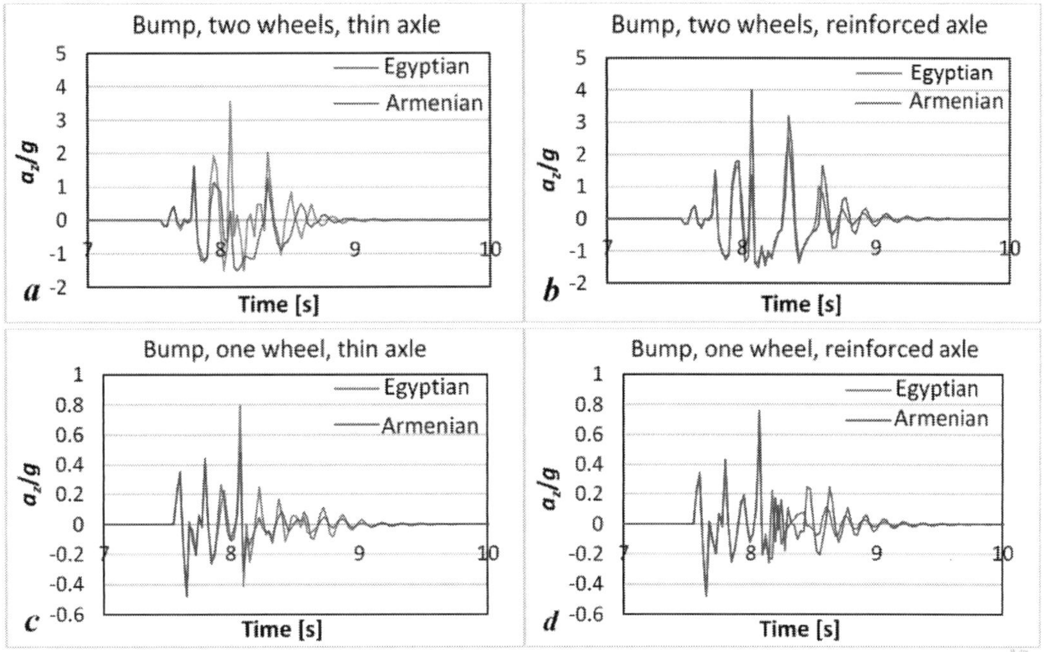

Figure 9. Vertical acceleration of the suspended mass with the Egyptian and Armenian chariots: a) double wheel bumping with thin axle; b) double wheel bumping with reinforced axle; c) single wheel bumping with thin axle; d) single wheel bumping with reinforced axle.

4. Conclusions

The plaster clast of the Mercurago wheel was acquired by laser scanner optical technology. The analysis of the acquired solid allowed obtaining some clues about the construction and the assembly technique. A solid model of the intact wheel was constructed and analyzed by finite element under vertical and lateral forces. The analyses highlighted mainly the weakness of the transversal rods in curve and allowed hypothesizing the anti-wear and anti-friction role of the inserted nave.

Dynamical simulations of the whole chariot in bumping at 40 km/h with single and double wheel were carried out, considering two different chariot configurations ("Egyptian" and "Armenian") and two axle design solutions. The analyses evidenced the better performance of the Egyptian one in terms of limitation of the vertical acceleration of the occupant, although the double wheel bumping could be a critical condition even for this chariot. Furthermore, the obtained loads on the wheel allowed concluding that the hypothesized solution with the thinner axle was likely unfit.

References

[1] Piggot S 1983 The Earliest Wheeled Transports – From the Atlantic Coast to the Caspian Sea (New York: Cornell University Press).
[2] Rossi S, Chondros T G, Milidonis K F, Savino S and Russo F 2016 *Front. Mech. Eng.* **11** 12–25
[3] Sandor B I 2004 *Oxf. J. Archaeol.* **23** 153–175
[4] Sandor B I 2004 *Fatigue Fract. Eng. Mater. Struct.* **27** 637–646
[5] Rovetta A, Nasri I and Helmi A 2000 *Mech. Mach. Theory.* **35** 1013–31
[6] Mazzù A, Gambari F M, Bodini I, Pasinetti S and Sansoni G 2017 *J. Archaeol. Sci. Rep.* **15** 138–149
[7] Green D W, Winandy J E and Kretschmann D, *Wood handbook-Wood as an engineering material* Gen. Tech. Rep. FPL–GTR–113. Madison, WI: U.S. Department of Agriculture, Forest Service, Forest Products Laboratory p 4-1

IOP Conf. Series: Materials Science and Engineering **364** (2018) 012017 doi:10.1088/1757-899X/364/1/012017

Automatic illustration of lithics from 3d scanned models

S. Barone[1], A. Paoli[1], P. Neri[1], A.V. Razionale[1], M.L. De Santis[2], F. Mailland[2]

[1] Department of Civil and Industrial Engineering, University of Pisa
[2] Centro Internazionale di Studi Preistorici ed Etnologici

a.razionale@ing.unipi.it

Abstract. The manual drafting of lithics artefacts could be a very time-consuming work, and it could be cumbersome on the archaeological site. In this case, a 3D digital model of the object could be very useful. Nowadays, several digitizing technologies are available to easily acquire information about the shape of an object. Virtual models could be used to create a digital museum or to share information between researchers. On the other hand, the manual drafting of a lithic object contains information about the technologies used to realize it. Information about the core setup, types of chipping surfaces, detach sequence of supports, and much more. In this work a method to easily obtain a hand-made-like draft of lithic artefacts is proposed. The method is based on the 3D acquisition of the object with a structured-light based scanner and a sequence of digital processing of the acquired data.

1. Introduction

Recent developments in digital 3D acquisition technologies have allowed innovative applications for recording, conservation, reproduction, study and fruition of sculptural, architectural and archaeological artworks [1.]. Originally developed for industrial application, 3D imaging techniques have been successfully extended to the field of artworks acquisition. In lithic studies there are several researches oriented to use of digital models instead of manual drafting [2, 3]. Of human made artefacts in prehistorical era, only the stronger, made of stone or flint are enduring to our days. The lithic manufacturing could be subdivided in chipping stone manufacturing and the smoothing stone one. In this work the chipping stone were considered. Lithic artefacts could be obtained by chipping it with another object (percussion hammer) from a raw stone. Final product can derive from the stone so chopped (double-sided, chopper, etc..), from one of the obtained splinter or from machining one of the splinter obtained from the core. The artefacts arising from a detachment usually have a ventral-face (i.e. the one facing the original core) and spinal-face (i.e. the one that preserves the original surface of the stone or nucleus). Moreover, in all the artefacts a proximal end (i.e. the one that received the chipping stroke) and a distal end opposed to it, are always present. Usually, the faces with manufacturing signs are the ones drawn by technicians [4].

The purpose of the lithic illustration is to represent as close as possible the manual manufacture of the object [5, 6]. The illustration could be of three main type: 1) the typological drafting, that is very schematic and gives information about principal used instruments (Figure 1 a); 2) the technical drafting, that is used to show simplified technical data (i.e. morphology, type of re-touching)(Figure 1 b); 3) the technological drafting, that is the most realistic one, it reproduces both technical and morphological aspects (Figure 1 c).

To obtain a good lithic representation analytical study of the artefact is required. Stones can have different type of surfaces to depict with different type of graphical signs (natural fractures, artificial fractures, re-touching, pseudo re-touching, dorsal face, ventral face, ...). Moreover, the orientation of the stone in the drafting is fundamental, also the reference system used has a technological meaning. The vertical axes should be the morphological one (the one with the main length of the piece), while the horizontal axes should be the principal chipping one, along the first fracture direction. Finally, also the light direction in the lithic drafting is very important. It should evidence, with the light and dark areas of the object, both its morphological and technical aspects. The light is usually placed in the top left spot of the artefact to be drawn.

Simple manual tools, like caliper and compass, are used to measure the length, the depth and the width of the stone, which is placed in an ideal bounding box. Ventral and dorsal profiles are so manually traced on the sheet by contouring the stone with a pencil, while for the depth profile a contour gauge is used [7]. Finally, the projection of the heel, if it is present, is manually traced.

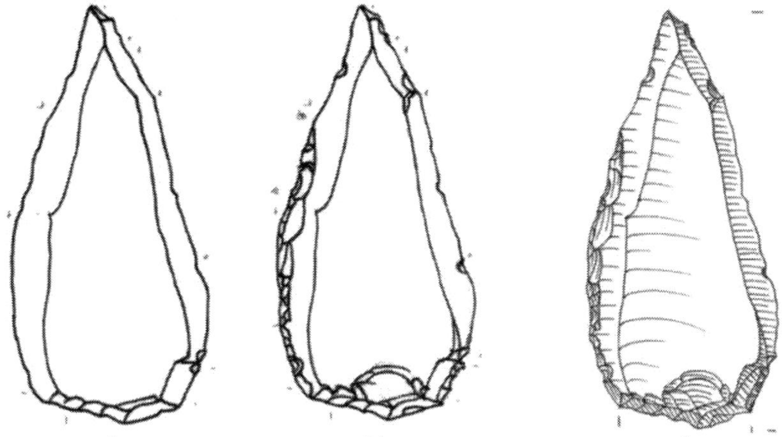

a) b) c)

Figure 1. Manual drafting of a lithic artefact: typological draft (a), technical draft (b), technological draft (c).

2. Materials and Methods

In this works a novel method to obtain manual-like illustrations from 3D scans of stones is proposed. The method is based on a semi-automatic segmentation of 3D data acquired by a structured light scanner. The segmentation is carried on by a cut algorithm based on the values of the crease angles between adjacent faces of the model, that is firstly decimated and smoothed.

2.1. High resolution 3D range camera system

To acquire the geometry of lithic artefacts a specifically developed 3D optical scanner was used. The device is a 3D range camera system composed of a DLP projector (1024 × 768 pixels) and one monochrome digital CCD camera (1280 × 960 pixels) (Figure 2). The acquisition method is based on a structured light approach, which uses binary patterns to capture 3D shapes. Projector is used as active device to scan the object with vertical striped light patterns. A calibration procedure is adopted to calculate distortion parameters and relative position of the optical devices. The projector is modeled like an inverse camera, and the calibration is exploited by using generated vertical and horizontal fringes [8]. A sequence of black and white vertical light stripes is projected onto the object (Figure 2). The crossing lines between black and white fringes define a pattern of planes that virtually slices the full object along a temporal sequence. Each pixel of the image acquired by the camera will be bright or dark depending on its location. A n-bit binary code could be assigned to each pixel, where n is the number of the projected stripe patterns, and values 0 and 1 are associated to the intensity levels, (i.e., 0 = black and 1 = white). The 3D coordinates of the measured points can then be computed by exploiting the intersection between optical rays from the camera with the projected planes. This methodology could provide the horizontal resolution from the projector while the vertical resolution from the camera. A high-density range map is hence obtained in a matter of seconds.

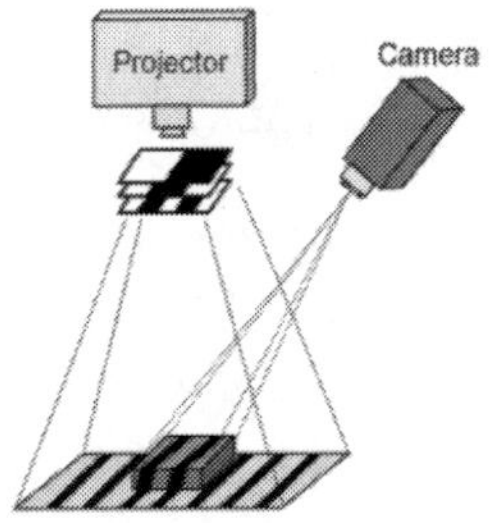

Figure 2. Scheme of the 3D range camera system.

The acquisition of a whole object requires several views of it from different directions. The alignment of the views could a cumbersome task and there are different methods that can be used. In this work a manual raw positioning of the object in front of the acquisition device has been used. The obtained range maps have been raw aligned by a 3-2-1 method and then processed with a standard pipeline to obtain the final model of the mesh (Figure 3).

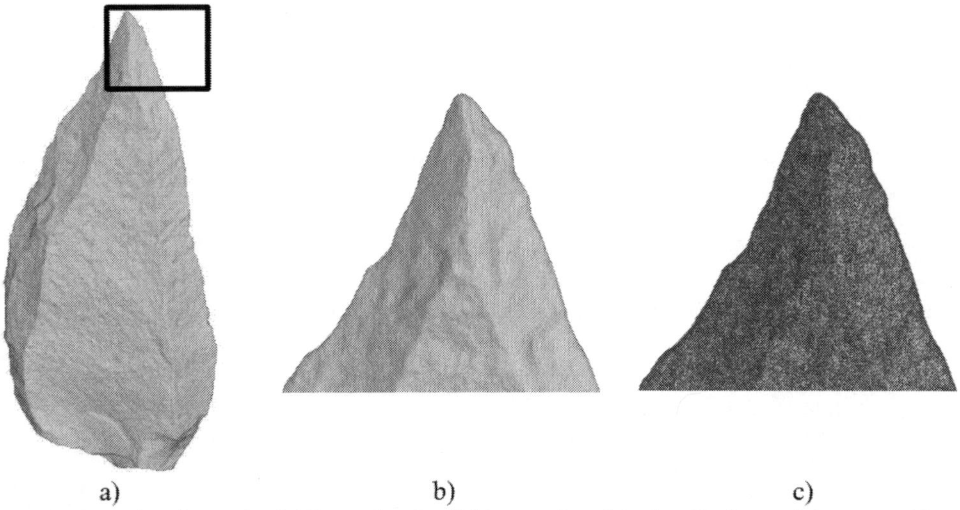

a) b) c)

Figure 3. Sample of acquired 3D mesh of a lithic artefact (a), detail view of the top without (b) and with (c) visualization of the triangle edges.

2.2. Automatic drafting from 3D model

To obtain a manual like drafting from the acquired 3D model a novel methodology has been developed. The methodology is based on a sequence of mesh processing algorithms that enhances first the feature lines of the object and automatically extracts the drafting as projection of these lines.

2.2.1. Model Decimation. The first step of the processing phase is the deçimation of the acquired model. Usually decimation is used in mesh processing to reduce the number of the faces with the aim at having light models (i.e. faster visualization and management). Several algorithms have been

IOP Conf. Series: Materials Science and Engineering **364** (2018) 012017 doi:10.1088/1757-899X/364/1/012017

developed for decimation of triangular meshes. In vertex-based decimation, meshes are reduced by removing vertices from them. Usually vertices are treated one by one, and the resulting hole after each removal operation are closed by a re-triangulation process. In edge-base decimation the number of faces is reduced by removing edges from them. The removed edge is replaced with a single vertex and again a re-triangulation process is needed to obtain the final mesh. In triangle-based decimation, faces that are nearly co-planar with their neighbouring are replaced by one face (or in a minor number of faces). The co-planarity between triangles can be evaluated by analysing local curvatures or face normal. The purpose of this step in the proposed methodology is to reduce the number of faces, without decrease the accuracy and the resolution obtained in the acquisition step. Moreover, the topology of the resulting mesh should be optimized in order to facilitate the later steps towards the automatic drafting of the object. A quadric edge collapse decimation algorithm has been used to reduce the number of faces acquired by the system.

2.2.2. Enhancement of feature lines. In this step the acquired 3D models are processed to enhance features areas (i.e. object boundaries) while smoothing noisy ones (i.e. homogeneous regions). These features could be lost with classical isotropic denoising algorithm. Anisotropic smoothing techniques, instead, could be used with the aim at holding corners and ridges (i.e. sharp edges) of the lithics artefacts (i.e. the lines to put in the technical draft). Different forms of anisotropic diffusion filters have been developed and could be used on meshes [9]. In this work an algorithm based on second-order properties of the surface has been used. Indeed, in the case of an edge between two areas of a mesh, the minimum curvature is zero along the edge, while the maximum curvature is perpendicular to this edge. A weighted mean curvature flow has been computed, this operator allows to penalize faces that have a large ratio between their two principal curvatures. Features like sharp edges could be held while noise, more symmetric by nature, could be reduced. Results of such algorithm has been shown in Figure 4.

Figure 4. Decimation and smoothing of acquired 3D mesh of a lithic artefact: original mesh (a), 10% decimated mesh (b) and smoothed mesh (c).

2.2.3. Drafting creation. The last step of the proposed method is the automatic creation of the draft from the processed 3D model of the lithic artefact. This task consists of 3 main sub-steps: model positioning, mesh cutting and 2D projection of feature lines. The first step consists in the orientation of the model in order to obtain a 2D drafting as much as possible similar to the manual drafting. For this purpose, an initial guess orientation is based on the principal axis orientation of the model, in which the bounding box of the model is placed along the principal direction. From this position a manual adjustment could be performed to obtain the desired final position. The mesh cutting is performed to obtain the edge of the artefact. The cutting of the mesh is created where the angle between two adjacent faces (crease angle) is greater than a specific value. Once the edges are created they could be used to automatic draw the desired views of the artefact (Figure 5).

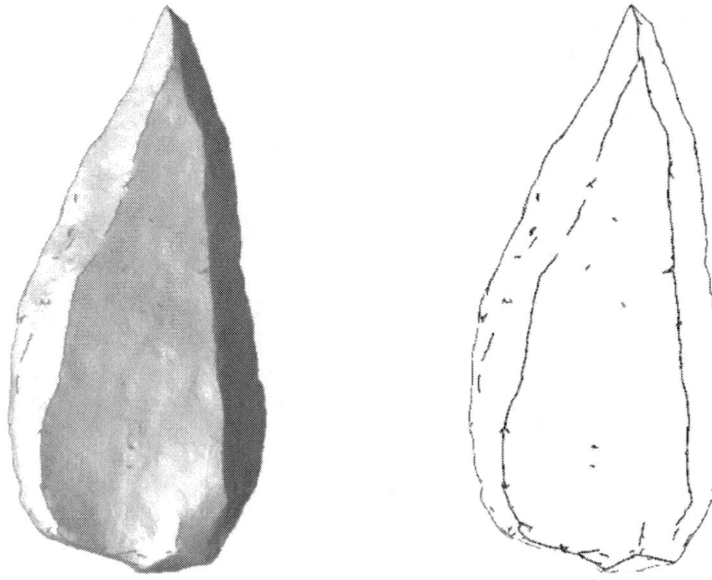

a) b)

Figure 5. Cutting step of the 3D model (a) and relative sharp edges view (b).

3. Results and discussion

Some lithic artefacts have been used to test the proposed methodology. In this works the result relative to one of the artefacts is presented. The artefact shown is a brown flint blade with heel's re-touches, obtained from a secondary chipping of a Palaeolithic detachment, with a height of 80 mm, a width of 30 mm and a depth of 10 mm (Figure 6). For the 3D acquisition of the artefacts 28 range maps (2.5M points) have been collected with a field of acquisition of 80x60 mm, a lateral resolution of 0.08 mm and an accuracy of 10 microns. The acquisition step along with registration phase and the data processing, has taken about two hours. The final model is a triangular mesh of 1.1 M triangles (Figure 7). The obtained model has been processed with the proposed methodologies and the obtained results are shown in Figure 8. The artefact has been also drafted by hand from a technician (Figure 9).

4. Conclusion

Even if there are several standards in manual drafting of lithic artefacts, the result could be variable because of the involved factors [2]. Indeed, the archaeological technician could evidence different characteristics of the stone based on his experience and on the purpose of the drafting [10, 11]. Moreover, when pictures are used instead of the technological drafting, these could not underline the most important details (e.g. due to colour or light reflection). 3D acquired models have been introduced in the last years in technical documentation of the artefacts and it could be of high potentialities in the archaeological field. The model could be used over time for looking at the object from different viewpoints or could be zoomed in to looking for specific details. Moreover, 3D model could be used to produce a replica by additive manufacturing techniques. Anyway, 2D technological drafting could be very useful, especially when printed documentation is needed, because of the additional information included.

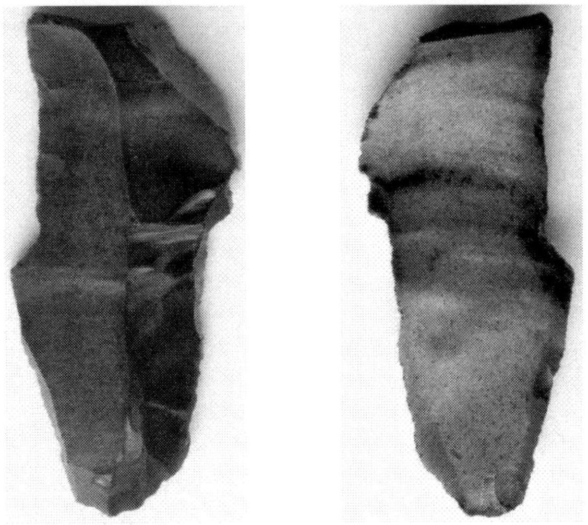

a) b)

Figure 6. Picture of the lithic artefact: frontal view (a) and rear (view (b).

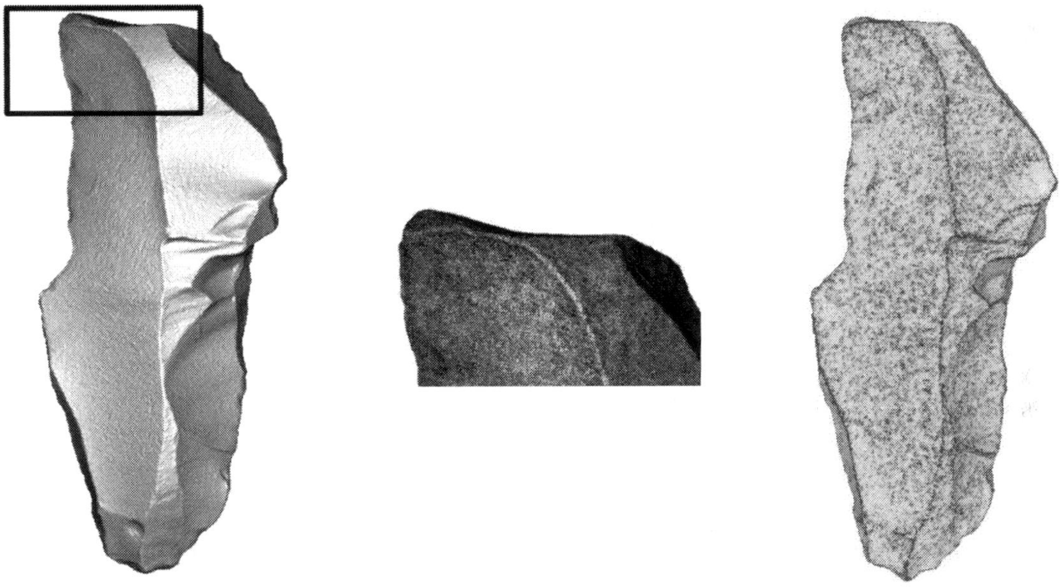

a) b) c)

Figure 7. Acquired 3D model of the lithic artefact: frontal view (a), detail view (b), curvature maps of the frontal view (c).

In this work 3D scanned data has been used to automatic create manual-like lithic technological drafting. The method is based on the processing of acquired data by algorithms of mesh decimation and smoothing. The obtained results show high potentialities of the proposed solution, especially for the technical drafting. Some little details miss from the drafting, anyway it is very interesting the mixed representation that could be easily obtained from 3D model. This, for example, could be used as a base from the technician for easily add technological details. From archaeological point of view, the typological and the technical drafting, digitally obtained from the proposed solution, are of higher accuracy with respect to the manual ones and, of course, faster.

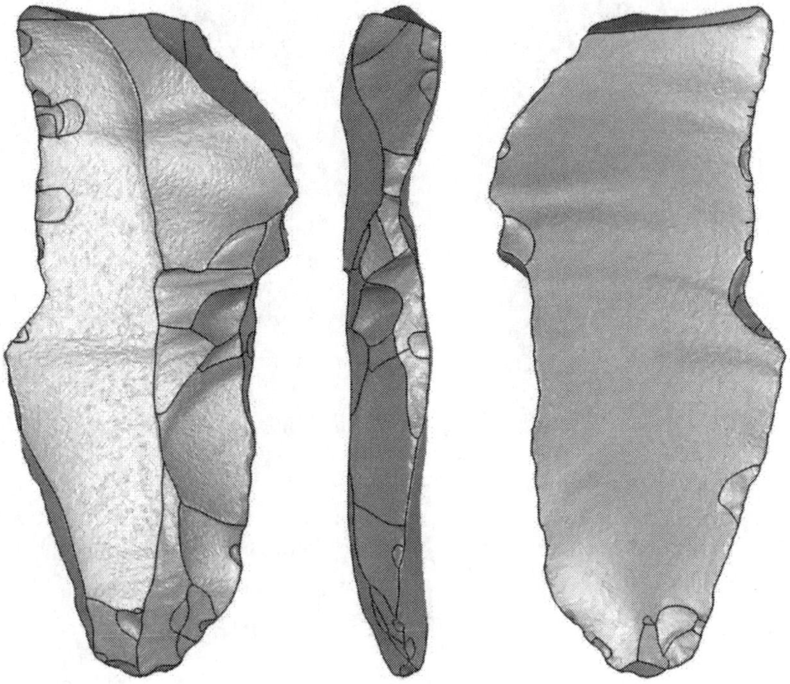

Figure 8. Automatic draft obtained for the lithic artefact.

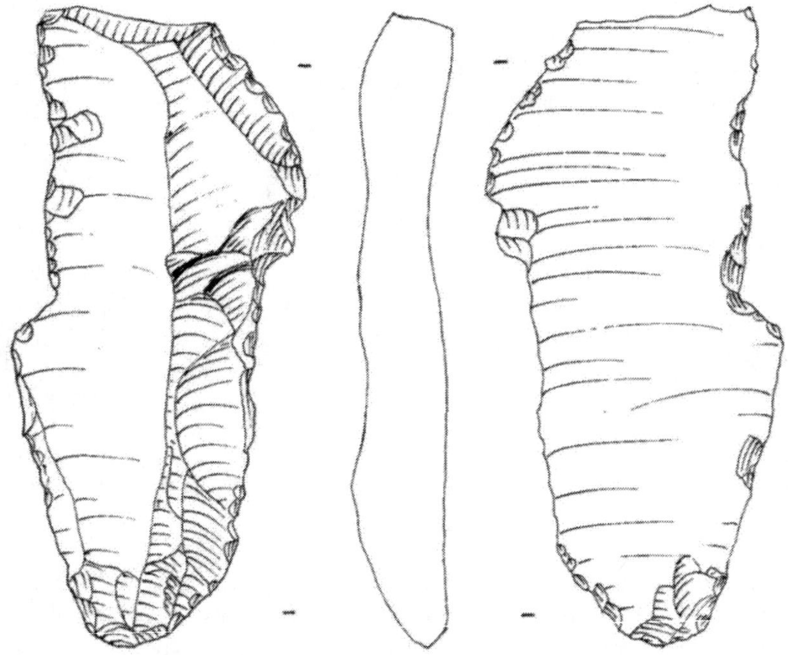

Figure 9. Hand-made draft of the lithic artefact.

References

[1] Abel, Richard L., Simon Parfitt, Nick Ashton, Simon G. Lewis, Beccy Scott, and Chris Stringer, *Digital Preservation and Dissemination of Ancient Lithic Technology with Modern Micro-CT.* Computers & Graphics, 2011, 35:878-884.

[2] Magnani, Matthew, *Three-dimensional alternatives to lithic illustration.* Advances in Archaeological Practice, 2014, 2.4: 285-297.

[3] Bruno, Fabio, Stefano Bruno, Giovanna De Sensi, Maria-Laura Luchi, Stefania, Mancuso, Maurizio Muzzupappa, *From 3D Reconstruction to Virtual Reality: A Complete Methodology for Digital Archaeological Exhibition.* Journal of Cultural Heritage, 2010 11:42-49.

[4] Broglio A., *Introduzione al Paleolitico,* Manuali Laterza, 2006.

[5] Dauvois M., *Precis de Dessin Dynamique et structural des Industries Lithiques Prehistoriques*, Fanlac a Périgueux: Pierre Fanlac, 1976.

[6] Tessaro C., *Preistoria e Protostoria – Tecnologia e Tipologia delle industrie litiche – Laboratorio – Disegno e Sintesi Grafica*, Università di Siena Dipartimento di Scienze Storiche e dei Beni Culturali, 2016.

[7] Medri M., *Manuale di rilievo archeologico*, G.O. Laterza, 2003.

[8] Barone, S., Paoli, A., Razionale, A.V., *A coded structured light system based on primary color stripe projection and monochrome imaging.* Sensors (Switzerland), 2013, 13 (10), pp. 13802-13819.

[9] Pelagotti, A., Del Mastio, A., Razionale, A.V., *Active and passive sensors for art works analysis and investigations*. Proceedings of SPIE - The International Society for Optical Engineering, 2007, 6491, art. no. 64910R.

[10] U. Clarenz, U. Diewald, and M. Rumpf., *Anisotropic Geometric Diffusion in Surface Processing*. In IEEE Visualization, 2000, pages 397–405.

[11] Addington, Lucile R., *Lithic Illustration: Drawing Flaked Stone Artifacts for Publication.* University of Chicago Press, 1986.

[12] Adkins, Lesley, and Roy Adkins, *Archaeological Illustration.* Cambridge University Press, 1989.

IOP Conf. Series: Materials Science and Engineering **364** (2018) 012018 doi:10.1088/1757-899X/364/1/012018

3D Virtual Reconstruction and Augmented Reality Visualization of Damaged Stone Sculptures

Francesco Gherardini[a]*, Mattia Santachiara[b], Francesco Leali[a]

[a]Department of Engineering "Enzo Ferrari", University of Modena and Reggio Emilia Via Pietro Vivarelli 10, 41125, Modena, Italy
[b]Santasco, Via Dardano Fenulli 19, 42123, Reggio Emilia, Italy

*corresponding author: francesco.gherardini@unimore.it

Abstract. This paper proposes the integration of photogrammetric reconstruction, 3D modelling and augmented reality application in order to achieve the complete visualization of a stone sculpture even if highly damaged or fragmentary. The first part of the research aims to the reconstruction of the original appearance of an incomplete sculpture, by using photogrammetric techniques based on standard resolution photos and, if possible, free software in order to obtain a digital model; then, we integrate this model with other 3D digital data (from other sculptures of the same period) or by 3D modelling in accordance with art-historical and archaeological studies. The second part of the research consists of embedding the 3D model in a custom Android application able to render in real-time the reconstructed parts of the damaged sculpture. Then, the rendering is overlaid on the video stream of the real scene and, as a result, a complete 3D digital model of the sculpture is achieved and could be visualized through a Virtual Reality viewer. As a case study, we focus on a Roman stone funerary sculpture of a male lion conserved in the Museo Lapidario Estense of Modena (Italy), which lacks of its muzzle and its four legs. The original appearance of the lion may be supposed by modelling and integrating the damaged sculpture with other photogrammetric details of funerary lion sculptures of the same period, according to art-historical sources. Finally, the lion is visualized through an augmented reality application which digitally overlays the reconstructed parts on the real sculpture.

Keywords. Photogrammetry, Augmented Reality, Virtual modelling, Real-time visualization, Cultural heritage.

1. Introduction

The role of digital technology in the cultural heritage field ranges from reconstructing to monitoring, from measuring to digitizing and visualization. Monitoring and investigating ancient structures or sculptures lead to use 3D models in a diagnostic perspective in order to study structures along with their actual dimensions, their stability, their modifications over time, etc. [1]. In measurement technology, data that were once extremely difficult and possibly dangerous to collect can now be collected and processed using only a digital camera and off-the-shelf software. Currently, digital close-range photogrammetry has made it possible to efficiently model a variety of structures in a budget-friendly manner [2]. Other photogrammetric applications deal with the reconstruction of digital models by means of images collected on the web [3,4], as in the case of strongly damaged or destroyed artefacts in order to reconstruct their appearance as they were before.

In [5], digital technologies are tested to assist stone carvers in restoring carved elements while also learning from the sculptors and masons to better understand their approach when conserving or restoring a sculptural element.

For most of these aims, the photogrammetric techniques have increased their central role in recent years, thanks to the support of better hardware and software, even open-source or free one, as tested by [6]. They focus on the accuracy and the metric content of the final result, which is globally positive even when low textured or reflective surfaces are present in the 3D scene.

Some technologies are used to reconstruct and show the complexity of the territory and its heritage, including 3D scanning, 3D printing and Augmented Reality (AR) [7]. Some museums and archaeological sites already exploit the potential of digital tools to preserve and spread their heritage but interactive services involving tourists in an immersive and more modern experience are still rare.

Modelling of complex figures using photogrammetric software is being increasingly used in the field of cultural heritage as an alternative to techniques based on 3D scans, due to its low cost and flexibility of use [8]: these authors applied this technique to the modelling of Mesoamerican sculpture, namely chacmool of Tenochtitlan, for obtaining a precise three-dimensional model, i.e. a complete record of its current state, used for the preparation of interactive museum models and for the production of stereoscopic videos.

The present paper focuses on the potentiality of integrating photogrammetry, for 3D reconstruction, with the AR benefits, in order to increase the interaction with the artefact.

As a case study, we focus on a Roman stone sculpture of a male lion conserved in the Museo Lapidario Estense (Estense Lapidary Museum) of Modena (Italy), which lacks of its muzzle, its four legs, and the end of its tail. The original appearance of the lion may be supposed by integrating the damaged sculpture with parts achieved by photogrammetric reconstruction of other sculptures of the same period and conveniently modified and re-modelled. Finally, the lion is visualized through an AR application which digitally overlays the reconstructed parts on the real sculpture.

The paper is structured as follows. Section 2 describes the general method, Section 3 presents the method application to a case study, which is a damaged Roman sculpture of a funerary lion, and finally Section 4 draws some final remarks.

2. Method

The first part of the research aims to the reconstruction of a damaged sculpture, by using photogrammetric techniques based on standard resolution photos and, if possible, on free software in order to obtain a first model; then, in order to integrate the missing parts of this model, we identify and digitally reconstruct other 3D sculptures (i.e. other sculptures of the same period) by photogrammetry, and by 3D modelling in accordance with archaeological studies and hypothesis from art-historians, aiming to achieve the original appearance of the damaged sculpture.

The steps are listed below:

- Photogrammetric reconstruction of the damaged sculpture and of the other sculptures that may provide some of the missing parts.
- Integration of these parts into the main model, if necessary by partially re-modelling them in order to reconstruct missing details.
- Finally the real model is "augmented" by the additional parts: in order to improve the interaction with the observer, a dedicated AR application is developed.

3. Case study

The statue selected as a case study depicts a male lion and is located inside the Museo Lapidario Estense (Estense Lapidary Museum of Modena), in the west span E, with number 52. It can be historically placed in the first half of the 1st century CE. In [9,10], the authors hypothesize that this lion could be part of a decorative element belonging to an ancient Roman funerary monument. The statue is realized in Vicentina stone, which was extracted from the Berici Hills (Vicenza, Italy) and sent to Modena by fluvial way, and often employed in the artistic field.

The lion statue (Fig. 1) appears, although damaged, in excellent condition: the lion is represented in frontal position, shows the muzzle slightly turned to the left side, compared to the observer. The lion is installed on a base whose dimensions are 620 x 1400 x (height) 345 mm. The locks of the lion mane are clearly visible and well maintained, extending over the entire front and back of the statue. On the sides it is possible to glimpse its muscular bundles and the ribs, and in the back part, starting from the right to the left it is possible to observe its tail, which lacks of its end, that embraces most of the lumbar area, while the muzzle and the four legs are missing. As regards the position of the legs, the hind ones may be thought to be similarly folded as if the lion was sitting, while the front ones appear to differ from each other, the left one (with respect to a frontal observer) seems to have a slightly inclined direction compared to the base, as the ribs still present may suggest that the lion was not with the legs perfectly perpendicular to the floor. As far as the other front leg is concerned, it is assumed that it was raised and taken forward, as if the lion was grabbing a prey or tending to someone or something.

Sculptures of this kind were widely used in Roman funerary structures of the Augustean era, similar elements are linked to other monuments such as the contemporary ones of Aquileia (Udine, Italy) and Sepino (Campobasso, Italy). The statue of this lion should have three other similar sculptures, placed in guardianship of a monumental sepulcher that had to rise along the ancient consular road "via Aemilia". Monuments of this type and size had to be very common in Modena, as in the case of the two Roman lions on the sides of the main door of the Cathedral (Duomo) of Modena (see next sub-section).

Figure 1. The funerary lion statue (first half of the 1st century CE, Museo Lapidario Estense, Modena, Italy). Photo credit: Marianna Grandi.

3.1. Photogrammetry of the damaged lion and the other sculptures

The first part of the research aims to the reconstruction of the original appearance of the incomplete sculpture of a funerary male lion, by using photogrammetric techniques based on standard resolution photos and, if possible, free software in order to obtain a 3D model. A first photogrammetry reconstruction is performed within the free version of software ReCap (Autodesk) by means of 95 photos (Standard compact camera Nikon P310, 16.1MP CMOS Sensor, Sensor size: 1/2.3" (~ 6.16 x 4.62 mm), Max. image resolution: 4608 x 3456), some of them are shown in Fig. 2. However, in this free version 100 is the maximum number of photos that can be processed so, because some details are missing in the achieved model, we further reconstruct the 3D model within the professional software PhotoScan (Agisoft) by means of 130 photos, which generates a more detailed textured model of the lion (Fig. 3), in particular in its back and bottom side.

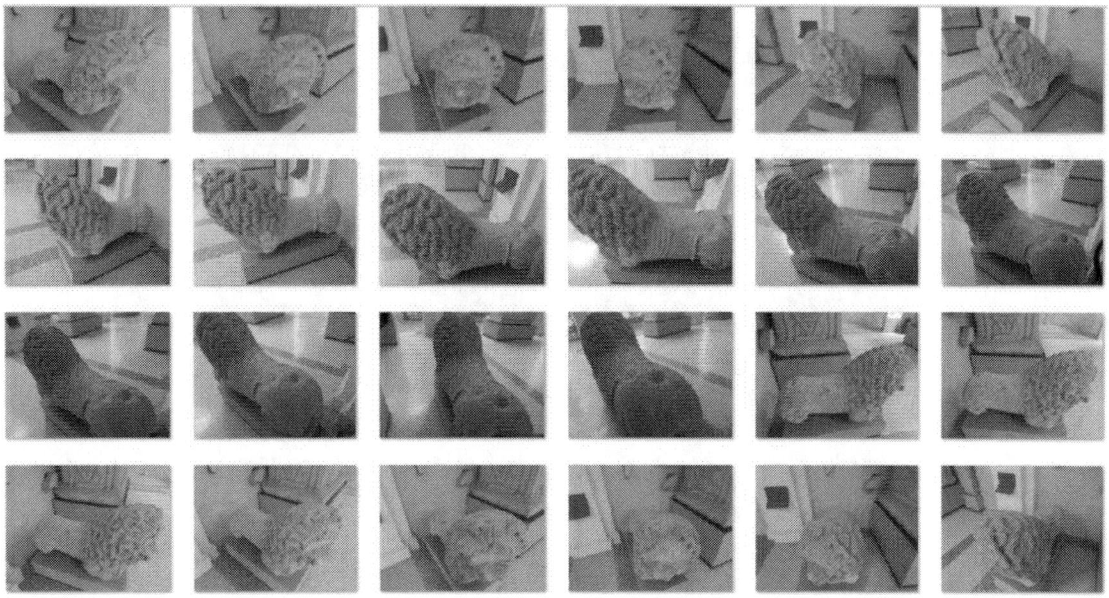

Figure 2. Examples of the photographic images used as inputs for the photogrammetric reconstruction.

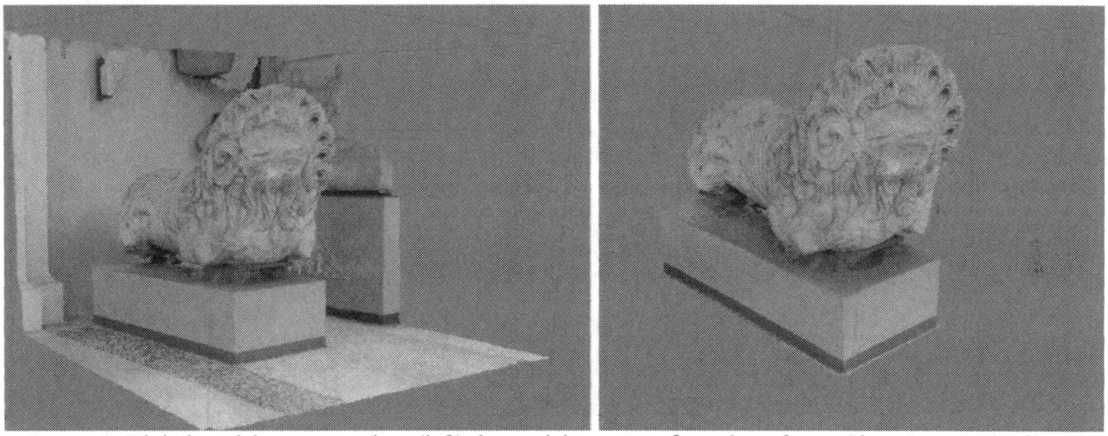

Figure 3. Digital model reconstruction: (left) the model as output from the software PhotoScan (Agisoft) and (right) after removing the background scene.

Similarly, we reconstruct: 1) the right one of the two Roman lions of the main door of the Cathedral of Modena (Fig. 4a), which are considered of the same age and originally with the same function (i.e. funerary lions), but re-used as building materials during the Cathedral of Modena construction [11,12]; 2) a crouching funerary lion of the 2nd Century CE (Fig. 4b), kept in the Capitoline Museum of Rome (Italy) [13], in order to extract some details.

3.2. Details integration, modelling and texturing.

The comparison between the damaged lion and the other two sculptures may suggest two different positions of the lion body: the lion may 1) have a crouching position or 2) stand on its four legs. Two different digital models of the funerary lion are then reconstructed in order to validate its integration with the parts achieved from the other models. For the crouching position, we extract the muzzle (Fig.

4c) from the Cathedral lion, according to [11, 12], while the front and hind legs (Fig. 4d e 4e) and the tail end (Fig. 4f) are achieved from the crouching funerary lion of the Capitoline Museum. For the standing position, we extract the muzzle (Fig. 4c) and the hind legs (conveniently remodeled as in Fig. 4g), while the front legs (with the standing one conveniently remodeled as in Fig. 4h) and the tail end (Fig. 4f). Then, we integrate these parts with the damaged lion sculpture model, by modifying the texture and by partially re-modelling them within the software Meshmixer (Autodesk). The additional parts need to be scaled, cut, refined, and smoothed: the two final 3D models are shown in Fig. 5 and will be used for the next phase of AR implementation.

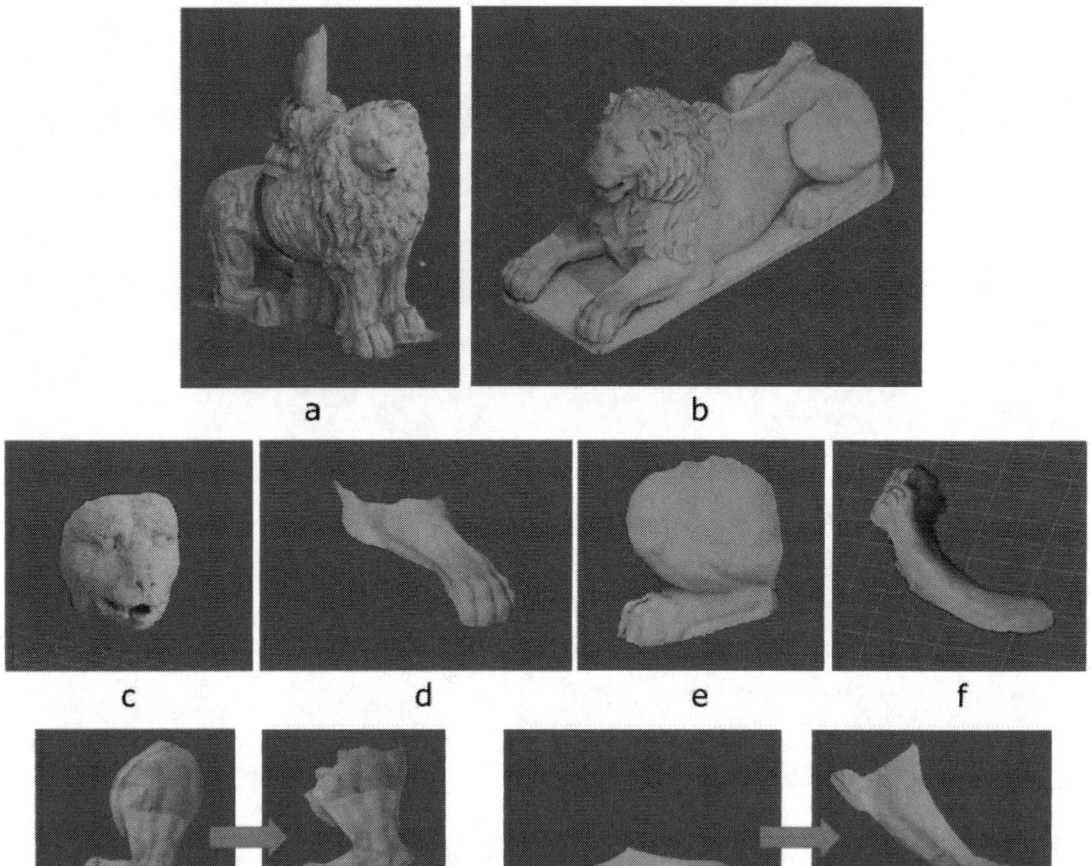

Figure 4. Top: The lion of the Cathedral of Modena (a) and the lion of the Capitoline Museum of Rome (b). Middle: Details from the other lion sculptures: muzzle (c), front leg (d), hint leg (e), tail end (f). Bottom: Re-modelling of the hind leg of the Cathedral lion (g) and the front leg of Capitoline lion (h).

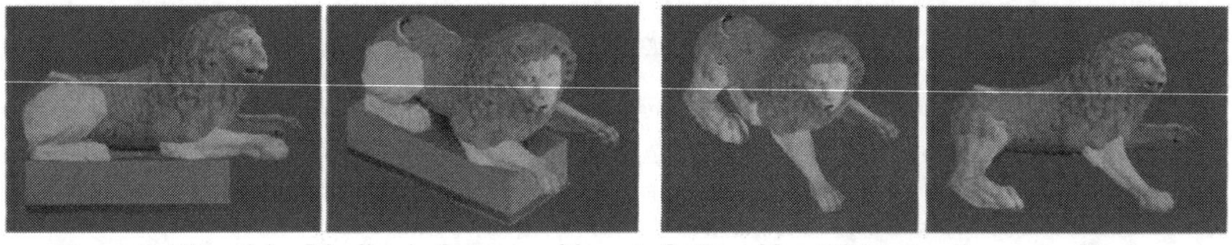

Figure 5. 3D models of the lion in the two positions: Left: Crouching; Right: Standing on the four legs.

3.3. Implementation of AR

The AR application is implemented within the Unity engine platform (Unity Technologies) by means of EasyAR tracking technology library. Its implementation requires the following steps:

1. The EasyAR plugin package is imported into the Unity platform.
2. The scene needs to be set up, whose essential components are already available (as prefab) in the plugin package. In the EasyAR, the following tools are available: a logical abstraction allows interfacing with the hardware device camera (CameraDevice Behaviour); a virtual camera of the scene allows the overlap between video flow and real-time 3D rendering (ARCamera Behavior); a tool for target tracking (Tracking ImageTracker); a component that detects and traces the ImageTarget (ImageTracker); a flat image that can be traced by an ImageTracker (ImageTarget).
3. The image to be traced required to be set: Once a new ImageTarget is inserted in the scene, the image (in .jpg or .png format) is copied in the subfolder of the "StreamingAsset" project, then set the X and Y proportions in the size of the ImageTarget in a manner consistent with those of the reference image. Assign an ImageTracker present in the scene in the Loader field. Then, at this point the application may recognize and trace the image.
4. The object to render is assigned to the tracker of the images: the object to visualize is set as a "child" of the ImageTracker. Specifically to the lion sculpture, which is a voluminous artefact, the parallelepiped base of the statue is considered as a reference marker, and an ImageTarget is generated for each face of the base. Each ImageTarget is positioned in such a way as to coincide with its reference face in the 3D model of the statue.
5. At this point we obtain a system of 3 ImageTargets (ignoring the upper and the back side of the statue). The reconstructed parts of the lion (Fig. 6) are then added as a child within each of the 3 ImageTargets. Then, we can trace the base and see the reconstructed parts which will be overlaid on the lion sculpture: nevertheless, only one ImageTarget at a time is required to be traced, in order to avoid disturbances and overlaps, by managing the priority based on the last ImageTarget detected. To this aim, a further part of the code is implemented, in order to let the observer turning around the lion for about 180° without losing tracking.
6. In order to use the AR application with the VR viewer, the adaptation for stereographic rendering requires that the display will be split in two parts: one will render for the left eye (left camera), the other one for right the eye (right camera). This is achieved by inserting in the scene two virtual cameras spaced apart by 0.6 on the horizontal axis and assigning to each a specific area on which to render in the viewport respectively $x = 0$, $y = 0$, $w = 0.5$, $h = 1$ for the right camera and $x = 0.5$, $y = 0$, $w = 0.5$, $h = 1$ for the left camera. This requires a double rendering pipeline: the scene is rendered twice. This could be a critical factor in the case of highly complex scenes, but negligible in the present case study.

The AR application is able to render in real-time the reconstructed parts of the damaged sculpture by overlaying them on the video-stream of the real scene (Fig. 7) and, as a result, the complete 3D digital model of the sculpture is achieved and could be visualized through a VR viewer.

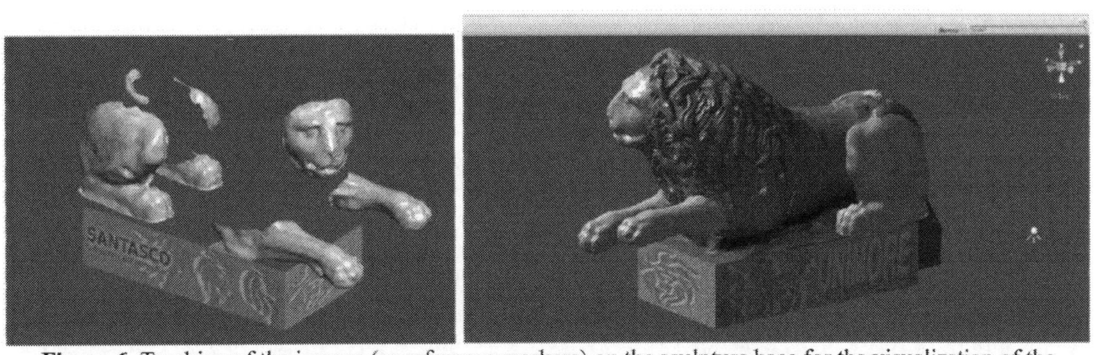

Figure 6. Tracking of the images (as reference markers) on the sculpture base for the visualization of the reconstructed parts (left) and their overlaying on the lion model (right).

Figure 7. Augmented Reality scene of the lion in crouching position.

4. Discussion and conclusion

This paper proposed a method for reconstructing and integrating the missing parts of stone sculptures by means of photogrammetric and AR techniques. The damaged sculpture is first digitally reconstructed and integrated with the missing parts, and then "augmented" by means of the digital models extracted and remodelled from other sculptures.

The method requires to reconstruct different models, which need to be compatible with the damaged one, in terms of surface finishing of the stone, colour, texture, or however, it requires to work extensively the parts and their feature. The selection of the "missing parts" requires the study and the investigation of models with similar characteristics, according to art-historical and archaeological studies. Due to photogrammetry, the lighting conditions may be a critical point in the method, which requires to post process the images in order to achieve better results. The use of a VR viewer requires a double rendering pipeline, so that the scene needs to be rendered twice: this could be a critical factor in the case of highly complex scenes, so the virtual model may require to be decimated and simplified.

On the other hand, the method is able to generate a tool for the assessment of hypothetical reconstruction of damaged sculptures, supporting the observer with AR visualization. As a further development, we will integrate the reconstructed lion within the scene of the whole funerary monuments, by means of AR.

Acknowledgments

The authors gratefully acknowledge the Gallerie Estensi, the Director Dr Martina Bagnoli, and Dr Chiara Marastoni; Lucio Cannito, B.Eng., for his support in the photogrammetric reconstruction; Aran Anderson, for sharing photographic images about the lion of the Capitoline Museum of Rome; the professional photographer Marianna Grandi, for her contribution in the realization of the images.

Photo credits

All images are by the authors unless otherwise noted.

References

[1] Capra A, Bertacchini E, Castagnetti C, Rivola R, Dubbini, M 2015 Recent approaches in geodesy and geomatics for structures monitoring *Rendiconti Lincei* 26(1) 53-61

[2] Leitch K R, M B Coon 2012 Noncontact Modeling of Structures Using Close-Range Digital

Photogrammetry *Practice Periodical On Structural Design And Construction.*
[3] Vincent M L, Flores Gutierrez M, Coughenour C, Lopez-Menchero Bendicho V.M., Remondino F, Fritsch D. 2015 Crowd-Sourcing the 3D digital reconstructions of lost cultural heritage. In: 2015 Digital Heritage, 1 pp 171–172
[4] Gherardini F, Leali F 2016 A Framework for 3D Pattern Analysis and Reconstruction of Persian Architectural Elements *Nexus Network J.* 18(1): 133–167
[5] Ouimet C, Gregga J, Kretz S, Chandler Ca, Hayes J 2015 Documentation and dissemination of the sculptural elements of Canada's parliamentary buildings: methodology development and evolution, a case study. *The International Archives of the Photogrammetry, Remote Sensing and Spatial Information Sciences*, Volume XL-5/W7, 2015
[6] Toschi I, Rivola R, Bertacchini E, Castagnetti C, Dubbini M, Capra A (2013) Validation tests of open-source procedures for digital camera calibration and 3D image-based modelling *ISPRS Archives* – Volume XL-5/W2, p. 647-652.
[7] Castagnetti C, Giannini M, Rivola R 2017 Image-based virtual tours and 3d modeling of past and current ages for the enhancement of archaeological parks: The visualversilia 3d project *Int. Arch. Photogramm. Remote Sens. Spatial Inf. Sci.* XLII- 5/W1 pp 639-645
[8] Modeling of an aztec sculpture with photogrammetry. Geneviève LUCET, Araceli CASAS. Facultad de Arquitectura, Universidad Nacional Autónoma de México. Ciudad Universitaria, Méx. D.F., Mexico 25th International CIPA Symposium 2015, 31 August – 04 September 2015, Taipei, TaiwanAnother reference.
[9] Malmusi C 1930 Museo Lapidario Modenese (Modena, Franco Cosimo Panini)
[10] Giordani N, Paolozzi Strozzi G 2005 Il Museo Lapidario Estense, Catalogo generale
[11] Rebecchi F 1984 Il reimpiego di materilae antico nel Duomo di Modena Lanfranco e Wiligelmo. Il Duomo di Modena.
[12] Sandonnini T Cronaca dei restauri del Duomo di Modena (1897-1925) Aedes Muratoriana (Modena, Italy).
[13] http://capitolini.info/scu00050/?lang=en (accessed March 15, 2018).

IOP Conf. Series: Materials Science and Engineering **364** (2018) 012019 doi:10.1088/1757-899X/364/1/012019

Smart Cultural Site: an Interactive 3d Model Accessible to People with Visual Impairment

V Rossetti[1], F Furfari[2], B Leporini[2], S Pelagatti[1] and A Quarta[1]

[1] Dipartimento di Informatica, Università di Pisa, Italy
[2] ISTI-CNR, Pisa, Italy

E-mail: susanna.pelagatti@unipi.it

Abstract. In our study, we proposed a prototype of "Piazza dei Miracoli" (Pisa, Italy) as an intelligent 3D interface system aimed at improving the accessibility and usability of cultural sites for all, including people with vision impairment. We combine tactile information with audio tracks in order to enable potential users to explore the artifact autonomously. Low-cost and partially open-source technologies are two important features leading our approach, creating a system easily replicable. First, we designed some 3D source-based models to be printed with an additive manufacturing technology. Next, the 3D prototype has been linked to a Raspberry for handling a set of detailed audio tracks to enrich the tactile exploration with an interactive audio guide. A preliminary evaluation was conducted to test the proposed system in order to better refine its design.

1. Introduction

In cultural heritage, technology is more and more exploited in order to grant everyone a concrete access to culture. The increasing development of modern technologies of rapid prototyping (RP), which allows reproducing an object from a 3D model, can offer new opportunities to experience for several purposes. Compared to traditional techniques, such as gypsum casts, thermoformed boards, etc., tactile patterns obtained with RP and 3D printing technologies speed up and improve the model reproduction process. Having 3D models of statues, buildings or plans is more suitable for users with special needs, such as blind people (BP), who rely on the sense of touch to perceive objects. In particular, tactile models allow a blind user to get information on fundamental details, such as 3D shapes and surface textures. However, as demonstrated by Hayhoe [6], tactile 3D models may be not enough to fully enjoy the content exploration and perception, especially for multifaceted objects. The correct understanding of a 3D reproduction is affected by many factors, such as personal skills, size (compared to the original), reproduction details and the overall quality of the tactile model. Details of some particulars are often missed by BP due to the limitation of model reproduction. Thus, a 3D reproduction of specific details of an artwork could be a valuable contribution for educational and an inclusive perception. In addition, some type of information cannot be reproduced in a tactile way. Thus, combining a tactile model with audio contents can help BP to build a better mental picture of the whole work.

In this paper, we propose a novel interface model which can provide a multimodal interaction to explore 3D objects. To exemplify our idea, we present a working prototype built up by of some 3D models developed using source-based modeling [4] and RP. The models are equipped with tangible buttons (placed aside). These are modelled in 3D format and allow the activation of descriptive audio tracks, in such a way that each shape means a different type of information. Tracks can also be interrupted and reactivated in a simple way. The prototype is built using low cost components and uses a Raspberry to connect each button to a Python script managing and playing the tracks. Some studies investigated interactive tactile prototypes to allow art works exploration by visually-impaired users. However, these solutions do not consider a 3D interactive modelling. For instance, [9] investigates a combination of vibration and speech feedback which can be used in order to make a digital map on a touch screen device more accessible. In [13], the authors created a prototype of a tactile-audio map based on a combination of tactile hardcopy and a SVG file used together to provide an interactive access to a map image through a touchpad. The result is a tactile-audio representation of the original input image.

These projects, along with other ones like the study presented in [1] and [12], propose interactive tactile approaches that are not based on truly 3D models, since they adopt a 2D touchable format in which the representation is basically 2D with some parts in relief. In our study, instead, we combined a 3D modelling approach with audio descriptions in order to enrich the BP's interaction. The contribution of this paper is about proposing and experimenting a working methodology to develop interactive interfaces of artistic objects or cultural sites, which is easily replicable, and can improve the cultural experience of BP while increasing their autonomy. In short, we aim to propose an enriched and interactive tactile model for BP with particular attention to details, which can be useful for touristic and educational purposes. The idea intends to face some issues often encountered when exploring tactile models available on the market, such as those reported in several projects [3], [15], [16]. The main issues observed can be summarized in: lack of details (granularity of some important particulars), replicability by specialized centers (more expensive), lack of choice in selecting information types, and free exploration by using both hands. Specifically, our approach is characterized by:

(1) Cost and replicability: develop a model which is easy to reproduce for other sites and objects at low cost.
(2) Avoiding unintentional selections: many BP appreciate the opportunity of exploring the model with both hands and no restrictions, that could not be feasible when sensors are embedded in the model itself.
(3) Semantic information and on-demand selection: buttons with different shapes are proposed for activating different types of information (historical, architectural and practical). An artefact is usually associated with a single audio track that describes all information and requires to be listened entirely. In our model, only desired information can be selected and listened.

The paper is organized as follows. Section 2 discusses some related work. Section 3 gives an outline of our approach. Section 4 details the methodology and technology used. Section 5 describes the interactive prototype and Section 6 reports a short description of the preliminary evaluation with BP. Conclusions (Section 7) end the work.

2. Related work

Several studies and projects investigated the opportunity to provide to BP a multimodal interaction to increase access to cultural contents. Tooteko [3] is based on a smart ring that allows the user to navigate a 3D surface with finger tips and get in return an audio content that is relevant to the part of the surface he/she is touching in that moment. The system is made out of three elements: a high-tech ring, a tactile surface tagged with NFC sensors, and an app for tablet or smartphone. When the ring reaches a NFC sensor it communicates with the app for activating the audio track. The 3D models are built using standard 3D printing. The hotspots are inserted inside the model, which needs to be reasonably large to accommodate them. Hotspots need also to be at a reasonable distance to be clearly detected by the reader. Hi-Storia [15] proposes a similar approach. In both these projects, sensors activating the audio tracks are integrated in the model itself. This has some important drawbacks. First, while listening to an audio track, the user can easily jam into another hotspot causing the stop of the old track and the activation of the new one. This is rather annoying since BP often use both hands when exploring a 3D object and listen to audio tracks [8]. Moreover, integrating sensors into the model is very costly, which

IOP Conf. Series: Materials Science and Engineering **364** (2018) 012019 doi:10.1088/1757-899X/364/1/012019

can refrain single small museums from adopting such an approach. Finally, these approaches assume an external person explaining the user how to move and what to do before starting to browse the model.

A different approach is proposed in 3D Photoworks [14]. Here, colour images are printed directly on the relief surfaces and infrared sensors are integrated into reliefs. Their interactive tactile prototypes accompanied by audio guides are based on motion capture [2], [10]. Depth cameras, placed above the tactile template detect the movement of the user's hands. The main disadvantage of hand tracking systems is their cost and complexity. Even though the technologies are continuously improving and becoming cheaper, the current low-cost trackers generally exhibit poor accuracy of tracking. In particular, while slow movements and limited rotations of the hand are usually traced in an effective manner, when speed increases the position of the hand is easily lost causing the user a deep feeling of frustration.

All these systems, although with the limitations outlined, represent an important first step towards a real access to cultural sites, using tactile interactive models. Our approach, explained in the rest of the paper, tries to overcome these problems by designing a low cost solution and moving the audio track activation outside the 3D model, still relying on sensors for their activation.

3. Approach

In this study our approach consists in reproducing an interactive 3D model by combining tactile perception with audio contents to explore the floor plans of monuments, rather than the main structure of the buildings. The reason of this decision is the lack of plans and indoor reproduction perceivable by touch. Likewise, the lack of particulars and details in a tactile reproduction led our study to investigate also these aspects. These types of reproduction (floor plans and particulars) can be, in fact, very useful for learning and perceiving details which are usually neglected for BP. However, this approach can be also exploited for reproducing the outside of the buildings.

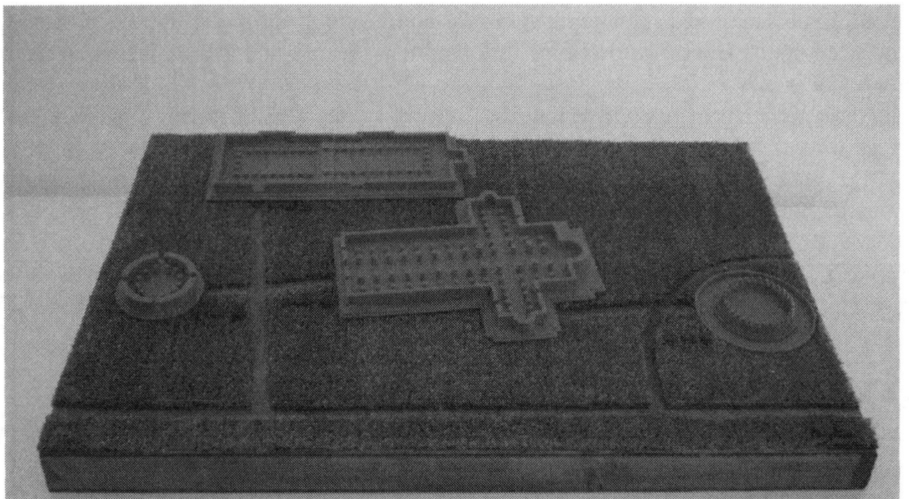

Figure 1. The 3D model of Piazza dei Miracoli

We applied our approach to a tangible 3D model of the four monuments placed in "Piazza dei Miracoli" in Pisa, Italy. We choose this case study because this site includes four cultural monuments with different floor plans: (1) the central plan of the Leaning Tower and the set of columns stacked outside; (2) the Latin cross plan of the Cathedral of Santa Maria Assunta; (3) the rectangular plan of the Monumental Graveyard; and (4) the central plan of the Baptistery. When visiting Piazza dei Miracoli, the plan of the square and of the pavements linking the four monuments is clearly visible for a sighted person. Our goal is to deliver a similar perception of the global structure available also for BP. To achieve this, we reproduced a 3D model of Piazza dei Miracoli, including the four monuments. The reproduction of each monument is built considering a two-dimensional plane which cuts the monument at height of approximately 1 m (3.28 feet) from the floor. In this way, we are able to show the section of the perimeter walls and of the columns which are important to understand the overall architectural setting (see Fig. 1). This tactile reproduction allows BP to perceive primary information, such as the thickness and position of the columns, the fact that the central nave is higher than the side naves, the position of different sets of arches, and the like.

4. Methodology

In building our prototype we followed guidelines and standards for tactile graphics [5]. However, these guidelines refer to the constructions of 2D and1D models, thus we have added some rules especially related to 3D. For instance, during participative design with a blind person we noticed that there is a minimal distance between two 3D objects (e.g. columns) which needs to be maintained in order to allow BP to perceive them correctly with the touch. Our system is built according to a modeling methodology for work of cultural sites based on the real reproduction [4]. The 3D model can be obtained using two different approaches to get the artwork (monument, statue, etc.) structure: (I) *reality-based*, such as via laser scanning or reflex camera; (II) *source-based*, via a digital representation of a work. Through this second approach, we can also model a work not existing anymore, or that has survived just partially, and whose reconstruction is based on historical sources. The approach actually used depends on the characteristics of the work and on the project goal. When using 3D modelling for huge monuments, as occurred in our case study, the work is simplified for a small scale reproduction (suppressing some uninfluential details). For this reason, we preferred to use a source-based approach, which is faithful to its original plan and to the historical facts, and we decided not to reproduce child elements that might not be easily perceptible by touch.

Starting from the digital source, the tactile models have been obtained through modern digital fabrication techniques. To model a source to be printed we: (1) got the plant in 2D; (2) model a 3D version based on the scale and the official sources (e.g. a circle on the plant is a cylinder in the 3D model); (3) select a simplified style plant, i.e. without architectural details; and (4) reproduce some architectural style details. The appropriate scale resolution was decided in a participative way involving BP during the whole design-development cycle (see Section 5).

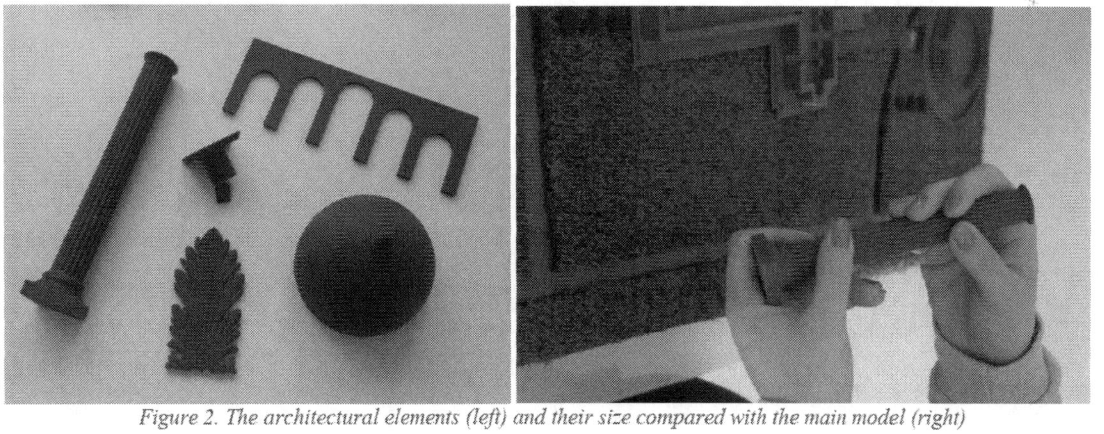

Figure 2. The architectural elements (left) and their size compared with the main model (right)

For the 3D printing step, an additive technical process has been adopted for our purposes: the object is produced by progressive accumulation of material [11]. The 3D prints were made with a Fused Filament Fabrication printer (FFF) that melts and extrudes a thermoplastic filament, in our case the Acrylonitrile-Butadiene-Styrene (ABS), an oil-based plastic. The models' size are appropriate to ensure a good exploration of the relevant details. We used Blender for developing the models and a proprietary software for slicing on the printer, a Zortrax M200 (available at the laboratory, cheaper printers can be

IOP Conf. Series: Materials Science and Engineering **364** (2018) 012019 doi:10.1088/1757-899X/364/1/012019

used instead). Because of technological limitations of the printer used, for some of the models, such as the Monumental Graveyard or the Cathedral, we had to print individual items in separate pieces glued together in postproduction. We also created 3D models for parts of the monuments that are relevant to understand the global structure, such as the dome and the nave profile of the Cathedral, a Corinthian column representing the main style of the Cathedral, the Baptistery and the Leaning Tower of Pisa (see Fig. 2). Reproducing these details is very important since they usually cannot be perceived by a BP.

Decorations were not reproduced on the 3D model of an element (e.g., acanthus leaves of column capitals) due to a scale limitation. However, some decorations are fundamental for the overall perception, such as the acanthus leaf in the Corinthian style representation. In this case, we preferred to produce a detailed model of the decoration, in a different scale, and print it. For instance, an acanthus leaf was printed on a larger scale separately from the capital. The architectural elements are located in a panel placed near to the main square model (Fig. 3). Each of these is introduced and described in the architectural audio description, whose reproduction is triggered by the button of circular shape.

5. The interactive 3D prototype

The interaction model proposed is conform to our aim of making the user entirely autonomous in the exploration of an artwork. To meet this aim, any piece of information should be provided using both the tactile and the audio channel. BP should be able to explore the model by touch and to activate an audio description related to a specific monument/detail. Spoken descriptions are organized into several audio tracks, which can be recorded by a person or via voice synthesizer through a specific audio content generation application, such as DSpeech**[11]** or Balabolka**[11]**. This allows to prepare various contents even into different languages. In addition, for each description more than one file can be available so that we can have different depth levels.

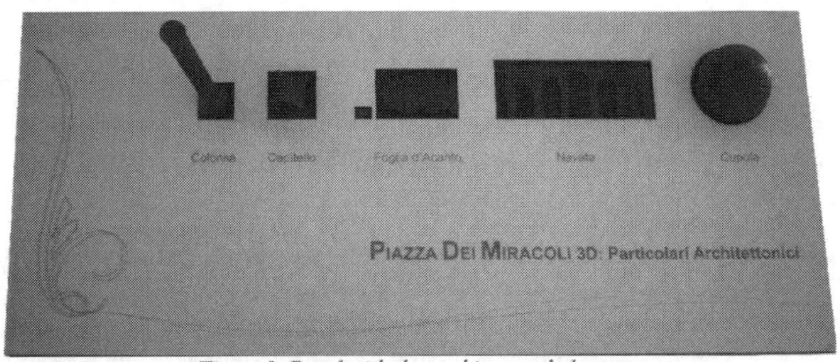

Figure 3. Panel with the architectural elements

The device used to manage the interaction user-3d model is Raspberry PI 3, a device single-board computer a few centimeters long, with general purpose input/output capabilities. The Raspberry is fixed below the prototype. Aural descriptions are activated by buttons placed aside each model (e.g. near the main entrance of each monument in Fig. 4). Buttons are connected to the Raspberry with cables. In our prototype each model has three buttons next to, and these are easily detectable via their geometric shapes: circular, triangular and squared. Each shape represents a specific type of information: (1) the circle is used for practical information, (2) the triangle for the historical overview and (3) the square for the architectural description.

When the user approaches to the model of the Piazza dei Miracoli, two proximity sensors activate a welcome audio track explaining how to interact with the model itself. The sensors, placed in front of the prototype, compute the distance between them and an obstacle in front of them. When a new obstacle (BP) appears in a distance between 0 and 50 cm, the sensors send to the Raspberry the information and the welcome audio track is activated.

The user is autonomous in deciding the times and the ways of interacting with the other audio tracks. A press on the button activates the correspondent track. Then the user can navigate the tracks pressing again the same button. The current track can also be suspended by pressing a pause button placed at the bottom right of the prototype.

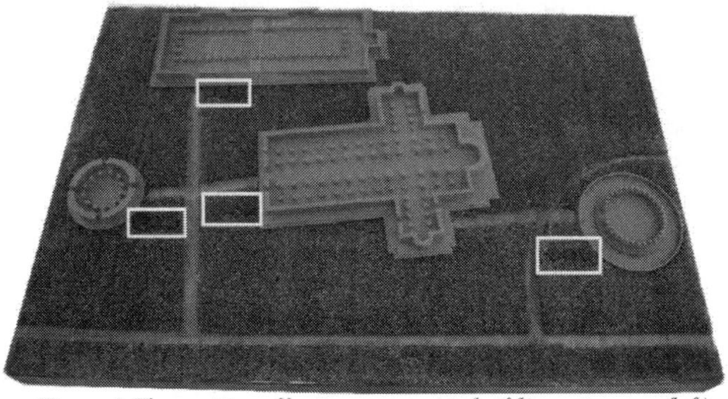

Figure 4. The position of buttons near main building entrances (left) and particular of Cathedral and Baptistery (right)

6. Evaluation

The aim of our approach is to support the user in an interactive autonomous exploration of a cultural site and/or an artwork through a 3D reproduction. The prototype has been thought as an interactive system usable by all, but useful especially for blind people who are able to 'see' a monument or a detail only via the sense of touch. In this perspective, we firstly evaluated our prototype with a group of blind and low vision users in order to evaluate if the approach of combining tactile interaction with structured audio descriptions is suitable and intuitive for a visually-impaired person. Thus, through the BP participation we aimed at evaluating:

(1) the appropriateness of the 3D models in terms of size, shape, texture and scale perceivable by touch;
(2) The suitability of the audio descriptions, i.e. the levels of the tracks, their activation, their structure, and so on;
(3) The difficulties encountered while exploring the models and using the audio tracks.

We tested the 3D models with the BP group during the design-development cycle in order to tune up the details of the prototype. Thus, we involved a group of three visually-impaired users who firstly familiarized with the model and expressed freely their impressions and perplexities. For our purpose we asked them:

- Are the monuments clearly perceivable by touch? This in order to understand if size and shape had been well designed and appropriately reproduced by the 3D printer.

- Are the audio tracks useful for the exploration, clear and easy to be listened to? These questions aimed at collecting impressions on how the tracks had been organized and proposed to the user.

- What about the interaction with the model? This type of question has been proposed to get opinions and feedback by the user with regards the interaction modality used to explore the prototype. Impressions and comments on audio descriptions activation, buttons styles, details reproduced in 3D, and so on were included in our investigation.

Two members of our research team observed the users while interacting with the prototype; this allowed us to collect information on those interaction aspects that were not mentioned by the people.

About the first goal, the BP were involved to set up the scale to use for reproducing elements and details. One user observed that some dimensions and spaces left between the columns were not easily perceptible by touch. Therefore, we consequently remodelled such part of the prototype. The users also suggested to reproduce some details of a piece (e.g. the arches) in a larger scale in order to improve their perception. So we modelled also that particular. One user observed that the base of the tower of Pisa had not a right gradient according to the leaning of tower. All the users suggested to reproduce more particulars into 3D format, since it is not easy to get information and 'see' a three-dimensional version of small objects, especially for perceiving the little architectural details. They declared that usually the 3D models reproduce monuments and artworks on the whole, but the particulars and the details are neglected.

With regard to the audio contents, the users expressed the preference for multiple tracks so that a blind person has the opportunity to listen to the amount and type of information according to the time and interests. Accordingly, we prepared three different types of information (architectural, historical and practical). Furthermore, each type of information has been recorded into different tracks in order to provide various levels of details which can be chosen by the user according to his/her preference.

With reference to the interaction modality, we observed that BP use to search for the buttons in specific positions; for example, we noticed that the users tried to find the buttons for activating the audio tracks in the same place for each artwork. The users clarified that if a certain consistency is preserved, for a blind person it is easier to localize the objects (e.g. all them are placed on the right of each monument). So we finally decided to place the three buttons near the entrance of each artwork. It was not possible to place them at the right of each artwork due to space limitation on the model.

Concerning the shapes of the buttons, the users declared to perceive them well, and to find very useful to listen to the audio tracks. Specifically, we discussed with them about the opportunity to have different levels of information to be listened to. This idea was immediately appreciated, in order to enable the user to better adapt the exploration according to his/her preferences and available time. Then the discussion moved on how to manage those types of levels. The proposed solutions were basically two: (1) two additional buttons to be placed in a fix position to perform 'previous' / 'next' of the current track (any it is); (2) the navigation among the tracks by pressing more than one times the same button which activated the current information. For example, the circle button when pressed actives the practical information (the first level), by pressing again it the second level, and so on. After a short discussion about this, all the blind users opted for the second proposal. So in the final version of our prototype the levels are handled through the same button activating the audio track.

Based on the comments and data observed by the users, we refined our prototype we will use for a more structured evaluation. As a result, we are currently working on a more extensive user testing with blind and non-blind people.

7. Conclusions

In this paper, we presented a novel approach to guide a blind user in an interactive exploration of a cultural artwork. The approach has been exemplified building a working prototype of Piazza dei Miracoli, the square in which is located the Leaning Tower of Pisa. Our approach considered indoor elements rather than the entire external monuments as well as particulars of the artworks in order to overcome the gap in this field. We worked to allow free model exploration with two hands and introduced a simple interaction model based on buttons with different shapes. At the moment, the prototype is thought for a single user exploration, a limit of most of technological devices used for the cultural fruition (e.g. Leap Motion or viewer like Oculus Rift, HTC Vive, etc.). For the future we would like develop a multi-user system. Moreover, our intention is to design an interactive system that is at the same time an emotional one.

References

[1] Brock A. 2013. *Touch the map! Designing Interactive Maps for Visually Impaired People*. In ACM SIGACCESS Accessibility and Computing (n° 105), pp. 9-14

[2] Buonamici F., Carfagni M., Furferi R., Governi L., Volpe Y. 2016. *Are We Ready to Build a System for Assisting Blind People in Tactile Exploration of Bas-Reliefs?*. In *Sensors*, vol. *16*, n. 9, p. 1361

[3] D'Agnano F., Balletti C., Guerra F., Vernier P. 2015. *Tooteko: a case study of augmented reality for an accessible cultural heritage. Digitization, 3D printing and sensors for an audio-tactile experience.* In Int. Arch. Photogramm. Remote Sens. Spatial Inf. Sci., volume XL-5/W4, pp. 207-213.

[4] Demetrescu E. 2015. *Archaeological stratigraphy as a formal language for virtual reconstruction. Theory and practice.* In *Journal of Archaeological Science* 57, pp. 42-55.

[5] Götzelmann T., Pavkovic A. 2014. *Towards Automatically Generated Tactile Detail Maps by 3D Printers for Blind Persons*. In 14th International Conference on Computers Helping People with Special Needs (ICCHP'14), At Springer International Publishing, vol. 8548, pp.1-7.

[6] Hayhoe S., 2008. *Arts, culture and blindness: studies of blind students in the visual arts.* Cambria Press, Youngstown, USA.

[7] Invitto S. 2013. *Neuroestetica e Ambiente percettivo: pensare strutture interattive a tre dimensioni.* In *SCIRES-IT,* Vol 3, Issue 1, pp. 35-46.

[8] Morrongiello B. A., Humphrey G. K., Timney B., Choi J., Rocca P. T. 1994. *Tactual object exploration and recognition in blind and sighted children.* In Perception, 23(7), pp. 833-848.

[9] Poppinga B., Magnusson C., Pielot M., Rassmus- Gröhn K. 2011. *TouchOver map: audio-tactile exploration of interactive maps*. In Proceedings of the 13th International Conference on Human Computer Interaction with Mobile Devices and Services, pp. 545-550.

[10] Reichinger A., Fuhrmann A., Maierhofer S., Purgathofer W. 2016. *Gesture-Based Interactive Audio Guide on Tactile Reliefs*. In Proceedings of the 18th International ACM SIGACCESS Conference on Computers and Accessibility, Reno, Nevada, Usa, pp. 91-100.

[11] Scopigno R., Cignoni P., Pietroni N., Callieri M., Dellepiane M. 2014. Digital Fabrication Technologies for Cultural Heritage (STAR). In 12th Eurographics Workshops on Graphics and Cultural Heritage, pp. 75-85.

[12] Senette C., Buzzi M.C., Buzzi M., Leporini B., Martusciello L.. 2013. Enriching Graphic Maps to Enable Multimodal Interaction by Blind People. In: UAHCI 2013 - 7th International Conference on Universal Access in Human-Computer Interaction (Las Vegas, NV, USA, July 21-26 2013). Proceedings, Constantine Stephanidis and Margherita Antona (eds.). Lecture Notes in Computer Science, vol. 8009, Springer, 2013. pp. 576 - 583

[13] Wang Z., Li B., Hedgpeth T., Haven T. 2009. *Instant tactile-audio map: enabling access to digital maps for people with visual impairment.* In Proceedings of the 11th international ACM SIGACCESS conference on Computers and accessibility, pp. 43-50.

[14] *3D Photoworks:* http://www.3dphotoworks.com. Accessed on November 02, 2017.

[15] *Hi-Storia:* https://www.hi-storia.it/. Accessed on November 02, 2017.

[16] *Touch Graphics:* http://touchgraphics.com/. Accessed on November 02, 2017.

[17] *DSpeech*: http://dimio.altervista.org/ita/. Accessed on January 18, 2018.

[18] *Balabolka* http://www.cross-plus-a.com/balabolka.htm. Accessed on January 18, 2018.

3D Reconstruction of rock paintings: a cost-effective approach based on modern photogrammetry for rapidly mapping archaeological findings

C Castagnetti[1,2], P Rossi[1] and A Capra[1]

[1] DIEF Department of Engineering 'Enzo Ferrari', University of Modena and Reggio Emilia, Via Vivarelli 10, 41125 Modena, Italy,

[2] Corresponding author: cristina.castagnetti@unimore.it

Abstract. The work describes a cost-effective methodology for the creation of 3D virtual models with both metric and photo-realistic content developed for archaeologists who need to rapidly map new findings during their field prospections. The potential of modern photogrammetry approach, based on Structure from Motion (SfM) and dense image matching algorithms, coupled with the use of low-cost cameras is evaluated for the creation of 3D models and orthophotos of rock paintings. The case study is located in Brealito valley, in the Calchaqui basin of the Province of Salta, Argentina. In these wide areas, actually known for the naturalistic value, a huge amount of archeological remains was found and surveyed for the first time in August 2013 within a research project funded by the Italian Ministry of Foreign Affairs. A methodological approach to record the findings in a cost-effective and reliable way is defined, acquisitions are performed with common use instruments such as digital cameras (off-the-shelf and single lens reflex - SLR) and measuring tapes. The novelty of the work lies in defining, experiencing and transferring to operators a simple methodology for digitizing and mapping the archeological heritage in order to provide 2D orthophotos and 3D models. Such methodology allows anyone, archaeologists or unprofessional personnel, to be trained and easily perform a survey in order to document, survey and study the investigated site.

1. Introduction

In recent years image processing and 3D modeling are finding great application in archaeology; the documentation of the heritage is becoming more and more digital and detailed thanks to innovative technologies such as photogrammetry, computer vision algorithms, laser scanning, remote sensing and topography based on satellites (GNSS - Global navigation Satellite System). Both traditional and satellite-based topography as well as remote sensing are well known in archaeology [1]. Laser scanning is also a widespread technique in cultural heritage documentation and monitoring [2, 3], although it has limitations in wild and mountainous areas because of the heaviness, the energy consuming and the encumbrance with respect to other systems. Traditional photogrammetry is widely used in archelogy for digitally document, usually in two dimensions, the excavations and the remains. The achievement of 3D products is also well known, there are countless publications about algorithms, procedures and potentialities of this technique [4], but it requires a deep knowledge of the procedures, the ability to properly design and carry out the survey and technical skills to accordingly process the data and deliver a reliable result. Thus, it is really hard to improve the method in terms of automation, rapidness of field activity and know-how transfer to unskilled personnel. On the contrary, the modern photogrammetry based on Structure from Motion (SfM) approach [5, 6] exploits digital images that are captured from different platforms (airplanes, unmanned aerial vehicles and from the ground) and sensors barely with no rigorous rules and provides automatic tools for data processing. Thanks to such an apparent easiness in obtaining a beautiful result, this technique rapidly spread the 3D virtual reconstruction of the digital heritage [7, 8], even if the actual challenge is still to produce a 3D documentation with an accurate metric content [9, 10]. Moreover, the recent technological improvement on software that are implementing SfM algorithms and hardware to speed up the image processing, allows a quite automatic method for 3D documentation.

Since 3D photorealistic models improve the representation of archaeological sites and enable post mission detailed studies and virtual enhancements of remote sites, the goal of the research is to develop, implement and test a cost effective methodology. User-friendly portable devices are exploited for data collection so that they can be carried out by archaeologists during prospections and do not need deep technical expertise. Moreover, such methodology is characterized by high automation so that the data processing is also carried out by personnel with a minimum training. The main steps of the research, as described in the paper, deal with methods for data collection and data processing as well as analysis and cross check of results in order to establish rules for the everyday field application of the methodology.

2. The case study: rock paintings in the Argentine Andes

The Argentine Northwest, and in particular two side valleys of the middle basin of the Calchaquí river (province of Salta), are the research sites of an archaeological mission based on the cooperation among Italian and Argentine scientists. The mission can boast two records: it is the first Italian archaeological research in Argentina and the interdisciplinary team combining both humanistic and technological skills has been driving to an innovative methodology to approach the digital 3D documentation of the Andean archaeological landscape. The specific context of the research is that of the lower Calchaqui river valley, southwest of the city of Salta (figure 1), in a densely populated region during the pre-Hispanic period. In particular, the studies took place in the areas of Luracatao, Brealito and Molinos where the archaeological prospection campaigns carried out since 2012 have allowed to identify several sites of cultural interest (fortified settlements, agricultural areas, caves with rock paintings, necropolis) [11]. Any findings and documentation in this area is of particular interest as it is unknown to archeology and it is located on the ancient transit routes between the Andean area and the low Argentine lands. The case study chosen to test the cost-effective methodology lies in the rock paintings of Brealito valley (figure 2).

3. Methods

3.1. Data collection

The archaeological prospection carried out in 2012 allowed to identify several sites of cultural interest that have been approximately positioned by means of portable low-cost *Garmin* receiver. In the following field campaign, that was carried out in 2013, some of them, such as the fortified settlements and agricultural terraces, have been recorded by means of GNSS static and kinematic surveys in order to provide detailed archaeological maps.

Some other sites, particularly the rock paintings, have been mapped by means of the modern photogrammetry approach in order to collect a suitable dataset to implement and test the cost-effective methodology for rapidly document in digital 3D way the archaeological heritage during field prospections. The investigated rock paintings are twofold (figure 2): the first one illustrates several white llamas with different sizes covering multiple faces of a rock within a cave; the second one, instead, paints animals (a monkey is particularly distinguishable) and figures with the colors of white, red and dark blue over a fractured rock.

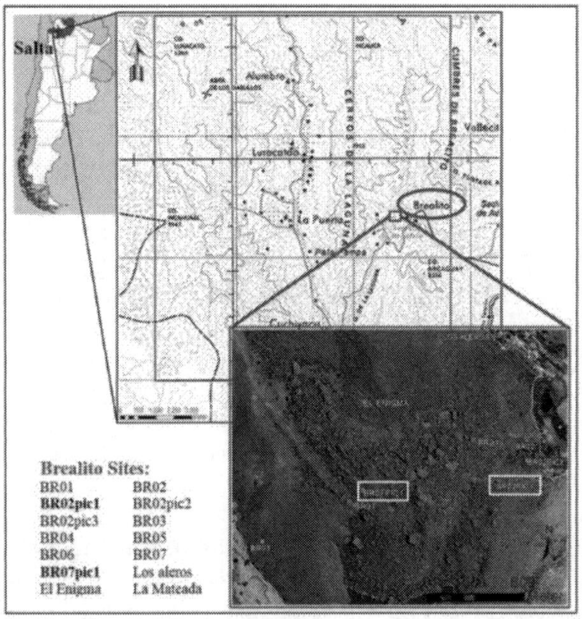

Figure 1. Location map of the study area: the Brealito valley (province of Salta – Northwest of Argentina) is included within the basin of the Calchaquì river where caves of cultural interest are found during 2012 field mission. Positions and identification codes are shown for all sites; the investigated rock paintings are BR02pic1 and BR07pic1.

Figure 2. Pictures of the investigated rock paintings with their name and identification codes. The white dots that stand out in any picture are the targets temporarily installed on the rock paintings with a double scope: to measure real distances in the field and then to properly process and give a geometric scale to the final 3D models.

In order to document the painting conservation, both the radiometric appearance and the 3D morphology of the rock are required; therefore, a 3D model with photorealistic content needs to be provided. The minimum suitable dataset consists of a proper number of images with good resolution (the higher the resolution, the better the final radiometric appearance) and a redundant set of geometric parameters. Usually, in photogrammetry, the geometric content comes from Ground Control Points (GCPs) whose coordinates are measured on purpose but this requires to carry on adequate technical instrumentation (for the specific environment, caves, the total station is the only feasible choice because GNSS receivers do not work unless in the open sky) and to own the proper competences to perform precise measurements, process the data and deliver accurate results. Very often archaeologists involved in field prospections cover wide areas by walking and examining the environment by means of only their eyes looking for evidences of past cultures; as soon as they identify any findings they have only the possibility to draw a sketch of the site on a paper and locate it by a *Garmin*. They come then back with technical equipment and topographers in case the site is of interest for excavations and full field missions. What if they could bring with themselves the knowledge and the ability to exploit mass market devices that are widely used and widely available such as a standard camera and a measuring tape? Such a minimum set of instruments is at the base of the data collection carried out in the August 2013 mission. Two different types of camera, a *Canon EOS 5DMarkII* and a *Panasonic DMC-FX60* have been tested in order to evaluate the performances of both Single Lens Reflex (SLR) and off-the-shelf cameras. Technical details of the cameras are listed in table 1.

Specific targets have been temporarily installed over the paintings in order to be included within the images and act as GCPs during data processing. Distances among the centers of targets have been measured by means of a simple measuring tape with an accuracy of few millimeters. These measurements provide the information about the real geometry and will then be used to process the data

IOP Conf. Series: Materials Science and Engineering **364** (2018) 012020 doi:10.1088/1757-899X/364/1/012020

and give to the final 3D model the proper geometry and an idea of the achieved accuracy. A redundant set of distances was collected for each investigated rock painting: 39 measurements for BR07pic1 and 34 for BR02pic1. The mean acquisition distance separating the camera and the painting was about 2.5 m when testing the Canon camera with the lens at 35 mm focal length and 1.5 m when testing the Panasonic camera and the Canon one with the lens at 24 mm focal length. Concerning the pictures, images have been collected from different points of view in a convergent way and with high redundancy, paying attention to guarantee an overlap of 70% or more between adjacent images (34 images with the Canon camera and 41 with the Panasonic one at the BR07pic1 test site; 31 pictures were capture at BR02pic1 site).

3.2. Image processing

The images have been processed by means of the software *Agisoft PhotoScan v. 1.1.6*, that implements the SfM algorithm and the dense image matching, in order to obtain the final 3D models of the rock paintings (figure 3 shows the workflow of the process). The process consists of aligning the images by means of common descriptors that are found in automatic way by the algorithm over multiple images; by this way a self-calibration of the camera is performed and a first sparse point cloud is achieved. In order to improve the alignment, the manual selection of targets pairs visible in different images can be used (by this way the targets are used as tie points). Then, the dense image matching allows to provide the dense point cloud and the 3D model after meshing procedure which looks like the real object thanks to the photorealistic appearance given by the radiometric content of photos (imaging texture) despite no geometric content is included yet.

Table 1. Technical details about the digital cameras used for data collection.

Camera	Canon	Panasonic
Model	EOS 5DMarkII	DMC-FX60
Type	SLR	Off the shelf
Lens	24 mm, 35 mm	4.5 mm
Resolution [np x np [a]]	5616 x 3744	4000 x 3000
Pixel size [um]	6.549 x 6.549	1.536 x 1.536
Pre-calibrated	No	No

[a] np number of pixel

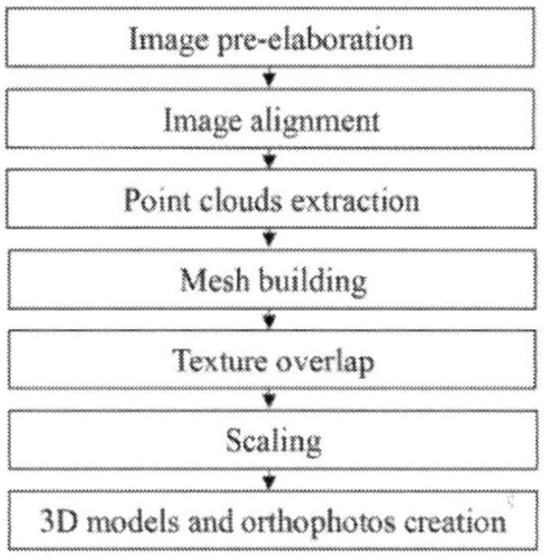

Figure 3. Workflow of the image processing procedure.

At this step, the target distances are used to scale the model and add the geometric content: the targets are identified over the 3D model and fixed to be the vertices of known segments, whose length is input in the software by the field measured distances (see figure 4). The validation of results as well as the evaluation of the final accuracy were performed by comparing the distances as measured over the obtained 3D model together with real distances measured in the field. In order to do that, the only usable field distances were the one that were not previously chosen for scaling the model. Several tests have been performed in BR07pic1 rock painting in order to evaluate the influence of the camera type and the specific focal length, the redundancy of the images dataset and the redundancy of the distances dataset. A summary of the various investigated dataset is described in table 2.

As soon as an accurate enough 3D model has been obtained, the methodology focuses on the creation of the orthophoto. Such a technical product is a 2D representation that is helpful to analyze in details the rock paintings conservation: the degradation of colours along time can be monitored thanks to the radiometric content whereas the geometric measurements that can be performed allows to check for deformations (such as thermal expansion, etc), fractures and further materic decay. The creation of orthophotos related to objects with a three-dimensional morphology requires the definition of multiple reference planes, where each one is orthogonal to the various faces of the rock in order to provide a comprehensive reliable representation. The automatic uncontrolled extraction of an orthophoto with a single plane, based on an average layout, provides a deformed 2D representation of the real geometry and furthermore any geometric analyses to monitor the conservation is meaningless. For complex morphologies a sort of unrolling of the cave has been implemented: projection planes as well as an internal coordinate system have been defined for the all painted surfaces.

Table 2. Technical parameters of data processing as modified in the various tests to evaluate the influence of camera type, lens and redundant dataset.

	BR07pic1					BR02pic1
Test ID	C35_34 [a]	C35_20 [a]	C35_10 [a]	C24_20 [a]	P4.5_41 [a]	C35_31 [a]
Camera	Canon	Canon	Canon	Canon	Panasonic	Canon
Focal length [mm]	35	35	35	24	4.5	35
N. of images	34	20	10	20	41	31
Mean capture distance [m]	2.63	2.63	2.63	1.23	1.37	2.5
Test area [m²]	~ 4.5	~ 4.5	~ 4.5	~ 4.5	~ 4.5	~ 1
N. of targets	24	24	24	22	24	16
N. measured distances	39	39	39	30	39	34

[a] Rule for the test identification name: 1st letter relate to the camera (C=Canon, P=Panasonic); 2nd number relate to the focal length; 3rd number relate to the number of images.

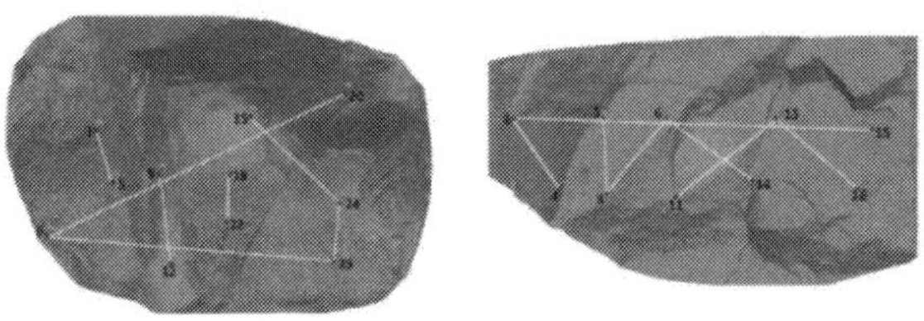

BR07pic1

Distance	Value [m]
6-20	2.441
1-5	0.44
9-12	0.69
22-18	0.377
6-25	2.165
24-25	0.422
15-24	0.887

BR02pic1

Distance	Value [m]
1-4	0.745
5-8	0.635
6-8	0.811
6-14	0.895
11-13	1.08
13-16	0.84
1-15	2.875

Figure 4. Scaling of the 3D model of BR07pic1 (left) and BR02pic1 (right) test sites. Top: the model with highlighted the know distances that have been selected and input to add the geometric content. Bottom: tables reporting the distances that have been measured during the field survey, for each test site respectively.

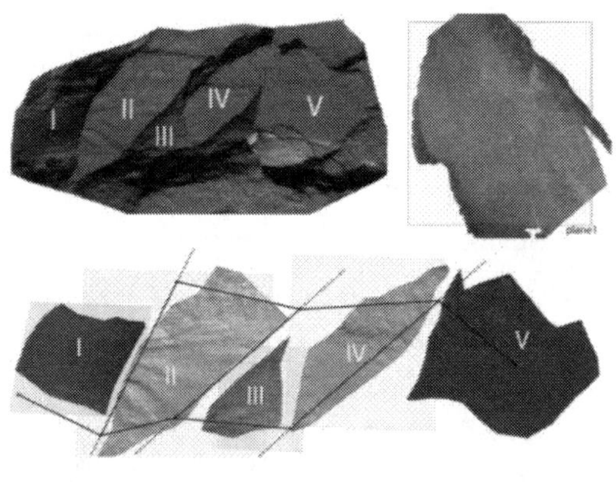

Figure 5. Orthophoto extraction: analysis of the 3D model and identification of sub-sectors where a single best-fit plane is able to describe the geometry layout (top left); computation of each best-fit plane (top right); planes and single coordinate systems to be used to perform the unrolling of the 3D model.

The area covered by paintings has been divided into as many sectors as the portions of rock that can be assimilated to a plane; each sector is then represented by an average best-fit plane (see figure 5). In order to compute the best-fit planes the 3D model has been processed with the modeling software *Rapidform XOR* (these tools are not available in *Agisoft PhotoScan*); the intersections between the planes provide the rotation axes to define the coordinate system. At this step, the coordinates of some points over these planes have been exported and introduced in *Agisoft PhotoScan* to create the coordinate system that is suitable to obtain the proper orthophoto, thus orthogonal to the surface, for the specific sector. The same procedure has been implemented for any sector, resulting in several portions of

IOP Conf. Series: Materials Science and Engineering **364** (2018) 012020 doi:10.1088/1757-899X/364/1/012020

orthophoto that are then combined to provide the final unrolled 2D representation of the rock painting. The main characteristic of such orthophoto is a full content both in terms of geometry and radiometry.

4. Results and discussion

The technical characteristics of the 3D models obtained by the various tests of image matching process are listed in table 3: the generated point clouds have a huge amount of points and a high resolution due to the low distance of cameras from the object (that has been especially designed with respect to the context) therefore a detailed 3D digital documentation of the rock paintings has been achieved. The multiple tests carried out on the BR07pic1 site by modifying the dataset and the processing parameters provide interesting results.

First of all the influence of the number of photos has been analyzed: by lowering the number of exploited images from the full set of 34 photos (100% of used images) to about a half, 20 photos (59% of used images), and to a third of them, 10 photos (29% of used images), the number of points of the dense point clouds slightly decrease (by using 59% or 29% of the total number of images, the resulting dense point clouds contain 88% and 68% of points respectively if compared to the full dataset - refer to dataset C35_34, C35_20, C35_10 in table 3). It is worth to conclude that the number of total point remains very high (tenths of millions for the dense cloud) and the mean resolution of the 3D models are really similar, less than 1 mm that represents an excellent result if one thinks to the rapid cost-effective way of collecting the data and the automation of the process.

Table 3. Results of the multiple tests: quality parameters of final 3D models obtained by varying the strategy of image processing to evaluate the influence of camera type, lens and dataset redundancy.

	BR07pic1					BR02pic1
Proc. ID	C35_34	C35_20	C35_10	C24_20	P4.5	C35_31
N. points sparse	145 185	101 468	48 742	104 465	190 540	141 268
N. points dense	16 625 786	14 696 760	11 409 307	14 950 825	14 352 837	16 193 554
Mean res. [mm]	0.9	0.9	0.9	0.6	0.8	0.9
N. GCP [a] / N. CP [a]	7/33	7/33	7/33	7/24	7/33	7/25
Mean [mm] [b]	3	3	3	3	3	2
Std.dev. 2σ [mm] [b]	4	4	4	3	4	2.1
Std.dev. 2σ [mm] [c]	--	0.5	0.5	0.7	1.1	--

[a] Number of measured distances used as known input for scaling (GCP) and number of measured distances used to validate the model distances (CP).

[b] Refers to the scaling error.

[c] Differences, computed as distances between meshes, with respect to the reference 3D model obtained by the dataset C35_34.

Moreover, the influence of the camera type has been addressed: the Panasonic dataset shows a comparable resolution with respect to the reference dataset (refer to P4.5 versus C35_34 in table 3), proving the success of exploiting mass market devices. The resolution about 1 mm, indeed, guarantees details in the documentation and performances that are unreachable by traditional manual drawings. The influence of the focal length has been tested with the Canon SLR camera type (refer to C35_20 and C24_20 in table 3): the comparison highlights that the performance of the final 3D model are quite similar, resulting again in great resolution (again about 1 mm) with the only disadvantage that it was not possible to properly represent the whole investigated area due to the lack of redundancy in the boundaries. Indeed, the shorter acquisition distance (1.23 m versus 2.63 m) and the different focal length (24 mm and 35 mm) did not allow a comprehensive detection of the scene unless a higher amount of images focusing on a larger area is accomplished, thus requiring more time.

In general, the comparison of all the tested scenarios for obtaining the 3D model suggests that the datasets provide 3D models quite similar and no significant differences are highlighted, as shown in figure 6; therefore the basic idea to use a low-cost and largely reachable camera to collect a redundant set of photos from an average distance to provide a good resolution (that can be also checked in the field by looking the first pictures on the screen) and from multiple perspectives allows to perform a powerful and successful digital documentation of the findings. Such images have to be coupled to a redundant set of distance measurements. At this point the reference 3D model (anyone else among the obtained 3D model could be used) is used to perform the unrolling of the multiple faces of the caves in order to provide the most reliable and accurate orthophoto of each rock painting (figure 6).

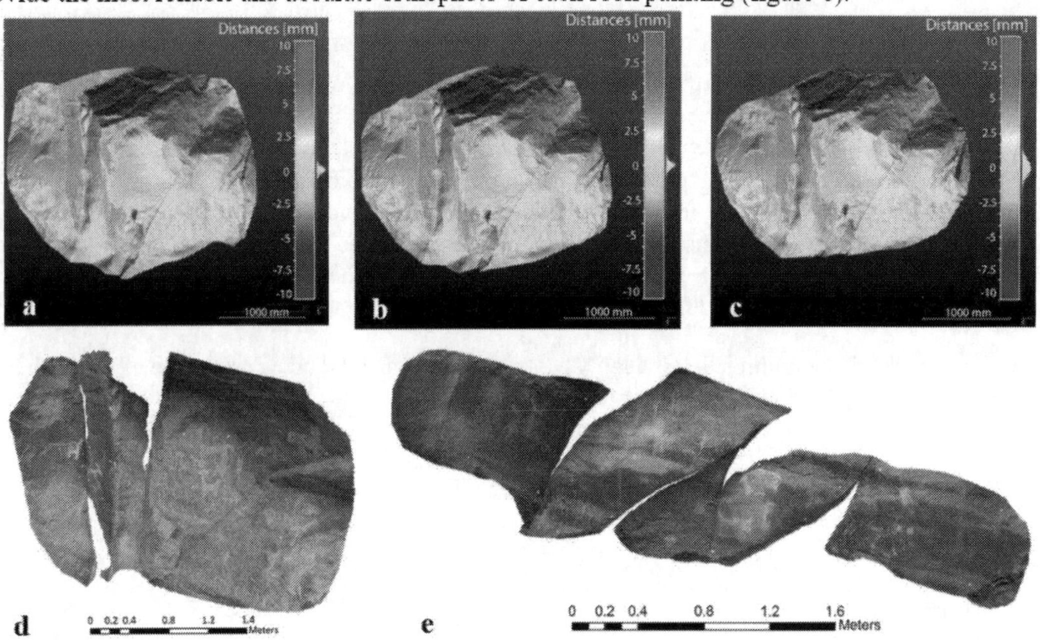

Figure 6. Comparison of the 3D models of BR07pic1 test site as obtained by modifying the image processing strategy: the reference solution is provided by the full dataset (C35_34) and is then compared to C35_20 (a),C35_10 (b), h_P4.5 (c). No significant differences exist within ± 10 mm. Then, final orthophotos have been computed with the reference 3D model and the unrolling procedure: from left to right, BR07pic1 and BR02pic1 rock paintings.

5. Conclusions

The creation of reliable 3D models with a metric content and a photorealistic appearance as well as 2D orthophotos allow the digital documentation of ancient rock paintings, thus providing archaeologists or scientists with technical data to be studied after the field mission and tourists or citizens with virtual reconstructions capable of enabling a remote visit. The photorealistic appearance combined to the geometric information allow to improve the knowledge and to monitor the evolution of rock paintings in an innovative way by exploiting the power of the third dimension. The use of modern photogrammetry based on image processing and SfM algorithm certainly simplifies both data collection and processing methods, but in spite of the simplicity it is still necessary to plan properly the image acquisitions and the ground measurements in order to provide the expected results in terms of accuracy, reliability and completeness.

Image matching is a cutting-edge research topic right now and new developments are being implemented. Ordinary image processing techniques, such as the classic photogrammetric approach, are not well suited for unknown archaeological findings located in wild areas, therefore the idea to develop, implement and test an innovative method for rapidly map new findings during archaeological prospections was pursued. The approach is evidently based on user friendly portable devices, such as a simple camera and a measuring tape; then more skills are required to process the data and be able to obtain a 3D virtual model and a suitable orthophoto but this step is leisurely carried out in office after the field mission is concluded. What is really of concern is to supply archaeologists with everything they need in the field, both devices and procedures, to be sure to be then able to fully document what they have been found. Then, on the basis of these preliminary data, future works can be planned. The present paper has shown the methodology and the results of several tests that allow to demonstrate the success of the methods and to give rules about the choice of devices, the way to collect the data and the way to process them for final helpful results. Moreover, these products enable the enhancements and the remote fruition of unmovable archaeological heritage, offering a modern way to attract tourists and the attention of the international community.

Acknowledgements
Authors would like to thank the Argentine team of archaeologists and particularly the Director, prof. Veronica Williams - University of Buenos Aires, for involving the whole Italian team in the great project "Archaeology and paleoenvironment of the middle Calchaquì valley". A special thank to the Italian team of archaeologists, dr. Carolina Orsini and dr. Elisa Benozzi, for promoting a productive and effective cooperation among scientists. The research has been funded by the Italian Ministry of Foreign Affairs – Archaeology Office, the Museum of Cultures - Municipality of Milan and the University of Modena and Reggio Emilia – DIEF Department of Engineering "Enzo Ferrari" (title of the project "The ecosystem of the Lagoon of Brealito", coordinated by dr. Carolina Orsini).

References
[1] Wiseman J R and El-Baz F 2007 *Remote sensing in archaeology* (Springer Science & Business Media) p 552.
[2] Yastikli N 2007 Documentation of cultural heritage using digital photogrammetry and laser scanning. *J. Cult. Herit.* **8(4)**, pp 423–427. doi: 10.1016/j.culher.2007.06.003.
[3] Capra A, Bertacchini E, Castagnetti C, Rivola R and Dubbini M 2015 Recent approaches in geodesy and geomatics for structures monitoring *Rendiconti Lincei* **26(1)** pp 53-61. doi:10.1007/s12210-015-0436-z.
[4] Toschi I, Rivola R, Bertacchini E, Castagnetti C, Dubbini M and Capra A 2013 Validation tests of open-source procedures for digital camera calibration and 3D image-based modeling. *Int. Arch. Photogramm. Remote Sens. Spatial Inf. Sci.* **XL-5/W2**, pp 647-652. doi:10.5194/isprsarchives-XL-5-W2-647-2013.
[5] Farenzena M, Fusiello A and Gherardi R 2009 Structure and Motion pipeline on a hierchical cluster tree. *Proc. of Computer Vision Workshops (ICCV Workshops) 2009 IEEE 12th International Conference (Kyoto, Japan)* p. 1489-1496. doi:10.1109/ICCVW.2009.5457435.
[6] Fonstad M A, Dietrich J T, Courville B C, Jensen J L and Carbonneau P E 2013 Topographic structure from motion: a new development in photogrammetric measurement. *Eearth Surf. Proc. Land.* **38** pp 421–430. doi:10.1002/esp.3366.
[7] Castagnetti C, Giannini M and Rivola R 2017 Image-based virtual tours and 3d modeling of past and current ages for the enhancement of archaeological parks: The visualversilia 3d project. *Int. Arch. Photogramm. Remote Sens. Spatial Inf. Sci.* **XLII- 5/W1** pp 639-645. doi:10.5194/isprs-Archives-XLII-5-W1-639-2017.
[8] Gherardini F and Leali F 2016 A Framework for 3D Pattern Analysis and Reconstruction of Persian Architectural Elements. *Nexus Netw. J.* **18(1)** pp 133–167. doi:10.1007/s00004-015-0287-z.
[9] Remondino F, Spera M G, Nocerino E, Menna F and Nex F 2014 State of the art in high density image matching. *Photogramm. Rec.* **29** pp 144–166. doi:10.1111/phor.12063.
[10] Capra A, Dubbini M, Bertacchini E, Castagnetti C and Mancini F 2015 3D Reconstruction of an underwater archaeological site: comparison between low cost cameras. *Int. Arch. Photogramm. Remote Sens. Spatial Inf. Sci.* **XL-5/W5** pp 67-72. doi:10.5194/isprsarchives-XL-5-W5-67-2015.
[11] Williams V I, Orsini C, Benozzi E and Castellanos M C 2014 Preliminary investigations in Brealito and Luracatao area: firsts results (dpt. Molinos, Salta). *Relaciones de la Sociedad Argentina de Antropologia* **XXXIX(2)** pp 539-549. ISSN 1852-1479 (online version).

IOP Conf. Series: Materials Science and Engineering **364** (2018) 012021 doi:10.1088/1757-899X/364/1/012021

Advanced Experience of Music through 5G Technologies

Adriano Baratè, Goffredo Haus, Luca A. Ludovico

Laboratorio di Informatica Musicale, Dipartimento di Informatica "Giovanni Degli Antoni", Università degli Studi di Milano, 20135 Milano, Italy

E-mail: {adriano.barate,goffredo.haus,luca.ludovico}@unimi.it

Abstract. This paper focuses on new models to enjoy music that will be implementable in a near future thanks to 5G technology. In the last two decades, our research mainly focused on the comprehensive description of music information, where multiple aspects are integrated to provide the user with an advanced multi-layer environment to experience music content. In recent times, the advancements in network technologies allowed a web implementation of this approach through W3C-compliant languages. The last obstacle to the use of personal devices is currently posed by the characteristics of mobile networks, concerning bandwidth, reliability, and the density of devices in an area. Designed to meet the requirements of future technological challenges, such as the Internet of Things and self-driving vehicles, the advent of 5G networks will solve these problems, thus paving the way also for new music-oriented applications. The possibilities described in this work range from bringing archive materials and music cultural heritage to a new life to the implementation of immersive environments for live-show remote experience.

1. Introduction

The domain of music and music-related information embraces heterogeneous forms of description. The goal of providing a comprehensive representation of a single music piece can be achieved by supplying:

- metadata – textual information about the piece, including its title, the list of involved artists, composition/publishing/release dates, etc.;
- logic information – a "virtual" description of score symbols (chords, rests, time indications, articulation signs, etc.) that does not depend on layout details;
- graphical content – notational instances referring to different score editions, each one with its own format and layout;
- audio/video information – recordings of different performances of the same piece, as well as partial takes of a single performance (like in multi-track audio and video);
- music structures – analysis- or musicology-oriented representations of musical entities (such as themes, subjects, etc.) and their relationships within the composition.

In scientific literature, the mentioned aspects are usually referred to as *layers*, as reported, e.g., in [1], [2], [3]. Currently, suitable computer representations of hierarchical multi-layer music descriptions make use of XML-based standard languages; among others, it is worth citing

IEEE 1599, MEI, and MusicXML formats. XML took foothold in the early 2000s thanks to its intrinsic characteristics, which include the possibility of structuring documents in a hierarchical way, extensibility, the availability of free tools for editing, both human and machine readability, and the support offered by most programming languages [4].

In the list above, we have mentioned a number of heterogeneous facets music information is made of. In addition, in the context of multi-layer representations, each layer can host multiple objects, e.g., different interpretations of the same piece, multi-angle and multi-track takes of the same performance, and differently notated and formatted versions of the same score.

The availability of a unique representation environment supporting multi-layer and multi-instance descriptions provides the user with the possibility to select, navigate, and interact with music contents according to their semantic relationships [5]. File formats and computer-based applications that reflect this approach are already in use, but the characteristics of current mobile-network technologies – in terms of bandwidth, availability, and reliability – have been limiting their use to off-line or wired contexts, so far.

In this sense, the imminent advent of 5G networks will introduce a paradigm shift: this new technology is expected to support reliable delivery of multiple and simultaneous high-quality streams. It will be possible to offer an advanced experience of music and music-related content anywhere and anytime on your own mobile device.

This approach will pave the way for a number of innovative – and sometimes unpredictable – applications, which can raise interest under different perspectives. Regarding entertainment, edutainment and cultural dissemination, users will be able to employ their own mobile devices to experience music in a more effective and engaging way. Also content providers should get an economic benefit, since ad hoc materials must be produced and released, and archive contents can be rediscovered and exploited in new ways. As stated in [6], applications dealing with integrated multi-layer content can even introduce new intellectual rights to be economically exploited, such as the *synchronization right*. Finally, hardware and software developers will have to keep pace with the technological progress and foresee user requests. In conclusion, the application of 5G technologies to musical heritage is expected to involve an entire cultural, multimedia and technological ecosystem.

This work is structured as follows: the next section will describe the state of the art, providing an overview of some relevant and already-available initiatives based on the concept of multi-layer representation of music; Section 3 will outline the characteristics of 5G wireless systems; finally, the last sections will give examples of possible 5G applications, and specifically Section 4 will address music archives, whereas Section 5 will discuss the case of live musical performances.

2. Examples of Multi-layer Applications for Music

In multi-layer formats, heterogeneous information is not only collected and organized within a single document, but also put in relationship so as to obtain intra-layer and inter-layer synchronization. Available formats may implement this concept in different ways, but they mainly rely on the concept of music-event identification: the logical description of a piece presents a number of anchors to which the different representations can be hooked.

A trivial but clarifying example is *score following*, namely the process of listening to a music performance and tracking the position in the score. This goal can be easily accomplished in the context of computer-driven performances, e.g., when music symbols are entered in a digital score editor with playback functions, able to produce a synchronized experience on the fly.

Thanks to multi-layer representation formats, such a concept can be considerably extended: first, it can be applied to other information sources (e.g., lyrics and videos); moreover, the audio layer can include any number of already available recordings, provided that synchronization data are present; the score itself can be any digital or digitized notational instance; finally, homogeneous materials (e.g., the current score or audio instance) can be switched and compared

IOP Conf. Series: Materials Science and Engineering **364** (2018) 012021 doi:10.1088/1757-899X/364/1/012021

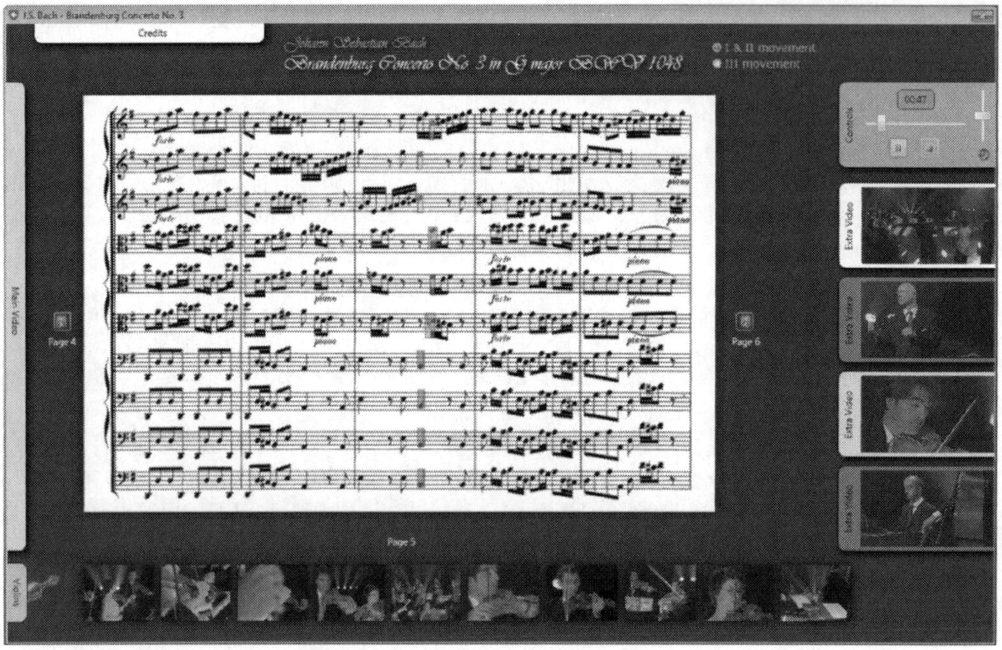

Figure 1. The application to navigate the score of the *Brandenburg Concerto No.3 in G major*. Its interface presents a main area for score following and a scrollable list of videos in the bottom, categorized by instrumental group. Multiple media contents can be played simultaneously in the right panels.

in real time, without affecting the score-following experience.[1]

For the sake of clarity, it is worth mentioning two use cases of multimedia products designed and implemented by the LIM staff and based on a multi-layer music format called IEEE 1599 [7]. The first use case is the prototype realized in 2007 for RSI Radiotelevisione svizzera, in cooperation with the Scuola Universitaria Professionale della Svizzera Italiana (SUPSI). The goal was to provide an advanced score follower for the *Brandenburg Concerto No.3 in G major, BWV 1048* by Johann Sebastian Bach. Its three movements were encoded into IEEE 1599 format with the finest granularity, i.e. allowing note by note synchronization. Music events were mapped onto a single printed score, but synchronized with multiple audio/video streams. The most advanced feature was the availability of about 40 simultaneous video takes, on average 8 per instrumental group; moreover, up to 5 videos could be loaded into ad hoc panels so as to offer a synchronized comparison, as shown in Figure 1. Thanks to this approach, the user could enjoy score following with a very detailed and customizable multi-angle view of the ensemble. Designed as a stand-alone off-line software, the need to keep a high number of video streams synchronized made it impossible – at that time – a web distribution.

Another example of highly demanding application is the multimedia product realized by the LIM in 2011 for the Orchestra sinfonica di Milano Giuseppe Verdi. It had a similar goal, namely advanced score following, but focused on the *Messa da Requiem* by Giuseppe Verdi and presented some additional features. Due to the length of the score and to the high number of orchestral parts to encode, the synchronization was performed at a less refined level of granularity, i.e. measure by measure. This product presented a single full-orchestra score, but 25 simultaneous

[1] Please note that the present work mainly focuses on *representation* issues, not on the way synchronized data are obtained, whether manually or automatically, in real time or through off-line processing, etc.

IOP Conf. Series: Materials Science and Engineering **364** (2018) 012021 doi:10.1088/1757-899X/364/1/012021

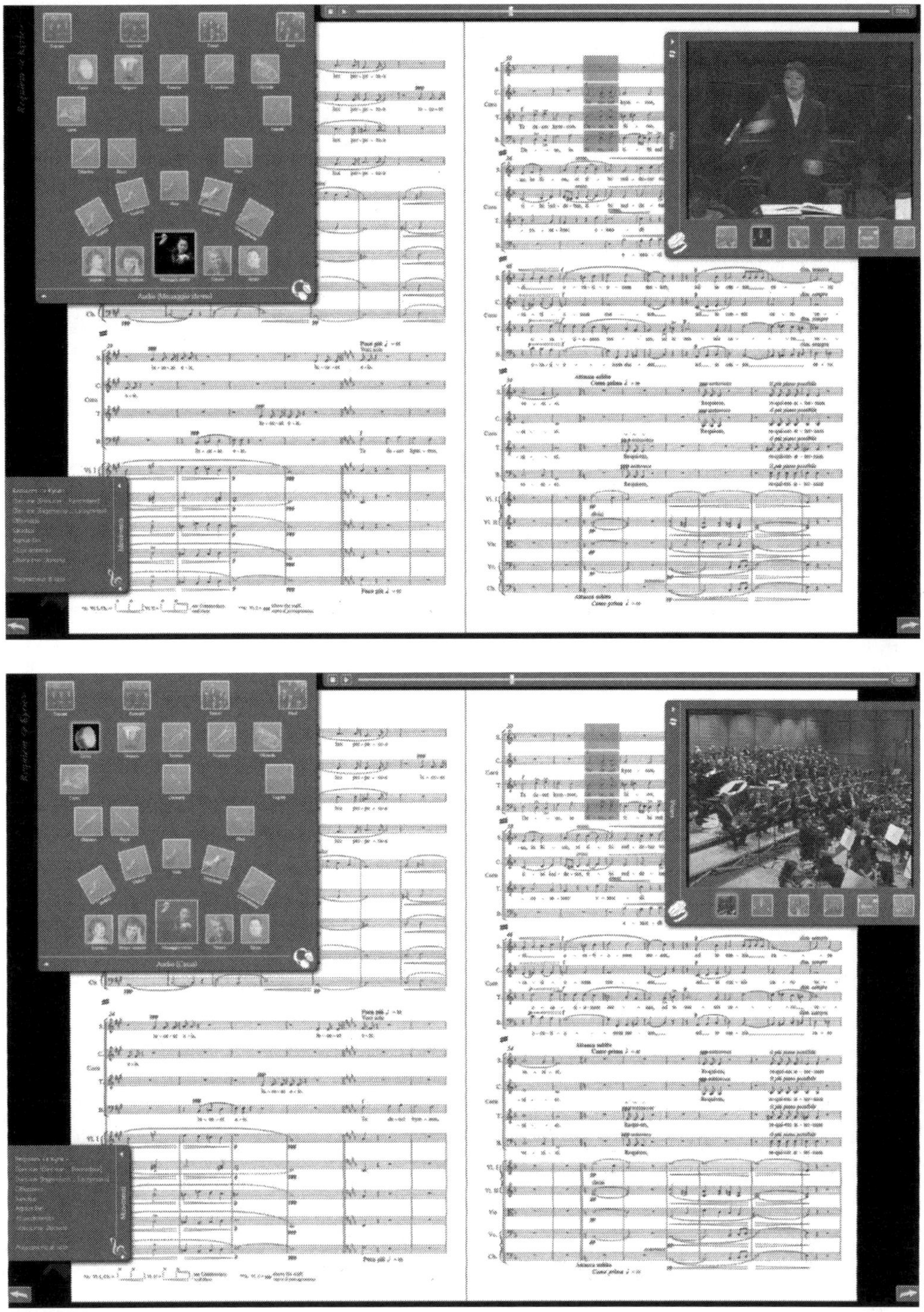

Figure 2. The interface of the *Messa da Requiem* application. The upper figure refers to the selection of the complete audio track and of the camera framing the conductor; in the lower one, the user chooses to listen to the audio from the bass drum microphone and to watch the orchestra from the left side.

audio tracks plus 6 video takes of the event. The most advanced function was the possibility for users not only to customize their visual perspective, like in the RSI prototype, but also to choose one of the multi-track recordings in order to listen to the performance as if they were immersed in the orchestra. The graphical interface of the application is shown in Figure 2.

In the timespan between the release of the two applications, some remarkable innovations were made in the field of network technologies. Defined in 1999, the 802.11b standard operated in the 2.4 GHz band and had a maximum raw data rate of 11 Mbps. The first commercial products supporting this standard appeared on the market in early 2000. An interesting work dating back to that period and addressing the problem of providing application-level quality of service in 3G/4G wireless systems is [8].

The evolution of the standard known as 802.11a was released in 2012, and – operating in the 5 GHz band – it allowed transmission and reception of data at rates of 1.5 to 54 Mbps. Considering congestion and data correction, it yielded realistic net achievable throughput in the mid-20 Mbps. For further details, please refer to [9], a work containing quantitative data about the performance of 3G mobile data offloading through WiFi networks.

Recalling the requirements of our applications, even the WiFi bandwidth would have been sufficient to support high-quality lossy-compressed audio (e.g., 25 audio streams at 320 kbps), but not multiple and simultaneous video content. For this reason, both the applications have been released as stand-alone and off-line products.

In that period, the LIM staff was working on solutions to optimize the delivery of multiple multimedia streams on wired networks, as reported in [10]. A similar fruition on mobile devices seemed far to be achieved. Now, technological advancements in this field are promising, and porting this kind of high-demanding applications onto mobile networks is no longer a chimera.

3. Key Features of 5G Mobile Networks

With respect to the 4G LTE standard, a major step forward is the publication of the "Draft new Report ITU-R M.[IMT-2020.TECH PERF REQ] - Minimum requirements related to technical performance for IMT-2020 radio interface(s)".[2] This document explicits the guidelines for the definition of the 5G standard, the new generation of mobile connectivity that should make its experimental debut in Italy in 2018.

The report establishes the minimum requirements for a mobile antenna to receive 5G certification and, therefore, comply with the specifications of the standard. In some months the draft should be reviewed and confirmed by the companies and research centers that are part of the ITU, the United Nations specialized agency for information and communication technologies. After that step, the draft will become standard and producers will start to release the first 5G-compliant devices.

The draft version of 5G technical specifications establishes at least 20 Gbps downlink and 10 Gbps uplink per single mobile cell. These values are purely theoretical: on one side, telecommunications operators could decide to install antennas with higher peak speeds (e.g., 100 Gbps or more) in order to guarantee better performances for their users; on the other hand, peak velocity refers to the entire cell, and the band – and therefore the connection speed – must be divided among all the users connected at the same time. Realistic transfer rates per single user are expected to be 100 Mbps in download and 50 Mbps in upload. These values are pretty close to the peak speed of some 4G LTE networks, but for 5G this is the average throughput.

Technical specifications fix relevant goals concerning massive connectivity and latency, too. The standard is expected to support at least 1 million connected devices per square kilometer, 1ms to 4ms latency, and the capability to allow up to 500 km/h access (suitable, e.g., for high-speed trains).

[2] `https://www.itu.int/md/R15-SG05-C-0040/en`

IOP Conf. Series: Materials Science and Engineering **364** (2018) 012021 doi:10.1088/1757-899X/364/1/012021

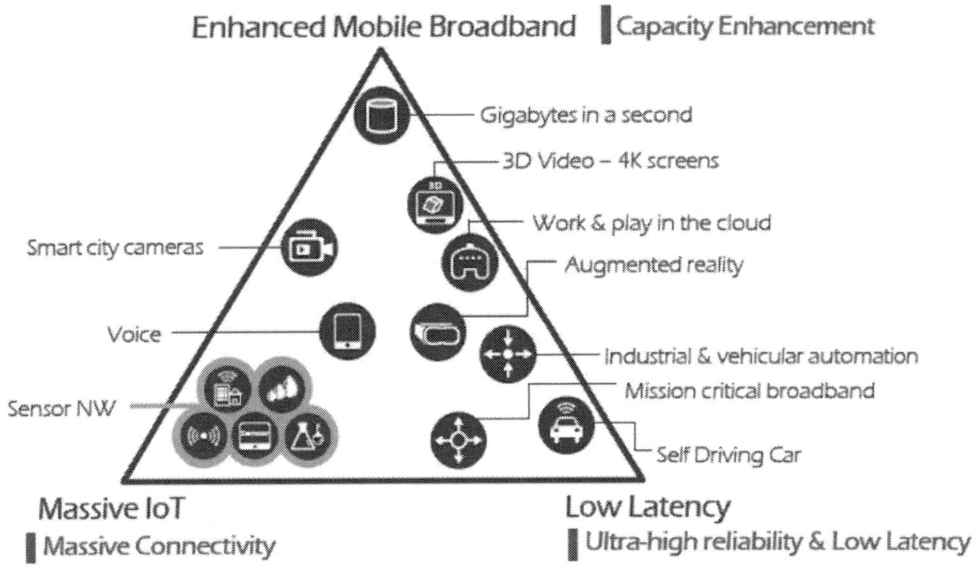

Figure 3. The triangle of 5G applications (source: ETRI graphic, from ITU-R IMT 2020 requirements).

Today, the main mobile services are related to human-centric communication, but the increasing number of devices associated to the Internet of Things (IoT) phenomenon produces forecasts of a total of about 30 billion connected devices by 2020 [11]. This scenario will lead to different communication characteristics, since the type of needs in terms of latency, data rates, density of devices, etc. varies among human-centric communication and IoT devices.

The 5G technology is expected to meet the needs of the next decade [12]. With respect to current network features, it will support:

- 1000 times higher mobile data volume per area;
- 10 to 100 times higher number of connected devices;
- 10 to 100 times higher user data rate;
- 10 times longer battery life for low-power massive machine communication (MMC);
- 5 times reduced end-to-end latency.

A number of scenarios not implementable today will be easily achievable in the future. For instance, Figure 3 outlines a number of possible applications, by arranging them along the axes that link the peculiar features of 5G systems: capacity enhancement, massive connectivity, and ultra-high reliability & low latency.

In [13] these scenarios are described as follows:

- *Amazingly fast* – Users can obtain very high data rates with instantaneous connectivity and low latency. This is crucial for multi-layer applications based on multiple high-quality media streams;
- *Great service in a crowd* – Currently, connectivity is limited when many users share the same area (e.g., stadiums, concert halls, etc.), but in the future also crowded places will permit a satisfactory experience;

IOP Conf. Series: Materials Science and Engineering **364** (2018) 012021 doi:10.1088/1757-899X/364/1/012021

Table 1. Video bit rates recommended by YouTube for standard-dynamic-range (SDR) uploads. Values for high-dynamic-range (HDR) videos are similar.

Type	Video Bit Rate Standard Frame Rate (24, 25, 30)	Video Bit Rate High Frame Rate(48, 50, 60)
2160p (4k)	35-45 Mbps	53-68 Mbps
1440p (2k)	16 Mbps	24 Mbps
1080p	8 Mbps	12 Mbps
720p	5 Mbps	7.5 Mbps
480p	2.5 Mbps	4 Mbps
360p	1 Mbps	1.5 Mbps

- *Ubiquitous things communicating* – The mix of IoT and human-centric communications tends to have different needs, and the 5G technology will efficiently handle these new requirements;
- *Best experience follows you* – Even when users are on the move (e.g., traveling by car or commuting), a high quality of service will be guaranteed;
- *Super real-time and reliable connections* – Future wireless systems will support new applications that take full advantage of very high reliability and low latency, thus allowing real-time fruition (e.g., augmented/virtual reality) as well as control (e.g., self-driving vehicles and industrial automation).

Another relevant consequence could be the possibility for some network operators to start deploying so-called *fixed wireless broadband*, i.e. delivering high-speed broadband to areas where there are few or no other broadband suppliers [14].

In order to evaluate the requirements of multi-layer applications, let us consider the bandwidth required by digital media formats commonly in use, focusing on audio and video.

Audio streams for Red Book audio CDs are two-channel signed 16-bit Linear PCM sampled at 44,100 Hz, whose bit rate is 1,411,200 bps, or 176,400 bytes per second. The DVD-Audio format presents a maximum permissible total bit rate of 9.6 Mbps. Concerning compressed audio, files typically present highly variable bit rates, depending on the encoding format, the expected quality, and the characteristics of media content. FLAC files usually have a bit rate that ranges from 220 to 1184 kbps, whereas for MP3 files acceptable values for music applications usually span from 128 to 320 kbps.

Also for video formats there is great variability. At the moment of writing, YouTube adopts an MP4 container with AAC-LC as the audio codec and H.264 as the video codec. Audio playback bit rate is not related to video resolution. Recommended bit rates for audio are: 128 kbps for mono, 384 kbps for stereo, and 512 kbps for 5.1; values for video are shown in Table 1.[3]

Before discussing innovative applications in the field of cultural heritage, let us reconsider the proposals mentioned in Section 2, and specifically a reimplementation that turns those stand-alone softwares into mobile apps interacting with a remote media server. Even leaving latency- and reliability-related issues out, what would the required bit rate be for multiple concurrent media streams?

The *Brandenburg Concerto* application was more demanding as it regards data transfer, due to the high number of selectable video tracks. Please remember that 5G users in metropolitan areas are expected to experience a limit of about 100 Mbps, while 40 full-HD videos encoded in

[3] Source: `https://support.google.com/youtube/answer/1722171`

H.264 format with a standard frame rate would require a total bit rate of 320 Mbps. Apparently, despite significant improvements on the bit rate, the 5G technology does not seem ready to support such complex and rich multimedia applications. First, the implications on content quality in order to arrange a given number of media streams may be unacceptable (in the mentioned case, videos could not have a resolution higher than 720p). Secondly, it would be inefficient to send all streams simultaneously to single devices, concerning both network occupation and server throughput.

Luckily, 5G wireless networks offer a solution. Thanks to their very low latency, the client-server request to send a new stream can occur on the fly, with no significant delay perceived in multimedia experience. In other words, it is no more necessary for the client to have a number of media streams simultaneously available and ready to be switched in real time. As a consequence, the band available to the user can be exploited in better ways, e.g. delivering on demand a 4K spherical video that requires 60 Mbps approximatively. Applied to the *Brandenburg Concerto* application, this approach lets the user not only watch the show choosing one of 40 fixed cameras, like in the original version, but also turn around as if he/she was sitting in the middle of the orchestra, in one of the 40 available virtual seatings, and listen to the performance accordingly.

In conclusion, thanks to the impressive features of 5G wireless systems, we can expect the design and release of musical applications not available today, or the reimplementation of already existing ones for personal mobile devices. In the next sections, we will propose two significant use cases focusing on music cultural heritage.

4. Exploitation and Revivification of Music Archives

Our first proposal addresses the exploitation and revivification of archived musical contents.

Until a few years ago, music archives preserved by research centers and international institutions offered very limited access to common users, and sometimes even to professionals. In recent times, thanks to digitization campaigns and web interfaces, many treasures hidden in archives and collections have been unveiled, thus giving new life to intangible cultural heritage.

This kind of cultural dissemination is expected to rise a great interest also in a non-specialized audience, thus becoming a potential source of income for musical institutions, too. Despite the great interest aroused by culture in our days, some critical factors are affecting the phenomenon. First, the user experience offered by specialized applications is often not adequately calibrated on the interests of a general public. In this sense, noteworthy examples are those portals that present lists of uncorrelated digital objects with catalog metadata, maybe useful for a specialized user but totally unappealing for a general audience. Besides, there are technological limitations that prevent a satisfactory fruition on portable devices (screen size and resolution, available memory, etc.), above all when a WiFi connection is not available (limited amount of downloadable data, slow transfer rate, etc.).

The adoption of multi-layer formats is a way to address the former issue, thus capturing the interest of a general public. In fact, by suitably combining, organizing and encoding different information sources, it is possible to achieve multiple results:

- offering a user-tailored experience;
- fostering cultural enrichment through engagement and edutainment applications;
- raising users' interest and extending the audience by launching new multimedia products;
- stimulating the production of ad hoc materials to meet these new needs;
- rediscovering and exploiting the past through novel approaches.

For instance, some historical audio performances have lost their commercial interest, since the availability of new artists (supported and promoted by record labels) and the advancements in recording techniques and formats make new products more appealing for the marketplace. But

IOP Conf. Series: Materials Science and Engineering **364** (2018) 012021 doi:10.1088/1757-899X/364/1/012021

in a multi-layer environment it is possible, and even desirable, to mix old and new performances into a unique multimedia product.

In the field of education, the revivification of old contents and the production of new ones can be the base for multi-layer applications with a strong valence. For example, the mentioned inter- and intra-layer forms of synchronization typical of a multi-layer environment can encourage learning approaches such as: score-following abilities for performers and conductors, the interpretation of music notations different from the Western traditional one (neumes, tablatures, etc.), the study of vocal repertoire by examples, the comparison of different harmonizations and orchestrations of jazz standards, and so on.

The other mentioned issue, related to technological limitations in the individual fruition of multimedia, will be solved by future advancements in network technologies. Specifically, the features of 5G networks will support any reasonable media payload in all contexts, even in very crowded places. For example, it will be possible to take musicology lessons in a university classroom where students have their own portable devices and anyone interacts autonomously with multiple synchronized media objects. Besides, the characteristics of 5G are fundamental to guarantee the delivering of high-quality multimedia streams with strict time requirements, namely in less than 100ms and aiming at a latency of 5ms, considered acceptable for real-time processing by Digital Audio Workstations.

The adoption of multi-layer formats and the technological support offered by 5G wireless systems could easily bring applications like the ones described in Section 2 into our portable devices. Going one step further, it is not necessary that the content to combine is already available in a single remote repository, rather it could be extracted on the fly from different archives, depending on the user needs and choices.

Let us recall the example of the *Brandenburg Concerto No.3* and re-implement it in an interoperable framework where an agent running on your personal device is able to crawl suitable information sources via the web. Instead of being stored in the limited local disk space, score versions could be provided on-demand by ad hoc repositories (such as IMSLP[4] and Bach Digital[5]), metadata could be retrieved from public digital libraries (e.g., from Europeana[6]), and audio tracks could be extracted from open-source archives (such as the Free Music Archive[7] and the National Jukebox of the U.S. Library of Congress[8]).

With this goal in mind, a relevant problem to solve is the one of standardization and interoperability among different sources. *Interoperability* can be defined as the compatibility of two or more systems such that they can exchange information and data and can use the exchanged information and data without any special manipulation, thus minimizing the loss of information due to technological differences [15]. Thanks to interoperability, users as well as software agents should be able to retrieve data by discovering through a single search what digital objects are available from a variety of collections, rather than having to search each collection individually [16]. From a technical point of view, this problem could be solved through suitable web services exposed by content providers. Unfortunately, another issue arises: collected data, potentially uncorrelated, should be automatically organized into a suitable multi-layer representations. Both goals at present are difficult to achieve.

So far, efforts have focused on the standardization of specific formats for "localized" interoperability. For instance, in 1975 the Cataloguing Commission of the International Association of Music Libraries, Archives and Documentation Centres (IAML)[9] and the

[4] `http://imslp.org/`
[5] `https://www.bach-digital.de/`
[6] `https://www.europeana.eu/`
[7] `http://freemusicarchive.org/`
[8] `http://www.loc.gov/jukebox/`
[9] `https://www.iaml.info/`

IOP Conf. Series: Materials Science and Engineering **364** (2018) 012021 doi:10.1088/1757-899X/364/1/012021

International Federation of Library Associations and Institutions (IFLA)[10] Committee on Cataloguing drafted the International Standard Bibliographic Description for printed music. The primary goal was to provide the stipulations for compatible descriptive cataloguing worldwide, in order to aid the international exchange of bibliographic records between national bibliographic agencies and throughout the international library and information community [17]. Similarly, in the context of score file interchange, MusicXML was explicitly designed and released to be a universal translator for programs that understand common Western musical notation [18].

Unfortunately, these initiatives – while going in the direction of interoperability – focused only on a subset of characteristics with respect to the potential richness of a multi-layer description. This issue has emerged in recent times in the context of the W3C Music Notation Community Group.[11] Their initial task was to maintain and update the MusicXML and SMuFL[12] specifications in order to develop a standard format for notated music. Starting from the problem of notation interchange over the web, the group soon had to face more general issues, such as enriching scores by arbitrary graphics and coupling them with audio resources such as MP4 and MIDI data, which is the core of the MNX draft.

In a multi-layer ecosystem, a further effort is required to support not only an interoperable exchange of homogeneous information (e.g., catalog metadata, or score files, etc.), but also the compatibility among heterogeneous media types. This problem is solved in multi-layer formats such as IEEE 1599 by providing a common data structure that serves as a dictionary containing the list of labels (ids) to identify each event to be described. After marking an event through an id, multi-layer instances have a non-ambiguous way to hook their descriptions to the event itself. Unfortunately, this approach is hardly feasible in a peer context where different information sources are independent and there is no shared authoritative list of ids.

In principle, synchronization could be achieved in real time by a software agent running on the local device, able to automatically parse logical and multimedia content referring to the same music piece, retrieve synchronization anchors, and feed the media player. Some of these aspects are still open research problems which stand at the intersection of artificial intelligence, pattern recognition, signal processing, and musicology. Examples include reliable optical music recognition [19], complex tasks of music information retrieval [20], and automatic polyphonic score following [21].

Nevertheless, this effort is justified by the potential impact on the revivification and dissemination of musical cultural heritage, and specifically the possibility to enjoy a user-tailored experience on your own device by choosing the multimedia content to combine from a huge set of sources.

5. 5G for Live Musical Performances

Relevant examples of intangible cultural heritages include music and theater, that sometimes merge into a single form of art, such as opera, musical theater, and ballet.

In 2012, a research group at LIM proposed an innovative approach to on-line experience of live performances [22]. The idea was to provide remote viewers, who were watching the show in real time via the web, with a browser application to let them select specific materials from a set of available media contents. For example, the user was allowed to choose the current camera (e.g., watching at the whole scene, focusing on a specific character, or even looking at the audience) and open ad hoc panels carrying additional information (e.g., the script in the case of theatrical performances, or the score for musical ones). A web interface was developed by the LIM to enjoy all the performances of "Prospettiva 09" festival[13] according to this paradigm

[10] `https://www.ifla.org/`
[11] `https://www.w3.org/community/music-notation/`
[12] Standard Music Font Layout, `https://www.smufl.org/`
[13] `http://archivio.teatrostabiletorino.it/collections/occurrence/detail/748/`

Figure 4. The interface of a web application for the multi-layer fruition of live shows.

(see Figure 4).

The extended interpretation of the concept of *multi-layer representation* implied the availability of multiple audio and video streams, several scripts translated into different languages for theatrical pieces, and different score versions for concerts. The strengths of that proposal were, once again, the concepts of multi-layer fruition, interactivity, and user-tailored experience.

Recalling the examples mentioned in Section 2, there were additional problems to solve, such as multi-cast management for multiple high-quality media streams, the support to real-time user interaction with remote content, and the need for a guaranteed minimum level concerning the quality of service. Due to technological limitations, also this application was intended for high-speed wired connections.

The characteristics of 5G wireless systems illustrated in Section 3 will solve network-related problems, thus supporting the fruition paradigm already explored in a web environment directly on individual devices. But turning a browser app originally conceived for traditional PCs into a mobile app offers further advantages, e.g. the possibility to remotely enjoy music in real time within an immersive virtual-reality (VR) environment. In this way, a plurality of participants may attend an entertainment and/or educational event simultaneously, both in the same physical location and remotely. The performance, delivered in form of spherical video, is intended to be viewed by participants through VR devices such as a head mounted display.

In the case of live musical performances, probably the audience will not be interested in changing the current media stream; nevertheless, additional content could greatly improve user experience. For example, even traditional opera settings are now commonly equipped with devices to follow libretto lyrics (big screens, individual displays, etc.). Through our proposal, it is possible to let the user choose among many translations, select one out of many score versions (all automatically synchronized with current audio), retrieve information about performers, book a new concert date, etc.

A satisfactory user experience implies the delivery of at least one high-quality 360-degree video and multi-channel audio, with a bit rate of about 60 Gbps. Besides, the connection must be reliable and content streaming must be continuous. As mentioned in Section 3, all these requirements will be met by 5G technologies.

Finally, let us mention the possibility to allow remote and geographically-distributed user interaction with the ongoing performance. For example, there are artworks where extemporaneous actions by the audience are encouraged so as to influence the evolution of the plot. It is the case of so-called *immersive theater*, that aims to create participative experiences where audience members give up their "observer" status to become co-actors and co-creators of the narrative and of the storytelling process [23]. The adoption of a multi-layer approach, coupled with the potential of 5G technologies, can foster not only an "augmented" passive experience, but also an interactive participation.

References

[1] Steyn J 2002 Framework for a music markup language *Proceeding of the First International IEEE Conference on Musical Application using* vol 1060 pp 22–29

[2] Haus G and Longari M 2005 A multi-layered, time-based music description approach based on XML *Computer Music Journal* vol 29 (MIT Press) pp 70–85

[3] Ludovico L A 2009 IEEE 1599: A multi-layer approach to music description *Journal of Multimedia* vol 4 (Academy Publisher) pp 9–14 ISSN 1796-2048

[4] Castan G, Good M and Roland P 2001 Extensible markup language (XML) for music applications: An introduction *The Virtual Score* vol 12 pp 95–102

[5] Baratè A and Ludovico L A 2016 Local and global semantic networks for the representation of music information *Journal of e-Learning and Knowledge Society* vol 12 (Italian e-Learning Association (SIe-L)) pp 109–123 ISSN 1826-6223 (print), 1971-8829 (electronic)

[6] Baratè A, Haus G, Ludovico L A and Perlasca P 2016 Managing intellectual property in a music fruition environment - the IEEE 1599 case study *IEEE MultiMedia* vol 23 (IEEE Computer Society) pp 84–94 ISSN 1070-986X

[7] Baggi D L and Haus G M 2013 *Music navigation with symbols and layers: Toward content browsing with IEEE 1599 XML encoding* (John Wiley & Sons)

[8] Fitzek F, Kopsel A, Wolisz A, Krishnam M and Reisslein M 2002 Providing application-level QoS in 3G/4G wireless systems: A comprehensive framework based on multirate CDMA *IEEE Wireless Communications* vol 9 (IEEE) pp 42–47

[9] Lee K, Lee J, Yi Y, Rhee I and Chong S 2013 Mobile data offloading: How much can WiFi deliver? *IEEE/ACM Trans. Netw.* vol 21 (Piscataway, NJ, USA: IEEE Press) pp 536–550 ISSN 1063-6692 URL `http://dx.doi.org/10.1109/TNET.2012.2218122`

[10] Baldan S, Ludovico L A and Mauro D A 2011 Managing multiple media streams in HTML5: the IEEE 1599-2008 case study *Proceedings of the International Conference on Signal Processing and Multimedia Applications (SIGMAP 2011)* ed Linares-Barranco A and Tsihrintzis G A (Seville: SciTePress) pp 193–199 ISBN 978-989-8425-72-0

[11] Ericsson 2015 *Ericsson Mobility Report - On the pulse of the networked society* white Paper URL `https://www.ericsson.com/assets/local/news/2016/03/ericsson-mobility-report-nov-2015.pdf`

[12] Osseiran A, Braun V, Hidekazu T, Marsch P, Schotten H, Tullberg H, Uusitalo M A and Schellman M 2013 The foundation of the mobile and wireless communications system for 2020 and beyond: Challenges, enablers and technology solutions *Vehicular Technology Conference (VTC Spring), 2013 IEEE 77th* (IEEE) pp 1–5

[13] Fallgren M, Timus B *et al.* 2013 Scenarios, requirements and kpis for 5g mobile and wireless system *METIS deliverable D1.1* URL `https://www.metis2020.com/wp-content/uploads/deliverables/METIS_D1.1_v1.pdf`

IOP Conf. Series: Materials Science and Engineering **364** (2018) 012021 doi:10.1088/1757-899X/364/1/012021

[14] Santo B Struggling towards 5G *EDN Network*

[15] Taylor A G and Joudrey D N 2017 *The organization of information* (ABC-CLIO)

[16] Tennant R 2001 Different paths to interoperability *Library Journal* vol 126 (Reed Business Information) pp 118–119

[17] Project Group on ISBD(PM) of the International Association of Music Libraries A and (IAML) D C 1991 *ISBD(PM): International Standard Bibliographic Description for Printed Music* 2nd ed (*UBCIM Publications New Series* vol 1) (K. G. Saur Verlag)

[18] Good M and Actor G 2003 Using MusicXML for file interchange *Web Delivering of Music, 2003. 2003 WEDELMUSIC. Proceedings. Third International Conference on* (IEEE) p 153

[19] Rebelo A, Fujinaga I, Paszkiewicz F, Marcal A R, Guedes C and Cardoso J S 2012 Optical music recognition: state-of-the-art and open issues *International Journal of Multimedia Information Retrieval* vol 1 (Springer) pp 173–190

[20] Hu X and Kando N 2017 Task complexity and difficulty in music information retrieval *Journal of the Association for Information Science and Technology* vol 68 (Wiley Online Library) pp 1711–1723

[21] Dorfer M, Arzt A and Widmer G 2016 Towards score following in sheet music images *Proceedings of the 17th International Society for Music Information Retrieval Conference (2016)*

[22] Baratè A, Haus G, Ludovico L A and Mauro D A 2012 IEEE 1599 for live musical and theatrical performances *Journal of Multimedia* vol 7 (Academy Publisher) pp 170–178 ISSN 1796-2048

[23] White G 2012 On immersive theatre *Theatre Research International* vol 37 (Cambridge University Press) pp 221–235

IOP Conf. Series: Materials Science and Engineering **364** (2018) 012022 doi:10.1088/1757-899X/364/1/012022

The bowed string instruments: acoustic characterization of unique pieces from the Italian lutherie

F Leccese[1], G Salvadori[1], G Bernardini[2], P Bernardini[2]
[1]Dept. of Energy Engineering, Systems, Territory and Constructions, University of Pisa, Italy
[2]Bernardini Lutherie, Grosseto, Italy
email: f.leccese@ing.unipi.it

Abstract. The Italian handicraft tradition in the construction of bowed string instruments is a very important element of the Italian cultural heritage, that has nothing to envy to other well-known traditions, such as painting and sculpture. The preservation of a handicraft musical instrument is fully incorporated in the preservation of the artworks of Italian heritage. In order to ensure suitable preservation, the systematic studies for the determination of the technical and acoustic characteristics of handicraft musical instruments are of particular interest. This study deals with the characterization of the bowed string instruments, with a particular attention to their acoustic behaviour. The idea of conducting such a research arises from the need to describe a systematic process aimed at the characterization of handicraft instruments and to collect in a single survey a description of different pieces from the Italian lutherie.

1. Introduction

The bowed string instruments owe their name to the fact that the sound is produced by rubbing the strings by means of horsehair stretched on a bow. The violin family as we know it today (violin, viola, cello and double bass) draws its origins from the oldest *viola da braccio*, an instrument with four strings tuned in fifths, and the *viola da gamba*, whose features remain mainly in reference to the handling position in the cello and the double bass [1-5]. Although different in size, the four components of the violin family consist of the same acoustic chain arranged in three groups (Fig. 1): 1- top plate, back plate and ribs; 2- top nut, tailpiece, bridge, scroll and strings; 3- sound post and bass bar. The representative of the bowed string instruments is historically the violin, invented by *Gasparo da Salò*, whose construction peak has been reached with the Cremona lutherie school with *Nicola Amati*, his pupil *Antonio Stradivari*, and others [1,4-7]. The realization of a violin involves the adoption of about 64 pieces; the operative mode, ranging from choosing the material to the recording of the instrument, is essential for the good quality of the violin itself. As a 'prince' of the bowed string instruments, the violin is the one of which the literature on acoustics mostly deals [8-15]. Viola, cello and double bass differ from the violin in terms of size and tuning, while the mechanisms of sound production are common: this makes that the results found for the violin can be easily extended to the other members of the bows family. However, the absence of a precise search for these instruments involves the lack of an acoustic optimization, especially with regard to low registers; what is mostly affected is the timbre of the instrument, significantly different from that of the violin.
This study deals with the characterization of the bowed string instruments, with a particular attention to their acoustic behaviour. The idea of conducting such a research arises from the need to describe a systematic process aimed at the characterization of handicraft instruments and to collect in a single survey a description of different pieces from the Italian lutherie. That process consists of a harmonic analysis and a temporal pattern analysis. In order to carry out a significant investigation, reference was made to the following unique pieces of Italian lutherie: a *Bernardini* violin (2006), a *Bernardini* viola (2010), a *Carletti* cello (1910) and a *Verdiani* double bass (2000). The conducted analyses enabled the processing of graphs, such as the frequency spectrum, which allowed to identify the main acoustic characteristics of the instruments, as well as to capture the differences. At the same time, it was also conducted the analysis of data and diagrams already in literature, searching for a direct comparison

with the musical instruments used and having, in this way, a global view of the characteristics of the bowed string instruments.

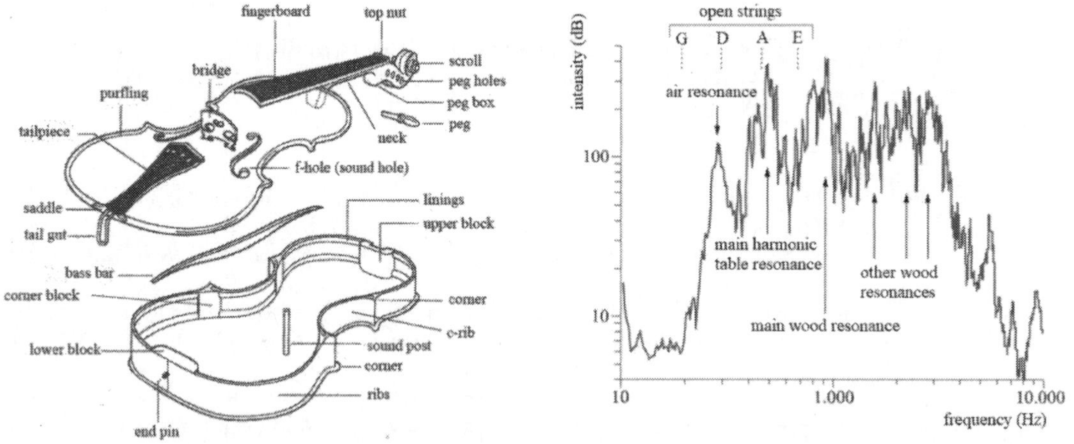

Figure 1. Components of a bowed string instrument (left); example of a sound spectrum of a violin (right).

2. Acoustic analysis of musical instruments: the ADSR envelope

The study of the acoustic characteristics of a musical instrument can be differentiated in two phases: the first is the one of harmonic analysis and is conducted by identifying a graph (the sound spectrum, e.g. Fig. 1) that returns the sound intensity (dB) according to the frequency (Hz); the second is the one that studies the time proceeding of sound by defining the ADSR envelope (Attack-Decay-Sustain-Release, e.g. Figs. 3-6), tracing the sound amplitude according to time. Both types of analysis are very important for a complete acoustic characterization of the various musical instruments. In this way, each instrument is unequivocally defined in order to make it recognizable even within large formations such as bands and orchestras.

With the expression 'envelope of the sound', is intended the timing of the magnitude of sound from the moment it is generated until it extinguishes. Each sound has its own dynamic evolution: there are sounds, such as the ones generated from percussion and piano, which have a flash attack, others, such as flute or violin, which have a sweeter and more gradual attack. Some sounds can be kept as long as the performer wants it (in the case of the violin this aspect is governed by the use of the bow, in the case of the flute instead of the musician's breath), while others disappear in more or less time without the performer may have a decisive influence on their behaviour. Generally, it is possible to say that all sounds start from a zero volume, they reach the maximum volume and then, after a time-varying evolution, they return to zero. In the case of the violin, and more generally of the bowed string instruments, execution is essential for the determination of the sound envelope: the notes may have an initial peak, or reach a maximum at the end, or maintain a constant volume without no peak in particular according to the will and needs of the musician.

These facts can be summarized saying that musical instruments are distinguished in two great classes that exhibit a very different sound envelope behaviour:

- 'free evolution' instruments, instruments in which the performer is limited to providing initial energy, and can not intervene significantly to influence the dynamic evolution of sound that gradually decreases until extinction (this category includes string instruments such as harp, guitar, lute, harpsichord, instruments with percussion strings such as the piano and percussions);
- 'controlled evolution' instruments, the performer provides energy continuously to maintain sound, when the performer ceases to power, the sound disappears more or less quickly (artificial instruments, wind instruments, voice, organs are part of this category).

It is now possible to define the dynamic variation of a sound over time as the set of four distinct phases articulated in the ADSR envelope (Tab. 1). It is important to point out that the four phases of enveloping are such a sketch of what is happening in reality, useful in studying the dynamic evolution of sounds. Also in the sustaining phase the sound is never perfectly firm (it would not be human) also because of executive practices such as vibrato or double strings of bowed string instruments. With regard to the specific case of bowed instruments, the ADSR diagram is generally found to be devoid of the decay phase, as the amplitude of the sound grows rapidly with the bow movement and reaches an uncontrolled sustain state. The performer can, however, change the trend of the envelope by inserting a decay phase with the execution of a *sforzato*, that is an accented sound for which, once the maximum volume has been reached, there is an initial decay followed by holding on a smaller volume.

Table 1. Sketch of an ADSR envelope and description of the four phases.

Sketch of an ADSR envelope	Phase	Description
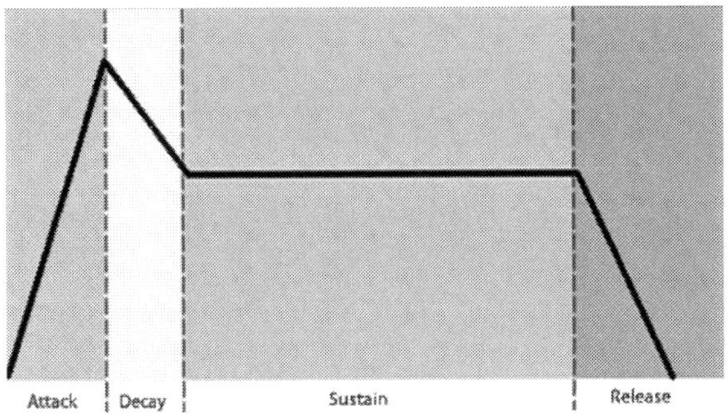	**Attack**	The phase from the moment in which the sound is generated until it reaches its maximum intensity; the beginning of the sound can be immediate (in the case of a piano or a percussion instrument this takes around 1/100 sec) or gradual (in the bowed and wind instruments the performer can create an attack in *crescendo* of the duration of various seconds).
	Decay	The short and rapid decrease in amplitude after the attack, which precedes the sound stabilization (also referred to as initial decay or first decay); this stage may also be completely missing to leave the place directly to the sustain phase.
	Sustain	The phase where the sound stays stable while the performer continues to provide energy; during the sustained phase the sound intensity does not undergo any major changes, remaining almost constant.
	Release	The phase that begins when the performer stops delivering energy and the sound declines more or less quickly (also known as the final decay), it is then shifted from the constant sound intensity of the sustain phase to a lower intensity even to the background noise; this stage can also be very long in free-evolving instruments, while it is usually short in the controlled-evolving ones.

3. Sampling of bowed string instruments

The interesting aspect of this work has been to analyse the results obtained by sampling the bowed string instruments. These have been sampled using the Cool Edit Pro 2.0 software, with the conduction of two types of analysis: with the former it has been obtained the acoustic spectrum of the four instruments; with the second one it has been evaluated the ADSR envelope of the four notes corresponding to the open strings of each instrument. The acoustic spectrum was obtained by playing *glissando* (that in music represents a glide from one pitch to another) on the four strings of the various instruments. The *glissando* was played by a professional musician. Through the Fourier analysis, the software returned the sound amplitude distribution in function of the frequencies emitted. The ADSR envelope was instead evaluated by playing the four open strings of each instrument individually. The results obtained for violin, viola, cello and double bass are reported. Finally, it is necessary to specify that the data collected are not to be considered as absolutely valid, but only about the instruments used in this particular survey and therefore related to the particular constructive features of them. The experimental characterization has been done in a special room, quite similar to a hemi-anechoic chamber, realized in the Bernardini Lutherie headquarter, being negligible the background noise level.

Figure 2a.
Bernardini violin

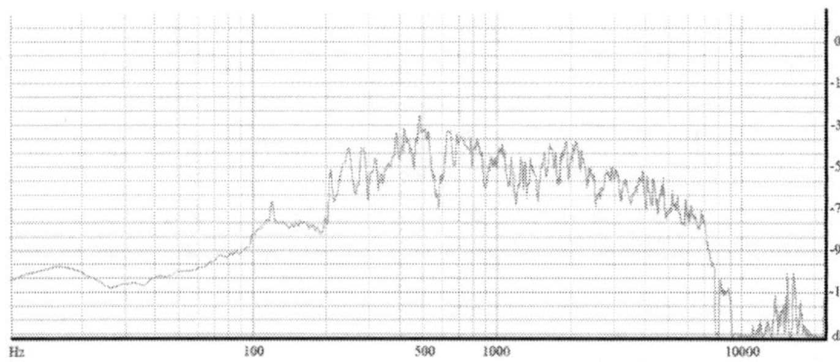

Figure 2b.
Bernardini viola

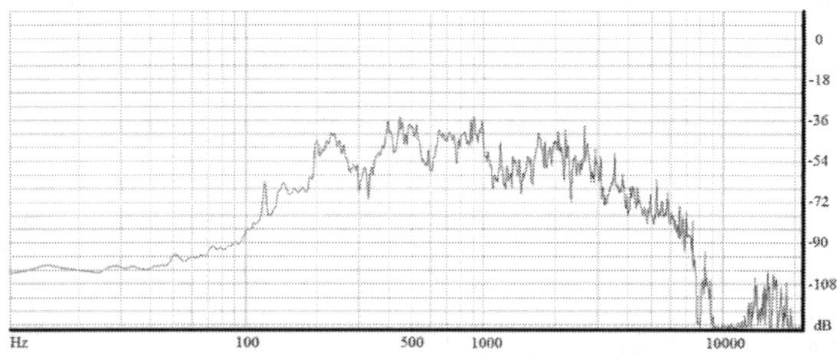

Figure 2c.
Carletti cello

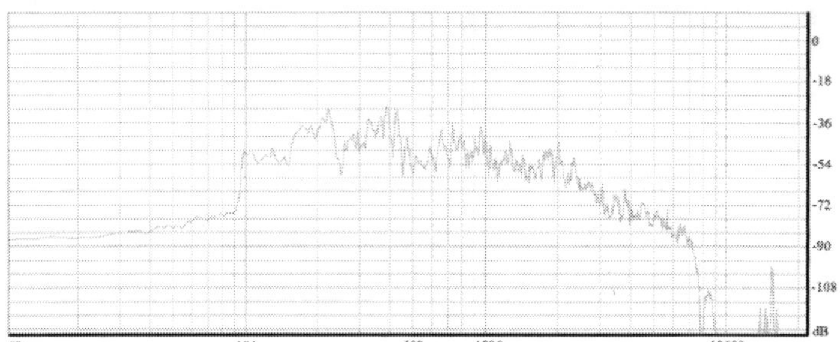

Figure 2d.
Verdiani double bass

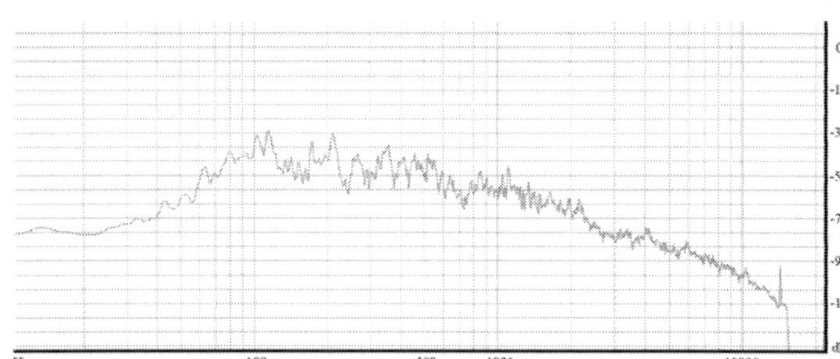

Note that the scale of intensity in the vertical axes returns negative values, characteristic of the particular mode of calculating decibel levels (logarithm of the relationship between a certain value and a value taken as a reference).

Figure 2. Acoustic spectra of the four bowed string instruments: (a) *Bernardini* violin, (b) *Bernardini* viola, (c) *Carletti* cello, (d) *Verdiani* double bass.

3.1 Violin

The measurements were conducted on a *Bernardini* violin of 2006. The frequencies commonly issued by a violin, taking into account only the fundamental notes, have lower and higher extremes of 196 Hz (G3) and 3136 Hz (G7); if we also consider the frequencies emitted by artificial harmonics then the field should be extended to more than 5000 Hz. As can be seen in Fig. 2a, the acoustic spectrum of a violin detected by the software occupies a much wider range than the one just quoted; whereas some peaks are also due to the presence of noises, such as strings rubbing or vibration of instrument parts, we can speak of a frequency range of approximately 60 Hz to more than 15000 Hz. The cause of this greater amplitude in the range of emissions is to be sought, of course, in the presence of natural harmonics from which one can not discern in the acoustic evaluation of an instrument (a violin is all the more rich and sonorous as many natural harmonics is able to emit). From the analysis of the spectrum of the violin considered there is a good correspondence with the spectra in the literature: the instrument has the maximum emission amplitude at 500 Hz (resonance of the harmonic table), maintaining a good sound intensity between 300 Hz and 2000 Hz. Before and after the maximum emission, there is a minimum of sound intensity: this is justified by the fact that at resonance the sound waves are of a constructive nature, therefore they must be preceded (and succeeded) by waves of destructive type; this is the case of the sound wave in phase and out of phase. After the sound spectrum and spectrograms analysis (not showed in this paper for the sake of simplicity), the ADSR envelopes were identified. This analysis was once again conducted for the notes produced by the four open string of the instrument: G_3 (196 Hz), D_4 (294 Hz), A_4 (440 Hz) and E_5 (660 Hz). In the graphs obtained through the Cool Edit Pro 2.0 software (Fig. 3), there are two coloured lines that approximate the envelope in a very schematic way: in particular, with the blue line (bottom line in the Fig. 3) it was attempted to follow more precisely the software redraw, with the orange line (top line in the Fig. 3) it was attempted to create an analogy with the standard lines with the ones that theoretically describe envelopes of this type. Note that only in the case of A and E notes the two approximations are almost coincident: for these two notes, therefore, the subdivision into the phases of Attack, Decay, Sustain and Release are fully recognizable; for the notes of G and D the trends of the two coloured lines do not coincide, emphasizing that the division into four phases is not always present in the envelopes of the notes emitted by the instruments. It is important to recall that the subdivision in four phases is a mere simplification to understand how the sound emission of the instruments works and how they produce sounds in a very different physical mode; it is not to be understood as a precise description of the physical phenomenon itself.

3.2 Viola

The instrument analysed in this case is a viola *Bernardini* of 2011, whose acoustic spectrum is reported in Figure 2b. It is still a quite uniform distribution in the range of frequencies that for this instrument has as extremes, referring to the fundamental notes, 131 Hz (C_3) and 1174 Hz (D_5); the forming frequency are those from 450 Hz to 1000 Hz. Anyway, as in the case of the violin, is from frequencies around 40 Hz to more than 10000 Hz. The analysis of the envelopes of notes corresponding to the open strings of the viola (Fig. 4) points out an aspect that has already been mentioned at the beginning of this paragraph and that has already found in the case of the violin. The four phases recognized in Attack, Decay, Sustain and Release are not always present or recognizable; especially in these measurements, the Decay phase was not detected for any of the four notes. Therefore, it was not possible to return the simplified blue line to the standardized one (orange line in violin envelopes, see Fig. 3). Notice how the sound attack is almost immediate and the release is very late in all four cases.

3.3 Cello

Measurements for the cello were performed on a *Carletti* instrument of 1910. The frequency range usually emitted, in relation to the fundamental notes, is recognized between about 65 Hz and 988 Hz. As it can be seen from Fig. 2c, the range of frequencies emitted is larger than the one just mentioned: this is due to the fact that the analysis software considers also the harmonics emitted by the instrument, such as in the cases of violin and viola. By comparing this spectrum with the ones of the violin and viola, it is obvious that in this case the decibel levels are lower and the sounds tend to extinguish below 10000 Hz. The instrument resonates at 400 Hz, reaching its highest sound level of -28 dB; before and after the resonance peak, there is an immediate decrease in the sound level, justified by the fact that the sound waves are out of phase. The cello has the same tuning of the viola, but at a lower octave; we therefore expect a greater concentration of harmonics at lower frequencies than those of viola and violin. The analysis of the envelopes for the notes corresponding to the cello open strings

(Fig. 5) show what has already been seen for the viola: in fact, even in this case, the four phases recognized in Attack, Decay, Sustain and Release are not always present or recognizable; only in the case of the A open string it was possible to refer the blue line characteristic of the note to the standardized line in orange, recognizing the presence of all four phases of the envelope. For the open string of C, G and D, this reconnaissance was not possible. Notice how the sound attack is almost immediate and the release is very late in all four cases.

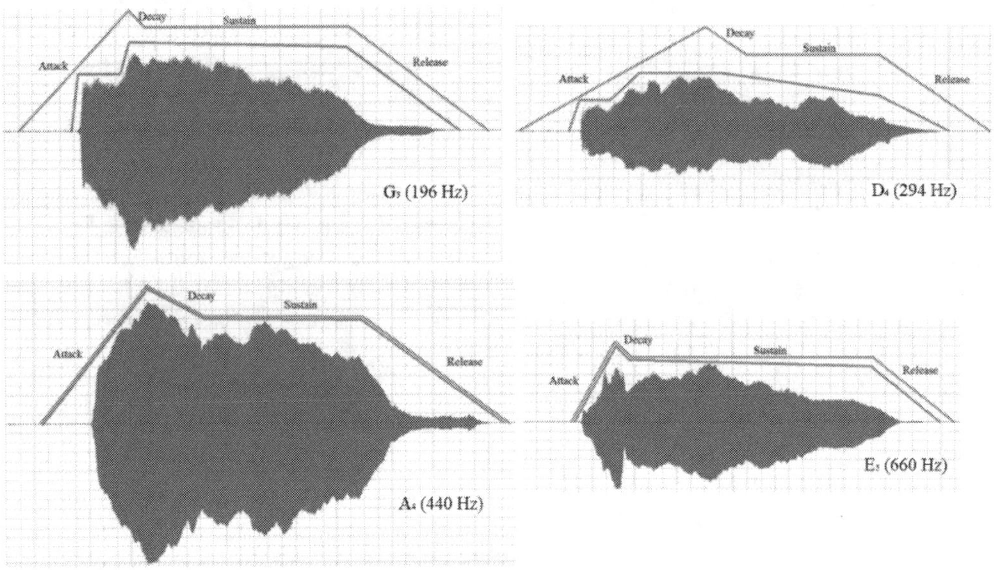

Figure 3. *Bernardini* violin (2006) open strings ADSR envelopes for fundamental notes of: 196 Hz (left, top), 294 Hz (right, top), 440 Hz (left, bottom), 660 Hz (right, bottom).

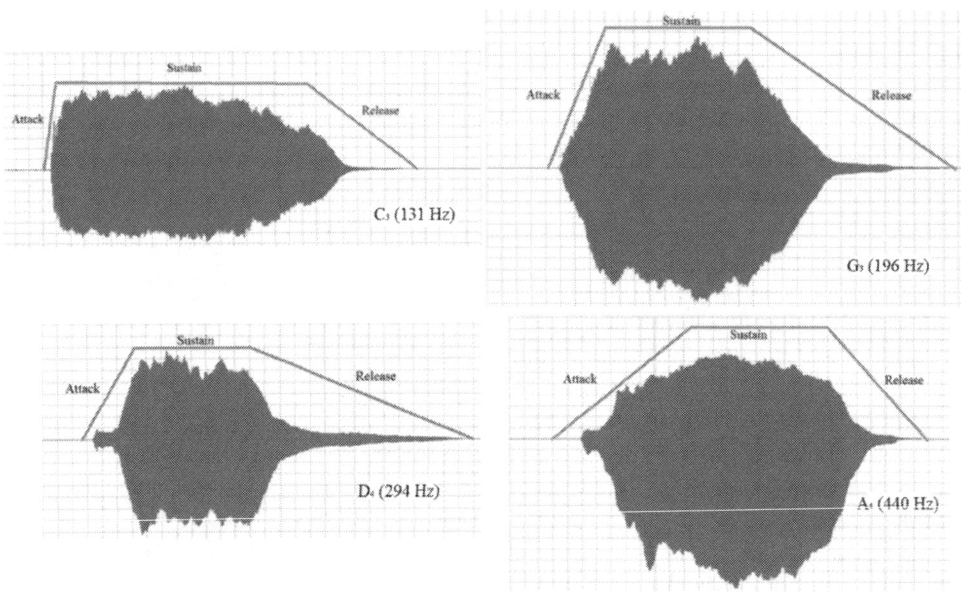

Figure 4. *Bernardini* viola (2011) open strings ADSR envelopes for fundamental notes of: 131 Hz (left, top), 196 Hz (right, top), 294 Hz (left, bottom), 440 Hz (right, bottom).

IOP Conf. Series: Materials Science and Engineering **364** (2018) 012022 doi:10.1088/1757-899X/364/1/012022

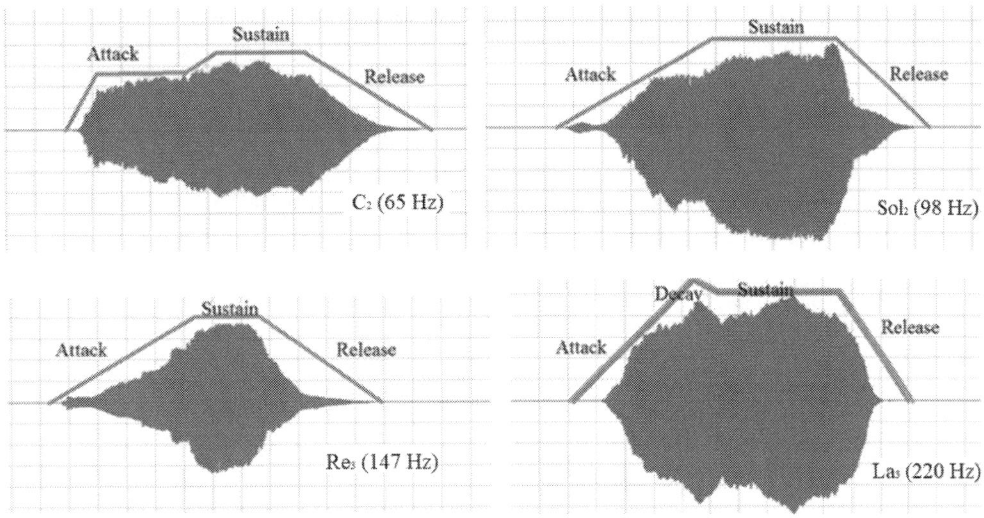

Figure 5. *Carletti* cello (1910) open strings ADSR envelopes for fundamental notes of: 65 Hz (left, top), 98 Hz (right, top), 147 Hz (left, bottom), 220 Hz (right, bottom).

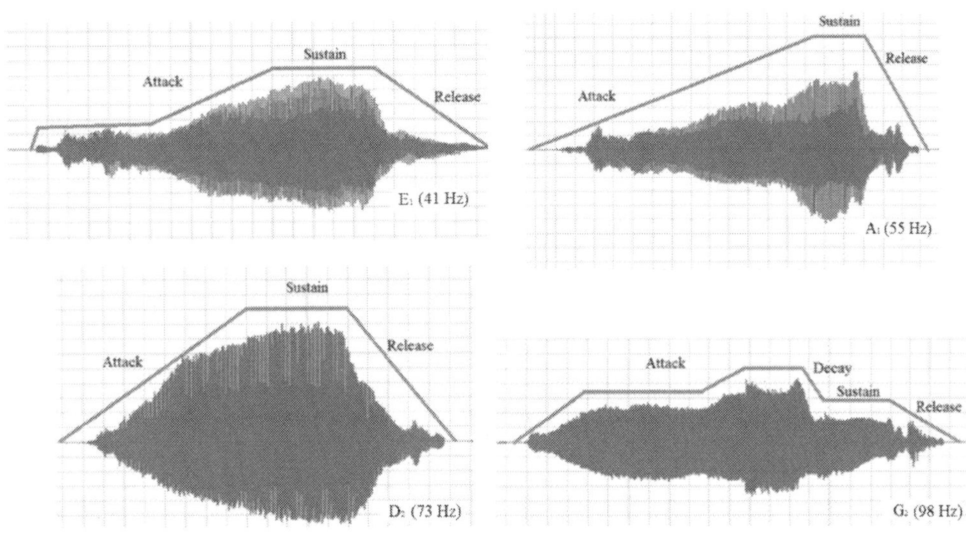

Figure 6. *Verdiani* double bass (2000) open strings ADSR envelopes for fundamental notes of: 41 Hz (left, top), 55 Hz (right, top), 73 Hz (left, bottom), 98 Hz (right, bottom).

3.4 Double bass

The double bass used is a *Verdiani* instrument of 2000. The particular double bass feature, in addition to having retained constructive traits typical of the *viola da gamba*, is that it has a different tuning compared to the others bowed string instruments. In fact, while violin, viola and cello are all tuned in fifth intervals, the double bass shows a fourth tuning; the succession of open strings, therefore, starting from the lowest note to the highest, is E, A, D and G. From a first analysis of the acoustic spectrum of the double bass shown in Figure 2d it can be observed that this does not reach the sound intensity levels detected for the other instruments: the highest level only stands at -36 dB at the frequency of 120 Hz while, for frequencies between 400 and 1000 Hz, where violin, viola and cello had resonance peaks at least -35 dB, the double bass shows an emission even below -45 dB. Even in the case of the double bass it is possible to consider a spectrum of emission larger than the one generally recognized in the literature, essentially between 41 Hz and 247 Hz. Taking into account of natural and artificial harmonics, it is legitimate to consider a spectrum of emission that starts from about 40 Hz and reaches at least 1000 Hz, although with very low intensity. Acoustic spectrum (and spectrograms) justifies the particularly gloomy nature of the double bass and the difficulty often encountered by luthiers in making instruments that have a full, clear and intense sound. The double bass study is also completed by the determination of the ADSR envelopes obtained once again for the open strings of the instrument (Fig. 6). As in the case of the viola, it was not possible to recognize in the envelopes founded the standard form, confirming again that the theoretical proposal respects the real situation only in some cases. In most cases, the phase division of Attack, Decay, Sustain and Release takes on various shapes and is not always present. For the double bass, the only one of the open string to present a four-phase subdivision is the one of G.

4. Conclusive remarks

The study of this limited number of instruments let us already understand that the family of the bowed string instruments is acoustically enough homogeneous, but at the same time is characterized by very different textures. These peculiarities are undoubtedly the ones that over the centuries have given these instruments a special attention by the composers, who have entrusted them with the most varied roles, from brilliant solos to orchestral filling. It is also clear how difficult it is to define the standards that characterize violin family in an objective way: the different types of wood used for their construction, the realization techniques of the various luthier schools, the real dimensions, the thicknesses of the various elements, they are all characteristics which give a certain degree of randomness in the acoustic behaviour of these instruments. That's why musicians often have to look for their ideal instrument for years to find the one that best suits their needs and that best suits their way of playing: they can not trust in a simple physical description of it, because it would not be able to fully represent its potential.

References

[1] Cingolani S., Spagnolo R. Acustica Musicale e Architettonica (in Italian). Torino (IT), *Città Studi Ed.* 2007. EAN 9788825173215

[2] Frova A. Fisica nella Musica (in Italian). Bologna (IT), *Zanichelli*. 1999. ISBN 9788808090126

[3] Pierce J.R. The Science of Musical Sound. New York (USA), *Scientific American Books*. 1983. ISBN 0-7167-1508-2

[4] Rossing TD. The science of string instruments. New York (USA), *Springer*. 2010. ISBN 978-1-4419-7109-8

[5] Woodhouse J. The acoustics of the violin: a review. *Rep Prog Phys*. 2014;77(11):115901.

[6] Dondi P, Lombardi L, Malagodi M, Licchelli M. Automatic identification of varnish wear on historical instruments: The case of Antonio Stradivari violins. *Journal of Cultural Heritage*. 2016;22:968–73.

[7] Zopf SR. Critical study of the use of a length unit in the design of 16th to 18th century Italian violins. *Journal of Cultural Heritage*. 2017;27:S26–33.

[8] Sterling M, Bocko M. Empirical physical modeling for bowed string instruments. *In: 2010 IEEE International Conference on Acoustics, Speech and Signal Processing*. 2010. pag. 433–436.

[9] [9] Vos R., Analysis and reproduction of the frequency spectrum and directivity of a violin, Institut de Recherche et Coordination d'Acoustique/Musique, Paris, January 2003, pp.1-37.

[10] Mitra AK. The violin arcs. *Applied Mathematics and Computation*. 1999;98(1):91–102.

[11] Jansson EV. Violin frequency response – bridge mobility and bridge feet distance. *Applied Acoustics*. 2004;65(12):1197–1205.

[12] Ravina E, Silvestri P, Montanari P, De Vecchi G. Spherical mapping of violins. *The Journal of the Acoustical Society of America*. 2008;123(5):3659–3659

[13] Park H, Lee B, Kim D. Violin Musical Tone Analysis Using Robot Finger. *Procedia Computer Science*. 2016;94:398–403.

[14] Aditanoyo T., Prasetiyo I., Ardhana Putra I.B. Study on vibro-acoustics characteristics of bamboo-based violin. *Procedia Engineering*. 2017, 170:286-292.

[15] Jang HS, Jeon JY. Acoustic characterization of on-stage performers in performing spaces. *Applied Acoustics*. 2016;114:159–70.

IOP Conf. Series: Materials Science and Engineering **364** (2018) 012023 doi:10.1088/1757-899X/364/1/012023

Assessment of thermo-hygrometric indicators for preventive conservation inside museums: in field monitoring and passive microclimatic control strategies applied to "La Specola" museum of Florence

Fabio Sciurpi*, Cristina Carletti, Leone Pierangioli

Department of Industrial Engineering, University of Florence, Via di San Niccolò 93, 50125 Florence, Italy

*fabio.sciurpi@unifi.it

Abstract. The University of Florence has equipped one of the most representative Florentine historical Museum, "La Specola", of a specific microclimatic monitoring system installed in some rooms. The paper presents the results of one-year record of temperature and relative humidity, regarding a room and its showcases. The elaboration and analysis of the microclimatic data have been carried out according to the Italian Standard UNI 10829 and the Performance Index of the most important thermo-hygrometric parameters has been calculated and assessed. Moreover, a dynamic thermal simulation to assess the effectiveness of some passive refurbishment strategies has been carried out in terms of annual and monthly Performance Index and dry bulb temperature variations. The results can be extended to several historical buildings, where poor envelope performances and large windows or skylights imply not acceptable indoor environmental conditions that can even cause damage to the artefacts.

1. Introduction

Indoor environmental conditions play a fundamental role in the deterioration process of the objects exhibited inside museums, and can be strongly affected by the values of temperature and relative humidity, as well as their time and space fluctuations [1]. As a matter of fact, the attitude is oriented to preventive conservation; it implies both "passive" techniques, aimed to minimize the potential damage to the object from the exhibition environment, and environmental monitoring and management directed to minimize variations of indoor parameters and to reduce impacts from outdoor. In the absence of a heating, ventilating and air conditioning system (HVAC), showcases are very important for the artefacts conservation; in general, they work as a filter against environmental attacks due to microclimate variations, chemical pollution and action of micro-organism; moreover, they offer a protection against vandalism, robberies and any direct damage that could come from visitors. Microclimatic monitoring of museum rooms and showcases is fundamental to assess the museum environment suitability to conserve exhibits and to plan any action required to reduce degradation risk [2] [3] [4]. Italian and European technical regulations about cultural heritage conservation [5] [6] establish guidelines and methods to measure indoor temperature, relative humidity and lighting level. Moreover, Guidelines for the Museums operation and management have been drawn up by the Italian

IOP Conf. Series: Materials Science and Engineering **364** (2018) 012023 doi:10.1088/1757-899X/364/1/012023

Ministry for Cultural Heritage [7], giving information about quality standard and technical equipment inside Museums.

In Italy, as well as in the Florence district, there are many important museums located in historical buildings that were built with a different destination, so that their restoration, improvement, safety and accessibility is difficult to achieve and maintain. Bad conservation conditions of the museums are mainly due to: high humidity levels, water infiltration from roof and windows, absence of lighting and solar control systems, poor thermal performance of the building envelope. As concern the microclimatic control systems in the museums, most of all have only a heating system, only some museums, or their specific sections have a HVAC system.

To choose suitable retrofitting strategies many aspects must be considered, and a. global comparative analysis has to be carried out to express a qualitative assessment associated to a particular strategy: effectiveness in terms of indoor climate improvement and reduction of the solar thermal loads, architectural integration, technical feasibility, management of the device in terms of easy maintenance, costs [8].

Aim of the research carried out by the University of Florence is to investigate microclimatic conditions inside museums in historical buildings in Florence by means of a long-term monitoring and the calculation of dedicated performance indices, such as PI_{θ} (%) and $PI_{\Delta\theta 24}$ (%). Moreover, the Research will outline the main problems for the exhibits related to poor climatic indoor conditions, assessing different strategies to refurbish them with a dedicated dynamic calculation code [9]. This research evaluated many different passive and active strategies to guarantee proper conservation on very peculiar and precious objects such as substitution of glasses and drapes, thermal insulation, ventilation.

In this paper the result of this Research applied to a room (Room XIX) of one of the oldest museum in Florence, “La Specola”, are presented and discussed. As conservative problems of the exhibits mainly refer to elevated temperature, the paper focus on the analysis of temperature variation and temperature performance indices both in the Rooms XIX as well as in its showcase.

To evaluate the effectiveness of different strategies, the performance indices of air temperature and relative humidity have been calculated and assessed by means of a dynamic simulation tool and are compared to the base case.

Among the different rooms of “La Specola”, Room XIX has been chosen for applying the methodology of the Research for its critical issues also related to an event that occurred in August 2014 when a cloudburst causes the windows to break and severe water infiltrations.

2. Description of “La Specola” Museum of Florence

“La Specola” is one of the oldest Museum of Florence and is situated at the second floor of Palazzo Torrigiani in the historic centre of Florence.

In about 1520, many thirteenth-century houses were merged by transforming the property into a building that was acquired in 1771 by the Grand Duke Pietro Leopoldo of Lorraine to turn it into the first European scientific museum. After extensive expansion and adaptation works the museum was opened to the public on February 21, 1775.

At present, Palazzo Torrigiani, which is one of the locations of the University of Florence, hosts the museum of Natural History La Specola, the Institute of Zoology and the Institute of Comparative Anatomy, General Biology and Genetics. “La Specola” zoological collections are rich, but only a small portion of the over three million specimens is visible to the public. Inside “La Specola” Museum are collected objects with different conservative requirements; the rooms contain showcases, of great historic and artistic value, in which very different specimen are exposed, such as: Diorama, Protozoa, Mollusca, Insects, Worms, Echinoderms, Carnivores, “Count of Turin” hunting trophies, Rodents, Cetaceans, Monkeys, Birds, Reptiles, Fishes, Anatomic Waxes, etc. The zoological exhibits are particularly sensible to high temperature values that can cause biological growth and activate putrefaction. Relative humidity variations are particularly important for the mammals exhibited that have an inner filling made of gypsum instead of straw; this filling presents a different behaviour then the animal skin and can cause its break due to high temperatures that determine a quick and uncontrolled reduction of relative humidity values in summer periods.

"La Specola" Museum is a building with a stone and brick masonry, wooden floors and turned ceilings. The many enlargements of the Palace determine a great variability of envelope structures about the wall thickness and materials. Shading devices combined with the windows and the skylights respectively consist of external shutters and internal drapes. The Rooms of the Museum have different orientations and dimensions as well as technological and thermal characteristics [10].

3. Microclimatic monitoring campaign

Environmental monitoring inside historical buildings can contribute to prevent damages of the exhibits due to inadequate thermal and hygrometric conditions also suggesting the Curator strategies to carry out in order to improve climatic indoor conditions.

To investigate and control the indoor microclimatic conditions, the University of Florence has equipped "La Specola" of a specific microclimatic monitoring system installed on 2011 in some rooms, with fixed data loggers to collect values of dry bulb temperature (θ) and relative humidity (RH). The monitoring system (table 1), has been described in [4]. In general, the internal measuring positions have been chosen considering the fruition needs of the Rooms, the operating requirements of the instruments (avoiding proximity to heat sources, direct light, direct contact with local disturbance causes that may affect the proper functioning of the instruments), as well as the geometric characteristics of the environments and the wireless connectivity. The loggers, placed in 16 rooms, inside 4 representative showcases and outside the museum, send the measure every 15 minutes to a master connected to a PC thanks to a LAN network. Recordings of each data logger are displayed in the bookshop's PC and in the museum manager's PC, to perform a real-time control of the measured environmental parameters.

Table 1. Main characteristics of the sensors used (http://www. giorgiobormac.com).

	Temperature Reading Resolution/ Accuracy	0,1°C/±0,5 °C
	Temperature Reading Range	- 30 to + 70 °C
	RH Reading Resolution/ Accuracy	0,1%/±2%
	RH Reading Range	0 to 100%

Optimal parameters for objects conservation can be defined both by the Conservators and by technical documents and UNI standards, such as DM 10.05.2001, UNI 10829, UNI EN 15757 and UNI 10969. Based on the climatic history of the exhibits, their material and structural characteristics, the recommended values for the artefacts by "La Specola" Conservators are the followings:

- for Anatomic Waxes: $20°C \leq \theta \leq 22°C$; $\Delta\theta_{24} \leq 1.5°C$; $40\% \leq RH \leq 60\%$; $\Delta RH_{24} \leq 5\%$
- for all other objects: $15°C \leq \theta \leq 24°C$; $\Delta\theta_{24} \leq 1.5°C$; $40\% \leq RH \leq 60\%$; $\Delta RH_{24} \leq 5\%$

4. Monitoring campaign analysis, results and discussion

The monitoring campaign in the Museum started on October 2011 and is still in progress. Data were collected, analyzed and discussed according to the methods for elaboration and analysis of data suggested by UNI Standard 10829. For each room and showcase monitored the temporal profile of temperature and RH, minimum, medium and maximum values of temperature and RH, daily gradients of temperature ($\Delta\theta_{24}$) and RH (ΔRH_{24}) were evaluated. Further on, the "Performance Index" (PI) was calculated; this index expresses a brief evaluation on the quality of the indoor environment in relation to the conservation of the object exhibited on it. It represents the percentage of time during which the measured parameters fall within their ranges considered as "acceptable" [1] [4]. At the moment, acceptable limits concerning different Performance Indices have not been established by Scientific Research on Cultural Heritage conservation.

IOP Conf. Series: Materials Science and Engineering **364** (2018) 012023 doi:10.1088/1757-899X/364/1/012023

Data monitored from a typical year have been analyzed and discussed for some air-conditioned and non-air-conditioned Rooms and for some showcases in a previous paper [4]. In the present paper, data monitored from March 2012 to March 2013 are analyzed and discussed for Room XIX and for the N/W showcase inside this room (XIXs), chosen on the basis of its importance and problems (exposition, great windows surfaces combined with not effective solar shadings).

Room XIX is without any HVAC system, has a not insulated opaque envelope, six large windows with internal drape and without external solar shadings (S/E, S/W and N/W oriented), and preserves relevant "large birds stuffed"; the showcases are placed against the three external walls. Tubular fluorescent lamps are used inside showcases and for artificial lighting of the room.

For one-year monitoring, the time profiles of mean daily temperature and RH measured in room XIX are presented and compared with the corresponding outdoor values (OUT), the showcase values (XIXs) and the accepted values (figure 1).

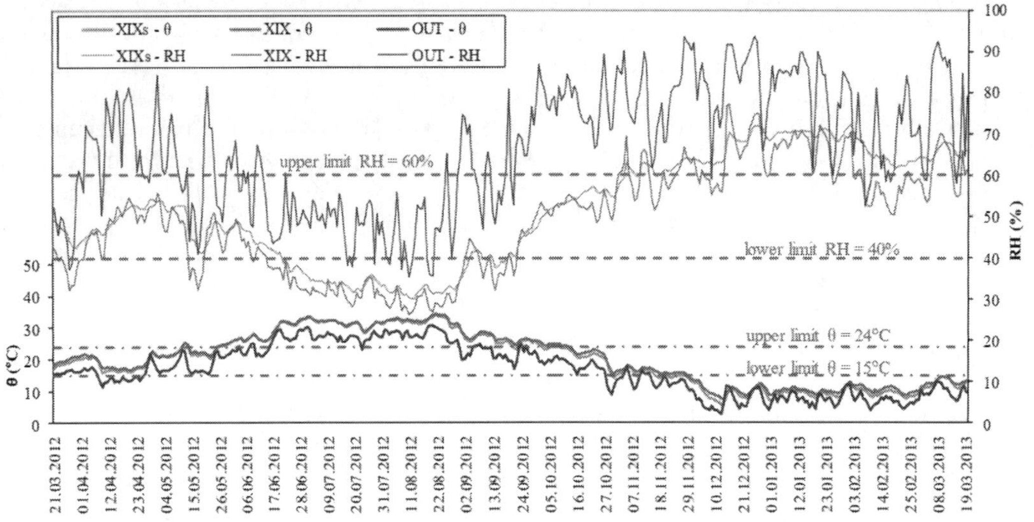

Figure 1. Time profile of mean daily temperature and RH values for one-year monitoring period for Room XIX, showcase XIXs and the outdoor compared with the accepted values.

From data collected it is clear that the temperature and RH trend for the Room XIX and the showcase XIXs follows the outdoor climate, although very dampened in sudden hourly and daily changes. In spring and autumn indoor temperature and RH are within the recommended range, whereas in summer and in winter temperature is respectively above and below the limits and RH is respectively below and above the limits. Particularly significant are the values of temperature inside the showcase that are maintained especially in winter below those of the room with a maximum value of about 3°C in February; as a matter of fact being this showcase placed against a not insulated and N/W oriented external wall, the lack of thermal insulation is reflected in lower temperatures inside the showcase. Faced with an external maximum daily gradient of temperature ($\Delta\theta_{24\ max}$) of 22.6°C, the maximum daily temperature variation in Room XIX is 4.7°C, while in the showcase XIXs is 3.6°C. The annual PI_θ is 32% for the room and 30% for the showcase, while the $PI_{\Delta\theta24}$ is very low for the environment monitored (15% for the room and 26% for the showcase). Faced with an external maximum daily gradient of RH ($\Delta RH_{24\ max}$) of 63%, it is clear the stabilization of this parameter inside the monitored environments (18% for the room and 7% for the showcase). The annual PI_{RH} is 47% for the room and 41% for the showcase, while the $PI_{\Delta RH24}$ is quite high (65% for the room and 99% for the showcase). Although the showcase is not sealed, it reduces RH variations respect those of the room and guarantees a damping of RH changes greater then temperature ones. The RH trend inside the case is related to the indoor temperature trend rather than to the outdoor RH variations. The percentage of time during which the temperature and RH simultaneously fall within their acceptable ranges ($PI_{\theta,RH}$) is only the 26%.

The analysis of the microclimatic parameters measured shows that especially the temperature conditions were not reasonably acceptable for the preservation of the kind of objects exhibited. The most problematic period for the kind of objects appears to be the summer, during which temperatures and their variations are always outside the range recommended by the conservators and can cause important and permanent damages.

5. Microclimate control strategies for museum retrofitting

Aiming "La Specola" retrofitting and to control the temperature values and their variations inside the Rooms, both passive and active strategies have been selected.

For Room XIX passive strategies referring to the improvement of the thermal performances of the envelope and of the solar shading device have been selected and assessed [11]. Wall thermal resistance has been improved with the application of an internal layer of a good insulating material positioned between the showcase itself and the wall. As the windows in Room XIX are single glazed with wooden frame, they have been successfully replaced with more performing glasses such as double glass or solar control glass. Moreover, to guarantee a good uniformity of illumination and avoid overheating effective, internal drapes have been evaluated.

Active strategies applied to Room XIX mainly refer to the improvement of the indoor climate as well as air quality, and consist of a ventilation and air filtration equipment.

With the purpose of evaluating the different actions, a model of the museum was made in Energy Plus®, trough the Design Builder® interface, using the conduction finite difference solution algorithm [10]. The hourly values of the main climatic parameters for the Florence location are from Test Reference Year (TRY) data set developed by CTI (Italian Thermotechnics Comitee).

The single actions are the follow:

- substitution of single clear glass (SG, thermal transmittance of the glass U_g=5.84 W/m^{2}K, total solar energy transmittance of the glass g=0.87, light transmittance of the glass τ_v=0.90) with a new double glass (DG, U_g=1.61 W/m^{2}K, g=0.62, τ_v=0.79) and with a new solar control glass (SCG, U_g=1.61 W/m^{2}K, g=0.21, τ_v=0.40). Thermal transmittance of the frame (U_f) is 2.2 W/m^{2}K;
- substitution of dark internal drape (ID-D, solar transmittance τ_e=0.3, solar reflectance ρ_e=0.1) with s light internal drape (ID-L, τ_e=0.70, ρ_e=0.15);
- introduction of a Demand Controlled Ventilation (DCV) system equipped with filter to prevent pollution due to dust and micro-organisms, to provide night free cooling (air change rate n=h^{-1} from 10 pm to 8 am);
- thermal insulation of external walls with pellets of cellulose blown into the air gap between the showcases and the external walls (ITI, thickness s=0.08m, conductivity λ=0.038W/mK, density ρ=55 kg/m^3, specific heat c_p=2544 J/kgK, vapour resistance factor μ=2). Thermal transmittance of the wall decreases from U=1,53 W/m^{2}K to U=0,36 W/m^{2}K).

The single actions have been combined in 6 different strategies compared with the performance indexes of the base case (BC, single glass, dark internal drape):

- DG/ID-L: double glass with light internal drape;
- SGC/ID-L: solar control glass with light internal drape;
- DG/ID-L/ITI: double glass with light internal drape and internal thermal insulation;
- SGC/ID-L/ITI: solar control glass with light internal drape and internal thermal insulation;
- DG/ID-L/ITI/DCV: double glass, light internal drape, internal thermal insulation and demand controlled ventilation;
- SGC/ID-L/ITI/DCV: solar control glass, light internal drape, internal thermal insulation and demand controlled ventilation.

Room XIX (13m x 13.5m x 6.9m high) has been modelled with 4 external surfaces: 3 walls (s=0.63m; U=1.53W/m^{2}K) and the roof (s=0.05m; U =2.99W/m^{2}K). Since the present analysis focused on the performance of the showcase adjacent to the external NW wall, the heat exchanges are mostly affected by Room XIX and the external environment. The other adjacent thermal zones were considered at the same thermo-hygrometric conditions of Room XIX. The Room (Figure 2) has 6 windows (2 windows on each external wall) oriented to NW, SW and SE. Each window is about 6.3m^2 and has a wooden frame and a single clear glass.

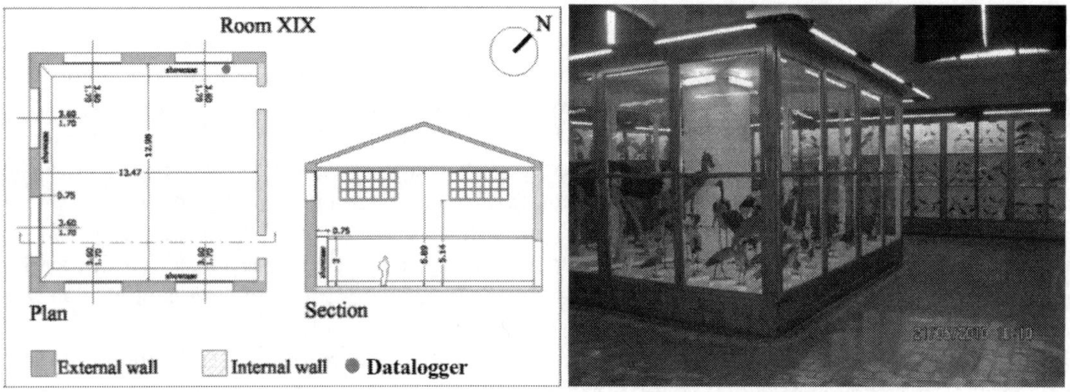

Figure 2. Plan and section of energy model and picture of the Room XIX

Each of the six strategies has been evaluated by means of the following indicators, referred to the air inside the N/W showcase: dry bulb temperature θ (°C), PI_{θ} (%) and $PI_{\Delta\theta 24}$ (%).

In figure 3 the monthly values of PI_{θ} for the analyzed strategies are reported. In figure 4 the monthly values of $PI_{\Delta\theta 24}$ for the analyzed strategies are reported. In figures 5 and 6 the hourly dry bulb temperature values respectively for a typical summer (19 June) and winter day (28 January) for the analyzed strategies are reported. From the analysis of annual values of the above-mentioned indicators, the following general observations can be made: the substitution of current single glasses with solar control glasses implies a significant improvement of $PI_{\Delta\theta 24}$, while the introduction of the Demand Controlled Ventilation system produces a light improvement of PI_{θ}. In particular:

- double glass combined with light internal drape (DG/ID-L) does not change annual PI_{θ} and makes $PI_{\Delta\theta 24}$ worse (from 48% to 39%); the introduction of internal thermal insulation further worsens $PI_{\Delta\theta 24}$ up to 25%;
- solar control glass combined with light internal drape (SCG/ID-L) does not change annual PI_{θ} and enhances $PI_{\Delta\theta 24}$ up to 78%); the introduction of internal thermal insulation does not modify the values of indicators referred to the Base Case (BC);
- double glass and solar control glass further combined with light internal drape, internal thermal insulation and demand controlled ventilation (DG/ID-L/ITI/DCV and SCG/ID-L/ITI/DCV) significantly enhance PI_{θ}, respectively up to 37% and 43%, and deeply worsen $PI_{\Delta\theta 24}$, respectively up to 18% and 26%.

Solar control glass combined with light internal drape is the most effective passive strategy in terms of enhancement of the percentage of time during which the daily variation of the temperature (Δθ24) inside the N/W showcase is less than 1,5°C. This strategy further combined with internal thermal insulation and demand controlled ventilation is the most effective passive action in terms of enhancement of the percentage of time during which the temperature inside the N/W showcase falls within the range of the acceptable values (from 15°C to 24 °C).

The results of the monthly simulations of PI_{θ} (%) and $PI_{\Delta\theta 24}$ reported in figures 3 and 4 agree with the results of the annual simulations.

The results of simulation in term of hourly temperature trend in a typical summer day (figure 5) and winter day (figure 6) confirm that solar control glass combined with light internal drape, internal thermal insulation and demand controlled ventilation (passive cooling active only in summer) is the most effective passive action in terms of summer dry bulb temperature reduction and winter and dry bulb temperature rise in comparison with the Base Case.

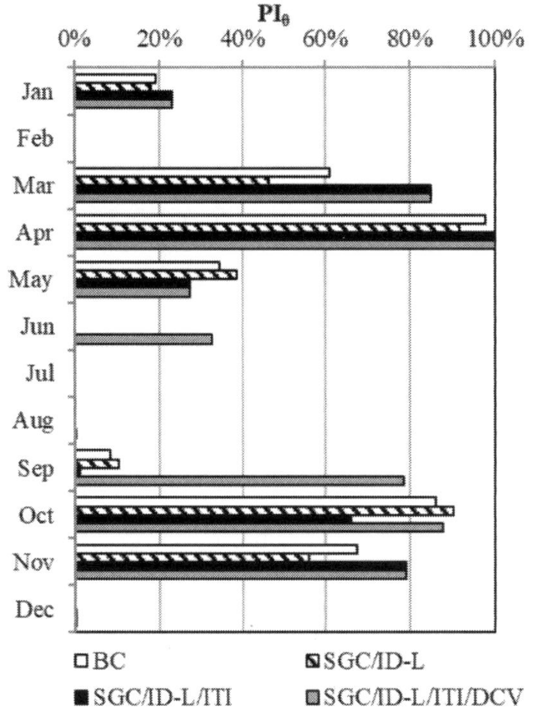

Figure 3. Monthly values of PI_θ for the analysed strategies.

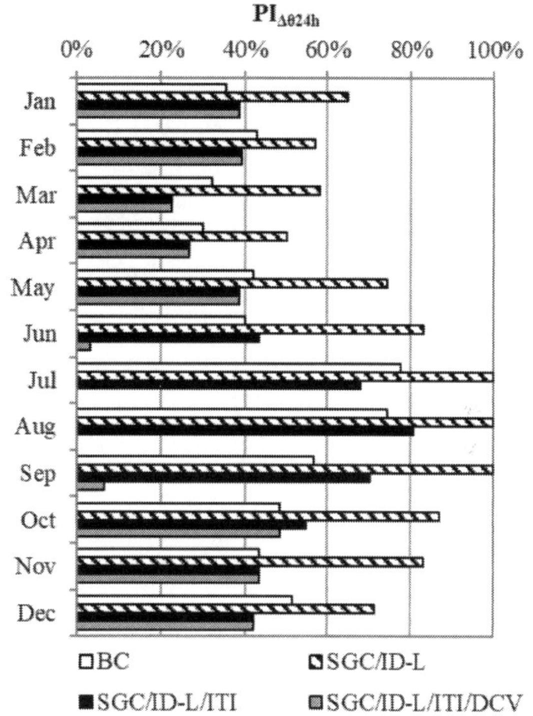

Figure 4. Monthly values of $PI_{\Delta\theta 24}$ for the analysed strategies.

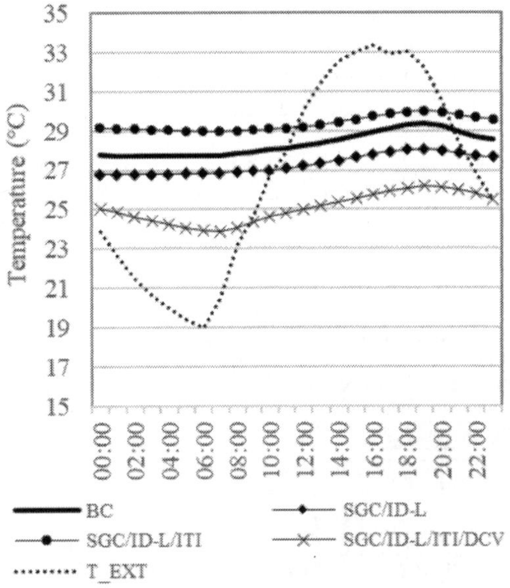

Figure 5. Hourly temperature values in a typical summer day for the analysed strategies.

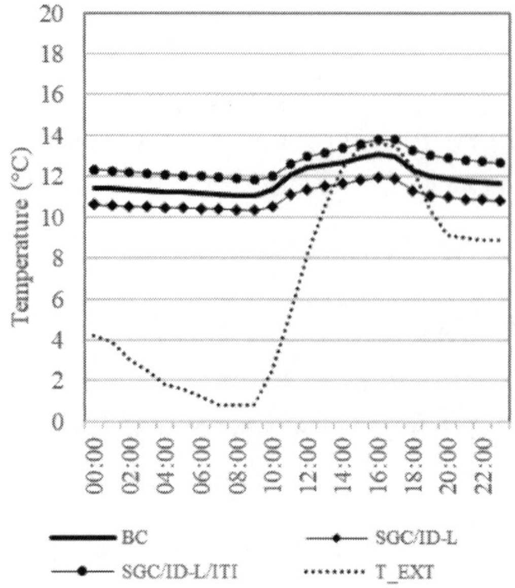

Figure 6. Hourly temperature values in a typical winter day for the analysed strategies.

6. Conclusions

One-year monitoring campaign carried out in "La Specola" Museum of Florence pointed out some critical aspects: temperature and relative humidity conditions were not reasonably acceptable for the preservation of the kind of objects exhibited (15°C<θ<24°C; 40%<RH<60%). Different refurbishment passive strategies have been selected and applied to a room of the Museum (Room XIX). Some combinations have been analyzed and discussed on the basis of the results of dynamic simulations in term of PI_{θ}, $PI_{\Delta\theta24}$ and dry bulb temperature inside the showcases. Solar control glass combined with light internal drape is the most effective passive strategy in terms of enhancement of $PI_{\Delta\theta24}$ while further combined with internal thermal insulation and demand controlled ventilation (passive cooling) is the most effective passive action in terms of enhancement of PI_{θ}. This methodology that combines long term environmental monitoring and dynamic simulation results expressed in terms of Performance Index seems to be suitable when the aim of the refurbishment is the optimal exhibits conservations.

References

[1] Corgnati S P, Fabi V and Filippi M 2009 A methodology for microclimatic quality evaluation in museums: Application to a temporary exhibit *Building and Environment* **44** 1253–60

[2] Pavlogeorgatos G 2003 Environmental parameters in museums *Building and Environment* **38** 1457–62

[3] La Gennusa M, Rizzo G, Scaccianoce G and Nicoletti F 2005 Control of indoor environments in heritage buildings: experimental measurements in an old Italian museum and proposal of a methodology *Journal of Cultural Heritage* **6** 147–55

[4] Sciurpi F, Carletti C, Cellai G and Pierangioli L 2015 Environmental monitoring and microclimatic control strategies in "La Specola" museum of Florence *Energy and Buildings* **95** 190–201

[5] UNI - Ente Nazionale Italiano di Unificazione 1999 UNI 10829. Beni di interesse storico artistico. Condizioni ambientali di conservazione. Misure ed analisi (In Italian). (Milan:

UNI)

[6] CEN - European Committee for Standardization 2010 EN 15757. Conservation of Cultural Property. Specifications for temperature and relative humidity to limit climate-induced mechanical damage in organic hygroscopic materials (Brussels: CEN)

[7] Ministry of Cultural Heritage and Activities and Tourism 2001 Ministero per i beni e le attività culturali. 2001. Ministerial Decree 10.05.2001. Atto di indirizzo sui criteri tecnico-scientifici e sugli standard di funzionamento e sviluppo dei musei (in Italian).

[8] Cellai G, Carletti C, Sciurpi F and Secchi S 2014 Transparent Building Envelope: Windows and Shading Devices Typologies for Energy Efficiency Refurbishments *Building Refurbishment for Energy Performance* Green Energy and Technology ed A Magrini (Springer International Publishing) pp 61–118

[9] Sciurpi F, Carletti C, Cellai G, Muratore V, Orsi A, Pierangioli L, Russo G and Schmidt E D 2017 Environmental monitoring and building simulation application to Vasari Corridor: preliminary results *Energy Procedia* **133** 219–30

[10] Carlucci S 2013 Thermal Comfort Assessment of Buildings (Milan: Springer)

[11] Carletti C, Sciurpi F and Pierangioli L 2014 The energy upgrading of existing buildings: Window and shading device typologies for energy efficiency refurbishment *Sustainability (Switzerland)* **6** 5354–77

IOP Conf. Series: Materials Science and Engineering **364** (2018) 012024 doi:10.1088/1757-899X/364/1/012024

Performance assessment of a heat and moisture dynamic simulation model in IDA ICE by the comparison with WUFI Plus

F Frasca[1*], C Cornaro[2] and AM Siani[3]

[1] Sapienza Università di Roma, Department of Earth Sciences, P.le A. Moro 2, 00185 Rome, Italy.
[2] Università degli Studi di Roma "Tor Vergata", Department of Enterprise Engineering, Via del Politecnico 1, 00133 Rome, Italy
[3] Sapienza Università di Roma, Department of Physics, P.le A. Moro 2, 00185 Rome, Italy.

* Corrisponding author: f.frasca@uniroma1.it

Abstract. Recently, in the field of preventive conservation, the use of accurate whole-building dynamic simulation models is becoming an effective approach for preventing degradation phenomena due to changes in indoor historic climate. Among microclimate parameters, the moisture plays a key role in the degradation of organic-hygroscopic artworks as well as in the durability of building components. Some simulation codes combine both heat and moisture transfer calculations, however their capability to accurately model the moisture transport is limited. The HMWall model coupled with IDA Indoor Climate and Energy (IDA ICE) software is one of those models.
This study aims at comparing the performance of the HMWall model with respect to WUFI Plus, developed by Fraunhofer Institute for Building Physics (IBP). Temperature (T) and relative humidity (RH) provided by both codes in the case of a building envelope with no infiltration, windows and incoming solar radiation, are compared. This allows to assess whether both models calculate the moisture transport throughout walls in the same way. *Dynamic simulations have been run over a year by using different T-RH outdoor conditions.* Even if both models are based on the same heat and moisture transport equations, RH behaviour simulated by HMWall is significantly different from that by WUFI Plus. This mainly depends on the calculation of saturated vapour pressure (p_{sat}) inside the material. Then, the Common Exercise 3 has been applied to test if HMWall were capable to affect indoor RH when cladding materials with different sorption behaviour are used.
The new HMWall implemented model is resulted more effective than the previous one, and in the case of simplified building, RHs modelled by both programs are highly correlated.

1. Introduction

In the last years, a commensurable interest on the use of accurate whole-building dynamic simulation models has been shown in preventive conservation studies, even though up to now this modelling approach has been extensively applied to determine energy efficiency of buildings [1-3] and thermal comfort optimization [4]. At the present time, the whole-building dynamic simulation model is also becoming an effective methodology to assess the microclimate risk on artworks when changes in indoor

IOP Conf. Series: Materials Science and Engineering **364** (2018) 012024 doi:10.1088/1757-899X/364/1/012024

historic climate occur due to refurbishment or climate change [5]. Physical measurements and simulation provide a complete evaluation of the indoor climate and interactions among object-environment and building-environment.

Both temperature (T) and relative humidity (RH) can induce degradation phenomena [6]. Specifically, RH is the main responsible of the deterioration in organic-hygroscopic artworks as well as in the durability of building components. For this reason, an accurate simulation and control of RH behaviour in these sites is of significant importance.

Most of simulation codes has been developed to model moisture exchanges between indoor and outdoor environments setting a specific moisture storage capacity to the interior of the building [7] and not to model the moisture flow between the air and porous surfaces, such as walls [8]. Some of them use two simulations distinctly but running together: one for building energy or envelope simulation and one for modelling the heat and moisture transport between the air and porous surfaces. The HMWall model coupled with IDA Indoor Climate and Energy (IDA ICE) and WUFI® Plus (hereafter called WUFI) belong to the latter group [9], and their performance will be compared in this paper.

WUFI® Plus is a holistic model based on the hygrothermal envelope calculation model developed by H.M. Kunzel's at Fraunhofer Institute for Building Physics (IBP) [10]. It takes into account sources and sinks of moisture inside a component, liquid water transport, diffusion and vapour ab - and de-sorption as well as the thermal parameters [11].Field and laboratory validation are performed for WUFI® Plus more than for other hygrothermal simulation tools [12]. Moreover, it has been widely compared with other hygrothermal software [13-15] and used within the European project Climate for Culture (CfC) [16].

The HMWall model coupled with IDA ICE has been developed in 1999 [17] and, then, updated in 2011. In the first version, the moisture transfer was modelled by one moisture-transfer potential (the humidity by volume), whereas the liquid water transport and the hysteresis of moisture transport were not taken into account. This version, coupled with IDA ICE 3.0, was used within the IEA Annex 41 [18] and to model the hygrothermal behaviour inside historical buildings [19-20]. In 2011, the code has been edited according to the same balance heat and moisture equations of WUFI [10] (section 1.1), mainly due to the lack of the hygrometric data of materials required to run the code. The latter has been validated to conform with the EN 15026:2007 standard validation test; however, no study about this HMWall model has been published yet.

Currently, the HMWall code differs from WUFI for the following features: the percentage of material porosity is not taken into account and hygrometric properties of materials, not available in the software library, have to be extracted from available database and time-by-time compiled in the model object. Hygrothermal curves can only be derived by mathematical relations and not built by using experimental data. Finally, heat or moisture sources (e.g. fraction of driving rain, pipe failures, etc.) cannot be added by the users within the wall component.

This paper aims at investigating for the first time the capability of the HMWall model, updated in 2011 and coupled with IDA ICE 4.7.1 released in 2015, to calculate the moisture transport across walls at increasing complexity of the building structures and boundary conditions. First, a simplified building envelope was modelled in IDA ICE and WUFI and, then, a comparison of time series of RH modelled by both models, was carried out. This has allowed to assess only the moisture transfer across walls, which is the main focus of this exercise. Then, the Common Exercise 3 (CE3) developed by the Fraunhofer IBP in the framework of IEA Annex 41 and also used within the CfC was applied to evaluate the capability of HMWall to simulate the influence of different cladding material in the rooms on the RH.

A successful performance of HMWall model coupled with IDA ICE will provide a validated improvement of IDA ICE packages to both users and experts of preventive conservation of historical buildings and artworks.

1.1. Basic governing equations

Both models are based on the following balance equations for heat (eq. 1) and moisture (eq. 2) transfers [10]:

$$\frac{dH}{dT}\cdot\frac{\delta T}{\delta t}=\nabla\cdot\left(\lambda\cdot\nabla\vartheta\right)-h_v\cdot\nabla g_v \qquad 1)$$

where dH/dT is the heat capacity of the wet material (J/m^3·K); $\delta T/\delta t$ is the change of temperature (T) in time (K/s),; λ is the thermal conductivity of the wet material (W/m·K); $h_v\cdot\nabla g_v$ is the latent heat source, h_v is the evaporation enthalpy of water (J/kg) and g_v is the vapour diffusion flux (kg/m^2·s).

$$\frac{dw}{d\varphi}\cdot\frac{\delta\varphi}{\delta t}=\nabla\cdot\left(g_w+g_v\right) \qquad 2)$$

where w is the equilibrium water content (w) and φ is the relative humidity; $dw/d\varphi$ is the moisture storage capacity of the material (kg/m^3); $\delta\varphi/\delta t$ is the change of relative humidity (φ or RH) in time (1/s); g_w is the capillary moisture flux (kg/m^2·s), i.e. $D\varphi\cdot\nabla\varphi$ with $D\varphi$ as the liquid conduction coefficient of water (kg·m/s); g_v as above, i.e. $\delta_p\cdot\nabla(RH\cdot p_{sat})$ with δ_p as the water vapour permeability of material (kg/m·s·Pa) and p_{sat} as the saturated vapour pressure of water (Pa).
In both codes, δ_p is calculated as the ratio between δ_a (the water vapour permeability of air given by Schirmer's equation) and μ (the dimensionless vapour resistance factor of the material).
Concerning $D\varphi$, the codes use two different equations both reported in [10]. In WUFI, $D\varphi$ is the product between the capillary transport coefficient (Dw), once the water absorption coefficient of the material (A_w) is known, and dw/dRH. In HMWall, $D\varphi$ is calculated taking into account both wet (μ_w) and dry (μ_d) cup vapour diffusion resistance factor. One of the main issues is that μ_w is not available in database and, if any, only for few materials.
The sorption curves are calculating as a function of RH and not dependent on T, since the equilibrium water content (w) is assumed to be little sensitive to T changes [10].
Equations 1) and 2) are strongly related when the number of variables in both equations are limited to T and RH. In fact, it follows that: λ in eq.1 is moisture-dependent; the enthalpy flux is related to g_v so that heat transfer considers the contribution of water vapour phase change; and g_v is dependent on T through p_{sat}.

2. Material and method

In this paper, the examination of the performance of the HMWall model coupled with IDA ICE 4.7.1 has been carried out as shown in the schematic workflow in Figure 1. The workflow consists of test cases at increasing complexity of the building envelope. In this way, it has been possible to assess: a) the capability of HMWall to calculate the heat and moisture transport across walls when the only difference between indoor and outdoor climate is given by RH (first test case); b) *the capability of HMWall to simulate the influence of different cladding material on the indoor RH (second test case). All simulations have been run in transient conditions using hourly step over a year.*

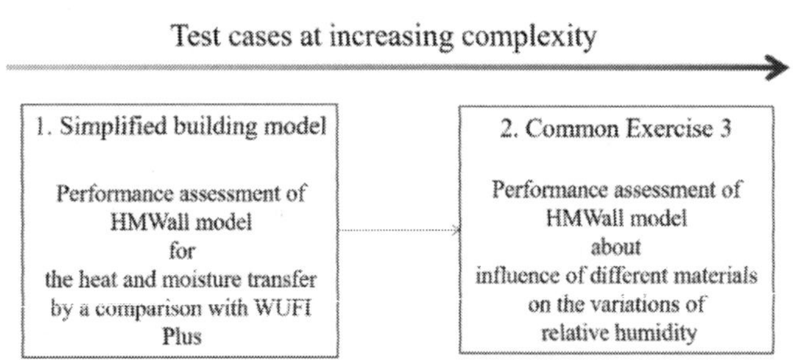

Figure 1. Schematic workflow of methodology applied in this study.

2.1. *A simplified building model*

The first test case is to perform the simulation considering a simplified building having the following characteristics: an envelope of a parallelepiped with a volume of 26.0 m^3, with height of 2.6 m and a length of 2.5 m oriented in east-west direction and of 4.0 m in north-south. It was designed without windows and with a flat ceiling. All opaque components are built by a monolayer of lime silica brick of 0.2 m, whose hygrothermal properties are reported in Table 1. Air changes by infiltration are set to 0.0 h^{-1} and no thermal bridges are taken into account. The initial values of indoor T and RH are 10°C and 50%, respectively. *The file of the external environmental conditions consists of T and RH values taken at 1-hour time slot, so that no contribution from solar radiation and ventilation can affect the heat and moisture transfer calculations. This is very important to avoid misleading with input parameters, such as wind, infiltration or radiation, whose setting varies from a code to another code. Two simulations are run with outdoor temperature T_{out}=10°C and relative humidity RH_{out}=60% (the first simulation) and T_{out}=10°C and RH_{out}=40% (the second simulation). A third simulation is run setting T_{out}=10°C and RH_{out} as a sinusoidal curve (RH_{min}=42.0% and RH_{max}=58.0%), in order to assess the response of both models when there is a seasonal variation of RH over a year. The simulations are performed over the period from January, 1st to December 31st at 1-hour step.* In this test case, variations of RH are mainly driven by water vapour partial pressure (p_v) transferred across walls and by the moisture storage capacity of envelope, since the heat transfer is constant and, consequently, the p_{sat} which is a function of T is constant.

Table 1. Hygrothermal properties of lime silica brick as provided by *MASEA Datenbank - Materialdatensammlung für die energetische Altbausanierung*

Hygrothermal properties	Value
Density (ρ)	1830.0 kg/m^3
Heat capacity (c_p)	850.0 J/(kg·K)
Thermal conductivity (λ)	1.0 W/(m·K)
Wet cup vapour diffusion resistance factor (μ_w)	18.0
Dry cup vapour diffusion resistance factor (μ_d)	27.0
Free water saturation (W_f)	257.1 kg/m^3
Equilibrium water content at 80% rel hum (W_{80})	27.5 kg/m^3
Water absorption coefficient (A_w)	0.059 kg/(m^2·s$^{1/2}$)

2.2. The Common Exercise 3 (CE3)

The Common Exercise 3 (CE3) consists of a double climatic chamber (test room and reference room) [7]. A detailed description of this exercise can be found in [18]. The CE3 aims at simulating the indoor climate of two rooms in order to assess the influence of the sorption of different material on the RH in the rooms. *Four boundary conditions around the rooms are set, so that internal walls are surrounded by controlled T and RH areas, whereas external walls are exposed to weather data of Holzkirchen, i.e. its TRY (Test Reference Year-type) weather data.* Indoor T is controlled by a small radiator (with maximum heat dissipation of 1000 W) at 20±2°C and the moisture production corresponds to 2.4 kg/day. Natural air changes by infiltration correspond to 0.09 h^{-1} for the reference room and 0.07 h^{-1} for the test room, whereas mechanical ventilations are 0.63 h^{-1} and 0.66 h^{-1}, respectively. The exercise consists in 3 steps, in which simulations are run with different cladding materials in the test room: 1) test room only with aluminium foil; 2) test room with gypsum boards on the walls and 3) test room with gypsum boards on the walls and roof. This paper will show the results from HMWall coupled with IDA ICE 4.7.1, to assess the influence of different materials on RH peaks [18], in the case of test room.

3. Results and discussion

3.1. A simplified building model

Figure 2 shows the scatter diagram (HMWall *vs* WUFI Plus) of indoor modelled RH values driven only by the moisture transport across walls. The initial indoor RH is 50% and the boundary RH is 60%. The WUFI RHs tend to increase towards the RH boundary value, whereas the HMWall RHs remain quite constant around initial condition (RH_{max}=50.3%). The slope of linear regression (red line in Figure 1) is 0.03%, demonstrating that the two codes have modelled indoor RH with significant differences. In fact, the RHs modelled by HMWall seem not be affected by the moisture transport across walls, suggesting that no difference between indoor and boundaries RH has been detected.

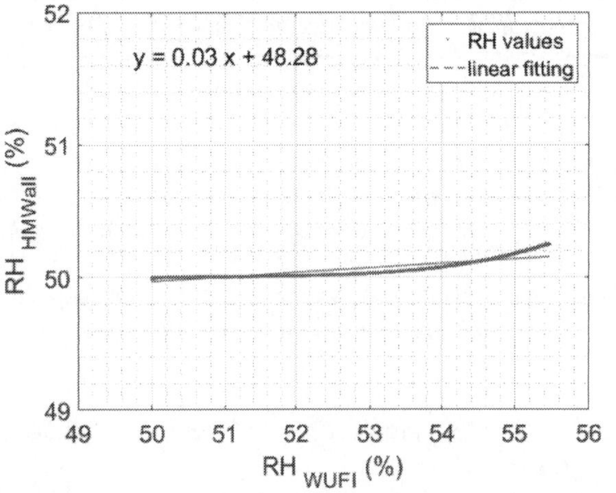

Figure 2. Scatter diagram of simulated RH values (HMWall *vs* WUFI Plus) over a year. The hygrometric boundary condition is set to 60%. Temperature is 10°C, constant inside and outside the envelope.

Since T was constant (T_{in} = T_{out} = 10°C), the p_{sat} was expected to be constant and equals to 12.30 hPa according to Magnus formula [21] for the pressure of water vapour at the saturation. It was observed that the p_{sat} calculated by HMWall was 13.08 hPa.

Moreover, it was found that the codes used a different calculation of the water vapour transfer (g_v) between the wall surface and the boundary air layer close to it. In WUFI Plus, g_v is determined in analogy to that used for the heat transfer, and water vapour transfer coefficient (β_p) is derived from the convective heat transfer coefficient. In HMWall, the calculation of moisture transfer at the boundaries of the building component is different from that in WUFI, since g_v is calculated as the discretized derivative of the vapour pressure between the last layer of wall material and the boundary air layer.

For the above reasons, we have modified this HMWall model (hereafter called HMWall_old) and in order to verify the sensibility of the code at g_v calculated as described by [10] and used in the WUFI Plus, two HMWall codes have been implemented as follows:

- HMWall (A), where only p_{sat} has been updated;
- HMWall (B), where p_{sat} is the same of HMWall (A) and g_v is also calculated as in the same as that in WUFI.

Figure 3 shows the scatter diagrams of the modelled RH values when the boundary conditions are set to RH_{out}=60%. RHs modelled by HMWall (A) and HMWall (B) tend to increase over time towards the equilibrium with the boundary condition in accordance with WUFI. Both HMWall (A) and (B) codes underestimate the maximum RH value modelled by WUFI, as if they buffer the moisture transport towards indoor over time. However, the coefficient of determination (R^2) is 0.994 for HMWall (A) and 0.996 for HMWall (B) (Table 2), showing a visible improvement of both codes with respect to the HMWall_old (R^2=0.771).

When the boundary conditions are drier (RH_{out}=40%) than the indoor climate (Figure 4), RHs modelled by HMWall (A) and HMWall (B) tend to decrease in accordance with WUFI. However, HMWall (A) (left panel in Figure 4) shows a better behaviour with respect to HMWall (B) with a slope of linear fitting close to unity. HMWall (B) (right panel of Figure 4) seems to go in equilibrium with boundaries more rapidly than the other two. The R^2 (Table 2) is 0.998 for HMWall (A) and 0.993 for HMWall (B).

Figure 5 shows the scatter diagrams of the modelled RH values when the hygrometric boundary conditions are defined as a sinusoidal curve. RH values modelled by both HMWall (A) and HMWall (B) are highly correlated with those modelled by WUFI. The slope of both linear fittings is 0.78, showing that both HMWall codes are in accordance to each other, when the hygrometric boundary conditions are more complex than the previous cases whose boundary conditions are constant.

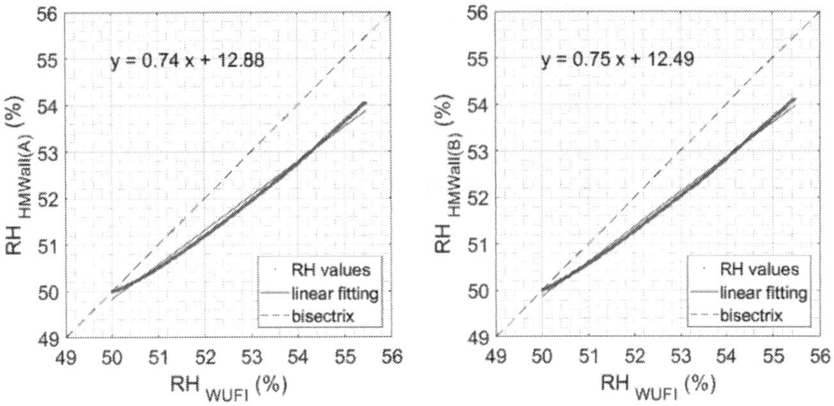

Figure 3. Scatter diagram of modelled RH values (HMWall *vs* WUFI Plus) over a year. The hygrometric boundary condition is set to 60%. T was constant ($T_{in} = T_{out} = 10°C$).

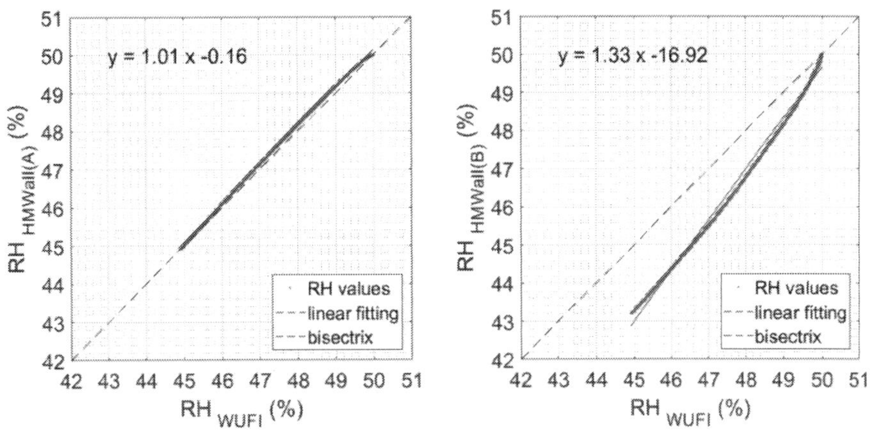

Figure 4. Scatter diagram of modelled RH values (HMWall *vs* WUFI Plus) over a year. The hygrometric boundary condition is set to 40%. T was constant ($T_{in} = T_{out} = 10°C$).

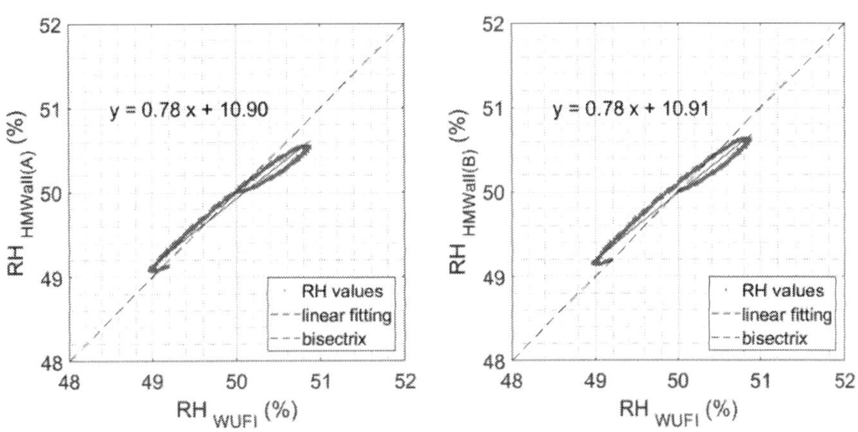

Figure 5. Scatter diagram of modelled RH values (HMWall *vs* WUFI Plus) over a year. The hygrometric boundary condition is a sinusoidal curve with a RH decrease in summer. T was constant ($T_{in} = T_{out} = 10°C$).

Table 2 summarizes the R^2, MBE (Mean Biased Error) and RMSE (Root Mean Squared Error) calculated for each case. Differences between HMWall (A) and HMWall (B) are at the third decimal place, demonstrating that both codes calculate the moisture transport in similar way. Especially, the HMWall (A) seems to be more in accordance with WUFI when boundaries are drier than indoors, since both the MBE and the RMSE are close to zero. Concerning the HMWall (B), it usually underestimates the RH modelled by WUFI except for the case 3 (MBE = 0.0%). For both the HMWall (A) and (B), the RMSE is close to the unity in the case 1 and close to zero in the case 3, whereas a significant difference is in the case 2 when RMSE is 0.1% and 1.2%, respectively.

Table 2. The coefficient of determination (R^2), the mean biased error (MBE) and the root mean squared error (RMSE) calculated for HMWall (A) and HMWall (B) in three different hygrometric boundary conditions.

Parameter	Case 1		Case 2		Case 3	
	A	B	A	B	A	B
R^2 (%)	0.994	0.996	0.998	0.993	0.979	0.977
MBE (%)	-0.8	-0.8	0.1	-1.1	-0.1	0.0
RMSE (%)	1.0	0.9	0.1	1.2	0.2	0.2

3.2. The Common Exercise 3

Three simulations have been run according to the three steps defined in the common exercise: 1) test room only with aluminium foil; 2) test room with gypsum boards on the walls and 3) test room with gypsum boards on the walls and roof. For this exercise, the maximum daily variations of RH modelled by HMWall (A) and HMWall (B) have been calculated in order to assess the influence of different cladding materials on the indoor RH.

Figure 6 shows that the HMWall (B) generally model higher daily span of RH values with respect to HMWall (A) in both rooms, which result to be very similar each other even though walls and ceiling of the test room were coated by aluminium foils, which means no sorption.

Scatter diagrams of the maximum daily RH variations computed for test2 are in Figure 7. In test room (right panel), the daily RH variations modelled by HMWall (A) are affected by gypsum boards covering walls provoking their reduction up to 6% with respect to HMWall (B). Results from test3 are not shown since no significant differences have been detected with respect to test2.

Generally, the maximum daily variations of RH modelled by HMWall (A) are lower than those modelled by HMWall (B).

4. Conclusion

A preliminary comparison between the HMWall model and WUFI Plus has been carried out in the test case of a simplified building. The difference between the two codes mainly depended on the calculation of saturated vapour pressure (p_{sat}) inside the layer material. This bug has been found and solved thank to the accessibility and adjustability of the HMWall code by the users. Moreover, it has been found that HMWall model did not calculate the water vapour transfer (g_v) between the wall surface and the boundary air layer close to it considering the effect of convection. For this reason, two HMWall codes, called HMWall (A) and HMWall (B), have been derived in order to verify the sensibility of the code at g_v calculated as that in WUFI. Both HMWall (A) and (B) show a visible improvement with respect to the previous HMWall code when RH_{out}=60%. The HMWall (A) seems more compatible with WUFI, especially when the hygrometric boundary condition is drier than indoor. In the case of a sinusoidal RH behaviour of boundaries, both HMWall (A) and (B) are quite similar with WUFI.

The main difference between codes can be related to the calculation of the liquid conduction coefficient of water ($D\varphi$). In HMWall, $D\varphi$ only varies as function of the saturated pressure (p_{sat}) and the water vapour permeability of material (δ_p), since μ_w and μ_d do not change over the calculation, whereas, in WUFI, $D\varphi$ varies according to the water content (w) and hence with RH inside the material.

From the Common Exercise 3, a comparison between HMWall (A) and (B) has been carried out. For both HMWall models, aluminium foils seem not to affect the indoor RH, whereas gypsum boards tend to reduce the maximum daily variations of RHs. However, the effect of gypsum boards is more effective in HMWall (A), suggesting that this model would allow a better simulation of sorption given by gypsum.

Further studies will be addressed in order to compare the results, modelled by modified HMWall models, with respect to the measured indoor data (not available at current state). The future goal is to validate the HMWall model coupled with IDA ICE, that better simulates indoor climate, in order to be definitively implemented in the software.

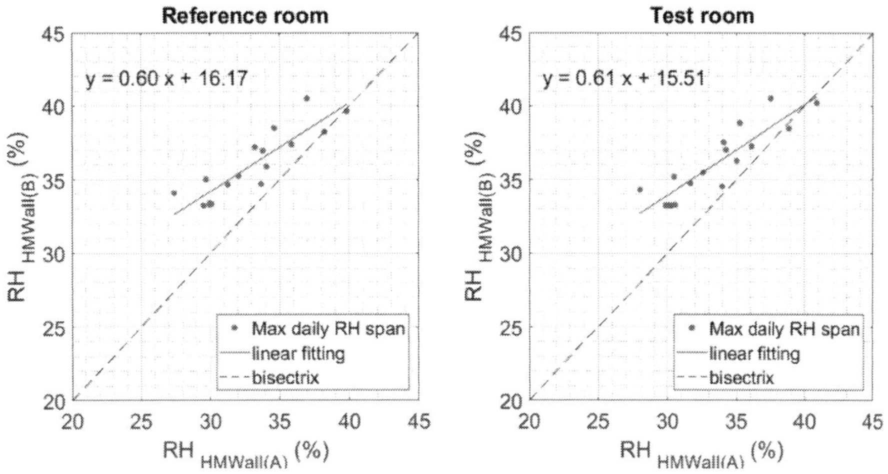

Figure 6. Scatter diagram HMWall (B) *vs* HMWall (A) of maximum daily RH variations modelled in reference room (left panel) and test room (right panel) when walls and ceiling are covered only with aluminium foil (test1). Simulation has been run from January 17th to February 2nd 2005.

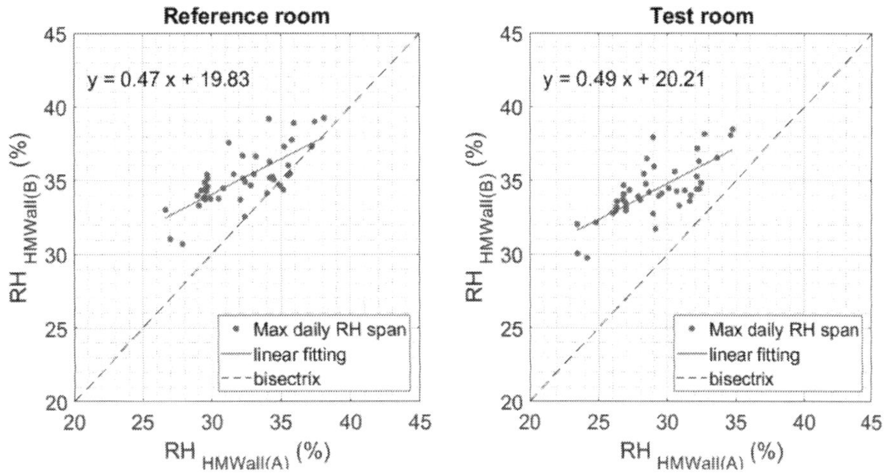

Figure 7. Scatter diagram HMWall (B) *vs* HMWall (A) of maximum daily RH variations modelled in reference room (left panel) and test room (right panel) when walls and ceiling are covered with gypsum boards (test2). Simulation has been run from February 14th to March 30th 2005.

Acknowledgements
Ph.D. student Francesca Frasca thanks Dr. Florian Antretter, Chief Group of Hygrothermic Department of Fraunhofer Institute for Building Physics (IBP), for supporting in simulations by WUFI® Plus.

References

[1] Gigliarelli E, Calcerano F and Cessari L 2016 *EuroMed Conf 2016, Part I, LNCS* Vol **10058** eds Ioannides M et al. (Springer, Cham) pp 91–103

[2] Tronchin L and Fabbri K 2017 Energy and Microclimate Simulation in a Heritage Building: Further Studies on the Malatestiana Library *Energies* **10** 1621

[3] Lucchi E 2016 Multidisciplinary risk-based analysis for supporting the decision making process on conservation, energy efficiency, and human comfort in museum buildings *J. Cult. Herit.* **22** 1079-89.

[4] Roberti F, Oberegger UF, Lucchi E and Troi A 2017 Energy retrofit and conservation of a historic building using multi-objective optimization and an analytic hierarchy process *Energ. Buildings* **138** 1-10.

[5] Leissner J, Kilian R, Kotova L, Jacob D, Mikolajewicz U, Broström T, Ashley-Smith J, Schellen H L, Martens M, Van Schijndel J, Antretter F, Winkler M, Bertolin C, Camuffo D, Simeunovic G and Vyhlídal T 2015 Climate for culture: Assessing the impact of climate change on the future indoor climate in historic buildings using simulations *Herit. Sci.* **3** 1–15

[6] Camuffo D 2014 *Microclimate for Cultural Heritage- Conservation, Restoration, and Maintenance of Indoor and Outdoor Monuments* (Amsterdam: Elsevier) pp 557

[7] Holm A, Kuenzel H M and Sedlbauer K 2003 The hygrothermal behaviour of rooms : combining thermal building simulation and hygrothermal envelope calculation *Build. Simul.* 499–506

[8] Rode C and Woloszyn M 2007 Whole-building hygrothermal modeling in IEA Annex 41. Thermal Performance of Exterior Envelopes of Whole Buildings, *ASHRAE*, Atlanta, 2007, pp. 1–15

[9] Delgado JMPQ, Barreira E, Ramos NMM and de Freitas VP 2013 *Hygrothermal Numerical Simulation Tools Applied to Building Physics* (Springer-Verlag Berlin & Heidelberg) pp 66

[10] Künzel H M 1995 Simultaneous Heat and Moisture Transport in Building Components One- and two-dimensional calculation using simple parameters. Doctoral dissertation. (Stuttgart: University of Stuttgart, IRB-Verlag) pp 65

[11] Antretter F, Sauer F, Schöpfer T and Holm A 2011 Validation of a hygrothermal whole building simulation software. Fraunhofer-Institut für Bauphysik, Holzkirchen , Germany *Ashrae Stand.* 14–6

[12] Corcoran L, Duffy A and Rouholamin S 2013 *2nd Central European Symp for Building Physics,* eds Mahdavi A and B Martens (Vienna: University of Technology - Faculty of Architecture and Regional Planning) pp 305–12

[13] Schmidt S, Lindauer M and Hoppe M 2012 *Proc 5th Int Building Physics Conf (IBPC).* (Kyoto: Kyoto University) pp 1143–48

[14] Ozolins A, Jakovics A and Ratnieks A 2013 *Comsol Conf.* (Comsol Incorporated). 1–6

[15] Barclay M, Holcroft N and Shea A D 2014 Methods to determine whole building hygrothermal performance of hemp-lime buildings *Build. Environ.* **80** 204–12

[16] Bichlmair S, Raffler S and Kilian R 2015 The temperierung heating systems as a retrofitting tool for the preventive conservation of historic museums buildings and exhibits *Energy Build.* **95** 80–5

[17] Kurnitski J and Vuolle M 2000 Simultaneous Calculation of Heat, Moisture and Air Transport in an Modular Simulation Environment *Est. Acad. Publ.* 25–47

[18] Woloszyn M and Rode C 2008 Tools for performance simulation of heat, air and moisture conditions of whole buildings *Build. Simul.* **1** 5–24

[19] Napp M and Kalamees T 2015 Energy use and indoor climate of conservation heating, dehumidification and adaptive ventilation for the climate control of a mediaeval church in a cold climate *Energy Build.* **108** 61–71

[20] Napp M, Wessberg M, Kalamees T and Broström T 2016 Adaptive ventilation for climate control in a medieval church in cold climate *Int. J. Vent.* **15** 1–14

[21] Parish OO and Putnam TW 1977 Equations for the determination of humidity from dewpoint and psychrometric data. *Nasa Technical Note, NASA TN D-8401* 1–24

IOP Conf. Series: Materials Science and Engineering **364** (2018) 012025 doi:10.1088/1757-899X/364/1/012025

Artworks in the spotlight: characterization with a multispectral LED dome

M Ciortan Irina [1], G Dulecha Tinsae [1], Giachetti Andrea [1], Pintus Ruggero [2], Jaspe-Villanueva Alberto [2] and Gobbetti Enrico [2]

[1]University of Verona, Italy [2] Center for Research and Development, Sardinia, Italy

E-mail: `irinamihaela.ciortan@univr.it`

Abstract. We describe the design and realization of a novel multispectral light dome system and the associated software control and calibration tools used to process the acquired data, in a specialized pipeline geared towards the analysis of shape and appearance properties of cultural heritage items. The current prototype dome, built using easily available electronic and lighting components, can illuminate a target of size 20cm x 20cm from 52 directions uniformly distributed in a hemisphere. From each illumination direction, 3 LED lights cover the visible range of the electromagnetic spectrum, as well as long ultraviolet and near infrared. A dedicated control system implemented on Arduino boards connected to a controlling PC fully manages all lighting and a camera to support automated acquisition. The controlling software also allows real-time adjustment of the LED settings, and provides a live-view of the to-be-captured scene. We approach per-pixel light calibration by placing dedicated targets in the focal plane: four black reflective spheres for back-tracing the position of the LED lamps and a planar full-frame white paper to correct for the non-uniformity of radiance. Once the light calibration is safeguarded, the multispectral acquisition of an artwork can be completed in a matter of minutes, resulting in a spot-wise appearance profile, that stores at pixel level the per-frequency intensity value together with the light direction vector. By performing calibrated acquisition of multispectral Reflectance Transformation Imaging (RTI), with our analysis system it is possible to recover surface normals, to characterize matte and specular behavior of materials, and to explore different surface layers thanks to UV-VIS-IR LED light separation. To demonstrate the system features we present the outcomes of the on-site capture of metallic artwork at the National Archaeological Museum of Cagliari, Sardinia.

1. Introduction

Reflectance transformation imaging (RTI) is widely used in the Cultural Heritage field, first and foremost for documentation purposes [1], as well as for enhancing visibility of details with small relief [2] and for the extrusion of the three-dimensionality of the real-word object. The technique is based on capturing multiple images of an object from a fixed viewpoint under different directional illumination.

The acquisition setup can be quite simple, with a camera on a tripod and a hand-held torch that is freely moved around the scene, recovering the illumination direction from highlights on a spherical target (H-RTI). This method is cheap and flexible, but makes light calibration hard and requires careful manual procedures. The counterpart of H-RTI is the dome-based system with higher stability and precision, being based on fixed lights equally sampled in the shape of a hemisphere. Dome solutions are, however, expensive and not too practical to be used for on-site acquisitions. In this paper we describe the design of a novel dome solution using low cost hardware and a reasonably portability making it suitable for both lab and on-site acquisitions. We describe our hardware and software choices, comparing them with prior work. We finally demonstrate the suitability of our system for Cultural Heritage applications, by showcasing an on-site acquisition carried out at the National Archeology Museum of Cagliari, Sardinia.

2. Related work

Several dome solutions have been introduced so far, with different cost requirements, dimensions and running environments [3, 4]. Schwartz et al. [5] presented a complex hardware design and the corresponding calibration and processing procedures for the "brute-force" sampling of a 6D approximation of Bidirectional Texture Functions (BTFs). In addition, their setup is capable of robustly and precisely reconstructing the mesoscopic material geometry, e.g. displacement maps, as well as capturing shape and reflectance of complete 3D objects. The system includes industrial digital video cameras, a rotation stage, LED lamps and projectors. The drawbacks of this arrangement are the high cost and high difficulty of handling outside the lab environment, as well as the need of complex calibration procedure due to many different hardware components. Hameeuw [4] proposed a portable solution extremely easy to use. It is almost a one-click solution that can provide a straightforward RTI acquisition pipeline. The dome is equipped with a 5 million pixel camera or a 29 million monochromatic sensor; it mounts 260 white LED light sources around a 80cm diameter dome. The main design purpose is to make it easy to assemble/disassemble the dome in less than half an hour. This requirement makes it operable both in museum collections or other in-situ scenarios. Recently, other domes have been presented that use both visible and invisible light wavelengths. One example is the Microdome with multispectral RTI capability presented by the RICH team [6]. It is equipped with 228 different LED light sources. Those LEDs are divided in five different spectra: ultraviolet (365 nm), blue (460 nm), red (523 nm), green (623 nm), and infrared (850 nm). A 28 megapixel monochromatic sensor is mounted on top of the dome.

In order to convert the raw image stack into shape and reflectance parameters or high quality relighted images, the calibration of light direction and intensity is required. Over the last decades a lot of methods have been published that deal with different kinds of acquisition setup, and several, non-ideal types of illuminants[7, 8, 9, 10]. Calibration should provide images and metadata suitable for the use with RTI processing/relighting pipeline.

In our work we propose an integrated dome solution coupled with our light calibration and processing software ([11]) able to recover reflectance parameters, relighted images and 3D models and we demonstrate its usefulness in CH applications.

3. Our system

Our dome (Figure 2) features 156 multi-spectral LEDs evenly distributed across 52 light positions over a 60cm diameter hemisphere. All the light sources have a wide aperture in order to stretch the homogeneity of the light intensity as much as possible. The light sources cover 5 bands (see Figure 1): two narrow in the ultraviolet (centered at 395nm) and infrared (centered at 850nm) regions and one broad in the visible to collect the RGB signals of the Nikon D810 DSLR camera with the infrared cut-off filter removed.

The camera is mounted on top of the hemisphere. The dome design has been thought in order to allow for its use in a fixed setup in an arbitrary orientation. This is achieved by integrating all the hardware components to the fixed dome structure, including the camera, as presented in Figure 3. It is possible to scan the objects horizontally, lying on a table by directly placing the dome on top of the acquisition surface (the most common setup for laboratory experiments or even on-site acquisitions, given that the cultural heritage objects allow to be moved) or

IOP Conf. Series: Materials Science and Engineering **364** (2018) 012025 doi:10.1088/1757-899X/364/1/012025

vertically, on a wall by mounting it on a tripod or other similar mounting accessory (appropriate for immutable objects, such as paintings, frescoes, etc.).

The control unit consists of some control boards (Arduino Mega 2560) and three constant voltage constant current programmable control supply power modules (DP20V2A). The boards drive the switch (on/off) of the single LEDs in the dome, while the power modules are used to tune the intensity of the three sets of LEDs (i.e., UV-VIS-IR). The Arduino boards are interconnected and can communicate with an external PC via USB cable using a serial protocol. This protocol allows a PC program to select and turn on one LED at a time; it is mandatory to wait until a LED is off before sending other commands.

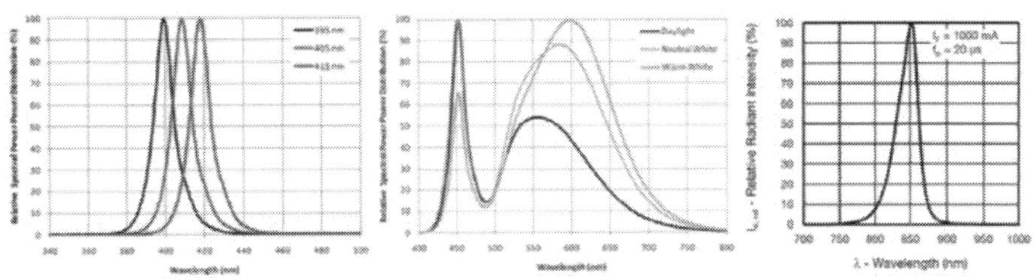

Figure 1. Spectral distribution of the LED spotlights in ultraviolet, visible and near infrared.

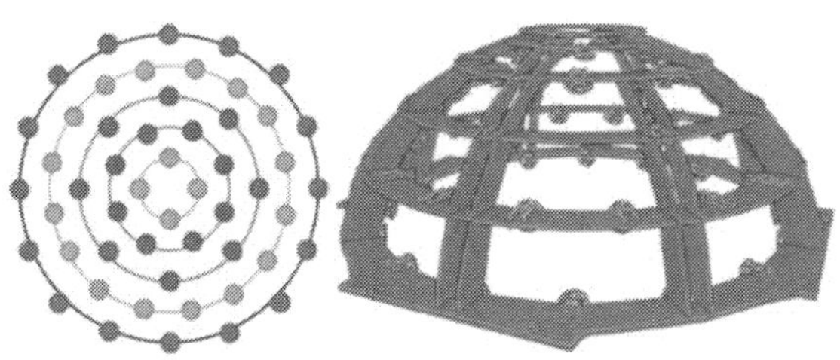

Figure 2. Light position distribution of the dome device.

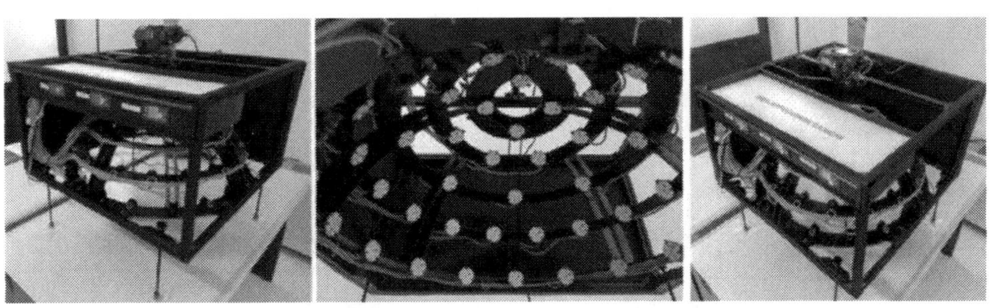

Figure 3. Several snapshots of the dome device.

3.1. Acquisition control software

An user friendly interface has been developed to control the image stack acquisition. The interface enables the easy control of both dome lights (sending commands to the Arduino according to the defined protocol) and of the digital camera by using a software library [12] with access to the camera settings. The GUI allows selection of predefined lights switch and synchronized camera

IOP Conf. Series: Materials Science and Engineering **364** (2018) 012025 doi:10.1088/1757-899X/364/1/012025

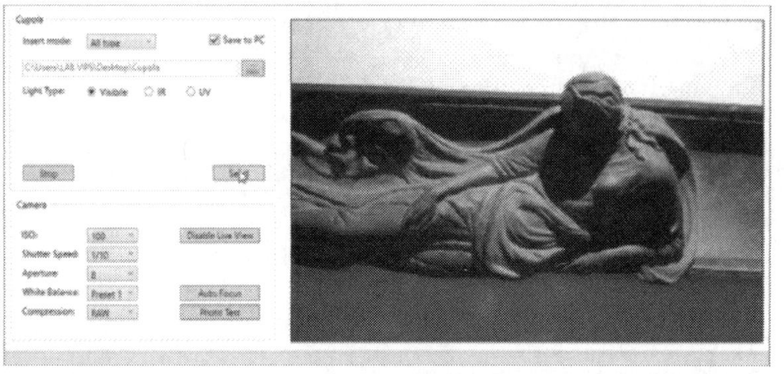

Figure 4. The graphical user interface of the dome system transmitting the live-view of a captured artwork, together with the camera settings.

capture sequences, or individual switches (Figure 4). The user can view in real-time the change of lights and the illuminated object during the acquisition thanks to the live-view window. The live-view feature is useful as well to visually check the image quality when adjusting the camera settings (ISO, shutter speed, aperture, white balance, image format and focus). The camera settings are optimally configured so as to capture sharp images of the object's surface.

3.2. Light calibration

The dome calibration method use methods and algorithms presented in [11], specifically adapted in order to obtain a preliminary estimation of per pixel light direction and reference illumination in the acquisition region that is then used to correct the image stack acquired during the actual object capture. This procedure consists in capturing two calibration targets at the same focal distance used for the object capture: the first includes four glossy spheres at the corners, and the second is a white Lambertian surface placed perpendicularly to the camera axis (Figure 5).

The image stack obtained capturing the first target is used to estimate per pixel interpolated light direction from the position of highlights on the spheres, as shown in [11]. The second image stack is used at the time of object image processing to correct original images with an intensity correction factors proportional to the measured illumination on the white target.

All these calibration procedures can be easily realized with a custom software tool (RTITool) freely available at *https://github.com/giach68/RTITool*. The tool allows semi-automatic sphere annotation and light direction estimate (Figure 6 (a)) as well as the intensity correction of the object images based on the reference images of the target (Figure 6 (b)). The tool features also

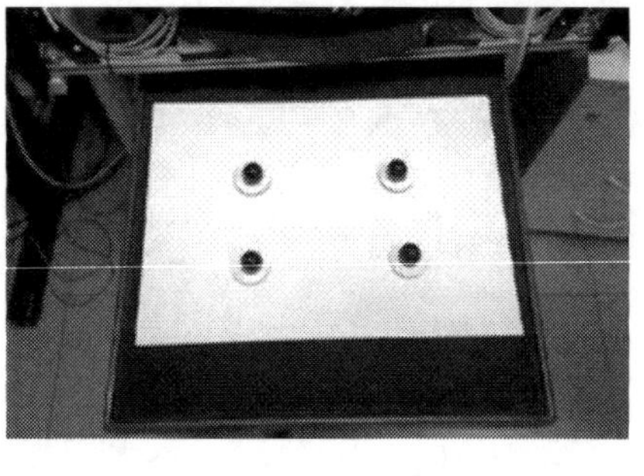

Figure 5. Target with reflective spheres being placed in the acquisition region for light direction calibration.

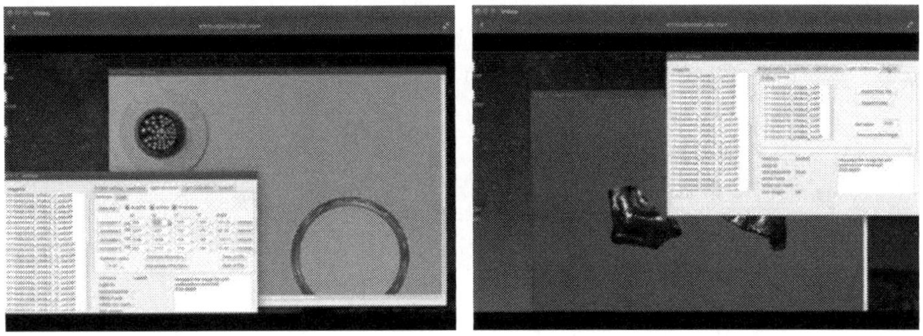

(a) (b)

Figure 6. Calibration software (RTItool) options for the dome system. (a) light direction estimation tab with the annotations of the four black reflective spheres; (b) intensity correction based on the loaded images of the white frame target.

several options for image preprocessing (e.g.undistortion) and on the fly calibration of RTI stacks based on targets directly added to the imaged scene (white frame and spheres).

The light calibration procedures and the interpolated light direction estimation proved to be effective in improving the accuracy of quantitative parameters that can be derived from the image stacks, e.g. normals and albedo [13].

4. Case Study

The dome has been tested on different artworks (paintings, bas-reliefs, coins). In this paper we present a particularly interesting case study related to an onsite acquisition of a metallic artwork at the National Archaeology Museum in Cagliari, Sardinia, Italy. The digitized object is an ancient gold lamina with historical inscriptions. Named "Lamina del Sulcis", it consists of two fragments of an inscription engraved on thin gold leaf "lamina" (a thin plaque or panel intended to be affixed to some other surface), that was found in the earliest stratum of the Sulcis tophet or infant cremation cemetery (West Sardinia). Only 1.4 cm x 1.5 cm x 0.05 cm in size, the whole lamina is thought to have once been attached to an iron object that has partially damaged its surface. The text has been dated to the 8th-7th centuries BCE on the basis of its paleographic features [14]. This makes the object extremely important for the Mediterranean archaeology [15].

In Figure 7 we show the acquisition setup. Before capturing the object images, the calibration procedure with the sphere target and the white target has been performed as described in Section 3. In the object acquisition setup, reflective spheres have been placed as well in order to allow a potential direct calibration from the acquired images. Different acquisition have been performed using different resolutions.

The RTI processing pipeline described in [11] has been used to extract albedo and normals of the lamina's surface, as shown in Figure 9.

Particularly valuable for the analysis of paintings, multispectral signal is also of great importance for metallic objects, and can increase the amount of information provided for a particular artwork. Looking at albedo and normal maps obtained with the three different lights it is possible to see that they capture slightly different information.

From the reflectance properties of the object it is possible to obtain useful information about it, in this case, for example, reading inscriptions. In [11] we have shown that measures like the Outlier Direction map (Figure 10), counting the number of incoming light directions causing non Lambertian reflection in the target object can reveal surface details. Estimating this map from

Figure 7. The acquisition setup: the dome is placed in order to have the object near the center of its sphere with its flatter part perpendicular to the camera axis.

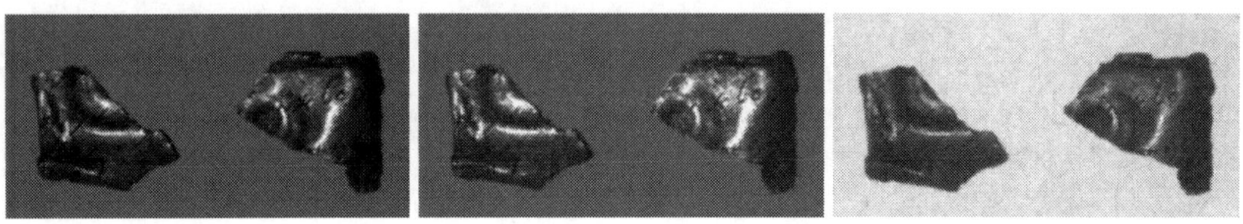

(a) (b) (c)

Figure 8. Example captured images with the different ligths. (a) visible. (b) IR. (c) UV.

(a) (b) (c)

(d) (e) (f)

Figure 9. Albedo and normals of the lamina, estimated with Photometric Stereo on Visible (a,d), IR (b,d) and UV (c,f) images.

PS fitting and mosaicking (manually) image patches in order to obtain an approximate object reconstruction it is possible to visualize the enhanced inscription as shown in Figure 10. It is possible to see that the map obtained with the IR images better enhances the characters, possibly due to the different shape of the specular highlight in the IR frequency reflectance function.

By relighting the image and simulating the appearance of the object from a virtual light direction, an enhanced visualization of the inscriptions is obtained (see Figure 11). In this case we performed relighting with Radial Basis Function interpolation as shown in [16], allowing a

(a) (b)

Figure 10. Outlier Directions maps enhancing local specularity of the surface can enhance useful detail. In this case the IR-based enhancement results in more readable characters.

good representation of the specular component of the reflectance.

Moreover, the recovery of the normals allows for a 3D reconstruction that further on leads to the identification of matching points between the two fragments of the lamina. The shape of the object has been obtained by integrating the input normal vector field. A mask identifies the object within the normal map images (Figure 9). Two types of integral have been performed. The first is a one-dimensional integral along the object/mask contour. Since the integral is defined up to a constant, a random point on the contour is chosen to have depth equal to zero. The boundary condition of the linear integral is the depth value of this point, which is the first and last point on the integration curve. Then a surface integral is computed, using the depth of the contour as the boundary condition. In both cases, an approximation of the second derivatives of the surface is employed to build a sparse Poisson based linear system, and it has been solved through a multidimensional Conjugate Gradient method. Figure 12 shows the re-attachment of the two parts into a whole piece, based on the corresponding points with highest matching probability.

5. Conclusion

The dome system presented in this paper is a flexible and reliable solution that can be extremely useful for the digitization of various types of artworks and underlying materials. By our multiple acquisition experiments both in laboratory and on-site, we have proved that our system is highly portable and this is thanks to the integrated hardware components that can be controlled by a single tool. Apart from the all-in-one controlling software, we have showed how the image stack delivered by the dome is easily compatible with image calibration and visualization tools developed in previous work for different RTI setups, by adding only minor tweaks. Above all, we

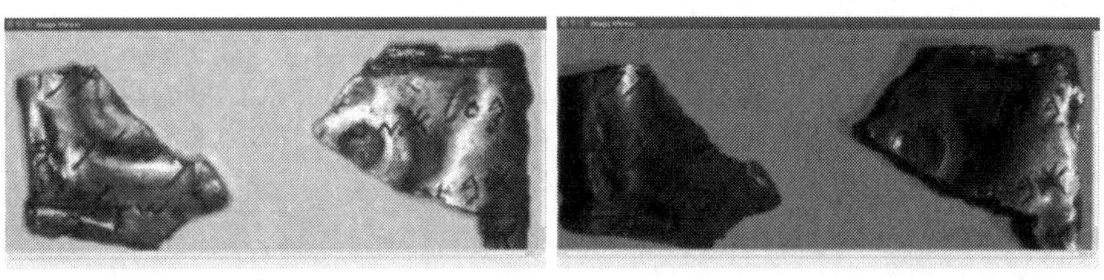

(a) (b)

Figure 11. Lamina images relighted from top center position (lx, ly) = (0,0) and with raking light from right (lx, ly) = (0,0.85) using Radial Basis Function interpolaton.

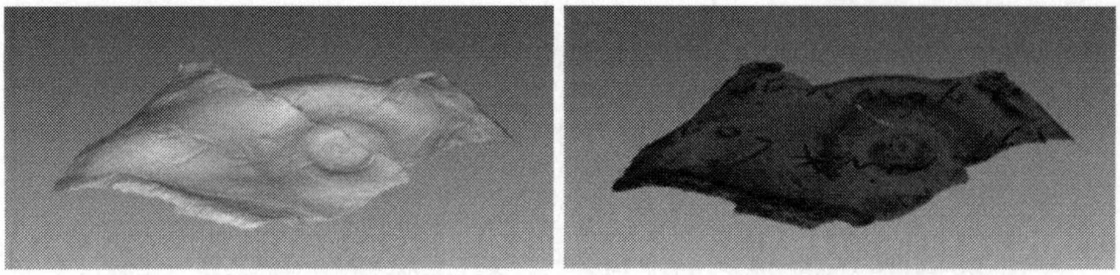

Figure 12. Shaded (left) and textured (right) rendering of the 3D model of the lamina obtained with 3D stitching of the two meshes recovered by normal map integration.

have seen that the multispectral dome in combination with reflectance fitting algorithms leads to revealing results for artworks, from detail visibility enhancement to 3D reconstruction and fragment reassembly.

As future work, we intend to enlarge the portfolio of acquisitions with the dome, by testing it on more artworks, with other properties than what was covered so far. At the same time, we are planning to continuously improve the supporting image processing tools, according to new challenges that may arise, and to extract more visual information based on the multi-light image stacks captured by the dome.

Acknowledgments

This work was partially supported by the Scan4Reco project funded by European Union's Horizon 2020 Framework Programme for Research and Innovation under grant agreement no 665091 and by the DSURF project (PRIN 2015) funded by the Italian Ministry of University and Research. We also acknowledge the contribution of Sardinian Regional Authorities. We thank Plumake SRL, Opificio delle Pietre Dure and the National Archaeological Museum of Cagliari for their collaboration.

References

[1] Mudge M, Voutaz J P, Schroer C and Lum M 2005 *VAST* vol 6 pp 29–40
[2] Palma G, Corsini M, Cignoni P, Scopigno R and Mudge M 2010 *Journal on Computing and Cultural Heritage (JOCCH)* **3** 6
[3] CHI 2017 Cultural heritage imaging URL `http://culturalheritageimaging.org`
[4] Hameeuw H 2015 *Bulletin van de Koninklijke Musea voor Kunst en Geschiedenis* **84** 5–48
[5] Schwartz C, Sarlette R, Weinmann M and Klein R 2013 *Material Appearance Modeling* pp 25–31
[6] LEUVEN K *URL: https://portablelightdome. wordpress. com/2015/04/29/rich-presents-the-new-multispectral-microdome* **2**
[7] Powell M W, Sarkar S and Goldgof D 2001 *IEEE Transactions on Pattern Analysis and Machine Intelligence* **23** 1022–1027
[8] Ackermann J, Fuhrmann S and Goesele M 2013 *VMV* pp 161–168
[9] Takai T, Maki A, Niinuma K and Matsuyama T 2009 *Computer Vision and Image Understanding* **113** 966–978
[10] Wong K Y K, Schnieders D and Li S 2008 *European conference on computer vision* (Springer) pp 631–642
[11] Giachetti A, Ciortan I, Daffara C, Marchioro G, Pintus R and Gobbetti E 2017 *Computer Vision and Image Understanding*
[12] Nikon 2017 Sdks for digital imaging products URL `https://sdk.nikonimaging.com/apply/`
[13] Giachetti A, Daffara C, Reghelin C, Gobbetti E and Pintus R 2015 *SPIE Optical Metrology* 95270B–95270B
[14] Barreca F 1965 *OA», IV* **1** 1–5
[15] Dixon H M 2013 *Phoenician Mortuary Practice in the Iron Age I-III (ca. 1200-ca. 300 BCE) Levantine" Homeland"* Ph.D. thesis University of Michigan
[16] Giachetti A, Ciortan I M, Daffara C, Pintus R and Gobbetti E 2017 *Eurographics Workshop on Graphics and Cultural Heritage* ed Schreck T, Weyrich T, Sablatnig R and Stular B (The Eurographics Association) ISBN 978-3-03868-037-6 ISSN 2312-6124

IOP Conf. Series: Materials Science and Engineering **364** (2018) 012026 doi:10.1088/1757-899X/364/1/012026

Technological improvements in creating controlled thermo-hygrometric conditions in sealed microenvironments:the Dew Point Climatic Generator

Ottaviano Allegretti[1] and Paolo Dionisi-Vici[2]
[1] CNR-IVALSA, via Biasi 70, I-38010 San Michele all'Adige (TN), Italy
email: allegretti@ivalsa.cnr.it , corresponding author
[2] Consultant, via G. Verdi 7, I-54100 Massa (MS), Italy
email: paolo.dionisivici@gmail.com

Abstract. In this contribution a new methodology of climate control, specifically designed for Cultural Heritage artifacts storage microenvironments, is presented. Standard conditioning systems work using the subtraction of vapor in the air by means of condensation on cold surfaces as a drying strategy and the addition of water vapor, obtained in different ways, to humidify the air. Temperature (T) and Relative Humidity (RH) are controlled by two different control loops.
Using this new system, currently in patent pending status, it is possible to obtain a wide range of T and RH values circulating the air flow (via a common air circulation pump) through a water gurgling device and a heating device, using the ΔT between the water temperature in the gurgling device and the set temperature of the heating section.
The system is very robust and straightforward: the sensible-latent transformation of the air flow in each one of the 4 possible directions (heating-drying, heating-humidification, cooling-drying, cooling-humidification) is obtained by two passages operated by only two devices (bubbler device + heating device) controlled by two temperature probes, one in the bubbler and one in the heating section.
The system has been successfully implemented in some different case-studies in Italy and in the United States.

1. Introduction

Climate stability is a benefit for the preservation of works of art made using hygroscopic materials, and especially wooden objects like wooden scuptures and panel paintings [1].

Even though the debate about relaxation of climate standards is ongoing in order to reduce the carbon footprint of the Museums all over the world[2], the need of sealed microenvironments with controlled climatic conditions is crucial for many reasons:

- a lot of objects are very sensitive and reactive to microclimate fluctuations, therefore needing to be exhibited in actively controlled display cases;
- the global size of loans for temporary exhibitions makes compulsory to reduce the climate induced stress to the objects, as specifically requested in the contracts between lending and borrowing institutions.

1.1 State of the art

Conventional air conditioning systems typically use a battery for cooling and condensing air, an adiabatic saturator and two heating batteries, one upstream and one downstream of the adiabatic saturator respectively.[3]

In various types of climatic chambers, vapor is directly produced by an atomizer.

Often these systems have modest energy efficiency and control precision.. Such systems can be improved if they are equipped with modulating valves, which allow to control the recirculation of the air and to partialize the air flows, so as to increase the energy efficiency and the precision of the control. Despite this improvement, energy efficiency and control accuracy remain non-optimal.

Moreover, in traditional dry-exchange heating coils, in many operating conditions the components of a traditional air-conditioning system work in opposite directions with negative consequences on the stability and efficiency of the system itself.

Another approach to climate conditioning system currently on the market is specifically used for conditioning display cabinets or museum display cases. In such system passive control systems use buffers of suitably conditioned hygroscopic substances (e.g. silica gels). Even if there are advantages due to simplicity and affordability, there are disadvantages due to the impossibility of controlling the temperature inside the display cabinet, and operating time limits requiring frequent interventions to recondition the hygroscopic substances are needed.

Furthermore, simplified active systems with a single cooling coil, for example a Peltier cell, are often used: in addition to having a very low efficiency they act only as cooling and dehumidification system with consequent evident operating limitations and a poor or non-existent temperature control.

1.2 Designing goals of the Dew Point generator

The described device[4] aims to provide a climate generator based on an innovative climate conditioning method, assuring an air flow with set Temperature and Relative Humidity conditions stable at the outlet, having less complexity, greater reliability, a lower production cost. and reduced maintenance requirements.

In summary, the device is based on the principle of controlling the T and RH parameters of a micro-environment through the production of an air flow having the target values of Temperature and Relative Humidity, in which the target values are achieved by bringing an air flow to be conditioned, whenever initial T and RH values are not known, at its dew point at a predetermined controlled temperature - different from the target temperature - and reaching the target RH by heating the supply air flow at the dew point in order to achieve the target T.

At the target T and RH values the mixture of air (air-vapor) thus defined has a determined dew temperature.

Therefore, in order to obtain the target values of T and RH of the air flow, the right amount of steam required is reached by saturating the air flow at the dew temperature through a latent/sensitive thermodynamic transformation. The saturated air (100% RH) and the dew temperature is then heated to the target temperature (undergoing a second sensitive thermodynamic transformation), and consequently the RH decreases to the target RH value.

The saturation step to get to the dew point takes place in a bubbler where water is kept at a controlled temperature in order to bring the air flow to the T corresponding to its dew point.

The control is then carried out simply by checking two temperatures: the temperature of the water in the bubbler and the target temperature. The target RH is a function of the difference between the dew-point temperature and the target temperature. The system therefore does not require RH sensors to control this parameter, as it does not require a condenser plus a saturator for regulating it, since the bubbler, whatever the characteristics of the incoming air (to be conditioned) are, allows to obtain the right amount of steam by condensing or releasing it.

The climate generator can be sized to work with latent or sensitive high thermal loads, or it can be miniaturized to operate in tight micro-environments such as display cases or museum display cases.

The climate generator also allows to obtain air free of dust and other pollutants in solid and gaseous phase.

With a simple construction variant, the climate generator can proficiently operate as a dynamic climate conditioning system which allows to perform very rapid, almost instantaneous humidity variations under isothermal conditions.

2. Detailed process description

With reference to figure 1, a first set up of the invention is illustrated in which a climate generator - indicated in the assembly with *G* - includes a bubbling device, adapted to receive a flow of air G_a to be conditioned and containing a predetermined volume of water *W* at a controlled temperature in a closed vessel or open vessel water circulation circuit. In the first case, the bubbler is a thermally insulated closed container, with pressure seal. The goal of the bubbler is to make an exchange of heat and humidity between the aforementioned volume of water at controlled temperature (thermostated) and the flow of air in transit from T and RH values of the incoming air flow not known to a value of dew T and RH of 100% of the outgoing air flow.

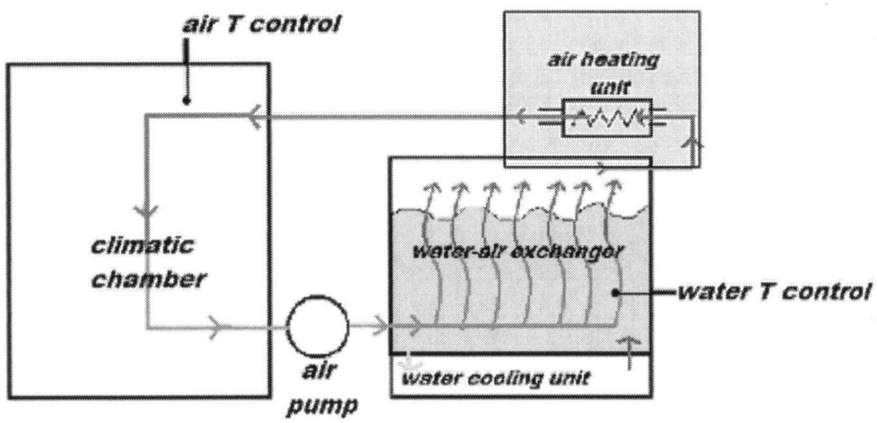

Figure 1. Design scheme of the Dew Point Climate Generator

The temperature of the water volume *W* in the bubbler is regulated according to the T and RH values required by the generator, as will be clearer in the following description, by means of heat exchanger, for example a heat exchange battery which can make a direct exchange if positioned inside the bubbler, or indirect exchange if positioned externally to the bubbler, when the latter is made with an open vessel water circulation circuit. The heat-exchanging devices are more generally responsible for a thermodynamic transformation of the volume of water *W* into the bubbler and of the flow of air to be conditioned through it, including the use of a latent amount of vaporization heat associated with the transformation from liquid to vapor phase in the volume of water *W* responsible for modifying the relative humidity characteristics of the air flow to be conditioned.

In order to obtain a good temperature control, the temperature is controlled by a Proportional-Integral-Derivative control, commonly abbreviated as PID. The climatic generator further includes parts for supplying the flow of air G_a to be conditioned to the bubbling device, which includes an air duct, having for example a tortuous path such as a coil provided with a high density of micro holes positioned on the bottom of the bubbler in such a way as to fractionate the air in transit in the volume of water *W* into bubbles which allows an optimal sensitive and latent exchange between air and water, or in order to optimally saturate the air and cool it to the dew temperature. It is connected to an air pump suitable for circulating the air in the duct and producing enough pressure to allow the air flow to overcome the pressure of the water column formed by the volume of water *W* present in the bubbler.

The air pump has also the purpose of providing air circulation, pushing the air up into the micro-environment to be conditioned.

More specifically, the pump has an inlet particulate filter and operates with a capacity and pressure sized to compensate for pressure losses and to overcome the pressure of the water column of the bubbler; for instance, the required minimum flow must be great enough to guarantee sufficient exchange and distribution of the air in the micro-environment to be conditioned and climatic stability, based on the latent-sensitive load in the environment and at the desired rate of achievement of these conditions . For example, an ultra-stable air-conditioning pump in a museum case is sized for around 100 air exchange/hour (determined by the ratio between pump flow and casing volume). Empirical tests showed acceptable performance around 1 air exchange / hour (using a pump flow rate of 1 m^3/h for a 1 m^3 case). For museum display cases placed in stable conditions and requiring specific restrictive effects on sound emissions, lower values may be sufficient.

The bubbler is designed and sized in order to optimize the height of the water column, of the water volume W and of the minimal size of the bubbles as a function of the air flow to be treated, so that the air-water exchange efficiency is maximized.

In the bubbler, the air flow G_a bubbles through the volume of water W present, saturating it with water vapor and reaching the temperature of the volume of water, an air flow collector is located in the upper part of the bubbler, suitable for canalize the flow of saturated air G_b towards the micro-environment to be conditioned by the climate generator.

Heating devices, such as a heating coil, are arranged downstream of the bubbler and suitable for heating the flow of air G_b at the outlet of the bubbler before being pushed into the climatic chamber.

In a cyclic setup, the deteriorated air flow G_a is taken from the climatic chamber to be re-conditioned.

In the currently preferred setup, shown in figure 2, the bubbler is a watertight, pressure-tight, thermally insulated container; its volume is filled with 3/4 of water thermostated to the dew point temperature (T_r) of the air flow. The container has a sealed cap on the upper side for topping up the water volume and the inspection. The container is connected on the upper side with two ducts, respectively for air supply and outflow. The air supply duct reaches the bottom of the container where it is connected to a porous tube. The air outflow duct is equipped with a goose neck trap to prevent any liquid being transported with air up to the heating device. The shape and volume of the container are sized so that the thermodynamic transformation of the incoming air is complete. Water volume should be oversized, in order to reduce the variations in water level.

The porous tube is preferably placed on the bottom of the container and connected to the air flow supply duct. The porous tube allows to introduce the air flow into the water as small bubbles, increasing the surface/volume ratio of the bubbles as much as possible in order to maximize the air/water exchange.

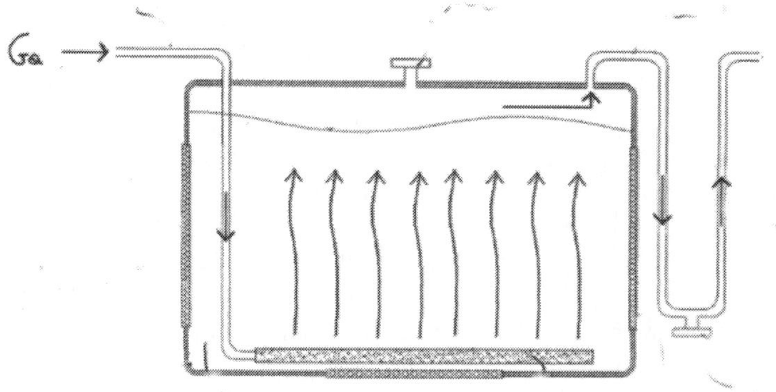

Figure 2. Design scheme of the bubbler

The heat-exchanging device is a system for thermostating the volume of water W of the bubbling device, and includes heat exchange batteries placed directly inside the container or externally in open vessel setup. The heat exchange batteries are made using a coil of thermally conductive material (such as copper) in which a refrigerated gas circulates from a refrigerating unit, or by means of peltier cells integrated on the walls of the container, which have a reduced efficiency but no working noise (which is a benefit in exhibiting locations).

The heating section is a thermostating system for the air aiming to heat the air flow coming out of the bubbler from the dew point temperature and at 100% RH to the target temperature value. It involves e.g. electrical resistances located inside a thermal insulated duct and air ducting towards the case, in which the electrical resistances are arranged near the display case or directly inside the case.

During the passage in the bubbling device, the incoming air flow G_a, which has an enthalpy value h_a, an absolute moisture content x_a,, temperature t_a and relative humidity i_a, undergoes a sensitive transformation, i.e. it is heated or cooled to a temperature t_b, and is saturated with water vapor making a latent transformation, whereby the output air flow G_b has enthalpy value h_b, absolute humidity x_b, temperature t_b and relative humidity $i_b = 1$.

Thereafter, the flow of air G_b is heated from the temperature t_b to the temperature t_c, to obtain the conditions of the air flow produced by the climate generator G_c conditioned to the desired values of enthalpy h_c, with an absolute humidity x_c, temperature t_c and relative humidity i_c, respecting the following proportions $xc = xb$, $tc \geq tb$ and $ic \leq ib$.

By setting the temperatures t_b and t_c, it is possible to obtain the desired climatic conditions at the set values of controlled temperature t_c and of controlled humidity i_c. In fact, once the desired controlled temperature t_c is established, the desired relative humidity condition is obtained by adjusting the water temperature of the bubbling device at the dew point temperature t_r of the air flow, i.e. by adjusting $t_b = t_r$, that is a function of the desired temperature t_c and of the desired relative humidity i_c.

The RH control i_c is then carried out by controlling the water temperature in the bubbling device, which is assumed for simplicity to correspond to the temperature of the air flow exiting from there at t_b. For this purpose, a function $t_b = f(t_c, i_c)$, in which the temperature value t_b is obtained as a function of the controlled temperature t_c and the controlled relative humidity i_c, can be directly implemented in the control. This function is expressed by the following formula, also known as Magnus equation[5]:

$$t_b = \frac{237.7 \cdot \dfrac{17.27 \cdot t_c}{237.7 + t_c} + \ln i_c}{17.27 - \dfrac{17.27 \cdot t_c}{237.7 + t_c} + \ln i_c} \tag{1}$$

where the symbols are previously specified and *ln* is the natural logarithm, and the calculated isotherms are displayed in the chart of figure 3.

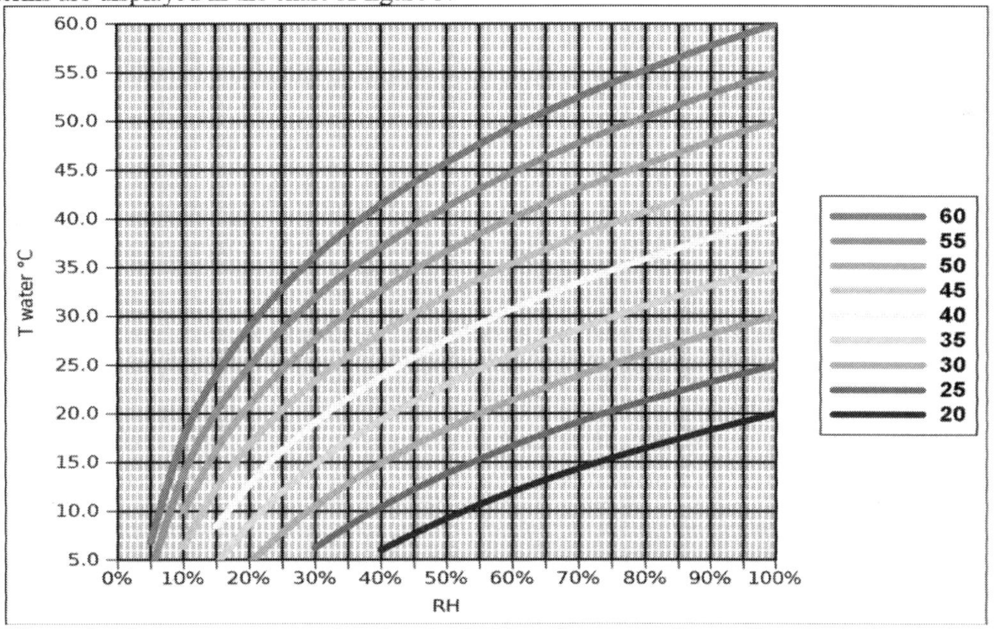

Figure 3. The curves representing the obtainable values according to formula 1. On x axis the i_c target values of relative humidity, on y axis the t_b values at the outflow of the bubbler. On the right legend the family of curves at constant t_c is displayed.

In an alternative setup as a dynamic climatic generator, the bubbling device includes a first bubbler and a second bubbler, the latter maintained at different water temperature conditions with respect to the water temperature of the first bubbler, e.g. through the use of the heat exchange coil which, by means of a modulating valve adapted to regulate a flow of water intake in the first and second bubbler respectively, as a function of the water temperature, serves both the bubblers. Means for diverting the flow of air to be conditioned, such as for example a three-way valve, are adapted to selectively deviate the flow of air to be conditioned to the first or to the second bubbler.

Using this setup, it is possible to obtain a system able to make sudden changes in relative humidity under isothermal conditions. Such a dynamic generator is necessary for some types of laboratory tests, as in the execution of some types of environmental simulation tests.

The advantage achieved by the climate generator and the corresponding climate conditioning process of the DPCG consists in the fact that only two transformation steps are necessary to control the climate of an environment. Thanks to this, a minor complexity, a greater reliability, lower production costs and reduced maintenance of the climate generator are obtained, despite of having a wide range of use and a stable and accurate control of the climatic parameters of T and RH.

Due to the need for dynamic systems where rapid and wide variations are required, the climate generator is highly efficient and has much lower manufacturing costs compared to known laboratory equipment, which are mainly constituted by a double climatic chamber served by a double climate control system .

Moreover, the treated air will be free of dust and pollutants thanks to the fact that the air passes into the water in the bubbler.

3. Case studies

The Dew Point Climatic Generator has been succesfully used during the restoration of a big panel painting by Jakob Seisenegger stored in Trento, Italy[6]. During the structural intervention aiming to re-design the crossbeams system, the technical need of measuring on site the response of the panel at different climate conditions and to follow the dynamics in achieving a new equilibrium with the surrounding microclimate has been possible thanks to the stability assured by the DPCG in keeping a setpoint and to its rapidity in passing from one setpoint to another.

Some tests (unpublished) carried out at the Metropolitan Museum of Art of New York on the response of dummies simulating panel paintings to temperature step variations at constant relative humidity were successfully completed thanks to the stability of the microclimate provided by the DPCG: the variations in moisture content of the dummies were so small to be otherwise unreadable using standard climate control systems because of the noise induced by the intrinsic micro-fluctuations.

4. Conclusions

An innovative climate control system specifically designed for museum display cases is described.

Its use of dew point as the starting point to produce an airflow at the desired T and RH conditions is intrinsically stable because not based on a feedback approach, which usually induces microfluctuations around the set point in traditional systems: using the traditional approach the required parameters are achieved by two systems working in parallel and often in opposition.

The tests carried out during the monitoring of a big panel painting in Trento and on dummies in a test climate chamber at the Metropoltan Museum of Art in New York showed how robust and reliable is the device and how tight can be the fluctuation range of the climate inside the display cases.

Its accuracy, long-term reliability and customizable size make it a challenging improvement for the best preventive conservation of works of art.

5. References

[1] Allegretti O, De Vincenzi M, Uzielli L, Dionisi-Vici P, 2013, *J Cult Herit,* **14**(3)S, 161, doi: 10.1016/j.culher.2012.10.022

[2] Klein L, Bermudez S, Schrott A, Tsukada M, Dionisi-Vici P, Kargere L, Marianno F, Hamann H, López V, Leona M, 2017, *Sensors* **17**(9), 1998, doi: 10.3390/s17091998

[3] EN: Italian patent application Nr. 102017000060109, filed on June 1, 2017

[4] Harriman L, Brundrett G and Kittler R, 2006, *Humidity Control Design Guide for Commercial and Institutional Buildings*, American Society of Heating, Refrigerating and Air-Conditioning, Inc., Atlanta, GA, 334-358, ISBN 1-883413-98-2

[5] Magnus G, 1844, *Ann. Phys.* 137(2), doi: 10.1002/andp.18441370202

[6] Allegretti O, Dionisi-Vici P, Bontadi J and Raffaelli F, 2017, *Appl. Phys. A*, **123**: 38, doi: 10.1007/s00339-016-0609-3

IOP Conf. Series: Materials Science and Engineering **364** (2018) 012027 doi:10.1088/1757-899X/364/1/012027

Diagnostics and Monitoring the quality of Dutch Heritage

Karl Kupka

Senior expert on Sustainable Urbanism and Reuse

Ministry of Internal Affairs – minBZK/DGBRW/B&E – Den Haag, NL
karl.kupka@minbzk.nl - Lorentzplein 11, Haarlem, the Netherlands

Abstract. Over the last century, housing has been one of the major concerns of government intervention in the Netherlands. Especially after World War II, housing supply for a fast growing population came in the first place. Natural demographic growth, immigration from the former colonies and from the Mediterranean called for a response in the production of dwellings - for the major part in the public sector. In the 21th Century there has been an effective shift in policy. Nowadays, financial engagement of Government in Housing is reduced, which does not mean that all problems have been resolved. Present challenges for the Dutch historic cities are: the effects of climate change and increasing vacancy of properties which could be used for housing. Recently, various methods have been developed and implemented to diagnose and monitor the situation in these fields. The results of these analyses are applied in operational policies for Urban Renewal. This review of policies and instruments is structured in two sections, each of which focused on the diagnostics and monitoring of foundation problems and the vacancy of old buildings:

- The period since the 1970s, in which Urban Renewal shifted from physical interventions (mainly demolition and substitution) towards reuse and changes of use;
- The upcoming decades, until a planning horizon of 2050, when consequences of climate change (raising water level, draught) and the growing vacancy of utility buildings have to be tackled.

This paper highlights the importance of monitoring as a preparation for public interventions in historic cities over time [1].1

[1] The author's principal experience with diagnostics and monitoring for Urban Renewal and Transformation regard the City of Amsterdam (1980-1990) and the Ministry of Housing (2005-2015). At present, he works on Urban policies for Climate Adaptation at the Ministry of Internal Affairs. His comprehensive theoretical approach and case studies have been published in the book *Redevelopment by Tradition – Urban Renewal in World Heritage cities* (Kupka, Cluva Venezia, 2012).

1. Urban Renewal: monitoring decay in constructions

In the First years after World War II, in the Netherlands great effort was made on housing production and the reconstruction of Historic Cities. In 1953, in addition to this construction task, came the flood disaster with devastated great part of the province of Zeeland. Both building quality as the quality of sea dikes asked for more efficient monitoring and reconstruction of infrastructures.

This paper highlights the importance of monitoring as a preparation for public interventions in historic cities over time.

In the post-war years, the quality of life in historic city centres was under growing pressure. This could be verified by social indicators, such as the concentration of inhabitants with lower incomes, as by the disappearence of economic activities. These phenomena were both a cause as a consequence of the worsening quality of historic buildings. Since the 1970s, Urban Renewal - as a public intervention - followed the diagnosis of the building stock at the level of properties and single houses:

1. The priority of intervention units was based on a district-wise monitoring of the external conditions of the properties. This was a global and external inspection of the buildings;
2. The first step was the individualisation of run down, vacant or under-utilised buildings;
3. The diagnostics of adjacent properties was necessary to decide the extent of the project;
4. If there are doubts about the underground construction (wooden piles, as used between 1880 and 1945) posed doubts, these foundations had to be inspected by partial excavation;
5. If the quality of the building remains ambiguous, the owner can be given a period to repair his property, otherwise the property can be included in an expropriation procedure.

Deterioration of wooden foundations is a problem in many old districts of Dutch cities. According a rough estimate about 750.000 dwellings (approximately 10% of the total stock) in the Netherlands are built on a wooden pile foundation. The number of residential buildings with foundation problems has been estimated between 70.000 and 200.000. Deterioration of foundations has negative effects on the capacity of the building for its proper use. It can be verified in damages of the brick facades, distortion or decay of the construction.

De Lange summarises the causes and consequences of foundation damage as follows: ''Wooden pile foundations with insufficient capacity and wood degradation due to fungi or bacteria. In case of carrying capacity problems the (draft) directive diagnosis describes a method to decide whether a foundation should be repaired or not based on the predicted future deformation of the building" [2]. Her thesis describes a draft directive method to calculate the residual capacity of the wooden piles in time. Based on this, one can assess the remaining life time of the construction and decide whether the foundation needs to be restored or not.

Analyses of the lifetime of wooden foundations has been applied as a diagnostic and monitoring method to decide the boundaries of urban renewal projects. Generally the reconstruction projects regarded new (social) housing between elder properties in good state in need of minor repair. De Lange's directive is consistent with this practice. It allows the delimitation and the determination of partial foundation repair of properties: "The directive can be used to test whether a recommended recovery solution meets the damage criteria. The directive provides a technical solution for foundation problems within low-rise residential blocks. In practice, whether the recommended recovery unit will be realised, depend on the attitude and financial situation of the building owners and municipal policy" [2].

Over the period 1970-2000 these criteria guided the processes of Urban Renewal, together with the substitution and reuse of vacant properties.

Figure 1. Historic Dutch (canal) house with wooden pile foundation

Figure 2. Example of distortion because of foundation decay (Amsterdam Centre)

1.1 . Actual situation and perspectives

By the year 2000 the Housing Ministry made a national scan of foundation quality among municipalities regarding the data of local government Building Inspection offices. The properties resulting from these data are divided into 4 categories regarding their expected maintenance period and cost of repair:

- Category 1: 25 years and longer;
- Category 2: up to 25 years;
- Category 3: within10 years due to foundation restoration;
- Category 4: within 5 years demolition and reconstruction.

The State provides subsidies to municipalities with a large stock of (old) properties in bad conditions.

As there are more than 7 million houses of which roughly 10% are built on wooden piles in the Netherlands, accurate diagnostics are difficult, also because the foundation quality is affected by changing climate conditions. It is estimated that in the upcoming decades about 10% of these wooden pile foundations will need repair. Problems with this type of foundation are mainly caused by wood decaying fungi, bacterial degradation and insufficient load-bearing capacity. If no measures are taken to deal with these problems, houses will undergo unacceptable subsidence and become uninhabitable over time. Foundation repair is essential to prevent this happens.

2. National Monitor of Vacant Properties

Since 1-1-2018 the National Vacancy Monitor is operative **[3]**. It consists of data from the Municipal Use Registers, the Register of Real Property, local taxes and building permits. All types of utility buildings are monitored and on the basis of this local authorities can incentive owners of vacant buildings to proceed towards transformation. The State provides some subsidies in this field, especially for buildings with Heritage Values. Municipalities can fine and ultimately occupy vacant properties according to the Housing Act (Wet Kraken en Leegstand). The monitor is based on 4 (municipal) data sets:

- BAG – Basis Registration of Uses for Locations and Buildings;
- WOZ – Value of the Properties for Taxation;
- BRP – Basis Registration for Persons (inhabitants at addresses);
- HR – Register of Commercial Activities.

Vacancy is recognised evident if there is neither an inhabitant (BRP), nor a registered used for taxation (WOZ), nor commercial activity registered. Vacancy shares vary in cities in and different parts of the country and by original use of the building:

(original) use	Amsterdam	Haarl.meer	Nijmegen	Rotterdam	Netherlands
office	12	42	37	23	17
industry	7	34	28	17	14
shops	5	10	18	12	10
public services	5	7	25	7	8
dwellings	0	2	3	4	3

2.1. Case: Vacancy of office space, 1991-2015 (PBL)

In the Netherlands, there is a surplus of office space. On the average, in 2015 more than 17% of office space was unoccupied and this percentage is still growing.

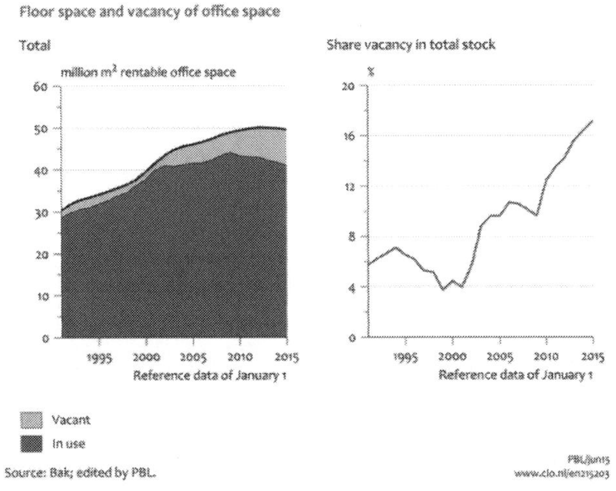

Figure 3. Floor space and vacancy of office space.

The office stock has gradually grown over the past two decades, just as the vacancy of other buildings, some with important heritage values like churches, barracks or farms. By the year 2000, growth of office vacancy accelerated as a result of the bursting of the internet bubble and the need for investors to look beyond shares and find alternative investment opportunities. PBL diagnosis: "The use of office space did not go hand in hand with stock developments. This led to higher vacancy rates. Since

2008, when the economic crisis set in, in 2015 vacancy rates have climbed to more than 17 percent. Other more structural factors also contribute to higher office space vacancy rates. This In has led to a reduced demand for business and office space. Businesses and organisations have adopted more flexible working methods, which means that fewer square metres (m^2) are required per individual worker. All these factors have contributed to an oversupply of office floor space. The largest volumes of office space are found in the Randstad region. More than half of total office space in the Netherlands is located in the provinces of North Holland (Great Amsterdam) and South Holland (Rotterdam-The Hague). The Randstad also has the highest vacancy rates, sometimes an excess of 20 percent. This is partly caused by the fact that many investors are active in this region, unlike in the peripheral regions where offices are often in the hands of owner-occupants. In addition to macro-economic, social, local and regional factors, geographical location and object characteristics are also important factors contributing to the attractiveness of office buildings. Within the office market of Amsterdam, for example, vacancy rates in the suburban regions are considerably higher than in the centre of Amsterdam" [4].

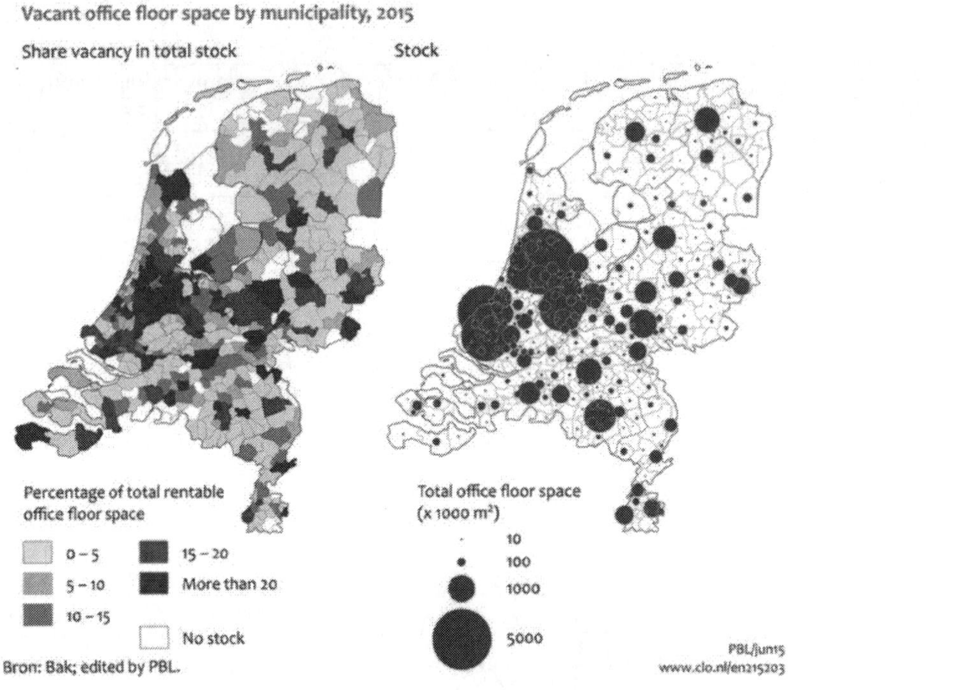

Figure 4. Vacant office floor space by municipality, 2015.

2.2. Transformation towards residential use

Over the past 10-15 years, a large area of new office space was added to the rental market. The amount of new additions was much higher than the amount of withdrawals due to demolition or renovation for other purposes. Since 2006 the State (Housing Ministry) facilitates transformation of empty utility buildings, especially towards residential functions. Several properties resulted to be adequate for student housing or hotels – especially in the larger cities. But also 'standard' dwellings have been accommodated in old factories, barracks etc. Compared to the pre-recession era, the total amount of m^2 annually added has been reduced to one fourth. Since 2013, more office premises are withdrawn from the stock than added, but as yet the vacancy rate is not falling because the use of office space is declining too rapidly.

The Netherlands are facing a problem of vacant buildings. There have been studies focusing on vacant office buildings and the possible solutions for their reuse.

The transformation or reuse of buildings is gradually growing. Statistics reveal a number of about 6.000 new dwellings in all kinds of utility buildings in 2012 and over 8.000 dwellings in 2016. Especially in the big cities like Amsterdam or The Hague an active reuse policy has been followed by the municipality. This has been a good start and an inspiring example for other municipalities.

Figure 5. Reuse as dwellings and ateliers of the former Philips factories in Eindhoven.

3. Climate Change: new challenges

Since the 1953 flood disaster (2.000 dead) onwards, Dutch government implemented large programmes for sea shore defence and inland water regulation. Since 2012, in view of expected developments in climate over the 21st Century, new policies have been developed. The Netherlands need to prepare for the consequences of the rising sea level, land subsidence and rising temperatures. This means looking further ahead and making effective plans for the long-term. It's not only about the future, however. Currently, flood protection falls short in some areas. The government states to "is carry out a number of projects to remedy this. These projects are included in the annual Delta Programme.

In January 2016, when the Netherlands commemorated the floods that affected the Zuiderzee area 100 years ago. These and other flood disasters that have hit the Netherlands in the past underline the urgency of continuing the Delta Works. With the Delta Programme, we do so in a new fashion that is unique in history" [5].

IOP Conf. Series: Materials Science and Engineering **364** (2018) 012027 doi:10.1088/1757-899X/364/1/012027

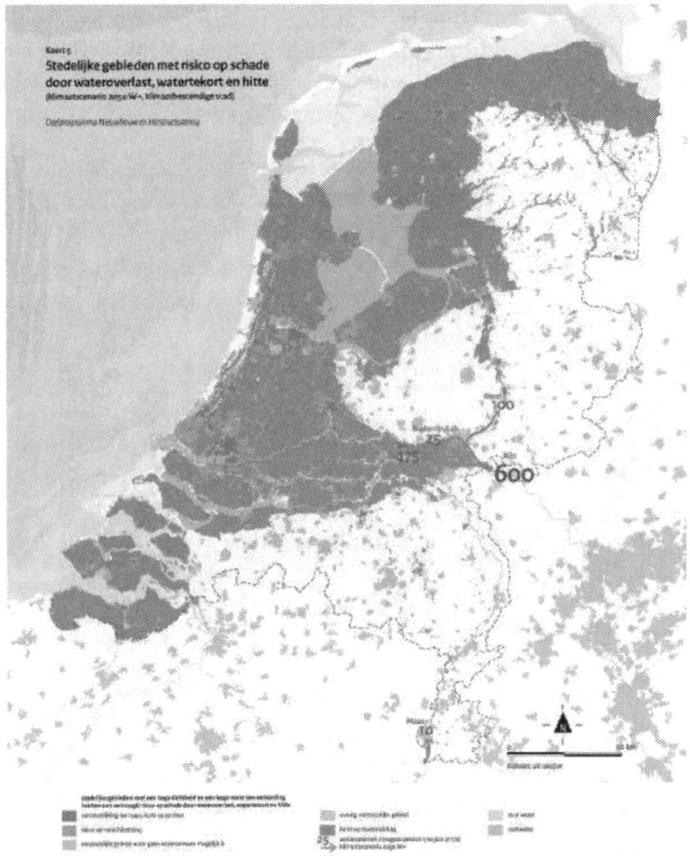

Figure 6. Map of the Delta Plan on Spatial Adaptation, (urban) areas affected by flooding (red); shortage of water (orange); heat (yellow); possibility of soil subsidence (gray).

The Delta Programme 2018 is the first one to include monitoring instruments on Spatial Adaptation. This sets out how municipalities, district water boards, provinces, and the national government intend to expedite and intensify the spatial adaptation process. The Delta Plan contains seven points of ambition to this end. It indicates the goals pursued by the parties, how they intend to attain these goals, and how they will visualise the results.

"The core of the Delta Decision on Spatial Adaptation is to have the Netherlands designed in a climate-proof and water-resilient manner by 2050. The public authorities will ensure that damage ensuing from heat stress, water logging, drought, and urban flooding will be kept to a minimum. This aim will be taken into account in the construction of new residential areas and business parks, the renovation of the existing built environment, sewer replacements, and road maintenance. Climate scenarios will be used to that end" [6]. Also *foundation repair* and monitoring is a part of this strategy. Deltares estimated the cost of foundation repair at almost 50% of total cost to reach a "Climate Sustainable City [7]. Therefore close collaboration of the authorities with the private sector and NGOs is essential. This will proceed in 3 steps:

- "Analysis": analysing the impact of climate change for the various functions in an area in the period up to 2050 (looking ahead to 2100);
- "Ambition": setting concrete goals for improving water-resilience and climate-proofing in the period up to 2050, and formulating a strategy to achieve these goals;
- "Action": setting down the goals and the strategy in plans, legislation, regulations and programmes pertaining to implementation, management and maintenance [6].

The central government, municipal authorities, provinces and district water boards conduct an *annual survey* to gauge the progress made. The monitors, conducted in 2015 (benchmark), 2016 and 2017, have revealed that all the government authorities have set to work on the topics; the district water boards appear to have made the most headway. All the parties are aware of the impact of climate change in terms of waterlogging, flood risk management and drought. Heat stress appears to be of lesser concern. The interim evaluation conducted in 2017 has shown that acceleration and intensification of the strategy is essential. For that reason, the parties have drawn up the Delta Plan on Spatial Adaptation.

A standardised *stress test* will be compiled to assist municipalities, district water boards, provinces, and the national government in mapping out the vulnerabilities of their area. The knowledge portal also features the revised Climate Impact Atlas, enabling the quick visualisation of various climate effects (heat, waterlogging, and drought). The effects may also be combined with information on, e.g., vulnerable groups, swimming water, or depths. The Climate Impact Atlas constitutes a sound basis for a stress test. In addition, the portal comprises an overview of ongoing adaptation projects.

Dutch central government ensures that "national vital and vulnerable functions (energy supply, telecom and IT facilities, public network and emergency communication), the wastewater chain, the supply of drinking water, healthcare, pumping stations and sluices, road transport, chemical companies and laboratories will be more resistant to flooding by 2050 [6].

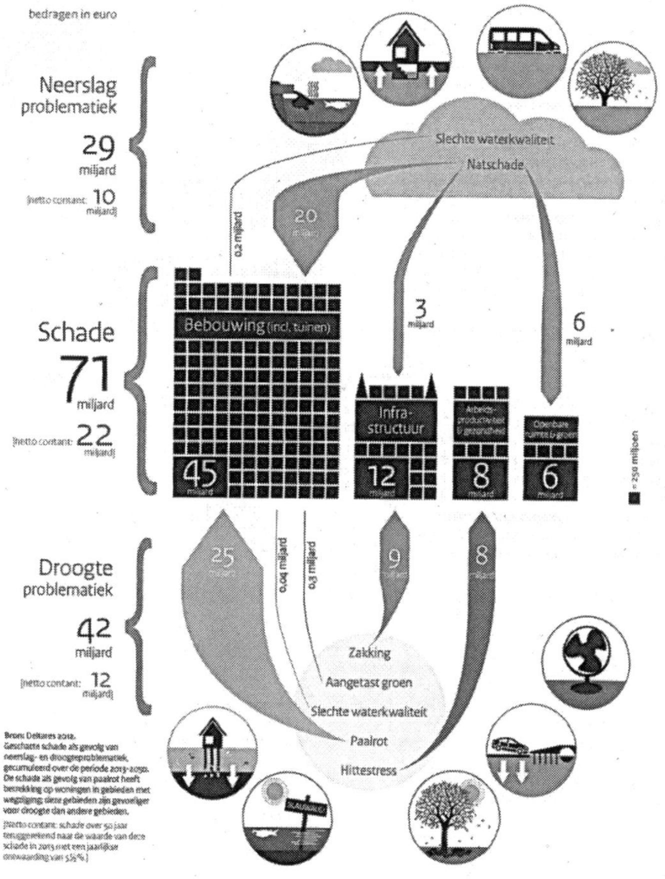

Figure 7. Conditions for "the Climate Sustainable City" (by Deltares, 2012)

Damage in billions of Euro's (if no precautions are taken):
Caused by rain/floods: €29 mrd. Nett: €10 mrd
Damage in properties: Buildings (45); Infrastructure (12); public space (14)
Caused by draught: €42 mrd. Nett: €12 (of which ca. 50% on foundations)

4. Local initiatives

An example of local elaboration is presented by the *Climate Change Adaptation Plan* of Rotterdam [8]. This municipal plan combines resilience and urban regeneration, it manipulates and reshapes the ground of the town, merging into a single network of urban spaces and hydraulic works into a desired landscape.

De Cristina and Massa compare the goals of the Rotterdam plan -controlling and slowing down floods by turning urban ground into a "sponge"- with those of the Venetian Republic as this "had done in the lagoon- pursued by adjusting the different types of urban fabric. Extended cross sections show the flow of water, through underground channels and surface paths, and its relationships with infrastructure, public spaces and building types. In order to distribute the costs of the actions, among the national government, public local institutions, businesses and private individuals, this strategy involves economic cooperation, and therefore encourages active social participation on vital targets recognized by all social subjects. The result of these actions are particular, hybrid and multipurpose, public spaces that did exist in old town but were almost forgotten in functional planning: squares that are flooded and become urban pools for a while; walking and cycling routes on the raised banks of rivers and waterways; flooding parks and gardens to be used as retention basins" [9]

Other Dutch authorities take action at a small scale. In 2018 the province of Brabant (2 million inhabitants) disposed of a subsidy budget of € 1,6 mln for (some 100) schools in order to convert (stone) school squares in green and water absorbing surfaces. Provincial government sees this – which is beside a climate measure – as a possible effect on education, reducing aggression among children.

References:

[1] Kupka, Karl, Redevelopment by Tradition – Urban Renewal in World Heritage Cities (Cluva, Venezia, 2012).
[2] Lange, Suzanne de, Partieel funderingsherstel van woningblokken (Partial repair of foundations under low-rise residential blocks) , TU Delft/Geo-Engineering, 2011.
[3] Centraal Bureau voor de Statistiek, Landelijke Monitor Leegstand, Den Haag 2017.
[4] Planbureau voor de Leefomgeving , Leegstand van kantoren 1991-2017, Den Haag 2017.
[5] Deltacommissaris, Deltaprogramma 2015, Den Haag 2015.
[6] Deltacommissaris, Deelprogramma Nieuwbouw en Herstructurering, Synthesedocument/ Achtergronddocument bij Deltaprogramma 2015.
[7] Deltares, The Climate Sustainable City, Delft 2012.
[8] Rotterdam Klimaatadaptatiestrategie (Climate Change Adaptation Strategy), City of Rotterdam, 2013.
[9] Di Cristina, Benedetto and Marco Massa, Protecting the earth from climate change: the "design of ground" notion in town planning and architecture, University of Florence, 2018.

IOP Conf. Series: Materials Science and Engineering **364** (2018) 012028 doi:10.1088/1757-899X/364/1/012028

Vulnerability assessment of cultural heritage sites towards flooding events

Alessandra Gandini[1], Aitziber Egusquiza[2], Leire Garmendia[3], José-Tomás San-José[4]

[1] Sustainable Construction Division, Tecnalia Research & Innovation, Derio, Spain
[2] Sustainable Construction Division, Tecnalia Research & Innovation, Derio, Spain
[3] Department of Mechanical Engineering. University of the Basque Country UPV/EHU, Bilbao, Spain
[4] Department of Metallurgical Engineering and Materials Science. University of the Basque Country UPV/EHU, Bilbao, Spain

E-mails: alessandra.gandini@tecnalia.com; aitziber.egusquiza@tecnalia.com; leire.garmendia@ehu.eus; josetomas.sanjose@ehu.es

Abstract. Historic sites are threatened by diverse weather patterns, mainly due to global climate change, such as sea-level rise and increasing frequency of storms and other extreme precipitation events. As climate change is becoming an increasing urban problem, heritage located in this context is considered as a sensitive and vulnerable element of the city. Adaptation should be oriented to a sustainable transformation of the historic city, leading to more resilient and safe environments. Risk-based approaches should incorporate an assessment of sensitiveness and capacity to adapt to these hazards. Vulnerability is often assessed on a large scale (e.g. regional, local) and buildings are not considered as part of the urban environment, while conservation is often developed on the operational scale of a monument or site. Management of cultural heritage requires therefore for an urban approach, which considers all the elements and buildings as part of the urban environment. Research presented in this paper describes a methodological approach (MIVES - Integrated Value Model for Sustainability Assessment) for vulnerability assessment of historic sites, supported by multilevel indicators (urban, building, element), in order to provide an informed decision-making. The solution proposed in based on an organised and structures decision tree, which provides a comparable and unique vulnerability index on the building level.

1. Introduction

Climate change adaptation, disaster risk reduction and cultural heritage preservation respond to the objectives of urban sustainable development [1] and aim at giving priority to people's quality of life improvement. Even if the topics are related to different fields of knowledge and competences, sustainable development can be reached only through a holistic approach, which considers all transformations and processes of change. In this context, the Sendai Framework [2] discusses the importance of understanding disaster risk, in all of its dimensions of vulnerability, adaptive capacity and exposure through the systematic evaluation of disaster losses and cultural heritage impacts, among others, in the context of event-specific hazard-exposure and vulnerability information.

The implementation of strategies related to cities' sustainable development, is a practice based on evidence based decision-making and information management. Data availability on the city level has been increasing in the last decades, leading to the development of diverse approaches. From one hand those requiring for a large amount of data and delivering highly accurate results and on the other hand those based in simple data but delivering generic results. Both approaches are not suitable for the strategic level of decision-making, which requires for a balance between data collection and accuracy of results in order to be cost-effective. Information should be organized and structured in order to be comprehensive and this can be supported by the use of proper data model. When dealing with vulnerability assessment, informed decisions are of the most relevant importance as the prioritization of interventions in the pre-disaster period is crucial to raise adaptation and decrease risk.

The use of value analysis methodologies can provide objective conclusions for establishing strategic priorities, in order to overcome the barrier of involving different stakeholders, with diverse profiles and needs.

2. Scope of the research

Main objective of the research presented is to develop a methodological framework for vulnerability assessment in historic areas against flooding events, through a decision-making methodology for the prioritization of adaptive and risk reduction interventions.

The methodology developed has the objective of evaluating the potential damage to historic buildings in case of flooding, where vulnerability is appraised and quantified by indicators, values functions and algorithms which delivers a vulnerability scoring based on a unique value index. The vulnerability assessment is supported by a multiscale urban model which standardize and synchronize geographic and semantic information. The research presented is therefore linked to different domains, namely climate change adaptation, disaster risk reduction and cultural heritage preservation.

The vulnerability assessment methodology applies the MIVES (Integrated Value Model for Sustainability Assessment) method through the definition of the decision-making process, the establishment of the information strategy for data acquirement and the development of a set of indicators sustained by the use of objective and justified calculation models.

Research is aligned with current scientific and political commitments, as climate change adaptation gains relevance and sets a new and emerging concern for public administrations. Some well-known organizations such as the European Commission, the United Nations, the World Bank, the IPCC, and UNESCO consider climate change impacts on urban areas as one of the strategic priorities which must be discussed internationally, leading to the launch of several initiatives at the local level [3], [4].

3. Methodological approach

Decision-making related to climate change adaptation at urban level should be based on an informed and evidence-based strategy implementation, especially if carried out on historic buildings. A building vulnerability assessment method is therefore the first step to be addressed. Flooding impacts should be determined in terms of the intrinsic conditions as well as social aspects of the area or building considered, as their characteristics makes them more susceptible to the effects of climate change or more capable of coping with them.

To be effective and in order to provide a solid basis for political and technical commitment, the method should find a balance between resources consume and results accuracy. Cities generate a large amount of heterogeneous data, at different scales, in different formats, and for different uses [5]. Many of these data can be used to determine to what extend a building is vulnerable to the effects of climate change, through the creation of representative typologies, by organising the building stock in categories with similar constructive elements and characteristics.

Once the building stock is organised and structured in a manageable number of categories, the value analysis method is applied and vulnerabilities are compared by using a unique index, thereby facilitating the prioritisation of interventions in a specific area or building of the historic city and providing greater objectivity in decision-making.

The vulnerability assessment method proposed in this research has been applied to the case study of San Sebastian and it was supported by the use of an interoperable and multi-scalar data model, which had the objective of organising required data and providing a visual understanding of results.

3.1. Building stock categorization

Vulnerability is mainly obtained by using macro-scale information, which results in the determination of which area is more vulnerable with respect to another one. On contrary, information needed to obtain a building diagnosis, is performed on a micro-scale, using a one by one approach through field surveys. When assessing the vulnerability of cultural heritage located in urban areas, it is important to define which buildings are more vulnerable in order to provide and define which adaptation measures are required. The micro-scale approach is often too expensive and time consuming for local governments and thus a balance between the two approaches should be seek. The compromise can be found by modelling the historic city through a statistical distribution of the buildings characteristics in a determined area, starting from the sample concept [6].

Information available from public sources and information systems is grouped according to the building characteristics, creating a limited number of samples [7], which reflect almost the entire building stock. These groups should reflect the flooding vulnerabilities, the historic value and the constructive characteristics of the buildings. A second and more detailed level of information is collected for sample buildings and vulnerability calculated. Results are then extrapolated to other buildings within the same category, thereby obtaining an overall vulnerability assessment for the whole historic district.

3.2. Integrated value model for sustainable assessment

Among the multi-criteria approaches that have been developed in the last decades [8], the Integrated Value Model for Sustainable Assessment (MIVES) has demonstrated its applicability in diverse complex scenarios related to sustainability [9], [10].

MIVES has been developed by the Polytechnic University of Catalonia (UPC), Tecnalia and the University of the Basque Country (UPV/EHU) and combines two different concepts, the Multi-Criteria Decision-making Theory and Value Engineering [11]. The model is used to compare and give equality to variables with different units of measurements, either quantitative or qualitative, by providing a dimensionless unit. The relative importance of the aspects considered is taken into account, giving as a result a unique and comparable index. As the methodology establishes the evaluation prior to the generation of alternatives, subjectivity in the decision-making is avoided [12].

4. Decision-making process

MIVES established different steps for the methodology, which are the following:

- Definition of the problem and decision to be taken;
- Definition of the requirements tree though the establishment of the information hierarchy;
- Set of the value functions which transforms all the aspects considered in a variable with values comprised between 0 and 1;
- Assignment of weights to compare the importance of one aspect compared to another one;
- Evaluation of the alternatives to obtain the value index.

Buildings, located in urban areas which are subject to climate change impacts, play an important role in the selection of the adaptive measures, which can be implemented in both the urban as well as building level. Knowing their vulnerability, in an objective way, is essential for the strategic planning approach.

4.1. The vulnerability assessment requirement tree

The requirements tree for heritage buildings vulnerability assessment has been built considering the connection with other disciplines, such as climate change adaptation and considers as main

IOP Conf. Series: Materials Science and Engineering **364** (2018) 012028 doi:10.1088/1757-899X/364/1/012028

requirements, elements which traditionally compose vulnerability: sensitiveness and adaptive capacity [13].

The requirements tree proposed follows a hierarchical structure based on three levels: requirements, criteria, and indicators. In the first levels, namely the requirements and criteria, general and qualitative aspects are defined, while in the last level, the indicators, concrete and measurable aspects are considered.

The sensitiveness requirement has the objective of assessing the degree to which a building is affected by an event. Depending on the conditions, typology and characteristics of the structure that is considered, its response to climate impacts varies. Criteria related to this requirement are therefore associated to the current state of the building, constructive critical elements, envelope characteristics, main use, and structural material.

The requirement of adaptive capacity refers to the ability of a system to assume the potential effects of an event, overcoming its consequences. In this case, criteria refer to interventions, socio-economic conditions and the cultural value of the buildings.

The requirement tree proposed is therefore defined by two requirements, eight criteria and fourteen indicators, as presented in the following Figure.

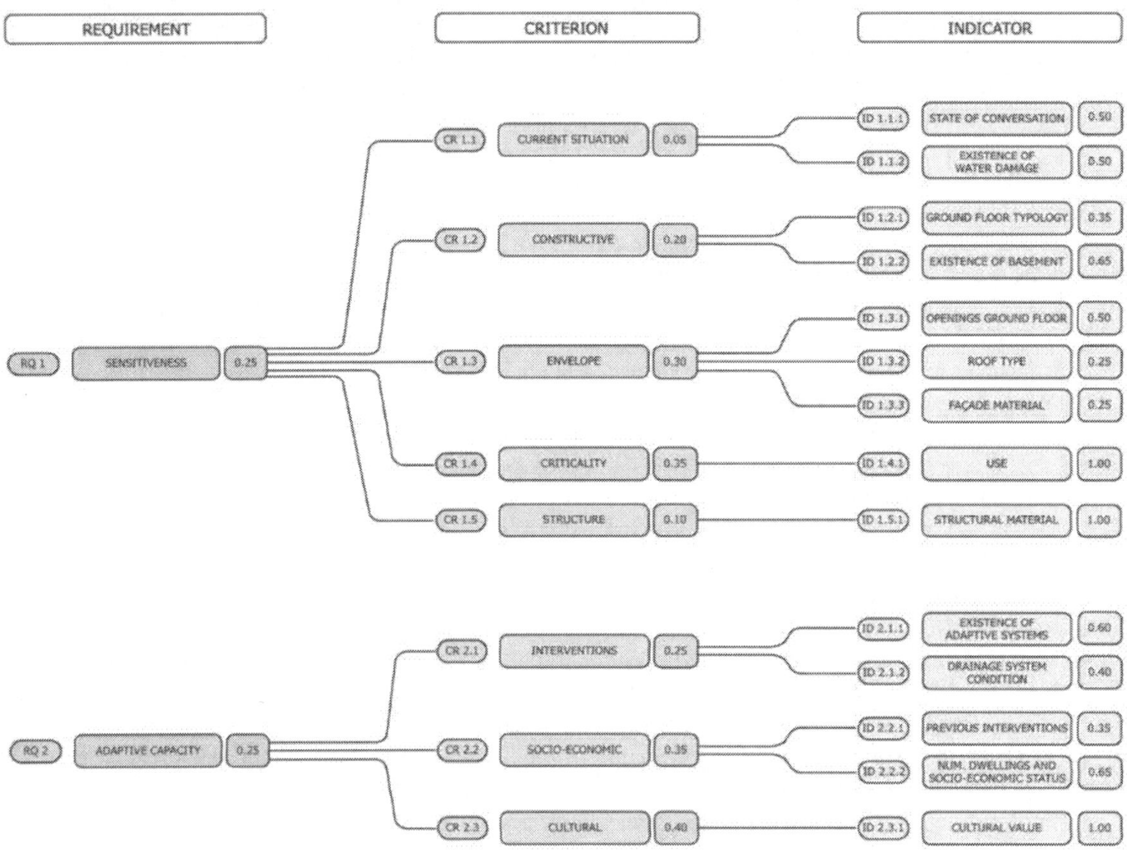

Figure 1. Requirements tree and overall weighting for the vulnerability assessment

Weights have been assigned by member of an expert panel, starting from the calculation of the γ weights of the indicators, followed by the β weights of the criteria, and lastly by the α weights of the requirements. Weight assignment is performed by comparing elements at the same level and in the same

IOP Conf. Series: Materials Science and Engineering **364** (2018) 012028 doi:10.1088/1757-899X/364/1/012028

branch of the requirements tree. Thus, the indicator weights are calculated according to other indicators belonging to the same criterion. In the same manner, a criterion weight is calculated by other criteria belonging to the same requirement.

Requirements, criteria and indicators have the objective of representing what we want to evaluate, avoiding the repetition of certain aspects or avoiding the use aspects which are out of scope. Indicators selected should therefore be representative, differentiating, complementary, relative, quantifiable and traceable [14].

Analytic Hierarchy Process (AHP) [15] was used for the weights assignment, by establishing the relative importance of each branch of the requirements tree. An adjustment was made of the final results, considering the opinion of each member of the expert panel.

4.2. Indicators value functions

Indicators are of different nature and present diverse units of measure. The different variables can be transformed into comparable and dimensionless units through the use of value functions, resulting in a value comprised between 0 and 1. Different value functions (concave, convex, linear, S-shaped) have been used for this purpose, where the vertical axis represents the minimum (0) or the maximum (1) level of satisfaction and the abscissa represents the variable of the indicator.

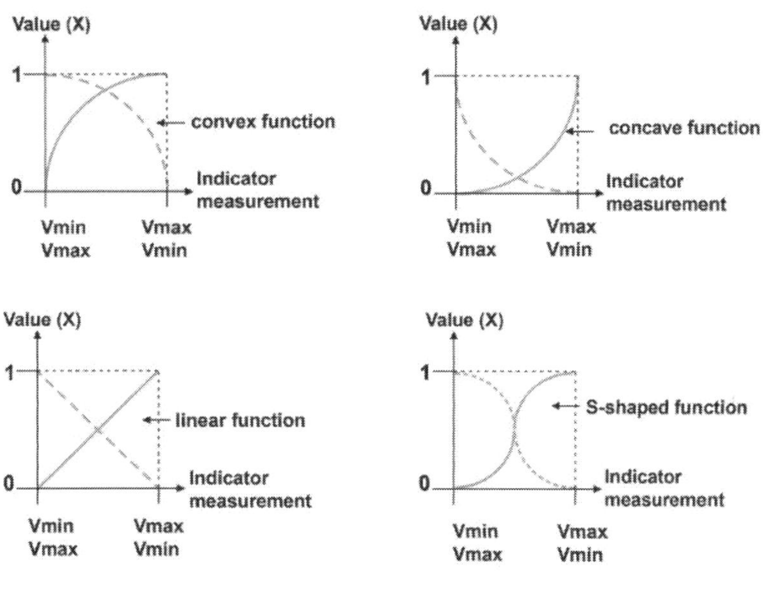

Figure 2. Different shapes of the value functions. Source:[16]

MIVES use the following equation as a mathematical model, in order to define the different value functions of each indicator,

$$V_{ind} = B * \left[1 - e^{-K*\left(\frac{|X-S_{min}|}{C}\right)^{P}}\right]$$

Where,

V_{ind} is the value of the indicator under evaluation.

B is a factor that allows the function to remain within the range from 0 to 1. It is assumed that the highest level of satisfaction has a value of 1.

For each indicator a value function has been created, to evaluate the different alternatives. In cases where the value function was not clear, it was defined by a working group. The following Figure shows the indicators selected for the vulnerability assessment and their relative values:

INDICATOR CODE	INDICATOR MEANING	VALUE MEANING	VALUE
ID 1.1.1	STATE OF CONSERVATION	GOOD	0.00
		FAIR	0.18
		POOR	0.73
		VERY BAD	1.00
ID 1.1.2	EXISTENCE OF WATER DAMAGE	NO EXISTING WATER DAMAGE ON THE BUILDING	0.00
		PRESENCE OF WATER DAMAGE ON THE BUILDING	1.00
ID 1.2.1	GROUND FLOOR TYPOLOGY	PORTICO STRUCTURE	0.00
		CLOSED STRUCTURE WITH NO ACTIVITY	0.50
		CLOSED STRUCTURE WITH ACTIVITY	1.00
ID 1.2.2	EXISTENCE OF BASEMENT	NO BASEMENT NOR SEMI-BASEMENT	0.00
		EXISTENCE OF BASEMENT OR SEMI-BASEMENT	1.00
ID 1.3.1	OPENINGS GROUND FLOOR	NO OPENINGS	0.00
		SMALL OPENINGS	0.49
		LARGE OPENINGS	1.00
ID 1.3.2	ROOF TYPE	PITCHED	0.00
		FLAT	1.00
ID 1.3.3	FAÇADE MATERIAL	BRICK/NON POROUS STONE	0.00
		MORTAR	0.38
		STEEL	0.62
		CONCRETE	0.82
		POROUS STONE	1.00
ID 1.4.1	USE	CULTURAL CENTRES, PUBLIC EQUIPMENT WITHOUT PRIORITY USE	0.00
		COMMERCE	0.22
		RESIDENCE	0.69
		EMERGENCY AND SANITARY	1.00
ID 1.5.1	STRUCTURAL MATERIAL	STONE	0.00
		BRICK	0.33
		STEEL	0.60
		CONCRETE	0.82
		WOOD	1.00
ID 2.1.1	EXISTENCE OF ADAPTIVE SYSTEMS	EXISTENCE OF ADAPTIVE SYSTEMS	1.00
		ABSENCE OF ADAPTIVE SYSTEMS	0.00
ID 2.1.2	DRAINAGE SYSTEM CONDITION	GOOD	1.00
		FAIR	0.78
		POOR	0.29
		VERY BAD	0.00
ID 2.2.1	PREVIOUS INTERVENTIONS	PREVIOUS INTERVENTIONS	1.00
		NO INTERVENTIONS MADE	0.00
ID 2.2.2	NUM. OF DWELLINGS AND SOCIO-ECONOMIC STATUS	X DWELLINGS, Y AVERAGE STATUS	Y*0.7+(X*0.0244)*0.3
ID 2.3.1	CULTURAL VALUE	GRADE I	1.00
		GRADE II	0.86
		GRADE III	0.61
		GRADE IV	0.27
		NONE	0.00

Figure 3. Values attached to each alternative of the sensitiveness and adaptive capacity indicators

5. Implementation

The methodological approach was applied to the city of San Sebastian, located on the northern coast of Spain. The area selected comprises 6 districts, each with its different characteristics, situated next to the boundaries of the Urumea river. The geometry of the 3D urban model was generated using the CityGML standard [17] and semantic properties for the categorization process added, as referenced in the Spanish cadaster. Following the statistical overview of the study area, it was considered that use, level of protection, existence of a basement and socio-economic status had to be taken into account, with regard to all their variables, as primary parameters and the threshold of minimum representation was established at 2%. The categorization process lead to a generation of 15 categories, representing the 76% of the total building stock. Sample buildings were then selected according to the representativeness of the parameters compared to the whole category and the availability of relevant information and semantic information completed and extrapolated to the whole category. As a result, the vulnerability index was calculated according to the values assigned to the variables of each indicator and the weights attached to each branch of the requirement tree.

Figure 4. Graphical representation of the lots' vulnerabilities

In order to verify the accuracy of the results of the methodology proposed, a survey campaign was carried out on 100 buildings with the objective of comparing results given by real data and the categorization method. The margin of error resulted in a 9% and the largest difference was appreciated in one of the districts, which is mainly characterized by single-family houses of diverse characteristics, while the methodology shows its highest potential on districts which have been characterized by a smooth development and present similarities.

6. Conclusions

In vulnerability assessment of historic buildings towards flooding, decision-making is a complex process in which several disciplines and interests intervene, leading to a difficult exercise of comparison and evaluation. Local governments are more and more keen to have accurate results with the minimum efforts. A balance among data acquisition and accuracy of results to include the relevant information in a unique model or platform is essential for the decision-making process. Furthermore, multiple criteria decision analysis (MCDA) process helps decision-maker in improving the objectivity and quality of results by providing a systemic and organised way of thinking. Research proposed has established a hierarchic structure based on a requirements tree, in order to provide decision-making with an objective intervention priority index, in which the characteristics of the vulnerability assessment are defined, displayed and organized. A method for the fine-tuning of the methodology, on the one hand, considering the sensitiveness index and, on the other hand, the adaptive capacity index, has been established, providing vulnerability levels defined by these parameters.

7. Acknowledgements

Authors would like to acknowledge the funding provided by the Basque Government through the ADVICE project and the research group IT781-13 at the UPV/EHU.

References

[1] I. Kelman, J. C. Gaillard, and J. Mercer, "Climate change's role in disaster risk reduction's future: Beyond vulnerability and resilience," *Int. J. Disaster Risk Sci.*, vol. 6, no. 1, pp. 21–27, 2015.

[2] UNISDR, "The Sendai Framework for Disaster Risk Reduction 2015-2030: the challenge for science," *R. Soc. Meet. Note*, p. 9, 2015.

[3] "European Climate Adaptation Platform." [Online]. Available: http://climate-adapt.eea.europa.eu/.

[4] "Covenant of Mayors for Climate and Energy." [Online]. Available: http://www.covenantofmayors.eu/Adaptation.html.

[5] A. Egusquiza, "Multiscale information management for historic districts′ energy retrofitting. A framework. A methodology. A model," Universidad Politécnica de Cataluña, 2015.

[6] I. Prieto, J. L. Izkara, and A. Egusquiza, "Building stock categorization for energy retrofitting of historic districts based on a 3D city model," *DYNA Ing. E Ind.*, 2017.

[7] I. Ballarini, S. P. Corgnati, and V. Corrado, "Use of reference buildings to assess the energy saving potentials of the residential building stock: The experience of TABULA project," *Energy Policy*, vol. 68, pp. 273–284, 2014.

[8] E. K. Zavadskas, T. Vilutienė, Z. Turskis, and J. Šaparauskas, "Multi-criteria analysis of Projects' performance in construction," *Arch. Civ. Mech. Eng.*, vol. 14, no. 1, pp. 114–121, Jan. 2014.

[9] I. Piñero, J. T. San-José, P. Rodriguez, and M. Losáñez, "Multi-criteria decision making for grading the rehabilitation of heritage sites. Application in the historic center of La Habana," *J. Cult. Herit.*, vol. 26, pp. 144–152, 2017.

[10] F. Pardo-Bosch and A. Aguado, "Investment priorities for the management of hydraulic structures," *Struct. Infrastruct. Eng.*, vol. 11, no. 10, pp. 1338–1351, 2015.

[11] J. T. San-José Lombera and I. Garrucho Aprea, "A system approach to the environmental analysis of industrial buildings," *Build. Environ.*, vol. 45, no. 3, pp. 673–683, 2010.

[12] B. Viñolas, F. Cortés, A. Marques, A. Josa, and A. Aguado, "MIVES: modelo integrado de valor para evaluaciones de sostenibilidad," in *Congrés Internacional de Mesura i Modelització de la Sostenibilitat*, 2009, pp. 1–24.

[13] IPCC, *Climate Change 2014: Synthesis Report. Contribution of Working Groups I, II and III to the Fifth Assessment Report of the Intergovernmental Panel on Climate Change*. 2014.

[14] A. Josa, "Introducción a MIVES y el análisis multicriterio." Derio, 2012.

[15] T. L. Saaty, "The Analytic Hierarchy Process," *Education*, pp. 1–11, 1980.

[16] J. P. R. J. Cuadrado, E. Roji, J.T. San-José, "Sustainability index for industrial buildings," *Proc. Inst. Civ. Eng. Struct. Build.*, vol. 165, no. 5, pp. 245–253, 2012.

[17] G. Gröger and L. Plümer, "CityGML - Interoperable semantic 3D city models," *ISPRS Journal of Photogrammetry and Remote Sensing*, vol. 71. pp. 12–33, 2012.

IOP Conf. Series: Materials Science and Engineering **364** (2018) 012029 doi:10.1088/1757-899X/364/1/012029

3d printing technologies: are their materials safe for conservation treatments?

D Cimino[1], G Rollo[2], M Zanetti[1] and P Bracco[1]

[1]Department of Chemistry, University of Turin, Italy
[2] Koinè, Conservazione Beni Culturali Scrl, Italy

Abstract. 3D printing technologies have been definitively introduced in conservation treatments. Despite the advantage of not requiring direct contact with the artwork, allowing the preservation of fragile objects, the printed item is located in direct contact with the object and the characterization of the filament used for the printing is not often taken into consideration. The following study was undertaken as an evaluation of filaments possibly employed for conservation treatments. The characterisation of the components was carried out through infrared spectroscopy, thermal and chromatographic analyses. Moreover, it was investigated whether such materials release volatile organic compounds (VOC) during their degradation process. Indeed, all of them released styrenic and alkyl compounds, all solvents for materials that can be found on artworks, employed by both, artist and conservator.

1. Introduction

In the last few years, 3D printing technologies have been introduced in conservation treatments as solution for different issues. Among 3D techniques, fused filament fabrication (FFF) method has found several applications in the field, in particular employed for the creation of reconstructions or integrations [1-7] and reinforcement structures [8] for fragmented art pieces. Surely, one of the advantages is that creating an object, starting from a CAD 3D model, does not require a direct contact with the artwork. In fact, the model can be drawn from artwork pictures reworking or from its laser scans. This allows preserving fragile objects, without inducing further damages as for the creation of a mold, which needs a direct contact with the artwork. On the other hand, in such cases, the printed item is located directly on the object and, despite that, the characterization of the filament used for the printing is not an aspect taken into consideration. Filaments are, mainly, based on polymers and are not tuned for conservation purposes, therefore their formulations might contain additives such as plasticizers, flame retardants, stabilizers, whose molecules, smaller than the macromolecular structure, can migrate towards the surface with time. These products can affect materials they are in contact with, changing their properties and their aspect, thus creating damage. Another issue could be the release of volatile organic compounds (VOC), derived from the degradation process of the polymer itself. Also in this case, these molecules may act negatively on other materials.

It is clear that on a large market it is not necessarily considered how the base-material will act with the ageing or will interact with artist materials, both questions to be considered though according to the contemporary conservation theories that articulate how conservation products must be reversible and chemically stable with time [9]. Such rules were at the base of a deep characterization of products more often employed in the conservation field or the research of specifically tuned formulations (e.g. Paraloid®B72 or BEVA®371) [10, 11].

Filament-producers tend to keep undisclosed the chemical composition of their products and their datasheets focus mainly on mechanical properties (e.g. deformation, elongation, tensile strength) and printing parameters, in particular temperatures of working plate and nozzle, together with the suggested printing speed. Nevertheless, missing parameters such as melting points and degradation-processes beginning temperatures are important information too.

Following this line of reasoning, the work here introduced aims at suggesting the elaboration of a list of potential materials that can be safely used on artworks. The project has started with the characterization of several filaments already used or ready to be used in conservation treatments. Only a multi-analytical approach, from infrared spectroscopy to thermal analysis and chromatographic techniques, has allowed to detect different chemicals, either components of the material or products released already in environmental conditions. Furthermore, some of them can potentially compromise restored objects. Indeed, results have underlined the need of having a full awareness of the composition of all involved materials in order to avoid further damages and continuous restoration treatments in a short time-frame on the same artwork.

2. Materials and methods

2.1. Filaments

Fourteen filaments were chosen on the base of the experience of three printing laboratories that had collaborated with conservators in the creation of integration of real artworks. The first material selection was based on the aesthetical aspect, since related with the final result other than on the ease of printing and following smoothing; base materials were not considered. Since the composition of some products is proprietary, production series names and manufacturers are not mentioned. Therefore, filaments are named as general samples, followed by their colour and producers are approximately indicated in order to gather filaments (Table 1).

Table 1. List of characterised filaments, together with their manufacturing; producers have been identified by abbreviation, due to confidential compositions

Filaments	Producers
Filament 1 – white	TF
Filament 2 – white	TF
Filament 3 – white	TF
Filament 4 – light red ochre	TF
Filament 5 – yellow ochre	TF
Filament 6 – light grey	TF
Filament 7– dark grey	TF
Filament 8 – dark red ochre	TF
Filament 9 – light wood	Unknown
Filament 10 – dark wood	CF
Filament 11 – bronze	CF
Filament 12 – transparent uncoloured	TO
Filament 13 – yellow	TO
Filament 14 – white	TO

Chemical characterization of printing materials was the focus of this first part of the work, especially changes induced already by natural ageing. All filaments were characterized through infrared spectroscopy, thermal analysis and gas-chromatography.

2.2. Analytical techniques

2.2.1. Thermogravimetric analysis (TGA)

TGA has been performed to assess the thermal degradation profile of each material and, more specifically, to identify the temperature at which components start to degrade (temperature degradation offset, Tdo), in order to verify the stability of the filaments at the suggested printing temperature. Also, the presence and the amount of inorganic filler in raw materials can be controlled.

Analyses were carried out with a Thermogravimetric Analyzer TGA 2050 (TA Instruments), in nitrogen atmosphere, with a temperature ramp from 40 to 600°C at a rate of 10°C/min. Materials were analysed directly, fragmenting the commercial reel and collecting an amount of sample of ca. 15mg, which was directly put in an alumina pan.

2.2.2. Fourier transform infrared spectroscopy (FTIR) in attenuated total reflection (ATR)

Infrared spectroscopy has been carried out to characterize their chemical nature.

ATR spectra were recorded on a Thermo Nicolet Nexus FTIR spectrometer, with a Smart Endurance ATR accessory with a diamond contact crystal and a ZnSe focusing element. The range investigated was from 4000 to 400 cm-1, with a resolution of 4 cm-1 and the result was the average of 64 scans.

2.2.3. Pyrolysis (Py) gas chromatography/ mass spectroscopy (GC/MS)

Aim of the analysis is a chemical characterization of the material which allows the identification of organic additives such as colorants, flame retardants or stabilizers.

The chromatograph is a 6890N Network GC system (Agilent Technologies) and the analytical capillary column employed has a solid phase made with methyl-phenyl-polysiloxane cross-linked with 5% phenyl methyl silicone (HP-5MS of Agilent Technologies) 0.25μm thick (30m x 0.25mm inner diameter). The coupled mass spectrometer is a 5973 Network Mass selective detector (Agilent technologies) set with the following parameters: interface temperature 280°C, ion source temperature 230°C, quadrupole mass analyser temperature 150°C, electron impact 70eV, scan range 40-600 m/z.

The instrument is connected with a CDS Pyroprobe 1000 with a platinum filament pyrolyzer, supplied by Analytical Inc.. The pyrolysis chamber temperature was 300°C, whereas the pyrolysis temperature was set at 600°C, reached with a ramp temperature of 5°C/ms and held for 10s. The chamber was also fluxed with helium.

Few milligrams of raw material were inserted in a quartz boat and kept in position with some quartz wool. The boat was then put in the pyrolyzer platinum coil and so inserted in the pyrolysis chamber. The pyrolysis and the chromatographic run began at the same moment and the temperature of the column, hold for 2 minutes at 50°C, was heated at 300°C at a rate of 10°C/min.

2.2.4. Solid phase microextraction (SPME) gas chromatography/ mass spectroscopy (GC/MS)

This technique allows the analysis of organic volatile components (VOC) in an enclosed environment. To have semi-quantitative information, ca 1g of sample, cut in small pieces, was sealed and left for one month at room temperature (RT) in a vial with silicone septum, suitable for the insertion of needles. The SPME fibre (65μm diameter, polydimethyl siloxane/divinyl benzene absorbent material, model 57310-U sold by Supelco) is located within a needle which has to be inserted in the vial to collect the volatile components in the inner atmosphere. Sampling was carried out for 30 minutes. The analysis was repeated heating up samples at 50°C for one hour and sampling was conduct in the meantime.

After sampling, the needle was removed and inserted for 2 minutes in the injector (280°C) of the chromatographer (previously described). A ramp temperature from 50 up to 250°C was set, at a rate of 10°C/min. Helium was used as carrier gas, at a constant flow of 1.0 ml/min. Analyses were carried out in split mode (ratio 1/20 of the total flow).

On the basis of the data obtained through FTIR and Py-GC/MS investigations, SPME analysis was limited to one sample per each different material (filament 1-3, 9-13).

3. Results and discussion

TGA, FTIR and Py-GC/MS allowed a first classification of selected filaments and data are introduced in table 2. Whenever the producer declared the composition of the product, analytical data added some additional details.

Table 2. Summary table of filaments main components, investigated through FTIR and Py-GC/MS. * data obtained by X-ray fluorescence (XRF), not introduced in the previous section due to the employment for this sample only

Filament	Polymer	Tdo [°C]	Filler	Flame retardant	UV-stabilizer	Antioxidant	Plasticizer
F 1	PS	245	$CaCO_3$ (10%)	Dicumene			
F 2	SAN	213	$CaCO_3$ (10%)		Tinuvin P		Dehydroabietic acid methyl ester
F 3	PA 12+PS	419	$CaCO_3$ (20%)		Tinuvin 770	» n.33 » SWP	
F 4	PS	234	$CaCO_3$ (15%)	Dicumene			
F 5	PS	234	$CaCO_3$ (15%)	Dicumene			
F 6	PS	234	$CaCO_3$ (15%)	Dicumene			
F 7	PS	280	$CaCO_3$ (20%)	Dicumene			
F 8	PS	234	$CaCO_3$ (15%)	Dicumene			
F 9	PLA+PS	225			Tinuvin 770		
F 10	PLA	258			Tinuvin 770		
F 11	PLA	290	Copper powder (80%) *		Tinuvin 770	Nocrac M17	Diethyl phthalate (DEP)
F 12	Acrylic polymers blend: P(S-MMA), PBA, PEA	250			Tinuvin P		
F 13	ABS + PTFE (or other fluorinated polymer)	350			Tinuvin 770		
Filament 14	ABS	230			Tinuvin 770		

Legend: PS polystyrene, SAN styrene-acrylonitrile, PA polyamide, PLA polylactic acid, MMA metylmetacrylate, PBA polybutyl acrylate, PEA polyethyl acrylate, ABS acrylonitrile-butadiene-styrene, PTFE polytetrafluoroethylene

TGA results, in terms of comparison of the T_{do} with the suggested printing temperature, highlighted how these two values are often close to each other and so how easily the printing process can be a kickoff of thermal degradation, endangering the workability of the object.

Polylactic acid (PLA), polystyrene (PS) and styrene-based polymers (acrylonitrile-butadiene-styrene, ABS, and styrene-acrylonitrile, SAN) were found as main components of most filaments. In some cases, more polymers are blended together, as for filaments 9, 12 and 13, where, it is assumed, this increases the processability of the material, more than its stability with time.

Filament 3 is mainly made of polyamide (PA). The characterization through Py-GC/MS allowed the identification of the main polymeric component as PA 12 thanks to the presence of laurolactam, which is the starting monomer. Furthermore, the technique highlighted the presence of PS, a component in lower concentration since it was not detected by IR spectroscopy, that it is probably added to provide stability to the formulation.

UV-stabilizers were often found in the formulations, especially Tinuvin P® and Tinuvin® 770, a UV absorber (UVA) and a hindered amine light stabilizer (HALS) respectively. Thus, it is possible to think of these materials as suitable for outdoor application. Other additives detected are flame retardants, antioxidants and plasticizers. In particular in filament 11, diethyl phthalate balances the high percentage of filler (copper, 80% w/w) added in the material. Beside the metal power, calcium carbonate was found as filler, in a concentration range between 10 and 20% (w/w).
Neither pigments nor colorants were identified. Organic colorants are added to polymeric formulations in low concentration, thus it is difficult to detect their presence with the techniques here involved. Only for white filaments the origin of the colour was linked to the addition of calcium carbonate, filler and pigment in the meantime. Wooden-aspect filaments reached their aesthetical aspects with the addition of wooden powder only in one of the two cases studied. Thanks to the VOC emission of furfural and other wooden derivates, it was possible to hypothesize the presence of lignin and cellulose in filament 10; on the contrary filament 9 did not released polysaccharide markers.

SPME-GC/MS results are listed in table 3.

Table 3. Summary table of SPME-GC/MS data

Filament	Polymer	VOC (RT)	VOC (50°C)
Filament 1	PS	Styrene derivatives	=
Filament 2	SAN	Styrene and acrylonitrile derivatives, acetone	=
Filament 3	PA 12+PS	Styrene derivatives and alkanes	=
Filament 9	PLA+PS	Styrene and lactic acid derivatives	=, terpenes
Filament 10	PLA	Lactic acid derivatives, furfural and other saccharide derivatives	=
Filament 11	PLA	Few lactic acid derivatives, in trace concentration	= in higher concentration
Filament 12	Acrylic polymers blend: P(S-MMA), PBA, PEA	Acrylate monomers (MMA, EA), aliphatic aldehydes	=
Filament 13	ABS + PTFE (or other fluorinated polymer)	Styrene derivatives, alkenes	=

Several studies about VOC released by polymeric filaments have been already reported in literature, but mainly with the aim of finding a correlation between 3D printing and human health, with respect to the inhalation of toxic organic volatile components released by the filament during the printing process. Therefore, tests have been carried out at temperatures set for the printing (ca 200°C) [12,13]. Here, the aim was verifying if the addition of printed items directly on an artwork might create further damage to the contact area. Considering that, the release at room temperature of remarkable amounts of styrene and its derivatives represents a danger due to the solvent action that these compounds have on several materials commonly employed by artists and conservators, e.g. natural terpenic resins or polymeric adhesives and consolidants.

4. Conclusion

For this project, 14 filaments for fused filament fabrication were selected among an ample group on the market. The choice of characterizing these materials is due to the always increasing employment of this 3D printing technique for creating integrations to be placed on fragmented artworks. Aim of the work was investigating the possibility of further damage to the artist object, induced by the insertion.

First classification of the samples was carried out with FTIR and Py-GC/MS: polystyrene based and PLA polymers are the most employed for the creation of such materials, as main component or copolymers in blend with other matrix. As filler, calcium carbonate is often added since its colour and opacity make this powder suitable as pigment or support for organic colorants as well. Organic colorants were not identified, due to the low concentration usually introduced in industrial formulations. Thanks to the chromatographic techniques, a good overview of additives was achieved. In particular antioxidants, UV-stabilizers and one flame retardant have been identified. Moreover, the VOC emission of furfural and other wooden derivates led to the identification of lignin and cellulose as filler and coloured component in one filament. SPME-GC/MS analysis have therefore revealed to be essential for a thorough characterization of the material, even though the aim of performing it was identifying VOC released by filaments components.

Data of VOC emission of the filaments showed for all of them the release of the unreacted monomers and their derivatives; moreover some of them showed also acetone and short-chain alkyl compounds, all solvents for materials that can be found on artworks, employed by both, artist and conservator. The compounds detected at room temperature were unexpected in such a short time of observation although documented in literature, both at room temperature and above. If safety of artworks wants to be preserved, these results should be taken into consideration under the workplace hazards point of view too.

TGA data underlined how the printing temperature is often close to the Tdo of the material in nitrogen atmosphere, therefore damages during this step cannot be excluded and should be verified.

Another issue related with these materials is that their formulation is often modified and characterising all of them is, in practice, impossible. Therefore, if conservators want to realise the conservation treatment with 3D printing techniques, the choice of the filament has to be made with attention and among a small group of selected materials, finding a fast method to check that the composition has not changed drastically. This aspect has then to be balanced with the needed time, the cost and the value of the treatment itself. Materials traditionally employed for such interventions have been deeply investigated and provide a safe and known answer to the integration issue. On the other hand, new techniques allow modelling pieces with specific software that permit to manage objects pictures in order to create a unique printed piece that perfectly fits on the area to be renovated.

4.1. Further developments

Since integrations on artworks might require a certain resistance to loads or particular forces due to their position on the object, mechanical properties should also be considered to have a complete view of the material wherewithal. Mechanical tests should be carried out on just printed and on aged specimens. Another aspect that was not investigated is the possible swelling and damage induced by water on some of the materials, especially PLA, as suggested by Proikakis [14]. These ageing investigations find reason considering that filaments, as in general conservation materials, should be suitable either for indoor and outdoor environments, or at least chosen with this awareness.

Furthermore, changes in crystallinity degree of just-printed and aged pieces could be investigated through DSC, thermal analysis not introduced here, but carried out to investigate deeper samples thermal transitions.

About VOC release, more information might rise from a calibration of the system in order to give an idea of real concentrations. Finally, tests on mock-ups are suggested, with the insertion of printed pieces on traditional materials (especially organic ones) and observation of the interaction, keeping in mind two of the fundamentals of conservation: reversibility and durability.

IOP Conf. Series: Materials Science and Engineering **364** (2018) 012029 doi:10.1088/1757-899X/364/1/012029

References

[1] Braghieri A et al. 2016 Utilizzo della stampante 3D per la ricostruzione di parti mancanti nel restauro di due sculture lignee. Oral presentation *Il Restauro nell'era della Fabbricazione Digitale*, Nov, 4, 2016, Parma (IT)

[2] Bigliardi G et al. 2015 **R**estauro e innovazione al Palazzo Ducale di Mantova: la stampa 3D al servizio dei Gonzaga *Archeomatica* **3** 40-4

[3] Neumüller M et al. 2014 3D Printing for Cultural Heritage: Preservation, Accessibility, Research and Education *3D Research Challenges*Eds M. Ioannides and E. Quak LNCS 8355, 119–34

[4] Scopigno R. et al. 2011 3D Models for Cultural Heritage: Beyond Plain Visualization *IEEE*, **0018-9162** 48-55

[5] Antlej K et al. 2011 Combining 3D technologies in the field of cultural heritage: three case studies Eds Dellepiane M et al. *Pro of The 12th International Symposium on Virtual Reality, Archaeology and Cultural Heritage* VAST 1-4

[6] Fantini M et al. 2008 3D restitution, restoration and prototyping of a medieval damaged skull *Rapid Prototyping J* **14(5)** 318-24

[7] La Trofa F 2017 Restauro 3D: la rinascita di Palmira *3D for everything*, website, http://3dstories.protocube.it/restauro-3d-la-rinascita-di-palmira/

[8] Hawotte S 2016 L'impression 3D au service du traitement de conservation-restauration d'une cruche en terre cuite gallo-romaine *CeROArt. Conservation, exposition, Restauration d'Objetsd'Art* **5** CeROArtasbl

[9] Brandi C 1981 *Teoria del restauro* (Torino: Einaudi)

[10] Amoroso GG Camaiti M 1997 *Scienza dei materiali e restauro. La pietra: dalle mani degli artisti e degli scalpellini a quelle dei chimici macromolecolari* (Firenze: Alinea)

[11] Berger GA 1971 Application of heat-activated adhesives for the consolidation of paintings *International Institute for Conservation of Historic and Artistic Works* **11(2)** 124-8

[12] Stephens B et al 2013 Ultrafine particle emissions from desktop 3D printers *Atmos environ* **79** 334-9

[13] Wojtyła S Klama P Baran T 2017 3D printing safe? Analysis of the thermal treatment of thermoplastics: ABS, PLA, PET, and nylon *Joccupenviron Hyg* **14(6)** D80-5

[14] Proikakis CS et al 2006 Swelling and hydrolytic degradation of poly(D,L-lactic acid) in aqueous solutions *Polym degrade stabil* **91** 614-9

IOP Conf. Series: Materials Science and Engineering **364** (2018) 012030 doi:10.1088/1757-899X/364/1/012030

Considering nonlinear properties of concrete in the design of reinforced concrete structures for torsion

N Jurkowska
Assistant professor, Tadeusz Kosciuszko Cracow University of Technology, Poland

E-mail: nadzieja.jurkowska@pk.edu.pl

Abstract. Restoration and maintenance of cultural heritage concrete and reinforced objects always require accounting for the torsional stress in the elements, as well as the tensile, compressive and shearing stress. It should be considered, that torsion always develops in the elements of spatial structures, including cross-ribbed systems, bridges and overlaps, elements of shells and domes, stairs, etc. This paper briefly systematizes substantive results of the last studies of an urgent problem of torsional stiffness of reinforced concrete elements, including theoretical investigations, experimental studies and numerical research. Stress redistribution in the statically indeterminate reinforced concrete structures' elements depends on both torsional and bending stiffness. However, while bending stiffness is considered in the design practice, the torsional stiffness is not aknowledged by engineers, nor used in building standards and software packages. This paper aims to substantiate the necessity of considering the torsional stiffness of reinforced concrete elements in design practice. This article is the result of numerous investigations conducted in the scientific school of professor Taliat Azizov. They certify that reduction of the torsional stiffness must be considered along with the bending stiffness in different stages of behaviour of concrete structures. Moreover, the research conducted for this paper proved that a change in the shear modulus of concrete also affects the deflected mode of structures and should be accounted for as the component of torsional stiffness. In order to provide engineers with specialized design techniques, they have been developed for the calculation of torsional stiffness of reinforced concrete elements with normal cracks of rectangular, triangular, t-sections, box-sections, and other cross-sections. The engineering bilinear shear stress-strain curve for concrete has also been proposed on the basis of experimental studies.

1. Introduction

Accounting for not only compressive and tensile stresses is essential during design of reinforced concrete structures. It has been proved, than neglecting shear capacity of the reinforced concrete elements is crucial [1,2]. However influence of torsion can also considerably influence on the strains' redistribution (in most cases in combination with the bending moment or / and shear force). In design or restoration, neglecting torsion on the strain redistribution in statically indeterminate structures results serious errors, affecting strength, stability, and deformability of such systems [3]. However international building standards don't consider it as an impact required a computation. According to the European design standard (BS EN 1992-1-1:2004, p. 6.3.1), a full torsional design for both ultimate limit state and the serviceability limit state must be provided for the structures, the static equilibrium of which depends on the torsional resistance of elements. For statically indeterminate structures it should not be provided in case, if torsion arises due to compatibility conditions only.

Meanwhile, in fact, torsion always arises in spatial structures' elements. For example, in the cross-ribbed systems, load is always unsymmetrical for some beams and results in the twist about their longitudinal axis. Any asymmetrical loading of plate causes torsion in the extreme beams. In order to identify the correct values of torsional moments of shells and domes' extreme elements, common deformations of the joint elements should be considered. Evaluation of the deflected mode of bridges must always include resistance to the torque, because elements often ruin exactly due to torsion, which is an important factor in the spatial structures' work.

While torsional strength is being widely researched nowadays, considering torsional stiffness is still mostly neglected. Neither building codes, nor engineering software (Ansys, Abaqus, Scad, Lira, Nastran, etc.) consider change of torsional stiffness in the design process.

Regularly, building standards set torsional stiffness as a constant value depending on bending stiffness, with no consideration of the shear modulus. According to the ACI code [5], the torsional stiffness should be considered for elements in equilibrium torsion and may be neglected for the case of compatibility torsion. The German standards do not provide torsional stiffness' evaluation, however in the design of frame structures the stiffness condition must be provided [6]. In Poland there are no strict regulations for the torsional stiffness of concrete [4,7]. Nevertheless, for reinforced concrete elements, which may affect essentially the stiffness and strength of the whole construction, it is recommended to design in torsion very carefully [8,9]. Complex sections are proposed to be divided into rectangular sections and set their resistance to the torsional moment to be proportional to the whole section's torsional stiffness.

Other European building standards [4], as well as Canadian, Australian, New Zealand and Hong Kong [10–15] do not consider torsional stiffness of reinforced concrete elements.

Speaking of post-Soviet countries, definition of the torsional stiffness is provided in Belarus. It is defined as a sum of torsional stiffness of individual rectangular parts of the section, so that the total stiffness of them must be maximum [16]. In Ukraine and Russia the torsional stiffness is neglected, and the accounting for only torsional strength is provided in special cases [17,18].

In 2001 the specific design techniques was proposed [3]. However till now the torsional stiffness is generally not considered in the design practice.

2. Experimental research of the torsional stiffness of reinforced concrete elements

2.1. Consideration of nonlinear properties of concrete in torsional stiffness' evaluation

This paper examines the nonlinear properties of concrete due to plastic deformations and cracks. The Ukrainian researcher Taliat Azizov derived the specialized theory of the spatial overlaps' behavior [3], which allows the consideration of torsion and torsional stiffness GI in the design process, as well as its change due to its significant influence on the strain redistribution. Furthermore, the scientist proved the serious impact of normal cracks, which arise in structural elements due to bending, on torsional stiffness [19].

The development of design techniques accounting for the nonlinearity of structures' deformation was carried out in two directions [20]: limit equilibrium and deformation models.

Most models for analytical determination of the torsional stiffness of reinforced concrete elements provide change of the initial shear modulus by considering the coefficient k_t, which is defined experimentally and might be defined through various relations (e.g. of torsional moments in the cracked and uncracked areas). There are also techniques for estimation of torsional stiffness considering nonlinear structure's deformation of particular sections for torsion with bending [21], at the elastic-plastic stage of work under action of bending and torsional moments, by changing Young's modulus depending on the level of loading and through the relation of elastic design deformations to the deformations in the elastic plastic stage [3,22].

First, on the basis of experimental data, scientists were tended to describe the concrete behaviour under pure torsion by the elasticity theory [23] or plasticity theory [24]. Data obtained by different researchers can be hardly compared due to significant differences between experimental dependences resulting dissimilarities of the factors, conditions and methods of experimental studies.

The author [25] accounts for the torsional stiffness as a function of the axial forces and states that according to N. J. Nielsen diagram the torsional stiffness doesn't depend on placement of reinforcement, but on the Young's modulus and slab's high for the uncracked element. For the cracked element one should use number of coefficients which characterize geometry of the section.

The scientist [3] recommends to define the torsional stiffness by comparing the displacements from torsion of the cracked and uncracked elements.

2.2. Research of destruction character of the cracked reinforced concrete elements of different cross sections

In the scientific school of the professor T. Azizov numerous experimental studies of pure and reinforced concrete of cracked and uncracked elements under pure torsion and torsion with bending have been conducted. The dependence "torque – angle of twist" was determined for all of the experimental series with the aim to explore the behavior of the cracked reinforced concrete elements.

Various cross-sections, including rectangular, triangular, t-sections and box-sections [3] have been explored. Those experiments proved the necessity of taking the torsional stiffness into account both experimentally and theoretically. The particular techniques and propositions for definition of the torsional stiffness and strength of the reinforced concrete elements of different sections with normal cracks under torsion have also been elaborated. The outcome of the proposed design technique helps to provide reinforcement of structures more precisely with accounting for necessary strength and stiffness of overlaps and beams. The artificial normal cracks simulated the cracks, which had already been formed in the concrete element due to bending.

There was the possibility to consider cracks of different height and distances between them allows evaluating the torsional stiffness of the entire element, not only of the specific section.

Conducted experiments for combined bending with torsion allowed making important conclusions, among which are:

1) diverse numerous investigations prove that the torsional stiffness decreases in the cracked elements. Sometimes significant reduction of stiffness results in destruction of the element;

2) reinforced concrete specimens with normal cracks, at stages closed to the destruction, have nonlinear dependence "torsional moment – angle of twist";

3) level of crack resistance increases, while considering torsional impact;

4) torsional displacements are larger, and cracks' number enlarges comparatively to the pure torsion;

5) the dowel force contracts deformations, and increase in reinforcement reduces the torsional displacements in both series of samples. Thus, increasing of the longitudinal bars' diameter and the height of stiffness of the compressed zone of the element with normal cracks results approaching of its stiffness to the stiffness of an uncracked element.

Besides of the geometrical parameters and, consequently, the moment of inertia I, the shear modulus of concrete G changes after cracking and effects the torsional stiffness GI. Due to the essential influence of the torsional stiffness of concrete on the stress' redistribution elaborating of the experimental technique for the complete stress-strain curve for concrete in shear was a vital task.

3. Research of the secant shear modulus of concrete

Similarly to the Young's modulus, the Kirchhoff's modulus of concrete can be defined by the analytical techniques or diagram methods. The differences of the curves are provided by the elastoplastic coefficients.

Deformation models allow tracing the change in the bearing capacity of the material considering variation in cross-sectional area in each load moment. The shear stress is generally taken either equal to the tensile stress, or depending on the tensile and compressive stresses (preventing margin of safety and due to absence of the relevant experimental data).

The exact stress-strain curve is difficult to obtain for concrete in pure torsion due to several reasons, one of which is the presence of significant differences between the experimental dependences obtained by various researchers due to diversity of the factors, conditions and methods of experimental research.

3.1. *Theoretical and bilinear shear stress-strain curve for concrete*

There is no option to change the secant shear modulus of concrete in the program software. It might be defined as a function of the Young's modulus. However, shear modulus and elastic modulus are linearly interdependent for the homogeneous and isotropic materials in the area of elasticity only, while after cracking the stress-strain relationship is not linear anymore due to plastic deformations and change of deformation characteristics of the material. That's why the iterative approach was applied.

The theoretical stress-strain curve for concrete in shear was elaborated on the bases on the theory of elastic plastic deformations [26], according to which the stress intensity in each moment of loading is correlated with the strain intensity for all types of stress condition, by analogy with the compression stress-strain curve used for bending [27]

The secant shear modulus of concrete is defined similarly to the secant elastic modulus [26]:

$$G_c = \frac{E_c^0 \vartheta_c}{2(1+\mu_c)} \tag{1}$$

where G_c – secant shear modulus of concrete; E_c^0 – initial elasticity modulus of concrete; ϑ_c – variation coefficient of the secant shear modulus of concrete [26], μ_c – Poisson's ratio.

3.2. *The bilinear shearing stress-strain curve for concrete*

In order to consider the secant shear modulus in design the bilinear stress-strain curve was implemented (Figure 1).

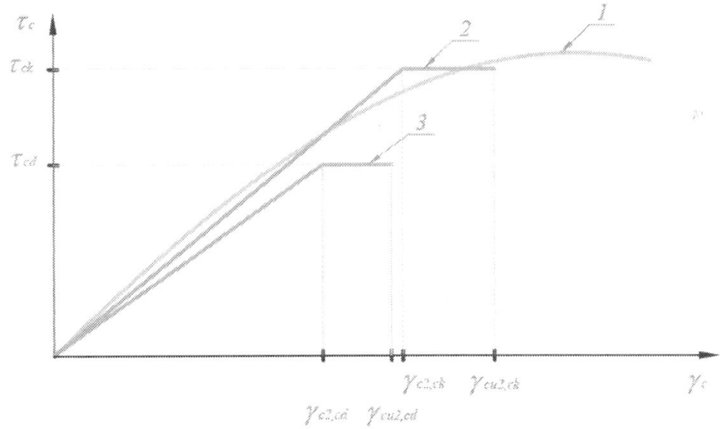

Figure 1. The shear stress-strain curve for concrete:
1. theoretical function; 2. bilinear characteristical dependence; 3. bilinear design chart

The theoretical shearing stress-strain curve has been simplified for the usage in the design practice by analogy with the compressive stress-strain curve for concrete [28].

$$\begin{cases} \tau_c = G_c\gamma_c \ for\ 0 \le \gamma_c \le \gamma_{c2} \\ \tau_c = \overline{\tau}_c \ for\ \gamma_{c2} \le \gamma_c \le \gamma_{cu} \end{cases} \tag{2}$$

where τ_c – tangent stress for concrete; $\overline{\tau}_c$ – ultimate tangent stress for concrete; G_c – secant shear modulus of concrete; γ_c – relative angular deformations for concrete in shear; γ_{cu} – relative ultimate

IOP Conf. Series: Materials Science and Engineering **364** (2018) 012030 doi:10.1088/1757-899X/364/1/012030

angular deformations for concrete under shear; γ_{c2} – relative angular deformations for concrete in shear correspondingly to the ultimate shearing stresses.

The design curve has been elaborated on the basis of the conducted experimental research [10].

3.3. Experimental shear stress-strain curve for concrete

Usage of the real deformation curves for concrete in shear in the building practice is constrained due to lack of reliable experimental data on its parameters [4,29]. Nonetheless, it should be used as a generalized characteristic of the mechanical properties of concrete, similar to the compression stress-strain curve.

The technique for obtaining shearing stress-strain curve for concrete is more complicated comparatively to the compression one due to difficulties in experimental definition of parametric points of the descending branch. The reason is in the sudden element's destruction while using traditional experimental setting due to the immediate realization of the ultimate potential deformation energy to the impact energy.

That's why the shear stress-strain curve of concrete in pure torsion considering plastic deformations has been examined, and dependence between stresses and angular deformations of concrete under short-term load has been determined.

Nine concrete cylinders of annular cross-sections have been tested for resistance to pure torsion. Thickness of the ring was varied for three series by three cylinders from 2 to 4.5 cm. The external diameter was 20 cm, and the length of the researched part was equal to 60 cm. The support areas were enlarged in order to avoid influence and specimens' destruction due to crushing. The fine-grained concrete was used due to thin annular sections of the specimen [28].

In the conducted experiments the load was transmitted to the concrete specimen of annular cross-section through a steel beam in order to control the external forces on the concrete cylinder. The special setting was elaborated to provide common deformations of the concrete model with the steel cross-beam at all stages of loading, for details see [10]. First there was obtained the complete stress-strain curve for concrete in shear and proved the influence of the shear modulus of concrete as a torsional stiffness' component on the strain redistribution in the spatial cross-ribbed systems. The correlation ratio, obtained by comparing of theoretical and experimental curves, is equal to 0.984.

Obtained data require more experimental research and verification for various concrete specimens of different classes for concrete.

4. Numerical investigations and consideration of the torsional stiffness when designing spatial structures

Influence of the torsional stiffness on the strains' redistribution of the cross-ribbed structures' elements was proved by professor Azizov and his students in the result of numerous numerical experiments, considering nonlinear properties of concrete in torsion.

The proposed research has been conducted for the noncracked elements in order to observe the change of nonlinear properties due to secant modulus' variation only.

Simulation of the cross-ribbed structures was accomplished in the SP "LIRA". Conducted calculations considered change of the torsional stiffness due to cracking. They were fulfilled by the iteration technique. First in the process of external iterations the torsional moments' values were obtained. Internal iterations consisted in calculation of the secant elastic and shear modulus by the deformational dependences [26,27]. According to the new values of stiffness the torsional moments were calculated anew, and the process repeated till convergence (the relative error between values of current and previous iterations did not exceed 10^{-7} %).

In such iterative calculations the least stiffness is taken at the initial iteration for the sections with maximum stresses. Therefore, in the next approximation the stiffness' ratio changes abruptly, and the design stresses increase significantly. With each subsequent iteration those values gradually converge. Thus, the technique allows to compare results of calculations with and without accounting for the torsional stiffness' change due to stresses' redistribution.

4.1. Accounting for the change of torsional stiffness due to secant shear modulus

Five 3 meter girders of section 25x25 cm were divided into 20 elements lengthwise. The span between them was 1 m. The slabs of section 15x5 cm are divided into 5 elements. Concrete class C 16/20, stiffness parameters of the ribs are: EF=1687500 kN, EI_y=8789 kNm^2, EI_z=8789 kNm ^{2}, GIt= 6196 kNm^2. Stiffness of the slab – EF=202500 kN, EI_y=379.69 kNm^2, EI_z=42.188 kNm^2, GI_t=63.0703 kNm^2.

First the uniformly distributed load q=50.8 kN/m was applied to the first beam, and only change of torsional stiffness as a function of the shear modulus of concrete was considered. Comparison of ultimate diagrams of the torsional moments for half of a beam is proposed on the Figure 2. The approximation area and the last meanings of torsional moments are pointed, and the numbers of the loaded beams are indicated above.

The ratio error between shear modulus, as well as between torsional stiffness, achieved 8-13% in different loading cases. The ultimate strain redistribution occurs in the extreme bar's elements when the last beam is under the load (Figure 3).

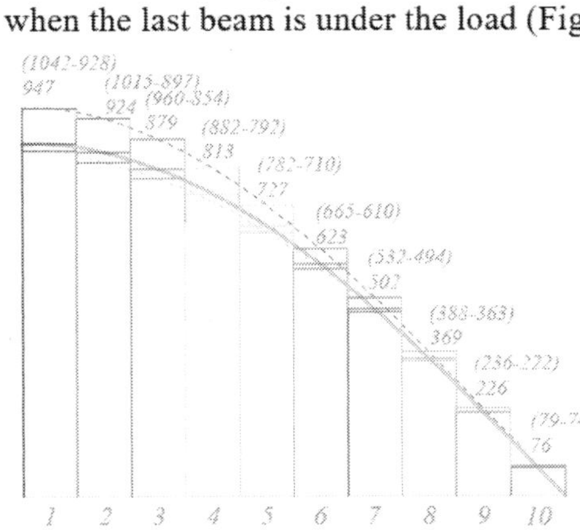

Figure 2. The torque (Mt) diagram for the element (in half lengthwise)

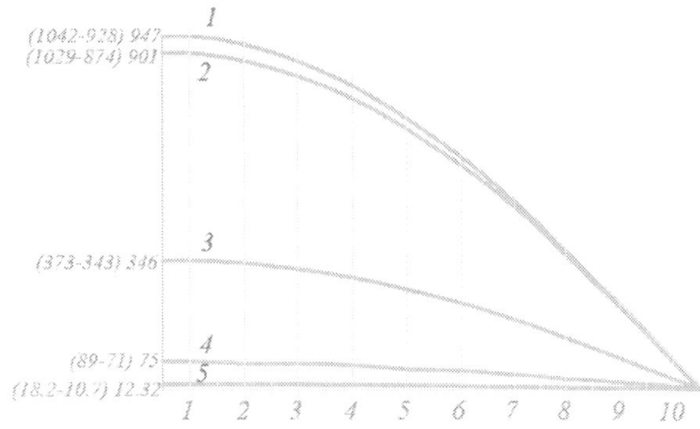

Figure 3. Comparison of ultimate Mt diagrams in 3 loading cases.

4.2. Considering of both secant Young's and Kirchhoff modulus of concrete as a torsional stiffness' component

In the second experiment the change of bending and torsional stiffness was researched along the length of the beam. Both secant modulus were evaluated and considered as a stiffness' components on each iteration for every finite element for every 5 beams (Figure 3). Uniformly distributed load q=45 kN/m was applied to the first beam.

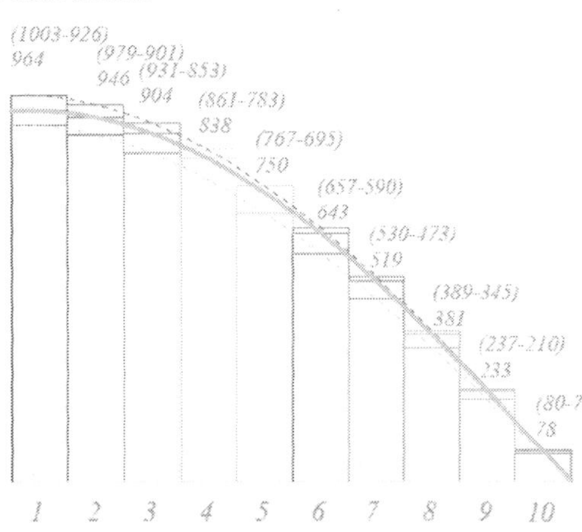

Figure 4. Approximation of the torsional moments M_t for the element (in half lengthwise).

In the result of comparison of the secant shear modulus' values G_c is 26,33% in the end element of the first beam, wherein the change of the secant elastic modulus E_c in the same element is 4,030%. In the central elements of the first beam change of E_c achieves 42,66%, and variation of G_c is 0,7896%.

Obtained data reaffirm the necessity of considering the shear modulus while determining the internal stresses and their redistribution in statically indeterminate systems, when considering torsional stiffness of reinforced concrete elements

5. Conclusions

1. Both torsional strength and stiffness should be considered during design of the new structures and restoration of the cultural heritage objects. Otherwise, neglecting torsion causes cracking and destruction of the elements (e.g. due to insufficient bending strength of the weakened by the spatial cracks elements, caused by torque, etc.).

2. The torsional stiffness should be researched equally with the bending stiffness of the reinforced concrete elements

3. The secant shear modulus of concrete, as the component of torsional stiffness, affects the deflected mode and the spatial structures' behavior should also be considered in the design techniques, building standards, as well as in the program software.

4. The parametric points of the complete stress-strain curve for concrete in torsion have been obtained, which allowed an observation of the secant shear modulus of concrete in torsion considering its nonlinear properties.

5. The bilinear engineering shear stress-strain curve has been elaborated on the basis of the experimental data in order to implement into building standards.

References

[1] Foti D 2014 Shear Vulnerability of Historical Reinforced-Concrete Structures *Int. J. Archit. Herit. Conserv. Anal. Restor.* 453–67

[2] Palmisano F 2017 A Preliminary Study on Shear Capacity of Historical Reinforced Concrete Beams *Int. J. Herit. Arch.* **1** 608–623

[3] Azizov T N 2006 *Prostranstvennaya rabota zhelezobetonnyh perekrytij. Teoriya i metody rascheta [Spatial work of concrete overlaps. The theory and techniques for calculation]* (Ukraine, Poltava)

[4] BS EN 1992-1-1:2004 2004 Eurocode 2: Design of concrete structures: Part 1-1: General rules and rules for buildings *London Br. Stand. Inst.* 230

[5] ACI 318-11 2011 *Building Code Requirements for Structural Concrete and Commentary (ACI 318M-11)* (Farmington Hills: American Concrete Institute)

[6] DIN 1045-4:2012-02 2014 Tragwerke aus Beton, Stahlbetoon und Spannbeton – Teil 4: Erganzende Regeln fur die Herstellung und die Konformitat von Fertigteilen *Ersatz für DIN 1045-42001-07* 378

[7] PN-EN 1992-1-1:2008/NA: 2010 2010 *Eurokod 2: Projektowanie konstrukcji z betonu. Część 1-1: Reguły ogólne i reguły dla budynków*

[8] Godycki-Ćwirko T 2006 *Skręcanie, w: Podstawy projektowania konstrukcji żelbetowych i sprężonych według Eurokodu 2* ed Sekcja Konstrukcji Betonowych KILiW PAN (Wrocław: Dolnośląskie Wydawnictwo Edukacyjne)

[9] Knauff M, Golubińska A and Knyziak P 2014 *Tablice i wzory do projektowania konstrukcji żelbetowych z przykładami obliczeń* (Wydawnictwo Naukowe PWN)

[10] Jurkowska N R 2016 *Kruchennya zalizobetonnyh elementiv v inozemnyh doslidzhennyah i standartah [Torsion of the reinforced concrete elements in the foreign researches and standads]: [monography]* (Kyiv: Interservis)

[11] AS 3600-2009 2009 *Australian Standard Concrete Structures* vol 2009(Australia, Sydney: Standards Australia International Ltd.)

[12] Hong Kong CP-04 2013 *Code of Practice for Structural Use of Concrete* (Hong Kong: The Government of the Hong Kong Special Administrative Region, H.K.B. Department)
[13] IS 456-2000 2000 *Code of Practice for Plan and Reinforced Concrete* (India, New Delhi: Bureau of Indian Standards)
[14] BC 2: 2008 2008 *Design Guide of High Strength Concrete to Singapore Standard CP 65 200. BCA Sustainable Construction, Series – 3* (Singapore: Building and Construction Authority)
[15] NZS 3101-06 2006 *Concrete Structures Standard* (New Zealand, Wellington: Standards New Zealand)
[16] SNB 5.03.01-02 2003 *Betonnye i zhelezobetonnye konstruktsii [Concrete and reinforced concrete structures]* (Belarus: Minsk)
[17] DBN V.2.6-98:2009 2011 *Betonni ta zalizobetonni konstruktsiyi. Osnovni polozhennya. [Concrete and Reinforced Concrete Structures. Substantive Provisions]* (Kyiv: Minregionbud of Ukraine)
[18] SNiP 2.03.01-84* 1989 *Betonnye i zhelezobetonnye konstruktsii [Concrete and reinforced concrete structures]* (Moscow)
[19] Azizov T 2009 Zhestkost' zhelezobetonnyh ehlementov pri kruchenii i ee vliyanie na prostranstvennuyu rabotu mostov [Stiffness of reinforced concrete elements in torsion and its impact on the spatial work of bridges] *Mech. Phys. Fract. Build. Mater. Constr. Collect. Sci. Work. H. V. Karpenko Physico-Mechanical Inst. Natl. Acad. Sci. Ukr.* 576–90
[20] Geniev G A, Kissock V N and Tyupin G A 1974 *Teoriya plastichnosti betona i zhelezobetona [The theory of plasticity of concrete and reinforced concrete]* (Moscow: Stroiizdat)
[21] Kasayev D H 2001 *Prochnost' ehlementov zhelezobetonnyh konstrukcij pri kruchenii i izgibe s krucheniem [Strength reinforced concrete structures' elements under torsion and bending with torsion]* (Rostov na Donu: Rostov university)
[22] Paramonov D 2012 *Zhestkost i prochnost zhb elementov z normalnymi treshinami pri izgibe s krucheniem [Stiffness and strength of reinforced concrete elements with normal cracks under bending with torsion]* (Ukraine, Odessa: Odessa State Academy of Civil Engineering and Architecture)
[23] Cowan H.J. 1972 *Kruchenie v obychnom i predvaritel'no napriazhennom zhelezobetone [Torsion in the ordinary and prestressed reinforced concrete]* (Moscow: Strojizdat)
[24] Kemp E L, Sozen M A and Siess C P 1961 Torsion in Reinforced Concrete. A report on a Research Project Sponsored by The University Research Board 128
[25] Gudmand-Hoyer T 2004 *Stiffness of Concrete Slabs. Report BYG – DTU R-092* vol 4(Danmark)
[26] Yaremenko O F and Shkola Yu.O. 2010 *Nesucha zdatnist' ta deformatyvnist' zalizobetonnykh sterzhnevykh elementiv v skladnomu napruzhenomu stani [Bearing capacity and deformability of the reinforced concrete strain elements in a compound stress].* (Odessa: Odessa State Academy of Civil Engineering and Architecture)
[27] Karpenko N I 1996 *Obshchie modeli mekhaniki zhelezobetona [General models of the mechanics of reinforced concrete]* (Moscow: Stroyizdat)
[28] Azizov T and Jurkowska N 2015 *The shear modulus of concrete considering plastic deformations* (Odessa)
[29] Bentz E C, Vecchio F J and Collins M P 2006 Simplified modified compression field theory for calculating shear strength of reinforced concrete elements *ACI Struct. J.* **103** 614–24

IOP Conf. Series: Materials Science and Engineering **364** (2018) 012031 doi:10.1088/1757-899X/364/1/012031

Knitting virtual tribes together: new audiences for cultural objects

Jane Malcolm-Davies
Aalto University Helsinki Finland & Centre for Textile Research Copenhagen Denmark

Jane.malcolm-davies@aalto.fi

Abstract. A focused re-analysis of knitted items in museum storage suggests a range of new approaches to researching the cultural biographies of objects and writing new chapters for them. The Knitting in Early Modern Europe (KEME) project examined more than 100 knitted caps held in European and North American museums. The project aims to bring a new scientific approach to the under-researched history of knitting as a textile craft and technology and to provide a protocol through which knitted objects which have been silent for too long can tell their stories. New chapters in these knitted objects' biographies are also being written using innovative research methodologies. Scientific analyses more usually applied to archaeological material (such as scanning electron microscopy) have revealed the sophistication of the raw materials required to make them. Attempts at reconstruction using methodologies drawn from experimental archaeology have provided clear evidence of the complexity of the processes required to turn fleece into aspirational faux fur or mock velvet. A range of audiences was invited to engage with the knitted caps as investigators in a Citizen Science project and as commentators on the evidence via an interactive online database. A tribe has gathered around the material and has an ongoing dialogue with it. Social media have created the opportunity for the knitted caps to continue their lives into the future as dynamic cultural objects as new and different people engage with them reconstruct them and reinterpret them for their own purposes. The KEME project has generated the possibility for mute objects lying in the dark drawers of museum storage to have multiple vibrant new lives among diverse tribes well beyond their geographical and traditional cultural communities. This paper explores how cultural biographies can be set free to change and grow without the need for physical examination or public presentation of the objects.

1. Introduction

Knitting is a widespread popular pastime which has experienced a renaissance in recent years - as serious leisure, as contemporary craft, and as a guerrilla art form among other activities [1]. It has well-known therapeutic effects and provides a strong sense of accomplishment [2]. This increased popularity has generated new interest in its history and a large potential audience for the material culture of knitted objects. Knitters' motivations include the pleasure of learning about the origins of their craft and the creative inspiration it provides. In addition, there is a demonstrable demand among dress historians and archaeologists, museum staff and wider publics for accurate information about the construction and reconstruction of historic dress, including knitted garments. Museum collections of knitted items are treasure troves of information from which new knowledge about textiles in the past may be generated. This paper explores how archaeological and historical knitted objects' biographies can be set free to change and grow by reaching large and diverse audiences without the need for physical examination or formal public presentation - via an interactive online database.

It has been argued that digital data collection, analyis and presentation in archaeology is a 'shift to the digital paradigm' and that in this communications revolution 'we are all digital archaeologists' [3, 4]. The study reported here combines textile analysis, dress history, experimental archaeology and the digital collection and presentation of archaeological and historical evidence for the history of knitting. The project explored ways in which a user group can be empowered to become digital archaeologists and suggests how an audience for a virtual collection may be cultivated while data is gathered and then contribute to the design and content of what is made available online.

2. The Evidence – the Problems and the Potential

Definitive evidence for the origin, development and spread of knitting is as yet unclear [5]. It was a key technological innovation of the medieval and early modern eras on which there is little published scientific scholarship. Unusually for a craft activity, European knitting has a short history. It appeared late and moved through Europe at speed compared to other textile crafts (such as weaving, netting and knotting which are millenia older) – a phenomenon almost unnoticed by historians. It appears in geographically diverse places from the 13th to 15th centuries, in some cases associated with 'magic' garments said to have been worn by the infant Christ which expanded as he grew [6-8]. One reason why this important development in clothing construction has gone unreported is that knitting is undervalued owing in part to its association with the domestic and the feminine [9].

Knitting produces fabric similar in appearance to other looping techniques such as netting, nalbinding and sprang. Close and careful study is often needed to determine whether an archaeological find or an historical object is knitted or constructed in another way. This has led to a lack of knitted evidence being securely identified in the archaeological and historical record. It is often incorrectly or minimally catalogued. A second problem is that those knitted items which are accurately identified have not been systematically studied because there are no agreed protocols for scientific interrogation of the physical evidence. This is in contrast to woven textiles for which there are a number of conventional systems for describing their characteristics (for example, the analysis framework developed by Centre International d'Etude des Textiles Anciens) [10]. As a result, much published work on the history of knitting pays little regard to the historical record, with a few notable exceptions [5, 6]. Some excavated items have been reported, although not always with reference to the key characteristics required for a comprehensive overview [11-13]. Knitwork has, somewhat surprisingly, lacked a common language in which to tell its tale. Knitted objects (if they are reported at all) are described in a contradictory collection of craft conventions (which are not agreed across geographical or cultural boundaries) and assumptions about how they were brought into being [14]. What is lacking are straightforward reports of the items which permit interested parties to draw their own conclusions about how they were made.

A recent research project brought a new scientific approach to the under-researched history of knitting as a textile craft and technology. Knitting in Early Modern Europe (KEME) has attempted to tackle these issues by recording a representative body of objects from the late fifteenth to the early seventeenth centuries. It has identified more than 100 knitted caps in museum collections of the kind worn by ordinary people in the early modern era. Despite this abundance of examples and the iconic longevity of the knitted hat, headwear has been of scant interest to scholars, with a few notable exceptions [15, 16]. The 100 caps are recorded as having been shipwrecked, deliberately concealed, preserved in peat bogs or discarded as beyond use. Despite the geographical spread of their discoveries, they present remarkable similarities in their materials and manufacture [17]. The knitted caps have, in most cases, doubtful provenance and little is known about the conditions in which they have been kept. This makes them highly representative of much material held in museums in general and of that which informs the history of knitting in particular. The knitted caps present a range of similar items from different places discovered at different times and under different conditions and kept in a variety of environments. Very few examples are on public display and those that are have taken on a significance beyond their representativeness. As more examples are made available online, the less influence these individual caps have on the popular perception of the genre [17]. The caps in storage offer an opportunity for extensive comparable data to be extracted and clues about the caps' construction and new insights into knitting's history to emerge. KEME aims to overcome the challenges to comparative analysis, the poor standard of reportageto date, and the caps' relative inaccessibility as physical objects by building a dynamic online digital collection.

3. The Data Collection

This rare evidence for lower-class dress has never before been systematically studied as a collection – partly because of its far-flung locations which range from Copenhagen to Croatia. Each of the identified caps (except three) were examined at the various museums where they are stored. A standard set of measurements, photographs, USB microscope images and (at those places where permission was granted) samples for further analysis were taken. How much of this evidence would prove useful to online visitors to the collection was not clear at this stage of the project.

The samples were taken in order to prepare new chapters in the knitted objects' biographies using innovative research methodologies. Most have not been studied in any detail although there are exceptions. Only a few early modern knitted caps have been reported in any detail [18-22, 13, 14]. Other work on knitted caps has been cursory or speculative [23-25, 5]. Most of it is descriptive rather than analytical, including a study identifying five distinct categories of caps of which the most common is the flat cap - a term which belies its constructional complexity [19].

Scientific analyses more usually applied to prehistoric and ancient textiles, such as scanning electron microscopy (SEM) have revealed the sophistication of the raw materials required to make them. [26] The fleece for the knitted caps must not only have been suitable for spinning into yarn but also suitable for fulling, napping and shearing into a silky pile after knitting. Results so far have included SEM images of fibres taken from the knitted objects. Further investigations will be undertaken using isotope analysis in the hope of pinpointing the items' provenance and radioncarbon 14 dating to see how closely the objects match their assumed timeframe for manufacture.

4. The Citizen Science KEME team

A series of research and outreach activities were designed which would assess how online users might interact with the material to provide guidance for the design of the digital interface. The intention was to discover what characteristics of the knitted caps would be of interest to online users. These discussions were also intended to inform the new protocol for reporting knitwork. Input from potential users was key to the identification of appropriate terminology for the continued scientific study of the material. This was motivated by the firm intention that the data be used by people beyond the academic and museum communities. The outreach activities included workshops at conferences and special events on the history of dress in the early modern era targeted at reenactors and other hobbyists interested in reconstructing and wearing historical clothing. A series of seminars were also held at the University of Copenhagen with a core of academic participants and a wider network of knitting and handicraft enthusiasts.

Key to the project was the recruitment of a representative audience for the digital collection before its content and design were finalised. The online knitting community is active and diverse [1]. Ravelry, the premier online hub for knitting, has more than 212,000 Facebook followers and more than four million members registered on its website. At least 25 of the subgroups within Ravelry pursue an active interest in historical knitting, including the study of sheep and yarn and how to knit reconstructions of historical objects. These groups are indicative of the online community of knitters at large for whom access to original archaeological and historical knitted objects is desirable. It was prudent to consider whether their requirements were in line with the plans for data collection and, if not, whether the digital evidence gathered could be interrogated to plug any gaps.

Recruiting a pilot audience followed the principles of Citizen Science, which began with the pioneering use of idle computers for identifying Mersenne prime numbers in 1996. It recruited 1.25 million participants (www.mersenne.org) to generate what has more recently been dubbed crowdsourced knowledge [27]. Facebook groups devoted to the discussion of sixteenth century clothing and its reconstruction were alerted to the need for volunteers and mailing lists of people who had expressed interest in this area of activity were contacted with an invitation to spread the word about participation in KEME. Many of the people who belong to these groups have an active interest in making historic garments and it seemed that their participation would be better secured by giving them a practical task relevant to the archaeological material.

An experimental archaeology project was already planned to explore a key variable in the construction of the knitted caps. This was developed into a participatory activity for which people could volunteer in order to stimulate interaction with the digital collection. This focused on the raw materials and processes required to manufacture and finish the caps. Unlike most modern knitted wool, which is fluffy and hairy, the knitted caps of the early modern era originally had a silky surface with smooth vertical fibres akin to velvet. The knitted wool fabric was fulled, napped and shorn to create this raised surface, which obscures the knitted loops with what looks and feels like velvet pile. It is likely that in the sixteenth century this nap was intended to mimic silk velvet or real animal fur – both luxury textiles usually reserved for the elite [28]

The choice of fleece from which the yarn is spun and knitted affects the quality of the nap produced. Modern European fleeces differ from those available in the sixteenth century and today vary across the world [29]. Sheep breeds have moved a long way from their early modern ancestors and it is difficult to know the precise differences in the yarns produced now compared to those of the past [30]. Volunteers were invited to participate by testing fleeces they could obtain locally in the hope of capturing a wide variety to compare and contrast. A small experimental flat circle of knitted fabric was designed (12.5cm in diameter) based on the circular linings found in some of the original caps. These circular swatches – or 'swircles' – became the evidence the experimental archaeologists produced by knitting, fulling and napping a variety of yarns. Details of the swircles were to be reported via an online questionnaire and then sent for physical evaluation according to the characteristics of the nap.

Workshops were also held at Jamestown, Virgina and Chicago, Illinois in the United States and in Copenhagen in Denmark. At these workshops, participants carried out fulling and napping experiments on swircles knitted from five different fleeces provided for the purpose and on examples they brought with them. Invitations were issued for the attendees to join the online KEME team with the opportunity to view and provide comment on the proposed digital collection of knitted caps. A series of eNewletters was circulated to the participants explaining how the project was progressing, promoting the workshops in Copenhagen, and encouraging people to visit and interact with the planned online material. These were not as regular or frequent as originally intended but, on average, one per month was sent out during the first eight months of the project.

A total of 164 people volunteered to participate online in the KEME fleece experiment which was well in excess of the 100 target for recruitment. Of these 164 volunteers 48% (78 people) were from the United States with 27% (45 people) from the United Kingdom. The remaining 25% were very diverse but when numbers were divided into language regions, English-speaking countries accounted for 85% (139 people) with the Nordic countries providing the bulk of the remaining volunteers at 8% (13 people). English was the only language used to promote the project, which explains the lack of diversity, although the representation of Nordic countries can be explained by the KEME project's base at the University of Copenhagen, Denmark which has close connections to its neighbours.

Volunteers were asked their primary reason for being interested in contributing to the project. There was a wide overlap in motivation with many volunteers expressing an interest in both knitting and archaeology. Interestingly, 25% (41 people) were primarily motivated for professional or tertiary education reasons, often citing qualifications in the field. A further 22% (36 people) were reenactors with a third of these specifically belonging to the Society for Creative Anachronism. Volunteers were also asked to rate their craft skill level. A large proportion (41%; 67 people) said they were expert knitters. Only two were non-knitters, both of whom stated that their interest lay in analysing and measuring the extant caps rather than providing swircles. The KEME volunteers had a wide range of professional and amateur knowledge relating specifically to experimental archaeology, which was something that had not been anticipated.

5. The KEME Database

A pilot database was developed using the data and images from 31 knitted cap linings, which provide a useful subset of the knitted caps. The pilot version of the database lacked a home page but provided attractive and easily navigable access to the archaeological objects with detailed information about them. The requirement to produce a nap on their knitted swircles encouraged many of the volunteers to register for the pilot KEME database, which was launched in May 2017. This provided photographs with overviews and measurements of the 31 knitted linings plus descriptions and dimensions.

The database was built using Filemaker by Jodie Cox, an illustrator and relational database specialist. The design drew on experiences with a previous visual database which provided online images of sixteenth century effigies in churches in England and Wales (www.tudoreffigies.co.uk). This database stimulated a great deal of feedback from local historians whose churches were featured, costume designers researching period clothing, and reenactors reconstructing historic garments [31] It was originally launched in 2007 and still generates considerable traffic. The current platform has been running since 2015: there were 34,307 page visits in that year followed by 21,410 in 2016 and 14,307 in 2017. The falling numbers probably reflect the lack of new material in the database.

It was crucial that the online collection be attractively presented with an intuitive interface to accommodate the anticipated international audience. An essential measurement to understand knitted fabric is the gauge – the ratio of knitted loops in the horizontal and vertical directions. These are conventionally recorded per inch or per 10 cm. Although it is relatively easy to convert from one to the other, British and North American knitters can visualise the former and mainland Europeans the latter very easily and find it tiresome to make the calculation. The database was designed with the option for each user to designate their preferred units.

Feedback on the data to be presented in the KEME database was also gathered in workshops held throughout the project. Participants had opinions of the usefulness of specific measurements – for example, the gauge of the knitting and the diameter of the yarn are essential for reconstructions. These discussions led to specific features being included in the database such as a 'knit this cap' icon, which displayed the gauge and yarn diameter when activated. A refined and expanded version of the database was launched in August 2017 with information on 37 split-brimmed flat caps in addition to the 31 linings. This provided opportunities to generate graphs of the data and to view a digital 3D reconstruction of a knitted cap. A series of curated collections was also offered on the home page to provide instant insights into the collection.

Each object's detailed record offers an invitation to comment via a detailed questionnaire asking the viewer to check the observations recorded in the database. Key questions about how the knitted caps may have been knitted were included – for example, knitters who wish to reconstruct the caps want to know whether they were knitted from the centre of the crown outwards or from the brim inwards. It is rare that any evidence is cited for one method or the other. One aim of the database interaction was to stimulate debate about how this method of working might be deduced with confidence. Visitors to the database were invited to give detailed responses to the objects including close inspection and measurements based on the photographs by answering ten questions. A few objects inspired respondents to provide very detailed observations, including tracings, which were upoaded to the questionnaire as jpgs. It was noteworthy that of the seven object-specific responses, four were from the same person.

People still continue to join the project. There are currently 179 people signed up for the experimental archaeology, which will officially end in August 2018. A total of 91 people continue to make regular visits to the database (not including the three people in the project development team). This represents more than half of the people who signed up for the project.

6. Discussion

The KEME project's aims were to provide a protocol through which knitted objects, which have been silent for too long, can tell their stories in a scientific and scholarly way, while providing access to the physical evidence for the history of knitting to a wide audience. The development of the online database employed Citizen Science to recruit a constituency of first adopters, whose engagement with it and feedback on it were crucial to its current format and content. As a result, a tribe has gathered around the material and has an ongoing dialogue with it. However, opportunities to respond to the material are currently too demanding and a lighter touch is required for users to feel they can interact easily and quickly.

The volunteers who signed up to the KEME team reported that they were more interested in analysing the online materials (41%) or the practical craft elements of spinning and knitting (51%). However, this distribution was not reflected in the contributions received from them. Despite 41% of volunteers expressing an interest in working with the database and a total of 339 visits to it being recorded, only eight observations of objects featured there were received from three people over the initial six months of the project. This was particularly disappointing as the database received traffic well in excess of the capacity for which it had been designed. This lack of direct engagement with the material was also reflected in the practical experimental archaeology. More than 200 people are now subscribers to the KEME eNewsletters and approximately 50% open them each time they are sent. Nevertheless, only 29 people reported their swircles data in the online questionnaire and just 15 actually sent processed swircles. Some of the participants sent more than one set of swircles bringing the total number received to 20. There may have been far more people who knitted and napped swircles but were not sufficiently inspired to report them or send them. Many people may have been satisfied with access to the database and to knit their own swircles without feeling the need to report their participation. This form of indirect engagement with an online community has been termed lurking and in some contexts it is criticised as parasitical behaviour [32, 33]. However, research into how people decide whether to reveal themselves online suggests that 'lurking is not free-riding but a form of participation that is both acceptable and beneficial to most online groups … all members of a group are part of a large social milieu, and value derived from belonging to a group may have far-reaching consequences' [34]. Guidance for encouraging more active online participation may be found in specialist research projects which have demonstrated how the fear of interacting online in inappropriate ways influences how active and visible a person will be [32].

The project has demonstrated that social media created the opportunity for the knitted caps to continue their lives into the future as dynamic cultural objects as new and different people engage with them, reconstruct them, and reinterpret them for their own purposes. But for this to be sustained, there must be regular social media activity to inspire and drive people to actively register their responses. Devising a reward for participation may help to build the active audience. Alternatively, tracking the number of people who log in and use the database provides a clear indication of its use, even if their relationship to the material is limited to lurking.

7. Conclusion

The knitted caps in the KEME project were for the most part a forgotten collection of objects which have lain in relative obscurity until the recent past. Very few have been on display and those that are have contextual information which merely hints at their complexity. A focused re-analysis of knitted items in museum storage suggests a range of new approaches to researching the cultural biographies of objects and writing new chapters for them. It is hoped that hobbyists can draw creative inspiration from online access to the archaeological material and specialists may use the evidence to build new knowledge.

The KEME project has produced a rich database of accessible evidence using an innovative blend of disciplines (archaeology, textile analysis, craft expertise, and dress history). An audience of early adopters was invited to engage with the knitted caps as investigators in a Citizen Science project and as commentators on the evidence via an interactive online database. The results suggested that imaginative effort is required to generate a tribe which wishes to engage in the material with the provision of simple pathways for interaction. There was evidence, however, that some individuals became very deeply engaged with the objects and were willing to put considerable time and thought into their interpretation of them.

This virtual collection, which is available online to both hobbyists and specialists, can now be viewed as a 'family' of objects despite the physical distance between them. As a result, their individual characteristics are available for scrutiny in new and different ways by a wide constituency of users. This paves the way for novel and exciting interpretations. It has been suggested that 'traditional expertise is seen as obsolete in the era of crowdsourced knowledge' and that it remains to be seen how 'digital community archaeology co-production public engagement and participatory technologies will affect the future of professional practise'. Further study of the KEME database and its tribe may prove instructive in this context. Nevertheless, at this stage, it may be confidently asserted that the KEME online interactive database has freed the cultural biographies of early modern knitted caps to change and grow without the need for physical examination or public presentation. The KEME project has generated the possibility for mute objects lying in the dark drawers of museum storage to have multiple vibrant new lives among diverse tribes well beyond their geographical and traditional cultural communities.

References

[1] Orton-Johnson K 2014 Knit purl and upload: new technologies digital mediations and the experience of leisure *Leisure Studies* **33** 3 305-321 DOI: 10.1080/02614367.2012.723730

[2] Riley J Corkhill B Morris C 2013 The benefits of knitting for personal and social wellbeing in adulthood: findings from an international survey. *British Journal of Occupational Therapy* **76** 2 pp50-57

[3] Roosevelt C Cobb P Moss E Olson B and Ünlüsoy S 2015 Excavation is destruction digitization: advances in archaeological practice. *Journal of Field Archaeology* **40** pp325–346 p339

[4] Morgan C and Eve S 2012 DIY and digital archaeology: What are you doing to participate? *World Archaeology* **44** pp521–537 p523

[5] Black S 2012 *Knitting: fashion industry craft* (London: V&A Publishing

[6] Rutt R 1987 *A history of hand knitting* (London: Interweave Press)

[7] Warburg L 1984 Den Strikkende Madonna I Syd og Nord *Cras* **39** pp79-90 Silkeborg Kunstmuseum

[8] Wyss R 1973 Die Handarbeiten der Maria: Eine ikonographische Studie unter Berücksichtigung der Textilen Techniken *Artes Minores: Danke aus Werner Abegg* (Bern: Stampfli)

[9] Turney J 2009 *The culture of knitting* (London: Bloomsbury)

[10] Centre International d'Etude des Textiles Anciens (CIETA) 2006 *Vocabulary of technical terms: fabrics* (Lyon: CIETA)

[11] Zimmerman H 1998 Textiel uit de beerput van de Latijnse School Hervonden stad: Jaarboek voor archeologie bouwhistorie en restauratie van de gemeente Groningen (Groningen: Stichting Monument en Materiaal)

[12] Walton P 1981 The textiles in ed P Harbottle and M Ellison An excavation in the castle ditch Newcastle upon Tyne. Archaeologia Aeliana 5th Series 9 pp190-228

[13] Crowfoot E Pritchard F and Staniland K 1992 *Medieval Finds from Excavations in London: 4 Textiles and Clothing c1150-c1450* (London: HMSO)

[14] Malcolm-Davies J Gilbert R. and Lervad S 2018 Unravelling the confusions: Defining concepts to record archaeological and historical evidence for knitting *Archaeological Textiles Review* **60** in print

[15] Zander-Seidel J. (2010) Haubendämmerung: Frauenkopfbedeckungen zwischen Spätmittelalter und Früher Neuzeit. In Schwinges Rainer C. Schorta R. Oschema K. (eds) *Fashion and Clothing in the Late Medieval Europe - Mode und Kleidung im Europa des späten Mittelalters* **1** 2 pp37-43

[16] Ribbert M 2003 *Auf Bäsler Kopfen: Kulturgeschichtliche Aspekte von Hüten Hauben Mützen* (Basel: Schwabe & Co)

[17] Malcolm-Davies J and Davidson H (2015) 'He is of no account … if he have not a velvet or taffeta hat': a survey of 16th century knitted caps eds Grömer K and Pritchard F *Aspects of the Design Production and Use of Textiles and Clothing from the Bronze Age to the Early Modern Era NESAT XII* (North European Symposium for Archaeological Textiles) Hallstatt Austria May 2014

[18] Botticello J 2003 *The conservation of two Tudor knitted woollen flat caps* undergraduate thesis Camberwell College London

[19] Buckland K 2005 Woollen caps ed Gardiner J *Before the mast* (Portsmouth: The Mary Rose Trust) pp31-35

[20] Flury-Lemberg M 1988 *Textile conservation and research* (Bern: Abegg-Stiftung) pp328-333 and 222-231

[21] Maeder E 1980 The restoration of a 16th century Swiss mercenary soldier's costume in the Historical Museum Bern Switzerland ed Pertegato F *Conservation and restoration of textiles: international conference Como 1980* (Milan: CISST) pp224-228

[22] Zimmerman H 2007 *Textiel in context: een analyse van archaeologosche textielvondsten uit 16e-eeuws Groningen* (Groningen: Stichting Monument & Materiaal)

[23] Levey S 1982 Glove cap and boot-hose *Crafts* 57 pp34-40

[24] Buckland K 2008/9 'A sign of some degree' - the mystery of capping *Text* **6** pp40-45

[25] Zimmerman H 2000 Barett: gestrickt und gewalkt – Replik eines Groninger Fundstückes aus der Zeit des Predinius*Ornamente* **2** pp44-4

[26] Malcolm-Davies J 2017 Shedding light with science: the potential for 21st century studies of 16th century knitting *Journal of Dress History* available at: https://www.dresshistorians.org/volume-1-issue-1 (last accessed 1 July 2017)

[27] Richardson L-J and Lindgren S 2017 *Open Archaeology* **3** 1 139–148 DOI: https://doi.org/10.1515/opar-2017-0008

[28] Malcolm-Davies J 2016 An Early Modern Mystery: A pilot study of knitting napping and capping *Archaeological Textiles Review* **58** pp57-74

[29] Ryder M 1984 Wools from textiles in the Mary Rose a 16th-century English Warship *Journal of Archaeological Science* 11 pp337-263

[30] Ryder M 1964 The History of Sheep Breeds in Britain *The Agricultural History Review* **12** 1 pp1-12 and 65-82

[31] Malcolm-Davies J 2007 'Wherein Taylors may finde out new fashions': Constructing the Costume Research Image Library (CRIL) eds Hayward M & Kramer E *Textiles and text: re-establishing the link between archival and object-based research* (London: Archetype) pp37-44

[32] Warburton S 2013 Space for lurking: a pattern for designing online social spaces In Warburton S. and Hatzipanagos S. *Digital Identity and Social Media* (Hershey PA: Information Science Reference) pp149-158

[33] Abramson T 2006 Learning styles lurkers and parasites *Journal of Instruction Delivery Systems* **20** 3 pp3-4

[34] Nonnecke, B and Preece, J (2000) Lurker demographics: Counting the silent Proceedings of CHI'2000 The Hague Netherlands pp73-80

IOP Conf. Series: Materials Science and Engineering **364** (2018) 012032 doi:10.1088/1757-899X/364/1/012032

The Revival of a Cultural Heritage: *Hangzhou Embroidery and its Development Today*

Chenxi Qi

Chinesisch-Deutsche Kunstakademie
China Academy of Art, P. R. China

Abstract. Hangzhou embroidery is a traditional embroidery handicraft in China. As a cultural heritage, it had a long history and faces a precarious situation today. The preservation of Hangzhou embroidery faces several challenges, including product marketing, business model and technological presentation. The leading role of preserving Hangzhou embroidery, Yi Hui Co.Ltd, are trying application of new technologies and organizational strategies. Nevertheless, it is not only the form, but also the content, the process that embedded the spirits of creating an artistic handicraft should also be preserved. The interaction between cultural heritage and technology and social context is key to its preservation and development.

1. Introduction

The correlation between preservation of cultural heritage and the interdisciplinary approaches to their future development has been a topic of intense discussion in China recently. Many institutes, universities, NGOs, private corporations and government agencies have been involved in this endeavor. The preservation of cultural heritage requires the application of new technologies and organizational strategies. Cultural heritage, both tangible and intangible, inevitably faces the challenge of regaining its own sustainability. It is not only the form, but also the content, that matters when we preserve our heritage. This paper will examine the current efforts for the preservation of Hangzhou embroidery – an ancient Chinese handicraft now in danger of extinction. What should be preserved is not just ancient skills and processes of a handicraft, but also its aesthetic and visual content transformed and implanted in new mass produced commodities.

2. Hangzhou embroidery

Hangzhou embroidery is one of traditional embroideries in China[1]. It first appeared almost 2000 years ago. Hangzhou, known in ancient times as Qiantang, has always been the capital of Zhejiang province, famous for its silk industry throughout the history of China. It provided the perfect environment for Hangzhou embroidery's development. When Hangzhou became China's national capital during the Southern Sung dynasty (1127-1279), Hangzhou embroidery reached its peak. Since that time, this special embroidery technique was available only to people inside the royal palace. For the centuries that followed that time, Hangzhou embroidery became increasingly popular among ordinary people as the civic market began to flourish. Another reason for its popularity among the merchant classes was the increasing demand for religious objects by local Buddhists. Hangzhou embroidery was renowned

[1] The most famous four types of traditional embroideries in China are: Su embroidery, Yue embroidery, Xiang embroidery and She embroidery, all named after the provinces where they are originated. And Hangzhou embroidery is name after the city of hangzhou.

for its use of Buddhist iconography as well as secular images.[2] (Figure.1)The decorative and ritual objects in local temples and private houses required lots of fabric embroidery handicraft.

Figure.1. Li Delian and Yao Dezhen, *Miao Fa Lian Hua Jing*, 44.1cm x 1953.3cm, 1355-1361, details.

At the beginning of 20th century, embroidery was mostly applied to clothing and accessories for the general commercial market. In 1950s, the local government started a factory to produce traditional opera costumes with Hangzhou embroider, in order to re-employ the embroidery artists who lost their careers during the years of warfare in the previous decade.[3]

According to the definition of 'intangible heritage' by UNESCO, the knowledge and skill to produce traditional crafts is an important aspect of cultural heritage. Preservation of heritage as a living tradition passed on from one generation to the next becomes far more important than the preservation of the objects themselves. The current situation of Hangzhou embroidery is quite similar to that of most ancient handicrafts all over the world. Less than ten Hangzhou embroidery masters are still working today. [4]There are many reasons attribute to its decline: inefficiency and complexity of the manufacturing process, lengthy apprenticeship but low payment afterward, massive changes in both environmental and social system of the region, shift from agricultural economy to industrial economy, the consumer market, etc.

---

[2] A piece of embroidered Buddhist scripture, *Miao Fa Lian Hua Jing*, 44.1cm x 1953.3cm, dated 1355-1361, made by Li Delian and Yao Dezhen, is now in Shanghai Museum, China, Figure.1.

[3] In 1956, Hangzhou government founded Hangzhou Jin Xiu Corporation, later developed into Hangzhou Costume Manufacture, recruited several Hangzhou embroidery maters, including Wang Guirao, Zhang Jinfa, Lou Baotu, Yang Bailin, Yang Rongsheng, etc.

[4] Among the masters of Hangzhou embroidery, Chen Shuiqin and Zhao Yijun are in their 70s, Jin Jiahong and Yu Zhiyin are from the younger generation. Jin Jiahong published several papers on Hangzhou embroidery and now is the secretary of China Embroidery Art Committee.

Figure.2 The family factory around 2012

However, the situation for another traditional branch of embroidery in southern part of China is completely different. Two hundred miles north of Hangzhou, a town called Zhenhu and its surrounding region is famous for its Suzhou embroidery industry. Suzhou embroidery has been well preserved and highly developed in the last few decades. With the support from Jiangsu province, Suzhou embroidery was identified as national intangible cultural heritage in 2006. It has become an important financial resource for that region. It is in mass production and there are over 10,000 women working as professional embroiderers there. Most of the orders are from the Internet. Even the Chinese government commissioned special pieces of Suzhou embroidery for diplomatic purposes. It seems that is the solution for reviving embroidery in a modern context requires reforming the industry, building a "brand" image and targeting the right clients. Will the solution for Hangzhou embroidery be the same?

3. Yi Hui Co.Ltd

One of the leaders of the movement to re-establish Hangzhou embroidery is a local company called Yi Hui Cultural Creativity Co. Ltd.. The owner had a family embroidery factory for over twenty years and formed the new company in 2013.[5] With the financial support from the family business, the new company has developed rapidly during last five years as so called a social company. Preserving Hangzhou embroidery is the explicit purpose of the company. The company combines multiple approaches to build an integrated system of industrialized embroidery production:

3.1. Cooperation with government, real estate group, non-profit organizations:

Yi Hui also works closely with government, real estate group, non-profit organizations to promote the Hangzhou embroidery industry. Their products were given as official presents by Chinese government in several important political events, such as G20 summit in 2016 and China-Arab States Expo in 2017. Hangzhou embroidery works, created by Yi Hui, were included in the collection of the permanent mission of the People's Republic of China to the United Nations office in Geneva. The company has developed new decorative products for the household market with a real estate group as

[5] The company is named after the CEO, Ms. Chen Yihui. Her grandmother used to be a embroider and founded an embroidery business, which was inherited by Chen Yihui's mother later. Chen Yihui was appointed to be the CEO of Yi Hui Cultural Creativity Co. Ltd. by her family.

IOP Conf. Series: Materials Science and Engineering **364** (2018) 012032 doi:10.1088/1757-899X/364/1/012032

partner. In 2015, International Silk Union (ISU) was founded in Hangzhou, the Yi Hui corporation joined it as a member of ISU History Culture Professional Committee.

Figure. 3 Yi Hui Co. Ltd. exhibited in Geneva

3.2. Exploring the new market both in and out of China.

Yi Hui is exploring new markets both in and out of China. The company is currently selling traditional embroidery objects as souvenirs to the tourist industry market which provides basic revenue for the company. It works in cooperation with Shangri-La Hotels and Resorts Group to expand its global trade. In August 2017, there was a business tour of Shangri-La Hotel and Resorts Group held in Hangzhou Midtown Shangri-La Hotel. The tour included 32 domestic hotels and 22 foreign hotels and over 200 guests attended the event. Yi Hui was invited to hold a charity embroidery show during the event.

Figure.4 Yi Hui Ltd. in promotion with Shangri-La Hotel

3.3. Develop a new system of production by updating design, manufacturing line and brand image.

Hangzhou is the center of e-commerce of China. With the support of big data and highly developed delivery service systems, Yi Hui is effectively connected to a complete system for the supply of raw materials, mass production lines, efficient delivery services, and point-of-purchase retail counters. Another challenge is to update its artistic style to meet new customers' taste. Even though the traditional visual elements in Hangzhou embroidery are most well known, such as Baoxiang flowers, birds, butterflies and images of Buddhism, etc., it is quite difficult to attract young buyers and high-end clients. The Yi Hui Ltd. works with young designers, trying to develop new brand image and new products. Besides typical items such as clothes, purses, bags, shoes and scarves, the owner of the company is also looking for opportunities to combine embroidery with different kind of products. For example, notebooks, memory stick, etc.

Figure 5. Yi Hui's new product line

3.4. Developing public education and market awareness.

Seeking new exhibition venues and greater access to technological resources for promotion and public education, the company has now launched a new gallery space in the historical Canal Region of Hangzhou. They are currently looking for opportunities to work with media artists and visual communication designers – searching for new ways to show the ancient story of embroidery, and let visitors experience the process of making embroidery with VR and interactive technology.

3.5. Engaging charity and education, especially recruiting deaf mutes to study Hangzhou embroidery.

Yi Hui has cooperated with local special education schools to train disabled people with embroidery skills. Qualified students could work for the company after graduation. It also founded a private fund in 2017 in order to promote its charity.

Figure.6 Yi Hui Ltd. in the exhibiting hall by the Grand Canal

4. The Challenges for embroidery as cultural heritage

One of the most unique techniques of Hangzhou embroidery is gold thread embroidery: the gold thread wound parallel to the surface of silk and red thread is used to fix them at certain points. Imagine

IOP Conf. Series: Materials Science and Engineering **364** (2018) 012032 doi:10.1088/1757-899X/364/1/012032

the shining gowns of antiquity with these embroideries on them. Now, in the exhibition hall, audiences can see the gold thread shining as if golden light is floating on the surface of the cloth under a moving light. This beautiful embroidered fabric exhibits a sense of richness and luxury that is unparalleled.

However, this specific pattern is still of a traditional aesthetic taste. It is likely to be buried among other embroidery patterns and forms, for most of the visual images of Yi Hui's products are flowers, birds, landscape. It represents the taste of the major body of Yi Hui's clients. Yi Hui is considering improving their brand and product design, but a bigger challenge is whether Hangzhou embroidery could be inherited as a rare handicraft or converted into a modern product.If the gold thread embroidery, for example, could be re-designed, it could certainly exist in a different context, including company brand logo, website, commercials and a variety of other products. As the core artistic element of the whole brand, both the historic cultural heritage and modern visual transformation of a traditional embroidery pattern might be easily recognized and embraced by customers, especially the younger ones.

The company developed new and more modern designs but didn't continue the practice. They were aware of the advantage of a transformative design but also worried about the result this may cause. There was something deeper in the understanding of this cultural heritage that made the company vacillate. First, from the company's point of view, using a new design meant a complete reshuffle of the product series, and its potential for the exploration of new markets. Secondly, the Yi Hui's CEO insists the best way of preserving this heritage is not only selling abundant of embroidered bags or fans but also retaining the process of hand making. According to UNESCO, the handicraft skills and techniques are integral parts of intangible cultural heritage. Is this traditional handicraft technique of Hangzhou embroidery able to re-establish itself in the marketplace, or must it be transformed into a mechanized process? Is there a new product that could present both its visual beauty and historic value? Would consumers be interested in such a product? What is the best way to preserve and propagate an authentic cultural heritage? How can the Internet be used for marketing a newly invigorated cultural tradition? What contemporary issues might influence the traditional designs of Hangzhou embroidery? To what extent can the visual effect of embroidery be retained? The company is actually in a turning point in its decision making process of how and what to preserve.

WISHOW

Figure. 7. Yi Hui's New logo

The Yi Hui company faces a dilemma: supporting the embroidery masters through the successful use of mass production leads to a possibility of losing the essence of the handicraft of embroidery, but reviving the cultural heritage of Hangzhou embroidery only by supporting those embroidery masters who keep small-scale studios is not economically feasible. The local government of Hangzhou once offered financial support to some of the embroidery masters, hoping they could recruit more apprentices to inherit their skills. Unfortunately, no more than five Hangzhou embroidery masters are still working now and they are known by few people. Some tried to work with local art schools, for instance, teaching some techniques in fashion design department or textile departments. Both of these strategies proved not to be viable in preserving Hangzhou embroidery. The success in preserving Suzhou embroidery industry is on account of numerous reasons including geography, cultural history and the differing economic structures of the two regions.

5. Form and content

There is an important question of how cultural heritage's development could be parallel to that of technology. Consumers' visual experience has been greatly changed by new exhibition technology and new forms of design and art, such as interaction design, integrated design, inter-media art, digital art, programming art, etc. Curating an exhibition is increasingly seen as another form of artistic creation. Yi Hui is currently developing a new exhibition plan for Hangzhou embroidery, which will apply interface touch screen, virtual reality and interaction design to show the process of making embroidery. There is mounting concern in the company that the delicacy and tranquility of hand making is easily overwhelmed by dazzling digital effects. The best way to experience embroidery is to hold the needle and to sew a pattern on a piece of cloth. Touch screens are for interaction between human and machines, but are not *real* materials. Visual communication design is effective for presenting information and history, essential aspects in public education programs, but do not, in and of themselves, lead to increased sales and revenue. Advanced technological tools for marketing and education involve a large capital investment for initial programming and installing digital devices. The technological expenses of mass production re well established and seem justified. The enhancement of marketing an important part of Chinese cultural heritage remains a central concern. The extent to which technology could be involved in the preservation of Hangzhou embroidery remains uncertain.

Figure.8. A piece of gold thread embroidery in detail.

Cultural heritage could be interpreted and effectively preserved even without any of these fascinating technologies. Mass media and social networking offer alternative methodologies to conventional marketing techniques and exhibition driven publicity. The film ***Jiro Dreams of Sushi***, directed by David Gelb, is the story of 85 year-old Japanese Sushi master Jiro Ono and his heir. Through this film, not only the 3 star Michelin Sushi bar but also the spirit of the master and his followers came to be viewed by a large public as the symbols of mastery of the art of one's work and a lifelong pursuing perfection – an important part of cultural heritage. Similar productions might capture the same spirit of Hangzhou embroidery and many other forms of handicraft. The use of current mass media technology raises questions about the necessity of using new technologies to present Hangzhou embroidery.

The relationship between the master and his son is a touching and compelling aspect of Gelb's ***Jiro Dreams of Sushi***. The connections and conflicts between the father and son, between human and nature, make the story even more complicated but add heightened interest and human emotion. If we look into the relationship between embroidery masters and their apprentices, we will very often find similar stories. A cultural heritage like Hangzhou embroidery, or Jiro's Sushi consists not only of the objects, but the process making-process, and a living physical, social and psychological human activity. How could technology preserve that? What connection could technology have to building the very human connection between ordinary viewers (consumers) and this craft in a short period of time? The sushi bar of Jiro Ono preserves its great cultural traditions, while becoming extremely successful financially. Its customers now need to make reservations at least one month in advance. The film obviously contributed to its popularity. To enjoy a splendid sushi service once in Jiro's bar is more than just a meal, but like a journey of experiencing the most precious spirit of mankind which rarely exists in our daily life. How could embroidery be interpreted in that manner? It is clear that film and social media can shape public awareness of its cultural heritage. Film is of course a technological innovation in itself, but it may be argued that it is the human expression – the art of film – rather than its technological spectacle that makes it a powerful tool for cultural awareness. The next wave of technology, that is to say, the emergence of artificial intelligence, challenges all forms of traditional culture, cognition and human behaviors. It seems essential that we struggle to retain our own sense of human spirit when we try to deal with it. With the ever expanding and increasingly rapid advance of technology, the most essential character of our own heritage is likely to be disseminated to the public. What ***is*** the essential aesthetic value of art of traditional embroidery – of any cultural heritage? How can it be adapted to the contemporary world in a cross-cultural and an interdisciplinary context?

6. Conclusion

By the end of 2016, there are 39 elements corresponding to China are included in UNESCO International Cultural Heritage lists. There are 52 World Heritage properties are presently located in China and over a thousand cultural heritage sites are recognized and supported by Chinese government. New technologies and successful business modes are being applied to the preservation of these national legacies. Many of these cultural heritage sites will face questions of the same kind that Hangzhou embroidery has been trying to answer. The question of whether technology would have a positive or negative impact on cultural heritage echoes our fear for the uncertainty of the role technology would play in the future of humanity. To revive the whole industrial production of Hangzhou embroidery, or any other cultural heritage, is an equally difficult question. Preserving cultural heritage not only protects the fading stories of our past but also reshapes our images for future generations. Even though the power of digital revolution is formidable, an ancient handicraft will be an important part of the definition of humankind, for we are not going to be digitalized in every way.

Reference

[1] Xu Jia 2016 Embroidered Painting: Research of Chinese Traditional Jiangnan Embroidery Ph.d paper
[2] Sun Jing 2013 Investigation Report and Inheritance Mode of Hangzhou Hand Embroidery MFA paper
[3] Kong Jia 2011 The Development of Hangzhou Embroidery and its Inheritance Pub. Arts.13 196
[4] Wang Haiqing 1955 The History of Hangzhou Embroidery and its heritage Zhejiang Art. Cra. 2 45
[5] Kranzberg M 1986 Technology and History: "Kranzberg's Laws" Tec.Cul. 27 544 (Baltimore: The Hopkins University Press and the Society for the History of Technology)

IOP Conf. Series: Materials Science and Engineering **364** (2018) 012033 doi:10.1088/1757-899X/364/1/012033

The future of Heritage Science and Technologies from a legal point of view

S. Segnalini
Assistant Professor PhD University of Rome, Italy – art lawyer Piselli & partners law firm, Rome, Italy
E-mail: silvia.segnalini@uniroma1.it and silvia.segnalini@piselliandpartners.com

Abstract. The development and the prospect of future cultural heritage should certainly be confronted with the development of new technologies applied to the arts which, from a legal point of view, implies mainly three important aspects.
1. ***Copyrights for software***: this seems to be the least interesting for the jurists as many aspects have already been addressed and solved, and therefore on will just quickly list some of them. In fact the copyright explicitely protects the software as work of wit and consequently it is clear that the author is the dominus of his creation in the sense that he is the only person who can decide how to allow other people use his software. So f.e . the relationship between the copyright legislation and open source is actually less difficult than it maybe thought. In addition open source use is fully regulated by Law no. 633/1941 (so called "legge sul diritto d'autore") which guarantees its proper use.
2. ***Legal issues related to the so-called BIM technology***: Building Information Modelling (BIM) is a software for plannig and implementing building projects based on 3D modeling of buildings or structures models. The BIM has already been adopted internationally for the realization of major project and one of the most exciting challenges for a lawyer is to develop a specific BIM contract for the particular field of art and cultural heritage.
3. ***Legal issues related to artificial intelligence***: this certainly represents a field yet to be discovered by jurists and here I will present a series of legal concerns that will engage jurists in the coming years. Such as responsibility and ethics of artificial intelligence. For the above-mentioned considerations, the European Parliament's Legal Affairs Committee approved a report calling on the European Commission to introduce a set of regulations on robotics.

1. Introduction

First of all, it's necessary to stress in which ground has been grown up the matching between the development of new technologies and arts. It all started with some interesting informations regarding robotics applied to the arts.

So if a robot can replace an orchestra conductor as it actually does in some places in South Asia, why does not it consider that the same thing can be done in the restoration field in the short run? Very soon the BIM technology is likely to be engaged to restore large archaeological sites (and to tell the truth I'm quite surprised this hasn't happened yet). At the hearth of all these future applications, there is the software and its legal problems, which unlike those of BIM and artificial intelligence, are now completely resolved.

However, it can be easily noticed that there are no references available regarding this specific topic, so the only way to proceed is by proposing some general ideas which represent the first step to be taken. For this reason, the paper lacks of scientific approach but it hopes to give a serious contribution over time by research and collecting evidence.

2. Copyrights for software

So this first concern represents a sort of "past" in this presentation and many of its important aspects have already been addressed and solved, therefore on will briefly list some of them.

Computer software is generally copyrighted just as a book is. Unlike a book, it is difficult to use just a small portion of it and call it fair use. Computer software can be sold, licensed or freely distributed. Just because software might be free, does not mean it is not copyrighted or that there is no license dictating terms of use.

A license is a legal agreement between you and your software publisher. The license spells out what you can and cannot do with the software. It might specify the number of computers on which the software may be installed or address resale rights. Software may include a validation feature as a check on proper use.

There are several categories of software and licenses:

Shared source - Microsoft developed "shared source." There is not one standard license for shared source, there are several. When using this software, read the accompanying restrictions on use.

Shareware is proprietary software that is distributed freely or at low cost as a way for users to test drive copyrighted software, they are interested in purchasing. Shareware will usually come with a license and registration. Shareware often distinguishes between commercial and non-commercial use.

Freeware – Although free, this software is copyrighted. It may have proprietary code or it may be open source software. It could even be in the public domain. Look for possible restrictions on use.

Open Source (OS) software is copyrighted, but the copyright holder ascribes to the OS license principles. OS comes with a license that uses special criteria known as Open Source Definition (OSD) established by the Open Source Initiative. See OSD for the ten principles. This is non-proprietary software. It can be downloaded, modified and redistributed. Commonly known OS software examples include Apache and Mozilla.

Implied License - This sort of license may appear on the box or in the printed information that comes with purchased software. There will be wording that states that your use of the software implies consent to the terms of use.

Shrink Wrap Licenses- These licenses are found after the buyer tears off the shrink wrap and opens the box. While "shrink wrap" licenses have been found to be unenforceable in some cases, one cannot assume this is will always be true.

Click Wrap License- Also known as "click through," this is similar to a Shrink Wrap license, but we find these all over the web or on CDs. The user is required to agree to terms before using the service.

Copyleft License - This liberal license retains copyright for the creator, but allows the user to modify and distribute the software with the understanding that "downstream" users will be equally generous.

GNU (General Public License) is a copyleft type of license.

Public Domain software – this is software that has no copyright restricting its use, no license, it's free for you to use any way you wish. The purchaser of an original copy of a computer program may "make or authorize the making of another copy or adaptation of the computer program provided:
(1) that the new copy or adaptation is created as an essential step in using the program on one's computer [an ephemeral copy] and that it is used in no other manner, or
(2) that such new copy or adaptation is for archival purposes only and that all archival copies are destroyed in the event that continued possession of the computer program should cease to be rightful".

The making of an archival copy permitted under this section of the law does not apply to software that is licensed unless the license allows it.

If a computer needs maintenance or repair, allows the owner or lessee of the computer to make a copy of a computer program in order to make the repair. This extra copy made in order to repair the computer must be destroyed immediately after the repair is made.

Original copies of software and copies made from originals may be "leased, sold or otherwise transferred…only as part of the lease, sale or other transfer of all rights in the program." Adaptations "may be transferred only with the authorization of the copyright owner." However, if your license speaks to the lease, sale or transfer of software, you must abide by those stated restrictions.

3. Legal issues related to the so-called BIM technology

This second concern looks at the "present", because the so called BIM, which has recently been received a first regulation in Italy as well, it's largely used and on can suppose it will be use soon also to restoration of large archaeological sites.

However many legal aspects have to be solved.

First of all, let's consider the meaning of BIM: acronym for Building Information Modelling is a software for plannig and implementing building projects based on 3D modeling of buildings or structures models.

Building Information Modelling is a set of technologies, processes and standards that allow many stakeholders to design, build and manage a structure in a collaborative way, within a virtual space. BIM is a methodology, a process of planning, planning, realization and maintenance that manages all the phases of life of the building, from the idea to the demolition. Also managing the 4D and 5D, times and costs All this is done by channeling all the disciplines into one model. In the traditional design we imagined a 3D object and represented through 2D and 3D elaborations with lines and volumes that were conventionally associated with objects.

Now, with the BIM methodology one imagines an object in 3D and through paramentamental objects a virtual 3D model is created from which all the documents and the documentation necessary for its presentation will be extrapolated.

The objects that make up the model are intelligent quinids, constructed using geometric parameters and technical information The model therefore has 3 characteristics: Parametric, since the aspects of the object are managed by parameters Informative, as it contains all the information on the technical data sheets Collaborative, because all the disciplines share all the information in a single model unlike the traditional design in which there was a file exchange intertwined with the consequent possibility of loss of data and errors. Nowadays, compared to BIM's level of maturity we find ourselves at Level 2 in which each discipline works on its model and then share the geometric information and not in a single federated model and updated in real time. One day we hope to get to the Shared BIM where all the disciplines work on a single model. The workflow according to the BIM methodology requires a high amount of work in the initial phases of the project for the organization of the successive phases during which the work will be halved.

In traditional design, on the other hand, it involves a lower initial commitment that substantially increases in the production phases of the works. A conscious adoption requires: Availability to invest in process innovation to be competitive in the market that requires it. Human resources specialized in the approach to the new methodology. Increase workload in the early stages and awareness of the advantages in the subsequent phases continuing with the BIM. In fact, changing methodology during construction would lead to a significant increase in work, time and therefore costs. The use of the methodology therefore brings important advantages: The disciplines convey in a single source from which all the graphic and non-graphic elaborations will be extrapolated; The model consists of intelligent objects that contain all the information of the real object allowing the study of interferenxe and thus reducing the variations during construction.

Moreover, when present, the variants will be managed with great ease since updating the model will be updated simultaneously all the processed From the model it will be possible to extrapolate also the metric calculations, which will therefore be automatic and dynamic as they are synchronized with the model In support of the BIM methodology, the most used tool is Autodesk's Revit software, which will help in the construction of a parametric model that can be shared in real time. For better interoperability we use the ifc export format supported by most of the software on the market. In conclusion, using the BIM methodology saves time and costs, reduces errors and simplifies management.

In Italy, a BIM Decree has been signed on last December 13, providing for the progressive introduction of digital information management models in the construction processes of buildings and civil engineering works. The obligation to use will start from 1 January 2019 for contracts for an amount equal to or greater than 100 million euro. Then there will be a progressive adjustment (year by year) up to the mandatory implementation of the BIM also for contracts of less than 1 million since 2025. The decree defines the methods and timing for the introduction of the compulsory electronic modeling methods and tools for building and infrastructure in the contracting stations and for the rationalization of the design activities and the related checks.

This is a useful intervention in the pursuit of a better quality of projects and works, as, through the digitization of the construction sector, benefits are made to public spending and real estate or infrastructure products and the work of insiders.

This of course doesn't consider all problems, and in my opinion the biggest challenge for a jurist is to imagine and build the perfect BIM international contract.

4. Legal issues related to artificial intelligence.

This last aspect represents a field yet to be discovered by both jurists and technologies experts and the day a robot will be able to replace a restorer is no doubt closer than we expect.

The history of Artificial Intelligence (hence A.I.) can be divided into three phases. The first sees the development of intelligent systems in the field of computation. The second is characterized by the imitation of problem solving processes and human reasoning. The third, finally, is distinguishes for the creation of "expert" systems and cc.dd. "connectionists" (Warwick 2012). More recently, with the progressive attention paid by legislators and policymakers to the phenomenon of industry 4.0, we begin to talk about A.I. even outside the laboratories where it had long been confined. It enters so silently, almost unnoticed, in the daily life of people, to mold, tame and often dull their minds, in the total absence of self-awareness and legal regulation.

First of all, a necessary clarification is necessary. A.I. in a broad sense it forms a circle within which numerous applications can be included (1). One thinks, above all, of the branch of cybernetics or that of robotics, which constitutes a sub-branch of the former and deals with those intelligent algorithms that operate in an external environment. Therefore, each approach to the topic must not fall into the error of confusing a part with the whole. A.I. in the strict sense it consists of three elements: one of the machine, one external to it and, finally, one indifferently internal or external to the machine. They are the algorithm, the information and the memory. This is true with reference to any type of intelligent application and is reflected in the most widespread definitions of A.I., which however do not always take into account all three elements (2).

In fact, from an empirical point of view, the A.I. it is the product of an algorithm able to process and synthesize a certain amount of data in a time inversely proportional to the computing power of the machine itself. Each algorithm of A.I. it is then designed to perform a certain function, to achieve a certain output pre-established by the programmer. However, the algorithm - rectius the language string that makes up the algorithm - constitutes only one component, perhaps the most important one, of A.I. The second component is given by the information processed. Needless to say, in fact, that the greater the amount of information processed, more refined and, one might say, 'intelligent', will be the answer that the machine will give to a specific external stimulus.

Thus, the solution given by an intelligent machine to a certain problem will be, in probabilistic terms, the more exact the information on which this solution is based. Finally, the third element is given by the memory, which allows the storage of information. This element should not be underestimated in the analysis of the phenomenon, since the greater the information to be processed, the greater will be the physical or virtual space to be engaged. The algorithm, under a legal profile, constitutes an invention, a work of ingenuity and, as such, subject to intellectual protection as an intangible asset.

On this point, legal science, even in the divergence of views about the type of protection to be granted to software, seems to be in agreement. Information in the broader sense, on the other hand, appears difficult to frame within a single legal definition. This is due to the composite nature of a term that includes data, public or private (sometimes even personal or sensitive), bilateral or plurilateral facts and interactions. Furthermore, it was stressed that the concept of information should not be confused with that of data. Data in itself is meaningless: they only assume it as processed by the algorithm (3).

Finally, the memory is neither more nor less than a warehouse, a place to record information. A place that, like any space (physical or virtual), lends itself to being illegally violated or altered. This circumstance is firmly rooted in the traditional problems that the nascent line of studies in the field of cyber security has been working for a few years now on the most careful doctrine. When the intelligent machine is equipped with a physical substrate, which allows it to move and interact with the outside world, a further element is added to the three that make up artificial intelligence. This further element is the body of the machine, which, in its four components, is thus defined as a robot.

The body of the machine, from a legal point of view, is on two levels: on the one hand, it extends the effects of the machine's actions in the external world; on the other, it allows the interaction of the machine with man and with phenomenal reality and above all, by virtue of this interaction, where the algorithm is so programmed, learning. In this last regard, it must also be taken into consideration the current tendency of the philosophy of science to broaden the research towards the c.d. IT'S AT. (Artificial Empathy), that is that area of robotics research engaged in the construction of synthetic agents with affective skills (4).

Balking (5) says: «when I was a boy, I read all of Isaac Asimov's stories about robotics. In Asimov's world, robots were gradually integrated into every aspect of society. They had various degrees of similarity to humans, but as the stories and novels progressed, the most advanced robots were very human in appearance and form.

The most famous feature of these robot stories is Asimov's three laws of robotics (6) that were to be built into every robot's positronic brain. The three laws are:

First Law: "a robot may not injure a human being or, through inaction, allow a human being to come to harm"; second Law: "a robot must obey orders given it by human beings except where such orders would conflict with the First Law"; third Law: "a robot must protect its own existence as long as such protection does not conflict with the First or Second Laws."

These three laws have been very influential, and even today people imagine what it would be like — or whether it would be even possible — to build them into robots, for example, into self-driving cars» (7).

But then, Balkin goes on, we are rapidly moving from the age of the Internet to the Algorithmic society.

On will diverge from Asimov at this point. Instead of focusing on laws directed at robots (or algorithms), I focus on laws directed at the people who program and use robots, A.I. agents, and algorithms. That is because what we need in the emerging Algorithm Society are not laws of robotics, but laws of robot operators.

The conceit of the Algorithmic Society is the harnessing of data and algorithms to govern society and make society better. The ambition of the Algorithmic Society is that of omniscience—to know all and to predict all—an ambition as old as humanity itself, but now seemingly ever closer to being within our grasp.

In the Algorithmic Society, the central problem of regulation is not the algorithms but the human beings who use them, and who allow themselves to be governed by them. Algorithmic governance is the governance of humans by humans using a particular technology of analysis and decisionmaking. Hence our need is not for robot-directed laws like Asimov's three laws of robotics, but laws directed at those who use robots to analyze, control, and exercise power over other human beings.

5. Conclusions

As asserts also by the most important living cosmologist, Martin Rees, baron of Ludlow, former Dean of Trinity College, the future commitment of mankind – and the same applies to jurists - is to work very close to the machines, watching over them, and mostly knowing that it is is more the individual or the small group that, by mistake of project or management, can cause a planetary disaster.

In such an interconnected global village, technology has become so effective that the action of the "village idiot" can have a global effect.

Over the last fifty years, we have had machines that are able to deal better than us.

Thanks to Moore's law (every two years the number of transistors of an integrated circuit doubles), there are machines that can learn by themselves to play chess and to recognize the human face. In some jobs which need to handle a huge amount of information - from traffic management to forecasting stock market trends - machines already work better than us. Luckily, there is still a gap between what machines can do and the versatility necessary to make them genuinely human reasoning. The computer that beats the Go world champion uses hundreds of kilowatts of energy. The human competitor only needs of 30 watts, the equivalent of a light bulb, and in the meantime he can do a lot of other things besides playing Go.

But even if the machines are still clumsy, they can already interact and damage our world through "Internet of things", with unintended but disastrous consequences.

The real challenge will be to instil ethics and common sense in machines and – for jurists – in the regulations behind them. In other words, it shouldn't be a real contrast between men and machines, but men ought to be the programmers, day by day, of their technological assistants, considering all side effects and implications from the legal point of view.

References

[1] Russel S J 2009
[2] Report of the European Parliament, containing Recommendations to the Commission concerning rules of civil law on robotics, doc. A8-0005 / 2017, 27 January 2017, p. 22, which however only deals with 'smart robots', according to which: "It is appropriate to establish a common European definition of intelligent autonomous robot, possibly including the definitions of its sub-categories, taking into account the following characteristics: - the ability to acquire autonomy thanks to sensors and / or through the exchange of data with your environment (interconnectivity) and the analysis of such data - the ability to learn through experience and interaction - the shape of the robot's physical support - the ability to adapt its behavior and its actions to the environment ". See also, Executive Office of the President, National Science and Technology Council Committee on Technology, Preparing for the future of Artificial Intelligence, October 2016, p. 6, where it is highlighted that: "There is no single definition of AI that is universally accepted by practitioners. Some define AI loosely as a computerized system that exhibits behavior that is commonly thought of as requiring intelligence. Others define AI as a system capable of rationally solving complex problems or taking appropriate actions to achieve its goals in whatever real world circumstances it encounters ". See also the recent discussion paper of the British Government's Information Commissioner's Office, Big Data, artificial intelligence, machine learning.
[3] Bateson G 1972
[4] Dumouchel M 2016
[5] Balking J M 2016 The Three Laws of Robotics in the Age of Big Data (the essay was originally presented as the 2016 Sidley Austin Distinguished Lecture On Big Data Law And Policy at the Ohio State University Moritz College of Law on October 27, 2016).
[6] Asimov I ROBOT 1950
[7] Deng B 2015 *Machineethics:Therobot'sdilemma*,Nature http://www.nature.com/news/machine-ethics-the-robot-s-dilemma-1.17881; Ulrike Barthelmess U and Furbach U 2014 Do We Need Asimov's Laws? https://arxiv.org/ftp/arxiv/papers/1405/1405.0961.pdf arguing that the laws reflect cultural anxieties about robots and it is unnecessary to build the laws into actual robots.

IOP Conf. Series: Materials Science and Engineering **364** (2018) 012034 doi:10.1088/1757-899X/364/1/012034

Digital restoration of the Last Supper by Leonardo Da Vinci and multimedia tools to experience edutainment.

Dott. Mario Taddei
Leonardo3, Italy

Email: mario@leonardo3.net
website: www.leonardo3.net
Google: mario+taddei+leonardo3

Abstract: Da Vinci's masterpiece can be understand in a totally new way by digitally exploring the painting and examining details that have been lost for centuries. A Virtual Reality set is able to bring us back like a "time machine" to interactively explore the Last supper room as never seen before. The research, the work and the exposition is actually on show in Leonardo3 Museum, Piazza Scala Milan. (www.leonardo3.net)
Today, The Last Supper is faded and cracked. Thus, these brilliant, saturated colors may appear shocking, but this reconstruction is the closest representation of how the fresco must have looked like when Leonardo painted it, with all the missing parts reconstructed. It is a sort of virtual archaeological reconstruction. The high-definition pictures allowed us to localize the original pigments," Pixel by pixel, we cloned da Vinci's original pigments, using virtual palette to restore areas where the color is irreparably lost.
In order to complete the fresco's missing parts we turned to contemporary copies of The Last Supper, such as the one by Giampietrino and others.
The reconstruction has indeed revealed some unknown details, such as an ideal city sketched in fragments to the left of Jesus' face, and a salt shaker that appears to have been knocked over by Judas. Multimedia software, and virtual reality has been realized to spread, history, science and heritage in the "edutainment" way, we believe is the future of knowledge.
The work will show the complete digital restoration of the entire arts, including parts that have never been seen before, and an interactive simulation will make everyone "paint" step by step the masterpiece. Virtual Reality as a time machine tool, to explore and understand history and heritage.

Keywords: Leonardo da Vinci, Art, Digital, Education, Edutainment, Virtual reality.

1. Last Supper: What we have today?

The Last Supper is a tempera and oil painting on plaster prepared with two layers of gesso. It measures 460 x 880 cm and was painted on a wall of the refectory of Santa Maria delle Grazie in Milan (Italy) in 1494. In February 1498, the mathematician Luca Pacioli referred to it as already completed. It is the largest of Leonardo's paintings and the only one executed on a wall that has survived. It represents the Eucharist and the meal during which Christ pronounces: "One of you will betray me". Leonardo froze the characters, expressions and gestures in the moment when Christ announces the presence of the traitor.
What we have today, after almost 500 years? The paint is no more the one Leonardo did and only a very little percentage of the original ideas of Leonardo survived. I believe that Leonardo will be ashamed of we have today, but of course is better not touch the original paint any more to make at least survive one little glimpse of the original idea (figure 1).

Figure 1. Las Supper as we find it today on books.

2. The Leonardo3 project

Always one to experiment, Leonardo executed this painting in oil rather than using classical fresco technique. He thus produced a masterpiece that several artists admired and copied immediately. Unfortunately, the painting began to disintegrate in the 16th century due to the oil technique and humidity. Today, despite restorations, it is only a faint shadow of what it was originally. In the year 2006 The private center of Leonardo Da Vinci studies, Leonardo3 has embarked upon an important virtual restoration that will bring the painting back to light as it originally appeared in the 16th century. In fact, initial interventions have already restored the colour in the missing areas, in turn revealing elements that have never been seen before, such as the bell tower to Christ's left, and all the tree. But the work of Leonardo does not stop on the square frame today we figure as the "Last Supper", all the surfaces around and the roof where painted ad decorated. The room was also a real monk's refectory full of tables, things and maybe also tapestries, like the virtual one Leonardo painted. So the real room and the real work of Leonardo is very far from the empty white, poor room we have today (figure 2).

Figure 2. Today only a part of the painting survive in an empty room

IOP Conf. Series: Materials Science and Engineering **364** (2018) 012034 doi:10.1088/1757-899X/364/1/012034

3. digital restoration of the Last Supper

According to the history of the Last Supper, what we see today is a far cry from what Leonardo actually painted, even after the excellent restoration that was finished in 1999. In fact, the oil and tempera paint Leonardo used began to flake off just a few years after its completion.
Numerous restorations and attempts to repaint it since then have further damaged the original. (figure 3-4).

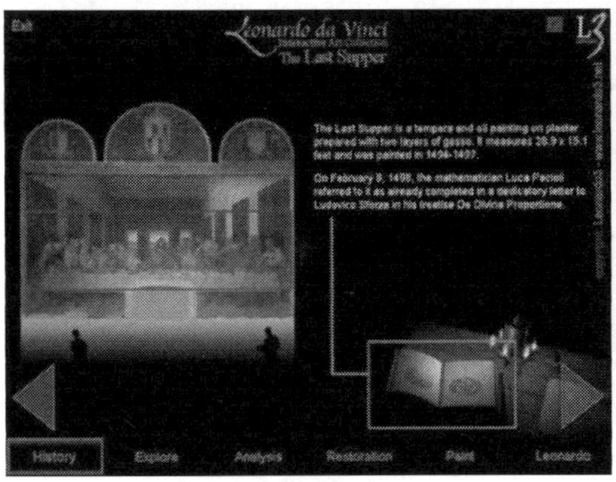

Figure 3. Paint studies

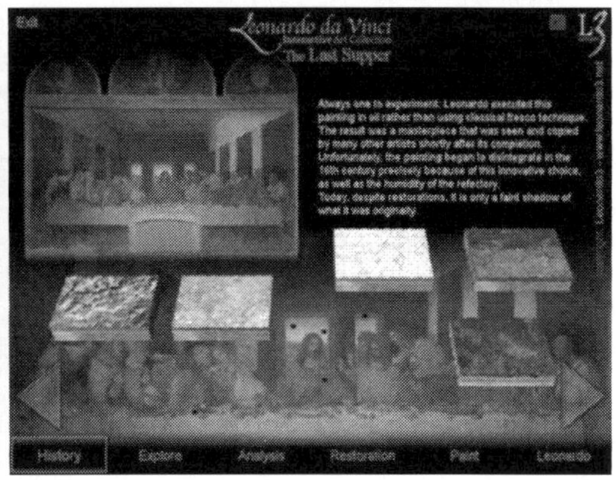

Figure 4. Surfaces today

Although the most recent restoration sought to cancel the effects of all those preceding it, it only succeeded in rescuing a ghost of a painting, over 50 percent of which had been destroyed (figure 5). Several parts of the painting have been lost and it's very difficult for us to realize how wonderful it must have looked to those who saw it shortly after its completion.
In 2007, the Leonardo3 research center began a process of digital restoration to bring the colors and forms that have been lost back to light. In addition to other paintings by Leonardo and our knowledge of the technique he used, the many copies made by other artists from 1497 onwards have been essential to our task. These paintings have allowed us to reconstruct not only the missing piece of the table and Christ's feet, but even more importantly the colors, which must have been a great deal more vibrant than they are today.

Figure 5. Less than 50% of surface restored

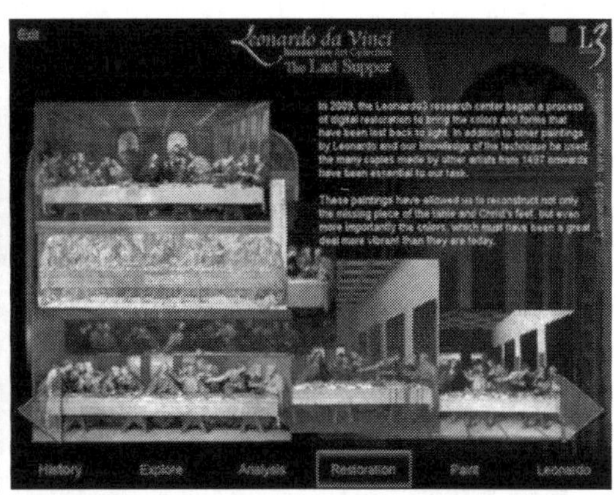

Figure 6. 1500's copies of the Paint

The restoration has uncovered several details that are difficult to see in the original painting, including plates of fish, slices of orange and a bell tower rising in the background to Christ's left.

IOP Conf. Series: Materials Science and Engineering **364** (2018) 012034 doi:10.1088/1757-899X/364/1/012034

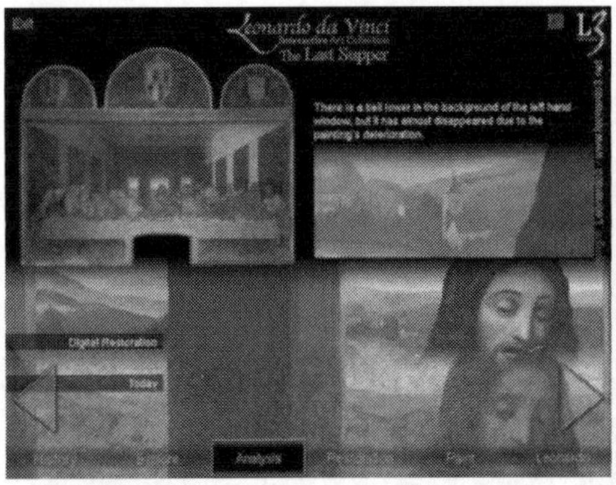

Figure 7. Background digital restoration.

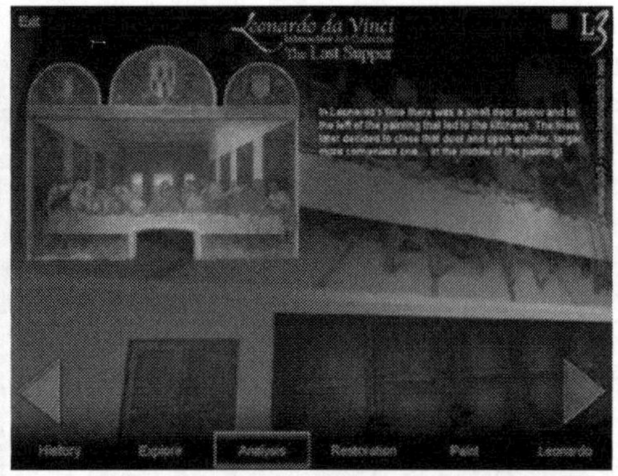

Figure 8. Original wall structure

In the background of the left-hand window is a bell tower, but the image has almost disappeared due to the painting's deterioration (figure 7). Following some poor restorations, in 1726 a hole was made in the wall to create a door, thus destroying the lower central part of the painting (figure 8).

We've also discovered that Judas has just knocked over a salt cellar with his right arm, which creates an important "movement" for representing a sudden, convulsive gesture. This salt cellar has totally disappeared from the painting (9-10).

We've been able to reconstruct several different elements of the painting by referring to copies made by other artists and similar details in the painting itself whose forms and colors we've "cloned". To reconstruct the piece of orange, for example, we cloned and distorted other painted details to cover the part of the slice that's been lost.

Figure 9. Giuda digital restoration.

Figure 10. Table digital restoration.

Figure 11. Pietro digital restoration.

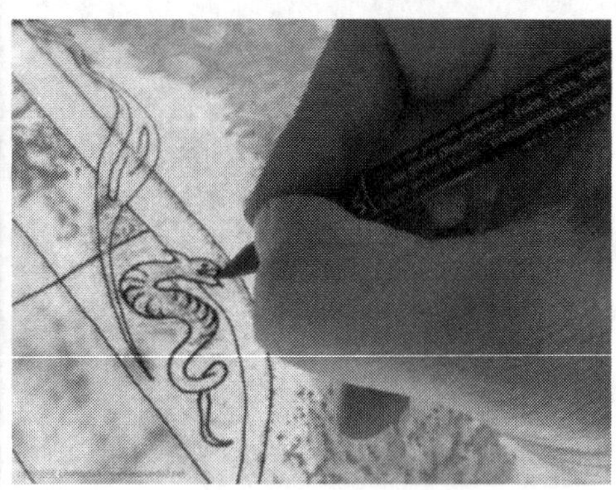

Figure 12. Reconstruction of lost detalis

We've also made use of Leonardo's preparatory studies for the digital restoration. In fact, some of these drawings can be perfectly superimposed onto the painting and have helped us to fill in the interplay of light and dark that's almost totally absent from the painting.

IOP Conf. Series: Materials Science and Engineering **364** (2018) 012034 doi:10.1088/1757-899X/364/1/012034

Although the lunettes above the painting seem less important, their restoration has also revealed some never-before-seen details. For example, the background, which is brown today, was originally blue to represent the sky and was covered with gold stars.
As for the three shields, which have almost completely disappeared, a parchment from 1498 is the most important document we have to reconstruct the upper lunettes. In fact, the three coats of arms and the letters are the same as those on the parchment.

MA[ria] M[a]X[imilianus] SF[ortia] AN[gliae] CO[mes] P[a]P[iae] - LU[ovicus] MA[ria] BE[atrix] EST[ensis] SF[fortia] AN[gliae] DUX [Mediolani] - SF[ortia] AN[gliae] DUX BARI (Massimiliano Sforza, Ludovico Sforza and Beatrice d'Este, Francesco Sforza the Second, Duke of Bari) (figure 13-14)

Figure 13. Center shield reconstruction

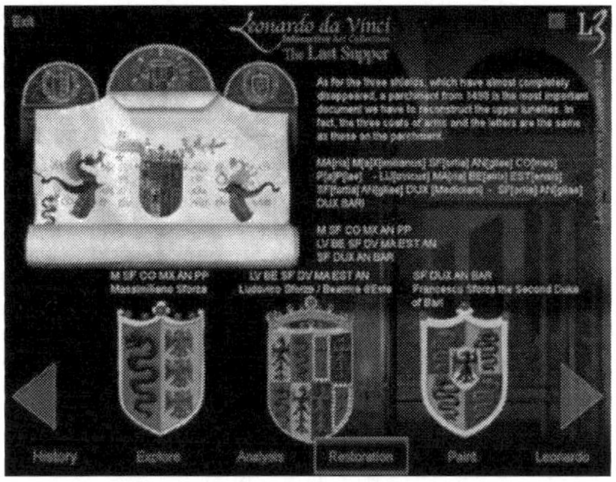

Figure 14. Document reference

An unexpected subject also emerged. We have reconstructed the image of a snake (or a dragon) that Leonardo drew with two pieces of fabric on top of the main coat of arms, which lines up directly above Christ. (figure 12) We're not talking about the stylized dragon on the Sforza coats of arms but a real serpent that Leonardo portrayed as though on the surface of the shield, wrapped around a piece of fabric. Perhaps Leonardo was poking fun at the patrons? After reconstructing the entire painted surface, we patiently calibrated the colours by referring to the parts that have survived, the copies made by other artists and the colours in other paintings by Leonardo.
Although the restoration continues (and could seemingly go on forever), our reconstruction is one of the first to bring lost details and colours back to life so that we can see what the painting might have looked like as soon as Leonardo completed it. Thanks to this digital restoration, which brings back to life all the details and colours that have been lost, people can examine and explore The Last Supper as it looked right after it was completed (figure 15-16).

Figure 15. Hi res digital restoration.

Figure 16. Hi res digital restoration.

IOP Conf. Series: Materials Science and Engineering **364** (2018) 012034 doi:10.1088/1757-899X/364/1/012034

4. Multimedia tools for heritage

The work done is fully described in details inside multimedia stations and videos that Leonardo3 has translated in many languages and exposed in several Da Vinci expositions around the world. In the 2017 A book and multimedia DVD , was published by Leonardo3 [1]. What is fascinating is the simple and direct approach of younger researcher of 5 years old, that feel technology as their own new languages. Of course adult people, simple visitors or scholars can have access to all informations, documents, animations and hi resolution images that are produced for this work with this stations. This provide a new way of spread history, science and art. Edutainment is "the" world, Education with Entrainment, a true powerful tool for knowledge (figure 17).

Figure 17. Last Supper multimedia room in the Leonardo3 museum, piazza Scala, Milan, Italy.

Figure 18. Last Supper multimedia station in the Leonardo3 exhibition in San Marino.

5. Next step: "The time machine"

During 2016 we worked on a interactive, real time 3d experience where the user can "walk" inside the Last Supper room as it was in the 1497. Using the last software techniques, used in Hollywood movies and video game industry we create a 3d environment that give the user the unique experience to "look" as it was. This software can be used also with Virtual reality systems.

Users are invited to wear the Oculus rift (VIVE is supported too) and walk in the Last supper room to see all details and items in a virtual environment. It is like a "time machine" to be able to experience history in an incredible way. The Virtual Experience of the Last supper is one of the

IOP Conf. Series: Materials Science and Engineering **364** (2018) 012034 doi:10.1088/1757-899X/364/1/012034

best attraction of the Leonardo3 museum (figure 19).

Figure 19. Last Supper Virtual Reality station, in the Leonardo3 museum, piazza Scala, Milan, Italy.

Figure 20. Screenshot from Discovery News Documentary, multimedia touch experience to explore the Last Supper paint, New York, Leonardoexposition.

Figure 21. Screenshot from Smithsonian Documentary, documents and paper of Leonardo

Figure 22. Interpretation of the Last Supper room as in 1497 , frame from the Virtual experience in the Leonardo3 museum.

References

[1] Taddei, L'Ultima Cena. Segreti, tecnica ed errori di un capolavoro come mai visto prima, Leonardo3 2007, ISBN 978-88-6048-017-0
[2] Taddei Laurenza Zanon, Le macchine di Leonardo da Vinci, Giunti Editore, 2005, ISBN 88-0904-363-4
[3] Taddei Lisa Zanon, Il Codice Atlantico di Leonardo da Vinci, Leonardo3, 2005, ISBN 978-88-6048-005-7
[4] Taddei Zanon Bernardoni, I Ponti di Leonardo, Leonardo3 2005, ISBN 978-88-6048-000-2
[5] Taddei Lisa Zanon, Il Laboratorio di Leonardo, Leonardo3 2006, ISBN 978-88-6048-002-6
[6] Taddei, I Robot di Leonardo. Leonardo3 2007, ISBN 978-88-6048-009-5
[7] Taddei Bernardoni Zanon, Leonardo, l'Acqua e il Rinascimento (catalogo mostra), Federico Motta, 2004, ISBN 88-7179-430-3
[8] Giuseppe Bossi, Del Cenacolo di Leonardo da Vinci, 1810
[9] Leonardo da Vinci, Trattato della pittura di Lionardo da Vinci, Firenze, 1792
[10] Leonardo da Vinci, IL CODICE ATLANTICO, Firenze, 1894
[11] Leonardo saggi e ricerche, Istituto poligrafico 1954
[12] I Manoscritti e i Disegni di Leonardo da Vinci Pubblicati dalla Reale Commissione Vinciana. 1928

Links

Main website: www.leonardo3.net , www.goo.gl/LTFGfT
Last Supper DVD: www.youtube.com/watch?v=Royl373MX0Q
Discovery video: www.youtube.com/watch?v=z2-njV2UtSs
Smithsonian video: www.youtube.com/watch?v=PgnY-MnxGys
Multimedia Last supper tool: www.youtube.com/watch?v=yp8Fmcf4m58
Rai5 il Cenalcolo: www.youtube.com/watch?v=NInk93Eu3fg
LS Leonardo3 museum and Last Supper: www.youtube.com/watch?v=MfFwLjWElZU

IOP Conf. Series: Materials Science and Engineering **364** (2018) 012035 doi:10.1088/1757-899X/364/1/012035

Immersive Virtual Reality As A Resource For Unaccessible Heritage Sites

Giovanni De Gasperis, Alessio Cordisco and Francesco Cucchiara
Department of Information Engineering, Computer Science and Mathematics
University of L'Aquila, ITALY

E-mail: giovanni.degasperis@univaq.it

Abstract. This contribution highlights the potential contribution that modern technologies offer for virtual reconstruction and immersive navigation of monuments that are unavailable due to catastrophic events or other causes. We propose a design and implementation methodology for the enjoyment of the Cultural Heritage sites through three-dimensional and photorealistic modeling, visualization and rendering in a immersive virtual world enriched with extra content. The project of virtual reconstruction of monuments damaged by catastrophic events arises from the need to continue to enjoy the architectural asset during the stages of safety-laying, reconstruction and finishing, large-scale interventions that usually take a long time, especially for buildings belonging to the Cultural Heritage. The experience of the earthquake in L'Aquila of 2009 has reflected on the importance of Heritage and collective identity that changes whenever a cultural asset is not used anymore. Santa Maria Paganica in Virtual Reality Project implements a freely navigable virtual model able to let discover and rediscover one of the monuments of L'Aquila, damaged by the 2009 earthquake, which during the past had a most important role for the social, economic and cultural life of L'Aquila people, and that stays still today, 2018, not reconstructed. Here we present the results and the methodology of the process which allowed us to get a model available in a immersive virtual reality environment. The chosen VR engine is the Unity software, which allows to implements models compatible with the two most common adopted VR headsets: HTC VIVE and OCULUS RIFT. The final VR artifact is an executable software like a videogame, playable on a PC; virtual lighting systems have been projected in order to get a better graphic performance avoiding to alter the mesh model, and the navigation system in VR has been implemented with specific C# scripts.

1. Introduction

This virtual reality experimentation is focused to a monument of the city of L'Aquila. On April 6, 2009 a Mag 6.3 Richter earthquake hit the city that destroyed most of the civil, residential and religious buildings. Even today, 2018, some areas of the historic center are still forbidden to the public and many monuments, which characterize the historical city center, are uninhabitable and inaccessible. The importance of the Heritage for the collective identity is undisputed; this link was reiterated in the Council of Europe Convention, adopted in Faro in 2005, which attributes to the Patrimony a strong social value recognizing it as the foundation of a democratic society, in which the community creates and reaffirms, through a common cultural heritage, a civic sense of belonging [1]. Nowadays modern technologies are commonly adopted by the scientific community and they constitute important tools for the reconstruction of memory, use and enhancement of inaccessible or no more existent Cultural Heritage sites destroyed by anthropic or natural transformation. Significant is the worldwide interest in the Syrian site of Palmyra, destroyed by ISIS in 2015, virtually rebuilt and now accessible via the web [2], being the subject of much research and experimentation [3; 4; 5]. A very interesting Italian example is the one carried out by the research team of the University of Bologna (Ravenna) on the church of Santa Maria del Porto outside, destroyed by the bombings of the Second World War and never rebuilt [6]. In the context of the post-seismic reconstruction activities carried out in the crater [1] territory, there are many research projects on Heritage that adopted modern technologies: starting from remote surveying in the emergency phase with UAV technologies and laser scanning [7; 8], up to the use of the thermal imaging camera [9], from digital photogrammetry of civil and religious buildings [10; 11; 12] up to the modern BIM trials [13; 14; 15]. These are all functional digital tools, generally aimed at the documentation of the damage and the state of the artifact at a specific time and the restoration projects of the heritage, where possible. However, there are still few technological applications for the enhancement and fruition of heritage, as the interesting projects that have been undertaken both by private individuals, as in the case of the enhancement of the Aterno valley [16], and by public bodies, such as the National Museum of Abruzzo, who digitized part of his collection making it public on the SketchFab online platform [17]. Therefore it was decided to carry on the VR experimentation on the church of Santa Maria Paganica, symbol of L'Aquila history and very dear to the inhabitants.

2. Case Study: The Church of Santa Maria Paganica in L'Aquila

The city head quarter of Santa Maria Paganica, one of the most significant monuments of the Aquilan area, is full of architectural superficies and stylistic stratifications that can be correlated with the many historical earthquakes that have hit the city over the centuries. Probably the structural interventions made in the past have compromised the stability of the building causing most of the damage by the earthquake of 2009: collapse of large part of the roof and the dome, the side wall and the facade [18] (Figure 1).

The earthquake, however, also brought to light unpublished architectural elements, reused in the walls, which document the past of this building, such as the Gothic inscription, found during the recovery of the rubble, which testifies the post-earthquake reconstruction of 1349 [19]. In fact, the original building, built in conjunction with the birth of the city at the behest of the inhabitants of Paganica village who first settled in this site, was very different from eighteenth-century (post-earthquake 1703) visible today characterized by a single-nave plant with noble chapels , a large transept in the presbyteral area surmounted by a majestic dome and semicircular apse. The building has a majestic facade, referable to the original plant as witnessed by the date 1308, in tanning stone with horizontal crowning and a mighty bell tower, attributed by many scholars to a pre-existing fortification that stood on the highest point of the city [20], which was later lowered by an edict of the Spanish rulers in the XVI century.

3. Methodologies and technologies for the Cultural Heritage

Before engaging in the virtual modeling of the church, we conducted studies aimed at the knowledge of the monument, as well as previous documentation for the retrieval of the graphic and photographic material prior to the 2009 earthquake; in addition to the study of the published material and private collections, we tried to record the historical memory of those who knew and visited the artifact in person; in this regard the parish priest Don Stefano Rizzo gave the most significant contribution making available, not only the photographic archive, but also its visual memory necessary for the reproduction of elements not photographed and/or digitized in the past. Furthermore, inspections were carried out on the outside and inside the building, indispensable for carrying out the photographic sampling and the measurements of the surviving

[1] the term refers to the area legally affected by the seismic event

Figure 1. Church of Santa Maria Paganica after the 2009 earthquake (photo available online).

environments with direct survey techniques. All the photographic material recovered has been analyzed and divided into two groups: monochromatic and color, using the black and white ones as the basis for the modeling, while the others reproduce the chromatic range of the walls, floors and furnishings. Subsequently, we carried out the virtual modeling in VR of the church, following the criteria proposed in the London Charter [21], that we translated into integrating different methodologies and technologies such as digital photogrammetry, three-dimensional modeling, rendering and assembly in the gaming editor. We have chosen to use photogrammetry to document and make usable only for those part of the architectural elements and liturgical furnishings that survived the earthquake; it was possible to recover a large amount of documentation material, but unfortunately not enough to reconstruct the whole building in detail. Particular attention was paid to the lateral portal and baptismal fountain base: we processed them with the Photoscan software by Agisoft, giving us respectively a digital model with a mesh of 1'801'062 faces, while the latter has a more accurate model with a mesh of 2'590'334 faces (Figure 2). The accuracy of the mesh needs to take into account the strict specification of the real time rendering during the VR session.

These meshes are generated from a set of images acquired in emergency conditions with suboptimal light exposure and noise elements; the raw results were very complex and not easily managed and therefore reworked within the Cinema 4D software to obtaining "clean" models that can be easily assembled with the 3D model of the church in a Unity scene. For the modeling phase, several software were used: SketchUp for basic modeling (Figure 3), Cinema 4D (Figure 4) for details and Lumion for video rendering. Adopting a 1: 1 scale plan and a side elevation representing the church as a base, we proceeded to create the perimeter walls, the openings, the noble chapels and the vaulted roof. The raw model created in this way has been modified several times until we reach a structure as faithful as possible to reality of the pre-earthquake situation. In some cases, using the photos as a basis, it was possible to reproduce models likely to be real, while in others, due to the absence of support materials, the elements are less accurate, but with

Figure 2. Photogrammetric models : (A) baptismal fountain base; (B) lateral portal.

Figure 3. Examples of elements modeled with SketchUp software: A - wireframe of the entire structure; B - wooden choir; C - capital; D - first left side chapel; E - transept; F - funerary tombstone

all the distinctive and distinguishable features.

Once the overall model was obtained, the modeling work focused on the details: capitals, pillars, cornices, altars, etc.; a very long process that led to the creation of 108 components collected in 12 groups, to which 211 images were used as surface textures. The extension of the building, the complexity and the richness of the furnishings led to a choice of representation, depicting some elements in 2D, such as the statues inside the niches and the decorative element of the top part of the three portals; to these two-dimensional objects, in the rendering phase, a normal map was applied to produce the relief effect and 3D. Attention was mostly reserved to the inner part of the building, unfortunately today destroyed; also to enrich the morphological context were to put the virtual church, we reproduced a minimalist square and the volumes

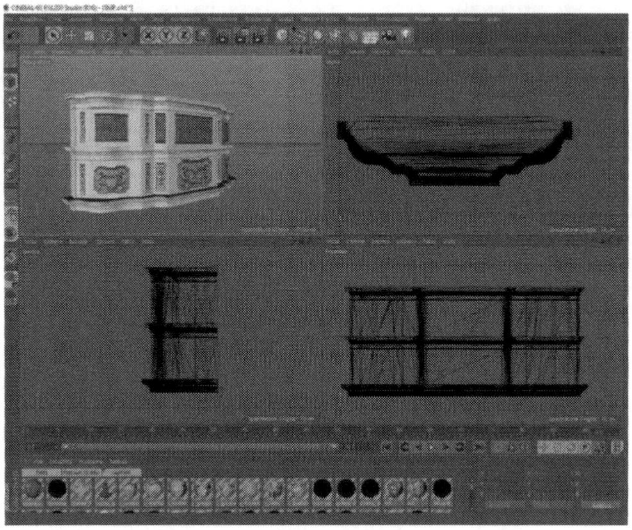

Figure 4. Modeling of liturgical elements in Cinema 4D software.

of the neighboring buildings, leaving important details for lack of time, such as the fountain and the terrain orography. As a final stage, we performed the rendering of the inside using the Lumion software, generally used for architectural visualization, which allowed us to obtain photorealistic rendering, 3D video and VR panoramas in a shorter time compared to the main competitors (Figure 5).

We adopted different tools for the further visualization of the model of the church of Santa Maria Paganica, from the conventional ones, like images and videos, to the most innovative ones like spherical 360' images linked together to form a virtual tour from specific landmark positions. However, these are rendering systems that constrain the user mobility in the model, not leaving much freedom of movement.

4. The Virtual Reality Implementation

4.1. Scene description

Geometry : the scene has approximatively 300 meshes, more than 1 million polygons and 130 PBR (physically based rendering and shading) materials using the standard shader in Unity that is quite computationally expensive. **Lighting** : since the geometry is very complex and the scene's objects are static, the best solution would have been to use a pre-calculated baked lighting system. But we decided not to do. Anyway baking such a large scene takes a lot of working time, so in this project we used two types of less computational intensive light models (ambient and directional) plus a spotlight to be used as a torch by the user only when the player needs it where she wants. **Post Effects** : the first notable thing in this VR scene was a strong aliasing effect. To avoid it we used a 8x multisampling antialiasing (MSAA), which is available only in forward rendering. Thanks to MSAA almost all the aliased edges were correctly smoothed and the users were not distracted any more by that unrealistic graphic behavior. Another useful post-effect we added to the rendering is the screen-space-ambient-occlusion (AO). The scene acquired a good realism as a result of this combined effects because the dynamic shadowing system only projected direct shadows on the scene's surfaces.

Figure 5. Virtual reality walkable model of the Santa Maria Paganica Church. A demonstration YouTube video is available at `http://youtu.be/G5Tf1Jv88gA`

4.2. User interaction

The user must be able to move freely and interact with the virtual environment. Since a monument can be wide and very high, it can show interesting points in usually inaccessible areas. For this reason the movement system makes sure that each interesting point can be easily visited. The navigation experience is composed by 3 main actions: physic movement, teleport and torch control.

Physic movement: recent VR headsets provides position tracking, allowing the user to move freely in the virtual space. Unfortunately in most cases the available physic space isn't enough wide to explore the whole scene, so it is necessary to expand the movement range. **Teleport**: the user should be able to instantly teleport herself to another area and explore it in the local physic space. The teleport system forces the user to first choose a destination point on the ground or on a floating platform and then activate the teleportation process. In this way the induced sickness side effect is completely avoided.

Teleport with controllers (tested on HTC Vive): we used the plugin Vive Teleporter by *Flafla2* (`http://github.com/Flafla2/Vive-Teleporter`) to create the teleportation system, similar to what is available in the *"The Lab"* VR game included with the headset. This plugin allows the user to choose destination points using a customizable parabola coming from one controller. The plugin uses a multi-raycast system to check collisions along a parabola. In order to setup a correct collision behavior it is necessary to create a navigation mesh that defines the areas in which the teleportation is allowed (we call these walkable areas). The system accepts the teleports on walkable areas and rejects the ones on non-walkable areas. Of course, to create a good navigation mesh it is essential to create a proper collider mesh. In our case, the original church model was very complex so we created a separated low-poly collider mesh. Using this method it is possible to move around an almost flat area, depending on the navigation mesh settings and on the parabola range, but sometimes it is not possible to reach a point of interest point because in the parabola may falls down before it can collide with it. Since one of the exposed parabola properties is the gravity to which it is subject, it is possible to reduce it to 0 and transform the parabola into a straight line that can potentially reach each point of the space. We modified the original script adding a straighten parabola mode to be used only when there is an intent of reaching a higher walkable area. The reason we did not use everywhere just the straight line is that the parabola is a much more natural way of selecting a destination point especially for large distances, because it does not allow the user to point towards areas that are too far from her.

Teleport without controllers (tested on Oculus Rift) : without space-tracked controllers the only useful device for the dynamic selection is the headset, so we implemented the teleport system to make it follow the head movement. In this case the only option possible is the straight line and the user has an always visible dot in the view centre that represents the gaze line's direction. All the inputs are from static controllers. In this case the valid/invalid areas are managed by a tag. The system checks whether the collider's tag corresponds to the ground or an obstacle and returns the user a yellow (valid) or a red (invalid) indicator. The line in this case is not visible and it virtually extends to the infinity, so there is no problem in reaching high interest points.

Dynamic torch behavior In some areas of the 3D environment the combination of ambient and directional light does not produce enough illumination. For instance, a crypt with a picture in it should be enlighten to let the user see the affresco that is inside of it. But having point lights near each interest point could be too expensive for the real time rendering of a complex scene, so we decided to create a torch system the user can activate to bring light to specific areas of the scene. The torch should be handled by a space-tracked controller when possible. If so, the light direction should follow the controller orientation but if there is no controller available the only way to illuminate what the user wants is to make the light direction follow the headset orientation. Also, the light should adapt its range in relation to the distance that is found from the object(s) which are under observation.

5. Conclusion

We demonstrated a quasi-philological virtual reality reconstruction of a church badly damaged by a strong earthquake using the most advanced consumer grade VR technology. We showed that it is possible to make the inaccessible, (virtually) accessible again. Such virtual reconstruction can only give back a living memory for Cultural Heritage sites that have a strong bond with a local community, while the real bricks and mortar reconstruction can get started and hopefully end by the life time of those who remember how it was.

6. Acknowledgments

This work was carried out with internal funding from University of L'Aquila. We thank Prof. Piero Tognolatti to encourage us to purse on this project and Prof. Silvia Mantini for insightful historical discussion about the Church of Santa Maria Paganica in L'Aquila.

References

1 Convenzione quadro del Consiglio d'Europa sul valore dell'ereditá culturale per la societá, Consiglio d'Europa - (CETS no. 199) Faro, 27.X.2005.

2 `http://www.newpalmyra.org` [Last accessed 22 Dec 2017]

3 Wahbeh W, Nebiker S and Fangi G 2016, Combining public domain and professional panoramic imagery for the accurate and dense 3D reconstruction of the destroyed Bel temple in Palmyra, ISPRS Annals of the Photogrammetry, Remote Sensing and Spatial Information Sciences, Vol. III-5:81-88.

4 Arkawi A 2017, Vision of the reconstruction of destructed monuments of Palmyra (3D) as a step to rehabilitate and preserve the whole site, The International Archives of the

Photogrammetry, Remote Sensing and Spatial Information Sciences, Vol. XLII-2/W5: 41-48.

5 Denker A 2017, Rebuilding Palmyra virtually: recreation of its former glory in digital space, Virtual Archaeology Review, Vol. 8-17: 20-30.

6 Volpe A 2017, Una mostra sulla distrutta chiesa di Santa Maria in Porto Fuori, Bollettino dantesco, 6 settembre 2017: 139-154.

7 Baiocchi V, Dominici D and Mormile M 2013, UAV application in post-seismic environment, International Archives of the Photogrammetry, Remote Sensing and Spatial Information Sciences, Vol. XL-1/W2: 21-25.

8 Pesci A, Loddo F, Casula G, Zampa F and Teza G 2010, Experience in mobile laser scanning by means of lynx system in L'Aquila city, INGV.

9 Cadelano G, Bortolin A, Cifan Gi, Ferrarini G, Mannella A, Petracca A and Bison P 2015, Ispezioni stutturali supportate dalla termografia IR, Act of XVI Convegno ANIDS, L'Aquila.

10 Bartolomucci C and Trizio I 2015, Dal rilievo del danno sismico al disegno per il progetto di restauro: un'applicazione di documentazione speditiva in condizioni di emergenza, DisegnareCon, Vol. 8 n.14: 24.1-24.8.

11 Giannangeli A, Marchetti A and Trizio I 2015, Dalla fotogrammetria digitale ai Pdf 3D: considerazioni sulle potenzialitá offerte dalle metodologie di rilievo speditivo basate sullo SfM, Act of XVI Convegno ANIDS, L'Aquila.

12 Dominici D, Alicandro M, Rosciano E and Massimi V 2017, Multiscale documentation and monitoring of L'Aquila historical centre using uav photogrammetry, The International Archives of the Photogrammetry, Remote Sensing and Spatial Information Sciences, Vol. XLII-5/W1: 367-371.

13 Brumara R, Della Torre S, Oreni D, Previtali M, Cantini L, Barazzetti L, Franchi A and. Banfi F 2017, HBIM challenge among the paradigm of complexity, tools and preservation: the basilica di collemaggio 8 years after the earthquake (L'Aquila), The International Archives of the Photogrammetry, Remote Sensing and Spatial Information Sciences, Volume XLII-2/W5: 365-371.

14 Continenza R, Trizio I and Tata A 2017, Riscoperta del palazzo della famiglia Branconio in L'Aquila. Applicazioni BIM a un cantiere della ricostruzione aquilana, in AAVV (eds) Territori e frontiere della Rappresentazione.

15 Continenza R, Redi F, Savini F, Tata A and Trizio I 2017, BIM for the Archaeology of Standing Buildings: Case Study of the Church of San Cipriano in Castelvecchio Calvisio (L'Aquila, Italy). Fogliaroni P, Ballatore A and Clementini E (eds) Proceedings of Workshops and Posters at the 13th International Conference on Spatial Information Theory (COSIT 2017). COSIT 2017 2017. Lecture Notes in Geoinformation and Cartography. Springer.

16 http://valledellaterno.it/torri-e-castelli [Last accessed 22 Dec 2017]

17 http://sketchfab.com/PoloMusealedellAbruzzo/collections/munda_museo-nazionale-dabruzzo-laquila [Last accessed 22 Dec 2017]

18 Carocci C, Cattari S, Lagomarsino S and Tocci C 2010, The Case Study of Santa Maria Paganica Church Damaged by 2009 L'Aquila Earthquake, Advanced Materials Research Vol. 133-134 (2010): 163-168.

19 Torrieri V 2010, L'Aquila, Santa Maria Paganica. Archeologia e rimozione delle macerie dopo il sisma del 2009, Quaderni di Archeologia d'Abruzzo 1/2009 (2010):65-69.

20 Antonini O 1999, Architettura religiosa aquilana, Edizioni del Gallo Cedrone, L'Aquila.

21 http://www.londoncharter.org [Last accessed 22 Dec 2017]

IOP Conf. Series: Materials Science and Engineering **364** (2018) 012036 doi:10.1088/1757-899X/364/1/012036

Efficiency and quality raising in preventive archaeology: work in progress of the project ARCHEO 3.0

S Rescic[1*], S Siano[2], R Manganelli Del Fà[1], I Cacciari[2], G Grosso[3], G Andreini[4], G Pocobelli[5], A Mencaglia[2], P Tiano[1], F Fratini[1] and C Riminesi[1]

[1]Institute for the Conservation and Valorization of Cultural Heritage (ICVBC), Via Madonna del Piano 10, 50019 Sesto Fiorentino (Florence -Italy)
[2] Institute of Applied Physics (IFAC), Via Madonna del Piano 10, 50019 Sesto Fiorentino (Florence- Italy)
[3]Aeffegroup s.rl., Via Dante Alighieri, 72, 50041 Calenzano (Florence -Italy)
[4]Studio Flu, Via Marche 10, 56123 Pisa (Italy)
[5]Cooperativa Archeologica, Via Luigi la Vista 5, 50133 Florence (Italy)

*) rescic@icvbc.cnr.it

Abstract. Preventive archaeology aims at assessing the archaeological interest of a given area, as well as discovering, interpreting, documenting, and protecting archaeological remains that might otherwise be destroyed by land development works. On request of private stakeholders or institutional bodies, archaeologists intervene on a given construction yard (of infrastructure or urbanisation works) in order to investigate the possible archaeological significance of zones subjected to excavation works. In many cases archaeological assays are carried out, which are associated with thorough stratigraphic documentations and definition of the final preservation conditions. The latter can involve suitable re-burying or musealization of some of the archaeological remains recovered. During these activities, the construction works are slowed down or completely interrupted up until the conclusion of the mentioned archaeological research, which can generate possible delays with respect to the planned time schedule and increase of the costs. Improvements are needed in order to minimise the duration of the archaeological assessment and, simultaneously, to guarantee quality standards even higher than tose stated by the competent superintendences. Within this framework, the project ARCHEO 3.0 "Integration of Key Enabling Technologies for rising the efficiency and quality of preventive archaeological excavations" (co-funded by Tuscany Region, POR-CReO 2014-2020) aims at speeding up the archaeological excavation and at improving its documentation. It foresees the development of a set of characterisation tools, their integration and validation in archaeological yards. Thanks to ICT solution such a multidisciplinary integration will be made possible by a dedicated web-based platform that will allow to manage 2D and 3D photogrammetry along with the data provided by electromagnetic, and acoustic sensors. The project aim at promoting a substantial step ahead of the preventive archaeology in order to overcome the traditional empiric approach. In this respect, data exchange and the ICT solutions for direct collection of information on archaeological evidences represent the major challenge. Here, we preliminarily focus on the general features of management protocols of the archaeometrical data, as well as on their suitable integration within the operative phases of the archaeological excavation.

1. Introduction

The topic of preventive archaeology has been discussed for decades both from the theoretical and methodological point of view and from the more strictly operative aspects. For a long time, in fact, archaeologists directly involved in the field have posed the problem of reconciling the needs of protection with the operational ones that involve excavation work in the construction sector, mining and in large infrastructural works. At the European level, experiences of preventive archaeology have taken place mainly in the countries of central and northern Europe starting from the second post-war period (excavations connected with the construction of the London underground). Also in Italy the first experiments in this sense took place with the intervention of English archaeologists followed, in the eighties of the XX century, by more systematic prevention activities throughout Italy, with interventions supported by both private and public clients. An opportunity to verify and consolidate the operational practices already tested on more limited territorial areas has been the construction of the high-speed railway lines, which have seen numerous preventive investigations, systematic and extensive excavation interventions from Lombardy to Campania, interventions still ongoing in some areas. Therefore, now it is common practice, by the Superintendences for archaeological heritage, to coordinate excavation interventions aimed at the construction of public and private works. At present these kinds of archaeological excavations are the overwhelming majority of those conducted in Italy. However, these excavations, after the scientific recovery of all the stratigraphic and structural data, necessarily require the removal of the archaeological context or modifications of the projects in order to allow the total or partial conservation of the remains found on site. In some cases, however quite rare, it was necessary to completely abandon the planned construction due to the discovery of archaeological assets of such importance that did not even allow scientific disassembly and relocation operations. The law 25 June 2005, n. 109 provides a legitimization for the interventions imposed by the same needs for the protection of archaeological heritage (established by the Code of Cultural Heritage and Landscape as reported in the Legislative Decree 22 January 2004, n. 42) and helps regulating a common behavior practice followed by all the archaeological Superintendences. Also in the new public procurement code (Legislative Decree 18 April 2016, n. 50) updated with the amendments introduced by the Corrective Decree (Legislative Decree 19 April 2017, n. 56, published in the Official Gazette 5 May 2017, n. 103), the article 25 of the legislative decree 18 April 2016, n. 50 establishes the need for prior verification of the archaeological interest carried out according to the usual practice [1-2].

2. Protocols and procedures in use for the preventive archaeology

Paragraph 13 of Article 25 establishes the need to adopt *".... guidelines aimed at ensuring speed, efficiency and effectiveness of the procedure referred to in this article. The same decree identifies simplified procedures... that guarantee the protection of the archaeological heritage, taking into account the public interest in the realization of the work. "*

This concept is explained in the indications reported in the Reference Practice UNI / PdR 16: 2016 "Guidelines for exploratory survey, archaeological assistance during construction and stratigraphic archaeological excavation"

The preliminary verification procedure of the archaeological interest requires that the contracting companies must transmit, before approval, to the territorially competent Superintendent copy of the preliminary project of the intervention including preliminary archaeological and geological investigations. This is a preliminary phase that involves four different types of operations, without involving excavation activities:

a) the collection of archive and bibliographic data, i.e. historical "knowledge;

b) surface surveys on the areas affected by the works: this is the so-called survey, which provides for the systematic collection of the findings brought to light seasonally during the ploughing or in sections exposed in the natural or artificial land exposures;

c) the "geomorphological interpretation of the territory" in relation to their settlement potential throughout the past ages;

d) the photointerpretation (provided exclusively for "network" works), i.e. the study of anomalies identified by reading aerial photographs available or realisable for the purpose.

The results of these operations must be "collected, processed and validated". In case of archaeological evidences in the areas affected by the works, supposed on the basis of the transmitted documentation and "further information available", the Superintendent may request a further investigation phase:

Phase 1) integrating of the preliminary planning:
1.1 - execution of cores;
1.2 - geophysical and geochemical prospecting;
1.3 - archaeological tests;
Phase 2) integrating of the final and executive planning:
2.1 - execution of surveys and excavations, even in extension.

The procedure ends with the drafting of the definitive archaeological report, approved by the territorially competent Superintendent. The report, satisfying the requirements established by the legislation, contains an analytical description of the investigations performed, qualifies the archaeological interest of the area, according to the following levels of archaeological significance of the site, and dictates the following requirements:

a) contexts in which the stratigraphic excavation directly satisfies the need of control and protection from archaeological point of view of the excavation area;
b) non-monumental contexts with a low level of conservation for which re-interventions or disassembly-reassembly and musealization are possible;
c) complexes of particular relevance, extension and historical-archaeological value that can be fully protected according to the Code of Cultural Heritage and Landscape.

The efforts of the project ARCHEO 3.0 are devoted to increase the speed and quality of these operations, overcoming the difficulty of storing of metadata and their management. In particular, it is focused on the Phase 2 of the planned procedure i.e. when excavation is required.

3. Materials and methods

The targets of the project ARCHEO 3.0 that we will discuss in this paper concern the following topics: development and techniques for a speditive approach for exploratory surveys; semi-automatic reconstruction of the SU (Stratigraphic Unit) during the excavation operations; management of the metadata collected during the excavation and their immediately processing. All these aspects are finalized to improve the speed and quality of the preventive archaeology yard.

3.1 Speditive approach for exploratory surveys

The purpose of exploratory surveys is to evaluate the presence of remains and archaeological evidences and their consistency. They are usually carried out by means of a forced section excavation with mechanical means provided with a smooth bucket. In presence of archaeological evidences, the excavation is stopped and the next operations are carried out in the framework of real archaeological yard, in according to stratigraphic criteria. The documentation drawn up during the operational phase in the yard must always include:

- description of the stratigraphy units found for each step of the excavation, the so-called stratigraphy diagram (Harris's matrix, or *matrix*);
- graphic documentation of the temporal succession of the excavation and sequence of deposition of archaeological evidences;
- list of the drawings, photos, videos, etc.;

The target of this topic has been achieved by using 3D technologies and digital instrumentations that allow improving the quality and speed of the survey in a compatible way with yard activities. In particular, the techniques of Structure from Motion [3] bring to new perspectives in the field of architectural and archaeological relief [4-5] and documentation; furthermore, Computer Vision algorithms are ready to be also managed by non-technical users.

3.2 Identification of the SU

The stratigraphic archaeological excavation is an activity that involves the removal of portion of terrain to bring out archaeological evidences like ruins of building or monuments, tombs and archaeological finds, and archaeological remains in general. The aim of this activity is the understanding of the relationship between remains and their ground layer, through a correct reading of the sequence of the ground layers it is possible to date the sequence of the events that interested the excavation. The stratigraphic archaeological excavation is carried out on the basis of the identified ground layers, distinguished according to *their different consistency, their composition and their colour*. The 3D modelling makes possible to realize a metrical restitution of each stratigraphic unit, giving priority to the correctness of its perimeter limits, relief and quotas after a properly graphic elaboration. To one specific area on the 3D model, or on the rectified photography (2D), can be referred the results obtained by different equipment and visual observation. Each stratigraphic archaeological excavation must include the following operations: identification of the stratigraphic units; manual cleaning of the stratigraphic units for suitable definition and graphic and photographic documentation; numbering and description; photographic documentation; graphic documentation.

These operations are achieved for each step of the excavation until to the archaeological deposit has been completely removed.

3.3 Management of metadata acquired on the yard

The exact and objective descriptive, graphic and photographic documentation of a stratigraphic archaeological excavation is a necessary condition for a correct recording of data and information, for possible future reviews of the investigated contexts and for the approach to restoration.

The proposed approach is based on the use of GIS to organize and analyze the acquired documentation (images, videos, results from analytic analysis or physical investigation, etc.) is developed by the following steps:

1. The topographic relief that gives the reference system on the official cartography and it is the reference for a right positioning of the archaeological targets inside the archaeological area;
2. The expeditious 3D relief of targets – objects or selected areas inside the excavation yard – that are positioned in the topographic relief (geo-referenced). The 3D model allows the extraction of digitalized sections of the area of interest improving the quality documentation;
3. The targets within the archaeological site are geo-referenced. Each target can be referred to the location of the archaeological find or to ruins, and it is possible associate every information (description, images, results of analysis, etc.) to it.

In such way, the proposed approach aims to increase the speed and quality of the information collected during the excavation phases. Then, a post GIS database linked to a QGIS project collects the aforementioned data through an internet server provided with REST services. Both services, Server API and Client API, use RESTfull data interchange protocol, such as in a classic client-server application in this way an easy and robust data transfer method is implemented.

The metadata analysis and management is made easier by procedures and protocols defined in the project and by a web-based platform properly designed, that allows to subsequent analysis and re-elaboration of data and metadata to drawing up the executive project. The drafting of the executive project will must comply with the following requirements: general report of the excavation; list of stratigraphic units; SU record cards; photographic documentation; graphic drawings including the positioning of the excavation area; list of boxes containing the findings and their description.

4. Improvements in the preventive archaeology yard

The ARCHEO 3.0 project aims to develop a series of innovative technologies to improve the surface and sub-surface investigation of the excavated area starting for the relief.

The proposed improvement on the relief procedure is based on the use of SfM technique applied on the frames acquired by a video recording of the excavation area. The SfM is a range imaging technique that allow us to reconstruct a 3D models starting from points extrapolated from 2D images. The 2D imaging can be automatically extracted from the video recording.

The video has been registered by GoPro Hero 3 camera equipped with stabilized rod. The camera was set with 1920x1080 pixels, and 50 fps as acquisition speed. The relief of the excavation area was just achieved in 11 min. To compare the performance with the standard procedure the same area was covered using a Nikon COOLPIX S3000 (setup: 4kx3k, f/3.2, ISO 200) the time used was 4 times longer. The modeling procedure consists of recovering 3D camera position related to pictures and 3D positions of the content of the images, this is obtained by identifying similar content between N views and solving 3D geometry problems. The test area for the speditive relief was the yard in the square of the Unità d'Italia, in Firenze (Fig. 1a). This is an emergency excavation carried out for the laying of the main sewage system in relation to the works for the construction of the tramway line in Florence. At the present, doesn't provide any musealization. At a depth of about 2 meters from the paved floor of the square, several remains have been found, maybe part of a farm linked to the exploitation of the fertile Florentine plain (Fig. 1b). These remains can be attributed at the end of the Republican age (end of the 1st century BC), when the Florentine plain was centuriated and divided among the veterans of the civil war. At the current state of knowledge, it isn't possible to establish with precision the chronology of the rustic settlement, of which we have found some levels of roof collapsed - made with tiles - and some masonry structures in *opus coementicium*, pebbles and limestone.

(a) (b)

Fig.1 - (a) Details of the excavation area of square of the Unità d' Italia in Firenze; (b) details of the remains found at 2 m depth respect to the paved floor

3D digital models were created using Agisoft PhotoScan. This software performs photogrammetric processing of digital images and generates 3D spatial data. Refinements, editing, cleaning and texturing operations on the obtained models were performed by MeshLab [6], an open source software for processing and editing 3D triangular meshes. 258 frames were extracted for the images alignment/matching that were achieved by the Scale Invariant Feature Transform (SIFT) algorithm through the identification of features in single images (grayscale converted). SfM estimates the relative camera position from the matched anchor points computed at the previous step and the sparse point cloud is built at the same time. In this way a sparse point cloud is obtained to roughly represent the scene. Approximately only 15 minutes has been sufficient to obtain the sparse point cloud by a notebook (equipped with Intel® Core™ i7-6700HQ @ 2.60GHz and RAM 16 GB).
The reconstruction of the 3D polygonal mesh was performed by using the PhotoScan software, in such way the geometry (mesh) is generated and the texture is applied using the photos initially acquired (Fig. 2). The realized model has been edited with MeshLab by performing a decimation of meshes, erase of detached components, close mesh holes, and other filtering operations. This process has been necessary to reduce the number of triangles that make up the 3D mesh so the 3D model is less "heavy" and easy to use on portable devices. This approach improves the management of the data acquired in the yard during all work stages by their direct insertion in a postGIS database that reduces the loss of information and allows changes as consequence of new archaeological evidence. Digital elaboration and procedures for the recognition of features useful for an automatic recognition and identification of SU are also aims of the project. A procedure for the automatic edge detection of the SU will be presented by I Cacciari et al. in this congress [7].

Fig. 2 - The textured model, rectified photography extracted from the 3D model

The mobile IT infrastructure aim is to integrates hardware and software devices to be able to acquire and store information in order to support the operational staff (archaeologist, technician and scientist) during their work, starting from the activity on site.

The platform is designed for an access through a tablet or other device capable of WiFi connection to the local server and to the central server in the headquarters. The infrastructure will allow to compile and store the documentation required directly online, sending the acquired information in real time to the specific server.

The principle scheme of the IT infrastructure is shown in Fig. 3.

The activity for the achievement of the IT infrastructure has been splitted into the following sub-activities:

1. Supply and appropriate configuration of the server that will host the database, the portion of the application that exposes the Application Programming Interface in order to interact with the database and a package of applications for a correct interface among the devices (server, mobile device, measurements systems, cameras, video cameras etc.).
2. Creation of a software structure in order to allow a safe interaction between the external world (applications that reside on device and server-side) and the database, with exposure of Application Programming Interfaces (APIs) with read/write access to the database;
3. Implementation of an executable app on the device (client-side application) that allows to the archaeologist, or end-user, to perform the cataloging and documentation of the finds directly on-site, and their processing.
4. Design of an on net software infrastructure to provide support to the dedicated app, and which allows the management and processing of data also through a PC.

The physical server was set up with Linux installation, allowing to integrates open-source software and advanced technology without license costs, Therefore, a Linux server was installed with Ubuntu distribution in the desktop 16.04.2 LTS (xenial), whose choice is motivated by two factors of fundamental importance:

- Desktop version so that the desktop version of a set of software can be run directly on the server, among which we include the geographic information system QGIS, Web GIS: QGIS and QGIS - Server.
- Ubuntu LTS because it is the long term support (up to 5 years) version of the Canonical desktop distribution.

This server runs the Apache 2.4.18 web server and the Tomcat application server, both open-source products, as well as having installed a series of desktop software for the project implementation. In particular, among the installed packages we include:
DBMS: PostgreSQL + PostGIS 9.5; PHP 7.0.15; Python 2.7.12; Java JDK 1.8; Jax-RS library for Java; Eclipse Oxygen; GIS: QGIS 2.18; WebGIS (MapServer); QGIS Server (2.18); QGIS Web client server; "pyArchInit" plugin for QGIS; Plugins for QGIS openLayers; Plugin for QGIS QGIS2Threejs.

IOP Conf. Series: Materials Science and Engineering **364** (2018) 012036 doi:10.1088/1757-899X/364/1/012036

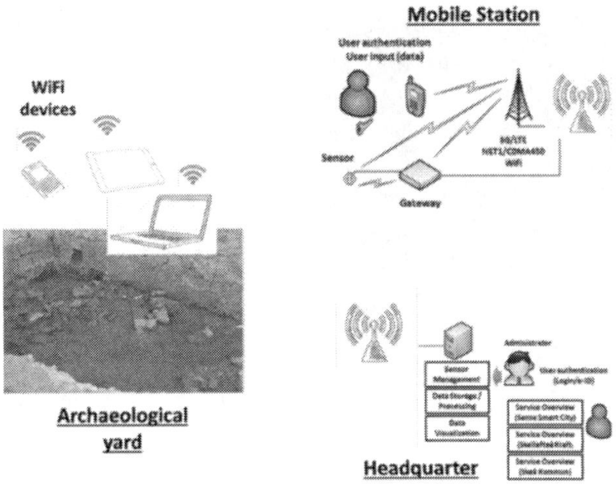

Fig. 3 - The IT infrastructure of the ARCHEO 3.0 project

For the second step, the software components are related to the interfacing on TCP / IP protocol, which allows to use the services also remotely with any device with Internet connection.

These products constitute the first-party software layer interposing between the database and the web app. These can be divided into two different APIs (API Server and API Client) both based on a data exchange protocol of the RESTFull type. The choice of standard restfull protocols both server-side and client-side is due to a simple access to data supply, typical of client-server applications.

In detail, with regard to the development of the Server API, it was decided to implement the latter in Java, using the Jersey library, with which it is possible to benefit from a set of tools able to create a REST infrastructure in a transparent way for the developer. For the realization of the WEB services, it was decided to implement, starting from the previously illustrated database, a three-level architecture as follows:

• Pojo: Objects that "map" (encapsulate) the tables present in the database, so that we can interact with these respecting the principle of encapsulation.

• Manager: Repository objects for entities mapped via Pojo, there are defined a number of methods that allow to perform queries in the database. Some of these operations, such as those identified by the acronym CRUD (Create, Read, Update and Delete) are common to all entities, while others will be specialized according to the entity considered (e.g. to recover a point of interest from a stratigraphic unit or vice versa). Among these operations we therefore include:

o GetById: Retrieve a record of a specific table in the database by mapping it as a Pojo object through its id;

o GetAll: Recovers all records of a given table by mapping them as a list of Pojo objects;

o GetAllPaginated: Same operation as the GetAll, but with the addition of the pagination for the results obtained;

o GetByFilter: Allows to select a certain number of records, mapping them as a list of Pojo objects, according to a given filter;

o DeleteById: Allows to remove the specified record via id;

o DeleteAll: Allows to remove all the records in the table;

o Insert: Insert a record in the database;

o Update: Performs the update of the selected record in the database;

Data exchange is expected both in XML format and in JSON format, the latter preferable for the conciseness in data representation. Therefore, based on the call made, the data will be directly displayed in either format. The OpenAPI tool to define the previously repoted services is shown in Fig. 4.

Fig. 4 - The IT infrastructure of the ARCHEO 3.0 project

Thanks to this tool, the documentation of the services and a skeleton have been realized on which it was then possible to continue with the development both for the server and client, since this tool allows server side and client side generation of the bees in almost all programming languages suitable for this purpose. In addition, OpenAPI, combined with the use of soapUI, has made it possible to directly test the services developed. In fact, by importing the documentation produced on soapUI, the unit tests are automatically generated.

6. Conclusion

In the case of preventive archaeology excavation often few time for survey and for the acquisition of information is allowed, therefore it is important to improve the rate of image and information to better define and understand the archaeological meaning of the site under investigation.

The acquisition through video of the excavation area has been proposed to reduce the time to achieved the 3D model, and to extract rectified photos, thanks to the implementation of SfM technique. The correct methodology for video capturing and the amount of frames to be extracted, is being studied in order to develop an effective pipe line of work.

And, the mobile IT infrastructure was realized to integrates hardware and software devices for acquiring and storing information in order to support the operational staff (archaeologist, technician and scientist) during their work starting from the activity on site.

Acknowledgements

The present study was carried out in the framework of the research project called ARCHEO 3.0 "Integration of Key Enabling Technologies for rising the efficiency and quality of preventive archaeological excavations" co-founded from Tuscany Region (POR-FESR 2014-2020).

References

[1] De Caro S 2009 Archeologia preventiva, lo stato dell'arte *Boll.ItaliaNostra*_num.444:1-3

[2] D'Andrea A & Guermandi M P 2008 *Strumenti per l'archeologia preventiva : esperienze, normative, tecnologie*. (Hungary-Budapest , Archaeolingua,): 125 p.

[3] Westoby M J, Brasington J, Glasser N F, Hambrey M J, Reynolds J M 2012 Structure-from-Motion' photogrammetry: A low-cost, effective tool for geoscience applications. *Geomorphology* **179:** 300-3014

[4] Fortunati M, Simone Zopfi L, Bishop J, De Rosa A, Malaspina F, Zanni S Archeomatica 4, 2014 *Applicazioni di Structure from Motion (SFM) - Tecniche open source di rilievo per interventi archeologici d'ermergenza*: 6-9

[5] Dellepiane M, Dell'unto N, Callieri M, Lindgren S, Scopigno R 2013 *Archeological excavation monitoring using dense stereo matching techniques*, Journal of Cultural Heritage 14(3): 201-210

[6] Cignoni P, Callieri M, Corsini M, Dellepiane M, Ganovelli F, Ranzuglia G 2008 MeshLab: an Open-Source Mesh Processing Tool. In 6th *Eurographics Italian Chapter Conference*: 129-136.

[7] I. Cacciari, G. F. Pocobelli, S. Cicola, S. Siano, 2018 Discrimination of soil texture and contour recognitions during archaeological excavation using Machine Learning. In press in Proceeding of Florence Heri-tech International Conference, Florence 16-18 May 2018.

Evaluation of visual descriptors for painting categorisation

Francesco Bianconi[1] **and Raquel Bello-Cerezo**[1]

[1]Department of Engineering, Università degli Studi di Perugia, Via Goffredo Duranti, 93 – 06125 Perugia, Italy

E-mail: bianco@ieee.org, bellocerezo@gmail.com

Abstract. The ever increasing availability of digital data from the Arts and cultural heritage calls for efficient methods to organise, categorise, and retrieve such information in an effective and reliable way. In this context, painting classification has attracted significant research interest in recent years. In this work we address the problem of style classification, which involves determining the school, period and art movement to which a painting belongs. Notably, this job is peculiarly different from other machine vision applications – such as material, object and scene recognition – since the concept of 'similarity' is much more difficult to define in this case. For this specific task we evaluate, in this study, the effectiveness of an array of hand-designed visual descriptors against a set of feature extractors based on last-generation convolutional neural networks. We also investigate the effect of pre-processing methods such as image split and pyramidal decomposition. The experiments are based on the open-access Pandora dataset. The results show that pre-trained models can significantly outperform hand-designed descriptors with overall accuracy surpassing 67%. This represents an improvement on the state-of-the-art by ≈12 percentage points.

1. Introduction

The Internet has long established itself as a common, open platform for cultural activities, and its role in shaping and factoring cultural interests can hardly be objected today [1]. Museums have not been oblivious to the potential of the Internet for reaching new users and, as a consequence, the availability of visual artistic data in digital format has increased dramatically in recent years. If, on the one hand, this calls for efficient methods to organise, categorise, and retrieve such information in an effective and reliable way, it makes, on the other, artistic data amenable of being processed by last-generation machine learning and data mining methods [2]. Within this context the present work is concerned with painting style classification. Notably, this job is peculiarly different from other machine vision tasks, as are for instance material classification, object and face recognition (see Refs. [3, 4] for a review), for the idea of *similarity* is more challenging to define in this context [5]. The very same concept of *style* is rather vague and imprecise: the Oxford English Dictionary defines this as the 'way of painting, writing, composing, building, etc., characteristic of a particular period, place, person or movement'. From a practical standpoint, however, few would object that among the visual elements that play a major role in making a style different from another are: the use of colour and its distribution across the painting, the shape and spatial layout of the geometric elements and the texture of the stroke.

Moving from these premises, various authors have tackled the problem of style categorisation using different techniques. Zujovic *et al.* [6] for instance proposed a combination of grey-scale features from steerable filters and colour statistics in the HSV space. Khan *et al.* [7] comparatively evaluated the performance of a set of global and local hand-crafted visual descriptors, and obtained the best results with a combination of Scale-Invariant Image Transform (SIFT) and colour names. Florea *et al.* [8] assessed a number of hand-designed descriptors for painting style classification and achieved the best performance with a combination of Local Binary Patterns (LBP) and Colour Structure Descriptor (CSD). Agarwal *et al.* [9] benchmarked five grey-scale local image descriptors and colour features obtaining the best performance with a combination of SIFT and colour histogram in the CIELab space. In recent years convolutional neural networks (CNNs) have dramatically changed the outlook in many computer vision tasks [10, 4] and therefore have been attracting much interest in computer analysis of visual arts too. Saleh and Elgammal [11] recently investigated the use of a pre-trained net along with metric learning schemes for large-scale classification of fine-art paintings. For the same task Tan *et al.* [5] proposed an end-to-end CNN trained from scratch. Of late, a deep multi-branch convolutional network for artist recognition and style classification has been proposed by Bianco *et al.* [12].

In this work we investigated the effectiveness of pre-trained convolutional networks as opposed to traditional, hand-crafted descriptors for painting style classification. We considered nine hand-designed visual descriptors combined with three pre-processing schemes on one side, and five pre-trained convolutional networks on the other. The pre-trained models surpassed the hand-designed descriptors by a large margin, allowing to reach overall accuracy in excess of 67%. This figure marks an improvement of ≈12 percentage points compared with the best result obtained in [8].

In the remainder of the paper we first describe the materials and methods used in this study (Secs. 2–3), then detail the experimental set-up in Sec. 4. We present and discuss the results in Sec. 5 and conclude the paper with some final considerations in Sec. 6.

2. Materials

We based our experiments on the recently released 'Pandora' dataset [8, 13]. This open-access, digital collection contains a total of 7724 images of paintings organised into 12 classes representing as many art movements: *Abstract expressionism*, *Baroque*, *Cubism*, *Fauvism*, *High Renaissance*, *Iconoclasm*, *Impressionism*, *old Greek pottery*, *Realism*, *Rococo*, *Romanticism* and *Surrealism*. The number of samples for each class is detailed in Fig. 1. The size, spatial resolution and shape factor (width/height) vary from one image to another; the imaging conditions such as illumination, camera/scanner type and acquisition settings are unknown.

3. Methods

We used a standard image classification pipeline based on three steps: image pre-processing, feature extraction and classification.

3.1. Image pre-processing

Image preprocessing consisted of either image split, pyramidal decomposition or no pre-processing whatsoever. No further preliminary operations such as colour normalisation, histogram equalisation, filtering or deblurring were applied.

3.1.1. Image split The original image is partitioned into s sub-images of equal size (Fig. 2). The visual descriptors are extracted from each sub-image separately and then concatenated into one single feature vector. As a result, the dimensionality of the original descriptor increases s-fold. In the experiments we used $s = 4$ and $s = 9$, which we indicate with subscripts '$4\times$' and '$9\times$' in the remainder.

IOP Conf. Series: Materials Science and Engineering **364** (2018) 012037 doi:10.1088/1757-899X/364/1/012037

Figure 1: Sample images from the 12 classes of the Pandora dataset. The figures in parentheses indicate the number of images for each class.

3.1.2. Pyramid decomposition A set of copies of the input image is obtained via recursive filtering and sub-sampling (Fig. 3). This enables a multi-scale analysis of the input image [14, 15]. Herein we used a Gaussian decomposition into four levels (including the original image – level 'zero'). The visual descriptors are extracted from each downsampled image separately and concatenated into one feature vector. The dimensionality of the original visual descriptor increases four-fold in this case. We shall indicate this option with with subscript '$p4$' in the remainder.

3.2. Visual descriptors for feature extraction

Visual descriptors can be classified as 'hand-designed' (also referred to as 'engineered' or 'hand-crafted') or based on deep learning [16, 17]. In the first group most of the mathematics that makes up a descriptor (i.e.: functional form and parameters) is established a priori. As a result such descriptors usually require little or no training. By contrast, deep learning relies on large sets of parameters the values of which need to be determined by training. The advantage, in

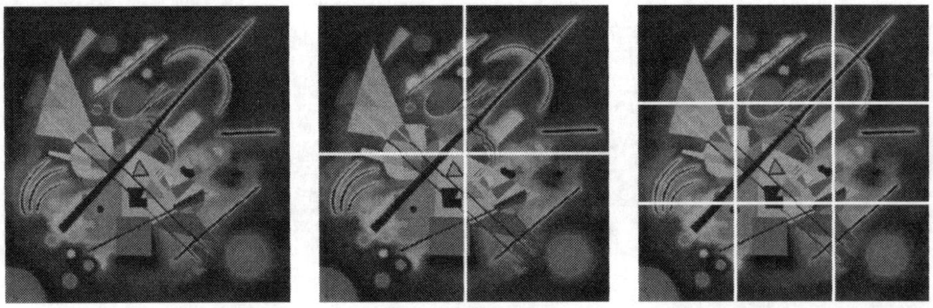

a) Original image b) Partition into four sub-images ('4×') c) Partition into nine sub-images ('9×')

Figure 2: Image split.

Figure 3: Pyramid decomposition ('*p*4'). From left to right: original image and downscaled images.

this case, is that a descriptor can be tailored to a specific application – and this can make it very accurate in that domain; the disadvantage that training requires large amounts of data and adequate computational resources. Fortunately, however, deep networks trained for specific tasks can be used effectively in completely different tasks as well (transfer learning [10]).

3.2.1. Hand-designed visual descriptors Hand-designed visual descriptors are generally classified into three main groups [3]: *spectral*, *spatial* and *hybrid* methods. Descriptors of the first group only consider the colour content of an image regardless of its spatial distribution. Conversely, those of the second disregard colour and only take into account the spatial variation of the image luminance (grey-scale intensity). Finally, hybrid methods combine spectral and spatial data in different ways. In this study we considered:

- Two spectral methods (joint colour histogram and marginal colour histograms)
- Three spatial methods (Histograms of Gradients – HOG, Local Binary Patterns and Improved Local Binary Patterns)
- Four hybrid methods (colour HOG, Local Colour Vector Binary Patterns, Improved Opponent Colour Local Binary Patterns and Opponent Colour Local Binary Patterns).

IOP Conf. Series: Materials Science and Engineering **364** (2018) 012037 doi:10.1088/1757-899X/364/1/012037

We briefly recall the basics of each descriptor here below and refer the interested reader to the references for in-depth explanations.

Colour histogram (ColHist) Joint, three-dimensional colour histograms in the RGB space [18]. Each colour channel was quantised into eight bins, which gives $8^3 = 512$ features.

Marginal colour histograms (MargHists) Concatenated marginal histograms of the intensity levels in each of the R, G and B channels [19]. In this case each colour was quantised into 256 bins, which results in a feature vector of dimension $256 \times 3 = 768$.

Histograms of oriented gradients (HOG) Histograms of the orientation of the local grey-scale gradient [20]. This was estimated using generalised Sobel filters at resolution 3px × 3px, 5px × 5px and 7px × 7px. The orientation was quantised into 64 bins. The number of features therefore is $3 \times 64 = 192$.

Colour HOG (ColourHOG) Histograms of oriented gradients extracted from each of the R, G and B channels. In this case the number of features is $3 \times 64 \times 3 = 576$.

Local Binary Patterns (LBP) Concatenation of directional, Local Binary Patterns histograms [21] computed over non-interpolated, eight-point digital circles of radius 1px, 2px and 3px (see [22] for details). This configuration generates $256 \times 3 = 768$ features.

Improved Local Binary Patterns (ILBP) Variant of LBP in which the local thresholding scheme is point-to-average instead of point-to-point [22]. The other settings were the same as LBP. The number of features in this case is $511 \times 3 = 1533$

Local Colour Vector Binary Patterns (LCVBP) Concatenation of Colour Norm Patterns (CNP) and Colour Angular Patterns (CAP) with the latter extracted from each of the R-G, R-B and G-B channel pairs [23]. The other settings were the same as LBP. Since CNP generates the same number of features as LBP and CAP three times as many, the final number of features is $256 + 256 \times 3 = 1024$.

Opponent Colour Local Binary Patterns (OCLBP) A colour variant of LBP obtained by applying LBP to each colour channel separately and in a pairwise manner between each of the R-G, R-B and G-B channel pairs [24]. This variant multiplies by six the dimensionality of LBP, therefore generates $256 \times 6 = 1536$ features.

Improved Opponent Colour Local Binary Patterns (IOCLBP) An extesion of ILBP to the colour domain based on the same scheme as OCLBP's [22]. The number of features is $511 \times 3 + 512 \times 3 = 3069$.

3.2.2. Deep learning We considered five pre-trained deep network models (details below) and, for each model, used as image features the L_1-normalised output of the last fully-connected layer. This is usually referred to as the 'FC' configuration (see also Refs. [25, 17] on this point). Pre-processing only involved resizing the input image to 224px × 224px – the size of the receptive field of all the models considered. The implementation was based on the pre-trained models available on MatConvNet [26] with no further modification whatsoever.

GoogLeNet (GoogLeNet) A deep network with 22 learnable layers based on the 'Inception' architecture as described in [27]. The dimension of the feature vector is 1024.

ResNet 50 (ResNet-50) A very deep network with 50 learnable layers based on the 'residual' architecture [28]. The dimension of the feature vector is 2048.

ResNet 101 (ResNet-101) A very deep network with the same overall architecture as ResNet-50's but with 101 learnable layers instead of 50 [28]. The dimension of the feature vector is 2048.

VGG very deep 16 (VGG-VD-16) A deep network for large-scale image recognition based on 16 learnable layers and small (3×3) convolution filters [29]. The dimension of the feature vector is 4096.

VGG very deep 19 (VGG-VD-19) A network with a structure very similar to VGG-VD-16's but with 19 learnable layers instead of 16 [29]. The number of features is 4096.

3.3. Classification
The feature vectors generated by the visual descriptors presented in Secs. 3.2.1–3.2.2 were fed to nearest neighbour and support vector classifiers. We used L_1 (cityblock) and L_2 (Euclidean) distance for the nearest neighbour (1-NN) classifier and Gaussian radial basis kernel for the support vector machine (SVM). Multi-class extension of SVM was based on the one-vs-one architecture; optimal box constraint (C) and kernel scale (γ) were determined via trial and error (final values were $C = 1000$ and $\gamma = 1$).

4. Experiments
We carried out a set of supervised image classification experiments to evaluate the effectiveness of each combination pre-processing/visual descriptor/classifier. To avoid excessively long feature vectors ('curse of dimensionality') pre-processing via image split and/or pyramid decomposition was limited to those descriptors that generate less than 1000 features. Accuracy estimation was based on four-fold cross-validation using the same partitions that come in bundle with the Pandora dataset. In each experiment a single fold was retained as the test set, and the remaining three were used as training data. The estimated accuracy was the fraction of paintings of the test set classified correctly. The results (Tabs. 1–2) were averaged over the four folds.

5. Results and discussion
As can be seen, the visual descriptors based on pre-trained CNNs outperformed the hand-designed ones by a large margin. This is remarkable, especially if we consider that: a) we used CNN models trained for other tasks (i.e. object recognition) with no adjustment whatsoever, and b) the input images had to be resized heavily to fit the visual field of each pre-trained net. In absolute terms the overall was in excess of 67%, which marks an improvement of more than 12 percentage points with respect to previous benchmarks [8].

Among the hand-designed methods, the best performance (45.7%) was obtained through joint colour histogram and preliminary image split ('9×'). This outcome is also interesting in that it shows that colour alone plays an important role in discriminating between styles. The results of the other hand-crafted descriptors (i.e. LBP, HOG and colour HOG) were otherwise in line with those available in the literature [8].

Table 1: Overall classification accuracy (in %) of the hand-designed visual descriptors. Boldface and underlined figures respectively indicate the best overall accuracy and the best accuracy for each group.

Descriptor	Classifier		
	NN_{L1}	NN_{L2}	SVM_{ovo}
Spectral methods			
ColHist	35.7	34.0	39.4
$ColHist_{\times 4}$	36.5	32.2	43.8
$ColHist_{\times 9}$	37.7	33.1	**45.7**
MargHists	30.9	30.9	27.7
$MargHists_{\times 4}$	32.8	30.8	35.0
$MargHists_{\times 9}$	33.8	31.6	38.7
Spatial methods			
HOG	27.6	28.0	15.9
$HOG_{\times 4}$	29.1	28.3	21.4
$HOG_{\times 9}$	26.8	26.5	28.0
HOG_{p4}	24.7	23.7	22.5
ILBP	39.5	35.8	20.8
$ILBP_{\times 4}$	39.3	36.5	31.7
$ILBP_{\times 9}$	36.5	35.3	39.0
$ILBP_{p4}$	<u>41.7</u>	38.9	37.0
LBP	38.2	35.0	21.3
$LBP_{\times 4}$	38.9	35.7	32.5
$LBP_{\times 9}$	36.3	34.6	38.2
LBP_{p4}	37.8	35.4	35.6
Hybrid methods			
ColourHOG	28.2	28.0	19.1
$ColourHOG_{\times 4}$	29.1	28.3	25.2
$ColourHOG_{\times 9}$	26.0	26.3	34.9
$ColourHOG_{p4}$	25.0	23.8	28.0
LCVBP	<u>40.8</u>	38.5	23.7
IOCLBP	36.5	26.7	25.7
OCLBP	36.4	29.1	26.6
Accuracy of random classifier (blind) = 8.3			
Accuracy of random classifier (with priors) = 9.9			

Table 2: Overall classification accuracy (in %) of the visual descriptors based on pre-trained convolutional neural networks (deep learning). The best accuracy is indicated in boldface.

Descriptor	Classifier		
	NN_{L1}	NN_{L2}	SVM_{ovo}
GoogLeNet	52.0	52.5	61.3
ResNet-50	59.7	60.7	**67.2**
ResNet-101	59.1	60.1	66.9
VGG-VD-19	51.2	51.9	60.9
VGG-VD-16	51.9	51.9	61.8
Accuracy of random classifier (blind) = 8.3			
Accuracy of random classifier (with priors) = 9.9			

6. Conclusions

In this work we have carried out an experimental evaluation of an array of visual descriptors for painting style categorisation. Compared with other machine vision tasks such as object, face, scene or material recognition this is a more challenging problem, for the concept of similarity is more difficult to define in this context. Besides, whereas humans can rely on previous background on art and art history for the job, the machine cannot [5]. In light of this, a success rate over 67% seems promising. Among the visual descriptors considered in this study, features from 'off-the-shelf' pre-trained CNNs emerged as the best strategy and largely outperformed hand-designed visual descriptors.

Acknowledgements

This work was partially supported by the Department of Engineering at the Università degli Studi di Perugia, Italy, under project *Machine learning algorithms for the control of autonomous mobile systems and the automatic classification of industrial products and biomedical images* (Fundamental research grants 2017), and by the Italian Ministry of Education, University and

Research (MIUR) within the framework *Finanziamento individuale annuale delle attività base di ricerca* (FFABR 2017).

References

[1] Karp C 2014 *Museum International* **66** 157–162
[2] Stork D, Coddington J and Bentkowska-Kafel A 2010 *Computer Vision and Image Analysis of Art (Proceedings of SPIE - The International Society for Optical Engineering* no 7531) (San José, USA) art. no. 753101
[3] González E, Bianconi F, Álvarez M and Saetta S 2013 *Advances in Optical Technologies* Art. no. 503541
[4] Nanni L, Ghidoni S and Brahnam S 2017 *Pattern Recognition* **71** 158–172
[5] Tan W, Chan C, Aguirre H and Tanaka K 2016 *Proceedings of the International Conference on Image Processing (ICIP)* (Phoenix, USA) pp 3703–3707
[6] Zujovic J, Gandy L, Friedman S, Pardo B and Pappas T 2009 *Proceedings of the IEEE International Workshop on Multimedia Signal Processing (MMSP)* (Rio De Janeiro; Brazil) art. no. 5293271
[7] Khan F, Beigpour S, Van De Weijer J and Felsberg M 2014 *Machine Vision and Applications* **25** 1385–1397
[8] Florea C, Condorovici R, Vertan C, Butnaru R, Florea L and Vrânceanu R 2016 *Proceedings of the 24th European Signal Processing Conference (EUSIPCO)* (Budapest, Hungary) pp 918–922 art. no. 7760382
[9] Agarwal S, Karnick H, Pant N and Patel U 2015 *Proceedings of the IEEE Winter Conference on Applications of Computer Vision (WACV)* (Waikoloa Beach, USA) pp 588–594 art. no. 7045938
[10] Razavian A, Azizpour H, Sullivan J and Carlsson S 2014 *Proceedings of the IEEE Conference on Computer Vision and Pattern Recognition Workshops, CVPRW 2014* (Columbus, United States) pp 512–519
[11] Saleh B and Elgammal A 2016 *International Journal for Digital Art History* **2**
[12] Bianco S, Mazzini D and Schettini R 2017 *Proceedings of the 19th International Conference on Image Analysis and Processing (ICIAP) (Lecture Notes in Computer Science* no 10484) (Catania, Italy) pp 414–423
[13] Image Processing and Analysis Laboratory 2016 Perceptual ANalysis and DescriptiOn of Romanian visual Art (PANDORA) available online at `http://imag.pub.ro/pandora/pandora_download.html`. Last accessed on Nov. 6, 2017
[14] Adelson E H, Burt P J, Anderson C H, Ogden J M and Bergen J R 1984 *RCA engineer* **29** 33–41
[15] González E, Bianconi F and Fernández A 2016 *Information Sciences* **361-362** 1–13
[16] Napoletano P 2017 *Proceedings of the 6th Computational Color Imaging Workshop (CCIW'17) (Lecture Notes in Computer Science* vol 10213) ed Bianco S, Schettini R, Tominaga S and Tremeau A (Milan, Italy: Springer) pp 259–271
[17] Bello-Cerezo R, Bianconi F, Cascianelli S, Fravolini M, Di Maria F and Smeraldi F 2017 *Intelligent Interactive Multimedia Systems and Services 2017. KES-IIMSS 2017 (Smart Innovation, Systems and Technologies* vol 76) ed De Pietro G, Gallo L, Howlett R and Jain L (Vilamoura, Portugal)
[18] Swain M and Ballard D 1991 *International Journal of Computer Vision* **7** 11–32
[19] Pietikainen M, Nieminen S, Marszalec E and Ojala T 1996 *Proceedings of the International Conference on Pattern Recognition (ICPR)* vol 3 (Vienna, Austria) pp 833–838 art. no. 547285
[20] Dalal N end Triggs B 2005 *Proceedings of the IEEE Computer Society Conference on Computer Vision and Pattern Recognition (CVPR)* vol 1 (San Diego, USA) pp 886–893 art. no. 1467360
[21] Ojala T, Pietikäinen M and Mäenpää T 2002 *IEEE Transactions on Pattern Analysis and Machine Intelligence* **24** 971–987
[22] Bianconi F, Bello-Cerezo R and Napoletano P 2018 *Journal of Electronic Imaging* **27** art. No. 011002
[23] Lee S, Choi J, Ro Y and Plataniotis K 2012 *IEEE Transactions on Image Processing* **21** 2347–2353
[24] Mäenpää T and Pietikäinen M 2005 *Handbook of Pattern Recognition and Computer Vision (3rd Edition)* ed Chen C H and Wang P S P (World Scientific Publishing) pp 197–216
[25] Cimpoi M, Maji S, Kokkinos I and Vedaldi A 2016 *International Journal of Computer Vision* **118** 65–94
[26] Vedaldi A and Lenc K 2015 *Proceedings of the 23rd ACM International Conference on Multimedia (MM 2015)* (Brisbane, Australia) pp 689–692
[27] Szegedy C, Liu W, Jia Y, Sermanet P, Reed S, Anguelov D, Erhan D, Vanhoucke V and Rabinovich A 2015 *Proceedings of the IEEE Computer Society Conference on Computer Vision and Pattern Recognition* (Boston, United States) pp 1–9
[28] He K, Zhang X, Ren S and Sun J 2016 *Proceedings of the IEEE Computer Society Conference on Computer Vision and Pattern Recognition* (Las Vegas, United States) pp 770–778
[29] Simonyan K and Zisserman A 2015 *Proceedings of the 3rd International Conference on Learning Representations (ICLR)* (San Diego, USA)

INTERACTION AND INFORMATION COMMUNICATION IN VIRTUAL MUSEUMS

Anastasios Latos
Dept. of Informatics, Ionian University, Corfu, Greece
E-mail: p11lato@ionio.gr

Vasileios Komianos
Dept. of Informatics, Ionian University, Corfu, Greece
E-mail: vkomianos@ionio.gr

Konstantinos Oikonomou
Dept. of Informatics, Ionian University, Corfu, Greece
E-mail: okon@ionio.gr

Abstract. This paper presents a virtual museum implementation designed in order to provide effective promotion of the represented cultural heritage objects and also to serve as a prototype of virtual educational museums. The considered virtual museum hosts 3D exhibits representing real cultural heritage artifacts and provides related to the exhibits information. User are able to navigate in it through the use of virtual characters and appropriate information communication methods are provided. The provided communication methods employ rich interaction techniques in order to engage users and enhance the learning process. An important aspect of this paper is the study of the navigational issues arising while navigating spaces of complex structure. Appropriate navigation assistance methods are deployed to reduce users disorientation incidents and a method that provides navigation recommendations is used to support their functionality. The paper also addresses issues related to the use of video game technology for cultural education purposes and provides guidelines on its effective deployment. Finally, tests including users evaluation are conducted in order to examine the effectiveness of the implemented information communication methods and the effectiveness of the provided navigation assistance.

Keywords: virtual museums, navigation methods, information presentation

1. Introduction

Three dimensional (3D) virtual environments, also called 3D virtual worlds, increasingly become a suitable solution for modern applications in the area of visual representation. Having many characteristics such as intuitiveness, easy of use, interactivity and immediacy, virtual environments are also applicable in promotion and preservation of cultural heritage [1]. Museums soon realized the potential and benefits of virtual environments, aiming to the development of educational content and services for their visitors. So they created their digital version, Virtual Museums, which are websites or applications that simulate a 3D world, in which the exhibits have been developed in virtual way. A virtual museum allows users tour into the virtual world, while interacting with exhibits and getting informed with related information about them. Virtual Museums aim to explore the benefits of using virtual reality in the premises of museums, in order to help both researchers and general public learn through an alternative, more entertaining way (Edutainment), while preserving and bringing to prominence cultural heritage exhibits. However, Virtual Museums indicate a lack of usability, which is usually due to non-comprehensible or too complicated tour techniques which are being used. In this paper two important aspects of virtual tours are studied: (i) *navigation assistance*; and (ii) *information communication*. The aforementioned aspects are considered important in order to enhance usability and information intake. A demonstrative virtual museum integrating navigation assistance and information communication methods is developed in order to evaluate their effectiveness.

2. Information Communication Methods in Virtual Museums

Information communication is considered to be an important aspect of both physical and virtual museums [2, 3]. Museums serve as information communication media by exhibiting their artifacts and by deploying appropriate information communication methods [4]. Multimedia content, e.g., video, images, audio, etc., is frequently used by virtual museums for information communication [5].

2.1. Video & Images

In this paper, videos and images, are studied as a single information communication method due to their visual form and the absence of interactivity. Videos are audiovisual means, which if they are prepared and processed in a right way, they can combine usefully many different types of information such as text, sound, stable or moving images, graphics, etc.

In addition, videos are able to present realistic 3D virtual exhibits, no matter their real location, size or age, giving the chance to visitors for learning about them through their virtual tour [6]. In this case, some of the exhibits are objects, buildings and places, which cannot be shown efficiently and completely during a virtual tour, due to their dimensions and the limited time that visitors provide for their tour. Using audiovisual means for providing information in virtual worlds, especially videos, is an ideal option for cases in which a lot of data and information have to be presented.

Additionally, visitors with audio-visual experiences during their tour in museums, were able to memorize, remember and analyze information and details about the exhibition they watched [7].

To summarize, using videos as a main information communication method in a virtual environment for the purpose of promoting and educate has many benefits which are concisely mentioned below: (i) they constitute direct and brief type of information, while they totally cover the presentation of the exhibit in a short time period; (ii) while creating them, the author has the opportunity to express personal feelings, opinions, viewpoints and bring out messages to the visitors; (iii) they can present every kind of data or information; (iv) they offer directness, as they visualize data information for the visitor; (v) they are able to create a special kind of atmosphere into the virtual world; (vi) they are able to provide users with realistic information about what they present; and (vii) they already stand as a known and friendly way of informing for every kind of visitor. Besides its advantages video, in its traditional form, lacks of interactivity.

Video projection in the presented virtual museum is described below. Into the virtual museum, videos play essential role for information communication and exhibit presentation. Videos are being projected on the surface of the walls, covering a fully presentation of virtual 3D exhibits, accompanied by audio. An interaction feature has been also integrated, as every video starts playing when visitors approach its surface and it pauses when they walk away from it. More interaction options are given to users, as they are able to start or stop playing the videos by giving keyboard entries.

2.2. Text

In addition to multimedia, text is probably the most widely used information communication method. Text contains information and details about presented exhibits but it cannot represent an exhibit by itself. However, it can be combined easily and efficiently with every other method existed, as it has the ability to provide information that cannot be displayed in different way. The main reason for choosing text to be combined with other techniques is because of the secondary role that text plays next to exhibits. In physical museums, text is used to accompany exhibits, providing additional information about them. Important details such as locations, age, dates, names, historical facts and more, can be easily provided. Moreover, information provided through text is direct and difficult to misunderstand by visitors.

Text as an information communication method provides the following advantages: (i) it is direct and contains only the needed information; (ii) it can be understood easily by everyone; (iii) it can be used in order to provide knowledge to every social kind of people, no matter their previous education level, age, experience, etc.; (iv) it is able to transmit author's personal messages, opinions, viewpoints; (v) it is hardly misinterpreted when is written, organized and presented in a right way.

Text integration in the presented virtual museums is described below. Text is also used into the museum for guiding visitors and provide them with additional information about the exhibits they are watching. Projected text is being read from text files, in order to be easily modified and organized. Text is integrated in both videos and 3D holograms, and it can be shown and hide when users want to, by giving keyboard entries. Users are also able to move on the next lines of the text while reading, by keyboard entries, too. Depending users position into the virtual environment, text is projected to them in order to guide them about their options, or provide information relatively with the exhibit they observe.

2.3. Audio

Audio is an important information communication method in both physical and virtual museums [8]. Audio data is usually used for providing additional information about virtual exhibits in virtual reality and in most cases accompany or replace text. As it is used for providing details about exhibits, audio usually has the form of narration. Narration and text have many similarities, as in most cases a narration can be printed in text. A well-organized narration during a virtual tour is able to transmit author's feelings, opinions and viewpoints about an exhibit in the museum. Moreover, it is easily understanding, direct and hard to misinterpreted. Narration is very helpful as a means of information communication used by people with vision impairment or a lower education level, which make it difficult or impossible for them to read.

2.4. 3D Holograms

3D holograms [9] is proposed in this paper as a method of information communication in virtual museums. 3D holograms are used to represent three-dimensional digital exhibits in a virtual environment. This kind of method has rose steeply during the last years, promising crucial changes in near future, at the field of information providing for both virtual and real-life environments. 3D holograms can be used to represent real world objects that have been shrinked in order to fit in place. 3D holograms can be passive as traditional museum exhibits or interactive. Holography as an Information Communication method for virtual exhibits has many benefits in contrast with other related techniques, while the same time is the best way to present the potential of 3d graphics in either virtual or real environments. Interaction features are incorporated enabling holograms to start rotating when visitors get closer to them, and stop when they move away. Moreover, users' keyboard entries are also acceptable for start and stop rotating holograms when they wish to.

2.5. Cave Automatic Virtual Environment

CAVE (Cave Automatic Virtual Environment) [10] is an interactive simulator of virtual environments which is used in real life world. In this paper virtual CAVE installations are proposed as a information communication method in virtual museums. CAVE is a room with the shape of a cube. In some cases, one wall of the block is missing, as this side is used for entrance. Outside the room there are projectors which synchronously light up every wall, creating an experience of a complete virtual world for the observer. Visitor uses 3D glasses which provide 3D-depth image. Surround speakers are also being used for achieving realistic sound effects. The system allows user to move and interact into the virtual world by using the control system in three ways: (i) Motion Joystick; (ii) Touch Pointer; (iii) Wireless Position Sensor Polhemus. CAVEs in virtual environments are quite different in the way they work. The concept of the projected room remains, but there is no need for special equipment such as projectors, 3D glasses etc. The visitor enters the room, into which a different virtual environment can be projected. In this way, many different topics with the form of virtual environments can be integrated into a single virtual tour. System controls can be managed by using simple peripherals, like keyboard or mouse, but also specialized technologies can be used, such as VR glasses, motion trackers, etc, for a better experience. While into the room, a whole virtual world is projected in visitors eyes, filled with available exhibits and related information. However, it remains a personal choice for the user about how he will move in the virtual world and with which exhibits is going to interact. CAVE is an alternative way of information communication as it uses interaction as its main feature in combination with the selected theme which is projected. In this way, it sets users free to tour themselves in the virtual world and interact into it, while a specific topic with a main idea is simulated into the room. This is why CAVE is currently being used in a variety of fields, such as education in universities, or just for entertainment by simulating games.

Koutsoumpidis et al. [11] propose an interesting method for cultural content communication as they use 3D avatars of historical personalities for information communication combining both visual and auditory stimuli.

3. Navigation and Interaction in Virtual Museums

Navigation is the process or activity of accurately ascertaining one's position and planning and following a route. During the last years, it has become an essential part of all Virtual Environments. While VR space grows, navigation becomes even more challenging, as its main goal is to be as simple as possible, and the same time preserving the elements of exploration and discovery [12]. This is what makes the technique which is used in VEs efficient. Due to its large scale and the number of included points of interest, users navigation problems [13] show up.

Effective users navigation is prerequisite as the virtual environment aims: (i) To cover extended areas; (ii) Having many points of interest; (iii) To provide users spatial knowledge useful in virtual and real visits too; (iv) Avoiding users disorientation that would cause frustration; (v) Recommend paths to follow. For an effective navigation, it is essential for users to be able to orient themselves in the virtual environment, observe navigation marks and obtain knowledge about the space. However, there are four different factors that constitute navigation and affect its efficiency in a virtual world. (i) Way finding; (ii) Travel; (iii) Users; (iv) Virtual Environment itself [14].

Komianos et al. [15] present the most common navigation techniques which are used into virtual environments and listed here (i) 2D maps; (ii) Location pointing using arrows; (iii) Oral

IOP Conf. Series: Materials Science and Engineering **364** (2018) 012038 doi:10.1088/1757-899X/364/1/012038

directions; (iv)Textual descriptions; (v) Orally-assisted destination recognition; (v) Destination recognition supported by appropriate illumination; (vi) 3D Tour Guide.

3.1. Static Indicative Arrows in Virtual Museums

The use of static arrows is an ideal option for navigating in not too complex environments such as the virtual museum, due to their low computing cost, in combination with the level of assistance they provide users with. Into the virtual museum, static arrows are being projected onto the floor, depending on the users space position. Arrows inform visitors in which direction they should move in order to visit every available exhibit. When the virtual tour starts, arrows are initially invisible, and they can be projected or hide again by visitors keyboard entry. Arrows created based on two essential characteristics, in order to avoid information loss or alteration. According to [15], arrows should be (i) evident and (ii) modest.

3.2. Interaction Borders in Virtual Museums

Interaction borders are proposed in this paper. Interaction borders are borders placed onto the floor of the building, marking the surface into which users are able to interact with the exhibits, as well which specific exhibit they involve. Those borders are projected depending visitor's position into the building, and can be shown or hide when users need to, after a keyboard entry.

3.3. Experimental Design & Main Idea

In order to evaluate the utility and need for navigation assistance in virtual environments, a user participation experiment has been designed. The navigation assistance method which was selected to be evaluated is static lines - which are similar to static arrows - places onto the floor of a building. More specifically, the experiment tries to simulate cases of virtual environments, in which users have to navigate under conditions of complicated space design, into environments with numerous possible points of interest. The evaluation is estimated relative to the time users need to navigate themselves into the virtual space. For this purpose, a large space virtual environment has been created, with sixteen similar rooms into it. The building has four entries in random positions, to its outer sides of walls. As for the entries to each room, they are also placed in random way. On the floor of the building, navigation lines has been designed, starting from each outer entry and finishing to each room. A teapot has also been placed into one of the rooms, which represents a possible exhibit. Moreover, each inner room represents a possible point of interest.

3.4. Procedure & Tasks

The experiment is conducted to random users who can participate as many times as they wish to. When the procedure begins, each user controls a virtual character who starts from one of the four outer building entries. The starting point is selected randomly. Into the building, one of the rooms, which is chosen also in random way, contains the teapot, while the rest of the rooms are empty. The users have to navigate themselves into the building with only one goal, to find the teapot among the rooms. In some cases, users are provided with assistance through the navigation lines on the floor of the building, but there are also instances, in which users have to navigate themselves alone and find the teapot without any help. During the procedure is in progress, an overseer observes the experiment and writes down the times users need to move in the virtual space until they find the requested teapot. The main concept of the experiment is to compare the recorded times, between cases with and without the navigation lines, in order to examine their utility, and the efficiency of navigation assistance.

3.5. Results Discussion

In our first attempt for collecting data, 11 random users participated in the experiment, with a total amount of 32 tries. Although time results during the procedure were interesting, the final results after analyzing data were more than informative. To be more specific, as the table below shows, the total average time users needed to find the teapot in cases without navigation assistance was 22,72 seconds, while in cases with navigation assistance they needed only 9,51 seconds.

Results prove that navigation assistance of static lines used in space complicated environments are able to reduce tour time up to about 58.14%. Of course, as it is expected, more data collected are able to lead on more representative results. Nevertheless, numbers still remain informative about the utility and efficiency of this specific navigation assistance method.

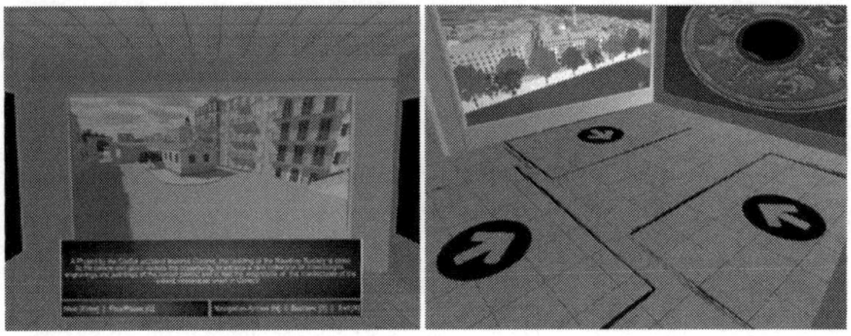

Figure 1. Video projection and virtual CAVE installation in virtual museum. The video is accompanied by text. The virtual CAVE installation uses static arrows and interaction borders in order to enhance its usability.

Figure 2. Holograms and interaction borders.

4. Conclusions

In this paper a set of methods providing information communication and navigation assistance in virtual museums is presented. Regarding the information communication, the proposed methods are mostly based on conventional methods which are appropriately integrated in virtual museums. Early tests show that video projection accompanied by narration and text descriptions have positive effect on users experience. According to users virtual CAVE is interesting and the authors consider that when equipped with appropriate scenarios can be further improved. Holograms are observed to attract users in early tests and increase their engagement. The provided navigation assistance

5. Bibliography

[1] Komianos, V., Kavvadia, E. and Oikonomou, K. (2014). Efficient and realistic cultural heritage representation in large scale virtual environments, *Information, Intelligence, Systems and Applications, IISA 2014, The 5th International Conference on*, IEEE, pp. 1–6.
[2] Schweibenz, W. (1998). The" virtual museum": New perspectives for museums to present objects and information using the internet as a knowledge base and communication system., *ISI* **34**: 185–200.
[3] Hooper-Greenhill, E. (1999). Museum, media, message, *Psychology Press.*
[4] Castells, M. (2001). Museums in the information era, Cultural connectors of time and space, *ICOM news*, International Council of Museums.
[5] Kravchyna, V. and Hastings, S. (2002). Informational value of museum web sites, *First Monday* **7**(2).
[6] Tsoumanis, G., Kavvadia, E. and Oikonomou, K. (2015). A v(irtual)-city implementation for promoting cultural heritage, *International Journal of Computational Intelligence Studies* **4**(2): 173–191.
[7] Allen, S. (2004). Designs for learning: Studying science museum exhibits that do more than entertain, *Science Education* **88**(S1).
[8] Keil, J., Pujol, L., Roussou, M., Engelke, T., Schmitt, M., Bockholt, U. and Eleftheratou, S. (2013). A digital look at physical museum exhibits: Designing personalized stories with handheld augmented reality in museums, *Digital Heritage International Congress* (DigitalHeritage), 2013, Vol. **2**, IEEE, pp. 685–688.
[9] Komianos, V. (2017). Content personalization approaches in cultural heritage (real, virtual and mixed) environments, Doctoral Thesis, Department of Informatics, Ionian University .
[10] Cruz-Neira, C., Sandin, D. J., DeFanti, T. A., Kenyon, R. V. and Hart, J. C. (1992). The cave: audio visual experience automatic virtual environment, *Communications of the ACM* **35**(6): 64–73.
[11] Koutsoumpidis, E., Komianos, V. and Oikonomou, K. (2017). Evaluation of virtual agents effectiveness in history class, *European Journal of Education and Applied Psychology* (Forthcoming).
[12] Chittaro, L., Ieronutti, L. and Ranon, R. (2004). Navigating 3d virtual environments by following embodied agents: a proposal and its informal evaluation on a virtual museum application., *Psychology Journal* **2**(1): 24–42.
[13] Darken, R. P. and Sibert, J. L. (1996). Navigating large virtual spaces, *International Journal of Human-Computer Interaction* **8**(1): 49–71.
[14] Dodiya, J., Alexandrov, V. N. et al. (2008). Navigation assistance for wayfinding in the virtual environments: Taxonomy and a survey.
[15] Komianos, V., Kavvadia, E. and Oikonomou, K. (2015). Cultural heritage recommendations and user navigation in large scale virtual environments, *International Journal of Computational Intelligence Studies* **4**(2): 151–172.

IOP Conf. Series: Materials Science and Engineering **364** (2018) 012039 doi:10.1088/1757-899X/364/1/012039

A Multi-Sensory Approach to Cultural Heritage: The *Battle of Pavia* Exhibition

Virginio Cantoni[1], Piercarlo Dondi[2,3], Luca Lombardi[1], Nahumi Nugrahaningsih[4], Marco Porta[1]*, Alessandra Setti[1]

[1] University of Pavia, Department of Electrical, Computer and Biomedical Engineering, Via A. Ferrata 5 – 27100 – Pavia, Italy

[2] University of Pavia, CISRiC, Arvedi Laboratory of Non-Invasive Diagnostics, Via Bell'Aspa 3 – 26100 – Cremona, Italy

[3] University of Pavia, Department of Civil Engeneering and Architecture, Via Ferrata 3 – 27100 – Pavia, Italy

[4] University of Palangkaraya, Department of Informatics, Palangkaraya, Kampus Unpar Tunjung Nyaho, Jl. Yos Sudarso – Palangkaraya 73112, Indonesia

* Corresponding author's e-mail address: marco.porta@unipv.it

Abstract. In the last years, several museums and exhibits have adopted new kinds of interactive installations that present artworks in more attractive ways, especially for young visitors. At the same time, new communication technologies have been introduced to allow vision and motion impaired people to visit arts centers. In this work, we present the multi-sensory solutions we have implemented for the "Battle of Pavia" Exhibition, a collateral event of Milan Expo 2015. The installation combined different interaction methods to achieve two main goals: providing visitors with engaging experiences and allowing blind and partially sighted people to appreciate the exposed artworks. The used technologies include gesture communication, gaze-based interaction, 3D character reconstruction, virtual avatars, and 3D tactile images. This work can be also viewed in the context of digital humanities for cultural heritage. To the best of our knowledge, this is the first exhibit to gather such a high number of interactive technologies in a single installation. The positive response from visitors is a great spur to continue our research in this direction.

1. Introduction

Digital technologies for enhancing the fruition of museums and exhibits are becoming an important research field, with the main aim to connect people, culture, and technologies [1][2].

Multimodal visits of exhibitions have the purpose to enhance the visitor's experience by inspiring self-motivated learning. This promotes creativity and engagement, but also requires new approaches to be developed for the design and implementation of digital spaces. Through different technologies, users can now interact with computers by means of gestures, voice, and eye input. Although the full potential of these Natural User Interfaces (NUIs) has not been achieved yet, their fast and steady enhancements are leading to cheap and reliable tools for content manipulation. NUIs, that can be also viewed in the context of digital humanities for cultural heritage, can offer new experiential ways to enhance the visit experience, and be an important support for disabled visitors.

IOP Conf. Series: Materials Science and Engineering **364** (2018) 012039 doi:10.1088/1757-899X/364/1/012039

In this paper, we present a case study related to the exhibition "1525–2015. Pavia, la Battaglia, il Futuro. Niente fu come prima" ("1525-2015. Pavia, the Battle, the Future. Nothing was the same again") held at the Visconti Castle of Pavia (Italy) from 14th June to 29th November 2015. Over 11,000 people visited the exhibition (10,556 tickets sold, including at least 400 family tickets).

The "Battle of Pavia Tapestries" are a collection of seven tapestries, each one describing a scene of the battle, permanently exposed at the "Museo di Capodimonte" in Naples (Italy). However, as a satellite event of Milan Expo 2015, one of them was presented in the exhibit at the Visconti Castle of Pavia. This paper describes our contribution to the exhibition, which, in addition to the real tapestry, included 3D printed characters, 3D "tactile images" for blind people, and virtual avatars. Visitors could also interact with virtual versions of the tapestries using gesture-based and gaze-based communication. Our work goes in the direction of NUIs, towards a more active and engaging involvement of visitors.

2. Gesture-Based Interaction

Interacting with computers through gestures and body movements has now become an almost frequent practice [3], and also museums and exhibitions can exploit gesture communication (both explicit and implicit [4]) to create more engaging experiences for visitors. To be really effective, explicit gestures should be as intuitive and comfortable as possible, so as to maximize the quality of the user experience.

2.1. Related Work

One of the first uses of gesture-based interfaces in the cultural heritage context was a finger tracking method proposed by Malerczyk [5]. The release of Microsoft Kinect in 2010 significantly increased the number of this kind of applications. For example, the combination of Kinect and RFID sensors was adopted by Cafaro et al. [6] to track and identify visitors in a museum. In the same year, an augmented reality application was developed which exploited the Kinect to allow users to gesturally manipulate 3D models of archaeological pieces from the Museo Regional de Huajuapan [7]. A Kinect-based application allowing visitors to perform a gesture interaction with virtual copies of historical relics was installed at the K2R International Expo in Rome, Italy [8]. Yoshida et al. [9] created an immersive environment, to teach paleontology to children. A similar approach was used by Soga [10] to allow visitors to "walk" through a 3D representation of the Shogo-in Temple.

2.2. Gesture Interaction at the Battle of Pavia Exhibition

For the Battle of Pavia Exhibition, we exploited the Microsoft Kinect sensor to allow visitors to control the computer through hand gestures (Figure 1). The system consisted of a personal computer, a 24" full HD monitor, and a Kinect v1 sensor. In its "slideshow stand-by mode", the program repeatedly displayed, in sequence, the seven tapestries. As soon as a visitor was detected in front of the device, the "interaction mode" could be started by simply holding a hand out, with the palm facing the sensor. By moving the screen cursor with the hand, visitors could select the tapestries and surf them. Besides tapestries, it was also possible to explore the map of the battle area. Visitors could select one of the seven tapestries or the map of the location of the battle in the screen shown in Figure 2.

By virtually "pushing" one of the eight rectangular areas in Figure 2, the corresponding enlarged version of the tapestry or of the map was displayed. Since pictures were big and only partially visible, the visitor could scroll them by clenching the fist and moving the hand in the four directions.

3. Gaze-Based Interaction

Human gaze can be exploited as an input channel to control the computer and other digital appliances. Eye input is the result of an eye tracking process carried out by eye trackers, i.e., devices that can detect the user's gaze direction [11].

Figure 1. A visitor browsing a tapestry with his hand.

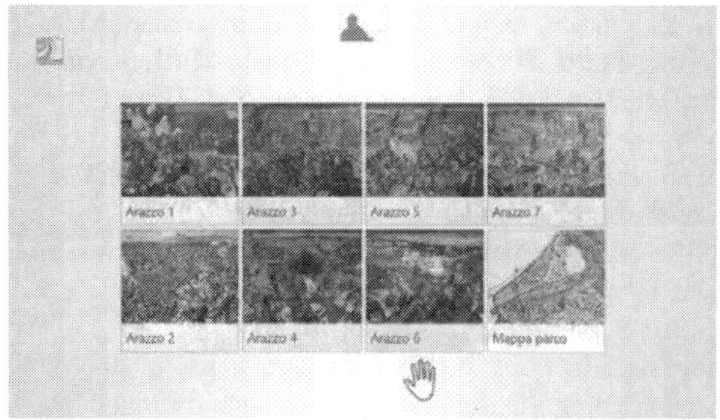

Figure 2. Screen for tapestry or battle map selection.

Basically, an eye tracker finds where the user is looking at and records the related gaze coordinates. Eye movements occur as sequences of saccades (lasting less than 100 ms) and fairly steady fixation periods (with durations normally between 100 and 600 ms). Eye tracking has been applied in a variety of contexts, from advertising (e.g., [12]) to accessibility (e.g., [13]) and biometrics (e.g., [14]). Among others, Milekic [15] provides an overview of gaze-based interaction and its potential applications in the context of museums and exhibitions.

In the Battle of Pavia Exhibition, we exploited eye tracking to allow visitors to expand and shrink high-resolution pictures of the tapestries. In addition, it was possible to scroll the images and see contextual descriptions displayed when specific characters or elements were observed. All these actions were performed without using the mouse or keyboard, but simply by means of gaze input. After the interaction with the virtual tapestries, visitors could also watch their "gaze replays" and download their "gazeplots".

3.1. Related Work

There are few examples of previous works in which eye tracking has been exploited within museums or exhibitions.

Buquet et al. [16] describe one of the first cases in which visitors' gaze behavior was assessed in a museum, specifically at the Cite des Sciences et l'Industrie de la Villette in Paris, in September 1986. Wooding et al. [17] employed an eye tracker in the "Telling Time" exhibition at the National Gallery of London, from 18th October 2000 to 14th January 2001. Mobile eye tracking was experimented by Wessel et al. [18] through a wearable device similar to glasses. Museum Guide 2.0 was an experimental project developed by Toyama et al. [19] as a virtual museum guide. Eghbal-Azar and Widlok [20] studied the use of a mobile eye tracker in two German exhibitions. More recently, Calandra et al. [21] employed E.Y.E C.U., an eye tracker to be placed below an artwork and tested in a museum of Naples (Italy).

3.2. Eye Tracking at the Battle of Pavia Exhibition

The eye tracking system for the Battle of Pavia Exhibition was developed to be robust and autonomous, as the interaction had to occur without any intervention from museum operators.

The employed eye tracking device was the 'Eye Tribe' (*theeyetribe.com*, company acquired by Facebook-Oculus in December 2016), a very cheap tool providing an accuracy of about 0.5 to 1 degrees and working at a 30 Hz sampling rate. The eye tracker was positioned below a 24'' Full HD screen (Figure 3). Unlike other eye tracking implementations, no further input tool was provided. The interaction could occur in Italian or English, and initial instructions were delivered both textually and through audio messages. Three eye tracking workstations were available at the exhibition (Figure 4). Gaze data was automatically synchronized with a remote server through an FTP module.

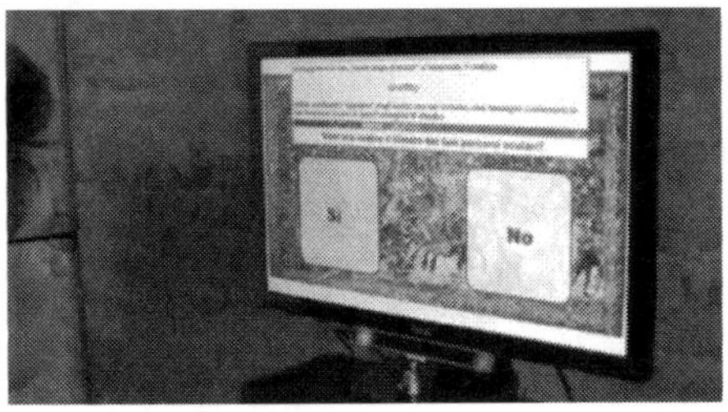

Figure 3. A visitor interacting with the eye tracking system.

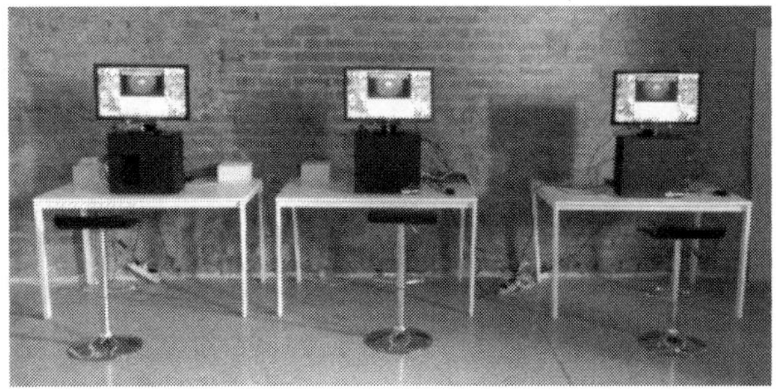

Figure 4. The three eye tracking workstations at the exhibition.

After a short initial calibration procedure, a Tutorial module allowed visitors to undergo a brief interactive training procedure explaining how to browse tapestries by means of the eyes (Figure 5). A zoomed-in tapestry image could be scrolled in the four directions by simply looking at the four screen edges. A visual menu composed of four "control icons" was displayed whenever a screen point was fixated for two seconds (Figure 5). Zooming operations (in and out) were performed by fixating the 'plus' or 'minus' lens icons of the menu for two seconds. Three zooming levels were available, which enabled visitors to closely observe the details of tapestries. To display another tapestry, the visitor had simply to look at the upper icon (the one with six squares). Each tapestry picture also contained four "sensitive areas", corresponding to "notable elements" (mainly characters). Sensitive areas were highlighted by means of a semi-transparent yellow rectangle as soon as the visitor's gaze was perceived on them. Immediately after that, a small rectangle containing a brief description of the watched element appeared. Looking at the 'Stop' control icon in the visual menu caused the interaction with the system to end. Visitors could then watch their own "gaze replay", i.e., an animation where gradually expanding circles represent fixations and straight lines between fixations indicate saccades. At the end of the interaction with the system, visitors could download their *gazeplots* (i.e., graphical representations of fixations superimposed on the tapestry images) from the website *http://vision.unipv.it/arazzi/*, by means of a code displayed once the interaction was concluded. In total, 1,053 visitors performed a "full interaction" with one of the eye tracking systems, going through the three steps of calibration, tutorial, and free tapestry surfing. About half of these visitors (51.02%) browsed only one tapestry, i.e., the first tapestry loaded when the visitor was detected in front of the eye tracker.

The gazeplot (Figure 6) that visitors could download at the end of the interaction (through the code they were provided with) is a graphical representation that shows the position of fixations (depicted with circles with areas are proportional to their durations).

Compared to previously developed eye tracking solutions for museums and exhibitions, the implemented system allows a much more complex interaction, purely based on gaze input.

Figure 5. A phase of the tutorial.

Figure 6. Gazeplot produced by a visitor while watching the first tapestry.

IOP Conf. Series: Materials Science and Engineering **364** (2018) 012039 doi:10.1088/1757-899X/364/1/012039

4. 3D Characters and Objects from the Tapestries

3D printing is a practical way to create copies of artworks that cannot be displayed directly, for example because they are not easily available, manageable, or accessible [22]. These 3D reconstructions can be also useful for partially sighted or blind people, who cannot enjoy the visual appearance of the treasures cultural heritage offers.

At the Battle of Pavia Exhibition, several 3D reconstructions of characters and items from the tapestries' scenes were on display. The models were created by the students of the Computer Vision course of the Master's degree in Computer Engineering of the University of Pavia using the *CINEMA 4D* and *Mixamo Fuse Character Creator* software tools. Figure 7 shows the 3D printing of five characters.

5. Virtual Avatars

Avatars are often exploited in historical exhibitions to provide computer-generated experiences in which visitors can play virtual roles. For example, the TOURBOT project [23] was one of the first works going in this direction.

In the Battle of Pavia Exhibition, some characters mimicked the visitor's facial expressions and head movements, in real-time (Figure 8). The application was developed by means of the facial capture and animation technologies provided by *Mixamo Face Plus*.

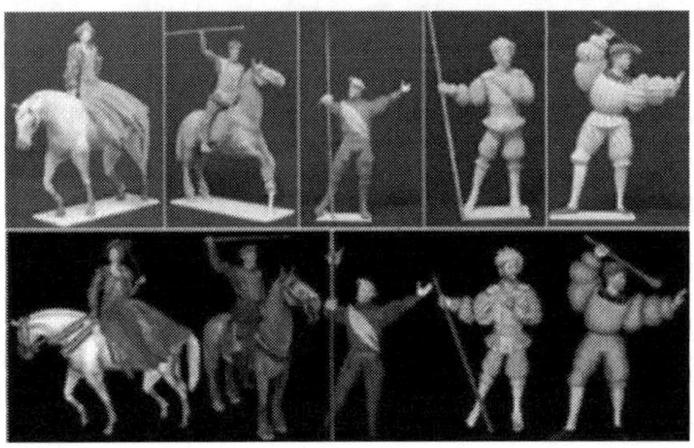

Figure 7. Some character models (bottom) and their 3D printings (top) made with a ProJet 460Plus (color) printer.

Figure 8. A user interacting with the avatar.

6. Tactile Images

The possibility to touch artworks or their reconstructions is very important for blind and partially sighted visitors of museums and exhibitions. *Tactile images* are 3D transpositions of pictures so that they can be read and interpreted by touch.

6.1. Related Work

Giving to visually-impaired people the possibility to visit museums and appreciate artworks is a complex task. Typical solutions involve the use of audio devices, which provide descriptions of the objects on display (e.g., [24]). However, also haptic interfaces have been deeply studied [25]. These implementations require technologies such as 3D scanning and modeling, sometimes combined with *ad hoc* devices. However, haptic interfaces are not directly applicable to two-dimensional artworks, such as paintings. In this case, it is necessary to create *tactile images*, sort of 3D representations of the original creations. Typical solutions can be summarized as variations of two main approaches: tactile diagrams, where only the main edges in the painting's elements are exploited [26], and bas-reliefs [27]. In the last years, the availability of relatively cheap 3D printing technology has definitively increased the interest in this field. Carfagni et al. [28] compared four different transformation methods from the original 2D artwork to its tactile representation. Furferi et al. [29] proposed a systematic method for semi-automatic generation of 2.5D models from paintings. Albertazzi et al. [30] tested for the existence of cross-modal visual and tactile associations in abstract art.

6.2. Tactile Images at the Battle of Pavia Exhibition

The Battle of Pavia Exhibition proposed tactile versions of the seven tapestries for partially sighted and blind visitors (Figures 9 and 10). The tapestry images were digitized, simplified, adapted, reconstructed as three-dimensional models, and printed in 3D so that they could be read through fingertips.

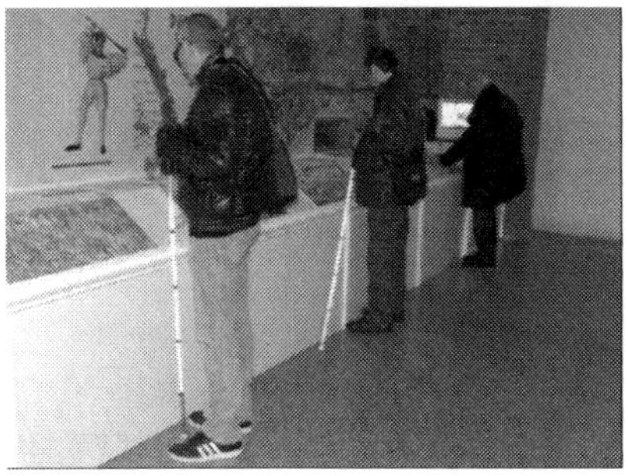

Figure 9. Tactile images at the Battle of Pavia Exhibition.

Figure 10. A tactile tapestry.

The tapestry modeling process for 3D printing was developed in several phases, namely: (1) A segmentation process was performed using the DoG filter to extract relevant parts [31]; (2) The extracted contours were then selected based on their relevance for proper scene comprehension. Segments could be either "full" (i.e., represented as high-relief areas) or "empty" (i.e., implemented as low reliefs). The CINEMA 4D software was used to process tapestries and create 3D models with three depth levels; (3) In each segment, a Braille alphabet letter (repeated in a regular pattern) was used to indicate the specific character or element — a Braille legend described such associations; (4) Finally, the tactile images were printed with a ProJet 460Plus full color 3D printer. Compared to other solutions, this kind of tactile images are much more informative.

7. Conclusions

In this paper, we have described the technological solutions we have implemented for the exhibition "1525–2015. Pavia, la Battaglia, il Futuro. Niente fu come prima" ("1525-2015. Pavia, the Battle, the Future. Nothing was the same again") held at the Visconti Castle of Pavia (Italy) from 14th June to 29th November 2015.

Our main goal was to study, also from a practical point of view, the opportunities offered by digital media for museums and exhibitions, in order to develop approaches, procedures, and tools that can maximize their benefits in such settings.

Although museums and exhibits increasingly exploit novel technologies, such as Natural User Interfaces, their number is still very low. It is only by promoting the use of new technologies in cultural settings that it will be possible to really assess their value, both as a way to create more engaging experiences for visitors and to provide disabled people (especially those who are visually and motion impaired) with alternative manners to access art.

As digital humanities point out, digital tools applied to the study of humanities foster collaborative and transdisciplinary research activities, enlarging the way of knowledge dissemination, and enriching the production of applications and techniques for cultural heritage.

The "Battle of Pavia" Exhibition can be considered a successful event, and the number of visitors went beyond expectations. The solutions proposed included gesture communication by means of the Kinect sensor (one workstation), gaze-based interaction (three workstations), 3D full-color element reconstruction and printing (six characters, a cannon, and two prints of the city of Pavia), a virtual avatar (one workstation), and seven full-color tactile images.

Our research goes in the direction shown by Natural User Interfaces, with the purpose to involve visitors more effectively and create positive experiences, also in relation to accessibility. To the best of our knowledge, no previous single art exhibition has been characterized by a so high number of multimedia and multimodal solutions.

References

[1] Bautista S S 2014 *Museums in the Digital Age: Changing Meanings of Place, Community, and Culture* (Altamira Press, USA)

[2] Din H and Wu S (Eds) 2014 *Digital Heritage and Culture: Strategy and Implementation* (edited by Din H and Wu S, World Scientific Publishing, Singapore)

[3] Dondi P, Lombardi L and Porta M 2014 Development of Gesture-Based HCI Applications by Fusion of Depth and Colour Video Streams *IET Computer Vision* Vol. **8** No. 6 pp 568-578

[4] Porta M. 2002 Vision-based User Interfaces: Methods and Applications *International Journal of Human-Computer Studies* Vol. **57** pp 27-73

[5] Malerczyk C 2004 Interactive Museum Exhibit Using Pointing Gesture Recognition *Proceedings of the 12th International Conference in Central Europe on Computer Graphics, Visualization and Computer Vision* pp 165–171

[6] Cafaro F, Panella A, Lyons L, Roberts J and Radinsky J 2013 I see you there!: developing identity-preserving embodied interaction for museum exhibits *Proceedings of the SIGCHI Conference on Human Factors in Computing Systems (CHI '13)* pp 1911-1920

[7] Ramos E, Ramírez M, Nila E, Figueroa D, Hernández J, García M and Pérez E 2013 Based Kinect Application to Promote Mixtec Culture *Procedia Technology* **7** pp 344-351

[8] Fanini B, D'Annibale E, Demetrescu E, Ferdani D and Pagano A 2015 Engaging and shared gesture-based interaction for museums the case study of K2R international expo in Rome *Proceedings of Digital Heritage*, Granada pp 263-270

[9] Yoshida R, Tamaki H, Sakai T, Nakadai T, Ogitsu T, Takemura H, Mizoguchi H, Namatame M, Saito M, Kusunoki F, Kamiyama S, Yamaguchi E, Inagaki S, Takeda Y, Sugimoto M and Egusa 2015 Novel application of Kinect sensor to support immersive learning within museum for children *Proceedings of 9th International Conference on Sensing Technology (ICST)*, Auckland pp 834-837

[10] Soga A 2015 Virtual Show, Go In!: Walk-Through System and VR Goggles of a Temple for Museum Exhibits *Proceedings of the International Conference on Culture and Computing (Culture Computing)*, Kyoto pp 199-200

[11] Duchowski A T 2007 *Eye Tracking Methodology – Theory and Practice* (2nd Ed. London: Springer-Verlag)

[12] Porta M, Ravarelli A and Spaghi F 2013 Online Newspapers and Ad Banners: an Eye Tracking Study on the Effects of Congruity *Online Information Review* Vol. **37** No. 3 pp, 405-423

[13] Porta M 2015 A Study on Text Entry Methods Based on Eye Gestures *Journal of Assistive Technologies* Vol. **9**, No. 1 pp 48-67

[14] Cantoni V, Galdi C, Nappi M, Porta M and Riccio D 2015 GANT: Gaze ANalysis Technique for Human Identification *Pattern Recognition* Vol. **48**, No. 4 2015 pp 1027-1038

[15] Milekic S 2010 Gaze-Tracking and Museums: Current Research and Implications *Proceedings of Museums and the Web 2010* Toronto: Archives & Museum Informatics (March 31)

[16] Buquet C, Charlier J R and Paris V 1988 Museum application of an eye tracker *Med. Biol. Eng. Comput.* Vol. **26** No. 3 pp 277–281

[17] Wooding D S, Mugglestone M D, Purdy K J and Gale A G 2002 Eye movements of large

populations: I. Implementation and performance of an autonomous public eye tracker *Behav. Res. Methods Instrum. Comput.* Vol. **34**, No. 4 pp 509–517

[18] Wessel D, Mayr E and Knipfer K 2007 Re-viewing the museum visitor's view *Proceedings of the Workshop 'Research Methods in Informal and Mobile Learning'*, Institute of Education, London, UK

[19] Toyama T, Kieninger T, Shafait F and Dengel A 2011 Museum Guide 2.0 - An Eye-Tracking Based Personal Assistant for Museums and Exhibits *Proceedings of Int. Conf. on Re-Thinking Technology in Museums*

[20] Eghbal-Azar K and Widlok T 2013 Potentials and Limitations of Mobile Eye Tracking in Visitor Studies Evidence From Field Research at Two Museum Exhibitions in Germany *Soc. Sci. Comput. Rev.* Vol. **31** No. 1 pp 103–118

[21] Calandra D M, Mauro D D, D'Auria D and Cutugno F 2016 E.Y.E. C. U.: an Emotional eYe trackEr for Cultural heritage sUpport (in *Empowering Organizations*, Torre T, Braccini A M and Spinelli R, Eds Springer International Publishing pp 161–172)

[22] Mazura M, Horjan G, Vannini C, Antlej K and Cosentino A 2015 eCult Vademecum - A Guide for Museums to Develop a Technology Strategy & Technology Providers to understand the Needs of Cultural Heritage Institutions (available *http://www.ecultobservatory.eu/sites/ecultobservatory.eu/files/documents/Vademecum_PDF_V2.0.pdf*, visited 19 December 2016)

[23] Tourbot 2000 TOURBOT – Interactive Museum Tele-presence Through Robotic Avatars (available *http://www.ics.forth.gr/tourbot/*, visited 10 January 2017)

[24] Picinali L, Afonso A, Denis M and Katz B F G 2014 Exploration of architectural spaces by blind people using auditory virtual reality for the construction of spatial knowledge *International Journal of Human-Computer Studies* **72**, 4 pp 393-407

[25] McLaughlin M L, Sukhatme G, Shahabi C, Hespanha J, Ortega A and Medioni G 2000 The haptic museum *Proceedings of 2000 EVA Conference on Electronic Imaging and the Visual Arts*

[26] Axel E S and Levent N S 2003 *Art Beyond Sight: A Resource Guide to Art, Creativity, and Visual Impairment* (Amer Foundation for the Blind Press)

[27] Oouchi S, Yamazawa K and Secchi L 2010 Reproduction of tactile paintings for visual impairments utilized three-dimensional modeling system and the effect of difference in the painting size on tactile perception *Proceedings of the 12th International Conference on Computers Helping People with Special Needs (ICCHP'10)* Vol. **6180** Springer Berlin pp 527–533

[28] Carfagni M, Furferi R, Governi L, Volpe Y and Tennirelli G 2012 Tactile Representation of Paintings: An Early Assessment of Possible Computer Based Strategies *Proceedings of Progress in Cultural Heritage Preservation, EuroMed 2012* (Lecture Notes in Computer Science, Springer, vol. 7616) pp 261-270

[29] Furfer R, Governi L, Volpe Y, Puggelli L, Vanni N and Carfagni M 2014 From 2D to 2.5D i.e. from painting to tactile model (2014) *Graphical Models* 76 (6) pp 706–723

[30] Albertazzi L, Bacci F, Canal L and Micciolo R 2016 The tactile dimensions of abstract paintings: A cross-modal study *Perception*, 45(7) pp 805-822

[31] Cantoni V, Levialdi S, Zavidovique B 2011 *3C VISION: Cues, Context and Channels* (Elsevier London)

IOP Conf. Series: Materials Science and Engineering **364** (2018) 012040 doi:10.1088/1757-899X/364/1/012040

ICT as a tool to foster inclusion: Interactive maps to access cultural heritage sites

Sebastiano Marconcini

Department of Architecture, Built environment and Construction Engineering (ABC), Politecnico di Milano, Via Ponzio 31, 20133 Milano, Italy

sebastiano.marconcini@polimi.it

Abstract. Within the European policies and strategies for a sustainable development, the role of cultural heritage as an economic, environmental and social driver has become increasingly significant. As an asset in people's lives, it's necessary to assure anyone the opportunity to access it, regardless of the physical abilities or cultural background of a person. In this sense, ICT could represent an important instrument in enhancing the level of inclusion in cultural heritage sites. This paper develops the discussion around interactive maps as a feasible solution to ensure a fully inclusive experience of a cultural heritage site. Thanks to the analysis of similar existing instruments, this operative tool has been conceived to perform two fundamental functions. First, provide an information system thanks to which everyone, particularly users with special needs, could be able to organize their movements and be aware of the proposed services (that also could be new ICT, like Virtual and Augmented Reality, implemented directly in the map) and opportunities to access cultural heritage. Second, offer a tool for the management of the projects related to the development of inclusion, therefore it is addressed to municipal administrators and other key actors involved in the governance of cultural heritage.

1. Introduction

From the 1972 UNESCO World Heritage Convention to the initiatives and events of the European Year of Cultural Heritage (2018), the role of cultural heritage as an economic, environmental and social driver has become increasingly significant.

To better understand its role, it is useful to recall the definition of World Heritage given by UNESCO, which states: "Heritage is our legacy from the past, what we live with today, and what we pass on to future generations. Our cultural and natural heritage are both irreplaceable sources of life and inspiration. [...] What makes the concept of World Heritage exceptional is its universal application. World Heritage sites belong to all the peoples of the world, irrespective of the territory on which they are located" [8].

The first part clearly underlines the importance of the cultural heritage in people's lives. In particular, with reference to Italian and European territories, its buildings and urban spaces house the main activities of the social life of a community, as well as the main cultural services. For this reason and for its architectural features, cultural heritage can play a key role in development strategies and policies. For example, it represents and reinforces individual and community identity and it can be an asset for new economic development opportunities, based on cultural tourism and the preservation of built heritage. Consequently, the real participation, in the social and cultural life of a community, is possible only through ensuring the accessibility to cultural heritage. In particular, as asserted in the second part of the previous definition, heritage has a broader value and it is meant for all, without any distinction based on a person's physical abilities or cultural/individual background.

2. Towards a role for ICT in fostering inclusive policies

According to these premises, the goal of this paper is to deepen the discussion about the role that ICT could play in fostering public participation in cultural heritage sites. In particular, once understood the needs that inclusive policies must address, the proposal of an operative tool will try to identify how its implementation represents a key resource for an inclusive design approach and which challenges still have to be faced.

2.1. Inclusive cultural heritage sites: the challenges of the current condition

Given the universal value of cultural heritage, to which everyone should have access, it is important to understand the meaning the term "inclusion" assumes in this work.

On the one hand, the attention is focused on the enhancement of the physical accessibility of the built environment, as a place that can enable or disable people in achieving their functioning and consequently their well-being.

In this regard, a fundamental document in understanding the influence of the built environment on everyone's life is the International Classification of Functioning, Disability and Health (ICF) by the World Health Organization (WHO). In relation to the topic of disability, the ICF represents the latest version of a tool introduced to assess a person's health and well-being. The importance of this classification system lies in the acknowledgment of another element, other than the physical conditions, that influences people's functioning and their ability to do personal activities and participate in life situations: contextual factors. The latter are also divided into environmental factors, that are the physical and social features of the surroundings in which people live their lives, and personal factors, which involve the particular background of an individual, like the cultural one [10]. This connection with the built environment has also been underlined in the United Nations Convention on the Rights of Persons with Disabilities, which defines disability as the result "from the interaction between persons with impairments and attitudinal and environmental barriers that hinder their full and effective participation in society on an equal basis with others" [7].

On the other hand, the focus is set on the importance of socio-cultural component in defining human well-being and how architecture should foster access to knowledge and public participation, regardless of the cultural background of a person.

Culture has appeared as another fundamental need for human beings. In their different forms, socio-cultural factors influence the possibilities of participation in the public realm and human well-being, but the key issue can be identified in the topic of identity. The American psychologist A. Maslow, in developing his "Theory of Human Motivations" [3], among the basic human needs pointed out those of belonging, self-esteem and self-realization. In our society, all people perceive the need for a stable, solidly based and generally high evaluation of themselves, through their own inclusion within a broader group. A person needs to develop and express, at the same time, his own personal identity and the belonging to some collective system. Since culture is the maximum manifestation of this identity, it is essential that the built environment conveys the latter in its physical features, but especially it has to offers spaces that house services that manifest the culture intangible assets and through which promote public participation and the acquisition of new knowledge; in other words, places that develop identity.

2.2. Foster inclusion: which role for ICT?

The previous sections have allowed to better understand how inclusion assumes a specific meaning in relation to cultural heritage sites. In addition to the importance to assure their physical accessibility, it has become clearer which potentialities they have, in turn, in fostering inclusion, especially from a socio-cultural point of view.

IOP Conf. Series: Materials Science and Engineering **364** (2018) 012040 doi:10.1088/1757-899X/364/1/012040

Despite the relevance of the topic is widely acknowledged, the current design practice reveals multiple problems in developing inclusion in the architectural project, especially in cultural heritage sites. In order to better comprehend the issues that will be shortly described, it's necessary to highlight how this paper is part of a broader research aimed at develop a fully inclusive view of the project, especially regarding historical contexts. Through the analysis of multiple European case studies and especially an on-site investigation applied to Mantua (a city located in Lombardy region, Northern Italy, entered in the UNESCO World Heritage List in 2008, along with the site of Sabbioneta), it has been possible to identify two main issues that can be addressed by the implementation of ICT in the design process.

First, the management of the applied actions to foster inclusion certainly represents one of the main problems. The high number of actors involved and the lack of a comprehensive and long-term planning vision have led to the realization of inadequate and incoherent accomplishments. In this sense, ICT can provide the proper tools for managing the necessary inclusive design processes.

A second fundamental matter, related to cultural heritage, is the coexistence of inclusion and conservation needs, which are frequently considered in contrast, although A. Bellini, among the first that have talked about architectural barriers in restoration, remembered "We heavy many difficulties, insurmountable difficulties, in imaging a monument that has not been produced for men, which is protected and preserved in itself, as an abstraction, and not for fruition. [...]. It is not an asset if it is not usable, pure contemplation does not belong to architecture" [1]. In those circumstances where the goal of accessibility can be in conflict with the needs of conservation, these new technologies can represent the only instruments able to guarantee everyone the possibility to access a particular cultural heritage site and, more in general, to have a better experience of it.

3. Interactive map as tool for inclusion

For the purposes of the work, the conception of an interactive map has seemed the proper solution in the perspective of producing a feasible operative tool.

At this stage of the research, where the focus is principally in defining the main objectives and features of the map rather than its software architecture, this instrument has seemed the ideal interface to be used. It's easy to think of how, even in the daily travel activities, we resort to the use of digital maps; not only they let us to organize our movements, but they are also one of the first tools through which we become aware of a place. In this sense, when the reference is to cultural heritage sites, digitals maps have a key role in collecting and sharing knowledge.

In addition to that, working through a cartographic base, this type of implement allows an easier integration with the existing documents and resources of urban planning, enabling to outline all the initiatives, whether they are architectural projects or not, necessary to allow inclusion to lead the transformation processes of the city.

3.1. Accessibility and interactive maps: background

In recent years, the success of digital maps has been widely recognized. This type of services, such as Google Maps, Bing Maps, Yelp and Waze, has contributed to the spread of an increasing number of information, although the data regarding the physical accessibility of the built environment are extremely poor.

For this reason, various initiatives focused on the issue of accessibility have been developed in parallel to these tools. One of the best-known examples is Wheelmap.org [9], an online map that collects accessibility information that has been reported by the users of its smartphone application, through on-site visits of the different points of interest within a city.

However, this kind of solutions still has several limitations. First of all, they suffer from a lack of records. If we consider the example of Wheelmap.org, a report on open accessibility data, made by Ding *et al*. in 2014, unearthed that only the 1,6% of the points of interest on the map had accessibility data [2]. In addition, this information refers only to punctual situations, measuring the accessibility of the single facilities without considering the paths that connect them. On the contrary, there are also some services that focus their attention only on the accessibility of the urban environment, like Project Sidewalk [5]. Developed in Washington D.C., this project has been aimed at mapping all the accessibility issues that the city sidewalks present, through virtual exploration of the streets.

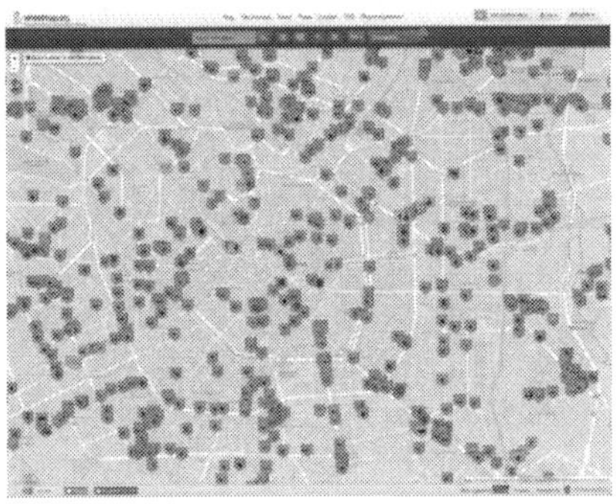

Figure 1. Screenshot of Wheelmap website. In the case of Milan (Italy), it's possible to see how many POIs still have missing information (in grey). Source: http://wheelmap.org

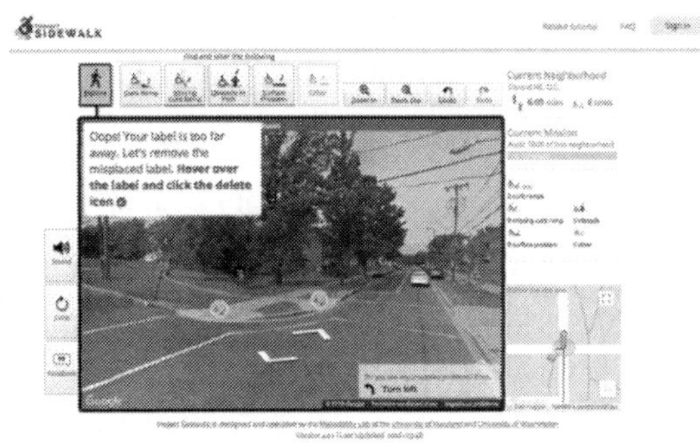

Figure 2. Screenshot of Project Sidewalk website. In this image is visible the user interface through which is possible to map the accessibility issues of the city sidewalks. Source: http://sidewalk.umiacs.umd.edu

Certainly, the data collected establish an important basis to develop inclusive policies, but the absence of information about facilities and other important "urban components", such as public transport, doesn't allow to have a complete vision of the city system and, in the case of end users, to be able to organize your movements. Moreover, these multiple tools also have the side effect of scattering information, since each service contains different one, making it complex for the final user to acquire thorough data about the accessibility of a place.

A further fundamental aspect that must be taken into consideration is that this kind of instruments are mainly based on crowdsourcing. This model surely allows a faster and more widespread data collection; however, the accuracy of the information can be strongly variable. On the one hand, it can be necessary to evaluate the survey methodologies and their effectiveness, as the experimentation developed by Salomoni *et al.* on three different mobile apps to crowdsourcing urban accessibility has shown [6]. On the other hand, it's very important to underline how the concept of accessibility can be interpreted differently from map to map and especially from person to person. In case the data have not been previously verified or no objective indications on the parameters used to define the level of accessibility of a place have been provided, the risk is to produce a strong discrepancy between users' expectation and their actual travel experience, aspect that has been recognized as one of the greatest difficulties disabled people have in traveling [4].

In addition to the previous limitations, it's necessary to draw attention on how these tools are mostly addressed to people with physical disabilities, without considering those affected by sensory or cognitive impairments. In this case, the reference is not only to this kind of parameters, completely missing when defining the level of accessibility of a place. In fact, the fundamental issue is to identify the proper technologies and solutions that allow visually and cognitive impaired users to obtain the same information provided by the map.

3.2. Interactive maps: objectives

On the basis of the presented limitations of the existing services and the unanswered needs identified through the case studies and on-site investigation, the following sections of the paper try to contribute to the discussions about the objectives and features that this kind of tools should fulfill. In particular, this interactive map has been conceived to perform two fundamental functions.

In accordance with the analysis conducted on existing tools, the primary purpose is to provide an information system for people. Especially for contexts of high tourist attraction, it has emerged how information is indispensable to ensure the ability to access and move in the built environment. In particular, people with disability need to receive as much data as possible, in order to know the level of accessibility of a place. The goal of defining a new interactive map is to offer only one and easy to use tool that makes everyone able to organize their movements, according to their personal needs and the possibilities offered to them through the proposed services and fulfilled projects from local institutions.

Furthermore, if properly designed, this map could become the appropriate tool with which citizens can participate in the process of fostering inclusion. In line with the model of crowdsourcing, users can use it to advise the administrative bodies of the critical situations located within the city. Although, for what have been said above, it is necessary a professional supervision, it is important to maintain this element because in this way, not only the sense of social participation can be fostered, but the planned interventions would be aimed at answering to real needs expressed by people, who could guide the design process, ensuring the effectiveness of the design actions.

In relation to what has just been said, there is another important function for which this tool has been considered suitable: the management of the projects related to the development of inclusion, therefore it is addressed to municipal administrators and other key actors, such as public or private institutions involved in the governance of cultural heritage. In this sense, the map should locate within the city all the situations representing a physical barrier that restricts people's public participation. This complete cognitive framework, which should be integrated with recommendations on how to overcome physical barriers, will be able to support the planning of the necessary interventions, in accordance with several parameters, such as priority and financial resources, which can be evaluated from time to time. This kind of tool should guarantee a general and long-term vision, capable of adapting to various external factor that could influence the fulfillment of the initiatives to be undertaken.

Another essential aspect is the possibility of coordination among the different actors involved in the development of inclusion. As it has been said, the latter is defined by multiple components, whose supervision is entrusted to many executives. For this reason, the interactive map should be aimed at be the only tool that everyone can interface with, so the planned actions can be combined and properly realized to reach the highest possible level of inclusion. In addition, if conceived as a single instrument, this managing system can offer the adequate supervision to data coming from crowdsourcing, ensuring the accuracy of the information provided. Finally, this tool should be conceived to allow the continuous monitoring of the achieved inclusion. In fact, after the first surveys, the map should be constantly updated, notifying the issues that have been resolved and the new criticalities that could emerge, always providing the most recent vision of the level of inclusion reached within the city.

3.3. Interactive maps: features

Once the two main aims have been defined, the focus of the discussion must necessary move on the features and functions the interactive map should possess.

First of all, this tool should be able to provide information on the presence of architectural barriers within the city, both at urban and building scale, if of public interest. As already mentioned above, this data could be the results of professional surveys led by local institutions or reports from crowdsourcing. For this reason, prior to the development of the map, it is necessary to designate a series of unambiguous parameters that can then be made available on a designated area of the website and smartphone application. This would allow both the records from crowdsourcing to be as objective as possible and the end users to be aware of how the level of accessibility has been defined. In addition, these data should be pointed out in different ways, depending on the identified type of criticality. As it has emerged through the analysis of different experiences and the on-site investigation of the case study of Mantua, it is possible to classify three sorts of physical barriers: the punctual situations, the ones with a linear profile and those that develop on an extended surface.

This distinction is critical both for end users to easily understand the extent of the limitations present and for the management purposes, because it already allows to express some preliminary evaluations on the interventions needed, the difficulty level of overcoming the architectural barriers and the relative costs to do so. The need to have this multiple information requires the map to be structured to provide an overall view of the city and, at the same time, more exhaustive and always updated data related to specific situations. In this sense, the proposed tool should overcome the approach of offering the assertion of the accessibility status, in favor of a more detailed information system that allows the end users to assess, according to their needs, the level of accessibility of a place. For this reason, in order to get this complete cognitive framework, a pop-up window must be associated with every situation identified. In this sense, also multimedia (like images, videos, etc.) should be integrated as an important objective resource for people to evaluate the existing possibilities and organize their movements.

Clearly, if it is essential to identify all the critical situations within the city, the final information that the map needs to offer is different in relation to the two main goals previously define for this tool. From the managing point of view, the key aspects are: the definition of the design proposals, the subdivision of the competencies of the single interventions, so, if necessary, it would be possible to coordinate different actors among themselves, and planning the projects that will be implemented.

Instead, as an information system for people, the data the map should supply are of different nature and more general interest. In fact, they should allow a person to identify the points of interest and the paths for reaching them. In particular, the tool should be accessed through a user profile, previously set according to the capabilities and consequent needs of the person. Therefore, this solution would allow to filter all the available data, in order to guarantee the best solutions to move within a place in complete safety and accordance with personal urgencies. In addition, this information should be provided with a simplified language, so everyone can be able to understand them regardless of their cognitive abilities.

Finally, in relation to physical accessibility issues, the information given not only has to focus on architectural barriers, but especially on those elements and services of the built environment, such as public transport, that could enable a better experience of social participation.

4. Conclusion: ICT implementation for a richer experience of cultural heritage

Through the discussions about the main objectives and features to which an interactive map should respond, it has been possible to highlight the new role that ICT can assume within an inclusive design approach. For the development of a fully inclusive environment, it is no longer enough to work on the component of physical accessibility, but also on the ones, like information, belonging to a managing dimension.

The discussion around the proposal of a new operative tool made it possible to transfer, in the ICT sector, the same measures that an inclusive design approach should implement, analyzing limitations and perspectives of these instruments. This isn't to be considered as the final step of a research. In fact, this paper is aimed at identify the opportunities that an instrument such an interactive map could offer.

From these premises, once defined a basic software architecture that allows the diffusion of this kind of instrument, it will be possible to assure an easier access to knowledge, especially in relation to cultural heritage. Particularly, this tool could be the user interface through which easily access additional ICT services, and not only related to accessibility issues. The reference is to new technologies, such as Virtual and Augmented Reality, which represent, in addition to being in some cases the only solutions to access/understand cultural heritage, a richer experience for everyone, fostering social participation and communities' involvement in heritage conservation.

IOP Conf. Series: Materials Science and Engineering **364** (2018) 012040 doi:10.1088/1757-899X/364/1/012040

References

[1] Bellini A 1998 La pura contemplazione non appartiene all'architettura *Tema* **1** 2–3
[2] Ding C Wald M and Wills G 2014 A survey of Open Accessibility Data *Web for All Conference*
[3] Maslow A H 1943 A theory of human motivation *Psychological review 50* **4** 370-396
[4] Packer T L McKercher B Yau M K 2007 Understanding the complex interplay between tourism, disability and environmental contexts *Disability and rehabilitation 29* **4** 281-292
[5] Project Sidewalk [online] Available at: http://sidewalk.umiacs.umd.edu [Accessed February 26, 2018]
[6] Salomoni P Prandi C Roccetti M Nisi V and Nunes N J 2015 Crowdsourcing urban accessibility: Some preliminary experiences with results *Proceedings of the 11th Biannual Conference on Italian SIGCHI Chapter* 130-133
[7] United Nations 2006 *UN Convention on the Rights of Persons with Disabilities* (United nations)
[8] United Nations Educational, Scientific and Cultural Organization, *World Heritage Definition*, [online] Available at: http://whc.unesco.org/en/about/ [Accessed January 17, 2018]
[9] Wheelmap.org [online] Available at: http://wheelmap.org [Accessed March 02, 2018]
[10] World Health Organization 2001 *International Classification of Functioning, Disability and Health: ICF* (World Health Organization)

IOP Conf. Series: Materials Science and Engineering **364** (2018) 012041 doi:10.1088/1757-899X/364/1/012041

Handling huge and complex 3D geometries with Semantic Web technology

Peter Bonsma[1], Iveta Bonsma[1], Anna Elisabetta Ziri[2], Ernesto Iadanza[3], Federica Maietti[4], Marco Medici[4], Federico Ferrari[4], Rizal Sebastian[5], Sander Bruinenberg[5], Pedro Martín Lerones[6]

1. RDF Ltd., Bulgaria

2. Nemoris srl., Italy

3. Consorzio Futuro in Ricerca, Italy

4. University of Ferrara, Department of Architecture, Italy

5. DEMO B.V., Netherlands, the

6. Fundación CARTIF, Spain

peter.bonsma@rdf.bg; iveta.bonsma@rdf.bg; annaelisabetta.ziri@nemoris.it; ernesto.iadanza@unifi.it; federica.maietti@unife.it; marco.medici@unife.it; federico.ferrari@unife.it; rizal@demobv.nl; sander.bruinenberg@demobv.nl; pedler@cartif.es

Abstract. In INCEPTION, a European collaborative research project, a Heritage BIM (H-BIM) ontology is being developed to store all relevant semantic data concerning cultural heritage objects. Similar to other projects dealing with storing semantics, one of the major questions is whether, and if yes, how should geometry be stored using semantic web technology. The INCEPTION cross-disciplinary research consortium chose to allow the storage of all relevant geometric information using semantic web technology. The alternative is to store geometry in a different way, or storing only the aggregated parts of geometry, for example through bounding box representations.

The geometry is originally generated by a CAD/BIM system and, as we are dealing with Cultural Heritage, in many cases it is derived from 3D point clouds. These result in a large amount of 3D data to be stored using semantic web technology. A well-known issue is that the performance of databases and inferencing engines for semantic web data drops considerably when the data grows to very large sizes. This paper explains how the performance issues on these large sets of geometric data can be solved while still being able to use the databases, inferencing engines, and the geometric data effectively.

1. Introduction

This paper describes the architecture and choices behind H-BIM approach within the INCEPTION project. It explains how the H-BIM approach can serve as the connection between all available knowledge, how it is populated and why it is structured the way it is. It also shows how all relevant BIM content (in the shape of IFC files) can be converted to both the H-BIM and the GEOM ontologies for respectively the semantic and geometric content of these BIMs. The complete set of geometrical information is processed even for large IFC files, overcoming performance issues as normally would be expected with the huge amounts of data that the INCEPTION platform should handle.

INCEPTION realizes innovation in 3D modelling of cultural heritage through an inclusive approach for time-dynamic 3D reconstruction of artefacts, built and social environments. It enriches the European identity through understanding of how European cultural heritage continuously evolves over long periods of time.

INCEPTION's Inclusive approach introduces novel solutions of 3D digital modelling:

- forever: INCEPTION "Time Machine" that represents an innovative use of time-scale for dynamic 3D reconstruction with emphasis on how the modelled cultural heritage evolves over time in association with its built and social environments;
- for everybody: portable, user-friendly and cost-effective hardware and software instruments for 3D capturing, modelling and analysis, as well as INCEPTION Semantic Web technologies and open-standard platform to creating and sharing understanding between various end-user group;
- from everywhere: INCEPTION's proposed standard procedures for data acquisition and open-standard format for cultural Heritage Building Information Modelling (H-BIM).

The INCEPTION project is funded by the European Commission under H2020-EU.3.6.3. - Reflective societies - cultural heritage and European identity.

The main technical challenge within this stage of the project is to move BIM models from a wide variety of existing commercial CAD systems towards a semantic web environment. One of the known issues is that CAD systems exports large amounts of both geometric and non-geometric data, both relevant for further use in the INCEPTION H-BIM platform. Although Semantic Web technologies are powerful, scaling towards large amounts of data is still an issue. This means that whatever the solution is, it has to be able to work with standard tools for Semantic Web technology on these large amounts of data without running into scalability issues.

The widely supported Open-BIM standard IFC is able to contain both geometric and non-geometric information like classification, relations and properties. Additionally, IFC is supported by all major CAD systems, scales very well and is very rich in the sense of BIM information.

The solution found and discussed within the project scope is to split semantic geometric knowledge from all other semantic knowledge, i.e. non-geometric semantic knowledge. Both parts are converted to semantic web technology. For the geometry the GEOM Ontology has been developed. All non-geometric knowledge is converted and stored in the H-BIM Ontology. Part of this H-BIM Ontology is directly generated from the IFC schema. This generation can be performed for all relevant IFC schema's, i.e. IFC 2x3 TC1, IFC4 ADD2 and IFC 4x1 Final. The generation is IFC-dependent but similar for all named IFC schemas and therefore different compared to existing Semantic Web BIM solutions like ifcOWL.

All generated semantic web content is loaded into the INCEPTION platform. To ensure performance, the geometric information stored in the GEOM Ontology can be used separately from the non-geometric information. This way it can be excluded from Fuseki-2. Visualization can be requested on demand by a dedicated service directly on top of the semantic web content. This dedicated service for geometry generation has limited functionality on top of the semantic web content and is therefore able to overcome the scalability issues intrinsic to all current developments for semantic web.

As a result in case, if there is a limited amount of information contained in the BIM it can be generated and transferred to the Fuseki 2 triple store. In the case that the amount of data is larger, all geometrically related data can still be converted but can be excluded from the Fuseki 2store. This structure still allows all benefits from using semantic web technology as having distributed data and connecting information from several different sources.

In addition to the conversion from IFC conversion from Collada, OBJ and DWG is also technically possible and is implemented step-by-step within the developed tools/applications and later on in services as part of the project. All converted data can be used in an integrated manner and visualization of 3D geometrical data is possible as sub-selection of all converted content in an integrated 3D view. Early tooling already available is able to visualize this integrated 3D data on virtually any modern mobile device.

2. IFC

IFC and BIM are often named in one sentence, although there are many BIM standards (both open and closed). The main idea behind IFC is to enable open exchange within and between all available disciplines within the Building & Construction sector. This holds for both non-geometrical data, i.e. relations, classification, properties, as well as for geometrical data. In practice, both geometric and non-geometric data are stored as a snap-shot of the data (IFC). Complex parametrical information is not included in the stored information. Since the geometrical information is used by different technical disciplines for different purposes, the definition of the geometry as an entity is essential. IFC has a strong focus on product information, storing this in an object based manner, although it has (limited) process related capabilities it is rarely used for supporting processes. The IFC 2x3 TC1 version of the standard already dates back a decennium, although it still is the main IFC exchange format used. Until very recently this 10-year-old IFC 2x3 TC1 standard was the only format software products could get certified for. While IFC 4 ADD2 has been available for a few years only a limited number of companies support this standard. The companies that do support the standard still struggle to export and import each other's data correctly. Within INCEPTION we found that although we were planning to use IFC4 for a few demonstration cases and the used CAD software was capable of exporting IFC4, the quality of the exported data was not good enough to be used even for use within the prototype software. We have to note here that the way CAD software is used is typical and different from normal architectural design work.

3. ifcXML / ifcOWL

There are two different ifcXML variants. The first variant is specific for IFC2x3 TC1 and although it really is 100% valid XML against a dedicated XSD schema, the style is very close to the original structure as found within STEP. The fact that it is valid XML / XSD enables many tools to be used for such a type of content. The amount of (developer) tools available, like XSLT, is much larger than the amount of tools available for SPFF (Step Physical File Format). The same counts for software developers that need to implement solutions.

One lesson learned from applying and using this XML serialization of IFC is that while it is technically valid XML / XSD, the fact that the structure still follows STEP ideas means the real practical use is somewhat limited compared to other XML standards. To overcome this for IFC4 a different XML serialization has been defined. The outcome is still 100% correct XML against a dedicated XSD, containing more than 99.9% of the knowledge as stored in SPFF, while the structure is much closer to what is expected by tooling in XML. This potentially enables the same XML enabled tooling to work more effectively on the content. The new XML serialization used for IFC XML serialization is called SimpleXML.

One of the problems with having these IFC dedicated different versions of XML serialization is that each new serialization will take considerable time to be implemented. Although IFC is a mature and widely supported open standard and officially promotes these new ifcXML serializations, the fact that the implementation cost is at the individual client and / or toolbox provider is often a reason not to support these serializations.

The Semantic Web serialization variant of IFC is called ifcOWL. First efforts were made over 10 years ago, however only the past few years it got serious interest and also an official serialization definition was proposed for BuildingSMART. While, like IFC, this first ifcOWL standard is perfectly following Semantic Web technology and can be processed by all available tools , however the solution still very close to the original structure of STEP. This means although technically working the content is semantically not in line wityh how somebody would define an ontology representing the same content.

In case of INCEPTION the official ifcOWL definition was not deemed to be a viable solution. Partly because of the architecture within INCEPTION where geometric knowledge is separated from non-geometric knowledge, and in part also because both parts are required to be closer to the original idea behind Semantic Web.

4. Cultural Heritage within H-BIM

Concerning the H-part of the H-BIM ontology, the peculiarity of 3D models representing historical buildings is that they are not conceived to be a footprint for construction and maintenance but more a representation of a valuable existing or reconstructed environment. These models has to represent all the complexity of the related context in term of ancient architectural techniques and components, important events related to the history of the building / site and all the data that define the actual value of the monument in term of historical heritage.

According to Historic England (2017) and Arayici et al. (2017) custom content creation is a requirement for H-BIM projects, since usual construction object libraries are related to new-build rather than historic building components. However, the inclusion of BIM objects from very different sources (creators, skills, purpose and context) is problematic for information consistency. Adopting standards is strongly advised and only content that has been checked for compliancy with these standards should be allowed in the project BIM object library and subsequently in the models. Thus to facilitating the international exchange of information ISO 128 (graphical representation of objects on technical drawings), ISO 16739:2013 (IFC for data sharing), ISO 1302:2002 (surface texturing in technical product documentation) and ISO 21127:2014 (reference ontology for the interchange of cultural heritage information) needs to be considered.

The H-BIM ontology is based on very well constructed and complete ontologies, while also integrating the BIM part of the H-BIM ontology.

5. The Architecture of the INCEPTION H-BIM approach

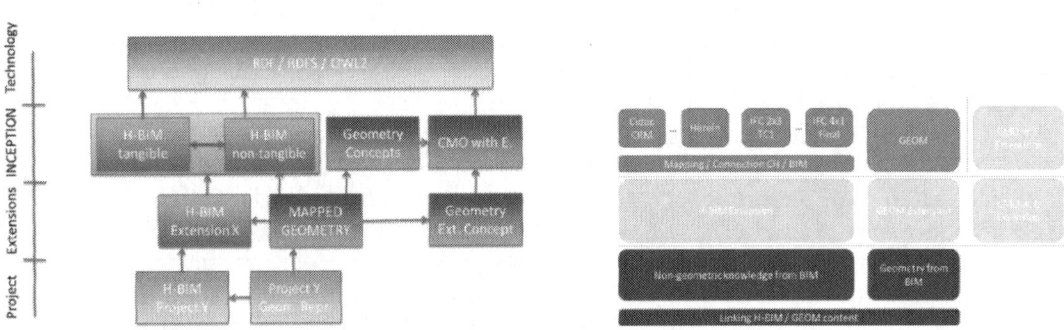

Figure 1, Architecture of the INCEPTION H-BIM approach

The image on the left represents the overall architecture of the INCEPTION H-BIM approach, while the image on the right is the detailed technical version of the same picture. In the technical picture the important parts for this paper are highlighted. In this diagram the blue and green parts are project independent, while the purple parts are project dependent.

6. Non-geometric Semantics from BIM

6.1. IFC Classification

The classification of BIM objects is a large task, with early estimations calculating over 25 person-years of work for a basic starting set of concepts. Within BuildingSMART, the organization behind IFC, bsDD and ISO 12006-3 (organization of information about construction works) an initial development in this direction was undertaken, are estimated to have used much more effort, while they still do not have a (nearly) finished classification available.

Within INCEPTION it is not viable to reinvent the wheel for BIM classification, and as such to keep things 'simple', the basic set of concepts as available in IFC and agreed on by most CAD vendors will be used. The classification is stored in the schema in the shape of different entities (read classes). Looking at 'instances' with a geometric representation we can restrict ourselves to any instance from entity IFCPRODUCT.

Making use of inheritance this adds up to approximately 100 different entities; depending on the IFC schema selected a few more or less. These entities include IFCWALL, IFCWINDOW, IFCDOOR, IFCCOLUMN etc. Within the conversion towards H-BIM the IFC part in the name will be removed and we will follow CamelCase naming conventions, resulting in Wall, Window, Door, Column, etc. This can be considered a good starting point to enrich classification for more specific historical definitions using the flexibility of Semantic Web approach. Some of the entities have a further refinement on IFC schema level by using enumerations. At the moment the conversions towards H-BIM do not contain these refinements, but if required it can be added easily in new versions of the INCEPTION H-BIM ontology.

6.2. IFC Relations

The IFC schemas donot only contain classification information, but also many different relations. Within the H-BIM ontology three relations are especially of interest:

- Decomposition
- Containment
- Grouping

Both decomposition and containment are defined on top of each IFCPRODUCT entity. Starting from an IFCSITE instance it is expected in a correct IFC file to find any IFCPRODUCT instance defined walking over these 2 relations (except for void instances). This knowledge is integrated into the H-BIM ontology, and as every available IFC schema works in the same way the implementation in the H-BIM ontology is the same for every IFC schema.

The grouping mechanism allows users of CAD systems to create their own grouping instances where every group is allowed to contain a subset of instances of the IFCPRODUCT entity. Allowing this specific information to pass through to the H-BIM ontology enables designers to create relevant groups, which are similarly represented in the H-BIM ontology.

6.3. IFC Properties

The object based structure of IFC is often used to 'carry' a vast amount of semantics (read information) in the form of property sets containing other property sets and / or properties. As IFCPRODUCT instances can be typed (completely in IFC4 and IFC4x1, partly in IFC2x3 TC1) different instances can share different property sets and properties.

Properties and property sets can be defined towards an agreed definition (containing +/- 3000 different properties), using naming convention PSet_xyz for the name of the property set. However every user can also create its own or application dependent properties and property sets.

Independent of the source of the properties and property sets, all available knowledge will be able to be stored against the H-BIM ontology. The structure of the properties is not defined in the H-BIM ontology other than that unit, and quantity type information can be stored.

7. Geometric Semantics from BIM - Alternative possible Geometric Ontologies

7.1. OntoBREP Ontology

OntoBREP is an ontology for CAD data and geometric constraints as a link between product models and semantic robot task descriptions. This ontology comes close to what is required within INCEPTION for being able to store geometrical data from BIM. One of the draw backs of this ontology is that it focusses on visualization exchange formats rather than geometrical definition exchange formats. In that sense it is much closer to open standards like Collada, X3D and older standards like VRML.

The real semantics within geometry representations like Boolean Operations, clipping, feature modelling within design trees cannot be stored using this standard. Another issue is that to be able to visualize any content defined against this ontology currently the results need to be converted back to STEP to be able to generate any type of visualization, which would make it complex to define a service generating geometric representations, enabling third party applications to visualize geometry.

7.2. ifcOWL Ontologies

Every ifcOWL ontology is generated by a piece of software that in itself is independent from the converted schema. This means also all geometric related entities within every IFC schema are converted towards classes almost 1-to-1. Already within IFC2x3 TC1 there are almost 200 entities somehow storing / describing geometrical information. It is important to note that IFCOBJECT entities, and its sub entities like IFCWALL, IFCDOOR etc. are not part of these 200 entities. Within IFC4 this grows to well over 200 entities where several of the entities are not 1-to-1 comparable with the entities in IFC2x3 TC1. IFC4x1 includes the alignment extension to support GIS related geometry and therefore is even larger than IFC4.

Looking at the 600+ entities in all relevant IFC versions there is a lot of redundancy and a large difference in aggregation level of the geometrical descriptions. Another issue is that there is no library available that understands the geometrical part of ifcOWL directly. Even if we would use ifcOWL there still would be no software components available to enable visualization and development of such components would be far too time consuming to develop it within the INCEPTION project.

7.3. CMO with Extensions

The open standard CMO with Extensions contains a small set of 'primitives', i.e. geometrical concepts to work with. While the architecture from this ontology has proven to be working, the geometry that can be represented with this set of geometrical concepts is limited. All typical geometrical concepts for GIS, such as Clothoids, vertical and horizontal alignments are missing as well as Bezier curves, B-Splines and NURBS are not available.

However this is still the most promising solution that could be found and its base is used and extended into a separate GEOM ontology. The GEOM ontology contains solutions for Splines, NURBS, 3D / 2D Boolean Operations, Boundary Representations, materials, textures and many other concepts.

8. H-BIM ontology Architecture

8.1. Technology Layer

Within this layer Semantic Web technology is represented. This means use of the RDF, RDFS and OWL2 as top layers of the H-BIM Ontology.

8.2. INCEPTION Layer

The INCEPTION layer of the H-BIM model contains the real knowledge from the specialists within the INCEPTION project. This is knowledge about Cultural Heritage but also knowledge from existing state-of-the-art open BIM and open GIS standards.

8.3. Extension Layer

Even with all the specialist and knowledge available within INCEPTION, there will always be structures missed, incorrect or incomplete. The Extensions layer allows any site, group or even

INCEPTION partners itself to extend the defined H-BIM ontology in a way appropriate. This way different views and/or more specialized knowledge can be integrated in the INCEPTION platform by every party, enabling optimal flexibility of the platform itself. As an example the INCEPTION H-BIM standard has embedded parts of the semantic structure of IFC (and therefore ifcOWL). This means classes Wall, WallStandardCase and CurtainWall exist. Adding a super class Walls and the knowledge that above named classes inherit from this new class Wall is an extension that works on all available content. A SPARQL query can be created to get all instances of new class Walls and it will directly have content for the majority of the Cultural Heritage H-BIM models stored in the INCEPTION platform.

8.4. Project Layer

Within the project layer the real content is defined, this content is arranged according to the layers above. All content can be queried according to the SPARQL queries defined on top of the INCEPTION layer. It is also possible to create solution specific queries as well as queries dedicated to certain extensions as defined in the extension layer.

9. Conclusion

One major challenge within the INCEPTION project is to create support for geometric representations using Semantic Web technology, taking into account the fact that in the area of Cultural Heritage geometry is often derived from point clouds coming from different 3D surveying techniques, and representing this information is therefore often even more space demanding than normal BIM models.

Within INCEPTION it was decided not to escape the issue itself but to look for a solution where this data rich content can really be converted towards content complying with Semantic Web technology. The distinction between geometric and non-geometric semantics as well as the structured setup to still keep all relevant knowledge allows a system in which data-heavy parts can be excluded from the core data stores, resulting in a minimal performance loss, even when handling vast amounts of data.

While such an approach is already interesting from a theoretical perspective, the INCEPTION project is making good progress and a working solution of this platform is already available and running in the cloud. It has proven to work on several of our demonstration cases and both performance and scalability seem to deliver the results expected from the theoretical solution. The amount of data in combination with the power of semantic web enables INCEPTION to deliver a platform that is working, scalable, powerful and useful.

References

[1] Arayici, Y.; Counsell, J.; Mahdjoubi, L.; Nagy, G. A.; Hawas, S.; Dweidar, K. (Eds.) (2017): Heritage Building Information Modelling. Routledge (UK).

[2] Historic England (2017): BIM for Heritage: Developing a Historic Building Information Model. Swindon, Historic England [accessed 25 January 2018]. Available at: <https://historicengland.org.uk/images-books/publications/bim-for-heritage/>.

IOP Conf. Series: Materials Science and Engineering **364** (2018) 012042 doi:10.1088/1757-899X/364/1/012042

Discrimination of soil texture and contour recognitions during archaeological excavation using Machine Learning

I Cacciari[1], G F Pocobelli[2], S Cicola[2], S Siano[1]

[1] Istituto di Fisica Applicata "Nello Carrara", Consiglio Nazionale delle Ricerche, v. M. del Piano 10, 50019 Sesto Fiorentino (FI), Italy

[2] Cooperativa Archeologia, v. L. La Vista, 5, 50133 Firenze, Italy

i.cacciari@ifac.cnr.it

Abstract. The most time-honoured tool for understanding the processes of the human past is represented by archaeological excavation. By examining an area at discrete temporal periods, archaeologists are literally able to look backwards in time: they can analyse incomplete material records in order to understand and reconstruct the cultural history of an area at particular moments in time. Since the digging process destroys the site forever, great care must be paid during both the excavation and the documentation. In general, after a stratum has been completely excavated, both the floors and walls are cleaned and made ready for documentation. Photos of both the sides and bedrock of a given excavation are collected, and several sketches of what the archaeologists have seen in the trenches are made. In these drawings are delineated the features and shapes of artefacts on the horizontal plane. In addition, depending on the colours and similarities of the textures, drawing are also made of the archaeological layers. This approach is time-consuming, is affected by human ability, and does not make possible a prompt digitization of the results. Within this context, the automatized identification of archaeological stratigraphy during excavation work is welcomed by archaeologists. Here, a k-means unsupervised machine learning algorithm has been used for colour clustering digital images of excavation sites. The algorithm that we have developed attempts to enhance the colour similarity while keeping the colours separate one from another as much as possible. The main idea is that pixels belonging to the same colour cluster are a part of the same layer. Once the layer has been identified, a statistical approach based on Haralick features is used to characterize each strata in terms of texture. Unsupervised machine learning combined with texture analysis could become a good practice in speeding up the documentation work of archaeologists and paving the way towards the creation of an "automated archaeologist".

1. Introduction

One of the key concept in modern archaeological theory and practice is stratigraphy. Although it derives from geology, it refers to the long-term build-up of consecutive layers of soil material that are due to both human and geological activities. Since archaeological findings are generally located below the surface of the ground, the most important tool for understanding the human past is represented by a controlled exploration of the layers below the surface: in other words, by means of excavation. After the said excavation has been completed, the archaeologist studies these layers in order to comprehend the historical processes of the site formation. In modern archaeology, lithological criteria are used to characterize the basic archaeological levels: these generally include colour, texture, hardness of the soil, etc. Since excavations come across different layers at various elevations under the surface, one of the main tasks of the archaeologist during an excavation is to distinguish the various layers. Once a layer has been completely excavated, both the wall and the floor of the trench are cleaned and then prepared for documentation. In general, archaeologists take photos of both the sides and the bedrock of the excavation, and then make a sketch of what they see. The extent, shape of the features, artefacts and layers in the horizontal plane are delineated by means of these drawings. However, this part of the archaeologist's work is not only considerably time-consuming and may be affected by human skill, but it also complicates the digitalization of the results. Therefore, any attempts at automatizing the identification of layers during excavation work are welcomed by the community. The aim of this work is to explore the possibility of creating an "automated archaeologist" [1] who can offer a prompt guide for "human archaeologists" in recognizing the archaeological layer during the excavation and in providing digital images of it. This, in turn, would simplify the drawing step and hence speed up the documentation of excavations. The "automated archaeologist" that we are imagining is based on image clustering created by Machine Learning (ML). This approach has been considered suitable for classifying regions (of excavation images) that may be considered similar (cluster identification). The potential of ML has been demonstrated in a variety of different fields [2-7]. However, very few examples of ML used in archaeology are known in the literature. Nonetheless, it has been applied to assessing the complementary value of geochemistry and machine-learning on predictive modelling in archaeology [8], in order to offer a complete methodology for the classification of archaeological data [9], as well as to a content-based image retrieval system for historical glass and to an automatic system for medieval coin classification [10].

In this work, ML has been used for speeding up layer identification that provides digital images of the archaeological excavation site. Among the different types of ML tasks we have considered unsupervised learning and, in particular, image clustering. The algorithm that we have developed is based on k-means. This makes it possible to enhance colour similarity and to keep colours separate from one another as far as possible: in other words, it divides the original image into k regions in which similar data points are grouped together, and these groups differ significantly one from the other. The work is divided as follows: section 2 introduces the images considered for colour clustering, the k-means algorithm, and the texture analysis performed in order to characterize the colour clusters obtained. Section 3 describes the experiments performed on the images and the results obtained. Lastly, Section 4 presents our conclusions as well as future perspectives.

2. Materials and methods

2.1. Images for colour clustering

In this work, two types of images have been colour clustered: namely, a mockup and several archaeological sites. The mockup was prepared in such a way as to simulate an excavation site characterized by different colours and textures. A green plastic dustsheet was spread under a wooden box; the box was then filled with soil (A) and peat (B1-B6) (figure 1).

Figure 1. Mockup prepared in order to simulate an excavation site characterized by different texture and colours. A) six pebbles were placed on an almost uniform soil layer. A peat layer was partially covered by: B1) sand and a circle of 8 pebbles, B2) sand and a large quantity of gravel, B3) sand and a small quantity of gravel, B4) randomly-placed pebbles, B5) a large quantity of gravel, B6) a small quantity of gravel.

A pebble circle was introduced on the soil layer (A) to simulate the case of a uniform soil background in which the anthropic environment was marked out by areas with different textures and

IOP Conf. Series: Materials Science and Engineering **364** (2018) 012042 doi:10.1088/1757-899X/364/1/012042

colours (pebbles). Moreover, six different areas were prepared on the peat background to simulate different combinations of texture and colour. A digital camera was used to image the scene with a 1280x960 resolution. Three actual archaeological sites were also considered for colour clustering. The images were photographed during excavation campaigns in central Italy. On the first site, a portion of ancient stone pavement (stratigraphic unit) emerges from the background of the excavation site. The second site represents a portion of a building, two orthogonal walls of which are recognizable. The third site represents another portion of that particular building: in the background are bricks that probably resulted from the collapse of the roof. The excavation limit is also visible.

2.2. *Archaeological texture*

In general, the archaeological surface may contain minute variations in colour, texture, composition and hardness. Some of them have a visual nature and others, a tactile nature. In archaeology, texture is considered a surface attribute that has a visual or tactile variety that may characterize its appearance. On the other hand, in image processing, texture describes the amplitude patterns and quantifies the spatial arrangement of colour or intensities in an image or in a selected portion of one. Within this framework, one of the most effective tools for quantifying the perceived texture of an image is based on the grey-level co-occurrence matrix (GLCM, [11]).

The elements of the co-occurrence matrix measure the number of times that different combinations of pixel pairs of a specific grey level occur in an image for various directions (ϕ) and different distances (δ) [12]. Given an $M\times N$ neighbourhood of an input image with G grey levels, let $f(m,n)$ be the intensity at pixel (m,n) of the neighbourhood. The element (i,j) of the GLCM is consequently defined as follows:

$$P(i,j|\Delta x,\Delta y) = \frac{1}{(M-\Delta x)(N-\Delta y)} \sum_{n=1}^{N-\Delta y} \sum_{m=1}^{M-\Delta x} A \qquad (1)$$

where

$$A = \begin{cases} 1 & \text{if } f(m,n)=i \text{ and } f(m+\Delta x,n+\Delta y)=j \\ 0 & \text{elsewhere} \end{cases} \qquad (2)$$

and $\delta=(\Delta x^2+\Delta y^2)^{1/2}$, $\phi=arctg(\Delta y/\Delta x)$. In this way, the matrix element P contains the second-order statistical probability values for changes between grey levels i and j at a particular displacement distance d and at a particular angle θ. For this study, we considered images with G=256 and δ=1. A normalized symmetrical matrix was computed by summing up the four matrices ϕ=0°, 45°, 90°, 135° and normalized by dividing each entry by the total number of pixel pairs. This made it possible to avoid a dependency of direction. The corresponding normalized co-occurrence values are in the [0, 1] interval, and this enabled them to be considered as probabilities. A number of textural features can be extracted from the co-occurrence matrices. Here, we considered energy to be defined as follows:

$$E = \sum_i \sum_j P_{i,j}^2. \qquad (3)$$

It returns the greyscale distribution homogeneity of images, and measures the textural uniformity of an image. Information on the random nature of the spatial distribution is then supplied. Energy assumes its highest values when grey-level distribution has either a constant or a periodic pattern. In a homogeneous image, very few dominant grey-tone transitions are expected, and the corresponding co-occurrence matrix has fewer entries of a larger magnitude, thus resulting in a sizable value for the energy feature. Because this feature is generally useful for highlighting continuity and geometry, it was considered suitable for this work [13, 14].

Moreover, the size of the neighbourhood partly determines the success of a texture-based image analysis. If the window size is too small, not enough spatial information can be extracted to distinguish between various different features; on the contrary, if it is too large, it could overlap different features and introduce spatial errors [15]. In this work, the same aspect ratio of the original image was used for

IOP Conf. Series: Materials Science and Engineering **364** (2018) 012042 doi:10.1088/1757-899X/364/1/012042

neighbourhood areas. For each neighbourhood size considered here, ten calculations chosen randomly from different areas of the same cluster were performed and their mean value was considered. The associated errors were estimated as the maximum deviations.

2.3. K-means colour clustering algorithm

ML is a powerful tool for solving a variety of problems, ranging from the visualization of high dimensional and cluster identification to pattern recognition. The basic aim of ML is to learn to identify automatically valid and potentially useful patterns and to make intelligent decisions based on data. In this work, we have considered unsupervised ML. This approach makes it possible to model a hidden structure or distribution in unlabelled data in order to learn more about given data. Clustering represents one of the main approaches of unsupervised learning: it assigns a set of inputs into subsets called clusters, so that each subset ideally shares some common characteristic and is able to place any new input within the appropriate cluster. Therefore, clustering can recognize different patterns in a dataset. In image processing, ML is used to divide a digital image into different regions for border detection or object recognition. In particular, it has been applied for different purposes in medicine [16-20], biology [21, 22], agriculture [23, 24], geophysics [25, 26], remote sensing [27, 28], security and crime detection [29], marketing and consumer analysis [30, 31], document clustering [32, 33], and automatic image annotation [34].

One of the simplest unsupervised algorithms that can be used to solve a clustering problem in digital images is represented by k-means. The algorithm developed in this work is based on k-means. It follows a simple and rapid way to cluster a digital image by means of a certain number of clusters established a priori. For each cluster, the main idea is to define a centroid (barycentre). The next step is to consider each point in the dataset and to associate it with the nearest centroid. This step is concluded, and an early groupage is completed, when no point is pending. New k centroids are then recalculated as the barycentre of the k clusters obtained in the previous step. The centroids are calculated according to the Euclidean distance between the colour dimensions and the centroids. Once new k centroids have been calculated, a new linking has to be established between the same dataset and the nearest new centroid (minimum distance). The iteration stops when the location of the k centroids no longer changes.

As described above, clustering enables dataset points to be grouped with some similarity along a dimension, while the points that differ from each other are kept further apart. In the case of digital images, the dimension used is generally colour, because the human vision system chooses colour, rather than shapes and texture, as its main discriminant feature. In this work, we have considered RGB, HSL and CieLab colour space. Images of excavation sites have been clustered using k-means in such a way that the different regions of the image are marked by k colours and the boundaries are revealed by separating the different regions. The outputs of the algorithm are k images in which all the non-zero pixels represent the object in the cluster. By assigning an 8-bit number to each pixel in a cluster, a composite image (in a false colour) is then produced: this helps the drawing of the layer contour using standard edge-detection techniques.

3. Results and discussions

The number of clusters (established a priori) is one of the main issues of the k-means algorithm: a poor colour clustering and, hence, a poor archaeological level identification may be the result of an underestimation of the k number. The best number of colours should be recognized by performing some preliminary tests. To this end, we performed colour (RGB) clustering on a portion of area A of the mockup, using different k values. In the case of k=2, it was observed that both the plastic dustsheet and the pebbles circles were considered as belonging to the cluster of the same colour, and the wooden box and soil, to the other. The wooden box in the corresponding composite was scarcely visible, and this underestimation of the colours number determined a mistake in the identification of the contour. The same held true for k=3 and k=4. The best colour clustering (for the portion of area A considered) was achieved with k=5. In this case, the anthropic environment (pebbles) was clearly separated from the background (dustsheet and soil). The contour of pebbles circle and of the wooden box could thus be drawn accurately.

In these preliminary tests we considered RGB colour space. In general, situations are likely to be found in which other colour spaces might be more suitable for colour clustering. To this end, we also considered HSL (Hue Saturation Lightness) and CieLab [35] for colour clustering of the portion of area A in the mockup. The results obtained with k=5 are presented in figure 2. It is interesting to note that, in this case, the best clustering was obtained with RGB colour space, because all the details were clearly separated. In both clusterings with HSL and CieLab colour spaces, the green dustsheet was considered to belong to the same cluster of the upper part of the wooden box, and the white pebbles, to the left part of the wooden box. The results of these preliminary tests led us to make two considerations: first, not only the k number, but also the colour space, may have impact on the cluster contour; secondly, since these two features should be tested before drawing the contour, this preliminary tests could be considered to be part of a good practice.

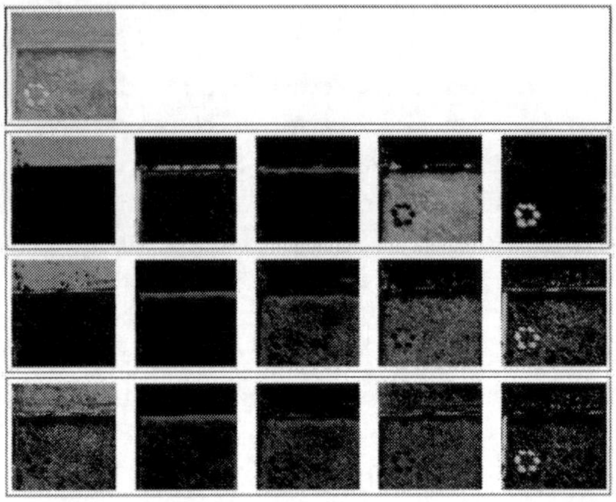

Figure 2. First row: portion of the mockup (original image), from second to fourth rows, RGB, HSL and CieLab colour clustering with k=5 respectively.

Analogous tests (number of colours and colour space) were performed on the other areas of the mockup: for the B areas, the best colour clustering was achieved with RGB and k=4. In particular, in the B1 case, the colour clustering has clearly separated the anthropic environment (pebbles) from the background. This represents a demanding case for colour clustering, because there is no strong colour difference between sand, gravel background, and the pebbles circle as there is in the A and B4 areas. This result may suggest that the algorithm could be useful also in cases in which the main differences are in the texture, rather than in the colour. In the B2 case, the background is fairly uniform (a sand layer partially covers the gravel); the texture is practically uniform, but is made with materials of two colours. The clustering brought out the two materials quite well, even if it was not possible to draw a contour. The same held true also for cases B3, B5 and B6, in which different combinations of sand, gravel and peat were considered for the background.

The texture clustering is considered by archaeologists to be equally interesting as colour clustering. Once the image of an archaeological excavation site has been colour clustered, it is remarkable for characterizing each cluster in terms of mean texture.

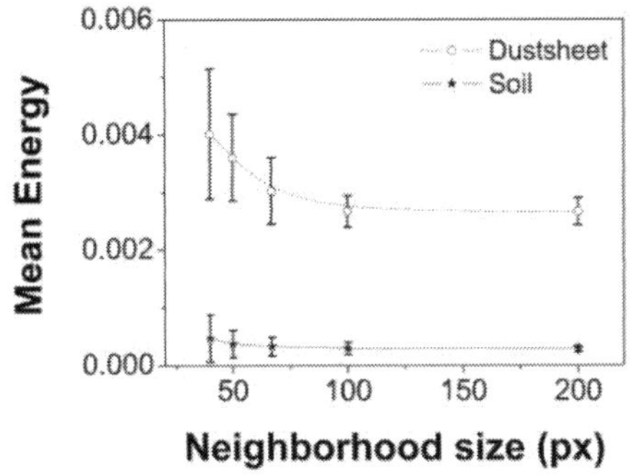

Figure 3. Mean energy calculated with different neighbourhood sizes in the dustsheet and soil areas in a portion of area A.

To this end, the energy (as one of the Haralick features that can be obtained from GLCM) was calculated with different neighbourhood sizes in the dustsheet and soil regions (in a portion of area A of the mockup). The results obtained are summarized in figure 3. With smaller neighbourhood sizes, the mean energy was greater than with ones of a larger size, and a downward trend was then established; then, with larger neighbourhood sizes, the mean energy values tended to be constant. In this work, we have suggested to use these values to characterize each colour cluster and, hence, the archaeological level. It is interesting to note that the dustsheet had greater energy than the soil. This corresponded to a greater degree of uniformity in the image and, therefore, to a smoother texture.

Moreover, colour clustering was also performed on images of two excavation sites. Figure 4 shows the composite image (RGB, k=3) and the original image (site 1) with, overlaid, the regional contours obtained by means of edge-detection. Even if the processing of the image is of poor quality, the stones of the pavement have been highlighted quite well. Analogously for site 2, the two portions of wall have been contoured quite well (figure 5).

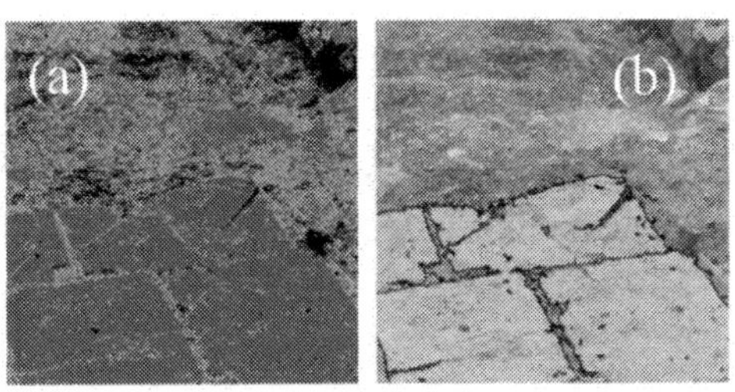

Figure 4. Excavation site 1: (a) composite image obtained with k=3 and RGB colour space; (b) original image with overlaid contours of the layer.

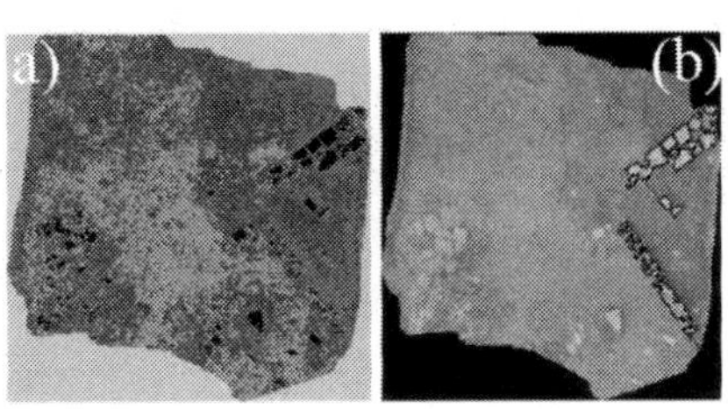

Figure 5. Excavation site 2: (a) composite image obtained with k=4 and RGB colour space; (b) original image with overlaid contours of the archaeological layer.

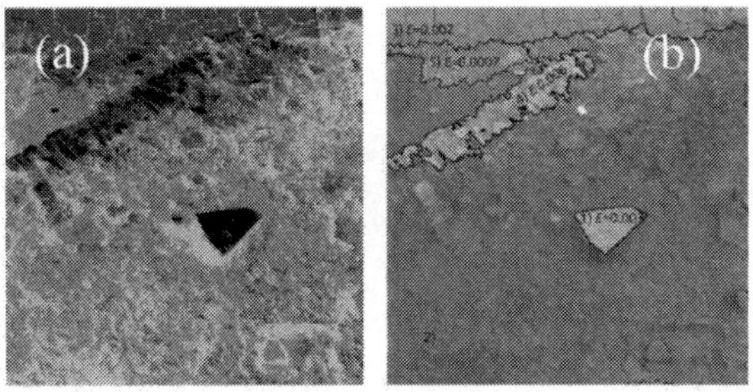

Figure 6. Excavation site 3: (a) composite image obtained with k=5 and RGB colour space; (b) original image with overlaid contours of layers and corresponding mean energy values.

The case of site 3 is particularly interesting. In figure 6 two layers are contoured: the excavation limit (3), a portion of a wall (4) and a region with a different texture (5) placed between (4) and (3). The stone (1) is also identified. This is interesting, since the algorithm may spread its potential to findings that are not actually levels, but are of natural interest to archaeologists.

Moreover, the colour clustering of site 3 highlights a layer (1) that seems to have a smoother texture than that of the background (2) (figure 6). The colour clusters have been characterized in terms of texture by calculating the corresponding mean energy. As each neighbourhood area should be included in the colour clusters, it is not possible to perform a texture analysis with larger-sized neighbourhoods. For this reason, for the image of site 3 (1100x1120 pixel) we have considered an area of 44x45 pixels. In addition, the results of the mean energy calculations are reported in figure 6. This site is extremely interesting, because different combinations of texture and colours can be observed. The results show that the two layers (labelled as cluster 4 and cluster 5) have approximately the same texture, even if they are chromatically different. Clusters 3 and 5 seem to be similar in terms of colour, but show different textures.

4. Conclusions

In this work, we aimed to explore the use of machine learning to automatize some tasks of archaeologists. This approach could help in identifying archaeological levels during an excavation campaign and in reducing the time spent for digitalizing the results. To this end, an unsupervised ML algorithm has been developed to colour-cluster digital images of archaeological excavation sites. The results obtained with a k-means algorithm and edge detection represent the first demonstration that layers can be readily and properly identified. The layers contoured have been characterized in terms of textural uniformity by calculating the corresponding energy. This proves that the combination of ML and texture analysis can serve to become a valid practice for speeding up the documentation work of archaeologists.

5. Acknowledgements

The present work has been carried out within the framework of the Archeo 3.0 project funded by the Tuscan Region (POR FESR 2014-2020).

6. References

[1] Barceló J A 2007 Proc. of the 35th International Conference on Computer Applications and Quantitative Methods in Archaeology, 413-17

[2] Dua S, Du X *Mining and Machine Learning in Cybersecurity*, ed. Auerbach Publications (Boston, USA)

[3] Gyorfi L, Ottucsak G, Walk H 2012 *Machine Learning For Financial Engineering*, ed. H. Walk (London: Imperial College Press)

[4] Clifton D A 2016 *Machine Learning for Healthcare Technologies*, ed D A Clifton (London: Institution of Engineering and Technology)

[5] Cracknell M J, Reading A M 2014 *Comput. Geosci.* **63** 22-33

[6] Harvey A S, Fotopoulos G 2016 *ISPRS-International Archives of the Photogrammetry, Remote Sensing and Spatial Information Sciences* **XLI-B8** 423-30
[7] Lary D J, Alavi A H, Gandomi A H, Walker A L 2016 *Geosc. Front* **7** 3-10
[8] Oonk S and Spijker J 2015 *J. Archaeol. Sci.* **59** 80-88
[9] Hörr C, Lindinger E and Brunnett G 2014 J. *Comput. and Cult. Herit.* **7** 1-23
[10] van der Maaten L, Boon P, Lange G, Paijmans H and Postma E 2006 *Computer Applications and Quantitative Methods in Archaeology* ed J. T. Clark and E.Hagemeister Proc. of the 34th Conf, United States
[11] Haralick R. M., Shanmugam K and Dinstein I 1973 *IEEE Trans. Syst. Man Cybern. Sys.* **3** 610-21
[12] Nixon M, Aguado A 2008 *Feature extraction & Image Processing* ed M. Nixon (Oxford: Oxford Press)
[13] Preethi G and Sornagopal V 2014 *Int. Conf. on Green Comput. Com. and Elect. Eng.* 1-6
[14] West B P, May S. R, Eastwood J. E and Rossen C 2002 *The Leading Edge* **21** 1042-49
[15] Toennies K D 2012 *Guide to Medical Image Analysis: Methods and Algorithms* (London: Springer-Verlag)
[16] Halberstadt W and Douglas T S 2008 *Comput. Biol. Med.* **38** 165-70
[17] Liao L, Lin T and Li B 2008 *Patt. Recogn. Lett.* **29** 1580-88
[18] Hill J, Corona E, Ao J, Mitra S and Nutter B 2014 *Advanced Computational Approaches to Biomedical Engineering* ed. P. Saha, U. Maulik, S. Basu (Berlin: Springer)
[19] Fouad S, Randell D, Hisham A G, Landini M G 2017 *Ann. Conf. Med. Image Understanding and Analysis* 767-79
[20] Govindaraj V, Vishnuvarthanan A, Thiagarajan A, Kannan M, and Murugan P R 2016 *J. Clin. Exp. Neuroimmunol.* **1** 1-10
[21] Hruschka E R, Campello G B and de Castro L N 2006 *Inf. Sci.* **176** 1898-1927.
[22] Xu Y, Wu J, Yin C.-C and Mao Y 2016 *PLOS*
[23] Papajorgji P, Chinchuluun R, Lee W S, Bhorania J and Pardalos P M 2009 *Advances in Modeling Agricultural Systems* ed. P. J Papajorgji, P. M. Pardalos (New York: Springer)
[24] Chinchuluun R, Lee W S, Bhorania J, Pardalos P. M 2008 *Advances in Modeling Agricultural Systems* ed. P. J Papajorgji, P. M. Pardalos (New York: Springer)
[25] Song Y-C, Meng H-D, O' Grady M J and O'Hare G M P 2010 *Computat. Geosci.* **14** 263-71
[26] Sidorova V S 2008 *Lect. Notes Comput. Sc.* **18** 693-99
[27] Bi H, Sun J and Xu Z 2017 *IEEE Trans. Geosci. Remote Sens.* **55** 3531-44
[28] Hassanzadeh A, Kaarna A and Kauranne T 2017 *Scandinavian Conf. on Image Anal.* 169-80
[29] Grubesic T H 2006 *J Quant Criminol* **22** 77-105
[30] Wang Y-J and Lee H-S 2008 *Inf Sci* **178** 1087-97
[31] Li J, Wang K and Xu L 2009 *Ann. of Operations Res.* **168** 225-45
[32] Cai X and Li W 2011 *Inf Sci* **181** 3816-27
[33] Carullo M, Binaghi E and Gallo I 2009 *Pattern Recognit. Lett.* **30** 870-76.
[34] Favorskaya Lakhmi M, Jain C and Proskurin A 2016 *New Approaches in Intelligent Image Analysis*, ed. R. Kountchev, K. Nakamatsu (Springer)
[35] Oleari C 1998 *Misurare il coloure* (Milan: Hoepli)

IOP Conf. Series: Materials Science and Engineering **364** (2018) 012043 doi:10.1088/1757-899X/364/1/012043

3D data integration for the digital reconstruction of cultural heritage monuments

Laura Micoli, Sara Gonizzi Barsanti, Umair Malik, Gabriele Guidi

Department of Mechanics, Politecnico di Milano, Via La Masa 1, 20156 – Italy
laura.micoli@polimi.it

Abstract. The ancient city centers often contain traces of monuments almost completely disappeared, whose few remains are hidden by modern buildings. This is the case for many monumental buildings of the Roman period in Italian cities, that have almost disappeared over time being gradually dismantled for the reusing their construction materials, leaving space to a dense urbanization. Knowledge of such monuments is often approximate and based on historical memories of various kinds. In such situations, three-dimensional surveys and digital 3D modelling reconstructions can be of great support for scientific-archaeological research and for providing a scientifically supported image of cultural heritage sites.
This paper shows the work on the emblematic example of the late Roman Circus of Milan, an impressive monument of about 450x85 meters, built when Mediolanum used to be capital of the Western Roman Empire (286-402 A.D). This fact is nowadays almost unknown to most citizens of Milan. The diachronic reconstruction approach is used in order to show the monument in his current state and in his presumed past aspect. The reconstruction is based on a proper mix of quantitative data originated by 3D surveys at present time, and historical sources like ancient maps, drawings, archaeological reports, archaeological restrictions decrees and old photographs. In this panorama, the georeferenced acquired data of the remains play a key role in order to validate the previous reconstruction of the building and to define some fixed points to hypothesize a reliable reconstruction of the area in different historical periods. The experimental activity is defined by the project Cultural Heritage Through Time - CHT2 (http://cht2-project.eu), funded in the framework of the Joint Programming Initiative on Cultural Heritage (JPI-CH) of the European Commission.

1. Introduction

For research purposes, documentation needs, preservation and conservation issues in the Cultural Heritage field, it is fundamental to appropriately record, document and survey artefacts and sites: indeed, an accurate and complete 3D digital documentation is a prerequisite for further analyses and interpretations. This digital representation is also important to properly communicate our Heritage to the public, often coupled with a 4D reconstruction, highlighting the real shape of a monument compared with what is still visible nowadays. This type of reconstructions can be based only on archival data but several works [1-4] have shown that a serious reconstruction work should start from the accurate digitization of the existing remains coupled with a deepen research of other sources of information. However, there are cases where the three-dimensional diachronic reconstruction is particularly complex, due to: i) the nearly total absence of remains to be surveyed; ii) few documentation on one or more periods of artefact's time life; iii) uncertainty of sources; iv) difficulty to correlate documents and data to a three-dimensional representation.

This paper presents a work done on the Late Roman Circus of Milan that, given the limited detectable findings, began with an in-depth philological analysis from the collection of historical and archival data, integrating the different contribution in the area started from the early 1939 [5-8]. The case study covers the south-west area of the city center of Milan that corresponds to the Roman Circus area [6, 9], where it is possible to see several traces of the different historical periods from ancient times until the densely urbanized structure of the present days. The Circus was the open-air venue for chariot and horseraces and also the place dedicated to the celebration of the Emperor's greatness and for this reason it was generally located near the Imperial Palace [10]. Archaeological studies were conducted mostly at the beginning of the '900 and after World War II, during the reconstruction of some private and public buildings, when it was possible to see the archaeological remains. Nowadays, only small portions of the monument remain visible and a lot of historical documentation has been lost during World War II. Starting from the '60s of last century, the northern area, corresponding to the to the complex of the Monastero Maggiore, was used as headquarter of the Archaeological Museum of the Municipality of Milan (Civico Museo Archeologico di Milano). The rest of the area is instead almost entirely occupied by residential buildings [11-12] and only some portion of circus remains is still visible in the basements.

2. Investigation methods

This work, given the limited detectable findings, began with an in-depth philological research started from historical and archival data that have been integrated to hypothesize a reconstruction of the monument, by referencing such documents to that specific location of the city.

One of the first step was a deep bibliographic research to collect information about the studies and the excavations done during the past centuries on the Milan's Circus and also on the others circuses of the same period to understand the design rules for that kind of monument [5, 10]. Another step of the work was the collection of historical maps and drawings showing the knowledge of the area and the extreme urbanistic changes during the centuries. In the Figure 1a) and 1b) for example it is possible to see how the northern area of the circus, converted into the ecclesiastical complex of the major monastery starting from VIII century BC, in the 19th century underwent a major transformation due to the opening of a new road axis of Via Bernardino Luini.

(a) b)

Figure 1. Detail of two historical maps in which the urban change in the 19th century is visible: a) 1837/1860, G. Pozzi; b) 1877,C. Fenghi (Courtesy of Archivio delle Stampe Bertarelli).

At the beginning of the activity an agreement was signed between the CHT2 project and the Milan Superintendence in order to obtain the permission to study the monument and to access to all the documents stored into the archive. Thanks to this, a long research was done taking into account: pictures, drawings, old digs diaries and archaeological constrain decrees.

The area of study, for example, was severely damaged during the II world war, and as a result of this, in the 50s and 60s, many buildings were demolished and rebuilt. During these works, some finds emerged and were subsequently removed, incorporated into new structures or covered and today they are no longer visible (Figure 2). Even for these peculiar reasons the analysis of the historical documents is particularly relevant.

(a) (b)

Figure 2. Photographs of the remains during the excavation of private buildings reconstruction after the WWII, actually not more visible: (a) Via Brisa 1951; (b) Via Medici-Circo (courtesy of Soprintendenza Archeologica di Milano).

3. Inspections and surveys

Parallel to the archival research, a capillary inspection was carried out in all the buildings in the area where the circus was expected to be. Thanks to this, it was possible to make a first check on the quantity and positions of the remains compared to the historical maps. In figure 3 it is possible to see a map of the circus with some portion filled in red corresponding to remains considered visible at that time [8], the overlapping portions filled with the red cross-hatching highlight the areas where currently there is nothing visible and detectable. As it can be noticed, there are large areas where historical information deviates from the real situation. Based on the obtained results of the inspection, the following activities were planned: i) comprehensive three-dimensional survey of all the remains currently accessible with different technologies depending on their peculiarities and the needs of study; ii) topographic survey of the area in order to correctly georeference each portion of the monument.

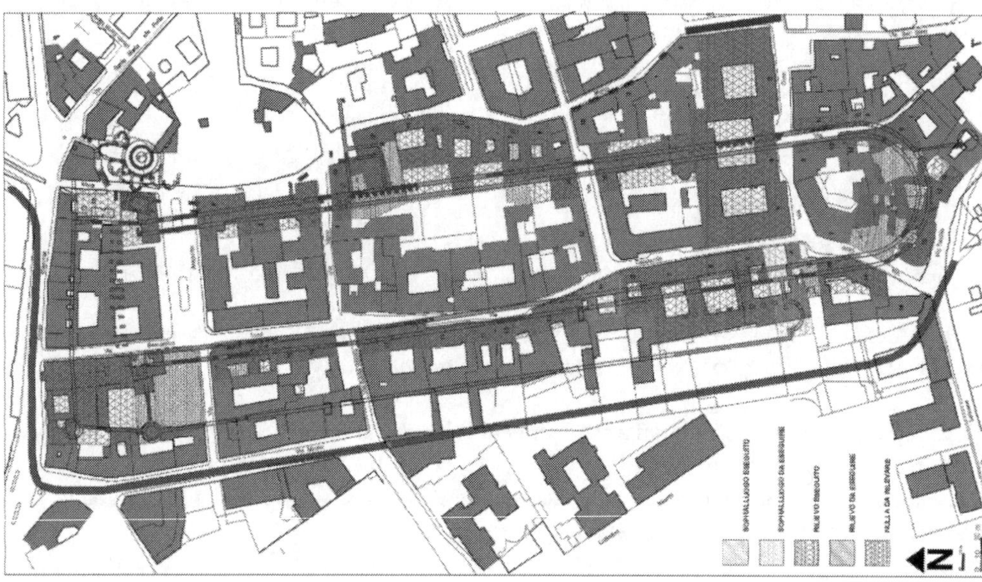

Figure 3. The map shows the state of the art of the survey campaign: cross-hatching in red the areas in which the archival data gave information of remains but now nothing is visible, in green the acquired data, in light blue the building not yet accessible.

3.1. 3D survey

The 3D survey regarded the data acquisition of all the archaeological remains still visible in the area. Different techniques were used: ToF, Phase Shift laser scanners, mobile mapping and photogrammetry, depending on the condition of the walls and the aim of the survey.

Laser scanning was used to acquire the geometric and dimensional information of the remains, maintaining a standard deviation lower than 3mm and a spatial resolution at least at 1cm. For the exterior parts a Faro Focus 3D 120 and LEICA HDS7000 were used.

While for the most of the inner parts and the connections to the exterior, in order to have the needed data for the georeferencing work, the hand-held mobile laser scanning GeoSLAM ZEB-REVO was used [13, 14] . This instruments is a portable mobile mapping systems (MMSs) consisting a 2D time-of-flight laser range scanner rigidly coupled to an inertial measurement unit (IMU) mounted on a motor drive; the data are combined by a 3D simultaneous localization and mapping (SLAM) algorithm is used to combine the 2D laser scan data with the IMU data to generate accurate 3D point clouds. The data is captured by simply walking through the area of interest drastically reducing the time consuming scanner set-ups and data registration associated with traditional terrestrial laser scanning methods. This aspect it was relevant because of some peculiar aspect of the survey: i) most remains are located in rooms and basements of private properties, often prestigious residences with problems related to privacy and security issues; ii) long and complex paths to go from the rest to the outdoor road position useful for georeferencing; iii) the need a short time work, both in the acquisition step, to stay as short as possible in residences, both in the post-processing step to minimize the alignment time.

A photogrammetric SfM/IM survey, was performed only in some case, where it is particularly important to enrich the geometric data with color information of the texture in order to be able to in order to better read the wall traces. It is the case for example of the wall of Via Vigna/Via Morigi, one of the few developed in high, where it is possible to analyze some structural issues and the stratigraphy of the monument, as well explained in a previous paper of the same authors [15].

Figure 4. The 3D textured model of the Via Morigi wall, from a photogrammetric survey realized by images captured with a Sony NEX 6 camera coupled with a 24mm F 1.8 Zeiss lens.

3.2. Topographic survey

To place all the remains in the same georeferenced system, a topographic survey with differential GPS was performed using 13 natural points for the exterior part of each portion of remains surveyed.

This step was a crucial stage for generating a reliable starting point for the reconstruction, allowing also to obtain georeferenced data and to correct some information about the monument that are uncertain as for example the total length of the circus. The two archival maps [5, 6] collecting the results of the excavations and used as basis for the reconstruction, show a difference in length of quite 10 m (Figure 5). With the results of the survey it will be possible to give a conclusive and correct data, that will be useful also for the Superintendence.

Figure 5. Map of the 13 GPS station on the circus area.

4. Results and future works

The knowledge of the Milan circus presents, as with many ancient monuments, some aspects still doubtful or unaddressed and in the specific of our case study there are three crucial issues. The first of these is the actual extension of the monument which, according to different historical sources, is indicated as 460 or 470 meters. Since in the North the carceres position is unequivocally identified by the tower conserved in the archaeological museum area it is fundamental to identify the correct position of the circus curve which, as visible from the geo-referenced historical maps shown in figure 6 (a), is not certain.

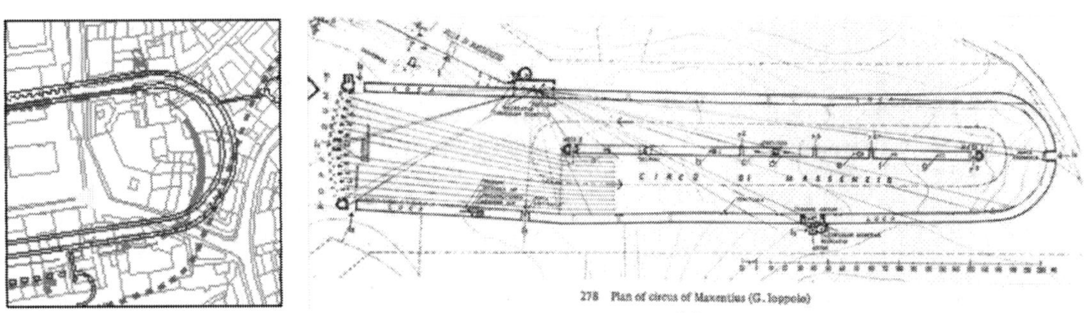

a) b)

Figure 6. Two imaging showing critical aspect of knowledge on the Milan circus: a) vector drawing of the archival maps showing the difference in length of the circus; b) the plan of the circus of Maxentium in Rome showing as in late roman circuses the walls were angled based on race rules.

An important contribute on this matter was given by the survey of the part of the wall belonging to the aps of the circus in the southern area. These impressive remains are the higher still visible of the monument, developed in four floors in a private house discovered during some restructuration works in the ’80. The 3D georeferenced survey realized will allow to right positioning the curve and calculate the building length (Figure 7 (a) and (b)).

IOP Conf. Series: Materials Science and Engineering **364** (2018) 012043 doi:10.1088/1757-899X/364/1/012043

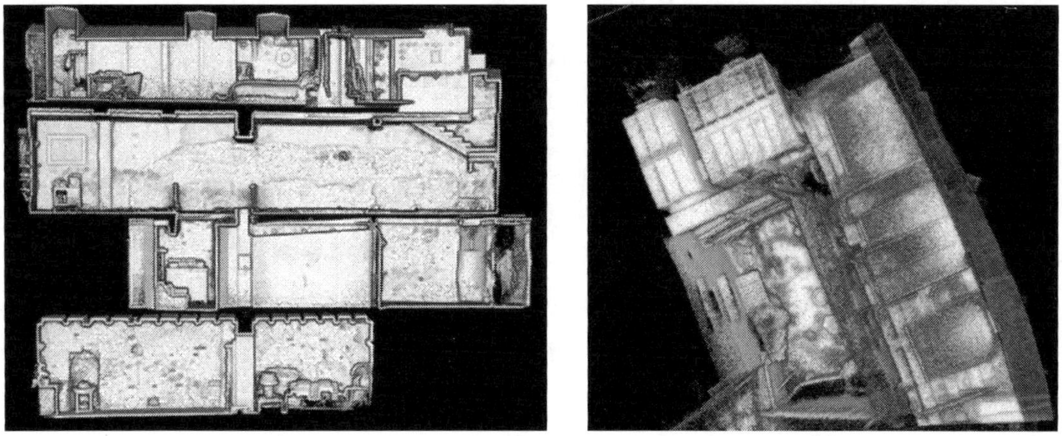

a) b)

Figure 7. The points of cloud of survey inside t private apartment of the remains of the apse of the Circus: (a) the extension of the wall along the 4 different floor of the apartment; (b) the curvature shape of the remains.

Another critical aspect is the orientation of the walls that in the circuses of the late Roman Empire, as evidenced by the studies of Humphrey [10] were not straight and parallel but had angles in order to allow all competitors to make an equal path and to avoid accidents. In the case of the Milan circus, due to the small remains and their distribution, the orientation of the walls to date is not certain and the studies of Mirabella Roberti have indicated plausible but not totally verified angles.

Thanks to the 3D capillary survey of all the remains and their georeferencing we will be able to perform an integrated reading of all the single portions, small and large, and to identify the actual course of the walls.

Finally, some doubts remain with respect to structural questions related to the support of the cavea. The recent survey provides us with new elements of analysis as we have been able to identify and acquire in the western area a site in which a large portion of the internal wall towards the arena and a small portion of the corresponding external wall are present. The remains of Via Cappuccio (figure 8 (a) and (b)), similarly to what has been previously at Via Vigna, this allows measurements to be made between the two walls and to make some hypotheses about the structures.

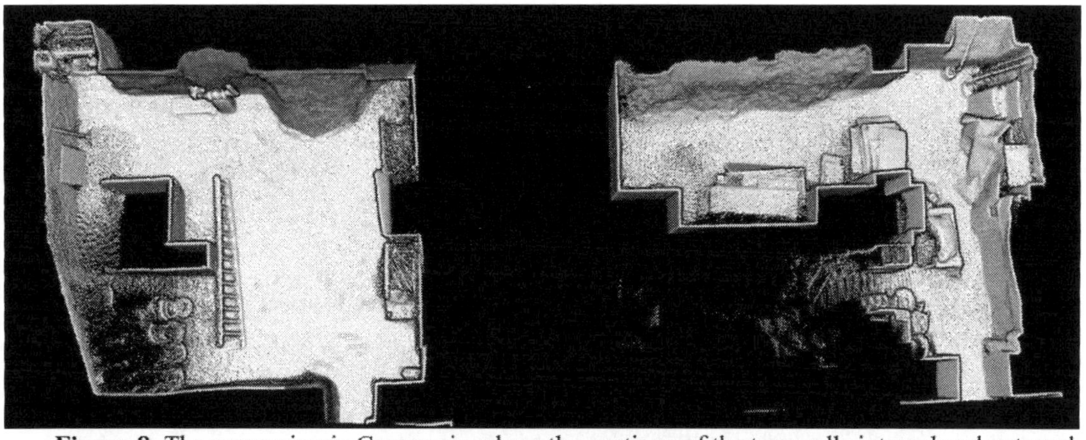

Figure 8. The survey in via Cappuccio where the portions of the two walls internal and external are visible (highlighted in green in the image).

All the data collected are crucial for hypothesizing and reconstructing the circus. In order to have a shared basis from which starting, it was decided to create a GIS platform (ArcMAP) to which a geodatabase will be connected to include all the information collected during the archival review and the surveys, in accordance also with the inspectors of the Superintendence.

On the basis of this georeferenced map, the reconstruction of the shape of the monument as it was, started as well descripted in [15], and will be updated with the results of the late 3D and GPS surveys. In this way, the reconstruction, done with the archaeologists in order to have a correct interpretation of the data, will be as much correct as possible.

5. Conclusion

The integration of the heterogeneous data collected will permit to provide a realistic reconstruction of a monument that is not visible anymore.

The so-called 4D reconstruction is usually intended as the process for obtaining the shape of real objects and its changes along a temporal dimension. This is based on a methodology [4] for integrating the different sources of non-uniform data. On the one hand the surveys georeferenced to the topographic network of the city, will help to identify the accurate positioning of the building and to obtaining precise measurements of the shape and length of the structures. By merging such data related with the current state of the monument with the archival material collected, a rearrangement of the historical representations will be made, like for example the normalization of historical plans in a uniform scale.

This reconstruction will be useful, not only for conservation and research purposes but also to let the citizens of Milan aware of the treasure hidden in their city and of their past.

6. Acknowledgments

The authors wish to acknowledge the support of the Joint Programming Initiative on Cultural Heritage (JPICH), for funding the project Cultural Heritage Through Time (CHT2).

The author would like to thank Anna Maria Fedeli from the Archaeological Superintendence of Milan, Donatella Caporusso and Anna Provenzali, from the Archaeological Museum of Milan for their precious cooperation in providing documentation on the area, helping also in the archival research.

We would also like to thank the colleagues of the FBK team working with us to the data acquisition and elaboration of the second survey campaign.

Finally, a special thank goes to all the citizens, owners and administrators of the private buildings involved in the inspection and survey activities of ours studies.

7. References

[1] Guidi, G.; Russo, M.; Angheleddu, D. 3D Survey and Virtual Reconstruction of archaeological sites. Digit. Appl. Archaeol. Cult. Herit. 2014, 1, 55–69, doi:10.1016/j.daach.2014.01.001

[2] Micoli, L. L.; Guidi, G.; Angheleddu, D.; Russo, M. A multidisciplinary approach to 3D survey and reconstruction of historical buildings. In 2013 Digital Heritage International Congress (DigitalHeritage); IEEE: USA, 2013; pp. 241–248.

[3] Guidi, G.; Russo, M. Diachronic representation of ancient buildings : studies on the "San Giovanni in Conca" Basilica in Milan. Disegnare Con 2009, 2, 69–80.

[4] Rodríguez-Gonzálvez, P.; Muñoz-Nieto, A. L.; Del Pozo, S.; Sanchez-Aparicio, L. J.; Gonzalez-Aguilera, D.; Micoli, L. L.; Gonizzi Barsanti, S.; Guidi, G.; Mills, J.; Fieber, K.; Haynes, I.; Hejmanowska, B. 4D Reconstruction and visualization of Cultural Heritage: analyzing our legacy through time. Int. Arch. Photogramm. Remote Sens. Spat. Inf. Sci. 2017, XLII-2/W3, 609–616, doi:10.5194/isprs-archives-XLII-2-W3-609-2017.

[5] De Capitani d'Arzago, A. Il circo romano; De Capitani d'Arzago, A., Calderini, A., Eds.; Ricerche della Commissione per la forma urbis Mediolani / Istituto di studi romani, Sezione lombarda; Ceschina, 1939;

[6] Mirabella Roberti, M. *Milano romana*; Rusconi: Sant'Arcangelo di Romagna, 1984; ISBN 88-18-33964-8.

[7] Blockley, P.; Caporusso, D. *Area del Monastero Maggiore in epoca romana / [testi di Paul Blockley ... et al.]*; Archeologia a Milano; Civico Museo archeologico: Milano, 2013; ISBN 978-88-97568-08-7.

[8] Caporusso, D.; Donati, M. T.; Masseroli, S.; Tibietti, T. *Immagini di Mediolanum - Archeologia e storia di Milano dal V secolo a.C. al V secolo d.C.*; Civico Museo Archeologico di Milano: Milano, 2014; ISBN 978-88-97568-00-1.

[9] Sena Chiesa, G. *Milano capitale dell'impero romano: 286-402 d.C.*; Silvana Editoriale: Milano, 1990; ISBN 978-88-366-0276-6.

[10] Humphrey, J. H. *Roman Circuses: Arenas for Chariot Racing*; University of California Press: Berkeley and Los Angeles, 1986; ISBN 978-0-520-04921-5.

[11] Fedeli, A. *Milano Archeologia. I luoghi della Milano antica*; Fondazione Cariplo: Milano, 2015; ISBN 978-88-86752-69-5.

[12] Capponi, C.; Ambrosini, A. *San Maurizio al Monastero Maggiore in Milano: guida storico artistica*; Silvana Editoriale: Cinisello Balsamo, Milano, 1998; ISBN 88-8215-095-X.

[13] Farella, E.M., 3D mapping of underground environments with a hand-held laser scanner, *Proc. SIFET Congress, Lecce, June 2016*

[14] Nocerino, E., Menna, F., Remondino, F., Toschi, I., Rodríguez-Gonzálvez, P., Investigation of indoor and outdoor performance of two portable mobile mapping systems, in: *Remondino, F., Shortis, M.R. (Eds.), Videometrics, Range Imaging, and Applications XIV; 103320I. International Society for Optics and Photonics, Munich, Germany, p. 103320I*

[15] Guidi, G.; Gonizzi Barsanti, S.; Micoli, L.L.; Malik, U.S. Accurate Reconstruction of the Roman Circus in Milan by Georeferencing Heterogeneous Data Sources with GIS. *Geosciences* 2017, *7*, 91.

IOP Conf. Series: Materials Science and Engineering **364** (2018) 012044 doi:10.1088/1757-899X/364/1/012044

3D Map Computation from Historical Stereo Photographs of Florence

Marco Fanfani, Fabio Bellavia, Giulio Bassetti, Fabrizio Argenti, Carlo Colombo

Department of Information Engineering (DINFO), University of Florence, Via S. Marta 3, 50139, Florence, Italy

E-mail: {marco.fanfani, fabio.bellavia, fabrizio.argenti, carlo.colombo}@unifi.it, giulio.bassetti@stud.unifi.it

Abstract.
The analysis of early photographic sources is fundamental for documenting and understanding the evolution of a city so rich in history and art as Florence. Indeed, by the 1860s several photographers used to work in town, and their images (often obtained through stereoscopic set-ups) can help us to reconstruct Florence in 3D as it was by the time of the Italian unification. The first and most delicate part of such reconstruction process is the computation of *disparity maps* from the historical stereo pairs. This is a very challenging task for fully-automatic computer vision algorithms, since XIX century photographs are affected by several problems—ranging from superficial damages to asynchronous acquisition—that are usually absent in modern images. In this paper we describe a semi-automatic method that, through minimal user input, allows the creation of coherent and realistic 3D maps of florentine scenes.

1. Introduction

Thanks to its immediate readability and faithful reproduction accuracy, photographic material from the last two centuries is an invaluable source of information for historians of art and architecture willing to track the changes occurred to the urban space and its monuments over the years. In the early days of photography (mid-19th century), several professional photographers were based in Florence, where they used to produce and sell various kinds of printed images, ranging from commissioned portraits to selections of remarkable city corners and works of art, which were particularly appreciated by tourists from both sides of the Atlantic. Anton Hautmann was one of the most active photographers in Florence between 1858 and 1862. He was a specialist in stereoscopic photography, a fashionable new technology at that time, by which an image pair of the same scene is obtained from two slightly different viewpoints. His "stereographs" were particularly sought-after by amateurs, who were thus able to recreate the depicted scene in 3D by watching the photographs through a simple stereoscope.

The project TRAVIS (Tecniche di Realtà Aumentata per la Visualizzazione di Immagini Storiche), currently carried out by the Computational Vision Group of the University of Florence and funded by Fondazione Cassa di Risparmio di Firenze, is aimed at developing effective computer vision techniques for the 3D reconstruction and rendering of (either single or stereoscopic) historical photographs of Florence. Its main goal is to develop a geo-referenced PC

(a) (b) (c)

Figure 1: (a) Digital scan of an original stereograph by A. Hautmann depicting a corner of Piazza Santissima Annunziata in Florence; (b) Disparity map obtained with a fully-automatic state-of-the-art software [1]; (c) Result obtained with the proposed method. (Better viewed on a computer screen.)

or mobile app allowing users to navigate through Florence as it was in the XIX century, being able to switch anytime to contemporary Florence, and mark the changes occurred in its urban structure.

While 3D reconstruction from images is a widely explored topic in photogrammetry and computer vision, its application to historical images is still challenging. Some published works describe solutions based on manual 3D modeling using commercial softwares [2, 3], others exploit projective constraints on manually selected features [4]. In [5] an automatic Structure from Motion (SfM) method is employed, leveraging on a huge data collection—800 photos from which the authors were able to select a small sub-set (44 images) on which computer vision algorithms work reliably. A similar problem is described in [6], where automatic methods can deal with images spanning only 1956 to 1975, while using manual input the authors are able to register images from 1864 to 2008.

Differently from other approaches, in this paper focuses on the dense stereo analysis of Hautmann's stereographs—see Fig. 1a. Being the first and most important step toward dense 3D reconstruction, dense stereo analysis is a well-known topic in computer vision, and many algorithms have been presented so far to solve the problem [7]. However, as SfM fails, so do even the best automatic dense stereo methods that are not able to deal satisfactorily with early stereographs (see Fig. 1b). These images are typically—and often simultaneously—affected by several problems not occurring in modern images: Low resolution, blurred areas, physical damages as scratches or stains, low contrast, asynchronous acquisition. For this reason, we developed a semi-automatic method that, by leveraging minimal user input, can obtain coherent and realistic *disparity maps* from historical photo pairs (see Fig. 1c). This is achieved by computing an initial disparity map, which is then improved through manual input and the use of planar projective transformations, according to the observation that urban man-made environments are mainly composed by planes.

The next Section briefly presents Hautmann's stereographs collection. Sect. 3 introduces basic dense stereo terminology and concepts. Section 4 presents the developed StereoDisparity App. Finally, in Sect. 5 conclusions are given and future work is outlined.

2. Hautmann's Stereographs Collection

Anton Hautmann (Munich 1821 – Florence 1862) was one of the most active photographers in Florence between 1858 and 1862. He left us with one of the largest photographic collections of Florence, characterized by both a high technical quality and a noticeable artistic skill. His

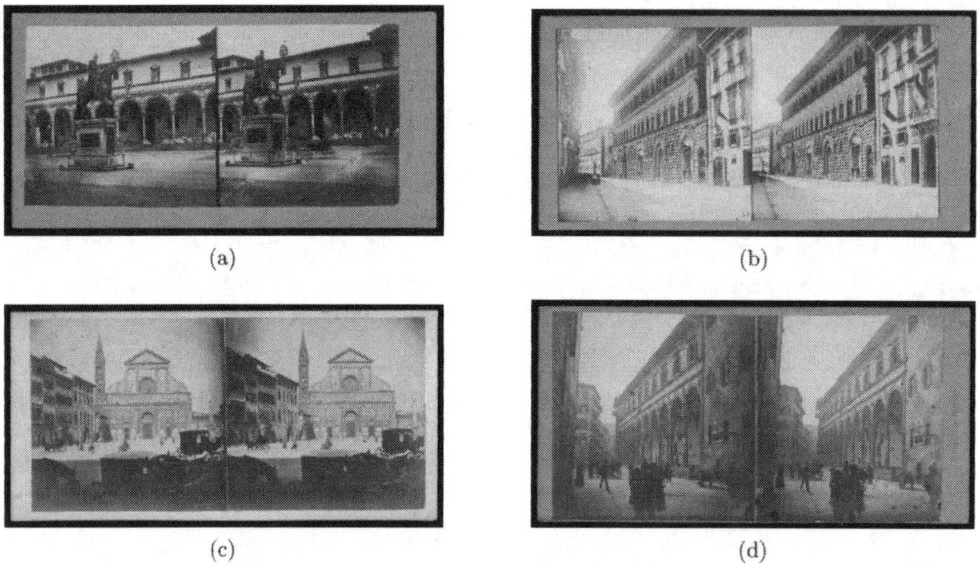

Figure 2: Examples of stereo photographs from H. Hautmann: note how all frames are affected by noise, presenting saturated areas, low contrast, different exposition between the left and right frames. Stains and scratches are clearly visible, as well as vignetting effects. In (a), a view of Piazza Santissima Annunziata; (b) Palazzo Medici Riccardi in Via Larga (today Via Cavour); (c) and (d) two views of Piazza Santa Maria Novella. Notice also that (a) is an example of asynchronous acquisition: in the right frame there is a carriage near the statue, absent in the left frame.

florentine views encompass a wide range of subjects (from palaces and squares to street life scenes), often taken from unusual yet effective viewpoints. Part of Hautmann's collection (123 stereoscopic photographs) is described in [8]. Thanks to Hautmann's living heirs, we were able to collate 174 stereographs of Florence and other places in Tuscany. High resolution digital scans of the original stereo pairs were then acquired in cooperation with the Kunsthistorisches Institut in Florence. Figure 2 shows some examples of Hautmann's stereographs collection.Observing the photographs, its easy to notice that the image quality is very poor. This is partly due to the limited photographic instrumentation available in the XIX century. Most photos are quite noisy and have poor definition, low contrast, saturated or blurred areas. Moreover, during the years, some photographs were partially damaged with scratches and stains. It is also worth noting that some stereo pairs were acquired with a single camera moved in two positions along a rail (asynchronous acquisition), instead than with a real stereo camera with two lenses: this can be easily detected by observing that some objects in the scene change their position in the two frames—this is a serious challenge for any automatic stereo matching algorithm.

3. Dense Stereo

Given two images of the same scene, it is possible to define a set of constraints, referred to as *epipolar geometry*, that can be exploited in order to compute the 3D structure of the scene [9, 10]. Let $\mathbf{X}_1$ be a 3D point in the scene, and $\mathbf{x}_1^l$, $\mathbf{x}_1^r$ be its projection onto the left and right stereo frames, respectively. Let $\mathbf{X}_2$ be a second 3D point, collinear with $\mathbf{X}_1$ and the left camera center $\mathbf{C}_l$. Under this setup, the projection of $\mathbf{X}_2$ will perfectly overlap with $\mathbf{x}_1^l$ on the left frame, while on the right frame $\mathbf{x}_2^r$ it will fall away from $\mathbf{x}_1^r$ (see Fig. 3a). This displacement between $\mathbf{x}_1^r$ and

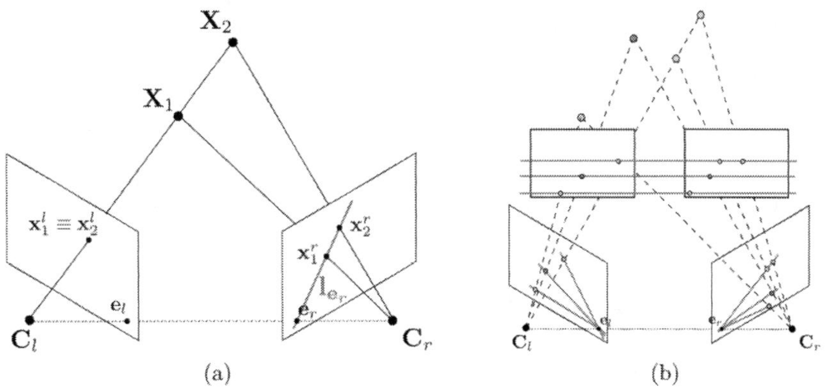

Figure 3: Epipolar geometry: (a) *Motion parallax.* $\mathbf{X}_1$ and $\mathbf{X}_2$ are collinear w.r.t. the left camera, so that $\mathbf{x}_1^l$ and $\mathbf{x}_2^l$ are coincident, while $\mathbf{x}_1^r$ and $\mathbf{x}_2^r$ belong to the same epipolar line $\mathbf{l}_{\mathbf{e}_r}$. $\mathbf{C}_l$ and $\mathbf{C}_r$ are the camera centres, while $\mathbf{e}_l$ and $\mathbf{e}_r$ are the epipoles (i.e. the projections of one camera centres onto the other image). (b) A graphical representation of the rectification transform that maps the epipolar lines in a parallel configuration, sending the epipoles to infinity.

$\mathbf{x}_2^r$ is called *motion parallax*: The closer the 3D points are to the camera, greater will be the motion of its projections on the left and right frame. The line connecting $\mathbf{x}_1^r$ and $\mathbf{x}_2^r$ (that is the projection of the line between $\mathbf{X}_1$ and $\mathbf{X}_2$) is called *epipolar line* and each 3D point projecting in $\mathbf{x}_1^l$ must fall on the right frame on the same epipolar line. So, for each point in the left image $\mathbf{x}_l$, there exists a line in the right image that passes through its corresponding point $\mathbf{x}_r$. This constraint is encoded in the *fundamental matrix* F, such that $\mathbf{x}_r^\top F \mathbf{x}_l = 0$.

Given a stereo pair, a matching problem has to be solved in order to obtain an estimate of the 3D structure of the scene. In particular, a given point in the left image has to match one-to-one with another point in the right image. To find these correspondences, it is possible to exploit the epipolar constraint to restrict the search space to a line. Better still, in order to ease the search problem, stereo frames are typically reported to a more convenient standard form using a *stereo rectification* technique (Fig. 3b). The aim of rectification is to transform the original frames in a way that all epipolar lines are overlapping, so that all matching points have the same row in both the left and right images. A rectified stereo pair can then be used to compute a *disparity map*, where each pixel value represents the displacement between the left and right views. If, in addition to epipolar geometry, the internal *calibration parameters* of the cameras are known, the disparity map can eventually be converted easily into a *depth map*, i.e., into a 3D representation of the scene.

4. The StereoDisparity App

Our StereoDisparity App, developed in Matlab, includes: (1) a point selection tool to identify correspondences between the left and right photos; (2) an automatic rectification algorithm to correct the original misalignment of epipolar lines; (3) enhancement algorithms to correct illumination and reduce noise; (4) a dense stereo correspondence software, based on block matching, with manual tuning of parameters; and (5) a densification procedure based on scene plane homographies. Details are provided in the following subsections.

As shown in the block diagram of Fig. 4, from a functional viewpoint StereoDisparity is composed by three main blocks: (i) a *MainGUI* that is used to call the other software modules as well as to compute the disparity map using automatic algorithms; (ii) the *RectificationTool*, used to rectify the stereo pair; and the *PlaneSelectionTool* for selecting corresponding planar

IOP Conf. Series: Materials Science and Engineering **364** (2018) 012044 doi:10.1088/1757-899X/364/1/012044

regions in the left and right images and to compute missing disparities according to planar homographies. At startup, the stereo pair is loaded in MainGUI. Then, RectificationTool is called to correct image misalignments.

4.1. Rectification

Even if the input stereographs seem to be already rectified (see again Fig. 2), they actually have some residual misalignments. To refine the input data, RectificationTool lets the user select point matches on the stereo pair with a few mouse clicks on a simple graphical interface. Once enough correspondences have been selected, the rectification algorithm described in [11] is run. To ease the matching selection, we also included two enhancement algorithms for improving the image quality: A color correction method that tries to mitigate the illumination differences between the left and right frame [12], and a de-noising method [13].

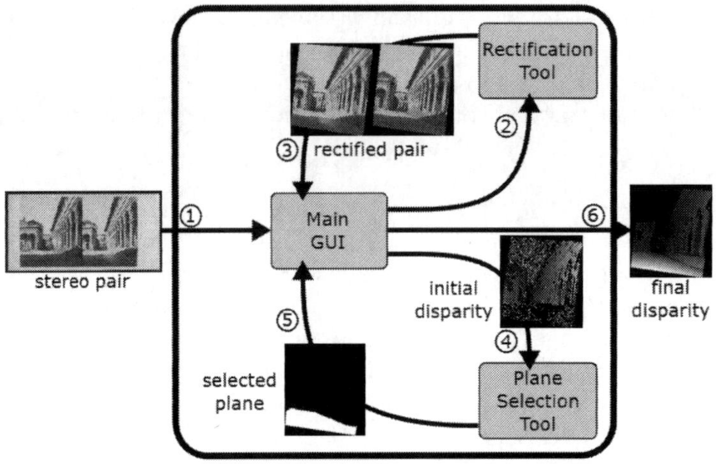

Figure 4: Block diagram of the StereoDisparity App. From the MainGUI, the Rectification Tool and the PlaneSelection Tool are called. Intermediate outputs are also shown.

4.2. Disparity map computation

Given the rectified pair, initial estimates of the disparities can be obtained with automatic stereo matching methods. We use three simple block matching methods: Two from the OpenCV library (the Block Matching and Semi Global Block Matching), and the Matlab implementation. For each of these solutions, StereoDisparity lets the user select appropriate parameters and observe the updated results (see Fig. 5a). However, the obtained disparity map, even after manual tuning, usually contains many missing areas and errors (see Fig. 5b) that can be fixed through PlaneSelectionTool, thus improving the result.

4.3. Disparity map improvement

PlaneSelectionTool is an interface—similar to Rectification Tool—for selecting matching points. In this case, the user has to pay attention on selecting only points belonging to corresponding planar surfaces. Once selected enough points, a *planar homography* transformation is estimated by the software. A planar homography is a perspective transformation. In particular, we are interested in homographies induced by a plane in the scene. Let $\mathbf{X}$ be a 3D point belonging to the plane π, and $\mathbf{x}_1$, $\mathbf{x}_2$ the projections of $\mathbf{X}$ on two different images. Then the homography H_π, induced by π, is a map such that $H_\pi \mathbf{x}_1 = \mathbf{x}_2$ (or $H_\pi^{-1}\mathbf{x}_2 = \mathbf{x}_1$). H_π has 8 degrees of freedom,

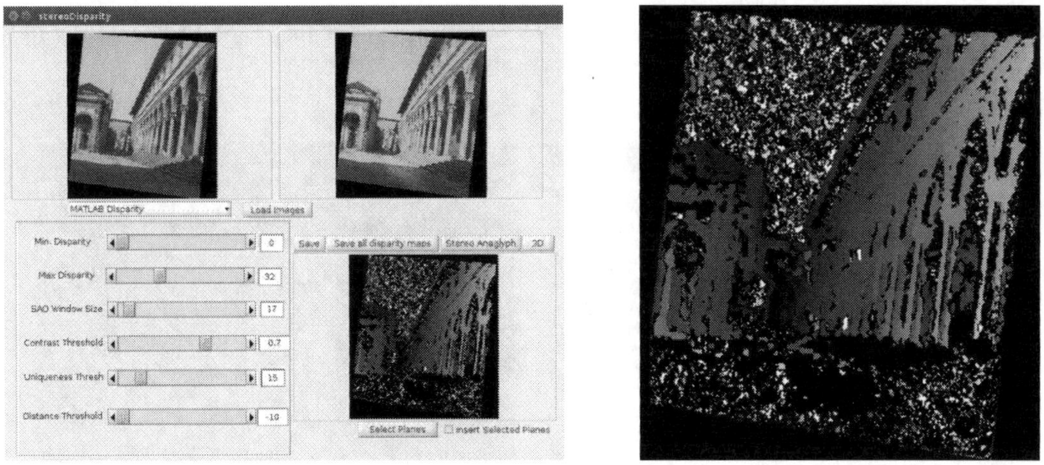

(a) (b)

Figure 5: (a) The MainGUI of the StereoDisparity App. The user can choose the stereo algorithm to use and tune its parameters. (b) Disparity map obtained with the automatic block matching algorithm of Matlab.

(a) (b) (c)

Figure 6: Interactive expansion scheme for planar surfaces: in (a) the initial planar surface detected. In (b) and (c) the region is progressively expanded by including super-pixels of areas belonging to the same plane (in this case the ground plane).

and each point correspondence produces two independent constraints, so four matching points are sufficient to estimate H_π. More details can be found in [9].

Manually selecting a whole plane region can be tedious and can take a lot of time. We simplify this step by implementing a planar expansion scheme exploiting super-pixels. After obtaining the initial plane region and the respective homography mapping (by selecting four o more correspondences), we segment the left frame into super-pixels, using EDISON that implements mean shift segmentation [14, 15, 16]. The initial planar region is superimposed on the image (Fig 6a), and the user can manually select single super-pixels to be included into the plane (Fig. 6b and 6c). In this way, the estimated homography can be applied to all the points of the selected regions, thus obtaining a dense disparity estimation. By repeating these operations, multiple planar regions can be selected to augment and improve the disparity map up to the desired level of detail.

Finally, all disparity values obtained by PlaneSelectionTool are fused with those of the initial diparity map, thus obtaining a richer description of the scene geometry (see Fig. 7).

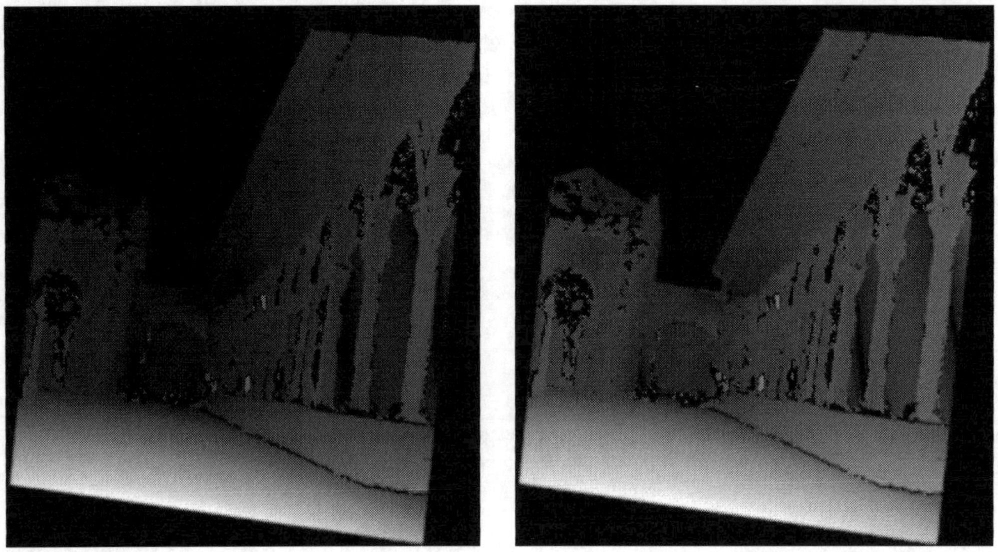

(a) (b)

Figure 7: (a): Disparity map obtained with the proposed method for the stereo pair of Fig. 1a. (b): The same after contrast enhancement, that emphasizes depth details. (Better viewed on a computer screen.)

This final dense disparity map can be used to achieve a 3D reconstruction of the scene. At the current state of the application, we limited our 3D output to a projective reconstruction, since calibration data (e.g. focal lengths) of the historical stereo pairs have not been estimated. Nevertheless, this 3D reconstruction it is already sufficient to appreciate the different depths of the scene, even if perspectively distorted (see Fig. 8).

5. Conclusions and Future Work

In this paper we presented StereoDisparity App, a computer vision-based graphical user interface to compute disparity maps and projective 3D reconstructions from stereo photographs, through limited manual user interaction. StereoDisparity can be used to work on historical images, that are practically intractable with automatic state-of-the-art stereo methods.

Future work will focus at first on obtaining a full metric 3D reconstruction. Camera intrinsics are missing for the historical photos: We will include in our StereoDisparity App a calibration tool exploiting self-calibration techniques, so that a metric 3D model can be finally obtained. Using such a reconstruction we will be able to register the old photos with the modern scene. This can be used, for instance, to implement augmented reality applications that superimpose the historical appearance to the contemporary scene.

Acknowledgements

This work has been carried out during the TRAVIS (Tecniche di Realtà Aumentata per la Visualizzazione di Immagini Storiche) project supported by Fondazione Ente Cassa di Risparmio di Firenze, within the call "Giovani Ricercatori Protagonisti" (University of Florence,

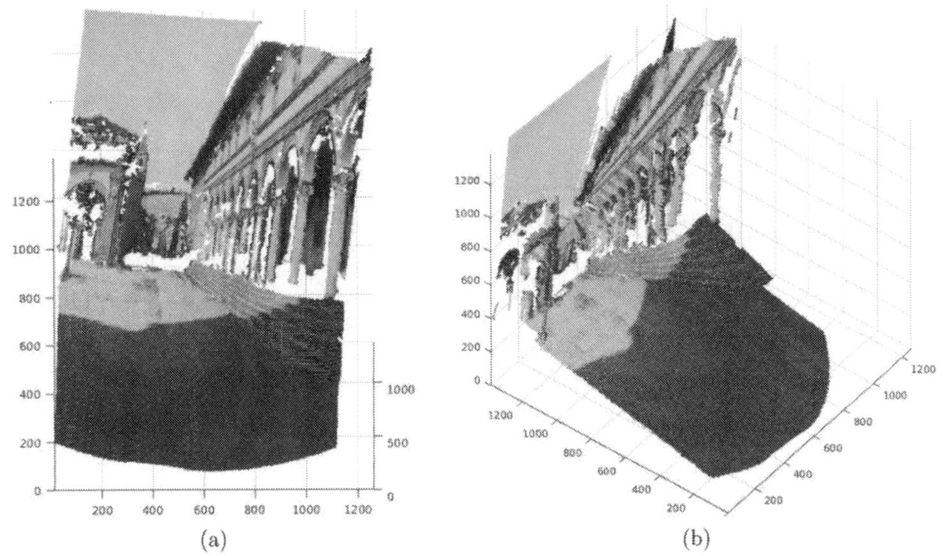

Figure 8: A 3D perspective reconstruction obtained from the enhanced disparity map.

D.R. N. 127804 (1206), 01/10/2015). We would like to thank Mrs Anna Pelleri Hautmann for allowing the reproduction of the photos in this paper. We also thank Drs. Costanza Caraffa and Ute Dercks at Photothek des Kunsthistorischen Instituts in Florenz—Max-Planck-Institut, for their kind support. Digital scans: ©Stefano Fancelli/KHI.

References

[1] Taniai T, Matsushita Y, Sato Y and Naemura T 2017 *IEEE Transactions on Pattern Analysis and Machine Intelligence (TPAMI)* (accepted) URL `https://github.com/t-taniai/LocalExpStereo`
[2] Dzwierzynska J 2017 *IOP Conference Series: Materials Science and Engineering* **245** 062015
[3] Bitelli G, Dellapasqua M, Girelli V A, Sbaraglia S and Tinia M A 2017 *ISPRS - International Archives of the Photogrammetry, Remote Sensing and Spatial Information Sciences* **XLII-5/W1** 113–119 URL `https://www.int-arch-photogramm-remote-sens-spatial-inf-sci.net/XLII-5-W1/113/2017/`
[4] Bräuer-Burchardt C and Voss K 2001 *Virtual and Augmented Architecture (VAA'01)* (Springer London) pp 159–170
[5] Maiwald F, Vietze T, Schneider D, Henze F, Münster S and Niebling F 2017 *ISPRS - International Archives of the Photogrammetry, Remote Sensing and Spatial Information Sciences* 447–452
[6] Schindler G and Dellaert F 2012 **7**
[7] Scharstein D and Szeliski R 2002 *International Journal of Computer Vision* **47** 7–42 URL `http://vision.middlebury.edu/stereo/`
[8] Fanelli G 1999 *Firenze in Stereoscopia* (Octavo Editore)
[9] Hartley R I and Zisserman A 2004 *Multiple View Geometry in Computer Vision* 2nd ed (Cambridge University Press, ISBN: 0521540518)
[10] Szeliski R 2010 *Computer Vision: Algorithms and Applications* 1st ed (New York, NY, USA: Springer-Verlag New York, Inc.) ISBN 1848829345, 9781848829343
[11] Fusiello A and Irsara L 2008 *19th International Conference on Pattern Recognition* pp 1–4 ISSN 1051-4651
[12] Reinhard E, Adhikhmin M, Gooch B and Shirley P 2001 *IEEE Computer Graphics and Applications* **21** 34–41
[13] Katkovnik V, Foi A, Egiazarian K and Astola J 2010 *Int. J. Comput. Vision* **86** 1–32
[14] Meer P and Georgescu B 2001 *IEEE Transactions on Pattern Analysis and Machine Intelligence* **23** 1351–1365
[15] Christoudias C M, Georgescu B and Meer P 2002 *Object recognition supported by user interaction for service robots* vol 4 pp 150–155
[16] Comaniciu D and Meer P 2002 *IEEE Transactions on Pattern Analysis and Machine Intelligence* **24** 603–619

IOP Conf. Series: Materials Science and Engineering **364** (2018) 012045 doi:10.1088/1757-899X/364/1/012045

A data mining approach for indoor air assessment, an alternative tool for cultural heritage conservation

D Leyva Pernia[1,4,5], S Demeyer[1], O Schalm[2], W Anaf [3]

[1] Department of Mathematics - Computer Science, University of Antwerp, Middelheimlaan 1, 2020 Antwerp, Belgium
[2] Conservation Studies, University of Antwerp, Blindestraat 9, 2000 Antwerp, Belgium
[3] War Heritage Institute, Renaissancelaan 30, B-1000 Brussels, Belgium
[4] Department of Physics, CEADEN, 502, Calle 30, 11300 Havana, Cuba

[5] Diana.LeyvaPernia@uantwerpen.be

Abstract. The exposure of cultural heritage to the environment has a significant impact on its degradation process and degradation rate. Consequently, managing the indoor air quality is vital to minimize further damage to historical artefacts and works of art. Despite its potential impact, the traditional assessment of the indoor air quality still represents a challenge for most collection guardians. This approach typically relays on the comparison of measured environmental parameters and corresponding acceptable values. However, determining the acceptable values and relative importance of the different environmental parameters turns out to be quite complex since it depends on the material types present in the collection and their preservation state. Furthermore, the significant amount of data generated during the measurements hampers the application of traditional methods of analysis.
Considering all these, we propose the use of data mining as an alternative method for the indoor air quality assessment in cultural heritage studies. Data mining can provide knowledge from vast volumes of heterogeneous data, through high-speed processing, detection, and analysis. Here we present its application to identify dynamics and patterns affecting the indoor air quality in a realistic case. Using data from a measuring campaign held at a late Gothic church in Belgium, we show that inappropriate periods can be identified without using standards. In addition, different types of periods can be identified by studying the relation between multiple parameters. For that we use the k-means clustering method, interpreting the results with both visual and statistical tools.

1. Introduction

Several studies have already demonstrated how the exposure to an aggressive environment has a significant impact on the degradation process and degradation rate of cultural heritage objects [1]. Based on these findings heritage guardians and institutions have now an increasing interest in managing the indoor air quality and provide stable environmental conditions to minimize, as much as possible, the occurrence of further damage to historical or valuable artefacts on their collections. However, monitoring and evaluating relevant environmental data is a challenging task for most heritage guardians, mainly due to the considerable amount of information generated, and the uncertainties behind the selection of the acceptable environmental values.

Continuous indoor air quality monitoring implies the collection of vast amounts of data that can quickly become overwhelming for specialists with limited experience in data science. This is aggravated when an increasing number of environmental parameters are analysed. Temperature and relative

IOP Conf. Series: Materials Science and Engineering **364** (2018) 012045 doi:10.1088/1757-899X/364/1/012045

humidity are the most common parameters monitored by the heritage community, but other parameters such as visible light, UV radiation or pollution also have a remarkable influence on the degradation rate of most objects. With the currently available technology these parameters can be monitored as well, and they should be considered to achieve an accurate assessment [2].

Traditional assessments usually rely on the comparison of measured environmental parameters with their corresponding acceptable values. Unfortunately, the target values are not precisely known, and they depend on variables such as the material type, preservation state, etc. This entails a strong dependency on the selected guidelines that define acceptable environmental values, and will be reflected in the accuracy of the analysis [3].

Given the preceding, we propose a complementary approach to aid indoor air assessment for heritage conservation based on the implementation of Big-Data related techniques, such as data mining and data analytics. The goal of these techniques is to extract relevant knowledge from data. In our specific context we focus on interesting patterns or atypical behaviours [4]. A similar approach has already been successfully implemented in indoor environment quality assessment for human health and comfort [5]. However, the application of this method for a comprehensive environmental characterization is still uncommon in the field of cultural heritage studies.

We present here our results from the implementation of the techniques mentioned above for recognizing inappropriate periods in the absence of standards or guidelines, and identifying dynamics or patterns affecting the indoor air quality in a specific case study.

2. Materials and Methods

2.1. *Case study*

To evaluate the viability of our approach, we analysed the data gathered from 23/07/2017 to 23/10/2017 in a measuring campaign held at a late Gothic church in the centre of a small Belgian city. The environmental information was collected with an innovative multi-sensor tool that registered every 15 minutes the values of temperature, relative humidity, illuminance, particulate matter (PM), nitrogen dioxide (NO_2) and ozone (O_3) concentrations among other parameters [2]. This resulted in a data matrix of 8928 measuring points (matrix rows) and 7 measurements per point including the above mentioned parameters (matrix columns). In addition, every measuring point was labelled with a timestamp and the corresponding weekday number in which the measurements were performed (e.g., Monday = 1).

2.2. *Identifying periods of elevated risk*

Our method for the identification of periods of elevated risk is based on three main suppositions.

1. A stable climate is more conducive for heritage conservation, especially in the case of temperature and relative humidity acting over hygroscopic materials [6].
2. Keeping the light exposure and the concentration of pollutants as low as reasonable achievable minimises the risk of material degradation [7].
3. Short-time fluctuations will not necessarily affect all objects, since the reaction time of the piece must be shorter that the fluctuation in order to have a potential impact on its conservation state. [8]

Considering the third supposition, we start the analysis by filtering out the short-time fluctuations. There are several classes of filtering methods, but the most widely used apply statistical measures over moving windows [9]. The mean value is often used as the statistical measure for the data in the window. Consequently, a moving mean is calculated creating series of averages of different subsets as the window shifts over the full data set.

Moving means are ordinarily used to smooth out short-term fluctuations and highlight longer-term trends or cycles in time series data. In our case, the use of the moving mean method filters out the fluctuations caused by the day and night variations exposing the regular seasonal variations or the anomalous behaviour that occurred during periods of elevated risk.

IOP Conf. Series: Materials Science and Engineering **364** (2018) 012045 doi:10.1088/1757-899X/364/1/012045

Considering now the first and second suppositions we can establish that the periods where the filtered data presented the higher fluctuations were the periods most likely to be linked to an elevated risk of degradation. For temperature and relative humidity, the relevant fluctuations can be found above or below the mean value, while for the rest of the environmental parameters, only the fluctuations above the mean would be noteworthy.

2.3. Clustering data, k-means clustering

Clustering is an unsupervised learning method that assigns labels to objects in unlabelled data that can be implemented by different types of algorithms [9]. For our research, we selected the k-means clustering, a data-partitioning algorithm based on iterative minimization of the sum-of-squares criterion [10].

Clustering validation was performed with two different methods: the Davies-Bouldin index, and Silhouette value. The Davies-Bouldin criterion is essentially a ratio of within-cluster and between-cluster distances. Based on its analysis the optimal clustering solution has the smallest Davies-Bouldin index value [11]. The Silhouette value is a measure of how similar a point is to points in its own cluster when compared to points in other clusters [12]. This value ranges from -1 to +1, where a high silhouette value indicates that the point is well-matched to its cluster and poorly-matched to neighbouring clusters.

3. Results and Discussion

The complete data set studied is presented in Figure 1. As mentioned in the previous section, seven different environmental parameters were monitored every 15 minutes during 93 days, resulting in data streams of 8928 elements each. This data set cannot be regarded as considerably large when compared to a typical climate characterization campaign that should cover, at least, one full year of data. However, it is more than enough to hinder a prelaminar assessment through visual inspection. There are certainly patterns in the data, but there is too much information to get a clear grasp of them. One characteristic that all the parameters share is the unremitting presence of short time fluctuations, mainly related to the day-night cycle.

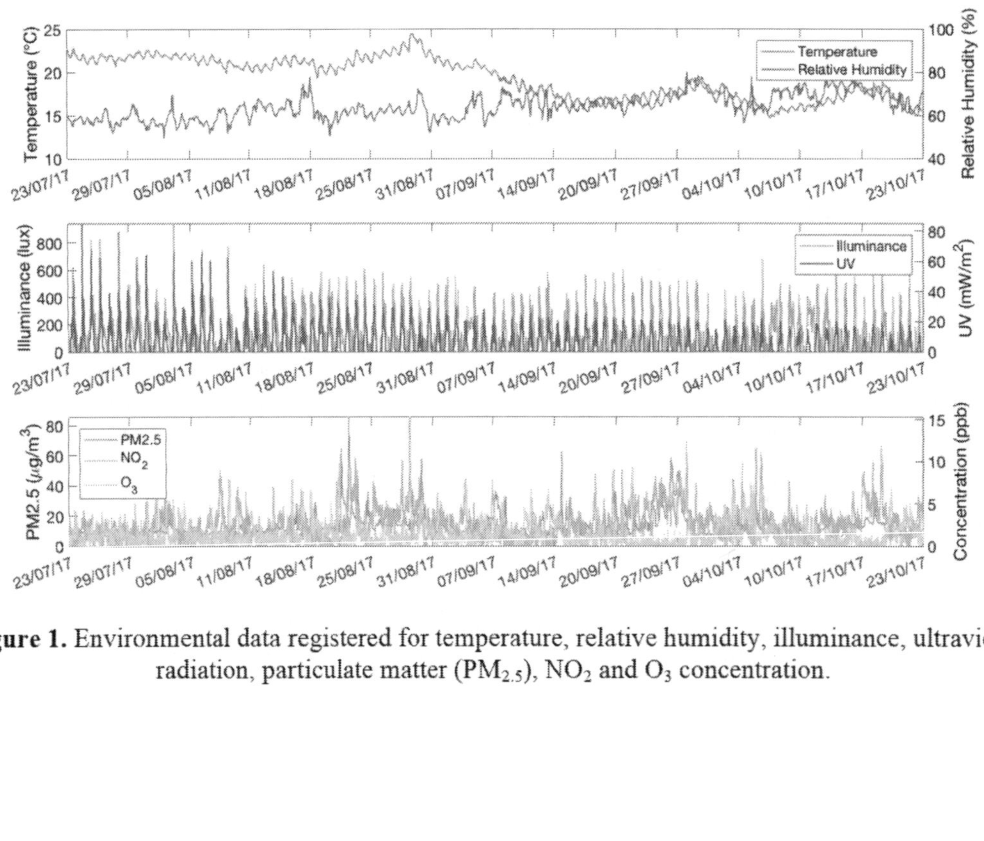

Figure 1. Environmental data registered for temperature, relative humidity, illuminance, ultraviolet radiation, particulate matter ($PM_{2.5}$), NO_2 and O_3 concentration.

In order to filter out the daily fluctuations we calculated the moving mean of the parameters with a symmetric window of 96 data points, which covers a period of 24 hours in the time scale of our case study. The result of its application on the case of relative humidity is presented in figure 2 (a). In this figure we plotted the measured data points for relative humidity, the mean value over the complete period analysed, the confidence bars delimited by the standard deviation (std) and the correspondent moving mean. By filtering higher frequency fluctuations, the moving mean shows now a gradual, but still appreciable overall increment during the studied period, which match the expectations for the seasonal transition from summer to autumn.

(a)

(b)

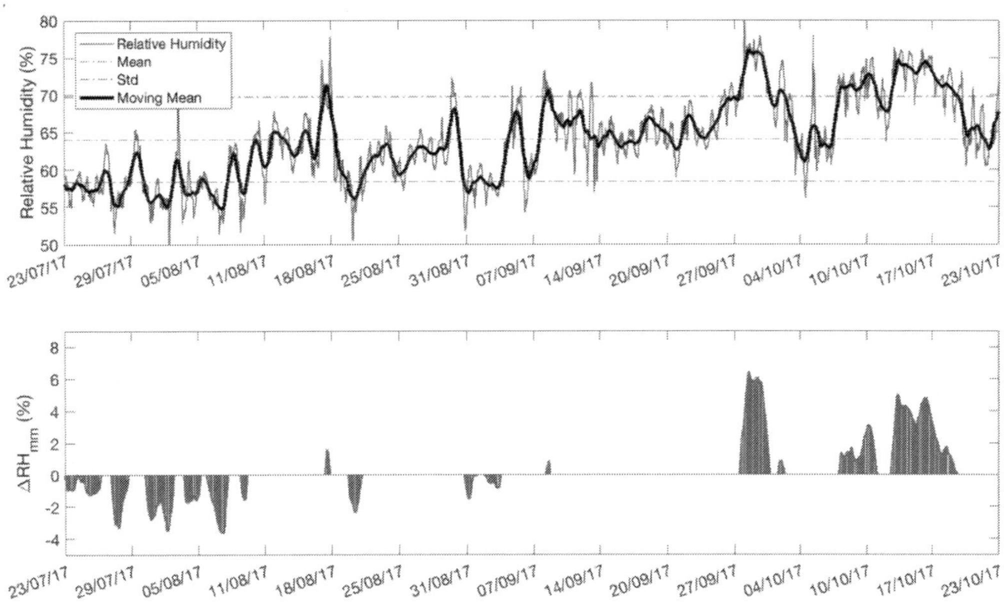

Figure 2. (a) Data registered for relative humidity, its mean value over the time analysed, confidence bar of ± the standard deviation, and moving mean for a symmetrical window of 24 hours. (b) Difference between the values of the moving mean outside of the confidence bar.

In figure 2(a) we can notice periods where the moving mean reached extreme values crossing the confidence bar. These periods are the most deviated from the typical behaviour, and consequently, more likely to be linked to unusual events or even to potentially dangerous occurrences. The magnitude of the difference between these points and the standard deviation can be used as a metric to characterize their anomaly. Plotting this magnitude would filter out the periods where the environmental parameter exhibits a standard behaviour, and therefore, directly guiding the analysis to periods of elevated risk. This analysis is illustrated in figure 2(b), where the difference between the values of the moving mean outside of the confidence bar and the standard deviation (ΔRH_{mm}) are plotted as a function of time.

This analysis facilitates the detection of extreme periods and can be implemented for all the environmental parameters on our dataset. Figure 3 shows the normalised values of these differences between the moving means and the corresponding confidence bars (y-axis) as a function of time for all parameters. The normalization was performed to establish an unbiased comparison unrelated to the measuring units of the parameters. For each case, the higher the difference value, the more atypical is the corresponding period. Due to the significant amount of data filtered out, it is still possible to identify the atypical episodes clearly, even when showing the information of all the parameters in the same graph.

IOP Conf. Series: Materials Science and Engineering **364** (2018) 012045 doi:10.1088/1757-899X/364/1/012045

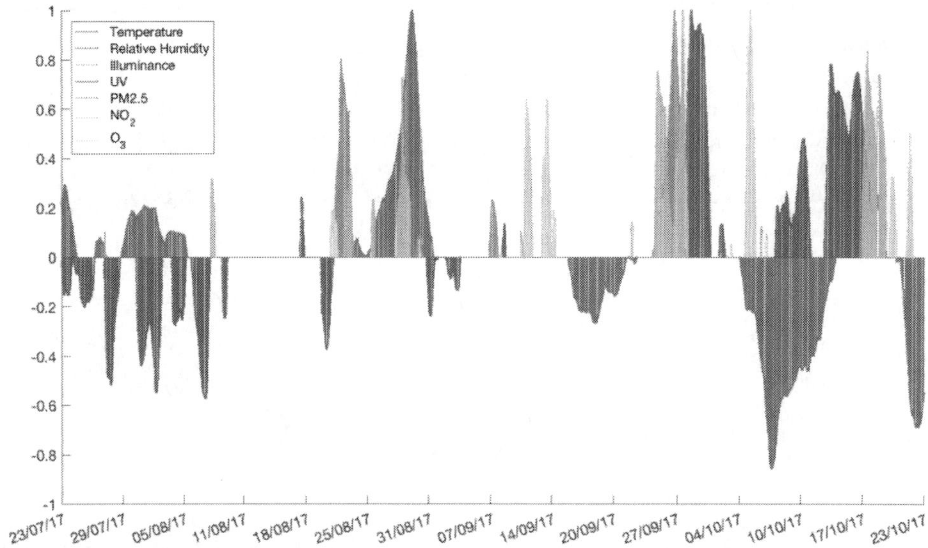

Figure 3. Normalised differences between the moving means of each environmental parameter and the corresponding confidence bars, identifying atypical periods.

For the implementation of the k-means clustering method, we analysed the entire data matrix of environmental data collected. Each column was normalised by rescaling in the range from 0 to 1 before applying the clustering method to avoid assigning artificial weights to the data. The clustering algorithm was implemented for k values from 2 to 15, each done in 5 separated runs. The resulting 13 cluster distributions were evaluated using the Davies-Bouldin and Silhouette criteria, and a detailed inspection of their characteristics. Based on these, we selected k=4 as the most relevant distribution for classifying our data set.

Table 1. Cluster interpretation data.

		Day	Temperature (°C)	RH (%)	Illuminance (lux)	UV (mW/m^2)	Pm2.5 ($\mu g/m^3$)	NO_2 (ppb)	O_3 (ppb)
Cluster 1	min	1	19.153	51.436	0.000	0.000	5.153	0.000	0.000
	max	7	24.282	74.018	357.674	22.470	64.358	15.323	7.999
	mean	**3.862**	**21.244**	**61.026**	**40.040**	**1.935**	**12.406**	**2.754**	**1.165**
Cluster 2	min	1	14.646	56.147	0.000	0.000	2.973	0.000	0.000
	max	7	21.237	79.006	680.531	25.351	64.605	11.779	7.358
	mean	**4.045**	**17.205**	**65.755**	**236.407**	**8.743**	**14.333**	**2.355**	**1.613**
Cluster 3	min	1	14.647	60.476	0.000	0.000	4.024	0.000	0.000
	max	7	20.950	80.067	279.758	18.022	62.399	12.423	7.389
	mean	**4.063**	**17.008**	**69.350**	**30.157**	**1.148**	**11.937**	**3.086**	**1.146**
Cluster 4	min	1	18.163	49.346	0.000	0.000	6.164	0.000	0.000
	max	7	24.462	74.655	942.825	85.150	85.777	10.240	9.895
	mean	**4.088**	**21.646**	**59.063**	**284.142**	**17.988**	**12.891**	**1.461**	**1.614**

The data interpretation of each cluster is presented in Table 1. We found two sets of patterns that provide physical meaning to the cluster distribution. One pattern is defined by the presence of lower values of illuminance and ultraviolet radiation (clusters 1 and 3) or higher values of these two

IOP Conf. Series: Materials Science and Engineering **364** (2018) 012045 doi:10.1088/1757-899X/364/1/012045

environmental parameters (clusters 2 and 4). The other pattern is set by the anti-correlation of temperature and relative humidity. Clusters 1 and 4 hold higher temperature and lower relative humidity (summer), while clusters 2 and 3 are the opposite (autumn).

Plotting the temporal distribution of the 4 clusters, figure 4(a), confirms our hypothesis on the seasonal influence. Comparing the graphs in figure 4(a) and 4(b) we can confirm that clusters 1 and 4 comprehend the periods where the temperature was above the mean value, while the temperatures in clusters 2 and 3 where below the average. From figure 4(a) we can also conclude that the distribution based on the illuminance and UV levels does not exactly reflect the day-night variations. The number of days during the analysed period does not match the number of repetitions of the clusters, therefore the difference comes from the distribution of brighter days (clusters 2 and 4) and nights or darker days (clusters 1 and 3).

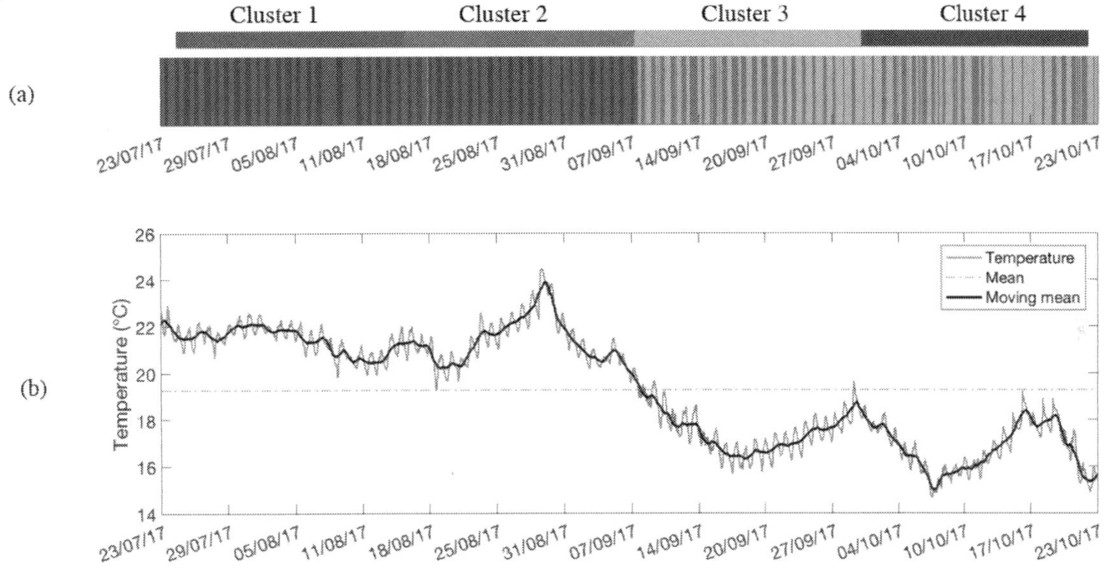

Figure 4. (a) Cluster distribution over time and (b) temperature distribution for comparison.

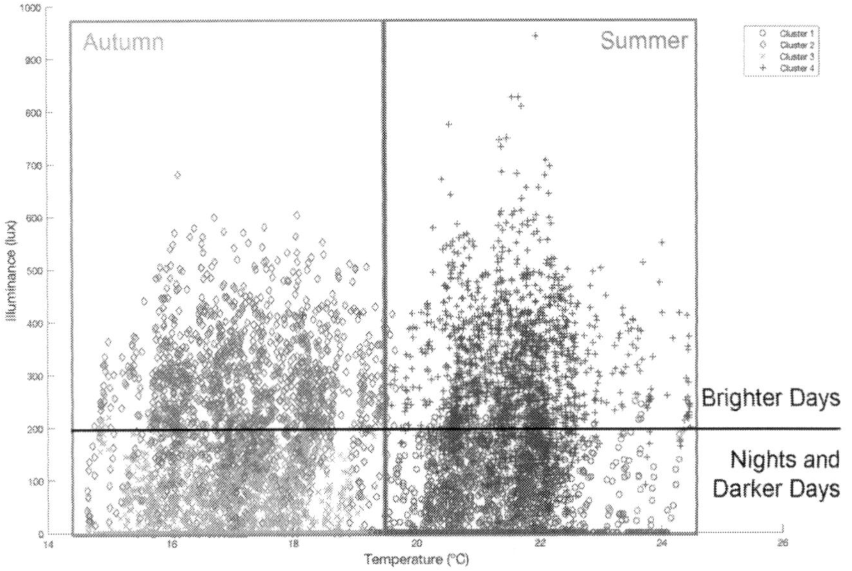

Figure 5. Cluster distribution for the plane temperature vs illuminance.

These distributions are also recognisable by plotting the clusters in the plane of temperature and illuminance (figure 5). In this graph the clusters can be found in 4 different quadrants representing nights and darker days of summer (cluster 1), brighter days of summer (cluster 4), brighter days of autumn (cluster 2), and nights and darker days of the same season (cluster 3); confirming our previous interpretation.

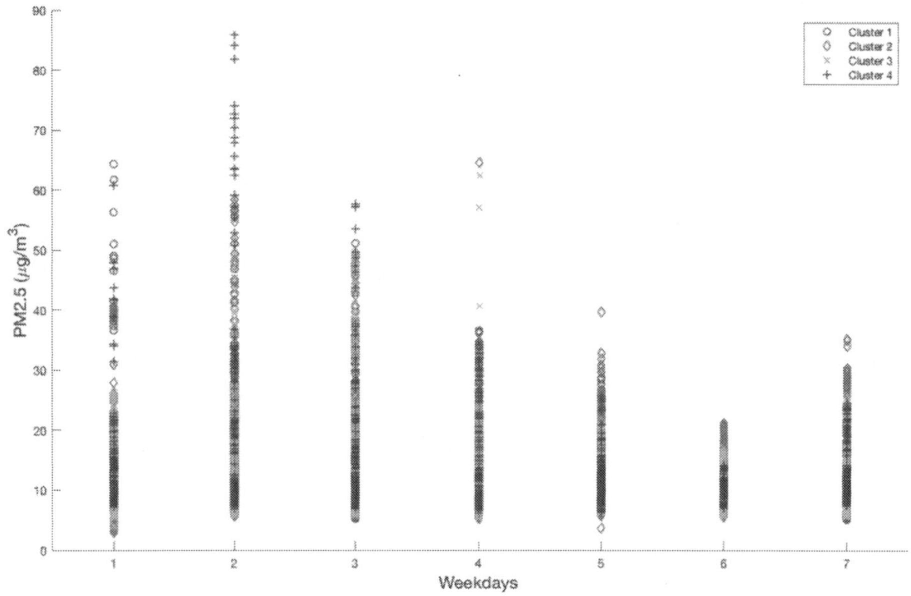

Figure 6. Cluster distribution regarding particulate matter and corresponding weekdays.

The analysis of the cluster distributions over different axes combinations can provide information about patterns in the environmental conditions. An example of this is presented in figure 6, where we plot the behaviour of particulate matter for each cluster as a function of the weekday number. In this figure is observed that, independently of the season, the lowest values of particulate matter were reached during the weekends, especially on Saturdays. In contrast, the higher concentrations, linked to periods of elevated risk, were reached on Tuesdays during brighter summer days (cluster 4). The source of this increment in the particulate matter concentration is most likely related to the construction works going on during the period. However, we cannot discard the influence of other sources, such as heavier traffic on workdays compared to weekends; or different internal sources, like a larger number of visitants during summer holidays. Nevertheless, being able to identify that this particular behaviour is connected to specific weather conditions can facilitate the selection of mitigation actions and alert heritage guardians about the potential risk during periods with similar characteristics.

4. Conclusions

This study revealed that, on the basis of three simple suppositions, it is possible to identify periods of elevated risk without the use of any specific guideline. The proposed approach detects periods of sudden change that deviates from the standard behaviour and filters out the information regarding stable periods. Despite this, the method would fail for periods of constant inappropriate environmental conditions. For that reason, we consider it as a complementary method, and not a substitution, to the use of norms or guidelines. However, in the cases where the application of direct guidelines is not a viable option, heritage guardians could still use our method to propose mitigation actions that would reduce the risk of degradation during the most extreme periods.

Our study also showed that the extended data matrix of environmental measurements does contain different types of patterns superposed on top of each other. By using clustering techniques, such patterns become easier to detect and interpret, facilitating the study of potential hazards.

Acknowledgements
The authors thank the financial support of the Belgian Federal Public Planning Service Science Policy (BELSPO) under project number BR/132/A6/AIRCHECQ.

References

[1] Thomson G 1986 *The Museum Environment*. Second edition ed.; Elsevier, Butterworth Heinemann: Oxford.
Caple C 2011 *Preventive Conservation in Museums*. Routledge, Taylor&Francis Group: London.
[2] Schalm O, Anaf W, Callier J and Leyva Pernia D 2018 New generation monitoring devices for heritage guardians to detect multiple events and hazards. *Proc. Int. Conf. The Future of Heritage Science and Technologies*, (Florence: Florence International Fair for Art and Restoration).
[3] Massimo A, Coppola F and Pavlovic A 2015 Application of the quality norms to the monitoring and the preventive conservation analysis of the cultural heritage. *International Journal for Quality Research*, 9: 299-308.
Leyva Pernia D, Demeyer S, Schalm O, Anaf W and Meert C 2016 New approach to indoors air quality assessment for cultural heritage conservation, *Proc. 14th Int. Conf. on Indoor Air Quality and Climate*, (Ghent: International Society of Indoor Air Quality and Climate); pp 490-497.
[4] Fayyad U M, Piatetsky-Shapiro G, Smyth P and Uthurusamy R 1996 *Advances in Knowledge Discovery and Data Mining*. AAAI Press, Menlo Park.
[5] Saarikoski M 2016 *A data mining approach to indoor environment quality assessment: A study on five detached houses in Finland*. MSc Thesis. Environmental Science. University of Eastern Finland; Department of Environmental and Biological Sciences.
[6] EN 15757:2010 *Conservation of Cultural Property – Specifications for Temperature and Relative Humidity to Limit Climate-Induced Mechanical Damage in Organic Hygroscopic Materials*, European Committee for Standardization, Brussels.
[7] Michalski S 1997 *The lighting decision*. In Textile Symposium 97, 97-104. Ottawa, Canada: Government of Canada.
ASHRAE 2011 Museums, Galleries, Archives, and Libraries. In *ASHRAE Handbook - Heating, Ventilating, and Air-Conditioning Applications (I-P Edition)*, American Society of Heating, Refrigerating and Air-Conditioning Engineers.
[8] Michalski S 1993 *Relative Humidity: A Discussion of Correct/Incorrect Values*. In ICOM Committee for Conservation.
[9] Runkler T A 2016 *Data Analytics*. Springer, Wiesbaden.
[10] Davies D L and Bouldin D W 1979 A Cluster Separation Measure. *IEEE Transactions on Pattern Analysis and Machine Intelligence*. Vol. PAMI-1, No. 2, pp. 224–227.
[11] Lloyd S 1982 Least Squares Quantization in PCM. *IEEE Transactions on Information Theory*. Vol. 28, pp. 129–137.
[12] Rouseeuw P J 1987 Silhouettes: a graphical aid to the interpretation and validation of cluster analysis. *Journal of Computational and Applied Mathematics*. Vol. 20, No. 1, pp. 53–65.

IOP Conf. Series: Materials Science and Engineering **364** (2018) 012046 doi:10.1088/1757-899X/364/1/012046

AbC Art blog Conservation

Blog Magazine about conservation and restoration of cultural Heritage

Marta Gómez Ubierna*, Eleonora Vittorini Orgeas**

* Freelance conservator, Via di Montefiesole 68, 50065 Pontassieve (Florence, Italy)

** Freelance conservator and scientist, Via Gianfrancesco Pagnini 23, 50134 Firenze (Florence, Italy)

* e-mail: gomubi@gmail.com

** e-mail: eleonora@vittoriniorgeas.it

Abstract. Recently, professionals in conservation have recognized the need to reach out to the public, or layperson, to communicate the contents and modalities of their work in order to create a more accessible culture.

At present, different forms of communication are used, and blogs have been identified as a very powerful tool, especially abroad. Furthermore, blogs offer a great opportunity to attract a broad audience of readers, and to invite commentary along with question and answer.

With the goal of forwarding shared knowledge across the conservation community and enhancing communication within the field we have created *AbC*, a conservation blog open to the public. *AbC* aims to be a web platform written in Italian, Spanish and English, with a multidisciplinary approach to the themes of conservation, restoration, and diagnostic investigation of artworks. It has been conceived as a popular and up-to-date heritage conservation blog that will to bring these themes into the center of public attention through the presentation of restorations with an understandable language and a captivating style through articles, photos, videos and other illustrations. The platform provides a *section* to inform readers on restoration and conservation works especially through the perspective of the work in progress which includes interviews and contributions by conservation professionals and a *training section* for keeping restorers technically up-to-date - beyond teaching to the common public on topics related to restoration, diagnostic and conservation of heritage.

In order to be one of the principal web platforms dedicated to conservation, restoration and diagnostic world of cultural heritage, *AbC Blog* takes advantage of collaboration with one of the most important magazines in conservation, *Kermes*. On the home of web-site of *Kermes, AbC* will be linked by a banner for an immediate and favourite connection. The blog themes will be complementary to web-site contents, proposing to have future links to other magazine web journals.

IOP Conf. Series: Materials Science and Engineering **364** (2018) 012046 doi:10.1088/1757-899X/364/1/012046

1. Introduction

In the digitization culture[1], museums and institutions are committed to satisfy the public's growing curiosity about art conservation using information technology.

A current recognition of the rights of the so-called "digital audience" has led to the development of technologies for content management and communication, in addition to the traditional website. A basic digital dialogue is formed by social media or communication platforms[2], through which museums and institutions can globally show new aspects of their content[3]. The Tate in London and the Metropolitan's digital departments, indeed, handle their own social media to engage a digital audience with their proposed culture events, encouraging a direct and immediate dialogue.

Global networks of heritage sites and museums are trying to involve communities also in the necessities of art conservation. Together with traditional programs, such as visits to the laboratories, other experiences have been created to exploit digital technologies. A recent example is the Monet exhibition, during which visitors could visualize the restoration process through a query-code[4]. Most of these initiatives highlight the fact that the public repeatedly requests more information on collection care and conservation researches[5].

The main question is how developments in Internet resources can communicate conservation issues? Due to the fact that most conservation research happens behind the scenes, the job of conservators needs to include more effort in making their activities and results more visible to the general audience, which should be able to know how the conservation process takes place in the laboratory[6].

2. Blogs and Conservation

Blogs have become an important popular tool for dissemination of information and raising public interest. Thanks to the recent acceptance of blogs as a widespread vehicle for web communication, the number of blogs related to cultural heritage has increased rapidly over the past few years[7].

A "blog"[8] is a web document created by software that allows material to be published on a web site in the same manner as log – or diary – entries are written in a journal. They are a format that emerged

[1] "Digital culture" is identified with a universe of communication technologies in which audience has an active role, thanks to which, information becomes bidirectional, quick and flexible.
[2] The use of web 2.0 technologies facilitate publication, edition or exchange of information created, also, by the users. Within this universe we can find collaborative tools (wikis, blogs), social networks (Facebook, Linkdln, Tuenti,...), microblogging (Twitter, Tumblr) or multimedia sharing (Flickr, Picasa, Youtube, Vimeo,...).
[3] The role of social media may have a positive impact in the democratisation of heritage but it requires resources, expertise and training (Taylor, 2017, p. 408).
[4] This experience has led to an interesting debate about increasing conservation curiosity (Gries, 2016, pp. 26-27).
[5] The positive impact of share information about conservation researches is highlighted in the project "Change or Damage?". The project allowing the public to feel involved and engaged in the research project of preventive conservation of house's museums in order to understand how environmental monitoring is conducted and the benefits for collections (Luxford, 2013, p. 66).
[6] Olcott, 2013, p. 15.
[7] A research on the web shows 257,000.000 "culture blogs" focus on art, technology, music, literature, etc.
[8] The term "blog" originated as an abbreviation of "web log".

in the 1990's - websites with chronological sorted entries with narratives built from text, images, videos, audio and links.

The more basic form of institutional blog is strictly a forum posting official communications, such as press or general information about hours, exhibits, events. Archives and historical museums are using blogs to share information about their collections and activities, highlight processing projects, publish date-based archival content and, in general, to publish news about events promoted by the institution.

The flexibility of blogging is used today to create innovative approaches for sharing information or to produce culture[9]. Processing blogs are a sort of outreach tool that share information about what experts find as they are processing a work. This capacity is used by University courses on art conservation which have incorporated "academic blogs" as a valuable source of shared and narrated projects carried out by students[10]. Several international museums are investing in blogs with more personality – for example – having staff write under their own names and including entries that discuss new acquisitions, often featuring digital images of the artworks[11]. Most of them convey restorations, which contain numerous and detail information with images or videos concerning conservation work, together with historic and artistic details on how they were carried out as a sort of a life-story able to attract a wide curiosity[12].

Thanks to these initiatives, arising above all within international museums, social media have acquired a sense of seriousness and authority as a highly qualitative and reliable means of information[13].

Based on the foundation provided by these examples, our initiative attempts to venture outside the laboratory and bring information on conservation to the public.

3. AbC project: toward a conservation web platform

The field of art conservation is one of the most interesting areas to develop a digital dialogue with new audiences in the field of cultural heritage. This is because reporting on conservation studies can reveal new and interesting information about an object. At the same time, object based narratives work best because a detailed and visual account of the process is more revealing than a mere description of the results.

---

[9] Dickinson College's blog and UMamot's blog are used to publish summaries of requests received and it serves for generating statistics about researches. Other experiences are based on a kind of "showroom blog" (Redondo, 2013, p. 35).

[10] Sunara, 2014 (online article).

[11] The Tate's blog, *Tate staff: behind the scenes,* offers first-hand news about acquisitions, interpretation of artworks, conservation or publication (www.tate.org.uk).

[12] *Collections Insights* is one of the Metropolitan of New York blogs aims to offer new perspectives on works in the collection, including restoration works (https://metmuseum.org/blogs/collection-insights). MoMA's website is equipped with a blog page - *Inside/Out* - within the section *Behind the scenes* presents conservation project on artworks of the collection (https://moma.org/explore/inside_out/category/conservation). More specific on conservation topics is the blog *Conservation*, of the Victoria and Albert Museum (www.vam.ac.uk/blog/section/conservation-blog), which present contributions from all departments to highlight current projects and researches. In addition, conservation staff offer and interpretation of the deterioration processes.

[13] About development of web technologies uses in the field of conservation, consult O'Connor, 2010, pp. 8-11.

IOP Conf. Series: Materials Science and Engineering **364** (2018) 012046 doi:10.1088/1757-899X/364/1/012046

With this purpose in mind, *AbC Art blog Conservation* is a kind of ongoing focus on artworks to provide conservation news, written with a closer and more informal style with an engaged first-person tone, in order to reach all kinds of audiences. Texts are short and include graphical documentation: photos, audio, videos and/or drawings. The blog presents a balance of different types of posts, classified into categories within specific sections and subsections.

The first section, entitled "News on Conservation", aims to address current topics on conservation news, from projects and studies in-progress to ongoing restorations and scientific plans. This section is divided in two specific subsections. The first one, “In the frontline”, where posts have an artworks-based approach that provides a more thorough explanation of conservation concepts and aspects such as documentation, collaborations or preventive conservation as well as protection policies for cultural heritage. The second one, “Share your work” is configured as a sort of "host structure" for restorers, that can use the blog as a platform to establish a more direct and continuous dialog with their visitors and gather opinions from a wider group of people. Posts on conservation treatments are presented in a chronological sequence such as a timeline with a sliding sequence of images, which are able to illustrate the procedural character of restoration. The audience can see the process behind the solution in order to acquire a high degree of understanding and perceive the excitement restorers feel during their work.

By integrating a multi-disciplinary approach, museum websites have yielded impressive results in generation audience satisfaction and positive publicity. For that reason, curators, educational experts and scientists have a space in another subsection dedicated to them, “Interview’s corner” where they are invited to talk about the history of the artworks, their concerns and expectations, or conservation issues of own collections. For this purpose, interviews of experts such as academic professors, conservation scientists or museum curators are included in this section to complete and deepen the conservation posts with their specific and professional point of view.

In addition, the blog contains a training section (“Focus On”) that include a subsection entitled “How we do it...?”, that provides didactic tools to illustrate the specific concepts or procedures of conservation and restoration treatments (laser using, consolidation and cleaning treatments, pictorial retouching, microclimatic monitoring etc.), experiments (cleaning tests, water absorption tests, scotch-tape test, evaluation of biological techniques for cleaning and consolidating, etc.) or scientific analyses (photo-documentation, monitoring, physico chemical analyses carried out in the lab or in situ, etc.). These are written as short progress reports to provide an example of how restorers and scientists conduct experiments and tests for conservation issues. For this purpose, articles written by experts in scientific fields are included. In this section, information is present in a simple and immediate way, including videos, photos or drawings with a clear and informal language. Articles aim not to be academic lessons, but rather educational posts with short and visual language as to be as attractive and light as possible a wider public audience. By reading posts, you can move from the explanation of basic topics, well known by experts, but of extreme curiosity for the general public, up to the presentation of conservation materials and methods of the last generation that mainly of importance to those working in the field of conservation. Another subsection, “Inside the lab...”, is dedicated to news originating in the laboratories and museums to highlight the technical solutions to the conservative problems regarding artworks where a different area of conservation (textiles, paper, paintings, sculptures) will periodically be the focus. The restorer's visual assessments regarding manufacturing techniques, materials and the results of analysis conducted by the scientific research and analysis laboratory are highlights of this section.

A future evolution of the blog is planned to facilitate the continued development of the technical expertise of restorers, the exchange of information, and the cross-disciplinary exchange with scientists through a “forum” for students and professionals where they can share information. The purpose of

IOP Conf. Series: Materials Science and Engineering **364** (2018) 012046 doi:10.1088/1757-899X/364/1/012046

this section will be to increase a collaborative learning and prevent the isolation of conservator by offering the blog for meeting and sharing knowledge and experience[14].

The last section, "News and events", is designed to keep conservation experts informed, both those who have been working for years, and those who have recently come to the world of restoration of cultural heritage. From cultural events dedicated to the last frontiers of restoration to scholarships or initiatives aimed at financing projects, with this section the blog will try to keep the public updated in real time on the news of the conservation world. In this way we aim to develop it also a practical tool that can be useful to experts in the most concrete way possible according to their needs.

3.1. AbC Blog structure

AbC blog is been created using a service which provides both online software and hosting, through siteground Wordpress.org[15]. The structure and template are developed by a webmaster[16].

Entries (or "posts") are classified into sections available to offer conservation and restoration news. When a new entry is published, it appears at the top of the web page, so in a reverse chronological order with a "jump" ("read more") that invites public to read the rest of the entry in another display. Each post will have identified by a title, author and data; also, in order to keep possible research, entries includes a tag (categories of information) to allow accessed by public. Another column of the web site contains news on conservation and a calendar of events related to art conservation.

Sidebar of the blog contains several sub-pages: these are used to describe the blog and its goals with a short-biography of the team, to display the list of lecture topics with assigned readings and links to references available on the internet, to present the projects and the professional literature they read.

There are many opportunities to reach new audiences through a blog. Overall, linking with social media channels (Facebook, Instagram and Youtube) provides a regular opportunity to improve visualization of the blog with brief messages and updates. Particularly, the aim is to improve the web use of the blog by exposing it to a less experienced public. In this way the *AbC* will have a strong social imprinting impact able to build a larger community[17].

Official websites of partners in the field of conservation are included in the blog. Soon, a direct link by a banner in the web-site home of *Kermes restauro, conservazione e tutela del patrimonio culturale* review will be one of the preferred channels for blog visitors[18]. This partnership with one of the most important Italian magazines in conservation of cultural heritage will allow an immediate connection with an extended public audience. This connection will take advantage of increasing the visibility and

---

[14] Recently European initiatives like *SHAKE in Conservation* have demonstrated the benefits of sharing information in the Web. It is a non-profit organization that aims mainly to connect conservators in Belgium, both students and working graduates, in all specialized fields, and is open to all heritage professionals (https://www.shakeinconservation.be/).

[15] Wordpress is the most known open-source content management software at the momento. It has an integrated user management system which allows more than one author with different administration rights to contribute blog posts.

[16] Webmaster services are in partnership with Spectra Entreprises srls (Florence, Italy) headed by Samantha Stout, Ph.D. in materials science and engineering applied to culture heritage conservation.

[17] The number of visitors of the Metropolitan museum's Facebook page has continued to grow rapidly since 2012.

[18] https://www.kermes-restauro.it.

IOP Conf. Series: Materials Science and Engineering **364** (2018) 012046 doi:10.1088/1757-899X/364/1/012046

the quality of the contents of both web magazines. A future development aims to have additional links to other conservation magazine web journals to improve the public reach of the blog.

4. Why a blog about conservation of cultural heritage?

Blogs and social media, as a part of Web 2.0, are useful internet experiences to create a community by engaging, interacting and, most importantly, creating new audiences[19]. The blog contributes to a new social practice of communication, widely available, that aims to aggregate people with common interests.

Publishing information in a chronological order make blogs very easy for most people to understand. Moreover, readers are often allowed to comment on the posts and join the discussion. As a part of the blog, the comments are a way of sharing conservation projects and their research findings. Thus, blogs can help to build social relations with readers that have a direct response on the blog posts.

Posts published in *AbC blog* are intended for several different audiences, whether for professionals in art conservation, those interested in practical treatments or analyses, or to a general public who are more interested survey posts, through which they can acquire a basic knowledge base. Most of the articles that will be published will be summaries of our visits to museums, institutions and restoration sites or laboratories, which are written in a journalistic style and can be quite interesting to a broader public audience. They will be translated in three languages: Italian, English and Spanish. Since young people especially are increasingly using Internet content to learn and share knowledge[20], we have considered technical and scientific blog posts specifically intended also for students. A kind of repository of tutorials, including test in situ, lab experiments and technical reports, to keep up to date and learn best practices.

In a future perspective we also think that reading posts that include links to other resources might motivate audience to do additional research. In this way the blog would be also functional to the knowledge distribution into the web-world being a part of a network where the sharing of ideas and information is at the service of an increasingly broad and motivated public within the field of conservation of cultural heritage. In addition, among the strengths of the blog, there is also the high quality of information so as to guide the user to specific areas and resources creating a cultural virtuous circle.

AbC blog intends to communicate the contents to the general public in an immediate, simple and above all visual way, aligning itself with the current trend of visual culture?[21]. Just as imaging has also become one of the privileged means of communication, it has changed conservation practice. One area in particular where conservators have adapted new and innovative technologies is in the use of imaging technology. Though photography has long been an important documentation tool, recent decades have seen a marked increase in the number of instruments that offer diverse ways of imaging material heritage. Various forms of imaging can not only improve the analytical evaluation of an object's materials but also reveal information about the object that is extremely interesting to the general public. The audience must be guided in the reading and interpretation of these images. In fact, blog posts are also a sort of a visual storytelling, where the images themselves can recount the story.

[19] An overview about this topic in Silberman, 2012, pp. 13-29.

[20] This trend seems not be temporary (O'Connor, 2017, p. 136).

[21] An example is the success of the "Wiki loves monuments" campaign, a competition to photograph monuments in 18 European countries started in 2011, which has inspired similar experiences (Pupe, 2017, p. 137).

Images are ordered in a specific way, either chronologically or as a series that should help the viewer understand the overarching steps of the conservation process or technique.

We believe that in a near future, the blog will become also a tool to explore how social media reframes our understanding and experience of art conservation. By examining the statistics of the site visitors, their visits and the comments, it will be possible to track the changes produced by the blog in the field of social practice and public information. Moreover, the possibility also exists to value the impact on the art network of the territories. In this sense, *AbC blog* could be a tool to increase and improve the use of cultural realities by the public at a local level. A collaboration with museums and institutions through surveys directed to the public is not excluded in the near future. These efforts would work to track and evaluate the sharing of art across the territorial network.

5. Conclusion

Although social media, making available by international museums and cultural heritage institutions, are showing that increased interest about art conservation is an irreversible process, restoration is still an exclusive topic, whose understanding is in the hands of a few specialists.

Based on this premise, our initiative aims to make conservation an accessible topic of culture heritage. The purpose is to communicate and increase understanding of the conservation issues, which contain a great deal of art historical, technical, conservation and scientific information on artworks. This makes conservation a complex matter that requires a direct a simple languages along with a stronger visual approach.

Based on the idea of "participatory culture", blogging is an opportunity to share our own knowledge more broadly and informally with a wide audience, discuss the intellectual content of the work process and contributing to tear down the physical barriers. If the public will be more conscious of the matter, in a simple and immediate way, the awareness and comprehension on the world of conservation will progressively increase. In this way, it is essential to maintain a high quality of content, in order to contribute to the circulation of high-level cultural ideas and knowledge.

Looking at the future we believe that by focusing on art conservation, it will be possible to create a community who will be able benefit and use the blog as a public service.

References

[1] Taylor, J.; Gibson, L.K., *Digitization, digital interaction and social media: embedded barriers to democratic heritage*, in *Digital Heritage and the public*, "International Journal of Heritage Studies", 2017, vol. 23, n. 5, pp. 408-420.

[2] Puppe, J.; Stadje, S., *Social Media in der Denkmalpflege*, in S. Winghar (ed.), *Oldenburg: Jahrestagung der Vereinigung der Landesdenkmalpfleger: 85. Tag für Denkmalpflege*, "Berichte zur Denkmalpflege in Niedersachsen", 2017, vol. 37, n. 2, pp. 136 -137.

[3] Gries, C., *Wie das digitale Museum restauratorisches Wissen vermittelt*, "Restauro: Zeitschrift für Restaurierung, Denkmalpflege und Museumstechnik", 2016, n. 2, pp. 26-27.

[4] Sunara, S.M., *Using Blogs to Teach and Promote Conservation*, in "CeROArt", 2014. Online

resource: http://journals.openedition.org/ceroart/4293.
[5] Redondo, M.; Maldonado, S., *La cultura se construye entre todos: el trabajo colaborativo a través del uso de blogs*. Caligrama, in C. Rodá (ed.), *Museos 2.0: hacia una sociedad conectada*, "RdM: revista de museología", 2013, n. 56, pp. 33-39.
[6] Olcott, L.; Stoehr, L., *Conservation curiosity: An Irreversible Trend*, in E. Williams (ed.), *The Public Face of Conservation*, London, Archetype, 2013, pp. 8-15.
[7] Luxford, N.; Thickett, D., *Change or Damage? Using Dissemination to Encourage Public Involvement in Conservation Research*, in E. Williams (ed.), *The Public Face of Conservation*, London, Archetype, 2013, pp. 66-75.
[8] Silberman, N.; Purser, M., *Collective memory as affirmation: people-centered cultural heritage in a digital age*, in E. Giaccardi, (ed.), *Heritage and social media: understanding heritage in a participatory culture*, Abingdon, Routledge, 2012, pp. 13-29.
[9] O'Coonor, A., *From collaboration to aggregation: Internet resources and the future of conservation*, "AIC news", 2010, vol. 35, n. 3, pp. 8-11.

Online resources

Blogs and forums

http://www.tate.org.uk/art (Tate Modern)

https://www.metmuseum.org/blogs (Metropolitan Museum)

https://blog.britishmuseum.org (British Museum)

http://www.vam.ac.uk/blog/ (Victoria and Albert Museum)

https://stories.moma.org (MoMA)

https://www.shakeinconservation.be (Shake in Conservation)

Reviews

https://www.kermes-restauro.it (Kermes)

IOP Conf. Series: Materials Science and Engineering **364** (2018) 012047 doi:10.1088/1757-899X/364/1/012047

Museum for All: Wearable Immersive Virtual Tours in Museums for People with Neurodevelopmental Disorders

Nicolò Messina, Vito Matarazzo, Daniele Occhiuto, Mirko Gelsomini, Franca Garzotto

Department of Electronics, Information and Bioengineering
Politecnico di Milano, 20133 Milano, Italy.

[name.surname]@polimi.it

Abstract. "Social stories" are used in educational interventions for subjects with Neurodevelopmental Disorders (NDD) to help them gain an accurate understanding of social situations, develop autonomy and learn appropriate behavior. Traditionally, a social story is a short narrative that uses paper sheets, animations, or videos to describe a social situation of everyday life, e.g., "going to school", "visiting a museum", "shopping at the supermarket". In our research, we exploit Wearable Immersive Virtual Reality (WIVR) technology to create a novel form of social story called Wearable Immersive Social Story (WISS). The paper describes the idea of transforming paper-based social stories, written for the project "Museo per tutti" (Museum for All), into WISS to better support persons with NDD in visiting museums and thus increasing their accessibility.

1. Introduction

"Social stories" are a storytelling approach used in educational and therapeutic interventions for people with Neurodevelopmental Disorders (NDD). NDD is an umbrella term for a group of disabilities that appear during the developmental period and are characterized by deficits and limitations in the cognitive, emotional, motor and intellectual spheres. A social story is a short narrative that uses visuals and in most cases also written text to describe a specific social situation, event, or activity in a clear and reassuring manner that is easily understood by the individual with disability. Social stories are used as learning material to promote the development of autonomy and appropriate behaviors, and to teach specific social skills, such as understanding rules, routines, and expectation, or taking another's point of view.

We exploit low-cost Wearable Immersive Virtual Reality (WIVR) technology to create a novel form of social story, i.e., Wearable Immersive Social Story (WISS). A WISS is composed of 360° videos enriched with superimposed interactive elements and visual clues (geometric shapes, images, audios, ...) that make the experience more fun and engaging, helping users gain a better understanding of the social situation. These videos are executed on a smartphone and viewed through a low-cost Head-Mounted Display (HMD) (e.g. Google Cardboard) that makes the user feel inside the virtual space.

The paper describes how we applied WISS within the project "Museo per tutti"[1], that aims at making museums more easily accessible for people with NDD. Specific paper-based social stories are written and designed for each museum. We translated them into WISSes made of 360° interactive videos recorded inside those museums, with the goal of making social stories more effective.

[1] Italian words meaning "Museum for all". https://museopertutti.it/

2. Related work

Social stories have been proposed as an effective intervention for persons with NDD, particularly children with ASD (Autism Spectrum Disorder) since the early 1990s. The term "Social Story" has been trademarked by its original creator to denote a narrative characterized by 10 detailed criteria [1], guiding the story creation in order to help the individual with disability understand the situation and develop social and practical skills. In most of the existing literature [2][3][4][5] the term "social story" has a broader meaning and corresponds to the definition given at the beginning of this paper. The research on the effectiveness of this instrument is limited and highlights highly variable effects in the learning process [6][7]. Still, social stories remain widely used in therapeutic and educational interventions for subjects with autism as well as with other forms of disability in the NDD spectrum.

Originally, social stories were based on paper-based visual and textual materials only. In today's practices, they often use digital media, like images and videos on computer displays [8][9]. In the research area of Virtual Reality (VR), there are some examples of applications created for subjects with NDD that focus on social situations and can be regarded as digital transpositions of social stories.

VR is thought appropriate for this target group to help learning about real life situations, because in the virtual space behaviors and responses can be practiced in a safe and repeatable environment, while interactivity promotes engagement and cause-effect understanding [10][11]. Strickland et al. (2007) [12] developed desktop 2D virtual environments to teach fire safety skills to young children with ASD. Josman et al. (2008) [13] used VR to teach autistic students aged 8-16 years to cross the road safely.

WIVR applications have been found to improve attention skill because the head-mounted display removes the distractions of the outside world, a feature that is important for subjects with NDD who often have severe attention deficits. Benefits in this area have been observed in the study reported in [3] where low-medium functioning children with NDD used a HMD to interact with the immersive digital transposition of paper-based fantasy tales.

Two important concerns related to the use of WIVR in interventions for persons with NDD are the acceptability of the headset and the risk of physical side-effects that are typical of wearable VR environments [14]: motion sickness (due to a disagreement between visually perceived movement in the simulated world and the vestibular system's sense of movement our body), double-vision (a particular condition under which the virtual elements are seen twice as overlapping copies instead of being perceived as one) and eye fatigue (the feeling that our eyes are burning, itchy and tired). These side effects were frequent in first generation VR headsets, characterized by poor viewing angles, high latency, and weight [15]. Today, HMDs are much more comfortable, including those commercially available at an affordable cost (e.g., Samsung Gear VR[2] and Google Cardboard[3]). In [3] the majority of study participants using WIVR on Google Cardboard had an enjoyable experience, were fascinated by the immersive experience, and manifested the willingness to play with it again.

VR headsets are increasingly being used to view 360° videos, e.g., in tourism, cultural heritage, professional training, and regular education (see for example the Google Expeditions Program[4] and the Immersive Education Initiative[5]). To our knowledge, the use of 360° videos has not yet been explored in learning interventions for persons with NDD, and the WIVR environments for this target group are usually created in computer graphics.

The device used in our research to view 360° videos is Google Cardboard (figure 1 – left), the cheapest WIVR headset in the market (5€ for the paper version, 30€ for more resistant plastic variants). This VR visor is composed of two biconvex lenses mounted on a plastic or cardboard structure. The smartphone positioned inside the visor displays the visual contents, splitting them into two near-identical bi-dimensional images (figure 1 – center). The illusion of space depth and immersion in the virtual environment is created by the stereoscopic effect generated by the viewer lenses (figure 1 – right).

[2] http://www.samsung.com/us/explore/gear-vr/
[3] https://www.google.com/intl/en/get/cardboard/
[4] https://www.google.com/edu/expeditions/#about
[5] http://immersiveeducation.org/

IOP Conf. Series: Materials Science and Engineering **364** (2018) 012047 doi:10.1088/1757-899X/364/1/012047

Interaction is achieved through gaze pointing and focus, assuming that the direction of the gaze focus is defined by head orientation (detected by smartphone sensors), and is always at the center of the screen. Users navigates the virtual world by rotating their head, which has the effect of rotating the virtual scene projected on the display.

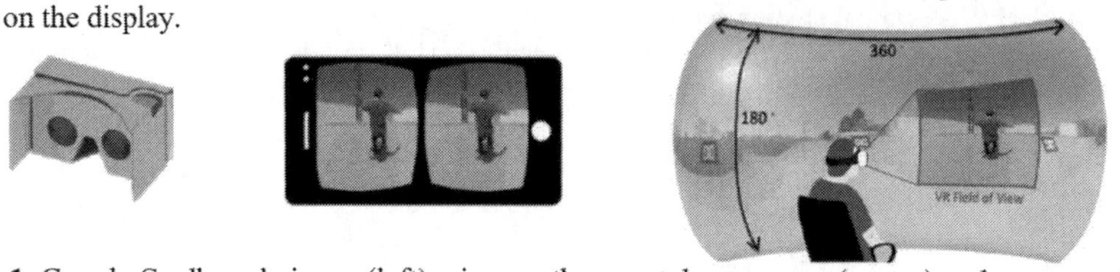

Figure 1. Google Cardboard viewer (left), view on the smartphone screen (centre) and conceptual view during the VR experience (right)

3. Approach

Knowing the benefits of social stories in educational and therapeutic practices for people with NDD, in our research we focus on mixing the power of social stories with that of WIVR. Transforming social stories into interactive immersive virtual narratives can increase the effectiveness of the traditional approach thanks to the feeling of immersion and engagement provided by HMDs.

The design of the concept of WISSes was a collaborative process between our university group and a team of 6 NDD specialists from two local care centers (special educators, neuro-psychiatric doctors, therapists). We did not involve individuals with NDD in the design activities because of the nature and severity of the disability of the persons attending the centers. This brought also to the development of XOOM, a tool that enables anyone to autonomously develop WIVR-based social stories, play them on any smartphone placed inside a VR headset, and supervise them from another connected device [16].

During these meetings, they had the idea of using this technology for the purposes of "Museo per tutti". In this project, they had created a set of paper-based social stories about the visits to different museums (see example in figure 2 – left) and wanted to transform these materials into WISS. The iterative co-design process comprised 4 workshops interplayed with development activities to create progressive prototypes of WISSes and the supporting technology.

Figure 2. Left: Paper-based social story about the museum visit; Right: script for the 360° videos (scenes 1 and 2) defined during workshops

4. Museo per tutti (Museum for All)

4.1. Paper-based tours in museums

Museo per tutti is an experimental project started in 2015, supported by Fondazione De Agostini[6] and designed and realized by L'abilità Onlus association[7], that aims at defining guidelines, tools and paths inside famous Italian museums to support persons with NDD in their visits. In the museum context, the NDD field is still not much explored. Therefore, this project aims at starting an important experimentation with the goal of improving the quality of life for people with NDD, through their participation and involvement in local and cultural contexts, that are usually common to all citizens.

The Archeological Museum in Cremona, La Venaria Reale in Turin and the World Cultures Museum – D'Albertis Castle in Genoa are the first three museums that, together with the National Gallery in Rome, which has joined the project later, were involved in the project working with a group from L'abilità Onlus that has specific knowledge in the fields of cultural heritage, art history and museums accessibility. In this way, experts in the NDD field and each museum's staff collaborated actively during all the project phases, to realize proposals and materials specific for each museum. In the future, two other museums will be involved in the project: Museo degli Innocenti in Florence and BEGO – Museum Benozzo Gozzoli in Castelfiorentino.

Since *Museo per tutti* offers mainly paper-based material, composed of words, pictures and illustrations, that helps people with NDD to familiarize with the museums before going physically there, the idea of the experts from L'abilità was to translate them into WISSes created through XOOM.

4.2. Wearable Immersive Virtual Tours in museums

The first workshop was devoted to identifying the contents and the technology for the social stories to be rendered in a virtual environment. We discussed the trade-offs between computer graphic contents and 360° videos, and between the different VR devices. We agreed to use 360° videos and the Google Cardboard headset, for several reasons. Story contents based on 360° videos were perceived as more immersive and realistic than computer graphic applications, enabling the person with NDD to experience the social space "as it really is". Inside the immersive digital space, the user is required to build and process a representation of the virtual environment "as a location" in order to successfully navigate it. Outside the virtual experience, the users are expected to capitalize on this generalization construct, linking their mental representation of the virtual environment to the real world in order to understand the physical environment and the social situation in it. Subjects with NDD have limited capability of generalization, and 360° videos of real contexts would facilitate this mapping. On the other side, high-end headsets (e.g. HTC Vive[8]) offer better quality VR experiences than cheaper devices, but at a much higher price. Still, the very low cost of Google Cardboard increases the potential for the adoption of WISSes among a larger number of people.

After this discussion, we selected the social story of a specific museum (La Venaria Reale) as case study, and the rest of the workshops were devoted to design the video scenes for its WIVR transposition. We explored pros and cons of recording each scene with a fixed camera (camera placed in a specific point of the environment) or a mobile camera (camera moving in the space suggesting the user the idea of walking through the scene). In the end, we generated a detailed video script, defining for each scene what the camera should record, in which mode (fixed or mobile) and which were the different possible effects and interactive elements to apply over the 360° videos as to facilitate the tour (see examples in figure 3).

Indeed, we designed XOOM as a customization tool for 360° videos, so that the same paper-based social story could be translated into different WISSes of increasing difficulty, to avoid the boredom due to the repetitiveness of the experiences. In this way, an individual with NDD can start by watching a

[6] http://www.fondazionedeagostini.it/
[7] http://labilita.org/
[8] https://www.vive.com/eu/

IOP Conf. Series: Materials Science and Engineering **364** (2018) 012047 doi:10.1088/1757-899X/364/1/012047

virtual tour full of suggestions and helpers that guide him/her through the museum, explaining everything; afterwards, when he/she has become confident with the virtual environment, he/she can switch to more difficult experiences made of the same previous videos, but with a less number of facilitators and possibly with the presence of some distractions.

Experts pinpointed the need for facilitators, i.e., visual clues superimposed on the video that attract the users' attention and help them focus on those elements in the virtual environment that would be more appropriate for understanding the current situation and the correct behavior. To make the experience more engaging and fun, some facilitators should be interactive, i.e., they generate visual (animated) effects when the user focuses the eye gaze on them. We defined the following types of facilitators: geometric shapes (e.g. arrow or circle); Highlight, that lights up a specific area of the scene "shading" the rest of the environment, to drive the user's attention on specific areas; PCS (Picture Communication Symbols), picture cards that represent objects, actions, people, or more abstract concepts like feelings and are widely used in Augmentative and Alternative Communication (AAC)[9]; sounds and textual popups, for instructions, comments and social clues.

On the other hand, reexamining the role of the stimuli acting as "facilitators", we realized that depending on the user, the contexts, and the learning goal, the same stimulus could have different meanings. For instance, the noise of a car in a city environment would act as facilitator, since users expect this sound as they see cars in the video. The same car roaring in a video of a natural landscape in the countryside would probably distract the user. An arrow pointing to a door could indicate that the user should interact with it to open it; but in case of an emergency door, the user should avoid interaction. So, we came out with the idea of "distractors", i.e., visual elements similar to facilitators in terms of attributes, but with the opposite role. Distractors could be introduced in advanced sessions to train selective attention and improve understanding of a social situation. Moreover, they contribute to reduce the gap between the quiet and safe virtual world and the chaotic and dangerous real world. In fact, the transition from the facilitated environment of WIVR to the real world, full of different stimuli and unknown events, can provoke confusion and unexpected reactions, above all in a person with NDD. Therefore, a tool that allows customization of the WIVR-based stories becomes fundamental, so that gradually more complex and verisimilar experiences can be created, adding various distractions to prepare users to face the real world. Nevertheless, we do not think that virtual reality should completely substitute experiences lived in the real world, but it can be a valid reinforcement tool to teach people with NDD correct behaviors in everyday life, if it is followed by a practical trial in real scenarios.

Finally, the experts recommended the introduction of "pause points". Suspending the video in some moments would give the user the time needed to explore the surrounding environment, to discover "facilitators" and to understand their meaning. Pause points would facilitate interaction with the interactive elements in a scene (interaction would be difficult while the video is running). At a pause point, the video could be suspended for a fixed amount of time, or could restart when the user focuses his/her gaze on an interactive element in the view.

Figure 3. Examples of distractor (left) and facilitator (right): the first is a shiny red sphere that doesn't belong to the real environment and, standing out, can distract the user from going through the hallway; the second is a Highlight area that draws user's attention to the relevant painting.

[9] Augmentative and Alternative Communication is used to describe various methods of communication that can 'add-on' to speech and are used to get around problems with ordinary speech. AAC includes simple systems such as pictures, gestures, pointing and more complex techniques involving powerful computer technology.

5. Software Platform and Video Recording

In parallel to the *Museo per tutti* project, we implemented the software platform for WISS creation, personalization and execution [16], and studied the best ways to record 360° videos, which brought us to build a tele-operated mobile robot for video shooting.

5.1. *The Software Platform*

The software platform – called XOOM – is *web-based* and integrates two main components: *Creator* and *Runtime Controller*. The first is an authoring tool for WISSes. It enables everyone to create a new WISS starting from a set of 360° videos, and to personalize an existing one (figure 4). The authoring functionalities include: allocation of videos on a timeline; definition of visual elements (facilitators and distractors) in terms of graphic and interactive properties and position in the videos; definition of pause points. The *Runtime Controller* manages the supervised execution of fully featured WISSes on a smartphone (they can also be played autonomously, without any supervisor, launching WISSes directly from a smartphone). This component also enables the visualization of the running WISS on an external PC and the manual control of its flow (launching a WISS, pausing/starting a video, launching a different WISS). This can be useful for a caregiver to follow a person with NDD in performing a WISS.

Figure 4. The Creator component of XOOM

5.2. *Studies about 360° videos*

We conducted some studies on how to record 360° videos in the best way. Existing guidelines[10] underline that in order to make the user feel present in first person in the virtual environment, the camera must be kept at almost 160 centimeters from the floor to simulate an average person's height, and all the key elements in the scene must be roughly 1 to 1.5 meters away from the camera. If they are too close, these elements look distorted, while if they are too far away they may be indistinguishable. Vibrations and unexpected or fast movements with the camera must be avoided and the camera's front side must point towards the movement direction in case of mobile camera videos, to reduce the risk of motion sickness. For this purpose, we built a simple remotely-operated robot – called Bob (figure 5 - left) – able to shoot 360° videos with a better quality than the ones recorded manually. Bob's body is composed by a plastic pipe inserted in a hard-plastic cone and a "hat" on top of the pipe where we placed a 360 camera. The cone has the goal of stabilizing the robot movements and reducing the vibrations of the camera. It is mounted on a mobile base that exploits the commercial iRobot Create[11] programmable platform. The

[10] http://www.samsung.com/ae/discover/how-to/how-to-film-the-best-360-videos; http://vrscout.com/news/avoid-motion-sickness-developing-for-vr/

[11] iRobot Create. http://www.irobot.com/About-iRobot/STEM/Create-2.aspx

camera height can be adjusted based on specific needs, e.g., reduced if we want to record the physical environment with a child's eyes or increased if we decide to capture the space as seen by an adult.

5.3. 360° videos recording

Bob was just used, together with the guidelines we studied, to record videos at a local supermarket (figure 5 – left) for another project involving WISSes. In collaboration with another therapeutic center called *Fraternità e Amicizia*, we wanted to create a WISS teaching how to shop in a supermarket, to choose the correct food based on a given dish to prepare, or on the moment of the day (e.g. breakfast or dinner), and to pay the bill and get change. With the recorded videos, we created a first demo of this WISS. Since it was one of our first attempts in recording 360° videos, we found something to improve and learnt new important aspects to take into account to obtain better results from recording 360° videos.

Figure 5. Left: Bob recording outside and inside the supermarket; Right: a scene in the "Shopping at the supermarket" WISS, with a PCS card superimposed to the video as facilitator

6. Discussion and conclusion

The main contributions of our work are in the definition of the concept of WISSes for people with NDD and its application as a tool to develop autonomies making museums more accessible to people with this kind of impairments.

During the workshops and XOOM's evaluation (Creator module has just been tested with 15 therapists and educators that found it easy to use), all the experts expressed a global positive judgment about usefulness of WISSes to teach their users how to behave in everyday life situations. They confirmed that an individual with NDD is expected to navigate easily within the 360° videos and to interact with its elements. Still, the experts also pinpointed some potential drawbacks in this technology. Individuals with NDD may have an initial resistance to wear the HMD and to explore a WISS, as in general these persons feel a strong need for routine and tend to get distressed when a situation or a pattern of behavior changes. Some educators suggested that preliminary familiarization activities without the visor would mitigate this resistance: for example, wearing a cardboard-based mask like the ones used in some physical games and looking at the 360° videos on a regular PC screen. The experts also said that a WISS, and WIVR technology in general, may be not suitable for users that suffer of psychosis or hallucinations, since they already live situations of detachment from the reality and the immersion in a virtual environment could worsen their condition. With the above caveat, the experts highlighted the motivational benefit of a WISS for persons with NDD, particularly children and young adults. Once the "distress" originated from novelty is overcome, a WISS offers a playful and enjoyable experience that can promote learning. The virtual story can act as a preparation to the reality, helping users in focusing attention towards the relevant elements and in facing unexpected events. Therapists declared the intention of adopting WISSes in their practices as a complement to existing tools. They pinpointed that these stories should be offered not only at the care center but also in other contexts of a person's life and in the real locations of the social story, e.g., the museum and the supermarket.

Finally, the experts were convinced that WISSes created with XOOM have many advantages. XOOM makes it possible to create and control visual stimuli inside a WISS in a way that is appropriate for each specific subject, and to create a progressively more complex story by changing facilitators and distractions at each repetition. XOOM offers a simple and smooth way to play the same story as many times as it is desired, or to change it when the subject is bored. Individuals with a persistent state of anxiety would receive particular benefit from the possibility of repeating the experience again and again.

7. Future work

The video script translating the paper-based social stories in WIVR virtual tours of La Venaria Reale is ready. As soon as possible, following the created script, we will record the required 360° videos directly in place. Aside from this, our future work in the short term envisions the execution of an empirical study investigating the learning benefits of WISSes also compared to more traditional forms of social stories. This long term empirical study will involve 40 subjects with NDD (most of them with high-functioning autism, plus some individuals with learning disability or psychomotor deficit) at the 2 care centers we collaborate with. As evaluation tools, we will use the WISSes created for the projects *Museo per tutti* and *Shopping at the supermarket.*

In case the study will give us positive results, firstly we would like to extend WISS to the other museums involved in the project *Museo per tutti.* Secondly, we desire to propose the same approach to the Malpensa Airport, that, together with L'abilità and other contributors, has developed paper-based social stories to prepare people with NDD in visiting the airport.

References

[1] Gray C A 2010 *The New Social Story Book* (Arlington: Future Horizons Inc.)
[2] Adams L, Gouvousis A, VanLue M and Waldron C 2004 Social Story Intervention: Improving Communication Skills in a Child with an Autism Spectrum Disorder. *Focus on Autism and Other Dev. Disabilities* vol 19 i2 pp 87-94
[3] Garzotto F, Gelsomini M, Clasadonte F, Montesano D and Occhiuto D 2016 Wearable Immersive Storytelling for Disabled Children *Proc. of the Int Working Conf Adv Visual Interfaces* pp 196-203
[4] Kuoch H and Mirenda P 2003 Social Story Interventions for Young Children With Autism Spectrum Disorders. *Focus on Autism and Other Dev. Disabilities* vol 18 i4 pp 219–227
[5] Swaggart B L, Gagnon E, Bock S J, Earles T L, Quinn C, Myles B S and Simpson R L 1995 Using Social Stories to Teach Social and Behavioral Skills to Children with Autism. *Focus on Autism and Other Dev. Disabilities* vol 10 i1 pp 1-16
[6] Nichols S L, Hupp S D, Jewell J D and Zeigler C S 2005 Review of Social Story interventions for children diagnosed with autism spectrum disorders *J. of Evidence-Based Practices for Schools* vol 6 i1 pp 90-120
[7] Reynhout G and Carter M J 2006 Social Stories for Children with Disability *J. Autism Dev. Disord* vol 36 i4 (Springer US) pp 445–469
[8] Hagiwara T and Myles B S 1999 A Multimedia Social Story Intervention. *Focus on Autism and Other Dev. Disabilities* vol 14 i2 pp 82-95
[9] Sansosti F J and Powell-Smith K A 2008 Using Computer-Presented Social Stories and Video Models to Increase the Social Communication Skills of Children With High-Functioning Autism Spectrum Disorders *J. of Positive Behavior Interventions* vol 10 i3 pp 162-178
[10] Andersson U, Josefsson P and Pareto L 2006 Challenges in designing virtual environments: Training social skills for children with autism *Int. J. Disability and Human Dev.* vol 5 i2 pp 105-112
[11] Lanyi C S and Tilinger A 2004 Multimedia and virtual reality in the rehabilitation of autistic children *Lecture Notes in Computer Science* vol 3118 (Berlin: Springer, Heidelberg) pp 22-28
[12] Strickland D, McAllister D, Coles CD, Osborne S 2007 An evolution of virtual reality training designs for children with autism and fetal alcohol spectrum disorders *Top. Lang. Dis.* vol 27 i3 pp 226-241
[13] Josman N, Ben-Chaim H M, Friedrich S. and Weiss P L 2008 Effectiveness of virtual reality for teaching street-crossing skills to children and adolescents with autism. *Int. J. of Disability Dev. and Human Dev.* vol 7 pp 49-56
[14] Sharples S, Cobb S, Moody A and Wilson J R 2008 Virtual reality induced symptoms and effects (VRISE): Comparison of head mounted display (HMD), desktop and projection display systems *Displays* vol 29 i2 pp 58-69
[15] Strickland D C, Marcus L M, Mesibov GB, Hogan K 1996 Brief report: Two case studies using virtual reality as a learning tool for autistic children *J. of Autism and Dev. Disorders* vol 26 i6 pp 651-659
[16] Garzotto F, Gelsomini M, Matarazzo V, Messina N and Occhiuto D 2017 XOOM: An End-User Development Tool for Web-Based Wearable Immersive Virtual Tours *ICWE 2017: Web Engineering* pp 507-519

IOP Conf. Series: Materials Science and Engineering **364** (2018) 012048 doi:10.1088/1757-899X/364/1/012048

Enhancing cultural tourism by a mixed reality application for outdoor navigation and information browsing using immersive devices

Federico Debandi, Roberto Iacoviello, Alberto Messina, Maurizio Montagnuolo

RAI, Centro Ricerche e Innovazione Tecnologica, Italy

E-mail: federico.debandi@rai.it, roberto.iacoviello@rai.it, alberto.messina@rai.it, maurizio.montagnuolo@rai.it

Federico Manuri, Andrea Sanna, Davide Zappia

Dipartimento di Automatica e Informatica, Politecnico di Torino, Italy

E-mail: federico.manuri@polito.it, andrea.sanna@polito.it, davide.zappia@studenti.polito.it

Abstract. In this paper a mixed reality application is introduced; this application runs on Microsoft Hololens and has been designed to provide information on a city scale. The application was developed to provide information about historical buildings, thus supporting cultural outdoor tourism. The huge amount of multimedia data stored in the archives of the Italian public broadcaster RAI, is used to enrich the user experience.

A remote application of image and video analysis receives an image flow by the user and identifies known objects framed in the images. The user can select the object (monument/building/artwork) for which augmented contents have to be displayed (video, text audio); the user can interact with these contents by a set of defined gestures. Moreover, if the object of interest is detected and tracked by the mixed reality application, also 3D contents can be overlapped and aligned with the real world.

1. Introduction

Augmented Reality (AR) provides an efficient and intuitive way to visualize computer generated information overlaid and aligned with objects in the real environment. AR technologies have been employed in a broad spectrum of applications: medicine, industry, military, and so on [1][2]. As real and virtual objects can coexist within the same augmented environment, AR solutions have been also used to enhance the user experience in cultural tourism contexts. Many solutions have been presented for indoor [3], where is quite easy to track the user position and light conditions can be controlled. Several museums provide users AR applications that framing well defined images (markers) recognize position and orientation of the user and overlap computer generated assets to the artwork the user is interested in. On the other hand, several challenges have to be tackled for outdoor applications [4]. First of all an accurate tracking of the user is not easy; often GPS-based solutions are provided, but they might be not able to accurately provide both the position and orientation. Moreover, GPS-denied environments should be also considered. The second issue is related to the impossibility to alter the environment by markers or target images. Finally, lighting conditions cannot be controlled.

If the term augmented reality refers to the possibility to overlap computer generated contents to the real environment, the term Mixed Reality (MR) denotes the possibility of the user to interact with these contents.

This paper presents a MR application for outdoor enhancing the user experience during cultural tours. With Mixed Reality, synthetic contents are anchored to some positions in the real space and the user can interact with them. In order to allow the user a full mobility, the Microsoft Hololens has been chosen as a Head Mounted Device (HMD); it is connected by a high-speed and high-throughput network (e.g, 4G/5G cellular networks) to a cloud infrastructure. The application gathers several times per minute images of what the user is watching; in a transparent way, images are sent to the cloud architecture where a visual search engine (MPEG Compact Descriptors for Visual Search - CDVS) is running. In this way, objects of interest framed by the user can be identified; a notification about a recognized object is sent back to the user, that can receive augmented contents about the object.

The application can display both textual information and movies about the framed monument/building/artwork and a semitransparent silhouette of the recognized item can be used to align the user with respect the target, thus enhancing the tracking robustness. When the AR application tracks the object, 3D virtual assets can be overlapped to the target.

The paper is organized as follows: basic concepts behind AR and MR are presented in Section 2 as well as some examples of outdoor augmented reality for tourism, Section 3 details the architecture of the proposed application and Section 4 shows tests and results gathered in the Turin's Archaeological Park.

2. Background

Several attempts of using AR and MR to support outdoor cultural tourism are known in the literature. For instance, Archeoguide [5] where a laptop computer is used to manage computer generated assets and track the scene. In particular, the tracking strategy is based on a set of reference images due to the impossibility to put any special markers onto the historical sites. For each viewpoint there is a set of reference images that are compared with the framed one. When a match is found, the transformation to fit the reference image to the video image is computed. This transformation allows to correctly align and overlap synthetic contents to the real world. A similar approach is also used in [6]. The project Archeoguide has been also extended in order to provide users a powerful mobile device for outdoor AR applications with standard off-the-shelf components [7]. At the date of [7] it was impossible to reach simultaneously the following design goals: intuitiveness, robustness, powerful and lightness. An example of MR application for the cultural heritage in outdoor environment is provided by the PRISMA project, which aim is to design, develop and implement a 3D visualization device based on Augmented Reality technologies [8]. Tourist AR binoculars display augmented contents to the user and the interaction with a pop-up menu allows to switch among different type of contents. Santos et al. proposed a similar approach, focusing on the technologies and approaches needed to enable outdoor mixed reality experiences in the cultural heritage domain [9].

Drawbacks and advantages of Mobile Augmented Reality (MAR) are presented in [10], whereas MAR using smartphones is considered in [11]. Many other examples of application of the MAR paradigm for tourism are known in the literature such as the LIFEPLUS system [12] (which aims to provide a guide for the archaeological site of Pompeii in Italy) and the work presented in [13], which improves the tracking phase on the basis of a combination between computer vision algorithms and device sensors.

Several works have been devoted also to investigate issue related to the expected user

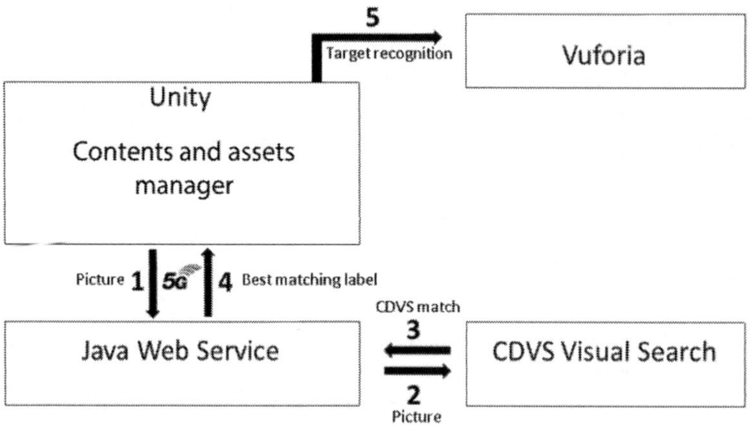

Figure 1. Software architecture and data flow exchange.

experience. For instance, in [14] it comes to light how users are interested in applications that are easy to use and highly customizable, whereas also the emotional point of view is analyzed in [15]: MAR services are expected to offer stimulating and pleasant experiences, such as playfulness, inspiration, liveliness, collectivity and surprise.

3. The proposed MR application

The device chosen for this project has to be a Head Mounted Display (HMD), as this kind of device provides users both an immersive experience and a full mobility. The Microsoft Hololens fulfills the above mentioned characteristics as it is a stand alone device that does not require an external computer for computation.

The detection step is the most important phase for a any AR-based application. After the recognition of a monument/building/art work of interest, information gathered in the RAI archives can be sent to the user. The users have to be able to select the wished kind of information (text, audio, video). Only when the recognized item is correctly tracked, 3D contents can be aligned and overlapped to the real object.

From the hardware point of view, the proposed architecture is very simple. The user wearing the Hololens is connected to the Internet by a high speed network (e.g., 4G/5G). The MR application is running on the glasses and it exchanges data in a bidirectional way with an application of visual content analysis and search running in cloud on a virtual machine (an Azure's node). For this purpose, the MPEG CDVS was integrated as a proof-of-concept. Any other machine learning tool may be seamlessly plugged into the system.

From the software point of view, the architecture is shown in Figure 1. The application is completely developed in Unity (https://unity3d.com/), but it is logically divided in two parts: the first part concerns the connection with the virtual machine and the visual search engine running on it, whereas the second part is in charge to track objects and display correctly aligned 3D models. The first part is composed by 5 different layers: Unity (a game engine) and UWP (Universal Windows Platform - a platform to develop applications for Windows 10, Windows Mobile 10, Xbox One and Hololens) for the application running on the Hololens; Java Web service, the Image recognizer software and Linux for the cloud side. Differently, the second part is composed by three layers Vuforia (https://www.vuforia.com/),a framework for developing AR application, Unity and UWP; all these three layers refer to the application running locally in the HDM.

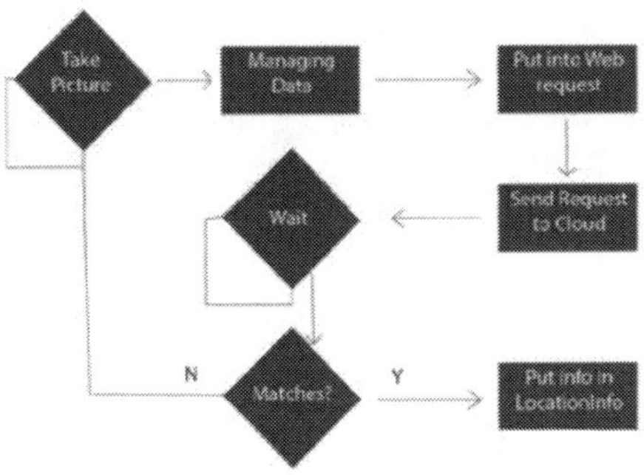

Figure 2. Flowchart of the first scene.

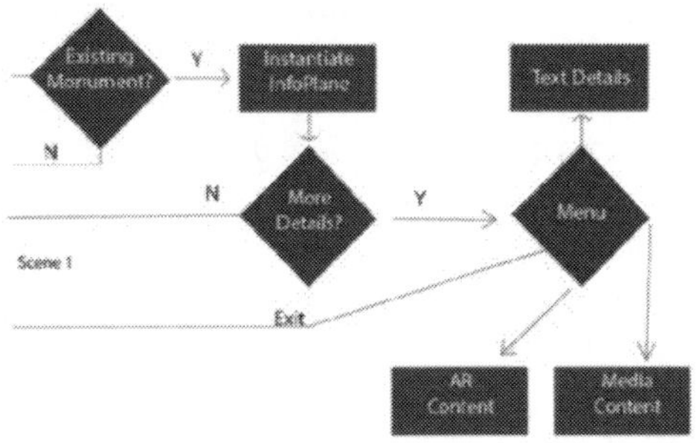

Figure 3. Flowchart of the second scene.

Unity provides many different components ready to use, like AudioSource, VideoPlayer and BoxColliders, but if it is necessary to develop a custom component, e.g., an Internet connection, this can be implemented only with the creation of a Script. Every particular behavior or action of a GameObject must be implemented by a Script. Moreover, in Unity, it is possible to divide the application in some scenes; in this project, there are three scenes: the first one for the cloud visual search, the second one for delivering object details and third one for the augmented contents.

3.1. The first scene

This scene is responsible for taking information from the camera and perform the connection with the cloud. This part is transparent to the user because for the recognition there are only functions that have to manage the application's behaviour. The camera is an important component in order to take the picture for the recognition and it is managed by the Microsoft Hololens SDK. In this scene, as shown in Figure 2, the application takes a picture every some seconds and performs the connection with the cloud while the user is walking freely in a city tour. Every picture is sent to the web service on the cloud that passes the image to the visual search service. CDVS compare the image with its database and return a result. After the computation, the web service get the result from the visual search service and send it to the application on the Hololens. If the result is a valid location, the application stores every important information and prepare the passage to the second scene.

3.2. The second scene

This scene, differently from the first one, provides many visible augmented elements. The application takes the location found and creates a first information hologram about the recognized monument. The plane shows a textual content that describes the recognized item and two buttons: Menu and Exit. If the user is interested to the given monument, he can select the menu button and then the general menu is shown. The menu plane contains 4 buttons that trigger the following actions:

- Details: a new plane is created and it shows particular information about the history of the monument;
- Media Content: a new plane is created and it shows additional multimedia contents such as pictures and movies;
- AR Content: Manager loads the next scene in order to show the virtual reconstruction of the monument;

IOP Conf. Series: Materials Science and Engineering **364** (2018) 012048 doi:10.1088/1757-899X/364/1/012048

The interaction model is a fundamental part for the selection of the contents and it works like a viewfinder: it consist of a cursor, which is controlled by the user gaze, and a gesture recognition module, managed by the Microsoft Hololens SDK. When the user looks at a menu button, he can perform a tap gesture anywhere in the camera range to activate the functionality associated to the targeted item.

3.3. The third scene

At the start of this scene all augmented reality components are initialized adding the Vuforia AR feature to the application. The application take the location info and creates the silhouette related. The user has to align the silhouette to the monument and sees it from the right perspective, this facilitate the vuforia's tracking. When the silhouette is perfectly aligned, the application makes the silhouette invisible and the appropriate 3D model is overlapped to the real monument. If the user move his gaze from the monument, the application loses the tracking, so 3D model is set to invisible and the silhouette appears again. Finally, when the user wants to stop the AR feature and to see the menu again, he has to perform a tap gesture.

3.4. The graphical user interface - GUI

In this application, the selection or interaction with objects is performed through the user's gaze and the "air-tap" gesture; therefore, users aims to something with their eyes and selects it performing a tap. In the first scene, the image recognizer is invisible to the user, so there is not any GUI. When an element is recognized, a GUI appears. InfoPlane is the first panel that the user can see. This is a very simple panel, as shown in Figure 4, with some general information and two buttons to make simpler the user choice. The MenuPlane is a simple panel with four buttons. It is the main menu and from this panel the user can see every available content. The buttons are always big to make them easily selectable. The others two panels are very similar to this, with one or two buttons and the same style.

4. Tests and results

This project aims to enhance tourists' experience during a city tour. Therefore, in order to evaluate a real usage of this application, some tests have been organized at Turin, Italy. Tests are focused on various type of parameters, they have been designed in order to gather both subjective and objective feedbacks. In particular, the usability and the simplicity of use have been the most interesting target of this tests. A questionnaire has been submitted after the test phase and users could rate from 1 (totally disagree) to 5 (totally agree) a set of claims.

Eight testers have been chosen randomly in order to get a representative set of samples. The group of testers is equally composed by women and men, the age ranges between 20 and 50 years, and testers have different knowledge of AR technologies. A half of participants never tried an augmented reality application and some of them have not any familiarity with computers. Testers were all volunteers and their participation was not remunerated.

The location chosen for the tests is the Archaeological Park in Turin [16], and the target monument is the Palatine Towers. This place is impressive, spacious and very popular for tourists. Palatine Towers was the Roman Age city gate, which provided access to the ancient city Julia Augusta Taurinorum. The users had to simulate a normal walking through the Archaeological Park, until the Palatine towers. When they arrived at the monument the application should recognize it and then users could navigate all information as shown in Figure 4. After a starting tutorial about the application, the users were left alone and they had to accomplish the following actions: reading initial information and tap on the menu button, tapping on the Details button and read some additional information, tapping on the Media Content button to see the video content about the monument and tapping on the Augmented Reality button and trying to examine the 3D model of the monument. Moreover, the users had

IOP Conf. Series: Materials Science and Engineering **364** (2018) 012048 doi:10.1088/1757-899X/364/1/012048

Figure 4. Initial information after monument recognition.

Figure 5. 3D model overlay of the monument.

to pay attention to some other factors in order to evaluate the usability and visibility of the system: ease to point, visibility, fluency, stability and speed of content switching. Before starting the tests, users were introduced to the pointing system of the Hololens and to the application interface. Furthermore, they were informed about the possibility to move freely in the park by using only their hands to interact with the system. At the end of the tests, the users filled a questionnaire, thus providing their comments about the application.

Before the results analysis, some technical and software limits have to be introduced. One of the most important issue is the limited field of view of Hololens, which is only $30 * 17$ degrees; therefore, the users cannot see large objects when they are too close the framed items. Moreover, the lenses brightness is limited, this makes the contents not really visible when the application is used in very sunny days. Another issue is the unstable tracking of the virtual reconstruction of the monument (see Figure 5). Nowadays, it does not exist a free and stable building recognizer to perform augmentation; furthermore, Vuforia is the only SDK supported by Microsoft to develop applications running on the Hololens. Vuforia recognizes an object by comparing the image framed by the camera with a set of pictures pre-loaded in a database; unfortunately, every light change or object movement with respect the images loaded in the database can compromise both the recognition and the tracking.

4.1. Results analysis

The questionnaire was made of 13 questions, as shown in Table 1, and they are related to different categories in order to have a good panoramic of the application's characteristics. Questions 1-4 discuss the likeability and usability of the application: it's important to know if the user likes to use the system and if he knows what to say or what to do. The next six questions focus on the response accuracy of the system and it's speed. The last three questions have the goal to inquire the usefulness of the application's.

More specifically, the first question concerns the system's usability and investigates the simplicity and the intuitiveness of the application commands. All participants think that the commands are simple and intuitive (see Figure 6). It is also important to know if the users find both the symbol and the text of menus and buttons intuitive or not: the majority of the users positively answered to this question. The following questions involve the fluency (Figure 7), which means the content switch's speed, the objects recognition's speed and the Vuforia's tracking stability.

From the Table 1 it is relevant that the 80% percent of the testers says that the application is quite fluent and the initial recognition is stable and fast. The only concern regards the Vuforia's tracking stability: even if there are many positive answers, the questions regarding the recognition's speed and the model overlapping precision provide an higher number of neutral

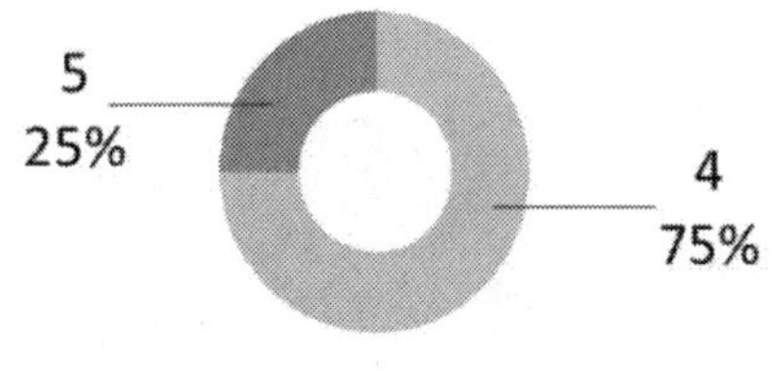

Figure 6. Answers to question 1: "Are commands simple and intuitive?".

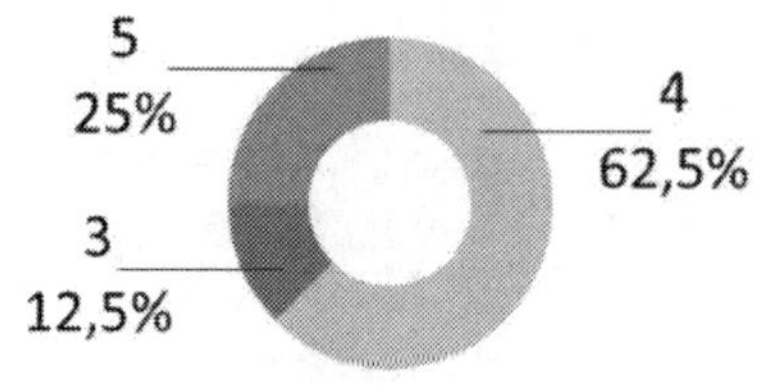

Figure 7. Answers about "Fluency of the application".

Question	**1**	**2**	**3**	**4**	**5**
Are commands simple and intuitive?	0	0	0	6	2
Instructions have been clearly exposed?	0	0	0	3	5
Can you easily come back to the menu?	0	0	0	7	1
Are text and symbology intuitive and clear?	0	0	2	3	3
Is the passage through contents fluid?	0	0	1	5	2
Is starting recognition fast?	0	0	3	1	4
Is 3D model overlay well executed?	0	1	2	2	3
Is the application well realized?	0	0	0	4	4
Is text content clear and visible?	0	0	1	3	4
Are video clips fluid and useful?	0	0	1	4	3
Do you think that read textual information is useful?	0	0	1	3	4
Do you think that see video clips is useful?	0	0	1	4	3
Do you think this application could helps cultural heritage?	0	1	0	2	5

Legend: **1**-Strongly Disagree **2**-Partially Disagree **3**-Neutral **4**-Partially Agree **5**-Strongly Agree

Table 1. Questionnaire's results.

answers. This variety of answers depends on the instability of Vuforia in recognizing images, as environmental conditions can change among tests. However, a noticeable feedback received from the users is that most of them appreciated the application and they think that AR/MR could really help and support the development of the national cultural heritage.

5. Conclusion

This work has been focused on the development of a smart guide able to enhances the experience of tourists during a city tour. The application enhances real monuments with textual and multimedia information; moreover, also an augmentation by 3D models can be provided. Tests show how users appreciate the application both from the usability point of view and from the information content point of view. The application is easy to use and it can really be a valuable tool for outdoor tourism. Indeed, the access to information stored in the RAI's archives is an added value.

On the other hand, several issues have to be tackled in order to move toward a production phase for these kind of tools. There are some hardware limitations that can impact on the user experience (e.g., limited field of view and low visibility in some lighting conditions), but the fast growth of AR technologies promises to limit soon these drawbacks. Another problem is the tracking instability. The fast changes in outdoor lighting conditions strongly limit the use of a image-based tracking. A possible solution could arise by a model-based tracking, which should

IOP Conf. Series: Materials Science and Engineering **364** (2018) 012048 doi:10.1088/1757-899X/364/1/012048

be much more stable than the image-based one. In this case, a CAD model of the object to be tracked should be available. Finally, the Hololens adoption rate from private citizens could be low and it could limit the diffusion of these applications due to its current cost. However, it is reasonable to believe that, likewise the audio guides usually available in museums, this kind of technology can be made available to the users as a service provided by a private society or even by city tourist offices.

Acknowledgments
This work has been co-funded by the H2020 European project 5GCity: http://www.5gcity.eu/ (GA No. 761508).

References
[1] Azuma R T 1997 A survey of augmented reality *Presence: Teleoperators and virtual environments* **6** 355-385.
[2] Sanna A and Manuri F 2016 A survey on applications of augmented reality *Advances in Computer Science: an International Journal*, **5** 18-27.
[3] Huey L C, Sebastian P and Drieberg M 2011 Augmented reality based indoor positioning navigation tool *In IEEE Conf. on Open Systems (ICOS)* 256-260.
[4] Azuma R T 1999 The challenge of making augmented reality work outdoors *Mixed reality: Merging real and virtual worlds* 379-390.
[5] Dähne P and Karigiannis J N 2002 Archeoguide: System architecture of a mobile outdoor augmented reality system *Proc. of the Int. Symposium on Mixed and Augmented Reality - ISMAR* 263-264
[6] Stricker D 2001 Tracking with reference images: a realtime and markerless tracking solution for out-door augmented reality applications *In Proc. of the conference on Virtual reality, archeology, and cultural heritage - VAST* 77-82.
[7] Gleue T and Dähne P 2001 Design and implementation of a mobile device for outdoor augmented reality in the archeoguide project. *In Proc. of the conference on Virtual reality, archeology, and cultural heritage - VAST* 161-168.
[8] Fritz F, Susperregui A and Linaza M.T. 2005 Enhancing cultural tourism experiences with augmented reality technologies *In Proc. of the conference on Virtual reality, archeology, and cultural heritage - VAST*
[9] Santos P, Acri D, Gierlinger T, Schmedt H and Stork A 2010 Supporting outdoor mixed reality applications for architecture and cultural heritage *In Proceedings of the 2010 Spring Simulation Multiconference 190-197.*
[10] Höllerer T and Feiner S 2004 Mobile augmented reality Telegeoinformatics: Location-Based Computing and Services Taylor and Francis Books Ltd. London, UK, 21.
[11] Yovcheva Z, Buhalis D, and Gatzidis C 2012 Smartphone augmented reality applications for tourism E-review of tourism research (ertr) **10** *63-66.*
[12] Vlahakis V, Pliakas T, Demiris A M and Ioannidis N 2003 Design and Application of an Augmented Reality System for continuous, context-sensitive guided tours of indoor and outdoor cultural sites and museums In Proc. of the conference on Virtual reality, archeology, and cultural heritage - VAST 155-164.
[13] El Choubassi M, Nestares O, Wu Y, Kozintsev I, and Haussecker H W 2010 An Augmented Reality Tourist Guide on Your Mobile Devices In Proc. of the Int. Conf. on Multiscale Materials Modeling - MMM 588-602.
[14] Han D I, Jung T and Gibson A 2013 Dublin AR: implementing augmented reality in tourism. In Information and communication technologies in tourism 511-523
[15] Olsson T, Lagerstam E, Kärkkäinen T, and Väänänen-Vainio-Mattila, K 2013 Expected user experience of mobile augmented reality services: a user study in the context of shopping centres Personal and ubiquitous computing **17**, *287-304.*
[16] The archaeological park in Turin web site: https://www.vimark.com/en/archaeological-garden-porte-palatine-torino/

The Digitized Old Books: First Grammar to Teach Romanian Language for Bulgarians

Velislava Stoykova

Institute for Bulgarian Language – BAS, 52, Shipchensky proh. str., bl. 17, 1113 Sofia, Bulgaria

E-mail: vstoykova@yahoo.com

Abstract. The paper offers analysis of digitized copy of the first printed school book to teach Romanian language for Bulgarians published in Bolgrad, Romania in 1865. The related study uses a digital copy of the book which was recently available on-line as a part of the initiative of the Central Library of the Bulgarian Academy of Sciences for digitizing old and rare printed books of its large collections. the Bulgarian Academy of Sciences for digitizing old and rare printed books of its large collections. The overview of the approach used for the digitization and the teaching methodology with related approaches are presented by analyzing the book's content including its bilingual Romanian-Bulgarian thematic dictionary. The book can be considered also as the first comparative grammar of Romanian and Bulgarian language, since it contains grammar and lexical examples of related parallel bilingual linguistic phenomena. The analysis, focused on the interpretation of both grammar and lexical phenomena of both languages given in the book is presented together with the related pedagogical implementations.

1. Introduction

Recent developments in Information and Communication Technologies (ICT) offer various approaches to manage, archive, and preserve cultural heritage objects of different kinds. The fact that contemporary ICT techniques offer systems and methods for ubiquitous and remote access not only to museum expositions but also to copies of old and rare printed books stored into the greatest world's libraries is giving unique opportunity for research, analysis and new findings on historical circumstances and their interconnections which have never been studied and discussed before.

The cultural heritage preservation of digitized old and rare printed books and their on-line availability throughout by offering the scanned copies is a fact that allows a multilingual interpretation of that books in a new cultural and nationally-specific contexts. For example, the research in education and history of foreign language teaching and related approaches may significantly benefits from analysing related old and rare school books interpreted in newly European context.

Further, we are going to present the overview of a multilingual approach used in foreign language teaching which was introduced and elaborated by the German linguist Johann Franz Ahn based on the digitized copies of related school books. Later, we shall analyze the digitized copy of the first printed school book to teach Romanian language for Bulgarians published in Bolgrad, Romania in 1865 using that approach. The book was recently offered on-line by the Central Library of Bulgarian Academy of Sciences' digitized initiative.

IOP Conf. Series: Materials Science and Engineering **364** (2018) 012049 doi:10.1088/1757-899X/364/1/012049

2. Johann Franz Ahn (1796-1865) and his works on foreign languages teaching

The works of the famous German linguist F. Ahn are worldwide recognized as to be the basics of what today we would say *methodology of learning foreign languages.* His *practical and easy method of learning foreign languages* follows the same approach as that for learning native language repeating the way the child is acquiring the native language from its mother, and the approach was considered to be with a multilingual application. The final aim is to speak that foreign language with a competence as the native one.

F. Ahn had published during his life an immense quantity of school books in many pedagogical genres and for learning various languages. For example, he prepared grammars, references, text readings and guides for foreign language learning. Also, his method was used additionally by great amount of authors to produce various school books to learn different foreign languages following the *practical and easy method.* The majority of that old and rare printed books can be found in various digital collections of several respectable libraries offering an opportunity to re-evaluate the impact of the author and presenting the analysis on the newly discovered applications of his method.

2.1. World's digitaized repositories and collections of F. Ahn's works

F. Ahn had published school books to learn French, German, English, Holland, etc. languages contributing to several pedagogical genres. At the same time, his *practical and easy method* was applied to almost all European languages (including Latin language).

Figure 1. The title page of F. Ahn's book *Praktischer lehrgang zur schnellen und leichten Erlernung der französischen sprache* from on-line sources of the LC.

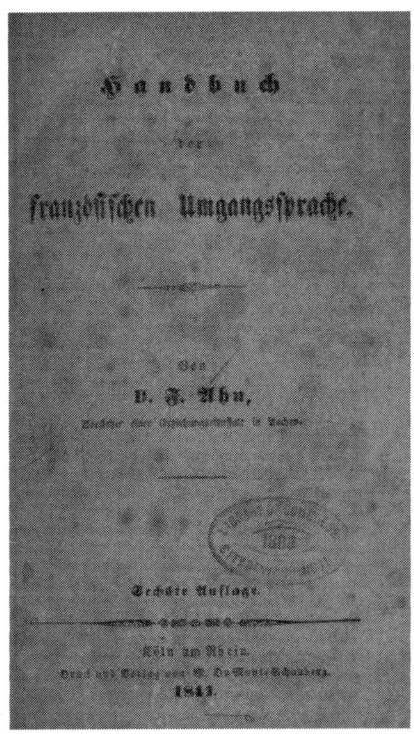

Figure 2. The title page of F. Ahn's book *Handbuch der französischen umgangssprache* from on-line sources of the LC.

One of the richest repositories of scanned copies of that books are the Library of Congress

(LC), the Harvard University (HU) and the Princeton University (PU). They offer on-line the scanned copies of ones on the first editions of the author. Generally, the beginning of F. Ahn's work was the publication of his school grammars of French [4] (a scanned copy is offered by the HU), English and Holland [1] language.

However, the most famous book of the author is offering a methodology to learn French for German readers [5] (a scanned copy of a later edition is offered by the LC (Fig. 1)). Later, F. Ahn applied the same methodology in his books to learn English and Italian [2] (a scanned copy is offered by the LC) language and to learn German language for French, English [9] (one of the latest editions) and Italian readers improving his methodology as to be with a multilingual application.

F. Ahn also contributed to the genre what today we would say *a guide for conversation*. He published the series of such books for French [3] (a scanned copy of a later edition is offered by the LC (Fig. 2)), German [8] (a scanned copy is offered by the HU), English and Holland language. Additionally, he published a *guide for commercial letter-writer* which contains German, French and English glossary [7] (a scanned copy is offered by the HU). The author also published the series of books with German poetry for English [6] and French readers.

At the same time, there are many authors which had applied the F. Ahn's *practical and easy method* to various languages. For example, C. Marchetti published a book to learn Italian language following that methodology [14], whereas P. Condigiorgi published a book [10] in 1871 in Constantinopole using that method as well. The cited works suggest that during the second half of the XIX c. the F. Ahn's method to learn foreign languages was known throughout the Europe including on the Balkans.

3. The application of F. Ahn's method for Bulgarian language readers

It was not until recently, that the process of digitization of the books published in Bulgarian language during the period of the National Revival has started. Thus, the National Academic Library and Information System (NALIS) organized the major Bulgarian libraries to start some years ago an initiative for digitizing their rich library collections of old and rare printed books allowing preservation and content management of that sources and offering on-line search.

3.1. The NALIS approach to digitization of old and rare printed books

Generally, the approaches to digitization of old and rare printed books allow to present their content (often written with different graphics for different books) in computer-readable format for on-line reading and searching. The techniques used start from simply scanning the book's content to structuring it by using meta-data annotation schemes as well as linked data for on-line search.

Contemporary library automation uses the electronic catalogues which present a searchable knowledge database. That electronic database uses the encodings and annotations establishing specific technical standards for presenting various types of book's information. Thus, the LC uses Machine-Readable Cataloguing (MARC) standards for electronic representation of bibliographic information [12]. The MARC standards of electronic data representation contain types of information fields which present book's specific numbers and code fields (for example, about the *Title*, *Physical Description*, *Linking Data Fields*, etc.) which allow on-line search.

At the same time, the EUROPEANA (European Commission's digital platform for Cultural Heritage) uses its own data model (EDM) for presenting Cultural Heritage digital content with specific meta-data format annotations combining text, image, video, etc. sources and allowing the use of Linking Open Data (LOD) and common search.

The NALIS uses the approach to annotate the old and rare printed books digital collections which also adopts EDM as well as LOD and is compatible with the standards for electronic representation of bibliographic information of the LC [11] and with the meta-data standards allowing common fast on-line web-search. The NALIS digital standards use content-oriented approach allowing options for *Navigation* (forward and backward) and *Display* (zoom in, zoom out), and options for *Search* the book content (within full text, chapters, tables, appendices, etc.) by web-based search. The meta-data annotation approaches are compatible to that of EUROPEANA and use common representation of book's *Title*, *Creator*, *Properties*, *Time*, *Provenance*, *References and Relations*, etc. and common on-line web-search.

As a result, some new studies connected to the chronological, geographical and linguistic aspects of the literature from the period of Bulgarian Revival were published [17]. At the same time, the main process of digitization was improved with the application of new standards which take into account the specific features of Bulgarian old and rare printed books from that period [21].

3.2. The F. Ahn's method and related grammars of learning French language for Bulgarians

The works related to F. Ahn's method and its application for Bulgarian readers to learn French language are spread through several Bulgarian libraries and are considered as old and rare printed books. They, basically, are included in the Repertoire of Books form Bulgarian Revival [18], which is the main and most representative bibliographic reference source about Bulgarian printed books of that period. However till now, only the Central Library (CL) of the Bulgarian Academy of Sciences is offering on-line electronic access to all of them. The resource includes two school books and one grammar to learn French language which are published by two authors following the F. Ahn's method adapted for Bulgarian readers.

Chronologically, the book by K. Tarnovsky [19] published in 1864 in Braila, Romania (in Bulgarian language) is a primary school book to learn French language for Bulgarians following the F. Ahn's method (a scanned copy is available by the CL). The translation of its title is *A Practical Method of Learning the French Language* and on its cover page is written that the book is a translation of F. Ahn. The book contains *Preface*, *First part*, *Second part*, *Short texts stories in French language*, and *Vocabulary for texts*.

Secondly, the book *A French Grammar for the Classical Civil Schools* [20] published by the same author in 1869 in Vienna (in Bulgarian language) is the first French grammar for Bulgarians following the F. Ahn's method (a scanned copy is available by the CL). The grammar is a translation of F. Ahn's book for which the German lessons were translated into Bulgarian language. The grammar contains a preface, fourteen chapters, table of content, and corrections. The preface is written by the author and contains methodological instructions how to teach the included grammar topics. Chapters from first to eleven include rules, exceptions, lessons and related exercises developed for the French language grammar categories of gender, number, case as well as for their related part-of-speech inflections. Chapters from twelve to fourteen include study materials on French syntax, orthography and exercises.

Finally, the book by S. Radulov [16] published in 1868 in Bolgrad, Romania (in Bulgarian language) with a title similar to that of K. Tarnovsky's book also uses the F. Ahn's method of learning French language for Bulgarians (a scanned copy is available by the CL). The book contains six parts: *Alphabet*, *Pronunciation*, *First part*, *Second part*, *Short texts stories in French language*, and *Guide for conversation*. The *First part*, the *Second part*, and the *Short texts stories in French language* are following the K. Tarnovsky's book (which is a translation from F. Ahn), however the *Alphabet*, the *Pronunciation* and the *Guide for conversation* are original author's contributions which follow methodology for learning and practice the French language also indebted to F. Ahn's pedagogical approach.

4. The first grammar to teach Romanian language for Bulgarians

The book by S. Radulov [15] *A Practical and Easy Method of Learning the Romanian Language* published in 1865 in Bolgrad, Romania (in Bulgarian language) is considered to be the first

IOP Conf. Series: Materials Science and Engineering **364** (2018) 012049 doi:10.1088/1757-899X/364/1/012049

school book to teach Romanian language for Bulgarians, and is an original implementation of its author's pedagogical approach which is indebted to that of F. Ahn's general methodology. The scanned copy of that book which is offered by the CL[1] is still the only available digital resource for on-line search and analysis.

The book contains: *Preface* (from F. Ahn), *Preface* (from the author), *Pronunciation*, *First part*, *Second part*, *Third part*, *Examples with the four conjugations*, *Reading exercises*, *Dictionary*, *Guide for easy conversation*, *List of subscribers* and *Afterword*. That content of the book is similar to the content of [16] published later by the same author and suggests a well established methodology for teaching foreign languages combining the classical pedagogical approaches to grammar and lexical rules acquisition with reading exercises, so to master the ability to speak and understand, which, in fact, follows the F. Ahn's general methodology of learning foreign languages. The Fig. 3 shows the title page of the book.

4.1. The two prefaces

The author offers two prefaces of his book (Fig. 4). The first one is a preface which summarize the general pedagogical approach of F. Ahn (a citation) to learn foreign language by repeating and combining language units (following the way the child learns the native language), so to acquire the ability to speak.

The second preface, written by the author, explain the need for a school book to teach Romanian language for Bulgarian readers. Among all other reasons, it points out that the book was intended to be used as a school book in all Bulgarian schools.

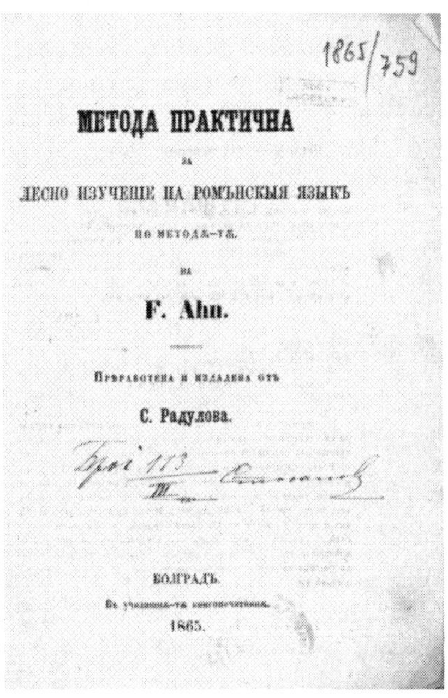

1865/759

МЕТОДА ПРАКТИЧНА

за

ЛЕСНО ИЗУЧЕНИЕ НА РОМЪНСКЫЯ ЯЗЫКЪ

по методѫ-тѫ

на

F. Ahn.

Прерàботена и издадена отъ

С. Радулова.

Болградъ.

Въ училищна-тѫ книгопечатница.

1865.

Figure 3. The title page of S. Radulov's book *A Practical and Easy Method of Learning the Romanian Language* from on-line sources of the CL.

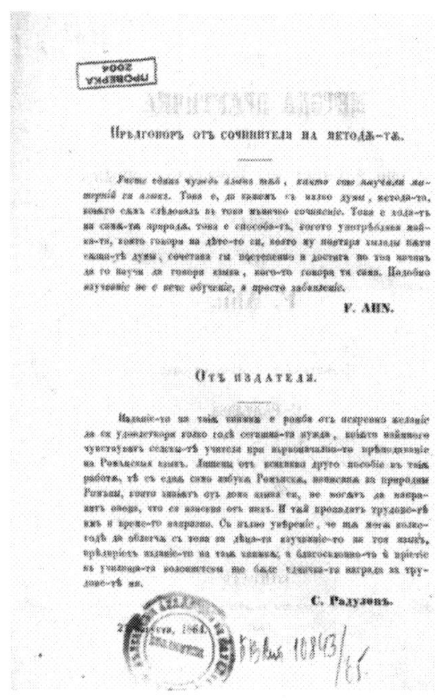

Прѣдговоръ отъ съчинителя на методѫ-тѫ.

F. AHN.

Отъ издателя.

С. Радуловъ.

Figure 4. The two prefaces of S. Radulov's book *A Practical and Easy Method of Learning the Romanian Language* from on-line sources of the CL.

[1] The copy is from the Library collection of Vasil Stoyanov – a Honorary member and one of the founders of the Bulgarian Academy of Sciences.

4.2. The Pronunciation
The first chapter of the book is the *Pronunciation* which contains the Romanian alphabet related to Bulgarian and accompanied by the pronunciation of all letters illustrated by explanations, rules and examples of all Romanian vocals and consonants.

4.3. The First part
The *First part* of the book contains 56 lessons which include the grammar category of gender, number, case, etc. introducing their part-of-speech inflections and supported by short text examples. The introductory texts are in Romanian language and they offer also the translations of unknown Romanian words. After the Romanian text, the text in Bulgarian language is presented, so to be translated into Romanian language.

4.4. The Second part
The *Second part* follows the same methodology as that presented in the *First part* but includes longer and more complicated texts in Romanian and Bulgarian language illustrated with the longer lists of unknown words and more verbs.

4.5. The Third part
The *Third part* also follows the methodology presented in the *First part* and *Second part* but introduces more verbs in related tenses illustrated with longer and more complicated texts in Romanian and in Bulgarian language.

4.6. The Examples with the four conjugations
The chapter *Examples with the four conjugations* includes example tables with the four conjugations and suitable examples of Romanian verbs (accompanied with their related translations in Bulgarian language). The verbs are presented according to the Romanian verb tenses and examples for: *present, imperfect, simple perfect, compound perfect, pluperfect, future* and *future perfect* tense relate them to the Bulgarian verb tenses' example translations with respect to the verbal features of *tense* and *aspect.* That presentation relates Romanian to Bulgarian verb also for all four Romanian verb moods: *indicative* (for all tenses), *subjunctive* (for *past* and *present* tense), *conditional* (for *past* and *present* tense) and *imperative* (for *present* tense). Also, the Romanian verbal forms for the *infinitive, past participles* and *gerund* are related to their Bulgarian counterparts.

Generally, the chapter presents for the first time an approach and related ideas and implementations for comparative grammar study of Romanian and Bulgarian verb outlining their semantic similarities and differences.

4.7. The Reading exercises
The chapter *Reading exercises* contains five short texts stories in Romanian language which are included for exercising the ability for reading and understanding the Romanian language texts by the Bulgarian readers.

4.8. The Dictionary
The chapter *Dictionary* presents what today we would say a *thematic* or a *conceptual* bilingual Romanian-Bulgarian dictionary which is an original author's contribution to the F. Ahn's methodology. It is interesting to note, that fifteen years after the publication of that book, similar dictionary of Latin Language [13] was published as an addition to F. Ahn's method.

The *Dictionary* is interesting in many aspects and generally, it presents an interesting genre of bilingual dictionary of synonyms which includes eleven cultural concepts (among them are *town*, *house*, *furniture*, *food*, *human*, *animal*, *birds*, etc.) described by their related hyponymic attributes for both languages.

That dictionary may be analyzed not only from the linguistic point of view (the meaning and the etymology of the included synonyms) but also from the anthropological point of view of the history of culture relating Romanian to Bulgarian concepts by revealing their new cultural and historical contexts.

4.9. The Guide for easy conversation

The chapter *Guide for easy conversation* also is thematically structured and includes twelve thematic areas (among which are *foods and drinks*, *arriving and departing*, *questions and answers*, *age*, *clock*, *time*, *greetings*, etc.) for which the related Bulgarian conversational phrases are presented together with their Romanian translations. That thematical structuring of the guide also is an author's contribution to F. Ahn's method since it did not appear in [16, 19]. The chapter also is very interesting from the point for view of inter-cultural study.

4.10. The List of subscribers

The chapter *List of subscribers* includes the names and families of subscribers of the book who lived in Bolgrad, Romania. It also presents a geographically related information (as a birthplace) about the origin of some subscribers as well as about the use of the book and its spreading within the related schools in Bolgrad, Romania.

4.11. The Afterword

The *Afterword* section includes acknowledgments of related persons to whom the author express his gratitude for their collaboration and support. That information is included also in the meta-data encoding and is specific to the digitization approach used.

5. Conclusion

The above description presents for the first time an analysis of content of the first grammar to teach Romanian language for Bulgarians by using its digitized copy from the on-line electronic resources of the CL of the Bulgarian Academy of Sciences. It considers the European tradition (the works of F. Ahn) and its implementation presented in the book by the use of related methodology for foreign language learning. At the same time, the author's original contributions are outlined with respect to the first comparative linguistic study of Romanian and Bulgarian verb, and to the inter-cultural aspects of analysis which consider the *Dictionary* and the *Guide for easy conversation* parts of the book as revealing new cultural contexts of the study.

References

[1] Ahn J F 1933 *Neue holländische Sprachlehre zum Selbstunterricht fur Deutsche: nebst einer Sammlung prosaischer und poetischer Stucke aus den besten Schriftstellern* (Koln: Pet Schmitz) (in German language)

[2] Ahn J F 1834 *Italienisches lesebuch in drei kursus, mit anmerkungen und einem wörterverzeichnis* (Leipzig: F Fleischer) (in German language)

[3] Ahn J F 1841 *Handbuch der französischen umgangssprache* (Köln am Rhein: M Du Mont-Schauberg) (in German language)

[4] Ahn J F 1845 *Französische Grammatik für Gymnasien und höhere Bürgerschulen* (Mainz: F Kupferberg) (in German language)

[5] Ahn J F 1847 *Praktischer lehrgang zur schnellen und leichten Erlernung der französischen sprache* (Köln: M Du Mont-Schauberg) (in German language)

[6] Ahn J F 1859 *The Poetry of Germany a Selection from the Most Celebrated German Poets of the Last Two Centuries Chronologically Arranged by Dr F Ahn* (Leipzig: F A Brockhaus)

[7] Ahn J F 1861 *German commercial letter-writer; a complete series of letters, circulars, and forms, suited to all the requirements of trade and commerce; with a German, French, and English glossary of all technical terms; and an appendix of suitable and business-like beginnings and endings* (London: Trbner)

[8] Ahn J F 1861 *Manual of German conversation or, vademecum for English travellers* (London: Trbner and Co)
[9] Ahn J F 2015 *A New, Practical and Easy Method of Learning the German Language* (New York: D Appleton and Co.)
[10] Condigiorgi P 1871 *Ahn's new practical and easy German method applied on the English language* (Constantinople: Ch Roth)
[11] Dikova E, Aneva S, Bogusheva E and Yavrukova B 2011 *Online Glossary of Bulgarian Library Terms* http://www.nalis.bg/library-standards/dictionaries (in Bulgarian language)
[12] Furrie B 2003 *Understanding MARC Bibliographic: Machine-Readable Cataloguing* (Follett Software Company, The Network Development and MARC Standards Office: Library of Congress)
[13] Henn P 1879 *Ahn's Latin vocabulary for beginners, methodical and etymological With a collection of Latin proverbs and quotations* (New York: E Steiger)
[14] Marchetti G 1857 *A new practical and easy method of learning the Italian language, after the system of F. Ahn* (London: Thimm)
[15] Radulov S I 1865 Метода практична за лесно изучение на ромънскыя языкъ по методата на *F. Ahn* Преработена и издадена от С. Радулова (Болград: В училищната книгопечятница) (in Bulgarian language)
[16] Radulov S I 1868 Практическа метода за изучавание на французскыий языкъ [Франц Ан] (Болград: Училищна книгопечятница) (in Bulgarian language)
[17] Stoykova V 2017 *Digitization of Bulgarian Natural Science School Books Published in Belgrade During the Period of National Revival (1806-1878) Review of the National Center for Digitization* vol. 31 pp 18-24
[18] Stoyanov M 1957 Българска възрожденска книжнина т 1 (София: Наука и изкуство) (in Bulgarian language)
[19] Tarnovsky K I 1864 Практическа метода за изучванието на френцкия языкъ, съчинител Ф Ан (Браила: Ромъно-българската книгопечятница на Хр Д Ваклидов) (in Bulgarian language)
[20] Tarnovsky K I 1869 Французска грамматика за класическите граждански училищта от д-р Ф Анн (Виена: у книгопечатн. Л Соммера) (in Bulgarian language)
[21] Zagorov V and Encheva M 2017 *Digitalization of the Repertoire of Books form Bulgarian Revival (1801-1878) Proceedings of ICERI2017 Conference* (Seville, Spain) pp 3545-3548

IOP Conf. Series: Materials Science and Engineering **364** (2018) 012050 doi:10.1088/1757-899X/364/1/012050

What is this painting about? Experiments on Unsupervised Keyphrases Extraction algorithms

M T Artese, and I Gagliardi

IMATI – CNR, Via Bassini 15, 20133 Milan, Italy
email: {teresa, isabella}@mi.imati.cnr.it

corrisponding author email: isabella@mi.imati.cnr.it

Abstract. A large number of cultural heritage archives are freely available on the web: they can be in Linked Open data format, or in any other format, such as databases, collections or archives, with some information for each object. To be really enjoyed and enjoyable by the users on the web, a set of scored keywords need to be associated with each item, manually or automatically. The overall problem here addressed is the automatic, unsupervised extraction of keywords/keyphrases from the items of cultural heritage archives, in different languages (English and Italian). The problem is very actual and in literature many papers are devoted to this topic and several approaches have been defined: we present here a work-in-progress, an experimentation done with the aim of automatically associating scored keywords/keyphrases to a painting archive. We have therefore tested five different methods present in literature, such as tf-idf, RAKE, TextRank, …, on two datasets, in English and in Italian, and evaluated the results - using recall and precision@n as the evaluation metrics.

1. Introduction

A large number of cultural heritage archives are freely available on the web: they can be in Linked Open data format, or in any other format, such as databases, collections or archives with pictures and some textual information for each object. To be really enjoyed and enjoyable on the web, users should be able to interact and navigate with simple and easy-to-use tools, using both the pictorial and textual information. Multimedia systems on the web, usually, provide facilities for searching, browsing, clustering and visualizing different kinds of visual data and related information. Many of their innovative tools are related to image searching and visualization functionalities in the context of cultural heritage web sites. To be effectively and efficiently searched, retrieval, and browsed, textual data need to be assigned (scored) keywords/keyphrases, manually or automatically.

Keywords / key phrases are single words or groups of words that characterize the content of the document and are useful for identifying the documents relevant to a given query and / or for "suggesting" something in some way related.

Keywords/keyphrases should, therefore, satisfy the following characteristics:

- have a good coverage: in that, all the keywords/keyphrases of an item are able to describe all its aspects;
- be relevant: general enough to represent more than a single item, as well as specific not to represent the whole set of items.

As well known, manually associating keywords to documents requires time, skills, specialized personnel, and more steps to ensure that the chosen terms are consistent, adequate, relevant, with a good coverage, sufficiently general and timely. To automate this activity, much work has been done over the years and researchers have explored both supervised and unsupervised techniques to address the problem of automatic keyphrase extraction. The problem is still very actual and in literature, many papers are devoted to this topic and different approaches have been defined. Furthermore, most of the studies and experiments have been conducted on texts in English, while there are few experiments concerning other languages, such as Italian.

In this paper we will consider the problem of automatic unsupervised extraction of keywords on texts, by presenting a work-in-progress, an experimentation done on a corpus of texts (in English and in Italian) related to paintings, to be integrated into a Multimedia Information system on the web.

We have applied five methods/algorithms present in literature, such as TextRank, RAKE, tf-idf, … to texts in English and Italian, to better understand the different characteristics of these methods and their use in a cultural heritage context in different languages, and present here the results, evaluating them in the context of our experimentation.

The paper is organized as follows: after the presentation of the overview of works related to automatic unsupervised keyword extraction, the data used and the experimentation runs are presented, together with the details of the solutions tested in the runs. The results obtained are then presented, compared and discussed.

2. State-of-the Art: Unsupervised Keyphrases Extraction algorithms

Although the problem of extracting keywords, keyphrases able to represent the content of pieces of text in natural language has been faced since the first Information Retrieval systems appeared, the advent of new tools and techniques makes it still very actual: in literature, many papers are devoted to this topic [1][2][3][4] and several approaches have been defined [5][6][7][8][9].

In the following, we present different methods that have shown to work well and have been tested in different domains, and that we have used in our experimentation.

2.1. TextRank

TextRank [7] is a graph-based ranking model for text processing, and is based on the same assumptions of PageRank [10][11], that is: i) important pages are linked by important pages, and ii) the PR values of a page is essentially based on the probability of a user visiting that page.

TextRank models a text document as a graph where each word (or sentence, according to granularity chosen) in the document is represented as a node, and the semantic relation between each node is represented by an edge. A scoring algorithm will then be run on the graph to assess the value of every word, which will be used to rank them, then top ranking words are selected as keywords. Keywords that are adjacent in the document are combined to form multi-word keywords. The authors [7] report that TextRank achieves its best performance when only nouns and adjectives are selected as potential keywords.

2.2. RAKE

Rapid Automatic Keyword Extraction (RAKE) [8] is an unsupervised, domain-independent, and language-independent method for extracting keywords from individual documents.

The algorithm is composed of the following steps: i) it extracts a set of candidate keywords (single words or sequences of contiguous words), ii) assigns a score to these candidate keywords, based on the co-occurrence graph: the score is based on the frequency of the words and their degree, iii) adjoins keywords and iv) extracts the top T scoring keywords.

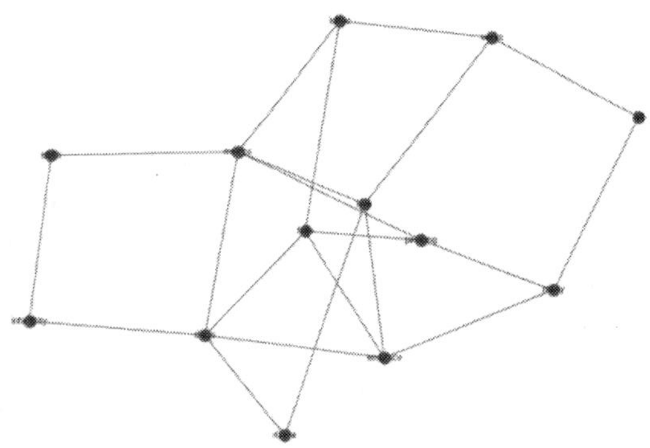

Figure 1. a graphical representation of TextRank algorithm for the painting of Frida Kahlo "Wounded Deer"

2.3. *Tf-idf*

Tf–idf [6][12], short for term frequency–inverse document frequency, is widely used in Information Retrieval and is used to measure the importance of a word in a document, within a collection or corpus. At a high level, a tf-idf score finds the words that have the highest ratio of occurring in the current document vs the frequency of occurring in the larger set of documents.

2.4. *Latent Semantic Analysis or Latent Semantic Indexing*

LSI [13] is a technique able of analyzing relationships between a set of documents and the terms they contain by producing a set of concepts related to the documents and terms.

LSI assumes that words that are close in meaning will occur in similar pieces of text. A term-document matrix, as that created by Tf-idf model, is the starting point for the technique used to reduce the number of rows in the matrix, while preserving the similarity structure among columns, thus converting terms and documents to points in a lower-dimension space.

2.5. *Other methods*

In literature, many other methods have been proposed and tested as, among others, SingleRank [14] and ExpandRank [15], that are essentially a TextRank expansion, or cluster-based approaches as KeyCluster [16].

3. Dataset

For our experimentations, we started from the 'paintings 91' [17][18] database, which consists of well-known paintings by famous artists ranging from Vermeer to Chagall, from Bosch to Matisse. The data set consists of 4,214 paintings from 91 different artists; is composed of images, and some basic information such as author, artist categorization, and style classification for each painting. No title nor painting description is available in the original data. We first associated to each painting its title (in Italian or in English), and then, in a completely automated way, we retrieved from the web a description of the painting, whose provenience is mainly from Wikipedia or the first result of a google search based on title + author query, in Italian and/or in English. To ensure that the webpages retrieved are semantically related to the paintings, a similarity score based on the co-occurrence of words in the web page title and painting title has been computed and assigned.

Table 1. Characteristics of Dataset used.

Dataset	No. items	Type of data	Language	Average length, max, min
Painting91 (original dataset)	4214	Authors, artist categorization, style	Title in Italian or English	-

		classification and title		
English	3794	Authors, Title and content	English	6348 chars (max 25926, min 500)
Italian	3088	Authors, Title and content	Italian	4047 chars (max 19436, min 502
Other languages	718	-	German, French, Dutch, Polish, …	-

Characteristics of the data:

- language: while data are available in different languages, such as German, French, Dutch, Polish, in this paper we present the results only for English and Italian data;
- length: informative content of each item in English language has an average length of 6348 characters, varying from 500 to almost 26000; while in Italian, texts are shorter, with an average of 4000 characters, (from 500 to less than 20000); in both languages, contents of less than 500 lengths were discarded;
- topics: all data refers to paintings, from Durer to Winslow, from Tintoretto to Boccioni, from Botticelli to Salvador Dalì, with a time span from 1400 to 2000, and with baroque, surrealism, pop art styles.

4. The experiments

With the aim of testing these methods and of identifying if (and, in case, which) method produces better results on these data with a set of test queries, and if there is any noticeable difference between English and Italian, a standard Information Retrieval approach [6] has been followed, shown in Fig. 2.

In the processing phase, the following steps, rather common, were carried out, with the aim to identify the most relevant keywords, to be used as an aid to perform a query and innovative ways to navigate the data:

- Pre-process data: remove stopwords from the text, perform stemming, tokenize, … ,
- Extract a list of candidate keywords /keyphrases using some heuristics,
- Score each candidate keywords/keyphrases, according to different criteria and methods,
- Select the first m keywords/keyphrases.

We run the following Unsupervised Keyphrases Extraction algorithms, described in section 2:

- TextRank
- Tf-idf
- RAKE
- Latent Semantic Indexing (LSI)

and the following python implementation:

- Automatic Keyword Extraction (AKE) (using gensim python package [21]): this method, based on the TextRank algorithm, has been applied "as is" with the default values as parameters, that is windows size = 2 and considering only nouns and adjectives. Because in this implementation pre-processing and stemming algorithms only works for English, we have disabled them, using texts as presented, in both languages.

We tested Tf-idf both using 1-gram and n-grams basis for word extraction; for LSI we performed tests with a different number of topics: here we report the results for 20 and 100 topics.

Some algorithms have to be performed document by document (TextRank, RAKE, AKE), while others have to be run on the whole collection of documents (tf-idf, LSI).

IOP Conf. Series: Materials Science and Engineering **364** (2018) 012050 doi:10.1088/1757-899X/364/1/012050

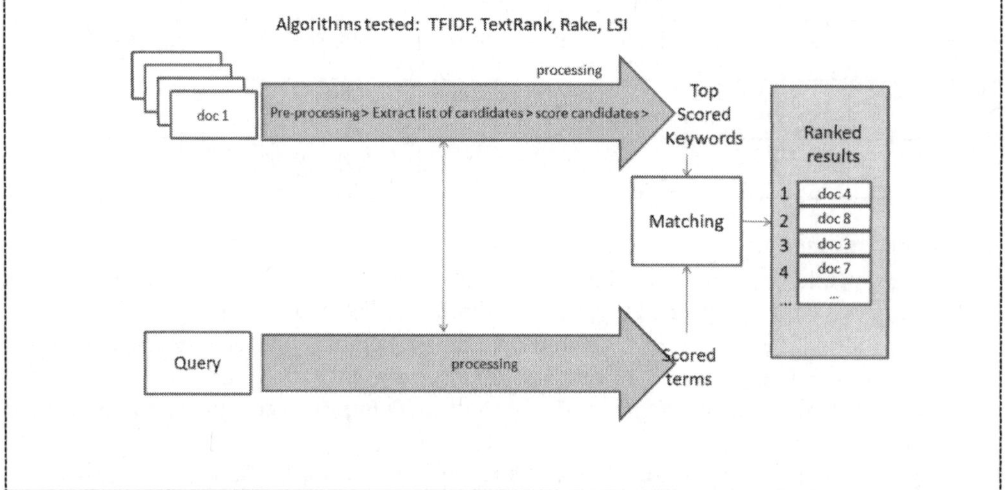

Figure 2. Information Retrieval architecture of our approach.

According to the diagram in Figure 2, the keywords / key phrases automatically extracted are then exploited for indexing documents, in a vector space model and the cosine similarity is the chosen metric used to retrieve and rank documents in response to the query.

Each run on an algorithm/dataset (English/Italian) gives rise to a different set of (weighted) keywords/keyphrases, associate to each document: so that the retrieved document list can be different both in the documents retrieved and in the order of the retrieval.

Table 2 shows information and some statistics on the results obtained from the processing phase, applying the 5 algorithms on Italian and English datasets, with tf-idf runs for 1-gram and n-grams, and LSI with 20 topics and 100 topics. In this phase the same parameters were applied to the 2 datasets, modifying, where necessary, stopword lists and stemming algorithms.

Table 2. keyphrases stats for the 5 algorithms tested.

Algorithm	No. different keyphrases	*Average length of keyphrases*	No. keyphrases with more than 10 occurrences (keyphrases10)	*Average length of keyphrases10*	No. keyphrases with more than 100 occurrences (keyphrases100)	*Average length of keyphrases100*
TextRank						
English	27.439	2,011	2237	1,581	133	1,083
Italian	14.674	1,755	958	1,280	63	1,015
Tf-idf						
English 1-gram	11.781	1	663	1	0	-
English n-grams	15.931	2,398	550	1,627	0	-
Italian 1-gram	11.599	1	486	1	0	-
Italian n-grams	16.254	1,755	1,28	1,280	1	2,0
RAKE	s					
English	8.511	1,963	549	1,828	16	1,375
Italian	6.300	1,838	405	1,476	15	1,133
LSI 20 topics						
English	98	1	89	1	39	1
Italian	98	1	95	1	21	1
LSI 100 topics						
English	314	1	286	1	125	1

IOP Conf. Series: Materials Science and Engineering **364** (2018) 012050 doi:10.1088/1757-899X/364/1/012050

Italian	330	1	320	1	66	1
AKE						
English	13362	1,106	1882	1,026	137	1
Italian	15562	1,355	1406	1,102	1333	1

Some considerations about the results:

- The number of different keyphrases varies greatly on the basis of the algorithm tested: from 98 of LSI with 20 topics to over 27.000 of TextRank.
- Except for LSI and AKE, the other algorithms consistently give lower values for Italian than English, due to the lower number of documents and shorter length of texts. Higher values in AKE are related to the lack of appropriate stemming algorithms for Italian;
- TextRank produces a relatively large number of keywords, using the recommended values: setting a threshold value, as the maximum number of keywords extracted per document or the weight of the nodes must be further tested and evaluated;
- TextRank, RAKE and AKE extract keyphrases longer than 1. Both in English and Italian. We tested tf-idf both for 1-grams and n-grams: in the latter case, the average length of the words extracted is more than 2 for English, and it is around 1.7 for Italian.
- LSI basically works as a clustering algorithm, associating each cluster with a set of keywords that represent it. The results reported were obtained by choosing 20 and 100 as the number of topics.

We tested the implemented algorithms on the two datasets, using the queries of Table 3. Since these methods and algorithms have already been tested and have proven to work well, in order to evaluate the results for our goal, i.e. the integration into a Multimedia Information system, we have assessed the results in terms of recall and precision. Recall is the fraction of relevant instances that have been retrieved over the total amount of relevant instances and its value is 1 when all the relevant instances are retrieved, decreasing up to 0 when none relevant instance is retrieved. Precision is the fraction of relevant instances among the retrieved instances, and the its value varies from 1 to 0, with 1 if only relevant documents are returned, and 0 if none relevant document is retrieved. In this paper, we use P@n, precision measured on the first n topmost results returned, with n = 10, 20 and 100.

Table 3. Queries used to test the algorithms

	Queries
English	art, artist, painter, braque, circle, man, people, landscape, nature, vermeer, green, museum, lissitzky, woman, rome, portrait, death, italian, paintings, mother, women, black, Picasso, Pablo Picasso, new york
Italian	arte, artista, pittore, braque, cerchio, uomo, persone, paesaggio, natura, vermeer, verde, museo, lissitzky, donna, roma, ritratto, morte, italiano, quadri, madre, donne, nero, picasso, Pablo Picasso, new york

TextRank, RAKE and AKE, another implementation of TextRank, index each document only with top ranked terms and when a query is performed, the matching algorithm retrieves and scores only those documents that have in common terms with the query; the other documents are scored 0; on the other hand tf-idf and (in some measure) LSI algorithms give a weight to each term in the document: thus in essence, documents are sorted according to the weight of the terms in the query.

For this reason, the accuracy of TextRank, RAKE and AKE is always very high and around 1, thus confirming the high quality of the algorithm results. On the other hand, recall values can be low because only those documents to which the searched terms have been associated are retrieved.

For the query 'woman', the dataset contains 149 paintings with woman in the title: TextRank retrieves 288 paintings, RAKE 52, and AKE 148. Rake has thus a recall value of 0,5 at the best.

Query 'Picasso': in the dataset are present 48 paintings by Pablo Picasso, for English TextRank retrieves 150 items, Rake 25 and AKE 73; and for Italian, the results are almost the same. Also in this case the recall of RAKE is about 0,5. No noticeable change for query 'Pablo Picasso'. In general, it can be noted that AKE results, for English, are comparable to TextRank, while for Italian, due to the lack of appropriate stemming algorithm, are less performing.

We evaluate the precision only for tf-idf and LSI. Tf-idf and LSI 100 give precision@10 and P@20 value almost close to 1, with lower precision on 100 items. Tests on LSI 20 have shown that 20 clusters for these 4000 items are too few: in fact, LSI 20 inserts documents that are very different from each other in the same clusters, so it fails to extract the keywords correctly.

Precision does not vary much comparing results for tests on Italian and English data sets, because the algorithms work on a single language at a time, thus identifying structure and words, adopting different stoplists when they process the two datasets.

The experimental setup has been implemented in Python 2.7, using NLTK, gensim, newspaper, ski-kit packages, together with some experimental packages in GitHub [23].

5. Conclusion and Future Works

We have presented here a work in progress for the automatic unsupervised extraction of keywords/keyphrases for a corpus of texts (in English and in Italian) related to paintings, to be integrated into a Multimedia Information system on the web. Five algorithms present in literature have been tested on two different datasets and the results of the experimentation have been reported here, to evaluate which algorithm works best, in terms of recall and precision, with respect to query tests. Preliminary results show that TextRank, RAKE methods have been proven to obtain the best values for precision, always around 1, while the recall, especially for RAKE is rather low. AKE results, for English are comparable to TextRank, while for Italian, due to the lack of appropriate stemming algorithm, are less performing. Tf-idf and LSI 100 provides good results, while LSI 20, in these tests fails to extract correct keywords.

Concluding, we can say that TextRank, tf-idf and LSI are the algorithms that provide the best results; TextRank and tf-idf do not need any adaptation, while, for LSI, the choice of the number of clusters is decisive and several tests may be necessary to obtain the best results.

The results of queries on English and Italian datasets perform almost the same, although English benefits from more studies, experimentations, and algorithms. Further tests, with real users, will be performed once these tools will be integrated in the Multimedia Information system on the web.

The results have shown that further works or investigations should be oriented towards:

- although some algorithms should automatically identify synonyms and homonyms, WordNet [19][20][24] could be used to further reinforce these aspects;
- test the algorithms using some controlled set of keyphrases against which match our results as the title of Wikipedia[25] pages or title of books or journal paper;
- test the algorithms on other multimedia archives;
- test other algorithms, such as for example SingleRank, ExpandRank or Keycluster.

References

[1] Hasan, K. S., & Ng, V. (2010, August). Conundrums in unsupervised keyphrase extraction: making sense of the state-of-the-art. In Proceedings of the 23rd International Conference on Computational Linguistics: Posters (pp. 365-373). Association for Computational Linguistics

[2] Hasan, K. S., & Ng, V. (2014, June). Automatic Keyphrase Extraction: A Survey of the State of the Art. In ACL (1) (pp. 1262-1273)

[3] Merrouni, Z. A., Frikh, B., & Ouhbi, B. (2016, October). Automatic keyphrase extraction: An overview of the state of the art. In Information Science and Technology (CiSt), 2016 4th IEEE

International Colloquium on (pp. 306-313). IEEE.
[4] Siddiqi, S., & Sharan, A. (2015). Keyword and keyphrase extraction techniques: a literature review. International Journal of Computer Applications, 109(2).
[5] Gollapalli, S. D., Caragea, C., Li, X., & Giles, C. L. (2015). Proceedings of the ACL 2015 Workshop on Novel Computational Approaches to Keyphrase Extraction. In Proceedings of the ACL 2015 Workshop on Novel Computational Approaches to Keyphrase Extraction.
[6] Christopher D. Manning, Prabhakar Raghavan, and Hinrich Schütze. 2008. Introduction to Information Retrieval. Cambridge University Press, New York, NY, USA.
[7] Mihalcea, R., & Tarau, P. (2004). Textrank: Bringing order into text. In Proceedings of the 2004 conference on empirical methods in natural language processing.
[8] Rose, S., Engel, D., Cramer, N., & Cowley, W. (2010). Automatic keyword extraction from individual documents. Text Mining: Applications and Theory, 1-20.
[9] Hulth, A., & Megyesi, B. B. (2006, July). A study on automatically extracted keywords in text categorization. In Proceedings of the 21st International Conference on Computational Linguistics and the 44th annual meeting of the Association for Computational Linguistics (pp. 537-544). Association for Computational Linguistics.
[10] Page, L., Brin, S., Motwani, R., & Winograd, T. (1999). The PageRank citation ranking: Bringing order to the web. Stanford InfoLab.
[11] Brin, S., & Page, L. (2012). Reprint of: The anatomy of a large-scale hypertextual web search engine. *Computer networks*, *56*(18), 3825-3833.
[12] Salton, G., & Buckley, C. (1988). Term-weighting approaches in automatic text retrieval. Information processing & management, 24(5), 513-523.
[13] Landauer, T. K. (2006). Latent semantic analysis. John Wiley & Sons, Ltd.
[14] Wan, X., & Xiao, J. (2008, July). Single Document Keyphrase Extraction Using Neighborhood Knowledge. In *AAAI* (Vol. 8, pp. 855-860).
[15] Wan, X., & Xiao, J. (2008, August). CollabRank: towards a collaborative approach to single-document keyphrase extraction. In *Proceedings of the 22nd International Conference on Computational Linguistics-Volume 1* (pp. 969-976). Association for Computational Linguistics.
[16] Liu, Z., Li, P., Zheng, Y., & Sun, M. (2009, August). Clustering to find exemplar terms for keyphrase extraction. In *Proceedings of the 2009 Conference on Empirical Methods in Natural Language Processing: Volume 1-Volume 1* (pp. 257-266). Association for Computational Linguistics.
[17] Khan, F. S., Beigpour, S., Van de Weijer, J., & Felsberg, M. (2014). Painting-91: a large scale database for computational painting categorization. Machine vision and applications, 25(6), 1385-1397.
[18] Artese, M. T., Ciocca, G., & Gagliardi, I. (2017). Evaluating perceptual visual attributes in social and cultural heritage web sites. Journal of Cultural Heritage.
[19] Fellbaum, C. (1998). *WordNet*. John Wiley & Sons, Inc..
[20] Miller, G. A. (1995). WordNet: a lexical database for English. *Communications of the ACM*, *38*(11), 39-41.

Websites (last consulted on 16 January 2018)
[21] Gensim topic modelling for humans https://radimrehurek.com/gensim/
[22] Newspaper: News, full-text, and article metadata extraction in Python https://github.com/codelucas/newspaper
[23] github https://github.com/
[24] WordNet: https://wordnet.princeton.edu/
[25] Wikipedia http://www.wikipedia.org

Review on new technologies in Art and Archaeology: The Vesuvian cities

A E Pérez Saborido[1,2], F Castanò[1], M Buono[1]

[1]Dipartimento di Ingegneria Civile, Design, Edilizia e Ambiente, Università degli Studi della Campania "Luigi Vanvitelli", Italy
[2]Departamento de Ciencias Historiográficas y Arqueología, Universidad Complutense de Madrid, Spain

anaelisp@ucm.es, francesca.castano@unicampania.it, mario.buono@unicampania.it

Abstract. The importance of Vesuvian archaeological sites goes back to the moment of their discovery in the eighteenth century, which since then has promoted a new perception of the antiquity that led to a systematic and scientific study of the cities. The continuity of the excavation work, as well as the interest which they have raised for more than three centuries, makes it possible to trace the evolution of the different initiatives through which technology has been exhibited to the service of archaeology. With the advent of digital technology, new means of study, diffusion, tutelage and accessibility are emerging which will certainly adopt this unique heritage as a subject of study. This paper analyses the different initiatives that have advocated the application of technologies that aspire to improve the visitor experience and to generate a greater understanding of the archeological sites and their global relevance. It therefore highlights the activities carried out in the last decade, both in situ involving the MAV (Museo Archeologico virtuale, Ercolano. 2008), The *Grande Progetto Pompei* (from 2011), as well as those carried out at the international level, such as the exhibition *Carlos III y la Difusión de la Antigüedad* (Naples, Madrid, Mexico, 2016-2017). The limitless possibilities that technology and interdisciplinary collaboration offer to Heritage, as well as a critical, estimable analysis for future proposals, will also be highlighted.

1. Technology at the service of Archaeology: a historical overview

The importance of the Vesuvian archaeological sites goes back to the moment of their discovery which, from the eighteenth century, promoted a new perception of the antiquity that led to the systematic and scientific study of the cities. The continuity of the excavation work, as well as the interest aroused by these for more than three centuries, makes possible to trace the evolution of the different initiatives in which the technology has manifested itself in the service of archaeology.

The multidisciplinarity, which might seem like a current demand, has always existed in an area as complex as archaeology. The discovery of the Vesuvian sites presented an unprecedented challenge: on one hand, the typology and nature of the terrain and, on the other, the unique circumstances that buried the cities. A process of excavation so difficult that continuous problems were emerging called for the collaboration of engineers, architects and restorers.

The toughness of the terrain, the excavation through tunnels under the city of Resina, and the vast number of finds preserved perfectly in their original context; both of great artistic value as well as those of daily life, assumed a revolution in the concept of archaeological heritage. Until that time it

IOP Conf. Series: Materials Science and Engineering **364** (2018) 012051 doi:10.1088/1757-899X/364/1/012051

was linked solely to art and collecting (which will never end to be eradicated). A material culture never before seen aroused the need to seek solutions never required until that time. This lead to the development of new restoration techniques, the creation of laboratories and specific museums and new instruments such as the famous "Macchina di Piaggio ", designed specifically to unwind the papyrus found in the Villa of the Papyri.

The systematic excavations included an archaeological record, the elaboration of a catalogue and an increasingly exhaustive control of all the activity related to the ancient cities; from the excavation, to the safeguarding of the different finds and their dissemination. The first reconstructions destined for these purposes, through the elaboration of plans, reconstructions and drawings by hand of great architects like Weber or La Vega, began to be carried out.

The exceptional nature of the discoveries soon acquired a didactic purpose, beginning in the first instance in a restricted and elitist character in Bourbon period and of a more democratic character in the time of Fiorelli. Publications, models and copies were made to enhance the study of sculptures and other findings. With the existing technology, plaster copies of the sculptures were sent to the academies of Madrid and Mexico in the eighteenth century which favored the development of the Enlightenment. Further models of the theatre of Herculaneum, made of wood and cork in 1808 by D. Padiglione, while models of Pompeii, were made by Fiorelli in the nineteenth century. This archaeologist also developed specific techniques that greatly increased the acquisition of archaeological information, such as the plaster casts of the victims of Pompeii and the establishment of the first initiative of an *in-situ* museum. Technology also collaborated during times when archaeology became a fundamental element of the foundations of new nationalism and cities were boosted as tourist attractions. In the nineteen thirties, not only were tunnels left behind to excavate Herculaneum, giving way to the assistance of mechanical excavators and electric arms, in addition, all kinds of technologies came into play in the promotion of the incipient mass tourism and the fruition of the cities. This included transport-related infrastructures as well as graphical and media documentation of the findings.

These past initiatives and technological advances which responded to the needs that the Vesuvian excavations required, are the precedents of what today's technology tries to contribute through 3d models, virtual reconstructions and *in-situ* immersive visits.

2. New needs, new technologies

Technologies have a fundamental role in the valorization of heritage. Previously, it has been presented that historically technology has been put to the service of archaeology as needs have arisen. In the same way, the technology today responds to the current needs of these sites, and the needs of our society in relation to culture and heritage. Some of these needs, despite the successive initiatives, have not changed. For example he conservation of the structures that remain in situ and the need to awaken an awareness of the importance of these cities to a society that is still far from assimilating the transcendence not only archaeologically, but also historically, linked to them. New needs arise however do arise which are typical of our time; such as the decontextualisation of objects and structures and the disastrous impact of human action as a result of mass tourism. Of course, the technological advances related to the diagnosis, conservation and restoration of heritage have acquired high levels of development increasingly specialized and that demonstrate a great advance in the time of safeguarding the remains of the antiquity, as shown by the Grande Progetto Pompei (from 2012) or the Herculaneum Conservation Project (since 2001), in collaboration with the Packard Humanities Institute.

With regard to the need to cultivate a society on the value of the Vesuvian cities from a didactic purpose, we see continuity through all the background evidence that the history of these sites allows us to trace; through plans, reconstructions, drawings and publications. Digital technology has allowed real and imagined reconstructions of the sites, with exact precision, each time with a wider scope for interactive possibilities, which allow individuals to immerse themselves in antiquity in a way that does not necessarily require them to visit the physical site.

Most European initiatives in relation to artistic and archaeological heritage seem to go in this direction. Exhibitions and museums use multimedia facilities to offer visitors an immersive experience that allows them to understand what they are observing in an interactive way. An illustrative example of this is the *Van Gogh Alive* itinerant exhibition, which tries to gather all the work of the artist in a single space and present it to different cities of the world for spectators to view the artists work simultaneously. Other examples of highly illustrative technologies offering an immersive explanation are the *Domus Romane di Palazzo Valentini* in Rome or *Chiesa de Santa Maria Antiqua* in Rome (via Videomapping). An increasing number of varied interactive devices integrated by the vast majority of museums, such as touch screens, holograms, motion sensors and apps for smartphones and tablets, make the content offered by 3D models, 3D printings, photogrammetry and virtual reconstructions available to the user.

3. The Vesuvian Cities in the era of digital technologies

Vesuvian cities have also called upon digital technology. As they are among the most emblematic of archaeological sites and the first to undergo systematic excavations, they offer a huge amount of information valuable for this type of purposes. Below the most representative cases that highlight the interaction of digital technology with this Vesuvian heritage will be presented as they have always been the object of interest for the application of cutting-edge technologies.

It is presented in the first place those centers related to the deposits in situ or in the vicinity that correspond to the first initiatives. Next, exhibitions and international projects that try to recompose the material and cultural dissemination that these sites have starred in through the new technologies are described. Finally, it is expounded the most recent proposals for valorization and fruition, which show the direction to which the new trends are directed.

3.1. The first 'virtual only' archaeological museum

While the Italian government declared the state of emergency after the deterioration of the ruins of Pompeii (visited by 2.5 million people a year) in Herculaneum, the first Virtual Archaeological Museum (MAV) of the world was inaugurated. After three years of work, in 2008 the public was offered the reconstructions of the cities of Pompeii, Herculaneum and Stabiae with technologies of last generation (laser, virtual reality, scanner, three-dimensional screens) without displaying any archaeological findings, with minimal scenery and the use of immersive, interactive and invisible technology. The project was designed by Gaetano Capasso, an electronics engineer and expert in virtual reality in the field of cultural heritage. The museum is the first example of an immersive structure: the experience is based on virtual reality and interactive electronic installations activated by tactile digital interfaces or motion sensors, similar examples of which can only be found in theme parks (Figure 2).

The MAV was also proposed as a didactic instrument to be used in parallel to a more traditional museum trying to overcome one of the limits posed by the latter with the difficulty in understanding by the general public. Virtual reality allows a re-creation of what has been lost, bringing the visitor closer to the world of archaeological culture by directly involving the visitor's senses, an application that may lead to new synergies between traditional museums and multimedia research.

Despite of the museum concept and technological innovation, the MAV presents the usual problems of this type of museums, like the fast obsolescence of the devices and others like certain incoherence in the narrative of the different topics from the different sites. In addition, with the exception of some devices such as the interactive table, the user traverses the different spaces being a mere spectator, since the reconstructions are being projected on screens (Figure 1).

However, the idea to use modern technologies was already born in 2006 to re-propose, in the noble plan of the palace, the original image of the Herculanense Museum and the cultural climate in which it was formed. The history of the excavations, their techniques and the procedures followed in the Bourbon age for the detachment of the frescoes were illustrated by multimedia projections or films. The Roman frescoes, once collected in the palace, were reproduced with the technique of the backlit paintings (Figure 3). A virtual reconstruction returns the theatre of Herculaneum, the first of the archaeological discoveries in the area. In the last multimedia room of the present museum stands the "Lanterna Magica" (Magic Lantern), a machine of vision that transports the visitor to the time of King Charles and his court. Inside are the wonders of the findings coming from the Vesuvian excavations (Figure 4).

Figure 1. Screen with virtual reconstruction inside the MAV.
Figure 2. Bourbon excavations hall, MAV.

Figure 3. Collection of backlit paintings, Herculanense Museum.
Figure 4. "Lanterna Magica", Herculanense Museum.

3.2. Breaking the geographical barriers

One of the most striking aspects in the history of the excavations has been the uncontrollable dispersion of the findings, which was a constant concern on the part of the entities in charge of their tutelage. Some projects have demonstrated the capacity that new technologies have acquired in providing context to these scattered findings, as well as the connections that the displacement of these pieces have caused, creating cultural bridges between countries and continents.

An illustrative example is the reconstruction of the frescoes from Boscoreale "*The Villa of Publius Fannius Synistor in reality and Virtual reality*" (2010), made by the Metropolitan Museum of Art, New York. It attempted to relocate virtually all the frescoes that were extracted from their original site and place them in their context, providing an ornamental coherence (Figure 5 and 6). The most interesting thing is that it gathers frescoes that are found not only in the MET but also in different geographical points: from the Musée Royal de Mariemont, Morlanwelz, Belgium, from the Musée de Picardie, Amiens, from the Villa Kérylos, Beaulieu-sur-Mer, France, Musée du Louvre, Paris, Allard Pierson Museum, Amsterdam, Musée Bonnat, Bayonne, Musées Royaux d´Art et d´Histoire, Brussels and the Museo Archeologico Nazionale of Naples.

Another example is one of the most innovative initiatives carried out in recent years: the commemorative exhibition organized on the occasion of the tercentenary of the birth of the monarch *Carlos III y la difusión de la antigüedad*. This exhibition was organized from December 2016 to March 2017 at the same time in Naples (National Archaeological Museum), Madrid (Royal Academy of Fine Arts of San Fernando) and Mexico (Academy of San Carlos). The exhibition aimed to highlight the role of the king in the dissemination of archaeological discoveries from Herculaneum, Pompeii and Stabiae and the impulse he gave to the knowledge of antiquities. Through virtual reality there were three stories that narrated three different moments of the eighteenth century: The excavations in the area of Vesuvius, the drawing room at the Academy of Fine Arts in San Fernando and the arrival at the Academy in San Carlos de Mexico of the plaster casts sent from Madrid.

The true innovation of this exhibition lies in the use of technology that allows the visitor to see the real scope of the work done three hundred years ago with the media today. This made it possible for the exhibition to be carried out simultaneously in three rooms, in three different museums and in three different countries, giving spectators the possibility of being in one of them and virtually visiting the other two at the same time. The rooms, connected by the theme of the work of the monarch, presented a similar installation being interconnected in real time via streaming and used technical means as virtual reality, augmented reality and spherical photography (Figure 7). All information about the content of the exhibition and access to 3d images and videos 360 ° (Figure 8) was made available for consultation through the APP *Carlos III*, downloadable from Play Store (Android) and on the Web page.

This exhibition aimed to be the first step in the design of a virtual exhibition model, which invited the interconnection of museums and their collections, allowing the accessibility to pieces that are thousands of miles away and inviting the interconnection of museums and their collections.

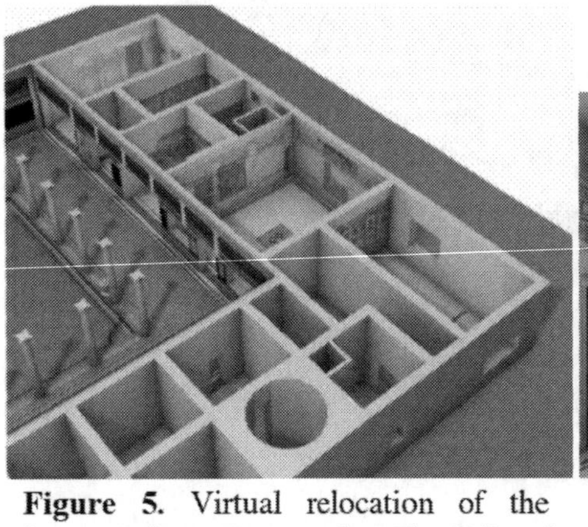

Figure 5. Virtual relocation of the frescoes from Boscoreale "*The Villa of Publius Fannius Synistor"*

Figure 6. Virtual relocation of the frescoes from Boscoreale "*The Villa of Publius Fannius Synistor"*

IOP Conf. Series: Materials Science and Engineering **364** (2018) 012051 doi:10.1088/1757-899X/364/1/012051

Figure 7. Augmented reality in *Carlos III y la difusión de la antigüedad*. Exhibition.

Figure 8. Access to 3d images and videos 360 ° in *Carlos III y la difusión de la antigüedad*. Exhibition.

3.3. Promoting awareness: Taking advantages of the possibilities of technology

The relationship between society and heritage has varied over time and education and understanding of the value of the latter has been a constant concern. The awareness of the value of the Vesuvian archaeological heritage is essential for its safeguarding, being on many occasions, despite the numerous initiatives, in deplorable conditions. In the last two decades, it has been tried to remedy through the two aforementioned important projects *Grande Progetto Pompei* and *Herculaneum Conservation Project*, with the collaboration of the Packard Humanities Institute. However, in parallel to the fruition of the ancient cities, the virtual era encourages projects of valorization that put, at the service of the society, a great quantity of accessible digital material that makes the understanding of the sites and its impact in the history of Archaeology easier.

The facilities arranged in the *Antiquarium* of Pompeii, re-opened to the public in 2016 are a good example of a natural integration of technologies to improve the understanding of visitors. Some screens were installed in a room that tried to show an immersive reconstruction of the moment of the eruption. Although the technology used is not the most innovative or revolutionary, it denotes the need for its use for any narrative in a museum context. Also, another good example of an attempt to integrate technological devices to attract attention to the Vesuvian archaeological remains of the museum is the video game produced by the Museo Archeologico Nazionale of Naples, "*Father and Son – the game*" (2017) (Figure 9), which tries to involve people in an exciting experience from the present to the past (https://www.museoarcheologiconapoli.it/it/father-and-son-the-game/) or the *Pompeii Touch app* (last updated 2018) (Figure 10) that provide the reconstructions of the ancient buildings in 3D with interactive GPS map in the smartphone or iPhone (http://www.pompeiitouch.com/). Users can choose a photo of a location and, with a touch or brushing the finger on the screen it appears correspondent and reconstructive hypothesis of places.

The *Estándares para la edición multimedia avanzada en yacimientos arqueológicos* (2017) by The Real Academia de Bellas Artes de San Fernando (Madrid). This project has replaced the concept of free navigation through the concept of orderly reading the content of the websites. This system gives access in a sequential way, from different chapters, to the databases, videos, specific texts and to all that multimedia content that can contribute to the better comprehension of the archaeological site that is presented. The first example created in this system is the Casa de la Diana Arcaizante in Pompeii (http://www.dianaarcaizante.com/), and apart of virtual reconstructions, it includes other technologies such as photogrammetry, 3D models of the findings and digital documents from diaries and archives.

Figure 9. *"Father and Son – the game"*, Museo Archeologico Nazionale of Naples

Figure 10. *Pompeii Touch app*

Figure 11. Casa de la Diana Arcaizante in Pompeii, *edición multimedia avanzada en yacimientos arqueológicos* by The Real Academia de Bellas Artes de San Fernando (Madrid).

4. Heritage and technology, an inevitable relationship

The repertoire of the exposed projects aims to analyse the links that unite them and highlight how virtual archaeology has been developed and consolidated around the Vesuvian cities. It has been widely verified that this discipline and new technologies are key allies for the documentation, interpretation and dissemination of heritage. The collaboration of different fields such as humanities and new technologies shows a natural consequence of the interaction with the past through the current language of society.

The potentiality of the Vesuvian cities not only concerns the material remains conserved, but encompasses a whole experience of valorization and dissemination of the heritage, from its discovery to the present day. Undoubtedly the Vesuvian sites are the most studied and those that have attracted more scholars, archaeologists and curious individuals who, over three centuries, have left a great amount of iconographic, photographic and documentary material that acquire new meaning to be exposed via the means that technology offers.

Research in this field should have as future objectives; the creation of new content and information and the increase of accessibility, facilitating the interaction with the remains of the past, parallelly alongside the development of new Digital devices that should make this possible. Through future technology, it will be possible to restore, not only how the current remains were in the past in their static state, but also a sense of time and history in such archaeological sites; allowing for

IOP Conf. Series: Materials Science and Engineering **364** (2018) 012051 doi:10.1088/1757-899X/364/1/012051

modifications, restorations, traces of the passage of time and the interaction with the environment, but also with society, of which the Vesuvian cities can certainly supply with an extensive source.

Seeing the stratifications of its phases, not only as a spectator but as participant in a immersive way will be possible. Devices will be able to satisfy not only visual but sensory experiences, awakening smell, touch, taste. It will be possible to establish a realistic link not with a single past, but with a sequence of past times and moments that have changed the state of the archaeological sites.

All the activity of valorization that happens around the cities Vesuvian shows a constant concern for improving its fruition, trying to remain one of the most important archaeological points, not only in Italy, but the world. This certainly positions these sites as strategic points where new initiatives can be carried out, allowing all the information which they can potentially offer to be extracted.

References

[1] Almagro Gorbea A and Maier Allende J 2010 *Corona y Arqueología En El Siglo de Las Luces* (Madrid, Patrimonio Nacional) 35–47; 202–35

[2] Benedetti B, Gaiani M and Remondino F 2010 *Modelli digitali 3D in archeologia: il caso di Pompei* (Pisa, Scuola Normale Superiore)

[3] Bergman B, De Caro S, Mertens JR and Meyer R 2010 *Roman Frescoes from Boscoreale. The Villa of Publius Fannius Synistor in Reality and Virtual Reality* (New York, TheMetropolitan Museum of Art) 11-33

[4] Boyko A 2012 *ICTs in museum education* in Information Technologies in Education (Moscow UNESCO, Institute)

[5] Burnette M 2011 *The Immersive Cultural Museum Experience – Creating Context and Story with New Media Technology* The International Journal of the Inclusive Museum **3**

[6] Cantinela R and Porzio A *2008 Herculanense Museum. Laboratorio sull'antico nella Reggia di Portici* (Milano, Electa) 285-97

[7] Camardo D, Court, S and Thompson J 2012 *Ten years of the Herculaneum Conservation Project* Papers of the British School at Rome **80** 360–62

[8] Camardo D and Notomista M 2017 *Ercolano: 1927-1961. L'impresa archeologica di Amedeo Maiuri e l'esperimento della città museo* (Roma, L'Erma di Bretschneider)

[9] Corti E 1957 *Ercolano e Pompei. Morte e rinascita di due città. Presentazione di Amedeo Maiuri* (München, Giulio Einaudi editor3) 237

[10] Croci V 2009 *Investigating Culture Through the Senses* Architectural Design **79** 128–31.

[11] MacDonald L 2006 *Digital heritage. Applying digital imaging to cultural heritage* (Oxford, Elsevier)

[12] Mattusch C 2005 *The Villa dei Papiri at Herculaneum. Life and Afterlife of a sculpture collection* (California, Getty Publications) 52

[13] Moormann E 2015 *Pompeii's ashes : the reception of the cities buried by Vesuvius in literature, music, and drama* (Boston, De Gruyter) 74-83

[14] O'Gorman K., Baxter I and Scott B 2007 *Exploring Pompeii: Discovering hospitality through research synergy* Tourism and Hospitality Research **7** 89–99.

[15] Pagano M 1991 *Metodologia dei restauri borbonici a Pompei ed Ercolano* Rivista di Studi Pompeiani. Associazione internazionale amici di Pompei 169-91

[16] Parslow C 1995 *Discovering Antiquity, Karl Weber and the Excavation of Herculaneum, Pompeii and Stabiae* (Cambridge, Cambridge University Press) 115-30

[17] Veronese L 2014 *L'invenzione dell'immagine turistica degli scavi di Ercolano. Contenuti e caratteri iconografici*. VI Convegno Internazionale di Studi, Università Degli Studi di Napoli Federico II 191–202

IOP Conf. Series: Materials Science and Engineering **364** (2018) 012052 doi:10.1088/1757-899X/364/1/012052

CHEOPS: Cultural Heritage Enhancement Over Cyber-Physical Systems

Yasmeen Abdrabou, Fatema Mohsen, Amr El Mougy, Slim Abdennadher

German University in Cairo, Egypt

E-mail: {yasmeen.abdrabou, fatema.mohsen, amr.elmougy, slim.abdennadher}@guc.edu.eg

Abstract.

The Internet of Things (Iot) field has been expanding in various domains with cultural heritage being no exception. Nevertheless, cultural heritage sites are still often managed in traditional ways and a prominent gap remains between tour managers and visitors in correctly identifying visitors interest to specific sites and monuments. Previous related works focused on tackling mood detection separately from particular touristic locations which lead to overly generalized assumptions. Accordingly, in this paper, we propose a comprehensive platform named CHEOPS to enhance the touristic experience by detecting the tourists interest in the sites. The proposed platform provides a customized tour through touristic sites by using indoor localization techniques coupled with information about smartphone activity. Accordingly, the visitor is directed to a more interesting location. In addition to interest detection, CHEOPS implements several techniques for interaction with the touristic site such as gaze detection, gesture recognition, and augmented reality. A prototype of the proposed platform is implemented and tested in real scenarios to detect the usability of the system. The output accurately detects the interest of the users, which can then be used to fill the gap between the site manages and the tourists.

1. Introduction

The cultural heritage industry has global renowned importance for many reasons that include national pride and economic value. In fact, the United Nations has recognized in its first resolution on culture and development the economic role of cultural heritage in sustainable development [1]. A study commissioned by Eurostat for cultural heritage [2] states that in certain countries some sites such as the Louvre in PAris and The Colosseum in Rome receive millions of visitors each year, and that trade in cultural heritage goods and services is worth 10s of billions of dollars per year. In addition, the same study states that the cultural heritage sector accounts for 2.9 percent of all employment in Europe. This means that the quality of the services offered in cultural heritage sites can have a significant effect on the experience of the visitors, which is likely to have a pronounced economic impact. Nevertheless, most cultural heritage sites offer traditional services that include static content (for example through audio guides) with minimum interaction. In addition, there is little consideration for the different interests that visitors may have towards the contents of the sites. If these interests are considered, the site tour can be customized accordingly. Furthermore, advanced interaction techniques can be used to engage tourists and provide attractive ways to enhance their experience.

Information and Communication Technology (ICT) can play a role in addressing these challenges. Interaction techniques such as Augmented Reality (AR), eye gaze detection, and gesture recognition can be used to enhance the interactivity with the site and provide a more attractive experience, especially for younger generations. In addition, indoor localization techniques are now inexpensive and mature enough to enable accurate location-based services. Moreover, the proliferation of smartphones can now leverage a wide range of smart services for tourists. Accordingly, we present in this paper a comprehensive platform named CHEOPS (Cultural Heritage Enhancement Over Cyber-Physical) for enhancing the services offered in cultural heritage sites, based on a set of ICT technologies.

2. Related Work

In this section, some pertinent research efforts that study the integration of technology with the world of tourism are explored.

2.1. Augmented Reality in Tourism

Augmented Reality (AR) is the augmentation of a physical world's view with digital content by the use of the camera of a smartphone or a tablet [3]. Some examples of smart tourism systems based on AR include the **Mobile AR system (MARS)** [4]. The prototype registered 3D graphical tour guide information with buildings and artifacts that the visitor sees. Another example,is [5] which used AR to display historical information to the visitors in the form of a 3D model that represents the site. Similarly, **Ancient Egypt ActiveLens** is a newly created AR app * that allows users to see Ancient Egypt through a simulated journey back in time. On the other side, in a collaborative AR scenario such as the one in [6], more than one user can share the same space through virtual objects while they are on the real world.

2.2. Eye Tracking and Pointing Detection in Tourism

Another area of interest to enhance tourism is by using special hardware, such as Eye Trackers. One example of eye tracking-based systems is **Museum Guide 2.0** [7], where the visitors enter the museum putting a wearable eye tracker. Then object detection mechanism runs on the area the tourist is standing in front of and gaze at then it plays a relevant audio file which provides information about the detected object. For pointing detection systems in museums, the mixed-reality system done by C.Malerczyk [8] comes to the foreground. The aim of this system was to allow the museum's visitor to explore the exhibits through interaction and give them the chance to explore some details in the paintings.

2.3. Interest-Based Systems

Interest detection has opened the door for several new applications as it provides much needed enhancements to already available systems to be more personalized to each user. In cultural applications, interest can be detected from the users' activity, specifically their average speed and step count during their visit. Step detection is done via activity tracking using Location-Based Services (LBS), where devices rely on GPS or other WiFi-based techniques [9]. Another prevalent approach is by using sensor fusion through combining accelerometer with gyroscope [10]. Due to researchers' growing interest in this field, several applications and services have been developed utilizing activity recognition and even more specifically step detection in Health [11], Entertainment [12] and Surveillance [13], but none of them was integrated in the tourism field. Thus, the purpose of this research is to enhance the touristic experience by employing the interest detection mechanism in the cultural heritage field aiming to fill the gap between the tourists and the site stakeholders.

*https://play.google.com/store/apps/details?id=com.computeam.ww1activelens&hl=en

3. CHEOPS Platform

CHEOPS is a comprehensive platform for enhancing the touristic experience inside cultural heritage sites. The platform is based on a set of technologies that are integrated to provide location-aware information services to the visitors. These technologies include Augmented Reality (AR), Computer Vision (CV), eye gaze tracking, and localization, among others.

3.1. Architecture

Figure 1 shows the CHEOPS system architecture, which proposes an e-Tourism system that is based on the visitor's interest detection on the site. The system first determines the interest by detecting the tourists' speed, duration spent in front of every monument, usage statistics of mobile applications and finally, the location of highest interest. Location information is demerited by two methods: using wireless devices such as Bluetooth beacons or by using computer vision to detect the distance from monuments on the site. If visitors are not interested, the system will redirect them to another nearby location and re-checks the interest. If they are interested, then info about the monument/statue will be shown. Displaying the info is supported by two modules: the first one is Augmented Reality(AR) which serves adults and children as well as individual and group visitors. The other module incorporates feature extraction, where the system displays info according to where the visitor is looking and pointing. Finally, the system will redirect the visitor to another area.

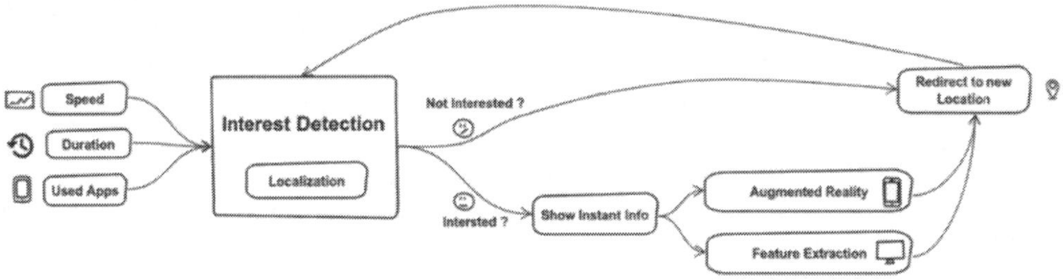

Figure 1. CHEOPS System Architecture

3.2. Interest Detection Methodology and Evaluation

The focus of this module is twofold: detecting the level of interest of the visitors and the level of engagement with the surrounding environment. In order for this system to be effective, the data needs to be easily accessed by the different stakeholders represented by the visitors and the tour guides/managers. The system relies on the visitors usage of their Smartphone to correlate their behavior to their level of interest. More specifically, this is done through two main methods; the first is measuring the users step count and walking speed during a visit. This is compared to the estimated average time it would take a person to visit the site. The second method is through detecting the users Smartphone activity where we correlate a high level of smart phone usage during a visit with a lack of interest in the site. However, the user's camera usage was excluded as it would indicate a higher level of engagement in the visit. After detecting the users' interest level regarding different artifacts, the data is used to redirect them around the site and provide them with information more tuned to their interest. This redirection module is location dependent, and its input detection mechanism is divided into two main parts.

- Motion detection: The system makes use of the Smartphone's accelerometer to record the tri-axial acceleration of the user. The obtained values are filtered to remove the effects of noise, gravity and changes in phone orientation. This is to ensure that only the user's walking speed is being measured without any external interferences. The filtered values can now be used to calculate user's speed, step count and distance traveled.

- Smart phone activity tracking: The system constantly checks the user's smartphone usage during their visit and calculates the percentage of interest. This is done through recording the start time of each application opened while on tour and then measuring the time spent using the application before the user closes it or switches to a different application. The system also tracks when the phone's camera is opened and the number of photos taken at each location.

This input is then sent to the server alongside the user's location, then it analyzes the acquired input in order to infer the user's interest in the surrounding site. First the user's speed, acceleration and time spent idle are examined. The system initially starts with a neutral interest i.e. 50% and according to the visitor's speed this value is increased or decreased on a percentage scale (0% - 100%). Here, high speed and acceleration indicate a low level of interest, and accordingly the interest percentage decreases. On the other hand, low speed, deceleration and an increase in time spent idle are all indicators of the user's increasing level of engagement with the site and hence, the interest percentage increases.

Second, we consider durations of application usage where we correlate a long duration of usage with declining interest and vice versa. To be able to adjust the overall interest, the change in percentage of the mobile apps usage is calculated and added or subtracted from the previously mentioned percentage. The camera usage is the only exception to the previous relation, where an increasing number of photos indicates a higher level of engagement in the monument.

An overall level of interest is then calculated from the previously calculated percentages and saved with the corresponding location. This data is later fed to a clustering algorithm (K-Means) that groups locations based on the user's percentage of interest, providing a more informative set of data. K-Means was chosen over other clustering approaches due to its scalability, especially in case of large data sets which is likely in our application. According to the clustering output, the locations with the highest percentages of interest are labeled and ranked for further suggestions to the visitors. These locations are used to redirect the visitors to other places by showing the rank of each location. Figure 2 shows the rankings of some locations. The X-axis represents the locations/halls and the Y-axis represents the percentage of interest. In the following section, we discuss the localization methods used in our platform.

3.2.1. Localization Location tracking is one of the main defining variables in interest detection. In this module, different approaches were tackled to get the user's location.

- Geo-fences
  A *Geofence* represents a virtual border that encompasses a certain location depending on a set radius. This can indicate when a user enters or exits the location using the Global Positioning System (GPS). The GeofencingApi ∗ is used in our system.
- Localization Using Beacons:
  Another technique is to get the visitor's location by using Bluetooth beacons. Once the location has been obtained from the geofences and the Smartphone's Bluetooth is activated, the service starts detecting active beacons in the vicinity and calculates location using triangulation †. This module also supports navigation to a specific location that can be entered by the user. In order to do this, a location map has to be pre-loaded to the module

∗https://developers.google.com/android/reference/com/google/android/gms/location/GeofencingApi/

†http://www.tothenew.com/blog/indoor-positioning-systemtrilateration/

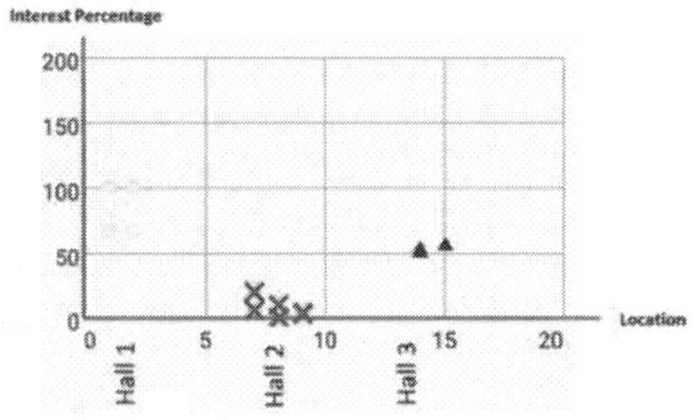

Figure 2. Locations Clustering

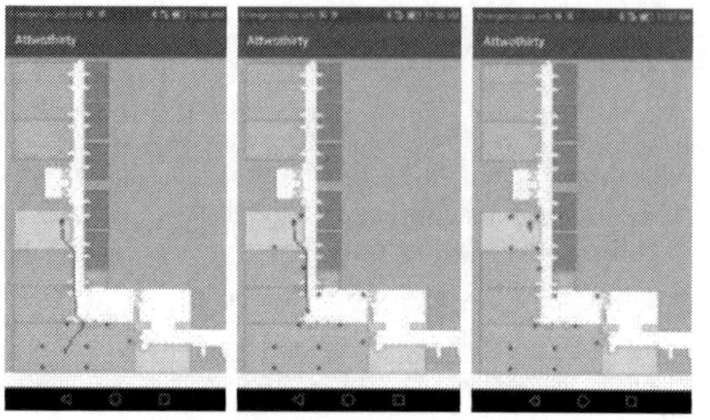

Figure 3. Highlighted location and updated path to destination

as a search space. Then, image processing algorithms are used to convert the image/map to gray scale and filter it to detect the obstacles/walls. Then, Gaussian filters were used along with erosion and dilation to filter the noise and to convert it to a binary image.
After constructing the 2D map, the A* algorithm * was used to get the shortest path. After the user chooses the destination, a green dot is drawn representing the destination point and a line is draw on the map as a guide representing the path. The location of the user and the path is recalculated every 10 seconds and the line path is redrawn as shown in Figure 3.

- Localization Using Computer Vision
In some cultural heritage sites, it is not possible to install wireless devices. Thus, this module presents an alternative solution for indoor localization using the Smartphone camera with the assistance of Open Computer Vision Library (OpenCV). Since cultural heritage sites are filled with artifacts, the application uses them as marker objects in the database to locate the user along with the real height of the object in meters, the pixel height of the object in the image, and the focal length of the mobile camera. By knowing these attributes and locations of such artifacts, the application would give an estimate to their location using depth calculations. The application uses ORB †, and SIFT ‡ as a feature detection algorithm to match artifacts and then calculates the distance from each artifact in the given image. The module takes into consideration the scaling and rotation factors of the captured artifacts. Afterwards, a modified trilateration algorithm [14] is used given the distances from the artifacts and successfully localizes the user.

Once the system detects a high level of interest in an object, an appropriate interaction technique is used to display proper information. For detailed monuments, eye gaze tracking and finger gesture recognition are used, while we use AR for other monuments.

3.2.2. Augmented Reality (AR) This module leverages the advantages of AR and encapsulates it in all modules mentioned in this section as a real-time interacting tool by displaying historical information of the place to visitors. The proposed AR methodologies differs from one module to another as explained below.

- AR for Adults

*http://theory.stanford.edu/~amitp/GameProgramming/AStarComparison.html
†http://docs.opencv.org/3.0-beta/doc/py_tutorials/py_feature2d/py_orb/py_orb.html
‡http://docs.opencv.org/trunk/da/df5/tutorial_py_sift_intro.html

In all AR modules, the user is supposed to point the device camera towards the showcased monuments. On detection of a monument, a button will be rendered on that monument to notify the user. When a button is pressed, an information panel appear presenting historical information about the detected monument. The panel has two options in case the user requires further information: the visitor can get more information either by pressing on the *Wikipedia* button or on *Google Search* button (see Figure 4). The module allows the user to take a picture of that monument and save it by a caption. If there is no info about that monument on the database, the module enables the user to search by image on *Wikipedia* and saves the info with the image for future users on the Vuforia cloud database * .

- AR for Children
  This module is dedicated to serve the particular interaction needs of children, mainly through more attractive interaction techniques and popular culture. Here, 3D cartoon models and animations are used as they are considered the most attractive visualization technique for the target audience. Animations like walking and moving the character's head are added, in addition to voice over audios with the character's tone containing the information given about the statues. The voice over audios, are added to help younger children who may not be able to read the information. Having the information presented in more than one language is an added value to reach more people with different cultures. Figure 5 shows the home page of the application and a screen-shot of the module. The child should first choose the language and then choose the cartoon character. The character chosen narrates the information about the monuments detected within the augmented reality scene. Even though these cartoon characters may not be related to the theme of the project, they are used for their popularity with children. Figure 5 on the right, shows the panel that is displayed with the statue's information displayed. The children can also share their tour and post it on social media.

Figure 4. Detected Targets and Information panel

Figure 5. Children module Homepage and 3D Cartoon Models Interface

- Collaborative AR
  This module extends the previous AR modules through collaboration in form of a *Treasure-Hunt* game. Uers play the treasure-hunt game to share with each other the detected monuments. The task of the treasure-hunt team is to detect specific monuments to complete all the letters of a word e.g."LUXOR", and thus be the *winning* team. The game is played simultaneously by different players and the flow is as follows: once a visitor chooses to play the game s/he chooses to join a team e.g.(A), a guide appears on the screen to help the player in finding the monuments that should be detected to get the letter "L", as shown in Figure 6 (a). Once a letter is detected, it appears on the screen of all players joining team (A) (without the need to detect the same monument again) and now it is time to collect the remaining letters. Once any of the users detect a right monument, the letter related to

*https://www.vuforia.com/

IOP Conf. Series: Materials Science and Engineering **364** (2018) 012052 doi:10.1088/1757-899X/364/1/012052

the monument appears to all users as shown in Figure 6 (b). When another visitor decides to play the game and chooses to be in another team e.g. (B), none of the letters will be displayed to this user, who then starts searching for the treasure. When a team completes all the letters of the word "LUXOR", this team wins and the other team loses, the winning team is saved in a *Firebase*∗ database where a message is sent immediately to all users of the winning team informing them that they won and a hard luck message is sent to the losing team as shown in Figure 6 (c) and 6 (d) respectively.

Figure 6. Team A: User 2 detects a monument and wins letter X

3.2.3. Feature Extraction In addition to AR, CHEOPS hosts a set of interaction techniques intended to provide fine-grained information about detailed scenes. Here, the interaction techniques are based mainly on two technologies: eye gaze tracking and finger gesture recognition. The module uses *Tobii EyeX* † for determining the user's eye position and inferring the user's interest. Moreover, it uses *Kinect version 2* (V2) ‡ for detecting the pointing gesture. The module involves taking the inputs from the eye tracker and the gesture from the Kinect and mapping the regions to get the interest area. This area is detected by extending the line of sight and mapping between the detected region from the eye tracker and the pointing region. Finally, the module shows info about the interest region displayed on the user interface. For testing, the painting § shown in Figure 7 (a) is used as a simulation for exhibitions and it is divided into sub regions. The testing scenario is as follows, if the eye tracker and the Kinect detect the same region, it means that this is a correct gesture. However, if they differ, this means that it is a false gesture and the visitor didn't mean to get information about this part.

A pilot study took place on a group of 8 users in Luxor museum to check the system accuracy, usability and to train the interest classifier. We received some comments on the usability of the user interface. However, they was pleased with the system.

4. Conclusion and Future Work

In this paper, we have presented CHEOPS, a comprehensive platform for enhancing services in cultural heritage sites based on ICT technologies. The proposed platform utilizes indoor localization techniques (based on Bluetooth beacons or computer vision) and information about Smartphone activity to detect the level of interest of the visitors, and accordingly provide a customized tour through the sites. In addition, several interaction technologies have been utilized to enable customized information display. Particularly, AR has been used to build interaction

∗https://firebase.google.com/
†https://tobiigaming.com/product/tobii-eyex/
‡https://developer.microsoft.com/en-us/windows/kinect/develop/
§https://goo.gl/YtSX77

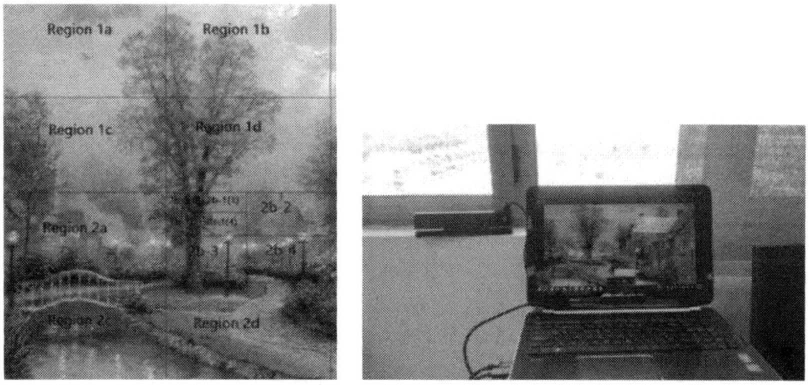

Figure 7. Accuracy Experiment Diagram and Setup

applications for different types of visitors, namely children, groups, and regular visitors. Furthermore, eye gaze detection and finger gesture recognition have implemented to allow fine-grained information display about specific parts of detailed monuments.The functionality of all the developed applications has been verified in realistic scenarios.

Acknowledgments
This work was in collaboration with Luxor Governorate and the ministry of antiquities. The authors would like to acknowledge the contribution of Ahmed Hazem, Safa Ads, Nadine Tarek, Nourhan Wael and Ahmed Elsayed in the implementation and testing phases.

References

[1] Rizzo I and Mignosa A 2013 *Handbook on the Economics of Cultural Heritage* Elgar original reference (Edward Elgar) ISBN 9780857930996 URL `https://books.google.com.eg/books?id=Sc3ENAEACAAJ`
[2] Eurostat 2016 Culture statistics - 2016 edition URL `http://ec.europa.eu/eurostat/documents/3217494/7551543/KS-04-15-737-EN-N.pdf/648072f3-63c4-47d8-905a-6fdc742b8605`
[3] Villarejo L, González-Reverté F, Miralbell O and Gomis J M 2014 *eLearn Center Research Paper Series* 06–14
[4] Feiner S, MacIntyre B, Hollerer T and Webster A 1997 A touring machine: Prototyping 3d mobile augmented reality systems for exploring the urban environment *Wearable Computers, 1997. Digest of Papers., First International Symposium on* (IEEE) pp 74–81
[5] Lee K 2012 *TechTrends* **56** 13–21
[6] Fuhrmann A, Löffelmann H and Schmalstieg D 1997 Collaborative augmented reality: Exploring dynamical systems *Proceedings of the 8th conference on Visualization'97* (IEEE Computer Society Press) pp 459–ff
[7] Toyama T, Kieninger T, Shafait F and Dengel A 2011 Museum guide 2.0-an eye-tracking based personal assistant for museums and exhibits *Proc. of Int. Conf. on Re-Thinking Technology in Museums* vol 1
[8] Malerczyk C 2004
[9] Stenneth L, Wolfson O, Yu P S and Xu B 2011 Transportation mode detection using mobile phones and gis information *Proceedings of the 19th ACM SIGSPATIAL International Conference on Advances in Geographic Information Systems* (ACM) pp 54–63
[10] Lee H h, Choi S and Lee M j 2015 *Sensors* **15** 27230–27250
[11] Long X, Yin B and Aarts R M 2009 Single-accelerometer-based daily physical activity classification *Engineering in Medicine and Biology Society, 2009. EMBC 2009. Annual International Conference of the IEEE* (IEEE) pp 6107–6110
[12] Abowd G D, Atkeson C G, Hong J, Long S, Kooper R and Pinkerton M 1997 *Wireless networks* **3** 421–433
[13] Lin W, Sun M T, Poovandran R and Zhang Z 2008 Human activity recognition for video surveillance *Circuits and Systems, 2008. ISCAS 2008. IEEE International Symposium on* (IEEE) pp 2737–2740
[14] Cotera P, Velazquez M, Cruz D, Medina L and Bandala M 2016 *International Journal of Advanced Robotic Systems* **13** 110 (*Preprint* `http://dx.doi.org/10.5772/63246`) URL `http://dx.doi.org/10.5772/63246`

IOP Conf. Series: Materials Science and Engineering **364** (2018) 012053 doi:10.1088/1757-899X/364/1/012053

From Alhambra to Digital Projection: Designing a Digital Media Cultural Heritage Project Based on the Transformative Qualities of Islamic Calligraphy

M J Khajavi

Associate professor
Faculty of Media, Volda University College, Norway

E-mail: khajavi@hivolda.no

Abstract. This paper traces how the design of a new media cultural heritage project was inspired by the transformative qualities found in many specimens of Islamic calligraphy. As certain scholars have pointed out, Islamic art is filled with examples of letters and words that appear to transform into things other than text. Indeed, such transformative qualities can be seen across the Islamic world and on a diverse range of artefacts, including the epigraphs on the walls of the Alhambra palace and the religious texts on Ottoman Turkish scrolls written in *ghubār* (which literary means "dust" in Arabic) script. In this paper, the author outlines two broad types of transformative qualities that can be distinguished in many specimens of Islamic calligraphic art, namely, the metamorphosis of the form and the transformation of the function of the letters and words. The paper then demonstrates how, inspired by such transformative qualities, the designers of a cultural heritage project created a calligraphic installation for museums and galleries in which letters are transformed from text to image and vice versa. Using a projection mapping system, words are transformed into images devoid of any semantic meaning, and then revert back into readable text again. Finally, the paper investigates how the transformation of the function of calligraphy renders readability a transitory process in this artwork, and how the technique used in the creation of this artwork can be applied to cultural heritage sites as an educational and entertainment tool.

1. Introduction

The Alhambra, a palatial city in Granada, Spain, contains examples of some of the more visually complex ornamentation found in Islamic art. Constructed in the 14th century on the remains of a Roman fortress by the Nasrids (r. 1238–1492), the last Arab Muslim dynasty that ruled parts of the Iberian Peninsula, its magnificent stalactite-like *muqarnas*[1] ceilings and its geometric tessellations—which famously were a source of inspiration for M. C. Escher (1898–1972)—exemplify Islamic artisanship at its finest. The carefully-designed details of the arabesques that decorate the interiors of the palaces and chambers at Alhambra are particularly mesmerizing to any visitor to this world heritage site. Moreover, visitors to the Alhambra may note that the interior walls of several of its chambers are decorated with calligraphic inscriptions.

The Arabic epigraphs in the Alhambra have been the subject of numerous scholarly investigations. These include detailed analyses of the signification and meaning of both the Qur'anic and the secular poetic inscriptions (see, for example, Grabar [1] and Bush [2]). In this article, I do not intend to discuss the religious and poetic content of these epigraphs. Rather, I take their style of writing as my point of departure. Central to the discussion in this article are some of the calligraphic inscriptions on the walls of the Alhambra, in which words and vegetal decorations appear to blend. Written in a style of calligraphy known as foliated *Kufic*, the shapes of the end parts of the Arabic letters engraved on these epigraphs intertwine and transform into foliage. Indeed, the Arabic phrases on these epigraphs seem to be in transformation, as they change into leaves and flowers. To an observer, these inscriptions appear as Arabic words for a moment, and then become visual decorations that blend into their vegetal backgrounds.

In fact, as scholars such as Grabar [3] and Marks [4-6] have pointed out, such transformative characteristics can be seen in many specimens of Islamic calligraphy [2] from different centuries and locations. Referring to the transformative quality of Islamic calligraphy, Marks polemicizes that: "throughout its history, from Konya to Khartoum, Granada to Samarkand, Islamic art is rife with text trying to become figure … as though the text could not be contained by its own bounds" [5]. In this article, I expand on Marks' discussion and distinguish two types of transformative quality in Islamic calligraphy. I then explain how a new media cultural heritage project, entitled *The Passion of Love* (2016), was designed on the basis of one of these transformative qualities.

2. The Letters' Desire for Transformation in Islamic Calligraphy

Let us now revert back to the Alhambra's epigraphs that are written in foliated *Kufic*. Kufic is one of the oldest styles used for writing in Arabic script [7], and different variations of it are found across the Islamic world [8]. These variations developed as a result of differences in regional tastes, or because of the purposes for which calligraphy was used. For example, some were more suitable for writing on paper, while others were developed solely for architectural decorations. One of these variations was foliated *Kufic*, which was most often used in the decoration of buildings. In this style of writing, the end parts of Arabic letters (the top of the vertical strokes and letter apices) break and branch out into vegetal motifs. In foliated *Kufic*, the shapes of the letters appear to metamorphose into flowers and plants. As Marks argues, this style of calligraphy seems to aspire to transformation. The words appear to be under "pressure from inside" to seek organic life and to transform into something other than text [6].

In other instances, calligraphy appears to transform into other things, such as birds and other animals. For example, on a pottery bowl from the city of Nishapur in Iran (probably created from the 10th to the 12th century), the letters of the Arabic word *baraka* (i.e. blessing) seem to have been transformed into birds (see Figure 1). Written in *kufic* style, the shapes of two of the letters have changed into bird-like creatures. Ettinghausen [9] calls such inscriptions "ornithomorphic" because the figures of the birds form integral parts of the words.

Figure 1- The word *baraka*, in ornithomorphic writing. The image is a drawn image of the inscription on a pottery bowl from Nishapur, Iran. The original pottery bowl is in the collection of Heeramaneck Galleries, New York.

Ornithomorphic inscriptions such as the one described above demonstrate that letters and words are not contained in their own contours in Islamic calligraphy. Rather, they desire to become alive. There are many other examples of Islamic calligraphy in which letters and words appear to transform into organic things, including humans, animals, and plants. Moreover, we can observe other specimens of Islamic calligraphic art in which transformation occurs in a different way.

In the Asian Civilizations Museum in Singapore there is an Ottoman paper scroll on which some Shi'i [3] invocation phrases are inscribed. The scroll is rather large (14 x 112 cm), as are the words of the invocation written on it. Looked at from the distance of a few meters, the Arabic phrases inscribed on this Ottoman scroll seem to have a texture. However, when viewed from a closer distance the magical aspect of this scroll becomes apparent. A closer inspection of the Arabic phrases in these Shi'i invocations reveals that what appeared to be the texture of the words is actually minuscule written text of the Qur'an. The Shi'i invocations on this Ottoman scroll actually contain the whole text of the Qur'an, written in the miniature-sized style of calligraphy known as *ghubār* (see Figure 2). Literally meaning dust-

IOP Conf. Series: Materials Science and Engineering **364** (2018) 012053 doi:10.1088/1757-899X/364/1/012053

like, *ghubār* script was probably originally used for written communications sent by pigeons [10], but it was later used for other purposes, including writing miniature Qur'ans that were sometimes carried by people as a form of protection.

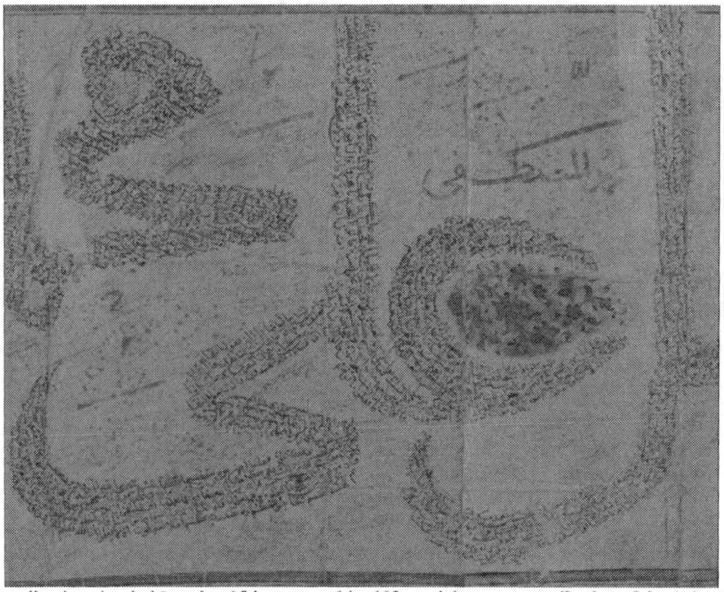

Figure 2- Ottoman Qur'an scroll written in *ghubār* script, 15th century, 14 x 112 cm, ink on paper, collection of the Asian Civilizations Museum, Singapore. Courtesy of the Asian Civilizations Museum in Singapore.

As the viewers move closer towards the object in this Ottoman Qur'an scroll, the Shi'i invocations reveal their hidden secret and suddenly transform into a container for the words of God, the text of the Qur'an. Both an outward and an inward transformation occurs here. The large-sized invocations become the outline for the small-sized Qur'anic text, while the smaller Qur'anic text becomes hidden (as if protected) inside the larger Shi'i phrases. This transformation reminds us of the Shi'i doctrine of *zāhir* (i.e. manifest) and *bātin* (i.e. latent). The doctrine of *zāhir* and *bātin*, which is believed by some Shi'i and Sufi[4] branches of Islam, holds that there is a deeper esoteric meaning behind the exoteric surface of things, particularly in religious texts and in the Qur'an itself [6]. According to this belief, the esoteric meaning is not available to everyone, but only to a few. While most people of faith see only a manifest layer of meaning, there is a latent hidden layer in the Qur'an that is not readily available to most people. Hence, in the Ottoman Qur'an scroll described above, the larger words and phrases comprise the exoteric layer, and they only transform and reveal their esoteric meaning to those who inspect them carefully.

While the transformation has a religious and almost mystical signification in the above mentioned Qur'an scroll, a secular transformation occurs in the calligraphy genre of *sīyāh-mashq*. Literally meaning "black-exercise" in Persian, *sīyāh-mashq* refers to a category of calligraphic works in which a whole page is almost completely covered with calligraphy. The practice was originally used as a form of exercise for calligraphers in order to warm-up their hands, but it gradually evolved into a high caliber art form and became a collectible category of calligraphic art. Figure 3 is an example of a work of *sīyāh-mashq*.

Figure 3- Bahman Panahi, *Never Got Tired of You*, 2001, page of *sīyāh-mashq*, ink on paper, 21 x 27 cm. Courtesy of Bahman Panahi.

As the figure demonstrates, letters and words are repeatedly inscribed on the page in *sīyāh-mashq* specimens, usually in an irregular manner. Different letters and words overlap and cross on sheets of *sīyāh-mashq*, rendering an arrangement of calligraphic forms that usually subsumes the semantic content of the calligraphy, and leaves little room for a reading of its literal content. While some of the letters and words on the pages of *sīyāh-mashq* are usually recognizable, and all of them are sometimes recognizable, it is rare for a meaningful phrase to be read from these sheets of calligraphic work. In fact, in many specimens of *sīyāh-mashq*, calligraphic forms appear to oscillate between being text and being an abstract image with little or no semantic signification. It can be argued that pages of *sīyāh-mashq*, such as the one in Figure 3, do not invite viewers to read them as text, but rather encourage an appreciation of the aesthetic quality of their image. Similarly to the previous examples, letters and words once more appear to transform into something other than text.

In all of the four examples presented above, calligraphy appears to fluctuate between being text and being image. Although the letters and words in these examples do not literally change over time, they demonstrate a desire for transformation. They point to a reversibility between word and image in Islamic calligraphy. As Marks [6] has argued, this internal force—the desire for transformation of the letters and words in Islamic calligraphy—can best be explored in time-based media such as animation.

In her article "Calligraphic animation: documenting the invisible", Marks claims that animation is a wonderful "playing field" for the transformative qualities of Islamic calligraphy [6]. Indeed, the transformative qualities of Islamic calligraphy, some of which have been seen in the above examples, can inform the temporal behavior of Arabic writing[5] in time-based media. All of the calligraphic inscriptions noted in this section suggest a form of temporal behavior, a transformation of text to or from something else. While transformation is only suggested in these specimens of calligraphy, and does not literally occur, a full transformation can occur over time in time-based media. I argue that specimens of Islamic calligraphy with transformative qualities, such as those discussed in this section, can inspire the design of time-based calligraphic artworks. Later in this article, I explain how these transformative qualities have inspired the design of a digital media cultural heritage project. However, I first distinguish between two types of transformative qualities that are observable in the examples discussed above and in many other examples of Islamic calligraphy.

3. Two Types of Transformation

The previous section presented four different examples of Islamic calligraphy in which the calligraphy appears to transform, and oscillates between text and image. In general, I argue that two similar but distinct types of transformative qualities can be observed in the above examples and in many other specimens of Islamic calligraphy. These are transformation of identity and transformation of the function of calligraphy.

In order to understand the difference between these two types of transformation, let us reconsider the examples discussed in the previous section. I began by referring to foliated *kufic* and ornithomorphic inscriptions. In these two types of calligraphy, one sees that it is indeed the shape of letters and words that appear to be in a process of transformation. In fact, it is the form of the letters that appears to change shape in these examples. This means that the contours of the letters are affected and are transfigured into organic shapes in these calligraphic styles. This clearly suggests a change or transformation in the shape of the calligraphic forms. It is because of this visual transformation in the shape of the letters that these calligraphic forms appear to oscillate between a verbal identity and a pictorial identity. In considering foliated *kufic* and ornithomorphic calligraphy, it appears that the identity of the calligraphic forms are oscillating between being text and being image in these types of inscriptions. Indeed, in such examples, the calligraphic forms seem to oscillate between having a verbal identity and a pictorial identity. Therefore, one can argue that such specimens of Islamic calligraphy suggest the transformation of the form and, consequently, of the identity of the letters and words.

However, in examples of Islamic calligraphy such as the Qur'an scroll from the Asian Civilizations Museum in Singapore and specimens of *sīyāh-mashq*, the shapes of the calligraphic forms are not in transformation. In such specimens, the shapes of the letters and words are fully preserved. For example, the shapes of the larger Arabic phrases (the Shi'i invocations) in the Ottoman Qur'an scroll do not change as one moves towards the scroll. Rather, the form of the words is intact in this specimen of Islamic calligraphy. Likewise, in *sīyāh-mashq* specimens such as that in Figure 3, while the calligraphic forms appear to oscillate between being text and being image, their shapes are preserved. I argue that in these cases it is not the form (or the identity) of the calligraphy that is in transformation, but rather its function. The term function here refers to the intended purpose for which something (in this case calligraphy) is used in a particular context. The purpose of written language is naturally expected to be the verbal communication of meaning. Therefore, we can say that the main function of writing is semantic communication. However, calligraphy transcends this as it is intended to add an aesthetic quality to writing. Nevertheless, in many contexts the function of calligraphy remains predominantly a semantic one. But in specific contexts, such as in some specimens of *sīyāh-mashq*, the balance between these functions of calligraphy may shift. In such cases, the semantic function of calligraphy may be subsumed under its aesthetic function. In some *sīyāh-mashq* specimens, for example, calligraphy is completely transformed from being text with a semantic function to being an abstract image with a purely aesthetic function. In such instances, one can say that a transformation in the function of calligraphy occurs. Such transformation is clearly different from a transformation in the form and identity of calligraphy. As is demonstrated in the following section, the digital media cultural heritage installation studied in this paper (*The Passion of Love*) was made based on the second type of transformative behavior, namely transformation of the function of calligraphy.

4. Transforming the Function of Calligraphy through Projection Mapping

As the above discussion indicates, the transformation of identity and the transformation of the function of calligraphy are the two distinct types of transformation suggested in many specimens of Islamic calligraphy that this author identifies. I have argued that the balance between the functions of calligraphy shifts in many specimens of *sīyāh-mashq*. Instead of having a predominantly semantic function (which is expected in written language), calligraphy comes to function primarily on an aesthetic level in these artworks, in which it therefore appears to become transformed from text into abstract image. Together with the art of *sīyāh-mashq,* this transformative quality informs the design of the digital cultural heritage installation considered in this paper. While such a transformation is only suggested and does not literally occur in *sīyāh-mashq* specimens, this transformative quality can unfold over time in time-based media, as we will see below.

The Passion of Love, the cultural heritage project considered in this paper, is a new media installation inspired by the transformative quality of Islamic calligraphy. It includes a calligraphic sculpture made from the words of a verse of a poem by the prominent Persian poet Rūmī (1207–1273). Figure 6 shows this calligraphic sculpture. Although the calligraphic forms (i.e. the letters and words from the poem) are legible (i.e. recognizable) in this sculpture, they are not readable due to the way in which the calligraphic forms are arranged in this composition. The letters and words of this sculpture overlap and collide. This specific visual structure of calligraphic forms impedes readability, if it does not make it totally impossible. In designing this sculpture, the emphasis was clearly on the aesthetic quality of the visual composition, rather than on the communication of the verse's semantic content. This visual structure of the calligraphic sculpture encourages the viewers to enjoy the aesthetic quality of the calligraphic forms and the composition they have comprised, rather than inviting them to read the semantic content of the text. It can be argued that, like a specimen of *sīyāh-mashq*, the calligraphic forms of this sculpture have a predominantly aesthetic function. In other words, the semantic function of calligraphy is subordinated to its aesthetic function in this art piece.

Figure 4- A photograph of *The Passion of Love*, 2016, digital media installation, 90 x 60 cm, exhibited at the photography studio of Nanyang Technological University, Singapore. Copyright © M Javad Khajavi.

This calligraphic sculpture was intended to be mounted on the wall of a gallery or museum. The installation was first exhibited at the photography studio of the School of Art, Design and Media of the Nanyang Technological University in Singapore.[6] Figure 5 shows the technical setup for this new media installation. A projector was installed in front of the sculpture and animated visual elements were then projected onto the calligraphic sculpture—using a projection mapping technique and the computer software VPT 7 (Video Projection Tools)— through a process that added notions of movement onto the otherwise static sculpture (see Figure 6 and 6).

IOP Conf. Series: Materials Science and Engineering **364** (2018) 012053 doi:10.1088/1757-899X/364/1/012053

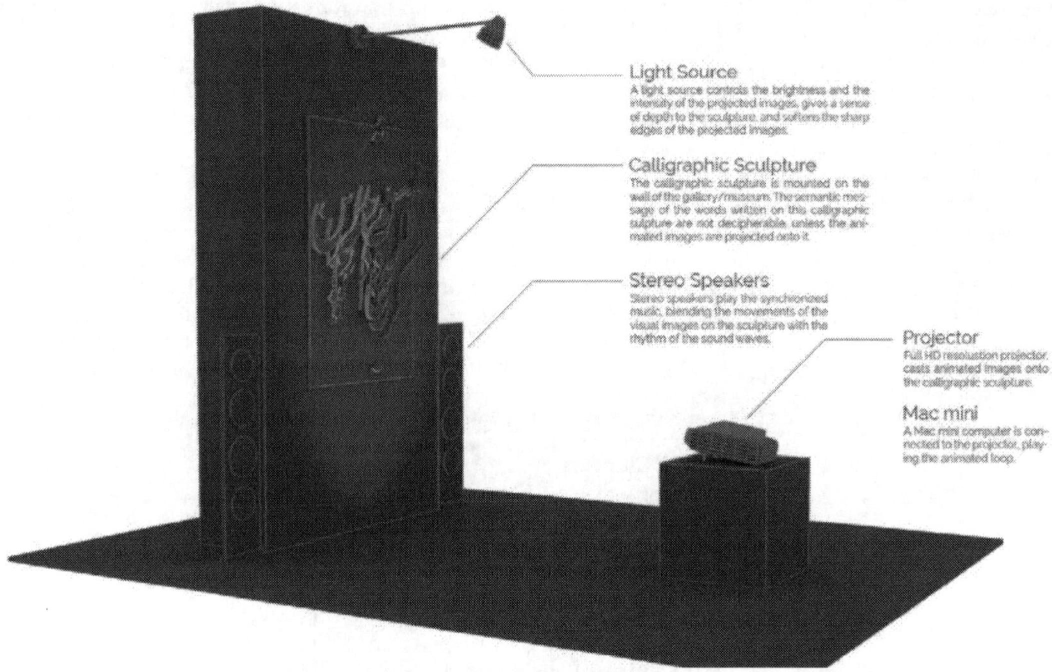

Figure 5- A diagram representing the technical setup of the digital media installation. Copyright © M Javad Khajavi.

The visuals projected directed the eyes of the viewers towards particular elements of the calligraphic sculpture (Figure 6). As the projected animation proceeded, it became clear to those viewers who could read the language that these animated elements were actually highlighting the words of the poem, one after another. They were clarifying the semantic message of the previously unreadable calligraphic forms. Thus, anyone who was able to read the language could follow the animated visuals and read the poem, which read: "This world would be engulfed in flames, if the lover's soul would speak".[7] Hence, the animated visuals that were projected onto the calligraphic sculpture led to a shift. In fact, the balance between the functions of the calligraphy shifted from a predominantly aesthetic function to a principally semantic one. In other words, a transformation occurred in the function of calligraphic forms due to the animated visuals that were projected onto the sculpture. This means that readability is a transitory process in this artwork. When there are no visuals projected onto the calligraphic sculpture, its semantic content is not readable. However, when the animated visuals are projected onto the sculpture, the calligraphic forms become readable and semantically meaningful. While the projection mapping technique used in this artwork is technically simple, it provides the opportunity for the viewers to see a relatively sophisticated process of shifting the balance between the functions of calligraphy.

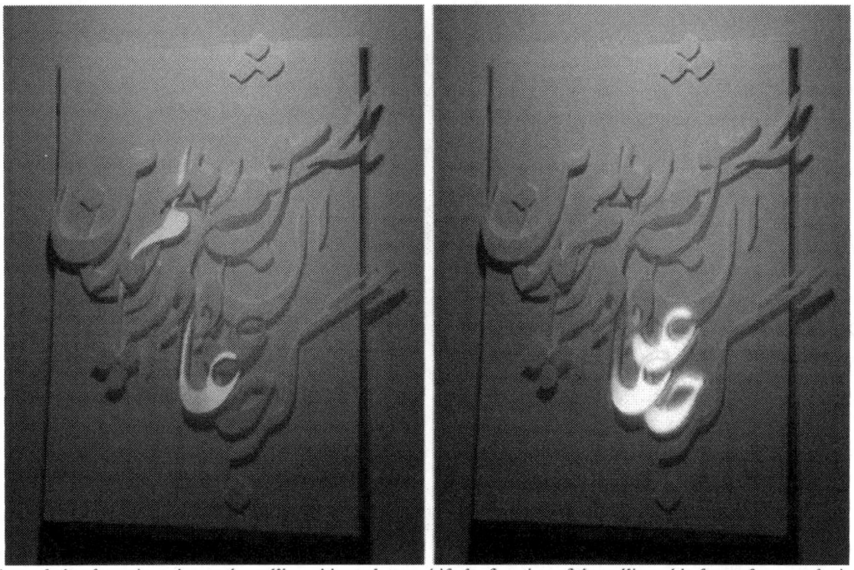

Figure 6- The animated visuals projected onto the calligraphic sculpture shift the function of the calligraphic forms from aesthetic to semantic, making the literal content of the piece available to the viewers. Copyright © M Javad Khajavi.

5. Conclusion

This article has explored how the transformative qualities of many specimens of Islamic calligraphy inform innovations in new media design. Ünlüer and Özcan rightfully argue that the "formal design solutions of past cultures can generate new innovative ideas" in new media design [12]. Their claim is particularly appropriate here, as this digital media installation received its inspiration from qualities that have existed for centuries in Islamic calligraphy. As we have seen, projection mapping was used in this digital media cultural heritage installation to activate and animate the transformative qualities that exist in many specimens of Islamic calligraphy. The design of this digital media installation was informed by examples of calligraphic art, such as *sīyāh-mashq*, in which calligraphy appears to oscillate between having a semantic and an aesthetic function.

The use of projection mapping to animate the otherwise static calligraphic forms is also of considerable importance here. Further applications of this technique can be explored in museum design and cultural heritage sites. For example, the technique can be used as an educational tool in museums to help viewers recognize the Arabic letters and words in epigraphs in which the calligraphy is written in styles that are not easily recognizable. Aside from projection mapping, other technologies such as Augmented Reality (i.e. superimposing computer-generated image on users' view of the physical world) can be used in Museums, galleries, or cultural heritage sites to transform the form or the function of epigraphs written in sophisticated styles. Clearly, such techniques can be used for educational as well as entertainment purposes. While some museums and heritage sites has started using Augmented Reality technologies to highlight Arabic words in epigraphs or to provide translation for such epigraphs, using such technologies are still limited.

Aside from these, the transformation of the form or the function of calligraphy through projection mapping can have applications in urban art. Currently, as part of a contemporary movement that experiments with Islamic calligraphy in new and innovative ways, there are many calligraphic sculptures decorating the streets, squares and public spaces in different countries, such as Qatar, United Arab Emirates, Kuwait, Iran, etc. The projection mapping technique mentioned in this paper can actually be used to add new aspects to these urban artworks, and animate such static calligraphic sculptures.

IOP Conf. Series: Materials Science and Engineering **364** (2018) 012053 doi:10.1088/1757-899X/364/1/012053

References

[1] Grabar O 1992 *The Alhambra* (Sebastopol, California: Solipsist Press)
[2] Bush O 2009 The writing on the wall: reading the decoration of the Alhambra. *Muqarnas* **26** 119-47
[3] Grabar O 1992 *The Mediation of Ornament* (Princeton: Princeton University Press)
[4] Marks L U 2010 *Enfoldment and Infinity: An Islamic Genealogy of New Media Art* (Cambridge: MIT Press)
[5] Marks L U 2010 *Variantology 4: on Deep Time Relations of Arts, Sciences, and Technologies in the Arabic-Islamic World and Beyond*, ed S Zielinski and E Fürlus (Köln: Walther König) p 45
[6] Marks L U 2011 Calligraphic animation: documenting the invisible. *Animation: an interdisciplinary journal* **6** 3 307-23, p 314
[7] Khatibi A and Sijelmassi M 1995 *The Splendour of Islamic Calligraphy* (London: Thames & Hudson)
[8] Grohmann A 1957 The origin and early development of floriated kūfic *Ars Orientalis* **2** 183-213
[9] Ettinghausen R 1957 The 'Wade Cup' in the Cleveland Museum of Art, its origin and decorations *Ars Orientalis* **2** 327-66
[10] Schimmel A and Rivolta B 1992 Islamic calligraphy *The Metropolitan Museum of Art Bulletin* **50** 1-56
[11] Lewis F D 2007 *Rumi: Swallowing the Sun* (Oxford: Oneworld Publications)
[12] Ünlüer A and Özcan O 2010 Sound and silence in the line: re-reading Turkish Islamic calligraphy for interactive media design *Leonardo* **43** 450-6

---

[1] Cellular honeycomb-like structures usually used in the design of domes, half-domes, and vaults in Islamic architecture.
[2] The term "Islamic calligraphy" refers to any sort of calligraphy written in Arabic script, and not exclusively calligraphy written for religious purposes.
[3] Shia and Sunni are the two main branches of Islam.
[4] Sufism is a mystical trend in Islam that promotes certain doctrines, values, and practices.
[5] It is important to note that Arabic script is used for writing many languages, including Arabic, Persian, Urdu, Kurdish, Ottoman Turkish, etc.
[6] A video ddocumentation of this exhibition can be found in the following link: https://vimeo.com/217946279.
[7] Translation by Lewis F D [11].

IOP Conf. Series: Materials Science and Engineering **364** (2018) 012054 doi:10.1088/1757-899X/364/1/012054

Lab4Dive Mobile Smart Lab for Augmented Archaeological Dives

D Scaradozzi[1,2], S Zingaretti[1], N Ciuccoli[1], D Costa[3], G Palmieri[3], F Bruno[4,5], G Ritacco[4], M Cozza[4], P Raxis[6] and A Tzifopanopoulos[6]

[1] Dipartimento di Ingegneria dell'Informazione, Università Politecnica delle Marche, Ancona, Italy
[2] LSIS, CNRS, UMR 7296, Marseille, France
[3] Dipartimento di Ingegneria Industriale e Scienze Matematiche, Università Politecnica delle Marche, Ancona, Italy
[4] 3D Research s.r.l., Rende (CS), Italy
[5] University of Calabria, Rende (CS), Italy
[6] ATLANTIS Consulting SA, Athens, Greece

d.scaradozzi@univpm.it, zingaretti.silvia@gmail.com, n.ciuccoli@pm.univpm.it, d.costa@pm.univpm.it, g.palmieri@univpm.it, fabio.bruno@unical.it, ritacco.gerardo@gmail.com, raxis@atlantisresearch.gr, tzifopanopoulos@atlantisresearch.gr

Abstract. Lab4Dive project aims to design, develop, and validate an innovative, marketable and competitive product for supporting underwater archaeologists in surveying, documenting and preserving the underwater cultural heritage. These activities could be improved considerably with the introduction of new technological devices and tools, helping underwater archaeologists to collect data from the marine environment. The project will provide a solution by equipping divers with portable, low cost and relatively small sized data-gathering systems (i.e. environmental sensors, acoustic localization system, HD cameras, etc.), so they can significantly improve both the amount of data collected in a single mission and the management of time and human resources at disposal. In this paper, the preliminary developments and the tests of the Lab4Dive system will be presented, which is mainly based on two modules: Cloud Server Module and the Underwater Module. The latter is based on two systems: an underwater tablet, that supports the diver during the immersion providing an augmented navigation interface that guides the diver along the selected targets; and a Docking Station, to acquire and exchange data coming from different environmental sensors, a high-resolution camera and an acoustic localization system, physically connected to it.

1. Introduction

Underwater archaeologists during their missions make extraordinary discoveries that expand the vision of humanity's past. Drowned settlements may contain well-preserved artifacts that can provide important new information about how people lived or when they arrived at various locations. These studies require long and repeated diving explorations in the interested sites ([1] and [2]). Until recently, the only sites accessible to underwater archaeologists were those that could be reached by

IOP Conf. Series: Materials Science and Engineering **364** (2018) 012054 doi:10.1088/1757-899X/364/1/012054

divers. In the last years, new set of tools incredibly helped them to survey submerged landscapes ([3] and [4]). In fact, once a place of interest has been located, divers investigate the small targeted area to collect samples and take notes on what they see, just like archaeologists do on land. Given the number of tasks for each diver and limited bottom times at deep depths, larger areas cannot be surveyed manually and this represents the biggest difference between surveying on land and surveying underwater.

Underwater survey technology has reached great advance. Remotely operated vehicles (ROVs) help scientists a lot to survey the sites at major depths and to cover bigger areas, thanks to the possibility of remotely piloting it to explore the seafloor, sending live video back to the crew on the research vessel ([5] and [6]). Moreover, ROVs are quite simply to use and even inexpert pilots are able to operate them easily ([7] and [8]).

Although this technology has improved a lot the research in the marine environment, there are a series of tasks and situations where the work of the underwater archaeologists cannot be substituted by a robot. Indeed, human intervention is often preferred when water depth makes the site of interest accessible by divers. Particularly regarding archaeological applications, great efforts have been focused on the documentation, preservation and conservation of our Underwater Cultural Heritage (UCH), which is an important part of the archaeologists' work. Therefore, the introduction of new technological devices and tools could considerably help divers to collect data from the marine environment, thus playing a significant role to the archaeological science.

This article presents the Lab4Dive project, co-funded by the EMFF programme of EU through the EASME and DG MARE call on "Blue Labs", and its preliminary results. The project aims to design, develop, and validate an innovative, marketable and competitive product for surveying, documenting and preserving UCH. Lab4Dive will provide a solution by equipping divers with portable, low cost and relatively small sized data-gathering systems, in order to hugely improve both the amount of data collected in a single mission and the management of time and human resources at disposal. The professional diver could use an underwater tablet equipped with environmental sensors, like high-resolution cameras and an acoustic localization system, where a dedicated data gathering system is accessible through its own application.

The rest of the article is structured as following: Section II describes in detail the project and the features of the Blue Lab concept; Section III presents the designed system architecture of the Lab4Dive device and Section IV presents the preliminary results obtained.

2. Project overview and Blue Lab establishment

Lab4Dive project has the aim of designing, developing and validating an innovative, marketable and competitive product for surveying, documenting and preserving underwater cultural heritage. It promotes the active cooperation of business experts, young researchers, tutors, end users and local stakeholders coming from different scientific disciplines and industries. In fact, it intends to encourage the multi-disciplinary cooperation, training young researchers in the design and development of new and innovative technologies for underwater survey and documentation, with the establishment of a Blue Lab.

The project aims also to provide a valuable solution for the protection and preservation of underwater archaeologies and cultural heritage, contributing with special events for the raising of public awareness regarding these thematic.

This system can be referred to as a "portable smart lab" to support archeological divers in the survey and documentation phases. In fact, the development of a cost effective, innovative and user-friendly solution could boost innovation capacities of underwater archaeologists. The smart lab will be based on a tablet coupled with an intelligent underwater case embedded with environmental sensors, a high-resolution camera and compatible with a number of commercial acoustic positioning systems, which will be integrated with a cloud data gathering system.

The Lab4Dive system is based on three physical devices:

- an underwater tablet (UT) that supports the diver during the immersion providing several functions;
- a docking station (DS) able to acquire and exchange data coming from different environmental sensors, a high-resolution camera and an acoustic localization system, that are physically connected to it;
- a remote server that hosts the cloud database and performs the 3D reconstruction from the photos captured during the survey.

The entire system will assist archaeologists before, during and after the dives on their daily on-site work to:

- Plan a survey;
- Show the area map based on acoustic bathymetry where points of interest are marked;
- Show the path followed by the diver during the immersion thus evidencing the areas already explored and the ones that needed to be investigated;
- Acquire geo-localized photos for marking interesting evidences or for the subsequent creation of orthophoto and 3D models of the explored areas;
- Acquire geo-localized environmental data from the available sensors;
- Acquire geo-localized notes for marking acquired samples and other interesting points.

Moreover, it offers the possibility of improving the safety of the diver: his position and movements can be shared with the other members of the diving team and with the support personnel that supervise the operation from the boat.

2.1. Blue Lab

The Lab4Dive project has invested resources and effort for promoting innovative "laboratories", called Blue Lab, to pilot novel way of working, where young scientists supported by researchers, industry and local stakeholders, team up to develop innovative solutions to support the development of a sustainable blue economy, while preserving marine resources and ecosystems.

The established Blue Lab is a distributed and virtual laboratory. It brings together professionals among different research and engineering groups, business experts, local stakeholders and under/post-graduate students, mainly from Università Politecnica delle Marche.

It provides the organizational space for collaborative and multidisciplinary activities involving people in specific design, development and lab-testing tasks of Lab4Dive system, under the technical supervision of tutors, leveraging young peoples' skills and creativity and increasing awareness of marine challenges and opportunities.

The implementation of the virtual laboratory for the support of the Blue Lab operation included the choice of:

- a cross-platform tool to target different O/S with the same basic code;
- an online platform for organizing collaboration among Blue Lab members;
- an online shared repository of documents to easily archive and exchange documents.

3. Architecture for data survey

The general scheme of the Lab4Dive distributed system derives from the typical configuration of a cloud system for the Internet of Things (IoT), where the underwater tablets used by the divers will be enriched with the capacity of gathering and presenting data on the field with a partial internet connection. It is composed of three main different elements, as shown in figure 1 and explained in detail in the following:

- Cloud Server Module
- Underwater Module

- Local Module.

The Local Module is an additional optional component which requirements have been analyzed but its implementation is left as a possible future development.

During the mission, the archeologists could collect a large quantity of data and images, for example for photogrammetric purposes; as soon as the diver is out of water and an internet connection is available, all the information acquired could be uploaded on an appropriate server in order to launch a 3D reconstruction process. This latter data-processing system is based on DocusScooter technology [3] and DiRAMa ([4]), a partially developed system developed by authors in previous years, and the results can be stored in the most common three-dimensional formats like PLY, OBJ + MTL, DAE and PDF. The possibility of choosing between low and high-quality results, which means different elaboration time, allows the researchers to analyze the data collected before the next dives or mission in order to plan it in the best way. Cooperation between man and technology can represent a winning solution to overcome the difficulties of operating in harsh but fragile environments.

The developed system will be tested at pilot sites in the Mediterranean Sea, at several locations and at varying depths in order to ensure that it meets the requirements of underwater archaeologists working in the rich UCH environment of the Mediterranean Sea.

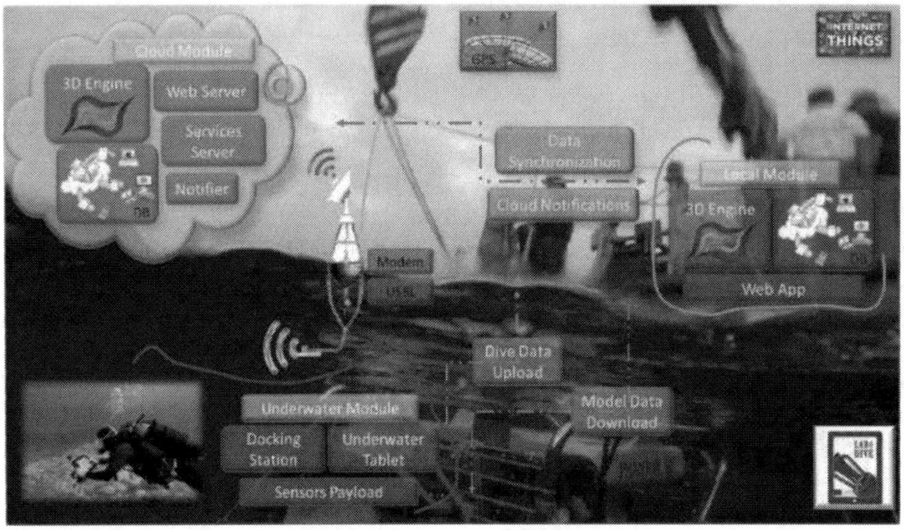

Figure 1 System architecture

3.1. Cloud Server Module

This module is composed of a Remote Server (RS) that hosts the database and manages the 3D reconstruction process. The Cloud Infrastructure is composed of:

- a database MySQL
- a web interface for backend CRUD operations
- a RESTful API to interact with the other devices
- the 3D Engine Module

The Cloud Infrastructure has the following functionalities:

- Manage the database
- Manage the communication with the other modules/devices of the system
- Perform a 3D reconstruction from the photos acquired during a mission through the 3D Engine Module
- Send a notification to the user about the 3D elaboration progress through the Cloud Manager

IOP Conf. Series: Materials Science and Engineering **364** (2018) 012054 doi:10.1088/1757-899X/364/1/012054

3.2. Underwater Module

The Underwater Module is based on two systems:

- an Underwater Tablet (UT), placed in an underwater case that ensures a fully functioning touchscreen, that supports the diver during the immersion providing several functions;
- a Docking Station (DS) able to acquire and exchange data coming from different environmental sensors, a high-resolution camera and an acoustic localization system, that are physically connected to it. It is provided with an embedded control board that manages all the electronics and that implements all the logic. The control board is integrated with a WiFi node for the communication with the tablet. This board provides also a bridge to the high-resolution camera connected.

The Underwater Module is composed of the following software components:

- Mission Planner (MP): an application developed using the framework Unity 3D. It enables a registered user to define a mission specifying all its properties like the bathymetry of immersion site and the targets list, i.e. the collection of points that the user wants to visit. All mission data are stored in the Cloud Server.
- Navigation App (NA): it provides an augmented navigation interface that guides the diver along the selected targets inside the underwater archeological site. Thanks to the 3D bathymetric model shown on the tablet display as a map, users can know their position in the underwater area, the path followed and the path to follow. Targets, which can be created and positioned directly by divers, are shown over the map as point to reach. This application can be easily turned into a drawing board where the divers can quickly and easily draw some sketches or take some notes about a specific point or area. Moreover, it allows users to control the camera in order to take photo in high quality format. Both the notes and the high-resolution pictures can be geo-localized. After the immersion, NA can send all the data collected over the internet to the Cloud Server or to a local one located on the boat. It can download new data like map, notes, sketches and targets generated in previous dives in order to use them as information for a new immersion.
- Docking Station Control Software (DSCS): it manages the communication between UT and DS through a WiFi channel, interacts with several environmental sensors and with different positioning systems to acquire in real-time the position of the diver.

4. Preliminary results

The preliminary tests of the Lab4Dive system have been conducted separately to verify the functioning of the Navigation and Mission Planner apps and of the Docking Station. The Underwater Module in fact is now being integrating.

4.1. Software components of the Underwater Module

The Underwater Module is composed of two main Unity software components: Mission Planner (MP) and Navigation App (NA).

The Mission Planner (MP) is a Unity 3D application that allows to create Mission Models. A Mission Model correlates a 3D bathymetric model and an ordered collection of targets to visit during a dive. The first prototype of the MP is released both as mobile and web application in order to ensure greater flexibility during the practical use of the system. The mobile application can be installed on a mobile tablet, including the Underwater Tablet, and used on the boat in order to plan the dive. In this case, the Underwater Tablet must be able to communicate to the Cloud Server, hence, an internet connection is required. The web based application can be integrated into the Cloud Server as a JavaScript program which uses HTML5 technologies and the WebGL rendering API to run Unity content in a web browser.

The workflow of the Mission Planner is given below and shown in diagram in Figure 2:

1. The user logs-in to the MP;

2. The user chooses the site of the mission;
3. The user imports an existing mission or chooses a 3D bathymetric model among the available ones;
4. The user adds targets, which can be related to a previously created content, such as textual notes, sketches, photos or 3D models;
5. The user is able to modify the position of already-added targets;
6. The user can choose an order for the targets;
7. The user edits the mission's information (name, description, etc.);
8. The user saves the Mission Model.

When the user saves the Mission Model, it is sent to the Cloud Server Module which stores it in the database. Then, it can be imported into the Navigation App in order to guide the diver during a survey.

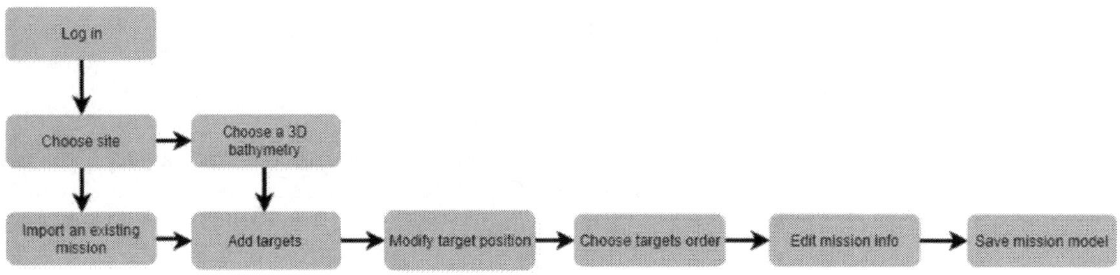

Figure 2 Mission Planner user workflow

The Navigation App (NA) is a mobile application that shows the diver's current position on a 3D bathymetric model representing the underwater archaeological site. The NA also shows the path to follow and the targets of a selected Mission Model. The Navigation App collects and stores the dive log, including geo-localized notes, sketches and photos taken by the diver during the survey.

The Navigation App implements three different communication protocols:

- One to communicate with the Docking Station Control Software, in order to get the diver's position and data from sensors;
- One to communicate with the Cloud Server, both before and after the dive. Before the dive, the NA requests information about the maps and the Mission Model. After the dive, the NA sends to the Cloud Server all gathered data, including the path followed, environmental data, notes, sketches and photos.
- One to communicate with the Underwater Camera, in order to control the acquisition of photos and to show a live-view.

Figure 3 Navigation App GUI

The Graphical User Interface (GUI) is shown in figure 2. The arrow represents the diver's position in the 3D representation of the underwater site. The diver can control the point of view of the map by using the left-side buttons. He can see the targets' list and the Underwater Camera live-view by using the right side buttons. He can also tap the button on the bottom right corner of the screen in order to add a new target (such as notes, sketches and photos). The GUI also provides a compass and an on-board computer which shows real-time information about the environment on the basis of the data received from the DS.

4.2. Hardware components of the Underwater Module

The overall architecture of the Docking Station consists of a NVIDIA Jetson TX1, which is connected to a Wi-Fi board through an Ethernet connection, to the battery, to the acoustic positioning system and to other sensors available. The general configuration is shown in figure 4.

The first prototype includes a depth sensor, an external IMU, a GPS and a battery in order to power all the system. All the components have been positioned in a custom-made underwater box to allow performing several tests. The first tests aimed to verify the connection and the communication between the components of the Docking Station in the underwater environment during a real mission. The system was indeed tested at a depth of about 10 meters and the immersion lasted about one hour. The second set of tests aimed to verify the tracking algorithm of the system. The diver held the device on the water surface during the test in order to have the GPS signal.

Figure 5 shows the tracking graph obtained by the algorithm. The GPS signal, which was captured every second, was lost twice during the test, about 10 seconds each time. During that period of time, the algorithm estimates the diver's position, with the mean square error shown in Figure 6.

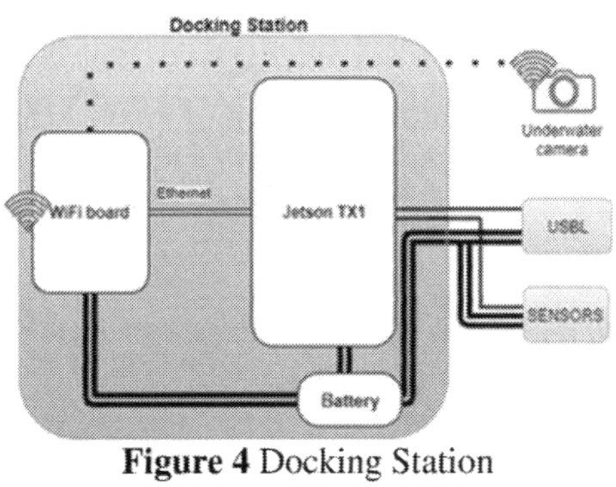

Figure 4 Docking Station

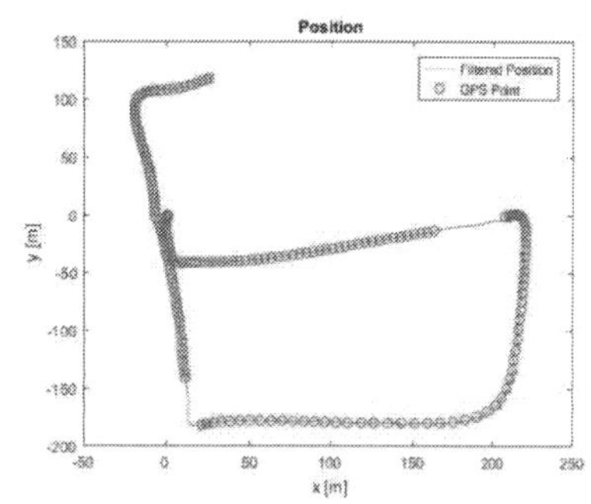

Figure 5 Diver's position

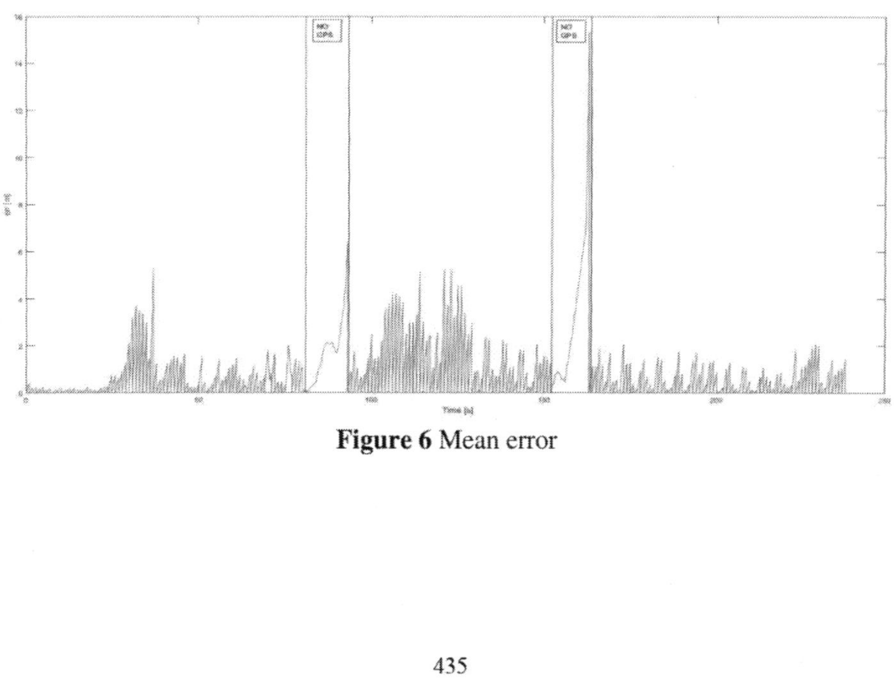

Figure 6 Mean error

5. Conclusions

In this article, has been presented the general overview of the Lab4Dive project, where partners from Greece and Italy are involved to design, develop, and validate an innovative, marketable and competitive product for surveying, documenting and preserving the Underwater Cultural Heritage of the Mediterranean Sea. The system architecture of the device that will be developed from this cooperation was illustrated, with the results of the preliminary tests performed in the first months of the project. The next steps will take into account firstly the integration of the two systems constituting the Underwater Module and then the integration with the Cloud Server Module.

6. Acknowledgements

The project is co-funded by the European Maritime and Fisheries Fund (EMFF) programme, supported by EASME and DG MARE, through the project id EASME/EMFF/2016/1.2.1.4/05/SI2.753831 - Blue Labs - innovative solutions for maritime challenges.

7. References

[1] Scaradozzi, D., Conte, G., De Capua, G.P., Sorbi, L., Luciani, C., De Cecco, P.G., Sorci, A., "Innovative technology for studying growth areas of Posidonia oceanica", Proceedings of 2009 IEEE Workshop on Environmental, Energy, and Structural Monitoring Systems, EESMS 2009, Crema; Italy; pp. 71-75, 2009

[2] Drap, P., Scaradozzi, D., Gambogi, P., Gauch, F., "Underwater photogrammetry for archaeology - The VENUS project framework", *GRAPP 2008 - Proceedings of the 3rd International Conference on Computer Graphics Theory and Applications*, pp. 485-491, 2008.

[3] D. Scaradozzi, S. Zingaretti, L. Panebianco, C. Altepe, S. Egi, M. Palma, U. Pantaleo, D. Ferraris and F. Micheli, "DocuScooter: a novel robotics platform for marine citizen science," in *ISOPE 2017,* San Francisco, 2017.

[4] Scaradozzi, D., Sorbi, L., Zoppini, F., Gambogi, P., "Tools and techniques for underwater archaeological sites documentation", *OCEANS 2013 MTS/IEEE - San Diego: An Ocean in Common*, art. no. 6741298, 2013.

[5] Conte, G., De Capua, G.P., Scaradozzi, D., "Designing the NGC system of a small ASV for tracking underwater targets", *Robotics and Autonomous Systems*, 76, pp. 46-57, 2016.

[6] Conte, G., Scaradozzi, D., Sorbi, L., Panebianco, L., Mannocchi, D., "ROS multi-agent structure for autonomous surface vehicles", MTS/IEEE OCEANS 2015 - Genova: Discovering Sustainable Ocean Energy for a New World, art. no. 7271543, 2015.

[7] Sorbi, L., Scaradozzi, D., Zoppini, F., Zingaretti, S., Gambogi, P., "Robotic tools and techniques for improving research in an underwater delicate environment", *Marine Technology Society Journal*, 49 (5), pp. 6-17, 2015.

[8] D. Scaradozzi, G. Conte and L. Sorbi, "Assisted guidance system for Micro ROV in underwater data gathering missions," *MED 2012 - Conference Proceedings,* pp. 1373-1378, 2012

IOP Conf. Series: Materials Science and Engineering **364** (2018) 012055 doi:10.1088/1757-899X/364/1/012055

Knowledge, conservation and technologies: a research project on the streets of Ancient Pompeii

Alberta Martellone, Massimo Osanna & Gianluca Vitagliano

Parco Archeologico di Pompei, Italy

alberta.martellone@beniculturali.it; massimo.osanna@beniculturali.it; gianluca.vitagliano@beniculturali.it

Abstract. The recent conclusion of a challenging project on surveying and monitoring the conservation status of the archaeological site of Pompeii was an opportunity to build a database of information about the city buried by the Vesuvius eruption in 79 A.D. The use of the most modern detection technologies (laser scanners, drones, etc.) has allowed to reach such degrees of accuracy as to open new perspectives of research on urban history.

The survey, in fact, has returned in detail for the first time the compositional geometries and the weaving of the blocks of the ancient roadmap of the site. In this perspective it seems interesting to systematise the new data and those from the archaeological excavations, also recent, on one of the most fascinating themes of the history of Pompeii: the forma urbis of the city, the chronology of the roads and the great work of monumentalisation consisting in their realization in stone blocks.

Therefore, on the basis of the study on the streets of Pompeii from a usage continuity perspective in the two phases of their life (the ancient, pre-eruption, and the modern, post-excavation), the research project that is illustrated aims to increase the levels of knowledge of the site, through the analysis of the techniques and the constituent materials, the definition of the relative chronology about the realization of the various stretches, the assessment of the conservation status. This also in order to identify appropriate management tools and appropriate strategies for the protection of such assets.

The project, for its implementation, will use: study of the construction typologies, surveys aimed to acquire knowledge of the constituent materials, observation of the stratigraphic relations between the different road sections, pre-eruption and post-excavation historical analysis.

The data that emerge will be integrated with excavation surveys, aimed at verifying chronologies and stratigraphic layers; they will be complemented by the results of the excavation of the alley between the *Insulae* 2 and 3 of the *Regio* V of Pompeii, to be executed, to return an undisturbed sample for study, both from the point of view of the geometries and of the techniques and the constituent materials.

1. Foreword

When, in 1748, the first ruins of Ancient Pompeii began to emerge, the interest of the excavators was primarily directed towards furniture, furnishings, works of art, frescoes and mosaics which were to be displayed in the rooms of the Royal Bourbon Museum: the excavations, whose locations depended on the will of their funders, were therefore geared towards maximising the recovery of finds, and thus keeping expenses under control. The situation changed under French domination, at the dawn of the nineteenth century, when excavations occurred in a more regular manner, proceeding towards unearthing the *domus* by advancing from the streets and the entrance halls. In those years the unearthing of the perimeter walls also began, thus clearly exposing the extent of the ancient city. The Pompeii which gradually emerged from the excavations held a particular charm for travellers of the Grand Tour, as borne witness to by the many views of ancient streets in the iconography of the era, as well as views of public buildings and private residences, even if for many years few people would be authorised by the Bourbons - even after the Restoration - to make drawings and conduct surveys *in situ* at Pompeii [1].

In 1861 Giuseppe Fiorelli became Inspector of the Excavations, and then Superintendent between 1863 and 1875, and his lucid and rational approach reigned supreme over the excavations of Pompeii: the liberalisation of access to the site, at the cost of 2 francs per person, and the techno-operative precautions taken for its systematic excavation are but a few of the initiatives which the indications of his proposal of "Legge organica del Real Museo e degli Scavi di antichità" would inherit. In it, articles 163 and 164 - which dictate the investigation methodology - set out that any new excavations must be carried out respecting the continuity of previous ones, in order to obtain uniformity over the areas yet to be unearthed [2]. The proposed law would remain as such, however the direction of the excavations of Pompeii would be based on the principles it specified beginning in 1863. Nevertheless, attention is still focused upon the building: the street, from which the excavation above all proceeds, represents solely the physical limit - the boundary - of the *insula* and, naturally, the preferred route of the tourist itinerary, "ormai vero fine dell'impresa di scavo"[3]. The excavation carried out by Fiorelli by then had a wholesale - and no longer leopard spot - restitution of the site as its objective and, to this end, the areas between the old excavation sites were excavated, so that by the end of his Pompeian mandate, approximately the western half of the ancient city had been excavated.

The excavation and restoration operations carried out between the end of the nineteenth and the beginning of the twentieth century continued in the wake of the innovations introduced by Fiorelli, at least until 1911, when Vittorio Spinazzola, Director of Pompeii until 1923, gave the go-ahead to a large scale excavation of the eastern section of Via dell'Abbondanza, going on to achieve the objective of reconnecting the Civil Forum district to the Amphitheatre. On closer inspection, this was an *ante litteram* attempt at an immersive experience: making it possible to follow "una strada pompeiana, e possiam dire di una cittadina campana o romana del Mezzogiorno d'Italia del I secolo dell'Impero, un aspetto che risulta in molta parte nuovo ed inatteso. Nuovo per la ricostruzione della sua vita, e nei rapporti dell'architettura e della decorazione architettonica: in ciò, anzi, che l'una e l'altra hanno di meno noto pei tempi grecoromani, la facciata, con il suo corteo di luci, aggetti, logge, coperture, ornamenti, colore"[4].

At the conclusion of the works in 1923, it was possible to traverse the entirety of Via dell'Abbondanza, from the crossroads with Via Stabiana to the *insula* of Octavius Quartio: the urban *facies* was now defined by the façades of buildings with upper floors and balconies [5], by the walling, by the colours of the frescoes and *tituli picti* preserved *in situ*; and then again by the pavements, by the public fountains and by the basalt paved streets. In short, everything refers to the street as a public social place, such as in the first life of Pompeii fixed to AD 79. Thus the experience of Spinazzola can be considered fundamental in the history of archaeology, which at Pompeii in the twentieth century, was immediately geared towards the understanding of social life in public spaces, well exemplified by the 'vision' which the Director pursued in his work [6].

Spinazzola documented the work carried out in a three volume work, *Pompei alla luce degli scavi nuovi di via dell'Abbondanza (anni 1910-1923)*, not published until 1953; in which he gives an account of the excavation methodology and finds - primarily the façades of the buildings along Via dell'Abbondanza - which allow us to re-examine the relationship between the private buildings and public spaces of the Roman city. Even if few words are given over to the role of the street as a significant element in its material substance [7], Spinazzola can be credited for having considered the excavation of the street as truly being *of* the street', in which the then followed approach appears to have been overturned: indeed for the first time, the objective of the excavation was the street itself, whose limit was fixed by the façade of the *insulae*, and not the other way round. Rarely did Spinazzola go further [8].

The street, therefore, came to exercise a particular charm, since it is configured as a system of elements, unique in the site, with the same function in Ancient Pompeii as today. Only the actors have changed: until AD 79, the street was the site of social interactions in which Pompeians played out their daily lives; in the modern age, travellers and tourists walk along it, living it as a place in which to sense life of the first century after Christ. In other words, even today the street system, with its pavements and fountains, continues to be a site of social interaction: tourists who stop near the fountains or entrances to the *domus* establish relationships among them neither more nor less than the inhabitants of two thousand years ago. Bearing this in mind, the streets are an element which strongly characterise the current site of Pompeii, and indeed in terms of its landscape, whose material integrity is to be protected.

However, every conservational action must imply an adequate 'knowledge operation'.

The 'Plan of Knowledge' of Pompeian streets arises from such premises, with the objective of shedding light on certain problematic areas related to aspects of material culture: it is not therefore a study of the urban development of Ancient Pompeii, but rather an investigation into the paving of the streets, which examines themes such as the construction techniques and production methods which, by necessity, had to have followed a diachronic logic before the eruption of AD 79. Awareness of the characteristics of the streets and verifying the state of conservation will allow us to specify protective operations, with a further view towards preserving material integrity in the face of a growing anthropic pressure due to the consistent flow of visitors [9].

2. The Project

The street, in this research project, refers to a system of elements in relation to each other which are only partially dedicated to the transit of vehicle, animals and people [10]. It has come to be characterised as a palimpsest upon which differing levels of understanding of the dynamics of urban development of Pompeii are arranged. Hence, as well as the street base, or more correctly the carriageway, whose role was - and remains - ensuring the outflow of water in the absence of a dedicated drainage system, components which contribute towards defining the Pompeian street are:

- the pavements, generally present to the sides of the carriageway, delimited by curbs and widely paved in Antiquity;
- the pedestrian crossings, raised above the level of the carriageway, which allowed the pedestrian to cross the street from one side to another in a straightforward way, keeping the walkway at the same level as the pavement;
- the guard stone blocks, found only on certain streets - mostly between Via di Nola, Via della Fortuna and Via di Vesuvio - and around a great number of public fountains;
- the public fountains, which were distributed along the paved arteries of the city, and were able to guarantee a water supply to the population;
- the other systems related to the water supply, including piezometric towers related to the aqueduct, which served the most important structures.

This project seeks to investigate this system of elements in its material components, with particular regard - at least in this first phase - to the carriageway. To begin with we proceeded to analyses of the data from the Plan of Knowledge [11], which in 2015 yielded the survey of the entire city at a scale of 1:50, utilising advanced technologies.

For the first time it is possible to understand the composite geometries and structure of blocks of the entirety of the ancient road network on the site. Previously, the most up-to-date maps of Pompeii, at a 1:500 scale, did not allow us to appreciate the details of the streets, such as the morphology and structure of the paving stones [12]. The results of the Plan of Knowledge are interwoven with the data which emerged during the recent archaeological excavations which formed part of the sites relevant to the Plan of Works of the Great Pompeii Project - the stabilisation of the *Regiones*, the new itineraries for improving accessibility and the restoration of certain structures - which provide additional data on the street network of the city.

For perhaps the first time, areal information will permit a comprehensive reading of the *forma urbis* of the city in AD 79, with a level of information that can be progressively implemented along with the results of other ongoing interventions.

In order to achieve the objectives outlined above, the project has been divided into phases.

The first phase, an account of the preliminary results of which are reported here, envisages mapping of the street paving and geometric characteristics.

The second phase consists of a vast study aimed at understanding the characteristics of specific elements of the infrastructure and any links with the context. In this phase, the potential use of a series of geognostic surveys carried out on all of the infrastructure will allow us to understand the thickness of the paving and alert us to the presence of any anomalies in the subsoil of the carriageway, to the additional end of putting into action any conservation work considered appropriate.

The complexity of the approach that a systematic study on the infrastructure network presents cannot disregard the preliminary definition of the situation regarding our understanding of the streets of Pompeii, integrated with information supplied by the Plan of Knowledge and *in situ* verifications.

3. Notes on recent excavations of the streets of Pompeii

As already hinted at, this project seeks to increase the level of understanding of the streets in their material consistency, with particular attention to the paving. It can not, however, ignore the study of urban development in Pompeii, a highly complex subject that is currently undergoing a renewed interest, both on the basis of information which is emerging from the conservational interventions being carried out on the site, and following stratigraphic excavations in various areas of the site and *extra moenia* which were launched in 2017, that will add data to our knowledge of the city's origins [13].

It is opportune to here recall the models of urban development which over time have animated the debate on the birth of the city. Among them, one well known model holds that the urban centre developed from the 7th century BC, starting from the *Altstadt*, corresponding to the Civil Forum and Triangular Forum [14], and continues in successive moments, until the colonial era, towards the eastern sector [15].

The alignments would already have been present at the beginning of the development of the city, and would have been accompanied by a 'tracing' operation, in which it is possible to recognise an urban planning project [16]: it is dated by certain recent studies to "at least the 6th century BC" and complied with at a later date [17]; others in the 4th or 3rd century BC [18]. The street paving must necessarily then be dated to moment after the earliest urban planning operation; in AD 79 it was not yet complete, and many streets were still *stratae terrenae*.

This research, which must necessarily take into account models of urban development which were developed in earlier studies, was born during the 'Pompeii for All' [19] intervention and amid the results of the preliminary investigations into the creation of new paving that primarily concerned the pavements, starting from the Amphitheatre, arriving at the Forum and spreading out towards the terrace of the Temple of Venus to the west and towards Via di Mercurio to the north. The archaeological investigations have made it possible to understand in detail the characteristics of the pavements and underground utilities, which are also known partially due to previous interventions [20], including: the private channels which occupy the base of pavement; the water system which distributes water in a capillary way via the lead *fistulae*, and the paving mixture created over an extended time sequence.

The 'Pompeii for All' operation was therefore a unique occasion to investigate these aspects, along the major arterial roads of Pompeii. It has allowed us to gather unexpected data on the history of the evolution of the ancient city, suggesting the notion that the street is a flexible system, able to metabolise the changes o urban construction over time: indeed, three experiments conducted in areas of the city somewhat distant from each other - along the northern side of the terrace of the Temple of

Venus [21], in Via di Mercurio [22] and in Via dell'Abbondanza, *Regio* III 7 side [23]- have clarified that the streets underwent modifications even after paving; these modifications consisted of a narrowing of the carriageway, in two cases in favour of widening the pavements.

The numerous experiments carried out along the arterial roads of Pompeii during the course of the most recent investigations suggest the absence of ballast or preparatory layers on which to lay the paving stones, which would be placed ad *solidum*; so the road foundation at Pompeii seems to consist of a compact layer of protohistoric volcanic eruptions [24].

Upon close inspection, the central zone, between *Regiones* VII and VIII - i.e the area corresponding to the *Altstadt* - is wholly paved with basalt flagstones. Certain sections of *Regio* I are *glareae stratae*; the rest of the streets are *terrenae*: including almost all of those of *Regio* II - including the area of the Amphitheatre and certain alleys between *Regiones* III, V, VI and IX; the work of paving the streets never reached there, and therefore another way had to be found to make the road surface compact, to additionally solve the problem of water flow.

Among the most interesting points of reflection that the project will face is the lack of paving in the sector corresponding to *Regio* II, which also includes the Amphitheatre. The latter lies in an area primarily crossed by pedestrian traffic: carts could only traverse Vicolo dell'Anfiteatro; the other alleys were inaccessible from Via dell'Abbondanza, where the pavement found no continuity [25].

4. The articulation of the project and the preliminary results

The framework of knowledge on Pompeian streets as it has been defined leaves certain problematic areas unexplored. These are areas of research whose examination is not possible without obtaining a vast and detailed understanding of the site.

The quantitative data has initially been specified: they allow us to understand the complexity of the street system of Ancient Pompeii within the city walls.

The city covers 66 hectares, of which 44 have been excavated. The street network runs for approximately 16.1km, of which 4.7 (equal to 29% of the total) are only theorised, since they have not been excavated. The remaining above-ground network is composed of streets paved with lava stone basalt slabs over a length of about 8.7km (54% of the total) and compacted earth streets along 2.7km (17% of the total).

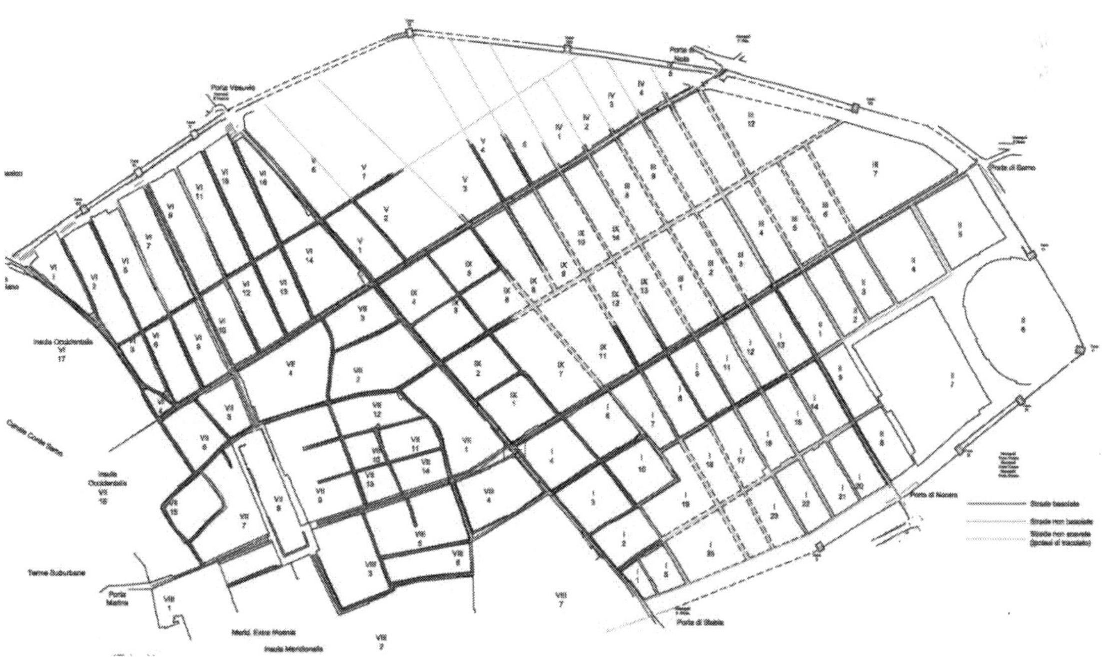

IOP Conf. Series: Materials Science and Engineering **364** (2018) 012055 doi:10.1088/1757-899X/364/1/012055

Regarding the basalt paved streets, the length of the network with an average section greater or equal to 4ml is equal to 2.7 km, or 31% of the excavated basalt paved roads: Via dell'Abbondanza, Via di Nola, Via della Fortuna, Via delle Terme, Via del Foro, Via Stabiana, Via di Vesuvio and Via di Mercurio. The total surface area occupied by paved carriageways is equal to 1.8 hectares.

The study of the streets dovetails with that of other infrastructure, including the pavements and public fountains. The pavements constitute a kind of system within the system, defined by the carriageway, whether paved or not, and by the kerb which defines its limit towards the street. In the carriageway run the *fistulae* for the capillary distribution of water - whose system in relation to the presence of piezometric towers needs to be better understood - and the small channels for private use, which carried water from buildings to the street. Furthermore, the stone blocks which mark the property limits are still documented on it.

The fountains are located along the paved streets, with a greater frequency along those which have a wider cross section (of the 42 fountains accounted for on the site, 21 are located along the principal streets of the city), most of them partially constructed on the street and partially on the pavement.

In this first phase, in addition to the survey of quantitative data, a sampling of road sectors was carried out on Via di Nola and Via di Vesuvio, two of the main arteries of the site, for which graphic illustrations were made available, being produced as part of the Plan of Knowledge. Road modules of 5 metre length have hence been identified; the measurement of these modules has been defined on the basis of assessing their feasibility in the space of a working week, including the moving and handling of the blocks and the surface treatment [26].

The sampling did not include those stretches of road which were tampered with in the post-excavation phase [27], or which suffered damage during the 1943 bombing. In the latter case, the survey allowed us to document traces of bomb damage, even where mapping carried out immediately following the bombing provided no information [28]. On the basalt slabs, indeed, one often finds traces of bomb explosions: one example is on Via di Vesuvio, between *Regiones* VI 14 and V 1, where one notes the absence of basalt slabs, corresponding to the burst crater filled with soil; around this block one see the splintering of blocks caused by the explosion of the bomb.

Traces of splintering can also be found on certain blocks on Via del Tempio di Iside, between *Regiones* VIII 4 and VIII 7, but their location must have been subject to some repositioning after the war, since the splintering lines do not seem to correspond.

Following the identification of the sections to be sampled, strategies were outlined which aimed to understand the distribution of the blocks according to their dimension, on the basis of the

IOP Conf. Series: Materials Science and Engineering **364** (2018) 012055 doi:10.1088/1757-899X/364/1/012055

documentation produced as part of the Plan of Knowledge. Specifically, blocks of a surface area greater than 0.20m² were selected, considering that stone elements of smaller dimensions can be moved by a single worker [29]. Such a distinction was thus considered ideal for verifying any differences between street sections which were attributable to different workers or time periods. The studies are still ongoing and will be completed at the end of the project, when the entire street network of Pompeii will be mapped out. However, an initial analysis already makes it possible to verify that the sampled street sections of Via di Nola exhibited a percentage of large blocks which averaged 57% compared to the total of the carriageway module [30], far from the 81% observed along Via di Vesuvio [31].

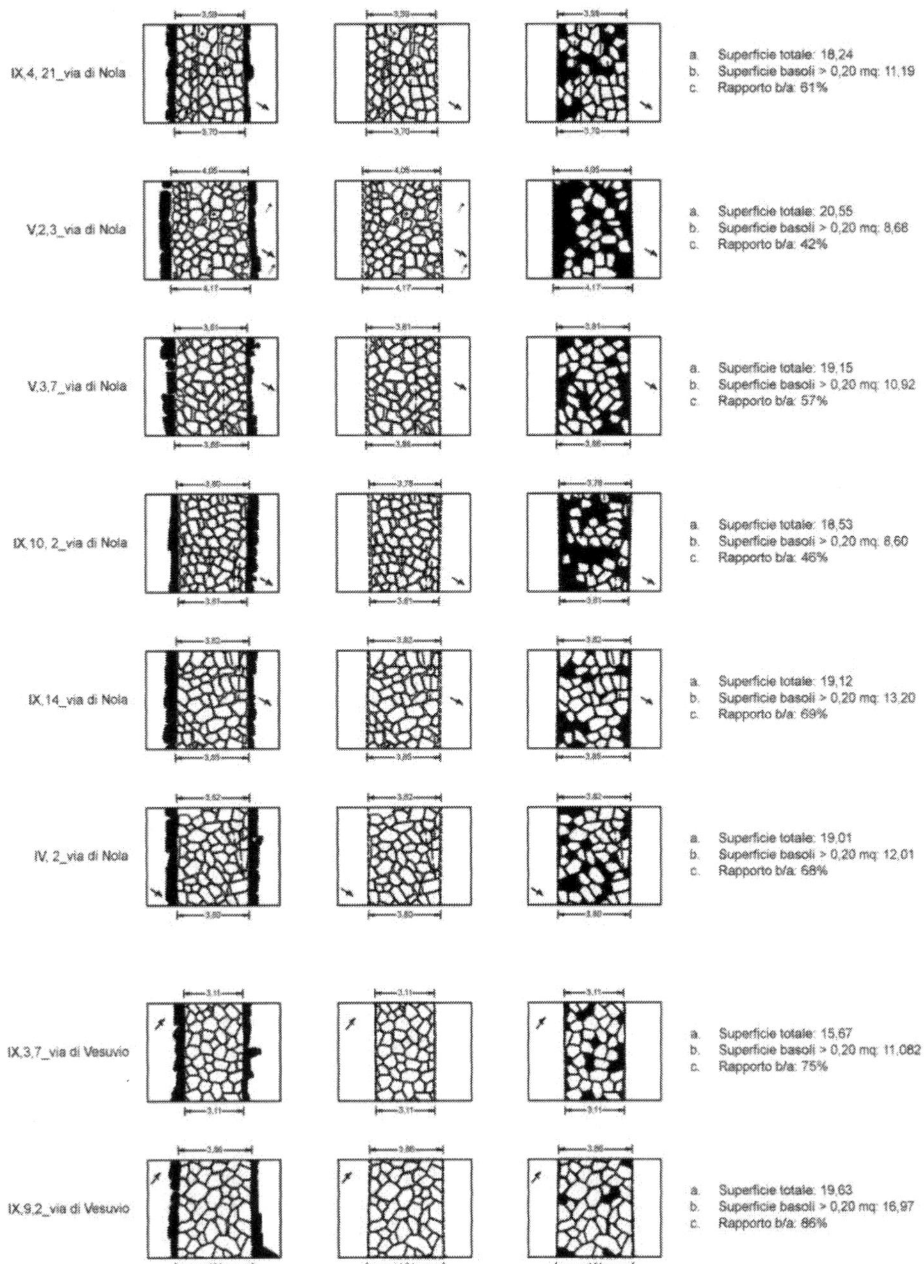

In this initial phase a study has been carried out on the pedestrian crossings. They can consist of a single block or a pair of juxtaposed blocks - corresponding with the alleys - or of an alignment of two, or three blocks in the case of wider streets. They take on a particular importance since they are linked to solutions of continuity in the structure of the basalt slabs, determined by a different modality or phase of laying the blocks compared to the creation of the rest of the paving.

Within the site, 163 crossing blocks have been identified, all on roads paved with lava stone blocks [32]. Some of them have been modified or removed due to particular conditions: for example, we know that two of the three crossing blocks at the crossroads of Via delle Terme and Via del Foro were removed before 1849, on occasion of the visit of Pope Pius IX [33].

Of further interest are the analyses of the distribution of these crossings in the site, according to demographic pressure: 103 crossings are located in *Regiones* VII and VIII and along Via delle Terme, della Fortuna and di Stabia which delimit the *Regiones* over an area of around 16.5 hectares. 65% of the crossings, therefore, are located on a little more than 37% of the excavated city.

Not only that: certain arterial roads, although paved, are devoid of crossings, such as in the case of Via dell'Abbondanza in the section in contact with *Regio* II, for a length of about 250 metres; or for the section of Via di Nola, in the direction of the port of the same name, which has no crossings for around 190 metres. This data confirms a distribution of the resident population with a lower incidence in the east of the city and a greater mobility of pedestrian flow in the western sector, around the Civil Forum and craft and commercial businesses.

5. First conclusions and perspectives

At the end of the surveys which were launched in the first research phase, a model will be created to survey the entire street network of Pompeii. The data gathered will yield information relative to solutions of continuity in the street sections, the types of cross section, the morphology and structure of the basalt slabs of the carriageway but also the types of surface working of the blocks and to a macroscopic reading of the lithotypes.

This survey will be aimed at understanding the specific aspects of Pompeian streets, such as the ability to identify temporal indicators for an absolute or relative dating of homogenous characteristics. Furthermore, the characterisation of traces of surface working of the blocks and the information regarding their 'wear' as a function of the period of construction and excavation, and of the lithotypes and current anthropic pressure, will allow us to put conservational measures into action. This last aspect, which does not only concern the lava stone blocks, also extends to analyses of the joints, and their possible construction typology, as well as evaluating the state of conservation and appropriate maintenance operations.

Finally, a series of non-destructive investigations utilising geognostic surveys across the entire street network will provide information on the thickness of the paving blocks and on any anomalies present in the subsoil.

In conclusion, the outlined project will yield an up-to-date reading of the role of streets in Ancient Pompeii, allowing an step forward in research in the field of archaeology, thanks to the use of innovative surveying and investigational technologies, as well as the results of the Plan of Knowledge drawn up under the Great Pompeii Project.

The vast amount of data which will be gathered, and the reflections which will result will allow us to define the most appropriate conservation strategies for the site, which itself represents a *unicuum* in the field of archaeology.

---

[1] On the role of Pompeii in European culture over the last two and half centuries, cf. Osanna M, Caracciolo T and Gallo L 2015 *Pompei e l'Europa 1748-1943 Catalogo della mostra* (Naples).

[2] "163. I nuovi scavi dovranno essere fatti in prosieguo degli antichi, né questi saranno intralasciati se prima non verrà tutto discoperto quello che erasi già cominciato a scavare…..

164. Al primo vestigio di edifizio si farà opera di scoprire le strade da cui può essere cinto; onde le città antiche dovranno essere dissotterrate per isole...", cf. Osanna M 2015 *"Tutto è stato riformato, moralizzato nella città morta". Giuseppe Fiorelli a Pompei* in Osanna M, Caracciolo T and Gallo L (eds) *Pompei e l'Europa 1748-1943 Catalogo della mostra* (Naples) p 230, In the articles of the proposal "Legge organica del Real Museo e degli Scavi di antichità" one reads (in a nutshell) of the successive reform of the nomenclature of Pompeii, with the identification of *Regiones* and *insulae*.

[3] De Caro S 1999 *Giuseppe Fiorelli e gli scavi di Pompei* in De Caro S and Guzzo P G (eds.) *A Giuseppe Fiorelli nel centenario della morte Atti del convegno Napoli 19-20 marzo 1997* (Naples) p 18.

[4] Spinazzola V 1953 *Pompei alla luce degli scavi nuovi di via dell'Abbondanza (anni 1910-1923) posthumous edition with foreword by S. Aurigemma* (Rome) p 11. Aurigemma credits Spinazzola for the innovative scope of his project, underlining that in 1923, at the end of the works, there were no longer "più strade pompeiane monotone, chiuse, deserte; ma finestre, balconi, tettoie e terrazze, che si seguono l'una all'altra, quasi che tutta la vita non avesse per iscopo se non la strada", cf. p XIII.

[5] It should be noted that the excavating from above and, therefore, the ability to reconstruct the upper floors, was already a widespread method under the direction of Fiorelli, who employed the wisdom gained from his experiences during the Bourbon era. The volumetric reconstruction of the House of the Hanging Balcony, for example, is due to Fiorelli (VII,12,28).

[6] F. Mangone, speaking of the works conducted at Pompeii since the last decades of the nineteenth century, observes that they tend to regulate the excavation areas according to the road network, "assecondando inconsapevolmente il nuovo interesse estetico per i vuoti urbani maturato nella Mitteleuropa di fine Ottocento", in Mangone F 2015 *Pompei, ginnasio dell'architettura europea, 1815-1914* in Osanna M, Caracciolo T and Gallo L (eds) *Pompei e l'Europa 1748-1943 Catalogo della mostra* (Naples) p 130. However, it is not by chance that the debate centred upon Pompeian city planning saw light in 1913, shortly after the beginning of new excavations on Via dell'Abbondanza by Spinazzola, with the contribution of Haverfield F 1913 *Ancient town-planning* (Oxford).

[7] On the other hand, interest in the materiality of the objects of use (which a street can be considered as) would come later than the historical period in which Spinazzola operated.

[8] Only in particular cases did Spinazzola go on to excavate rooms which faced onto the street (such as for example, the Fullonica of Stephanus or the House of the Moralist); in the majority of cases work proceeded by highlighting the façade and creating retaining walls for holding back earth behind the ancient walls. Such containment walls were also used alongside the alleys perpendicular to Via dell'Abbondanza, conveniently antiquated so as not to hinder their overall perception.

[9] In 1950 the site welcomed 300,000 visitors; in 2017 visitor numbers exceeded 3,400,000. For data regarding the last fifteen years, see http://www.pompeiisites.org/Sezione.jsp?titolo=Dati+Visitatori&idSezione=9 (last access 12.01.2018). Data on visitors to the site throughout the twentieth century can be found in Longobardi G 2002 *Pompei sostenibile* (Rome).

[10] "In campo urbanistico il termine definisce uno spazio di percorrenza con una misura prevalentemente allungata, delineato con intenzionalità costruttiva da parte dell'uomo", source *Via (Strada)* from Wikitecnica.com, F. Angelucci (ed.), http://www.wikitecnica.com/via-strada/ (last access: 14.01.2018). Regarding the widespread presence of vehicle grooves along the street network of Pompeii and traffic routes, see the recent Poehler E 2017 *The Traffic Systems of Pompeii* (Oxford).

[11] As part of the Great Pompeii Project, between 2015 and 2016, the "Plan of Knowledge. Diagnostic and monitoring services of the state of conservation at Pompeii" was carried out, which was an organised program of surveying and monitoring of the state of conservation at the archaeological site. With laser scanning technology, it surveyed the entire city and organised this data into an open multimedia 'container', the 'Information System'. This is a new GIS which forms the basis for both present and future studies and research which allows us, through periodic monitoring of the conservational state of ancient structures, to plan in an effective and timely manner, consolidation and restoration works which are necessary for the security and protection of the archaeological site of Pompeii. It is also becoming the digital archive for information resulting from archaeological and geognostic investigations conducted at the site.

[12] The "Nuova Cartografia Pompeiana Georiferita", edited by R. Morichi and R. Paone, after a long work project which began at the end of the 1980's and lasted about 25 years until 2012, for the first time granted Pompeii a digital map on a topographical basis at the conventional scale of 1:500, cf. Morichi R, Paone R, Rispoli P, Sampaolo F and Sodo A 2011 *Sulla nuova Cartografia Digitale di Pompei* in Rivista di Studi Pompeiani (XXII) pp 133-143.

[13] In 2017, the Archaeological Park of Pompeii began the "Scavi e ricerche a Pompei e siti della Soprintendenza" intervention which will be completed by 2018. However, the stabilisation interventions of entire *Regiones* and enhancement works which form part of the Great Pompeii Project have often provided for excavations which have increased the level of understanding of the site, particularly regarding the street network. Of particular interest is the intervention regarding *Lavori di messa in sicurezza dei fronti di scavo interni alla città antica, messa in sicurezza del fronte sud della Regio VIII e mitigazione del rischio idrogeologico delle Regiones I-III- IV-V- IX*, which was recently launched, and which envisages the excavation and opening up of the alley between *insulae* 2 and 3 of *Regio* V.

[14] Cf. Haverfield F J 1913 *Ancient town-planning* (Oxford). During the twentieth century A. von Gerkan and H. Eschebach entered the debate regarding the development of the city, perfecting the model proposed by Haverfield, cf. von Gerkan A 1940 *Der Stadtplan von Pompeji* (Berlin); Eschebach H 1970 *Die staedtebauliche Entwicklung des antiken Pompeji mit einem Plan 1:1000 und einem Exkurs: die Baugeschichte der Stabianer Thermen* (Heidelberg).

[15] However, it is not possible, under the current state of awareness, to substantiate this record with certain archaeological data, which is based on the observation of the shape of blocks; the attempt to reinforce it with stratigraphic data from archaeological investigations in a building remains weak, cf. Guzzo P G 2006 *Pompei. Storia e paesaggi della città antica* (Verona) p 48.

[16] Guzzo identifies the phase of 'tracing' of the original orientation in the western sector, and also underlines that it is "ben arduo immaginare che i tracciati siano stati formati altro che da battuti, sia che da essi abbiano, già entro il VI secolo, tratto forma e orientamento tutti gli isolati, per non parlare della rispettiva edificazione", in Guzzo P G 2006 *Pompei. Storia e paesaggi della città antica* (Verona) pp 48-49.

[17] For a summary of the wide debate, see the recent contribution of Giglio M 2016 *Considerazioni sull'impianto urbanistico di Pompei. Teorie sull'evoluzione urbanistica del settore orientale di Pompei* in *Vesuviana* (8) (Pisa-Rome) pp 11-48, with the relative bibliography.

[18] De Caro S 1992 *Lo sviluppo urbanistico di Pompei* in *Atti e Memorie della Società Magna Grecia* (3-1) pp 67-90; F. Coarelli F 2008 *Il settore nord-occidentale di Pompei e lo sviluppo urbanistico della città dall'età arcaica al III secolo a.C.* in Guzzo P G and Guidobaldi M P (eds.) *Nuove ricerche archeologiche nell'area vesuviana (scavi 2003-2006)* Atti del convegno internazionale (Roma 1-3 febbraio 2007) (Rome) pp 173-176. In the discussion forum of the conference "Nuove ricerche archeologiche nell'area vesuviana (scavi 2003-

2006)", F. Coarelli states that "lo strato più antico [della strada] è di fine IV - inizi III a.C.", in Coarelli F 2008 *Discussione*, in Guzzo P G and Guidobaldi M P (eds.) *Nuove ricerche archeologiche nell'area vesuviana (scavi 2003-2006)* Atti del convegno internazionale (Roma 1-3 febbraio 2007) (Rome) p 514.

[19] The 'Pompei per tutti. Percorsi per l'accessibilità ed il superamento delle barriere architettoniche' intervention is part of the Plan of Works of the Great Pompeii Project and was completed at the end of 2016. It involved the creation of easily accessible routes within the site over a length of 3 kilometres. The origins and results of the intervention in detailed in Sirano F 2016 *Pompei per tutti* (Naples).

[20] Nappo S C 1994 *L'impianto idrico a Pompei nel 79 d.C. Nuovi dati* in De Haan H and Jansen G (eds.) *Cura Aquarum in Campania Proceedings of the Ninth International Congress on the History of Water Management and Hydraulic Engineering in the Mediterranean Region*, Pompeii (Leida) pp 37-45.

[21] There was a reduction in the width of the functional carriageway upon the construction of the boundary wall on the northern side, which involved the removal of basalt slabs, which were never reintegrated. The investigations carried out as part of the "Pompeii for All" on the terrace of the Temple of Venus, north side, are currently being published (L. D'Esposito, publication underway).

[22] In this case, the trace in negative of a pair of removed basalt slabs, flush with the residues of an old pavement, suggest that there was a narrowing of the carriageway in favour of a greater width in the pavement (D'Esposito L, Galeandro F, Iadanza M and Martellone A *Il contributo dell'archeologia alla conoscenza* in Osanna M and Picone R (eds.) *Restaurando Pompei*, proceedings of the conference 3rd-4th April 2017, publication underway).

[23] From the pavement base of *Regio* III 7, a section of paving emerged whose basalt slabs have dimensions inferior to those of the central section; even the level of placement appears to be incompatible with the central paving of Via dell'Abbondanza: both of these elements suggest an interesting hypothesis, namely that the basalt slab identified predates the known one of AD 79, belonging therefore to a phase in which the carriageway was wider (this has also been encountered in Via di Mercurio, where it has however not been possible to verify the typology of the basalt slab and placement height of the blocks).

[24] However, precisely in the absence of ballast, a potential cause of localised subsistence of the street foundation can be identified.

[25] Guzzo P G 2006 *Pompei. Storia e paesaggi della città antica* (Verona) p 139.

[26] This data is taken from the analysis of the *in situ* relocation of certain paving blocks along Via dell'Abbondanza, near the entrance to the *Praedia* of Julia Felix, which was carried out as part of the 'Pompei per tutti. Percorsi per l'accessibilità ed il superamento delle barriere architettoniche' in 2016. On that occasion, it became necessary to proceed with the relocation of certain basalt slabs which the explosion of wartime ordnance of 1943 had damaged; the explosion of the bomb had disintegrated some of them. During the archaeological excavation, the margins of the crater were identified as being at least fifty centimetres deep compared to the street level. it is likely that during the removal of the rubble, the blocks were placed along the pavements. In 2016 they were therefore detected and, on the basis of their geometry and the alignment of the cart grooves, they were repositioned in the carriageway, which however still had a substantial gap due to the lack of other blocks. The relocation operation was conducted without the assistance of machinery, by a team of three workers, and took place during normal working hours from the 26/05/2016 to the morning of the 30/05/16; the surface covered by the end of the operation was equal to $10m^2$; hence, for the laying of the lava stone blocks for a paved surface equal to $20m^2$ (the 5 metre long module considered for an average width of 4 metres), one can reasonably consider a working week to be necessary. Due once again to the explosion of

IOP Conf. Series: Materials Science and Engineering **364** (2018) 012055 doi:10.1088/1757-899X/364/1/012055

munitions, a reconstruction intervention similar to that on Via dell'Abbondanza was conducted in Vicolo di Tesmo where, however, a substantial number of blocks was preserved, so that it was possible to restore the street paving in its entirety.

[27] Those sections in which evident disconnections of the paving due to maintenance interventions or the crossing of modern subservices were rejected: consider for example the anomalous repositioning of blocks, in which the cart grooves are not aligned with the main directions - one can find a clear case of tampering along Via di Stabia, corresponding to the entrance to the Theatres; along with cases in which the lithotypes clearly trace back, due to their conservational condition, to recent interventions.

[28] For the sections of street damaged by 1943 bombing, reference was made to the map *Pompeii - Bomb damage. 1943* attached to the *Final Regional Report Campania Dec 1945-May 1946*, in the Central State Archives, Allied Commission Control fund, Headquarters Allied Commission-Monuments & Fine Arts (and Archives).

[29] The specific weight of the lava stone is around 2800 kg/m3. If one assumes an average block thickness of around 20m, 1m^2 of lava stone block weighs around 560kg. However, blocks with a smaller surface area have smaller thicknesses: a block of 0.2m^2 paving surface can have an average thickness of 15cm, even weighing less than 70/80kg, and is therefore movable by a worker. The calculations reported show the distribution of blocks with a surface area less than 0.20m^2 (marked areas) compared to larger blocks.

[30] In the 6 sampled sections of Via di Nola, the percentages of surfaces occupied by large blocks compared to the total area are, respectively: 61%; 42%; 57%; 46%; 69%; 68%. The selected samples are indicated according to a west-east direction.

[31] In the 2 samples sections of Via di Vesuvio, the percentages of surfaces occupied by large blocks compared to the total area are: 75%; 86%. The selected samples are indicated according to a north south direction.

[32] The sole exception consists of three crossings (two near the intersection of Via di Castricio with Vicolo dell'Efebo, and one in Vicolo di Paquius Proculus) in which there is scant evidence of lava stone block paving. At present, the crossing on Vicolo dell'Efebo has a higher surface compared to the same walking level of the street, which was recently repaved (2016).

[33] To allow the visit of the Pontiff by carriage within the site, some other crossing blocks were likely removed along Via della Fortuna. The news, supplied by Vincenzo Sabini, technical assistant to the Archaeological Park of Pompeii, needs however documentary confirmation, which will be awaited during the course of research. For the visit of Pius IX on the 22nd October 1849, cf. Ascione G C and Conticello B 1987 *Pio IX a Pompei. Memorie e testimonianze di un viaggio* catalogo della mostra (Naples). It is conceivable that the absence of a series of three crossing blocks near the House of Marcus Lucretius on Via di Stabia can be attributed to the same explanation.

IOP Conf. Series: Materials Science and Engineering **364** (2018) 012056 doi:10.1088/1757-899X/364/1/012056

New generation monitoring devices for heritage guardians to detect multiple events and hazards

O Schalm[1,2], W Anaf[1,3], J Callier[3,4] and D Leyva Pernia[5,6]

[1] Conservation Studies, University of Antwerp, Blindestraat 9, B-2000 Antwerp, Belgium
[2] Antwerp Maritime Academy, Noordkasteel Oost 6, B-2030 Antwerp, Belgium
[3] War Heritage Institute, Renaissancelaan 30, B-1000 Brussels, Belgium
[4] Royal Museum of Fine Arts of Belgium, Museumstraat 9, B-1000 Brussels, Belgium
[5] Computer Science, University of Antwerp, Middelheimlaan 1, B-2020 Antwerp, Belgium
[6] Department of Physics, CEADEN, 502, Calle 30, 11300 Havana, Cuba

E-mail: olivier.schalm@uantwerpen.be

Abstract. Environmental parameters such as temperature, relative humidity, visible light, UV radiation and pollution influence the deterioration rate of heritage items. To judge on the environmental appropriateness for heritage conservation, it is therefore important to monitor the environment. Often, an incomplete set of environmental parameters is measured, or sporadic or time-averaged measurements are performed. As a result, a wide range of undesirable situations and hazards remain unnoticed. This might lead to an underestimation of environmental dangers (i.e., inaccurate judgement) or to inappropriate mitigation measures (i.e., inaccurate decision making).

We present an innovative and user-friendly monitoring device that simultaneously and continuously measures (1) environmental parameters and (2) material behavior. An extended combination of off-the-shelf sensors for temperature, relative humidity, air speed, CO_2, NO_2, O_3 and particulate matter are connected to a multipurpose datalogger. In-house developed sensors for the shrinkage and expansion behavior of wood, as well as sensors for metal corrosion rates are connected to the same datalogger. Such extended monitoring shows the identification of a wider range of undesirable situations, and it facilitates the search for correlations between such situations and the sources that cause them, i.e., the hazards.

1. Introduction

Scientific studies on heritage degradation and preventive conservation often focus on the (mathematical) relation between the object's degradation rate and its environment [1]. Although precise relations vary from object to object, an average relation could be assumed. This means that monitoring the environment gives substantial information on the degradation rate of heritage objects. Usually, the environmental appropriateness is estimated from temperature and relative humidity measurements. However, it is known that many other parameters such as visible light, UV radiation or pollution affect the degradation rate as well. Since many relevant parameters remain invisible with the commonly used monitoring systems, underestimation of environmental dangers might occur.

Studies in preventive conservation often fit within a well-defined framework of risk management [2]. Although risk refers to the future, examples of past damage are considered as they illustrate what might happen in future. The risk approach supposes that the accumulation of harm is governed by a sequence of randomly occurring incidents. Up to the authors' knowledge, the application of that approach on environmental data streams has not been used to its full potential. To apply the approach on environmental data streams, 3 types of undesirable situations are distinguished: dangerous occurrences, events and incidents. All three concepts define periods with elevated risk but with increasing risk level.

- **Dangerous occurrence:** Reportable period with elevated though limited risk for accelerated accumulation of harm;
- **Event:** Reportable period where at least 1 environmental parameter exceeds the corresponding guideline and where the risk for accelerated accumulation of harm is significant;
- **Incident:** Reportable period where the environmental conditions induce harm to at least 1 object.

Periods with elevated risk might remain invisible when only temperature and relative humidity are monitored. For example, housekeeping or moving actions in a museum environment may cause a sudden rise in particulate matter (PM). NO_2 and black carbon concentrations may significantly increase inside buildings due to traffic rush hours. In order to detect periods of elevated risk, a multi-sensor tool that combines off-the-shelf sensors for temperature, relative humidity, light, UV radiation, CO_2, NO_2, O_3, particulate matter, air speed, and human activity is developed. To directly monitor material behavior, in-house developed sensors for the shrinkage and expansion behavior of wood surfaces, as well as sensors for metal corrosion rates were incorporated in the monitoring system. The possibilities of the multi-sensor tool to identify undesirable situations are explored with a pilot study. By identifying undesirable situations, actions could be taken to avoid hazards to occur in future. The hazard is the potential source of harm that causes the period of elevated risk.

2. Materials and method

2.1 Sampling location

Monitoring has been performed in a late Gothic church in the center of a small Belgian city. The multi-sensor tool was located at the organ loft at a height of around 7 meter. Data collection started on 03/07/2017 and will continue until the summer of 2018 in order to collect one year of data. For the current article, the period from 23/07/2017 to 23/10/2017 was selected, including the transition from summer to autumn. This period was dominated by the installation of a new 'cold air' heating system. This means that there was no heating, that doors were regularly open and that the construction works generated dust. All precious objects of art were protected behind plastic foils with openings covered with Tyvek®.

2.2 Multi-sensor tool

The multi-sensor tool's backbone is a multi-purpose data logger (DataTaker DT85, Thermo Fischer scientific, Australia) to which sensors for a wide range of environmental parameters are connected. The total cost of the multi-sensor device was around 7000. Data of temperature (T), relative humidity (RH) and CO_2 were collected with a GMW90 (Vaisala, Finland). The intensity of visible and UV light were monitored with the upward positioned sensors SKL310 and SKU421, respectively (Skye Instruments, UK). Air speed was measured with a HD403TS omni-directional hotwire sensor (Delta Ohm, Soest, The Netherlands). Concentrations of NO_2 and O_3 were collected with NO2-A43F and OX-A431 sensors from Alphasense (Essex, UK). Particulate matter was collected with a DC1100 Pro Air Quality Monitor (Dylos Corporation, CA, USA). The measured concentration in number of particles m^{-3} has been converted into $\mu g\ m^{-3}$ using an empirical formula provided by the supplier. Human activity was measured with a passive infrared motion sensor with a detection range of 10 m (Panasonic Electric Works, Osaka, Japan).

Besides the environmental parameters, the behavior of two materials was monitored as well. The shrinkage and swelling of the surface of a cube of 16th century oak with a volume of 10 dm^3 was monitored by adhering a PFLW-30-11 strain gauge on the wood using the corresponding PS adhesive (TML, Tokyo). The wood originates from a beam of a historic house and was dated by means of dendrochronology. The beam was not in its original position anymore, but kept in the basement of the house. A small piece was sacrificed to be used as a sensor. The metal corrosion sensor is an atmospheric electrical resistance (ER) probe Model 610 consisting of Ag with a thickness of 250 nm (Cosasco Corrosion Monitoring and Chemical Management Systems, Santa Fe Springs, CA, USA). The resistance of this sensitive probe increases with progressing corrosion [3]. The wood and silver sensors were directly attached to the data logger using a Wheatstone bridge.

All sensors were read out in phase with a frequency of 1 minute, and 15 minute averages were saved by the data logger. The air speed sensor is the exception, for which the maximum speed within each 15 minutes time interval is stored. Data could be downloaded wirelessly using a 4G router.

2.3 Data processing

The stored data consists of a matrix of measuring points that are organized as rows. Each measuring point consists of the timestamp and the series of sensor readouts. The parameters measured are organized as columns. In-house developed software written in MATLAB combines the readouts of the different environmental parameters from the same measuring point and converts them in an overall indoor air quality index (IAQ-index). The IAQ index is calculated by (1) comparing each measured parameter with published target values, and (2) considering the importance of each measured parameter, since not all parameters endanger a collection to the same extent. The time dependency of the IAQ-index is visualized in a color bar. For this, the IAQ-values are associated to the jet color map used in MATLAB. A dark blue color is assigned to an appropriate environment, while a red color indicates an environment with elevated risk. More information about this data processing can be found elsewhere [4].

IOP Conf. Series: Materials Science and Engineering **364** (2018) 012056 doi:10.1088/1757-899X/364/1/012056

3. Results

All environmental data are plotted as scatter plots where measuring points are connected with lines (Figure 1a, with Figure1b a one week detail). Dangerous occurrences, events and incidents could be identified as sharp peaks that arise on top of a relatively constant background level. This is clear for all plotted parameters except for temperature and relative humidity. For the latter parameters, undesirable situations are more difficult to identify. This is mainly due to the fact that the fluctuations of ambient temperature and relative humidity cannot be described as a deviation from a baseline. For all other environmental parameters deviations could be considered as deviations from a baseline (e.g., a perfect dark situation for visible light, average CO_2 concentration on earth of around 400 ppm).

At certain moments, remarkable peaks are identified in the parameter-specific plots. Such an exemplary undesirable situation is the one on Sunday 20th August. On this day, a remarkable increase in CO_2 is observed (Figure 1b). This peak attracts attention due to its height and width. Therefore, this period is studied into detail. Information was collected on what happened that day. The church was open to the public and attracted around 1350 visitors. In the same time frame, an elevated air displacement was detected by the air speed sensor. In combination with the high visitors' amount, it is hypothesized that open church doors highly increase the air speed. At the same time, a reduction in RH of almost 10% occurred. This RH-drop was also recorded with other monitoring systems present in the church. This is counterintuitive when large numbers of people are present in the same building. Due to the open doors, outdoor influence was expected. However, daily averaged outdoor RH-values indicate a higher outdoor level compared to the indoor levels for this specific day. The indoor level of illumination is high until 11:00 PM. This is much later compared to sunset, and indicates the use of additional artificial illumination. Despite the large visitor number, hardly any rise in particulate matter is observed. The case study demonstrates that many more dangerous occurrences (e.g., several peaks in wind speed) can be seen in the graphs. Some of these occurrences would remain unnoticed when only temperature and relative humidity information is available. With the known periods in which they appear and the known parameters where sudden increase occur, substantial information is available to search for additional hazards in a targeted way.

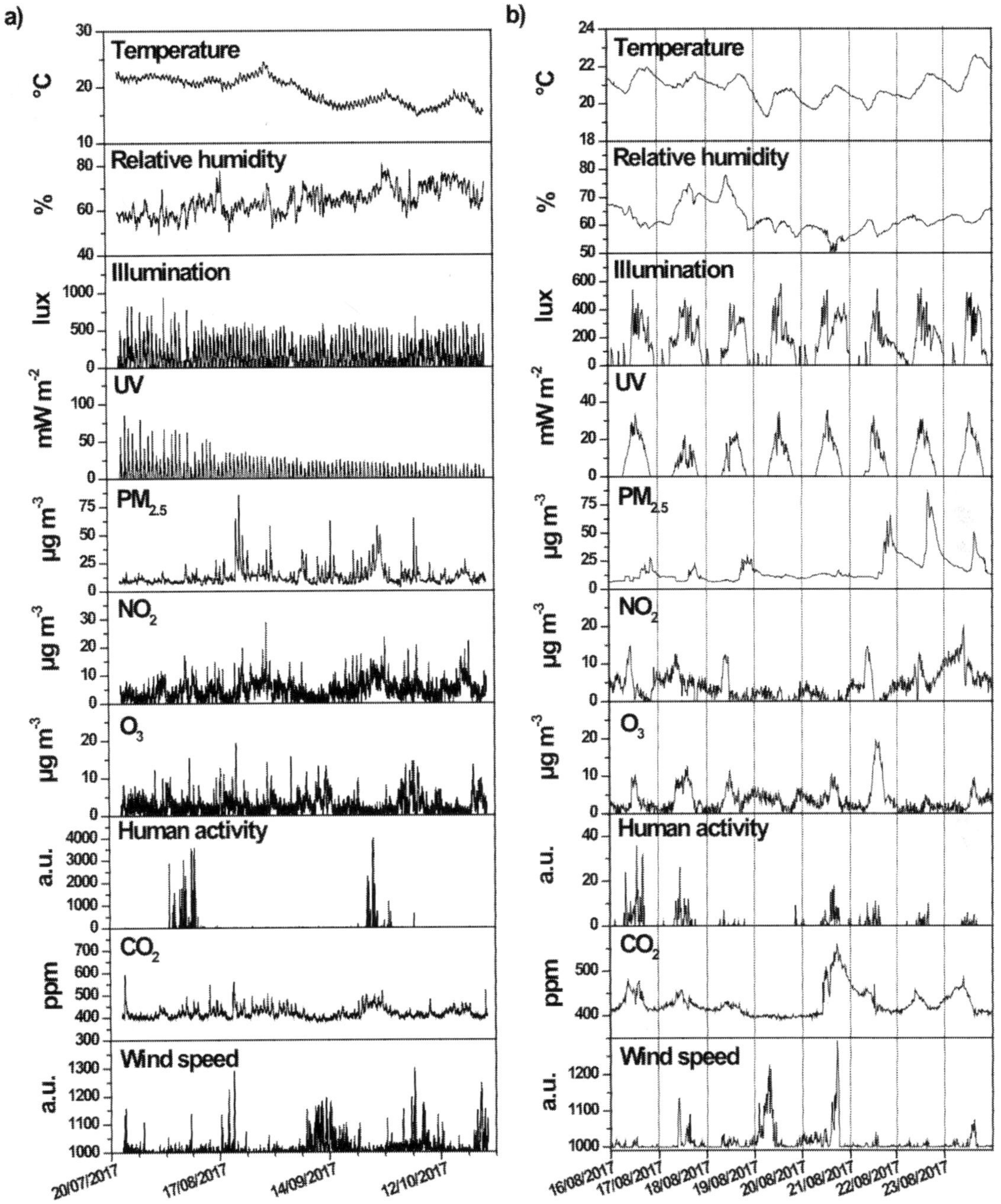

Figure 1. (a) Plots of environmental data over time from a period of 3 months (between 23/07/2017 and 23/10/2017); (b) One week detail (between Wednesday 16/08/2017 and Wednesday 23/08/2017).

As demonstrated in the example of the public day, periods with elevated risk could be visually identified in the parameter-specific plots, i.e., peaks on top of a background. Such periods could often be related to specific hazards. Not every peak has the same height. A clear example are the peaks in particulate matter in Figure 1b. On 16th, 17th and 18th August, small peaks appear, whereas on 21st, 22nd and 23th the PM-peaks are remarkably larger. This implies a difference in risk magnitude.

To guide the heritage guardians in detecting undesirable situations, in-house developed software generates color bars that visualize the dataset. Events are defined based on thresholds. When a period deviates from the thresholds, that period is colored in shades from cyan, yellow to red. Thus, all colors except of blue should be considered as events. The more the color reaches the red hues, the larger the level of risk. Figure 2 shows the result for a general collection, i.e., a mixed collection without any particularly sensitive objects. The overall conditions are classified in the blue color range, which means they could be considered as acceptable. However, several shorter periods show orange to red colors and should be checked in detail. The events in the current dataset are mainly due to inappropriate RH-conditions. The red periods in the IAQ-index reflect periods where the RH-levels exceed 75%. The cyan/greenish periods mainly reflect daily RH-fluctuations larger than 10%. This data processing and corresponding visualization demonstrates that the precise moments when events take place could be easily identified and the main parameter responsible for elevated risk could be found as well.

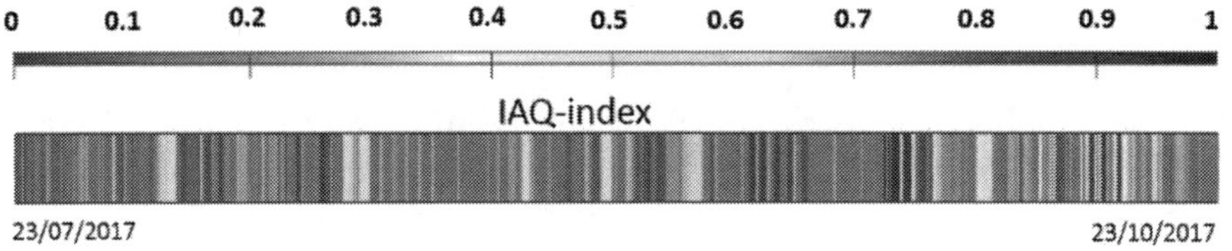

Figure 2. Summary of the environmental data collected in the period 23/07/2017 to 23/10/2017. The IAQ-index visualizes the environmental appropriateness for a general collection. Dark blue colors indicate appropriate conditions, while red colors are periods with significant environmental risks.

The data streams of the environmental parameters give meaningful information about the environmental dangers to which the heritage collection is exposed to. However, synergetic effects remain difficult to evaluate on the basis of the parameter-specific plots. This could be covered by directly monitor the material behavior. The wood and silver sensors monitor the material behavior of two totally different materials. Wood is hygroscopic and therefore sensitive to changes in RH. Silver, on the other hand, is mainly sensitive to pollution, and more specifically to sulfide reducing compounds.

The behavior of both material sensors is illustrated in Figure 3. Fig 3a shows the overview of the whole sampling period, while Figure 3b represents a detail of the same week as depicted in Figure 1b. The material behavior demonstrates that degradation of materials should not be considered as a constant over time. The fast response of the wood surface has a clear correlation with the relative humidity. The overall increase in wood strain is related to the overall increase in RH. The latter is less clear in the RH-plot due to the large fluctuations on top of an increasing trend. However, both parameters have a linear correlation coefficient

R^2 of 0.75. The short drop in RH during the open day (20th August) is not reflected in the wood behavior, which means that the four-hour lasting drop in RH is too short for the wood surface to react.

The silver sensor is subjected to progressing corrosion, but the corrosion rate is not constant. This is reflected in different slopes. For example, in September, the slope and thus the silver corrosion rate is lower compared to the summer months July and August. In October, the corrosion rate increases again. From the zoom in Figure 3b, it becomes clear that certain peaks in the silver corrosion signal correspond to RH levels, e.g., on the 21th August. This could be explained by an increase in surface conductivity at increasing RH levels. This means that the sensor read out is affected by the relative humidity. On the 20th August, there is no clear relation with the RH. Therefore, the number of visitors on that day had a (limited) impact on the silver corrosion rate.

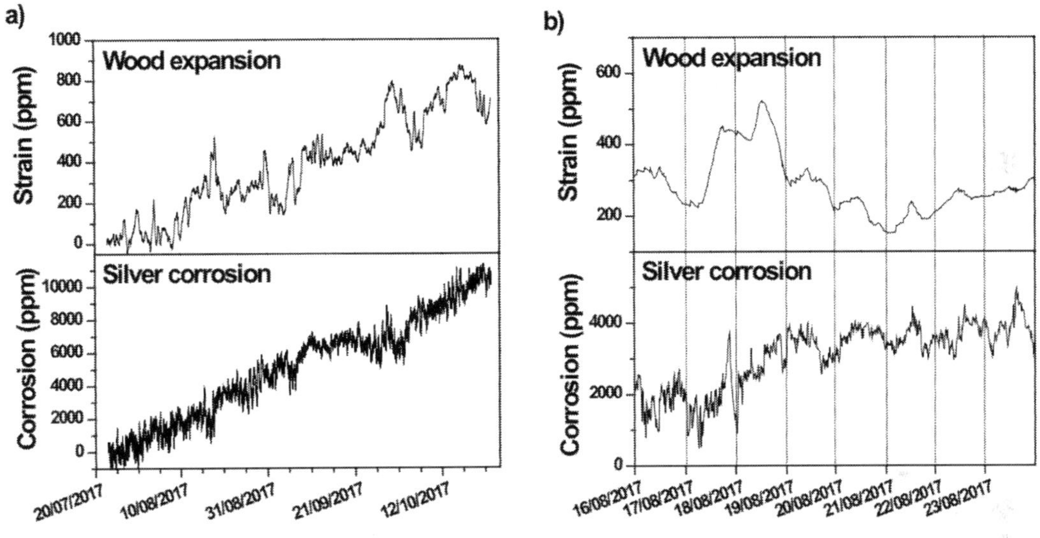

Figure 3. Continuous expansion and shrinkage information of the wood surface relative to a fixed resistance of 120 Ω and to the starting day and continuous corrosion depth information of silver relative to the reference electrode and relative to the starting day, both expressed in parts per million. (a) Plots of material behavior over time from a period of 3 months (between 23/07/2017 and 23/10/2017); (b) One week detail (between Wednesday 16/08/2017 and Wednesday 23/08/2017).

4. Discussion

This study demonstrates that current technology enables the monitoring of additional relevant environmental parameters and material behavior with a limited budget, on top of temperature and humidity measurements. The information obtained from the multi-sensor tool is relevant to determine the overall indoor air quality. The large number of parameters monitored suggests that the environmental conditions are more variable than when only temperature and relative humidity are considered. Several parameters consist of a background with peaks on top. These peaks reflect periods with elevated risk and are considered as undesirable situations. When identifying undesirable situations based on peak detection, caution should be exercised since their identification is affected by several factors:

- **Imperfect measurements:** Although many of the measured parameters are scalars (e.g., temperature, relative humidity, concentration of gaseous pollutants and particulate matter) other parameters are clearly vectors (e.g., illuminance, UV, air speed). Due to the use of a limited number of sensors in one position, not all directions could be monitored. Moreover, the environmental parameters are not uniform throughout a room but show gradients.
- **Imperfect sensors:** The availability and performance of continuous pollution sensors is relatively new and developing fast. It is expected that the technology will highly improve in near future. The gas sensors suffer from cross-sensitivity and the measured concentrations are rather close to the detection limits of the sensors. For certain gases such as sulfide reducing compounds, currently available sensors do not reach the low concentrations needed for heritage applications. The currently available inexpensive sensors to monitor particulate matter are generating a signal that is proportional to the number of particles per m^3. The guidelines, on the other hand, report target values in µg per m^3. Therefore an average density has to be used to convert the measurements in an appropriate quantity. This could result in large errors on the estimated PM-concentration. However, the results of these imperfect sensors could still be used to make decisions because the presence of peaks are undeniably the result of suddenly changing environmental conditions.
- **Imperfect target values:** Periods with elevated risk are easily identifiable in graphs through peak detection. However, to evaluate whether these periods should be classified as events, threshold values are needed. Such thresholds are not always well defined and are often material specific. Moreover, not all measured parameters have widely accepted threshold values (e.g., air speed). Other parameters do not affect heritage degradation in a substantial way (e.g., CO_2) but help in identifying undesirable situations.
- **Imperfect peak shapes:** Undesired situations are not always reflected as peaks in the graph. For example, a constantly too high RH-level is an event, but will not result in a peak when plotting the data. Our in-house developed software tackles this, since it detects undesirable situations based on published thresholds and not on peak detection. On the other hand, the software detects periods of undesired situations as it were sharp peaks while the actual values can be more gradually in reality.

5. Conclusion

The developed monitoring device demonstrates that a large range of relevant environmental-related parameters can be monitored simultaneously, and that also material behavior can be followed-up continuously. Many of the measured parameters show numerous peaks. These peaks are considered as undesirable situations because they are associated with periods of elevated risk. The case study demonstrates that monitoring a large number of environmental parameters allows the identification of many more undesired situations when compared to the monitoring of temperature and relative humidity only. Moreover, due to the continuous measurements the precise moments at which undesirable situations occur could be determined. This facilitates the search of the hazards causing these situations. The height of the peaks give an indication of the level of risk. Based on that height and the corresponding guidelines, the peaks could be classified as dangerous occurrences, events or incidents.

Since hazards could reoccur in future with a possible increased level of risk, it is advised to perform mitigation actions to avoid or reduce the identified hazards to occur in future. As a result, heritage collections will be exposed to a reduced level of risk in the future. This means that identifying undesirable situations contain valuable information and should not be neglected, even when they did not cause any noticeable harm so far.

This contribution integrates risk management vocabulary in environmental monitoring. This combination becomes relevant when multiple parameters are monitored simultaneously, and when well-resolved peaks could be identified as undesirable situations.

Acknowlegdements

The authors thank the financial support of the Belgian Federal Public Planning Service Science Policy (BELSPO) under project number BR/132/A6/AIRCHECQ.

References

[1] Luxford N, Thickett D and Wyeth P 2010 *Applying Preventive Conservation Recommendations for Silk in Historic Houses* (Rome: ICOM-CC Joint Interim-Meeting of five Working Groups) p 12
Strlic M et al. 2015 *Herit Sci* **3(33)**
Tétreault J, Cano E, van Bommel M, Scott D, Dennis M, Barthés-Labrousse M-G, Minel L and Robbiola L 2003 *Stud Conserv* **48(4)** 237-50
[2] Pedersoli Jr J L, Antomarchi C and Michalski S 2016 *A Guide to Risk Management of Cultural Heritage* (Sharjah: ICCROM-ATHAR) p 118
Michalski S, Pedersoli Jr J L 2016 *The ABC method: a risk management approach to the preservation of cultural heritage* (Ottawa: Canadian Conservation Institute)
Brokerhof A W and Bülow A E 2016 *J Inst Conserv* **39(1)** 18-28
Waller R R 2003 *Cultural Property Risk Analysis Model. Development and Application to Preventive Conservation at the Canadian Museum of Nature* (Ottawa: Acta Universitatis Gothoburgensis) p 108
[3] Dubus M, Prosek T 2012 *e-PS* **9** 67-71
[4] Leyva Pernia D, Demeyer S, Schalm O, Anaf W and Meert C 2016 *Proc. 14th Int. Conf. on Indoor Air Quality and Climate* (Ghent: International Society of Indoor Air Quality and Climate)

IOP Conf. Series: Materials Science and Engineering **364** (2018) 012057 doi:10.1088/1757-899X/364/1/012057

Seismic Assessment of Artefacts: The Case of Juno's Fountain of The National Museum of Bargello

G Cerri[1], G Pirazzoli[1], M Tanganelli[1], G Verdiani[1], T Rotunno[1], B Pintucchi[2] and S Viti[1]

[1] Department of Architecture (DiDA), University of Florence
[2] Department of Civil and Environmental Engineering (DICeA), University of Florence

giada.cerri@alumni.imtlucca.it

Abstract. Earthquakes represent huge risks for cultural heritage. Although there is a large literature about prevention on historical buildings, the seismic assessment on collections and exhibition designs is usually underestimated. "RESIMUS" long running project involves different knowledge and fields; its goal is to prevent seismic damages to museum collections and artefacts, combining both museographical solutions and anti-seismic methods. This contribution focuses on the "Fontana di Giunone", designed in 1555 by Bartolomeo Ammannati and currently kept at the Museo Nazionale del Bargello. The museographical and historical researches have been combined with the seismic performance of the reconstructed Fountain. In this paper the first results provided by the seismic analysis of this artefact are shown. A 3D structural model has been set, based on the three dimensional digital model provided by an detailed laser scanner survey. An elastic dynamic analysis has been performed, by representing the seismic input through an ensemble of seven ground motions expressing the seismic hazard of the area. The analysis has provided the seismic response of the sculptures, evidencing its sensitivity to the dynamic properties of the considered ground motions.

1. Introduction

RESIMUS is a multidisciplinary research group, whose purpose is to define a new methodology of analysis applied to museums so to measure the capacity, the so-called efficiency, of a site in case of dangerous events. The research focuses on an outstanding case-study: the National Museum of Bagello. Specifically, the goal is combining knowledge on museography and anti-seismic methods –in order to prevent risk to the museum collections and artifacts in case of earthquakes. In this phase, the research focuses on the seismic performance of the Fountain of Juno. First, the paper illustrates the artistic value of the masterpiece, its particular history, recent studies, and its set up. It is well known that cross-disciplinary studies facilitate the correct interpretation and exhibition of art works in the museum settings; in seismic areas, museum practices have to include anti-seismic methods to prevent risks. A brief literature review [1-8] illustrates the growing interest that the scientific community is addressing to the seismic assessment of artefacts in recent years. The research program is aimed at assessing the reliability of a case-study subjected to seismic actions. The case-study under examination is the statue of Venus, the most vulnerable (i.e. slender) sculpture of Ammananti's Juno Fountain, currently located in the courtyard of the Bargello Museum in Florence. In this paper, some preliminary results of the research, still in progress, are shown. A dynamic elastic analysis has been performed through a Finite Element (FEM) code , by assuming the ensemble "statue-pedestal" as a continuum body standing on the courtyard floor, therefore neglecting the interface [9,10] between both soil and pedestal and pedestal and statue. A finite element model has been set, on the basis of a detailed laser scanner survey. The seismic input assumed in the analysis has been defined according to the Italian Technical Code, by selecting an ensemble of seven ground motion spectrum-compatible to the seismic hazard of the area for the Collapse Prevention (*CP*) limit state. The analysis has been provided the elastic response of the case-study under the assumed seismic input, and pointed out the sensitivity of such response to the dynamic properties of the assumed ground motions, as a consequence of the relationship between the main frequency of the system and the ones of the ground motions.

The paper shows that multidisciplinary approaches are needed for a comprehensive interpretation and exhibition of artifacts. The paper inaugurates a series of studies, whose final aim is to include anti-seismic methods and relative technological and museographic solutions inside the museum environments.

2. The case study: the Fontana di Sala Grande

2.1 The Fountain history

Juno's Fountain (Figure 1), also known as "Fontana di Sala Grande", is one of the sculptural masterpieces by Bartolomeo Ammannati [11]. Currently exhibited at the National Museum of Bargello [12,13], the fountain attracts the interest of scholars and curators for its interesting layout, original destination, and uncertain history.

In the middle of the XVI Century, Cosimo I de' Medici moved his residence to Palazzo Pitti and transformed Palazzo Vecchio in a sort of headquarter of the Ducato, with new spaces and offices (Uffizi). He converted a Medieval fortress in a modern palace, so to host and impress foreign visitors. The first works in the throne room were designed by Baccio Bandinelli and Giuliano di Baccio d'Agnolo and coordinated by del Tasso [15]. The works concentrated on the North wall of the hall. When Vasari succeeded to Tasso in the job direction, the Sala Grande's yard was suspended for ten years, until Francesco I's wedding (1555). Vasari sponsored his friend Ammannati to the Duke, and he became the designer of the South wall of the Sala Grande.

In continuity to Bandinelli's work, Ammannati's design foresaw a rich architectural articulation, an internal fountain, and many statues in marble and bronze. The architectural arrangement of the sculptures is one of the main features of the unique composition. The so-called *Concerto di Statue* (Concert of Statues) fascinates for its original conception: a magnificent fountain to be placed inside the Sala Grande (now Salone dei Cinquecento) of Palazzo Vecchio.

The fountain is a tribute to the Medici's magnificence. The feminine statue at the top of a rainbow alludes to both the goodness and the duchess Eleonora di Toledo; the two peacocks are her emblems. A rainbow originates from two sources of water: Arno, a powerful man sitting on a lion, and Spring of Parnasus, a gentle feminine figure laying on a wing-horse. Inside the water, Cerere (Venus), pressing her breasts, embodies life and richness. Two of the sculptures, Fiorenza and Prudenza, are placed laterally to the group. According to the original project [14], these two pieces should have been positioned in the niches of the South wall of Sala Grande. When Vasari presented Ammannati's project of the Sala and of the Fountain to Michelangelo, the latter suggested raising the height of the room so to harmonize the global proportions. Vasari, as head of the *Fabbriche Medicee* (Medici's Yard) applied Michelangelo's recommendations: the Sala Grande became the Salone dei Cinquecento (see Figure 2), and the fountain works (1560) stopped; meanwhile, Ammannati started working on the Neptune Fountain in *Piazza della Signoria*, the so-called *Biancone*.

There are no drawings describing Ammanati's design; the only information we have about the project is the sculptures themselves and some writings reporting the project as a "concerto di statue" (concert of sculptures). After the project interruption, archive documents testify the presence of the marbles in the Medici's properties through time [16], others [14] indicate the presence of some statues inside Ammannati's *bottega* (atelier). Borghini [17], for example, recognizes the statues of Arno and Parnasus at Pratolino Villa. Probably, Ammannati finished the back of the statues when they were set up in the Pratolino's garden. In 1588, the sculptures moved to Palazzo Pitti; there, Ammannati himself curated the arrangement in the courtyard. Juno's Fountain changed location once again in 1635, when the pieces of the fountain were scattered between the Casino di San Marco and the Boboli gardens.

Figure 1. Current reconstruction at the National Museum of Bargello

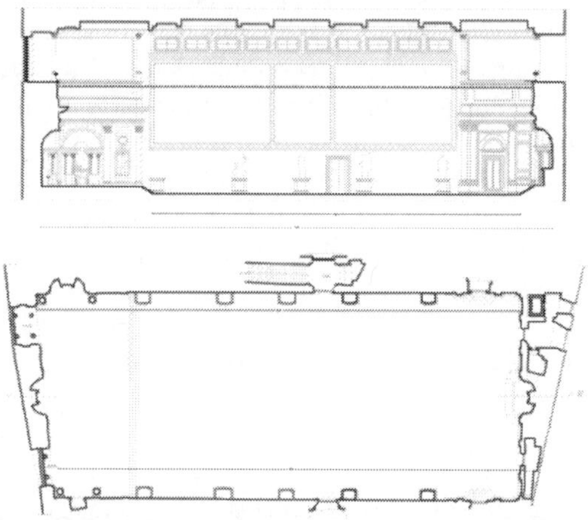

Figure 2. Salone de' Cinquecento (The red line indicates the old height of Sala Grande)

2.2 The set up. Behind the scene: documentation and survey campaigns

The final set up is the result of a long process, which involved the curators of the museum and the University of Florence. Today, Juno Fountain is considered a unique sculptural group, testimony of the greatness of the Medici's family, which can finally be admired in one of the most important museum of the world.

The first intuition about the Fountain's original layout is by Detlef Heikamp [18]. The scholar conducted a rigorous study on Ammannati as sculptor, and he finalized a reconstructive proposal for the group of sculptures (Figure 3). Years later, the traditional study combined to the new technologies helped finding a liable solution, which, at the end, is not far from Heikamp's suggestion. The current reconstruction suffers the limit of the missing historical documents about the original project (how was the South wall configuration project? Where were displayed the sculptures? Which was the relation among water, light, and volumes?) and of the subsequent wandering phase [19]. During the XX century, the statues were reunited at the National Museum of the Bargello.

The current set up has been realized in occasion of the 500 years from the birth of Ammannati, celebrated in 2011 [13]. It has been possible thanks to the combination of different methodologies and expertise. The curators' knowledge has been fundamental for the interpretation of the sculpture iconology and, therefore, the coherency between positions and meanings [16], the restorers analyzed the statues, the 3D modeling (subsequent to the three 3D scanner laser survey campaigns) allowed the correct documentation and the preview of the set up [20], the museographers designed the reconstruction [12] and, with the structure team, calculated the new structure (in particular the new arch holding the cast of Juno and of the peacocks). The sculptures of Cerere and Fiorenza are now placed beside the group. The two sculptures were part of the challenging sculptural program. Their position is a quotation of Guerra's sketch (Figure 4). The original marbles of Juno and the peacocks are placed near the recomposed fountain.

The preparation works involved the uses of new technologies and specific professionals. The survey campaigns and the treatment of the data set, allows the realization of detailed virtual models. Such models served to multiple purposes: global virtual reconstruction, 3D print, caster and restoration, exhibition proposals, documentation and preservation. Such cross-disciplinary work is now the invaluable support for the continuation of the study on the Fountain. The research is currently focusing on the safety and conservation of the Fountain, in particular, on the seismic risk.

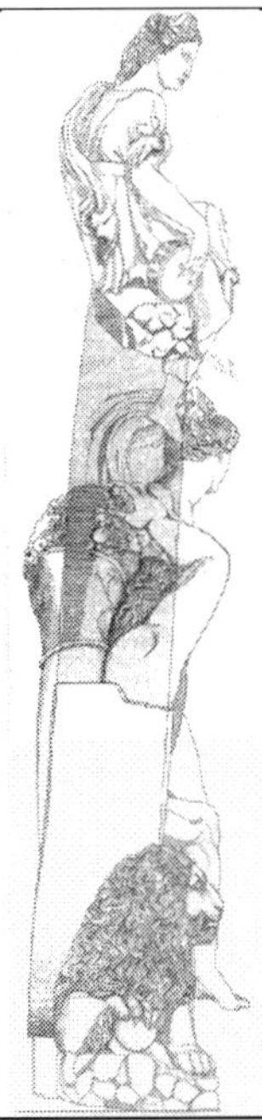

Figure 3. Heikamp's reconstruction.

Figure 4. Guerra's sketch in Pratolino.

3. The seismic analysis

The vast documentation about the current arrangement of fountain and its fame bring the RESIMUS research group to use Ammannati's masterpiece for this seismic analysis. In a preliminary analysis [21] the five sculptures constituting the Fountain have been checked in terms of slenderness on the basis of their geometry; since Venus came out to be the slenderest statue of the complex, the dynamic analysis has been performed on the model representing Venus only. The following Sections shows the model adopted for analysis, the selected ground motions represented the seismic input and the obtained results.

3.1 The model

The structural analysis has been performed on a 3D Finite Element (FE) model consisting of four-nodes tetrahedral isoparametric elements. Such model has been set on the basis of a 3D laser scanner survey. The survey was made using a Cam/2 Faro Photon unit, based on phase shift measurement technology; to complete some parts of the back of the statue, a Nextengine unit was used, since it better fits with the narrow space between the statue and the wall. All the scans were later aligned on morphological similarities, creating a pointcloud with all the information about shape and detailed characteristics of the statue. The FE model used for the analysis has been set from geometrical survey by making a series of changes. As a first step, the virtual model, rendered after the survey campaign, has been reduced in size and definition through the software MeshLab [22] in order to be used for the analysis. Furthermore, the surface model provided by the laser-scanner survey has been transformed in a volume one (second step), by introducing a set of nodes in the volume inscribed by the statue surface, and completed by additional nodes representing the pedestal (third step). In Table 1 the number of polygons referred to each model is listed.

Table 1. Number of polygons of the geometrical models.

Model	Initial surface model	Simplified surface model	Simplified volume model	Simplified volume model including the pedestal
Number of polygons or tetrahedrons	900,000	10,000	34,592	43,929

3.2 The seismic input

The seismic input has been represented through an ensemble of seven ground motions, spectrum-compatible to the elastic spectrum provided by the Italian Code, NTC 2008 [23] for the site. The Museum area, such as the rest of the Florence basin, consists of plio-pleistocene palustrine and alluvial deposits, followed by two sedimentary cycles related to the paleo-Arno river and the holocene geomorphic evolution [24]. As a consequence, as the investigation on the foundation soil is still in progress, a soil-type B, according to the NTC 2008 classification, has been assumed. The considered seismic intensity refers to a Return Period of 949 years, i.e. the *Collapse Prevention* limit state, with a class of use (c_u) equal to 2 (strategic buildings). The ground motions have been selected by the database Itaca [25] through the software Rexel [26] In Figure 5 the elastic spectra of the ground motions are shown and compared to the corresponding spectrum provided by NTC 2008. As can be noted, the seven ground motions differ very much from each other both in the maximum acceleration and in the frequency content.

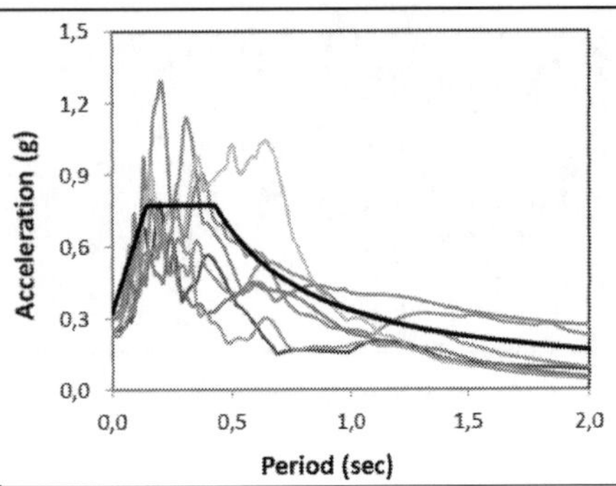

	Name	Station	Magnitude	PGA (g)
E_01	SRCO_HNE	S.Rocco	6.0	0.250
E_02	STR_HNE	Sturno	6.9	0.316
E_03	STR_HNN	Sturno	6.9	0.225
E_04	TLM1_HNE	Tolmezzo	6.4	0.315
E_05	GMN_HNE	Gemona	5.8	0.298
E_06	GMN_HNE_01	Gemona	6.0	0.255
E_07	GMN_HNN_01	Gemona	6.0	0.252
NTC 2008				

Figure 5. Elastic spectra of the assumed ground motions.

3.3 *The analysis*

The analysis has been performed through the software Strauss 7 [27]. An elastic time-history analysis has been performed on the FE model described in § 3.1. In the analysis, the marble of the statue has been described through a Young's modulus equal to 50 GPa, a Poisson's coefficient equal to 0.2 and a density equal to 2700 kg/m^3. The pedestal, made in masonry, has been represented through a Young's modulus equal to 5 GPa, a Poisson's coefficient equal to 0.2 and a density equal to 1700 kg/m^3. No link components have been introduced between the statue and the pedestal or between the pedestal and the floor. Therefore, the system "statue+pedestal" has been assumed as a single body, simply standing on the floor. A viscous damping equal to 4% has been assumed for the system.

A preliminary modal analysis, whose results are shown in Figure 6, has been made on the FE model to check its dynamic behavior. As can be seen from the Participation Factor (*PF*) shown in Figure 6, the first and the third eigenmodes refer to the x-axis (front to back of Venus), with a *PF* of about 57 % and 23%, respectively; the second mode occurs along the y-direction (lateral bending), while the fourth mode is a torsional one. Since the system is very rigid, its fundamental frequency is high (over 17.62 Hz, corresponding to a period of 0.058 sec).

Mode	Frequency (Hz)	PF-X (%)	PF-Y (%)	PF-Z (%)
1	1.762E+01	57.1	0.1	0.1
2	2.031E+01	0.1	58.0	0.0
3	1.014E+02	22.3	0.2	5.6
4	1.100E+02	0.2	13.2	0.0
5	1.303E+02	0.1	10.3	0.1
6	1.780E+02	2.0	0.2	77.8
7	2.637E+02	6.7	0.0	0.3

Figure 6. Results of the modal analysis.

IOP Conf. Series: Materials Science and Engineering **364** (2018) 012057 doi:10.1088/1757-899X/364/1/012057

By comparing the period of the first mode to the elastic spectra of the ground motions representing the seismic input (Figure 5), it can be seen that their peaks in acceleration occur for higher periods compared to the fundamental one of Venus. It can be expected, therefore, that no resonance phenomena will occur in the dynamic response of the system. As a consequence of the obtained eigenmodes, the time-history analysis has been performed along the x-direction only, that is the most sensitive to the statue dynamic response. The response parameter assumed in the analysis is the top displacement of the system, i.e. the displacement of the top of the sculpture.

4. Results

Figure 7 shows the stress distribution found for the case-study subjected to its own weight only. Even when no horizontal actions are applied, the statue experiences a slight horizontal displacement (0.05 mm), due to its asymmetry. As regards the stress level, in the pedestal it is almost uniform, as well as in the top part of the sculpture. In the bottom part of the statue, instead, due to the reduction in cross section (see Figure 7), there is a peak of stress, which achieves 0.18 MPa in tension and 0.82 MPa in compression. Therefore, the maximum stress arising in the material under gravitational loads are well below its elastic capacity.

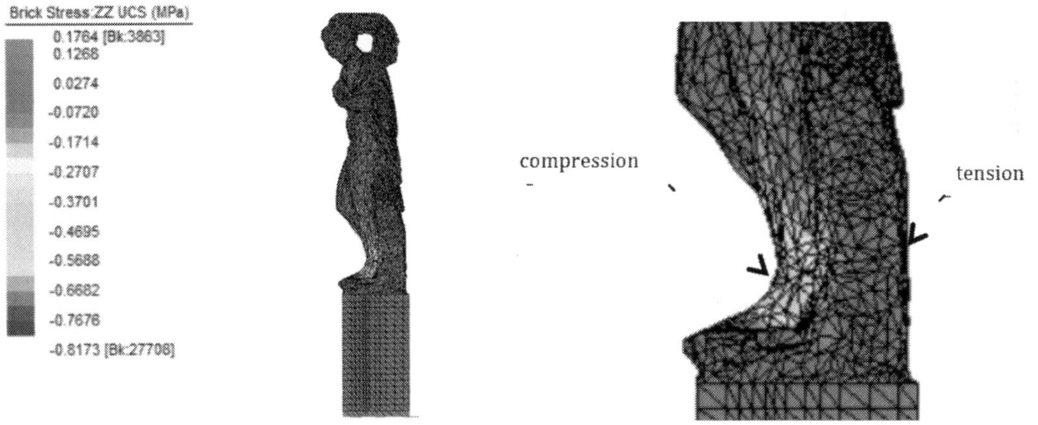

Figure 7. Stress distribution under gravitational loads.

In Figure 8, the time-histories of the top displacement provided by the seven ground motions are shown, together with their maximum value and the first (higher) peak in frequency of each record. The dynamic response of the case-study could be related both to the content in frequency of the considered ground motions and to their accelerations at the fundamental Period of the system. The maximum stress arising in the statue in the dynamic analysis is equal to 1.10 MPa and 3.14 MPa for tensile and compressive stress, respectively. The tensile stress in the material under seismic excitation, therefore, results to be more than 6 time larger than the one due to gravitational loads. It should be reminded, however, that the analysis has been performed by assuming a linear elastic behavior for the system, and neglecting any possible interaction between the statue and the pedestal and the pedestal and the soil.

5. Conclusive remarks

This paper presents the preliminary results of a multidisciplinary research concerning the seismic vulnerability of Ammananti's Juno Fountain, currently located in the courtyard of the Bargello Museum in Florence. More precisely, the analysis focuses on Venus, the main and slender statue of the sculptures complex. A 3D Finite Element (FE) model, consisting of four-nodes tetrahedral isoparametric elements, has been set on the basis of a detailed laser scanner survey.

IOP Conf. Series: Materials Science and Engineering **364** (2018) 012057 doi:10.1088/1757-899X/364/1/012057

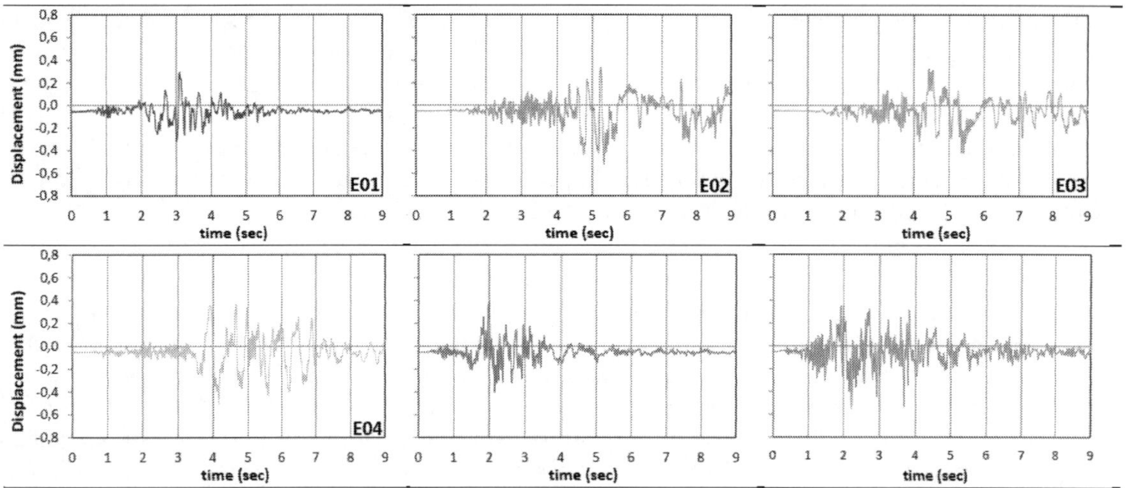

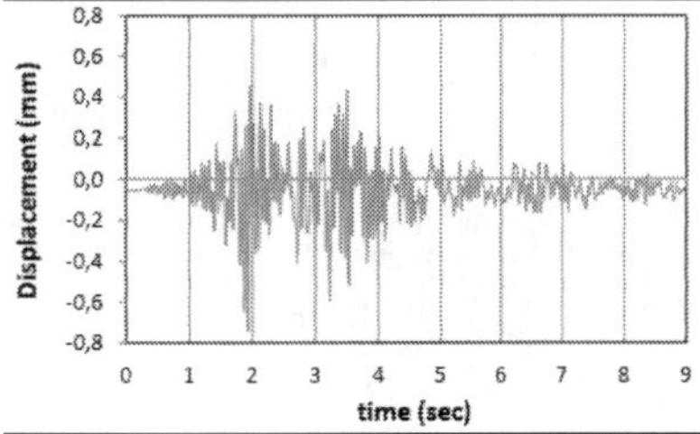

Record	First peak in frequency (Hz)	Max Displacement (mm)
E01	7,69	0.321
E02	7,69	0.526
E03	6,67	0.418
E04	6,25	0.505
E05	14,29	0.405
E06	11,11	0.550
E07	14,29	0.744

Figure 8. Top displacement provided by the time-history analysis

An elastic time-history analysis has been performed, by representing the seismic input through an ensemble of seven ground motions spectrum compatible to the elastic spectra as provided by the Italian Code for the Collapse Prevention limit state of the sculpture site. The top displacement time-histories of the case-study have been compared to the frequency content of the seismic input, compared to the sculpture's one. The maximum stress, both tensile and compressing, arising in the statue due to gravitational loads and to seismic excitation does not seem to exceed the capacity assumed for the material. The performed analysis neglects *i)* the possible nonlinearity of the seismic response of the case-study, and *ii)* the possible sliding between the statue and its pedestal. Therefore, further analyses should be performed, by representing the statue through more refined models.

6. References

[1] Parisi F, Augenti N (2013). Earthquake damages to cultural heritage constructions and simplified assessment of artworks. *Engineering Failure Analysis* **34**, 735–760.

[2] Spyrakos CC, Maniatakis CA, Taflampas IM (2008). Assessment of seismic risk for museum artifacts. *XIV World Conference on Earthquake Engineering*. October 12-17, 2008, Beijing, China.

[3] Spyrakos CC, Maniatakis CA, Taflampas IM (2017). Application of predictive models to assess failure of museum artifacts under seismic loads. *Journal of Cultural Heritage* **23**, 11–21.

[4] Lowry MK, Farrar BJ, Armendariz D., and Podany J. (2007). Protecting Collections in the J. Paul Getty Museum from Earthquake Damage. *WAAC Newsletter* Volume 29.

[5] Wittich CE, Hutchinson TC, Wood RL, Seracini M, Kuester F (2016). Characterization of Full-Scale, Human-Form, Culturally Important Statues: Case Study J. *Comput. Civ. Eng.*, **30**(3).

[6] Borri A, Grazini A (2006) Diagnostic analysis of the lesions and stability of Michelangelo's David, *Journal of Cultural Heritage* **7**, 273–285.

[7] Berto L, Favaretto T, Saetta A, Antonelli F, Lazzarini L (2012). Assessment of seismic vulnerability of art objects: The "Galleria dei Prigioni" sculptures at the Accademia Gallery in Florence, *Journal of Cultural Heritage* **13**, 7–21.

[8] Pascale G and Lolli A, 2015. Crack assessment in marble sculptures using ultrasonic measurements: Laboratory tests and application on the statue of David by Michelangelo. *Journal of Cultural Heritage* **16**: 813–821.

[9] Monaco M, Guadagnuolo M, Gesualdo A. (2014). The role of friction in the seismic risk mitigation of freestanding art objects. *Nat Hazards* **73**: 389–402.

[10] Pintucchi B, Rotunno T, Tanganelli M and Viti S (2018). Bartolomeo Ammannati's Fountain: comparisons among different numerical models. *Proc. XI International Conference on Structural Analysis of Historical Constructions (SAHC 2018)*. Cusco (Peru), September 11-13.

[11] Cerri G (2014). From the traces to definition of the monumental space. The case of Bartolomeo Ammannati's "Fontana di Sala Grande," *Proceedings of the CHNT 18*, 2013, Wien.

[12] Pirazzoli G (2011). "L'acqua, la Pietra, il fuoco. Bartolomeo Ammannati Scultore. Note sull'allestimento della mostra per il V centenario," in Paolozzi Strozzi and Zikos (eds.), *L'acqua, la Pietra, il fuoco. Bartolomeo Ammannati Scultore,* Firenze: Giunti.

[13] Paolozzi Strozzi and Zikos (eds.), 2011. L'acqua, la Pietra, il fuoco. Bartolomeo Ammannati Scultore, Firenze: Giunti.

[14] Ferretti, 2011. Bartolomeo Ammannati, la Fontana di Sala Grande e le trasformazioni del Salone dei Cinquecento da Cosimo I a Ferdinando, in "Paolozzi Strozzi and Zikos (eds.), L'acqua, la Pietra, il fuoco. Bartolomeo Ammannati Scultore", Firenze: Giunti.

[15] Morolli G (2006). Il palazzo del giovane duca: Giuliano di Baccio d'Agnolo, Baccio Bandinelli, Giovan Battista del Tasso. In: Carlo Francini, Palazzo Vecchio. Officina di opere e di ingegni. Mailand 2006, pp. 130–147.

[16] Zikos, 2011. Fontana di Sala Grande 1556-1561, in "Paolozzi Strozzi and Zikos (eds.), L'acqua, la Pietra, il fuoco. Bartolomeo Ammannati Scultore, Firenze: Giunti

[17] Borghini, 1584. Il riposo, p. 59.

[18] Heikamp, 1978. Bartolomeo Ammannati's Marble Fountain for the Sala Grande of the PalazzoVecchio in Florence, in MacDougall and Miller (eds.), Fons Sapientiae: Renaissance Garden Fountains, Washington: Dumbarton Oaks, Trustees for Harvard University.

[19] Cerri 2011. Thesis (honours), Bartolomeo Ammannati E Giorgio Vasari, Appunti per un dialogo di acqua e di luce,.

[20] Verdiani, Pirazzoli, Cerri, 2012. The Reconstruction of the "Fontana di Sala Grande" and some hypothesis about its original layout, in Act of counsil Virtual Systems and Multimedia (VSMM) Virtual Systems in the Information Society 2012, Milano.

[21] Cerri G, Pirazzoli G, Verdiani G, Tanganelli M and Viti S (2018). The Bartolomeo Ammannati's Fontain: an artifact in progress. Proc. of the International Conference on Structural Analysis of Historical Constructions (SAHC 2018).

[22] Cignoni P, Callieri M, Corsini M, Dellepiane M, Ganovelli F, Ranzuglia G (2008). MeshLab: an Open-Source Mesh Processing Tool. *Sixth Eurographics Italian Chapter Conference*, pp 129-136.

[23] NTC, 2008. Norme tecniche per le costruzioni. D.M. Ministero Infrastrutture e Trasporti 14 gennaio 2008, *G.U.R.I.* 4 Febbraio 2008, Roma (in Italian).

[24] Coli M & Rubellini P (2013). Geological anamnesis of the Florence area, Italy. Z. Dt. Ges. Geowiss. (German J. Geosci.), 164 (4), p. 581–589.

[25] Itaca (2008). Database of the Italian strong motions data. http://itaca.mi.ingv.it

[26] Iervolino I, Galasso C, Cosenza E (2009). REXEL: computer aided record selection for code-based seismic structural analysis. *Bull Earthq Eng* **8**:339–362. doi:10.1007/s10518-009-9146-1.

[27] Engisoft, 2013. *Straus7: User manual* (http://www.enginsoft.it/software/straus/index.html)

Acknowledgement

This research belongs to the project "*RESIMUS: la valutazione della resilienza e del rischio in un caso di rilevanza internazionale: il Museo del Bargello a Firenze*", supported by the Foundation "Fondazione Cassa di Risparmio di Firenze".

IOP Conf. Series: Materials Science and Engineering **364** (2018) 012058 doi:10.1088/1757-899X/364/1/012058

Improvement of the visibility of concealed features in misregistered NIR reflectograms by deep learning

Jan Blažek[1], **Oldřich Vlašic**[2], **Barbara Zitová**[1]

[1] Institute of Information Theory and Automation of the CAS, Pod Vodárenskou věží 8, Prague 8, Czech Republic

[2] Faculty of Nuclear Sciences and Physical Engineering, Břehová 7, Prague 1, Czech Republic

E-mail: blazek@utia.cas.cz

Abstract. Features of Old Master paintings hidden under the upper layer of a painting are often studied using NIR reflectograms; however their interpretability can be reduced due to the visible content. In our previous work [3] we described the possibility of increasing the visibility of concealed features in NIR reflectograms from the painting surface. The method output, enhanced NIR reflectogram, is produced by extrapolating the VIS data to a NIR range reflectogram and subtracting it from the acquired data in the NIR spectral subband. As a result, separated information from the NIR domain is obtiained. This method has a severe limitation, because it requires precise image registration of the VIS and NIR spectral bands. This is often hard to achieve, because DSLR cameras or multiple devices with various optical systems are used for data collection, and the mutual spatial relation of the images is often unknown. Thus, in the original form ,the algorithm was applicable only for data acquired using special scanners producing spatially registered images (as in [4]). In this work, we present an extension of the previous algorithm inspired by deep learning. The new concept allows processing of images only partially registered with pixel precision; subpixel accuracy is no longer needed. We suggest an extension of neural network input with neighboring pixels and allocation of extra ANN layers for translation compensation. The results are demonstrated on misregistered images captured by DSLR camera in VIS and NIR.

1. Introduction

The study of underdrawings and pentimenti is a standard procedure of artwork investigation. However, the quality of scanning devices is often not sufficient, in the sense of acquisition of spatially registered dataset. For this reason, reflectograms from various spectral bands are in most case studies processed separately. This is a serious limitation because some relevant information can remain unnoticed. The information gain of a spectral subband is often covered with high intensity information which is shared among modalities and in this way remains unnoticed. In our previous research, we suggest an algorithm for separation of information which is shared among the VIS and NIR part of the spectra and the information gain of NIR [3]. The prerequisite for such separation is the accurate registration of all relevant images with subpixel precision.

1.1. Registration

Automatic registration of multimodal images of artworks is still an unresolved problem. This is a difficult problem becasuse there is no universal similarity measure for various modalities.

IOP Conf. Series: Materials Science and Engineering **364** (2018) 012058 doi:10.1088/1757-899X/364/1/012058

The set of captured modalities contains images (ordered according to their availability) in these bands:

(i) Visible light (VIS)
(ii) Near infrared (NIR and also SWIR)
(iii) Ultraviolet fluorescence (UVF)
(iv) X-ray
(v) X-ray fluorescence (XRF)
(vi) Far infrared or "terahertz" (THz)

While one scanner can be used for both the VIS and NIR bands (with the same optical elements) [4, 6], for THz or X-ray images multiple scanners are necessary. The use of one scanner means that acquired raw data can be spatially registered, or that a registration function is measurable. On the other hand, the use of more than one scanning device requires spatial normalization according to the optical elements used (lenses, prisms, monochromator, etc.), position of the scanned artwork, and physical properties (index of refraction) of the acquired wavelengths. An efficient estimation of the parameters of such a transformation from an acquisition setting is significantly harder than their estimation from the collected dataset.

Precise registration of a multimodal dataset is required for further digital image processing (image comparison [9], false-color images [12, 7], spectral analysis [10, 11], pigment classification, etc.). The registration techniques from the referenced papers are based on control point selection or on minimization of mutual information [1]. As the authors in [1] described, registration of this type is not precise: *"..., the attenuation correction is not always accurate. While errors in registration are a possible cause, ..."*. The same problem occurs when registration based on control points is used, in spite of the precise registration of control points (precise position of control points is achieved by the application of the similarity measure on roughly placed points [5]). The problem here is that the present transformation function is not polynomial, thus its approximation from a finite number of control points may be impossible. The registration precision varies over the data set in this case.

The most relevant question is "Why is there no relevant similarity measure?"

In our previous research [3] we were looking for the answer. There were also various approaches regarding similarity estimation: intensity difference variance [9], mutual information [2], and coupled dictionaries [8]. All of them worked with a matrix of corresponding intensities (ICM), which contains the frequency of correspondent pixel intensities in two different modalities (2D histogram of intensities). The number of rows and columns corresponds to the intensity levels recognized in each modality and greatly affects the estimated separation. If we use a correspondence matrix as a visualization of mutual information (or entropy), we recognize two extreme patterns:

- The highest entropy/lowest mutual information is reached when the correspondence matrix is uniform with zero total variance.
- The lowest entropy/highest mutual information is reached when there is only one non-zero column for each row, and vice versa.

The first case leads to a very smooth registration error function, which means, that global minima is searched quickly, but often does not reflect the present transformation function, because correct mapping is not a function (multiple values in one modality can correspond to one value in the second one).

The second case leads to a precise definition of global minima without any information gain for each modality.

The similarity of two modalities (without the context of neighborhood pixels) describes well the ICM and, according to the previous description, such a measure is limited to "good" cases. In [3] we quantify the minimal correspondence required for separation of spectral subband information gain, which is about 90%. This threshold was empirically measured on numerous ANNs with a perfectly registered dataset. When these conditions are satisfied, subband information gain can be computed and concealed features are uncovered in the subband reflectogram.

The question is now a little bit shifted. Let's assume a suitable dataset for VIS/NIR comparison. We would like to use ICM according to [3]. However, "What happens when the dataset is not registered well?"

We try to use the same similarity measure for automatic registration.

2. Extension of the comparison algorithm

The use of ANN is advantageous due to its modularity. The main idea is to extend ANN input with the intensities of neighboring pixels. In this case, the ANN can also estimate transformation, necessary for registration of the input and output images of our original ANN. In the following paragraphs we present a proof of this concept on a "dummy architecture", estimating the image shift.

The achitecture of ANN is proposed in Figure 1. The original algorithm contained only white and ochre neurons. These neurons play the role of an extrapolation function from the information content of VIS spectral bands into NIR spectral bands on the registered dataset. The inputs of the original ANN are VIS pixel intensities, and the outputs are intensities taken in NIR. Extrapolation was computed by two hidden layers.

When misregistration is present, we extend the input layer to accept not only one pixel but the whole neighborhood of a specified size. The reason behind this is that we don't know which pixel corresponds to the one at the output. To prevent the curse of dimensionality, we limit neighborhood size according to the estimated maximal pixel shift. An effective estimation of maximal pixel shift gives us faster computation and more precise results. After the input layer, the registration layers follow. In our case we use two layers with several neurons for shift estimation. We expect that estimation of more complex transformations will require more neurons (and probably more sophisticated architecture), but for pixel shift two layers are more than sufficient. We assume that such architecture of the registration layers (controlled by the extrapolation error function) will identify a pixel from the pixel neighborhood corresponding to the expected output pixel.

The improvement achieved by this new concept compared to the previous work [3] is demonstrated in Figure 2.

3. Hypotheses and Experiments

In evaluation of the proposed concept, we track several targets. First of all, we need to confirm or disprove our expectations. Second, we would like to tune ANN architecture for fast training and efficient data processing. We postulate three hypotheses which can disprove the proposed concept or indicate a direction for further research. Moreover, to prevent an over-parametrization of the proposed system, we evaluate effect of each relevant parameter separately. As a result, we would like to define the limits of registration layers.

The atomic operation for evaluating of our hypotheses was the processing of phantoms. These phantoms were created from a precisely registered VIS-NIR dataset (obtained from real artworks — source INO-CNR in Florence) where the NIR bands were slightly shifted. The shift was specific for each phantom and goes from 1px up to 40px. If the shift amount was a non-integer, the NIR intensities were computed by cubic spline approximation.

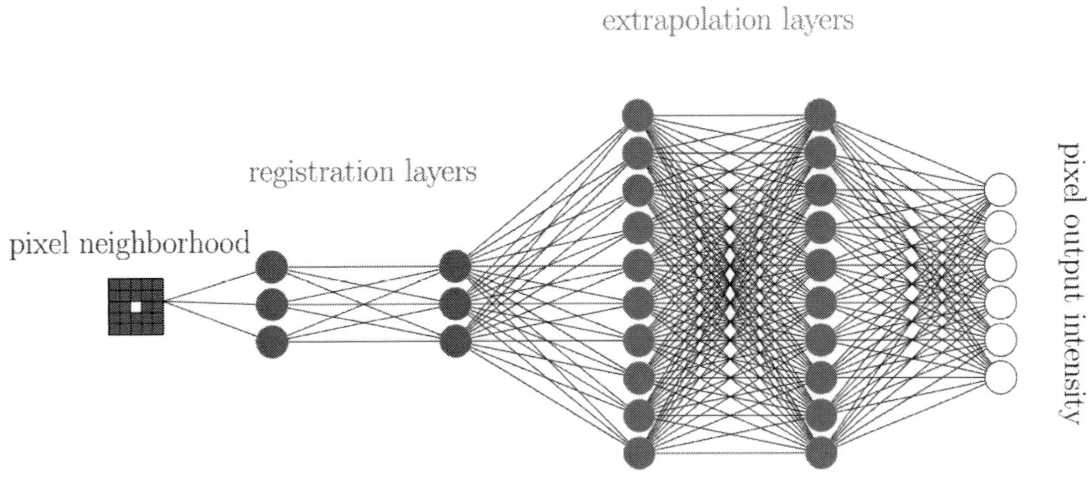

Figure 1. Extension of original algorithm with new registration layers and input pixels (marked with red color).

3.1. Hypothesis 1: The absolute amount of misregistration affects performance

We expect that input images must be partially aligned. The reason for this hypothesis is the statistical principle of an ANN training. Reduction of the space of possible transformations can improve the speed and accuracy of the ANN training. And the partial registration decreases the number of relevant input pixels.

For the evaluation, we prepare phantoms with a shift going from 1 to 15 pixels and then measure, on several runs of the algorithm, the error per pixel. Results are shown in Figure 3. The first observation confirms our hypothesis: **An increasing shift amount produces a more erroneous output**. The second observation is the convergence of the histogram curve with the increasing shift amount.

The explanation for the second observation is not clear, but in general we can say that information relevant for the ANN training drops under the ratio where training is efficient.

3.2. Hypothesis 2: Layer(s) for estimation of modality misregistration can improve the achieved results

A logical split of the ANN in registration and extrapolation layers is a part of our concept but should be supported by another experiment. There is a possibility that without extra layers, the ANN will be similarly effective. In Figure 4 we present a comparison of the results of ANN with extra layers versus ANN without extra layers. Various layer widths have been tested as well, going from $3 - 3 - 25 - 25$ and $5 - 5 - 25 - 25$ to $40 - 40 - 25 - 25$.

According to our metric (histogram of errors) we see that small amount of neurons in *registration layers* (i.e. $3 - 3$ or $5 - 5$) is worse than ANN without these layers. This fact poses a question: After the training, are the *registration layers* focused on selecting one of the input pixels? The answer seems to be negative, as observation on trained ANN weights shows that ANN needs extra regularization parameters to achieve this behavior. Consequently, we reject this hypothesis, with the promise of further tests on more suitable ANN architectures.

Here we propose the requirements for *registration layers*. For a shift transformation, the ANN *registration layer* has to identify the most relevant pixel in the input set. This will be represented as the highest weight for the corresponding pixel and decreasing weights with growing distance.

IOP Conf. Series: Materials Science and Engineering **364** (2018) 012058 doi:10.1088/1757-899X/364/1/012058

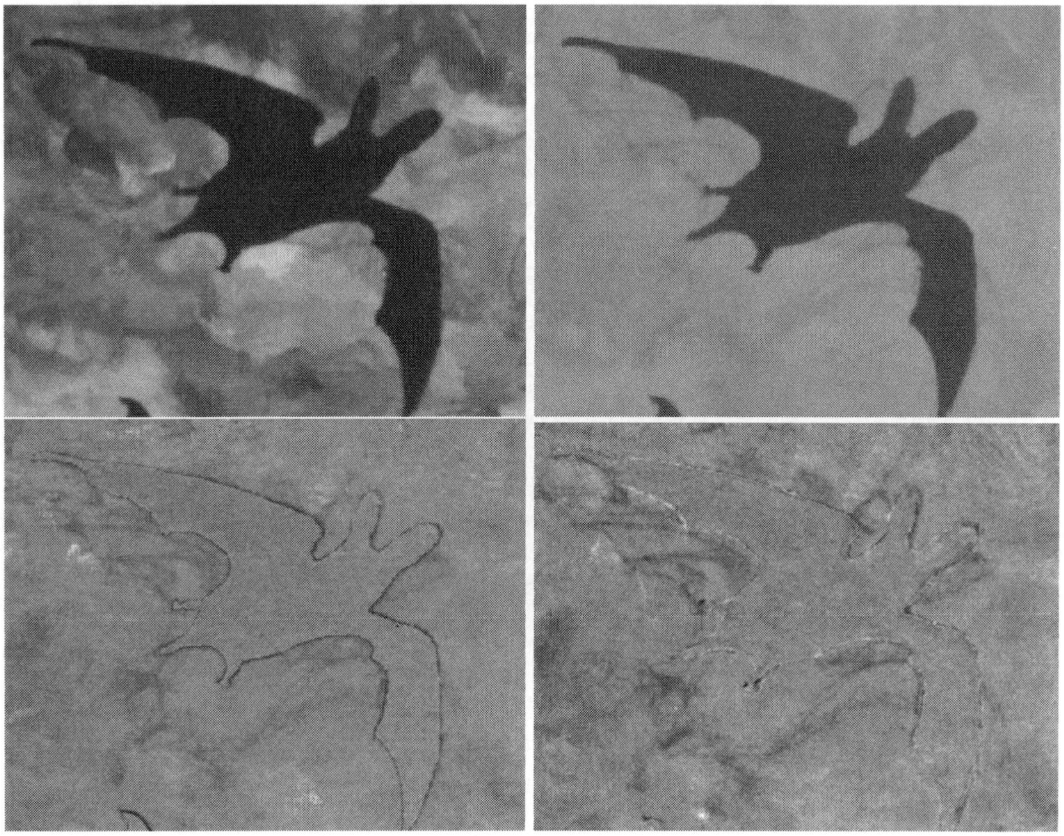

Figure 2. Comparison of the result of the original algorithm, working with registered images, with the improved one. (a) VIS subband, (b) misregistered NIR subband, (c) output of original algorithm (without pixel neighborhood), (d) improved algorithm (input contains neighboring pixels). The visibility of underdrawings on image (d) is much better. ANN has efficiently estimated the transform function, while the original algorithm works as an "edge detector".

3.3. Hypothesis 3: A higher number of output bands processed at one time decreases the performance of a single band

In practice, it is very time consuming to manually go through numerous images and select the best one for processing. Therefore, the efficiency of "processing simultaneously" should be measured. Such an approach can lead to the failure of ANN training, and therefore it is relevant to ask whether the number of output layers negatively affects the results. Our dataset contains 15 bands in NIR, which allows us to construct an experiment comparing the error achieved for ANNs with from 1 to 15 output neurons, see Figure 5.

According to Figure 5, the algorithm performs better for a smaller number of bands. However, a deeper analysis of the results shows that for low shifts of modalities (from 0 to 10 pixels), the histograms are almost the same. The hypothesis was therefore neither proved nor disproved. We conclude here that a better metric of *quality* is necessary.

4. Conclusion

We suggest an improved variant of our algorithm for improvement of the visibility of concealed features in a misregistered NIR modality. We demonstrate the efficiency of the proposed algorithm on a real dataset and demonstrate its limits on manually shifted NIR images. We

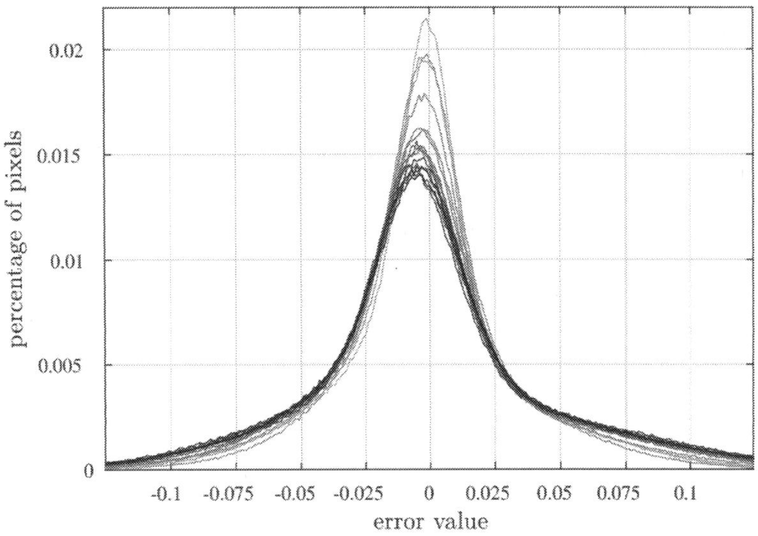

Figure 3. Histogram of error per pixel. ANN architecture $5 - 5 - 25 - 25$. The colors go from the lightest 1px shift to the darkest 15px shift.

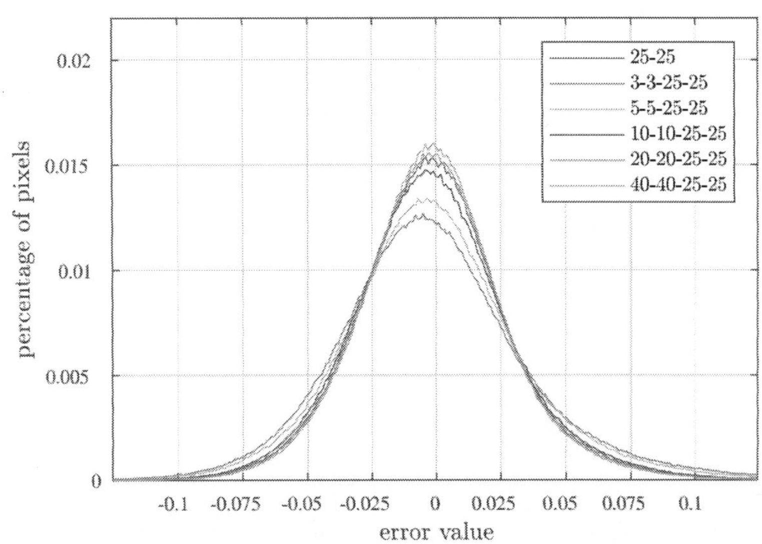

Figure 4. Comparison of histograms of error per pixel for different ANN architectures. Layer widths are given in the graph legend.

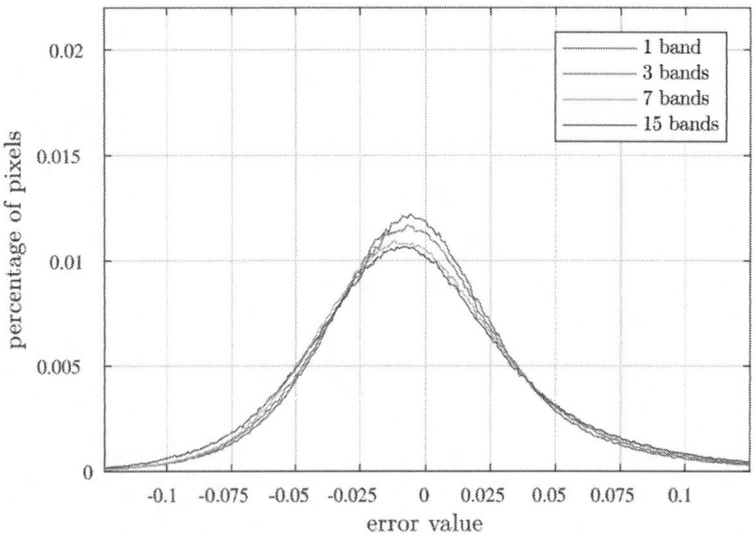

Figure 5. Histogram of error per pixel with respect to the number of target NIR subbands, namely 1, 3, 7, and 15. Errors were evaluated over a single band common to all configurations. ANN architecture $3 - 3 - 25 - 25$.

identify that a crucial part of the improvement is the addition of neighboring pixels into ANN inputs. We also evaluate several configurations and define limits of the suggested ANN architecture.

Our concept joins two tasks into one. This approach is rare because the complexity of the problem can get out of control. However, we have shown that registration can be joined together with extrapolation using the same similarity measure. A logical future work is the preparation of stronger rules for ANN training to achieve splitting of the registration and extrapolation steps.

This preliminary research raises a lot of questions which should be answered in future work:

- How can we regularize *registration layers* of the ANN to obtain a transformation function?
- Is there an ANN architecture that prevents the curse of dimensionality? And if so, is it possible to process images without a preceding rough registration?

These questions should be answered, and we will focus on them in subsequent work.

Acknowledgments

Access to computing and storage facilities owned by parties and projects contributing to the National Grid Infrastructure MetaCentrum, provided under the programme "Projects of Large Infrastructure for Research, Development, and Innovations" (LM2010005), is greatly appreciated. Without them this research would not be possible.

We would like to acknowledge the Czech Science Foundation, which has supported this work (project no. GA18-05360S).

References

[1] Anila Anitha, Andrei Brasoveanu, Marco F Duarte, Shannon M Hughes, Ingrid Daubechies, Joris Dik, Koen Janssens, and Matthias Alfeld. "Virtual underpainting reconstruction from X-ray fluorescence imaging Data". In: *19th European Signal Processing Conference*. Eusipco. 2011, pp. 1239–1243.

[2] Anila Anitha, Andrei Brasoveanu, Marco Duarte, Shannon Hughes, Ingrid Daubechies, Joris Dik, Koen Janssens, and Matthias Alfeld. "Restoration of X-ray fluorescence images of hidden paintings". In: *Signal Processing* 93.3 (2013). Image Processing for Digital Art Work, pp. 592–604. ISSN: 0165-1684. DOI: `10.1016/j.sigpro.2012.09.027`. URL: `http://www.sciencedirect.com/science/article/pii/S0165168412003659`.

[3] Jan Blažek, Jana Striová, Raffaella Fontana, and Barbara Zitová. "Improvement of the visibility of concealed features in artwork NIR reflectograms by information separation". In: *Digital Signal Processing* 60 (Jan. 2017), pp. 140–151.

[4] Claudio Bonifazzi, Pierluigi Carcagnì, Raffaella Fontana, Marinella Greco, Maria Mastroianni, Marzia Materazzi, Enrico Pampaloni, Luca Pezzati, and Davide Bencini. "A scanning device for VIS–NIR multispectral imaging of paintings". In: *Journal of Optics A: Pure and Applied Optics* 10.6 (2008), p. 064011. URL: `http://stacks.iop.org/1464-4258/10/i=6/a=064011`.

[5] Damon M. Conover, John K. Delaney, and Murray H. Loew. "Automatic registration and mosaicking of technical images of Old Master paintings". In: *Applied Physics A: Materials Science and Processing* 8790 (2015), pp. 1–8. ISSN: 14320630. DOI: `10.1007/s00339-015-9140-1`.

[6] Costanza Cucci, Andrea Casini, Marcello Picollo, Marco Poggesi, and Lorenzo Stefani. "A hyper-spectral scanner for high-quality image spectroscopy: Digital documentation and spectroscopic characterization of polychrome surfaces". In: *Art '11: 10th International Conference on Non-Destructive Investigations and Microanalysis for the Diagnostics and Conservation of Cultural and Environmental Heritage* (2011), pp. 1–7.

[7] John K. Delaney, Jason G. Zeibel, Mathieu Thoury, Roy Littleton, Kathryn M. Morales, Michael Palmer, and E. René de la Rie. "Visible and infrared reflectance imaging spectroscopy of paintings: Pigment mapping and improved infrared reflectography". In: *Proceedings of SPIE*. Ed. by Luca Pezzati and Renzo Salimbeni. Vol. 7391. Springer International Publishing, July 2009, p. 739103. ISBN: 978-0-8194-7674-6. DOI: `10.1117/12.827493`.

[8] N. Deligiannis, J. F. C. Mota, B. Cornelis, M. R. D. Rodrigues, and I. Daubechies. "Multi-modal dictionary learning for image separation with application in art investigation". In: *IEEE Transactions on Image Processing* PP.99 (2016), pp. 1–12. ISSN: 1057-7149. DOI: `10.1109/TIP.2016.2623484`.

[9] Amy Gooch and Jack Tumblin. "Visualizing pentimenti: Revealing the hidden history of paintings". In: *Journal of Mathematics and the Arts* 00.00 (2007), pp. 1–11. ISSN: 17513472. DOI: `10.1080/1751347YYxxxxxxxx`.

[10] Anna Pelagotti, Luca Pezzati, Natalia Bevilacqua, Veronica Vascotto, Vincent Reillon, and Claudia Daffara. "A study of UV fluorescence emission of painting materials". In: *Art '05:8th International Conference on Non-Destructive Investigations and Microanalysis for the Diagnostics and Conservation of Cultural and Environmental Heritage* (2005), A97.

[11] Paola Ricciardi, John K. Delaney, Michelle Facini, Jason G. Zeibel, Marcello Picollo, Suzanne Lomax, and Murray Loew. "Near infrared reflectance imaging spectroscopy to map paint binders in situ on illuminated manuscripts". In: *Angewandte Chemie - International Edition* 51.23 (2012), pp. 5607–5610. ISSN: 14337851. DOI: `10.1002/anie.201200840`.

[12] Chad Weiner. *Improved Acquisition of Underdrawings in Oil Paintings Using IR-Reflectography*. Tech. rep. Rochester Institute of Technology, May 1998.

IOP Conf. Series: Materials Science and Engineering **364** (2018) 012059 doi:10.1088/1757-899X/364/1/012059

Application of coatings on silver studied with punctual and imaging techniques: from specimens to real cases

Vittorio Barra[1], Claudia Daffara[1], Simone Porcinai[2] and Monica Galeotti[2]

[1] University of Verona, Dept. of Computer Science, Strada Le Grazie, 15, 37134 Verona (Italy), [2] Opificio delle Pietre Dure e Laboratori di Restauro, via Alfani, 78, 50129 Firenze (Italy)

Corresponding author: monica.galeotti@beniculturali.it

Abstract. The protection of silver artifacts with a coating of organic material isolating the surface from oxygen and S-bearing gases is still one of the most used techniques for the preservation of silver surfaces. A variety of materials suitable to this purpose are being used in the conservation practice. Among them there are acrylic polymers, nitrocellulose lacquer, microcrystalline wax and combination of them in superimposing layers. They are applied either by brush or, in some cases, by spraying. The first stage in a conservation project is often the cleaning of residuals of old protectives, aimed at making the silver surface free from remnants of yellowish materials and enabling the application of new coatings. Unfortunately, removing aged coatings and applying new ones are "blind" processes, since the techniques available for not invasively monitoring the surface of silver artifacts are punctual and do not allow an overall overview of the surface. This is the case of reflectance FTIR and thickness measurement with the Eddy Current technique, which, apart from being punctual, are tricky to apply to uneven surfaces. In this paper, we aim to show the benefits of using an imaging technique, UV induced VIS luminescence, for monitoring the presence and the distribution of protective coatings on silver surfaces and of combining it with reflectance FTIR and thickness gauge. At first, our study was focused on specimens treated with nitrocellulose, acrylic emulsion and microcrystalline wax. The application of the treatments was studied with UV-induced VIS luminescence as well as with the abovementioned punctual techniques to cross-validate the outcome of each method and to assess their suitability for application on organic protective layers on silver artifacts. In a further stage, UV induced VIS luminescence was applied to a real case, a silver chalice that was treated with nitrocellulose lacquer after cleaning.

1. Introduction

In the field of the conservation of historical silver artworks, the very first step to properly plan a preservation project is the identification of varnishes and their remains from previous restoration treatments. Indeed, it is not easy to detect the presence of such varnishes and to characterize their spatial distribution, not even by means of optical microscopy, except for those made up of materials that have turned yellow over time. Applying the coating after the cleaning treatment is another stage of the silver conservation process in which checking the homogeneity in the spatial distribution of organic layers plays a key role. This is particularly true in the cases of more complex shaped artwork or when reliefs, folds and curves are present, as there is no possibility for the conservator to visually control the process and to prevent the development of thickness inhomogeneity or even gap in the varnish layers. Therefore, there is a clear need to have an available method for monitoring the varnish layer even when applying the coating after the cleaning treatment

To this aim, further analyses, in addition to the visual examination of the surface, are recommended either by microinvasive methods (namely by sampling a minimal amount of varnish) or through noninvasive ones, using anyhow FTIR spectroscopy for the molecular characterization of old materials [1]. Although these materials in most cases will be removed during the cleaning treatment, it is important to keep the sample amount for analysis as much as limited as possible in order not to remove old patinas or not to damage the metal surface. For this reason, there is an increased preference towards a non-invasive approach by means of reflectance FTIR spectroscopy. This method proves particularly effective for the molecular characterization of thin layers of organic materials on surfaces that do not adsorb the mid IR wavelengths, such as metals [2, 3]. Thanks to its noninvasiveness, reflectance FTIR allows analysis in different areas of the artwork to gain information on the presence and the relative amount of a coating. Nevertheless, it has to be taken into account that the methodology is based on a punctual probe and, owing to constraints in the acquisition setup of the currently available tools, is not suitable to effectively map materials over a surface with a complex shape and with either embossed or chiseled details.

Homogeneities in the coating can be monitored by means of thickness measurements with the Eddy Currents technique [4, 5]. This method allows the measuring of the thickness of non-conductive films, such as organic films, on conductive substrates, such as silver: even in this case, however, the measurement provides point-wise data rather than an image. Further limitations of the Eddy Currents technique are that, although sampling is not required, a contact between the probe and the surface under measurement is necessary and, in some cases, unacceptable indentations or impressions are left on the surface, if the film itself is not interrupted. Besides that, the Eddy Current technique is sensitive to surface roughness, curvature, and distance from an edge. Most recently implemented imaging techniques such as terahertz [6] and ultrasound [7,8] have provided promising results regarding the investigation of coatings on metal, although the difficulty in their employment, transportability, and data interpretation may result in a limitation for a widespread use in conservation practice. Indeed, in a conservation project it is often necessary to carry out quick checks, even to the detriment of the completeness of information that could be gathered by means of more complex methods. For this reason, in recent years, several attempts have been made to employ imaging techniques for the analysis of metal artworks, in order to investigate both alteration products and transparent varnishes [9].

This work is focused on the optimization of the acquisition setup for the photographic technique in order to obtain high-quality reflectance and photo-induced luminescence imaging of silver artifacts in the UVA-VIS regions. We propose an approach for the investigation of varnishes on silver artworks that combines UV-induced VIS luminescence imaging with the Eddy Current thickness measurement and reflectance FTIR spectroscopy to relate differences in luminescence intensity to both thickness and IR band intensity and shape.

2. Materials and methods

2.1 Samples preparation.

Some 7 cm x 2.5 cm x 0.1 cm sterling silver alloy plate specimens (95% Cu-5% Ag) have been prepared with smooth and with chiseled surfaces. On the smooth specimens, an incision with the sharp point of a compass was made with the aim of creating a discontinuity on the surface similar to the ones found on objects with various decorations. The specimens were polished with sandpaper and finished with abrasive pastes with different grain sizes. In the end, surfaces were degreased with acetone. As a coating, three varnishes commonly used in the conservative practice were chosen:

- Zapon (a nitrocellulose lacquer supplied by Lechler) 70% in a mixture of 20% isopropyl alcohol - 40% isothane - 40% acetone;
- Incral44 (an ethyl methacrylate and methyl acrylate based blend, supplied by CTS Europe) 70% in butyl acetate;
- R21 (microcrystalline wax supplied by CTS) 2% in cyclohexane.

The coatings were applied by brush. A strip on the short side of the sample was left uncoated as a reference area. The wax applied hot and later polished with a cloth. Table 1 sums-up the treatments applied on the specimens (smooth and chiseled).

IOP Conf. Series: Materials Science and Engineering **364** (2018) 012059 doi:10.1088/1757-899X/364/1/012059

2.2 UV-induced VIS luminescence imaging.
UV-induced VIS luminescence was collected employing xenon-based sources. The UV component was selected using optical glass band-pass excitation filters with 65.8% transmittance at 360 nm and very low transmittance in the violet (0.22% at 390 nm). A red component reflected by metal surfaces, clearly visible in the images shown below, was due to the very low transmittance in the VIS region between 690 and 750 nm (with a 2.2% peak at 710 nm) of the excitation filters employed. It was not deemed necessary to subtract the red component, neither in the acquisition nor in post-processing phases, because it was not considered a disturbing element for the data interpretation. The UV component in outgoing was blocked with a long-pass filter with almost no transmittance (10-5%) at 400 nm and 50% at 418 nm. The low luminescence signal was amplified during both the data collection and post-processing ensuring an optimal signal-to-noise ratio thanks to the employment of a high performance CMOS sensor. More specifically, we used a commercial camera for astronomical photography equipped with a CMOS Full Frame sensor (35.9x24.0 mm, pixel pitch of 4.9 μm, 14 bit dynamics).

2.3 Thickness measurements.
A Karl Deutsch Thickness Gauge, based on the Eddy Currents method, was used to non-destructively evaluate the coating thickness; the measurement area is 1mm diameter and the zero calibration procedure was adopted on uncoated silver surface. Measurements were checked on the same surface used for calibration with thickness standards. On each area, a number of 20 evenly distributed measurements were repeated.

2.4 Reflectance FTIR spectroscopy.
The composition and amount of the coating material were characterized with a portable FTIR spectrometer (Alpha Bruker) operating in total reflection mode. The instrument is equipped with a camera shooting the region around the measuring spot, which is 3 mm diameter Spectra were acquired with resolution of 4 cm^{-1} in the 7500-375 cm^{-1} range, collecting 100 scans.

Sample	Surface finishing	Surface coating
13	Smooth with engraving	1 layer of Zapon+ 1 layer of R21
15	Smooth with engraving	1 layer of Incral44+ 1 layer of R21
18	Chiseled	1 layer of Zapon
22	Chiseled	Half sample with one layer of Zapon, half sample with two layers of Zapon

Table 1

3. Results and discussion
Figure 1 shows the visible photo and the UV-induced VIS luminescence image of four flat samples. Samples 13 and 15 are smooth and were respectively coated with a layer of Zapon (13) and Incral (15) with a further layer of wax on top. Although the photo in visible light shows some inhomogeneity in the coating, with accumulations distributed at the end of the samples (short sides), the UV-induced VIS luminescence image clearly highlights the brush strokes and the filling of small holes of the surface.

Thickness measurements show that the average thickness of the two types of coating is very similar (around 8 μm for the specimen with Incral+ wax and around 9 μm for the specimen with Zapon+wax). In the sample 13, three zones are differentiated according to the thickness: one near the engraving (average thickness 5.9 μm with 13% standard deviation), one central (average thickness 7.8 μm with 17% standard deviation) and one close to the untreated strip (average thickness 14 μm with 14% standard deviation).

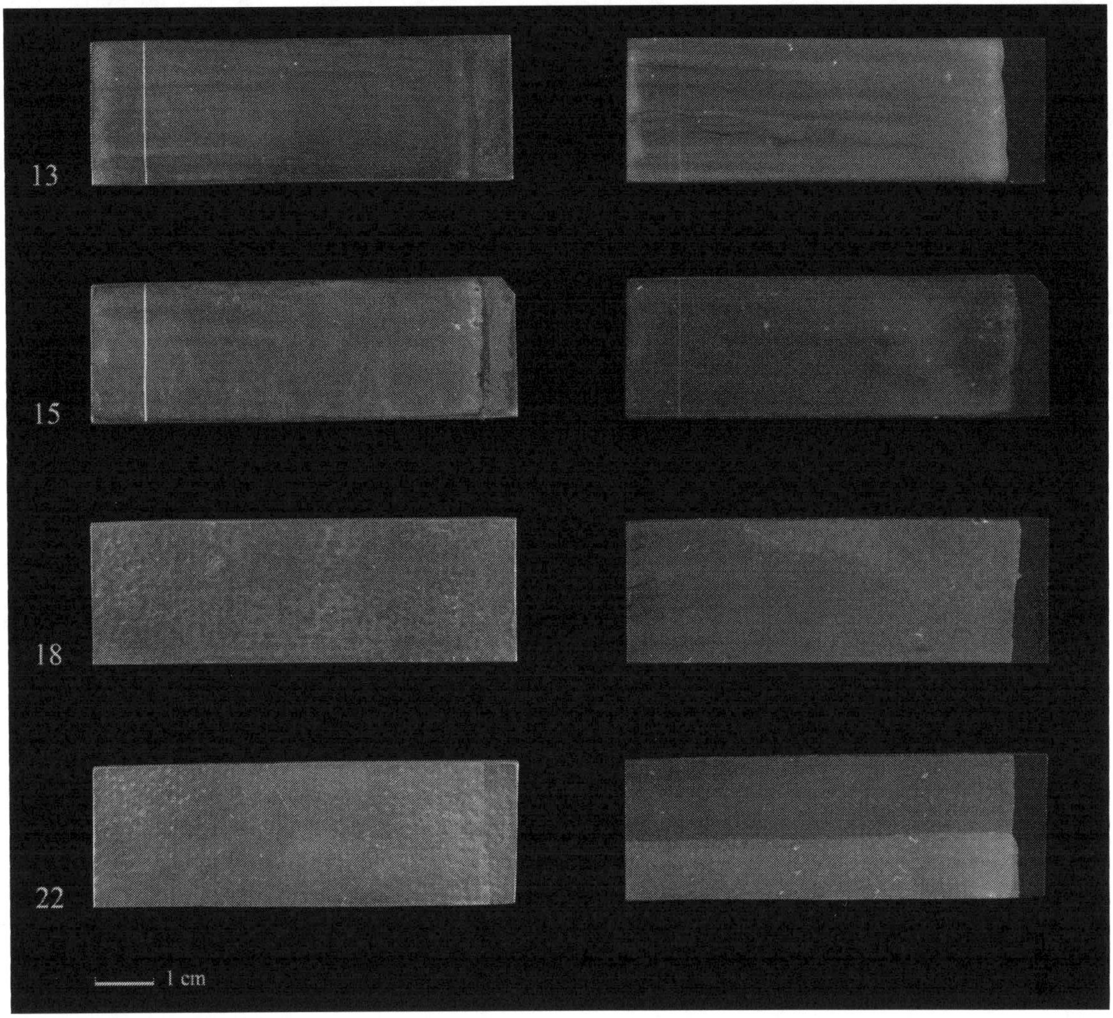

Figure 1. Flat Sterling silver specimens with different coatings, in reflected VIS light (left) and UV induced VIS luminescence (right).

Reflectance FTIR spectra show different features in correspondence of different thickness and different luminescence response. Near the untreated strip, where the thickness is higher, the spectrum of Zapon (not shown here) exhibits both derivative-like bands and not distorted bands, while near the incision, where both luminescence response and thickness are lower, the spectrum features no distortions at all, being the coating so thin that the Fresnel component of the reflected beam is negligible. Sample 15 shows an overall less intense luminescence response. Thickness measurements give an area near the untreated strip with a smaller thickness (3.6 μm with 30% standard deviation), an area in the center with a greater thickness (15 μm with 17% standard deviation) and an area near the engraving with thickness of 6 μm (23% standard deviation), all in agreement with luminescence distribution shown in the image. The probe sensing small width (about 2 mm in diameter) accounts for the observed pretty high values of standard deviation, being the film thickness very variable on a extremely short scale, as observed when luminescence images are magnified. Samples 18 and 22 are instead chiseled and have been coated with Zapon alone. In Figure 2, we show the reflectance FTIR spectra of three areas of sample 18. Area 1 is close to the untreated strip, where more luminescence is detected, Area 2 refers to the central portion of the coupon with a luminescence of medium intensity, while Area 3 is near the edge, where little luminescence is shown. The area of the bands that do not show distortions (like the signal due to C-H stretching in the 2830-3030 cm^{-1} region) increases from Area 3 to Area 1. Other bands (like the one at 1286 cm^{-1} associated to nitrate group [10]) show changes from a derivative-like shape to mixed character for Area 2 and 3, when luminescence decreases. One half of specimen 22 was coated with a second layer of the same nitrocellulose varnish.

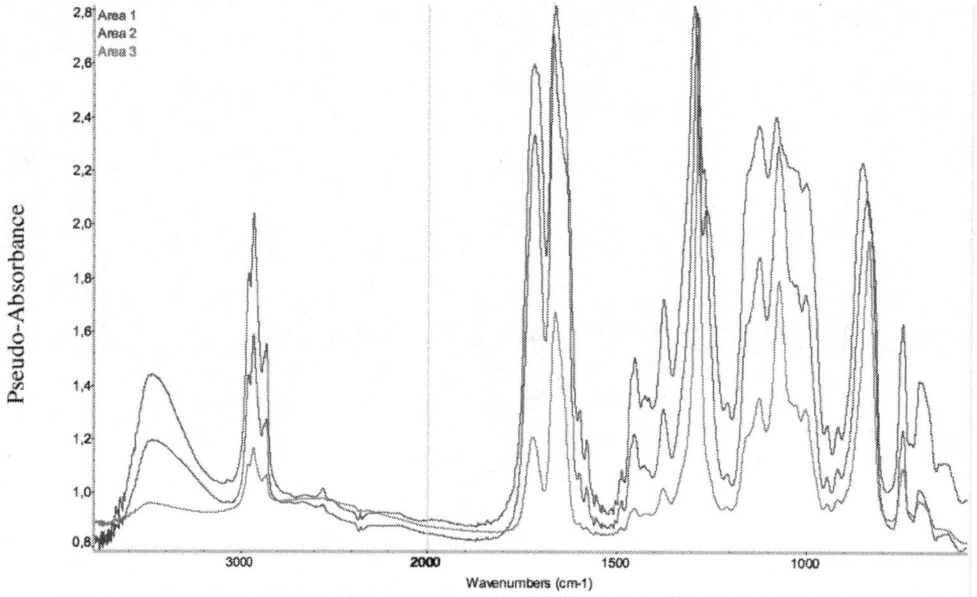

Figure 2: Reflectance FTIR spectra of three areas of sample 18 (Area 1 corresponds to high luminescence, Area 2 to luminescence of medium intensity and Area 3 to scarce luminescence).

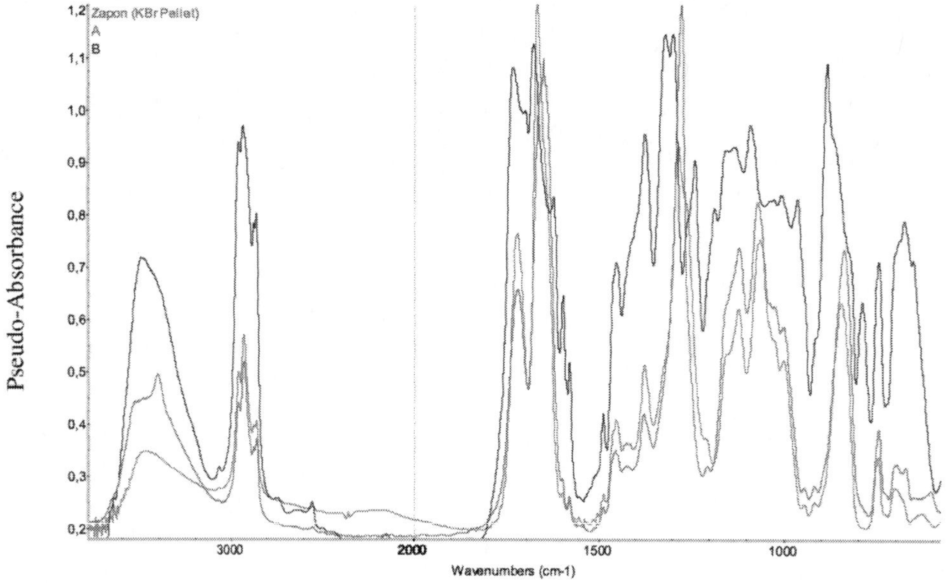

Figure 3: Reflectance FTIR spectra of sample 22, spectrum A: area coated with one layer of Zapon, spectrum B: area coated with two layers of Zapon. Transmittance spectrum of Zapon is shown as reference.

The luminescence image clearly shows the difference in response of the two parts. Thickness measurements provide a value of 11.6 μm on the single layer side and of about 23 μm on the double layer side. In both cases, the poor planarity of the surface has an influence on the accuracy of the measurement and accounts for the high standard deviation, around 30%. Changes of bands shape and intensity are observed in reflectance FTIR spectra, as reported in Figure 3, where spectra of the two half of sample 22 are shown, together with reference transmittance spectrum of Zapon. The overall intensity of bands of spectrum B (two coatings of Zapon) is higher than those of spectrum A and, in particular, the peak area for the 2830-3030 cm^{-1} bands (not distorted in both spectra) is 67.5 in spectrum B, nearly twice than for the same bands of spectrum A (peak area 34). Less intense peaks (1380, 1071 and 840 cm^{-1}) are not distorted in A, while exhibit a derivative-shape for B. Intense, distorted bands for both spectra, on the other hand, are distorted also for the one layer half but show a change from a derivative-like to a mixed aspect when the thickness increases, owing to the increasing contribution of the Restrahlen character [11]. Thus, the behavior of IR reflectance signals is in agreement with the changes of the luminescence response and with the measured thickness.

The effectiveness of the UV-induced VIS luminescence technique in characterizing the spatial distribution of the protective was tested, as a significant case study, on a tarnished silver chalice with engravings, chiseled and embossed parts, dated back to the first half of the 18th century (a detail is shown in Figure 4). The Tuscan manufactured chalice comes from the Museo di Arte Sacra di Incisa val d'Arno (Italy), has a height of 22.5 cm and a base diameter of 11.5 cm. The chalice was coated a few years ago with Zapon by students during a conservation class. In the case of a real case study, the distribution of the coating, partly observed even with visible light, is less evident to the naked eye than in the case of the samples shown above, partially due to differences in the color of the metal between zone and zone (such as the yellowish tone of the upper edge). The UV-induced VIS luminescence (Figure 4), instead, documents a quite uneven application with crossed brushstrokes in the upper part. The lower part, exhibiting an elaborate decoration, shows areas with a more intense UV-induced VIS luminescence and areas where the response is nearly absent. Thickness values were measured on the cup of the chalice (Figure 5). The thickness data shown in the figure are consistent with the intensity of fluorescence: thickness on the central and left areas above the embossed decoration is higher than on the right. The upper, less luminescent, portion of the cup, visible in figure 4, shows a thinner layer of varnish of about 13 μm. The average standard deviation of thickness on the whole upper part is quite high (30%), confirming the dispersion of thickness values. In the lower part of the chalice, owing to the complex shape of the worked details, only the UV-induced VIS luminescence was able to provide information on the presence and homogeneity of the organic film.

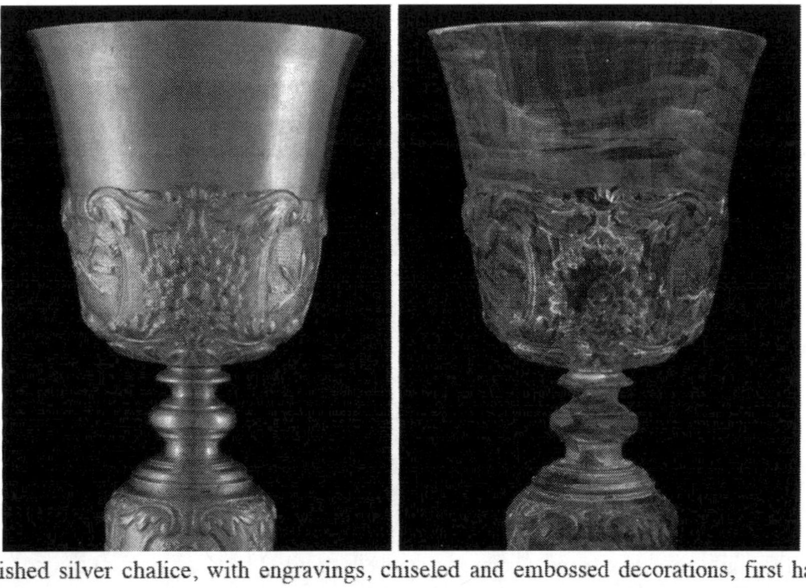

Figure 4. Tarnished silver chalice, with engravings, chiseled and embossed decorations, first half of the 18th century, detail. The chalice in reflected VIS light (left) and UV-induced VIS luminescence (right).

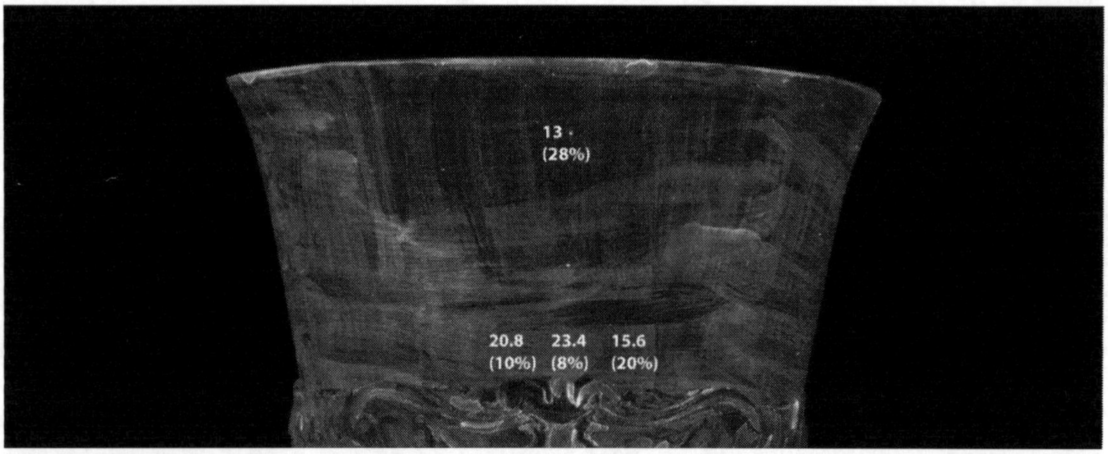

Figure 5. Detail of the cup of the chalice with average thickness (in ìm) measured on areas with different luminescence intensity. Standard deviation in % is also shown.

4. Conclusions

This work has explored the possibility of using UV-induced VIS luminescence imaging for monitoring the application of protective organic coatings on silver artworks. The method was tested on Sterling silver samples with different features that mimic some of the working techniques usually encountered on historical silver objects, like engravings and chiseled surfaces. The coating materials include the most widespread used protectives for historical silver, i.e. acrylic resin, nitrocellulose lacquer, wax and combination of them.

The particular set-up used in this work enabled us to get an enhanced UV-induced VIS luminescence response and to reveal different evenness of the protective coatings. Measurements of coating thickness made with the Eddy Current methods show a good agreement with the intensity of luminescence of the different areas. Intensity differences in each sample match also the trend of reflectance FTIR spectra, since those collected in areas with greater luminescence intensity exhibit distorted signals and higher peaks respect to those from less bright areas.

After validation trials on samples, the method was tested on a silver chalice treated with nitrocellulose lacquer. Notwithstanding the difficulties of imaging an object with such a complex shape, UV-induced VIS luminescence provided evidences of not uniform coverage of the surface, in particular on areas with embossed details, like in the chalice foot. Again, thickness variability is in agreement with luminescence differences. Therefore, the imaging of UV-induced VIS luminescence is recommended as a hands-on and powerful method to check the application of the most common protective coatings on historical silver objects. Another point is that the combination of three techniques offered the possibility of an interesting cross-validation on very thin layers of organic materials on metal. Despite their substantial different physical principles and their different acquisition approaches (imaging, punctual, with contact and contactless), the achievements are basically consistent and let us use each of them in a more confident way as tools to support conservation projects on silver artworks. Further work is planned to test the sensitivity of UV-induced VIS luminescence, together with that of reflectance FTIR and Eddy Current gauge, to very small amount of protective materials and to check the suitability of this approach to follow the phase of the removal of old protective films on silver and to detect even tiny residuals of films upon cleaning.

Acknowledgements

The specimens used in this work were prepared for the Horizon2020 EU project Scan4Reco (Grant n. 665091, www.scan4reco.eu), which includes activities aimed at the improvement of characterization of coatings on metal surfaces. The silver specimens were prepared and treated by Alessandro Pacini, Mary Nesi, Giulia Basilissi, Stefania Agnoletti. A special thank to Cinzia Ortolani for providing the chalice treated with nitrocellulose lacquer.

References

[1] Aldrovandi A., Cagnini A., Celi C., Erbetti M., Galeotti M., Lalli C. G., Lanterna G., Porcinai S., Rizzi M., Salvadori B., Tosini I., 2012, Applicazioni della spettroscopia infrarossa portatile nella diagnostica e monitoraggio dei Beni Culturali: vantaggi e limiti, 'OPD restauro', n.24, pp.57-70

[2] B. Salvadori , A. Cagnini, M. Galeotti, S. Porcinai, S. Goidanich, A. Vicenzo, C. Celi, P. Frediani, L. Rosi, M.Frediani, G. Giuntoli, L. Brambilla, R. Beltrami, S. Trasatti,, Traditional and innovative protective coatings for outdoor bronze: Application and performance comparison, J. APPL. POLYM. SCI. 2017, DOI: 10.1002/APP.46011

[3] B. Salvadori, E. Cantisani, C. G. R. Tognon, M. Nesi, A. Cagnini, M. Galeotti, S. Porcinai. Single- and multiple-layer coatings for bronze statuary: preliminary study and in situ evaluation after five-year outdoor exposure. Proceedings of the European Corrosion Congress EUROCORR 2014, Pisa, 8-12 Sept. 2014, O-7464

[4] S. Agnoletti, T. Bruni, A. Cagnini, M. Galeotti, S. Porcinai, B. Salvadori, A. Santagostino Barbone, Le applicazioni della tecnica delle correnti indotte (Eddy-Current) per la conservazione e lo studio di manufatti metallici di valore storico-artistico, 'OPD restauro', 2016, n.28, pp. 162-173

[5] Eddy Current testing at level 2 Manual for the syllabi Contained in iaea-TeCDoC-628.rev. 2 "Training guidelines for non Destructive Testing Techniques", IAEIA Training Course Series, 48, 2011

[6] J.Dong, A. Locquet, D.Citrin, Terahertz reflective imaging of damage mechanisms in the coating on metal substrate, atti del convegno '41st International Conference on Infrared, Millimeter, and Terahertz waves' (IRMMW-THz), Copenhagen, 2016

[7] J.Hoffmann, S.Sathish, E.B.Shell, S.Fassbender, N.Meyendorf, Acoustic Imaging Techniques for Characterization of Corrosion, Corrosion Protective Coatings, and Surface Cracks, 'Nondestructive Materials Characterization', 2004, pp 294-322

[8] G. Karagiannis, G. Apostolidis, A. Cagnini, M. Galeotti, S. Porcinai, Nondestructive investigation of metallic objects varnish using acoustic micro tomography, Proceedings of ART17, Torino

[9] A. Mignemi, O. Caruso, A. Cagnini, S. Porcinai, M. Galeotti, A. Aldrovandi Le indagini fotografiche nello studio delle superfici metalliche in leghe di rame:prime applicazioni, 'OPD restauro', 2014, n.26, pp. 202-216

[10] T. Learner, 2005, Analysis of modern paint, Getty Conservation Institute, Los Angeles p. 91

[11] PR.Griffiths, J.A. de Haseth, Fourier transform infrared spectrometry, Chapter 13, Specular Reflection, p.277, 2007, John Wiley & Sons, Inc.

IOP Conf. Series: Materials Science and Engineering **364** (2018) 012060 doi:10.1088/1757-899X/364/1/012060

Revisiting Reflectance Transformation Imaging (RTI): A Tool for Monitoring and Evaluating Conservation Treatments

Rosa Boute[2], Marjolein Hupkes[2*], Noa Kollaard[2], Sepha Wouda[2], Kate Seymour[1], Lauren ten Wolde[1]

1 Stichting Restauratie Atelier Limburg, Avenue Céramique, 6221KX Maastricht
2 University of Amsterdam

* Author of correspondence: Marjolein Hupkes, marjoleinhupkes@gmail.com

Abstract. Reflectance Transformation Imaging (RTI) has already been used for about twenty years in the field of Conservation and Restoration of Cultural Heritage for non-invasive condition assessment and documentation. However, its potential can be expanded. This paper will discuss the benefits of applying this readily available and low-tech technique to evaluate and monitor conservation treatments. Since the RTI software calculates the exact angle of the light for each RTI image taken, lighting conditions can be reproduced, leading to a standardisation of oblique light imaging. As a result, images produced before and after a conservation treatment can be compared. This allows for an objective evaluation of the treatment carried out and enables monitoring the condition of the painting over time.

1. Introduction

Reflectance Transformation Imaging (RTI) is a technique that is increasingly being used in the field of conservation. It is a computational photographic method that creates an interactive image, which highlights the object's topography (the object's surface shape and texture). It was first developed at Hewlett Packard Laboratories in 2001 and the open-access software has been further developed by Cultural Heritage Imaging (CHI) for its use for cultural heritage objects [1].

The technique has found applications for a broad range of objects for non-invasive investigation, condition assessment and documentation of both two- and three-dimensional objects [2]. Studies have included illuminated manuscripts, archaeological objects, ancient sea scrolls and coins. It has been used to give insight into minor surface details, typically not visible to the naked eye. It also aids in the identification of tool marks, materials and layer build-up, manufacture techniques and in stylistic comparison of objects. In several studies, RTI has proven to be a useful tool in the examination of panels and canvas paintings as it emphasises the three-dimensionality of the paint layer. These case studies also demonstrated the suitability of RTI in capturing gilded and tempera painted surfaces of icons [3, 4]. The majority of these studies to date have been diagnostic in nature.

The application of this readily available technique can extend beyond the diagnostic. This paper will present two case studies in which the authors intend to show the possibilities of using RTI as a tool for evaluating and monitoring conservation treatments, at specific moments during the conservation process and in the long-term.[I] In this context, evaluation refers to the assessment and validation of results during and immediately after a treatment. The monitoring of a restoration treatment entails checking whether the structural stability of the object changes over time after a treatment is carried out. The case studies described will focus on the conservation of painted panels treated at Stichting Restauratie Atelier Limburg (SRAL). The applicability and effectiveness of RTI will be discussed in relation to its use to aid conservators within the conservation process.

2. How RTI works

Traditionally, to obtain information about surface textures - such as impasto, cracks, raised paint, damages, canvas weave and wood grain - raking light photography is used.[II] This technique uses a low angle light source almost parallel to the surface. The light source emphasises the surface topography by creating shadows of the highpoints. Images captured are typically converted into black-and-white as any colour registration is distorted. More advanced raking light investigations may involve moving the light source to different sides of the object, adjusting the angle of the raking light to provide the most visibly effective result. This system is, however, limiting because each image taken is of a static moment and of a single angle. Additionally, shadows cast over the surface may mask adjacent raised details. RTI provides many advantages compared to this technique.

RTI is "a computational photographic method that captures a subject's surface shape and colour and enables the interactive re-lighting of the subject from any direction" [1]. As the CHI website mentions, "the enhancement functions of RTI reveal surface information that is not disclosed under direct empirical examination of the physical object" [1]. RTI has a distinct advantage over traditional observations made using raking light. A single and unregistered angle of light is replaced by taking multiple high-resolution digital photographs of the surface of an object lit from different angles. The light creates different highlights and shadows for each individual image at each specific defined angle. The images are mathematically synthesised by free software, available from CHI. The RTIBuilder software is able to calculate the surface normal for each pixel of the image, in order to reconstruct the surface topography of the object.[III] Synthesised light angles are created for the 'gaps' in the image. The captured images are stacked and registered so that they are overlaid exactly on top of one another. These images are merged into one virtual reconstruction of the surface and can be viewed in RTIViewer, a specific, separate programme. The images captured are in colour but can also be viewed in black-and-white. This system, thus, provides a photo documentation in visible light of the current condition of the object.

2.1.1. *Light and set-up*

The best images are taken in a darkened room and the light source needs to be direct rather than diffuse. The light source can be either a flashlight or a continuous light source. Different light angles can be achieved by constructing a dome, with numerous fixed positions for the lights, that is placed over the object. Alternatively, an upright stand that is moved around the object can be used with a light positioned at different heights. The dome system requires multiple fixed light sources on a rigid structure, whereas the second system can be constructed using a simple vertical stand and a single light source [6]. Both the dome and stand lighting rigs are more easily used with the object placed on the floor. The object can also be mounted on a vertical support, such as an easel when using the second system.

The background on which the object is placed for photography needs consideration. It should not contrast against the image, and thus it needs to be non-reflective and of a neutral colour. For the best result, a photographic backdrop can be used.

[I] Terminology of conservation and restoration treatments can be found on the ICOM-CC website [5].

[II] Raking light is a term used in conservation and is commonly known as oblique light.

[III] Surface normal: 'the vector at right angles to the surface at that particular point' [7].

IOP Conf. Series: Materials Science and Engineering **364** (2018) 012060 doi:10.1088/1757-899X/364/1/012060

2.1.2. *Registration*
The light angle for each image can be registered using a small, reflecting sphere. With this sphere as a reference point, the surface normals are calculated. The sphere should have a reflective, glossy surface and is preferably black. Literature mentions the use of different sizes, from very small ball-bearings to snooker balls. The size of the sphere depends on the size of the object and the distance of the camera to the object as it has to be well visible in each image. The cast shadow of the sphere should not cover the object. The sphere must not be moved during the capture of images within one RTI session. The reflection in the sphere is recorded and thereby the angle of the light can be registered. Since the sphere is static, the RTIBuilder software uses its position to stack and overlay the images captured.

A photographic colour checker is also included in the image capture to ensure adjusted colour registration for each image. This allows for an accurate colour adjustment in the capture and post-processing of the digital image.

2.1.3. *Camera and software*
The camera has to be placed in a fixed position, usually mounted on a tripod or a stand, that does not move from one image capture to another. Typically, a digital single-lens reflex (D-SLR) camera with high resolution capacity is employed. The higher the resolution of the camera, the better the final results will be. For best results, the camera can be tethered to a computer and controlled by a software programme such as Adobe® Lightroom [8]. This ensures that the camera is static while the image capture occurs, aiding the image processing during the next step. The image capture software allows the white balance to be set to the lighting conditions, providing a colour corrected image. Full colour information is therefore also obtained with each shot. Captured images are manipulated to remove extraneous information and ensure image consistency using a software programme such as Adobe® Photoshop. These images are further processed in the RTIBuilder software.

A minimum of 24, but better between 40-60 images, should be taken to obtain a good quality RTI image. The more images used, the less normals need to be constructed by the RTIBuilder software to fill the 'gaps' between the images taken. The captured images are then registered and stacked using the RTIBuilder programme creating an interactive, digital display of the object, which includes a movable light source. The processed image is viewed via the RTIViewer programme (fig. 1).

Fig. 1. Digital display of the RTIViewer showing one of the case studies in the rendering mode 'Specular Enhancement' (Image: SRAL).

IOP Conf. Series: Materials Science and Engineering **364** (2018) 012060 doi:10.1088/1757-899X/364/1/012060

The user can view the effect of different lighting positions on the surface of the object by moving the curser around the dynamic green mouse-pad, found on the right of the viewer screen. Full colour and raking light conditions can be viewed. Further digital processing manipulations of the image can be carried out. Features, such as 'diffuse gain' and 'specular enhancement', can be used to further highlight textural information.[IV]

3. Methodology

RTI captures have become more regularly used at SRAL in the last few years. The technique is used to aid the conservator determining the condition of the object, such as the extent of flaking paint or deformations of the support, and to evaluate and monitor the results of conservation treatments. Two case studies will be described below. Both used a standardised system for taking the RTI captures (fig. 2).

Fig. 2. RTI capturing of one of the case study objects, *The Dream of Jacob.* (Photo: SRAL).

The set-up for the RTI images taken at SRAL is as follows:

- The entire process is carried out in a dark room to eliminate light reflections other than those from the light source. Large objects are placed on the floor, which has a neutral grey linoleum covering. The smaller objects are placed on a green cutting mat, also placed on the floor. The grid pattern on the cutting mat aids the cropping of the image in the post-processing phase.
- The camera is mounted on a photographic stand with an extendable arm or attached to the ceiling. For larger objects, partial RTI images can be taken.
- The camera used is a NIKON D7100 with a zoom lens (AF-S NIKKOR 18-70mm, 1:3.5-4.5G ED). Image capture is set at the highest resolution in RAW format. Forty-eight images are taken, which is a number determined by the set light positions.
- An LED light source producing a direct beam is used. The LED lights were designed for SRAL by Van Kan and Willems, and consist of a 6x6 array set to daylight colour temperature via computer software.
- The position of the painting within the centre of the circle is recorded.

[IV] Definitions of these features can be found in the glossary on the Cultural Heritage Imaging website [1].

- The positions for the light direction are determined by first creating a circle around the object and marking twelve clockwise divisions. At each of the clock positions, four height angles are calculated using an upright stand on which they are marked. The light is positioned at each determined height for each position on the clock, which ensures consistency. Thus, 48 images are captured. The light angles can be reproduced by using the same lighting positions for future RTI captures.
- The reflective sphere consists of a small copper spherical ball painted black. The diameter of the sphere measures 2 cm and is placed adjacent to the object on washers to ensure they remain fixed in place during RTI image capture. Care is taken not to allow the cast shadow to cover the object.
- A colour checker (Qpcard 203 Camera Color Calibration) is placed in view frame of the camera [9].
- The camera is tethered to the computer using Adobe® Lightroom. This allows remote viewing of the image, which is especially important when the camera is mounted at ceiling height. The software allows for image manipulation prior to image capture, as aperture, exposure time, and white balance can be pre-set and standardised. Following instructions in the RTIBuilder manual, the images are processed and stacked. This programme also calculates the normal angles, which are used to synthesise the final RTI file. The RTI file is displayed in the RTIViewer programme, which allows an interactive viewing experience of the reflectance transformation images.

The most expensive element of the RTI tools described above is the camera. Other materials used in the system are relatively inexpensive and can be purchased easily. The Adobe® software package is also relatively inexpensive. This makes the system readily available to a wide audience. The set-up of the photographic studio for RTI captures takes time, but does not need specialised expertise. Instructional manuals and videos are available not only from CHI, but also from various other sources such as Cultural Heritage Science Open Source (CHSOS) [1, 10 and 11].

4. Case studies

The following section will evaluate the RTI process as described by using two case studies. The first case study used RTI captures to compare and evaluate the development of the aesthetic treatment of a fifteenth-century panel painting. The second case study used before and after treatment RTI captures to assess the structural treatment of a sixteenth-century panel painting. The after treatment RTI capture will be used to monitor future changes in the structure of the panel that may occur under the influence of uncontrolled environmental conditions.

4.1.1. *Case study 1: Spanish panel*

The first case study concerns a Spanish panel painting by an unknown artist, dated between 1450 and 1475 (fig. 3). It belongs to the Suermondt-Ludwig Museum in Aachen. The artwork has gilded areas and is painted with both tempera and oil-resinous paint [12]. The five-hundred-year-old painting has had a long history of damages and related treatments. Recently, old varnish layers, repaints and fillings were removed. Losses in some areas extended as deep as the ground layer and required filling to bring the surface to a uniform level.[V] It was important that the fillings would be well integrated into the original paint surface and that they would not be visible under museum lighting conditions. In this case, the RTI captures were used as a means to evaluate the quality of the applied fillings (fig. 4). The possibility of changing the light direction in the RTIViewer programme enabled the assessment of the height and texture of the fillings from all directions. Any fills that appeared disturbing in the RTI images were further adapted by the conservator to ensure their integration into the surface (fig. 5 and 6). In this way, a better result could be achieved.

The before and after RTI captures provide the conservator with two specific data sets that can be directly compared and assessed. In this case, the highlights present in the before treatment RTI

[V] The filling treatment was carried out by Cerys Fry, Fellow at SRAL (2017-2018).

capture are more pronounced than the RTI capture after the filling treatment. The reduction of the number of highlights shows that an improvement was made by the treatment. This allows the conservator to evaluate the treatment and improve the fillings further prior to integrating the fills to create a unified comprehensive image.

Fig. 3 and 4. Unknown master, *The Beheading of St. John the Baptist and the Feast of Herod* (GK84b). Circa 1475 Mixed media on panel, ~92 x 48 cm, Suermondt-Ludwig-Museum, Aachen. Left image is a direct light photo before treatment. The right image is an RTI capture before filling treatment (Photos: SRAL).

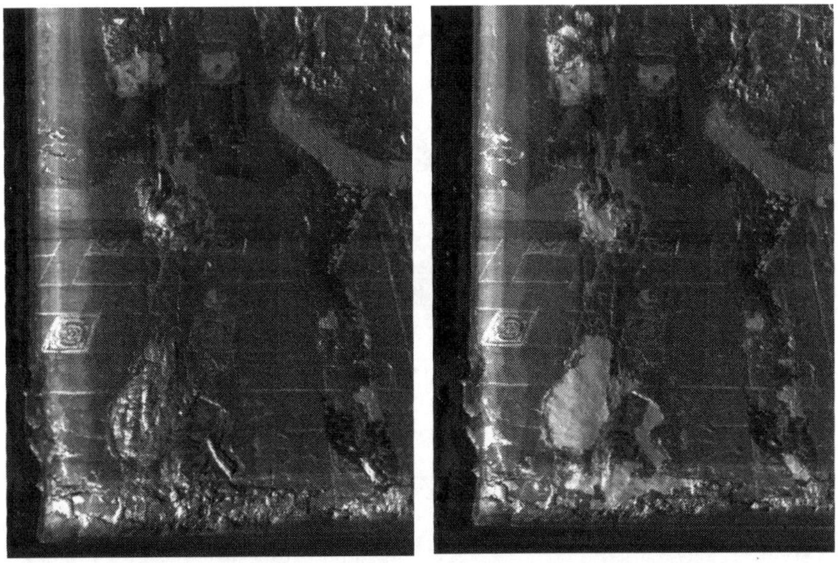

Fig. 5 and 6. Left image shows the RTI capture before treatment and the right image shows the RTI capture after the application of fillings. Both images were taken with the exact same light angle coordinates. (Photos: SRAL).

4.1.2. *Case study 2: The Dream of Jacob*

The second case study involved a sixteenth-century, Southern-Netherlandish panel painting, entitled *The Dream of Jacob* by an anonymous artist (fig. 7) [13]. The painting belongs to the Basilica of Our Lady (Onze-Lieve-Vrouwebasiliek) in Maastricht. It required a structural treatment of the support as the original butt joints between the three wooden boards of the support had failed over time, and splits had occurred in the upper and lower boards. Further damage was induced by the application of a secondary cradle support.[VI] This had restricted the natural movements of the wooden panel in response to changes in environmental relative humidity. Tension build-up in the wood caused deformations in each individual board of the support to develop. In the field of paintings conservation such deformations are generally described as 'wash boarding'.

The challenge of this treatment was to re-align the individual boards and create a levelled paint surface with a continuous curvature.[VII] RTI captures were carried out before and after rejoining the panel in order to assess the result of the treatment. Comparison of the RTI captures showed that the wash-boarding effect had strongly diminished after rejoining the boards (fig. 8, 9, 10 and 11).

In the future, the after treatment RTI capture can be used as a reference to assess if the support develops new deformations. The wooden support will continue to react to fluctuations in relative humidity as it is a hydroscopic organic material. Monitoring the degree of curvature of the panel over time will allow conservators to determine if compression tension builds up in the support. This can happen if the movement of the support is restricted as relative humidity varies. To monitor alterations in the condition of the panel, new RTI captures can be taken and compared with the existing after treatment RTI image. In this way, damage caused by factors such as adverse environmental conditions can be demonstrated and perhaps even prevented.

Fig. 7. Artist Unknown/Flemish, *The Dream of Jacob* (SRAL inv. nr. 0.2016.071), first half 16th-century, oil paint on oak panel, 79 x 109.2 x 0.7 cm, Parish of the Onze-Lieve-Vrouwebasiliek. (Photo: SRAL, direct light, before treatment).

[VI] A cradle consists of attached wooden members adhered to the reverse of the support in the same direction as the wood-grain of the support boards. Perpendicular sliding battens are inserted into half-lap inserts into the fixed members. The intention is that these flex as the panel support responds to fluctuations in relative humidity. The sliding battens often become blocked as the wood swells when absorbing moisture from the environment, negating their intended function. The aim of the cradle is to support the painting and maintain its flatness.

[VII] The treatment of *The Dream of Jacob* was carried out by Lauren ten Wolde, Fellow at SRAL (2017-2018).

Fig. 8 and 9. The top image shows the RTI capture before treatment and the bottom image shows the RTI capture after structural treatment. Both images were taken with the exact same light angle coordinates. (Photos: SRAL).

IOP Conf. Series: Materials Science and Engineering **364** (2018) 012060 doi:10.1088/1757-899X/364/1/012060

Fig. 10 and 11. The top detail shows the RTI capture before treatment and the bottom detail shows the RTI capture after treatment. Both images were taken with the exact same light angle coordinates. In the top image the light source emphasises the deformation of the lowest board, whereas in the bottom image the light source illustrates the continuation of the curvature. (Photos: SRAL).

IOP Conf. Series: Materials Science and Engineering **364** (2018) 012060 doi:10.1088/1757-899X/364/1/012060

5. Discussion

The first main advantage of using RTI as a tool for evaluating conservation treatments is the flexibility of this technique. In contrast to raking light photography - in which only one light angle is captured - RTI captures allow the light angle to be changed in a virtual reality. The light angle can be adjusted at each specific area, to obtain the most information about the surface texture of the painting. This is especially useful for strongly deformed wooden supports. In these cases, a single raking light photograph will not contain information about the whole surface of the panel as the shadow cast will extend over adjacent deformations, masking their severity. This problem does not occur when RTI captures are made.

A second advantage of RTI is the reproducibility of the imaging technique. If a consistency is kept in the lighting situation, it is possible to use the same set of lighting coordinates in the RTIViewer in the two captures. This allows for an accurate and reliable comparison between the different captures at different points in time. Differences visible between the captures can therefore be ascribed to actual differences in the surface shape and texture of the painting and not to a difference in the angle of the light. For example, the improvement to the curvature of *The Dream of Jacob* is substantial and is demonstrated by comparing the two RTI captures.

RTI also has drawbacks since the captures may be difficult to read. This was the case when interpreting the RTI capture of the Spanish panel. Here the surface is extremely irregular and comparing the overview RTI images taken before and after treatment did not show a distinct change or improvement. Zoomed images could be studied to provide more local information and allow for more detailed observations. These subtle details may be more easily determined by the eye of the trained conservator than that of a lay audience.

Although it is possible to create high-tech RTI captures using advanced equipment, the authors experienced that good quality RTI captures can be produced with low-tech tools. It requires time to arrange the whole set-up, however, once the preparation is done, making the actual captures can be done quickly. Therefore, it is possible to implement RTI in typical conservation studio practice.

6. Conclusion

RTI can be used beyond diagnostic application, as shown by the two case studies in this paper. The imaging technique proved to be a useful tool to evaluate the specific goals of the described treatments. Furthermore, the images that were obtained with RTI are valuable for documentation and as a tool for communication with owners and other stakeholders.

The flexibility of the RTI technique proved to be useful when evaluating highly irregular surfaces because the light can be placed at the best angle for each specific area in the virtual viewer programme. Moreover, because of the reproducibility of the RTI set-up, the exact same area can be viewed with the same light angle in a next RTI capture. This enables accurate comparison between the different stages within a conservation treatment, creating a valuable record for monitoring, evaluation, documentation and communication. The authors recommend that a more extensive use of this low-tech, easily affordable system could be wider implemented in the field of conservation and restoration.

7. Acknowledgements

The authors thank Cerys Fry and Inês Carreira for providing information on the case studies. Also, we are grateful for the help of Madeleine Vaudremer and Chiara Manfriani.

References

[1] Cultural Heritage Imaging, Inc.: http://culturalheritageimaging.org/Technologies/RTI/ (accessed on: 18-1-2018)

[2] Manfredi M, Bearman G, Williamson G, Kronkright D, Doehne E, Jacobs M and Marengo E 2014 *A New Quantitative Method for the Non-Invasive Documentation of Morphological Damage in Paintings Using RTI Surface Normals* Sensors pp 12271-84

[3] Kotoula E and Earl G 2014 *Integrated RTI approaches for the study of painted surfaces* (University of Southampton Archaeological Computing Research Group) CAA: Paris. Text and PowerPoint URL: https://caa2014.sciencesconf.org/45911/document and http://www.academia.edu/7078660/Integrated_RTI_approaches_for_the_study_of_painted_surfaces

[4] Kotoula E 2015 *Reflectance Transformation Imaging Beyond the Visible: Ultraviolet Reflected and Ultraviolet Induced Visible Fluorescence* CAA pp 909-18

[5] Website ICOM-CC: Terminology for Conservation http://www.icom-cc.org/242/about/terminology-for-conservation/#.WmBq7ainE2w (accessed on: 18-1-2018)

[6] Padfield J, Saunders D and Malzbender T 2005 *Polynomial texture mapping: a new tool for examining the surface of paintings*. 14th Triennial Meeting The Hague 12-16 September 2005 Preprints Volume 1 (ICOM Committee for Conservation). Ed. Isabelle Verger. London: Earthscan ltd. pp 1-7

[7] Gabov A and Bevan G 2011 Recording the weathering of outdoor stone monuments using reflectance transformation imaging (RTI): the case of the guild of all arts (Scarborough, Ontario) (*J. CAC, volume 36, Canadian Association for Conservation)* pp 1 3-14

[8] Website Adobe Creative Cloud: http://www.adobe.com/nl/creativecloud.html (accessed on: 18-1-2018)

[9] Website Qpcard: www.qpcard.com (accessed on: 19-1-2018)

[10] Website Cultural Heritage Science Open Source: https://chsopensource.org/ (accessed on: 19-1-2018)

[11] Duffy S M, Bryan P, Earl G, Beale G, Pagi H and Kotoula E 2013 *Multi-light Imaging for Heritage Applications* (English Heritage)

[12] Albrecht M, Daugherty M, Van Oudheusden S, d'Hont L, Seymour K, Rief M, Marchant R and Uffelman E 2016 *Unraveling the History of Two Fifteenth-Century Spanish Panels* The Noninvasive Analysis of Painted Surfaces: Scientific Impact and Conservation Practice. Ed. Austin Nevin and Tiarna Doherty. Washington D.C., Smithsonian Institution Scholarly Press, pp 37-51

[13] Ten Wolde L M A 2018 *Research and Conservation Report: The Dream of Jacob – artist unknown – 16th century. Maastricht: Stichting Restauratie Atelier Limburg* (Unpublished report)

IOP Conf. Series: Materials Science and Engineering **364** (2018) 012061 doi:10.1088/1757-899X/364/1/012061

Intrinsic limits of reflectance spectroscopy in identifying pigments in paint layers

L Pronti[1], A Pelagotti[2], F Uccheddu[3], L Massini Rosati [1], A C Felici[1]
[1] Department of Basic and Applied Sciences for Engineering, LANDA-Laboratory of Non Destructive Analysis and Archaeometry, University of Rome Sapienza, via A. Scarpa 16, 00161, Rome, Italy
[2]National Institute of Optics, (INO) CNR, L.go E. Fermi 6, 50125 Firenze
[3]Departiment of Industrial Engineering, University of Florence, via Santa Marta 3 50123 Firenze

Corresponding author: lucilla.pronti@uniroma1.it,

Abstract. The study of pigments is of fundamental importance for authentication and restoration purposes of painted artworks. To this aim reflectance spectroscopy has been widely studied on samples and in a number of real cases. A number of valuable efforts have been dedicated to this quest; however, the identification of a pigment on a painting surface by means of reflectance spectroscopy appears to be possible only in very few cases also due to the fact that artist materials are often mixed. In this paper we review a number of relevant works and present results on a new set of test samples especially produced for exploring the possibilities of reflectance spectroscopy on pigment mixtures. We also include a case study of a real painting.

1. Introduction

The chemical composition of materials used to create an artwork is crucial information. Materials for art production changed over time or between a production place and another, an artist and its competitors. Assessing them is thus vital for e.g. uncovering forgeries.
Historically, chemical analysis of paint layers has been among the first scientific tests performed on artworks. This is still an invasive technique and the results are representative only of a limited area.
For quite some time scientists have been trying to replace these analyses with non-destructive methodologies. Reflectance spectroscopy has been often proposed. The underlying assumption is that each material displays a specific spectral signature in the ultraviolet (UV), visible (Vis) and infrared (IR) regions, which carries more information compared to what the eye is able to discriminate.
However, while it proved to be very useful for monitoring the presence of alteration products and for the analysis of colour changes on paintings (see eg. [1], [2]), we argue that it can be severely misleading when used for the identification of pigments. In fact, although some pigments present a specific spectral behaviour, in practical applications pictorial layers are hardly ever composed by a single pigment (or colorant). Paint mixtures or paint overlaying are very common practices, but a model for how the different components contribute to the final reflectance spectrum has not been conclusively defined yet, thus unmixing the signals is not trivial. In order to characterize spectral signatures, mostly peaks and inflection points (or flexes, i.e. points of the curve where the curvature

changes its sign and the second derivative equals zero) are described, giving the max or min values and making approximation with e.g. Gaussian forms (see e.g. [3], [4]). In this work we analyse some of the critical issues regarding the use of the reflectance spectroscopy for the identification of pigments or dyes, making reference to existing literature, providing evidence using samples realized ad hoc, and presenting an exemplificative case study.

2. Reflectance spectroscopy in identifying pigments in paint layers. A brief review

Reflectance spectroscopy from its debut in 1938 [5] has been extensively used [6], [7]. Soon, it became apparent that if the technique was to be used for pigment identification, a database of spectra of artist's materials was needed for comparison. A number of such databases have been developed in the years for several pigments, binders and paint methods (see e.g. [8], [9], [10]) although it is probably intrinsically not conceivable to cover the full spectrum of paints and paint mixtures used by artists over several centuries, including the most recent years. For this reason, a method to recognize single pigment in mixtures would be needed [11], [12].

Several methods were researched in order to characterise the spectral signature of each given material, , i.e. its typical variation of reflectance or emittance. Some methods have been proposed for spectra classification [13], [14]. They however show how reflectance spectroscopy on its own does not generally permit pigment identification even for a single material [4]. The implications are very relevant. For example the spectra of *Naples Yellow* and of *Cadmium Yellow* are very alike but the use of the first is compatible with an Old Master painting while latter is a modern compound.

In the analysis of real paintings the possibility of identification of pigments by reflectance spectroscopy is moreover complicated by the presence of several additional elements like the geometry of painting surfaces [11], the fact that pigments are generally present in mixtures [13] and/or as overlapping layers [15] of varying thickness [16]. Moreover also aging plays a role, with the intense yellowing of the binder or of the overlapping varnish, due to oxidation processes occurring with natural ageing [17] and/or other environmental agents like light, humidity, and temperature [12] that can alter some pigments. Moreover other intrinsic factors like the nature and grain size of pigments [19], the differences of provenance [20], [21], [22] and of manufacturing processes, also play a significant role, generally making pigment identification with reflectance spectroscopy not conclusive. Moreover, when extending reflectance spectroscopy to the near IR region, the transparency of most thin layers layer of pigments to the IR radiation complicates the matter. In the red and/or NIR region the depth of the penetration of the radiation into painted layers can reach the preparation layer, and this will affect the acquired spectrum depending of the reflectance properties of the preparation layer.

3. Materials and methods

3.1. Samples

In order to specifically investigate the spectral response of complex paint structures, a number of samples were produced and investigated. The set covers 31 pictorial layers consisting of two or three pigments mixed with linseed oil or egg yolk, 32 oil colour tubes, 30 home-made oil based paints and 15 red-glazes spread on two different pictorial layers.

- **31 pictorial layers composed by two or three pigments mixed with linseed oil or egg yolk:**
  - MIX SL:10 samples mixtures of Lead White and Raw Sienna Earth at different volume percent, 5 of which diluted with linseed oil and the other with egg yolk;
  - MIX SLG: 6 samples are mixtures of Lead White, raw Sienna Earth and Green Earth with linseed oil. 3 of them are mixtures of Lead White and raw Sienna Earth at 50:50 volume percent with an increase of the quantity of Green Earth (0.10g, 0.20g and 0.30 g) at each sample and the others are mixtures of Lead White and raw Sienna Earth at 10:90 volume percent with the same increase of the quantity of Green Earth (0.10g, 0.20g and 0.30 g);
  - MIX LZn: 6 samples are mixtures of Lead White and Zinc White at different volume

percent, 3 of which diluted with linseed oil and the other with egg yolk.
- Zn_H: 1 sample of a mixture of Zinc White and Hematite with linseed oil;
- Pb_H:1 sample of a mixture of Lead White and Hematite with linseed oil;
- Zn_Pb_H: 1 sample of a mixture of Lead White and Zinc White at 50:50 volume percent with linseed oil; 1
- MIX LZH: 6 samples are mixtures of Lead White and Zinc White at 50:50 volume percent with an increase of the quantity of Hematite (0.05g, 0.10g, 0.15g), 3 of which diluted with linseed oil and the other with egg yolk

- **32 oil colour tubes and 30 hand-made oil colours**:
  - Oil colour tubes from several manufacturers: Maimeri, Ferrario, Van Dick, Pēbēo, Williamsburg
  - Home-made oil colours with pigments from several manufactures: Maimeri, Zecchi, Poggi.The pigments are diluted with linseed oil (Lefranc & Bourgeois). all the oil colour were spread on a microscope slides.
- **15 red-glazes spread on two sets of pictorial layers.** Red-glazes produced by mixing commercial alizarin based-pigments with linseed oil (Lefranc & Bourgeois) were spread on microscope slides and on two different "tempera" pictorial layers (Red Ochre mixed with egg yolk and Cinnabar mixed with egg yolk). Moreover two different thicknesses were produced: one more covering than the other. Both the two "tempera" blends were spread on a commercial-prepared wood panel. The commercial alizarin-based lakes were purchased by Kremer Pigmente and Zecchi.

3.2. Reflectance spectroscopy equipment
Measurements were performed by means of an Avantes AvaSpec spectrophotometer equipped with a halogen lamp (HL-2000 FHSA, Avantes and a diffraction grating of 300 lines/mm, blazed at 500 nm, with (300 - 1100 nm spectral range), and a CCD linear sensor (2048 pixels). The spectral resolution is 0.8 nm. We illuminated the samples at 45° with a bifurcated fiber (diameter of 600 μm. The reflected light was collected with a 200 μm diameter fiber at 90° respect to the surface. When using microscope slides, we had a white background.

4. Reflectance spectroscopy applied on case studies
Given the limited space, we present here only very few of the spectra we acquired, although similar conclusions could be drawn from many others.
The reflectance spectrum of a mixture of two or more pigments is a combination of their spectra. The way the spectra combine is however not clarified. But, even adding white pigments, (having reflectance practically constant over the visible range), to a mixture, it generally modifies at least the flexes position, generally without retaining any information about the pigment added. Which white pigment has been used in a painting is however of crucial importance. Lead or Zinc White could completely change the dating of a painting, being the first an ancient pigment and the latter a modern one. Unfortunately, when Lead White is mixed with other pigments it is impossible to recognize it, as shown in **Figure 1**. In the case of Zinc White, although it has a very specific behaviour in the UV region, when in combination with other pigments, the peak at 390 nm is hardly visible. See e.g. the spectrum Pb_H and the mix LZE: the two spectra in the range 400-900 nm are practically not easily distinguished, especially considering that in real cases the peaks and the absorption bands will be less pronounced [4]. The situation is not dramatically better even when the pigments have distinctive spectral signatures e.g. for blue pigments. For example Smalt is characterized by an intense absorption in the 550-650 nm range, with three sub bands at 540 nm, 590 nm and 640 nm [23]. However it is extremely transparent and the spectral behaviour is strongly influenced by the underlayer [14]. Reflectance spectra in **Figure 2A** exhibits different ratios between the reflectance maximum in the blue range and the red-NIR range. Minima values disappear when the layer thickness is high and when smalt is spread on a black layer.

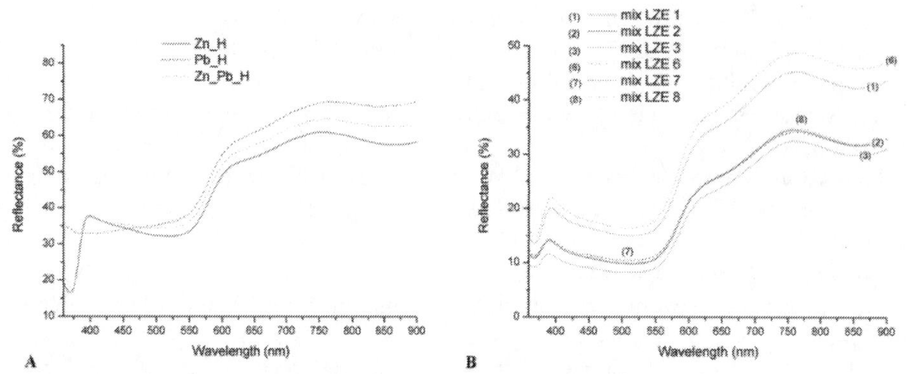

Figure 1: A) Reflectance spectra of the samples of Hematite mixed with: Lead White (Pb_H), Zinc White (Zn_H) and a mixture of Zinc White and Lead White 50:50 (Zn_Pb_H) in linseed oil. B) Reflectance spectra of a mixture of Lead White and Zinc White 50:50 with increasing quantities of Hematite in linseed oil (solid lines) or egg yolk (dash lines).

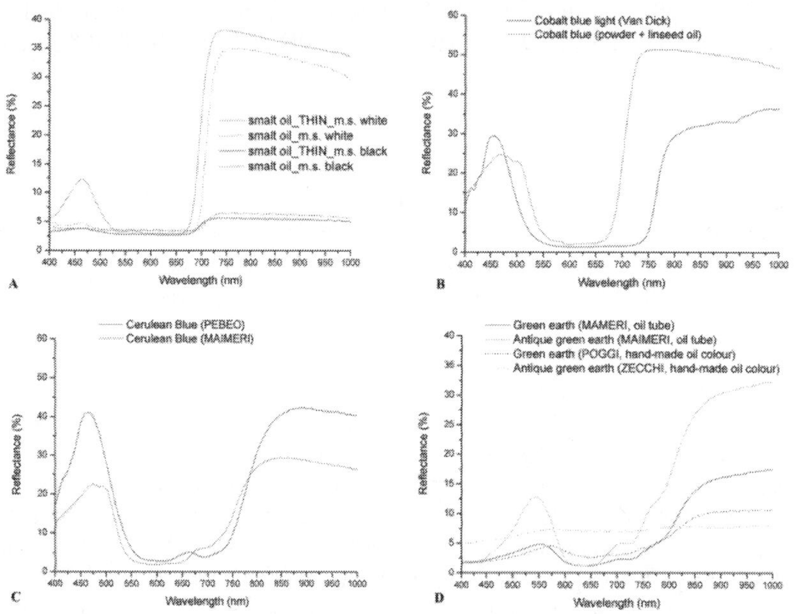

Figure 2: A) Reflectance spectra of Smalt in linseed oil. The terms "m.s. white" refers to microscope slide laid on a white support, "m.s. black" on a black support. The addition of "THIN_" refers to the thickness of the thin pictorial layer. B) Reflectance spectra of modern oil tube of Cobalt Blue (black line) and a powder of Cobalt Blue dispersed in linseed oil (gray line). C) Reflectance spectra of two modern oil tubes of Ceruleo: PĒBĒO (black line) and Maimeri (gray line). D) Reflectance spectra of green oil colours.

Cobalt blue dispersed in linseed oil (home-made oil based paint) presents two absorption bands at about 480 nm and 495 nm, as shown in **Figure 2B**, however this spectral features are linked to the specific manufacturer recipe. Ferrario Cobalt blue shows e.g. two minima at about 417 and 430 nm. The same we can say for many Blue pigments (like Cerulean), **Figure 2C**, and green pigments, **Figure 2D**.

Yellow pigments, except Cadmium Yellow, that contain Zinc sulphide as Lithopone [9], are generally identified by the position of flexes and shoulders. However, the inflection points could shift when the pigment is mixed with others and the position of the shoulders could be modified due to the increase of transparency at longer wavelengths. Although some yellow pigments present a characteristic reflectance curve, mixing them with a white pigment, such as Lead White and a green pigment, such as Green Earth (MIX SLG sample), we obtain different spectral features, as shown in **Figure 3A**. The two blends 50:50 and 90:10 volume percent of Raw Sienna Earth and Lead White, mixed with green earth and linseed oil as a binder, show different inflection points at about 525 nm and 550 nm, respectively. The reflectance spectra of the samples with the highest quantity of Raw Sienna Earth, and lower of Lead White, are weaker at long wavelengths and show a second well-pronounced band at 750 nm. Despite the low quantity of Green Earth, all the reflectance spectra show weak bands between 650 nm and 800 nm of the Green Earth pigment, **Figure 3B**.

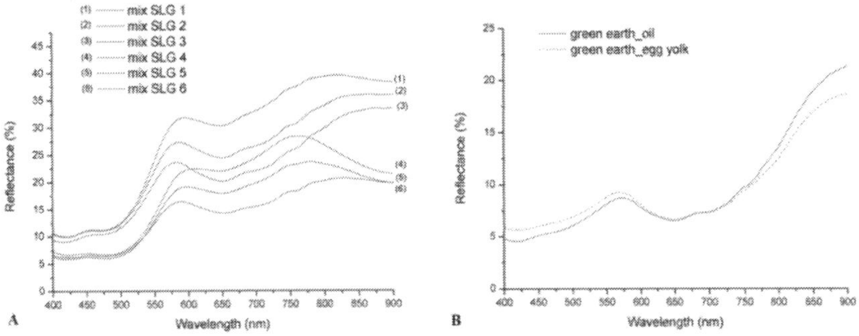

Figure 3: A): Reflectance spectra of the samples of raw Sienna Earth, Lead White and Green Earth in linseed oil or egg yolk. Samples 1 to 3 are mixtures of raw Sienna Earth and Lead White 50:50 and increasing quantities of Green Earth mixed with linseed oil, while 4 to 6 are mixtures of raw Sienna Earth and Lead White 90:10 and increasing quantities of Green Earth mixed with egg yolk. B) Reflectance spectra of the samples of Green Earth with linseed oil (solid line) or egg yolk (dash line).

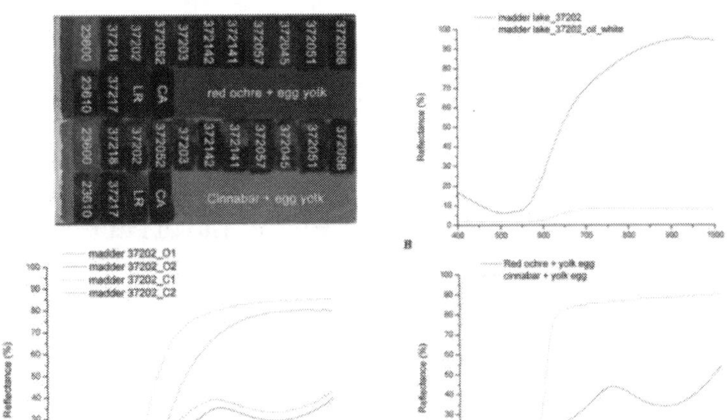

Figure 4: A) Reflectance spectra of madder lake with linseed oil spread on Red Ochre bonded with with egg yolk (O) and on Cinnabar in egg yolk (C). The number 1 and 2 is referred to the thickness of the layer: thin (1) and thicker (2). B) Reflectance spectrum of madder lake in powder. C) Reflectance spectra of red ochre mixed with yolk egg (gray line) and Cinnabar with yolk egg (black line).

Also the identification of red inorganic pigments are mainly associated to the position of inflection points and shoulders, except for the iron-based pigments that show a characteristic shape [9], [24]. The reflectance spectroscopy is used also to recognize red dyes e.g. [20], [25], [26], however this approach can be misleading when the lakes are spread above another layer, for example Vermillion, as it generally occurs in real paintings. In this case, the spectral features of the tempera layer of red dyes could be recognized if it is spread as dilute mixture or as thin layer, but the positions of the absorption bands could be shifted [20]. To study the lakes, we prepared 15 red-glazes spread over two different pictorial layers. The reflectance spectra of one lake (37202_Kremer) can be considered representative of the entire set of samples depicted in **Figure 4**. The absorption bands of this Alizarin-base lake (Madder Lake), at about 505 nm and 550 nm, is completely indistinguishable in linseed oil, **Figure 4B**. When the red lakes are spread above Red Ochre or Cinnabar layers, the spectral features of the red lakes are completely modified, as shown in **Figure 4A**, as far as, the presence of a superficial lake layer modifies the spectral behaviours of the Red Ochre and the Cinnabar pigments, **Figure 4C**. In the case of real paintings, the interpretation of the reflectance spectra could be extremely difficult and could lead to a wrong identification of the artist's palette. To this aim we report a case study of "Virgin and Child and Saint Catherine of Alexandria" painting by Matteo di Giovanni (or an imitator), in storage at the Accademia di San Luca (Rome), **Figure 5A**.

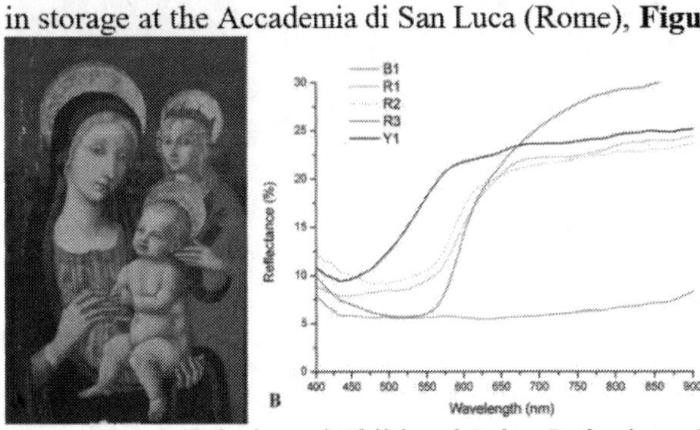

Figure 5: A) "Virgin and Child and Saint Catherine of Alexandria" painting. B) Reflectance spectra of three areas of the Virgin and Child and Saint Catherine of Alexandria painting. B1: Virgin's mantle, R1: Virgin's dress (flower), R2: Virgin's dress (cuff), R3: Saint's dress Y1: Saint's hair

The reflectance spectrum of the Virgin's mantle shows a flat trend, as shown in **Figure 5B** suggesting the use of a black pigment, however XRF[1] analysis revealed the presence of copper, as reported in **Table 1**, probably to be attributed to a copper-based pigment, generally appearing green or blue. These results could suggest a multilayer system or a blackening of the surface that could have been caused by a high pigment/binder ratio or to an alteration [27]. The reflectance curves relative to vest, flowers and cuff show approximately the same inflection point at 590 nm. This could suggest the presence of Cinnabar or Vermillion (these two pigments are anyway difficult to discriminate with non-invasive techniques). However XRF analysis, a very sensitive technique, reveals this pigment only in the Saint's dress through the detection of mercury (Hg). For the others areas we can suppose the presence of Red Ochre and pigments at low atomic number. The Y1 curve relative to the Y1 area (Saint's hair) shows two minima at about 430 nm and 630 nm (weak), an inflection point at about 540 nm, a shoulder at about 580nm, and a weak band at 670 nm. The absence of the typical S-shape could indicate the absence of Iron-based pigments, although we detect a moderate count of Iron by means of XRF. Naples Yellow presents a shoulder at about 580 nm [9] but XRF did not detect Antimony, also

[1] X-Ray Fluorescence Spectroscopy performed with an X-ray tube (EIS srl) with an anode target of palladium and a beryllium window of 150 μm of thickness, powered with an accelerating potential difference of 40 kV and an electronic current of 0.4 mA. The detector is a Peltier cooled silicon drift with amplifier and integrated multi-channel analyser (Amptek 123-SDD) which provides for every X photon revealed a current pulse of amplitude proportional to the photon energy. The detector has a surface of 25 mm^2 and a thickness of 500 μm, a beryllium window of 12.5 μm and its energy resolution is 140 eV full width half maximum of the manganese Kα line at 5.9 keV.

Cobalt yellow presents a shoulder at about 580 nm and a second one at about 670 nm, similar to those of the Y1 curve [9], but XRF did not find Cobalt. We may suppose the use of a mixture of more pigments and/or dyes, however we cannot identify which ones could possibly have been employed.

	S	Cl	K	Ca	Ti	Cr	Mn	Fe	Cu	Zn	As	Sr	Hg	Pb
B1	112	71	395	5892	56	-	232	2536	287157	3822	2045	378	-	17128
R1	372	-	271	1875	-	-	332	1417	597	627	-	792	175	102271
R2	417	-	327	2295	49	-	149	1126	352	501	-	259	147	115636
R3	1063	58		1560	51	90	85	756	543	300			56767	58387
Y1	560	48	53	597	-	-	53	3616	394	271	-	642	115	53317

Table 1. X-Ray Fluorescence counts.

5. Conclusions

In this paper we discussed the inherent limitations of reflectance spectroscopy for the purpose of material identification on a painting. We reviewed the existing literature and presented results on a set of test samples especially produced for exploring results on pigment mixtures, with an increased degree of complexity compared to the single pigment swatches that are often used. We also discussed the application of this technique to a real case. Results show how this technique is often not conclusive for several reasons. The databases used for comparison cannot be exhaustive, as many pigments come in different formulas, showing different spectra, also depending on the manufacturer. The fact that in most cases pigments are not used pure but mixed with others, in the absence of a way of modelling interactions, leads often to the impossibility to find the components of the mixture. Moreover aging and other factors change the reflectance of pigments in a not always known or predictable way. We can therefore conclude that currently this technique should not be proposed for the purpose of non-invasive pigments identification. A correct analytical approach should at least involve the integration of complementary techniques for the analysis of chemical and physical features of the materials employed.

References

[1] S. Acquaviva, E. D'Anna, M. L. De Giorgi, A. Della Patria, and L. Pezzati, "Optical characterization of pigments by reflectance spectroscopy in support of UV laser cleaning treatments," *Appl. Phys. A Mater. Sci. Process.*, Vol. 92, no. 1, pp. 223–227, 2008.

[2] L. Bullock: "Reflectance Spectrophotometry for Measurement of Colour Change", National Gallery Technical Bulletin, Vol. 2, pp. 49-55, 1978

[3] http://www.brera.unimi.it/istituto/archeo/schede_Vicenza.html (retreived 5-1-18)

[4] Tesi di Dottorato di ricerca in Fisica, Astrofisica e Fisica Applicata di: Davide Pandini, Ciclo XXIV, Coordinatore: Prof. Marco Bersanelli, Tutore: Nicola Ludwig, "Studi per la modellazione della riflettanza spettrale negli strati pittorici", Università degli Studi di Milano, Facoltà di Scienze Matematiche, Fisiche e Naturali, Anno Accademico 2010-2011

[5] N. F. Barnes: "A Spectrophotometric Study of Artists' Pigments", Technical Studies in the Field of the Fine Arts, Vol. 7, pp. 120-138, 1938

[6] M. Bacci, F. Baldini, R. Carlà, and R. Linari: "A Color Analysis of the Brancacci Chapel Frescoes", Applied Spectroscopy, Vol. 45, no. 1 pp. 26-31, 1991.

[7] Bacci, M., Baronti, S., Casini, A., Lotti, F., Picollo, M., and Casazza. O.: "Non-destructive spectroscopic investigations on paintings using optical fibers", Materials Research Society Symposium Proceedings, Vol. 267, pp. 265-283, 1992.

[8] A. Aldrovandi, M. L. Altamura, M. T. Cianfanelli, P. Riitano: "I materiali pittorici: tavolette campione per la caratterizzazione mediante analisi multispettrale",OPD Restauro, Vol. 8, pp. 191-210, 101-103, 1996.

[9] "Fiber Optics Reflectance Spectra (FORS) of Pictorial Materials in the 270-1700 nm range."

[Online]. Available: http://fors.ifac.cnr.it/index.php
[10] "Database of diagnostic analyses". [Online]. Available: http://webimgc.inrim.it/Hyperspectral_imaging/Database.aspx
[11] T. Cavaleri, A. Giovagnoli, M. Nervo: "Pigments and mixtures identification by Visible Reflectance Spectroscopy", Procedia Chemistry, Vol. 8, pp. 45-54, 2013
[12] G. Dupuis, M. Menu: "Quantitative characterisation of pigment mixtures used in art by fibre-optics diffuse-reflectance spectroscopy", Applied Physics A, Vol. 83, no. 4, pp. 469-474, 2006
[13] L. Pronti, A. C. Felici, M. Ménager, C. Vieillescazes, M. Piacentini: "Spectral Behavior of White Pigment Mixtures Using Reflectance, Ultraviolet-Fluorescence Spectroscopy, and Multispectral Imaging", Applied Spectroscopy, Vol. 71, no. 12, pp. 1-10, 2017
[14] L. Pronti, P. Ferrara, F. Uccheddu, A. Pelagotti, and A. Piva, "Identification of pictorial materials by means of optimized multispectral reflectance image processing," in 2015 IEEE International Workshop on Information Forensics and Security (WIFS), pp. 1-6, 2015.
[15] D. Comelli, A. Nevin, G. Valentini, I. Osticioli, E. M. Castellucci, L. Toniolo, D. Gulotta, R. Cubeddu: "Insights into Masolino's wall paintings in Castiglione Olona: Advanced reflectance and fluorescence imaging analysis", J. of Cultural Heritage, Vol. 12, no. 1, pp. 11-18, 2011
[16] L. Bonizzoni, S. Caglio, A. Galli, G. Poldi: "A non invasive method to detect stratigraphy, thicknesses and pigment concentration of pictorial multilayers based on EDXRF and vis-RS: in situ applications", Applied Physics A, Vol. 92, no.1, pp. 203-210, 2008
[17] G. Antonioli, F. Fermi, C. Oleari, R. Reverberi: "Reflectoscopic analysis of cultural goods for knowledge and preservation", MIDECH'05, Convegni Aracne 13, Aracne Editrice srl Roma, 2005, ISBN 88-548-0029-5 175–181
[18] G. Poldi, S. Caglio: "Phthalocyanine Identification in Paintings by Reflectance Spectroscopy. A Laboratory and In Situ Study", Optics and Spectroscopy, Vol. 114, no. 6, pp. 929-935, 2013
[19] S. P. Best, R. J. H. Clark, M. A. M. Daniels, C. A. Porter, R. Withnall: "Identification by Raman Microscopy and Visible Reflectance Spectroscopy of pigments on an Icelandic Manuscript", Studies in Conservation, Vol. 40, no. 1, pp. 31-40, 1995
[20] C. Bisulca, M. Picollo, M. Bacci, D. Kunzelman: "UV-VIS-NIR Reflectance Spectroscopic of red lakes in paintings", 9th International Conference on NDT of Art, Jerusalem Israel, 25-30 May 2008, no. May, pp. 1-7, 2008
[21] M. Picollo, M. Bacci, D. Magrini, B. Radicati, G. Trumpy, M. Tsukada, D. Kunzelman "Modern White Pigments: Their Identification by Means of Noninvasive Ultraviolet, Visible, and Infrared Fiber Optic Reflectance Spectroscopy", Proc.Modern Paints Uncovered, Maggio 16-19, Londra, 2006. Editori: T.J.S. Learner, P. Smithen, J.W. Krueger, M. Schilling. The Getty Conservation Institute, Los Angeles, pp. 118-128, 2007
[22] M. Bacci, M. Picollo, G. Trumpy, M. Tsukada, D. Kunzelman: "Non-invasive identification of white pigments on 20th-century oil paintings by using Fiber Optic Reflectance Spectroscopy", JAIC, Vol. 46, no. 1, pp.27-37, 2007
[23] M. Bacci, M. Picollo: "Non-destructive spectroscopic detection of cobalt (II) in paintings and glass", Studies in Conservation, Vol. 41, no. 3, pp. 136-144, 1996
[24] M. Elias, C. Chartier, G. Prévot, H. Garay, and C. Vignaud, "The colour of ochres explained by their composition," Mater. Sci. Eng. B Solid-State Mater. Adv. Technol., Vol. 127, no. 1, pp. 70–80, 2006.
[25] H. Schweppe and J. Winter, "Madder and Alizarin," Artists pigments. A Handbook of their history characteristics. Volume 3, E. W. FitzHugh, pp. 295–355, Ed. 1997
[26] M. A. Maynez-Rojas, E. Casanova-González, and J. L. Ruvalcaba-Sil, "Identification of natural red and purple dyes on textiles by Fiber-optics Reflectance Spectroscopy", Spectrochim. Acta - Part A Mol. Biomol. Spectrosc., Vol. 178, 5 Maggio, pp. 239–250, 2017.
[27] J. Dunkerton, A. Roy: "The Materials of a Group of Late Fifteenth-century Florentine Panel Paintings", National Gallery Technical Bulletin, Vol. 17, pp. 20-31, 1996

IOP Conf. Series: Materials Science and Engineering **364** (2018) 012062 doi:10.1088/1757-899X/364/1/012062

Field microscopy applied to the understanding of the technology and conservation of wall paintings

E Geddes da Filicaia[1], G Verri[2] and Y Goren[3]

[1] The Courtauld Institute of Art, Florence, Italy
[2] The Courtauld Institute of Art, London, UK
[3] Ben-Gurion University of the Negev, Beer-Sheva, Israel

E-mail address: eugeniagdf@gmail.com

Abstract. Wall paintings are complex, heterogeneous structures open to the environment and composed of multiple layers. Additionally, they are frequently located in remote areas, where basic infrastructure is not available. The causes and mechanisms involved in the complex deterioration phenomena, which wall paintings are often subjected to, are therefore difficult to identify and address. Developments have been made to target conservation issues in situ with non-invasive and portable investigative techniques, however standard practice to investigate stratigraphic information still involves analysing samples in a laboratory setting. In remote sites this approach is inefficient, as in many cases decisions on treatment cannot wait for the results of the laboratory examinations. This may have serious consequences on the integrity of the painting. This paper describes adaptation of classical methods for the in-situ preparation of cross-sections, thin-sections and dispersions utilising custom-made equipment. The methodology allows for a direct examination of samples *in-situ*, using a portable polarizing microscope (field microscopy). The system was adapted and developed to not only capture incident and transmitted light, but also to capture multispectral images (infrared-, ultraviolet-reflected and ultraviolet- and visible-induced luminescence), using external radiation sources. Comparisons with benchtop equipment are made to demonstrate the efficacy and usefulness of the process.

1. Introduction

Incident light microscopy is one of the most useful tools employed in art conservation. A great deal of information may be gained quickly, with a relative small capital investment and without the need for specialist technicians. Microscopy, defined by McCrone as 'the application of any enlarged-image process enabling visualization of objects nominally invisible to the unaided human eye' [1], permits a skilled and trained conservator to identify aspects of original technology or of condition that would otherwise be missed. The polarised light microscope (PLM) takes this process forward, allowing visualisation, interpretation and chemical identification of microscopic objects through their examination between polarising filters. Optical properties not visible in a standard incident light microscope may thus be observed. Materials identifiable with this micro-analytical tool include pigments, fibres, plaster components and alteration products [2].

In the conservation of wall painting, samples are collected from a site, which is often remote and only accessible for investigation for a limited period of time [3]. In broad terms, two types of samples for microscopy are commonly collected:

1. a first sample typology is collected to investigate the original technology (to observe the composition and complexity of paint and plaster layers, for example) and to understand past conservation/restoration interventions (such as the presence of a later varnish or repainting); and
2. a second typology is researched to scrutinize conservation issues and the effectiveness of current conservation interventions (for instance to observe freshly consolidated plaster, or the composition of experimental grouts or of salt species).

Under standard working conditions, and following appropriate visual observations and non-invasive investigations [4], microscopic samples are taken back to a laboratory for examination. Considerable time may pass between the collection of samples and their analysis, with an inevitable lack of continuity in the conservation process. In addition, as the conservation process is iterative in nature [5], questions raised during the analysis cannot be addressed before a new campaign takes place, and new samples are collected [6]. Moreover, in transport friable samples may suffer from mechanical damage due to vibration.

1.1. Cross-sections, thin-sections and dispersions

A wall painting, in the field of conservation, is considered a complex, heterogeneous structure made up of multiple layers. In order to study such an intricate object, it is common practice to collect a minute sample and mount it as a cross-section. This enables a conservator to view the stratigraphy of the wall painting under a microscope, and to gain crucial information on the technology and state of conservation. To obtain a cross-section, the sample is embedded into a mounting medium, commonly an organic resin, and polished on one side until the paint stratigraphy is exposed. Alternatively, the sample may be cut using a microtome. By observing various characteristics of the particles in the paint layers, such as colour, size, it may be possible to preliminarily characterise the pigments [7], as well as to obtain important information on optical and physical properties related to their change and ageing [8]. Histochemical tests may be used to investigate organic binders [9][10].

A thin-section may be obtained from a cross-section adhered to a glass slide and polished to a standard thickness of 30 μm [11][12]. In conservation literature, a range of thicknesses between 5 and 50 μm is reported [13]. This thickness is estimated by comparing the birefringence colours of known minerals within the section to the Michel-Lévy chart. Quartz grains are often added next to the section to be used as a control. Samples prepared in such a way are more commonly analysed under PLM, and have traditionally been used in petrography and micromorphology studies [14], to identify inorganic materials and locate their source, to group items with a common source, or, less frequently, to provide information on conservation treatments and on the state of deterioration [12]. In wall painting conservation, thin-sections are usually prepared for plaster analysis. Observed microscopically and coupled with XRD analysis, mineralogical components and their origin may be identified [15]. Moreover, thin-sections of grouts allow to observe their morphology, particle distribution and, when impregnated with coloured resin, their porosity [16].

Pigment dispersions are created by immobilising pigment particles on a microscope slide with a resin of a known refractive index [2]. Their microscopical observation may aid in the identification of the pigments present and their source, through the study of the combination of their optical properties visible in plane- and crossed-polarised light [17][18]. Pigment dispersions may be especially useful for the differentiation between natural and synthetic pigments [8].

2. Embedding resins for in-situ preparation of samples: characteristics and field constrains

Many embedding materials and equipment for sample preparation are available. In order to select the most appropriate ones, required properties for resins used in the preparation of cross-sections and thin-sections in situ need to be identified. These are based on a literature review of the most frequently used materials and on the field constraints considered below.

2.1. Definition of field constrains and of resin properties

Preparation of samples in situ may be hindered by field constraints, outlined below:

- **Availability of materials**: materials involved should be easily available (dependant on countries);
- **Portability**: equipment must be portable, lightweight, not bulky, and durable;
- **Transportability**: equipment and materials must comply with relevant health and safety regulations for transport. This must be checked with the competent authorities on a case-by-case basis;
- **Ease of use**: the methodology should be relatively simple and reproducible, without the need for highly specialized skills and resources. Ideally, it should be possible for the methodology to be implemented even with scarce resources, such as when water or electricity are not available; and
- **Cost**: cost should be reasonable and necessary material and equipment should be affordable by small private conservation firms/individuals. Ideally, equipment and materials can be sourced in the country where the measurements take place.

The characteristics that should be kept in mind while choosing an adequate embedding material pertain both to its immediate use, so to its behaviour as soon as it interacts with the sample (working properties), and to its performance in the future with respect to the sample, after its initial use and manipulation (performance characteristics). Working properties include:

- **High transparency**: allows visibility of the sample during preparation;
- **Low viscosity**: maximises penetration within the pores of the sample;
- **Fast curing time**: ensures that samples can be analysed within tight time constraints;
- **Simple preparation methodology**: facilitates the preparation of the samples in adverse conditions;
- **Long shelf life**: minimizes costs and resources; and
- **Health and safety**: the resin must be safe to use for the users and the environment.
- Performance characteristics include:
- **Hardness**: a hard resin protects the sample, and allows polishing without sample deformation;
- **Uniformity**: colour uniformity increases reproducibility of the results;
- **Low shrinkage**: minimizes the risk of sample deformation;
- **Cold-curing mechanism**: prevents overheating and subsequent damage of the sample;
- **Low luminescence**: maximises the detection of weakly luminescent compounds in the sample; and
- **High stability**: minimises colour change and deterioration over time, allowing future analysis of the sample.

2.2. *Common embedding resins*

The sensitivity and fragility of samples of cultural heritage materials require the use of cold-curing embedding resins. The process of curing, however, even in cold systems, usually produces heat to different extents associated with the polymerization process. Generally, slow-setting resins produce less heat than fast-setting resins, which can reach temperatures as high as 100 °C [22]. Cold systems can be further classified as thermosetting, (including epoxy and polyester resins), and thermoplastic, (Canada balsam, acrylics and cyanoacrylates) polymers.

Thermosetting polymers are more resistant to solvents, abrasion and temperature, so they are generally considered to perform better [13]. For thin-sections, the resin functions as an adhesive—to hold the sample on the glass slide and to allow manipulation and storage—as well as a consolidant, especially in cases of powdery samples, to preserve its structure during polishing [23]. Epoxies are frequently deemed the best choice due to general performance characteristics [22][13], although polyester resins are often preferred due to their minimal shrinkage, low cost, moderate toxicity, ease of preparation and availability [24]. Both these types are suitable for luminescence imaging [13].

IOP Conf. Series: Materials Science and Engineering **364** (2018) 012062 doi:10.1088/1757-899X/364/1/012062

Identification of pigments in dispersion relies on comparison with known references, frequently prepared with Cargille Meltmount™ , which is by far the most commonly used medium [19][20][21].

3. Method for in-situ sample preparation

Much of the equipment needed for the preparation of cross-sections, thin-sections and dispersions is already portable. For the few exceptions, compact alternatives were sought. For cross- and thin-sections, the main points to consider are the way in which the sample is embedded in the resin, the grinding, polishing and vacuum-impregnation processes.

Preparation of cross-sections in situ follows standard procedures [7], except for the grinding and polishing which will have to be done manually (and evenly) on wet polishing cloths of various grades (120, 500, 1000, 2400, 4000, 8000 and 12000). If the section will be further polished to a thin-section after examination and imaging, some quartz grains must be placed in the sample-holder adjacent to the sample. The grains must be exposed when the cross-section is finalised, then the exposed side may be glued to a microscope glass-slide and further polished on the other side. The section must be frequently checked under cross-polarised light, until the quartz grains reach the colours in accordance with the right thickness (according to the Michel- Lévy birefringence chart). For 30 µm, the quartz should appear beige-grey. The adaptation of laboratory equipment to on-site equipment for dispersion was very straightforward, with the only point to consider being the heating process for the resin. A cup warmer, able to heat above 65 °C in order to melt the resin, was selected as a lightweight and compact alternative to a standard hot plate.

The quality of samples prepared is comparable to that obtained using standard laboratory equipment. The benefits of preparing cross-sections, thin-sections and dispersions anywhere with minimal equipment, allowing a conservator to gain information quickly and effectively, greatly outweigh the longer time required to prepare samples with this method, compared to laboratory preparation.

3.1. Custom-made sample holders

Manual polishing of sections requires more time than with the use of polishing wheels or microtomes [7]. It follows that samples should be embedded in a reduced amount of resin, to decrease the amount of polishing required. Moreover, smaller resin blocks will also reduce the cost, the amount of resin and hardener to be transported and the health and safety concerns[1]. For this study, sample-holders were thus designed and converted in a CAD file using AutoCAD software. They were then 3D printed in acrylonitrile butadiene styrene (ABS), a general-purpose, low-cost plastic material. Finally, they were replicated with silicone moulds, and reproduced with resin (figure 1).

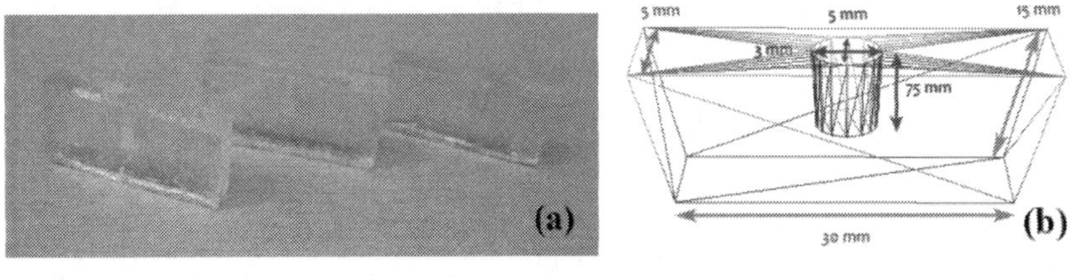

Figure 1. Sample holders in resin (a) and their measurements (b).

The sample-holders are thin rectangular blocks with an elliptical well reaching the middle of the block. The sample is placed in the well, which is then filled with resin. The rationale behind this design is the following:

- its thinness (5 mm) minimises the amount of resin needed for embedding and polishing the sample. The sample-holder is at the same time sufficiently large to be conveniently handled;

[1] Valentine Welsh [VW Fecit] also designed ready-made sample holders (easysections.com). The holders designed for this study allow for the positioning of the sample in the middle of the holder, minimizing the risk of overpolishing.

- the shape and width of the well (3 mm) allows placement of quartz or calcite crystals alongside the sample, needed for the preparation of thin-sections;
- crucially, it allows for the positioning of the sample in the middle of the block, minimising the risk of over-polishing when manually preparing thin sections.
- the sample holders are small and lightweight and can be prepared in the laboratory and carried ready-made on site.

3.2. Vacuum pump and chamber
Thin-sections might require thorough embedding of the sample in the resin, especially for decohesive or very porous samples. This serves to stabilise the sample during its preparation as a thin-section [25]. Samples are thus frequently embedded under vacuum [14]. Whether a vacuum is desirable, or required, also depends on the fluidity of the resin chosen. Vacuum removes air from the pores in the sample, and also the bubbles that might have formed within the resin. The more viscous the resin, the easier it is for air bubbles to remain trapped. If present, they may obstruct the observation of the sample: with resins of lower hardness than epoxy, air bubbles may be enlarged during grinding due to the pressure created within them by loose resin particles, which may also remain trapped in the air bubbles. For our purposes, a high-level of vacuum such as that achieved by a rotary vane pump used in laboratories is not required, thus it was deemed suitable to use a hand-held vacuum pump (used to bleed breaks in cars and motorcycles). This pump is able to reach maximum 760 mmHg pressure, about 1 bar. A small, modified Tupperware container was initially used as a vacuum chamber [14], but an economical mini-vacuum desiccator was later found to perform better (figure 2).

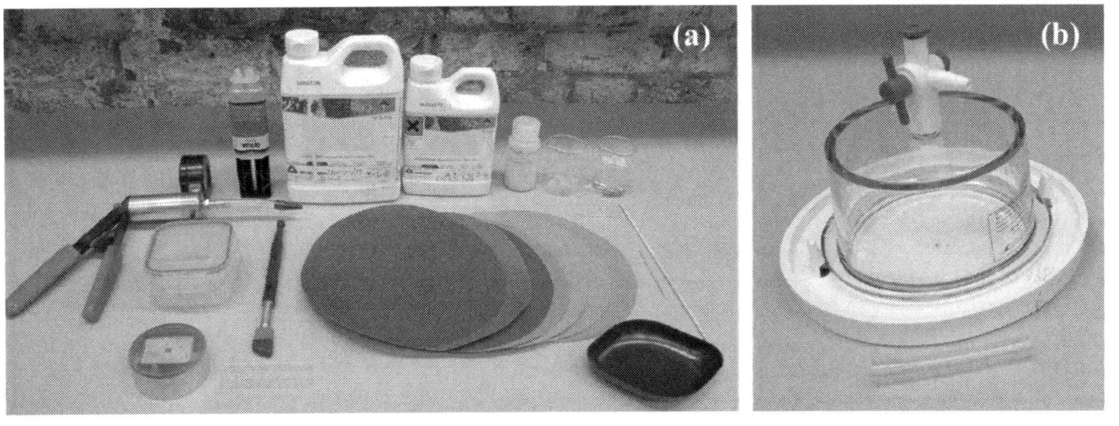

Figure 2. Materials required for in-situ preparation of cross- and thin-sections (a), including from left and clockwise: vacuum-pump and Tupperware used as a vacuum-chamber, epoxy glue to adhere the cross section to the polishing slide, resin to embed the section, quartz grains to determine the thickness of the thin-section, polishing cloths, glass-cutter to cut regular glass slides to size to fit into the thin-section holder. In (b), a mini-vacuum desiccator, which connects to the vacuum pump in (a).

4. Examination of samples: comparison between portable and benchtop microscopes
Once the samples are prepared, they can be examined in situ with the Goren-Pol microscope. This portable polarising microscope is principally designed for observation of samples in transmitted light; however, incident illumination can easily be provided externally by placing a torch on a stand. A high-end benchtop Leica DMRD was utilized to compare the quality of the examination and imaging of samples obtainable with the Goren-Pol. By removing the left eyepiece, a camera may be fitted on the Goren-Pol, to allow capturing of images both in incident and transmitted light configurations.

IOP Conf. Series: Materials Science and Engineering **364** (2018) 012062 doi:10.1088/1757-899X/364/1/012062

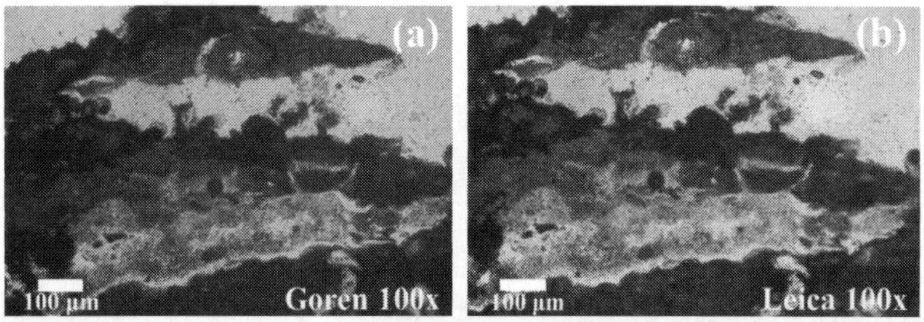

Figure 3. Plane-polarised light image of a thin-section (a and b). Degradation of plaster due to salts. The sample was first mounted as a cross-section (c).

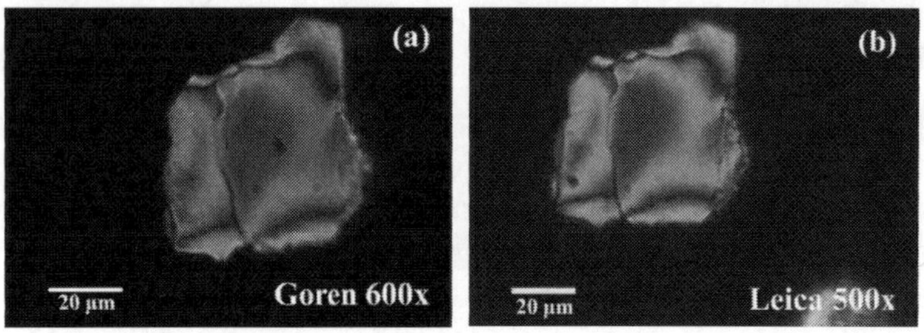

Figure 4. Particle of azurite imaged in cross-polarised light with the Goren-Pol (a) and Leica (b), at 600x and 500x respectively.

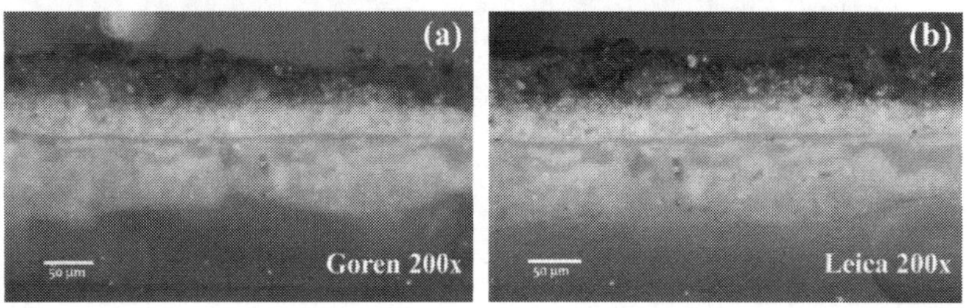

Figure 5. Incident light images of a cross-section of ultramarine.

Transmitted light: In general, examination with the Goren-Pol yields excellent results, comparable in quality with the Leica, especially in transmitted light (figures 3 and 4). The light source of the Goren-Pol (LED) is whiter than that of the Leica (halogen) and therefore provides a better colour rendition of the various phenomena used for the identification of materials. Incident light: Examination with both microscopes in incident light is limited to about 200-400x. While very flat samples yield better results, in general, images in incident light captured with the two microscopes can be different due to the direction of the incident light and the difference in type of source. The light for the Goren-Pol, being external, reaches the sample at a different angle than it does with the Leica, equipped with and internal source (figure 5). This may result in a more raking view of the sample, depending on the working distance of the objective. Nonetheless, incident light examination of thin- sections and dispersions is often clearer with the Goren than with the Leica, especially at high magnifications, as images appear less hazy and better defined, even if slightly raking.

5. Adaptation of the Goren-Pol to Multispectral Imaging (MSI)

The Goren-Pol microscope was adapted for multispectral imaging of microscopic samples, easily done by using different radiation sources, filters and modified cameras. Set-ups for imaging of cross-sections in infrared-reflected (IRR), visible-induced infrared luminescence (VIL) and ultraviolet-induced visible luminescence (UVL) are briefly described below (Figure 6). Further avenues of investigation include using more advanced and sensitive cameras and a more specific system of filters.

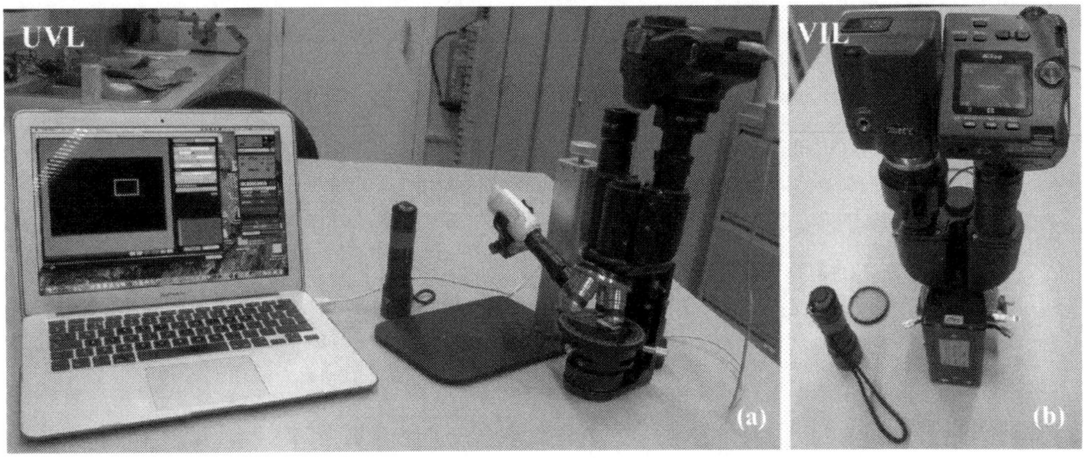

Figure 6. Set-up with the Goren-Pol: (a) a filtered UV torch (5W, 365 nm, Schott UG11 filter) illuminates the sample; the camera, equipped with a B&W420 filter, is connected to a computer through tethering software. In (b) a standard filtered visible torch (400-700 nm, 240 lm, Schott BG38 filter) excites the sample with visible light, re-emitted in the IR range as seen on the camera screen. The camera is equipped with a Schott RG830 filter.

5.1. Ultraviolet-induced luminescence imaging in the visible range (UVL)

The Goren-Pol requires an external UV light source for UVL observation and imaging, while the Leica is equipped with an internal one. A 5W LED UV torch with an emission centred at 365 nm was used for this research as the external UV radiation source. A Schott UG11 excitation filter was positioned in front of the torch, to remove the unwanted violet-blue radiation. The UV radiation excites luminescence from the sample, which can be observed through the eyepiece. For visual observation, the eyepieces must be equipped with an emission UV-blocking filter, in order to block the reflected UV radiation and avoid damage to the unprotected eye. During image capture, a B&W420 emission filter was placed between the relay lens and the camera sensor. Difficulties in focusing weakly fluorescent samples can be overcome by using a visible LED torch.

Up to 200x, excellent results were obtained with the Goren-Pol, only requiring slightly more adjustment of exposure to obtain accurate images comparable to those of the Leica (figure 7).

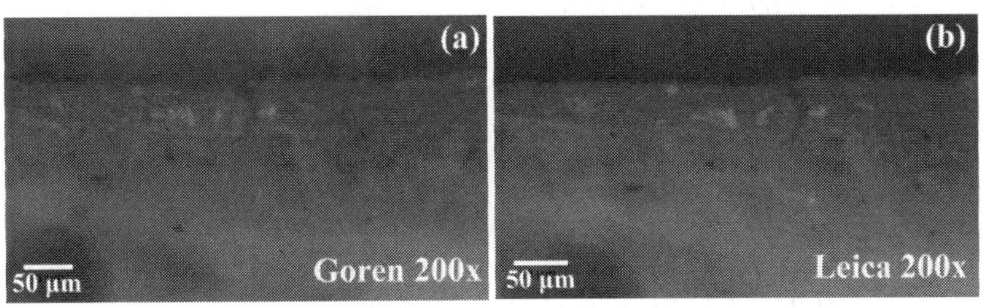

Figure 7. UVL image of a cross-section with a fluorescent paint layer containing particles of bright yellow (Indian yellow) [26].

5.2. Infrared-reflected (IRR) and visible-induced infrared luminescence (VIL) imaging

When equipped with a set of filters and a modified CCD camera, sensitive in the IR range, the Goren-Pol microscope can capture images in the IR range. A Nikon Coolpix 995, following the removal of the inbuilt UV&IR-blocking filter, and equipped with an IR cut-on filter (Schott RG715 or RG830), was used to capture images in the IR range. The function of this filter is to allow IR radiation to reach the sensor and to block any visible light.

For IR-reflected imaging, a 5W IR torch, centred at 850 nm, was used as an external radiation source. For visible-induced luminescence imaging, a visible LED torch, equipped with an excitation IR-

IOP Conf. Series: Materials Science and Engineering **364** (2018) 012062 doi:10.1088/1757-899X/364/1/012062

blocking filter Schott BG38, was used as a source of visible light. The visible light is absorbed by luminescent compounds and re-emitted in the IR range.

Figure 8 shows the IRR and VIL images of a cross-section of a sample from a 4th Century BCE temple in Cumae, Naples, Italy, and consists of a calcium carbonate ground and a paint layer containing particles of Egyptian blue. This pigment is known to emit IR radiation when excited in the visible range [27][28]. Focusing, for both reflected and luminescence images, was achieved using the 5W IR torch.

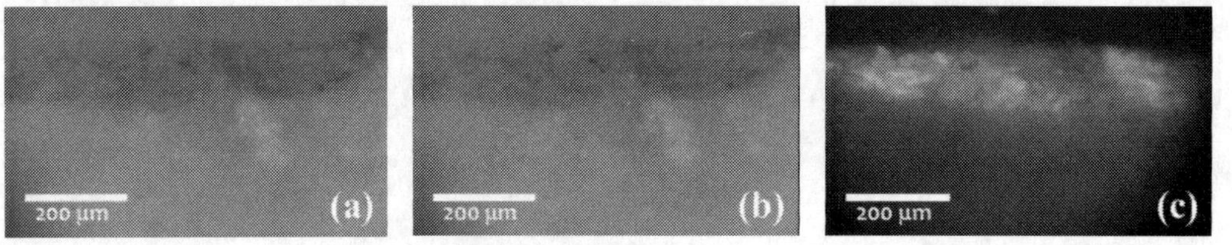

Figure 8. Visible (a), IRR (b), and VIL (c) images of a cross-section containing Egyptian blue.

6. Conclusion

Preparation and examination of cross-sections, thin-sections and dispersions for the conservation of wall paintings is more often carried out off-site, in a laboratory with state-of-the-art, but expensive equipment. However, this discipline would greatly benefit from the possibility of carrying out this type of examination in situ, using portable, low-cost and versatile equipment that may be tailored for specific needs. The present paper illustrates the possibility of achieving this by modifying standard preparation procedures to suit the examiner's needs, designing custom-made equipment and by using a suitable portable microscope, such as the recently developed Goren-Pol. Comparatively to a state-of-the-art Leica DMRD, the results obtained with the Goren-Pol are excellent, bearing in mind its compact size and much lower cost. The set-up could easily and inexpensively be altered, to examine samples in incident light, UV and IR radiation.

Acknowledgements

The authors wish to thank Professor Sharon Cather at The Courtauld Institute of Art, London, for her support and for suggesting to 3D print the sample holders.

References

[1] McCrone W C 1994 *Journal of the American Institute of Conservation* **33** (2) 11, 14

[2] McCrone W C 1982 *Journal of the International Institute for Conservation - Canadian Group* **7** (1-2) 11-4

[3] Cather S 2004 *2nd Int. Conf. on the Conservation of Grotto Sites (Dunhuang)* ed N Agnew (Los Angeles: The Getty Conservation Institute) pp 22-32

[4] Pique' F and Verri G 2015 *Project Report: Organic Materials in Wall Paintings* (Los Angeles: The Getty Conservation Institute) p 13

[5] Cather S 2004 *Proc. Int. Sem. Diagnosis, Conservation and Restoration of Far Asian Mural Paintings (Ravenna)* ed R Mazzeo pp 28-29

[6] Hunt P N and Griffiths D R 1989 *World Archaeology Ceramic Technology* **21** (1) 165-72

[7] Plesters J 1955-6 *Studies in Conservation* **2** 110-13

[8] Silva C L 2006 'A Technical Study of the Mural Paintings of the Interior Dome of the Capilla de la Virgen del Rosario, Iglesia San José, San Juan, Puerto Rico', *Theses (Historic Preservation)* pp 16, 40

[9] Messinger J M 1992 *Journal of the American Institute of Conservation* **31** (3) 267

[10] Magrini D, Bracci S and Sandu I C A 2013 *Procedia Chem.* **8** 194

[11] Gribble C and Hall A 1995 *Optical Mineralogy: Principles and Practice* (London: UCL Press) p 32

[12] Reedy C L 1994 *Journal of the American Institute of Conservation* **33** (2) 115

[13] Sandu I C, Schafer S, Magrini D, Bracci S and Roque C A 2012 *Microsc. Microanal.* **18** (4) 862-3
[14] Goren Y 2014 *New York Microscopical Society Newsletter* pp 2, 3, 7, 10
[15] Zizola C 2005 La conservazione e il restauro delle pitture murali della Madrasa Amiriya, Radà, Yemen *Centro di Conservazione Archeologica* (Roma) 2
[16] Griffin I 1997 'Pozzolanas as additives for grouts: an investigation of their use in wall painting conservation' *Unpublished dissertation* (London: The Courtauld Institute of Art) p 37
[17] Santopadre P and Verita' M 2006 *Studies in Conservation* **51** (1) 32
[18] Gaetani M C, Santamaria U and Seccaroni C 2004 *Studies in Conservation* **49** (1) 16
[19] Berrie B H, Leona M and McLaughlin R 2016 *Heritage Science* **4** (1) 8
[20] Yong L and Shiwei W 2014 *Studies in Conservation* **59** (5) 315
[21] Stoner J H and Rushfield R 2013 *Conservation of Easel Paintings* (Oxford: Taylor & Francis) p 315
[22] Bousfield B 1992 *Surface preparation and microscopy of materials* (Chichester: Wiley) pp 21, 167
[23] Pouyet E, Lluveras-Tenorio A, Nevin A, Saviello D, Sette F and Cotte M 2014 *Anal. Chim. Acta* **822** 53
[24] Derrick M, Souza L, Kieslich T, Florsheim H and Stulik D 1994 *Journal of the American Institute of Conservation* **33** (3) 228
[25] Sabatini G, Giamello M, Pini R, Siano S and Salimbeni R 2000 *J. Cult. Heri.* **1** S13
[26] Tamburini D, Martin de Fonjaudran C, Verri G, Accorsi G, Acocella A, Zerbetto F, Rava A, Whittaker S, Saunders D and Cather S 2018 *Microchemic. J.* **137** 242
[27] Verri G 2009 *Anal. Bioanal. Chem.* **394** 1011
[28] Aramini F, Sidoti G and Santopadre P 2013 *Bollettino ICR* **27** 20

IOP Conf. Series: Materials Science and Engineering **364** (2018) 012063 doi:10.1088/1757-899X/364/1/012063

A versatile system for *in-situ* speckle and thermography-based diagnostics of artifacts

L Perlini[1], D Ambrosini[2,3] and C Daffara[1,3]

[1] University of Verona, Dept of Computer Science, Strada Le Grazie 15, Verona (VR), Italy
[2] University of L'Aquila, DIIIE, Piazzale Pontieri 1, Monteluco di Roio (AQ), Italy
[3] CNR - Institute of Applied Science & Intelligent Systems, via Campi Flegrei 34, Pozzuoli (NA), Italy

E-mail: `luca.perlini@univr.it`

Abstract. Holographic nondestructive techniques already showed their effectiveness in detecting structural defects on or beneath the surface of historical artifacts, not emerging from conventional visual inspections. Unfortunately, such tools are difficult to be operated out of the stability conditions guaranteed by an optics laboratory. The peculiar features of many works of art, however, make often impracticable their transport to dedicated facilities. In this work, a compact, stable, portable holographic system was designed and implemented: the resulting apparatus allows *in-situ* acquisitions, even in non-laboratory conditions, and provides rapid and clearly interpretable diagnostic information by the completely non-invasive procedure of speckle pattern photography (SPP). Complementary data are furnished by an integrated infrared thermography system. A map of the artifact surface can be obtained, representing the distribution of defects and their extension. The technique will hopefully help or replace conventional methods, more invasive or less systematic (visual inspections, surface tapping), still largely used for the same purposes.

1. Introduction

Historical artifacts undergo, with time, structural alterations induced by aging and by the interaction with the environment [1, 2]. Daily and seasonal variations of temperature and humidity expose every artwork to cyclic stresses which, associated to the progressive weakening of the constitutive materials, can lead to irreversible damage on or beneath the surface of the specimen.

Artistic supports presenting highly layered structures, such as paintings or *frescoes*, are especially susceptible to the onset of detachments, voids, microcracks and delaminations, in consequence of the decay process. A detailed evaluation of aging-induced defects and a monitoring of their evolution are required to guide the restorers' work and to prevent major damages, especially in the presence of possible drastic triggering events, like earthquakes.

Beside conventional investigation techniques (visual inspections, surface percussion, etc.), still largely used to this aim, a number of systematic, reliable and non invasive tools are provided by optical techniques [3, 4].

Surface holography, and in particular speckle-based methods, already showed their effectiveness in detecting sub-superficial defects – thus invisible to the naked eye – on different types of supports [2, 5]. In addition to the advantageous whole-field, high resolution and non-contact characteristics,

IOP Conf. Series: Materials Science and Engineering **364** (2018) 012063 doi:10.1088/1757-899X/364/1/012063

such techniques include algorithms for the digital analysis of images [6] providing a clearly interpretable description of the state of conservation of the artwork in rapid times.

The greatest obstacle left to the systematic application of such techniques is represented by their high sensitivity to external disturbances, like vibrations. The optimal conditions of mechanical stability are hardly reached and maintained, except that in laboratory situations [2]. Since, however, the nature of many artifacts makes inadvisable or impossible their transport to dedicated facilities, the research in this field is now addressed to the problem of allowing *in-situ* acquisitions.

Speckle digital methods, like electronic speckle interferometry (ESPI) [6, 7] and speckle photography (SPP) [7, 8], are able to meet this need.

In this article, a particularly simple, compact, easy-to-transport and accessible speckle-based setup was implemented, characterized and optimized for the research of sub-superficial detachments on paintings and *frescoes*.

Considering the complex structure of works of art, a number of studies [9–12] suggested the integration of non-invasive, complementary and independent tools, in order to reach complete diagnostic information. The proposed setup was thus integrated with a compact infrared thermography system, acquiring data from the sample at the same time as the speckle system.

2. Basic Principles

A detached region beneath the surface of a painting does not emerge from a simple visual inspection, but can reveal its presence when the artwork is thermally stressed. In correspondence of the detachment, the lack of adhesion reduces the heat dispersion rate and slows down the return to equilibrium conditions. This translates into two observable surface effects: an anomalously high temperature in that location; and an irregular deformation field, both in intensity and in time persistence.

2.1. Speckle Methods

Speckle effect arises when a coherent radiation field hits a rough diffusing surface [8, 13]. Many secondary spherical wavelets are back-scattered, and the optical interference between different contributions creates a granular pattern on the surface. The geometry of the speckles strongly depends on the imaging system used for the observation [7, 8].

A relation holds between such pattern – which can be described in statistical terms [13] – and the micro-morphology of the surface. In particular, every deformation of the latter results in a modification of the corresponding speckle pattern. Such correspondence is maintained down to displacements in the order of multiples of the radiation wavelength, namely to the micrometer scale. Moreover, a "locality" principle is satisfied [14]: a local deformation of the surface causes a local deformation of the speckle pattern in that precise location.

Such properties constitute the basis of speckle metrology, which investigates shape and deformations of a surface by acquiring and comparing speckle patterns in sequence. Two are the principal approaches, both based on digital acquisition and analysis procedures [6].

Speckle interferometry (ESPI) involves a two-beam configuration analogue to the traditional holographic interferometry, and found advantageous application in the conservation field [15–20]. However, it requires some stability conditions and the presence of an optics-skilled specialist to be operated [21].

The information about some particular components of the surface displacement field is accessible even in the absence of the reference beam [8]. In this case, the speckle pattern generated by the object beam alone, which is called a specklegram, is acquired. Comparative correlation analyses of successive specklegrams from the same surface, before and after an external *stimulus*, allow to measure the involved deformation field. Such one-beam option, called Speckle Pattern

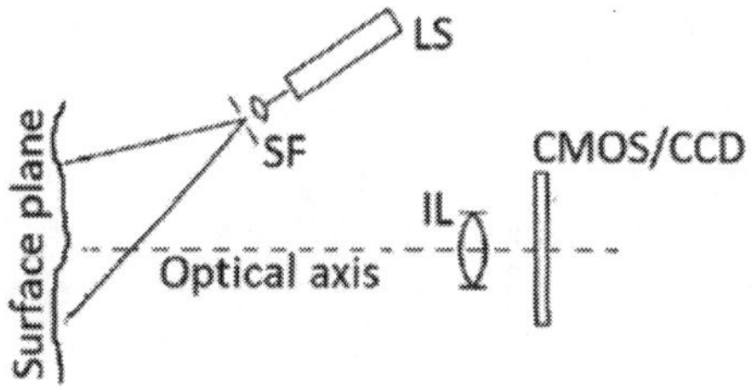

Figure 1: Schematic SPP setup. LS: laser source; SF: spatial filter; IL: imaging lens.

Photography (SPP), is especially sensitive to in-plane displacement components and local tilting [8], within the range $10^1 \div 10^3$ µm.

Despite these limits, the final setup, the acquisition procedure and the stability requirements turn out to be very simplified, if compared to ESPI, making conceivable a compact dedicated apparatus, suitable for *in-situ* campaigns and operable by non-scientists too. A scheme of a typical SPP setup is reported in Figure 1.

Laser speckle is not limited to reflected light geometry: the same effect arises when coherent light is transmitted through a diffuser medium, such as a ground glass [13]. In the image of such medium, projected on the inspected surface, strong fluctuations in intensity are observed, caused by the superposition of secondary wavelets generated and dephased inside the diffuser.

2.2. Infrared Thermography

A radiometric system measures the energy of the electromagnetic radiation emitted by a surface and correlates it to its temperature, following the relationship between the two quantities provided by Plank's blackbody theory. In the case of objects around the room temperature – the ones which conservation deals with – the theory provides the emission peak to fall in the far-infrared (FIR) interval ($7 \div 14$ µm).

The employment of such technique as a diagnostic tool is made possible by the fact that the actual thermometric response of a real object, like an artifact, is influenced by a number of properties: spectral (emissivity, absorption, reflection), thermal (conductivity, diffusivity, specific heat) and general physical conditions, including but not limiting to porosity, water content and surface treatment. Ultimately, the presence of a structural defect alters locally one or more of such properties, and hence the emission regime. It is thus possible to detect and map the thermal variations of a work of art and correlate them to the material conditions, to its microstructure and to the morphology of its surface [1].

In the so-called *active* approach [22], an external source, generally in the form of a pulsed heat *stimulus*, is used to create thermal contrast. Again, a detached region is associated to a lower heat transfer rate and a higher temperature with respect to the surrounding regions, and it is revealed as a thermal discontinuity on the artifact surface.

Among the advantages offered by IRT, non-contact and full-field features must be mentioned, together with rapid acquisition times and clearly interpretable results [22]. On the other hand, the difficulty to obtain a uniform thermal solicitations of wide areas and the uncertainty about the emissivity and reflectance of composite materials are the most critical points to deal with.

3. Experimental Setup

The proposed setup is schematically reported in Figure 2. Its design favors compactness and modularity, and allows fine components adjustments without affecting the alignment of non-involved parts. In the present study, the system was optimized for speckle photography measurements (in one-beam geometry) together with thermography.

Speckles produced in transmission configuration are largely exploited in the present work, since they lead to some practical advantages: the speckle field is projected onto a very wide area (in

IOP Conf. Series: Materials Science and Engineering **364** (2018) 012063 doi:10.1088/1757-899X/364/1/012063

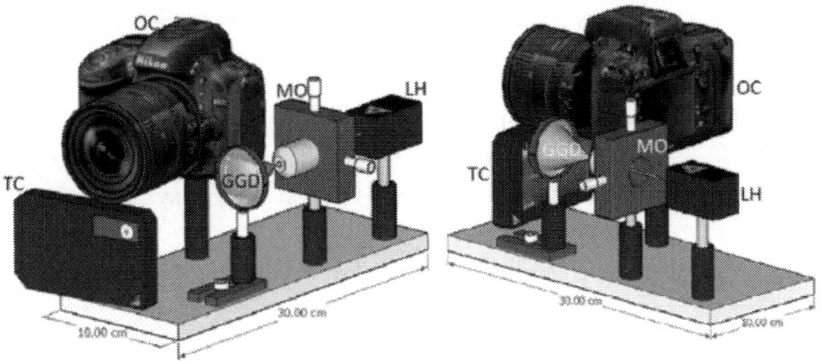

Figure 2: 3D model of the proposed apparatus. LH: laser head; MO: microscope objective; GGD: ground glass diffuser; OC: optical camera; TC: thermal camera. Approximate total weight: 3 kg. The supporting optical breadboard can be mounted on a stiff photographic tripod.

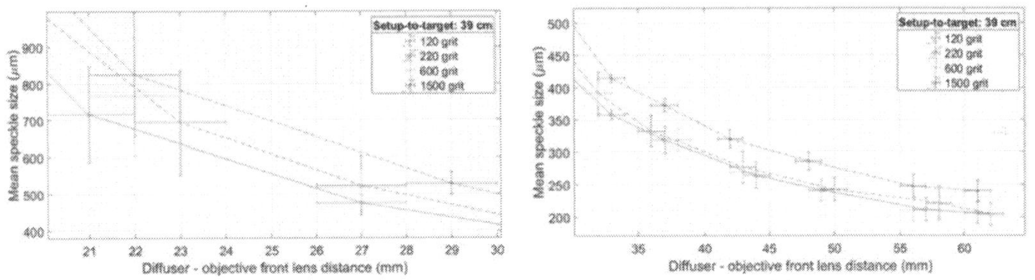

(a) (b)

Figure 3: Mean speckle dimension as a function of the diffuser grain and of its distance from the front element of the microscope objective. The calibration curve also depends on the distance between the sample and the diffuser: in this case, such parameter is 39 cm.

the order of $1\,\mathrm{m}^2$) in a uniform way and it is barely subject to distortions or aberrations induced by the beam-expander. Moreover, the mean speckle size is generally increased, decreasing the sensitiveness to displacement but relaxing the resolution requirements for the acquisition phase.

The coherent light source is provided by a DPSS laser with output wavelength of 532 nm and maximum power of 125 mW. The chosen operative power depends on the background illumination conditions and on the nature of the sample: a number of tests suggested powers in the range $60 \div 90$ mW.

Sensitiveness to displacements is reached even if the laser beam impinging on the surface is divergent [8]. Therefore, an Olympus Plan-N 10x/0.25 microscope objective is used as a beam expander: this solution guarantees very high enlargement and allows to illuminate a wide area.

A ground glass diffuser is placed after the objective to operate in transmission geometry. The distance between these two components determines the speckle dimension (and thus, for the greatest part, the minimum spatial resolution): in order to have such dependence under control, the mean speckle size was measured for varying diffuser grain level and distances from the objective front lens. As observable from the graphs in Figure 3, both factors concur to the final speckle size, which is included, for this particular setup, in the range $150 \div 700\,\mu\mathrm{m}$.

The three components addressed to surface illumination are fixed on a 10×30 cm optical breadboard, together with a CMOS camera (Nikon D810, $7360 \times 4912 \times 14$ bits) mounting a 50 mm imaging lens with aperture in the range from f/9 to f/11. A green band-pass filter on the front part of the lens limits the background light collected on the sensor. The camera, used to acquire the specklegrams of the surface before and after the thermal stress, is remotely controlled by computer – as well as the laser source – in order to avoid direct interventions of the operator during the acquisition.

A bolometer-based thermal camera (FLIR C2) is mounted on the breadboard, just in front of the optical camera (see Figure 2). This model of infrared camera is characterized by a very compact design, which makes it suitable to be integrated in the system without affecting its overall geometry. The device provides access to the raw radiometric data, allowing the correct estimation of temperature also in the complex situation of surfaces with heterogeneous emissivity or in presence of reflections. Moreover, it can be operated in streaming mode with a computer via USB connection.

The whole setup is protected from mechanical disturbances and vibrations by a set of sorbothane feet.

The resulting system can be laid on a scaffolding plane; alternatively it can be mounted on a stiff photographic tripod, at variable distance from the inspected surface. After some tests, the optimal distance for acquisitions of projected speckles was estimated about 40 cm, corresponding to a 20 cm diameter ROI.

The thermal *stimulus* is provided by a 750 W infrared tungsten lamp pointed toward the front surface of the artifact (the inspected one) from a distance of 40 cm. The collected data are referred to long-pulsed solicitations of variable duration, between 5 and 30 s.

4. Digital image processing

The local surface displacement is deduced from that of the two speckle patterns, acquired before and after the external solicitation. The comparison between them is performed by computer algorithms, usually based on cross-correlation calculations. Basic assumption of the diagnostic process is that defective subregions are identified by local irregularities in the direction or in the intensity of the average displacement field. In particular, since a SPP setup is especially sensitive to in-plane deformations, such regions correspond to places in which the initial speckle pattern undergoes an anomalously high translation, when compared to the surrounding. Within this region, the correlation between the two specklegrams drops.

Our analysis relies on the adaptation of two Matlab scripts, based on quite different approaches. In both cases, the raw speckle images are pre-processed for contrast enhancement, exposure fluctuations and vibrations compensation and for cutting any residual contribution from background radiation. Since we operate in green laser light illumination, the red and blue channels of every image are put to zero.

The Speckle Correlation (SC) method consists in the evaluation of a local parameter, estimating the correlation of the speckle patterns after the modification of a surface element: it was applied, in the past, during similar studies on different artistic supports (wooden paintings, mosaics and *frescoes*). Calling, respectively, $u_{\text{ref}}(x,y)$ and $I_{\text{ref}}(x,y)$ the field amplitude and intensity relative to the surface in its initial state, and $u_{\text{mod}}(x,y)$ and $I_{\text{mod}}(x,y)$ the same quantities after the solicitation, a local correlation coefficient is defined as [13, 23]

$$\rho(x,y) = \frac{|\langle u_{\text{mod}}(x,y)u^*_{\text{ref}}(x,y)\rangle|^2}{\langle I_{\text{mod}}(x,y)\rangle\,\langle I_{\text{ref}}(x,y)\rangle} \tag{1}$$

Under some statistical assumptions about the signal and about the linear response of the camera, it was demonstrated [13] that $\rho(x,y)$ can be obtained from the spatial average of the squared arithmetic difference of the two specklegrams, according to [23]

$$\rho(x,y) = 1 - \frac{\langle[I_{\text{ref}}(x,y) - I_{\text{mod}}(x,y)]^2\rangle}{\langle I^2(x,y)\rangle} \tag{2}$$

where the angle brackets denote a spatial local average, executed over an area containing many speckles. Eq. (2) ensures that subtracting the displaced specklegram from the reference one, squaring the result, averaging over many speckles and dividing by a normalization factor, the complement-to-one of the correlation parameter is obtained. Thus, in the resulting image, bright areas represent regions of higher decorrelation, associated to strong in-plane displacement fields, and thus to the presence of detached layers or other types of defect. Practically, local averaging is effected by a discrete convolution between the gray-scale image and a matrix of ones, dimensioned such to contain a statistically significant number of speckles.

A quantitative geometrical reconstruction of displacement can instead be reached by adapting Particle Image Velocimetry (PIV) tools [9]. One example is provided by MatPIV, developed in the Matlab environment [24]. It receives as input the two specklegrams, which are divided into sub-images of dimensions set by the user. Every sub-region from the undisplaced image is associated to the corresponding one from the displaced image through a cross-correlation calculation, and an average in-plane displacement vector is constructed in each window. The FFT algorithm is exploited in the calculation to speed up the running time, which remains in the order of some minutes even for inexpensive computers. A detailed description of the software functioning is provided in its documentation (see [24]). In our case, cross-correlation is calculated with one iteration through the images and a 50% overlap of the interrogation windows.

5. Results and Discussion

In order to evaluate the effectiveness of the proposed setup in detecting hidden detachments, a plausible diagnostic procedure was carried out on a specimen with known structure. A sample model of $60 \times 40\,\mathrm{cm}$ was realized, reproducing the layered structure typical of Renaissance paintings. The choice of materials and fabrication methods was guided by *Il libro dell'arte* by Cennino Cennini, one of the main treatises about the sixteenth-century paintings techniques. At various levels of the pictorial support – based on linen stripes fixed on a poplar table by a series of successive hands of Bologna chalk and rabbit-skin glue mixture – a number of plastic leafs were inserted, in order to impede adhesion between adjacent layers and simulating the effect of aging-induced detachments. Insertions differed from each other in dimension, position and depth.

The intensity of the external thermal *stimulus* was also varied. The long-pulse duration of the IR source was progressively increased from 5, to 15, to 30 s: the superficial temperature of the sample was measured by the thermal camera in real-time. As recommended for these type of solicitation on very fragile artifacts [3, 16], the temperature gradient did not exceed values of $4 \div 5\,^{\circ}\mathrm{C}$. The post-*stimulus* specklegrams were acquired both at the top of the heating and during the cooling phase, respectively after 5 and 65 s from the lamp switching-off.

Some results are shown in Figure 4. When comparing the reference specklegram with the one acquired during the cooling phase (65 s after the lamp switching-off), the SC-based software gives an account of a bright area – and thus very decorrelated – correspondent in position and dimension to the defective one (Figure 4c). Other partially decorrelated regions are detected, not associated with the presence of underlying plastic sheets: it is possible for such "false positives" to arise as a consequence of a non-uniform external solicitation or of an unexpectedly unstable part of the sample. No comparable effectiveness seems to be obtained when acquiring the displaced specklegram at the top of the heating phase, as shown in Figure 4b, in which no contrast enhancement could be performed.

Figures 4d and 4e show the results of the MatPIV-based analysis. The script was adapted to visualize only the local displacement outliers, i.e. the subregions in which the deformation underwent abrupt direction or intensity changes with respect to the surrounding and to the average of the ROI. The same area as before is recognized as defective, presenting a very high concentration of anomalous displacement vectors. In general, a wide irregular field is reconstructed all over the bottom part of the investigated area: the most likely explanation is analogue to that

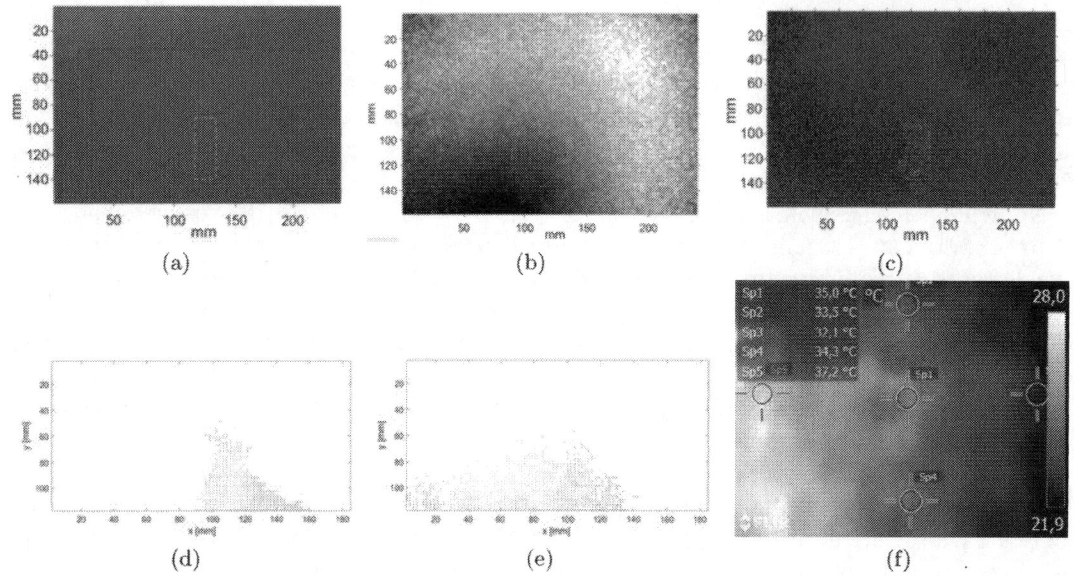

Figure 4: (a): Investigated ROI. The dashed rectangle locates the detachment, the solid red one indicates the region analyzed by MatPIV. Laser output power: 60 mW; exposure time: 1/5 s; aperture: f/11.0; surface temperatures (pre-*stimulus*, 5 s post-*stimulus*, 65 s post-*stimulus*): $T_0 = 23.5 \pm 0.5\,°\mathrm{C}$, $T_1 = 26.7 \pm 0.5\,°\mathrm{C}$, $T_2 = 25.5 \pm 0.5\,°\mathrm{C}$. (b)-(e): SC and MatPIV speckle displacement data relative to a thermal *stimulus* of 5 s duration. Displaced specklegram acquisition: (b),(d): 5 s and (c),(e): 65 s after the lamp switching-off. (f): Thermographic data relative to a 30 s-lasting *stimulus*.

provided for the SC analysis.

Two indications can be extracted from the data. First, that the presence of a defect is most likely to emerge when a "response time" is waited after the end of the *stimulus*, before acquiring the displaced specklegram. This fact, already established by similar studies [3], is explained by recalling that, during the redistribution of thermal energy from the surface to the bulk, detached areas undergo anomalous heat accumulations, and thus deformations more persistent in time. Maximum contrast is then generated, during the cooling phase, with the regular regions, which instead return faster to the equilibrium situation. Second, that a more intense *stimulus* is not necessarily associated to higher defects detectability; a fact that is useful to further reduce the potential invasiveness.

Thermographic data from the same ROI are presented in Figure 4f. The detachment emerges, as expected, as a region with higher temperature than the surrounding.

6. Conclusions

In this work a speckle imaging setup for the diagnostics and evaluation of historical paintings was proposed, characterized by portability, modularity and suitability for *in-situ* measurement campaigns. The adopted hardware and software solutions provided a tool sensitive to the presence of hidden sub-superficial detachments, but at the same time barely influenced by the main noise sources (vibrations and fluctuations in background radiation).

Placing side by side a speckle and a compact thermographic system optimized the measurement procedure: in fact, by exploiting an external thermal stress it was possible, during a unique cycle, to obtain complementary and independent diagnostic data from the two techniques, as well as to monitor in real time the surface temperature.

IOP Conf. Series: Materials Science and Engineering **364** (2018) 012063 doi:10.1088/1757-899X/364/1/012063

Preliminary tests proved the system to be characterized, making use of some medical terminology, by good diagnostic sensibility when applied to the research of detachments on painting-like supports. Specificity can instead be improved, for example by ensuring a very uniform thermal solicitation all over the inspected surface. A full understatement of the potentialities of the system and a characterization of the optimal measure protocol will however require other data, for example from the application to artistic supports other that paintings.

References

[1] Moropoulou A, Labropoulos K C, Delegou E T, Karoglou M and Bakolas A 2013 *Construction and Building Materials* **48** 1222–1239
[2] Ambrosini D and Paoletti D 2004 *Studies in Conservation* **49, sup. 1** 38–48
[3] Fotakis C, Anglos D, Zafiropulos V, Georgiou S and Tornari V 2006 *Lasers in the preservation of cultural heritage: principles and applications* (CRC Press)
[4] Zhu Y K, Tian G Y, Lu R S and Zhang H 2011 *Sensors* **11** 7773–7798
[5] Hinsch K D 2007 *Oscillation, Waves and Interaction* 259
[6] Rastogi P K 2000 *Digital Speckle Pattern Interferometry and Related Techniques, by PK Rastogi (Editor), pp. 384. ISBN 0-471-49052-0. Wiley-VCH, December 2000.* 384
[7] Cloud G 1998 *Optical methods of engineering analysis* (Cambridge University Press)
[8] Jones R and Wykes C 1983 *Holographic and Speckle Interferometry* second edition ed (Cambridge University Press)
[9] Ambrosini D, Paoletti D and Galli G 2008 *Lasers in the Conservation of Artworks, Proceedings of the International Conference LACONA* vol 7 pp 399–405
[10] Hinsch K D, Gülker G and Helmers H 2007 *Optics and Lasers in Engineering* **45** 578–588
[11] Ambrosini D, Daffara C, Di Biase R, Paoletti D, Pezzati L, Bellucci R and Bettini F 2010 *Journal of Cultural Heritage* **11** 196–204
[12] Daffara C, Parisotto S and Ambrosini D 2017 *Optics and Lasers in Engineering* (In Press, Corrected Proof)
[13] Goodman J W 2007 *Speckle phenomena in optics: theory and applications* (Roberts and Company Publishers)
[14] Shih Y C, Davis A, Hasinoff S W, Durand F and Freeman W T 2012 *Computer Vision and Pattern Recognition (CVPR), 2012 IEEE Conference on* (IEEE) pp 33–40
[15] Albrecht D, Franchi M, Lucia A C, Zanettaa P M, Aldrovandi A, Cianfanelli T, Riitano P, Sartiani O and Emmony D C 2000 *Journal of Cultural Heritage* **1** S331–S335
[16] Tornari V 2007 *Analytical and bioanalytical chemistry* **387** 761–780
[17] Hinsch K D, Zehnder K, Joost H and Gülker G 2009 *Journal of Cultural Heritage* **10** 94–105
[18] Lucia A C, Zanetta P M and Facchini M 1997 *Optics and lasers in Engineering* **26** 221–233
[19] Paoletti D and Spagnolo G S 1996 *Progress in Optics* **35** 197–255
[20] Krzemień L, Lukomski M, Kijowska A and Mierzejewska B 2015 *Journal of Cultural Heritage* **16** 544–550
[21] Memmolo P, Arena G, Fatigati G, Grilli M, Paturzo M, Pezzati L and Ferraro P 2015 *Journal of Display Technology* **11** 417–422
[22] Maldague X 2001 *Theory and practice of infrared technology for nondestructive testing* (Wiley)
[23] Spagnolo G S, Paoletti D, Ambrosini D and Guattari G 1997 *Pure and Applied Optics: Journal of the European Optical Society Part A* **6** 557
[24] Sveen J K 2004 *Preprint series. Mechanics and Applied Mathematics http://urn. nb. no/URN: NBN: no-23418*

IOP Conf. Series: Materials Science and Engineering **364** (2018) 012064 doi:10.1088/1757-899X/364/1/012064

Multispectral Reflectance and UV Fluorescence Microscopy to study painting's cross sections

Anna Pelagotti[1], Lucilla Pronti[2], Emanuela Massa[3], Monica Galeotti[4], Anna Candida Felici[2]
[1]National Institute of Optics, CNR, L.go E. Fermi 6, 50125 Firenze
[2]Department of Basic and Applied Sciences for Engineering, LANDA-Laboratory of Non Destructive Analysis and Archaeometry, University of Rome Sapienza, via A. Scarpa 16, 00161, Rome, Italy
[3]Art-Test, via S. Spirito 11, 50123 Firenze
[4]Opificio delle Pietre Dure, via Alfani 78 50123 Firenze
Corresponding author: anna.pelagotti@ino.it

Abstract. UV Fluorescence microscopy is a powerful mean for visualizing and identifying structures in a specimen, and it has long been used in many applications, including pigment identification in artworks. Typical painting cross section samples are the result of multiple layers and especially for paintings produced from the thirteen to sixteen centuries, it is possible that for different pigments a different binder has been used, resulting in a complex stratification of materials. In this paper we propose a novel multispectral microscopy imaging system, which is specially designed for cultural heritage application, capable of acquiring both reflectance and fluorescence images in different wavebands, so to be able to document UV fluorescence emission and reflectance spectra.

1. Introduction

In the cultural heritage field, since the discovery of the Wood filter [1], UV fluorescence has been applied both for imaging the artwork's surface and for microscopy of paint cross-sections. On the painting surface, UV fluorescence imaging can be used to detect varnishes and retouches [2]. The first ones, in fact, are typically strongly fluorescent. Moreover, the emission of most painting materials commonly used, increases with aging. Thus retouches, more recent that the rest of the paints, would appear darker under the UV light. Multispectral imaging techniques also for UV fluorescence imaging, have been developed to introduce new assessment possibilities [3] [4].
At the same time, also fluorescence microscopy has been widely used to analyse cross-sections of micro-samples, often complemented by SEM microscopy with EDS.

In fact, UV Fluorescence microscopy is a powerful mean for visualizing and identifying structures in a specimen. However, like for UV fluorescence imaging, in order to be able to properly analyze the fluorescence emission of a sample, and not to stop at a qualitative examination, the microscopy system needs to be calibrated. Moreover, in order to extract the fluorescent signature, the emission at various wavelengths needs to be recorded, together with the corresponding reflectance signal. Currently, to the best of our knowledge, there is no proposed method that implements such requirements.

In this paper we propose a novel multispectral microscopy imaging system, which is specially designed for cultural heritage application. Although reflectance multispectral imaging microscopy have already been investigated [5], the novel microscope presented is capable of acquiring both reflectance and fluorescence images in several wavebands, chosen ad hoc to highlight the most relevant features, which are generally in the blue region, of the fluorescence spectra of paint samples. In addition, the method developed is designed to provide calibrated images. This is particularly useful when, like most of the times, samples exhibit multiple layers in cross sections. The underlying layers are in fact not visible on the painting surface. However, their understanding is extremely useful both for studying the artist's technique and for conservation purposes. Especially for paintings of the thirteen till sixteen centuries, it is possible that for different pigments a different binder has been used, leading to a complex stratigraphy that cannot be easily studied otherwise. This is especially true for what concerns the organic matter of each layer. In fact, if inorganic components can be quite successfully investigated by SEM-EDS and microRaman spectroscopy, only micro-reflectance FTIR might provide some information about the organic compounds like binders and varnishes. Limitations owing to low signal-to-noise ratio from layers as thin as few tenths of micrometers, interference of strong overlapping peaks of inorganics and shape and intensity distortions in the reflectance spectra make micro-reflectance FTIR not suitable for a widespread use when applied to a painting stratigraphy. The immuno-fluorescence assay and staining tests [6] does not permit the identification of binders different from proteins and it leads to irreversible changes of cross-sections.

Documenting the reflectance and UV fluorescence emission for each layer in different wavebands on the other side, enables a detailed study of the painting's inner structure. Binders, like egg tempera or linseed oil exhibit typical UV fluorescence emissions in the visible range, however displaying different peaks and absorption bands, while most pigments are generally not fluorescent [7].

In this paper we analyze the multispectral UV fluorescence microscopy images acquired with this novel technique.

2. Test samples

2.1. Paint samples from the OPD collection

A set of samples has been prepared at the Opificio delle Pietre Dure to be used as reference for testing diagnostic techniques. The samples have been made according to the techniques described by Cennino Cennini and other ancient treatises on arts and reflect also the evidences collected from actual paintings studied at the OPD [8].

Two preparation coatings made with rabbit glues and calcium sulphate dihydrated were applied on wood. An "imprimitura" layer, made of rabbit glue and water, overlaps this ground layer. Pictorial layers with different pigments, pure or in mixture, as thin as few tenths of micrometers were applied over the abovementioned ground layer (preparation + "imprimitura"). Pigments were supplied by Zecchi (Italy) and checked by FTIR spectroscopy and/or SEM-EDS as for their actual composition. For some samples, "stand" linseed oil (Zecchi, Italy) was used as binding medium, while for other ones tempera was prepared by mixing egg yolk, egg white and vinegar with a 2:1:1 volume ratio. On top of such samples, a glazing layer with a semitransparent organic colorant, like madder lake or saffron, with linseed oil, was also applied.

We have chosen this set of samples because it has been widely studied and analysed, and we could access a variety of data on it (see eg. fig. 2).

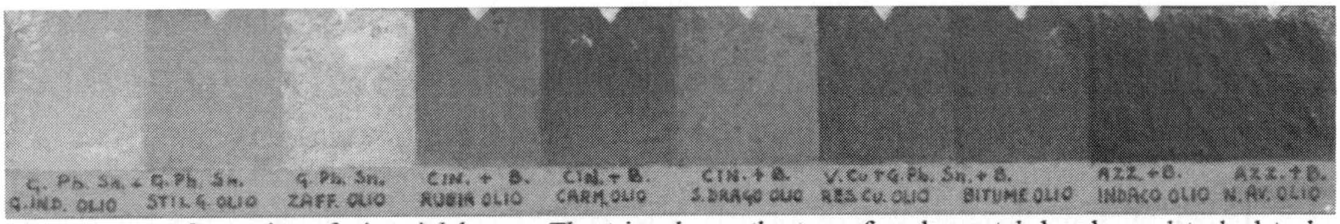

Fig. 1 A set of samples of pictorial layers. The triangle on the top of each swatch has been detached to be analysed under the standard optical microscope, SEM, and with the novel multispectral microscope.

IOP Conf. Series: Materials Science and Engineering **364** (2018) 012064 doi:10.1088/1757-899X/364/1/012064

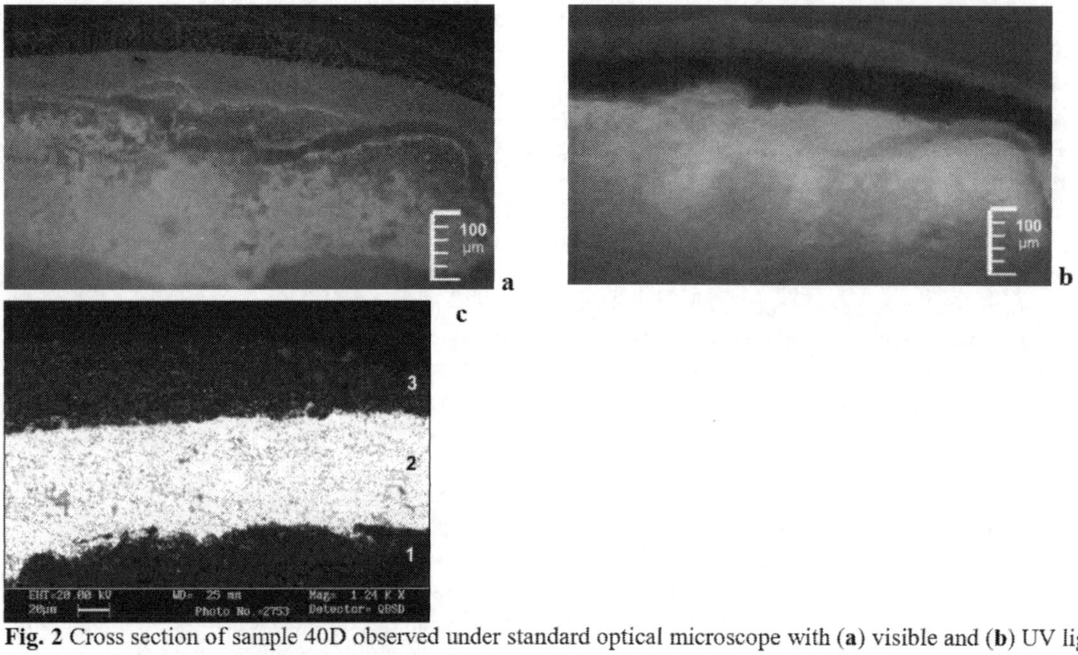

Fig. 2 Cross section of sample 40D observed under standard optical microscope with (**a**) visible and (**b**) UV light and (**c**) under scanning electron microscope (SEM-EDS). Not all the materials used can be detected using the current analytical techniques. The EDS spectrum of rubia lake shows the presence of aluminium, related to alum in the lake. The larger grains of layer 2(in **c**) are composed by cinnabar, whereas the smaller ones consist of lead white. No information can be retreivd on the binder used

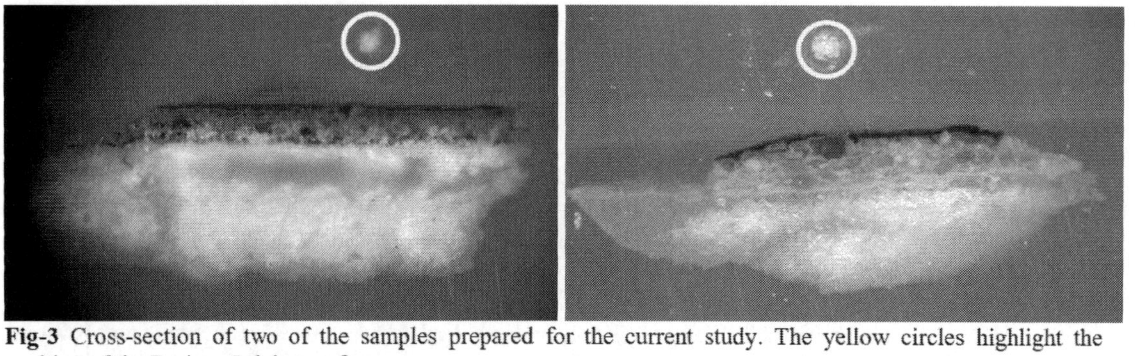

Fig-3 Cross-section of two of the samples prepared for the current study. The yellow circles highlight the position of the Barium Sulphate references.

2.1.1. Sample preparation

We took micro-fragments from near the edge of the sample (see fig. 1 and fig. 2). The sample were then embedded in an alkyd resin and polished with abrasive paper. A micro hole was made in the resin where Barium Sulphate powder was pressed (see fig. 3). Barium Sulphate exhibits a very uniform reflectance from UV to Near IR. Moreover it does not exhibit any UV fluorescence emission.

3. Novel multispectral microscopy

3.1. Multispectral microscope

The acquisition set up is composed by two UV lamps, a fiber optic illuminator with a ring-light guide (PL3000, A-1170, Photonic), an optical system, a CCD camera and a set of band-pass filters, as shown in Fig. 4a. Each UV lamps has inside 3 UV LEDs (High Power LED M365D1, Thorlabs, characterized by a spectral emission peaked at 365 nm with a FWHM of 10 nm, and an optical output power of 190 mW each LED. The sources were placed at 45° with respect to the normal samples.

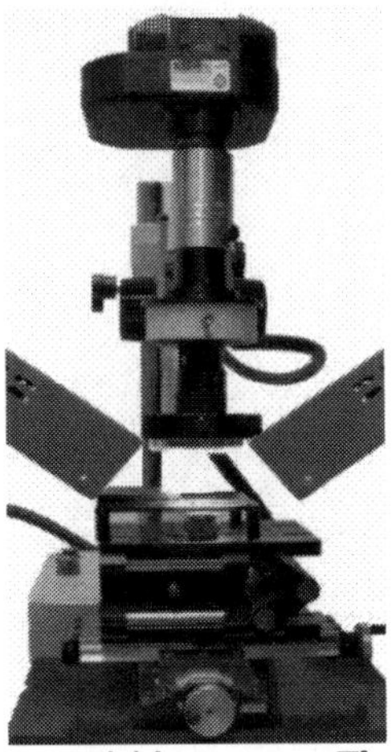

Fig. 4a The acquisition set-up. The microscope is coupled with a multispectral CCD camera

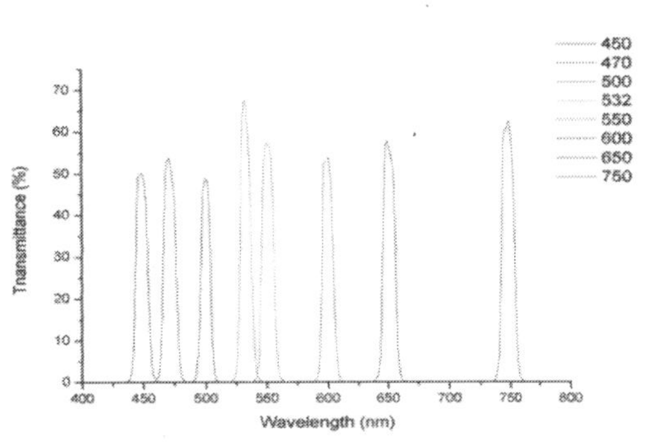

Fig. 4b Transmittance curves of the interferential filters used for the acquisition of the multispectral images

The camera is a monochromatic QSI 583w camera with a CCD full-frame sensor (KAF8300) with 3326x2504 pixels (pixel size of 17.96mm x 13.52mm). The CCD camera is equipped with an internal filter wheel. The filter selection reflects the known characteristics of paint reflectance and fluorescence [7] They are narrow bandpass filters with a FWHM of 10 nm with central wavelengths of 450, 470, 500, 532, 550, 600, 650, 750 nm (Thorlabs) (Fig. 4b). The optical system is composed by a zoom body tube (with 6.5x magnification, 1-60135, Navitar), a lens attachments (with 2x magnification, 1-600113, Navitar) and an adaptor for the CCD camera (1-6015, Navitar).

4. Calibration procedures

We aimed at acquiring reflectance images, thus images that at every pixel present the reflectance value of the paint sample, and calibrated UV fluorescence emission images, thus images that, at every pixel, record an emission value independent from the system used to acquire them, and from environmental conditions. In order to reach this goal, several steps are needed.

4.1. Image registration and calibration

The registration procedure removes geometrical misalignments among the images. These are mainly due to small changes in the optical path, between the object and the CCD sensor for the various filters. The algorithm is based on mutual information maximisation [9].

All acquired images have to be corrected for uneven illumination and for stray light. For fluorescence images it is also mandatory to un-mix the contribution of UV fluorescence emission in the visible range from the stray light due to visible environment radiation (although the measurements were taken in a dark room, some stray light is unavoidable).

For each filter, we acquired with the novel microscopy system a "raw" fluorescence image ($\mathbf{I}_{R\_UV}$) of the cross section. We recorded in this portion of the image, values directly proportional to a known percentage of the incident stray-light.

Two "flat" images i.e. image of a uniform surface placed in the same position as the cross section were also acquired for each wavelength, one using UV lamps ($\mathbf{I}_{F\_UV}$) and another using white light ($\mathbf{I}_{R\_R}$). The camera automatically subtracted a "dark" image for each band.

We computed the Corrected Fluorescence+StrayLight Image ($\mathbf{I}_{C\_UV+S}$), and the Corrected Radiance Image ($\mathbf{I}_{C\_R}$) according to the following:

IOP Conf. Series: Materials Science and Engineering **364** (2018) 012064 doi:10.1088/1757-899X/364/1/012064

$\mathbf{I}_{C\_UV+S} = (\boldsymbol{a}_{\_F\_UV})*(\mathbf{I}_{R\_UV})/(\mathbf{I}_{F\_UV})$
$\mathbf{I}_{C\_R} = (\boldsymbol{a}_{\_F\_R})*(\mathbf{I}_{R\_R})/(\mathbf{I}_{F\_R})$

where $\boldsymbol{a}_{\_F\_UV}$ is the average value of $\mathbf{I}_{F\_UV}$ over the whole image, and $\boldsymbol{a}_{\_F\_R}$ is the average value of $\mathbf{I}_{F\_R.}$ over the whole image.

4.1.1. Reflectance images

The Reflectance Image ($\mathbf{I}_{Ref}$) can then be computed as

$\mathbf{I}_{Ref}$=$\mathbf{I}_{C\_R}$/***s_lr*** x ***(100/R)***

where ***s_l*** is the average of the pixels' values in $\mathbf{I}_{C\_R}$ over the part where the reflectance standard was imaged and (***R***) is the reflectance percentage of the reflectance standard used.

4.1.2. Calibrated UV fluorescence images

The Corrected Fluorescence Image ($\mathbf{I}_{C\_UV}$) can be obtained subtracting the Stray-Light image ($\mathbf{I}_S$) from the $\mathbf{I}_{C\_UV+S}$ image. $\mathbf{I}_S$ can be computed as

$\mathbf{I}_S$=$\mathbf{I}_{Ref}$ x ***s_luv*** x (***100/R***)

where ***s_luv*** is the average of the pixels' values where reflectance standard is in $\mathbf{I}_{C\_UV+S}$.

Signal acquisition was performed by a CCD array. Its response depends not only on the signal energy but also on the system behaviour for each filter, in particular: the spectral transmittance of the optical systems in front of the detector array $\boldsymbol{o(\lambda)}$, the spectral transmittance of the ***k***-th optical colour filter $\boldsymbol{\Phi_k(\lambda)}$ and the spectral sensitivity of the CCD array $\boldsymbol{\alpha(\lambda)}$.

To determine the calibration factors $\boldsymbol{W_k}$ (k=1,2...7) that need to be introduced in order to derive the calibrated images $\mathbf{I}_{Cal\_UV}$ from $\mathbf{I}_{C\_UV}$ images, which are independent of the specific camera and filters used, we proceeded with the indirect method as described in [4].

5. Image analysis

Two samples, 20D and 40D, of the set described in [7] were chosen to illustrate how the method works. The visible image of sample 20D, together with the UV fluorescence image, reconstructed starting from the multispectral images acquired, are shown in fig. 5 and fig. 6 respectively. The single reflectance images in the 8 bands chosen and the images of the UV fluorescence emission are presented in fig. 7. The visible image of sample 40D together with the UV fluorescence are shown in figures 9 and fig. 10, respectively. The reflectance images in the 8 bands chosen and the images of the UV fluorescence emission are presented in fig. 11.

Both samples exhibit, over the white preparation, a red layer made of a mixture of cinnabar and lead white, but the binder is tempera for sample 20D and linseed oil in case of sample 40D. Above the red layer with cinnabar, a glaze layer with madder lake and linseed oil is applied. Since the fragments were taken near the edge of the samples, in some cases the paint dropped down along the end side of the stratigraphy and to a certain extent entered below the layer underneath.

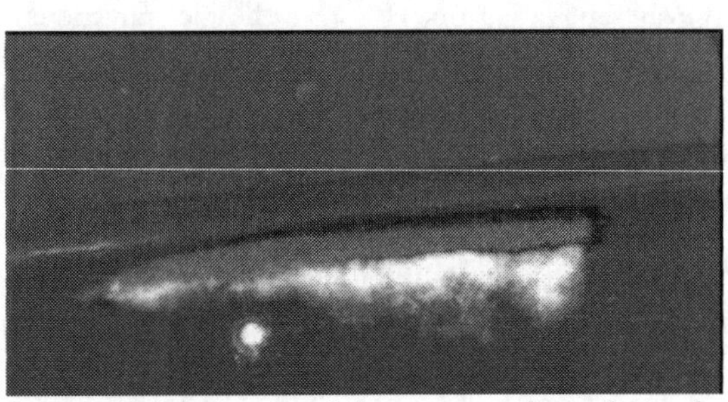

Fig. 5 Visible image of the cross-section of sample 20D

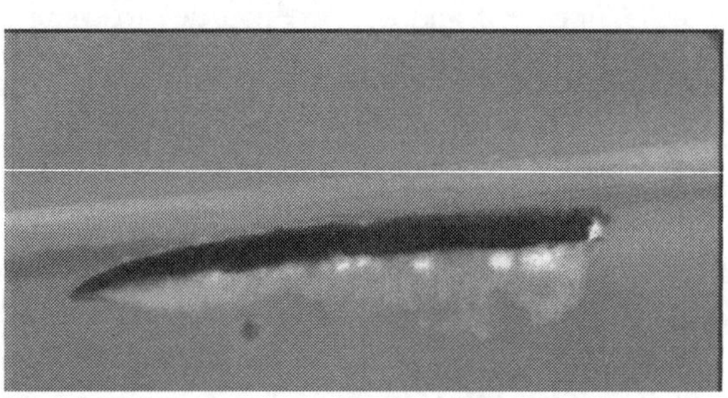

Fig. 6 UV fluorescence image of the cross-section of sample 20D

IOP Conf. Series: Materials Science and Engineering **364** (2018) 012064 doi:10.1088/1757-899X/364/1/012064

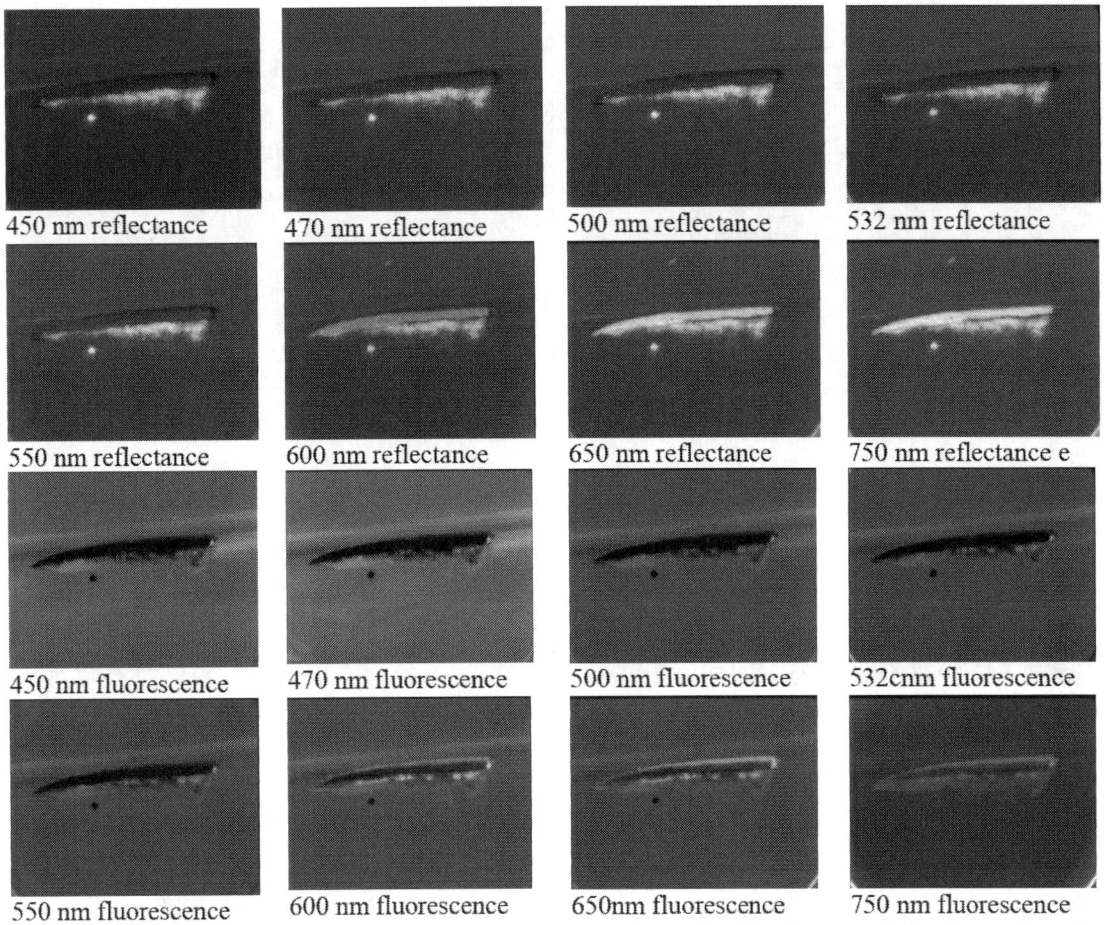

Fig. 7 Reflectance and UV fluorescence multispectral images of Sample 20D. The numerical reference is to the peak of the interferential filter used (see fig 4b)

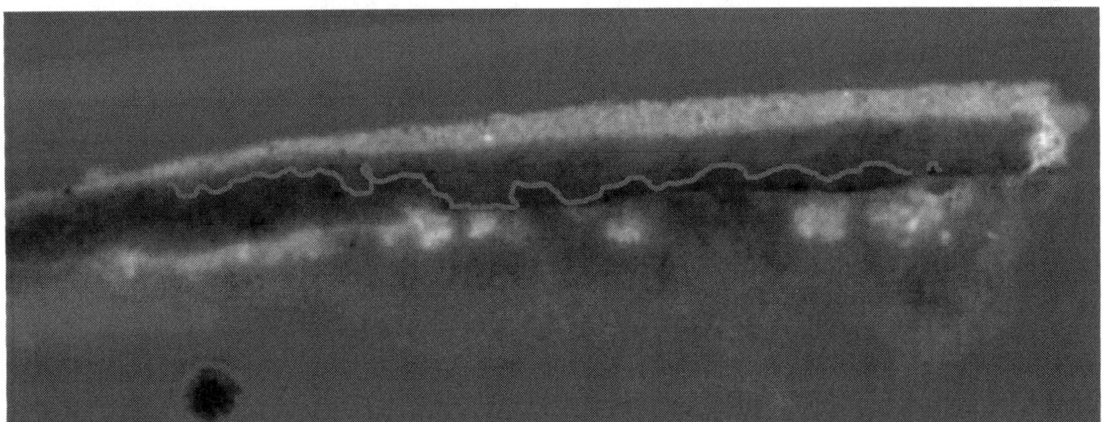

Fig. 8 UV fluorescence image at 650 nm of sample 20D. The cross section exhibit a complex fluorescence emission. The Cinnabar layer in particular shows rather difference emission within the layer (highlighted by the blue line), probably due to a penetration of the binder of the top layer (linseed oil) in the Cinnabar layer, which was originally a tempera. Strong fluorescence emission is also displayed by some areas in the preparatory layer. The reason for this needs further investigations

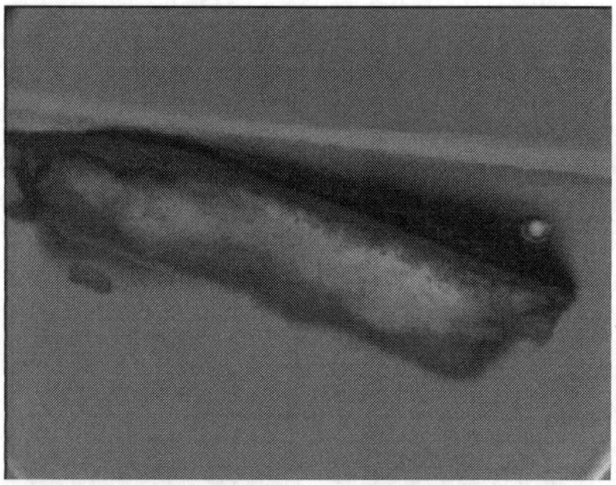

Fig. 9 Visible image of the cross section of sample 40D

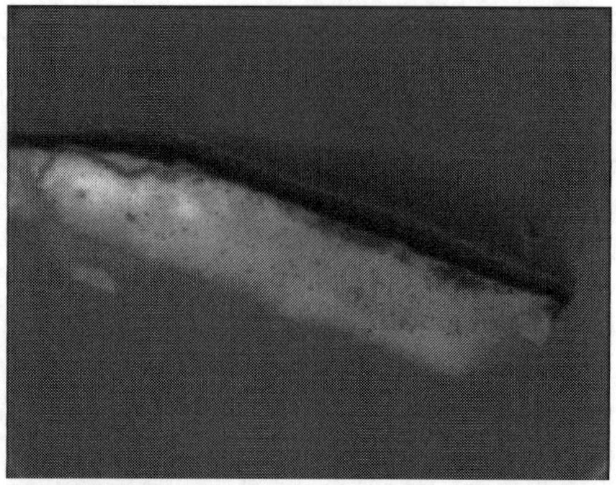

Fig. 10 UV fluorescence image of the cross section of sample 40D

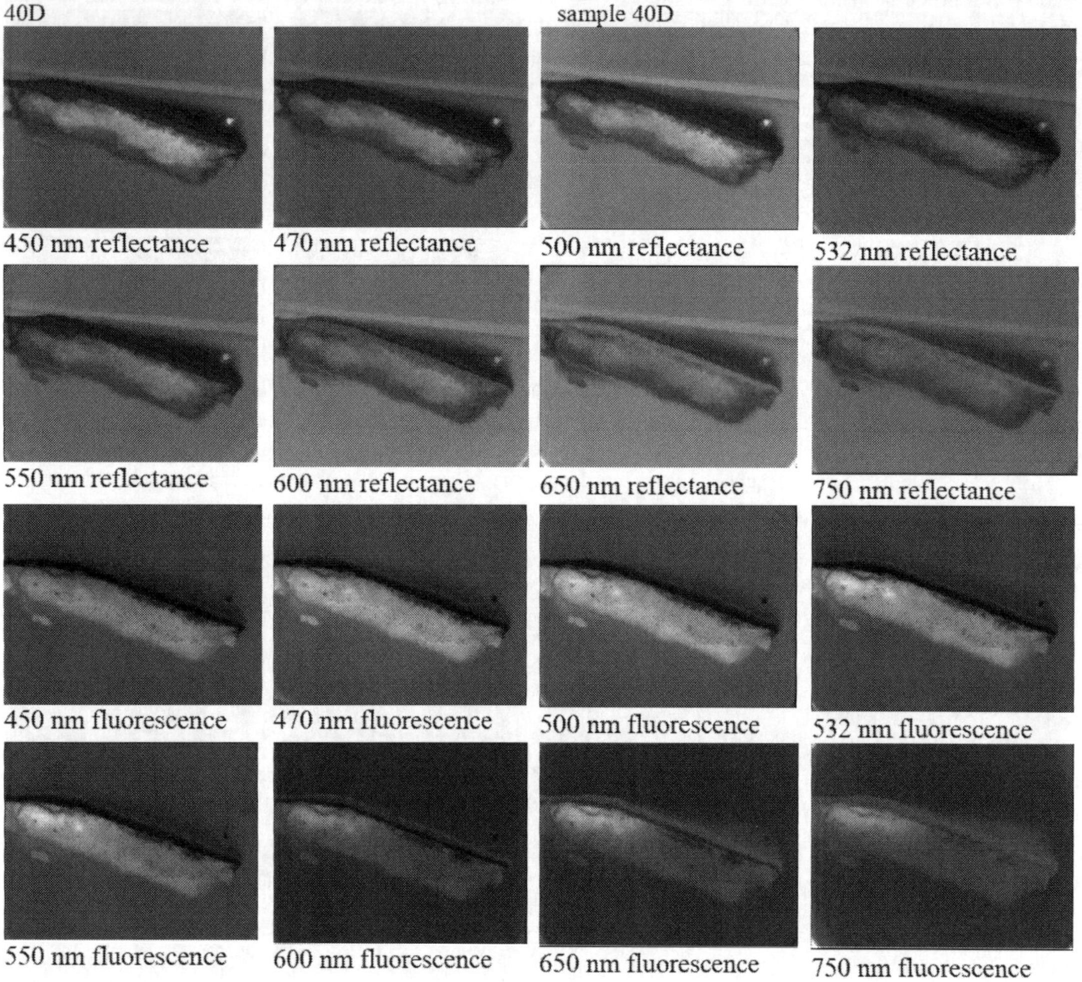

Fig. 11 Reflectance and UV fluorescence multispectral images of Sample 40D. The numerical reference is to the peak of the interferential filter used (see fig 4b)

5.1. Spectral features

The multispectral images acquired document the inner structure of the samples, showing the reflectance and the UV fluorescence emission in different wavebands. We were expecting to find a

IOP Conf. Series: Materials Science and Engineering **364** (2018) 012064 doi:10.1088/1757-899X/364/1/012064

clear distinction in the emission peaks of each layer, depending on the paint and on the binder used, corresponding to the known behavior of the reference materials.

However, in reality, the cross sections displayed a much more complex scenario. Different emissions are also present within each layer, possibly because of occurred interactions. For example one binder could have migrated from one layer to the next, as it seems to have happened e.g. for the case illustrated in fig. 8. Such cross-sections had also been coated to be analysed with SEM technique, and then uncoated. This could also have introduced unwanted.

6. Conclusions and future work

To the best of our knowledge, the method presented is novel and addresses an unmet goal in the analysis of paint materials. Sample microscopy has been used in CH field for many years, however not in a multispectral fashion, and generally providing only qualitative and not quantitative data. Multispectral reflectance and UV fluorescence emission microscopy presents several clear advantages for the understanding of a complex stratigraphy. For example organic matter, like that of binders and varnishes, cannot be easily studied with conventional techniques. In fact, only micro-reflectance FTIR might provide some information about the organic compounds, however, the signals of layers few tenths of micrometers thin, generally display a low signal-to-noise ratio, interference of strong overlapping peaks and shape and intensity distortions. On the other side, documenting the reflectance and calibrated UV fluorescence emission for each layer in different wavebands, enables a detailed study of the painting's inner structure. Understanding the different emissions and correlating them with the presence of some materials will be extremely useful both for studying the artist's technique and for conservation purposes. We plan to further develop the technique proposed, increasing the magnification and possibly the number of interferential filters, to reach its full potential.

References

[1] R.A. Lyon, "Ultra-violet Rays as Aids to Restorers." Technical Studies in the Field of Fine Arts 2 (1934): pp 153-157.

[2] R. de la Rie, "Fluorescence of paint and varnish layers (part I- II- III)", Studies in Conservation, Vol. 27, pp 1-7, 65-69,102-108,1982

[3] A. Pelagotti, L. Pezzati, A. Piva, and A. Del Mastio, "Multispectral UV fluorescence analysis of painted surfaces", 14th European Signal Processing Conference (EUSIPCO 2006), Florence, Italy, Sept 2006

[4] A. Pelagotti ; A. Del Mastio ; A. De Rosa ; A.Piva, "Multispectral imaging of paintings", IEEE Signal Processing Magazine, Vol.25 Issue: 4, 2008, pp. 27 – 36

[5] J. van der Weerd, M. K. van Veen, Ron M. A. Heeren, and J. J. Boon, "Identification of Pigments in Paint Cross Sections by Reflection Visible Light Imaging Microspectroscopy", Anal. Chem. 2003, 75, pp. 716-722

[6] M. Vagnini, L. Pitzurra, L. Cartechini, C. Miliani, B. G. Brunetti, A. Sgamellotti, "Identification of proteins in painting cross-sections by immunofluorescence microscopy", Anal Bioanal Chem, 2008, 392: pp. 57–64.

[7] A. Pelagotti, L. Pezzati, N. Bevilacqua, V. Vascotto, V. Reillon, C. Daffara "A Study Of Uv Fluorescence Emission Of Painting Materials", Proc. Of ART2005, May 2005, Lecce, Italy.

[8] A. ALdovrandi, M.L.. .Altamura, M.T.Cianfanelli, P.Riitano, I materiali pittorici:tavolette campione per la caratterizzazione mediate analisi multispettrale, OPD restauro, 1996, Centro Di, Firenze, pp.191-210

[9] V. Cappellini, A. Del Mastio, A. De Rosa, A. Piva, A. Pelagotti, and H. El Yamani, "An automatic registration algorithm for cultural heritage images," in IEEE International Conference on Image Processing, ICIP 2005, vol. 2, Sept 2005, pp. II–566–9.

IOP Conf. Series: Materials Science and Engineering **364** (2018) 012065 doi:10.1088/1757-899X/364/1/012065

Comparative analysis of two self-portraits by Igor Vasiljev using non-invasive methods

M Djuric[1], V Jovanovic[1], Z Velikic[2], M Jovanovic[1]

[1]ArtLab, Bulevar Kralja Aleksandra 391, Belgrade, Serbia
[2]Institute of Physics, Pregrevica 118, Belgrade, Serbia

E-mail: vmaja74@gmail.com

Abstract. This paper presents a research on materials and techniques used in two paintings dating from October 1952, signed by Igor Vasiljev. The aim of the research was to provide important answers from the art history standpoint and explain the existence of two almost identical self-portraits. Standard non-destructive diagnostic methods for painting examination were applied. First, a qualitative analysis of recorded visible reflectance spectra in 350-1100 nm domain was performed, and then its results were regarded along with the information obtained applying imaging techniques such as UV induced fluorescence, VIS-NIR-IR reflectography and radiography. Since the paintings do not have a varnish layer, the UV fluorescence gave especially interesting results, showing difference in used paints. Although the comparative colorimetric spectra indicated the use of same pigments but different mixtures in most points, the infrared reflectography showed very clearly the existence of underdrawing and significant changes in composition (*pentimenti*) on only one painting. In addition, the details of that painting observed under the great enlargement indicated great spontaneity of paint application. While a visual inspection of these two self-portraits suggested the same author - a very talented and creative Igor Vasiljev, a comparative analysis of the results obtained by spectroscopic and imaging techniques showed that there is a significant difference between them.

1. Introduction

The existence of two paintings "Self-portrait in Rubashka[1]" signed by Igor Vasiljev, dating from the same period, has drawn the attention of connoisseurs and admirers of his art. The application of scientific methods in the analysis of these paintings, primarily the painting process, materials and techniques used, has opened some new questions, but also showed the potentials of various nondestructive methods. The accent has been put on the use of imaging techniques, since their results have indicted a different creation process of these two self-portraits. Inclusion of other non-destructive methods has significantly expanded the field of research and contributed to a better understanding of the artistic process of Igor Vasiljev, especially considering the fact that his work has never been explored in that way before.

Painting A (figure 1) - Self-portrait in Rubashka, dimensions 70 x 50 cm, was executed in oil on kraft paper that has been laminated to cardboard. It has been signed in the lower right angle: Igor X 52. Already at a first glance, before going into any detail, the painting possesses distinct marks of the portraits painted by Vasiljev during that year. The line of his drawing is strong, Van Gogh-like exiting

[1] from Russian рубашка - a type of shirt or tunic traditionally worn by Russian peasants

but precise, with no hesitation. His stroke is clear and spontaneous, consisted of rich brush deposits and a transparent background.

Painting B (figure 2) – The paper carrier is made of transparent paper, dimensions 71 x 50 cm, laminated to a cardboard, dimensions 71,5 x 53 cm.

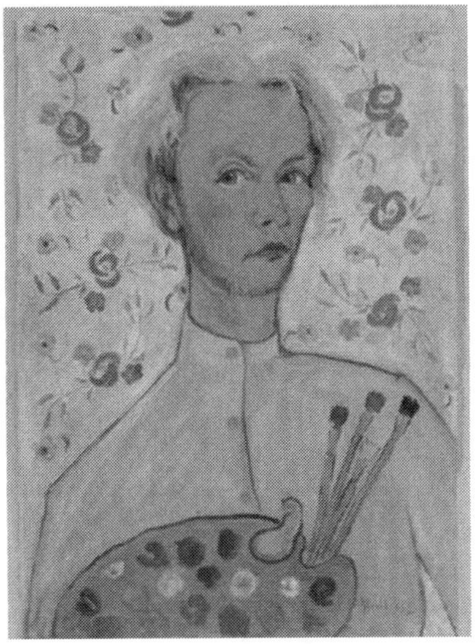

Figure 1. Self-portrait in Rubashka, painting A, visible light

Figure 2. Self-portrait in Rubashka, painting B, visible light

Igor Vasiljev (1928-1954) was the descendant of Russian immigrants and intellectuals, born in Belgrade. He painted intensely since 1948 and in 1953 he organizes his first solo exhibitions in Belgrade and Zagreb. The same year he gets accepted in the Association of Fine Artists of Serbia as its youngest member. He died in an accident in 1954. The cause and circumstances of the accident are still a mystery and a subject of speculation. Utterly distinctive, consistent and charismatic, he left a signature in Serbian painting after the Second World War. Nine solo exhibitions have been organized posthumously, the last one in March 2018 at the Gallery of Serbian House of Army in Belgrade. The "Self-portrait in Rubashka", in the ownership of Canvas Art Gallery, has drawn the attention of the numerous visitors from all generations on that occasion.

2. Experimental

The chosen set of non-destructive techniques has provided ultra-high resolution digital images at multiple wavelengths, including ultra-violet fluorescence, visible, infrared reflectography and X-ray radiography. Besides these, UV-VIS-NIR and XRF spectroscopy were applied in selected points as well.

2.1. Visible (VIS) examination (400-700nm)

The first step in the analysis of paintings was surface examination under uniform illumination and raking light. The high resolution digital images were acquired with a Nikon D700 (figure18) and CANON EOS 6D (figure 1, figure 2); camera lighting source was TUNGSRAPAR-SPOT4 lamp.

The high resolution digital photography showed clear traces of pencil made preparatory drawing, along with a significant difference in painting technique.

In both paintings, A and B, a cardboard of similar type and age has been used as support, but the difference is in the type of paper which served as a carrier of the drawing and the painted layer. The paint application of both works is very expressive, yet it is of a much less uniform thickness in painting A. Painting B on the other hand, is characterized by a uniform paint layer.

IOP Conf. Series: Materials Science and Engineering **364** (2018) 012065 doi:10.1088/1757-899X/364/1/012065

2.2. UV induced VIS-NIR fluorescence

Besides the information on the areas of retouch or the type of varnish, UV induced fluorescence can also be a method for obtaining data about pigments, especially in case of the examined paintings [1].

The UV light source was UV-Handleuchte uf LED 5000 lamp (Uniflux Germany). The UV induced fluorescence was recorded in visible (CANON EOS 6D) and IR region (Samsung S1050). Since there is no varnish layer, differently colored fluorescence is evident [2].

A difference between the two paintings is noticeable on the UVF photos in the visible area (figures 3 and 4). Pigments from painting A fluoresce much more intensely and in different colors. The fluorescence of one part of the yellow paint in orange is particularly visible only on this painting, suggesting that different mixtures of pigments were used in A and B.

Figure 3. Painting A, UVF - visible region

Figure 4. Painting B, UVF - visible region

Figure 5. Painting A, UVF - NIR region

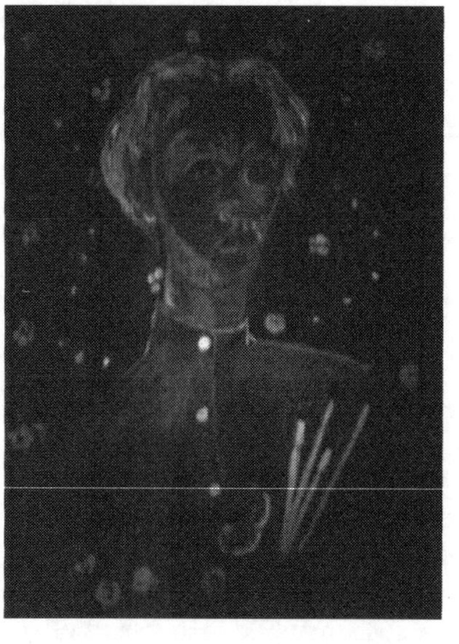

Figure 6. Painting B, UVF - NIR region

The UVF photos in NIR region (850 - 1100nm) (figures 5 and 6) have shown fluorescence on both paintings not only of yellow, but also of other paints for which it can be assumed to contain the same yellow pigment, most probably Cadmium yellow [2,3]. The only visible difference in yellow parts is in the fluorescence of the yellow brush which doesn't fluoresce on painting B in the IR region as well, indicating the use of another pigment.

2.3. IR reflectography (NIR-SWIR)

Comparison of paintings at different wavelengths in the IR region can provide additional information about the material and the painting technique. IR reflectography has been done in the NIR and the SWIR area using Samsung S1050 camera with 850nm - 1100nm filter; Nikon D600 with 950nm - 1100nm filter and LR-SWIR InGaAs PhotonicScience camera with 1600nm filter (bandpass 50nm).

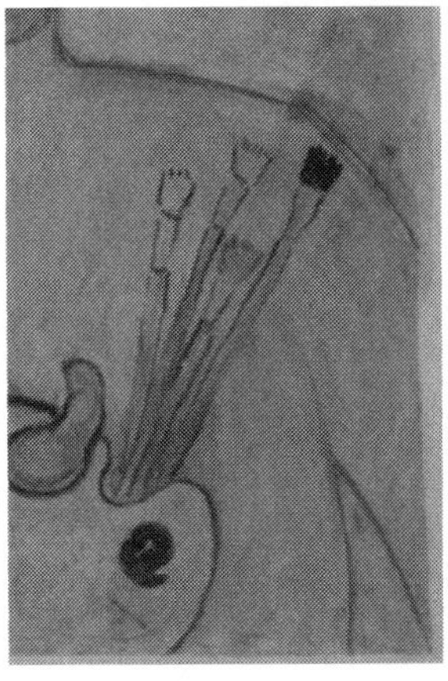

Figure 7. Brushes, painting A, 1600 nm

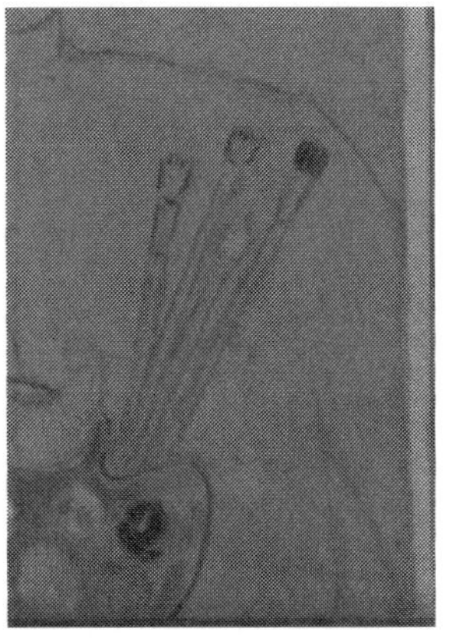

Figure 8. Brushes, painting B, 1600 nm

Figure 9. Head, painting A, 1600 nm

Figure 10. Head, painting B, 1600 nm

Besides showing the preparatory drawing, obtained images were also used for determination of certain pigments, considering the differences in their transparency in the IR region.

Observation of the recordings in visible and IR region (figures 7 to 10) led to a conclusion that there is a clearly visible and solid preparatory drawing executed in graphite pencil, with a very dynamic and rich tone colors gradation and thickness of lines on painting A. The drawing on painting B was made in such manner that colored areas are linearly confined, i.e. each colored space is precisely contoured.

IOP Conf. Series: Materials Science and Engineering **364** (2018) 012065 doi:10.1088/1757-899X/364/1/012065

Also, the change in the position of brushes and figure can be noticed on painting A, which is not the case on painting B. A different transparency in the IR region is especially noticeable at 1600nm, which indicates the difference in strokes. The yellow colored brushes of painting B show evident reflectivity in the IR region. Moreover, higher reflectivity and lower contrast can be noticed on this painting, probably because the support is less covered with paint, which has already been noticed on the photographs taken in the visible area.

2.4. X-ray radiography

All of the noted differences have been confirmed by digital radiography as well. The imaging has been performed using Roentgen scanner HD-CR 35 NDT Plus by DURR, Germany. Several recordings have been made with different exposition periods and different cathode tube voltages.

X-ray radiography can provide a wealth of information not obtainable with other techniques, for its capability to cross all layers of a painting. With this method it is possible to discover underlying paintings, investigate author's *pentimenti*, previous conservation treatments, hidden inscriptions and signatures, structural anomalies, etc. [4].

No *pentimenti* have been noticed on the X-ray recordings (figures 11 and 12), but the difference in the painting technique and in the background has been spotted: painting A has thicker coating and most of the colors are mixed with white. It is also clear that the yellow paint is absorbing the X-rays. Painting B shows a uniform background, with most prominent details painted in yellow which is also non-transparent for the X-rays.

Figure 11. Painting A, radiography

Figure 12. Painting B, radiography

2.5. UV-VIS-NIR reflectance spectroscopy

This technique enables the recognition of spectral characteristics of pigments by comparing them with a standard base. Performing comparative analysis in the UV-VIS-NIR region it is possible to make presumptions about the spectral features of the background as well.

Spectroscopic analysis has been done in several spots (figures 13 and 14). Considering the results obtained with previously described techniques, only spectra of green and yellow paints will be presented. Spectra were obtained with AvaSpec 2048-usb2-VA-50 Avantes specrometer in 400 - 1100nm area

IOP Conf. Series: Materials Science and Engineering **364** (2018) 012065 doi:10.1088/1757-899X/364/1/012065

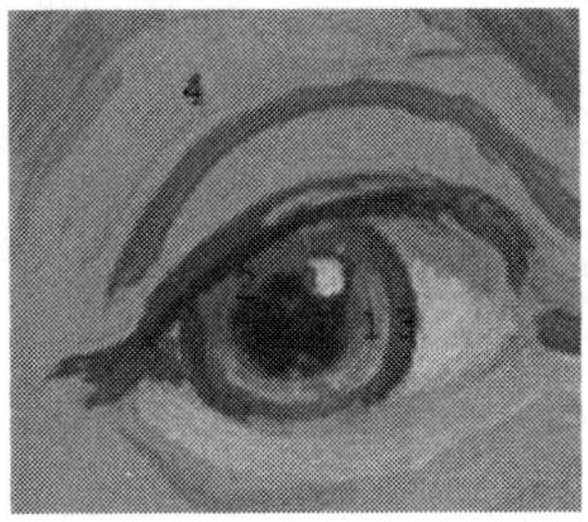

Figure 13. Spectroscopy spots of the green paint

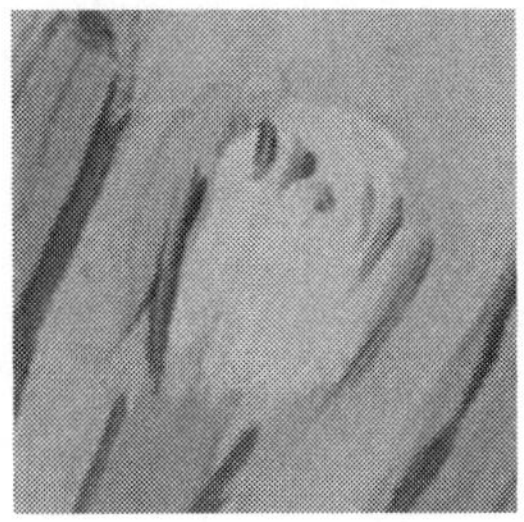

Figure 14. Spectroscopy of the yellow paint

2.5.1. Green paint. Spectra of the spot number 3 (figure 15) are different in shape and indicate most probably the use of different green pigments when colouring the outer edge of the iris, while the ones of the spot number 4 (figure 16) clearly show presence of yellow in the green paint. Spectra of painting B show much higher reflectivity in the NIR region.

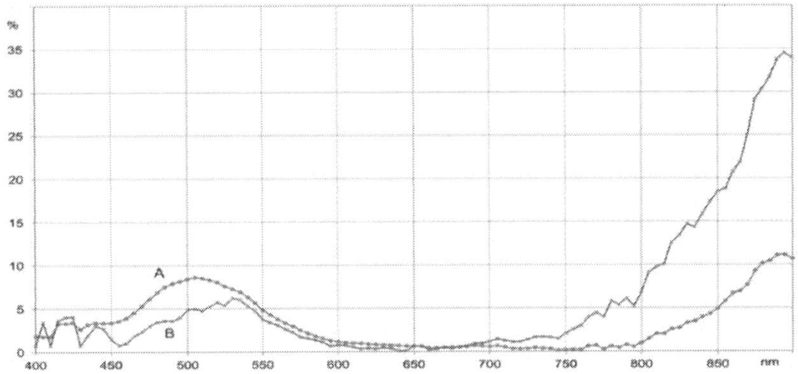

Figure 15. Reflectivity vs wavelength of green paint in spot number 3

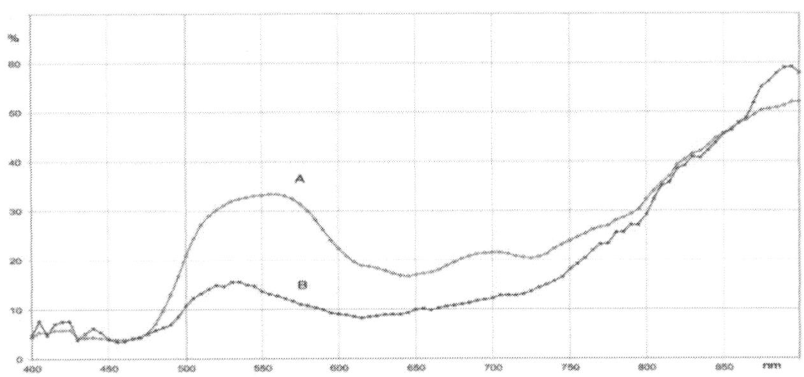

Figure 16. Reflectivity vs wavelength of green paint in spot number 4

2.5.2. Yellow paint. The largest difference between the paintings has been noticed in the yellow paint spectra, especially of the yellow brush (figure 17). Already the UV images have shown completely different properties of the yellow paints from the two paintings. Spectral analysis indicates different mixture of pigments in the yellow paint: the spectra of the spot on painting A is moved much more towards the orange, while the one on painting B has a higher content of green.

IOP Conf. Series: Materials Science and Engineering **364** (2018) 012065 doi:10.1088/1757-899X/364/1/012065

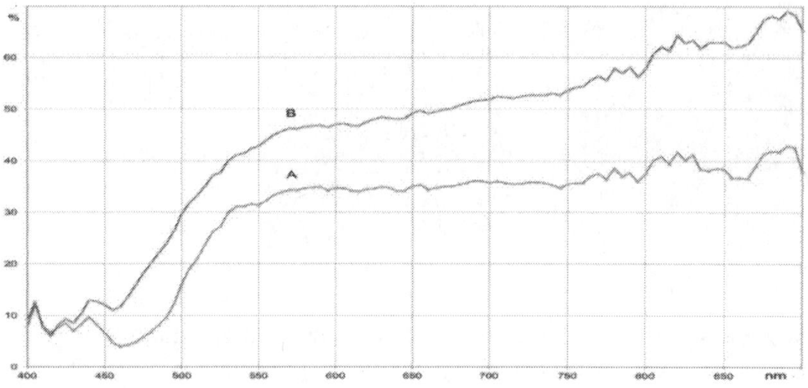

Figure 17. Reflectivity vs wavelength of yellow paint in a spot on the brush

2.6. XRF spectroscopy

The XRF spectra were obtained by means of the ARTAX X-ray analysis device model 200, from Bruker Microanalysis GmbH, spot 0,65mm.

The analysis depth depends mostly on the type of material that has been analyzed. The limitation of the technique when it comes to examination of paintings is that it is not possible to know for sure in which layer the detected element is situated.

The analysis has been performed at the same spots of both paintings in which VIS-NIR spectra were made: the yellow brush as well as the green at the outer edge of the iris. Obtained results were compared to the available database [5]. Based on this data, the paint composition in the selected points has been determined (table 1).

Table 1. Results of the XRF analysis.

	Elements detected	Pigment
Yellow brush, painting A	Cd ,S ,Zn, Ba, Se, Pb	Cadmium yellow, Cadmium orange
Yellow brush, painting B	Pb, Sb, Zn, P	Naples yellow
Green brush, painting A	Zn, S, Ba, Cl, Cu, Ca	Phthalo green
Green brush, painting B	Zn, S, Ba, Cr, Fe, Ca, Cu	Chrome green
Background	Zn, Ba, S	Lithophone

3. Discussion

Observation of the two self-portraits with the naked eye has led to the notion that the paintings are similar in terms of pigments that have been used, but quite different when it comes to painting technique, stroke, etc. Technological and stylistic characteristics of painting B give an impression that tracing/copying of the drawing from painting A could have taken place. This is also indicated by the fact that the drawing of painting B has been made on a transparent paper, its tone colors gradation is much poorer and the line has quite a uniform thickness.

Information obtained by means of the imaging techniques continued to point to a different way of the occurrence of these two paintings, but have also indicated the use of different pigments.

Very few pigments, such as: Zinc white, Cadmium yellow, Cadmium orange, Cadmium red and Madder Lake, have strong fluorescence [3]. The fluorescence of the yellow paint is very interesting: in some parts it fluoresces green and in some parts it fluoresces orange (figures 18 and 19). A similar case has been presented in an article which explains it as the degradation of Cadmium yellow pigment

IOP Conf. Series: Materials Science and Engineering **364** (2018) 012065 doi:10.1088/1757-899X/364/1/012065

[6], but in our case the XRF analysis has shown the presence of Selenium which indicates the possibility of the use of Cadmium orange (which also fluoresces orange) in the yellow paint. Cadmium-based pigments are highly absorbing for the X-Rays, which is also visible in the radiography images. Images in the NIR have shown strong fluorescence not only of the yellow, but of the most part of the green as well, hence it can be assumed that it has been mixed with the yellow pigment. Since some details from painting B fluoresce only in the IR region, it is possible that it is a cadmium pigment, because the fluorescence of cadmium pigments falls in the red and the infrared part of the electromagnetic spectrum [3], unlike the yellow brush which has been done in the Naples yellow and it doesn't fluoresce. The latter is also highly absorbing for the X-Rays, hence it is not possible to use radiography in order to distinguish those pigments.

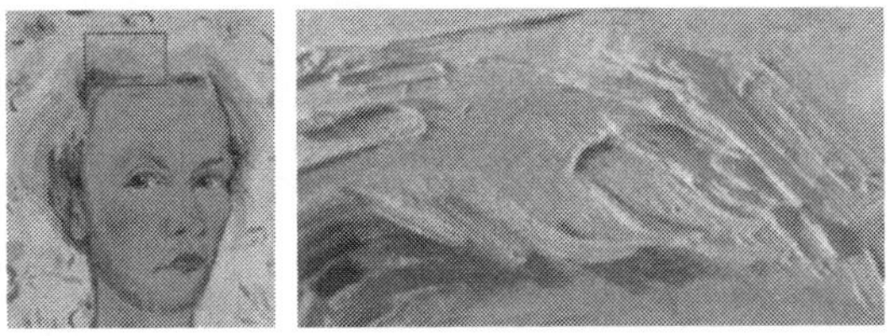

Figure 18. Yellow painted detail, visible light **Figure 19.** Yellow painted detail, UVF

These images have determined the spots in which to perform the spectroscopy and the XRF analysis: the yellow and the green painted areas.

The difference between the yellow pigments has been confirmed by the XRF analysis. The VIS-NIR spectra (figure 17) show good superposition with the spectra from the database [7], only that our spectrum of the spot on painting A is more shifted toward the red, which corresponds to the detected Selenium, typical for the Cadmium orange [8].

It has been confirmed by the XRF analysis that there is no Cadmium yellow in the non-fluorescing green paint. Comparison of its spectrum with the standard spectra of the green pigments [7] has led to the conclusion that the most probable pigment in the spot 3 of painting A is Phthalo green, keeping in mind that this technique shows a huge dependence of the support, which has also been shown in the IR image. Pigment Chrome green has been found in spot 3 of painting B, again by combining the data of the appropriate spectra and the XRF analysis. Its reflectivity in the NIR region is independent of the type of support [8]. X-Radiography supports this by showing higher X-ray absorption in the area of spot 3 in painting A, which is explained by the higher atomic mass of Copper versus Cadmium.

4. Conclusion

Applied techniques and examinations have shown some differences in the used pigments as well as in the production process of these paintings such as: different preparation of support, different preparatory drawing, change in position.

Two pigments that have most probably been used are Cadmium yellow and Cadmium orange (red), which produced different effects in UVF image, although it may also be due to the degradation of Cadmium yellow. This certainly requires further examination, for example the application of FTIR. The same Cadmium pigments have been used on painting B, except on the brush, which was clearly visible on the IR recording. The XRF analysis has confirmed it is Naples yellow, its spectrum matches the one recorded in that spot.

Green pigments are completely different on both paintings: Phthalo green has been used in painting A, while Chrome green has been used in painting B. This was clearly indicated by the recorded spectra and the IR images, which confirmed the transparency of these pigments in the IR region [9]; the final identification was made by XRF.

The information that can be obtained with imaging techniques overcome their usual purpose, however more research is necessary in order to precisely determine them, but even then a multitechnical approach in diagnostic research should be preferred, because each technique has its own limitations due to a high number of influencing variables.

In the light of a newly discovered self-portrait by Igor Vasiljev from the same year, but different month (November), the research will be expanded in order to make further technological and stylistic comparisons.

Acknowledgments

We would like to thank Prof. Dr. Jonjaua Ranogajec, Dr. Bojan Miljevic, and the team of the Laboratory for Examination of Materials in Cultural Heritage from Technological Faculty, University of Novi Sad (Serbia), for the preformed XRF analysis and consultations, as well as Mr. Nebojsa Filipovic and the Welding Institute, Belgrade (Serbia) for the radiography imaging. We would also like to thank Zeljko Radovic for the high-quality photographs.

References

[1] Consentino A, 2015 Practical notes on ultraviolet technical photography for art examination *Cons. Patrim.* **21** ARP 53-62

[2] Pelagotti A, Pezzati L, Bevilacqua N, Vascotto V, Reillon V and Daffara C 2005 *8th Int. Conf. on Non-Destructive Testing and Microanalysis for the Diagnostics and Conservation of the Cultural and Environmental Heritage* (Lecce) A study of UV fluorescence emission of painting materials

[3] de la Rie R, Fluorescence of paint and varnish layers, parts I, II, III *Stud in Cons*, **27**

[4] Thoury M, Celaney J K, de la Rie R, Palmer M, Morales K and Krueger J, 2011 Near-Infrared Luminescence of Cadmium Pigments:In Situ Identification and Mapping in Paintings *Appl. Spectrosc.* **65**(8) 939-51

[5] Larsen R, Coluzzi N, Cosentino A, 2016 Free XRF spectroscopy database of pigments checker *Int.Journ. of Cons. Sci.* 7(3) 659-68

[6] Mass J, Uffelman E, Buckley B, Grimstad I, Vila A, Daleney J, Wadum J, Andrews V, Burns L, Florescu S and Hull A, 2016 *The Noninvasive Analysis of Painted Surfaces,* ed A Nevin and T Doherty (Washington, D.C: Smithsonian Institution Scholarly Press) pp 53-64

[7] FORS database A. Consentino

[8] LAWRENCE Berkly National Laboratory Pigment Database

[9] Cosentino A, 2014 Identification of pigments by multispectral imaging; a flowchart method *Herit.Sci.* 2:8

IOP Conf. Series: Materials Science and Engineering **364** (2018) 012066 doi:10.1088/1757-899X/364/1/012066

Development of a multi-analytical approach to investigate the fading of eosin in painting matrices

F Sabatini[1], R Giugliano[1], I Degano[1], A Lluveras-Tenorio[1], R Sokolová[2], M Thoury[3] and MP Colombini[1]

[1] Dipartimento di Chimica e Chimica Industriale, Università di Pisa, Via Giuseppe Moruzzi 3, I-56124 Pisa, Italy
[2] J. Heyrovský Institute of Physical Chemistry of the Czech Academy of Sciences, Dolejskova 3, 18223 Prague, Czech Republic
[3] IPANEMA, CNRS, Ministère de la Culture, UVSQ, USR3461, Université Paris-Saclay, F-91192 Gif-sur-Yvette, France

E-mails: ilaria.degano@unipi.it, f.sabatini4@gmail.com, ro.giugliano@gmail.com, anna.lluveras@unipi.it, romana.sokolova@jh-inst.cas.cz, maria.perla.colombini@unipi.it

Abstract. Eosin is a xanthenic synthetic organic dye/pigment synthesized by Caro in 1873. It was precipitated with lead and aluminum salts and commercialized as geranium lake and it was one of the most brilliant hues in the palette of the Impressionist painter Vincent van Gogh. Unfortunately, the high photosensitivity of eosin has caused the evident fading of the intense pink colorations of his canvas paintings into lighter ones.
This work aims at underlying the contribution of several complementary different analytical techniques in disclosing a very complex and still little-known topic such as the degradation process of eosin.

1. Introduction

Eosin, also called Pigment Red 90 or Acid Red 87, is a xanthenic synthetic organic dye/pigment synthesized by Caro in 1873 (Figure 1). Although widely commercialized after precipitation with lead and aluminum salts as geranium lake, its identification in historical artworks has been rarely achieved due to its high photo-instability [1,2,3]. Moreover, the description of the involved mechanisms in its loss of color is still missing. Characterizing eosin degradation products is rather challenging, not only due to eosin strong pH sensitivity and photosensitivity, but also to the presence of impurities in commercial products. When dealing with historical materials, the complexity of the heterogeneous painting matrix increases the number of variables involved in the rate and the mechanism of organic pigments ageing [4]. Thus, for shading light on the phenomena involved in the degradation of eosin, the application of a multi-analytical approach is fundamental. In this work, a multi-analytical approach was applied to study eosin both in solution, and in oil media. The potentialities and limitations of the different approaches are herein discussed. First, colorimetry was used for monitoring the fading. Then, artificially aged paint replicas where characterized using luminescence spectroscopy and an optimized HPLC-DAD method.
Infrared imaging techniques based on Synchrotron Radiation were also applied to establish the chemical distribution of eosin-based compounds within the micrometric layers of paint model systems.

IOP Conf. Series: Materials Science and Engineering **364** (2018) 012066 doi:10.1088/1757-899X/364/1/012066

2. Materials and Methods

2.1. Chemicals
Standard sodium eosin (EO-Na) was purchased from Sigma Aldrich (USA). Stock solutions were prepared in dimethyl sulfoxide (DMSO; J.T. Baker, Holland) and kept in the dark. All the solvents were HPLC grade and were used without any further purification. The eluents for the HPLC-DAD analyses were: bi-distilled water and acetonitrile both HPLC grade (Sigma-Aldrich, USA). All eluents were added with 0.1% v/v Trifluoracetic acid (TFA; 99% purity, Sigma Aldrich (USA)). PTFE filters (4 mm thickness and 0.45 μm pore diameter) were used for the purification prior to HPLC injection.

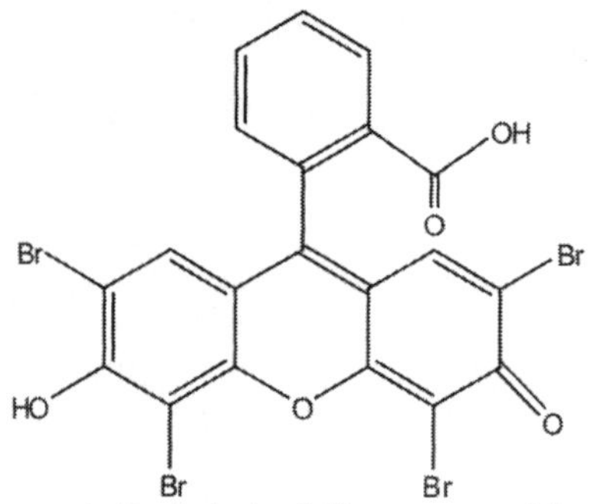

Figure 1: Eosin in its fully protonated form.

2.2. *Reference samples*
Paint replicas were prepared by admixing eosin with linseed oil, and eosin, linseed oil and lead white (Kremer Pigmente, Germany).

2.3. *Colorimetric measurements*
The colorimetric measurements were performed with a Konica-Minolta Mod. CM-700d, with a spot of 8 mm (diameter), with a wavelength range of 360-740 nm and D65/10° for the illuminant/observer, specular component excluded (SCE). The data were acquired in triplicates on two different spots of the sample. The standard deviation of the three replicates in terms of ΔE was below 0.1.

2.4. *Chromatographic analyses*
HPLC-DAD: a HPLC consisting of a PU-2089 Quaternary Pump with degasser, equipped with an autosampler AS-950 and coupled to a spectrophotometric diode array detector MD-2010 was used (Jasco International Co, Japan). The data were processed with ChromNav software. The working conditions were: DAD spectra acquisition in the 200-650 nm range every 0.8 sec with a resolution of 4 nm; liquid chromatographic separation at room temperature (30 °C) on an analytical reverse phase column TC-C18 (2) (4.6 x 150 mm, particle size 5 μm, Agilent) with a precolumn TC-C18 (2) (4.6 x 12.5 mm, particle size 5 μm, Agilent), flow rate 1 mL/min and injection volume of 50 μL. The two eluents were 0.1% (v/v) TFA in water (A) and 0.1% (v/v) TFA in ACN (B). The elution program was 85% A for 5 min, then to 50% A in 25 min, then 30% A in 10 min, then to 0% in 1 min and hold for 4 min, and re-equilibration took 7 min.
LC-ESI-Q-ToF: a HPLC 1200 Infinity, coupled with a Quadrupole-Time of Flight tandem mass spectrometer 6530 Infinity Q-ToF detector by a Jet Stream ESI interface (Agilent Technologies, (USA)), was used for confirmation of the chromatographic peak assignments [5].
The sample pre-treatment was the same for both the instrumentations. C.a. 0.1 mg of each paint model system was submitted to extraction with 700 μL of dimethylsolfoxide (DMSO). The solution was sonicated for 10 min at 30° C, filtrated using a PTFE filter and then injected in the chromatographic system.

IOP Conf. Series: Materials Science and Engineering **364** (2018) 012066 doi:10.1088/1757-899X/364/1/012066

2.5. *Micro-FTIR analyses*
Synchrotron radiation micro-Fourier Transform Infra-Red Analysis (micro-FTIR) was carried out at the IRIS beamline (BESSY II) at Helmholtz-Zentrum, Berlin (Germany) on a Thermo Nicolet Continuum™ microscope equipped with a mercury–cadmium–telluride (MCT) detector. The sample was mounted on a diamond cell on a motorized microscope stage and raster scanned through the synchrotron beam with a diameter of 15 μm collecting a grid-like pattern of IR spectra spaced in 10-μm increments. Measurements were performed in transmission mode at a magnification ×32 using confocal objectives. Infrared spectra were registered between 4000 and 650 cm−1 with a spectral resolution of 4 cm−1. An accumulation of 128 scans per point was used. The spectrum and mapping acquisition was performed by using the OMNIC Atlμs™ software. The sample preparation protocol consisted in the microtoming at 10 μm of the samples with an embedding-free approach as described in the literature [6].

2.6. *Luminescence measurements*
Luminescence measurements were performed at IPANEMA laboratory (Saint Aubin, France). A spectroradiometer SPECBOS 1211 UV (JETI Technische Instrumente GmbH, Jena) that allows collecting measurements in irradiance mode was used. The spectroradiometer was equipped with a focusing optics allowing a collection area of 0.5 mm diameter. A CoolLed model PE-4000 (Andover, UK) was used as excitation source. The detection range was 350-1100 nm and integration time was 626 ms.

3. Results and discussion

The study of the reference materials and aged specimens through the plethora of analytical techniques available allowed to evaluate the fading of eosin and to assess the presence of possible degradation products.
In particular, the application of colorimetry to the paint replica prepared with linseed oil allowed to quantify the color change after artificial ageing. Figure 2 reports the color changes observed in the course of accelerated ageing, including the shift in the values of lightness/darkness (L*), redness/greenness (a*), yellowness/blueness (b*) and total color (E*). It is important to highlight that in theory, only $\Delta E^* > 0.8$ are noticeable by the human eye [7, 8]. In this case, the lightness sharply increases after the first few hours of accelerated ageing, and the ΔE changes abruptly, as a* and b* slowly approach zero [4].
The paint replicas were analysed, after solubilisation in DMSO. The chromatographic profiles obtained before and at the end of the artificial ageing experiments for the paint replica containing eosin and linseed oil are reported in Figure 3. The presence of degradation products can be appreciated in the chromatogram of the aged paint replica (d1 and d2), whose DAD spectra features absorbance in the range of yellow compounds. Moreover, it is worth noticing that the chromatogram of standard eosin (Eo) presents a peak due to tribromofluorescein or dibrominated eosin (DBEo), possibly as a synthesis by-product. The nature of the synthesis by-product and of degradation products was confirmed by mass spectrometric detection and is the object of a forthcoming publication.
The decrease in the concentration of eosin in the chromatogram of the extract of the aged sample correlated with the increase observed for L*. Nonetheless, the loss of eosin was not as dramatic as expected from the evaluation of the ΔE change, suggesting that the analysis of the bulk sample by chromatographic analysis only yields partial information. Thus, experiments aimed at locating eosin in the paint cross section were performed.

IOP Conf. Series: Materials Science and Engineering **364** (2018) 012066 doi:10.1088/1757-899X/364/1/012066

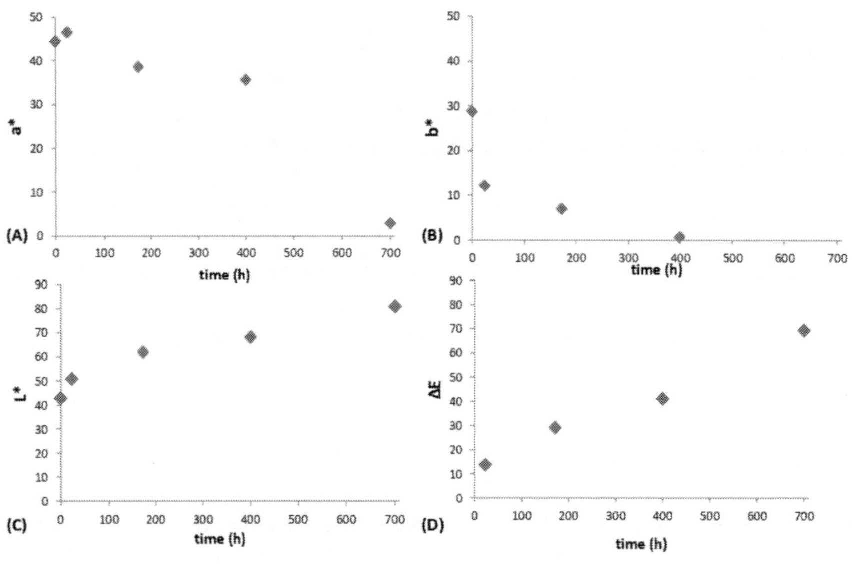

Figure 2: Results of the colorimetric analysis of the paint replica subjected to increasing doses of UV-Vis light (accelerated ageing) in CIELAB coordinates: values of a* (**A**); b* (**B**); L* (**C**) and ΔE (**D**) with ageing time.

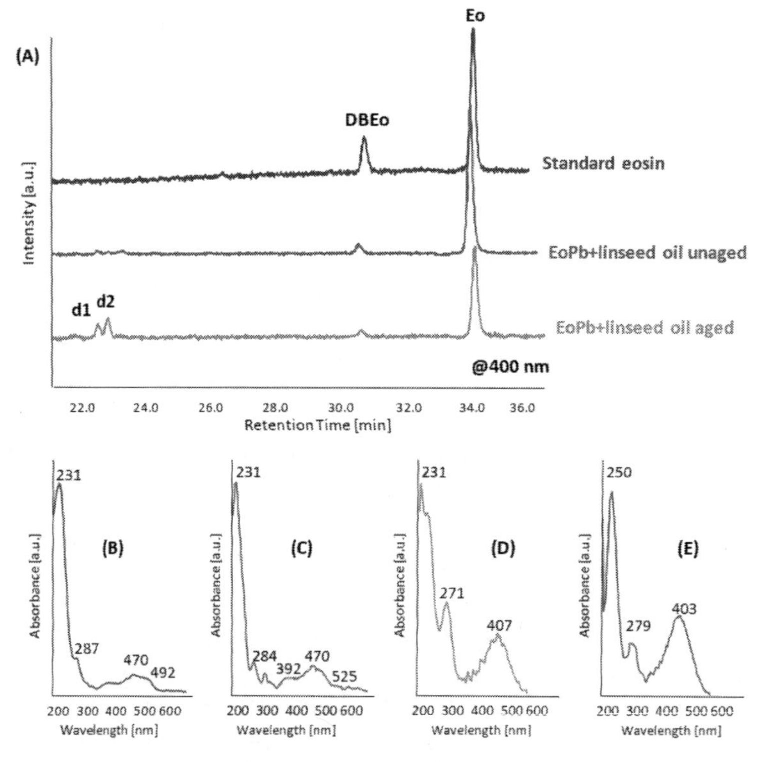

Figure 3: (**A**) Chromatographic profiles at 400 nm of the DMSO extracts of the reference paint replicas before and after artificial ageing, along with the profile obtained by the analysis of a DMSO solution of standard eosin; UV-Vis spectra of (**B**) eosin, Eo; (**C**) debrominated eosin, DBEo; (**D**) ageing product d1; (**E**) ageing product d2.

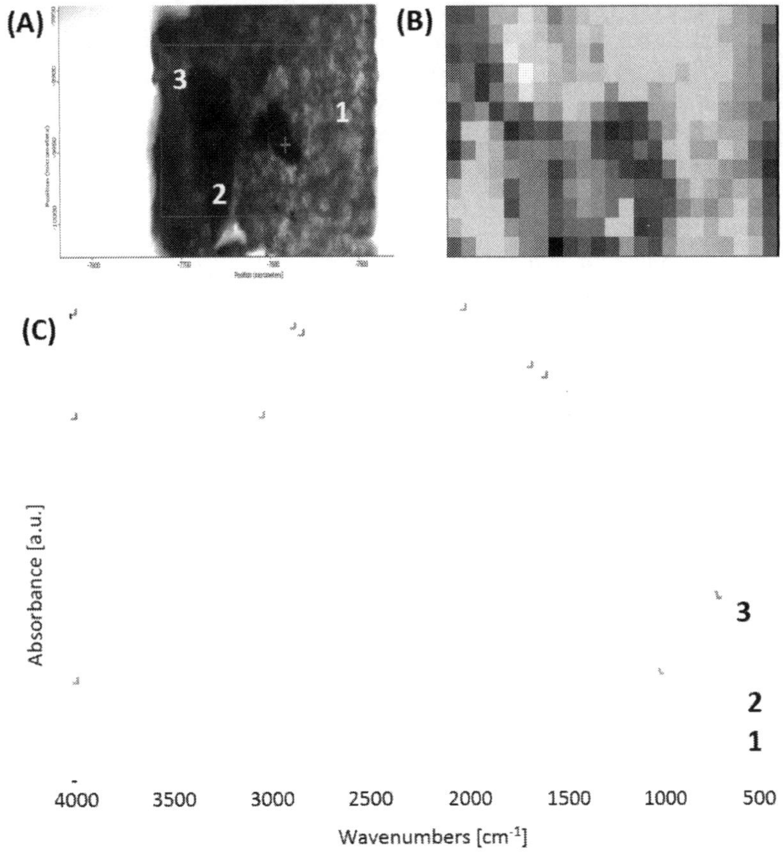

Figure 4: SR μ-FTIR study of aged PMS admixed with lead white: (**A**) microphotograph of the thin section analyzed. The area mapped (red square) shows the red paint layer with big red grain of eosin and the faded superficial layer; (**B**) chemical maps of a carbonyl group stretching at 1739 cm^{-1} corresponding to the oil as binder (green areas), eosin (1550 cm^{-1}, red areas), band at 1660 cm^{-1} from an unknown compound (blue area) (**C**) SR μ-FTIR spectrum of the three points labeled in (A).

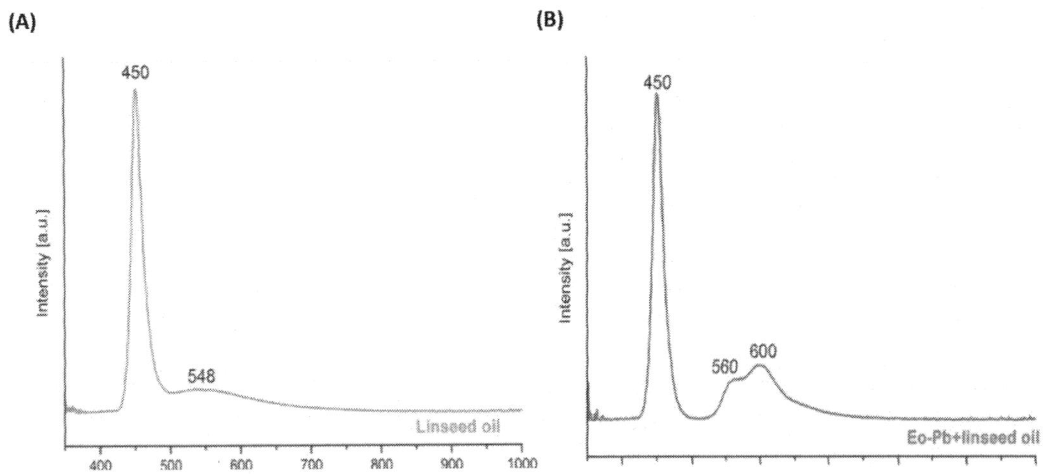

Figure 5: Luminescence spectra of the reference paint replicas: (**A**) linseed oil, used as blank; (**B**) eosin in linseed oil. (λ_{ecc}= 450 nm).

IOP Conf. Series: Materials Science and Engineering **364** (2018) 012066 doi:10.1088/1757-899X/364/1/012066

The application of synchrotron radiation-based techniques allowed us to identify different areas in the paint cross-section, namely a superficial layer where eosin was almost completely faded, and an inner layer where particles of the organic pigment were still visible. The results are depicted in Figure 4.
In order to increase our ability to detect eosin in very degraded samples, both by microdestructive (i.e. chromatographic) and non-destructive (i.e. spectrophotometric) techniques, the applicability of luminescence related methods was investigated (Figure 5).

4. Conclusions and perspectives

The application of a multi-analytical approach to the understanding of the behavior of eosin lakes in paintings highlighted the need for sensitive and selective methods. Most important, it was proved that the necessary following step will entail the use of mapping techniques to establish the chemical distribution of eosin-based compounds within the micrometric layers of paint model systems before and after ageing and to evaluate the influence and interaction of the whole paint matrix in eosin degradation process. Work in progress comprises the interpretation of the fading mechanism in eosin lakes through spectroscopic and spectrometric detection, in order to elucidate the nature of the degradation products. Moreover, the application of fluorescence and luminescence based analyses is ongoing.

Acknowledgements. Part of this research has been developed in the framework of the IPERION-CH project (Task 7.2) in collaboration with Eva Eis from Kremer pigmente (Germany), Industria Maimeri a Fila Group Company (Milano, Italy), (Berlin, Germany), Ancient Materials Research platform IPANEMA (France). We also thank HZB (Berlin) for the allocation of synchrotron radiation beam time (proposal 15102267-ST) and Ulrich Schade from the Institute for Material Development (Berlin, Germany) for her invaluable help while working at the beamline station.

References

[1] de Keijzer M and van Bommel M 2017 *The diversity of Dyes in history and archaeology* (London/Archetype Publications) p 326
[2] de Keijzer M, van Bommel M and Geldof M 2009 *EU-Artech Symposium on Vincent van Gogh and Contemporaries* (Amsterdam)
[3] Geldof M, de Keijzer M, van Bommel M, Pilz K, Salvant J, van Keulen H and Megens LL 2013. *Van Gogh's studio practice,* Vellekoop M, Geldof M, Hendriks E, Jansen L, deTagle A, editors (Brussels/Mercatorfonds)
[4] Ghelardi E, Degano I, Colombini MP, Mazurek J, Schilling M, Khanjian H and Learner T 2015 *Dyes Pigm* **123** 396–403
[5] Degano I, Tognotti P, Kunzelman D and Modugno F 2017 *Heritage Sci* **5** 7–22
[6] Pouyet E, Lluveras-Tenorio A, Nevin A, Saviello D, Sette F and Cotte M 2014 *Anal Chim Acta* **822** 51–9
[7] Leon K, Mery D, Pedreschi F and Leon J 2006 *Food research international* 39 (10), 1084–1091
[8] Furferi R and Governi L 2011 Color Research & Application **36** (3) 179–191.

IOP Conf. Series: Materials Science and Engineering **364** (2018) 012067 doi:10.1088/1757-899X/364/1/012067

Determination of material parameters of thermal insulation boards for the application on interior side of historical walls

Lukáš Fiala, Miloš Jerman and Robert Černý

Department of Materials Engineering and Chemistry, Faculty Civil Engineering, Czech Technical University in Prague, Czech Republic

fialal@fsv.cvut.cz

Abstract. At the reconstruction works on historical buildings, considerable financial means are spent. Therefore, it is desirable to assess the durability of applied materials in the particular conditions of a specific building. This cannot be done effectively without the knowledge of their hygric and thermal properties which can be used as input parameters of computational models. In this paper, hygric and thermal properties of several types of materials which can be used as interior thermal insulation layers in reconstructions of historical buildings on Czech territory, among them calcium silicate, hydrophilic mineral wool, and sheep wool, are investigated. Experimental results show that all analyzed materials can be considered as suitable for interior thermal insulation systems of historical building envelopes, in general. Their hygrothermal performance in specific applications should though be verified using computational simulations, taking into account the complex character of some historical construction systems.

1. Introduction

The interior thermal insulation of exterior walls is not an easy task but it can be considered as an effective solution how to save energy and increase thermal comfort for historical constructions. The application of interior thermal insulation systems on historical buildings should be analyzed from the point of view of their composition and hygrothermal performance because there are many risks, such as interstitial condensation or frost damage. There are two main possibilities how to build an interior insulation system, either it can be tight or it can be based on a capillary active material. The main risk of a tight system is that the external walls are not able to evaporate accumulated moisture; it can lead to decreasing of freeze/thaw resistance [1–3] and increasing the moisture content in the interior. If a capillary active system is used, capillary active thermal insulation is able to redistribute the condensed water. It can lead to higher moisture content in the thermal insulation material and to an increase of its thermal conductivity [4]. As there are substantial risks for both systems, the interior thermal insulation must be designed very carefully.

Computational models present a very effective tool for the hygrothermal assessment of renovated structures, which can then be used for the service life assessment and possible planning of subsequent reconstructions, but computer simulations should be performed with care because the hygrothermal analysis of any historical building can involve quite a few uncertainties which may affect negatively the accuracy of obtained results. The quality of computer simulations depends on the quality of input parameters. One of them are external climatic data, such as temperature and relative humidity, solar radiation, and driving rain over one year, and prescribed indoor conditions like temperature and relative humidity. Very important input parameters are initial conditions as the initial moisture content in a building construction. Other important input parameters are material characteristics of all components of the studied system. The basic physical properties like bulk density, open porosity and matrix density are fundamental parameters in that respect. Water vapor transport and storage is expressed by water vapor diffusion coefficient and sorption isotherm. Among liquid water transport parameters, water absorption coefficient and moisture diffusivity in dependence on moisture content are utilized in the models. Thermal properties are represented by thermal conductivity and specific heat capacity in dependence on moisture content. In this paper complete sets of heat and moisture transport and storage parameters of selected thermal insulation materials in dependence on moisture content are determined.

2. Investigated materials

Two modern synthetic materials and one natural were investigated. Calcium silicate (CS) is a material which often serves also as an insulation for high temperatures. The second material is mineral wool (MW) which is primarily used for plant growing and green roofs. Both materials have a high ability to transport liquid water, therefore they can be utilized in interior thermal insulation system. The third material is sheep wool (SW).

3. Experimental methods

3.1. Basic physical properties

As fundamental physical material characteristics, the bulk density ρ [kg·m^{-3}], the open porosity ψ [Vol.-%] and the matrix density ρ_{mat} [kg·m^{-3}] were determined. For synthetic material calcium silicate and mineral wool the water vacuum saturation method [5] was used. Because the sheep wool is a deformable material, its bulk density was determined using the mass and linear dimensions. The matrix density of sheep wool was measured by water vacuum saturation method.

3.2. Water vapor transport properties

The water vapor permeability characterizes the material ability to transport water vapor under a vapor pressure gradient. In this study the dry cup method was employed [5]. The sealed cups containing silica gel were placed into a controlled climatic chamber with 50% relative humidity and weighted periodically. Measurements were carried out for two weeks at temperature 25 °C. The water vapor diffusion permeability δ [s] was then calculated according to the equation

$$\delta = \frac{\Delta m \cdot d}{S \cdot \tau \cdot \Delta p_p}, \qquad (1)$$

where Δm [kg] is the amount of water vapor diffused through the sample, d [m] is the sample thickness, S [m^2] is the specimen surface, τ [s] is the period of time corresponding to the transport of mass of water vapor Δm, and Δp_p [Pa] is the difference between partial water vapor pressure in the air under and above specific specimen surface.

The water vapor diffusion coefficient D [m^2·s^{-1}] and the water vapor diffusion resistance factor μ [-] were determined on samples with dimensions of 100 x 100 x 20 mm and calculated according to Eq. 2 and Eq. 3, respectively.

$$D = \delta \cdot \frac{R \cdot T}{M}, \qquad (2)$$

where R [J·mol^{-1}·K^{-1}] is the universal gas constant, M [kg·mol^{-1}] is the molar mass of water and T [K] is the absolute temperature.

$$\mu = \frac{D_a}{D}, \tag{3}$$

where D_a [m^2 s^{-1}] is the diffusion coefficient of water vapor in the air.

3.3. Liquid water transport properties

The water absorption coefficient A [kg·m^{-2}·s$^{-1/2}$] and the apparent moisture diffusivity κ [m^2·s^{-1}] were measured using water sorptivity experiment [6, 7]. The specimens were water- and vapor-proof insulated on four lateral sides and the face side was immersed 1-2 mm into water. Increase of the mass was then recorded by automatic balance. Measurements were conducted on samples with square cross sectional area of dimension 100 mm and the water absorption coefficient A [kg·m^{-2}s$^{-1/2}$] was then calculated by formula

$$i = A \cdot \sqrt{t}, \tag{4}$$

where i [kg·m^{-2}] is the cumulative water absorption, t [s] is the time of the suction experiment. The water absorption coefficient was then employed for the calculation of the apparent moisture diffusivity in the form [6]

$$\kappa \approx \left(\frac{A}{w_c - w_0}\right)^2, \tag{5}$$

where w_c [kg·m^{-3}] is the saturated moisture content and w_0 [kg·m^{-3}] the initial moisture content. Inverse analysis of moisture profiles using Boltzmann transformation was used for determination of the moisture diffusivity dependence on the moisture content. For such calculations, basically just one moisture profile is needed. However, the more measured profiles, the higher accuracy of the method. Such method consists in application of Boltzmann transformation which converts all measured moisture profiles in various times into a single profile. The method is supposed to be used only when the boundary condition on the dry end of the specimen is not yet effective [8].

$$\kappa(u_x) = \frac{1}{2t_0\left(\frac{du}{dx}\right)_{x_0}} \int_{x_0}^{\infty} x \frac{du}{dx} dx, \tag{6}$$

where u [kg kg^{-1}] is the moisture content by mass, u_x [kg kg^{-1}] is the value of moisture content at the position x = x$_0$. The moisture profiles were measured by the gravimetric method. The moisture diffusivity was calculated from 3 moisture profiles within this paper.

3.4. Sorption isotherms

The water vapor adsorption isotherms were measured using the desiccator method [5]. The samples were placed into the desiccators with different salt solutions to simulate different values of relative humidity. The initial state was dry sample. The mass of samples was measured in specified periods of time until steady state value of mass was achieved. Then, the moisture content by mass was calculated.

3.5. Thermal properties

The thermal conductivity λ [W·m^{-1}·K^{-1}] and the specific heat capacity c [J·kg^{-1}·K^{-1}] were measured using the commercial device Isomet 2104 (Applied Precision, Ltd.). The measurement is based on

IOP Conf. Series: Materials Science and Engineering **364** (2018) 012067 doi:10.1088/1757-899X/364/1/012067

analysis of temperature response of the analysed material to heat flow impulses. The heat flow is induced by electrical heating using a resistor heater having a direct thermal contact with the surface of the sample. The measurement of sheep wool was conducted by needle probe on samples with dimensions of 70 x 70 x 300 mm. Calcium silicate and mineral wool were measured by using surface probe.

4. Experimental results

4.1. Basic physical properties

The basic physical properties are given in Table 1. The lowest bulk density achieved sheep wool, the highest calcium silicate. The porosity of all materials was very high.

Table 1. Basic physical properties

Material	ρ	ρ_{mat}	ψ
	[kg m^{-3}]	[kg m^{-3}]	[%]
MW	126.5	2608	95.1
CS	264.5	2508	89.5
SW	15.0	1155	98.7

4.2. Water vapor transport properties

The water vapor transport properties are shown in Table 2. All materials were able to transport water vapor very easily.

Table 2. Water vapor transport properties of studied materials

Material	5/50%		
	δ	D	μ
	[s]	[m^{2}s^{-1}]	[-]
MW	4.90E-11	6.74E-6	3.43
CS	6.15E-11	8.44E-6	2.83
SW	1.44E-11	1.32E-5	1.93

4.3. Liquid water transport properties

The water absorption coefficient and apparent moisture diffusivity are given in Table 3 and Figure 1. The lowest ability to transport liquid water exhibited sheep wool. The highest water absorption coefficient and apparent moisture diffusivity achieved hydrophilic mineral wool. The high ability to transport liquid water had also calcium silicate. The liquid water transport parameters are very important for interior thermal insulation materials. Therefore, the application of sheep wool should be considered with care.

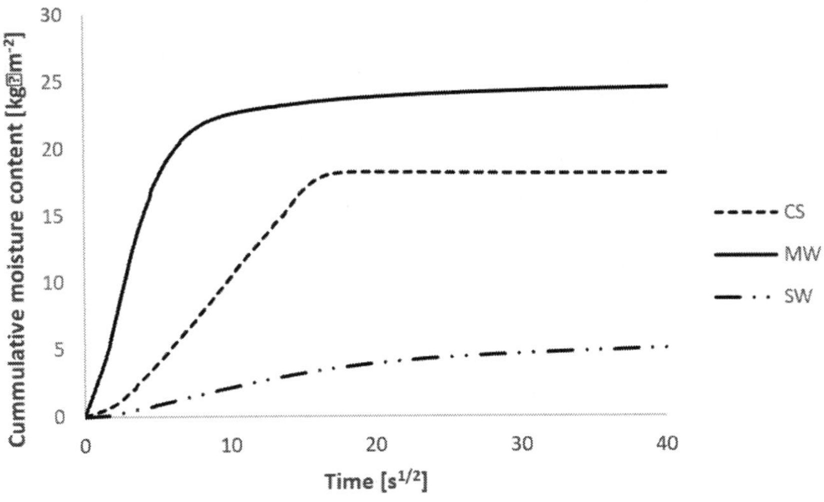

Figure 1. Water absorption as function of time

Table 3. Water absorption coefficient and apparent moisture diffusivity

Material	A	κ
	kg m^{-2}·s$^{-1/2}$	m^2·s^{-1}
Calcium silicate	1.256	2.31 E-6
Mineral wool	2.004	4.43 E-6
Sheep wool	0.07	4.42 E-9

Moisture diffusivity in dependence on moisture content is shown in Figure 2. The highest moisture diffusivity had mineral wool, for calcium silicate it was only slightly lower. The lowest ability to transport liquid moisture exhibited sheep wool. The results were in the agreement with the measurements of water absorption coefficient and apparent moisture diffusivity.

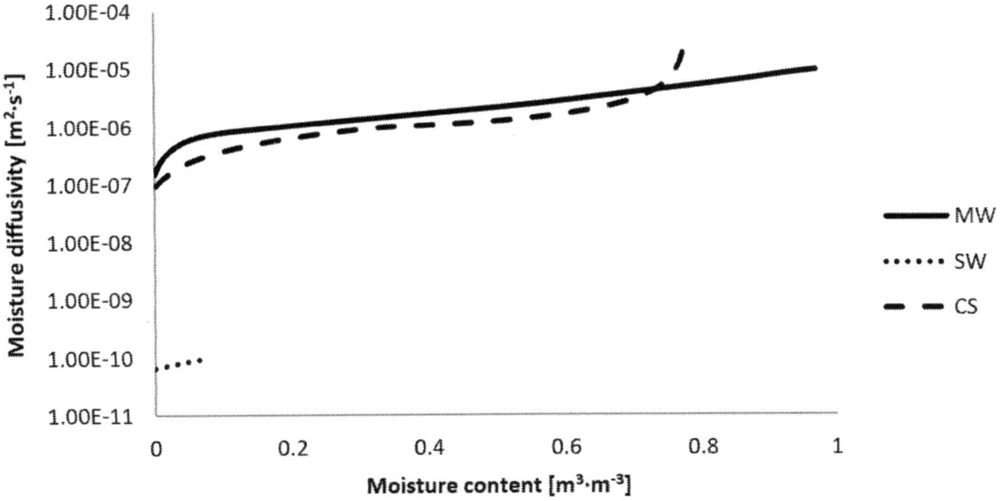

Figure 2. Moisture diffusivity in dependence on moisture content

IOP Conf. Series: Materials Science and Engineering **364** (2018) 012067 doi:10.1088/1757-899X/364/1/012067

4.4. Sorption isotherms

The sorption isotherms of all studied materials are shown in Figure 3. The highest capability of adsorbing gaseous moisture showed sheep wool. On the other hand, mineral wool adsorbed almost negligible amount of water vapor.

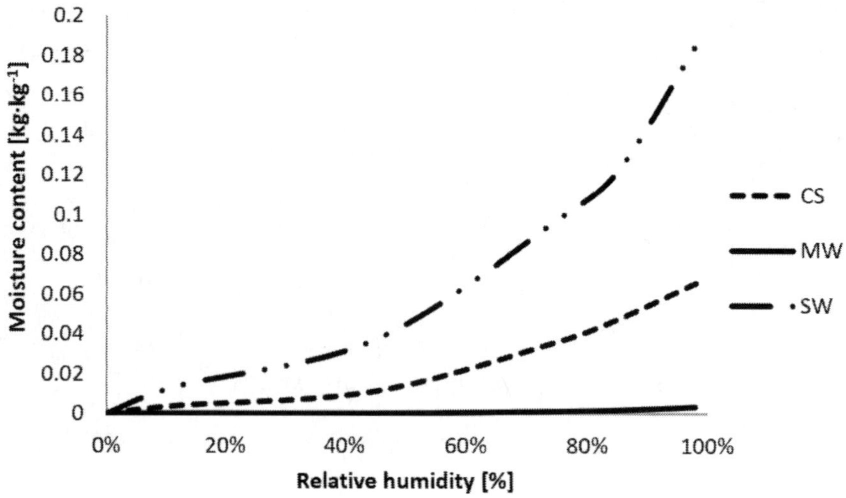

Figure 3. Sorption isotherms

4.5. Thermal properties

The values of thermal conductivity and specific heat capacity in dry state are shown in Table 4, thermal conductivity and specific heat capacity in dependence on moisture content are presented in Figures 4 and 5.

The highest thermal conductivity in dry state achieved calcium silicate; mineral wool and sheep wool exhibited almost 50% lower values. The effect of moisture content on thermal conductivity of all materials was very significant; they lost their insulation capabilities already for moisture contents as low as 5-10%. This underlined the high importance of their liquid water transport parameters.

The highest specific heat capacity in dry state achieved sheep wool, mineral wool and calcium silicate exhibited significantly lower values. The specific heat capacity as a function of moisture content was calculated according to the theoretical procedure described in [9], under the assumption that this heat storage parameter is an additive quantity in the sense of the linear theory of mixtures:

$$c_{wet}(w) = \frac{\rho_{dry} c_{dry} + \rho_w c_w w}{\rho_{dry} + \rho_w w}, \tag{7}$$

where ρ_{dry} [kg·m^{-3}] is the bulk density of dry material, ρ_w [kg·m^{-3}] is the density of water, w [m^3·m^{-3}] is the moisture content by volume, c_{dry} [J·kg^{-1}·K^{-1}] is the specific heat capacity of dry material, and c_w [J·kg^{-1}·K^{-1}] the specific heat capacity of water. The specific heat capacity of all three materials increased rapidly with increasing moisture content, which was due to high specific heat capacity of water.

IOP Conf. Series: Materials Science and Engineering **364** (2018) 012067 doi:10.1088/1757-899X/364/1/012067

Table 4. Thermal conductivity and specific heat capacity in dry state

Material	Thermal conductivity [W·m^{-1}·K^{-1}]	Specific heat capacity [J·kg^{-1}·K^{-1}]
Calcium silicate	0.0654	972
Mineral wool	0.0421	718
Sheep wool	0.0420	1940

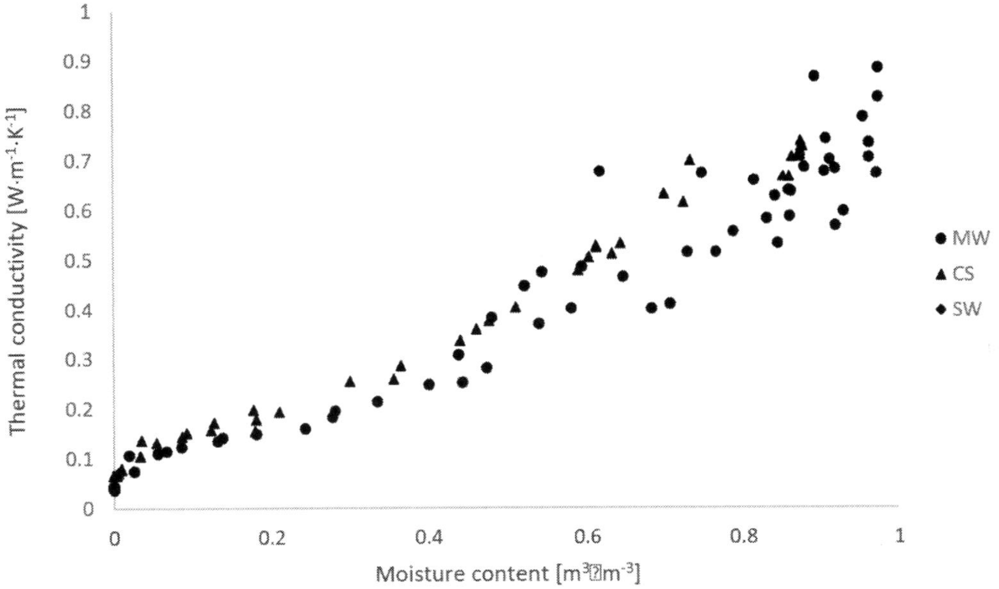

Figure 4. Thermal conductivity in dependence on moisture content

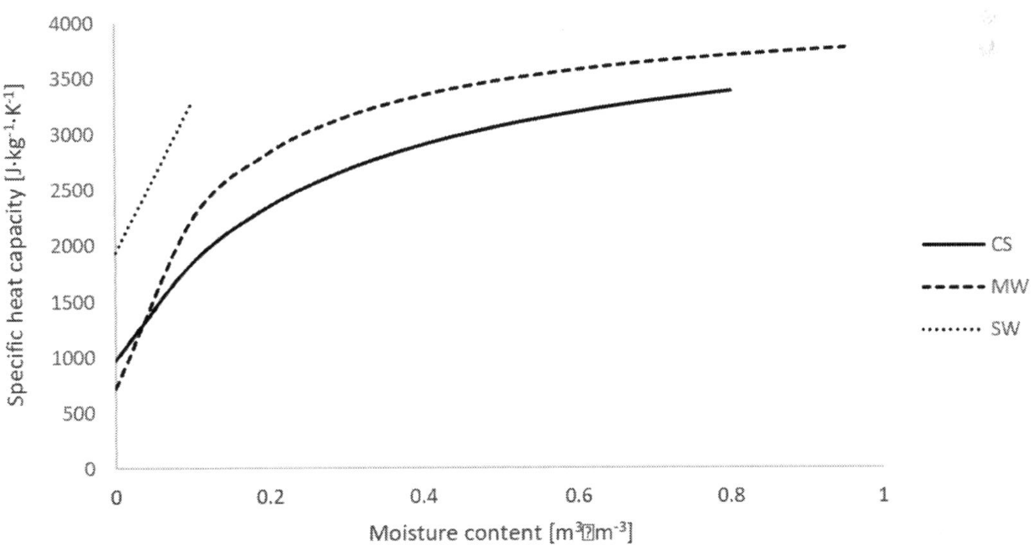

Figure 5. Specific heat capacity in dependence on moisture content

5. Conclusions

Three types of thermal insulation materials, namely calcium silicate, hydrophilic mineral wool and sheep wool were investigated in the paper. The main results can be summarized as follows. The lowest bulk density showed sheep wool, the open porosity of all materials was very high. All three thermal insulations were very open to water vapor transport. The highest ability to transport liquid water exhibited hydrophilic mineral wool, calcium silicate was only slightly worse. The lowest liquid water transport parameters had sheep wool. On the other hand, sheep wool presented a very high capability of water vapor adsorption, even higher than calcium silicate. Thermal conductivity of all three materials was sufficiently low for low moisture contents only. Their ability of transporting fast the liquid water was thus imperative for their proper function in building envelopes.

Acknowledgments

This research has been supported by the Ministry of Culture of the Czech Republic, under project No DG16P02H046.

References

[1] Zagorskas J, Zavadskas E K, Turskis Z, Burinskiene M, Blumberga A 2014 Thermal Insulation Alternatives of Historic Brick Buildingsin Baltic Sea Region, *Energy and Buildings* ***78*** pp 35-42

[2] Johansson P, Geving S, Hagentoft, C E, Jelle E B, Rognvik E, Kalagasidis A S, Time B 2014 Interior Insulation Retrofit of a Historical Brick Wall Vacuum Insulation Panels: Hygrothermal Numerical Simulation and Laboratory Investigation, *Building and Environment 79* pp 31-45

[3] Vereecken E, Gelder L V, Janssen H, Roels, 2015, Interior Insulation for Wall Retrofitting – A Probalististic Analysis of Energy Saving and Hygrothermal Risks, *Energy and Buildings* ***89*** pp 231-244

[4] Vereecken E, Roels S, 2015 Capillary Active Interior Insulation: Do the Advantages Really Offset Potential Disadvantages?, *Materials and Structures* ***48***, pp 3009–3021

[5] Roels, S., Carmeliet, J., Hens, H., Adan, O., Brocken, H., Černý, R., Pavlík, Z., Hall, C., Kumaran, K., Pel, L. & Plagge, R 2004 Interlaboratory Comparison of Hygric Properties of Porous Building Materials *Journal of Thermal Envelope and Building Science 27*, pp 307-325

[6] Kumaran, M K, 1999, Moisture diffusivity of building materials from water absorption measurements *Journal of Thermal Envelope and Building Science* ***22*** pp 349-355

[7] Vejmelková E, Pavlíková M, Jerman M, Černý, R, 2009, Free Water Intake as Means of Material Characterization *Journal of Building Physics,* ***33*** pp 29-44,

[8] Z. Pavlík, R. Černý, 2012, Determination of moisture diffusivity as a function of both moisture and temperature *International Journal of Thermophysics* ***33*** pp 1704–1714

[9] Jerman, M, Černý R, 2012 Effect of moisture content on heat and moisture transport and storage properties of thermal insulation materials *Energy and Buildings* ***53*** pp 39–46

Basic physical, mechanical and hygric properties of renders suitable for historical buildings

E Vejmelková, M Keppert, M Jerman & R Černý

Department of Materials Engineering and Chemistry, Faculty Civil Engineering, Czech Technical University in Prague, Czech Republic

eva.vejmelkova@fsv.cvut.cz, martin.keppert@fsv.cvut.cz, milos.jerman@fsv.cvut.cz, cernyr@fsv.cvut.cz

Abstract. Several renders suitable for historical buildings were designed and their basic physical, mechanical and hygric properties were tested. Lime was used as the main binder, three different pozzolanic materials, namely metakaolin and metashale as commercially produced admixtures and brick dust as a waste material from precise brick blocks manufacturing, were utilized as secondary binders instead of cement. Metakaolin and metashale showed a very good performance, as for the mechanical and water transport properties of renders, significantly increasing the compressive strength and decreasing the water absorption coefficient. On the other hand, brick dust was not found effective in that respect.

Keywords: renders, historical buildings, lime, pozzolans, properties

1. Introduction

It is a high concern of every nation to preserve the value of historical buildings as a part of its cultural heritage. Therefore, the restoration of such monuments is provided under the control of competent authorities, which specify the conditions of the restoration process and the used materials. Since the use of cement as a component of plasters dates from the twenties of the 20th century, it is common, that the historical buildings are covered with lime plasters and so a renewal plaster must be of as much as possible similar composition. However today, the lime plasters suffer greatly from the increasingly more polluted environment. Calcium carbonate, as the product of lime plaster hardening, carbonation, is easily decomposed by exhaust acid gases [1]. On the other hand, it is known, that from the ancient times pozzolanic materials have been used as lime plaster admixtures to enhance its properties in both fresh and hardened state [2,3,4]. These were mostly volcanic materials, which are not available in the Czech Republic, but there are many other materials with pozzolanic properties on the market, either in the form of wastes from industrial processes or manufactured intentionally for the use as admixture to binders.

The pursuit is to understand better the action of the available pozzolanic materials in the hardening process of plasters and their influence on the plasters´ final properties. These are profusely used in concrete and cement composites taking part in the hydration process [5]. When incorporated into lime plaster, they hydrate to form calcium silicates and aluminates, similar to products of Portland cement hydration. The altered microstructure of the plaster is then responsible for better resistance to harsh environment and longer durability [6,7].

IOP Conf. Series: Materials Science and Engineering **364** (2018) 012068 doi:10.1088/1757-899X/364/1/012068

In this paper, a comparison of the action of three different pozzolanas in the lime plasters is presented. Brick dust is a by-product of precise brick blocks manufacture, which is generated in high amounts and need to be utilized somewhere. Due to its fineness and good pozzolanic properties it is a prospective material for use as an admixture in binders. Metakaolin and metashale are burnt and grinded intentionally in order to achieve specific qualities typical for binder admixtures. The comparison is made on the basis of basic physical properties, pore characterization of the plasters, mechanical properties and hygric properties.

2. Materials and methods

2.1. Materials

Figure 1 gives the schematic composition of studied plasters. The main binder was hydrated lime completed with 20% of a pozzolan, except from the reference mixture, which was bonded just by the lime hydrate. Silica sand of the 0-4 mm fraction was used as filler, its granulometry is shown in Figure 2. Water was dosed in order to achieve good consistency, i.e., the flow of the fresh mixture 165 mm in diameter. Three types of pozzolans were used, brick dust represented a waste material, while metakaolin and metashale were representatives of intentionally manufactured admixtures. Figures 3 and 4 show the distribution curves of the raw binder materials – lime, metakaolin, metashale and brick dust and Table 1 gives their chemical composition. After manufacture the plasters were marked as LR, LK, LL and LB, which stands for lime-reference, lime-metakaolin, lime-metashale and lime-brick dust plaster, respectively.

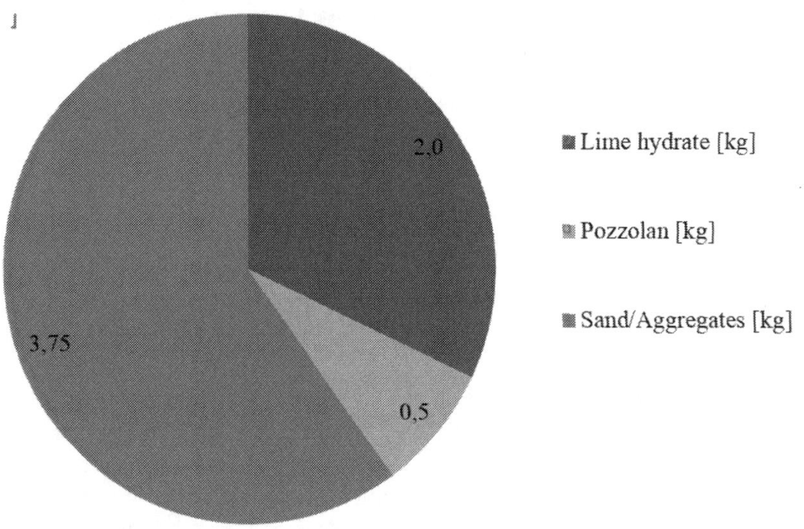

Figure 1. Composition of studied plasters.

IOP Conf. Series: Materials Science and Engineering **364** (2018) 012068 doi:10.1088/1757-899X/364/1/012068

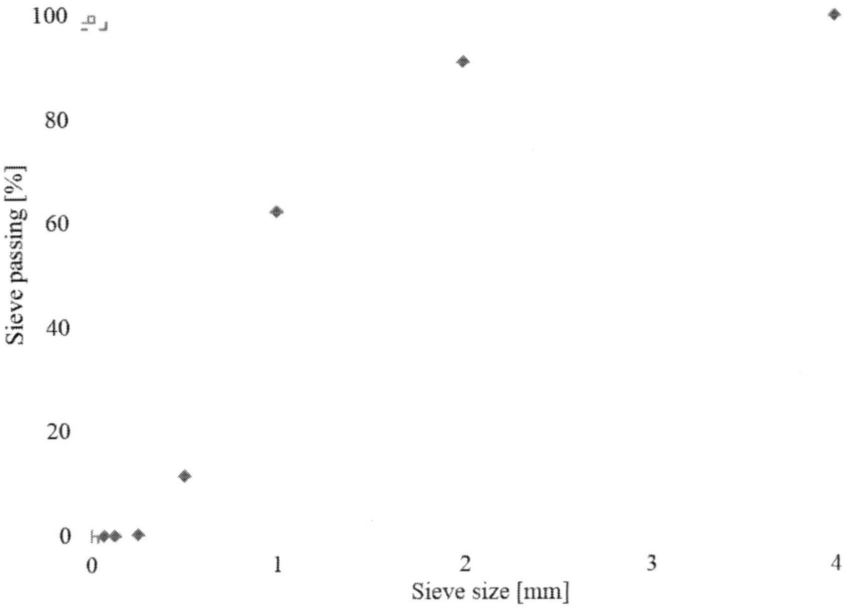

Figure 2. Granulometric curve of silica aggregate.

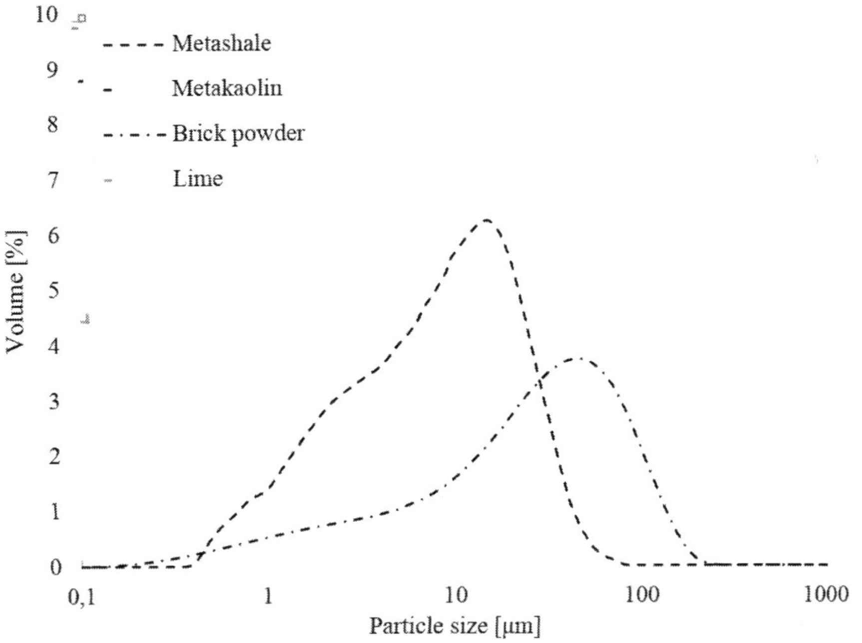

Figure 3. Particle size distribution of binders.

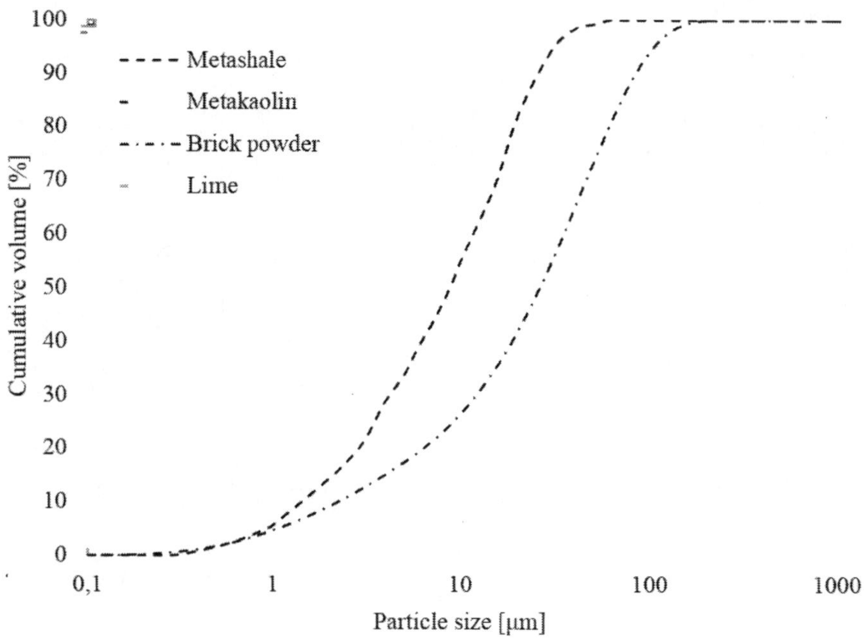

Figure 4. Cumulative curve of particle size distribution of binders.

Table 1. Chemical composition of binder components.

Component	Metakaolin	Metashale	Lime hydrate	Brick dust
Al_2O_3 [%]	38.5	41.9	-	20.0
SiO_2 [%]	58.7	52.9	-	51.3
K_2O [%]	0.9	0.8	-	3.2
Fe_2O_3 [%]	0.7	1.1	-	6.0
TiO_2 [%]	0.5	1.8	-	0.8
MgO [%]	0.4	0.2	0.5	4.5
CaO [%]	0.2	0.1	99.3	11.5
SO_3 [%]	-	-	0.1	1.0
Na_2O [%]	-	-	-	1.3

2.2. Methods

2.2.1. Basic physical properties and pore system characterization. Bulk density ϱ_v [kg m^{-3}], matrix density ϱ_{mat} [kg m^{-3}] and total open porosity ψ_0 [%] were measured as the basic physical properties. The used method was water vacuum saturation method [8].

The pore system characteristics were measured by means of mercury intrusion porosimetry. The experiments were carried out by Pascal 140 and Pascal 440 devices.

2.2.2. Mechanical properties. To characterize mechanical properties, compressive and flexural strength were determined on 40x40x160 mm samples according to ČSN EN 1015-11[9]. Firstly, the three-point flexural strength test was carried out and then on the left fragments the compressive strength test was performed.

2.2.3. Hygric properties. Absorption experiment was performed to assess the water transport properties characterized by the water absorption coefficient A [kg $m^{-2}s^{-1/2}$] and moisture diffusivity $\varkappa$ [m^2s^{-1}]. The 50 x 50 x 50 mm samples were immersed to water to the depth of 1 – 2 mm and their mass increases were recorded by a digital scale and saved automatically by a computer program. The generated curve of cumulative mass increase served for the determination of both parameters [10].

Water vapour transport parameters were measured by dry and wet cup methods. The cups were placed in a controlled climatic chamber with 50% relative humidity. In the dry cup method silica gel was used, while in the wet cup method water was placed in the cup, which induced partial pressures causing the transport of water vapour through the material. From the records of mass changes of studied samples, the water vapour diffusion resistance factor μ [-] was calculated [11].

3. Experimental results

3.1. Basic physical properties

Table 2. Basic physical properties of studied plasters.

Plaster	ρ_v [kg m^{-3}]	ρ_{mat} [kg m^{-3}]	ψ_o [%]
LR	1805	2879	28.8
LK	1767	2638	33.2
LL	1741	2688	32.2
LB	1761	2598	32.0

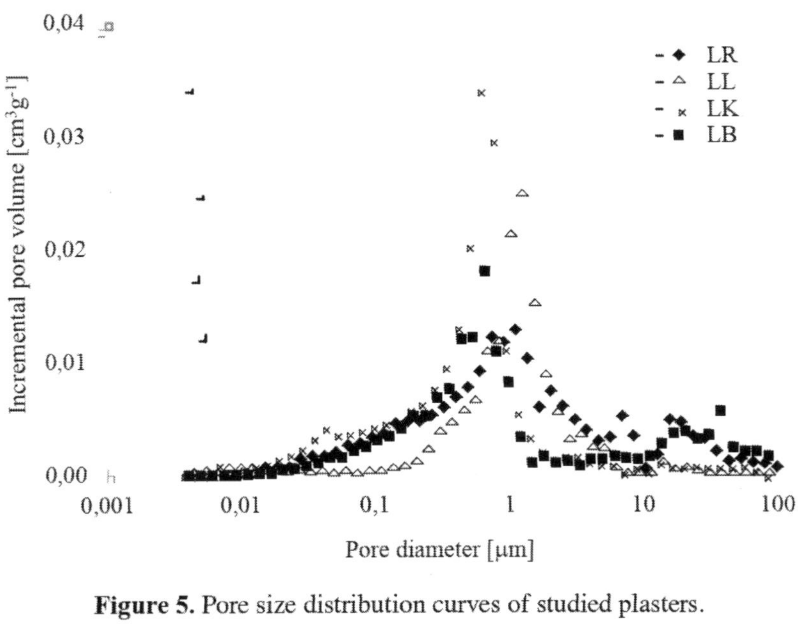

Figure 5. Pore size distribution curves of studied plasters.

Table 2 provides the values of basic physical properties, i.e., bulk density ρ_v [kg m^{-3}], matrix density ρ_{mat} [kg m^{-3}] and total open porosity ψ_0 [%]. Bulk density went down with the addition of all pozzolanas, most pronouncedly for plaster with metashale – LL, which was within 6%. The changes in matrix density were slightly higher, reaching the lowest value for plaster containing brick dust, which was by 10% lower compared to the reference plaster. It can be seen, that the presence of a pozzolana increases the total open porosity, in the worst case by 13%, which was recorded for metakaolin plaster.

Figure 5 shows the pore size distribution curves of studied plasters. While the LK and LL systems are strictly unimodal with majority of pores around 1 μm, the reference and brick containing mixtures LR and LB contain also certain volume of larger pores of diameter 15-100 μm. One observes re-fining effect of metakaolin and metashale in the plaster. It is caused by finer microstructure of ITZ zone compared to just lime-sand mixture due to the presence of CSH and CAH. The porosity results also indicate that brick dust pozzolan was not efficient enough to refine the ITZ.

3.2. Mechanical properties
Figure 6 gives values of mechanical properties – both compressive and flexural strength. the extreme value of compressive strength was recorded for lime plaster containing metakaolin – LK, which was almost 5 times higher compared to the reference mixture. Also metashale incorporation into the lime plaster mixture caused an increase in compressive strength, but in lesser magnitude – about 2.3 times higher than the reference. Plaster with brick dust showed comparable values of strength to reference mixture.

The values of flexural strength follow the same trend but with lower differences. The highest value for LK plaster was 2.7 times higher than reference, while metashale plaster showed 1.4 times higher value. The mechanical properties show that metakaolin and metashale are – at least in the studied system – more effective pozzolans than brick dust.

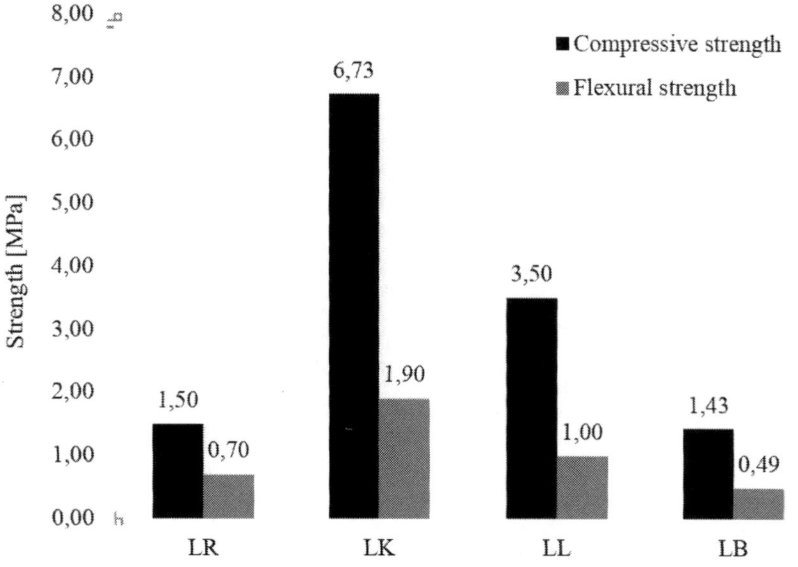

Figure 6. Compressive and flexural strength of studied plasters.

3.3. Hygric properties
Results of the water absorption test characterizing water transport properties are displayed in Table 3. The water absorption coefficient A [kg m^{-2}s$^{-1/2}$] and moisture diffusivity $\varkappa$ [m^{2}s^{-1}] follow the same trend, showing that the reference mixture demonstrated the highest ability to water transport. By 35% lower value was recorded for plaster with metashale admixture, metakaolin caused decrease by 27% and brick

dust only by 6%. These results are in good agreement with the porosimetry, where the ability of metakaolin and metashale to refine the large – capillary active – pores was documented.

Water vapour transport properties were characterized by the water vapour diffusion resistance factor [-] obtained by wet and dry cup methods (Figure 7). It can be seen, that reference plaster values are comparable with plaster containing metakaolin and metashale, only wet cup method showed an increased value for metakaolin plaster. However, brick dust caused an increase in measured values obtained by both wet and dry cup methods, in case of dry cup the water vapour diffusion resistance factor reached 2 times higher value. As it is favourable for plasters to be open to diffusion, but the liquid transport should be limited, it is not easy to reach these two somewhat contrary qualities in one layer plaster. Nevertheless, the low water absorption coefficient and satisfactory diffusion properties favour metakaolin and metashale plasters.

Table 3. Water transport properties of studied plasters.

Plaster	A [kg m^{-2}s$^{-1/2}$]	κ [m^{2}s^{-1}]
LR	0.206	8.80E-07
LK	0.151	2.02E-07
LL	0.133	1.71E-07
LB	0.194	4.47E-07

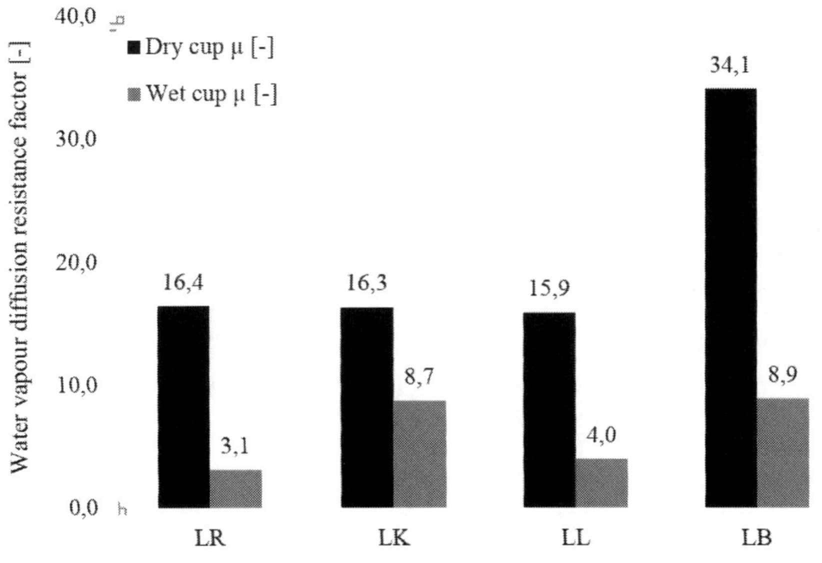

Figure 7. Water vapour transport properties of studied plasters.

4. Conclusions

Renovation plasters should maintain the cultural value of historical buildings and thus it is required to reach as much as possible similar composition as the original plaster had. That is the reason, why lime is used as the main binder component. However, lime suffers from pollution in today´s environment and thus its lifetime is limited. Therefore, it can be favourable to incorporate pozzolanic admixtures in the plaster, which has been used in plasters since ancient times, although with different origin and composition.

This paper investigated the properties of lime plasters containing pozzolanic materials – brick dust as a waste product, metakaolin and metashale as commercially manufactured admixtures. Studied properties were basic physical properties, pore system characterization, mechanical properties, liquid water and water vapour transport properties. Although all pozzolanic admixtures caused an increase in open porosity, the investigated properties were mostly enhanced or at least at the same level which was probably caused by their action in the hardening process, creating calcium silicate- and calcium aluminate hydrates as the reaction products. The best mechanical properties were found for plaster containing metakaolin in the means of both compressive and flexural strength, while water transport properties showed the most promising values in case of metashale incorporation into the plaster. Brick dust plaster demonstrated neither enhanced performance nor significant worsening of investigated parameters.

Acknowledgments
This research has been supported by the Ministry of Culture of the Czech Republic, under project No DG16P02H046.

References
[1] Bakolas A, Aggelakopoulou E and Moropoulou A 2008 *J. Therm. Anal. Calorim.* **92** 345–51
[2] Cabrera J and Rojas M F 2001 *Cement Concrete Res* **31** 177–82
[3] Černý R, Kunca A, Tydlitát V, Drchalová J and Rovnaníková P 2006 *Constr. Build. Mater.* **20** 849–57
[4] Charola A E, Rodrigues P F, McGhie A R and Henriques F M A 2005 *Rest. Build. Monum.* **11** 111–118
[5] He C, Osbaeck B and Makovicky E 1995 *Cement Concrete Res.* **25** 1691–1702
[6] O'Farrell M, Sabir B B and Wild S 2006 *Cement Concrete Comp.* **28** 790–99
[7] Valipour M, Pargar F, Shekarchi M and Khani S 2013 *Constr. Build. Mater.* **41** 879-88
[8] ČSN EN 1015-10: 2000 *Methods of test for mortar for masonry - Part 10: Determination of dry bulk density of hardened mortar* (Prague: Czech Standardization Institute)
[9] ČSN EN 1015-11: 2000 *Testing methods for mortar for masonry - Part 11: Determination of compressive and flexural strength of hardened mortars* (Prague: Czech Standardization Institute)
[10] ČSN EN 1015-18: 2003 *Methods of test for mortar for masonry - Part 18: Determination of water absorption coefficient due to capillarity action of hardened mortar* (Prague: Czech Standardization Institute)
[11] ČSN 1015-19: 2005 *Methods of test for mortar for masonry - Part 19: Determination of water vapour permeability of hardened rendering and plastering mortars* (Prague: Czech Standardization Institute)

IOP Conf. Series: Materials Science and Engineering **364** (2018) 012069 doi:10.1088/1757-899X/364/1/012069

The San Giovanni Baptistery in Florence (Italy): characterisation of the serpentinite floor

AP Santo, E Pecchioni, CA Garzonio

Università degli Studi di Firenze, Dipartimento di Scienze della Terra, Via La Pira 4, 50121 Firenze (Italy)

E-mail: alba.santo@unifi.it

Abstract. The San Giovanni Baptistery in Florence (Italy) is an outstanding worldwide monument, representative of Romanesque architecture, dating back at least to the XII century. The whole edifice is externally revetted by white marble of different provenance and, subordinately, of serpentinite, commercially known as "green marble". This latter is a rock formed from the transformation of ultramafic rocks during low-grade metamorphism or hydrothermal processes; it has been used for building and ornamental purposes since ancient times. In the San Giovanni Baptistery, in addition to the facade, the "green marble" was used on the floor, which consists of white and green tarsias forming an oriental motif. In this work, we studied, through a multi-analytical approach, the green tarsias to compositionally characterise the Baptistery's floor. At the same time, serpentinite samples from different ancient Tuscany quarries were collected and studied to compare their characteristics with those of the "green marble" from Baptistery floor.

1. Introduction

Serpentinites are commercially referred to as "Green Marble", but this name has no correspondence to their mineralogy, geochemistry and physical properties. Serpentinites are rocks forming, at low (100°C) to intermediate (700°C) temperature [1], in many different geologic settings, by the hydration of olivine-rich ultramafic rocks; they consist mostly of serpentine-group minerals (approximate formula $Mg_3Si_2O_5(OH)_4$) and subordinately by magnetite, brucite, talc, calcite and magnesite.

During the serpentinization process, the primary Mg-rich minerals (olivine, orthopyroxene, and clinopyroxene) are replaced by serpentine-group minerals. The typical combination of mineral products results in a fine-grained, dark green to black rock (commonly crossed by veins of variable composition), whose density is significantly lower than that of the primary peridotite [2]. The most common serpentine-group minerals are lizardite, chrysotile, and antigorite, characterised by different structures [3-5]. Lizardite, which displays a flat crystal structure, i.e. a triangular plate truncated at each corner [2], is the most common pseudomorph of olivine. Chrysotile, very poor in Al and Fe^{3+}, is characterised by cylindrical fibre even several centimetres in length, which generally fills secondary microfractures crosscutting the serpentinite; it is the main constituent of commercial asbestos, a material showing good outstanding electric, thermal, and phonic insulation properties, but its fine dust is thought to be deleterious to human health [6]. Antigorite, with its interlocking texture, show difficult cleavage and enhanced hardness and seismic velocities compared to those of lizardite and chrysotile; it commonly occurs in partially recrystallised and metamorphosed serpentinite.

The serpentine minerals are stable under a wide range of temperature and pressure; in particular, lizardite and chrysotile are stable at low temperature (0-300°C) and pressure (<1GPa) conditions but, at high water/rock ratio, chrysotile becomes more stable than lizardite [7]. Antigorite is considered to be the high-temperature high-pressure stable phase of serpentine [1]. The serpentinization process produces variation and deformation of the original rock structure with the consequent weakness of its mechanical properties. The resulting rock display relatively high porosity and absorption coefficient, and a high tendency to shear and failure. Also, the serpentinite can undergo a carbonation process trough precipitation of carbonates in its shears and fractures or transformation of serpentine depending on the availability of magnesium and calcium in the precursor rock. The carbonation process will obviously influence the serpentinite behaviour.

Due to its compositional and physical-mechanical characteristics, serpentinite displays moderate weathering resistance which also depends on the environment conditions in which it is placed and whether it is carbonated or not. Despite its fragile behaviour, this rock has been worldwide used since ancient times for ornamental and building purposes. In this paper we studied the "green marble" used on the floor of San Giovanni Baptistery in Florence; we present the first compositional results of a wider study realised in collaboration with the Opera di Santa Maria del Fiore aimed to the complete characterisation of the materials prior than their use during restoration and replacement work.

2. The San Giovanni Baptistery and the "green marble"

The San Giovanni Baptistery is an outstanding worldwide monument, representing the oldest religious monument in Florence (Italy). Due to the lack of primary sources and specific dating issues, the Baptistery origins have been debated for a long time and are still controversial: different hypotheses, covering seven centuries, have succeeded. Among the hypotheses, it was believed built over the ruins of a Roman temple dedicated to Mars, dating back to the 4th-5th century A.D., or edified during the Longobard period [8]. The archaeological excavations effected in the period 1895-1915 seemed to strengthen the thesis of a medieval origin [9]. Recently, Bernabei et al. [10] studied the timber structure of the dome and obtained a radiocarbon dating of the chestnut from the 11th-12th centuries; also, they dated, through the dendrochronological analysis, a silver fir element to 1268. This date, possibly, represents a replacement coincident with the period, between 1270 and 1300, when a substantial part of the mosaics of the dome was positioned [10].

Fig. 1 - Detail of Baptistery floor, showing green (serpentinite) and white (marble) elements

The Baptistery, a representative of Romanesque architecture, has a typical octagonal plan layout, inscribed in a circle with a diameter of 36 m. Walls are doubled: an external wall of about 1.8 m width and an inner structure with columns at the ground level and squat pillars on the second and third Matroneo (Women's Gallery) levels at 9 m and 15.6 m from the ground, respectively. An inner octagonal dome surmounts the Baptistery, covered on its outside by an octagonally shaped masonry. The whole edifice is externally revetted by white marble originating from Italy and Greece [11] and, subordinately, of serpentinite that, in addition to the facade, was used on the floor, which consists of white and green tarsias (Fig. 1) forming an oriental motif.

2.1. The Tuscany serpentinites

In Italy, serpentinite outcrops occur in the Alps, and in the Northern, Central and Southern Apennines. The Baptistery serpentinite has been historically referred to the Ophiolitic Complex (Northern Apennine), outcropping in Tuscany in different localities such as Florence, Impruneta, Prato, Pistoia,

Siena and Volterra. It originated, during the Jurassic period, through a serpentinization process of ultrabasic rocks in a seafloor environment [12]. The sequence of Tuscany Ophiolitic Complex mainly consists of two serpentinite masses separated by a gabbro lens. The ophiolitic sequence includes: a dark serpentinite, sometimes deeply altered, a granular metagabbro, and a bluish-green diabase, minor lenses and veinlets of ophicalcite, cherts, and ophilolite breccias.

From a variety of historical sources [13 and references therein] it results that the supply quarries of the Baptistery serpentinite, locally known with different name such as Verde di Prato, Nero di Prato [14], Marmo Nero [15, 16], Paragone [17,18], were those located in the area of Monte Ferrato (Prato) and Impruneta (Florence); in particular, the oldest quarries were on Piano di Gello, on the eastern slope of Monte Piccioli (Monte Ferrato area), which were already exploited since the eleventh century, and currently closed. Among the old descriptions of these rocks, the most complete is that of Targioni Tozzetti [19], who referred to the different lithologies as "gabbro". At those times, the Italian term *serpentino* and gabbro were considered synonymous. Von Buch [20], introduced the term gabbro in the scientific literature with his modern meaning in 1809, in a misinterpretation of the Tuscan use whereas the term ophiolite was used, for the first time, by Brongniart [21] in 1813, as Hellenism for the Italian term *serpentino (serpente* = snake), used in Tuscany for many centuries for a snake-skin-like ornamental stone [22]. The serpentinite rocks display a characteristic green colour, from light to almost black, with a bluish reflection at times; in some cases, whitish veins crosscut the rock. A variety of this rock, light green with a dense network of dark veins, is known by the local name of "Ranocchiaia" (frog rock) [17].

3. Sampling and Analytical Methods

During this work, serpentinite samples from the Baptistery floor and from some ancient quarries in the area of Monte Ferrato and a small outcrop at Impruneta were collected and analysed for their mineralogy and bulk major element composition. The mineralogical characterisation was obtained at the Dipartimento di Scienze della Terra and CNR-IGG (Firenze) by using polarising microscope, XRD and EPM; the bulk major and trace element composition was obtained by ICP at the Activation Laboratories, Ancaster (Canada).

The textures of the serpentinites were studied on thin sections (30 microns thickness) using a Zeiss Axio Scope.A1 polarising microscope, equipped with a camera (resolution 5 megapixel). XRD analysis was conducted using a PANalytical diffractometer with radiation CuKα1 (λ=1.545 Å) and graphite monochromator, operating at 40 kV, 20 mA, investigated range 2θ=5–70°, software X'PertPRO and High Score for data acquisition and data interpretation of the mineralogical composition.

The mineral chemistry composition of major elements was performed by using a JEOL JXA-8600 microprobe equipped with four wavelength-dispersive spectrometers (WDS), under the following operating conditions: accelerating voltage at 15 kV, beam current at 10 mA and beam diameter at 3-5 um. Counting time was 10s for Na, 15s for all other major elements and 40s for minor elements (such as Mn, Sr and Ba). Matrix effects were corrected by PAP algorithm, following the Pouchou and Pichoir method [23]. Analytical data on international reference standards measured during the analytical session show a good precision, with a variation coefficient lower than 3% on major elements, in accordance with the long-term accuracy and reproducibility of the lab as given in Vaggelli et al. [24].

For whole rock, major element analyses, samples were prepared in a batch system. Each batch contains a method reagent blank, certified reference material and 17% replicates. Samples were mixed with a flux of lithium metaborate and lithium tetraborate and fused in an induction furnace. The molten melt was immediately poured into a solution of 5% nitric acid, containing an internal standard, and mixed continuously until completely dissolved (~30 minutes). The samples were run on a combination simultaneous/sequential Thermo Jarrell-Ash ENVIRO II ICP (inductively coupled plasma). Calibration was performed using seven prepared USGS and CANMET certified reference materials. One of the seven standards was used during the analysis for every group of ten samples. The detection limit is in the range 0.01 and 0.001 (%).

4. Results and Discussion

Serpentinite samples from the San Giovanni Baptistery and from serpentinite outcrops in the areas of Monte Ferrato and Impruneta display a general similar aspect; the collected Impruneta rock samples show a darker green colour imputed to the uniform distribution of magnetite small crystals.

The studied rocks display an overall similar texture. The primary Mg-rich minerals, olivine, clinopyroxene and orthopyroxene, are completely replaced by serpentine (only chrysotile is recognisable in cross-fibre veins), fine-grained (20-80 μm) magnetite and spinel. Ghosts of pyroxenes, displaying a coarse particle size (2-4 mm), are easily distinguishable. The serpentine minerals are present as a lattice of cross-fibre veins (100-250 μm), that give rise to a mesh texture (serpentinisation of olivine) and as a bastitic texture (serpentinisation of pyroxenes), in variable percentages (Fig. 2 a). In the Impruneta rock sample, the mesh texture is subordinate to the bastites and shows a dominant fine lattice (50-100 μm). All the samples are affected by the presence of a variable abundance of late fractures (40-100 μm), filled with talc, which also occurs as small colourless grains scattered in the groundmass of serpentinite minerals. The studied rocks do not exhibit evidence of carbonisation processes.

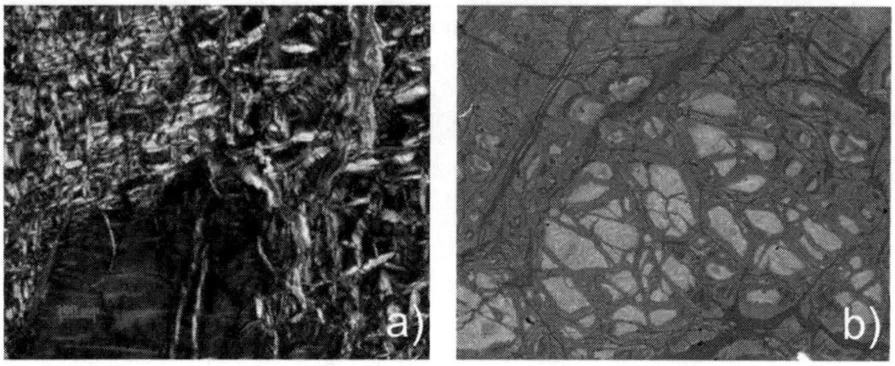

Fig. 2 - a) Polarised light micrograph showing mesh-textured and bastite from BPT1 sample; b) Back-scattered electron image (100 X) of BPT1 sample showing mesh texture

The diffractometric analyses permitted to distinguish two varieties of serpentine, not resolvable trough the optical microscope. All the studied serpentinites show the same mineralogical composition, characterised by lizardite and chrysotile and subordinate presence of magnetite.

Based on the texture and mineralogy, the studied serpentinites can be ascribed to the pseudomorphic petrological type [5, 25]. These results are in agreement with the data obtained through electron microprobe analyses (in Table 1 are reported representative analyses of the serpentine-group minerals). The different analysed rock samples do not display significant compositional differences: in each sample two different serpentine-group minerals (Fig. 2b), whose composition corresponds to a different texture is present. According to previous studies [5, 26], serpentine pseudomorphs after olivine, orthopyroxene and clinopyroxene are indistinguishable as groups by their Fe, Mg, Mn and Ni contents.

Table 1 - Representative microprobe analyses of serpentine-group minerals from the studied rocks

Sample	SF1-1	SF1-3	SF1-6	SF1-9	SF1-10	SF3-1	SF3-2	SF3-3	SF3-5	SF3-7	SF3-8	SF3-9	SF3-11
Texture	*bastite*	*bastite*	*bastite*	*mesh*	*mesh*	*mesh*	*mesh*	*bastite*	*bastite*	*bastite*	*bastite*	*mesh*	*bastite*
SiO_2	39.2	41.9	39.7	42.6	41.4	39.4	41.9	38.2	39.4	38.9	36.7	39.4	40.0
TiO_2	0.004	0.040	0.040	-	0.069	0.015	0.012	0.094	0.017	0.075	-	0.128	-
Al_2O_3	1.88	1.36	1.57	0.48	0.26	1.72	0.35	2.65	2.22	2.64	5.55	1.92	1.47
FeO	6.56	1.32	5.82	3.54	4.63	8.28	4.23	6.74	7.18	4.42	4.69	7.84	5.46
Cr_2O_3	1.550	1.510	0.853	0.064	0.017	0.060	0.017	1.163	1.012	1.002	1.348	0.053	0.420
MnO	0.111	0.064	0.101	0.098	0.054	0.192	0.054	0.157	0.168	0.100	0.008	0.166	0.092
MgO	37.3	40.4	37.9	38.5	39.0	36.2	40.2	36.1	35.6	37.8	37.6	37.4	37.8
CaO	0.086	0.098	-	0.010	0.088	0.050	-	0.031	0.093	0.061	0.127	0.085	0.229
Na_2O	-	-	-	-	0.011	0.078	-	0.004	-	-	-	0.048	0.153
K_2O	0.019	-	-	-	0.003	0.040	-	-	0.013	0.006	0.014	0.009	-
Total	86.7	86.7	86.0	85.3	85.6	86.0	86.7	85.2	85.6	85.0	86.0	87.0	85.6
H_2O*	13.29	13.26	14.00	14.68	14.41	13.96	13.31	14.80	14.36	15.00	14.00	12.98	14.42
			Impruneta							Baptistery			
Sample	SI1-1	SI1-3	SI1-5	SI1-6	SI1-9	SI1-11	BPT1-1	BPT1-4	BPT1-6	BPT1-10	BPT1-11	BPT1-12	BPT1-13
Texture	*bastite*	*bastite*	*bastite*	*bastite*	*bastite*	*bastite*	*bastite*	*bastite*	*bastite*	*mesh*	*mesh*	*bastite*	*bastite*
SiO_2	42.7	37.7	36.5	38.3	39.4	38.7	40.4	37.8	39.8	41.3	41.0	37.7	39.5
TiO_2	0.137	0.060	0.094	0.069	0.064	0.037	0.188	0.048	0.087	0.080	0.047	0.091	0.090
Al_2O_3	1.47	4.07	3.40	3.27	4.71	4.87	1.47	5.30	2.76	1.19	4.05	4.95	1.92
FeO	5.43	5.78	7.66	6.68	6.43	5.60	6.97	2.71	5.41	8.06	2.59	4.18	7.01
Cr_2O_3	0.964	0.890	0.942	0.695	0.850	1.064	0.694	1.539	0.877	0.014	0.130	1.253	0.968
MnO	0.100	0.072	0.112	0.144	0.138	0.078	0.172	0.028	0.137	0.146	0.066	0.086	0.202
MgO	33.3	35.0	35.7	34.2	35.6	34.5	35.9	38.6	35.6	34.5	36.8	36.8	35.7
CaO	0.319	0.095	0.065	0.071	0.008	0.009	0.064	0.000	0.166	0.048	0.068	0.150	0.056
Na_2O	0.159	0.062	-	0.083	0.080	0.185	0.062	0.196	0.019	-	-	0.070	0.010
K_2O	0.019	-	0.013	-	0.020	-	-	0.007	0.006	-	0.032	0.002	0.094
Total	84.6	83.8	84.6	83.5	87.4	85.1	85.9	86.2	84.9	85.4	84.7	85.3	85.6
H_2O*	15.37	16.20	15.44	16.49	12.63	14.90	14.08	13.75	15.15	14.60	15.26	14.75	14.44
*Estimated by difference													

However, in the comparison of mesh-textured serpentine with bastites, significant differences, with respect to Cr, were observed; indeed, this element migrates very little during the serpentinisation process because of its low solubility in low temperature aqueous solutions [27]. Mesh-textured pseudomorphs, analysed in this study, contain only minor amounts of chromium (0.014-0.06 % Cr_2O_3). The Cr_2O_3, contents of px-bastites range from 0.40 to 1.5 %.

The chemical composition of the studied serpentinites from the three different provenances (Table 2) shows very small variability; the rocks have high MgO% (37.6–39.1) and SiO_2% (~38) contents with a considerable amount of iron (Fe_2O_3% = 7.6-8.3) as well as variable values of Al_2O_3% (0.8–1.7). The considerable amount of Fe_2O_3 may be attributed to the presence of magnetite and chromite minerals whereas the low CaO contents may be attributed to the absence of dolomite and calcite. All samples have high loss-on-ignition (LOI) values (13.5 to 15 wt.%) indicative of the extensive serpentinisation of the rocks.

Table 2 - Major element analyses (mean) of the studied serpentinites

	Monte Ferrato	Impruneta	Baptistery
Sample	SF1	SI1	BPT1
SiO_2	37.7	37.8	37.8
TiO_2	0.025	0.038	0.035
Al_2O_3	0.85	1.70	1.60
Fe_2O_3	8.30	7.80	7.60
MnO	0.071	0.099	0.117
MgO	39.1	37.6	38.6
CaO	0.02	0.10	0.03
Na_2O	<0.01	0.01	0.02
K_2O	<0.01	<0.01	<0.01
P_2O_5	<0.01	<0.01	<0.01
LOI	13.6	15.0	13.9
Total	99.7	100	99.7

5. Concluding remarks

The data obtained in this work revealed only small differences among the studied rock samples; these differences are not sufficient to assert with confidence that the serpentinite used in the Baptistery floor came from the Monte Ferrato or Impruneta quarries. Additional compositional analyses and physico-mechanical measurements are in progress on a wider number of rock samples. However, on the basis of: a) documents conserved at the Archivio dell'Opera di Santa Maria del Fiore; b) documents present in the Archivio di Stato di Firenze and Prato; c) old literature data [e.g.14-18, 27] we can assume that the greatest part of material supplies came from the Monte Ferrato quarries. This fact does not exclude that small quantities of material, possibly used only for decoration [27], were obtained from the Impruneta quarries.

From the preliminary obtained data, we conclude that the followed multy-analytical approach is a reliable one and permits to obtain details of textural and mineralogical characteristics of serpentinitic rocks; the availability of these data play a relevant role in any restoration and replacement work. In fact, according to several authors [e.g. 28] the main factor favouring the resistance of serpentinite to weathering seems to be the mineralogical composition of this type of rocks and, in particular, the nature of the serpentine-group minerals. Our research is, thus, focused on obtaining documentation of the details of the serpentine-mineral assemblage that characterises various commercially available serpentinites which allows the best possible choice of materials to be used for any architectural purpose.

Acknowledgments

Authors wishing to acknowledge the Opera di Santa Maria del Fiore and the head of Area Restauro e Manutenzione Monumenti, Dr. Arch. Beatrice Agostini, for the collaboration and financial support.

References

[1] Guillot S, Schwartz S, Reynard B, Agard P, Prigent C 2015 Tectonic significance of serpentinites *Tectonophysics* **646** 1–19
[2] Evans BW, Hattori K, Baronnet A 2013 Serpentinite: What, Why, Where? *Elements* **9** 99-106
[3] Whittaker EJW, Zussman J 1956. The characterization of serpentine minerals by X-ray diffraction *Mineral. Mag.* **31** 107-126
[4] Wicks FJ, Whittaker EJW 1975 A reppraisal of the structures of the serpentine minerals *Can. Mineral.* **13** 227-243
[5] Wicks FJ, Whittaker EJW 1977 Serpentine textures and serpentinization *Can. Mineral.* **15** 459-488
[6] Fubini B, Fenoglio I 2007 Toxic potential of mineral dusts *Elements* **3** 407-414
[7] Evans BW 2004 The serpentinite multisystem revisited: Chrysotile is metastable *Int. Geol. Rev.*

IOP Conf. Series: Materials Science and Engineering **364** (2018) 012069 doi:10.1088/1757-899X/364/1/012069

46 479-506
[8] Degl'Innocenti P 1994 *Le origini del bel San Giovanni. Da tempio di Marte a Battistero di Firenze* Del Girasole Press Firenze Italy
[9] Racano L, and Tarlini E 2014 La datazione del Battistero di Firenze: Metodi, teorie, problemi. Memorie e contributi *Proc. Int. Conf.* on *"Il Battistero di San Giovanni, Conoscenza, Diagnostica, Conservazione,* ed F. Gurrieri Mandragora Press Firenze Italy pp 11–12
[10] Bernabei M, Bontadi J, Quarta G, Calcagnile L, Diodato M 2016 The Baptistry of Saint John in Florence: The Scientific dating of the Timber Structure of the Dome *Int. J. Arch. Her.* **10(6)** 704-713
[11] Garzonio CA, Cantisani E, Coli M, Cuzman O, De Luca D, Lubritto C, Ricci M, Vettori S, Sibilia E 2014 I materiali costitutivi del Battistero *Proc. Int. Conf. on "Il Battistero di San Giovanni, Conoscenza, Diagnostica, Conservazione,* ed F. Gurrieri Mandragora Press Firenze Italy pp 181-191.
[12] Principi G, Bortolotti V, Chiari M, Cortesogno L, Gaggero L, Marcucci M, Saccani E, Treves B, 2004 The pre-orogenic volcano-sedimentary covers of the western tethys oceanic basin: a review *Ofioliti* **29 (2)** 177-211
[13] Corti M 2006 *Il verde di Prato: una risorsa naturale e storica dell' Area protetta del Monteferrato* Prato Centro di Scienze Naturali Press pp 158
[14] Repetti E, 1839 *Dizionario Geografico Fisico Storico della Toscana* Anastatic copy Off. Grafiche F.lli Stianti Press San Casciano Firenze 1972.
[15] Villani G 1345 Cronaca libro VIII 3 Firenze
[16] Guasti C 1887 *S. Maria del Fiore-La costruzione della Chiesa e del Campanile secondo i documenti tratti dell'archivio dell'Opera secolare e da quelli di Stato* Anastatic copy A. Forni Press Bologna Italy 1974
[17] Vasari G 1568 *Le vite dè più eccellenti pittori, scultori et architettori* Firenze Milanesi Press 1878-1885
[18] Del Riccio A 1597 *Istoria delle Pietre.* Anastatic copy of document MS 230 of the Riccardiana Library care P. Barocchi S.P.E.S. Press Firenze Italy 1979
[19] Targioni Tozzetti G 1768 *Relazioni d'alcuni viaggi fatti in diverse parti della Toscana* Anastatic copy A. Forni Press Bologna Italy 1972
[20] Von Buch L 1809 Reise über die Gebirgszuge der Alpen Zwischen Glarus und Chiavenna in Aug 1803 Berlin Ges. Naturf. Freunde Magazine Entdeck ges Natk Jahrg **III**
[21] Brongniart A 1813 Essaie d'une classification minèralogique des roches mèlangès *Journal des Mines* **199** 5-48.
[22] Bortolotti V, Principi G, Treves B 2001 Ophiolites, Ligurides and the tectonic evolution from spreading to convergence of a Mesozoic Western Tethys segment Eds Vai G.B. Martini I.P *Anatomy of an Orogen: the Apennines and Adjacent Mediterranean Basins* Kluwer Academic Publishers Dordrecht pp 151-164
[23] Pouchou JL, Pichoir F 1987 Basic expressions of PAP computation for quantitative EPMA *Proc. Int. Conf. on ICXOM 11* Ontario Univ. Western Ontario Press pp 249-253
[24] Vaggelli G, Olmi F, Conticelli S 1999 Quantitative electron microprobe analysis of reference silicate mineral and glass samples *Acta Vulcanol.* **11 (2)** 297-303
[25] Wicks FJ, Zussman NJ, 1975 Microbeam X-ray diffraction patterns of the serpentine minerals *Can. Mineral.* **13** 244-258
[26] Dungan M A 1979 A Microprobe study of antigorite and some serpentine pseudomorphs *Can. Mineral.* **17** 771-784
[27] Sartori R 2002 *Pietre e Marmi di Firenze* Alinea Press Firenze pp 89
[28] Ismael IS, Hassan MS 2008 Characterization of some Egyptian serpentinites used as ornamental stones *Chin. J. Geochem.* **27** 140–149

Self-cleaning and antifouling nanocomposites for stone protection: properties and performances of stone-nanomaterial systems

M Roveri[1], F Gherardi[1], S Goidanich[1], D Gulotta[1], V Castelvetro[2], R Fischer[3], L Winandy[3], J Weber[4] and L Toniolo[1]

1. INSTM – Department of Chemistry, Materials and Chemical Engineering "Giulio Natta", Politecnico di Milano, Italy
2. INSTM – Department of Chemistry, University of Pisa, Italy
3. KIT – Karlsruhe Institute of Technology, Germany
4. University of Applied Arts Vienna, Austria

marco.roveri@polimi.it

Abstract. The development of nanocomposites combining photocatalytic, antifouling and protective features has provided interesting and promising results in the last years. However, few data about the behaviour of the nanomaterials applied on stone surfaces are available in the literature. In the framework of the EU-Horizon 2020 project "Nano-Cathedral", nanostructured protective treatments have been designed with different nanoparticles (TiO_2, Ag, ZnO), solvents and silane/siloxane-based polymeric matrices. The innovative formulations have been applied on 6 lithotypes, selected among the stones used in five medieval cathedrals (Vitoria-Gasteiz, Ghent, Cologne, Vienna and Pisa) and a contemporary theatre (Oslo Opera House), which are emblematic of different European geological and environmental areas. The treated stone specimens have been fully characterized to evaluate the surface optical and morphological compatibility, the reduction of water absorption by capillarity, the change in wettability and water vapour permeability properties. The selected treatments fulfil all these requirements and exhibit good photocatalytic and antifouling properties once applied on stone specimens. Different accelerated ageing procedures have also been performed in order to evaluate the stability of the polymeric matrices in the presence of photoactive TiO_2.

1. Introduction

The use of nanomaterials in the conservation of natural and artificial stone materials has been an active research topic over the last fifteen years, showing a potential for tackling different aspects of conservation, from consolidation to protection against chemical and biological decay agents.

In particular, the development of chemically stable nanocomposites based on the addition of inorganic TiO_2, ZnO or Ag nanoparticles (NPs) to silanic or acrylic matrices has provided an effective means of combining the water repellency features of siloxane, fluorinated and acrylic polymers with the photocatalytic activity of nano-TiO_2 [1], able to promote the solar-driven degradation of many pollutants, and the biocidal properties of Ag and ZnO [2].

"Nano-Cathedral", a Horizon 2020 project currently in progress, is devoted to developing a systematic approach to the use of TiO_2, ZnO and Ag-based nanomaterials in the protection of European architectural heritage, by evaluating the performance of an array of innovative research products applied on natural stones with highly different properties.

In this work, three innovative nanocomposites based on different silicon-based matrices and containing different types and concentrations of NPs were set up with the contribution of companies and their protective performance was assessed on different stone substrates.

As stone substrates, three lithotypes with largely complementary properties were selected, i.e. Apuan marble, Balegem limestone and Schlaitdorf sandstone. These natural stones were employed in relevant historical buildings involved in the project: Apuan marble is used in the Cathedral of Pisa (Italy, XIII cent.) and in a contemporary high-standing building, the Oslo Opera House (Norway, 2008), Balegem stone is present in the Cathedral of Ghent (Belgium, XIII cent.) and Schlaitdorf stone in the Cathedral of Cologne (Germany, XIII cent.).

The performance of the selected nanocomposites was evaluated on fresh stone substrates in laboratory conditions according to the recent EN Standard Protocol for the assessment of protective treatments in the field of Cultural Heritage [3] and by assessing their photocatalytic/biocidal activity.

Moreover, three different artificial ageing protocols that take into account the separate effects of heat, UV light irradiation and meteoric run-off are currently being conducted in laboratory conditions, with the aim of investigating the durability of nanocomposites once applied on stone substrates and predicting their long-term response towards environmental exposure. Partial results from this investigation will be presented, including those related to the effects of heat and UV light irradiation.

As a final step of the Project, the nanocomposites were applied on deteriorated surfaces in pilot areas of the respective historical buildings and the evaluation of their performance is currently being performed in the real environmental conditions, which are very different from site to site. Results from this onsite experimentation will be complementary and determinant for a thorough evaluation of the effectiveness of the considered new protectives and will be available at the end of the Nano-Cathedral project.

2. Materials and methods

Three nanocomposites were selected, according to a preliminary screening protocol, among those prepared by the companies involved in the project. Owing to the confidential non-disclosure agreement signed within the Project, only partial information can be given about the products, as shown in table 1.

Table 1. Description of the nanocomposites.

Product	Composition	Solvent	Property
WNC	alkyl alkoxy silane oligomers (15% w/w) with TiO_2 NPs (0.96% w/w)	water	photocatalytic
ANC	alkyl alkoxy silane monomers (40% w/w) with TiO_2 NPs (0.12% w/w)	2-propanol	photocatalytic
TNC	hydrophobized silica with AgO and ZnO nanoparticles (0.2% w/w)	2-propanol	anti-bacterial

The stone substrates belong to three different lithotypes, i.e. Apuan marble, Balegem limestone and Schlaitdorf sandstone, whose main characteristics are summarized in table 2.

Table 2. Main characteristics of lithotypes (porosity data are from mercury intrusion porosimetry measurements).

Lithotype	Provenance	Description	Open porosity (%vol)
Apuan marble	Carrara region, Italy	Metamorphic rock with almost purely calcitic composition and highly compact texture	0.7±0.1
Balegem limestone	Ghent region, Belgium	Sedimentary rock with large siliceous clasts in calcitic matrix; compact texture	9.9±0.8
Schlaitdorf sandstone	Neckar area, Germany	Sedimentary rock with chiefly siliceous composition and coarse texture	16±1

5x5x1 cm and 5x5x2 cm specimens (6 for each size) of fresh lithotypes were gently polished with abrasive paper (P180 carborundum paper), washed and kept in deionized water for 1 hour in order to remove possible soluble salts and then dried in oven at 50 °C until constant weight (a minimum of 48 hours). The nanocomposites were applied by capillary absorption for 6 hours, using a filter paper pad saturated with the treating material, according to EN 16581:2014 standard [3]. The time of application was defined after checking the time necessary to allow for the penetration of the products inside the depth of lithotypes. A commercial protective treatment based on silanes and siloxanes, Silres BS 290 (Wacker Chemie, dil. 8 wt.% in white spirit), was used as reference material for comparison. In order to determine the amount of absorbed dry matter, all stone specimens were weighed before the treatment and then after solvent evaporation until constant weight was achieved (a minimum of 48 hours). For all products, the amount of dry matter per unit area is reported in table 3.

Table 3. Dry matter (mg/cm^2) absorbed by the fresh stone specimens.

Treatment	Apuan Marble	Balegem	Schlaitdorf
WNC	0.3±0.2	16±1	26±4
ANC	1.3±0.1	10±3	18±10
TNC	4.5±0.2	7±1	7.7±0.5
Silres BS 290	2±1	12±8	18±7

The colour change of stone specimens after the application of the products was measured with a Konica Minolta CM-600D Vis spectrophotometer with a D65 illuminant at 8°, in the 360-740 nm wavelength range. 25 measurements were performed on each specimen before and after the treatment. The results were expressed in the CIE L*a*b* colour space and the average values of L*, a* and b* were used to calculate the colour change ΔE^*. A threshold value of $\Delta E^*=5$ was set as requirement for aesthetic compatibility [4].

Static contact angle test was performed on 15 points for each sample, according to EN Standard Protocol [5], using an OCA (Optical Contact Angle) 20 PLUS (DataPhysics, Germany) equipment, with a drop volume of 5 µl, after 10 seconds. The test was carried out before and after the application of the treatments.

The capillary water absorption of the stone specimens was measured following the EN Standard Protocol [6] on untreated and treated specimens. The data were elaborated according to the literature [7]. In order to evaluate the reduction of water absorption, two parameters were considered: the relative capillary index (CIrel) and the ratio of the absorption coefficients of treated and untreated specimen (ACt/ACnt).

Water vapour permeability tests were performed according to the EN Standard Protocol [8] on the specimens before and after the application of the treatments, using the "wet cup" system described in [9]. In order to evaluate the retention of permeability, the ratio of the permeability values of treated and untreated specimen ($\delta t/\delta nt$) was calculated.

The photocatalytic properties of the products were assessed through the Rhodamine discolouration test, according to the laboratory procedure and data elaboration described in [9]. The extent of discolouration (D*) of specimens treated with WNC and ANC was divided by that of specimens treated with Silres BS 290, which has no photocatalytic properties and can be used as blank.

In order to determine the antibacterial activity of TNC, sterilized stone samples of the three chosen lithotypes were treated with the products and, after the complete evaporation of the solvent, were incubated with fresh Bacillus cereus or Pseudomonas putida cultures for 16 h at 28 °C. As controls, untreated stone samples were also incubated with the bacterial strains. After the incubation, the stone samples were washed once with fresh medium to remove unattached cells. Remaining bacteria were collected by scratching an area of 5 cm^2 with a sterile swab. The swab was then transferred to 3 ml fresh medium and vortexed to release the bacterial cells in the medium. Serial dilutions of this suspension were placed on solid medium plates and incubated at 28 °C for another 16 h. Afterwards, single colonies could be observed and were counted to determine the number of CFU (Colonie Forming Units).

The artificial ageing of stone specimens in laboratory conditions was conducted for 750 hours through two different protocols: one thermal ageing in oven at 65 °C and one UV light ageing in an irradiation chamber (Suntest XLS$^+$, URAI S.p.A) equipped with a Xenon arc lamp (cut-off filter for λ<295 nm, 765 W/m^2 irradiance). Two specimens per lithotype and treatment were selected for each protocol, among the six specimens that were characterized in the former part of this investigation. At the end of the first ageing cycle, the stone specimens were characterized through the static contact angle and capillary water absorption tests, according to the previously described methodologies. A complete testing procedure including all the tests will be carried out in the next few months, after another ageing cycle of 750 hours has been finished.

3. Results

The results of the colour measurements on fresh stone specimens are summarized in table 4.

Table 4. Total colour difference (ΔE*) between treated and untreated specimens.

Treatment	Apuan Marble	Balegem stone	Schlaitdorf
WNC	1.5±0.5	3.4±0.4	1.5±0.2
ANC	2.8±0.8	1.7±0.5	7.1±0.6
TNC	4±1	4.5±0.8	2.6±0.5
Silres BS 290	1.8±0.9	3±2	8±2

It can be observed that all treatments preserve the original surface colour on Apuan marble and Balegem stone. On Schlaitdorf stone, WNC and TNC show excellent colour compatibility, whereas ANC and Silres BS 290 produce a colour change higher than the threshold value of 5, due to a darkening and yellowing effect. In both cases, the application of a lower amount of product would probably ensure a lower chromatic alteration.

The results of water absorption measurements on treated and untreated specimens are reported in table 5.

Table 5. Relative capillary index (CIrel) and relative absorption coefficient (ACt/ACnt) of treated specimens.

Treatment	Apuan Marble		Balegem stone		Schlaitdorf	
	CIrel	ACt/ACnt	CIrel	ACt/ACnt	CIrel	ACt/ACnt
WNC	0.75±0.05	0.4±0.4	0.35±0.07	0.1±0.1	0.44±0.05	0.11±0.05
ANC	0.13±0.02	0.19±0.03	0.07±0.01	0.02±0.01	0.08±0.02	0.028±0.009
TNC	0.5±0.2	0.03±0.02	0.19±0.09	0.01±0.01	0.09±0.01	0.031±0.007
Silres BS 290	0.2±0.3	0.05±0.07	0.5±0.3	0.6±0.3	0.11±0.07	0.02±0.01

On Apuan Marble, ANC shows the best results in terms of water absorption reduction. Good results were obtained by Silres BS 290 but the very high standard deviation values of both CIrel and ACt/ACnt is an indication of the heterogeneity of the treatment applied on different marble samples. TNC proves to be a good water repellent product especially after a short time interval of contact with water (30 min), whereas WNC is not effective in water absorption reduction. On Balegem stone, the best effectiveness was obtained by ANC both after 4 days (CI rel) and after 30 min (ACt/ACnt) compared to the other treatments. Good results were obtained by TNC and WNC, whereas Silres BS 290 shows poor effectiveness in the reduction of the water uptake. Finally, on Schlaitdorf stone, excellent results in terms of water absorption reduction were obtained from specimens treated with ANC, TNC and Silres BS 290, with CIrel values of about 0.1. WNC shows rather good reduction as well, especially within 30 min of contact with water.

The results of water vapour permeability measurements on treated and untreated measurements are reported in table 6.

Table 6. Ratio of water vapour permeability values of treated and untreated specimens (δt/δnt).

Treatment	Apuan Marble	Balegem stone	Schlaitdorf
WNC	1.2±0.4	0.9±0.3	0.8±0.3
ANC	0.9±0.2	0.6±0.3	0.9±0.2
TNC	0.5±0.3	0.2±0.1	0.4±0.2
Silres BS 290	0.3±0.2	0.3±0.2	0.436±0.2

On Apuan marble, both WNC and ANC do not affect the original vapour permeability, since the δt/δnt values are about 1. Further investigations would be necessary to explain, in the adopted treatment conditions, why WNC increases the mean values of vapor permeability of the stone. However, a similar effect has been reported in the literature concerning other coating materials [10]. TNC leads to a reduction of about 50% of permeability, whereas Silres BS290 significantly reduces it, indicating that probably the product is partially occluding the pores of the stone. On Balegem stone, WNC does not significantly affect the original water vapour permeability (reduction of about 10%), while ANC reduced the permeability of about 40%. Silres BS 290 and TNC show the highest reduction of permeability (about 70-80%). It is important to notice that, except for TNC, high values of standard deviation could be observed, indicating a high heterogeneity of the treated stones. On Schlaitdorf stone, WNC and ANC do not significantly change the original water vapour permeability; Silres BS 290 and TNC show comparable results with a reduction of the permeability of about 60%.

The results of static contact angle measurements on treated and untreated specimens are reported in table 7.

Table 7. Static contact angle (θ) of treated and untreated specimens (expressed in degrees); on untreated Schlaitdorf stone the contact angle could not be measured due to the very fast absorption of the drop.

Treatment	Apuan Marble		Balegem stone		Schlaitdorf	
	untreated	treated	untreated	treated	untreated	treated
WNC	66±10	113.8±0.7	71±8	116±4	-	128±2
ANC	72±18	140±10	37±7	143±3	-	137±3
TNC	55±7	129±12	54±10	149±3	-	133±7
Silres BS 290	58±11	101±5	32±8	125±12	-	127±6

In the case of Apuan marble, the application of all the new products reduces the surface wettability compared to the result obtained from Silres BS 290, with static contact angle values higher than 100°. In particular, ANC and TNC show a significant increase in the values of static contact angle (about 130-140°). On Balegem stone, all the products change the original surface wettability. In particular, specimens treated with TNC and ANC reach values of static contact angle of about 140-150° and the low standard deviation proves the homogeneity of this surface property of the treated stone. Finally, all the treatments applied on Schlaitdorf stone show water repellent properties, with contact angle values higher than 100°. In particular, ANC and TNC lead to a significant reduction of the surface wettability (about θ=140°), with low standard deviations proving the homogenous distribution of the products on specimens.

The results of the Rhodamine discolouration test are reported in table 8.

IOP Conf. Series: Materials Science and Engineering **364** (2018) 012070 doi:10.1088/1757-899X/364/1/012070

Table 8. Ratio of D* values of specimens treated with WNC/ANC and specimens treated with Silres BS 290 at 150 min (D^*_t/D^*_{SILRES}).

Treatment	Apuan Marble	Balegem stone	Schlaitdorf
WNC	1.34	2.08	3.04
ANC	1.30	1.40	3.90

It can be observed that, on Balegem stone, WNC produces a higher discolouration compared to ANC. This is readily explained by its higher TiO_2 concentration. However, when it comes to Apuan marble, the photocatalytic performances of the products are comparable and the performance of ANC is significantly better in the case of Schlaitdorf stone, indicating that this product, despite a lower nanoparticle concentration, has a better performance compared to WNC once applied on certain substrates.

Concerning the results of the test for the antibacterial activity of TNC (table 9), it can be observed that Apuan marble specimens show minor reduction of the CFU number for both Bacillus cereus and Pseudomonas Putida. This could be explained either by the small amount of protective product absorbed by the marble, or by the small amount of bacteria able to attach to the marble surface. On the other hand, the antibacterial activity of Balegem and Schlaitdorf specimens prove the effectiveness of TNC in the reduction of CFU compared to the respective untreated specimens, especially regarding Bacillus cereus.

Table 9. CFU values of untreated specimens and specimens treated with TNC (/10^6 counts/mL).

Treatment	Apuan Marble		Balegem stone		Schlaitdorf	
	Bacillus cereus	Pseudomonas putida	Bacillus cereus	Pseudomonas putida	Bacillus cereus	Pseudomonas putida
TNC	1.9±0.4	4.4±0.4	7±1	10.4±0.3	5.7±0.1	11.2±0.7
untreated	1.9±0.4	4.9±0.4	21±3	12.0±0.9	57±8	17±1

In conclusion, the laboratory tests conducted on fresh stone specimens showed that on Balegem and Schlaitdorf stones, ANC and TNC prove to be the best products regarding both protection efficacy and photocatalytic/antibacterial activity. In the case of Apuan marble, ANC has a comparable performance, while TNC shows only poor effectiveness in antibacterial activity.

Consequently, ANC and the reference Silres BS290 were chosen to undergo the ageing protocols on all three lithotypes, while TNC was only selected for Balegem and Schlaitdorf stones.

The results of the water absorption tests after 750 hours of thermal and UV light ageing are reported in figure 1. By comparing the values of CIrel before and after the ageing, it can be observed that thermal ageing does not produce negative effects on the protective performance, while in the case of UV light ageing, some differences can be highlighted in the behavior of the products. Indeed, no effect is observed in the case of TNC, while the effectiveness of Silres and ANC in the reduction of water absorption is slightly reduced, but the change is not remarkable. The results prove that the addition of nanoparticles in the formulations does not negatively affect the protective properties of the treated specimens.

The results of the static contact angle tests after the ageing are reported in figure 2. It can be observed that in all cases thermal ageing does not significantly affect the surface water-repellency of the treated stones. On the other hand, UV light ageing leaves the specimens treated with TNC unaffected, but it produces a visible effect in the case of Silres BS 290 and has a very strong impact on specimens treated with ANC in the case of Apuan Marble and Balegem, where substantially the surface wettability of untreated stone is reverted, while on Schlaitdorf stone the loss of hydrophobicity is much less pronounced. This significant decrease observed on specimens treated with ANC may be related to the photocatalytic activity of TiO_2 nanoparticles towards the polymeric matrix in which they are embedded. These results indicate that, although photocatalytic treatments applied on stone are predictably affected by a certain extent of degradation under UV light, this degradation remains largely a surface effect and the ability of the treatment to reduce water penetration is not significantly impaired.

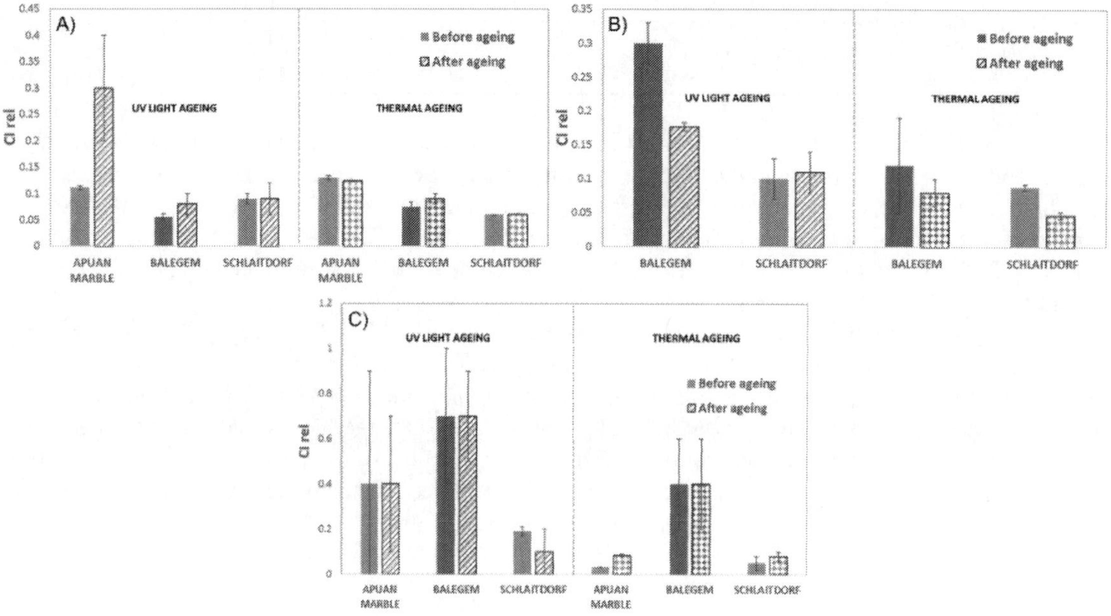

Figure 1. CIrel values of specimens treated with A) ANC, B) TNC and C) Silres BS 290 *before* and *after* the two ageing protocols.

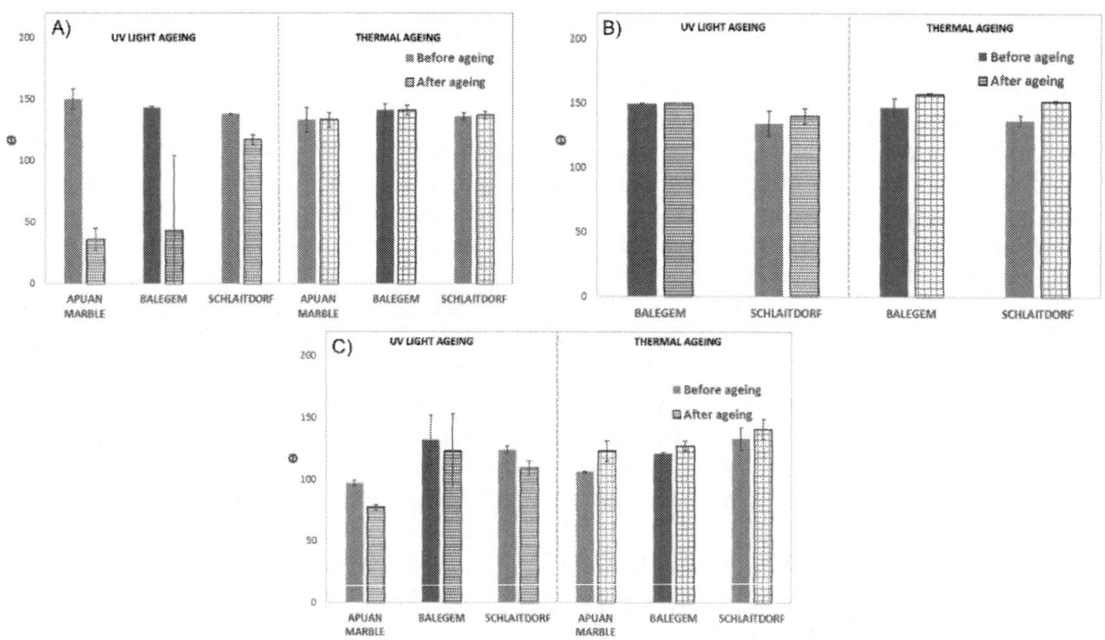

Figure 2. Values of static contact angle Θ (°) of specimens treated with A) ANC, B) TNC and C) Silres BS 290 *before* and *after* the two ageing protocols.

4. Conclusions

In this study, three innovative nanocomposites with photocatalytic or antibacterial properties, developed in the framework of the Nano-Cathedral H2020 Project, have been applied for protection purposes on three different lithotypes. The performance has been assessed through a set of laboratory tests and the durability has been evaluated by means of two different protocols of artificial ageing. The results indicate that two of the three products, one antibacterial and one photocatalytic product, show very high and promising protective effectiveness. The antibacterial product also showed excellent response in the durability tests, while the photocatalytic product proved to be able to retain much of its effectiveness in the reduction of water absorption but contact angle tests indicated a loss of hydrophobicity after the UV ageing protocol. This suggests the fact that, upon prolonged exposure to UV light, the product undergoes a degradation process, but this degradation, being a surface effect, does not significantly affect its effectiveness in reducing water absorption.

Acknowledgments

The Nano-Cathedral Project (Nanomaterials for conservation of European architectural heritage developed by research on characteristic lithotypes) is supported by the European program Horizon 2020 (GA n. 646178). The AA. wish to acknowledge the kind collaboration of the companies ChemSpec Srl (Italy), Colorobbia Srl (Italy) and Tecnan-Nanomat (Spain). Matea Ban from Technical University of Wien is gratefully acknowledged for Mercury Intrusion Porosimetry measurements.

References

[1] Quagliarini E, Graziani L, Diso D, Licciulli A and D'Orazio M 2018 *Journal of Cultural Heritage* **30** 81–91
Toniolo L and Gherardi F 2018 The protection of marble surfaces: the challenge to develop suitable nanostructured treatments *Advanced materials for the conservation of stone* ed M Hosseini and I Karapanagiotis (New York: Springer International Publishing)
Cappelletti G, Fermo P and Camiloni M 2015 *Progress in Organic Coatings* **78** 511–6
[2] Van der Werf I D, Ditaranto N, Picca R A, Sportelli M C and Sabbatini L 2015 *Heritage Science* **3** (29) 1-9
Serra C L, Tulliani J M, Sangermano M 2014 *Macromolecular Materials and Engineering* **299** (11) 1352–61
[3] EN 16581 2014 *Conservation of cultural heritage - surface protection for porous inorganic materials - laboratory test methods for the evaluation of the performance of water repellent products* (European Committee for Standardization)
[4] García O and Malaga K 2012 *Journal of Cultural Heritage* **13** (1) 77-82
[5] EN 15802 2009 *Conservation of cultural property - Test methods - Determination of static contact angle* (European Committee for Standardization)
[6] EN 15801 2009 *Conservation of cultural property - Test methods - Determination of water absorption by capillarity* (European Committee for Standardization)
[7] Peruzzi R, Poli T and Toniolo L 2003 *Journal of Cultural Heritage* **4** 251-4
[8] EN 15803 2010 *Conservation of cultural property - test methods – determination of water vapour permeability* (European Committee for Standardization)
[9] Gherardi F, Roveri M, Goidanich S and Toniolo L 2018 *Materials* **11** 1
[10] Kronlund D, Bergbreiter A, Meierjohann A, Kronberg L, Lindén M, Grosso D and Smått J H 2015 *Progress in Organic Coatings* **85** 159-67

Cleaning and protection of historic objects – biotechnology and nanotechnology approach

Snežana Vučetic[1], Jonjaua Ranogajec[1], Helena Hirschenberger[2], John Milan van der Bergh[1], Ana Vidaković[1] and Siniša Markov[1]

[1] University of Novi Sad, Faculty of Technology, Laboratory for Materials in Cultural Heritage, Bul. cara Lazara 1, 21000 Novi Sad, Serbia, janjar@uns.ac.rs;
[2] University of Novi Sad, Faculty of Technical Sciences, Trg Dositeja Obradovića 1, 21000 Novi Sad, Serbia

Abstract. Biocleaning procedure has been tested successfully on many materials and has proved to be more selective than mechanical and chemical cleaning techniques. Despite its success in treatment, biocleaning is not well-established method for the reduction of salt crystals. Therefore, the innovative approach combines biocleaning with the traditional (poultices) cleaning procedure. It is a new technology for salts reduction, starting from the design of clay poultices and joining their effects with targeted microorganisms. The chosen bacterial cultures increase the desalination capacity of the designed clay poultices, as a consequence of their metabolic activity, providing the mobilization of salts and salt reduction by the advection mechanism. Cleaned with this technology, architectural historic materials are properly prepared for long-term protection with photocatalytic coating based on TiO_2 and LDH suitable for fragile heritage materials. This protective material, designed under the FP7 HEROMAT project, has been already proved in laboratory and in-situ to be effective in maintaining aesthetic and functional properties of the mineral substrates and ensuring their long-term resistance to environmental conditions. This promising two-stage treatment presents a method which is both environmental friendly and cost/time effective for the reduction of nitrate salts and continuous self-cleaning protection of the surfaces of historic buildings.

1. Introduction

Building materials are exposed to wind, sunlight, rain, snow, atmospheric pollution and temperature variations during their exploitation and consequently, they degrade over time. Even the most durable materials are modified by deposition of ambient dust which enables the colonization of biological organisms such as cyanobacteria, fungi and algae [1]. One of the principal outcomes of the interaction between the environment and historical buildings is the deterioration of materials and the formation of soluble salts [2]. Nitrates, chlorides, oxalates and sulphates are known to be among the most harmful soluble salts for porous materials. Under real environmental conditions these salts are subjected to cycles of crystallisation-dissolution. The growth of salts crystals within porous building materials can cause serious surface damage or even structural collapse [3].

The two-step method of cleaning and protection presents a promising way to restore and improve the characteristics of deteriorated constructive materials and objects. Currently, the most used method for desalination is based on the application of poultice [4]. The use of water based poultices for the extraction of salts is a well-established technique in conservation. The desalination poultices available on the market are predominantly either cellulose powder based or mixtures consisting of cellulose fibres, clay materials (bentonite, kaolin, sepiolite) and sand or light aggregate. In practice, the composition of commercial products and self-made recipes do not undergo any modification in order to target individual properties of the substrates. However, the research findings have revealed that some variations of the textural properties of the poultices are needed to enhance the desalination efficiency [5]. In recent years, biocleaning (microbial cleaning) has been tested successfully on many materials and has proved to be more selective than mechanical and chemical cleaning techniques [6]. Biocleaning procedure with the use of non-pathogenic and non-sporulated microorganisms is a safe and environmental friendly way for cleaning the surfaces of architectural cultural heritage objects [7]. However, only about 30 articles related to biocleaning of building materials (www.sciencedirect.com) can be found, which leaves room for further investigation [8]. In addition, publications exploring the usage of conventional water-based poultices as support for bacteria cultures are difficult to find.

After the cleaning, the surface should be properly protected, since many deterioration processes start from the surface. For this, the application of a suitable surface coating is required. Surface treatments such as water repellent coatings [9] and antigrafiti coatings [10, 7] (based on silicones, fluoropolymers, polyurethanes and alkoxysilanes) strongly reduce water and vapour transport and consequently are not considered compatible with most historical building materials. A very promising approach has been found in the use of nano-photocatalyst based material. By applying inorganic-inorganic coatings, the compatibility of these materials with the mineral substrates (built-in materials) increases, compared to the organic coatings available on the market. Different types of coatings with additional functions have been developed so far. A specific problem of these applications is the lack of compatibility of photocatalysts with the surface of building materials. This difficulty could be solved by the appropriate immobilization of TiO_2nano-particles onto the photocatalyst support, layered double hydroxides (LDHs) [10]. Since the degradation processes start on the material surface, the deposition of suitable protective coating might be a very useful tool for surface protection.

Considering all above mentioned, the aim of this work was to study the possibility of coupling good properties of the designed cleaning procedure (combination of the traditional poultices and bio-cleaning) with a long-term protection using the photo-catalytic coating based on TiO_2 and LDH. The chosen cleaning procedure was performed using the already proven methodology based on effective bioactive denitration poultices [8]. Combining the effects of targeted designed water-based poultice and already verified bacterial culture for the removal of nitrate (*Pseudomonas stutzeri ATCC 17588*), an effective bioactive system was prepared. In this cleaning system, the poultice ensures the transport of soluble salts from the interior of the mineral substrate to the surface. Based on our previous studies, poultice was developed in order to suit the properties of the targeted mineral substrate (carbonate stone material). It possesses a wide pore size distribution incorporating large pores that can act as reservoirs for wetting, and small pores to ensure advection from the substrate to the poultice. At the same time, the poultice needed to take the role of a good microbiological carrier (water activity and appropriate pH values). When both requirements were achieved, the salt cleaning procedure (denitration) was tested. Following the denitration, the treated stone substrates were protected with a photocatalytic coating system designed under the FP7 project HEROMAT [11]. This protective coating (doped clay material) was used directly on the stone substrates surface, with or without prior cleaning. It enabled the monitoring and examination of the joint effects of the used cleaning and protection procedures on the treated stone substrate.

2. Materials and methods

The experimental procedure includes three main groups of experiments: 1 - contamination and examination of stone mineral substrates; 2 - preparation and application of bioactive poultice and characterization of treated samples; 3 – protection and examination;

2.1. Contamination and examination of stone mineral substrates

The uniform samples of carbonate stone (marble) were prepared in the form of cubes of 5 cm x 5 cm x 2 cm (length x width x height). The obtained samples were contaminated with potassium nitrate at a level of about 500 mg ml^{-1}. The desired contamination level was achieved using the following protocol: (1) drying samples in the oven at 105°C for 2 h and (2) complete immersion of the samples in a saturated solution of potassium nitrate for 22 h. The described steps were repeated in 5 cycles.

Textural characteristics, pore size distribution and total porosity values, were measured using Mercury Intrusion Porosimetry (MIP). MIP is a well-known and often used method for the determination of pore size distribution in stone substrates. It is commonly used in the investigation and assessment of degradation caused by salt crystalization. The prepared samples were investigated using Hg-porosimeter Autopore IV 9500 (Micromeritics, USA) with mercury intrusion pressures ranging up to 228 MPa, which enabled the determination of pores with diameters ranging from about 390 μm to 0.005 μm. The assumed contact angle between mercury and stone was standard 141°.

2.2. Preparation and application of bioactive poultice and characterization of treated samples

Based on our previous study [8], the poultice material was prepared using the easily available industrial components - kaolin:quartz sand:talc = 1:1:1. The composition of the prepared poultice material was designed to be of adequate textural properties required for the advection mechanism of desalination and at the same time, to retain the water necessary for a normal bacterial activity. The usage of talc was beneficial due to its surface adsorption capacity [12], while kaolin was used for its moderately shrinking properties. The particle sizes of the used components were as follows: kaolin – particles less than 43 μm, quartz sand –particles less than 710 μm, with 45mass. % of the particles with diameter of 180 μm and talc particles less than 6,5 μm. All used components were firstly mixed in dry state. In order to achieve consistency, bacterial suspension (for **bioactive poultice**) was added after mixing in the following manner - dry components: bacterial suspension =1:0.5. For the preparation of bioactive poultice the strain *Pseudomonas stutzeri ATCC 17588* was used. The selected strain was stored in Nutrient Broth (HiMedia, Mumbai, India), with the addition of glycerol (Lach-ner, Neratovice, Czech Republic) as a cryoprotectant and kept in the freezer for ultra-low temperatures (Snijders Lab, Tilburg, The Netherlands) at -80 °C. The strain was grown overnight at 37 °C in Nitrate Broth (DifcoTM, Becton, Dickinson and Company, Le Pont-de-Claix, France). After centrifugation (3 x 4200 rpm for 10 min), the obtained pellet was washed twice and suspended in sterile distilled water. The final cell concentration was adjusted to approx. $3x10^8$ CFU ml^{-1} corresponding to McFarland No.1 standard.

After the application of the prepared bioactive poultice **the denitration process was monitored** through only one face of the carbonate stone substrate. The other five were completely covered with silicon, to ensure unidirectional diffusion. The freshly prepared poultices were applied on the surface of contaminated models (1 cm thick layer). Denitration process was monitored after 24 h, 48 h and 6 days. In the case of the samples that were used for monitoring of the denitration process for 6 days a new fresh layer of poultice was applied after 48 h (two-stage application).The poultice was removed from the surface prior to nitrate content determination. In order to determine the nitrate content in the bulk, the stone samples were powdered. The nitrates from the powdered substrate were extracted with distilled water (1:1) and the nitrate content was determined by Quantofix® test stripes (Macherey-Nagel, Düren, Germany).

2.3. Protection and examination

The used protective coating was developed under the FP7 HEROMAT project and its self-cleaning properties had already been proven on different mineral substrates (brick, mortar, render, sandstone…) in laboratory, as well as in real environmental conditions [11]. The synthesized suspension is based on the layered double hydroxides (LDH) and photocatalyticaly active TiO_2. The LDH structure was selected in order to prevent the aggregation of the TiO2 nanoparticles and to provide adequate porosity of the deposited coatings. Another reason for its selection was the compatibility with the chosen mineral substrates: mortars, renders, bricks.

Four types of stone substrate samples were investigated: 1) B - blank sample (non-contaminated and non-protected); 2) NC+P - non-contaminated and protected sample, 3) C+P – contaminated and protected; 4) C+CL+P - contaminated, cleaned and protected. All sample groups were prepared for the colourimetric tests using the following procedure: non-polished side of the stone substrate sample (dimensions 5 cm x 5 cm) was covered with 1.5 ml of the protective suspension, evenly distributed over the whole surface using a brush. Protected samples were then dried in a laboratory dryer for 30 min at 105°C. After that, the protective suspension was again applied on the same side of the samples and dried using the same methodology as described before. Dried samples were then left to cool down to room temperature.

2.3.1. Aesthetic appearance test. The influence of the photocatalytic coating on the visual appearance of stone substrates was examined by measuring the colour change. All samples were recorded before and after the application of the coating with Konica-Minolta CM700-d spectrophotometer (2 cm diameter spot size) in CIE LAB 76 colour space. Total colour differences (ΔE) before and after the application of the coating was calculated as follows:

$$\Delta E = \sqrt{\Delta L^{*2} + \Delta a^{*2} + \Delta b^{*2}}$$

where ΔL*, Δa* and Δb* are the differences in the respective values of colourimetric coordinates before and after the application of the coating.

2.3.2. Contact angle measurements. Contact angle measurements between the experimental fluid (glycerol) and the coated surfaces of the investigated stone samples were performed by the *Surface Energy Evaluation System (Advex Instruments)*. The liquid drops, of about 5 µl in volume, were gently deposited on the substrate surface using a micro syringe. The measurements of the initial contact angle (θ_{ci}, after 1s) were performed at five different points of the examined samples. The examined samples were the same ones used for photocatalytic activity measurements.

2.3.3. Rhodamine B colorimetric test. In order to examine the photocatalytic activity of the coated stone samples, a modified Rhodamine B colourimetric test was performed [13]. Rhodamine B is a fluorone dye that exhibits a characteristic red-pink colour and is commonly used as a test dye for photocatalytic activity. The chemical structure of this dye is related to polycyclic aromatic hydrocarbons, which are often found among soiling agents in real urban conditions.

The 100 µmol/L Rhodamine B water solution was applied to previously prepared dried samples– 1.5 ml of solution (per sample) was uniformly distributed with a brush on the sample surface covered with the protective suspension. The samples were then left to dry in the dark at room temperature for 24 h.

The dyed samples were put in a chamber with a UV-light source (2 Sankyo Denki 20 W UV-lamps, 368 nm)located 20 cm above the samples. The UV-A light intensity absorbed by the samples was 0.80 mW/cm^2. The samples were irradiated during 24 h, and their colour was measured using Konica-Minolta CM700-d spectrophotometer (0.6 cm diameter spot size) in CIE LAB 76 colour space. Even though CIE LAB 76 colour space uses 3 colourimetric coordinates (L*, a*and b*) to define colour, only coordinate a* was considered, since Rhodamine B solution dyed the sample surface reddish-pink. The measurements of a* coordinated were averaged over the entire sample surface and were performed after irradiation periods of 30 min, 1.5 h, 2.5 h, 3.5 h, 4.5 h and 24 h. The photocatalytic activity (A) was calculated using the following equation (1):

$$A = \frac{a_0^* - a_x^*}{a_0^*} \cdot 100\% \qquad (1)$$

where a_0^* is coordinate a* that corresponds to the substrate before UV irradiation (0 hours), and a_x^* is coordinate a* that corresponds to the substrate after x hours of UV-irradiation.

IOP Conf. Series: Materials Science and Engineering **364** (2018) 012071 doi:10.1088/1757-899X/364/1/012071

3. Results and discussion

3.1. Cleaning: reduction of nitrate content in stone substrates (biotechnology approach)

3.1.1. Textural characteristics of the stone substrate and of the applied poultice. The total open porosity and the pore size distribution of poultice and contaminated stone sample measured by MIP are presented in figure 1 and table 1.

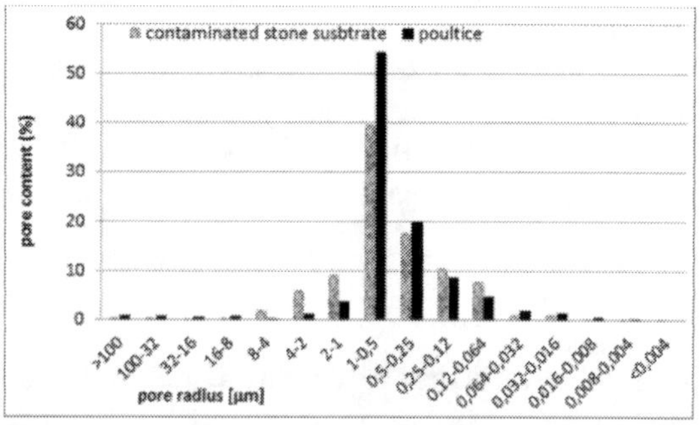

Figure 1. Pore size distribution of the contaminated stone sample and the used poultice

Table 1. Textural properties of poultice and contaminated stone sample

Sample	Total porosity (%)	Total cumulative pore volume (ml/g)
Contaminated stone sample	1.73	0.019
Poultice	47.79	0.347

In the case of the advection mechanism of desalination, the dissolved salts are transported from the mineral substrate into the poultice by capillary forces during the drying process [14]. The saline solution travels from large pores of the mineral substrate into the fine pores of the poultice. Accordingly, the poultice must have a larger quantity of fine pores than the stone substrate, but for the wetting phase (the first phase of desalination) the poultice has to contain large pores.

Based on the results presented in figure 1 and table 1, the advection mechanism of desalination could be expected. Namely, the measured values of pore size distribution in the developed poultice are in very good correspondence with the pore size distribution of the contaminated stone samples, allowing denitration by the advection mechanism, figure 1. The total porosity of the developed poultice is 40 times higher than the same value measured for the contaminated stone, thus a very good desalination capacity of the developed poultice could be expected, figure 2. Moreover, the developed poultice has a higher amount of both small and large pores compared with the contaminated stone samples. It also has a very well developed network of fine pores, which are necessary for the advection mechanism of denitration.

3.1.2 Application of the designed bioactive poultice. Based on the obtained results, figure 2, it could be concluded that the application of the developed cleaning system led to the significant nitrate reduction in the contaminated stone samples (initial concentration was 500 mg/L). Namely, after one day of application, the nitrate concentration was 100 mg/L, which is 350 mg/L lower compared with the initial concentration. Moreover, after six days of the bioactive poultice application, the final nitrate concentration was only 50 mg/L. According to the obtained results, it may be assumed that the bioactive poultice presented in this work has a great potential in the decreasing procedure of nitrate content in stone materials, even up to 90%. The results have a good correlation with our previous research of the application of bioactive poultices for the removal of nitrate from brick samples [8].

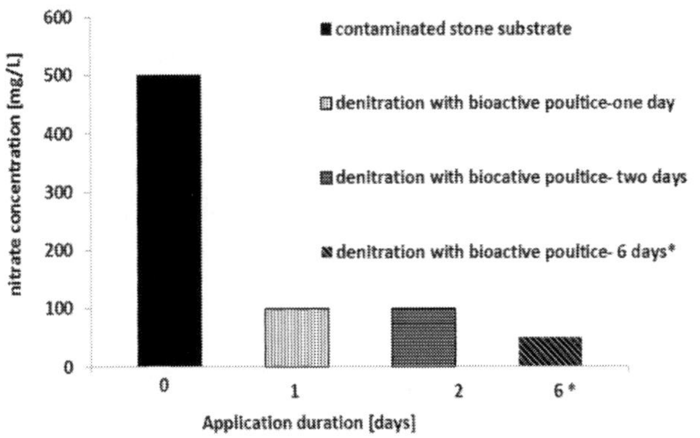

Figure 2. Nitrate concentration in contaminated stone substrates before and after cleaning procedure

*a new fresh layer of poultice was applied after 48 h (two-stage application)

The efficiency of the designed biocleaning has been proved in laboratory and already tested in situ on two case study objects in Serbia (Bač Fortress and calcite stone monument in urban area of Novi Sad). The long-term monitoring is being performed, while the possible side-effects of residual bacteria cultures within the stone pores should also be investigated.

3.2. Protection: nanotechnology approach

The total colour change of the investigated stone samples is given in table 2.

Table 2. Total colour change of the protected stone samples (ΔE)

non-contaminated and protected sample	NC+P	1.00
contaminated and protected sample	C+P	1.95
contaminated, cleaned and protected sample	C+CL+P	3.38

Table 3. Contact angle values [°]

blank sample	B	101.00
non-contaminated and protected sample	NC+P	80.55
contaminated and protected sample	C+P	86.50
contaminated, cleaned and protected sample	C+CL+P	82.64

The results presented in table 2. show that there is no significant colour change for the stone samples after the application of the developed photocatalytic suspension. Namely, the values of ΔE^* are very low (< 5), signifying that the TiO_2/ZnAl LDHs coating deposited onto surfaces of the stone samples does not affect their aesthetic appearance. This is a necessary criterion for the application of protective suspensions both in cultural heritage buildings and modern ones.

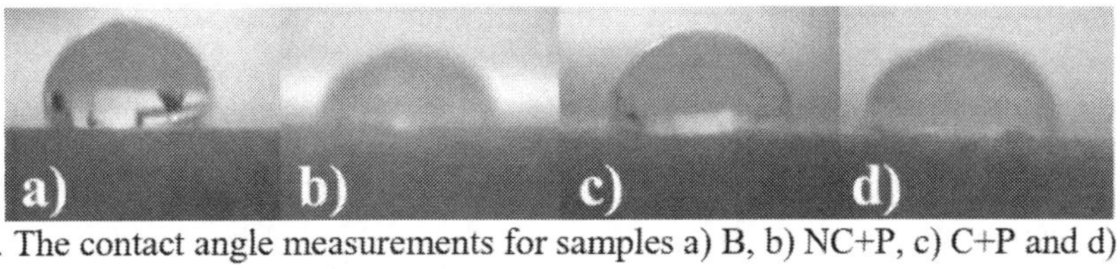

Figure 3. The contact angle measurements for samples a) B, b) NC+P, c) C+P and d) C+CL+P

The results of the contact angle measurements, table 3 and figure 3, indicate that the application of the developed coating leads to a decrease of the contact angle. These results proved the hydrophilic effect of the developed material. Moreover, the identified decrease is evident in all the examined

IOP Conf. Series: Materials Science and Engineering **364** (2018) 012071 doi:10.1088/1757-899X/364/1/012071

cases. Nevertheless, the contact angle values are lower in the case of previously cleaned samples and converge to the non-contaminated and protected samples, the fact that speaks strongly in favour of the biocleaning method used. The results of the Rhodamine B colorimetric test are shown in table 4 and figure 4. In the first 3.5 h of irradiation no significant difference can be observed among the photocatalytic activities of the samples NC+P (34%), C+P (35%) and C+CL+P (39%). This result implies that the presence of the nitrate salts does not negatively influence the photocatalytic function of the coating. Additionally, the described cleaning procedure has no negative impact to the photocatalytic properties of the coating either.

Table 4. Results of photocatalytic activity measurements (Rhodamine B colourimetric test)

		B	NC+P	C+P	C+CL+P
		photocatalytic activity (%)			
Irradiation time (h)	0.5	-4.43	21.36	21.95	19.39
	1.5	-8.33	28.00	34.55	26.97
	2.5	-6.69	33.68	28.49	31.23
	3.5	-5.27	34.10	35.16	39.29
	4.5	-3.06	36.70	37.01	51.44
	24	4.85	42.35	51.50	57.71

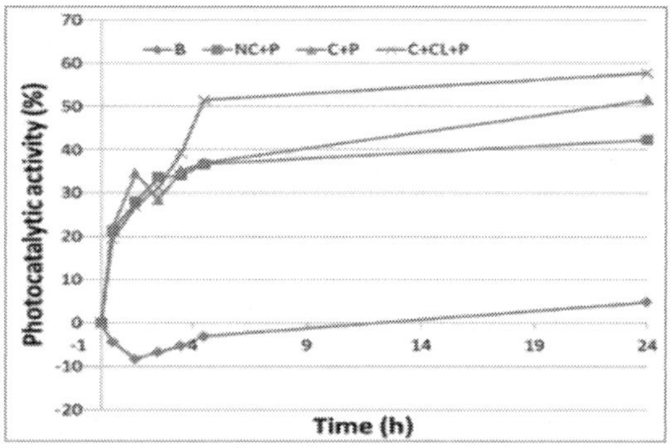

Figure 4. Results of photocatalytic activity measurements (Rhodamine B colorimetric test).

*blank (B); non-contaminated + protected (NC+P); contaminated + protected (C+P); contaminated + cleaned + protected (C+CL+P);

The photocatalytic activity results obtained after 24 h of UV-irradiation, figure 4, suggest that the photocatalytic function of the coating is more expressed on the cleaned and protected surface. Moreover, the high photocatalytcic activity values for the protective coating were gained in the case of previously cleaned samples (C+CL+P: contaminated + cleaned + protected), which leads to the conclusion that there are no negative effect due to the application of the proposed procedures.

4. Conclusion

In this study, a two-step methodology for the cleaning of soluble salts and protection with photocatalytic coating is presented. The already confirmed methodology for the development and application of biocleaning poultices for the reduction of nitrate salts on the brick substrate is also confirmed on stone substrate. Moreover, the necessity to design the most appropriate bioactive poultice (a combination of the traditional poultice as a support for bacterial culture) is once more emphasized. Namely, the comparative study of the pore sizes and distribution of the contaminated stone substrates and bioactive poultice already promised the optimal conditions for advection mechanism of desalination and high desalination efficacy. The obtained results confirmed this hypothesis - after the application of bioactive poultice, the measured reduction of nitrate salts was very high. After one day of bioactive poultice application, the quantity of nitrates fell from 500 mg/L to 100 mg/L, while after six days of two-stage application, the nitrates content was only 50 mg/L, which approximately presents 90% efficiency of the applied cleaning (desalination) procedure.

Following the successful cleaning procedure (nitrate salts reduction), the protection of the stone substrates was performed with the photocatalytic material designed in the FP7 project HEROMAT. A TiO_2-layer double hydroxide (TiO_2-LDH) suspension was applied onto the surface of stone samples. In order to examine the coupling effect of the performed biocleaning and protective treatments, the contact angle measurements and photocatalytic activity test were performed. The results indicate good photocatalytic activity and hydrophilicity of the examined samples, emphasizing better values of photocatalytic activity in the case when the protective suspension was applied on previously contaminated and cleaned samples (C+CL+P).

The combination of this newly designed desalination method based on biocleaning and the application of developed protective coating could replace a group of traditionally used procedures in the maintenance of built cultural heritage. The proposed innovative techniques (bulk desalination and surface protection based on self-cleaning phenomenon) could have highly positive environmental and economic impacts: the salt crystallization process and damaging microbiological growth will slow down. As a consequence, the time between the two interventions will increase leading to the reduction of long-term maintenance costs. The outcomes of this study could be tested on other mineral substrates further increasing the value of the obtained results and possibilities of their utilization in practice.

Acknowledgments
The authors acknowledge the financial support of the Ministry of Education, Science and Technological Development of the Republic of Serbia within the Project III 45008.

References

[1] Berdahl P, Akbaria H, Levinsona R, Millerb W 2008 Weathering of roofing materials – An overview *Constr. Build. Mater.* **22** 423-33
[2] Nasraouia M, Nowik W, Lubelli B 2009 A comparative study of hygroscopic moisture content, electrical conductivity and ion chromatography for salt assessment in plasters of historical buildings. *Constr. Build. Mater.* **22** 1731–51
[3] Marcello L.D, Bitossi M.G 1198 Characterisation of salt efflorescences in cultural heritage conservation by thermal analysis, *Thermochim. Acta* **317** 133-40
[4] Martins J, Dionisio A and Neves O 2017 Agar gel for Ançã limestone desalination *Procedia Earth Planet. Sci.* **17** 754-7
[5] Vergès-Belmin V and Siedel H 2005 Desalination of masonries and monumental sculptures by poulticing: A review *B. und Baudenkmalpflege* **11** 1-18
[6] Cappitelli F, Toniolo L, Sansonetti A, Gulotta D, Ranalli G, Zanardini E and Sorlini C 2007 Advantages of using microbial technology over traditional chemical technology in removal of black crust on historical stone surfaces of historical monuments *Appl. Environ. Microbiol.* **73** 5671-5
[7] Kohli R and Mittal K L 2013 *Developments in Surface Contamination and Cleaning* vol 6 (Amsterdam: Elsevier)
[8] Vučetic S, Ranogajec J, Markov S, Vidakovic A, Hirsenberger H, Bera O 2017 Development and modeling of the effective bioactive poultices for reducing the nitrate content in building materials *Constr. Build. Mater.* **142** 506-13
[9] Licchelli M and Marzolla S J 2008 Enhanced protection against mural writings by crosslinked fluorinated polymers *Proc. 11th Int. Cong. on Deterioration and Conservation of Stone (Torun)* (Torun: Nicolaus Copernicus University) pp 967-74
[10] Rudić O, Ranogajec J, Vulić T, Vučetić S, Čjepa D and Lazar D 2014 Photo-induced properties of TiO_2/ZnAl layered double hydroxide coating onto porous mineral substrates *Ceram. Int.* **40** 9445-55
[11] Vidaković A, Ranogajec J, Markov S, Lončar E, Hiršenberger H and Sever Škapin A 2017 Synergistic effect of the consolidant and the photocatalytic coating on antifungal activity of porous mineral substrates *J. Cult. Herit.* **24** 1-8
[12] Hao D and Miler J D 2007 A molecular dynamics simulation study of water structure and adsorption states at talc surfaces *Int. J. Miner. Process.* **84** 172-84
[13] Ruot B, Plassais A, Olive F, Guillot L and Bonafous L 2009 TiO_2-containing cement pastes and mortars: Measurements of the photocatalytic efficiency using a rhodamine B-based colourimetric test *Sol. Energy* **83** 1794-801
[14] Lubelli B, van Hees R. P. J. 2010 Desalination of masonry structures: Fine tuning of pore size distribution of poultices to substrate properties *J. Cult. Herit.* **11** 10–8

IOP Conf. Series: Materials Science and Engineering **364** (2018) 012072 doi:10.1088/1757-899X/364/1/012072

Detecting metal elements inside the dome of Santa Maria del Fiore in Florence using cosmic ray muons

C Blasi[1], L Bonechi[2], R D'Alessandro[2], E Guardincerri[3], V Vaccaro[4]

[1] University of Parma, Italy
[2] University of Florence an INFN section of Florence, Italy
[3] Los Alamos National Laboratory, USA
[4] Board of the Opera di Santa Maria del Fiore

E-mail: blasi.unipr@gmail.com

Abstract. As part of the ongoing study of Brunelleschi's great dome in Florence, the Opera di Santa Maria del Fiore has approved a proposal to solve the age-old problem of the possible presence of hoops or iron stirrups within the dome's inner wall by exploiting the flux of natural muons traversing the dome. Ascertaining presence of a metal hoop in the dome would obviously be relevant to assess the static and seismic safety of the monument. The project, proposed by Dr. Elena Guardincerri from LANL and by Prof. Carlo Blasi, began in 2016 at LANL with the construction of special prototype-panels for the tracking of muons, and has expanded this year to include researchers from the Department of Physics of the University of Florence and INFN-Florence.

1. The Opera di Santa Maria del Fiore and Brunelleschi's Dome

1.1. The activity of the Opera di Santa Maria del Fiore to preserve the dome of the Florence cathedral.

According to the tradition, the first stone of the new Florence Cathedral, designed by Arnolfo di Cambio, was laid on September 8, 1296. On the same day the institution of the Opera di Santa Maria del Fiore [1] was born. After a first phase of ecclesiastical and municipal co-management of the cathedral, the institution went, after a few decades, to a laical conduction according to schemes already in use in Florence, with its assignment to the guilds of trades. In the XIV century Francesco Talenti modified and enlarged the cathedral, adding a fourth span to the three planned.

The body of the Basilica was completed in 1380, and in 1420, after the construction of the chapels, the construction of the great dome began, which was finally entrusted to Filippo Brunelleschi. In 1934 a ministerial decree modified the organization of the Opera and sanctioned the institutional form still in force today, according to which the institution is managed by a Board of Directors made up of seven members – in part elected by the Bishop of Florence and in part by the Italian Ministry of the Interior - among which a President is elected.

Today the Opera di Santa Maria del Fiore is a non-profit organization, run according to its own Statute promulgated in 2001 and listing among its institutional goals "the promotion of culture and art". The Opera manages, besides the Cathedral of Santa Maria del Fiore with its majestic dome, Giotto's Campanile, the Baptistry, the new museum of the Opera, recently enlarged, the center for the Arts and Culture at piazza S. Giovanni, the palace in via della Canonica, where its offices are, and many more buildings around the Cathedral. It recently expressed its support for a project aiming at finding metal elements inside the dome by measuring the deflection of muon particles travelling through its wall.

1.2. The Dome and the argument about its hoop

Brunelleschi built the dome without explaining the rules he followed to check its geometry and to grant it stability; he did not even write memoires that could explain his beliefs and his modus operandi. For this reason the expression "*secrets of the dome*" has been in use for a long time. What is known today

IOP Conf. Series: Materials Science and Engineering **364** (2018) 012072 doi:10.1088/1757-899X/364/1/012072

was deduced from the reports from the construction period, from the documents in the Opera's archives and from studies performed during the past centuries by many researchers.

In particular, the studies following the 600-year anniversary of Brunelleschi's birth in 1977 have been crucial, especially those by Pasquale e Andrea Chiarugi. In those years, in fact, the layout of the bricks was precisely determined: these are laid on conical beds and on spirals according to a "herringbone" pattern.

Some core samples were taken, and they proved that the masonry was carefully built throughout the entire thickness of the wall, not as in a dry stone wall; the cracks were observed and their evolution with time was studied; the system to trace and cross check the geometry by rampant centering was explained and the current monitoring system was installed, which is regarded today as the most accurate monitoring system for a monument. Lastly, the reasons for the cracks were understood. It can be argued

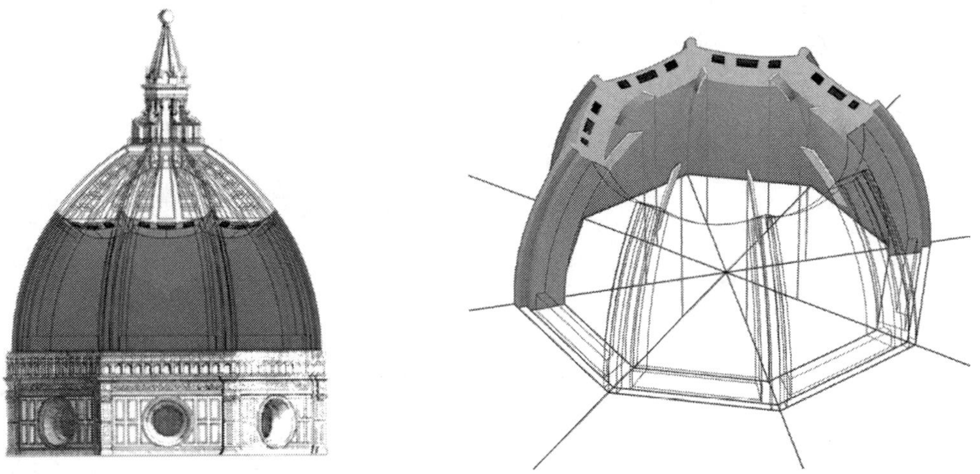

Figure 1: Drawings showing the conical beds upon which the bricks are laid and the tracing system based on rampant centering.

that nowadays there are no longer big "secrets" about the construction and layout of the dome, but some important questions to be answered still remain. Among them, the most important is probably whether there is an effective hoop. Knowing whether an iron hoop or a "*macigno chain*" (made of large stones connected by metal clamps) are present is crucial to evaluate the safety of the dome and to determine the interventions that could be performed to grant it an adequate static and seismic stability. Such interventions would stop the slow but inexorable widening of the largest cracks, today almost 7 cm wide, affecting the entire thickness of the dome and threatening its stability.

In order to oppose the centrifugal thrust and to provide an adequate hoop the Florence dome, contrary to all the major masonry domes in history, is only reinforced by a wood chain located by Brunelleschi in the space between the two shells. The chain cracked and was refurbished in the past by the Opera, whose experts were aware of its importance. The very same fact that the chain broke where the major cracks are shows its usefulness and effectiveness. The chain, made by Brunelleschi with great care and many iron clamps and nails, crosses the ribs and was probably laid based on the model of that already present in the Baptistry. In the Baptistry, however, due to the 4-5 cm wide cracks that had formed, an iron chain was added in 1514 which prevented further problems. Conversely nothing was made for Brunelleschi's dome, despite the opinions of famous experts (among them, in primis Vincenzo Viviani) and the wooden chain has not been mantained in the last decades and fares now in a state of extreme degradation.

A survey recently performed by the technical staff of the IVALSA-CRN laboratory from San Michele all'Adige, led by Prof. Ario Ceccotti, showed how the chain is subject to a traction of about 200-300 kN in the section considered. This proves that the chain, despite its current state of decay, still has a crucial static purpose.

IOP Conf. Series: Materials Science and Engineering **364** (2018) 012072 doi:10.1088/1757-899X/364/1/012072

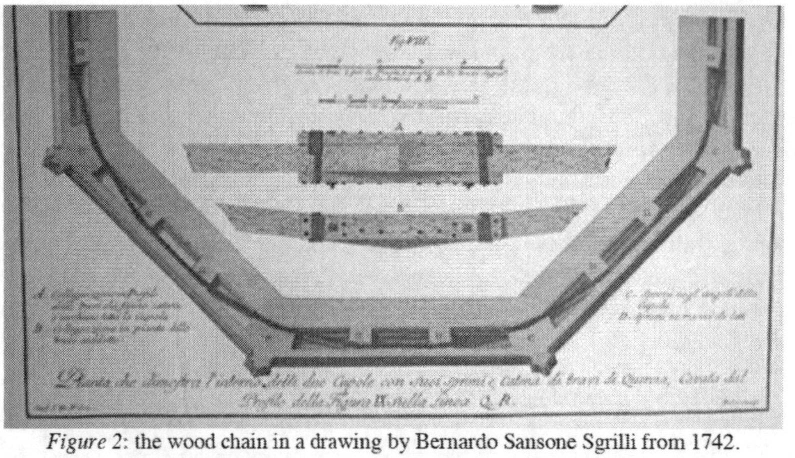

Figure 2: the wood chain in a drawing by Bernardo Sansone Sgrilli from 1742.

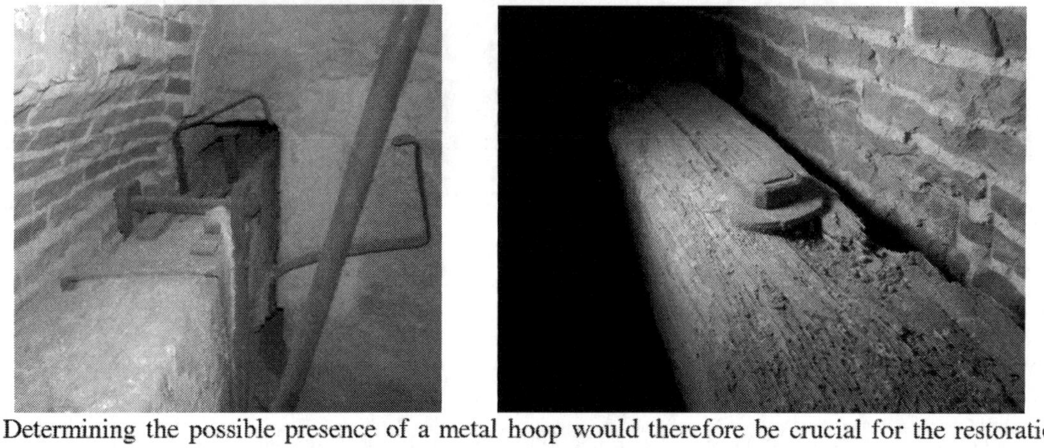

Determining the possible presence of a metal hoop would therefore be crucial for the restoration

Figure 3: Four pictures showing the wood chain and its status of decay

programs of the Opera; many investigations have been performed in the past with metal detectors and GPRS with unsatisfactory results, given the remarkable thickness of the dome.

The images that can be obtained using cosmic ray muons traversing the dome could help solve the problem. Two different non-invasive methodologies could be used; both are based on the detection of muons, elementary particles similar to electrons, but with a mass about 200 times larger, generated in the upper layers of the atmosphere by the interaction of cosmic radiation with the Oxygen and Nitrogen nuclei. These particles reach the ground naturally at speeds close to the speed of light and, thanks to their high mass, are able to penetrate the earth's crust like infinitesimal projectiles, reaching depths up to hundreds of meters, depending on their energy.

The former technique exploits the fact that muons are absorbed when travelling through matter, and the absorption increases with the density of the object traversed. The first use of cosmic ray muons for radiography was in 1955, when George measured muon attenuation to determine the overburden of rock above a tunnel [2]. This was followed by Alvarez et al., who used this method to confirm the Second Pyramid of Giza did not contain any undiscovered chamber above the Pharaoh's burial chamber [3]. More recently, several groups have used muons to examine structure and geophysical features of different sites [4][5][6][7][8][9][10] and for archaeological studies [11][12].

A different method, developed at Los Alamos National Laboratory, uses measurements of the multiple scattering angle of individual muons passing through an object to create tomographic images of the object's interior structure [13][14]. This technique was originally developed to inspect cargo containers for illicit trafficking of nuclear material [15][16][17], and has since been applied to detection of Special Nuclear Materials [18], industrial corrosion [19], nuclear reactors [20][21][22][23] and high level nuclear waste [24][25].

The two aforementioned methodologies, both proposed to be used within this project, and the measurement devices currently available are described in the following two paragraphs.

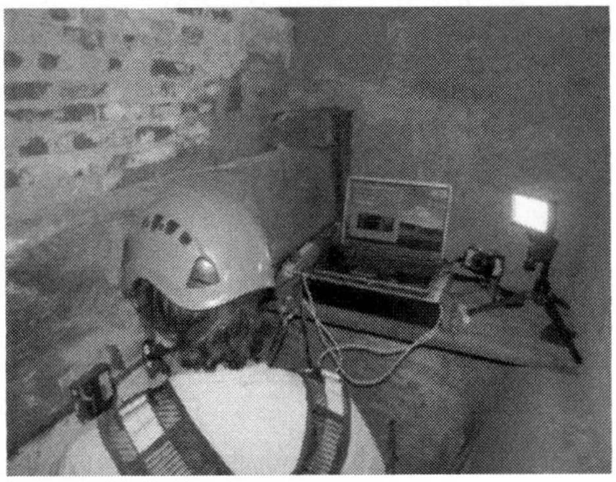

Figure 4: Picture taken during the survey performed by the IVALSA laboratory of CNR to measure the force applied by the wooden chai

2. Imaging techniques using cosmic muons

2.1. Multiple scattering muon tomography – Los Alamos National Laboratory (LANL), USA

Multiple scattering muon tomography is a 3D imaging technique, developed at the Los Alamos National Laboratory, which exploits the measurement of the multiple scattering angle of the individual muons traversing an object to create tomographic images of its internal structure.

The method is based on the fact that muons, when travelling through materials, undergo many single Coulomb scatterings with the charged atomic nuclei. As a result, they are deviated from their initial trajectory and exit the material with a different direction from the original one. By measuring the distribution of the cosmic muons' scattering angles after they traverse an object of interest, an image of the object itself can be reconstructed as described in [20]. The measurement of the deflection of muon trajectories requires the determination of the muons' trajectories before and after they traverse the object of interest. This is done using two "trackers" operated in coincidence upstream and downstream of the object to be studied.

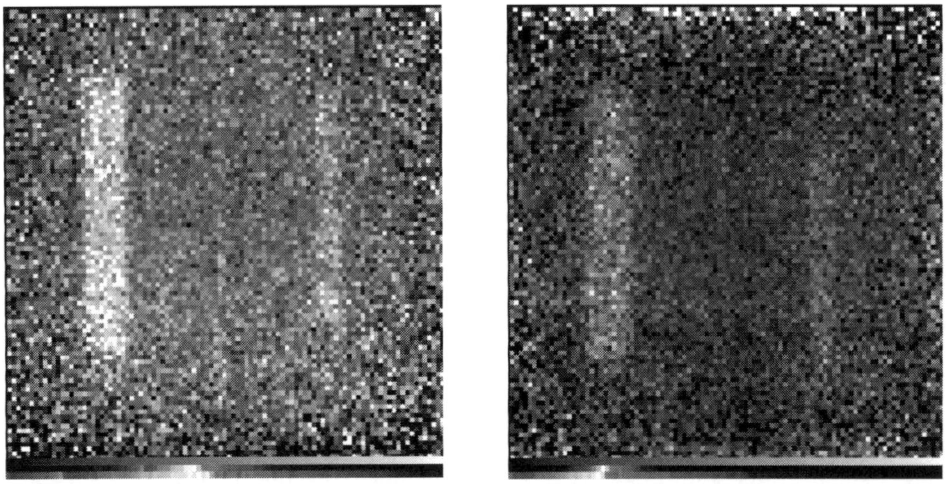

Figure 5: Left: Image of the 3 iron bars inside the concrete mock-up wall, obtained (35 days of data collected with the MMT). Right: Same image obtained from a Monte Carlo simulation.

IOP Conf. Series: Materials Science and Engineering **364** (2018) 012072 doi:10.1088/1757-899X/364/1/012072

By using multiple scattering muon radiography on the inner wall of the Florence Dome we plan to identify and image iron elements inside the masonry of the dome. In order to prove the feasibility of this measurement, a demonstration measurement was performed at LANL during the summer of 2015. If any reinforcement structure is present in the masonry of the Cupola's wall, it most likely has to be located inside the inner shell. This shell is much thicker than the outer shell, and thus plays a more important role in sustaining the whole structure.

A concrete wall having the same equivalent thickness (in terms of the effects of the material on the muon flux) as the inner wall of the Dome was built. More details about this measurement can be found in [26].

Three iron bars were placed inside this mock-up wall. The bars had rectangular or square cross-sections with different cross sections, the thinnest was 2 cm x 3 cm. The LANL Mini Muon Tracker (MMT [27]), was used to perform the demonstration measurement. The detector consists of two trackers made of aluminum drift tubes. The two trackers were deployed on opposite sides of the wall and collected data for 35 days. An image of the vertical plane passing through the iron bars was reconstructed and the resulting image is shown in the left insert in Figure 5. All the three iron bars are visible, including the thinnest one. For comparison, a Monte Carlo simulation of the same experimental set-up was performed and its data was analyzed with the same analysis technique. The areal density image obtained from the Monte Carlo simulation is also shown in Figure 5. The good agreement between the data and the simulation demonstrates that the technique is well understood and that the performances of future measurements can be confidently predicted. This measurement furthermore demonstrates that iron elements inside the inner wall of the dome of the Florence Cathedral could be located and imaged using a pair of muon trackers similar to those used to perform the aforementioned measurement.

We are therefore assembling two trackers made of carbon fiber drift tubes closed by conductive plastic endcaps: the trackers are modular, made of elements small and light enough to be hand carried to the base of the dome through the narrow spiral staircases leading there. Each tracker is made of 576

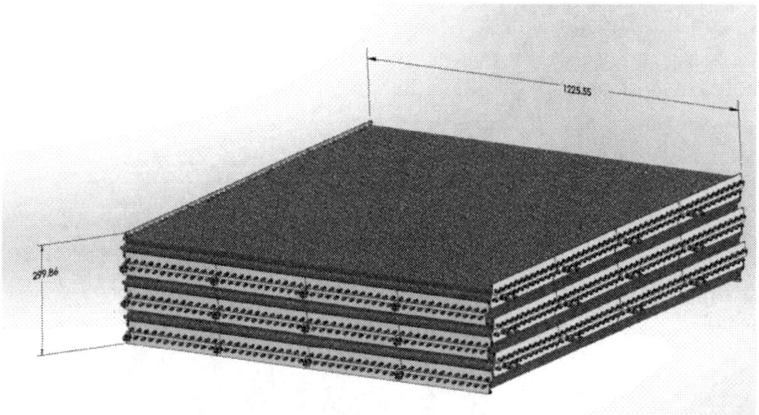

Figure 6: Mechanical drawing showing one of the two LANL trackers and its constituent modules

drift tubes, each one having a diameter of 2.54 cm and a length of 122 cm. Each tube has a 20 μm diameter anode wire at its center. The tubes are grouped in modules of 24. Each module has been assembled and tested at the Los Alamos National Laboratory. The modules will be filled with a mixture of Ar, CF_4 and ethane and operated at 1 bar of pressure. The overall size of each tracker, including the electronics, will be 137 cm × 137 cm × 30.5 cm and its weight will be around 130 kg. A mechanical drawing of a single tracker is shown in Figure 6 and a picture of one tracker, showing also the aluminum frame that holds together the tube modules, is shown in *Figure 7*.

The timing of the tiny (few fC) charge pulses produced at the anode wire of each tube will be read out by custom electronic boards, currently being designed by the Department of Physics and Astronomy at the University of Pennsylvania, and the information used to determine the trajectory of each muon crossing the trackers. The existing image reconstruction software developed by LANL will then be used to obtain radiographic images of the volume between the two trackers. Once the trackers are fully assembled and tested they will be shipped to Florence and deployed and operated on two opposite sides of the inner wall of the dome of Santa Maria del Fiore.

Figure 7: One of the two LANL trackers assembled

2.2. Muon absorption radiography - Department of Physics and Astronomy and INFN, Florence, Italy
Unlike the tomographic technique presented in the previous paragraph, based on the measurement of the deflection of muon trajectories, the muon absorption radiographic technique is based instead on the measurement of the degree of absorption of the cosmic ray muons crossing the material under examination. The Department of Physics and Astronomy of the University of Florence and the INFN section of Florence have started since 2009 a new development of the methodology, initially declining it to the sole volcanological area. This activity, stimulated by similar studies proposed by the Japanese group of Prof. H. Tanaka [28], took shape in the INFN Mu-Ray R&D project and in the subsequent "Progetto Premiale" (reward project) MURAVES [29], funded by the Italian Ministry of Education, University and Research (MIUR). The main purpose of this project is the validation of muon radiography as a standard prospection tool for Volcanology, by means of a detailed study of the internal structure of Vesuvius. The project involves the Italian INFN and INGV research institutes. Starting from 2012, the Florence research group has started a parallel study aimed at the possible application of this methodology in other research fields, in particular focusing on case studies of archaeological and geological interest. Concerning the dome of the Florence cathedral a particle detector of the latest generation (analogous to a classic optical telescope) will be used to measure the natural muon flux in two different experimental configurations: 1) with the detector installed inside the dome, pointing to the zone of interest and 2) with the detector placed outside the dome and pointing to free sky along the same direction of the previous case (reference measurement). The comparison of the two measurements makes it possible to highlight cavities or dense areas (like metal structures) hidden inside the predominant material constituting the volume under examination. The Florentine proponent group has already constructed a detector suitable for the use within this project, called MIMA (Muon Imaging for Mining and Archeology).

The MIMA detector consists of a cube-shaped hodoscope, about 50 cm in length, housed on an altazimuth mount to be easily oriented in space, thus defining precisely the measurement direction. The approximate total mass is 70 kg. In *Figure 8* the instrument is shown in different phases of its development and use for measurements performed in real sites of interest. The MIMA detector consists of six detection planes, each of them made of 21 bars of scintillating material, optically shielded with respect to each other by means of a reflective film. The bars, 40 cm long and with an isosceles triangle section with a 4 cm base, are arranged alternately with the right angle directed upwards or downwards, so as to form practically homogeneous flat structures, 2 cm thick, as shown in Figure 9. The crossing of these planes by a muon gives rise to the production of photons along the muon paths inside the struck bars. The light signals thus generated propagate within each affected bar and are partly collected by SiPM photosensors (silicon photomultipliers), low-consumption optical devices of the latest generation, installed at both ends of each bar (see Figure 10).

IOP Conf. Series: Materials Science and Engineering **364** (2018) 012072 doi:10.1088/1757-899X/364/1/012072

The information provided by the SiPMs for each bar crossed by a muon allows reconstructing a coordinate along the muon trajectory. Thanks to the particular spatial arrangement of the scintillating bars, inherited from the MINERVA experiment [30] at Fermilab (USA), the trajectories of the muons cross always at least two adjacent bars for each plane of the detector.

This allows reconstructing a coordinate along the muon trajectory with a spatial resolution of about 3 mm, exploiting the information on the relative intensity of the signals released by the two adjacent

Figure 8: Left: the MIMA muon hodoscope at the INFN-Florence laboratories. Center: installation at the Galleria Borbonica in Naples. Right: measurement at a river bank in collaboration with geologists of the University of Florence.

bars (a sort of barycenter of the measured signals). The MIMA hodoscope consists of a total of three pairs of x-y scintillating planes, i.e. three planes with parallel bars, arranged for the measurement of the coordinate along the x-axis, and three planes with the bars rotated of 90° compared to the previous ones, for the measurement of the coordinate along the y-axis, orthogonal to the previous one. The measurement of three points of passage of the muon allows uniquely and reliably identifying the real straight trajectory traveled by the particle.

On the basis of previous experiences, the study of the dome (of the order of a few meters thick wall) with the detector located 1-2 meters from the internal surface requires approximately an exposure time of at least a few weeks, with the currently available apparatus (MIMA).

The spatial resolution of 3 mm obtainable with a single scintillating plane allows reconstructing the

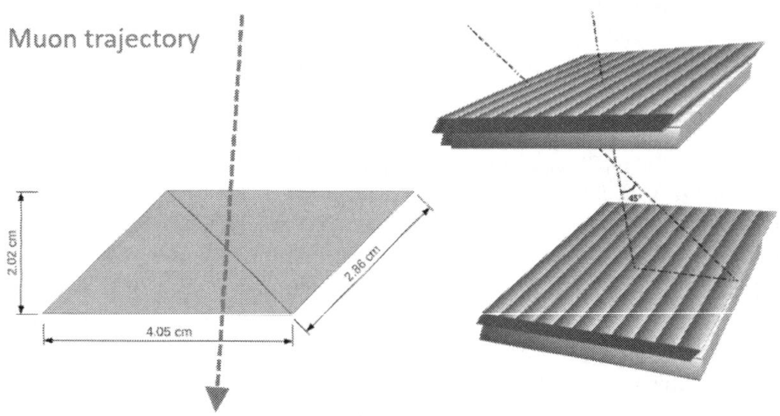

Figure 9: Details of the structure of the detection planes of the MIMA odoscope.
Left: passage of a muon through two adjacent bars.
Right: arrangement of the scintillating bars in pairs of x-y planes (in the example only two pairs of three are represented).

muon trajectories with an angular resolution of about 10 mrad. Considering a structure placed at 2 meters from the detector, this angular resolution allows in principle to carry out a measurement of the structure's details with a spatial resolutions of about 2 cm. By moving the detector away from the volume under

Figure 10: SiPM optical sensors. Left: a SiPM sensor used for the MIMA experiment. Right: SiPM sensors glued at the ends of two reflective bars wrapped in a reflective film.

examination, the spatial resolution obtainable increases (worsens) linearly with the distance. We believe that the installation of the MIMA hodoscope close to the internal surface of the dome, does not present critical problems. Once installed the detector can be safely left running for a few weeks without the need for any on-site maintenance.

References

[1] Fabbri L, Archivio dell'Opera di S. Maria del Fiore di Firenze. Quaderni di Archimeetings n.15, Firenze 2007.
[2] George E P, *Commonwealth Engineer*, 1955: 455-457.
[3] Alvarez L W et al., *Science* 167 (1970): 832-839
[4] Carloganu C et al. (TOMUVOL Collaboration). *Geoscientific Methods, Instrumentation and Data Systems* 2, no. 1 (2013): 55-60.
[5] Ambrosino F et al., *Journal of Instrumentation* 9 (2014).
[6] Nagamine K, et al. , *Nuclear Instruments and Methods A* 356, no. 2 (1995): 585-595.
[7] Lesparre N et al. ,*Geophysical Journal International* 183 (2010): 1348-1361.
[8] Guardincerri E et al., *Pure and Applied Geophysics* 174 (2017): 2133–2141.
[9] Nishiyama R et al., *Journal of Geophysical Research: Solid Earth*, 2014: 699-710.
[10] Olah L et al. *Advances in High Energy Physics* 2013 Article ID 560192 (2013).
[11] Morishima K. et al., *Nature*, no. 552 (2017): 386–390.
[12] Saracino G. et al. *Nature, Scientific Reports* **7**, Article number: 1181 (2017)
[13] Priedhorsky W et al., *Revue of Scientific Instruments* 74, no. 10 (2003).
[14] Schultz L et al., *Nuclear Instruments and Methods A* 519, no. 3 (2004): 687-694.
[15] Borozdin K et al., *AIP ADVANCES* 2 (2012): 042128 .
[16] Borozdin et al., *Nature* 422, no. 277 (2003).
[17] Morris C L et al. , *AIP Advances* 3 (2013): 082128.
[18] Guardincerri E et al., *Nuclear Instruments and Methods A* 789 (2015): 109-113.
[19] Durham J M et al. , *AIP Advances* 5 067111 (2015).
[20] Borozdin K et al., *AIP ADVANCES* 2 (2012): 042128 .
[21] Miyadera H et al., *AIP Advances* 3, 052133 (2013).
[22] Perry J O et al., *Journal of Applied Physics* 113, 184909 (2013).
[23] Morris C L et al., *Applied Physics Letters* 104, 024110 (2014).
[24] Durham J M et al. *Journal of Nuclear Materials Management* 44, no. 3 (2016).
[25] Poulson D et al. *Nuclear Instruments and Methods A,* 842 (2017): 48-53.
[26] Guardincerri E et al., *AIP Advances* 6 (2016): 015213.
[27] Perry,J O, *Ph.D. Thesis*. Albuquerque: The University of New Mexico, July 2013.
[28] Tanaka H et al., Nucl. Inst. and Meth. A507, 657-669 (2003)
[29] Saracino et al., Annals of Geophysics 60, 1 (2017) S0103

[30] Aliava L et al., Nucl. Inst. and Meth. A743, 130 (2014)

Photocatalytic hydroxyapatite-titania nanocomposites for preventive conservation of marble

Enrico Sassoni[1], Eros D'Amen[2], Norberto Roveri[2], George W. Scherer[3], Elisa Franzoni [1]

[1] Dept. Civil, Chemical, Environmental & Materials Engineering (DICAM), University of Bologna, Italy

[2] Chemical Center S.r.l., Via S. Donato 5, Granarolo dell'Emilia (BO), Italy

[3] Dept. Civil and Environmental Engineering (CEE), Princeton University, USA

E-mail: enrico.sassoni2@unibo.it

Abstract. Soiling of external surfaces is an increasing issue for conservation of architectural heritage. Providing architectural surfaces with self-cleaning ability is one of the most promising routes of preventive conservation. To this aim, several methods have been proposed in the literature, based on the use of photocatalytic TiO_2 nanoparticles, either directly applied onto the surfaces or incorporated in protective coatings. However, when nano-TiO_2 is directly applied onto architectural surfaces, the particles are easily removed by rain. When TiO_2 is incorporated in polymeric coatings, durability issues arise as well, because the photoactivity of TiO_2 nanoparticles can promote degradation of the polymer. Here, we present an innovative alternative method, based on combination of TiO_2 nanoparticles and hydroxyapatite (HAP). The incorporation of nano-TiO_2 into an HAP coating protects the nanoparticles from leaching by rain, thanks to the chemical bonding between TiO_2 and HAP, without diminishing their photoactivity. As a result, marble treated with HAP-TiO_2 composites exhibits high self-cleaning ability and high durability, with results superior to those achieve by direct application of nano-TiO_2 onto the surface, as frequently performed on site.

1. Introduction

Soiling of external surfaces by particulate matter present in the atmosphere is extremely important for cultural heritage conservation. To prevent soiling, the use of photocatalytic TiO_2 nanoparticles has been proposed, with the aim of providing existing surfaces with self-cleaning properties [1]. Two main routes have been pursued: (1) applying suspensions of TiO_2 nanoparticles directly onto the architectural surfaces, so that particles are retained on the surface after solvent evaporation; (2) incorporating TiO_2 nanoparticles into protective coatings. The first approach has the advantage of leading to high photocatalytic efficacy, because TiO_2 nanoparticles are directly exposed to solar radiation. However, because of the weak adhesion between the particles and the substrate, nano-TiO_2 directly applied onto historic surfaces is easily removed by rain [2]. This led to the second approach, i.e. incorporating the particles into coatings aimed at protecting them from leaching by rain. Coatings of various types have been proposed, including inorganic, organic and hybrid coatings. This approach has the advantage that particles are protected from leaching and mechanical strengthening can also be achieved by a single treatment. However, when organic matrices are used, the polymer binding the TiO_2 particles might be

IOP Conf. Series: Materials Science and Engineering **364** (2018) 012073 doi:10.1088/1757-899X/364/1/012073

deteriorated by exposure to UV radiation and might even undergo accelerated degradation caused by the photocatalytic activity of the particles themselves [3].

With the aim of developing photocatalytic composites with enhanced durability, in the present study we investigated the combination of nano-TiO_2 with hydroxyapatite (HAP, $Ca_{10}(PO_4)_6(OH)_2$). HAP coatings have been proposed for protection of marble from dissolution in rain, by formation of a superficial layer of HAP (having very low solubility) over calcite [4]. To form HAP, marble is treated by an aqueous solution of diammonium hydrogen phosphate (DAP, $(NH_4)_2HPO_4$) [5]. HAP-TiO_2 composites were here developed by introducing TiO_2 nano-particles into the DAP solution, so that particles are embedded by HAP formation. The photocatalytic activity and the resistance to simulated rain of such composites were evaluated and compared, on the one hand, to those of HAP alone and TiO_2 alone and, on the other hand, to a combined treatment simply involving sequential application of TiO_2 over an already formed HAP coating.

2. Materials and methods

2.1. Materials

Carrara marble was used for the tests, as this lithotype has been widely used in historic architecture and sculpture and because the issue of nano-TiO_2 leaching by rain is particularly relevant in the case of smooth surfaces such as marble. Moreover, the good efficacy and compatibility of the HAP-treatment on marble has been previously assessed [6,7]. Samples with 30×30×20 mm^3 size were sawn from a slab supplied by Imbellone Michelangelo s.r.l. (Italy). Before any treatment, all the samples were preliminarily artificially weathered by heating in an oven at 400 °C for 1 hour, to reproduce the micro-cracking typically affecting naturally weathered marble [6]. Part of the samples was left untreated and used as reference, while the remaining samples were treated as described below.

2.2. Treatments

Samples subjected to the hydroxyapatite-treatment alone (labelled "HAP") were treated by application of a 3 M aqueous solution of DAP (Sigma-Aldrich, assay ≥ 98%), applied by brush until apparent refusal (8 brush strokes). After curing for 48 hours wrapped in a plastic film, samples were rinsed with water and dried at room temperature. A limewater poultice was then applied, to further promote HAP formation and to remove unreacted DAP during drying [7]. Samples were finally rinsed with deionized water and dried at room temperature.

Samples subjected to the TiO_2-treatment alone (labelled "TiO_2") were treated by application of a 2 wt% suspension of nano-TiO_2 in 80-20 wt% water-isopropanol, applied by a single brush stroke.

Samples subjected to the sequential application of the HAP and TiO_2 treatments (labelled "HAP then TiO_2") were treated by applying the two single treatments described above, in sequence.

Samples subjected to the combined treatment (labelled "HAP with TiO_2") were treated by application of a 3 M DAP solution also containing 1.5 wt% of the same suspension used for the "TiO_2" treatment. In this case, the treatment was applied until apparent refusal (8 brush strokes), and then the samples were subjected to application of the limewater poultice, as described above.

2.3. Characterization

The coating composition was analysed by Raman spectroscopy (Renishaw Raman Invia spectrometer, linked to a Leica DMLM optical microscope). In particular, the possible formation of Ti-O-P bonds in the "HAP with TiO_2" treatment was investigated.

The surface morphology was analysed by observing the samples with an environmental scanning electron microscope (FEI Quanta 200 FEG ESEM), after carbon coating. The microstructure of the "HAP" and "HAP with TiO_2" coatings was investigated by cutting a cross section using a focused ion beam SEM (FEI StrataTM DB 235 FIB), after carbon coating. In particular, the possible effect of TiO_2 particles on HAP crystallization was investigated.

The aesthetic compatibility of the coatings was evaluated by determining the colour change (ΔE^*) after treatment, calculated as $\Delta E^* = [(\Delta a^*)^2 + (\Delta b^*)^2 + (\Delta L^*)^2]^{1/2}$. The CIELAB colour parameters L^* (black÷white), a^* (red÷green) and b^* (yellow÷blue) were measured by an optical fibre spectrophotometer Ocean optics USB2000+ in reflective mode.

The photocatalytic activity of the coatings was evaluated by the methylene blue degradation test [1]. Samples were first stained by a drop of methylene blue solution (10 g/l) and the colour parameters were determined. After exposure for 2 hours to UV radiation (50 W, λ=365 nm, distance from the lamp to the samples of 20 cm), the colour parameters were measured again.

To evaluate the resistance to nano-TiO_2 leaching by rain, samples (previously not used for any other test) were subjected to dripping of deionized water at pH 5.6 onto a 9 cm^2 surface, by means of a custom-designed apparatus [8]. After dripping a volume of water corresponding to the amount of rain that falls in ~6 years in Bologna (Italy), samples were dried and then subjected to the methylene blue test, carried out as described above. To evaluate the TiO_2 leaching from the marble surface, the runoff water was collected and the Ti content was analysed by inductively coupled plasma optical emission spectrometry (ICP-OES), using a Horiba ULTIMA 2 spectrometer.

3. Results and discussion

3.1. Coating composition, microstructure and aesthetic compatibility

In the case of the single treatments and the sequential treatment, Raman spectroscopy detected bands ascribable to HAP (959, 594 and 434 cm^{-1} [9]) and anatase (638, 520, 406 and 147 cm^{-1} [11]), as expected (Figure 1). Alongside HAP, formation of octacalcium phosphate (OCP, $Ca_8(PO_4)_6H_2 \cdot 5H_2O$)) cannot be completely excluded, as the detected bands are also compatible with OCP [9]. In any case, OCP formation is not expected to negatively affect the coating performance, because OCP is much less soluble than calcite [4]. In the case of the "HAP with TiO_2" treatment, in addition to the HAP and TiO_2 bands, a new band appeared at 1007 cm^{-1}, which suggests formation of a Ti-O-P chemical bond [11]. Formation of such a bond is expected to guarantee good adhesion between the TiO_2 nanoparticles and the HAP coating, with a positive outcome in terms of durability to simulated rain.

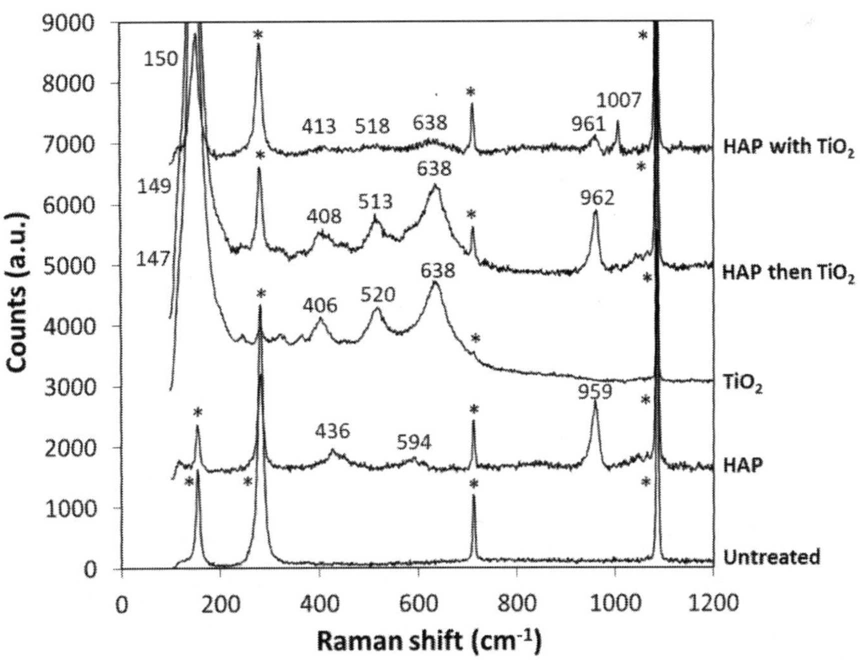

Figure 1. Raman spectra of untreated and treated samples. Calcite bands are indicated by a star.

IOP Conf. Series: Materials Science and Engineering **364** (2018) 012073 doi:10.1088/1757-899X/364/1/012073

The surface morphology of untreated and treated samples is illustrated in Figure 2. Although the coatings are all affected by diffused cracking (Figures 2-5), the "HAP with TiO_2" coating exhibits less cracking and apparently denser microstructure (Figures 2 and 6). This is thought to be a positive consequence of the presence of TiO_2 nanoparticles during HAP formation, because: (i) the particles might act as "seeds" during HAP crystallization [12]; (ii) the particles might limit shrinkage and cracking during drying (providing a similar action as they do in particle modified consolidants) [13]; (iii) the isopropanol contained in the TiO_2 suspension added to the DAP solution might favour HAP formation [14]. Accordingly, the FIB cross sections of the "HAP" and "HAP with TiO_2" coatings reveal that, for the latter treatment, a refined microstructure is present (Figure 2, dashed box).

In terms of aesthetic compatibility, all the treatments are suitable for application on white marble, because in no case did the colour change (ΔE^*) after treatment exceed the threshold commonly accepted for conservation treatments ($\Delta E^* = 5$ [13]) (Table 1).

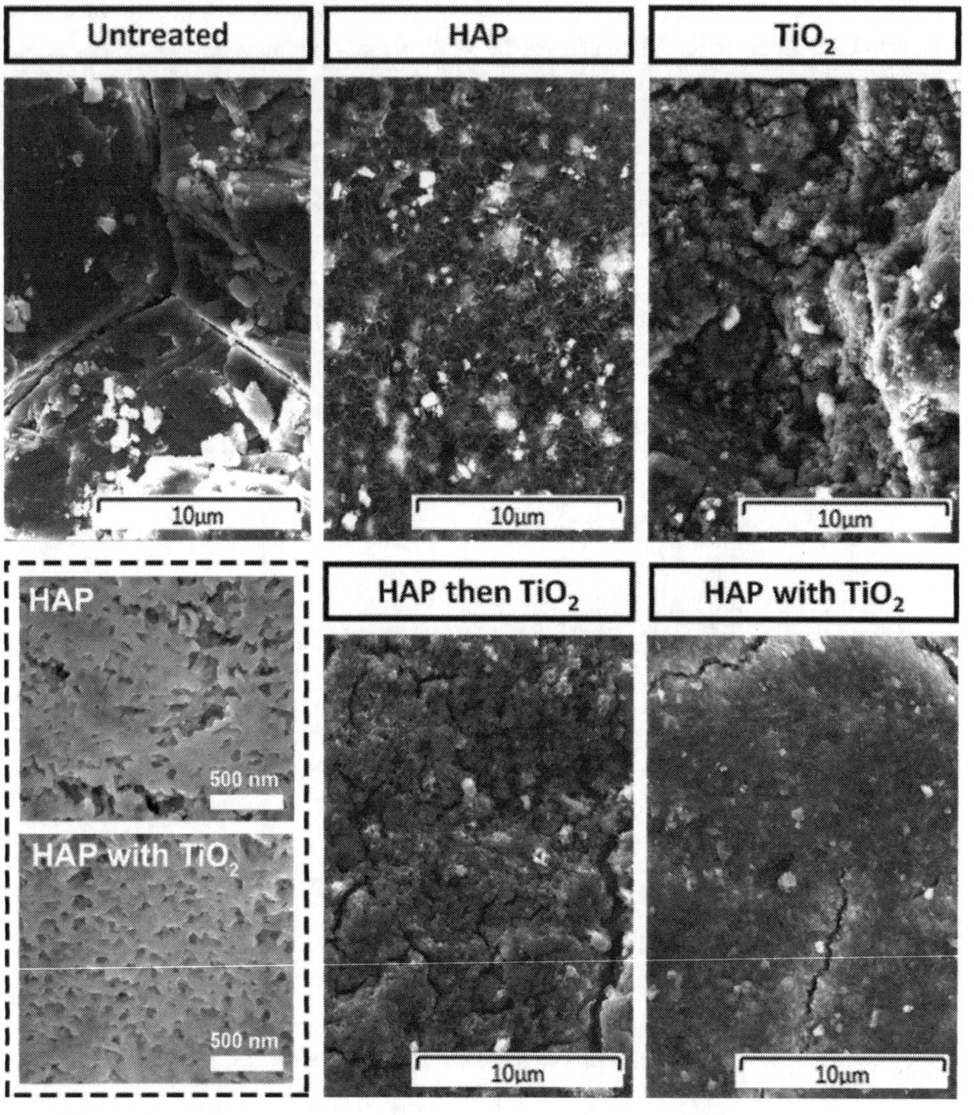

Figure 2. Surface morphology of untreated and treated samples and FIB cross sections of the "HAP" and "HAP with TiO_2" coatings (dashed box).

IOP Conf. Series: Materials Science and Engineering **364** (2018) 012073 doi:10.1088/1757-899X/364/1/012073

Table 1. Colour variation (ΔE^*) after treatment and after the methylene blue discoloration test, carried out before and after exposure of the samples to simulated rain.

	Compatibility	Self-cleaning ability (degradation of methylene blue)	
Condition	ΔE^* after treatment	ΔE^* pre-simulated rain	ΔE^* post-simulated rain
Untreated	-	1.3	1.6
HAP	2.2	4.9	3.4
TiO_2	1.9	20.9	13.2
HAP then TiO_2	1.2	20.7	17.1
HAP with TiO_2	0.6	17.6	16.3

3.2. Photocatalytic ability

The results of the methylene blue test are reported in Table 1. Whereas the untreated reference exhibits very little discoloration and the "HAP" sample only limited self-cleaning ability, as expected, all the treatments containing TiO_2 exhibit high discoloration ability, fully comparable to values reported in the literature for alternative coatings [1]. In detail, the "TiO_2" and the "HAP then TiO_2" coatings provided the highest self-cleaning ability, consistent with the fact that high and comparable amounts of TiO_2 are present over the surface of the respective coatings (Figures 4 and 5, "PRE" condition). Notably, the self-cleaning ability of the "HAP with TiO_2" coating was a little lower (17.6 *vs* 20.7-20.9, Table 1), but still remarkable, considering that in this case the amount of TiO_2 detected on the surface is very low (0.1 *vs* 31.9-32.4 wt%, Figures 4-6, "PRE" condition). Such a low value is due to the low amount of nano-TiO_2 suspension added to the DAP solution (cf. § 2.2) and to the application of the limewater poultice at the end of the treatment (cf. § 2.2), which removed all loosely bonded particles.

3.3. Durability

After exposure to simulated rain, the self-cleaning ability of the various treatments was altered as reported in Table 1. Minor changes were observed in the case of the untreated reference and the "HAP" coating, as expected. The "TiO_2" coating experienced the highest reduction (-35%), as a consequence of abundant removal of TiO_2 nanoparticles from the sample surface. This is explained by SEM observation of the sample surface (Figure 4, "POST" condition), which reveals a drastic reduction in Ti content (from 31.9 to 1.1 wt%), and by ICP analysis of the runoff water, containing the highest amount of Ti (0.256 ppm).

The "HAP then TiO_2" coating exhibited a lower reduction in photocatalytic activity (-17%), even though the TiO_2 nanoparticles had been simply applied over the HAP coating, with no encapsulation in the coating and no apparent chemical bonding (Figure 1). Nonetheless, TiO_2 removal from the surface was minimal (Figure 5, "PRE" and "POST" conditions) and a much lower Ti content was detected in the runoff water, compared to the "TiO_2" coating (0.085 ppm). The scarce removal of particles in this coating and the consequent good residual self-cleaning ability are thought to be a consequence of the improved adhesion, although physical in nature, between the particles and the HAP surface. Compared to the marble surface, the flower-like morphology and the high specific surface area of HAP allow for better adhesion of the particles, which are better retained during exposure to simulated rain.

In the case of the "HAP with TiO_2" coating, the chemical bonding between the particles and the HAP coating ensured that almost no removal of TiO_2 occurred (Figure 6, "PRE" and "POST" conditions), with consequent very low decrease in photocatalytic activity (-7%) and very low amount of Ti detected in runoff water (0.025 ppm). These results point out the high potential of the composite coating obtained by incorporating nano-TiO_2 into HAP, characterized by high photocatalytic activity and high durability at the same time.

IOP Conf. Series: Materials Science and Engineering **364** (2018) 012073 doi:10.1088/1757-899X/364/1/012073

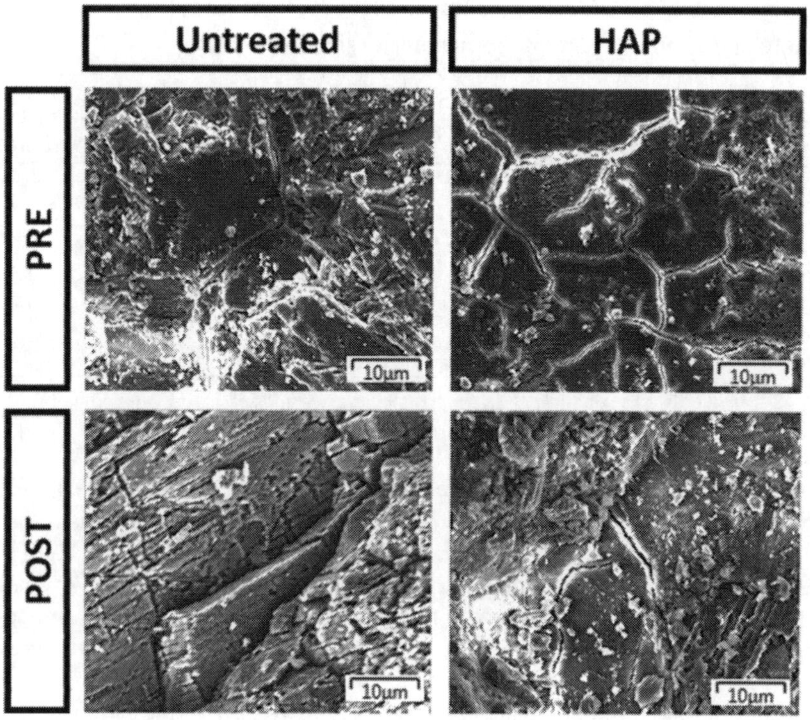

Figure 3. Surface morphology of untreated and "HAP" samples, before (PRE) and after (POST) exposure to simulated rain.

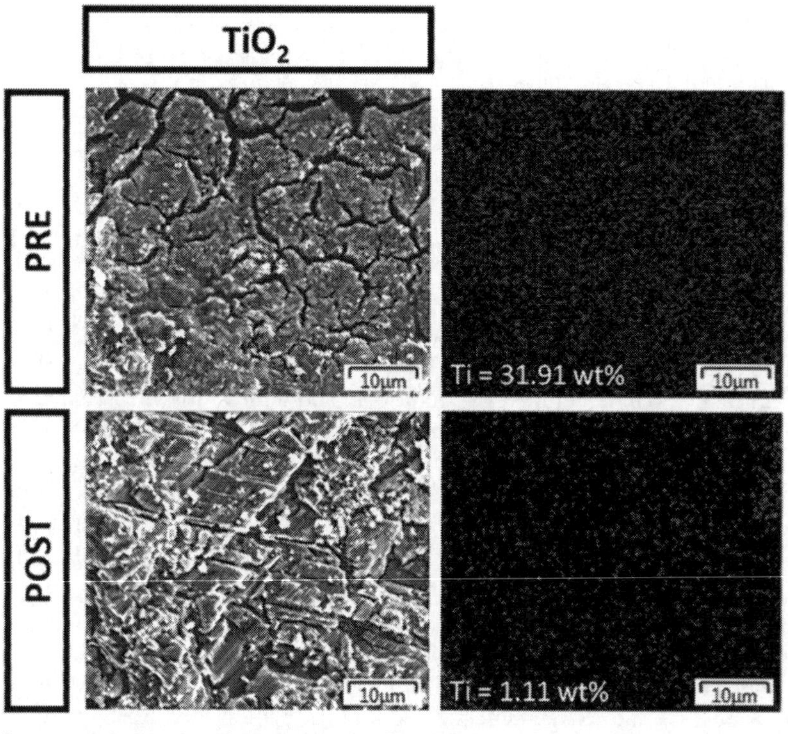

Figure 4. Surface morphology and EDS maps of "TiO_2" samples, before (PRE) and after (POST) exposure to simulated rain.

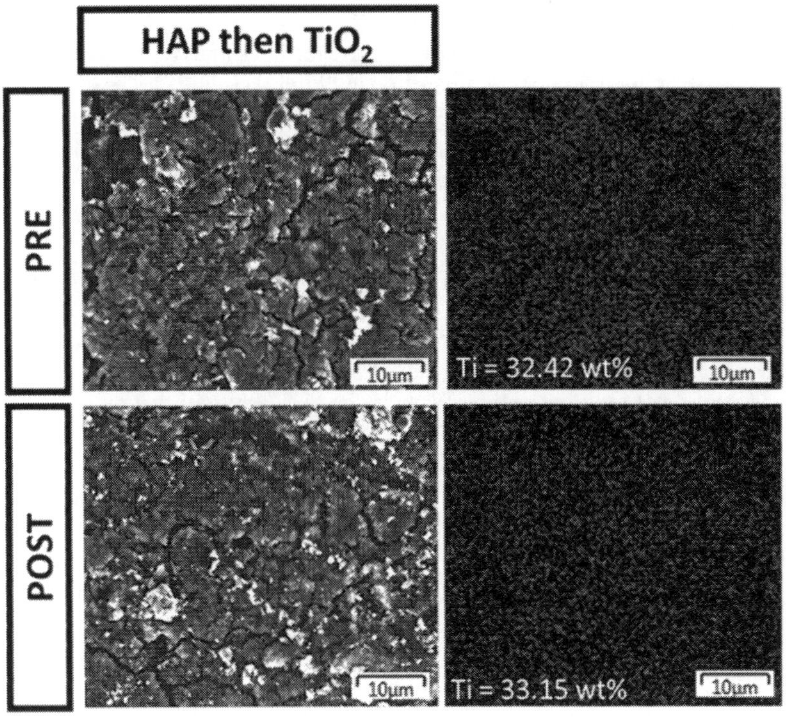

Figure 5. Surface morphology and EDS maps of "HAP then TiO_2" samples, before (PRE) and after (POST) exposure to simulated rain.

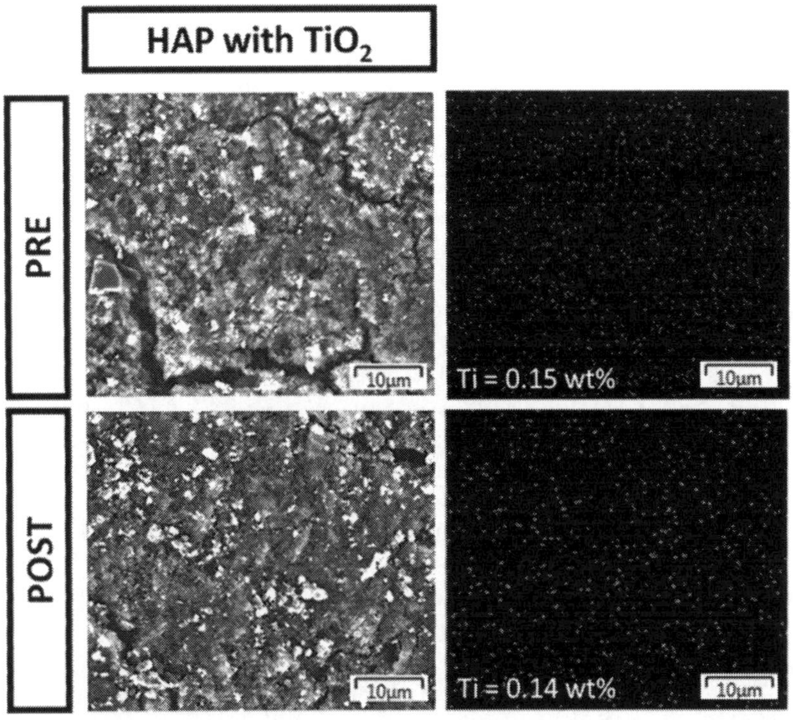

Figure 6. Surface morphology and EDS maps of "HAP with TiO_2" samples, before (PRE) and after (POST) exposure to simulated rain.

IOP Conf. Series: Materials Science and Engineering **364** (2018) 012073 doi:10.1088/1757-899X/364/1/012073

4. Conclusions

In this study, we investigated the combination of TiO_2 nanoparticles and hydroxyapatite as an innovative route to obtain coatings with high self-cleaning ability and high resistance to TiO_2 leaching by rain. The HAP-TiO_2 composite outperformed simple application of TiO_2 nanoparticles directly onto marble surface, which is currently the most common practice in the field. Very good results were obtained also in the case of sequential application of nano-TiO_2 over an already formed HAP coating, because the high specific surface area of HAP allows for a better physical adhesion between the particles and the substrate, compared to calcite. Although results of mechanical tests were not reported here for brevity's sake, both the combined and the sequential treatments were also able to fully restore the cohesion of artificially weathered marble, which is a further important aspect. Because the HAP-based treatment is attracting increasing attention and it has been explored for a variety of different substrates (from other stone types to concrete), the development of HAP-TiO_2 composites looks like a very promising route for preventive conservation of architectural surfaces in the built environment.

Acknowledgements

This project has received funding from the European Union's Horizon 2020 research and innovation programme under the Marie Sklodowska-Curie grant agreement No 655239 (HAP4MARBLE project, "*Multi-functionalization of hydroxyapatite for restoration and preventive conservation of marble artworks*"). M.Eng. Tobia Scurani is gratefully acknowledged for useful collaboration.

References

[1] Munafò P, Goffredo GB and Quagliarini E 2015 TiO_2-based nanocoatings for preserving architectural stone surfaces: An overview *Constr. Build. Mater.* **84** 201–18
[2] Munafò P, Quagliarini E, Goffredo GB, Bondioli F and Licciulli A 2014 Durability of nano-engineered TiO_2 self-cleaning treatments on limestone *Constr. Build. Mater.* **65** 218–31
[3] La Russa MF, Ruffolo SA, Rovella N, Belfiore CM, Palermo AM, Guzzi MT and Crisci GM 2012 Multifunctional TiO_2 coatings for Cultural Heritage *Prog. Org. Coat* .**74** 186–191
[4] Naidu S and Scherer GW 2014 Nucleation, growth and evolution of calcium phosphate films on calcite *J. Colloid. Interf. Sci.* **435** 128-37
[5] Sassoni E, Naidu S and Scherer GW 2011 The use of hydroxyapatite as a new inorganic consolidant for damaged carbonate stones *J. Cult. Herit.* **12** 346-55
[6] Sassoni E and Franzoni E 2014 Sugaring marble in the Monumental Cemetery in Bologna (Italy): characterization of naturally and artificially weathered samples and first results of consolidation by hydroxyapatite *Appl. Phys. A-Mater.* **117**, 1893-906
[7] Sassoni E, Graziani G and Franzoni E 2015 Repair of sugaring marble by ammonium phosphate: Comparison with ethyl silicate and ammonium oxalate and pilot application to historic artifact *Mater. Design* **88** 1145-57
[8] Graziani G, Sassoni E, Franzoni E and Scherer GW 2017 Resistance to simulated rain of hydroxyapatite- and calcium oxalate-based coatings for protection of marble against corrosion *Corros. Sci.* **127** 168–174
[9] Koutsopoulos S 2002 Synthesis and characterization of hydroxyapatite crystals: A review study on the analytical methods *J. Biomed. Mater. Res.* **15** 600-12
[10] Balachandran U and Eror NG 1982 Raman spectra of TiO_2 *J. Solid State Chem.* **42** 276-82
[11] Maeda H, Tamura T, Kasuga T 2017 Experimental and theoretical investigation of the structural role of titanium oxide in CaO-P_2O_5–TiO_2 invert glass, *J. Phys. Chem. B* **121** 5433–5438
[12] Moreno EC, Zahradnik RT, Glazman A and Hwu R 1977 Precipitation of hydroxyapatite from dilute solutions upon seeding *Calcif. Tiss. Res.* **24** 47-57
[13] Miliani C, Velo-Simpson ML and Scherer GW 2007 Particle-modified consolidants: A study on the effect of particles on sol–gel properties and consolidation effectiveness *J. Cult. Herit.* **8**, 1-6
[14] Lerner E, Azoury R and Sarig S 1989 Rapid precipitation of apatite from ethanol-water solution *J. Cryst. Growth* **97** 725-730

IOP Conf. Series: Materials Science and Engineering **364** (2018) 012074 doi:10.1088/1757-899X/364/1/012074

Innovative metagenomic approaches for detection of microbial communities involved in biodeteriorattion of cultural heritage

Brunella Perito[1], Duccio Cavalieri[1]

1. Department of Biology, Florence University, Via Madonna del Piano 6, 50019 Sesto Fiorentino, Firenze, Italy.

duccio.cavalieri@unifi.it

Abstract. Microorganisms can colonize any known environment and modify it due to their extremely wide and versatile metabolism. From the microorganisms point of view, cultural heritage (CH) objects represent a series of very heterogeneous habitats (from historical stones to paints or archaeological wood) where they can grow as well as in natural habitats. Aesthetical and/or structural modifications of CH materials due to microbial growth and activities are referred as biodeterioration. In most cases biodeterioration is not due to a single species of microorganisms, but is the results of the activities of a composite microbial community at some stage of its development on the artwork. The knowledge of microbial community composition and its metabolic potential is essential to individuate the microorganisms mainly responsible for deterioration and take the appropriate measures to control their growth. Microbial communities can be investigated by two different methods: culture-dependent (based on the cultivation of microorganisms) and culture-independent (based on the analysis of diagnostic molecules, mainly nucleic acids). Even if culture-dependent methods allow to study the physiology and metabolism of the isolated microorganisms, they provide limited information on the diversity of complex microbial communities since the vast majority of microorganisms (approximately 99% for environmental samples) are not yet culturable. Microbiology applied to Conservation is a relatively young branch of environmental microbiology. For many years methods used to detect microorganisms inhabiting CH objects have been culture-dependent while molecular techniques developed from microbial ecology have been adopting only in recent years. Here we discuss the potential application of innovative metagenomics approaches, to investigate microbial communities on CH. Metagenomics approaches can be divided in targeted and untargeted methods. The targeted methods amplify and sequence marker regions of the ribosomal DNA to obtain a genus level picture of the microorganisms present in different substrates. The untargeted approaches can provide strain level information of the genes and the biochemical functions present in a microbial community. This review emphasizes the importance of applying metagenomics based, culture-independent techniques, in combination with the culture-dependent ones, to assess the biodiversity of microorganisms colonizing CH objects and take proper actions for their conservation.

1. Introduction

Microorganisms can colonize any known environment and modify it due to their extremely wide and versatile metabolism. They play a key role in almost all the biogeochemical processes involved in transformation of organic as well inorganic matter. If such transformations occur in materials made and/or used by human beings, they are very often considered as negative. Biodeterioration can be defined as "any undesirable change in a material brought about by the vital activities of organisms" [1].

From the microorganisms point of view, cultural heritage (CH) objects represent a series of very heterogeneous habitats (from historical stones to paints or archaeological wood) where they can grow as well as in natural habitats. Bacteria, archaea, fungi, algae and lichens are the microorganisms commonly causing problems in the conservation of CH because of their biodeteriorative potential since they inhabit and penetrate into the materials. Even if microbes are invisible, they cause not only visible aesthetical damages (such as patinas or discoloration), but also severe structural modifications of CH materials such as fractures and material loss, due to acid corrosion, enzymatic degradation and mechanical attack [2]. This occurs on (and in) all types of cultural artefacts (organic or inorganic materials, historic or modern ones, art objects and historical buildings, paintings, costumes, ceramics, mummies, books and manuscripts, human remains and burial-related materials) and in indoor (public museums, private art collections, churches, caves and catacombs) as well as in outdoor environments (stone monuments, architectural surfaces).

For a long time abiotic chemical and physical processes were believed the dominant factors of decay of CH materials. Only in recent decades the role of microorganisms in causing deterioration of CH has been recognized, also due to the increase of knowledge about the ecological role of microorganisms in natural environments, and since the late 1980s significant effort was applied to ascertaining the biodiversity in the component materials of works of art [2].

Microbiological investigation is then a younger scientific discipline applied to Conservation than the more traditional chemical, physical and geological disciplines and now working alongside them.

It is now generally agreed that the variety of biodeterioration phenomena observed on CH materials are determined by chemical, physical and biological factors, and their interactions, linked to the nature of the material itself and to the environmental conditions of the object exposure and keeping (Ecology of biodeterioration).

In most cases biodeterioration is not due to a single species of microorganisms, but is the results of the activities of a composite microbial community at some stage of its development on the artwork. The knowledge of microbial community composition inhabiting the CH objects and its metabolic potential is essential to identify the microorganisms and mechanisms mainly responsible for deterioration and take the appropriate measures to control their growth, and has become an important research field both for conservators and for microbiologists [3].

Microbial techniques for investigating microbial communities can be divided into culture-dependent and culture-independent methods. Culture-dependent methods are based on the cultivation of microbial strains, thus allowing to study the physiology and metabolism of the isolated microorganisms. The drawback is that they provide limited information on the diversity of complex microbial communities. In fact it has been estimated that the vast majority of environmental microorganisms occurring in a viable state (approximately 99%) cannot be cultured according to standard procedures [4]. This is partly due to the lack of appropriate methods to grow environmental microorganisms, in particular those belonging to unknown or poorly known species or the anaerobes.

Moreover, very often culture conditions favor fast-growing opportunistic species and hinder the development of microorganisms playing a significant role in the environment. Microbiology applied to Conservation is a relatively young branch of environmental microbiology and has adopted many techniques developed from other sciences, in particular microbial ecology and clinical microbiology, to detect microorganisms. For many years methods used to investigate microorganisms inhabiting CH objects have been culture-dependent, often associated to Microscopy. This has allowed Microbiology to achieve important results in demonstrating the deteriorative role of microorganisms. Nevertheless, the limitation of traditional cultivation techniques and the value of culture-independent methods to assess the biodiversity of complex microbial communities were recognized by microbiologists working on CH at the beginning of the current century [5].

Culture-independent methods are mainly based on DNA level analyses to assess microbial community structure, function and dynamics regardless of the requirement to cultivate microorganisms.

The application of these techniques to CH has been reviewed in detail in [6] and in [7]. The most common techniques used so far have been genetic fingerprinting analyses such as DGGE/TGGE, ARDRA, t-RFLP, SSCP, ARISA, despite of interest their limitation is to fail to assign a taxonomic identification of the different members of a given microbial community. The most promising molecular techniques are those based PCR amplification of specific target genes, mainly rRNA genes, that serve as evolutionary clock and allow taxonomic assignment of the sequences obtained from a microbial community. Initially these techniques were applied to single colonies or libraries from environmental samples, following cloning of the single PCR products, yet in recent years, the application to microbial communities of next generation sequencing techniques has revolutionized microbiology, enabling massive sequencing of DNA samples, independently of cultivation and cloning. This field has been named metagenomics, the term refers to a study of genetic material isolated directly from an environmental sample, whereby it is possible to analyze all of the microorganisms from the sample, including the non-culturable ones [7]. Many studies on microbial communities carried out by sequencing of DNA libraries, carried out over the first decade of our century can be considered as metagenomic studies. Nevertheless, the depth of the investigation is being now enormously enhanced by the fast pace advancement of high throughput next-generation sequencing (NGS) technologies, which allow in-depth sequencing and data analysis of various types of environmental samples [3]. The NGS based methods are increasingly being employed to successfully characterize microbial community composition and functional diversity in different environments, from soil to human gut [8]. In the field of CH these methodologies are being adopted from few years and only few publications are available in the literature.

We will describe these state-of-the-art methodologies in Section 2 and review very recent studies that used them in assessing microbial diversity in CH objects in Section 3.

2. Metagenomics approaches and tools

Metagenomics offers a powerful lens for microbial microorganisms directly in their natural environments bypassing the need for isolation and cultivation of individual species. Before the coming of Next Generation Sequencing (NGS), microorganisms analysis were only based on cultured studied and sequencing with Sanger method [9]. The consequent advance offered by NGS is the ability to produce an enormous volume of data rapidly and cheaply. The ability to sequence the whole genome of many related organisms has allowed large-scale comparative and evolutionary studies that were unimaginable just few years ago [10]. NGS, and in particular Pyrosequencing analysis, "allows to provide cost-effective, time-consuming steps, rapid and highly parallel sequencing of large numbers of fragments from complex samples of DNA or even transcriptomes" [11].

This field of study is constantly evolving and its application can be divided mainly in two different approaches, targeted and untargeted metagenomics. Targeted metagenomics, the most common, sequences in parallel and massively target genes, serving as evolutionary clocks. The evolutionary clock of election is ribosomal RNA (rRNA). The success of this biomarker is based on the enormous database of rRNA gene sequences (more than 200,000) that have been collected for reconstructing the universal Tree of Life and that is exponentially increasing both thanks to targeted and untargeted seqeucning methods [12]. All organisms have rRNAs similar enough to each other that they can be recognized as the "same molecule" but different enough that the differences are a good measure of evolutionary distance.

The 16S rRNA gene has become an accepted standard in studies of Bacteria and Archea, in which the discrimination of different species occurs when two organisms have less than 97% identity of this gene [13]. 16S gene consists of about 1500 pairs of base [10], composed of conserved regions and nine short hypervariable regions, numbered from V1 to V9 (fig.1.5) that characterize the bacterial taxonomy. From literature emerges, that the nine-hypervariable regions show different capacities in the identification of bacteria [14]. In order to obtain a meaningful microbial identification, genus-or species-level classification is important. Using 16S rRNA gene sequence data, genera and species are typically distinguished at levels of 95% and 97% pairwise sequence identities, respectively [15]. The HMP Consortium, basing on NGS data, found out that 454-generated data describe microbial community better than those generated with 3730 Sanger sequencing and determined that the statistically significant most accurate window region was the V3-V5, with the V6-V9 giving the worse results (Group Jumpstart Consortium Human Microbiome Project Data Generation Working Group, 2012 and [16]). In the case of metagenomics analysis of fungi, the region analysed was the ITS1 region. The ribosomes of eukaryotic organisms are composed of two structural subunits, 60S and 40S. The 60S subunit contain three different types of rRNA fragments: 28S, 5.8S and 5S rRNA. The 40S has a single 18S RNA fragment. The genes coding for the different fragments of rRNA are repeated genes, constituting the nucleolus organize region (NOR). Within the chromosome are clustered into structural units separated by non-coding spacer sequences called NTS (non-transcribed spacer) and IGS (intergenic spacer). Each unit consists of a structural and external transcribed spacer (ETS), located at the 3 'end of the 18S gene, two internal transcribed spacers (ITS), situated respectively: the ITS1 between the gene 18S and 5.8 S gene, the ITS2 between the 5.8S 3 'end and the 28S gene and 5S gene is separated from the 28S gene from another non-coding spacer sequence. Currently the ITS1 region, such V3-V5 region of the 16S, is a hypervariable regions, allowing identification of fungi at the species or genus level, and in the case of a metagenomic analysis, to understand the composition of the fungal community present in a particular habitat [17].

From a technological point of view, initially the metagenomics revolution was led by, Roche/454 pyrosequencing Life Sciences (Roche), used for the identification of ribosomal RNA and for the investigation of entire genomes. Pyrosequencing generates a large number of rDNA sequence tags by amplifying select variable regions within 16S or ITS rRNA gene that are currently regarded as the most versatile phylogenetic markers.

In parallel to pyrosequencing Illumina has recently developed a number of instruments enabling both targeted and untargeted sequencing of 16s or ITS hypervariable regions. The chemistry and chip design are different from Pyrosequencing, yet the potentials are greater, since the throughput and costs are exponentially decreasing.

Illumina is also the technology of choice for untargeted massive sequencing [18]. Untargeted metagenomics is significantly more expensive than targeted, and is based on generating large amount of short reads on the entire DNA community. Assignment of sequence to an OTU is a challenge dependent on the number of sequenced reference genomes available, thus this approaches are viable for the human metagenome, but are quite challenging for environmental samples. Yet the big advantage of these technologies is to provide unique insight on strain level variation [19] and on the microbial functions present [20].

The future of next generation sequencing technologies is now the ability to obtain long reads longer than 10kb, using single molecule sequencing approaches, that do not require amplification, allowing entire assembly of small genomes or organelles [21].

The fast pace at which the NGS is evolving promises to further reduce the costs and expand the potential, nevertheless the decision on the best technology to be used really depends on a careful definition of the biological question [22]. The main challenge being the bioinformatics development and the ability to extract meaningful information from the large datasets generated.

Other paragraphs are indented (BodytextIndented style).

3. Recent studies on CH biodeterioration based on Metagenomics and NGS Methods

Authors [28] characterized algal and fungal epilithic communities living on the surface of stone heritage structures in Belfast, UK, by using molecular techniques, namely TRFLP and high-throughput 454 pyrosequencing, to investigate the relationship between eukaryotic community composition and a variety of substrate characteristics. They found that community structure varied in association with major differences in stone geochemistry (i.e. limestone vs. sandstone). By 454 pyrosequencing a number of algal species that have not been associated with building stone were identified. However, algal communities were found relatively homogeneous and largely unaffected by minor differences in sandstone type or by morphological features such as surface profile or surface texture. In contrast, fungal communities were much richer and more spatially heterogeneous than the algae. According to the

IOP Conf. Series: Materials Science and Engineering **364** (2018) 012074 doi:10.1088/1757-899X/364/1/012074

authors, the relative simplicity of epilithic algal communities, coupled with their spatial invariance, should make these assemblages amenable to efforts to model the impacts of future climate change (e.g. through the use of manipulative experiments).

Authors in [23] investigated the diversity and abundance of microbial communities thriving in degraded mural paintings, selected for representing different environmental conditions in Portugal, by combining culture-dependent methods and culture independent methods, such as denaturing gradient gel electrophoresis (DGGE) and pyrosequencing. The coupled use of culture-dependent methods and DGGE did not give enough information to investigate the diversity and abundance of microorganisms present in wall paintings. Pyrosequencing, used for the first time for this kind of materials, allowed the identification of a large number of microorganisms, confirming some already identified by the cultivation-dependent methods (such as fungi of the genera Penicillium and Cladosporium), but also providing a great contribution in the identification of several genera and species not previously identified in these artworks. The results obtained on several mural painting samples showed a strong relationship between the most deteriorated areas of the paintings and higher microbial contamination. According to the authors, the application of pyrosequencing provided an important and exhaustive description about the microbial population that develops on murals paintings and allowed to expand the knowledge about them, giving a detailed overview of contaminants which was not possible with the other techniques.

Authors in [8] studied the phylogenetic diversity and activated metabolic pathways of microbial communities colonizing construction materials, such as wood, brick, mortar, and stone in historic buildings located in the former Auschwitz II–Birkenau concentration and extermination camp in Oświęcim, Poland. For this purpose molecular, microscopic and chemical methods were combined. Selected specimens were examined using Field Emission Scanning Electron Microscopy (FESEM), metabolomics analysis by ultra-high performance liquid chromatography (UPLC) coupled to high-resolution mass spectrometry (HRMS) and high-throughput Illumina sequencing. UPLC/HRMS- based metabolomics was used for the first time in the study of biodeterioration of heritage materials. DNA sequences documented the presence of 15 bacterial phyla representing 99 genera and nine fungal classes represented by 113 genera, confirming NGSM as a superior tool for interrogating the diversity of the microbial populations associated with heritage materials compared with culturing methods and clone library approach (previously used by the same authors on historic building materials). Most of the identified sequences were characteristic of organisms implicated in deterioration of wood and brick. Metabolomic data allowed the detection of numerous active metabolic pathways present in heritage materials, including those regulating the production of primary and secondary metabolites. The study demonstrated that a combination of FESEM imaging with metabolomics and genomic techniques allows to link the phylogenetic information and metabolic profiles of microbial communities, to shed new light on biodeterioration processes and to develop novel conservation strategies.

Authors in [29] carried out a high-throughput investigation of the prokaryotic communities colonizing the exterior walls of the medieval church of San Leonardo di Siponto (Italy) by Illumina-based deep sequencing.

The metagenomic analysis of sequences revealed the presence of Archaea, Bacteria, and photosynthetic Eukarya. The microbial diversity was explored at qualitative and quantitative levels, evaluating the richness and the abundance of reads associated with each OTU. The results highlighted a structured community, showing low diversity, made up of extremophile organisms adapted to desiccation and UV radiation. The authors concluded that molecular tools, and in particular the easy-to-run next-generation sequencing, are powerful to perform an accurate microbiological investigation and diagnosis in order to plan better strategies for restoration and protection of historical and artistic monuments.

In [24] the microbial communities colonizing the stone monuments of biodeteriorated ancient Buddhist statues in Hangzhou, China, were identified by high-throughput Illumina sequencing (performed on a MiSeq platform) of 16S rDNA for bacteria and ITS sequences for fungi. The diversity and distribution of the microbial communities in samples collected from three different environmental conditions with signs of deterioration were analyzed by means of bioinformatics software and diversity indices. In addition, the relationship between these communities and environmental factors, including light intensity, air humidity and concentration of NO2 and SO2 were evaluated. The authors showed a positive association between the diversity and distribution of microbial communities and the considered environmental parameters and considered their data useful for the evaluation of effective control measures.

In a following study [25], the same authors characterized the microbial (prokaryotic and eukaryotic) communities colonizing other biodeteriorated Buddhist statues with signs of deterioration from the same site through a combination of high-throughput sequencing by Illumina MiSeq and culture-dependent techniques. The latter were used to test the ability of the isolates to degrade synthetic materials used to protect and reinforce the stone monuments. They found that the diversity of the microorganisms varied with the environmental conditions where samples were collected from, while a large proportion of the isolates had the ability to degrade protective materials.

In [3], 16S rRNA amplicon NGS using the Ion Torrent™ sequencing technology platform was applied for the first time to investigate the microbiota, Bacteria and Archaea, involved in the biodeterioration of building materials, as brick and paint coating used in historical buildings in Lodz (Poland). To evaluate the sensitivity of taxon detection depending on the type of primer pairs, amplicon sequencing of the 16S rRNA gene was performed using three sets of well-known primers (universal, bacteria- and archaea-specific primers). Results showed that the Ion Torrent™ sequencing platform has an enormous potential to evaluate the microbiome of historic building materials in a fast and easy way. The brick sample displayed a total of 1178 genera belonging to 27 bacterial and 3 archaeal phyla, while the paint coating showed 145 genera belonging to 11 bacterial phyla. Nevertheless, analyses showed to be influenced by the primer choice, and the authors recommend using, in addition to the universal primers, bacteria- and archaea specific primers in order to cover a higher taxonomic diversity.

Since deteriorating monuments, especially salt weathered ones, can provide suitable habitat for the proliferation of halophilic microorganisms, a special focus was given to detect them. The analyses revealed halophilic and halotolerant bacterial sequences, represented by genera which biodegradation potential was described and discussed in the literature. Moreover, the list of halophilic microorganisms obtained in previous work from cultivation and molecular techniques, comprising mainly DGGE and cloning analysed, applied to the built cultural heritage was remarkably expanded by the implementation of next generation sequencing using the Ion PGM™ System in metagenomics studies.

Authors of [26] employed and compared culture-dependent and culture-independent, using high-throughput sequencing on Illumina MiSeq platform, approaches for the first time to analyze the deteriorating microflora of book samples. It is worth noting that rRNA genes were amplified from total RNA extracted from the microflora sampled, which allows to detect metabolically active microorganisms. The comparison of results obtained from culture-dependent and culture-independent methods showed few similarities. High-throughput sequencing allowed for the detection of higher biodiversity than the culture-dependent approach, giving a deep view of the microflora contaminating different book portions. The authors compared their results with those coming from other few studies based on other culture-independent methods to analyse the bacterial communities in archival documents. Although bacteria members of the phyla Actinobacteria, Proteobacteria and Firmicutes very frequently contaminating archival objects were detected also by different approaches applied to the analysis of cellulose materials in previous works, the Illumina investigation allowed a more accurate analysis of this kind of microbial community. For fungi, Illumina results confirmed the findings on detected genera of Ascomycota, known to be plentiful in paper documents, of previous investigations and allowed to detect new taxa contaminating these materials, although their presence was in the range of 1-4%. Moreover, it evidenced the presence of members of Basidiomycota, not considered as the major fungal contaminant of archival documents, with a large diversity of this fungal group not been detected until now on archival items.

4. Conclusions and Perspectives

The high throughput NGS methods are increasingly employed to successfully characterize microbial communities occurring in different environments, from soil to human gut, since their fast-growing output per cost ratio [8, 3] and the advancements in appropriate bioinformatics tools to analyze the amount of phylogenetic data obtained. Nevertheless, these state-of-the-art methodologies are being adopted only very recently in the field of CH and only few publications are available. These studies, here reviewed, used targeted Metagenomics methods based on sequencing of partial rDNA amplicons to describe prokarytic and eukaryotic communities colonizing deteriorated heritage materials, mainly stone structures, monuments and construction materials. The NGS sequencing platforms used were different, with the Illumina Miseq platform being the most adopted. Independently on the platform and the rDNA target region, a general agreement emerges among the authors on the indisputable advantage of using NGS methods as a powerful tool to deep analysis of microbial populations over conventional microbiological or other molecular techniques adopted in the study of biodeterioration. Moreover, these methods can be combined with a non-invasive (or micro-invasive) sampling, an essential requirement for working on historic or artistic products.

In order to get a deeper understanding of biodeterioration processes and to design effective and case-specific conservation strategies, it is of outmost importance to know not only the taxon composition of microbial communities, but also how microbial populations interact with each other and with heritage materials causing damage.

Metagenomics approaches on airborne microbial communities is also showing that climate change and pollution are changing the structure of microbial communities and their transport [27]. Thus it is expected that microbial communities involved in biodeterioration could also be affected and reflect this global change.

It is then absolutely necessary to develop appropriate methods for monitoring both composition of microbial communities and their physiological activity and combining different methods [2, 8, 26]. Monitoring methods must be optimized in order to be able to assess the function and possible community shifts, the effects of conservation treatments, climate change or biocide application [2, 26].

While cultivation methods are still widely used to investigate the physiological role of microorganisms in biodeterioration, studies attempting to investigate physiology and quantify microbial activity without cultivation in CH, i.e. studies on RNA or metabolic products, are still rare [2]. Among the studies here reviewed, only one used a metabolomic approach [8].

Metagenomics, metatranscriptomics and metabolomics technologies offer more powerful tools to understanding the activity and function of whole microbial communities and allow to generate unprecedented amount of genetic and chemical data on biodeterioration processes. These state-of-the-art technologies can be still combined to cultivation methods. NGS techniques can give an exhaustive picture of the microbial communities (richness and abundance) inhabiting CH objects allowing to identify microorganisms not identified by cultivation but probably involved in biodeterioration if detected with high percentages of reads and/or with a metabolic potential compatible with the investigated damage. These findings can direct efforts towards developing and improving cultivation strategies able to isolate selected groups of microorganisms with a potential role in deterioration. This is a challenge for next investigations in order to enhance the knowledge on cultivation of microorganisms from a specific material and their real deterioration effects [26].

Meta-omics approaches applied to the study of biodeterioration of CH are still in their infancy. Nevertheless, considering their deep detection power, the huge amount of data produced in a short time and the availability of bioinformatics tools to process them, and that the costs for molecular analysis are also going down, in the next future they should become a routine tool, also combined with implemented more conventional techniques, for monitoring structure and activity of microbial communities associated to biodeterioration.

References

[1] Allsopp D. 2011. Worldwide wastage: the economics of biodeterioration. Microbiol.

Tod.,38:150-153.

[2] Sterflinger K, Piñar G. 2013. Microbial deterioration of cultural heritage and works of art — tilting at windmills? Appl. Microbiol. Biotechnol., 97(22):9637-9646.

[3] Adamiak J, Otlewska A, Tafer H, Lopandic K, Gutarowska B, Sterflinger K, Piñar. 2017. First evaluation of the microbiome of built cultural heritage by using the Ion Torrent next generation sequencing platform. Int. Biodeter. Biodegrad. DOI: 10.1016/j.ibiod.2017.01.040.

[4] Rappé MS, Giovannoni SJ. 2003. The uncultured microbial majority. Annu. Rev. Microbiol. 57:369 –394.

[5] González JM, Saiz-Jiménez C. 2005. Application of molecular nucleic acid-based techniques for the study of microbial communities in monuments and artworks. Int. Microbiol., 8:189–194

[6] Dakal TC, Arora PK. 2012. Evaluation of potential of molecular and physical techniques in studying biodeterioration. Rev. Environ. Sci. Biotechnol., 11: 71–104.

[7] Otlewska A, Adamiak aJ, Gutarowska B. 2014. Application of molecular techniques for the assessment of microorganism diversity on cultural heritage objects. Acta Biochimica polonica 61(2):217-225.

[8] Gutarowska B, Celikkol-Aydin S, Bonifay V, Otlewska A, Aydin E, Oldham AL, Brauer JI, Duncan KE, Adamiak J, Sunner JA, Beech IB. 2015. Metabolomic and high-throughput sequencing analysis-modern approach for the assessment of biodeterioration of materials from historic buildings. Front. Microbiol., 6:979.

[9] F. Sanger, S. Nicklen, A. R. Coulson, Proc. Natl. Acad. Sci. U.S.A. 74 , 5463 (1977)

[10] Weinstock G.M., (2012). *Genomic approaches to studying the human microbiota*. Nature, 489 (7415), 250-256.

[11] Lozupone C. A., Stombaugh J. I., Gordon J. I., Jansson J. K., Knight R. *Diversity, stability and resilience of the human gut microbiota*. Nature. Sep 13, 2012; 489(7415): 220–230. doi: 10.1038.

[12] Ramazzotti M., Berna L., Donati C., Cavalieri D. riboFrame: An Improved Method for Microbial Taxonomy Profiling from Non-Targeted Metagenomics. Frontiers in Genetics, 17 November 2015

[13] Whitman, W. B., D. C. Coleman, and W. J. Wiebe. 1998. Prokaryotes: the unseen majority. Proc. Natl. Acad. Sci. USA 95:6578–6583

[14] Petrosino J.F., Highlander S., Luna R.A., Gibbs R.A., Versalovic J., *Metagenomic Pyrosequencing and Microbial* Identification. Clinical Chemistry. (2009). 55 (5), 856-866.

[15] Peterson, D. A., Frank, D. N., Pace, N. R., Gordon, J. I. (2008). Metagenomic approaches for defining the pathogenesis of inflammatory bowel diseases. *Cell host & microbe*, *3*(6), 417-427.

[16] Albanese D, De Filippo C, Cavalieri D, Donati C. Explaining diversity in metagenomic datasets by phylogenetic-based feature weighting. PLoS Comput Biol. 2015 Mar 27;11(3):e1004186.

[17] Strati F., Di Paola M., Stefanini I., Albanese D., Rizzetto L., Lionetti P., Calabrò A., Jousson O., Donati C., Cavalieri D. and De Filippo C. Age and gender affect the composition of fungal population of the human gastrointestinal tract. Front. Microbiol. 7:1227. doi: 10.3389/fmicb.2016.01227.

[18] Quince C, Walker AW, Simpson JT, Loman NJ, Segata N. Shotgun metagenomics, from sampling to analysis. Nat Biotechnol. 2017 Sep 12;35(9):833-844.

[19] C. Donati, M. Zolfo, D. Albanese, D. T. Truong, F. Asnicar, V. Iebba, D. Cavalieri, O. Jousson, C. De Filippo, C. Huttenhower and N. Segata. Uncovering oral Neisseria tropism and persistence using metagenomic sequencing. Nat Microbiol. 2016 May 27;1(7):16070.

[20] De Filippo C., Ramazzotti M., Fontana P., Cavalieri D. Bioinformatic approaches for functional annotation and pathway inference in metagenomics data. Briefings in Bioinformatics 2012 Nov;13(6):696-710.

[21] Ferrarini M, Moretto M, Ward JA, Urbanovski N, Stevanovi V, Giongo L, Viola R, Cavalieri D, Velasco R, Cestaro A, Sargent DJ. An evaluation of the PacBio RS platform for sequencing

and de novo assembly of a chloroplast genome. BMC Genomics. 2013 Oct 1;14(1):670.

[22] Vincent AT, Derome N, Boyle B, Culley AI, Charette SJ. J Microbiol Methods. Next-generation sequencing (NGS) in the microbiological world: How to make the most of your money. 2017 Jul;138:60-71.

[23] Rosado T, Mirão J, Candeias A, Caldeira AT. 2014. Microbial communities analysis assessed by pyrosequencing—a new approach applied to conservation state studies of mural paintings. Anal. Bioanal. Chem., 406:887-895

[24] Li Q, Zhang B, He Z, Yang X. 2016. Distribution and Diversity of Bacteria and Fungi Colonization in Stone Monuments Analyzed by High-Throughput Sequencing. PLoS ONE 11(9):11-17

[25] Li Q, Zhang B, Wang L, Ge Q. 2017. Distribution and diversity of bacteria and fungi colonizing ancient Buddhist statues analyzed by high-throughput sequencing. Int. Biodeter. Biodegrad., 117:245-254.

[26] Kraková L, Šoltys K, Otlewska A, Pietrzak K, Purkrtová S, Savická D, Puškárová A, Bučková M, Szemes T, Budiš J, Demnerová K, Gutarowska B, Pangallo D. 2017. Comparison of methods for identification of microbial communities in book collections: Culture-dependent (sequencing and MALDI-TOF MS) and culture-independent (Illumina MiSeq). Int. Biodeter. Biodegrad. DOI: 10.1016/j.ibiod.2017.02.015

[27] Weil T, De Filippo C, Albanese D, Donati C, Pindo M, Pavarini L, Carotenuto F, Pasqui M, Poto L, Gabrieli J, Barbante C, Sattler B, Cavalieri D, Miglietta F. Legal immigrants: invasion of alien microbial communities during winter occurring desert dust storms. Microbiome. 2017 Mar 10;5(1):32.

[28] Cutler NA, Oliver AE, Viles HA, Ahmad S, Whiteley AS. 2013. The characterisation of eukaryotic microbial communities on sandstone buildings in Belfast, UK, using TRFLP and 454 pyrosequencing. Int. Biodeter. Biodegrad. , 82:124-133.

[29] Chimienti G, Piredda R, Pepe G, . Dorothé van der Werf, Sabbatini L, Crecchio C, Ricciuti P, D'Erchia AM, Manzari C, Pesole G. 2016. Profile of microbial communities on carbonate stones of the medieval church of San Leonardo di Siponto (Italy) by Illumina-based deep sequencing. Appl. Microbiol. Biotechnol., 100:8537-8548.

BIONANOSCULP, an ongoing project in biotechnology applications for preventive conservation of outdoor sculptures

P R Moreira[1,2,3],N Silva[1], R C Pullar[4], B A E Ben-Arfa[4], J G Abreu[2], F Henriques[2], M M E Pintado[1] and E Vieira[1,3]

[1] Center for Biotechnology and Fine Chemistry (CBQF), Escola Superior de Biotecnologia, Universidade Católica Portuguesa, Rua Arquiteto Lobão Vital Apartado 2511, 4202-401 Porto, Portugal
[2] Research Center for Science and Technology on Arts (CITAR), Escola das Artes, Universidade Católica Portuguesa, Rua Diogo Coelho, 1327, 4169-005 Porto, Portugal
[3] School of Arts, Universidade Católica Portuguesa, Rua Diogo Coelho, 1327, 4169-005 Porto, Portugal
[4] Department of Engineering of Materials and Ceramics / CICECO - Aveiro Institute of Materials, Campus Universitário de Santiago, 3810-193 Aveiro, Portugal

Corresponding author: prmoreira@porto.ucp.pt

Abstract.
The objective of this paper is the presentation of the research strategies adopted and results of the ongoing BIONANOSCULP research project that is aiming to develop solutions in the area of sustainable nanomaterials, which are non-invasive and high-performance in their preventive conservation approach.
An integrated methodology was designed as a holistic strategy to the characterisation of the microbiota present on the surface of public outdoor sculptures. Gathering objective data in the characterization of the surface microbiota of public outdoor sculptures is important, in order to design strategies for the preventive conservation of these objects that make use of biotechnology innovative coatings. Such is one of the objectives of the project BIONANOSCULP. Methodologies applied include conservation reports, surface sampling methodologies using gels, 3D modeling, SEM, flow cytometry and metagenomics.
The project is already significantly contributing to create a bridge between the experts from different areas: the skills of biotechnologists, microbiologists, materials scientists, art historians and conservators-restorers to assess the state of conservation, biodeterioration and biocontamination of a selected number of sculptures, and to design the appropriate materials to pursue preventive conservation through coatings with anti-microbial activities.

1. Introduction

Urban outdoor public sculptures are among the most vulnerable cultural objects that can be found in a modern city. Subjected to extreme atmospheric variations and pollution, the conservation of such objects is, most of times, problematic. The variety of materials and the chemical and physical alterations suffered, as well as a wide range of microbiological and biological organisms that thrive on their surface increases the complexity of conservation treatments, and even more, definition of preventive conservation actions.

IOP Conf. Series: Materials Science and Engineering **364** (2018) 012075 doi:10.1088/1757-899X/364/1/012075

Conservation treatments for this type of object usually aim at eradicating the responsible biological agents, however there is the need to aim further than devising restorative treatments that must be repeated every few years. This project aims expanding the traditional boundaries of conservation science to emergent scientific areas such as biotechnology and nanotechnology, changing the preventive conservation paradigm, taking the concept beyond climate control.
There are already several research efforts and reviews on biodeterioration of art objects and many on the specific microorganisms that are responsible for that process in stone and metallic sculptures [1]. They usually reveal important relationships between specific microorganisms and materials, but the variety of materials employed (from stone, to metal and human-made such as concrete) makes generalizations difficult. Since these objects often represent the past and validate the present of the inhabitants of the city, materializing their common cultural base, it is very important to find innovative solutions for conservation problems that affect them.
The need for preventive conservation in such cases is self-explanatory as it is impossible to control the environment that surrounds these objects and the sources for biocontamination responsible for the colonization, most of the time visible as biological patinas that grow into mature ecosystems comprising a large variety of microorganisms, bacteria, fungi, microalgae, lichens and small plants.
The research approach is based the development of innovative materials for coatings whose effectiveness will be tested in a real context, a limited number of public outdoor sculptures with selected material specificities from the metropolitan region of Oporto. The specific aims of the project are: (a) assessment of the state of biodeterioration and conservation state of a selected number of sculptures; (b) design the appropriate materials to pursue preventive conservation thru coating with anti-microbial and/or anti-UV activities as well as consolidation properties; and (c) to test their effectiveness over time.
A partnership with the Division of Culture of Oporto City Council, the Contemporary Sculpture International Museum of Santo Tirso (MIEC, http://miec.cm-stirso.pt) and the Museum of the Faculty of Fine Arts of Oporto (MFBAUP, http://museu.fba.up.pt) allowed access to seven selected sculptures from many different materials and forming part of the cultural heritage of the metropolitan region.
In a second development phase of the project, new coatings will be tested for their effectiveness in preventing the growth of the different agents involved, and also for the requirements for applications in this area. This includes interaction with the object (in this case, the sculpture) from the physical, chemical and aesthetic point of view, as well as reversibility, durability, and applicability to different types of materials. The clear advantage over other products is the low level of toxicity in our material, due to the use of biocompatible biomaterials such as chitosan [2].
One of the challenges of BIONANOSCULP is the interaction between researchers from such different scientific backgrounds such as biotechnologists, microbiologists, materials scientists, art historians and conservators-restorers as the proposed project is by nature collaborative and inter-disciplinary. As a result, it also aims at improving the dialogue between research and cultural heritage stakeholders.
Furthermore, expected products and protocols developed could be adopted by Municipalities, cultural heritage stakeholders, conservation and restoration institutions and companies. The results might also in the future be applied to other cultural heritage objects and collections both in a national and international dimension.

2. Methodologies

2.1 Sculptures selection, description and conservation status evaluation

The seven sculptures represent an ensemble with the common feature of being urban outdoor sculptures of the Oporto metropolitan area.
An initial survey of the catalogued public outdoor stone and metal sculptures in the Metropolitan Area of Porto was performed. The evaluation of the sculptures to integrate the research plan was achieved by observation of the materials present in the sculptures, as well as by visual recognition to the naked eye of the microbial contamination and their general conservation state. The accessibility and surrounding environmental conditions were also evaluated and taken into account.
From all the sculptures analyzed, the following objects were selected to integrate the research project:

Stone- and modern materials-based sculptures: *Rosalía de Castro* (1951) by Salvador Barata Feyo, pink granite, in Praça da Galiza, Massarelos, Porto; *Sol, Lua e Vento* (1997) by Satoru Sato, grey granite, in MIEC; *Afonso de Albuquerque* (1930) by Diogo de Macedo, limestone, in Largo de D. João III, Lordelo do Ouro, Porto; *Movimento* (1994) by Augusto Jorge Ulisses, marble, in MFBAUP; *Repouso* (1953) by Gustavo Bastos, cement mortar, in MFBAUP; *O guardador do Sol* (1953) by José Rodrigues, bronze, in MFBAUP (Fig. 1.) and *Eu espero* (1999) by Fernanda Fragateiro, stainless steel, in MFBAUP.

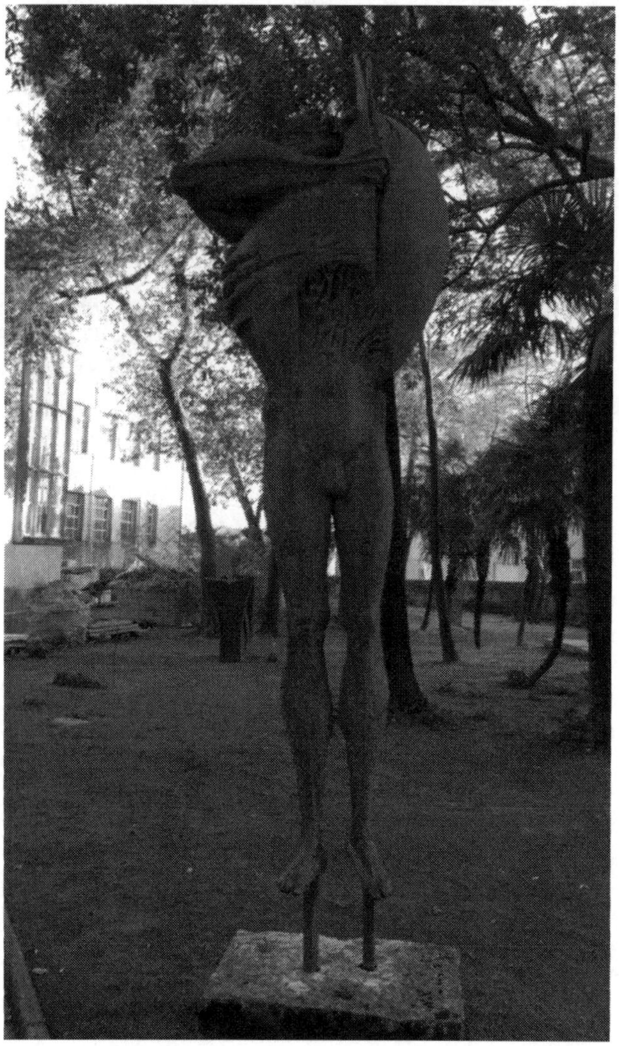

Figure 1. *O guardador do Sol* (1953) by José Rodrigues, bronze, in MFBAUP, one of the selected sculptures studied in BIONANOSCULP.

The selected sculptures are either constituted by stone or metal and all represent important Portuguese artists with the exception of *Sol, Lua e Vento*, a grey granite sculpture by Satoru Sato (MIEC). Their selection for the project was based not only on the materials present, easy access for sampling and (apparent) conservation state but also on their artistic and historical importance, that being reflected on the fact that two of the selected sculptures are also part of the Public Art Map rotes of the Oporto City Council that underlines the Public Art Program of the city (http://www.porto.pt/noticias/mapa-de-arte-publica-hoje-lancado-propoe-5-rotas-pelo-museu-a-ceu-aberto). The evaluation of conservation status was performed by the CITAR team researchers making use of an evaluation form specially designed since no previously defined form was available for outdoor sculptures. The art history specialist in the research team was able to gather important information about the artist, production details of the object, present and past locations, its relation to the institution that currently is responsible for its safeguard, as well as other relevant details that make up the history of the sculpture as detailed as can be described. Such information is often found disperse and is difficult to obtain but of incomparable importance for tracing the relationship between environment conditions and conservation status for outdoor objects.

2.2 3D Modeling and mapping analysis points

Nowadays it is also crucial to establish a knowledge database that includes 3D virtual models that can help explaining the objects [3]. These modeling process is important for mapping the state of biodeterioration and biocontamination, allowing the aquisition of details by the viewer that with a 2D image is not possible, namely in interfaces of materials and different surfaces. The data acquisition technique used is a type of photogrammetry, the structure from motion (SfM) technique. The photogrammetric 3D model was generated using Agisoft Photoscan® (http://www.agisoft.com/). This type of acquisition has the advantage of producing a photorealist 3D models of the sculptures. This characteristic is very important for mapping with accuracy the several locations where the nanomaterials will be tested. The photogrammetric models are then post-processed with Blender, a free and open source 3D creation suite software with GNU General Public License (www. blender.org). Figure 2 depicts a 3D model of *Repouso* (1953) by Gustavo Bastos.

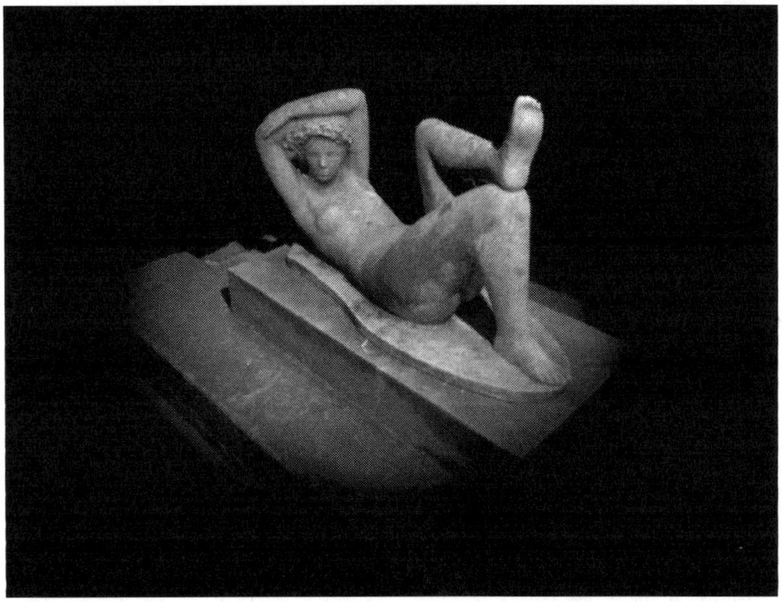

Figure 2. 3D model of *Repouso* (1953) by Gustavo Bastos, in MFBAUP, one of the selected sculptures studied in BIONANOSCULP.

2.3 Surface sampling and characterization methodologies

After testing several gels including mixtures of PVA, HEMA, PVA/PVP and agar and agarose, a sampling procedure was defined that included swabbing with a sterile cotton swab and pressing a sterile disk of a HEMA/MBAm cryogel [4] onto the sculpture's surface followed by a protocol consisting of several centrifugations and filtrations to eliminate debris (adapted from [5]). Samples for SEM were collected with double-sided carbon adhesive tape and sputter-coated with gold and Fungi-tape™ (Scientific Device Laboratory, USA) was used to collect samples for optical microscopy at 100x amplifications. Two granite fragments were also collected from the sculpture with sterile tweezers and a chisel and used for flow cytometry, SEM and DNA extraction for metagenomics analysis. Figure 3 depicts an example of a SEM microcopy image obtained from sampling the surface of *Sol, Lua e Vento* sculpture in MIEC.

Figure 3. SEM image (2000x amplification) resulting of sampling the surface of *Sol, Lua e Vento* sculpture in MIEC, one of the selected sculptures studied in BIONANOSCULP.

2.4 Biocontamination quantification

Classic microbiology methodologies were performed from samples described in section 2.3 using Plate Count Agar (PCA) and Potato Dextrose Agar (PDA) by the spread plate method and incubated at 25 °C in the dark up to 7 days. For the growth of microalgae, BG-11 liquid media supplemented with amphotericin B was used. For aerobiology studies, PDA and PCA Petri dishes (in duplicate) were left open for 10 minutes in selected sites and incubated in the same conditions. Colorimetric measurements were performed using a CR-400 chroma meter (KONICA MINOLTA, Japan) on the sculpture's surface in the five zones previously selected, and L*a*b* values were recorded. ATP was detected by swabbing the sculpture's surface and determining the emitted light after reaction of the samples with a luciferin/luciferase mixture, using a HY-Lite®2 luminometer (Merck, Germany). BD Accuri™ C6 flow cytometer (BD Biosciences, USA) was used to acquired data without probes in order to calculate the number of autofluorescent microorganisms. Cell viability was determined after addition of thiazole orange (TO) and propidium iodide (PI) (BD™ Cell Viability Kit, BD Biosciences, USA).

2.5 Organism identification

To identify all culturable and unculturable microorganisms, total genomic DNA extraction was performed with the Ultraclean® Microbial DNA Isolation kit (MO BIO Laboratories, Inc., USA). DNA was quantified by spectrophotometry and amplified by PCR reaction using the bacterial and fungal universal primers 27F/1492R and ITS4/ITS5, for metagenomics analysis.

3. Discussion and conclusions

It was possible to devise a universal characterization protocol for outdoor sculptures that allows the integration of the results obtained by very different techniques, namely artistic and historical description, conservation reports, 3D modelling, surface characterization, biocontamination quantification and organism identification as described in sections 2.1 to 2.5. We believe this is the first time such a complete and interdisciplinary protocol is devised and performed for outdoor sculptures with the possibility to be applied independently of their material constitution, format, location and size. Gathering of data from the seven selected sculptures using previously described methodologies is still ongoing. The combination of the data obtained will allow the throughout

characterization of the sculptures regarding their biocontamination and conservation status and surely allow for reaching insightful conclusions regarding their similarities and dissimilarities.
Within some of the methodologies, there were already interesting specific results, that result in innovation within the characterization of biocontamination. It was described for the first time, the use of a sampling hydrogel of HEMA/MBAm for cells and DNA in order do identify surface contamination on cultural objects.
It was also possible to evaluate the potential of using either swabs or hydrogels for samples that were later analysed by flow cytometry, which allowed for a quick and effective determination of the total number of microorganisms. Flow cytometry also allowed the determination of the total number of microorganisms that was confirmed by detection of ATP. Highest levels of ATP were observed in accordance with the results obtained with flow cytometry and growth on culture media.
The presence of autofluorescent microorganisms was detected, being higher in the samples collected with swabs than those collected with hydrogels. It was detected a great variation in the number of autofluorescent microorganisms detected, according to the methodology used to collect the samples (swab or hydrogel).

The development of the new coatings from chitosan is currently ongoing and in the near future they will be tested for their effectiveness in preventing the growth of the different biological agents identified on the sculptures; different requirements for application in preventive conservation will also be tested, including the coatings interactions with the object's materials from a physical, chemical, and aesthetic point of view, as well as reversibility, durability, and applicability to different types of materials. The clear advantage over other products is its low level of toxicity, which is very important from the perspective of the conservator-restorer or archaeologist.

4. References

[1] Warscheid T, Braams J 2000 *Int. Biodet. Biodeg.*, **46** 343-368.
[2] Kumar M N V R 2000. *React. Funct. Polym.* **46** 1-27.
[3] Historic England 2017 Photogrammetric Applications for Cultural Heritage. Guidance for Good Practice. Swindon. Historic England.
[4] Keles B, Inanan T, Tüzmen N, Denizli A 2015 *Croatia Chemica Acta* **88** 139-149.
[5] Anvarian A H P, Cao Y, Srikumar S, Fanning S, Jordan K 2016. *Front. Microbiol.* **7** 1-15.

5. Acknowledgments

This research work is supported by national funding from FCT- Fundação para a Ciência e a Tecnologia, I.P. for project BIONANOSCULP- Desenvolvimento de BIONANOmateriais para revestimento anti-microbiano de eSCULturas exteriores metálicas e de Pedra PTDC/EPH-PAT/6281/2014. This work was developed within the scope of the project CICECO-Aveiro Institute of Materials, POCI-01-0145-FEDER-007679 (FCT Ref. UID /CTM /50011/2013), financed by national funds through the FCT/MEC and when appropriate co-financed by FEDER under the PT2020 Partnership Agreement. R.C. Pullar thanks the FCT for funding under grant IF/00681/2015.

IOP Conf. Series: Materials Science and Engineering **364** (2018) 012076 doi:10.1088/1757-899X/364/1/012076

SOLUBLE SALTS EXTRACTION WITH DIFFERENT THICKENERS: MONITORING OF THE EFFECTS ON PLASTER

M Bertasa[1-2], F Bandini[3], A Felici[3], M R Lanfranchi[3], R Negrotti[1], C Riminesi[4], D Scalarone[2], A Sansonetti[1]

[1] ICVBC-CNR, Via Roberto Cozzi 56, Milano, Italy
[2] Department of Chemistry, University of Turin, Via Pietro Giuria 7, Turin, Italy
[3] Opificio delle Pietre Dure di Firenze, V. le F. Strozzi 1, 50122 Firenze, Italy
[4] ICVBC-CNR, Via Madonna del Piano 10, Sesto Fiorentino, Italy

Corresponding author e-mail: mbertasa@unito.it

Abstract

Soluble salts content and their crystallization play an important role in the deterioration of porous substrates; the common approach to solve this conservative problem is focused on the identification of a proper thickener and application protocol able to reach a good level of desalination. Here we report the results of a systematic study on a late Renaissance plaster aimed at reducing the content of gypsum and sodium nitrates comparing five different type of thickeners. The extractive process has been monitored by Evanescent Field Dielectrometry (EFD) by means of the evaluation of salinity index (SI) and humidity content (HC), whereas the quantification of the salinity level, before and after desalination, has been performed with ionic chromatography (IC). Data coming from the field, have been integrated with laboratory measurements aimed at evaluating the water release trend, according to the micro-structural features and the pore size distribution of both the thickeners and the substrate. This step has the purpose to identify an effective correlation among the characteristics of the desalination systems and the porosity in order to estimate a proper trend of effectiveness, suitable for the procedure of conservators.

1. Introduction

Scientific literature underlines how soluble salts content and their crystallization are still an open conservation problem[1-5]; they could be considered among the primary causes of irreversible damage in artistic objects, such as sculpture, masonry and mural paintings. The common approach adopted to front this problem and minimize the damage, is aimed at reducing the salts content by means of water-based systems composed by different thickeners. Water has the role to solubilize the salts, while the extraction mechanism is strongly dependent to the typology of the thickener and its microstructure. Both diffusion and capillary suction are involved in the extraction mechanism. As regards diffusion, salts ions are transported from the artwork bulk to the poultice thanks to a concentration gradient; for what concern capillary suction the size of capillaries could be the driving force of the transport phenomena[2]. Pel et Al. report about the physical working principles of ions salt and moisture transport phenomena; starting from the penetration depth of the saline damage, they demonstrate an important dependency in salts extraction efficiency with the time of treatment[6].
Water evaporation links the migration of soluble salts to crystallization on the outer surface (efflorescence) or in the underneath region (sub-efflorescence), this latter being the most potentially damaging, due to both the lack of ease in visibility and the huge crystallization pressures, which could rapidly act on cohesion forces[7]. The most common soluble salts are composed of chlorides, nitrates, sulfates of sodium, potassium, calcium and magnesium and their mixtures; generally, their chemical composition is strongly dependent by the composition of the host object.

IOP Conf. Series: Materials Science and Engineering **364** (2018) 012076 doi:10.1088/1757-899X/364/1/012076

In this paper we report the results of a systematic study on a late renaissance plaster, located in the Sforza Castle in Milan recently object of a conservation work. As a first step the soluble salts were characterized: a mixture of gypsum ($CaSO_4$ $2H_2O$), magnesium sulphates (both hexahydrite $MgSO_4$ $6H_2O$ and epsomite $MgSO_4$ $6H_2O$) and nitratine ($NaNO_3$) were detected by X-Ray Diffraction[8]. In the second step the conservator team choose a set of different thickener in order to identify the most adequate poultice for a proper desalination. Five different systems were compared: agar (applied by brush and as a rigid gel), cellulose pulp (arbocel), cellulose pulp + sepiolite and cellulose pulp + sepiolite + sand; each one was applied to the plaster with different contact time (2 ½ h, 5h and 24h). A multi-analytical approach was applied to study the extractive process: Evanescent Field Dielectrometry (EFD)[9,10], was used to monitor the desalination in terms of salinity index (SI) and humidity content (HC) and ionic chromatography (IC) was used to quantify the salts content into the plaster, sampled by drilling as powder, but also extracted from the plaster, before and after treatment.
The plaster showed both efflorescence, removable by brush, and sub-efflorescence. This latter decay product could be quantified only with the aid of sampling, but the procedure is often not allowed. For this reason, the application of a non-destructive testing as the Evanescent Field Dielectrometry was crucial and proved to be a winning strategy.
This study suggested to explore another important issue related to the efficiency of the thickeners such as the water release. To reach this goal, a set of model samples composed by specimens of different lithotypes characterized by decreasing porosity, has been assessed. Noto calcarenite, Manciano sandstone and Bergamo limestone have been selected and weighted before and after the application of a water based poultice composed by the same thickener used in the field, measuring accurately the water release at a given time (30 min). At each step, before and after the application of the poultice, also the SI and HC measurements were performed by EFD.

2. Materials and Methods

2.1 Poultices and their formulations

Five different formulations of water based poultices were prepared:
1. agar at 4% in liquid form and rigid (AgarArt, CTS srl)
2. cellulose pulp (Arbocel BC 200)
3. cellulose pulp + sepiolite (Bresciani S.r.l. Milan)
4. cellulose pulp + sepiolite + quartz sand (quarried in Ticino river deposits)

Each system was applied to the plaster with different contact time (2 ½ h, 5h and 24h). The proportions used are reported in table 1. The amount of water in poultices mixes C, D and E, was accurately defined via preliminary tests in order to apply them on a vertical surface.

Table 1 - Formulations of poultices.

System code	Materials	Proportions
A	agar gel	4 % in water
B	agar gel applied by brush	4 % in water
C	arbocel	/
D	arbocel + sepiolite	3:1
E	arbocel + sepiolite + sand	3:1:1

2.2 Evanescent Field Dielectrometry (EFD)

The EFD system used here (called SUSI©) consists of a resonant probe operating in the microwave frequency range (1-1.5 GHz), of a Vector Network Analyser (VNA) for measuring the probe response

IOP Conf. Series: Materials Science and Engineering **364** (2018) 012076 doi:10.1088/1757-899X/364/1/012076

and of a PC portable that driven the previous device by a dedicated software and elaborate achieving the diagnostics results (HC and SI) (Figure 1a).

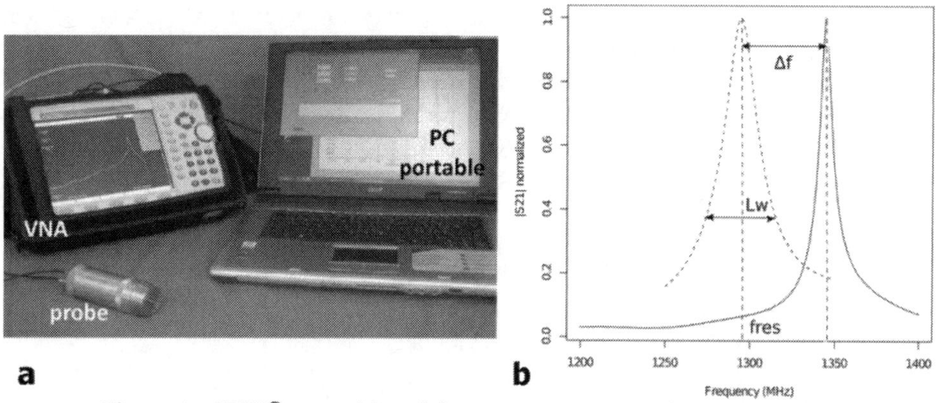

Figure 1 - SUSI© setup (a) and frequency response of resonant probe (b).

The electrical parameters of interest of the resonant probe response are the resonance frequency *fr* and the quality factor *Q* (or resonance bandwidth *Lw*) shown in Figure 1b. Δfr corresponds to the resonant frequency shift of the probe between the loaded behaviour, which is in contact with the material under test (*fr*), and for unloaded behaviour, adressed on air (*fo*).

These resonant parameters are related to the diagnostics parameters HC and SI[9,10]. In first instance, the frequency shift is related to the HC while the spreading in bandwidth is due to the combination of the dielectric losses in the operation microwave region of the system and conductivity losses, both related to the presence of salts, or better related to electrolytes in solution. The HC of the examined material can be related to Δfr, according to the simplified equation shown in reference 10 and 11, where a preliminary calibration, performed by of several measurements at the dry state and wet state, is necessary for the determination of the correlation constants.

The range of humidity content that can be investigated is between 0% to 20% (on dry basis), and for the salinity index from 1 to 10. The SI is a semi-empirical parameter[11].

2.3 Ionic chromatography (IC)

The soluble salt content of the poultice materials (F^-, Cl^-, NO_2^-, NO_3^-, PO_4^{3-}, $SO_4^{=}$, $C_2O_4^{=}$) was determined by a Dionex DX100 ion chromatography system equipped with Dionex AS4A/CS12 columns.

2.4 Stone specimens

Three different stones characterized by different lithology and decreasing porosity have been chosen in order to measure the specific water release using each specific poultice formulations: Noto calcarenite, Manciano sandstone[12] and Bergamo limestone[13], each one sized 5 cm × 5 cm × 2 cm.

2.5 Water release measures

Each specimen was used for the determination of water release from the top to the bottom. The poultices have been shaped on the top of the specimen to cover precisely the upper surface. Each poultice has been sized exactly 1 cm thick, in order to give the same "reservoir effect", and the measure has been prolonged for 30 minutes. At the end of the test each specimen has been weighed on a precision scale, where the weight of water absorbed has been determined by subtracting the weight of the dry specimen. Each test series has been averaged on the basis of 5 specimens. The tests have been carried out at room controlled temperature (25°C) avoiding the evaporation during the 30 minutes, by wrapping the system in a polyethylene film.

3. Results and Discussion

The late Renaissance plaster

The late renaissance plaster has a worrying state of conservation, with an important presence of efflorescence and sub-efflorescence causing a general whitening of the surface[8]. The presence of the soluble salts (gypsum, hexahydrite, epsomite and nitratine) as reported in introduction is possibly due both to the inappropriate use of the room as a stable till the XIX century and to the high level of sulphation involving calcium and magnesium carbonates.
These salts produce a series of whitish rings, some of which visible in Figure 2, together with a general overview of the poultices applied; hence the conservator needed to plan a removal of the soluble salts avoiding as much as possible a large release of water. The conservation problem focused on the correct balance in between desalination effectiveness and low amount of water release.

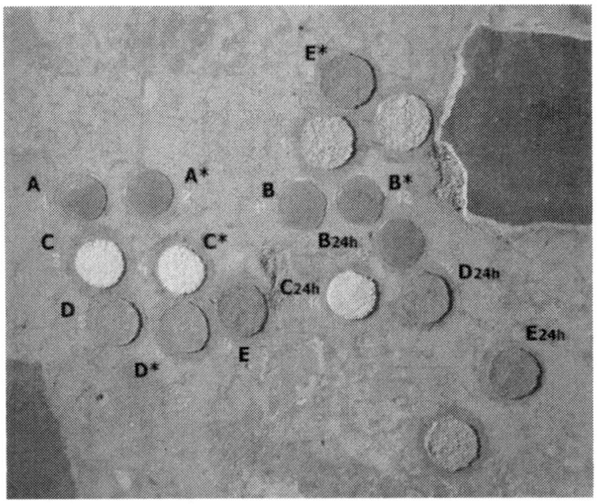

Figure 2 - A general view of different desalination systems applied on plaster. A = Rigid Agar gel; B = Fluid Agar Gel applied by brush; C = Arbocel; D = Arbocel + Sepiolite, E = Arbocel + Sepiolite + sand; * indicates the poultices kept in contact for 5h. In Figure are showed some other poultices not included in the present study.

SUSI

The characteristics of SUSI system, non-destructive measurement and ability to perform the measurement in real-time, are addressed to be the ideal tool for checking the extraction efficacy of poultice of salts. With the aid of SUSI is possible to control the status of the support (HC and SI parameters) before the application of the poultice and immediately after its removal; in this case the monitoring continued up to 48 hours after the removal of the poultice, following the re-equilibrium process of the support with the environment (monitoring of the evaporation process where the efflorescence risk is greater).
The results obtained by SUSI system are presented as to kind of poultice (see Table 1) and time of application: 2½ h, 5h and 24h. The values of HC and SI are plotted in the same graph with a double *y*-axis; the legend in the graph allows distinguishing the marker for the respective parameter.
Since the volume of the material investigated by the probe is a hemisphere (2 cm in diameter) inside the support, it is needed to perform several measurements on the area interested by the poultice, in

order to obtain an average information on the salts contamination of the substrates. Hence, at least 3 measurements were performed by SUSI, for each contact area of the poultice.
In Figure 3 the results obtained for each poultice (Table 1) at time contact of 2½ h, have been selected as the most significant. The first measurement series, at wet state, are quite similar both in terms of HC (about 1-2%) and SI (3-4), this confirms the homogeneity of the areas under comparison. After the application of the poultice, we observed an increase in the salinity index SI (contact time 2½h) for each system, probably due to salts moved by water released by poultice. The effect of the increasing in salts amount, is still measurable after 48 h, and it is more relevant for system C, D, E, where the amount of water released in the support is probably greater.
Analogous results were obtained as to the other contact times, without a significant variation, except for a visible trend in increasing the index of salinity SI.

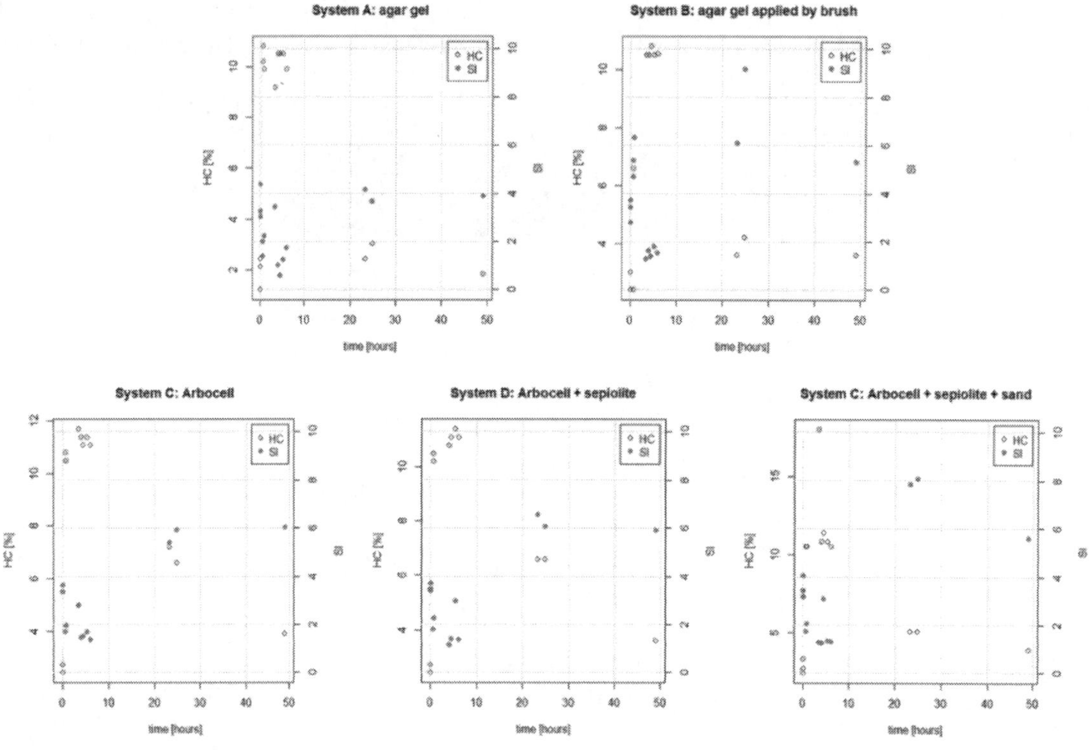

Figure 3 – Plots of HC and SI parameters by SUSI system on the areas where applied the poultices described in Table 1.

Ionic Chromatography

The desalination level performed by the different systems (see 2.1 and Table 1) were estimated by ionic chromatography, focusing on the fluctuation of nitrate and sulfate content (Table 2a-b). At the first glance, all systems applied for 5h exhibit a good effectiveness compared to the ones applied for 2 ½; in particular nitrate ion enhance its content in the poultice material doubling and tripling at least in the 5 hours extraction. This result is probably due to the high solubility of the salt, that is around 91.2 g/100 mL at room temperature in the case of sodium nitrate. On the contrary the removal of sulphates does not increase importantly, if the systems are kept in contact for 5 hours, respect to the values obtained for 2½. it is worth noting that the desalination capacity enhances greatly up to the contact time 24h for all the poultices both for nitrate and sulfate groups. Unfortunately, it was not possible to keep in contact the rigid agar gel for 24 hours because of its own weight and its low adhesion to the vertical surface due to the water evaporation. The same increasing trend with the contact time is observable with chloride, despite its lower values. Agar gel both as a rigid gel and applied by brush seems to be the only system effective in removing phosphates.

Table 2a - Desalination value detected by ionic chromatography on agar gel systems for different contact time

System Code	Time of contact	F^-	Cl^-	NO_2^-	NO_3^-	PO_4^{3-}	$SO_4^=$	$C_2O_4^=$
A	2 ½ h	0.000	0.062	0.086	0.049	0.108	0.094	0.000
A	5 h	0.004	0.042	0.069	0.135	0.048	0.090	0.000
A	24 h	-	-	-	-	-	-	-
B	2 ½ h	0.000	0.119	0.000	0.646	0.038	0.144	0.000
B	5 h	0.000	0.215	0.071	1.147	0.085	0.150	0.000
B	24 h	0.000	0.318	0.000	2.756	0.024	0.439	0.000

Table 2b - Desalination value detected by ionic chromatography on traditional systems for different contact time

System Code	Time of contact	F^-	Cl^-	NO_2^-	NO_3^-	PO_4^{3-}	$SO_4^=$	C_2O_4=
C	2 ½ h	0.000	0.05	0.000	0.008	0.000	0.007	0.000
C	5 h	0.000	0.006	0.000	0.022	0.000	0.008	0.000
C	24 h	0.000	0.010	0.000	0.076	0.000	0.012	0.000
D	2 ½ h	0.006	0.011	0.000	0.028	0.000	0.030	0.000
D	5 h	0.007	0.020	0.000	0.118	0.000	0.039	0.000
D	24 h	0.000	0.094	0.000	0.899	0.000	0.212	0.000
E	2 ½ h	0.000	0.010	0.000	0.029	0.000	0.021	0.000
E	5 h	0.000	0.017	0.000	0.103	0.040	0.027	0.000
E	24 h	0.000	0.064	0.000	0.725	0.000	0.211	0.000

Indeed, agar gel extract the major amount of nitrates and sulfates compared to the other systems: based on the ionic chromatography results, the desalination systems ordered in terms of effectiveness could be listed as follows: agar gel applied by brush > agar gel in rigid form > arbocel + sepiolite > arbocel + sepiolite + sand > arbocel.

The water release on model sample

Thickeners and gels could be chosen in function of the control of the liquid released in the surface and near-surface region; as concern agar gel according to the concentration of the polymer, they show a certain degree of flexibility, which helps in modulating some characteristics. The water release is also strongly dependent with the porosity of the substrate on which the poultice is applied. In this work water release was measured by weight comparing all poultices tested on site, on several stones specimen in the lab; stone was chosen with progressive higher porosity (Bergamo Limestone, Manciano sandstone and Noto Calcarenite).

Figure 4a shows the amount of water released by agar gels at 3, 4 and 5% w/v: it decreases proportionally with the increase of the agar concentration on Noto Calcarenite (R2 =0.999) and on Manciano sandstone (R2 =0.928), as well. The case of Bergamo Black Limestone is problematic, due to the compactness of the stone: the series measured at 4% w/v is not linear with the ones at 3 and 5 % w/v.

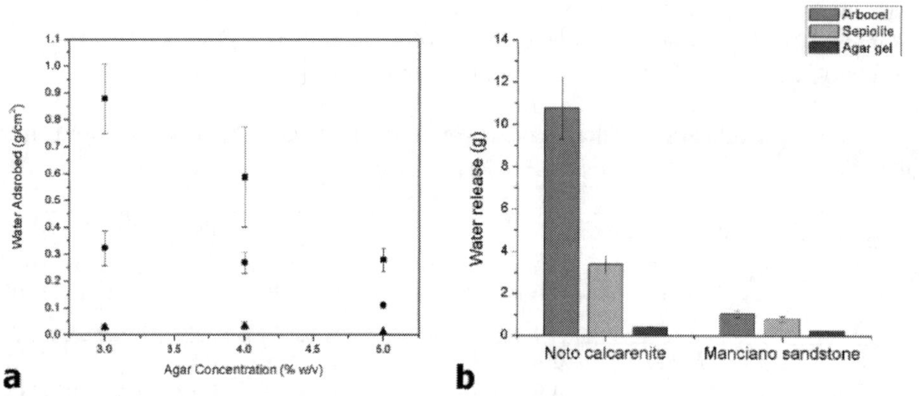

Figure 4 - a) Water absorbed at 30 minutes by agar gel at different concentration on Noto calcarenite (■), Manciano sandstone (•) and Bergamo limestone (▲); b) Water released by paper pulp, sepiolite and agar gel at 3% w/v, on Noto calcarenite and Manciano sandstone.

As it is possible to observe by Figure 4b, in the case of Noto Calcarenite the water released by Arbocel is more or less ten times greater the one released by agar gel; for what concern sepiolite it is the triple. The situation is similar for what concern the other stones, even if is the total amount of water released is lesser than the one measured for Noto Stone; this is certainly due to the total open porosity value of the calcarenite.
Moreover, the preliminary result about the percentage of the water absorbed is estimated and it is around 2.39 % for Noto stone, 1.44 % for Manciano sandstone and 0.12 % for Bergamo limestone.

4. Conclusions

In this research, a systematic study on desalination systems demonstrates that agar gel applied by brush, records the best results in saline extraction capacity if compared to the traditional systems and to agar gel in rigid form too. In general, the agar gel is considered as a *peelable* system, but this statement could be considered correct if the gel is applied in a rigid form and not applied by brush, where some inconveniences about rips have been observed. Moreover, the other systems exhibit a lower extractive capacity and the Japanese paper shall be applied on the treated surface to avoid unwanted residues.
Finally, only one application of the extractive poultices didn't bring in a reduction of the SI on the plaster, so the applications should be repeated several times until to observe an effective reduction in the SI parameter.

Acknowledge

Authors would like to acknowledge *Raccolte Artistiche* del Castello Sforzesco in Milan.

REFERENCES

1) Rodríguez-Navarro C. and Doehne E. 1999 Earth Surf. Process. Landforms, **24** pp 191-269
2) Sawdy A. *et al.* 2008 Proc. Inter. Conf. on Salt Weathering on Buildings and Stone Sculptures (Denmark) pp 1–28.
3) Vergès-Belmin V. and Siedel H. 2005 Restoration of Buildings and Monuments, **11** pp 1–18.
4) Ahl J. 2004 J. Mat. Sci., **39** pp 4247–4254.
5) Petkovic J. *et al.* 2007 Mater. Struct., **40** pp 475–490.

IOP Conf. Series: Materials Science and Engineering **364** (2018) 012076 doi:10.1088/1757-899X/364/1/012076

6) Pel L. *et al.* 2010 J. Cult. Herit., **11** pp 56-67
7) E.Winkler. Stone in Architecture. Properties and Durability. Springer, 1997.
8) Rosina E. *et al.* 2017 Construction and building Materials, **136** pp 643-652.
9) Olmi R. *et al.* 2006 Meas. Sci. Technol., **17**, pp 2281-2288.
10) Olmi R. and Riminesi C. 2008 Italian Physical Society DOI 10.1393/ncc/ic2008-10298-0, 14p.
11) Riminesi C. and Olmi R. 2017 Proc. SWBSS 2017 Fourth International Conference on Salt Weathering of Buildings and Stone Sculptures, pp 49-55.
12) Ludwig N. *et al.* 2017 Measurement, in press, https://doi.org/10.1016/j.measurement.2017.09.002.
13) Marinoni N. *et al.* 2002 J. Cult. Herit., (2002), 3(4): 241-249.

IOP Conf. Series: Materials Science and Engineering **364** (2018) 012077 doi:10.1088/1757-899X/364/1/012077

Atmospheric pressure plasma as a tool for decontamination and functionalization of wooden surfaces

Constanze Roth, Sven Gerullis, Oliver Beier, Dr. Andreas Pfuch, Dr. Annett Hartmann, Dr. Bernd Grünler

INNOVENT e.V. Technology development, Pruessingstr. 27b, 07745 Jena, Germany

cr1@innovent-jena.de

Abstract. Possible applications of plasma technology on wooden surfaces have been investigated by new research projects. There are different initial situation: The residuals of several timber preservatives, based on organo-chlorine compounds, are a high challenge for human protection, storage and wood restoration. Among the most common organo-chlorine substances that have been used as pesticides in the past are Dichlorodiphenyltrichloroethane (DDT), Lindane and Pentachlorphenol (PCP). Lindane and PCP are toxic and/or carcinogenic and all are considered to constitute several environmental hazards. The pesticides were widely used for the treatment of trusses in private homes, churches, castles as well as of furniture or wooden artefacts. This creates the necessity to develop efficient and economically viable methods for removing the harmful biocides from the contaminated timbers without damaging the original materials. For the gentle removal of biocide residuals (so-called blooming of mainly DDT crystalline forms) atmospheric pressure plasma was used as a cleaning tool for the first time. For selected working conditions the removal of up to 75% DDT contamination could be achieved on plain timber surfaces.

Using atmospheric pressure plasma jets (APPJ) it is also possible to create thin films (~ 50 – 200 nm) with different functionalities such as hydrophilicity, hydrophobicity or antimicrobial properties. To create antimicrobial effects active agents were embedded into silicon dioxide films matrices. Antimicrobial tests showed high effects of these films towards bacteria strains such as *Escherichia coli* and *Staphylococcus aureus*. An effect towards mold fungi was low using films containing only one active agent. However, significant improvements in activity were achieved by the combination of active agents. This paper will address the production of samples, analytic methods and open questions about decontamination and functionalization of wooden surfaces by using atmospheric pressure plasma.

1 Introduction

Motivation

In order to protect wooden surfaces against microbial attack preservatives based on organochlorine have been used in the past. According to a current study a high number of contaminated objects can be found in museums and collections in Germany [1]. So, there is a need to clean precious goods on the one side and to find alternative methods to protect the material on the other side.
Wood and wood-based materials as natural products are degradable by microorganism such as bacteria, fungi and/or alga. The biological degradation and the associated infestation by microorganism can be adverse, because performance characteristics such as aesthetic appearance or mechanical properties can decrease. Further critical disadvantages caused by microorganism can be health hazards such as bacterial infections [2]. Common routes to prevent microbial colonization are the insertion of active agents by impregnation procedures or the application of active agents as additive in paints or varnishes. Disadvantages of these kinds of wood treatment can be the location of the biocides over the whole cross section. Furthermore, large amounts of biocides are often need to achieve antimicrobial properties. Another point to consider is that biocide product regulations (Regulation EU No 528 2012) demand the reduction of embedded amounts of biocides. In accordance with it, a negative impact on the customer's perception can caused by embedding biocides.
This paper is based upon two research projects to decontaminate wooden surfaces and create antimicrobial films on wood by using atmospheric pressure for the first time.

Atmospheric pressure plasma

Atmospheric pressure plasma systems, especially those creating cold plasmas, are very interesting for a wide range of applications. Depending on process parameters, non-thermal plasma sources such as dielectric barrier discharges and atmospheric pressure plasma jets (APPJ) generate plasmas with temperatures down to room temperature.
Operating at atmospheric pressure, such plasmas are composed of charged species (positive ions and electrons), high-reactive species such as O^{2+}, O_3, and NO_x as well as photons from ultraviolet to infrared [3, 4]. High-reactive species with energy up to 20 eV can cause polymer chain scission and promote the formation of free radicals and ions. This can further result in a fragmentation, isomerization, cross-linking and the introduction of functional groups to the surface of numerous kinds of materials [5, 6].

The wood components cellulose, lignin and extractives, contain groups such as C-H, C-C, C=C, C-O and C=O. Considering the structure of biocides such as dichlorodiphenyltrichloroethane (DDT), possible groups are C-H, C-C, C=C and C-Cl. These groups have bond energies less than 20 eV and with respect to this, plasmas are intense enough to dissociate almost all chemical bonds occurring in wood and DDT and to create free radical species. These free radicals can further react with oxygen-based polar species and form hydroxyl, carbonyl and carboxyl groups. Hence, these reactive plasma environments can be apply to improve surface characteristics such as wettability and surface energy [7, 8], to clean and decontaminate material surfaces [9] or even to create functional films [10, 11]. Plasma modification effects depend on the substrate, gas flow and nature of the working gas, exposure time, electrical power and others and needs to be adjust to the desired application. These technologies could also be interesting for wood processing due to their in-line capability and therefore the possible implementation in industrial applications and scalability.

Wood preservatives

One of the most common timber preservative in the past is the liquid preservative Hylotox 59®. It contains chlororganic compounds: 3,5% DDT and 0,5% Lindane in a solvent mixture. Until the prohibition of fabrication in the end of the 1980s this very efficient but toxic preservative was used very often in the former German Democratic Republic and in the Federal Republic of Germany [1].

IOP Conf. Series: Materials Science and Engineering **364** (2018) 012077 doi:10.1088/1757-899X/364/1/012077

Wood preservatives usually have a long-lasting effect. Objects treated with Hylotox 59® are also a health risk. It is of great importance to protect employees and visitors of archives and museums by reducing the toxic components from the contaminated materials. The chlororganic residues appear as sparkling crystals on the wooden surfaces and strain the atmosphere. Hence, the removal of the unwanted crystals is desired for two reasons: first to protect the health of employees and visitors and second. to reverse the visual changes of the artefacts.

2 Experimental Methods

2.1. Materials (samples and sample preparation)

The tests for the decontamination researches were carried out by using old contaminated original samples and samples produced in laboratory scale.
The original samples were provides by the Thuringian Schlossmuseum Sondershausen. The most important piece of the museums collection is the so called "Golden Carriage" from the early 18th century. It is an artful emblazed and gilded carriage, made of beech wood, and one of the last five existing parade carriage in Europe [12]. The Golden Carriage served a several princes of Schwarzburg-Sondershausen as a representative state car.
Based on restorer's documentations, the treatment of constructions parts of the Golden Carriage with *Hylotox-59®* against vermins, was firstly reported in 1969. [12]. Due to the uniqueness of this collection piece, the Schlossmuseum provided other contaminated samples of the depot like a probably fifty years old pine wood strip which served as backing protection of an easel painting (Fig. 1).

Fig. 1: The old contaminated pine wood strip, dimensions: 18 x 7 cm². Foto: Birgit Schmidt

It has to be mentioned that for these samples no reproducible depletion quantities could be determined since the original chlorine loading by the wood preservative was not documented or directly known. GC/MS (gas chromatographic-mass spectronomy) measurements revealed here from sample to sample significant differences in DDT concentrations allowing no clear statements for the different treatment conditions.

Since the construction corpus of the Golden Carriage is made of beech wood and for the statistical evaluation of the examinations, new samples were prepared from beech veneers. The specimens were about 20 x 15 mm² in size, about 0.5 mm thick and get a defined chlorine loading. After extensive analysis and comparing measurements INNOVENT prepared a liquid of DDT, mineral oil and solvent. The composition of this mixture was very similar to the aged Hylotox 59®. To receive a similar depletion quantity with Hylotox 59® treated old samples, every new beech veneer sample was loaded with a defined amount of DDT of 30 µg. More than 450 samples in total have been prepared to compare the results and to determine exactly depletion rates.
For the evaluation of the new test specimens six pieces per parameter set were plasma treated and analyzed in order to obtain a statistically reliable mean value.

2.2 atmospheric pressure plasma system

The temperature of the used cold atmospheric pressure plasma is between 30 to 250 °C. The plasma intensity can be adjusted across the entire spectrum and investigations of plasmas itself were already made in restoration projects [13]. INNOVENT compared two different sources namely the TIGRES Plasma MEF (TIGRES GmbH, Marschacht, Germany, see Fig. 2a) and kINPen 09 (neoplas tools GmbH, Greifswald, Germany, Fig. 2b). The last mentioned source is very easy to handle, has a gentle performance and is already in use for dermatological treatments [14] while the TIGRES source offers more variations in parameters due to its technical design.

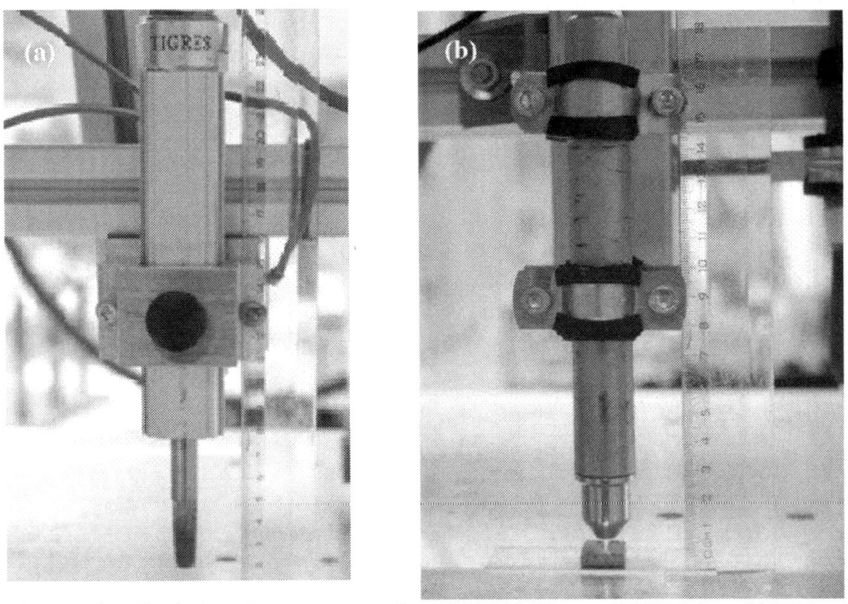

Fig. 2: The single jet plasma system by TIGRES Plasma MEF (a) and kINPen 09 by neoplasm Tools GmbH, Photo: INNOVENT e.V.

The parameters gas type, number of treatments (run), the distance between the wooden surface and the plasma as well as the distance between the treatment lines (pattern distance) were varied. The process gases during the experiments for the TIGRES plasma MEF were nitrogen, air or a nitrogen-hydrogen-mixture (5% hydrogen), for kINPen 09 argon, air and the nitrogen-hydrogen-mixture. Other parameters were kept constant; the velocity of the substrate relatively to the stationary plasma jet was set to 50 mm/s for both sources.
The results for TIGRES MEF single jet with different types of gas, a pattern distance of 1 mm or 3 mm and the number of treatment runs from 1 to 10 times are shown in Tab. 1. For every parameter six samples have been plasma treated and analyzed by GC/MS measurements. The depletion rate of DDT peaks were compared to untreated reference samples measured on the same day to minimize environmental effects.

2.3 system modifications for film deposition

In order to create active agents containing thin films the plasma system was modified in two ways. The first way contains the extension of the process gas module to introduce gaseous hexamethyldisiloxane (HMDSO). HMDSO was provided using the dosage system Pyrosil® STS 10.0 (SURA Instruments GmbH, Jena, Germany). The second way is based on the use of a dosing system, built to spray liquids or liquid dispersed particulate materials through a modified plasma nozzle into the active plasma discharge. To form active agent-containing composite films besides the HMDSO secondary precursors such as silver nitrate, zinc nitrate or copper nitrate solutions were sprayed into the plasma zone, thereby generating the corresponding silver, zinc or copper (-oxide) nanoparticles *in situ*, which are embedded uniformly over the whole growing SiO_x film. The active agent-containing solutions were sprayed directly into the plasma as a nebulized aerosol using an atomizing nozzle. Moreover, by using an x-y positioning system, which realizes a computer-controlled meandering movement of the substrate holder relative to the plasma jet nozzle, it is possible to create areal coatings. A more precise description of the used plasma system can be found elsewhere [8].

2.4 Analytical methods

An evaluation of the chlorine residues was performed with headspace GC/MS for DDT peak areas. The samples were heated in headspace vials and the evaporated phase was analysed before and after the plasma treatment.
Additionally, a documentation and characterization of the sample surface is performed before and after plasma treatment using optical microscopy.
Temperature measurements during the plasma treatment have been investigated by using a type K thermocouple. The data depends on the choosen type of gas, the number and speed of plasma treatments and the chosen power of the plasma source.
To determine antimicrobial effects of thin film modified wood, materials were tested applying for example the so-called agar-diffusion test (Fig. 4). In this test method the sample were embedded into a bacteria (*Staphylococcus aureus* DSM 799 and *Escherichia coli* DSM 1576)-containing nutrient agar. Following an incubation step (storage at 37 °C, 24h), agar plates were visual evaluated according to the areal growth around the wood material. Furthermore, modified materials were tested to evaluate the resistance towards fungi (mold) by applying an assay in accordance to the standard DIN EN 60068-2-10 (2005). Samples (50 x 50 mm²) were prepared with suspensions containing spores of *Aspergillus niger* DSM 1957, *Paecilomyces variotii* DSM 1961, *Penicillium funiculosum* DSM 1944 and *Trichoderma viride* DSM 1963. Samples were then stored at 95% rel. humidity and 26 °C over a time of four weeks. A visual evaluation according to the areal mold growth was conducted weekly.

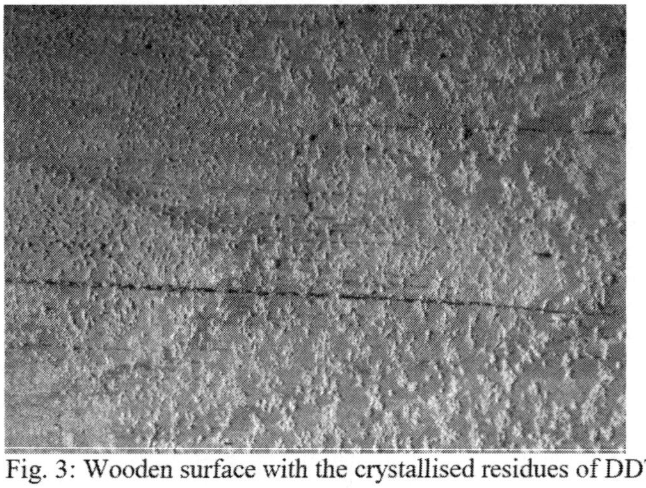

Fig. 3: Wooden surface with the crystallised residues of DDT. Photo: Wikimedia Commons.

3 Results and Discussion

3.1 Decontamination

In the first series of tests, appropriate parameters for cleaning of wood surfaces were investigated. The prerequisites were: minimal loading of the surface by the plasma treatment and a significant (damage-free) cleaning effect. In general, different DDT depletion rates could be achieved by variation plasma treatment parameters. All tested types of gas did reduce DDT peak amount. Tab. 1 shows the results for TIGRES MEF single jet:

Process gas	Pattern distance	Runs	Mean value decontamination rate / total volume DDT	Standard variation of decontamination rate / total volume DDT
N_2	3	3	50%	14%
N_2	3	5	35%	14%
N_2	3	7	37%	10%
N_2	3	10	65%	8%
N_2	1	5	60%	6%
N_2	1	7	65%	8%
N_2	1	10	70%	9%
N_2 H_2	3	1	29%	10%
N_2 H_2	3	2	52%	12%
N_2 H_2	3	3	51%	6%
Air	3	3	52%	10%
Air	1	3	56%	21%
Air	1	5	51%	14%
Air	1	7	47%	18%

Tab.1: Evaluation of the DDT peak areas of the specimens after GC / MS analysis of MEF single jet.

Using TIGRES MEF single plasma jet with nitrogen as type of gas ten times on one sample and a pattern distance of 1 mm, the highest depletion rate of DDT could be measured. With these parameters a reducing of the DDT total value about 70% is possible. Under these conditions, a maximum surface temperature of 120 °C was measured one millimeter below the substrate surface with the Type K thermocouple. These results are reproducible.
Using the gas mixture nitrogen-hydrogen, a DDT reduction of over 50% was achieved after only three runs and a pattern distance of 3 mm. So, this type of process gas is suitable for implementing a DDT reduction with a shorter treatment time. Even by using compressed air comparable DDT degradation rates can be achieved. However, the use of compressed air was not pursued because the presence of oxygen during the plasma treatment should be avoided to protect sensitive surfaces of wooden artefacts.

The second plasma source, kINPen 09, did not reached similar high depletion rates of DDT. Using a gas mixture of nitrogen / hydrogen, 1 mm distance to the samples surface and a plasma treatment of ten runs, only a depletion rate of DDT about 60% could be reached. It is positive to mention, that the maximum surface temperature 1 mm below the substrate surface did not exceeded 70 °C.

No results could be achieved about the penetration of the plasma treatment, the duration of the removal of DDT crystals from the surface and metabolites by using atmospheric pressure plasma on biocidal treated wooden surfaces. These topics have to investigate in further research projects.

3.2 Antimicrobial films

Most antibacterial modified materials (SiOx films containing silver, copper or zincoxide on wood) showed strong effects towards bacteria. Figure 4 demonstrates the antibacterial effect: the reference

IOP Conf. Series: Materials Science and Engineering **364** (2018) 012077 doi:10.1088/1757-899X/364/1/012077

material on the left side shows a bacterial growth all over the agar plate and around the material. However, by applying a thin active agent-containing SiO_x thin film (right part of this figure), a growth-free zone (inhibitory zone) around the material was clearly visible.
In contrast to the strong antibacterial effect, most of the surfaces were covered by mold even after short incubation times and only small mold resistance effects could be achieved. In general, copper containing films showed the highest impact. However, this film forming procedure also allows the embedment of different kinds of active agents in one films (combinations of active agents) These tests showed that an increased resistance towards mold growth can be achieved especially by combinations of active agents silver, copper and zinc oxide in one film [11].

A point needs to discuss is the applicability of the here presented methods. This technology allows the creation of thin films on nearly every material without the use of organic solvents. Hence, there is no need of curing processes such as tempering or UV-treatment and films can also be realised on materials which are incompatible with organic solvents. Another advantage of this technology is, besides the possibility to create areal coatings, the possibility to treat chosen parts of a material, for example areas frequently colonized by microorganism in sensitive sectors. Using the single-jet systems, the realisation of larger areal coatings can be time-consuming. But recent system development, using a linear array of several plasma-jets [15], allows the treatments of larger areas and consequently makes this process less time-consuming and more economical.

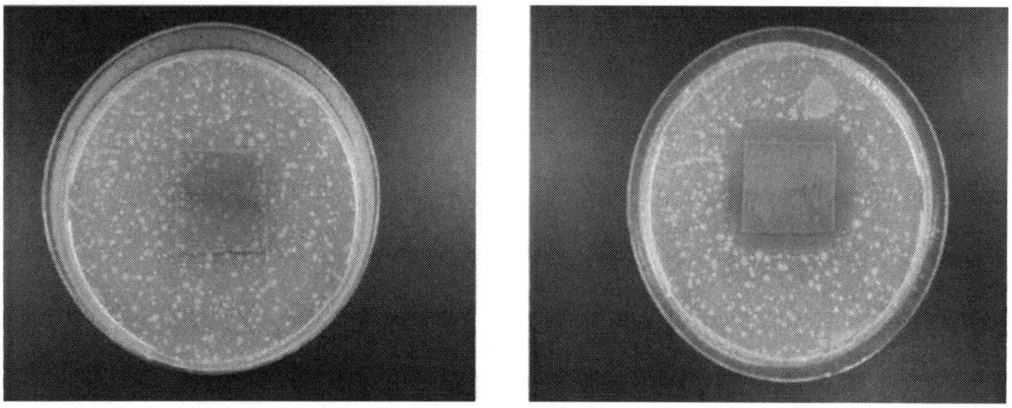

Fig. 4: Antibacterial effect on Escherichia coli using agar-diffusion test: uncoated wood slide (left), active agent-containing SiOx film on wood slide (right) [16]. Photo: INNOVENT.

4 Conclusion:

The removal of DDT crystals from wooden artworks as a result of a biocidal treatment with cold atmospheric plasma was demonstrated. Using the TIGRES plasma MEF with nitrogen as working gas on new contaminated samples depletion rates of 70% could be achieved with a relatively high number of treatment runs. Using nitrogen-hydrogen gas mix a depletion rate of 70% could be achieved with half of treatment runs. The low energy plasma source kINPen09® achieved similar results as TIGRES plasma MEF by using nitrogen as working gas (ca. 60% depletion of DDT).
Open questions about decontamination of wooden surfaces are left about the penetration of the plasma treatment, the duration of the removal of DDT crystals from the surface and metabolites by using atmospheric pressure plasma on biocidal treated wooden surfaces. Also the matter of "patina", an aged surface of a wooden artefact, has to be interdisciplinary discussed between restorers and scientists to improve the technical development of plasma sources. Suitable applications for cultural heritage should be considered in further plasma systems.

By the modification of plasma jet systems, the introduction of precursors into the active plasma zone is possible. Using this approach, the formation of thin, active agent (silver, copper, zinc oxide) containing films on wood materials was achieved. Furthermore, this technology is a suitable and economical way to create antimicrobial properties, especially in comparison to conventional methods, which incorporate much larger amounts of biologically active components in the whole bulk material. Applying the agar-diffusion test, a strong antibacterial effect was observed. More challenging to achieve was the creation of an effect towards mold fungi. Here, films containing only one active agents showed poor antimycotic properties and sample were strongly covered by mold even after short incubation times. Significant improvements were obtained by using combinations of active agents.

Acknowledgement:
The authors would like to thank restorer Birgit A. Schmidt, Dr. Jörg Krüger, Simone Pentzien and Andrea Conradi (BAM Berlin) for open minded discussions and Karolina Barnikol (INNOVENT) for GCMS measurements. Thanks to the Schlossmuseum Sondershausen for providing old contaminated samples.
This work was financed by the Deutsche Bundesstiftung Umwelt DBU (AZ 32953) and The German Federal Ministry of Economic Affairs and Energy, ZIM "PlaCoW".

References

[1] H. Tello, B. Paz, Über den Einsatz von Bioziden in naturkundlichen, botanischen und musealen Sammlungen – Rückblick und Ausblick, in: P. Zalewski (Ed.), Biozidbelastete Kulturgüter- Grundsätzliche Hinweise und Texte zur Einführung in die Problematik. Bericht über das EU-/ ESF-Projekt „Kleine und Mittlere Unternehmen und Wissenschaft im Dialog. Dekontamination von Kulturgütern“ , Frankfurt (Oder), 2014, pp. 107 – 124.

[2] Hankammer, Gunter, and Wolfgang Lorenz, "Schimmelpilze und Bakterien in Gebäuden." Erkennen und Beurteilung von Symptomen und Ursachen. Rudolf Müller, Köln (2003), Regulation (EU) No 528/2012 of the European Parliament and of the Council of 22 May 2012 concerning the making available on the market and use of biocidal products Text with EEA relevance

[3] A. Schutze, J.Y. Jeong, S.E. Babayan, J. Park, G.S. Selwyn, R.F. Hicks, The atmospheric-pressure plasma jet: a review and comparison to other plasma sources, IEEE transactions on plasma science 26 (1998): 1685-1694.

[4] Jamali, A., and P. D. Evans. "Etching of wood surfaces by glow discharge plasma." *Wood science and technology* 45.1 (2011): 169-182.

[5] Denes, Ferencz S., L. Emilio Cruz-Barba, and Sorin Manolache. "16 Plasma Treatment of Wood." *Handbook of wood chemistry and wood composites* (2005): 447.

[6] Zhou, Xiaoyan, et al. "Glass transition of oxygen plasma treated enzymatic hydrolysis lignin." *BioResources* 7.4 (2012): 4776-4785.

[7] Noeske, Michael, et al. "Plasma jet treatment of five polymers at atmospheric pressure: surface modifications and the relevance for adhesion." *International Journal of Adhesion and Adhesives* 24.2 (2004): 171-177.

[8] Cheng, Zhang Liye, and Ru-Juan Zhan. "Surface modification of polymer fibre by the new atmospheric pressure cold plasma jet." *Surface and Coatings Technology* 200.24 (2006): 6659-6665.

[9] Iwasaki, Masahiro, et al. "Nonequilibrium atmospheric pressure plasma with ultrahigh electron density and high performance for glass surface cleaning." *Applied Physics Letters* 92.8 (2008): 081503

[10] Beier O, Pfuch A, Horn K, Weisser J, Schnabelrauch M, Schimanski A (2013) Low temperature deposition of antibacterially active silicon oxide layers containing silver nanoparticles, prepared by atmospheric pressure plasma chemical vapor deposition. Plasma Processes and Polymers 10: 77-87. doi: 10.1002/ppap.201200059.

[11] Gerullis, Sven, et al. "Thin antimicrobial silver, copper or zinc containing SiOx films on wood polymer composites (WPC) applied by atmospheric pressure plasma chemical vapour deposition (APCVD) and sol–gel technology." European Journal of Wood and Wood Products (2017): 1-13.

[12] B. Bärnighausen, H. Bärnighausen, Die Goldene Kutsche von Schloß Sondershausen. Amtlicher Führer Special, Deutscher Kunstverlag GmbH München, Berlin 2001, p. 7.

B. Bärnighausen, H. Bärnighausen, Die Goldene Kutsche von Schloß Sondershausen. Amtlicher Führer Special, Deutscher Kunstverlag GmbH München, Berlin 2001, pp. 37.

[13] Alessandro Patelli, Monica Favaro, et al., PANNA Project – Plasma and Nano for New Age Soft Conservation. Development of a Full-Life Protocol for the Conservation of Cultural Heritage, in: Progress in Cultural Heritage Preservation, 4th International Conference, EuroMed 2012, Limassol, Cyprus, October 29 – November 3, 2012. Proceedings, 793-800.

[14] T. von Woedtke, K.-D. Weltmann, Grundlagen der Plasmamedizin, MKG-Chirurg 10-2016, Springer-Verlag Berlin-Heidelberg, 2016. Published online.

[15] Ramm M, Pfuch A, Beier O, Spange S, Schimanski A ,Taylor-made surfaces by means of inline plasma coating at atmospheric pressure. Melliand International 4, 2013, 232-234.

[16] Gerullis, Sven et al. "Antimikrobielle Beschichtungen auf Holz mittels Atmospheric Pressure Plasma Chemical Vapour Deposition (APCVD)" Holztechnologie 57.5 (2016): 16 – 25.

IOP Conf. Series: Materials Science and Engineering **364** (2018) 012078 doi:10.1088/1757-899X/364/1/012078

The laser as a tool for the cleaning of Cultural Heritage

A Zanini and L Bartoli

El.En. S.p.A., Via Baldanzese, 17, 50041, Calenzano (FI), Italy

l.bartoli@elen.it

Abstract. Laser ablation is one of the most important irradiation effect which can be induced on optically absorbing materials. Laser cleaning is a particular case of laser ablation where a specific substrate is uncovered through the removal of undesired layers. The idea of using laser ablation for the selective removal of encrustations from the surface of artifacts dates back to the 70s. However, only in the last years, thanks to studies on the ablative process and on the development of dedicated laser systems, laser cleaning has become a daily routine in the world of conservation. The development of brand new devices optimized for the cleaning of artworks is here reported: the innovative machines exploit the versatility provided by the presence of different pulse durations, which turned out to be a key parameter in the laser- target interaction. In addition to the well-known laser cleaning of stones, novel applications are discussed such as the cleaning of metals, wall paintings, and the novel use of the Er:YAG laser for the removal of aged varnishes from easel paintings. The growing flexibility and versatility of the new laser systems dedicated to the cleaning of artworks enlarge the panorama of successful application of this cleaning method.

1. Introduction

The properties of monochromaticity, collimation, and coherence of laser light and the associated interaction features have favoured the development of a variety of laser applications in several fields, such as for examples industrial , biomedical, and cultural heritage.

Laser cleaning is a particular case of laser ablation where a specific substrate is uncovered through the removal of undesired layers.

The application of laser cleaning in the conservation of artworks was proposed by J.F. Asmus and L. Lazzerini since the beginning of 1970s [1, 2, 3] through a set of practical tests carried out in Venice on encrusted stone artefacts: the novel approach, though, did not overcome the experimental stage for several years mainly because of the technological limits of the pulsed laser sources available at that time. During the 1990s several research centres, conservation institutions, and restoration enterprises started constructive synergies aimed at developing tailored laser systems and methodologies to face different classes of materials and deterioration problems.

The gradual acceptance within the scientific and conservation communities of laser cleaning as a restoration tool is documented by the high number of studies published in various applied physics, applied chemistry, and interdisciplinary journals and by the spread of dedicated conferences such as LACONA (Laser in the Conservation of Artworks) which has become a fundamental reference for the development and dissemination of the laser techniques in the conservation field.

IOP Conf. Series: Materials Science and Engineering **364** (2018) 012078 doi:10.1088/1757-899X/364/1/012078

In Italy, where laser cleaning is nowadays a common practice in restoration, the conference APLAR (Applicazioni Laser nel Restauro – Laser applications in conservation) is also a very interesting discussion forum and meeting point for restorers who daily use laser as a cleaning tool.

2. Laser cleaning features

Laser ablation occurs when the laser irradiation fluence (pulse energy per unit area) overcomes a critical threshold, which is an intrinsic property of the material structures under irradiation.

Laser ablation process involves optical, photothermal and photomechanical phenomena depending on the laser parameters and material properties. The main important parameters in laser cleaning are: wavelength, pulse duration and pulse energy.

The process of laser cleaning can be explained in a simplified way by evaporation (photo thermal ablation mechanism) and ablation/spallation (photo mechanical and photo chemical ablation mechanism) effects. The simplest case is a white stone covered by a black dark crust. The laser action causes the complete evaporation/ablation of the black crust; as the laser beam interacts with the white surface of the stone there is no more dark material absorbing the laser light and the laser will be reflected away without producing any damage on the white stone (see figure 1).

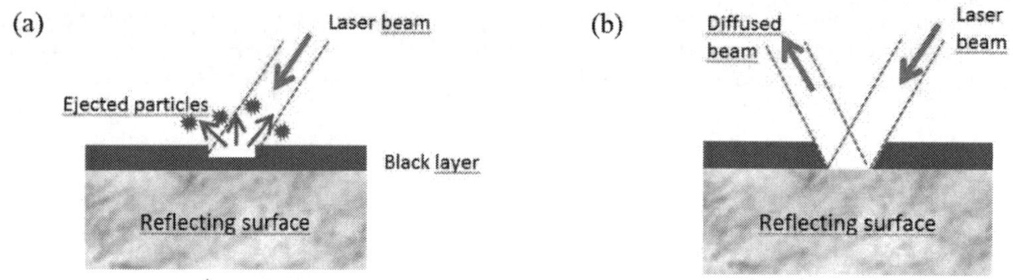

Figure 1: (a) the laser beam interacting with the black crust causes the ejection of particles, (b) as the black layer is removed the laser beam does not interact with the reflecting surface and it is diffused away.

However, in reality such clear simple cases are rare due to the extreme variety of substrates and encrustations found in Cultural Heritage conservation, therefore the process of laser cleaning is actually much more complex. A fundamental parameter of the laser cleaning process is the pulse duration. The material removal, indeed, follows different "channels of ablation" depending on the pulse duration and on the pulse energy: spallation, thermal explosion, fast or slow vaporisation are just some of this channels. Only pulsed lasers are actually used in the cleaning of art and historical objects. Up to a few years ago, the most employed systems in stone cleaning were Q-Switched (QS) Nd:YAG lasers emitting at the fundamental harmonic (1064 nm) with pulses of typical duration of 8-20 ns . Usually, also the Free Running (FR) regime is available on these commercial laser systems providing pulse durations of 200-500 μs. QS systems could cause photomechanical stress on the target, whereas FR laser can cause photothermal damages on the substrate: to avoid these problems a novel class of Intermediate Pulse Duration Nd:YAG lasers was proposed. These systems are based on Short Free Running (SFR) [4] and Long Q-Switching (LQS) [5] regimes providing pulse duration of 100ns and 30-120 μs, respectively, intermediate between the QS and the FR regimes, reducing hence both the thermal- and mechanical damages to the substrate [6, 7].

3. The evolution of laser cleaning in Cultural Heritage field

3.1. Laser cleaning of stones

Since the second half of the 1980s laser cleaning was widely applied in stone artefacts restoration in Italy, France [8, 9], England [10, 11] and other countries, mainly for the removal of black crusts produced by environmental pollution but also for the removal of intentional dark layers applied in the past. Laser cleaning was applied massively on restoration of stone reliefs, historical façades [12, 13], ancient archaeological artworks such as the West Frieze of the Parthenon [14] and famous Renaissance masterpieces such as Profeta Abacuc, San Marco and Pulpito by Donatello, the capitals of the leaning Tower in Pisa, panels of the Giotto's tower of the Florence Cathedral by Andrea Pisano [15].

In the last years the laser has been used also in the cleaning of the St. John Baptistery in Florence to remove the crust from the very fragile friezes of the windows and from some capitals where chemical and mechanical traditional methods proved to be too aggressive. Still in Florence, the laser has been lately used for the cleaning of the worldwide known statue of Neptune by Ammannati in Piazza della Signoria for the removal of the thick calcareous crust covering the artwork (see figure 2)

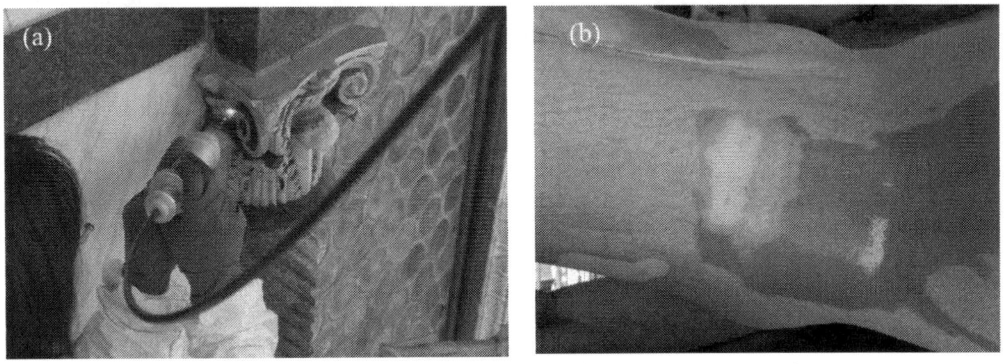

Figure 2: (a) cleaning of a capital of the St.John's Baptistery, (b) detail of some laser cleaning trials on the Neptune's statue by Ammannati (courtesy of R.A.M. restauri).

In the last years, the use of laser cleaning overcame the European borders to land in Asia and the USA where important stone conservation works has been performed by laser such as at the Agra Temples in India, the Venus Temple in Balbeek (Lebanon), the Metropolitan Museum Cloisters in New York (USA).

This extensive application of laser cleaning was accompanied by basic studies on the phenomenological characterization of irradiation effects, diagnostic of the material removal and physical modelling which allowed the definition of operative fluence ranges ensuring discrimination between encrustation to be removed and the substrate to be preserved [16, 17] and solving the well-known problem of yellowing [18, 19, 20] . The first case can be addressed by using SFR [4, 20] and double wavelength QS [21, 22] lasers which permit the control of the chromatic hue of the stone surface. Laser ablation of stone surfaces provides many advantages with respect to mechanical and chemical methods in terms of gradualness, self-termination, selectivity, environmental impact and safeguard of the so-called "age patina".

3.2. Laser cleaning of metals

The application of laser cleaning on metals concretely started with the case study of the gilded bronze panels of the Porta del Paradiso by Lorenzo Ghiberti: a careful optimization of laser parameters was performed and led to the introduction of the already mentioned Long Q-switching laser system [23]. The effectiveness and safety of the new temporal regime for the cleaning of mercury amalgam gilding, gold laminas, silver and related alloys was proved during the years. After this first successful study laser cleaning was in fact applied to other gilded artworks such as the Amore Attis and the David by

Donatello [24, 25], and the North Door of the Baptistery in Florence by Lorenzo Ghiberti. The South Door of the Baptistery, the oldest one of the three, manufactured by Andrea Pisano in 1330 is currently under restoration at the Opificio delle Pietre Dure in Florence and the laser is, once again, widely used for the cleaning of gilded parts.

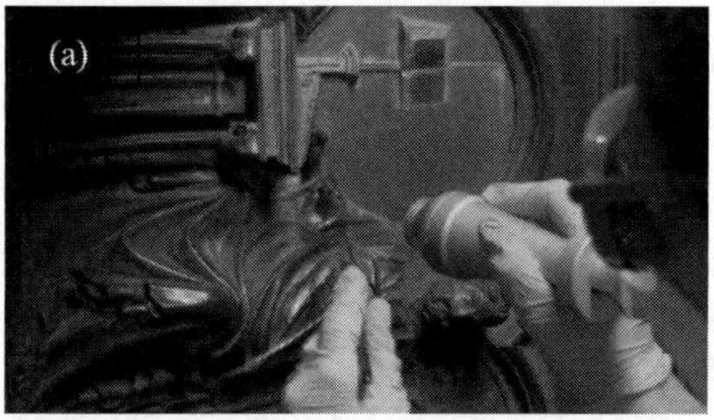

Figure 3: (a) laser cleaning of the North Door of the Baptistery of Florence, (b) cleaning of the David by Donatello. Both artworks are made of gilded bronze: the North Door is amalgam gilding and the David is gold leaf.

The restoration of silver included the underwater cleaning of a Roman Treasure (around 300 silver alloy Roman coins) [26] and the application of laser ablation to solve the problem of silver tarnishing [27] . A step forward was represented by the cleaning of oil gilding decorations by using the SFR laser: the restoration of the David by Verrocchio [28] and of the Santi Quattro Coronati by Nanni di Banco [29]demonstrated the effectiveness of the SFR laser to safely uncover gold leaf decorations.

Large bronze sculptures were also lately laser cleaned such as the Etruscan statue "Arringatore" or the sculptural group "Decollazione del Battista" by Vincenzo Danti [30] from the Baptistery of Florence. Moreover, a more recent statue as the bronze Napoleon as Mars the Peacemaker [31] by Antonia Canova (1811) in the courtyard of the Brera Palace in Milan has been cleaned thanks to the LQS laser in combination with more traditional mechanical methods.

3.3. Laser cleaning of wall paintings

The application of laser cleaning on wall paintings concretely started with the case study of the frescoed walls of the Old Sacristy and the Chapel of the Mantle in Santa Maria della Scala, Siena [32,20]. The painted walls and the vaults of the Old Sacristy and of the Chapel of Mantle were coated with layers of whitewashing and lime applied in the past.

Preliminary cleaning tests were carried out with SFR and LQS lasers. These two intermediate-pulse systems proved to be extremely effective and safe; used together or one by one, they resulted in the successful removal of the whitewashing, revealing the frescoes resting underneath (figure 4).

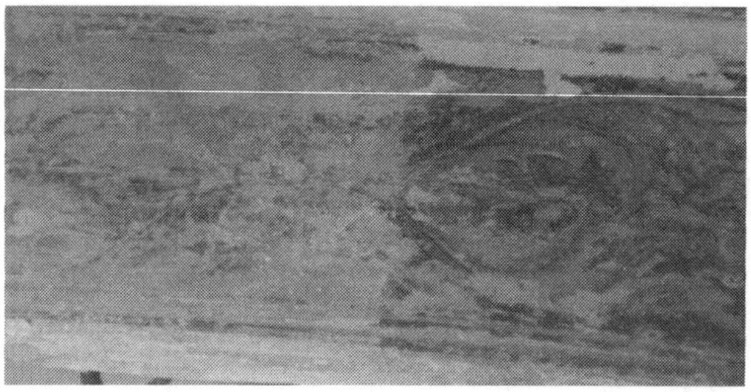

Figure 4 – Uncleaned (left) and laser-cleaned (right) fresco's detail in the Old Sacristy in Siena.

After these successful results in Siena, the laser started to be used also on extreme environments such as catacombs. One interesting example is the cleaning of the wall paintings in the "Portico dei Fornari" in the Domitilla's catacombs in Rome [33]. During the last twenty years, the removal of the dark film covering the vaults has been carried out mostly manually but the results obtained with this method were unsatisfactory as they did not ended up in the complete cleaning of the surface. Thanks to the laser the dark film was completely removed from all the shades of colours (white, green, ochre, etc) and the wall paintings were completely uncovered by the black layer that concealed them.

Another example of successful laser cleaning in hypogean environment is the cleaning of the decorated stuccoes of the vault of the Pythagorean Basilica of Porta Maggiore in Rome. The stuccoes in the vault were covered by a thick and strong layer of calcium carbonate whereas the decorations on the walls were covered by earthy residues. A powerful QS laser was used to safely remove the encrustations bringing back to light the delicate decorative motifs.

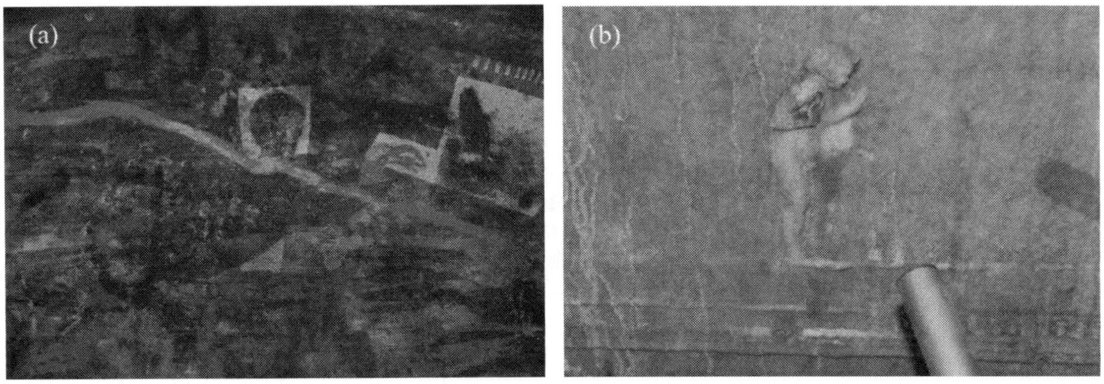

Figure 5 - (a) laser cleaning of a detail of a fresco in the Domitilla's catacomb and (b) removal of thick calcareous crust from the decorated plaster of the Basilica of Santa Maria Maggiore (courtesy of R. Mancinelli – C.S.R. Restauri).

In some particularly complicated cases the combined use of different laser systems is required. One of this cases is the restoration of the Carracci Gallery in Palazzo Farnese, Rome. The laser has allowed the recovery of inscriptions and drawings in sanguine and charcoal, made between the late '600 and 800 by artists who visited and studied the Gallery. The drawings were concealed by a thick layer of lime and, only thanks to the synergistic action of laser systems with different operating modes, it was possible to remove the lime and bring to light the beautiful sketches that were hidden under it. The restorers, depending on the thickness of the crust and on the nature of the underlying drawing, employed both the Er:YAG laser than the Nd:YAG LQS and SFR systems.

A set of different lasers is currently being employed also in the Sforzesco Castle in Milan at the Sala delle Asse on the walls painted by Leonardo Da Vinci. The decoration is based on a dense interweaving of branches and foliage, which develops in a sort of pergola and ideally breaks through the wall of the closed hall, transforming it into a piece of open-air landscape. Particularly interesting is the fragment on the east wall, brought to light in 1954, covered by a black-white monochromatic preparatory drawing, where large roots can be seen penetrating some rocky stratifications: this is the base of the tree stem, from which originally departed all the green leaves that cover the vault of the room. On the east wall the laser beam is used to remove the layers of whitewash in search of the original preparatory drawings, while on the vaults the laser helps the restorers to look for the authentic colours of Leonardo below the different repaintings. In this conservation project the restorers are using Nd:YAG systems with QS, LQS and SFR pulse durations.

IOP Conf. Series: Materials Science and Engineering **364** (2018) 012078 doi:10.1088/1757-899X/364/1/012078

Figure 6 - The laser acting on a drawing in the Carracci Gallery in Palazzo Farnese in Rome (courtesy of Paolo Pastorello-CRC Restauri).

3.4. Laser cleaning of easel paintings

The cleaning of easel paintings i.e. the removal of thin films of varnishes, resins, repaintings or deposits from the pictorial layer, represents the main challenge of laser cleaning.

Recently, the interest turned to the Er: YAG laser systems with a wavelength of 2940nm which proved to be particularly effective in the removal of repaintings and aged, yellowish varnishes from paintings on canvas and board [34, 35]. The wavelength of 2940nm is highly absorbed by OH bonds therefore, the efficiency of the laser ablation is directly proportional to the amount of OH groups present in the materials. When these OH bonds are not present on the original material, they can be obtained by the addition of hydroxylated liquids, which can also help to limit the heat penetration.

The application of Er:YAG laser cleaning to easel painting is the newest frontier of this technique. There are quite few case studies reported in literature [36] and a scientific, rigorous approach is still at its initial stage but the interest has enormously increased in the last period thanks to some successful cleaning results obtained on the removal of mastic and dammar varnishes, linseed oil/ sandarac varnishes, Paraloid B72 and animal glue.

Restorers, conservators and scientists are working together to fully exploit the huge potential of this brand new approach that will surely bring a new important tool in the restorer's toolbox: the laser.

Figure 7 - Duke University professor Adele De Cruz using the Er:YAG laser for the removal of aged varnish from a Borrassa's Panel Painting (15th Century).

References

[1] Asmus J F, Murphy C and Munk W 1973 *in: Wuerker, R.F. (Ed.) Developments in Laser Technology II, Proc. SPIE 41, Bellington WA,* Studies on the interaction of Laser Radiation with art artifacts 19-30

[2] Asmus J F, 1973 *Technol. Conserv* **3** Light Cleaning: laser technology for surface preparation in the arts 14-18

[3] Lazzarini L, Asmus J F and Marchesini M 1972 *In: Proceedings of the 1st International Symposium on the Deterioration of Building Stones, La Rochelle* Lasers for the cleaning of statuary, initial results and potentialities 89-94

[4] Margheri F, Modi S, Masotti L, Mazzinghi P, Pini R, Siano S and Salimbeni R 2000 *J. Cult. Heritage 1*, **No 1S** Smart Clean: a new laser system with improved emission characteristics and transmission through long optical fibers 119-123

[5] Salimbeni R, Pini R and Siano S 2003 *J. Cult. Heritage* **4-S1** A variable pulsewidth Nd:YAG laser system for conservation 72s-76s

[6] Bartoli L, Pouli P, Fotakis C, Siano S and Salimbeni R 2006 *Laser Chemistry*, **Article ID:81750** Characterisation of stone cleaning by Nd:YAG lasers with different pulse duration 6 pages

[7] Siano S, Giamello M, Bartoli L, Mencaglia A, Parfenov V and Salimbeni R, 2008 *Laser Physics* **18** Laser cleaning of stone by different laser pulse duration and wavelength 27-36

[8] Orial G and Gaufillet J P Nice 1989 *Technologie Industrielle Conservation Restauration di Patromoine Culturel, Colloque AFTPV/SFIIC* Nettoyage des monuments historiques par désincrustation photonique des salissures 118-25

[9] Vergès-Belmin V, Pichot C and Orial G 1993 *In: M.J. Thiel (Ed.), Conservation of stone and other materials: Proceedings of International RILEM/UNESCO Congress, Paris* Elimination de croutes noires sur marbre et craie: à quell niveau arreter le nettoyage? 534-41

[10] Cooper M I, Emmony D C and Larson J H 1992 *Proocedings of the 7th International Congress on Deterioration and Cleaning of Stone, Lisbon* A comparative study of the laser cleaning of limestone 1305-07

[11] Cooper M I, Emmony D C and Larson J H 1993 *Proceedings of conference on Conservation Science in the UK, Glasgow, UK* Laser cleaning of limestone sculpture 69-73

[12] Calcagno G, Koeller and Nimmrichter J, 1997 *Restauration Blaetter, LACONA I* Laser based cleaning of stoneworks at St. Stephen's cathedral 39-43

[13] Pini R, Siano S, Salimbeni, Piazza V, Iannucci A M, Giamello M, Sabatini G and Bevilacqua F 2000 *J. Cult. Heritage 1*, **No 1S** Application of a new laser cleaning procedure to the mausoleum of Theodoric based on a 50m long fiberoptic delivery system 93-97

[14] Pouli P, Totou G, Zafiropulos V, Fotakis C, Oujja M, Rebollar E, Castillejo M, Domingo C and Laborde A 2007 *In the Proceedings of the 6th International Conference on Lasers in the Conservation of Artworks (LACONA VI), Series Eds. J. Nimmrichter, W. Kautek and M. Schreiner (Springer Proceedings in Physics* **116** Comprehensive Study of the Coloration Effect Associated with Laser Cleaning of Pollution Encrustations from Stonework 105-14

[15] Giusti A 2000 Donatello restaurato. I marmi del pulpito di Prato, Maschietto&Musolino, Siena

[16] Siano S, Fabiani F, Pini R, Salimbeni R, Giamello M and Sabatini G 2000 *J. Cult. Heritage* **1** Determination of damage thresholds to prevent side effects in laser cleaning of pliocene sandstone of Siena 47-53

[17] Salimbeni R, Pini R and Siano S 2001 *Spectrochimica Acta Part B- Atomic Spectroscopy*, **56** Achievement of optimum laser cleaning in the restoration of artworks: expected improvements by online optical diagnostics 877-85

[18] Klein S, Fekrsanati F, Hildenhagen J, Dickmann K, Uphoff H, Marakis Y and Zafiropulos V 2001 *Appl. Surf. Sci.* **171** Discoloration of marble during laser cleaning by Nd:YAG laser wavelengths 243-51

[19] Dickmann K, Klein S and Zafiropulos V 2001 *In: Salimbeni R, editor, Laser Techniques and Systems in Art Conservation* **vol. 4402** *Washington: SPIE 4402* Laser cleaning of marble: discoloration effects using various Nd:YAG laser wavelengths (ω, 2ω, 3ω) 54–67

[20] Siano S and Salimbeni R 2010 *Acc. Chem. Res.* **43** Recent advances in laser cleaning of artworks and objects of historical interest 739-50

[21] Marakis G, Pouli P, Zafiropulos V and Maravelaki-Kalaitzaki P 2003 *J. Cult. Heritage* **4** Comparative study on the application of the 1st and the 3rd harmonic of a Q-switched Nd:YAG laser system to clean black encrustation on marble 83–91

[22] Pouli P, Fotakis C, Hermosin B, Sáiz-Jiménez C, Domingo C and Oujja M 2008 *Spectrochimica Acta Part A: Molecular and Biomolecular Spectroscopy* **71** The laser-induced discoloration of stonework; a comparative study on its origins and remedies 932-45

[23] Siano S and Salimbeni R, 2001 *Studies in Conservation*, **46**, **N.4** The Gates of Paradise: physical optimization of the laser cleaning approach 269-81

[24] Burrini F and Gennai S 2005 in Il ritorno d'amore: l'Attis di Donatello restaurato, ed. by B. Paolozzi Strozzi (Museo Nazionale del Bargello, Firenze,) 100–16

[25] Nicolai M L 2008 in Donatello, il David restaurato, ed. by B. Paolozzi Strozzi (Giunti editore, Firenze,) 160–67

[26] Siano S, Miccio M and Palleschi P 2004 Il tesoretto di Rimigliano: indagini analitiche preliminari e pulitura, in: un tesoro dal mare, Bandecchi&Vivaldi, Pontedera, 87-92

[27] Belluzzo P, Siano S, Pieri G, Lanterna G and Innocenti C 2007 *Kermes- La rivista del restauro* **67/68** Pulitura laser di manufatti in argento 59-64

[28] Siano S, Nicolai M L and Porcinai S 2003 in: B. Paolozzi Strozzi et al (Ed.), Verrocchio's David restored, Giunti, Firenze Verrocchio's David: characterization and conservation treatments, 97-108

[29] Siano S, Salimbeni R, Pini R, Matteini M, Porcinai S and Giusti A M 2003 *J. Cult. Heritage* **4** The Santi Quattro Coronati by Nanni di Banco: cleaning of the gilded decorations 123-28

[30] Agnoletti S, Brini A et al 2008 I grandi bronzi del Battistero: l'arte di Vincenzo Danti discepolo di Michelangelo, Giunti editore 307

[31] Sansonetti A, Colella M, Letardi P, Salvadori B and Striova J 2015, *Studies in Conservation*, **Volume 60**, **Number S1** Laser cleaning of a nineteenth-century bronze sculpture In situ multi-analytical evaluation 28-33

[32] Siano S, Brunetto A, Mencaglia A, Guasparri G, Scala A, Droghini F and Bagnoli A 2007 in: Nimmrichter J., Kautek W., Schreiner M. (Eds.), LACONA VI, Proceedings Springer, Heidelberg Integration of laser ablation technique for cleaning the wall painting of the Sagrestia vecchia and Cappella del Manto in Santa Maria della Scala, Siena 191-201

[33] Patrizi M G, Espunes R S, Mazzei B, Mascalchi M, Agresti J, Osticioli I and Siano S 2013 in A. Brunetto (ed.), APLAR 4. Applicazioni laser nel restauro, Saonara (PD) Il cubicolo "dei fornai" nelle catacombe di Domitilla a Roma. Considerazioni a conclusione della pulitura laser 143-54

[34] de Cruz A, Wolbarsht M L and Hauger S A 2000 *J. Cult. Heritage* Laser removal of contaminants from painted surfaces 173-80

[35] Bracco P, Lanterna G, Matteini M, Nakahara K, Sartiani O, de Cruz A, Wolbarsht M L, Adamkiewicz E and Colombini M P 2003 *J. of Cult. Heritage* **4** Er:YAG laser: an innovative tool for controlled cleaning of old paintings: testing and evaluation 202–08

[36] Bordalo R, Morais P J, Gouveia H and Young C 2006 *Laser Chemistry*, **Article ID 90279** Laser Cleaning of Easel Paintings: An Overview 9 pages

IOP Conf. Series: Materials Science and Engineering **364** (2018) 012079 doi:10.1088/1757-899X/364/1/012079

A customised atmospheric pressure plasma jet for conservation requirements

A. Patelli[1], E. Verga[2], L. Nodari[3], S. M. Petrillo[4], A. Delva[2], P. Ugo[2] and P. Scopece[5]

[1] Dipartimento di Fisica e Astronomia, Università di Padova, via Marzolo 8 35131 Padova, IT
[2] Università di Venezia, via Torino 155 30172 Venezia, IT
[3] ICMATE–CNR, Corso Stati Uniti 4 35127 Padova, IT
[4] SMP International Photo Conservation Studio, Roma, IT
[5] Nadir srl, via Torino 155 30172 Venezia, IT

E-mail: alessandro.patelli@unipd.it

Abstract. In the last years atmospheric plasma has gained an increasing interest in the field of Cultural Heritage. Isolated experimentations that yet generated interesting results have drawn the attention of a wider community working on coating removal, oxide reduction and sterilization applications. Thanks to this growing multidisciplinary interest, the EU funded in 2011 the PANNA project, which includes the atmospheric plasma as a novel cleaning tool for cultural heritage. Plasma cleaning has some advantages compared to standard techniques because it avoids the use of solvents, it is contactless and differently from laser it works through a surface chemical action and not with mechanical shock or local heating. Moreover, atmospheric plasma has already shown good results in reducing mode for the conversion of corroded metal surface layers and in oxidation mode for biological cleaning and organic layer removal.

In order to fulfil the Cultural Heritage requirements a novel plasma torch has been developed with a new design which allows to overcome the actual limits of commercial atmospheric plasma devices. The new device permits fast and effective treatments at temperatures lower than 50 °C, even at room temperature, without any electrode material deposition. These performances will be presented in relation to the removal of epoxy coatings and daguerreotype cleaning as examples.

1. Introduction

1.1. The novel atmospheric plasma jet

Plasma processes have drawn the attention of cultural heritage restoration since the 80s, mainly because of their low temperature. Since his first article on the topic Daniels [1] identified the potential applications for plasmas: when the gases used are argon and oxygen, organic compounds can be removed from many substrates, while argon-hydrogen mixtures can be used in our apparatus to reduce several corrosion products back to their respective metals. Starting from this intuition, different experiments have been carried out by means of vacuum plasmas during the 80s and 90s. Daniels' preliminary work inspired other scientists to investigate this topic. Daniels tested the removal of some organic coatings in argon-oxygen plasma at a rate in the range of few micrometers per hour, while on metals only lead, copper and silver showed surface oxidation. On the other side argon-hydrogen plasma led to lead oxides and carbonates conversion to sub-oxides and to silver reduction. Silver reduction was identified as the more

IOP Conf. Series: Materials Science and Engineering **364** (2018) 012079 doi:10.1088/1757-899X/364/1/012079

efficient process, in particular when corrosion is confined in a thin surface layer. For this reason the more striking results can be observed on daguerreotypes [2]. The use of reducing plasmas has been tested also on archaeological artefacts in order to remove the corrosion layer [3,4]. For iron artefacts reduction plasma treatment at temperatures higher than 300 °C are needed for chlorides and oxides conversion after a preliminary mechanical cleaning [5]. These protocols are now in use for example at the Swiss National Museum. This protocol however shows the drawback of a potential surface cracking due to volume reduction from oxide to metal and potential phase transformations. Papadopoulou [6] tested the reduction process at lower temperature on bronze archaeological artefacts verifying that chlorides still remain on the surface. Therefore, the stabilization of objects with extensive Cl attack is not feasible without preliminary chemical treatment. The plasma vacuum processes offer also the possibility to deposit at room temperature protective layers such as silica and to improve their resistance to oxidation [7].

Vacuum plasma has been tested also on organic media such as photographs for oxides or sulfides removal or the etching of dust and greasy substances [8]. Soot has been also successfully removed on gesso, paint and paper [9]. On paper substrates these treatments can affect cellulose, therefore oxygen and hydrogen plasma effects have been widely tested for decontamination purposes [10,11]. Since plasma treatments seem not to de-polymerise cellulose, consolidation processes have also been evaluated through the deposition of a thin super-hydrophobic siloxane coating [12] or a polymethylmethacrylate plasma polymer film [13]. Despite the interesting results obtained by vacuum plasmas, their application is still limited because of some intrinsic technological features: the size of the treated artefacts is limited to the vacuum chamber, the whole surface must be treated and it is expensive, but the most important limit is that it cannot be continuously controlled by the restorers in-situ.

The atmospheric plasma technology allows to remove all these barriers to plasma processes. Its first application has been the improvement of surface wettability. Documented successful case studies are the activation of polypropylene wall tapestries for a following coating [14] or of paper surfaces for a following chitosan [15] or $Ca(OH)_2$ nanoparticles [16] treatment. The use of atmospheric plasma treatments has been suggested for fungal decontamination on wood [17] or for varnish removal [18,19]. The efficiency of these processes is however still challenging because it leads to low organic coatings removal rate. At the same time, the use of more efficient plasma such as the arc torch devices could lead to high temperatures on the substrates and to the deposition of particles eroded by the electrodes [20].

An easier and faster application is represented by the silver surface cleaning by the atmospheric plasma in reduction mode, therefore using noble gas and hydrogen mixtures, which shows the most impressive results on daguerreotypes. The daguerreotype is the first example of photographic product that, around year 1850, reached a worldwide circulation. Each daguerreotype constitute an unique image which cannot be reproduced. The image is formed by silver-mercury amalgam particles produced by photosensitization and following development of a plate made of copper and silver, sensitized with silver iodide and developed with mercury vapours and eventually stabilized by gilding. Resulting artefact is particularly fragile and, if in contact with the environmental atmosphere, can be easily oxidized leading to the formation of silver oxides and silver sulphides. Typical cleaning techniques include the use of chemical reagents, of electrochemical processes or laser equipments [1,21,22]. However, such techniques present several drawbacks: they can leave residues which disfigure the image (chemical methods), require the use of electrolyte solutions (electrochemical methods) or may cause the partial removal of the outermost layer of the surface of the daguerreotype (laser methods). The atmospheric plasma jets can offer an interesting alternative as can be seen by results are already available in literature [23,24]. On silver alloys the effects are more complicated since they involve also the corrosion processes but some results in remote conditions have already

IOP Conf. Series: Materials Science and Engineering **364** (2018) 012079 doi:10.1088/1757-899X/364/1/012079

highlighted the potentialities of atmospheric plasma [25,26].

The use of atmospheric plasma is still a promising topic for Cultural Heritage applications but it needs a deeper experimentation through case studies in order to fully validate the technique, understand its effects and define protocols. However, in order to achieve this investigation stage the right tool must be available for the CH community. The plasma device needed for CH has specific requirements that standard commercial torches available on the market do not meet. In general, high efficiency torches like arc devices show the drawback of high temperatures and deposition of particles derived by electrodes erosion. On the other hand, dielectric barrier discharge (DBD) devices are developed for use on humans and are not efficient. Here we present an atmospheric plasma jet that has been developed following all the requirements of the CH field: it is applicable to any surface, including those temperature sensitive ($T < 50$ °C); it does not generate electrodes deposition on the substrate; removal rate and cleaning costs are compatible with restoration works; it is suitable for localized treatment, it can be stopped at any time by the restorer and it is suitable for in-situ operation (portability).

The performances of this novel device are then shown in the two main applications: the organic layers removal in oxidizing atmosphere and the daguerreotype cleaning in the reducing atmosphere.

2. Experimental

2.1. The novel atmospheric plasma jet

The patented atmospheric pressure device [27] is a DBD plasma jet where no electrodes are in direct contact with plasma (Fig. 1). This configuration with electrodes hidden from plasma ensures clean processes with no undesired deposition since they can not be eroded. The electrodes are positioned externally and coaxially to an alumina tube where argon is fluxed and plasma is ignited. Two other ducts are present: an inner duct for precursors inlet in the vapour or aerosol phase and an outer duct for atmosphere control at the exit of the torch, where air or nitrogen are generally used. The main feature that characterizes the device is the use of a double couple of electrodes: a first upstream couple powered with an high voltage (HV) supply in the kilohertz regime ($\sim$ 17kHz) and a downstream couple in radio frequency (RF) at about 27MHz.

The choice to work with argon noble gas, which avoids energy thermalization by molecular roto-vibrational motion and by dissociation/recombination reactions, allows to keep the temperatures of plasma as low as possible. The use of helium is not considered due to its cost. Furthermore, the RF regime with its fast oscillations avoids the formation of high current carrying streamers and enhances the generation of plasma in the volume and not on the surfaces leading to a more glow regime. The RF choice therefore goes further in the direction of reducing the thermal load on the treated surface without losing in plasma density. This is also the main design of different plasma jets used for human body applications [28]. The general drawback of this design is its ignition and stability in atmospheres when reactive species are added to the plasma. The upstream couple of electrodes solves this problem assuring ignition and stability also when high voltages are needed, without the need to increase the RF power. Basically the configuration allows to decouple plasma needs: the HV controls ignition and stability with low power, while the RF allows the control of plasma density. The coupling in fact is not just the overlapping of two different plasmas but they also interact with each other. The major effect is that the streamers created by the HV electrodes are blown up by the RF field that rapidly changes the polarity. As a consequence, if hundreds of mA streamers can be measured on a conductive substrate if HV is switched on when both power supplies are active, only a small current bumps of few tens of mA can be measured (Fig. 2b). Streamers that are current carrying channels are warmer because of their high charged species densities and fields gradients. The new design allows to avoid their propagation on treated surfaces reducing local heating. The

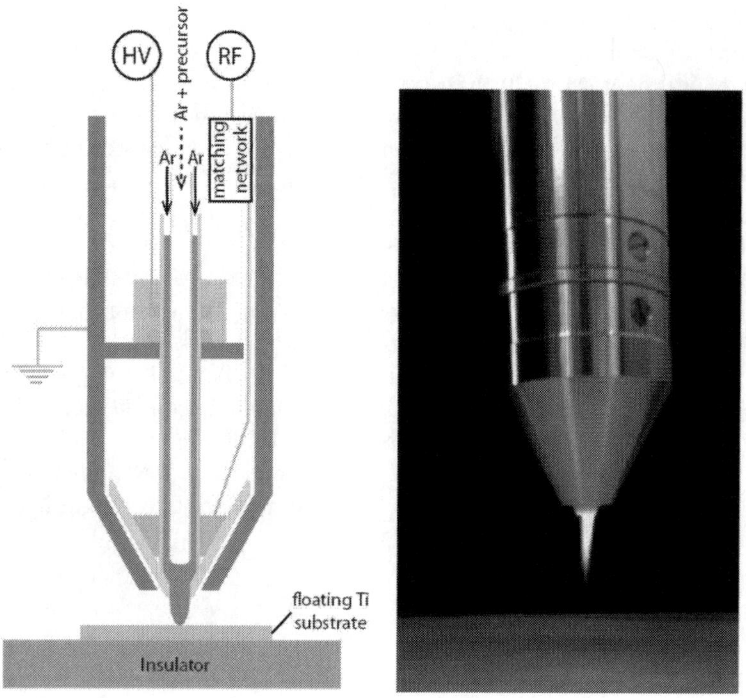

Figure 1. Scheme of the novel atmospheric plasma jet characterized by its three coaxial gas ducts and the double couple of electrodes RF an HV (on the left). On the right a photograph (time averaged) of the device.

low temperature of the novel jet design can be observed in Fig. 2a presenting a thermographic camera image after 5 minutes of treatment on a plastic substrate at 15W of RF power with argon as process gas. The image clearly shows that plasma has no emissions in the infra–red spectral region since it is not thermal. On the substrate the mirrored image of the jet can be observed.

2.2. Experimental conditions for the organic removal

An Istria limestone substrate sample was prepared by polishing and an epoxy coating (Araldite®) was applied on the surface [20]. The coating underwent photo-oxidative artificial aging according to the Italian standard UNI 10925. Photo-oxidation was carried out with a SolarBox 1500e apparatus (COFOMEGRA,Milano I), equipped with a xenon light source filtered for λ=295 nm. Irradiation was kept at a constant power of 550 W/m^2 during the lamp operating life. The maximum temperature of the samples, during irradiation, was 45 °C. Total exposure was fixed at 2000 MJ/m^2 (corresponding to 1000 h of exposure).
For the epoxy removal the plasma device has been used in oxidizing atmosphere with a mixture of Ar and O_2 at the 0.008% at a flux of 10 l/min. RF power was kept at 20 W and the HV at 8 W. Working distance was 3 mm, while in the outer duct there is fluxed compressed air. The treatment was performed in punctual mode without moving the jet on the surface as a function of the time. The results are characterized by optical imaging, FT–ATR spectroscopy (Perkin Elmer Spectrum One) and profilometry.

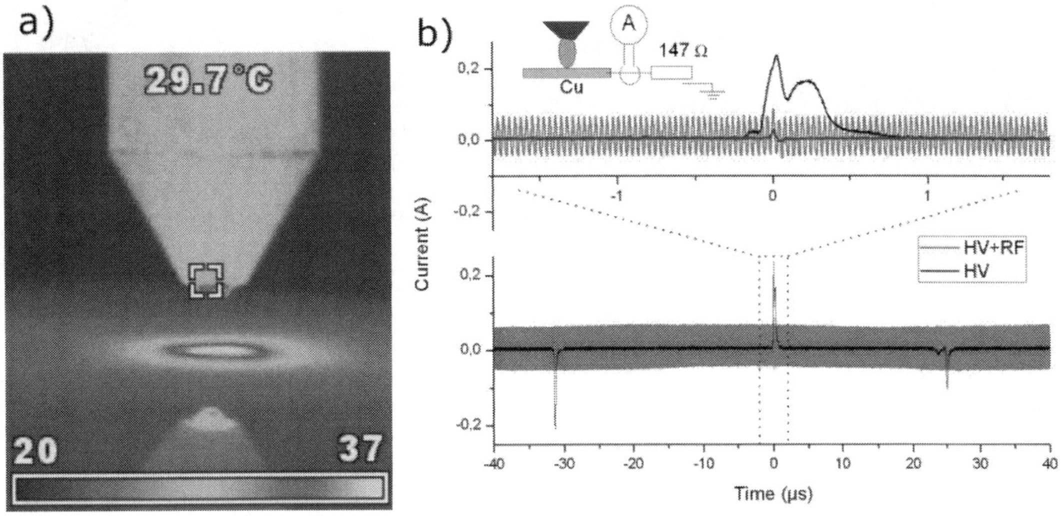

Figure 2. a) Image obtained with a thermographic camera of the plasma device on a plastic substrate in stationary conditions (after 5 minutes). b) Electrical characterization on a substrate of the coupling of the HV and RF plasmas.

2.3. Experimental conditions for the daguerreotype cleaning

The studied daguerreotype (6.7×4.8 mm) can be presumably dated around 1850–1860 and was provided by an American collector. According to the American style for daguerreotypes production, it was kept in a leather case, with a brass preserver and brass mat with an oval window, everything covered by a glass plate. Before treatment, the case was carefully dismounted, according to good restoration practice, in order to study and operate on the daguerreotype plate only, and remounted after the treatment.

The daguerreotype was in bad conservation state with scratches, deformations, imprints and surface chemical oxidation and sulfuration. The chemically damaged area appeared in different colours from blue to green, yellow or brown darkening as the thickness of the corroded layer increased. The cleaning process was performed gradually dividing the area in quarters. For each quarter the treatment time was set to 60 seconds and it was operated by hand. The distance of the jet from the surface was set to 5 mm, RF power was kept at 15 W and the HV at about 8 W. A mixture of argon and hydrogen at the 0.008% was used as process gas . The surface of the daguerreotype had been visually inspected before and after the treatment.

3. Results and Discussion

3.1. Aged epoxy layer removal

The coating applied on the sample surface has a thickness of about 50 μm. After the aging process the surface appears yellowish because of the reticulation process and C=C bonds formation. This aged layer is in fact only few microns thick and it is related to the penetration depth of the UV radiation. The surface does not appear smooth as the substrate but the coating deposited by aerosol presents a drop–like structure. After 25 minutes of treatment, as it can be observed in Fig. 3 it is fully removed under the plasma jet spot (⊘ 8 mm) and the shiny stone surface is revealed without damages. The thickness profile of the eroded spot is also highlighted in Fig. 3. As it can be observed the treatment consists in a low temperature process where the epoxy is not carbonised or melted and no liquid epoxy displacement is visible on the border of the crater. The border is instead characterized by a gradual erosion by the creation of volatiles

IOP Conf. Series: Materials Science and Engineering **364** (2018) 012079 doi:10.1088/1757-899X/364/1/012079

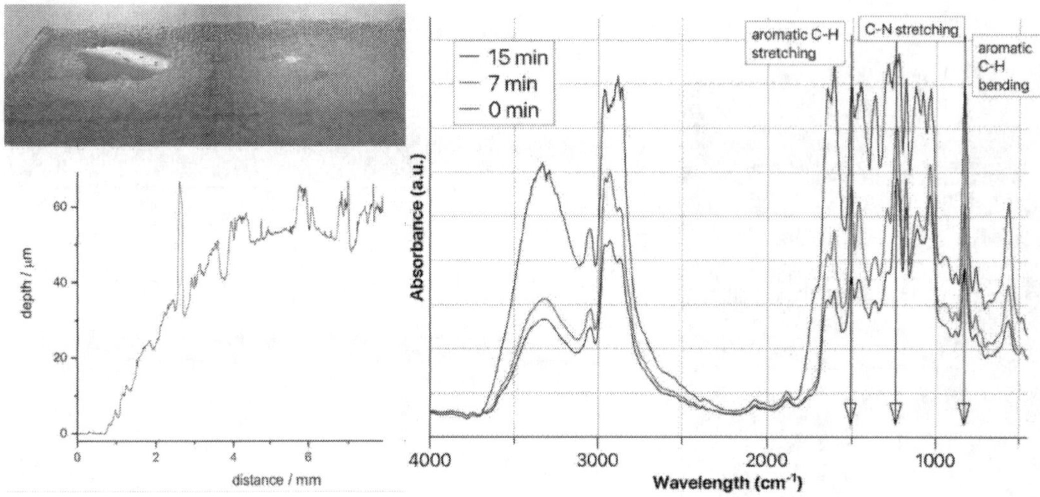

Figure 3. (top left) Photograph of the removed coating spot on the polished Istria limestone (left spot 25 min, right spot 10 min). (bottom left) Thickness profile of the removed spot. (right) FT–ATR spectra before and after 7 and 15 minutes of treatment.

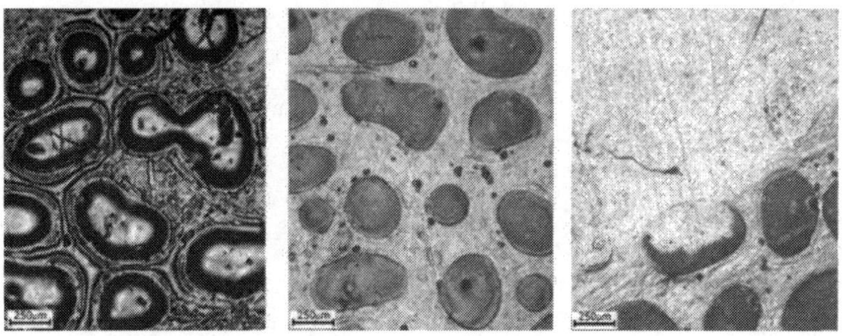

Figure 4. Optical microscope images of the epoxy coating at different stage of removal: before the treatment, after 10 min and after 25 min.

products. As the few microns thick aged layer is removed the not yellowed original epoxy is revealed. Plasma can often be used not to remove the whole thickness of the aged polymers but only to remove the first layer, allowing a subsequent easier action of the solvents. The removal of the coating is mainly a chemical oxidation. From FT–ATR spectra it can be observed that at first C–H bending out of plane of the aromatic ring are removed; then the bridge among the epoxy chains, made by the harder molecules, is broken; at the end, the fragments are oxidized, leading to H_2O, CO_2 and mono-nitrogen oxides (NO_x) (Fig. 3).

Optical microscopy clearly shows the gradual chemical etching of the surface that does not alter the morphology in a micrometer scale (Fig. 4). The droplets of the aerosol deposition are uniformly and gradually thinned. At a middle stage of the erosion process only the thicker parts of the droplets remain on the surface and after 25 minutes the whole coating is removed.

3.2. Daguerreotype cleaning

As shown in Fig. 5A, the studied daguerreotype presents some degraded areas. In particular, two types of oxidation can be noted, the first one is localized along the edges of the slab, characterized by an iridescent colouring, the second surrounds the oval window of contact with

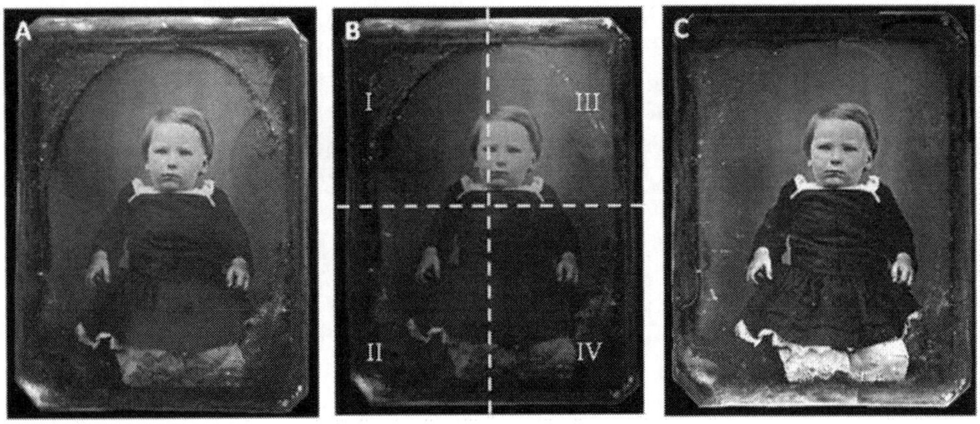

Figure 5. Photographs of the studies daguerreotypes, removed from the original case: (A) before, (B) during treatment optimization and (C) after final cleaning with plasma.

the brass mat. There are also alterations of a physical nature: the two corners on the right side show a detachment of the superficial layer, where the underlying silvery layer can be watched, while the lower left corner shows an obvious deformation. In the central area there are stains with jagged edges tinged to yellow, probably caused by previous cleaning operations, as supported by the presence of a pencil writing on the cardboard on the back of the plate with the words "Re sealed 17 March 99 L.M. Hodde ". In order to optimize the cleaning, different plasma conditions were tested in four different zones, indicated as I-IV in Fig. 5B. Between the different sectors, the high voltage percentage was changed between 60% and 70%, with treatment time ranging from 60 to 120 seconds. On the basis of these preliminary tests, it was finally chosen to clean all the plate by setting the high voltage percent to 70%, using a treatment time spanning from 40 seconds, for less degraded areas, to 80 seconds for the highly oxidized zones. The final result is shown in Fig. 5C.

An evident attenuation of the degradation can be observed, particularly at the edges of the image, where rather thick and iridescent stains of oxidation products were initially present. The central spots, probably due to sulphurization phenomena, are also significantly attenuated after the treatment. Many details, particularly in the highlight zones, that previously were barely readable, are now very well defined and clear. In particular, the lower part of the picture, corresponding to the child's outfit, is finally well defined, the iridescent oxidation spots that previously overlapped the dress being no longer visible.

Preliminary surface characterization analyses performed by scanning electron microscoy, optical profilometry, macrophotography in grazing light, optical and metallographic microscopy indicated that the treatment didn't cause any major change in the morphology of the amalgam particles which constitute the daguerreotype image. Further detailed analyses are in progress.

The obtained results confirm that the novel proposed APPJ device can be used as a promising cleaning tool, particularly respectful of the delicate daguerreotype surface and of the silver and amalgam micro- and nanoparticles that compose the image, allowing the localized tuning of the cleaning procedure, depending on the degree of degradation of different objects or of different areas of the same object.

4. Conclusions

The newly developed APPJ offers safe tool for CH applications and with its patented design fully suits the CH requirements. It has a demonstrated room temperature operation mode, does not deposit undesired material on the surfaces, it suits efficiency rates ad costs of the restoration works and it is portable and can be used also in-situ. Its action is gradual, acts only on the topmost surface by chemical oxidation or reduction locally, therefore offers to restorers the possibility to control its action and to stop the process as desired. It can be used on organic and inorganic, insulating or conductive materials independently from their colour. Due to these features it is mainly focused on small areas or artefacts treatments where a soft and respectful treatment is needed. It can find interesting applications also in combination with other technologies, as for example in the presented case study on the removal of an organic coating: it is unbeatable for the removal of aged layers and after this stage solvents can act faster and are cheaper on the not aged polymer below the surface.

This atmospheric plasma technology is therefore ready to the CH community a new tool for deeper experimentation in case studies to fully validate the technique, understand its effects and define protocols. As potential applications, for the selection of case studies, the APPJ would fit better on sensitive materials as plastics as in modern art or as natural organics like for example silk like in textiles in particular where different materials are present or as support technique to clean metallic artefacts.

Acknowledgements
The research leading to part of these results has received funding from the European Community's Seventh Framework Programme (FP7/2007-2013) under grant agreement n. 282998, project acronym PANNA.

References
[1] Daniels V D, Holland L and Pascoe M W 1979 *Studies in Conservation* **24** 85
[2] Daniels V 1981 *Studies in Conservation* **26** 45–49
[3] Patscheider J and Veprek S 1986 *Studies in conservation* **31** 29–37
[4] Sjogren A, Mathiesen T, Van Janschott J, Turgoose S and Hawkins C 1997 *Zeitschrift für schweizerische Archäologie und Kunstgeschichte = Revue suisse d'art et d'archéologie* **54** 34–40
[5] Schmidt-Ott K and Boissonnas V 2002 *Studies in conservation* **47** 81–87
[6] Papadopoulou O, Novakovic J, Vassiliou P, Filippaki E and Bassiakos Y 2013 *Applied Physics A: Materials Science and Processing* **113** 981–988
[7] Grassini S, Angelini E, Mao Y, Novakovic J and Vassiliou P 2011 *Progress in Organic Coatings* **72** 131–137
[8] Ioanid E G, Ioanid A, Rusu D E, Popescu C M and Stoica I 2011 *Journal of Cultural Heritage* **12** 399–407
[9] Rutledge S K, Banks B A, Forkapa M, Stueber T, Sechkar E and Malinowski K 2000 *Journal of the American Institute for Conservation* **39** 65–74
[10] Vohrer U, Trick I, Bernhardt J, Oehr C and Brunner H 2001 *Surface and Coatings Technology* **142-144** 1069–1073
[11] Laguardia L, Vassallo E, Cappitelli F, Mesto E, Cremona A, Sorlini C and Bonizzoni G 2005 *Applied Surface Science* **252** 1159–1166
[12] Laguardia L, Ricci D, Vassallo E, Cremona A, Mesto E, Grezzi F and Dellera F 2007 *Macromolecular Symposia* **247** 295–302
[13] Totolin M I and Neamu I 2011 *Journal of Cultural Heritage* **12** 392–398
[14] van Oosten T B, Fundeanu I, Bollard C, de Castro C and Laganà A 2008 *Plastics: looking at the future and learning from the past: papers from the conference held at the Victoria and Albert Museum, London: 23-25 May 2007* pp 97–105 ISBN 978-1-904982-43-2 (pbk.)
[15] Mikula M, Trnovec B, Černáková , Šutý Š and Jančovičová V 2009 *Acta Chimica Slovaca* **2** 62–69
[16] Li Q, Xi S and Zhang X 2014 *Journal of Cultural Heritage* **15** 159–164
[17] Leclaire C, Lecoq E, Orial G, Clement F and Bousta F 2010 *Wood science for conservation of cultural heritage: Braga 2008: proceedings of the international conference held by COST Action IE0601 in Braga (Portugal), 5-7 November 2008* 101–105
[18] Fricke K, Steffen H, Von Woedtke T, Schröder K and Weltmann K D 2011 *Plasma Processes and Polymers* **8** 51–58
[19] Pflugfelder C, Mainusch N, Hammer I and Viöl W 2007 *Plasma Processes and Polymers* **4** 516–521
[20] Voltolina S, Nodari L, Aibéo C, Egel E, Pamplona M, Simon S, Falzacappa E V, Scopece P, Gambirasi A, Favaro M and Patelli A 2016 *Journal of Cultural Heritage* **22** 940–950

IOP Conf. Series: Materials Science and Engineering **364** (2018) 012079 doi:10.1088/1757-899X/364/1/012079

[21] Golovlev V V, Gresalfi M J, Miller J C, Romer G and Messier P 2003 *Journal of Cultural Heritage* **4** S139–S144
[22] Barger M S, Krishnaswamy S and Messier R 1982 *Journal of the American Institute for Conservation* **22** 13–24
[23] Boselli M, Chiavari C, Colombo V, Ghedini E, Gherardi M, Martini C, Rotundo F, Mater A and Bologna S u 2013 *Plasma Science (ICOPS) 2013 IEEE International Conference* 1–8
[24] Grieten E, Schalm O, Tack P, Bauters S, Storme P, Gauquelin N, Caen J, Patelli A, Vincze L and Schryvers D 2017 *Journal of Cultural Heritage* **28** 56–64
[25] Schalm O, Crabbé A, Storme P, Wiesinger R, Gambirasi A, Grieten E, Tack P, Bauters S, Kleber C, Favaro M, Schryvers D, Vincze L, Terryn H and Patelli A 2016 *Applied Physics A: Materials Science and Processing* **122**
[26] Schalm O, Storme P, Gambirasi A, Favaro M and Patelli A 2018 *Surface and Interface Analysis* **50** 32–42
[27] Patelli A, Verga Falzacappa E, Scopece P, Pierobon R and Vezzu S 2014 *WO2015071746*
[28] Tiede R, Hirschberg J, Daeschlein G, von Woedtke T, Vioel W and Emmert S 2014 *Contributions to Plasma Physics* **54** 118–130

Art restoration, protection and education in China -Murals, materials study in humid areas in South China

Shan Zeng

School of Art Management and Education, China Academy of Art, Nanshan Road 218. Hangzhou 310002,China

art65@126.com

Abstract. This paper introduces the museum development history and cultural heritage education in China. The number of museums and precious collections increases tremendously, the professionals specially trained to perform Art administration, repair and restoration are next to non-existence. Facing the challenge of well-qualified teaching and researching staff, China Academy of Art (CAA) in Hangzhou China strived to offer graduate and undergraduate programs in Art Administration, Art Education and Art Appreciation in 2015. The efforts are expected to train students in an interdisciplinary approach so that they will be equipped with art knowledge, scientific techniques and practical skills. It also discusses an interesting recent studies and findings performed by the professor and his students. This paper describes special historical materials of southern Murals in China, and analyzes the traditional materials, production techniques and artistic characteristic of these Murals. The plan is to rebuild an old wall with murals as a laboratory for future studies in CAA.

1. Museums and Cultural Heritage Education

1.1. Museum and Collection

China is an ancient cultural country with 5,000 years of civilization. In the treasure house of Chinese heritage, a rich and colorful cultural heritage has been left behind, including a large number of tangible cultural heritage such as ancient ruins, tombs, architecture, books, artifacts, gardens, literature, paintings, sculptures and arts and crafts. They are not only China's treasures, but also the most precious legacy of all humanity. In recent years, more and more museums, art repair and protection education has also received more and more attention.

Chinese Modern Museum was founded in the second half of the 19th century. In 1868, French priest Pierre Heude first established the Xujiahui Museum in Shanghai, mainly collecting specimens of plants. Subsequently, the British, Japanese, Americans, Russians and other successively in Shanghai, Jinan, Chengdu, Lushun, Harbin and other places to build museums. The first public museum founded by the Chinese themselves was Nantong Museum built by Zhang Jian in 1905.

Social changes in 1911 promoted the activity of science, culture and education in China and started a number of museums in Beijing, Tianjin, Taiyuan, Nanchang, Shenyang, Kaifeng, Nanjing, Lanzhou, Hangzhou and Guangzhou. In 1925, the Forbidden City of China changed to the National Palace Museum with nearly 1 million items. At present, there are 15 museums with over 100,000 pieces, such as Tianjin Natural History Museum and Nanjing Museum. 20 to 30 years of the 20th century, the Chinese Museum had had a more significant period of development. On the eve of the Anti-Japanese War in 1936, there were 77 museums, 56 art galleries and 98 antiquities stores. The Japanese invasion of China in 1937 severely damaged the Chinese museum. From the early 1950s to the late 1950s, the number of Chinese museums soared from 24 to 480, forming another significant period of development in the history of Chinese museums. The "Cultural Revolution" started in the mid-1960s caused a large number of Chinese museums to be closed, the collections of cultural relics and specimens to be seriously damaged, and the museum project suffered huge losses. Since the 1980s, the development of Chinese museums has accelerated significantly. By the end of 2008, there were 2,970 museums in the country, including 2,161 state-owned museums, 490 state-owned museums and 386 private museums, 8 times those in 1978 and 138 times those in 1949. Chinese Museum collection of more than 2000 million pieces.

Forbidden City repair department also admitted that with the current restoration staff to repair the existing artifacts at the Forbidden City at least a hundred years to be more. According to the scale and other conditions, Chinese museums are divided into four levels, namely, state pavilion, provincial (autonomous region and municipality) level pavilion, prefectural (city) level pavilion and county (district) level pavilion, and are managed by the relevant departments at different levels. March 23, 1982 Chinese Museums Association Chinese Museums Association (CMA) to create. From January 2008 onwards, 1,447 museums across the country have been gradually liberalized. As of the end of 2009, the museum received 820 million visitors in two years.

The national museum not only develops rapidly in quantity but also continuously improves the quality and standard of museums. However, the overall professionalism of museums is relatively limited, which is far behind that of developed countries. Many large and medium-sized museums are mostly under-researched, and their ability to interpret and interpret collections needs to be improved urgently. Some small-scale museums lack hardware and the environment for preserving and displaying collections is grim. Many private museums have weak public awareness, lack of professional awareness in museums and relative social functions limited. There are many problems related to the training of professionals.

1.2. Program and academic education

Cultural relics protection technology is a comprehensive professional knowledge, including the production of cultural relics, protection and prevention and control related to science and technology, material properties, operating techniques and various survey and testing. The professional is a combination of science and technology, science and technology infiltration, modern science and technology and humanities science combination of emerging edge discipline. In 1989 in colleges and universities began to establish cultural heritage conservation technology Preservation Techniques of Cultural Relics undergraduate. At present, there are 23 institutions in the country that set specialties related to the protection of cultural relics. The main course for the major of cultural relic protection is chemistry. Therefore, major science students are enrolled. Such as Peking University, Jilin University, Northwestern University, and other colleges and universities in recent years, there are several professional courses opened in this area, but because of China's institutional reform with the recent shortage of less than 2% of graduates accepted and digestion. After 2007, the appraisal and restoration of cultural relics (code: 660113) was canceled. The professional training of arts colleges is different from the professional goals of University. Its goal is to cultivate high-level technical and practical talents with a certain degree of cultural and artistic accomplishment, mastery of cultural relics identification and restoration of knowledge and skills. Professional core competence is to cultivate the skills and skills of rehabilitation and rehabilitation of the cultural relics with the basic knowledge and ability of cultural relics appraisal. Professional Core Courses: Ancient Chinese Foundation, Ancient Chinese History, Modern and Contemporary Chinese History, Ancient World History, Modern and Contemporary World History, A Brief History of World Archeology, Museology, General Archeology, Aesthetics, Chinese Art History, Introduction to Cultural Relics, Fine Arts Archeology, the world's ancient art, ancient

IOP Conf. Series: Materials Science and Engineering **364** (2018) 012080 doi:10.1088/1757-899X/364/1/012080

Chinese architecture, Chinese painting, sculpture, ceramics, Chinese four treasures, Chinese seal, Chinese coins, cultural relics identification, cultural relics, painting and calligraphy identification training, Four treasures and seal identification training.

The art of Chinese cultural relics protection has a long history. In the Tang Dynasty (618-907), there was a record of the wedges being placed on the crooked ancient building beams. According to Huang Hu-fu's Record of Famous Masters in Yizhou, Chengdu had migrated three Wall murals, after 200 years is still intact.

On the statue, murals, paintings and other ancillary works of art can not be free to repair. Most of the destroyed painting and calligraphy use the traditional method of exposing and framing; the protection of the framed and framed art of calligraphy and painting is quite mature to the Tang Dynasty. "Paint sticky stones, sticky for wood" is a long-standing tradition of restoration of stone artefacts and wood artefacts. With the progress of science and technology later, polymer materials, physical detection technology has gradually been introduced to the work of cultural relics protection. Silk screen ancient silk fumigation available sterilization; fragile textiles available mesh reinforcement or traditional mounting method of reinforcement; ceramic fracture with shellac or epoxy resin and other polymer materials bonding. Chinese painting mounting technology is basically the master apprentice mode, the general situation to go through 5-10 years of study before it is possible to independently host art repair, for example, the quality of paste production directly affects the mounting quality. To handcraft training, did not form a complete theoretical system. The academic education has not been closely linked with the practical application.

Professionals can not keep up with the development of art galleries. The collection department is a department set up by most art galleries. The collections are the foundation of the museum. However, the preservation and restoration department is unusual. Even if the capital equipment is complete, but professionals are difficult to hire. In order to break the bottleneck of craftsmen, in 2015, the Forbidden City and Beijing International Vocational Education School jointly opened the first professional restoration and protection of cultural relics. The first batch of 20 people, so far, the profession enrolls 35 students. On August 31, 2016, a two month training program for the first national art collection protection and restoration was conducted by the Chinese Ministry of Culture and the National Art Fund, sponsored by the National Art Museum of China. The number of new art students in Chinese universities is 530,000 each year, with 2 million students enrolled. There are less than 300 art students specializing in the protection and restoration of artwork.

The China Academy of Art is the first Academy of fine art in China and exists since 90 years. The Academy was founded in 1928. and has more than 10000 students now studies are organized in 10 colleges of Academy. With altogether 84 degrees (art, design, architecture, art management, art history and art restoration)

The college of Art Management and Education was established in 2015. it's the first institute focus on Art restoration, protection and Art Management in China. And includes 3 Departments: Art Management, Appreciation and Restoration also Art Education.

The Department of Art Management has as main teaching content in art administration, art communication and art economy, and as well The Department Appreciation and restoration focus of Material Research, Art Studies, Theory Studies. Main teaching method is theory and practice combination that the appreciation and collection, protection and restoration of room ware, traditional calligraphy and painting in China, both practical foundation and theoretical accomplishment. Aimed to a high level of professional and practical teaching and practicing the academic concept applied project. Currently 2015 China Academy of Art in Art Management and Education College has 340 students, of whom 72 graduate students and undergraduates in the Department of Art Kam learning. In addition, Beijing and Shanghai Institute of Fine Arts also set up such professional courses, but this highly practical college education focuses on theory, less opportunities, even to have the opportunity to participate in the specific restoration work, but also to trainee the Lord. It is hard to cultivate and train outstanding students without great opportunities to participate in the rehabilitation. The purpose of the training process is not very clear, post-graduate job is not clear. The education of oil painting restoration and

IOP Conf. Series: Materials Science and Engineering **364** (2018) 012080 doi:10.1088/1757-899X/364/1/012080

protection is almost blank in China. Currently only the Shanghai Institute of Visual Art opened a painting repair professional. Weak teachers is a very real problem.

2. Study and research on mural technology and protection in humid areas in southern China

China has a vast territory that from northern to southern and from western to eastern about 5000 kilometers, the climate, soil and vegetation are very different. In winter between north and south is 50 degrees Celsius with difference Temperature. The climate in northern is dry, From Jin, Tang, Song, Yuan, Ming, Qing dynasties, like Cave mural in Dunhuang, Kizur xinjiang region, temples Mural in Yonglegong, Jincheng in Shanxi, or Beijing and other places more (Annual average humidity 43-50). but in south area like province Zhejiang Fujian, Hubei, Hunan Sichuan the climate is hot and humid (Annual average humidity 77-80). This paper researches on special historical materials of southern Murals in China and analyzes the traditional materials production techniques and artistic characteristic (eg such like one of the future projects or a professional level will be the rebuilding of an old wall with murals as a laboratory for future studies.

At present, the research and protection of traditional murals in China are mostly concentrated in the northern areas mainly based on Dunhuang. Due to the prevailing stereotypes in the south that "there is a paucity of remnants of the climate due to damp weather, it has seriously lagged the relevant research." For a long time, many well-made and historically high value traditional murals in the humid southern areas have not received the attention and attention they deserve. They are located in Longquan, Jinhua, Quzhou, Tonglu, Sichuan, Chengdu, Xinjin, Guanghan, The traditional murals in the wet areas of the south are notoriously inadequate, under-researched and poorly preserved.

There are a large number of frescoes in Zhejiang Province, and there are 94 frescoes of the Qing Dynasty in the Palace of the King of Jinhua with an area of 294 square meters. The murals reflect the high standard of ink and wash murals in wet southern areas. , Pines crane (Songhe) , cats and butterflies, cypress deer, bee monkeys, watchtower barracks, tower pavilions, landscape flowers and seasons fishing chart, mountain mining fireworks, etc., its style is different from the northern murals fine detainery, with South Sect murals The typical characteristics of art and heritage value is higher.

In addition, the survey found that Lin Yueqing ancestral hall in Yueqing, Lei Gong Temple Yongjia, Yongjia Green House Taiyin Gong, Yongjia Jingzhou Taiyin Gong, Ninghai Yu ancestral hall, Hu ancestral hall, Pan's ancestral hall, Lee's ancestral hall, Jiangshan Mao's ancestral hall, Wenchang Palace, Shengzhou Wang ancestral hall, bamboo ancestral temple, New Town Temple 13 ancient buildings have 253 color murals, with a total area of about 583.8 square meters.

Some murals in the open air of the Ming and Qing dynasties are still clear after years of hard work. Anhui shexian Shengzeng An Temple mural is the painting by master Huang Zhu, created in the Ming Dynasty Wanli years (1573 -1620). The murals are also works of ink drawings, the outline of the brushwork is more bold and full of passion, character and clothing outline to a fine level of the Ming Dynasty called first-class level. Sichuan Pengxi Bao Fan Temple murals open character and outline the color are extremely refined, is a typical style of the Ming Dynasty, although the era and style is different, but both in terms of sophistication and cultural value comparable to the Yuan Dynasty in Paradise in Shanxi mural.

The hot and humid weather in southern China, the southern mural protection does exist problems, resulting in improper protection measures. Mural easily damp moldy, picture peeling, the picture is damaged. There is a lack of aesthetic appreciation in the repair, lack of understanding of the traditional materials and techniques, and inadequate research on the artistic style, tool materials and painting techniques, resulting in improper patch full color.

According to the present situation, the art management and education institute of China Academy of Art has made important achievements in the in-depth study of Professor He Shiyang and his team in the Department of Art and Design.

The research team first conducted its work on materials and workmanship, conducting research from mural supports, poles and layers. South traditional building wall rammed earth wall or brick wall, black brick firewood tunnel kiln, the temperature 1000 to 1100 degrees firing more than 20 days, and then boring, carburizing, etc., about 30 days to go out of the kiln. Bricks are brighter than red bricks and better in firmness. Winter temperatures are low, easy to crack red brick, brick will not. Sichuan is also useful bamboo as the skeleton, and on the glutinous rice paste, brown silk pounded mud and lime, paper fiber, bamboo (bamboo fiber), hemp rope.black brick structure for the wall mainly in the vicinity of Zhejiang Province, Zhejiang Province, the brick wall of the ancient method of construction, the surface to the ancient law approved the lime wall for painting; the southern coast of the ancient method of building brick walls, the surface of the ancient method of scratch Oysters shell gray wall for painting; South Fujian and Guangdong area to ancient law rammed earth wall rammed, smooth polished wall for painting; Sichuan area to bamboo structure, painted brown silk pounded mud and lime walls for painting

Southern wall murals aesthetic, follow the Tang Dynasty Wu Daozi, Li Gong-lin in the Song Dynasty heavy line sketching a light stroke with a pulse of painting techniques to strengthen the use of pen to mention, pay attention to the performance of line drawing brushwork, color simple and simple, with mural "South Sect "aesthetic characteristics

China needs a lot of research and work on art protection, and the protection and restoration of traditional materials and methods are fundamental to education in this area. At the same time, the application of modern science and technology and new materials is also necessary, and more cross-disciplinary and intercultural Cooperation and exchange.

References

[1] The Yearbook of Beijing Museums1988-1991,1992-1994.1995-1998.1999-2003.2004-2006.2004-2008,2009-2012, ISBN:978-7-5402-3570-3, Beijing Yanshan Press 2015.

[2] Material Studies of traditional painting ,He Shiyang ,reports of China Academy of Art.

IOP Conf. Series: Materials Science and Engineering **364** (2018) 012081 doi:10.1088/1757-899X/364/1/012081

Study on Removal of Iron Rust from Paper Objects with Thiourea Dioxide

Huiping Xing*, Ouya Ma, Qi Yong, Chun Yang, Xiaolian Chao and Yuhu Li
Engineering Research Center of Historical Cultural Heritage Conservation Ministry of Education, Material Science and Engineering School, Shaanxi Normal University, 710019 Xi'an, China

*E-mail: xhp@snnu.edu.cn

Abstract. There are many paper-based documents, archives and books contaminated by iron rust beacuse they were fixed with iron staples or clips. Iron rust not only disfigures the paper, covers the writtings or images of paper, but also makes paper brittle. Iron rust is very difficult to remove due to its insolubility in water and organic solvents. Thiourea dioxide(TDO), an agent with strong reducibility, easy accessibility, stability and safety was applied to remove iron rust in this work. The concentration of TDO solution, iron rust removal effect and removal rate were studied. The results showed that TDO was an effective and safe agents to remove iron rust from paper documents. The concentration was higher, the removal speed was faster. But to get high concentration, TDO need to be dissolved at higher temperature which is damage to paper. In view of the above facts, the proper treatment bath was 5% at 40°C. Results also showed that the combination of 5% TDO and EDTA can remove iron rust from paper completely in 4 hours.

1. Introduction

There are many paper-based documents, archives and books fixed with staples or clips. The composition of these staples and clips are iron usually. So the paper is easy to be rusted under high humidity (Fig. 1). The iron rust not only disfigures the paper, but also makes paper become brittle.

IOP Conf. Series: Materials Science and Engineering **364** (2018) 012081 doi:10.1088/1757-899X/364/1/012081

Therefore, the effect removal of the iron rust is always an interesting problem for the paper conservators.

Iron rust, $Fe_2O_3 \cdot H_2O$ (or FeOOH), is difficult to remove due to its insolubility in water and organic solvents. As reported in literatures[1-3], the effective way to remove it from paper is using the reducing agents to reduce the insoluble ferric iron (III) compound to a soluble ferrous iron (II) compound. Then the colorless ferrous iron is washed away with chelating agent in case it be reoxidized by atmospheric oxygen back to the colored ferric compounds. And it also mybe catalyze oxidation of cellulose, which has been mentioned in many literatures about the iron gall ink corrosion [4-6].

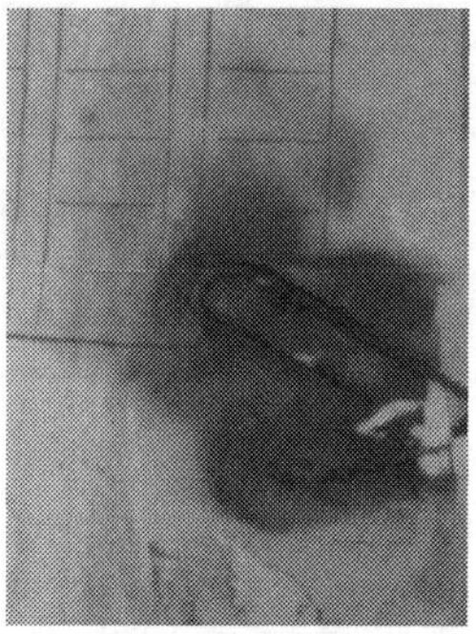

Fig. 1 The documents contaminated with iron rust

The reduction agent most frequently mentioned in the literatures for iron rust removal is Sodium Dithionite ($Na_2S_2O_4$), also called Sodium Hydrosulfite. It has been proven to be effective for iron rust removal but has significant practicality problems. It is extremely flammable and prone to spontaneous combustion. According to the problems of Sodium Dithionite, another reducing agent, Sodium Metabisulfite ($Na_2S_2O_5$), was brought forward to remove the rust from paper [7]. Sodium Metabisulfite has three advantages than Sodium Dithionate (1) one more oxygen atom,(2) not labeled as a hazardous substance according to Material Safety Data Sheet (MSDS), and (3) more cheaper than Sodium Dithionate. It was observed to be effective at removing rust stains especially for light-to-medium stains, but it took a long time from 24 to 48 hours because of its weak reducibility. So the reducing agent with strong reducing and safety to the environment is an urgent need for the paper conservator.Thiourea dioxide (TDO, $CH_4N_2O_2S$), due to its strong reducibility, easy accessibility and stability, has been widely used in the field of textile printing, paper, photographic and leather processing industries. The reduction process generates sodium sulfite and urea which are easy to manage since these are common commercial wastes. Although TDO has been widely used as strong inducing and environmentally friendly regent in industry, it has never been reported to be applied in paper restoration. This paper will study on the effect of TDO on removal iron rust from paper.

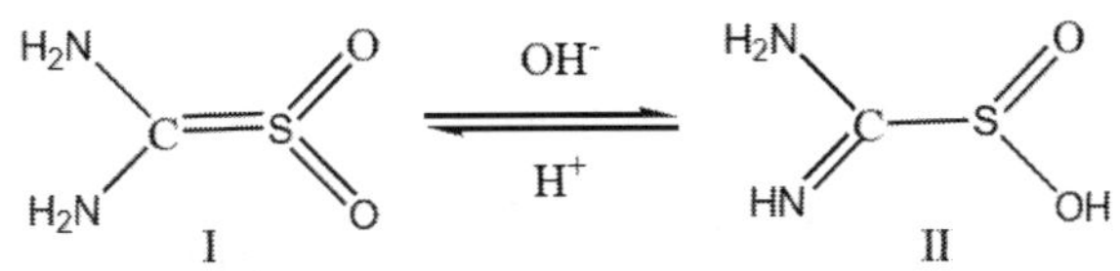

a

$$NH_2(NH)CSO_2H + H_2O \longrightarrow NH_2CONH_2 + H_2SO_2 \qquad b$$

$$H_2SO_2 + Fe_2O_3 + H_2O \longrightarrow Fe^{2+} + SO_4^{2-} + OH^- \qquad c$$

Fig. 2 Thiourea dioxide reaction at alkali or high temperature (a,b) and reaction with iron rust (c)

TDO has two kinds of tautomers in the experimental solution (Fig. 2 a). Under acidic conditions, TDO mainly exists in the form of isomeri. Under basic conditions or at high temperatures, TDO exists in the form of isomer II [8]. In fact, TDO appears terrifically stable at 20–30°C in solution. However, when it is heated or catalyzed by alkali, it decomposes rapidly to produce sulfoxylic acid and urea(Fig . 2 b). Sulfoxylic acid, act as a strong reductant, can react with iron rust-Fe_2O_3 to reduce the insoluble ferrici iron (III) to soluble ferrous iron (II) (Fig.2 c). This paper will try to find the effect of TDO removing iron rust from paper, and the proper concentration to be used.

2. Experimental

2.1. Samples preparation

The paper from a modern notebook was selected for the preparation of paper samples with iron stain. The paper clips, bought from a local stationery shop (Fig. 3), were used to fix the paper. A desiccator with some water at the bottom was used to form high relative humidity (Fig.4 c).

The paper samples were divided in the size of 2.5cm to 7.5cm and fixed with the clips (Fig.4 a). Then they were put on the plate of the desiccator(Fig.4 b) and the desiccator was sealed(Fig.4 c). One month later. heavy iron rust were formed on the papre samples. Then the clips were removed away from the samples and the samples were dried in the atmosphere (Fig.4 c).

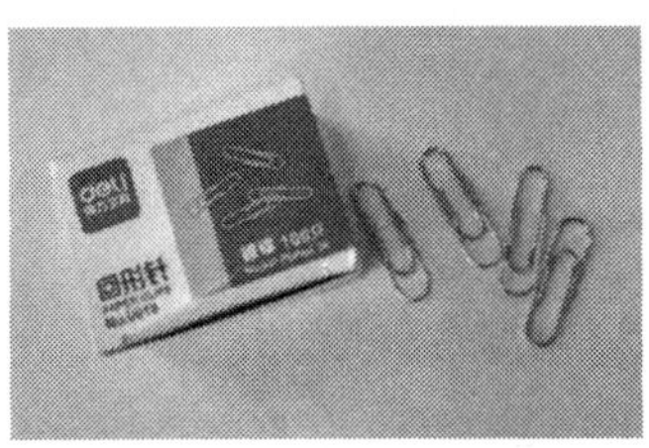

Fig. 3 Paper clips

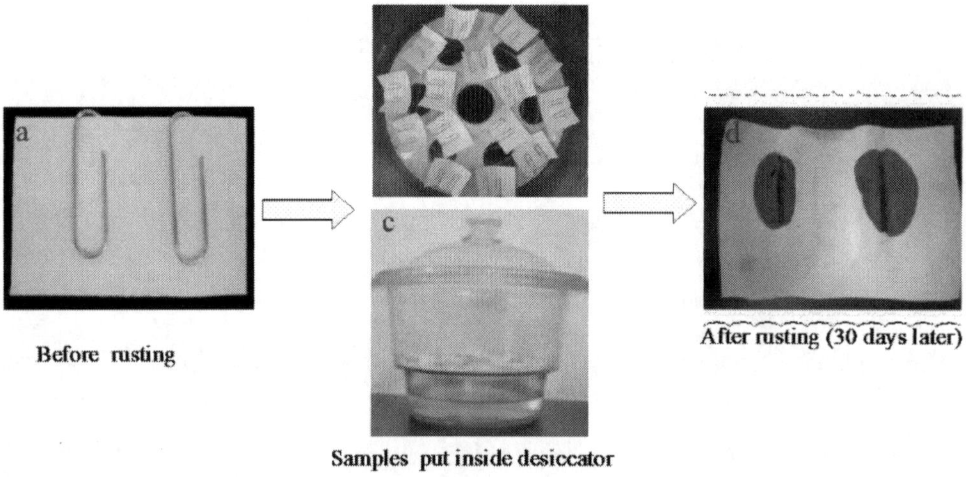

Fig.4 The process of samples rusted

2.2. TDO solutions preparation

The solubility of TDO in water is poor. At 20°C, the solubility is approximately 3%. But with the increasing of temperature, the solubility is increasing too. 2.5%, 5.0%, 7.5% and 10.0% TDO solutions were prepared at 25°C, 40°C, 60°C and 70°C separately.

2.3. EDTA+TDO solution preparation

2.5g EDTA was added in the 5.0% TDO solution.

2.4. Iron rust removal

The paper samples were kept in the above TDO solutions and TDO+EDTA solution at 40°C, to avoid the precipitation of TDO in high concentration. Then they were washed by water.

2.5. Determination of iron content

The Thermo Scientific Niton XL3t handheld X-ray fluorescence (XRF) analyzer was used to determine the amount of iron in the paper before and after treatment with TDO+EDTA solution.

2.6. The tensile strength of paper

The tensile strength of samples before and after with TDO+EDTA solution was tested by the tensile test machine.

3. Results and discussion

3.1. The influence of TDO concentration on the iron rust removal

Fig. 6 shows the process of samples treated with different concentration TDO solutions. Within10 mins, the iron rust removal speed of the samples immersed in the higher concentration was faster than the samples immersed in the lower concentration. But after 30 min, the above TDO solutions, except 2.5%, had almost the same effect on removal of iron rust. The removal speed of all the samples slowed down after 30 min. There are three possible reasons: (1) The unremoved iron rust after 30 mins was on the tightest touching part of clips and paper, so the amount of iron rust on which part was the highest. It was also the most difficult part to remove. (2) The concentration of TDO solutions was decreased after 30 mins due to their reaction with iron rust. (3) The high concentration of TDO solutions did not improve the removal speed because some amount of TDO was precipitated during the process due to low temperature.

So if the samples are immersed into the unused TDO solutions for several times, the removal consuming time will be shorten greatly. For example, it has been proved that it only took 1.5 hours to remove iron rust from the sample with 5% TDO bath when the TDO solution was changed every 30 minutes.

3.2. The influence of EDTA on the iron rust removal

Fig.7 shows the samples before and after treatment with EDTA+TDO solution. EDTA can chelate Fe^{2+}, which produced during the reaction of TDO and iron rust, and keep most of the Fe^{2+} retain in the solution. It also showed that EDTA can increase the removal speed. The iron rust can been removed completely in 4 hours with the combination of TDO and EDTA, which shorten the removal cosuming time greatly. The possible reason is that EDTA chelating with Fe^{2+} reduced the concentration of iron in the samples and promoted the reaction remove to right direction faster (Fig.4 c).

3.3. The effect of TDO removal iron rust

The iron content in the samples was detected by portal-XRF (Table 1). The results show that the content of the iron in the samples before treatment was about 90%. It decreased to around 30% after treatment with EDTA+TDO solution, which was about the same concentration of the iron in the samples without iron rust. This indicates that the iron rust has been removed completely.

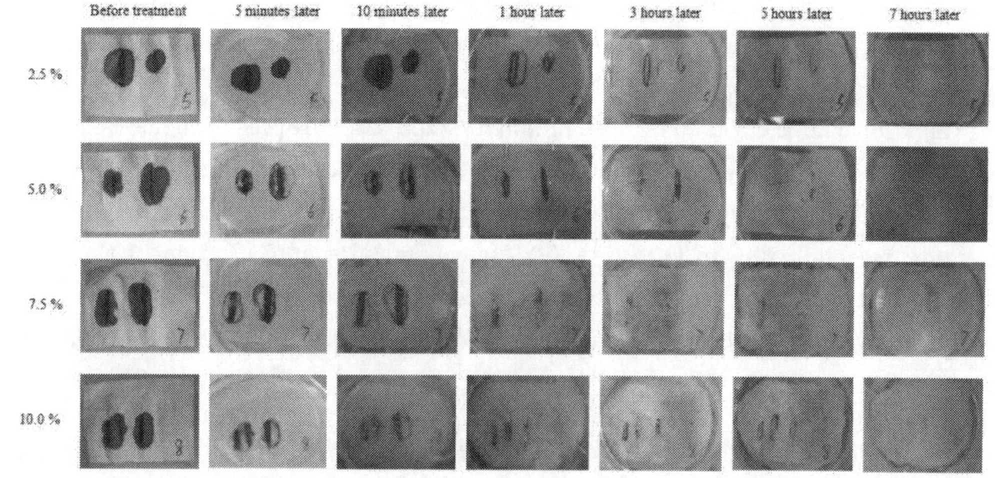

Fig. 5 The process of samples treated with different concentration TDO solutions(from up to down2.5%, 5.0%, 7.5%, 10.0%) in 7 hours (from left to right: before treatment, 5 minutes later, 10 minutes later, 1 hour later, 3 hours later, 5 hours later and 7 hour later)

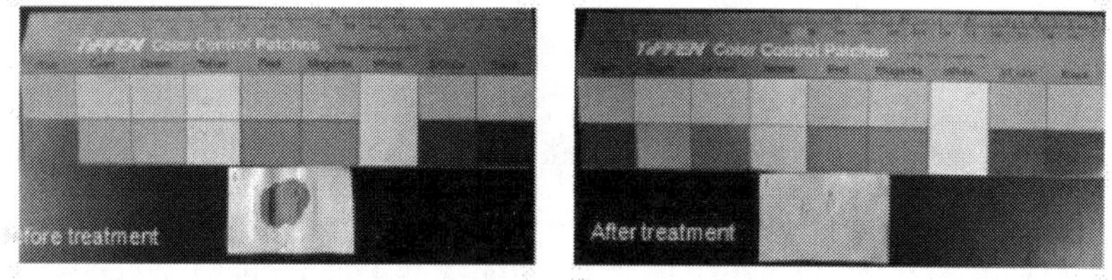

Fig.6 The samples before and after treatment with EDTA+TDO solution

3.4. The tensile strength of paper treated with TDO+EDTA

Table 2 shows the tensile strength of samples before and after treated with EDAT+TDO solution. There was a slight decrease of the sample treated with EDAT+TDO compared with the untreated samples.

Table 1 The iron content in the samples before and after treatment

Sample	Before treatment(%)	After treatment (%)
Iron rust sample	89.388	31.706
Control sample	34.830	32.596

Table 2 The tensile strength of the samples before and after treatment

	Before treatment	After treatment
Tensile strength (kN/m)	1.036	1.014

4. Case study

The documents were written in 1949 by some famous artists from a private collection. Since these documemts were fixed with clips, iron rust was formed on the surface of the paper seriously (Fig. 7 a).

A small area of iron rust (Fig. 7 b) was treated with 5% TDO+EDTA solution. After treatment, the iron rust was removed completely (Fig. 7 c).

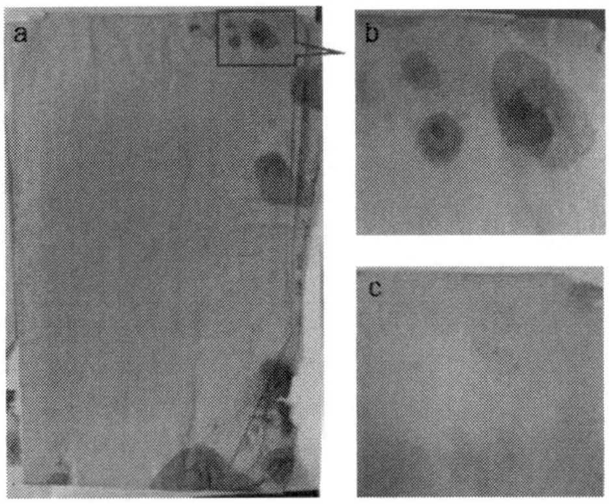

Fig. 7 (a) The original document, (b) before treatment, (c) after treatment

5. Conclusion

In this work, Thiourea dioxide has been employed to remove iron rust from paper objects.The results show that TDO was an effective and safe reducing agents to remove iron rust from paper documents. The TDO concentration was higher, and the removal speed was faster within 30 mins. But the higher concentration needs to be dissolved at higher temperature, which is damage to paper. Low temperature caused the precipitation of TDO and reduced the effect of removal. The low concentration of TDO solution (2.5%) could not remove the rust completely. So the proper TDO concentration solution was 5% at 40°C. The combination of TDO and EDTA improved the removal speed greatly. In addition, the tensile strength of sample treated with EDTA+TDO had a slight decrease, so the treatment of brittle paper with EDTA+TDO should to be very carefully.

Acknowledgements

This work was funded by National Natural Science Foundation of China (51403121) and the Fundamental Research Funds for the Central Universities (GK201603060).

References

[1] The American Institute for Conservation of Historic and Artistic Works Book and Paper Group:*Paper Conservation Catalog.In: Foxing.*1992: 22.

[2] Malešič J, Kojc M and Šelih VS 2008 Assessment of the Effect of Various Bleaching Agents on Papers with Foxing Stains *Restaurator* 29 142-54.

[3] Suryawanshi DG and Bisaria SK 2005 Removing Metallic Stains from Paper Objects Using Chelating Agent EDTA *Restaurator* 26 276–85.

[4] Neevle JG 1995 Phytate: a Potential Conservation Agent for the Treatment of Ink Corrosion Caused by Irongall Inks *Restaurator* 16 143-60.

[5] Calvini P and Gorassini A 2002 The Degrading Action of Iron and Copper on Paper A FTIR-Deconvolution Analysis *Restaurator* 23 205-21.

[6] Poggi G, Giorigi R, Toccafond N, Katzur V and Bglioni P 2010 Hydroxide Nanoparticles for Deacidification and Concomitant Inhibition of Iron-Gall Ink Corrosion of Paper *Langmuir* 26 19084-90.

[7] Irwin S 2011 Comparison of the Use of Sodium Metabisulfite and Sodium Dithionite for Removing Rust Stains from Paper *The Book and Paper Group Annua*l 30 37-46.

[8] Wang HY, Li QM and Gao CZ 2014 Preparation of Nanometer Nickel Powder from Spent Electroless Nickel Plating Baths by Using Thiourea Dioxide as a Green Reductant. J*ournal of Cleaner Production* 84 701-06.

IOP Conf. Series: Materials Science and Engineering **364** (2018) 012082 doi:10.1088/1757-899X/364/1/012082

Phosphate-based treatments for consolidation of salt-bearing Globigerina limestone

G. Graziani[1,2], E. Sassoni[2], G.W. Scherer[3], E. Franzoni[2]

[1] Rizzoli Orthopaedic Institute (IOR), Laboratory of NanoBiotechnology, Bologna, Italy

[2] Dept. Civil, Chemical, Environmental & Materials Engineering (DICAM), University of Bologna, Italy

[3] Dept. Civil and Environmental Engineering (CEE), Princeton University, USA

E-mail: gabriela.graziani2@unibo.it

Abstract. Phosphate-based treatments have proven to be effective for consolidation of porous limestone. However, application on salt-laden stone must be evaluated, as salt-weathering is among the main deterioration phenomena for this lithotype and salts present in the stone can interfere with the outcome of consolidation. For this reason, here, the phosphate-based consolidant was tested on a salt-bearing porous limestone (Globigerina limestone). To achieve salt contamination and to induce weathering, stone samples were subjected to crystallization cycles in sodium sulfate, which is among the most common and damaging salts that affect limestone on site. Phase formation and distribution of the consolidant were evaluated in the presence of salts. The treatment efficacy was investigated in comparison with ethyl silicate, which is currently the most used consolidant for porous limestone. Promising results were obtained, indicating that the presence of sodium sulfate does not hamper penetration of the phosphate-based consolidant. By penetrating deeply into the stone and sealing cracks caused by weathering, the consolidant enhances strength and stiffness. Moreover, the presence of salts does not interfere with the formation of calcium phosphates, hydroxyapatite being the only phase to be detected. These results indicate that the phosphate-based treatment is suitable for consolidation of limestone, even when contaminated with salt.

1. Introduction

Consolidation of porous limestone and other carbonate substrates by phosphate-based treatments has yielded promising results in terms of efficacy, compatibility and durability [1-12]. These treatments aim at forming insoluble hydroxyapatite (HAP, $Ca_{10}(PO_4)_6(OH)_2$) inside the stone, by reaction of a water-based solution of diammonium hydrogen phosphate ($(NH_4)_2HPO_4$, DAP) with calcium ions (deriving from the stone itself or externally added). Notably, because the precursor is applied in solution and no particles are present, even at the nanoscale, the consolidant exhibits remarkable penetration depth [3]. In addition, it is capable of successfully binding cracks formed by weathering, hence guaranteeing efficacy while maintaining good compatibility [3].

The efficacy of the treatments depends on the formation of hydroxyapatite, or, at least, calcium phosphates having low solubility, such as octacalcium phosphate ($Ca_8H_2(PO_4)_6{\cdot}5H_2O$, OCP). However, different outcomes in terms of amount, morphology and composition of phases that form as a result of the treatment could be experienced, depending on the substrate, the treating solution and procedure and on the presence of substrate contaminants, such as soluble salts [6,13].

In particular, salts in the stone might affect the outcome of consolidation by two main mechanisms:

- by blocking pores, thereby inhibiting penetration and homogeneous distribution of the consolidant in the substrate;
- by interfering with HAP-formation reactions. This can be detrimental if soluble calcium phosphates, or other phosphate compounds (not including calcium) form instead of HAP [14-16]. On the other hand, increased phase formation in the presence of salts has also been reported [6,13], indicating that the substrate lithotype, weathering level and surface roughness, as well as the salt type and its amount, can all have an impact on phase formation.

Salt contamination of porous limestone is very common in the field, as salt weathering is among the most frequent and severe deterioration phenomena for this lithotype [17-22]. In addition, it is rarely possible to completely remove salts from the stone, even when desalination procedures are carried out.

For these reasons, here, application of phosphate-based treatments is tested on Globigerina limestone, purposely contaminated by sodium sulfate crystallization cycles. The effects of the treatments on mechanical properties are investigated and the impact of the presence of sodium sulfate on phase formation is evaluated in terms of their nature, morphology and distribution.

2. Materials and Methods

2.1. Materials

Tests were carried out on Globigerina limestone, an organogenic limestone, mainly composed of calcite with fractions of silicates (kaolinite) and quartz [23]. Cores (5 cm in height and 2 cm in diameter) were used for all the tests, all cut from one single quarry slab.

2.2. Contamination by sodium sulfate crystallization cycles

Samples were contaminated by 3 cycles of sodium sulfate crystallization, performed as follows:

1) Wetting phase (24 h): samples were immersed to a depth of about 1 cm in a 14 wt% solution of sodium sulfate decahydrate ($Na_2SO_4 \cdot 10H_2O$) in deionized water as recommended in the European Standard EN12370;

2) Drying phase (22 h): samples were dried in a ventilated oven with forced convection at 50°C;

3) Cooling (1 h) and measuring of weight and dynamic elastic modulus (total: 2 h).

The number and duration of the cycles were selected based on preliminary tests, with the aim of contaminating the stone with salts and cause weathering, while not causing rupture and conserving a morphology suitable for mechanical testing. To reproduce the conditions on site, where treatment of stone exhibiting abundant efflorescence is uncommon and the substrate is normally desalinated by poulticing before the application of the consolidant, the cores were desalinated after contamination, purposely leaving a fraction of the salts in the stone ("samples DESALT"). This allows evaluation of the effects of salt contamination and simulates the conditions on site, where it is often impossible to completely remove salts from the substrate, even when desalination is performed. Desalination was achieved by applying a deionized water poultice (water:cellulouse pulp ratio 1:5, thickness ~1 cm). The poultice was kept sealed for 24 hours to allow solubilization of the salts, then left to dry in contact with the samples to allow advection of the salts towards the external surface.

2.3. Treating procedure

Two HAP-based treatments were tested that were previously developed for the consolidation and protection of limestone and marble [2,8]:

1) Treatment "3M": a 3 M DAP solution is applied by brushing (10 strokes) followed by the application of a limewater poultice (1.7 g/L $Ca(OH)_2$, limewater:poultice ratio 1:6). Limewater poultice is applied with the aim of providing further calcium ions to boost HAP formation and to remove possible residual DAP from the stone after treatment;

2) Treatment "0.1M": a 0.1 M DAP solution also containing 0.1 mM $CaCl_2$ and 10 vol% ethanol is applied, followed by the application of a 0.1 M DAP + 0.1 mM $CaCl_2$ solution without the addition of ethanol. Both solutions are applied by brushing (10 strokes). This procedure was developed for the consolidation of marble, for which it was found that ethanol promotes HAP formation (and thus efficacy), while using lower amounts of precursors [16,24].

Ethyl silicate (ES) was also tested for comparison's sake, as it is currently the most widely used product for consolidation of Globigerina limestone. For ES treatment a commercial formulation (Estel 1000, CTS, Italy, composed of 75 wt% ethyl silicate and 25 wt% white spirit D40, also containing 1 wt% dibutyltin dilaurate as a catalyst) was applied by brushing (10 brush strokes). After treatment, samples were left to cure for 1 month in room conditions prior to testing.

For all the procedures, stone cores were treated on the whole external surface.

2.4. Post-treatment characterization

All characterizations were performed on partially desalinated ("DESALT") cores, before and after treatment.

Mechanical properties after treatments were evaluated in terms of dynamic elastic modulus (Ed, measured by ultrasonic tests) and tensile strength (σ_T, measured by tensile splitting tests). Ultrasonic tests were carried out by a Matest instrument with 55KH transducers. Rubber couplants and vaseline were used to ensure proper contact between the samples and the transducers, while avoiding contamination of the sample. Briefly, an ultrasonic pulse is transmitted across the sample by piezoelectric transducers. The time required by the wave to cross the sample is recorded, that is linked to dynamic elastic modulus by the equation: $Ed=\varrho v^2$, v being the ultrasonic pulse velocity. Tensile splitting test consists in subjecting cylindrical specimens to compression along two lines parallel to the axis. σ_T is calculated with the equation $\sigma_T = 2P/(\pi DL)$, P being the maximum load, D the diameter and L the length of the cylindrical specimens, respectively. Three samples were taken for each consolidation treatment (3M, 0.1M and ES) and for the untreated reference (UT). Variations in dynamic elastic modulus and tensile strength caused by the treatments were also calculated, according to the formulas:

$$\Delta Ed= (Ed_{,after\ treatment}-Ed_{,before\ treatment})/Ed_{,after\ treatment};\ \Delta\ \sigma_T = (\sigma_{T\ after\ treatment}-\sigma_{T\ before\ treatment})/\ \sigma_{T\ after\ treatment}$$

Salt content after salt contamination, desalination and consolidation treatment was evaluated by ion chromatography (IC), performed on the ground stone after salt extraction and filtering. Salt extraction was performed by grinding the samples and boiling the obtained powder in deionized water, to favor solubilization of the salts. Samples (~5x5x5 mm^3) were collected from the tops of the cores (i.e. the area opposite to that immersed in the sodium sulfate solution), where salt concentration is expected to be maximal.

Phase morphology and distribution were evaluated by SEM/EDS, to understand whether the presence of salts prevents consolidants from penetrating and homogeneously distributing into the stone and to evaluate the treatments' capability of sealing cracks caused by weathering.

Phase formation after 3M and 0.1M treatments was examined by FT-IR to verify which phases form in the presence of salt contamination. FT-IR (Perkin Elmer Spectrum 2, ATR mode, resolution 2 cm^{-1}, 32 scans) was performed on ground samples, to maximize the intensity of the bands. Powder was collected from the surface of the samples (0-2 mm depth). FT-IR was also performed on ES-treated samples to verify whether the presence of salts interferes with ES curing, according to the very same procedure.

IOP Conf. Series: Materials Science and Engineering **364** (2018) 012082 doi:10.1088/1757-899X/364/1/012082

3. Results

As a result of salt contamination and desalination, sulfate concentrations of 1.4 wt% and 0.3 wt% are achieved, respectively.

Dynamic elastic modulus and tensile strength after salt contamination, desalination and consolidation are reported in Table 1. HAP-based treatments cause a minor change in dynamic elastic modulus, but a significant increase in tensile strength, indicating good efficacy for both treatments. The different behavior in terms of Ed and σ_T can be explained as follows. σ_T is essentially determined by the capability of the consolidant to increase stone cohesion by bridging cracks caused by weathering. For the consolidant to be effective, it must seal at least the tip of the cracks caused by weathering, so as to bind the surfaces together and prevent them from propagating through the boundary. However, some porosity in the binder can be beneficial, because it allows volume variations in calcite grains from temperature changes to be accommodated without causing stress [25]. Ed, on the other hand, is raised by pore occlusion and by large amounts of precipitates, even if they don't bind strongly to the substrate. As a consequence, if the consolidant bridges cracks without significantly filling them, the effect can be much more visible in terms of σ_T than it is in terms of Ed. ES, in contrast, causes a significant increase in terms of Ed and, especially, σ_T. Such a large increase in tensile strength can cause over-strengthening, as it is generally believed that increases lower than +50% are to be preferred [26,27]. In fact, the final dynamic elastic modulus and tensile strength are significantly higher than those of unweathered stone (Ed= 14.2 GPa, σ_T = 3.1 MPa). Both treatments with DAP exhibit similar increases in both dynamic elastic modulus and tensile strength, indicating that the treatments have similar performance despite significant differences in the amount of DAP used as a precursor, pointing out the efficacy of the addition of ethanol to the treating solution.

Table 1. Mechanical properties after treatment.

	Ed (GPa)	ΔEd (GPa)	σ_T (MPa)	$\Delta\sigma_T$ (MPa)
UT	12.8 ± 0.2	/	2.3 ± 0.4	/
3M	13.1 ± 0.1	+2%	3.3 ± 0.2	+ 39%
0.1M	13.1 ± 0.2	+2%	3.3 ± 0.3	+ 40%
ES	15.7 ± 0.1	+23%	5.5 ± 0.9	+ 134%

SEM images of samples 3M and 0.1M (Figure 1) show that phosphate phases formed both on the exterior surface in the form of a coating (~1-2 μm thick) and in the interior of the samples. This indicates that the presence of the salt did not prevent the penetration and curing of the consolidant due to pore blocking.

Phase formation (Figures 1 and 2) is more abundant for samples treated by 3M compared with 0.1M, which is consistent with the much higher amount of precursor used. However, for both treatments, the phosphate phases that form are capable of bridging cracks caused by weathering, including in the areas far from the surfaces (see red arrows in Figure 1). This explains the similar increase in the mechanical properties caused by the two treatments, in spite of the different amount of phases formed. In ES-treated samples, agglomerates of salts, probably incorporated in the glass network constituted by cured ethyl silicate, are visible across the whole sample. This indicates that significant pore clogging occurs, which can be responsible for the large increase in mechanical properties. For this reason, evaluation of compatibility and durability of the treatments is in progress.

IOP Conf. Series: Materials Science and Engineering **364** (2018) 012082 doi:10.1088/1757-899X/364/1/012082

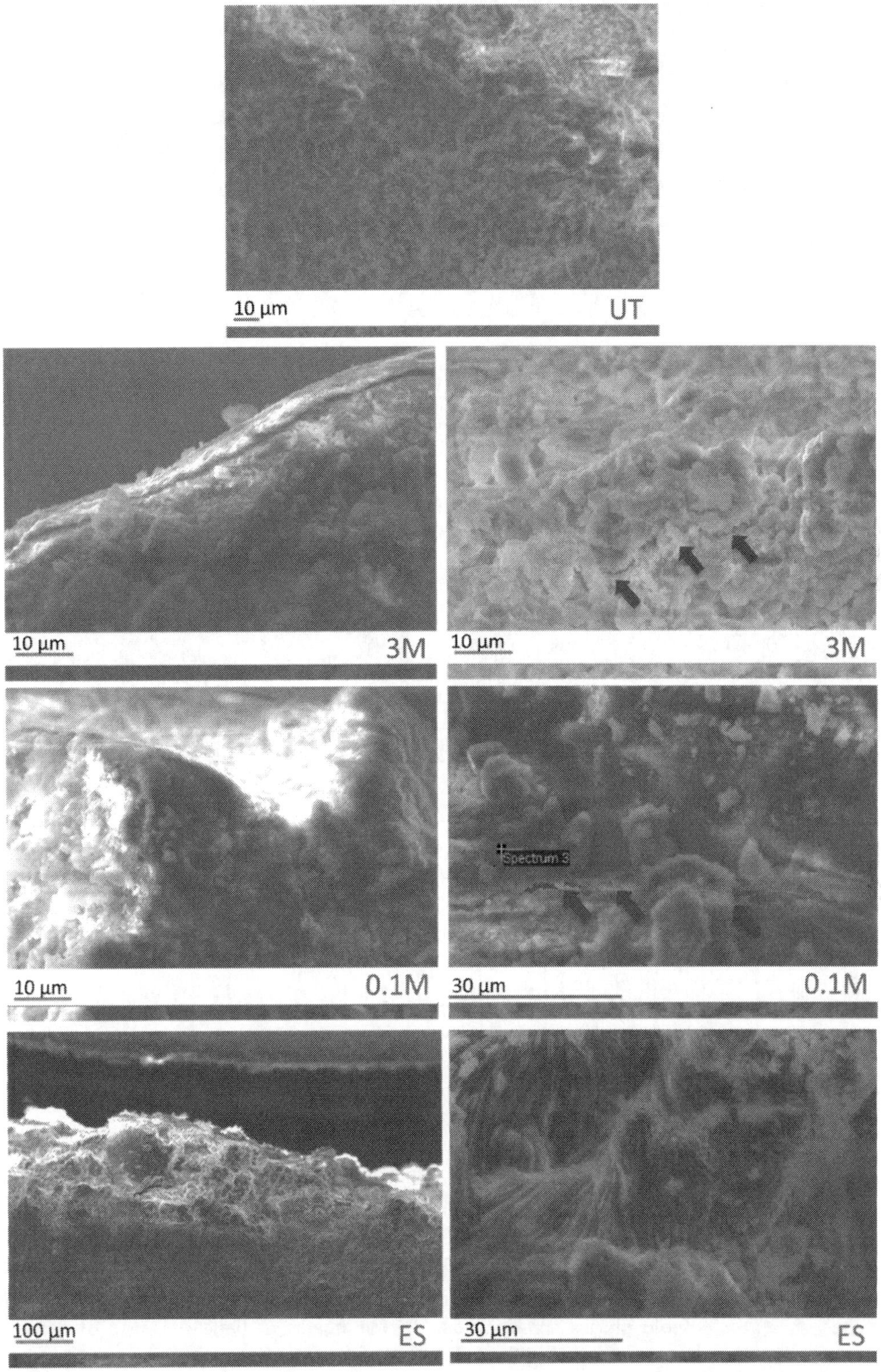

Figure 1. SEM images of cross sections of treated samples, at the surface (left column) and at a depth ~1 cm from the treated surface (right column). Red arrows indicate cracks bridged by HAP.

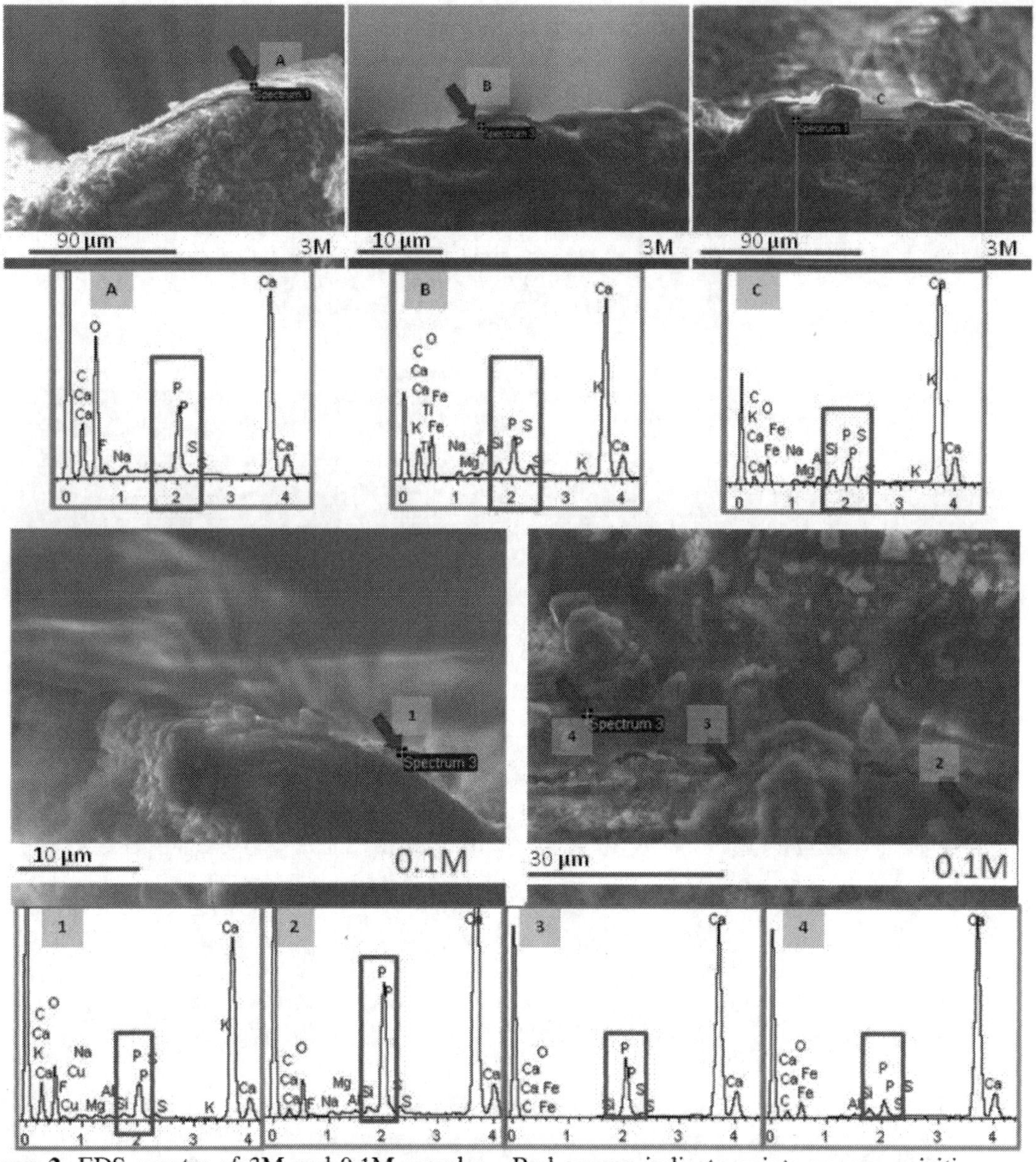

Figure 2: EDS spectra of 3M and 0.1M samples. Red arrows indicate points were acquisition was performed. As can be seen by the spectra, abundant formation of phosphate phases can be detected on the exterior surface and around cracks caused by weathering.

FT-IR spectra are reported in Figure 3. Bands detected are characteristic of calcite (873, 712, 472 cm^{-1}) [28,29] and kaolinite (1035, 525, 470 cm^{-1}) [29,30] of Globigerina limestone, sodium sulfate (1138, 1110 cm^{-1}) and either hydroxyapatite (1030, 600, 560 cm^{-1}) [28,31] or ethyl silicate (1080, 1001 cm^{-1}) [32-34]. FT-IR data indicate that all phosphate phases, formed as a result of 0.1M and 3M treatments and observed by SEM/EDS, correspond to HAP. No bands characteristic of other calcium phosphates phases or soluble phases are detected. This indicates that the presence of salts did not interfere with HAP nucleation by causing the formation of less stable compounds, which would have been detrimental for the treatment success and could represent an issue for the application of the consolidant on site. This was confirmed by IC results, which indicate that no soluble phosphates were extracted when treated samples were crushed and rinsed with water. Phase formation is more abundant for treatment 3M compared to 0.1M, as already shown by SEM. Samples 3M also exhibit a negligible content of residual sodium sulfate, which was successfully removed by the limewater poultice, together with unreacted DAP. This is confirmed by ion chromatography, indicating a much lower sulfate concentration for samples 3M (0.07 wt% sulfate vs. 0.31 wt% for ES and 0.1M). The possibility of achieving desalination during consolidation can be regarded as a further advantage of the 3M treatment. Residual salt content is greatest in samples ES, for which the main bands are those of sodium sulfate.

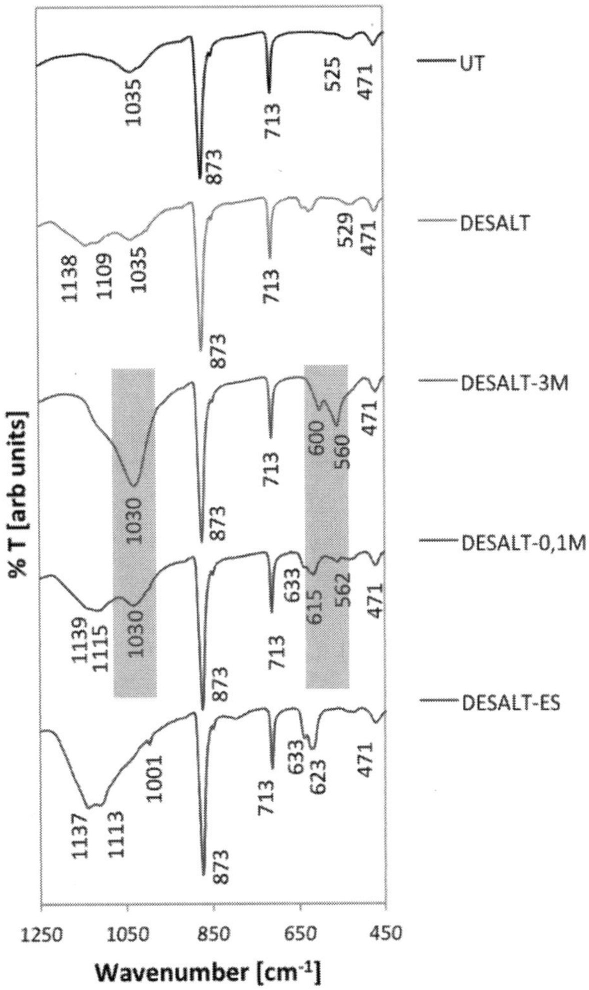

Figure 3. FT-IR spectra of salt contaminated samples. Bands characteristic of HAP (1030, 600, 560 cm^{-1}) are highlighted in blue.

4. Conclusions

The application of two HAP-based treatments was tested on Globigerina limestone contaminated by sodium sulfate. The following conclusions can be derived:

- The presence of salts in the stone does not prevent penetration of the HAP-based consolidants, nor cause formation of soluble phases. In fact, HAP is the only phase to form.
- Because no formation of soluble phases, nor insufficient or inhomogeneous HAP distribution were found, even for high levels of contamination, it can be concluded that the treatments can be safely applied to sodium sulfate-containing substrates.

- 0.1M and 3M treatments result in a good increase in stone tensile strength as they lead to the formation of HAP that binds cracks caused by weathering. Both treatments have comparable efficacy, even though 3M results in more abundant calcium phosphate formation, as treatment 0.1M still guarantees effective crack-sealing.
- ES provides greater, but possibly excessive, increases in Ed and σT. Durability tests will be carried out to better evaluate the long-term performance of all treatments.

References

[1] Sassoni E, Naidu S and Scherer GW 2011 The use of hydroxyapatite as a new inorganic consolidant for damaged carbonate stones *J. Cult. Herit.* **12** 346–355

[2] Franzoni E, Sassoni E and Graziani G 2015 Brushing, poultice or immersion? Role of the application technique on the performance of a novel hydroxyapatite-based consolidating treatment for limestone *J. Cult. Herit.* **16** 173–184.

[3] Sassoni E, Graziani G and Franzoni E 2016 An innovative phosphate-based consolidant for limestone. Part 1: effectiveness and compatibility in comparison with ethyl silicate *Constr. Build. Mater.* **102** 918-930

[4] Possenti E, Colombo C, Bersani D, Bertasa M, Botteon A, Conti C, Lottici PP and Realini M 2016 New insight on the interaction of diammonium hydrogen phosphate conservation treatment with carbonatic substrates: A multi-analytical approach *Microchem. J.* **127** 79-86

[5] Sassoni E, Graziani G and Franzoni E 2016 An innovative phosphate-based consolidant for limestone. Part 2. Durability in comparison with ethyl silicate *Constr. Build. Mater.* **102** 931-942

[6] Graziani G, Sassoni E and Franzoni E Hydroxyapatite-based treatments application to salt-bearing Porous limestones: A study on sodium sulphate-contaminated Lecce Stone, Fourth International Conference on Salt Weathering of Building Stones and Sculptures, Potsdam (G), 20 -22 September 2017

[7] Matteini M, Rescic S, Fratini F. Botticelli G 2011 Ammonium phosphates as consolidating agents for carbonatic stone materials used in architecture and cultural heritage: Preliminary research *Int. J. Archit. Herit.* **5** 717-736

[8] Yang F, Zhang B, Liu Y, Guofeng W, Zhang H, Cheng W, Xu Z 2011 Biomimic conservation of weathered calcareous stones by apatite *New J. Chem.* **35** 887-892

[9] Ma X, Balonis M, Pasco H, M. Toumazou, Counts D, Kakoulli I 2017 Evaluation of hydroxyapatite effects for the consolidation of a Hellenistic-Roman rock-cut chamber tomb at Athienou-Malloura in Cyprus *Constr. Build. Mater.* **150** 333–344

[10] Barriuso BC, Botticelli G, Cuzman OA, Osticioli I, Tiano P, Matteini M 2017 Conservation of calcareous stone monuments: Screening different diammonium phosphate based formulations for countering phototrophic colonization *J. Cult. Herit.* **27** 97-106

[11] Snethlage R, Gruber C, Tucic V, Wendler E 2008 Transforming gypsum into calcium phosphate - a better way to preserve lime paint layers on natural stone? *Proc. Int. Symp. Stone consolidation in cultural heritage* 1-13, eds. J. Delgado Rodrigues and J. Manuel Mimoso ISBN 978-972-49-2135-8

[12] Balonis-Sant M, Ma X, Kakoulli I 2013 Preliminary Results on Biomimetic Methods Based on Soluble Ammonium Phosphate Precursors for the Consolidation of Archaeological Wall Paintings, *Archaeological Chemistry VIII. Washington DC: American Chemical Society, ACS Symposium Series* 420-47

[13] Sassoni E, Graziani G and Franzoni E 2015 Repair of sugaring marble by ammonium phosphate: Comparison with ethyl silicate and ammonium oxalate and pilot application to historic artifact *Mater. Design* **88** 1145-57

[14] Dorozhkin SV 2012 Biphasic, triphasic and multiphasic calcium orthophosphates *Acta Biomater.* 8 963–977

[15] Dorozhkin SV 2011 Calcium orthophosphates *Biomatter.* **12** 121–164

[16] Graziani G, Sassoni E, Franzoni E and Scherer GW 2016 Hydroxyapatite coatings for marble

protection: Optimization of calcite covering and acid resistance *Appl. Surf. Sci.* **368** 241-57

[17] Cardell C, Delalieux F and Roumpopoulos 2003 K Salt-induced decay in calcareous stone monuments and buildings in a marine environment in SW France *Constr. Build. Mater.* **17**

[18] Cardell C, Benavente D and Rodríguez-Gordillo J 2008 Weathering of limestone building material by mixed sulfate solutions. Characterization of stone microstructure, reaction products and decay forms *Mater. Charact.* **59** 1371-1385

[19] Flatt RJ 2002 Salt damage in porous materials: how high supersaturations are generated *J.Cryst. Growth* **242** 435-454,

[20] Tsui N, Flatt RJ and Scherer GW 2003 Crystallization damage by sodium sulfate *J. Cult. Herit.* **4** 109-115,

[21] Benavente D, Garcı´a del Cura MA, Bernabéu A and Ordóñez S 2001 Quantification of salt weathering in porous stones using an experimental continuous partial immersion method *Eng. Geol.* **59** 313-325

[22] Benavente D, Garcı´a del Cura MA, Garcı´a del Cura J, Sánchez-Moral S and Ordóñez S 2004 Role of pore structure in salt crystallisation in unsaturated porous stone *J.Cryst. Growth* **320** 532-544

[23] Cassar J 2002 Deterioration of the Globigerina limestone of the Maltese Islands *Geol. Soc. Spec. Publ.* **205** 33-49

[24] Sassoni E, Graziani G, Franzoni E and Scherer GW 2018 Calcium phosphate coatings for marble conservation: Influence of ethanol and isopropanol addition to the precipitation medium and performance *Corros. Sci.*, DOI: 10.1016/j.corsci.2018.03.019

[25] Sassoni E, Graziani G, Ridolfi G, Bignozzi MC, Franzoni E 2017 Thermal behavior of Carrara marble after consolidation by ammonium phosphate, ammonium oxalate and ethyl silicate, Mater. Design 120 345-353

[26] Maravelaki-Kalaitzaki P, Kallithrakas-Kontos N, Korakaki D, Agioutantis Z and Maurigiannakis S 2006 Evaluation of silicon-based strengthening agents on porous limestones *Prog. Org. Coat.* **57** 140–8.

[27] Maravelaki-Kalaitzaki P, Kallithrakas-Kontos N, Agioutantis Z, Maurigiannakis S and Korakaki D 2008 A comparative study of porous limestones treated with siliconbasedstrengthening agents *Prog. Org. Coat.* **62** 49–60.

[28] Tao J 2013 FTIR and Raman studies of structure and bonding in mineral andorganic–mineral composites *Method. Enzymol.* **532** 533–556

[29] Gunasekaran S, Anbalagan G and Pandi S 2006 Raman and infrared spectra of carbonates of calcite structure *J. Raman Spectrosc.* **37** 892-899

[30] Saikia BJ and Parthasarathy G 2010 Fourier Transform Infrared Spectroscopic Characterization of Kaolinite from Assam and Meghalaya, Northeastern India *J. Mod. Phys.* **1** 206-210S.

[31] Koutsopoulos S 2002 Synthesis and characterization of hydroxyapatite crystals: areview study on the analytical methods *J. Biomed. Mater. Res.* **62** 600–612

[32] Franzoni E, Graziani G and Sassoni E 2015 TEOS-based treatments for stone consolidation: acceleration of hydrolysis–condensation reactions by poulticing *J. Sol-Gel Sci. Technol.* **74** 398

[33] Xu F and Li D 2013 Effect of the addition of hydroxyl-terminated polydimethylsiloxane to TEOS-based stone protective materials *J. Sol-Gel Sci. Technol.* **65** 212–219

[34] Rubio F, Rubio J and Oteo JL 1998 A FT-IR study of the hydrolysis of tetraethylortosilicate (TEOS) *Spectrosc. Lett.* **3** 199–219

IOP Conf. Series: Materials Science and Engineering **364** (2018) 012083 doi:10.1088/1757-899X/364/1/012083

On the generation of numerical models from point clouds for the analysis of damaged Cultural Heritage

Gabriele Bitelli[1], Giovanni Castellazzi[1], Antonio Maria D'Altri[1], Stefano de Miranda[1], Alessandro Lambertini[1] and Ilenia Selvaggi[1]
[1] Department of Civil, Chemical, Environmental, and Materials Engineering (DICAM), University of Bologna, Viale Risorgimento 2, Bologna 40136, Italy

E-mail: (gabriele.bitelli, giovanni.castellazzi, antoniomaria.daltri2, stefano.demiranda, alessandro.lambertini, ilenia.selvaggi2)@unibo.it

Abstract. Nowadays, a growing necessity for preservation and documentation of the Cultural Heritage is related to the precise 3D surveys as the input for multidisciplinary analysis. An efficient tool that meets the aforementioned requirement is the HBIM approach, linked to numerical models for structural analysis. In this context, the innovative CLOUD2FEM procedure, which semi-automatically generates numerical models from point clouds, has been applied to the surveyed point clouds – obtained by laser scanning – in a complex case study of a damaged building: the San Felice sul Panaro Fortress (Italy), hit by a severe earthquake in 2012. The numerical model generated is employed in the framework of advanced seismic nonlinear analysis. The promising results show the effectiveness of the method.

Keywords: *Cultural Heritage, point cloud, geometric modelling, HBIM, structural analysis, FEM*

1. Introduction

The field of documentation and conservation of Cultural Heritage, in particular of architectural interest, is experiencing a strong innovation in these years [1], linked primarily to the digitisation of information. This topic can be found above all in the phase of data acquisition and integrated data management.
With regard to the first, thanks to the rapid development of new techniques and technologies, first of all terrestrial laser scanning (TLS) and new methods of close-range photogrammetry (CRP), the geometry of objects can be described in a complete and very detailed way through high density point clouds. On the other hand, this new type of data, which was normally not available until twenty years ago, sets the basis for creating new approaches to the management of data and objects, not only in terms of documentation but also envisaging an integrated knowledge management, bringing great flexibility of use and new application possibilities.
An obvious example is given by the possibility of being able to derive from point clouds the necessary data for the management of the asset and its study in structural terms. In the first area, interest is growing in applying a Building Information Modelling (BIM) approach to existing assets, which is described here as Historical Building Information Modelling (HBIM). HBIM can be described as a multidisciplinary system composed by parametric library parts mapped onto the point cloud, in order to allow the modelling of complex or also irregular shapes that characterise historic buildings [2]. While BIM approach can be considered consolidated, there are still open issues and limitation in HBIM [3]. Nevertheless, various case studies are devoted to identifying best practises for the use of BIM tools in

ancient context [4]. Furthermore, recent methodologies support the creation of more detailed structural analyses based on objective high-precision data, thereby increasing the reliability of the results [5].
The connection between Geomatics, the modern scientific discipline that deals with surveying with the most recent digital methodologies, and the sciences that deal with structural verification and control, has become tighter. These modern developments can stimulate new collaborative and interdisciplinary contributions towards a better knowledge and management of the assets, especially when dealing with ancient buildings which deserve and require special attention.
This paper presents some considerations derived from such a scientific collaboration, specifically devoted to compute a complete 3D model useful for a Finite Element Method (FEM) analysis.

1.1. Case of study and historical information

The case of study, here presented and discussed, is based on *San Felice sul Panaro Fortress* (Latitude: 44.8388; Longitude: 11.1412), the symbol of the homonymous city (Figure 1) in the province of Modena (Italy), that was hit by a severe earthquake in 2012. The historic fortress is particularly interesting for the intrinsic complexity of the building and this peculiarity has influenced all the phases of the survey and the following data processing. The final goal is to obtain a complete model with high precision and geometric detail.

Figure 1. Geographic location of the case study, from left to right: Emilia-Romagna region overview (satellite imagery), San Felice sul Panaro Fortress overview (aerial imagery AGEA, 2011).

On 20 May 2012 San Felice sul Panaro was hit by the Emilia earthquake with a magnitude peak of Mw=5.86 and an epicentre located between *Finale Emilia* and *San Felice sul Panaro* municipalities. Nine days later another magnitude peak of Mw=5.66 followed. Further minor peaks were recorded in the following days. [6]
The roofs of the different towers have suffered a considerable damage that has led, in some cases, to the collapse of the roof itself. Furthermore, the external perimeter walls have suffered a serious decay and, in some cases, partial collapses. Also, the interior rooms have been damaged by numerous collapses, involving their walls and their vaults. As a first measure of response to the earthquake, the Municipality of San Felice sul Panaro installed large plastic tarpaulins to prevent infiltration of rainwater into the towers with collapsed roof and all the cracks have been filled with mortar. In the following months, steps were taken to ensure safety by means of interventions aimed at preventing further deterioration [7]. Debris, rubble and the any unsafe elements have been subsequently removed as shown in Figure 2 in order to avoid further damages.

Figure 2. Perspective view of the North side, from left to right: before the earthquake, after the first event, condition at the epoch of survey.

IOP Conf. Series: Materials Science and Engineering **364** (2018) 012083 doi:10.1088/1757-899X/364/1/012083

2. Geomatic techniques

2.1. Data collection and data processing

A survey of the building was planned through the use of geomatic techniques, mainly TLS, CRP and total station for the alignment network. The Fortress had suffered numerous and serious damages and at the epoch of the survey it was in a state of heavy degradation. The survey, performed by Abacus srl, was aimed at obtaining a detailed representation of the condition of the building and its output is a dense point cloud. All the metric information have been gathered, georeferenced and linked in a relation within a database towards a BIM system, from which it will be possible to derive subsequent further results as necessary. In this context a BIM approach is a convenient and efficient tool for the storage of multidisciplinary data, which can be used as a support for restoration procedure or as collection of the actual condition. The geomatic survey resulted in 163 different point clouds from different scanning positions, composed by over 40 million points (Figure 3).

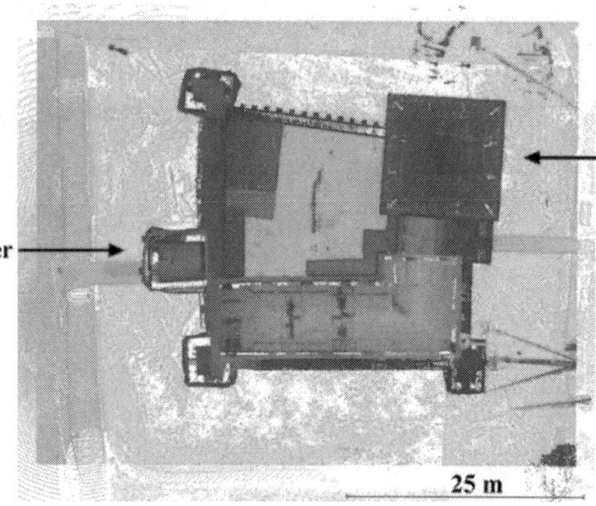

Figure 3. Initial raw point cloud of the entire Fortress, gradient colour based on elevation, highlighting the principal elements object of this work.

In order to obtain a complete and coherent geomatic model, targets (as visible in Figure 4) were placed on both the interior and exterior of the whole building, acquiring data for each room and each façade. The point cloud is initially reduced to 3.2 million with a regular spatial sampling of 5 centimetres in order to obtain a model that could be processed without losing any relevant detail for further analysis. Moreover, any elements outside the surveyed perimeter of the Fortress was manually removed from the point cloud in a brief cleaning phase. The final dataset is then composed by only 1.9 million points.

2.2. Innovative approach on critical issues

The CLOUD2FEM [8] procedure has been applied to transform the point cloud dataset into a model suitable for FEM analysis. This allows to transform through a semi-automatic workflow any three-dimensionally surveyed historical monumental buildings into a sequence of two-dimensional slices of containing all the relevant geometrical data in a pixel grid format. Finally, the aligned pixel grids are stacked into a voxel model. Therefore, the output of the geomatic techniques becomes the input of the structural analysis.

The mentioned procedure is now stressed analysing the whole Fortress, discretized through a sequence of slices perpendicular to the Z-axis (horizontal). Critical aspects and peculiarities emerge from this analysis and offer suggestion for further investigations, regarding both the geometric and structural perspective, as the Mastio, the tallest and most stressed tower, composed by six floors [8] and the North Tower, involved by a partial collapse of the perimeter walls. Furthermore, an extension and adaption of

the CLOUD2FEM procedure has been applied to the detailed analysis of the different floors and their vaults at a higher resolution [9].
The TLS survey difficulties derived from the presence of debris, discontinuities, furniture, stairs and railings. The point cloud was also predictably affected by these problems, due to foreign elements not belonging to walls, floors and structural elements in general terms. In a traditional geometric approach, the creation of a complete and watertight model is time-consuming. Nevertheless, the structural analysis needs a filled model and the proposed method guarantees its creation despite the critical issues.
An explanatory example is given by the analysis of the Giulio II Hall, in the Mastio tower. This room is characterized by a great amount of debris as a consequence of the earthquake. Its vault is also compromised by a complex crack pattern as show in the picture captured inside the room (Figure 4). It is possible to observe the details captured from the cross-section of the point cloud, where the real geometry of the floor is not directly visible and therefore measurable. This happens not only in Giulio II Hall, but also in several other spaces due to different causes: presence of furniture or stairs (Figure 5).

Figure 4. Picture from the interior of the Giulio II Hall in the Mastio.

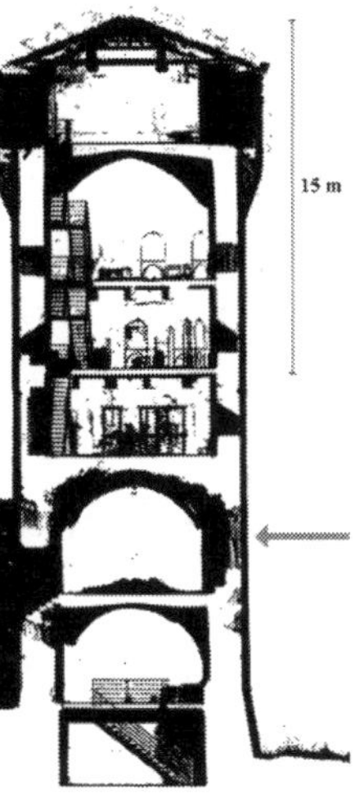

Figure 5. Cross-section of the Mastio, with Giulio II Hall highlighted.

Concerning the North Tower, it has been possible to measure the rotation of the perimeter walls along the whole elevation of the tower. In particular, the study is then focused on the vault at the second level, deformed due to the rotation of the walls. It has been possible to measure the displacements between the theoretical geometric model of the original vault, starting from the real model obtained from the surveyed point cloud.
By comparison of the two models it is clear the displacement between the surveyed vault and its geometrical model. Furthermore, on both models isolines have been overlapped at a constant step. In particular, on the peak of the surveyed model an irregular trend can be clearly observed (Figure 6).
Following the results of this analysis, the distance between subsequent slices in correspondence of each transition between floors was decreased in order to increase the resolution of the final model, down to 0.1 meter (Figure 7). This new proposed approach allows a more accurate geometrical analysis for the vaults and also the structural analysis will benefit from the improved accuracy [10].

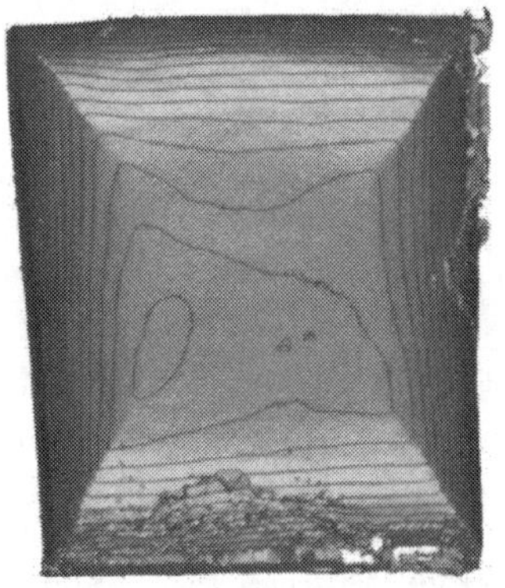

Figure 6. Mesh of the vault with isolines.

Figure 7. Polygonal model of the North Tower. Studied vault highlighted in red.

3. FE modelling and analysis

3.1. Generation of the FE model

Following the CLOUD2FEM procedure [8], the pixelized two-dimensional slices, which contain all the significant geometrical data, are stacked to generate a voxel model. Figure 8 shows an example of stacking of the slices. Successively, the voxels are automatically transformed into solid hexahedral FEs. Consequently, the output of the CLOUD2FEM procedure is a FE solid model ready to be used in the framework of numerical analysis.

In the case at hand, the FE model generated refers to the damaged condition of the structure, see Figure 8. Therefore, this model could be used to investigate the current condition of a damaged building, which can be beneficial for the knowledge of its actual structural behaviour and, consequently, for the design of repairing and strengthening interventions. To this aim, the updating of the FE model by means of ambient vibration measurements could improve the reliability of the model. A vibration-based FE model updating procedure for historical masonry structures which have suffered severe damage due to seismic events has been recently developed in [11].

However, numerical models are often employed to understand the causes of the occurred damage [12]. Thereby, the undamaged configuration of the structure (before the earthquake) appears to be in the engineers' interest.

The rational discretization of the model generated through the CLOUD2FEM procedure also allows to simply process the slices to obtain the model of the undamaged structure, see Figure 9. Indeed, by manually processing the 3D voxel model slice by slice, the collapsed portions of the structure can be added in the voxel grid. Additionally, by following the same process, the material properties can be defined along the whole model domain (Figure 9). This operation, although manual and, in general, time consuming, allows to obtain a very accurate description of the materials distribution in the structure (e.g. different kinds, textures and degradation states of masonry, etc). In Figure 9, the materials of the undamaged model have been characterized in according to the material definition collected in [13].

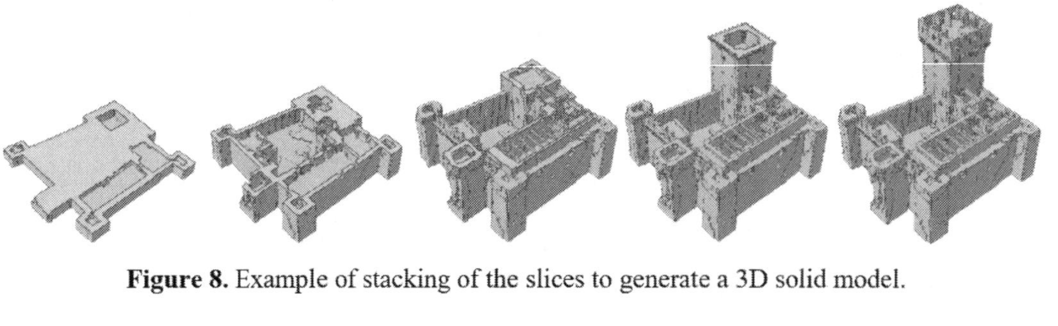

Figure 8. Example of stacking of the slices to generate a 3D solid model.

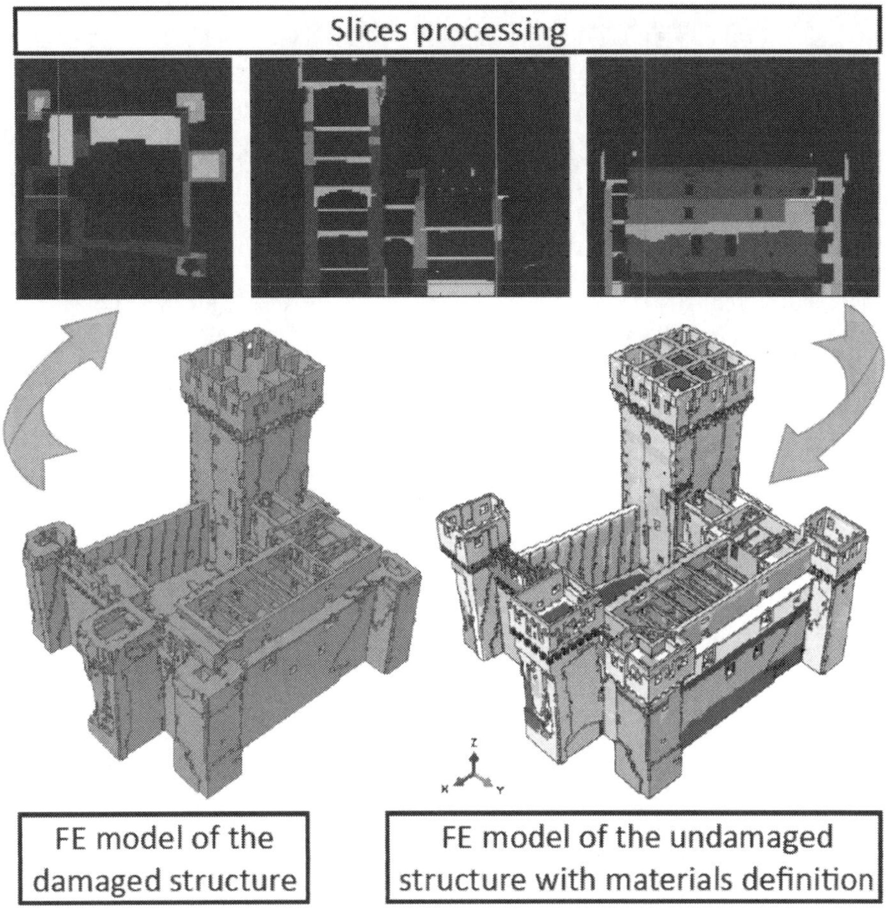

Figure 9. FE models generated from the point cloud: example of slices processing to obtain the model of the undamaged structure, as well as the material definition.

3.2. FE nonlinear analysis

To show the effectiveness of the approach, a nonlinear dynamic analysis has been carried out employing the undamaged model of Figure 9. The actual accelerogram, recorded on May 29th, 2012 close to the fortress, has been applied at the base of the structure. Particularly, a plastic-damage constitutive law has been adopted for masonry, following the setting of the mechanical parameters given in [14]. Additionally, a Rayleigh damping model has been used, as adopted in [15].

Figure 10 shows the accelerogram applied at the base of the structure (top) and the nonlinear dynamic analysis results (bottom), in terms of tensile damage contour plots at subsequent instants. The obtained cracking pattern, predicted by the numerical model (Figure 10), is in good agreement with the damages experienced by the actual structure: among all consider for instance the severe damage computed for the surrounding towers, localized at the upper part (Figure 10), and the experienced damage that ended with the collapse of the towers upper part (Figure 2).

IOP Conf. Series: Materials Science and Engineering **364** (2018) 012083 doi:10.1088/1757-899X/364/1/012083

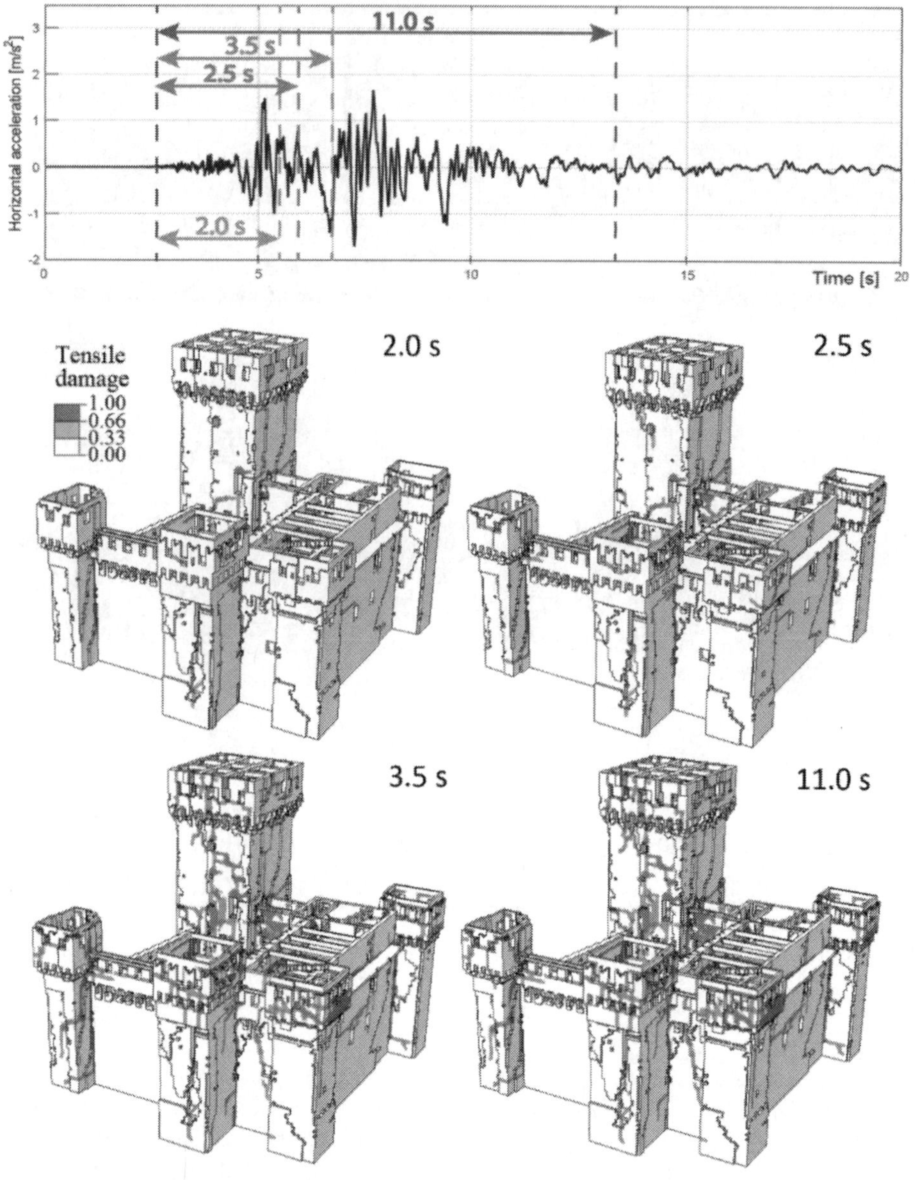

Figure 10. Nonlinear dynamic analysis: accelerogram applied at the base of the structure (top) and tensile damage contour plots at subsequent instants (bottom).

4. Conclusions

The innovative CLOUD2FEM procedure has been applied to a whole historical building damaged from an earthquake in the 2012: the San Felice sul Panaro Fortress. Several surveyed point clouds are an efficient support for multidisciplinary studies aimed to the conservation of the Cultural Heritage, towards a HBIM approach. The precise 3D model was completely analysed. Moreover, the model has been exploited in detail to investigate all the peculiar elements of the building from different perspectives: geomatic and structural. Complex elements, such as vaults or damaged structures, have been further investigated in a comprehensive geometric analysis. Finally, the 3D numerical model semi-automatically generated has been employed in the framework of advanced seismic nonlinear analysis. The promising results showed the effectiveness of the method.

References

[1] Bitelli G, Balletti C, Brumana R, Barazzetti L, D'Urso M G, Rinaudo F and Tucci G 2017 Metric Documentation of Cultural Heritage: Research Directions from the Italian GAMHer Project *ISPRS - Int. Arch. Photogramm. Remote Sens. Spat. Inf. Sci.* **XLII-2/W5** 83-90

[2] Murphy M, McGovern E and Pavia S 2009 Historic building information modelling (HBIM) *Struct. Surv.* **27** 311–27

[3] Dore C and Murphy M 2017 Current State of the Art Historic Building Information Modelling *ISPRS - Int. Arch. Photogramm. Remote Sens. Spat. Inf. Sci.* **XLII-2/W5** 185–92

[4] Bitelli G, Dellapasqua M, Girelli V A, Sanchini E and Tini M A 2017 3D Geomatics techniques for an integrated approach to cultural heritage knowledge: The case of San Michele in Acerboli's church in Santarcangelo di Romagna *ISPRS - Int. Arch. Photogramm. Remote Sens. Spat. Inf. Sci. - ISPRS Arch.* **42** 291–6

[5] Barazzetti L, Banfi F, Brumana R, Gusmeroli G, Previtali M and Schiantarelli G 2015 Cloud-to-BIM-to-FEM: Structural simulation with accurate historic BIM from laser scans *Simul. Model. Pract. Theory* **57** 71–87

[6] Scognamiglio L, Margheriti L, Mele F M, Tinti E, Bono A, De Gori P, Lauciani V, Lucente F P, Mandiello A G, Marcocci C, Mazza S, Pintore S and Quintiliani M 2012 The 2012 Pianura Padana Emiliana seimic sequence: Locations, moment tensors and magnitudes *Ann. Geophys.* **55** 549–59

[7] Cattari S, Abbati S D, Ferretti D, Lagomarsino S, Ottonelli D and Tralli A 2014 Damage assessment of fortresses after the 2012 Emilia earthquake (Italy) *Bull. Earthq. Eng.* **12** 2333–65

[8] Castellazzi G, D'Altri A, Bitelli G, Selvaggi I and Lambertini A 2015 From laser scanning to finite element analysis of complex buildings by using a semi-automatic procedure *Sensors* **15** 18360–80

[9] Bitelli G, Castellazzi G, D'Altri A M, de Miranda S, Lambertini A and Selvaggi I 2016 Automated voxel model from point clouds for structural analysis of cultural heritage *ISPRS - Int. Arch. Photogramm. Remote Sens. Spat. Inf. Sci.* **XLI-B5** 191–7

[10] Selvaggi I 2017 Surveying and three-dimensional modeling for preservation and structural analysis of Cultural Heritage *PhD thesis in Civil, Chemical, Environmental and Materials Engineering* (University of Bologna) XXIX cycle

[11] Bassoli E, Vincenzi L, D'Altri A M, de Miranda S, Marianna F and Castellazzi G 2018 Ambient vibration-based Finite Element model updating of an earthquake-damaged masonry tower *Structural Control and Health Monitoring* (in press)

[12] Castellazzi G, D'Altri A M, de Miranda S, Chiozzi A and Tralli A 2018 Numerical insights on the seismic behavior of a non-isolated historical masonry tower *Bull. Earthq. Eng.* **16** 933–961

[13] Ferrari L and Goldoni G 2014 Behind the sign. Inside and Outside the San Felice sul Panaro Fortress restoration work *Master thesis* (University of Parma)

[14] Castellazzi G, D'Altri A M, de Miranda S, and Ubertini F 2017 An innovative numerical modeling strategy for the structural analysis of historical monumental buildings *Engineering Structures* **132** 229–248

[15] D'Altri A M, Castellazzi G, de Miranda S, and Tralli A 2017 Seismic-induced damage in historical masonry vaults: A case-study in the 2012 Emilia earthquake-stricken area *Journal of Building Engineering* **13** 224–243

Acknowledgments

The authors would like to thank the municipality of San Felice sul Panaro (MO), Cineca (www.cineca.it), Abacus s.a.s. (www.arcoabacus.it), Jacopo Ponti. This work has been partially supported by the GAMHer project: Geomatics Data Acquisition and Management for Landscape and Built Heritage in a European Perspective, PRIN: Progetti di Ricerca di Rilevante Interesse Nazionale – Bando 2015, Prot. 2015HJLS7E.

IOP Conf. Series: Materials Science and Engineering **364** (2018) 012084 doi:10.1088/1757-899X/364/1/012084

Duality between BIM and GIS: an example related to the medieval bridge Azzone Visconti

L Barazzetti

Department of Architecture, Built environment and Construction engineering (ABC), Politecnico di Milano, via Ponzio 31, Milan, Italy

Email: luigi.barazzetti@polimi.it

Abstract. The paper presents the work carried out on the medieval bridge Azzone Visconti in Lecco, Italy. The project required the use of both BIM and GIS technologies to assess the loading capacity of the bridge. First, a detailed geometric survey was carried out with laser scanning and photogrammetric techniques. The aim was the generation of a BIM. Displacements were then measured via geometric levelling during loading tests carried out in different days. Such data were integrated and stored in the BIM. Finally, the simulation of traffic conditions was used to investigate the usability of the BIM for a geographic area around the bridge. Results highlighted that the BIM used in the first stage of the work had to be replaced with another integrated BIM-GIS solution more suitable for analysis based on geospatial data.

1. Introduction

This paper presents the different geospatial data collected and processed during the project related to the medieval bridge Azzone Visconti, in Lecco (Italy). A multi-disciplinary team of specialists carried out several analysis to assess the loading capacity of the bridge with both experimental and numerical analyses. More details are described in [1].

The work described in this paper wants to underline the contribution of Geomatics techniques (surveying, modelling monitoring, etc.) and the different types of geospatial data acquired and processed. Different instruments and techniques were integrated to (i) measure very small displacements during loading tests (precision of about ±0.1 mm), (ii) create a detailed BIM [2] and drawings at a scale 1:50 (precision of about ±2-5 mm) for condition assessment and finite element analysis, and (iii) produce a new integrated BIM-GIS model for advanced traffic simulations (metric precision of about ±200-400 mm). In the case of analysis (i) and (ii), a Cartesian reference system was established, which became suitable for most analysis at the local scale (the scale of a construction). The integrated BIM-GIS model required a cartographic reference system (UTMWGS84-ETRF89) and preliminary georeferencing via GNSS of specific information collected in the local reference system. Additional existing geospatial data, such as DTM, shape files, and orthophotos, were added as well.

A clear definition of the work in complete and integrated BIM-GIS projects is still not very clear at the present. In some cases, processing is carried out in independent software with a continuous import/export, which usually results in loss of information. Another approach consists in a BIM imported in a GIS environment, which is also able to perform basic BIM processing, and vice versa.

Nowadays, several authors have described different experiences for a variety of projects where both technologies were discussed (e.g. [3-15], among the others).
It could be said that Geographic Information System (GIS) and Building Information Modelling (BIM) can be very useful when the work requires different scales. For instance, BIM sw are mainly designed for the scale of the building, and GIS for the district, city, regional, national and international scales. On the other hand, such distinction is not always valid since many examples with GIS sw were carried out also for small objects. Another distinction could be based not only on the type of data, but also on the type of operations. GIS engines have advanced functions for geospatial analysis, whereas BIM sw have advanced parametric modeling tools. A common feature of both technologies is the database connected to the graphic representation of specific objects.
The previous similarities and differences are just an overall indicator to understand why BIM and GIS could be intended as different technologies. It is the author's opinion that (at the present), BIM and GIS sw have to be simultaneously exploited to solve related but different problems. An integrated approach in data processing requires a continuous transition through specific software packages. It could be said that BIM and GIS can be used together as they are complementary.

2. The first model of the bridge: working in a local reference system

The project started with the acquisition of photogrammetric and laser scanning data to create a BIM of the bridge in Autodesk Revit. In addition, a set of 2D measured drawings (plans, sections, elevations) at the scale 1:50 was requested. The survey was carried out with these basic requirements in mind. A geodetic network was measured with a Leica TS30. The network was adjusted via least squares obtaining an average point precision of about ±1.5 mm. Points of the network provide the local reference system used for the generation of the model.
A set of 77 scans was acquired with a Faro Focus 3D laser scanner. Scans were registered in the local system with an average accuracy of about ±3 mm. Then, the project was integrated with several photogrammetric blocks. Most instruments and techniques allowed us to achieve a metric accuracy better than ±5 mm, which is more than sufficient for the creation of deliverables at the scale 1:50.

The creation of the BIM was carried out dividing the bridge into its constructive elements. Specific objects were generated and assembled. They include stones, ashlars, sprinters, decks, floor beams and stringers. The complexity of the bridge, with irregular shapes not available in existing object libraries, required the development of specific procedures for parametric modelling. This was necessary to overcome the lack of commercial software able to provide detailed geometric reconstructions which preserve the level of detail encapsulated into laser scanning point clouds [16-17]. For this reason, a procedure based on surfaces made up of NURBS curves and NURBS surfaces [18] was used to create new parametric BIM objects. Some images of the model are shown in Fig. 1.

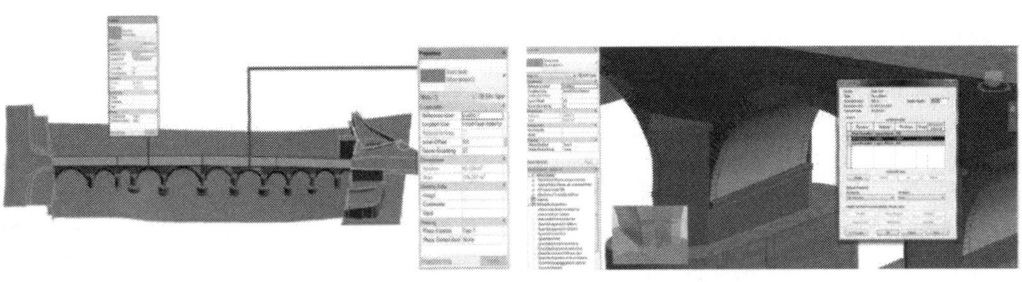

Figure 1. The first BIM in Autodesk Revit.

A condition assessment was carried out using the orthophotos generated via photogrammetry, vector data were instead used to run structural simulations. As mentioned in the introduction, it is not the author's intention to describe such operations, for which the reader is referred to [1]. On the other hand, the work carried out to measure the vertical displacements of the bridge deserves to be mentioned. Vertical movements were measured via geometric levelling, after mounting a set of benchmarks on the bridge. Different loading tests were carried out and least squares adjustment provided the displacements of benchmarks with an average accuracy of about ±0.15 mm (Fig. 2). Specific "benchmark" objects were added to the Revit Model to identify the location of the measured points. Then, graphs and tables were connected to the model, which enriched the existing database.

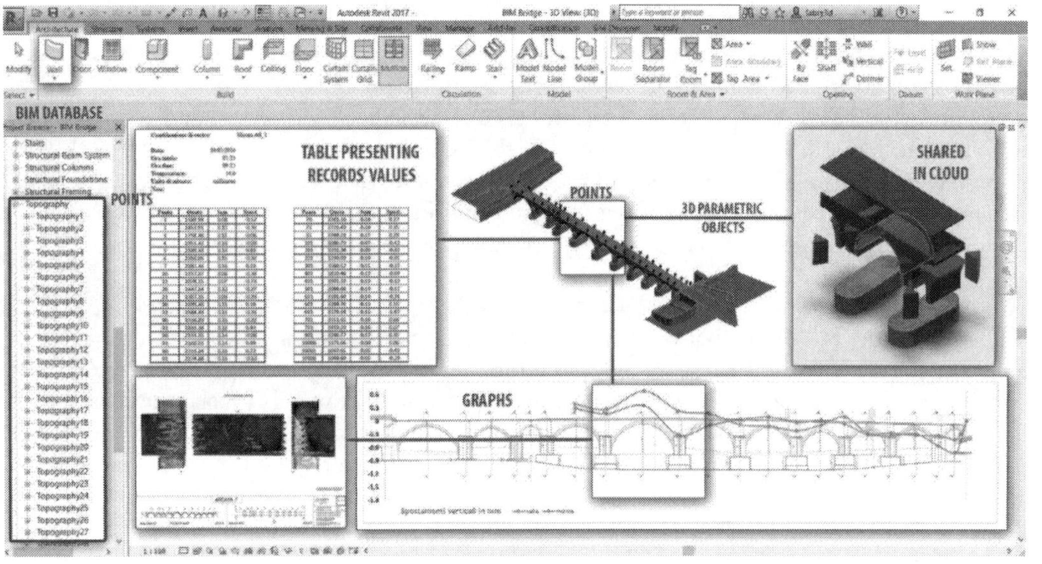

Figure 2. Integration of monitoring data in the BIM.

3. The second model of the bridge, towards BIM-GIS

The second phase of the work was a traffic flow simulation in the area around the bridge, assuming different configurations:

(i) Bridge with 2 lanes;
(ii) Bridge with a single lane;
(iii) Bridge closed.

Configuration (i) reflects the conditions before the beginning of the work. Configuration (ii) is the actual conditions. In fact, after completing the work described in section 2, it was decided to (permanently) close 1 lane. The last configuration (iii) was the case of a limited time period (1 week + another week later), when the bridge was temporarily closed to allow specialists to collect all the technical data to assess the loading capacity.

It is clear that the different configurations have direct impact on the traffic in the area around the bridge. In such case, data analysis must be moved from the local scale of the bridge to a wider area, which includes also the existing road network.

Traffic analysis tools are available in GIS software and could be used to simulate traffic flows at specific hours of the day. On the other hand, we decided to carry out the traffic simulation via BIM to reuse (as much as possible) the data already processed in section 2. The BIM of the bridge as well as the raw data (geodetic network, laser point clouds) were georeferenced in the cartographic system UTM-WGS84 ETRF89 and integrated with additional geospatial data. Georeferencing was carried out with some points measured with GNSS receivers in RTK.

Existing geospatial data include both raster and vector layers from the geospatial database of the province and municipality. An additional LiDAR (aerial) point clouds was also integrated. Such

IOP Conf. Series: Materials Science and Engineering **364** (2018) 012084 doi:10.1088/1757-899X/364/1/012084

dataset was acquired in 2005 using the Optech ALTM3033 altimeter. The theoretical density of the points was 0.2–1 points/m^2 (without taking into account the overlapping area of different strips) while the real density in the considered area is about 1.7-1.8 points/m^2. Such resolution is significantly lower than the point cloud acquired via terrestrial laser scanning, but it provides a coverage for the entire city.

Fig. 3 shows a detail of the vector layers used for traffic simulation: the existing road network, buildings, and water bodies. As can be seen, the bridge was removed from GIS data, since the idea is to replace the corresponding shapefile with the model in Revit.

Figure 3. The different shapefiles used to simulate traffic conditions. The bridge was removed and replaced with the BIM in Revit.

The software used for traffic simulation is Autodesk InfraWorks, which is designed for BIM of infrastructures. It has instruments able to simulate traffic conditions starting from a BIM that encapsulate geospatial data. The road shapefile was imported in Infraworks and projected on the digital terrain model. Additional information about intersections (e.g. stop signs) was added manually. Buildings were also integrated (the height of each building is available in the shapefile) since they provide visibility occlusions during traffic simulations. Buildings are basically represented as boxes. Then, data from OpenStreetMaps were superimposed to the available cartographic layer to check their consistency. Such operations were carried out in QGIS, where a spatial subset in the considered areas was extracted to limit the amount of data. Since the road networks from OpenStreetMaps was consistent with the geospatial database, we decided to use such dataset because it is more recent.

Then, the georeferenced BIM in Revit was added to the InfraWorks project. On the other hand, such BIM was not recognized as a valid bridge object for traffic simulation. A careful simplification would be needed to provide a model usable for this kind of analysis. We therefore decided to create a simplified BIM directly in Infraworks. Such bridge is still built parametrically and is able to interact with other objects such as roads and water bodies. It is curios that the mediaeval bridge is here replaced with a modern one, which proved to be sufficient for a detailed traffic simulation. After setting specific parameters for the simulation (such as speed, number and type of vehicles, period of the day), a numerical estimate of the traffic flow was carried out. Some example images are shown in Fig. 4.

Figure 4. Some images of the traffic simulation in InfraWorks with different lane configurations: a single lane (top) and two lanes (bottom).

4. Considerations and outlooks

This sections tries to compare the work carried out in the previous sections. Two models were generated and handled in different software: Revit and InfraWorks. The first software is surely more suitable for analysis at the local scale of the building, whereas InfraWorks can integrate different geospatial data for projects at a different scale, which could be defined as the scale of the infrastructure.

It is possible to share data between the two software, although specific analyses are feasible only using a distinct workflow. At the same time, data import/export revealed some loss of information.

Fig. 5 shows the different data and techniques for both projects, which are not completely uncorrelated. In the first case, a Cartesian reference system was sufficient for the modelling phase. In the second phase, it is necessary to move to a cartographic reference system. The connection between the two systems is available through some benchmarks measured with total station and GNSS receivers. The transformation used is based on a 2D roto-translation with variation in scale (4 parameters) estimated via least squares, starting from and a set of corresponding points $(X, Y)_{LOCAL} \leftrightarrow$ (East, North). Given the size of the bridge (less than 150 m), transformation of coordinates along the vertical line $Z_{LOCAL} \leftrightarrow H$ was carried out with a simple translation. H is the orthometric elevation value, estimated via GNSS and corrected for geoid undulation.

Some operations on the geospatial data have been carried out using a GIS software. Indeed, InfraWorks lacks some instruments for geospatial data analysis available in GIS packages. InfraWorks is more oriented to infrastructure design, where geospatial data have to be incorporated. It provides instruments to model specific typologies of objects such as roads and railways, which becomes parametric objects and their attributes can be modified by working on the database. Here, the work can start from geospatial datasets in both vector and raster formats. Then, addition information has to be added after importing each file in InfraWorks. For instance, information about the typology has to be assigned. A line shapefile that represents a road has to be identified as a road object. Although this is quite obvious, file typologies have to be defined manually resulting in a time-consuming operation. Then, the software provides different basic geometry for roads (such as highway roads, arterial roads, collector roads or local roads), with specific speed limits. Additional parameters are the number of lane forwards and backward, creation date, lifespan, etc.

IOP Conf. Series: Materials Science and Engineering **364** (2018) 012084 doi:10.1088/1757-899X/364/1/012084

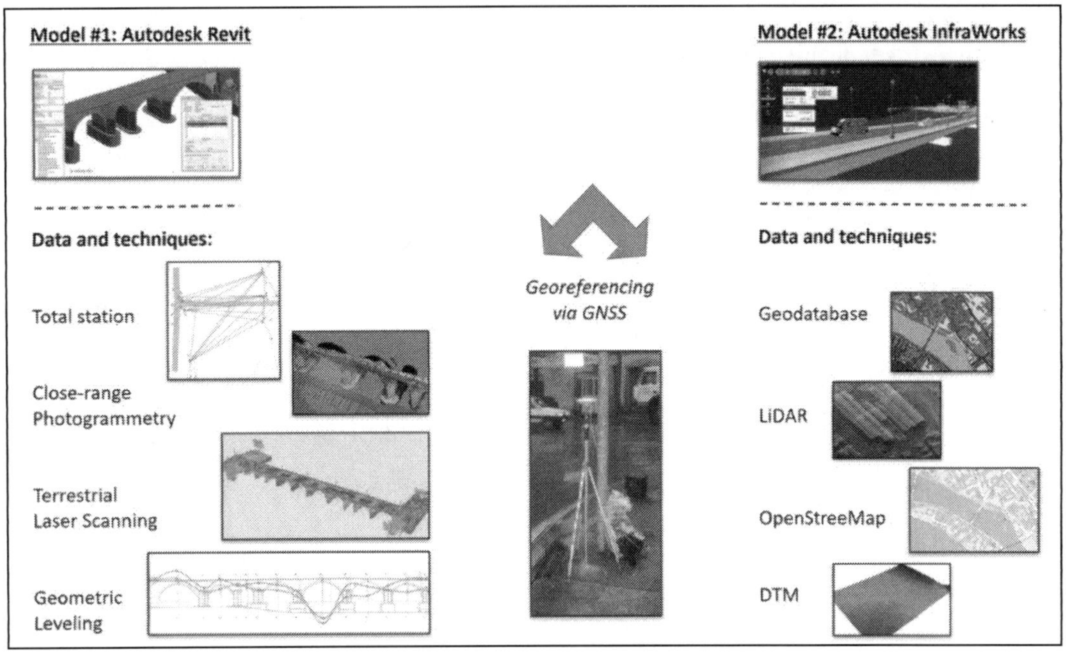

Figure 5. The different data collected and processed in this project.

BIM software like Revit are instead mainly designed to represent the building and its elements. The typical workflow for the generation of a new (as-designed) BIM requires the assembly of predefined objects already subdivided into categories (walls, columns, floors, roofs, windows, doors, ceilings, ramps, etc.). Objects are placed in levels (e.g. foundation, floor, ceiling, roof, etc.), which provide a sort of reference system. Levels can be intended as horizontal planes for the different storeys of the construction. The generation of the BIM of a medieval bridge in Revit was not a simple task because of the lack of parametric objects able to represent the geometry of the bridge. Here, additional tools and functions were developed and used to represent the geometry captured by photogrammetry and laser scanning, as described in [19].

GIS software cannot represent the complexity of the building with this level of detail. A similar consideration is valid also for InfraWorks. It is therefore rather clear that BIM, GIS and BIM-GIS are still complementary solutions, which have to be exploited together. On the other hand, this paper reflects the work achievable with the commercial solution available on the commercial market in 2017. The integration between BIM and GIS is a field of research of primary importance that is developing were fast. For example, Autodesk and Esri have announced a partnership to "put GIS and BIM at the centre of construction projects, driving smarter decisions, optimized designs, accelerated project approvals, reduced cost, creating smarter cities and resilient infrastructure. Building the bridge between GIS and BIM will change the way we plan, design, and build our future" (text copied from http://www.esri.com/landing-pages/autodesk-esri-strategic-alliance). It is therefore expected that substantial changes will be available soon.

5. Conclusion

One of the outcomes of this research was the creation of "two BIM" for the local survey and the work in the cartographic reference system. Two software were used to handle the models: Autodesk Revit and Autodesk InfraWorks. In the first case, the model of the bridge was generated considering the requirements of the project: geometric survey, condition assessment, destructive and non- destructive testing, finite element analysis, estimation of the loading capacity. InfraWorks was instead used to integrate some geospatial data and to run traffic simulations with the bridge in different configurations. Although some data can be integrated, processed, and visualized in both software, specific tools and functions were available only in a specific package. This required a set of import/export operations to take advantages of both packages. At the same time, the use of a "pure GIS software" was mandatory to handle the geospatial data with the typical geospatial tools only available in such packages. In fact, although InfraWorks can exploit geospatial information, the available tools are more oriented to preliminary design of infrastructures. This underlines the dual role between BIM and GIS at the infrastructure level, which makes the simultaneous use of both solutions still mandatory for structure and infrastructure projects based on geographic information.

Acknowledgments
This work has been supported by the GAMHer project: Geomatics Data Acquisition and Management for Landscape and Built Heritage in a European Perspective, PRIN: Progetti di Ricerca di Rilevante Interesse Nazionale – Bando 2015, Prot. 2015HJLS7E.

References
[1] Martinelli, P., Galli, A., Barazzetti, L., Colombo, M., Felicetti, R., Previtali, M., Roncoroni, F., Scola, M., Di Prisco, M., 2018. Bearing capacity assessment of a XIV century arch bridge in Lecco (Italy). International Journal of Architectural Heritage, Volume 12 - Issue 2, pp. 237-256.
[2] Eastman, C., Teicholz, P., Sacks, R., Liston, K., 2008. BIM Handbook - A guide to Building Information Modeling for owners, managers, designers, engineers, and contractors. John Wiley & Sons, Inc.
[3] Amirebrahimi, S., Rajabifard, A., Mendis, P., Ngo, T., 2016. A BIM-GIS integration method in support of the assessment and 3D visualisation of flood damage to a building, Journal of Spatial Science, 61 (2), pp. 317-350.
[4] Baik, A., Yaagoubi, R., Boehm, J., 2015. Integration of Jeddah historical BIM and 3D GIS for documentation and restoration of historical monument. International Archives of the Photogrammetry, Remote Sensing and Spatial Information Sciences - ISPRS Archives, 40 (5W7), pp. 29-34
[5] Borrmann, A., Kolbe, T.H., Donaubauer, A., Steuer, H., Jubierre, J.R., Flurl, M., 2014. Multi-scale geometric-semantic modeling of shield tunnels for GIS and BIM applications. Computer-Aided Civil and Infrastructure Engineering. Article in Press.
[6] Del Giudice, M., Osello, A., Patti, E., 2015. BIM and GIS for district modeling. eWork and eBusiness in Architecture, Engineering and Construction - Proceedings of the 10th European Conference on Product and Process Modelling, ECPPM 2014, pp. 851-854.
[7] El Meouche, R., Rezoug, M., Hijazi, I., 2013. Integrating and managing BIM in gis, software review. International Archives of the Photogrammetry, Remote Sensing and Spatial Information Sciences - ISPRS Archives, XL-2/W2, pp. 31-34.
[8] Fan, D.-K., 2016. Research on the fusion technology of BIM and GIS applied in railway information deployment. Journal of Railway Engineering Society, 33 (10), pp. 106-110.
[9] Kang, T.W., Hong, C.H., 2015. A study on software architecture for effective BIM/GIS-based facility management data integration. Automation in Construction, 54, art. no. 1887, pp. 25-38.
[10] Karan, E.P., Irizarry, J., 2015. Extending BIM interoperability to preconstruction operations using geospatial analyses and semantic web services. Automation in Construction, 53, pp. 1-12.
[11] Liu, X., Wang, X., Wright, G., Cheng, J.C.P., Li, X., Liu, R., 2017. A state-of-the-art review on the integration of Building Information Modeling (BIM) and Geographic Information

IOP Conf. Series: Materials Science and Engineering **364** (2018) 012084 doi:10.1088/1757-899X/364/1/012084

System (GIS). ISPRS International Journal of Geo-Information, 6 (2).

[12] Vilgertshofer, S., Amann, J., Willenborg, B., Borrmann, A., Kolbe, T.H., 2017. Linking BIM and GIS models in infrastructure by example of IFC and CityGML. Congress on Computing in Civil Engineering, Proceedings, pp. 133-140.

[13] Wang, Q.K., Li, P., Xiao, Y.P., Liu, Z.G., 2014. Integration of GIS and BIM in metro construction. Applied Mechanics and Materials, 608-609, pp. 698-702.

[14] Wu, W., Yang, X., Fan, Q., 2014. GIS-BIM based virtual facility energy assessment (VFEA) - Framework development and use case of California State University, Fresno. Computing in Civil and Building Engineering - Proceedings of the 2014 International Conference on Computing in Civil and Building Engineering, pp. 339-346.

[15] Zlatanova, S., Isikdag, U., 2016. The need to integrate BIM and geoinformation. GIM International, 30 (10), pp. 27-29

[16] Barazzetti, L., Banfi, F., Brumana, R., Previtali, M., 2015. Creation of Parametric BIM Objects from Point Clouds Using NURBS. The Photogrammetric Record, Vol. 30(152), pp. 339–362.

[17] Barazzetti, L., 2016. Parametric as-built model generation of complex shapes from point clouds. Advanced Engineering Informatics, 30, pp. 298-311.

[18] Piegl, L. A., Tiller, W., 1999. Computing offsets of NURBS curves and surfaces. Computer-Aided Design, 31(2):147–156.

[19] Banfi, F., Barazzetti, L., Previtali, M., Roncoroni, F., 2017. Historic BIM: a new repository for structural health monitoring. The International Archives of the Photogrammetry, Remote Sensing and Spatial Information Sciences, Volume XLII-5/W1, 2017, 22–24 May 2017, Florence, Italy, 6 pages.

IOP Conf. Series: Materials Science and Engineering **364** (2018) 012085 doi:10.1088/1757-899X/364/1/012085

Advanced procedure for documenting and assessment of Cultural Heritage: from Laser Scanning to Finite Element.

Grazia Tucci[1], Gianni Bartoli[1], Michele Betti[1], Valentina Bonora[1], Mustafa Korumaz[2] and Armağan Güleç Korumaz[2]

[1] Department of Civil and Environmental Engineering (DICEA), University of Florence, Via di Santa Marta 3, Firenze, Italy

[2] Department of Architecture, Selçuk University, 421030 Campus, Konya, Turkey

E-mail: (grazia.tucci, gianni.bartoli, michele.betti, valentina.bonora)@unifi.it; (mkorumaz, ulec.armagan)@selcuk.edu.tr

Abstract. This paper discusses an approach for identification of historic buildings that combines Terrestrial Laser Scanning (TLS) survey, Deviation Analysis (DA) and Finite Element (FE) numerical modelling. The methodology is presented through the application to an illustrative case study: an early medieval period brick minaret located in Aksaray (Turkey). Precise direction of inclination, leaning angle, local deviations from circular building shape, deflections from vertical planes, local curvatures and related maps were obtained with high accuracy by DA, based on detailed point cloud 3D mesh model. In addition, differently from traditional approaches in FE analysis, a method for direct transfer of high accuracy TLS based 3D model to FE structural analysis is introduced. The FE model is subsequently employed to interpret and verify structural health of the historic building.

Keywords: *Terrestrial Laser Scanning (TLS); Deviation Analysis (DA); Cultural Heritage (CH); Finite Element (FE) modelling; Structural analysis; Health assessment.*

1. Introduction

The process of preservation of architectural heritage introduces different issues concerning documentation, conservation, restoration and reusing activities: it needs specific analyses based on accurate broad-spectrum surveys (especially when structural aspects are concerned). In general, the structural analysis of a monumental building requires the development of an interconnected series of operations aimed at obtaining a satisfactory knowledge of the construction where in-situ investigations are performed together with advanced numerical analyses [1], [2]. The structural assessment of architectural heritage is consequently composed of several steps that can be summarized as follows: i) geometric and topographic surveys with the identification of main constructive steps (through the analysis of historical documents) and damage survey including its evolution over time; ii) assessment of the actual damage through essays and surveys defining the variables that characterize the structural behaviour of the structure; iii) measurement and interpretation of existing stresses in some significant areas of the structure by means of static tests and identification of the monument dynamic behaviour; iv) development of numerical models (with gradually increasing complexity) able to reproduce the experimental evidence; v) numerical simulation of the structural response with respect to several load conditions (e.g. horizontal movements induced by earthquakes, etc.) that could affect the monumental building. This research consists of an iterative approach aimed at obtaining an adequate level of knowledge

where the typology and the extension of the experimental survey must be combined with the results of numerical models of the building that are iteratively updated [1].
The accuracy of measured data directly affects decision-making and analysis process [2], [3]. Nevertheless, in documentation of cultural heritage, the needs to preserve the integrity of the historic construction make difficult to develop an extensive experimental investigation and/or to apply traditional techniques. However, local authorities and agencies devoted to Cultural Heritage (CH) preservation need complete and updated databases on territorial scale in order to plan interventions and to prevent possible damages to CH. For these reasons, there is a great interest in expeditious and non-destructive methods [4]. In this respect, Terrestrial Laser Scanning (TLS) technology allowing high accuracy measurements can be used as a versatile and powerful tool for rapid no-contact survey. TLS is a remote sensing technique offering accurate geometric characterization of the historic building, providing key information for structural and historical purposes [5] and preserving the integrity of historical constructions.
Finite Element (FE) analysis has become in last decades a predominant tool, and it is used in a majority of analysis and simulation applications. One of the key challenges in applying FE analysis to realistic model is the process of converting the acquired geometric data into a suitable format for FE analysis software. Concerning FE analysis, automated 3D mesh model generation is an old problem but great advances have been made over last years and most commercial tools have automated the process for large classes of geometries. However, it can be observed that laser scanning technologies have not been effectively used in structural engineering research fields, and only few recent researches discusses such integrated approach [6], [7]. Mesh models used in FE analysis are generally created in a CAD software and they cannot be strictly considered as "reality based" 3D models since they come from a manual user-performed remodelling approach. The adopted industry-wide solutions are to help for simplifying complex shaped models, with removing some geometric non-structural features. However, in general, accurate and detailed models allow the study of the relationship between the shape, size and state of conservation of buildings and available technical and historical information for architectural and engineering studies.
This paper discusses these issues presenting an approach for identification of historic buildings that combines Terrestrial Laser Scanning (TLS) survey, Deviation Analysis (DA) and Finite Element (FE) numerical modelling. The approach is aimed to straightforward connect architectural survey and FE modelling, and the discussion is developed by an elucidating a case study; a minaret located in the city of Aksaray (Turkey) [8] [9] (Figure 1). The case study is particularly significant since the structure has lost its vertical configuration in past time and the assessment of its actual structural behaviour asks for an exhaustive evaluation of its current geometrical configuration.

Figure 1. Photos of Minaret: (a) General view; (b); Steel cables and clamped part of minaret.

2. Methodology

Data accuracy and measurement workflow employed as methodology within the case study are shown in Figure 2. As first step, topographic survey was carried out in a local coordinate system. Afterwards, five scans were performed just around the minaret. Because of the security measurements, accessing to the inside of the minaret was not allowed. Interior of minaret was modelled based on previously available manual survey drawings, and integrated with point cloud based mesh model. Then post-processing procedure was followed and a mesh model of minaret was created. This model was subsequently converted in a watertight mesh model to be imported into the FE software employed to perform structural analyses. Some deviation analyses were carried out using point cloud based mesh model. In this analysis, the geometrical property of minaret, the inclination direction and some deviations were determined. The presented methodology is intended to provide benchmark guidance for subsequent applications with point cloud data for deviation and structural analysis.

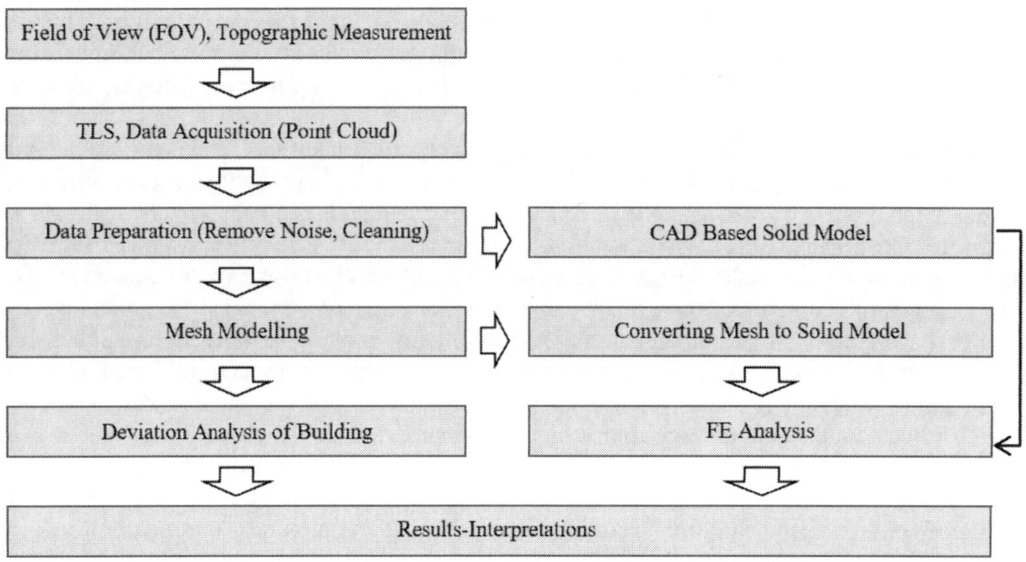

Figure 2. Methodology.

3. Minaret case study

Eğri (Leaning) Minaret is an important CH of Seljuk Empire in Aksaray (Turkey). Because of its red bricks used as construction elements, it is also referred to as Reddish Minaret. The construction period of minaret is estimated between 1221 and 1236 A.D. This minaret is an early example of Turkish buildings in Anatolia which were constructed with exactly the same Iranian and middle Asia region architectural style, construction techniques, materials and decorations as well. The Minaret has a cylindrical main trunk, with a squared shape stone basement. The distinguishing features of Eğri Minaret are the zigzag ornaments (Figure 1), some relief inscriptions on the trunk and blue and green tiles on the top of building. The height of the minaret is about 30.6 m. An internal spiral brick staircase with 92 steps leads to the balcony.

Today the minaret leans towards North-West direction. For safety reasons, in 1973 the minaret was strapped down from the upper part and was tightened to the ground with steel cables. In recent years, some additional investigations have been carried out to determine possible reasons for the movement and the inclination of the minaret. Eğri Minaret was employed as representative case study to test a methodology where the actual geometric anomalies and the structural state of the minaret are assessed and evaluated in a diagnostic perspective with point cloud based 3D models and data.

4. Architectural survey

TLS acquires high resolution data sampling of the observed object surface with an accuracy of few millimetres. It allows the acquisition of high-resolution point cloud data in relatively short time and gives the possibility of reducing the in-situ survey time, and the cost on human employment [10]. The need for detailed survey of complex historic structures highlights that traditional surveying techniques are usually cumbersome and time consuming when applied to historic constructions. The experiences of last decades proved that traditional survey techniques are not able to provide the required accuracy,

to limit the in-situ survey time and to reach the level of detail needed for the digitalization of cultural heritage. Today, the necessity of providing data for digital databases has become more essential, compared to the past. With the help of the fast development and improvement of the information and communications technologies (ICT), to ensure data for digital databases has become more easy and strong, but also demanding. Data acquisition with TLS provides detailed 3D models capable to describe complex architecture for different purposes, such as: photorealistic representations, historical documentation, FE analysis aimed at checking the safety of the structures or risk assessment analyses aimed to verify the effects resulting from progressive damages or from the onset of new damages [11]. TLS based point cloud 3D models allow many type of geometrical check and comparison (horizontal or vertical deviations, rotations and inclinations of walls, relative or absolute displacements, etc.). TLS can also provide precise geometrical survey in order to know the buildings in details with all deformations and anomalies (crack movements, surface deformations, material, etc. [12]).
In this study the FARO Focus^{3D} S-120 terrestrial laser scanner was employed and high resolution and quality settings were adopted in order to obtain good quality data. The acquired data set was approximately 7 million points. After simplification, 3 million points were used for triangulated surface model of minaret.

5. 3D Modelling process

With rapidly increasing availability of laser scanner data, and the growing pressure to use it, engineers who need to work on spatial data have tried to find optimum solutions for obtaining 3D models based on point cloud data [7], [13]. However, it can be observed that still the engineer's community commonly do not use directly point cloud data as basis to build the computational models needed their FE analysis. Tognaccini [14] discussed that the development of structural models based on point cloud data can be performed substantially by three different techniques: the first one is to prepare model in CAD software, the second one is to create a mesh model and convert it to a structural model using sections. Last one is to create non-uniform rational B-splines (NURBS) surfaces with some approximations of point positions. Extending Tognaccini, classification to different techniques for data acquisition structural models can be obtained with five different methods [9]: 3D modelling procedures from a first (CAD modelling) to a last typology (creating mesh convert mesh to NURBS) are growing in complexity. Correspondingly, as long as techniques for 3D modelling getting more advanced, 3D models accuracy increases.

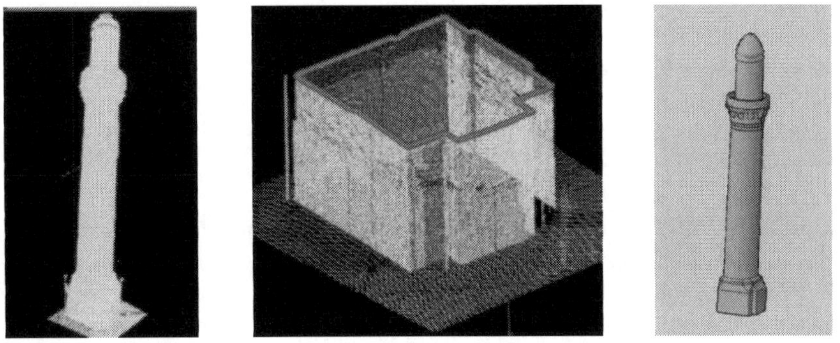

Point Cloud in CAD — Creating section in CAD — Creating solid model using sections.

Figure 3. CAD base solid modelling process.

CAD base models are widely used for FE analysis because both do not require having complex software and easily allow to obtaining results. This technique, the most commonly employed methodology for generating FE models, is based on regular 3D modelling in a CAD software combining primitive solid shapes with some unions, subtractions and intersections commands, without using point cloud data. A CAD model based on point cloud data requires almost the same process than a manual or topographic survey based modelling, but in this process needed distances are taken directly from point

IOP Conf. Series: Materials Science and Engineering **364** (2018) 012085 doi:10.1088/1757-899X/364/1/012085

cloud data (Figure 3). At the end of these processes, a 3D model can be obtained but many characteristics of the original geometry can be lost. A positive aspect of this approach is that processing and converting watertight 3D models to numerical models for FE analysis are relatively easy for simple shape geometries.

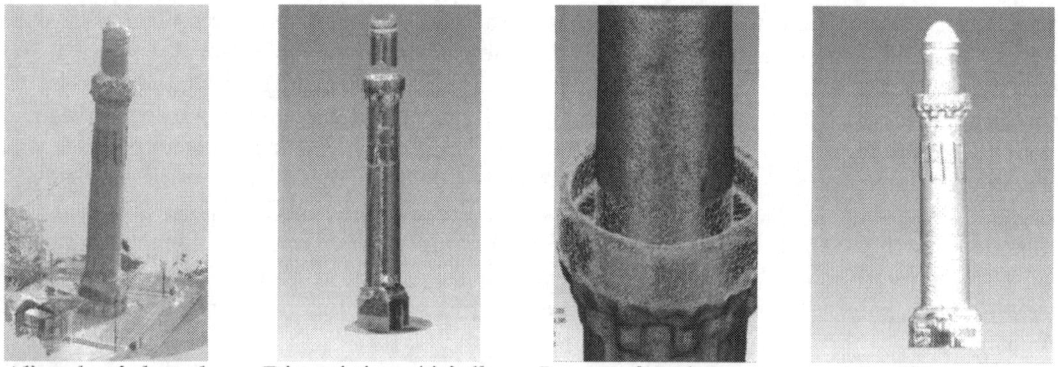

Aligned and cleaned point cloud data | Triangulation with halls and un-scanned parts. | Process of repairements filling of halls and gaps. | After filling and repairements.

Figure 4. Mesh model process.

Although there are some commercial solutions for the transformation of laser scanning point cloud data into mesh and solid models, these tools have significant limitations that are mainly related to the management of big amount of data. These mesh models need to be further converted into suitable numerical models. The 3D model for FE analysis was created by using these techniques (Figure 4). The sections can be defined with regular steps (as contour lines) up to upper part of the building, and can be exported to CAD software to create 3D mesh / solid model. A second limitation in the models transformation process is the need of overcome spars and missing data and to limit the loss of accuracy in the conversion process, that is a challenging task.

It can be observed that while the methodology for acquiring 3D data of objects using TLS is well-established with different approaches, data analysis procedures still need to be improved. In order to obtain reliable results, high accuracy metric survey becomes more essential in order to determine the vulnerability of buildings, especially in case of ancient structures that are characterized by irregularities [15]. It is finally noteworthy to observe that since Eğri Minaret was scanned from the ground level, there were several "holes" in the point cloud 3D model especially at the top of the balcony. In this case, it is possible to develop a hybrid workflow to obtain some data at points which laser scanner beams cannot reach, such as aerial photogrammetry with UAV (Unmanned Aerial Vehicle) [16].

6. Deviation Analysis

Deviation Analysis (DA) method allows recognition of possible anomalies that can be related to non-uniform material distribution, localized defects, basement and/or terrain instabilities and past seismic effects. Besides regular contact and contactless structural analysis, DA method can provide some data about the structural problems of the building, and it can detect the deformations which have to be considered in modelling process. Different deviation patterns can be correlated with different type of deformations and damages, and they can be used to identify possible external causes. The DA method is composed by three main steps: i) computing deviations of the model with respect to a reference object usually considering Euclidian distance, ii) visualizing deviations or creating deviations maps, and iii) analysing these maps for the description of problems [17].

6.1 Determination of inclination direction

Studies on verticality of high buildings can be carried out by cutting point cloud or mesh models and analysing the obtained sections. The trend of vertical axis can be computed by identifying the centre of each section and joining them. This method can provide data related to local leaning, tapering angle, radius of building, and local deviations from local curvatures [5]. Reconstruction of vertical axis may be refined through the extraction of more sections by a smaller interval without complicating the analysis [15]. This method was applied to the minaret, and several curves were used for determining the leaning angle which resulted -70.04° North. The vertical inclination trend was described by connecting the centre points of consecutive curves, and comparing it with a vertical line. At the end of this process, the deviation between the vertical reference and the average leaning angle was measured as 2.79° North. In a similar way, it was possible to compute the distances from the centre of every curve with respect to the vertical axis. The maximum linear deviation from the vertical axis was obtained on the top of the minaret and resulted 1.28 m (Figure 5).

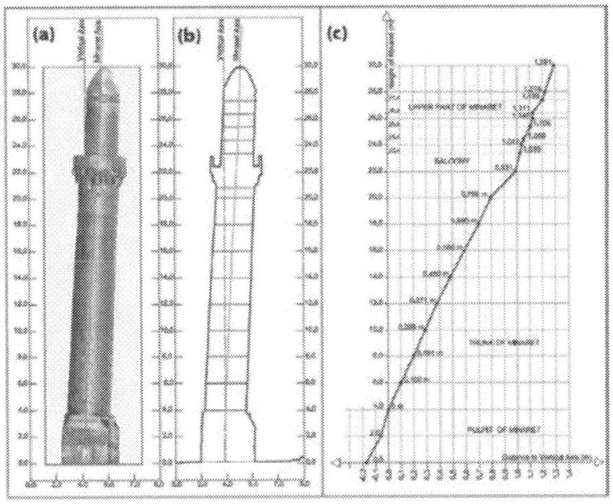

Figure 5. (a) Side elevation of minaret with central axis and inclination direction; (b) Section of minaret with central axis and inclination direction; (c): Diagram of inclination direction and distances from centre of the curves to vertical axis.

6.2 Deviation analysis with respect to orthogonal planes

Deviation analysis is not only useful for documentation of tall buildings; it can be also employed for documentation of all interiors, internal walls, celling and floors of all typology of buildings. Identifying of overhang, progressive changes of inclination, differential movements of the structures and detailing the study of structural elements are just some examples of many fundamental information that can be obtained and useful for analysing the structure behaviour [15]. This data can be obtained by some analyses based on mesh or point cloud models of structures. A DA is not just measurement of an object, it is exactly a comparison of a precise mesh model based on point cloud data with a reference plane or created 3D shapes. Thanks to this analysis some geometric differences could be defined by carrying out detailed comparisons of deflection from verticality with respect to a vertical plane orthogonal to the one (which is also vertical) containing the main inclination direction.

The reference plane was set up tangent to the minaret in the lower part of the trunk since the inclination starts from there, and this point considered reference point of the minaret (Figure 6). Deviations maps respect to this plane were obtained: the range of the mapped distances was between +1 m and -1 m. Maps of the colours from blue to red represent different distances between references plane and minaret surface. It can be observed that the deviation of the upper part of the minaret exceed the maximum value (1.281 m from vertical axes). Deviation of main trunk upper part reaches to maximum value (0.931 m). Deviation of base of minaret was observed less than 0.2 m from the vertical axis. Similar to this comparison, main trunk of minaret was compared with a best fit cylinder which was created using minaret horizontal sections (from lover and upper part of the minaret). It was observed that, because of the inclination, the maximum distance was measured on the North-East side of the main trunk (0.058 m). Some deflations can be observed as dark blue colour areas in deviation map on lower and upper part of trunk especially on South-West and South-East directions. These deviations can be observed mainly in the opposite site of bulged part of minaret (Figure 7).

IOP Conf. Series: Materials Science and Engineering **364** (2018) 012085 doi:10.1088/1757-899X/364/1/012085

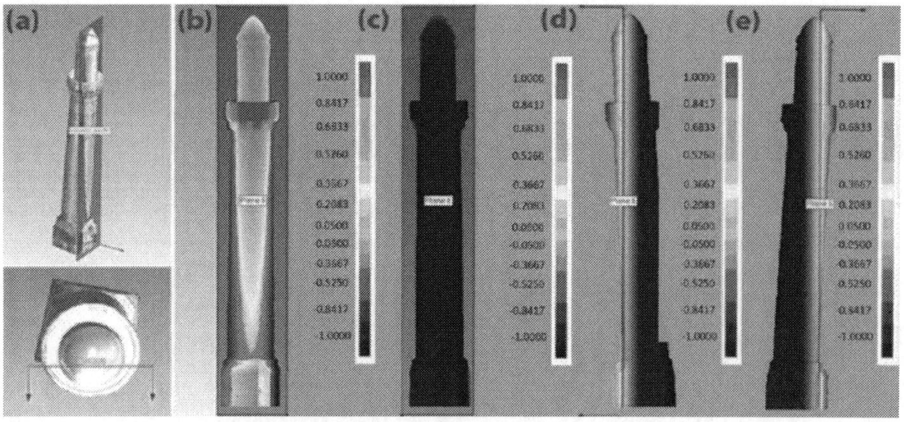

Figure 6. (a) Position of reference plain; (b) Leaning direction deviation; (c) Opposite site of leaning; (d) West site of leaning; (e) East site of leaning.

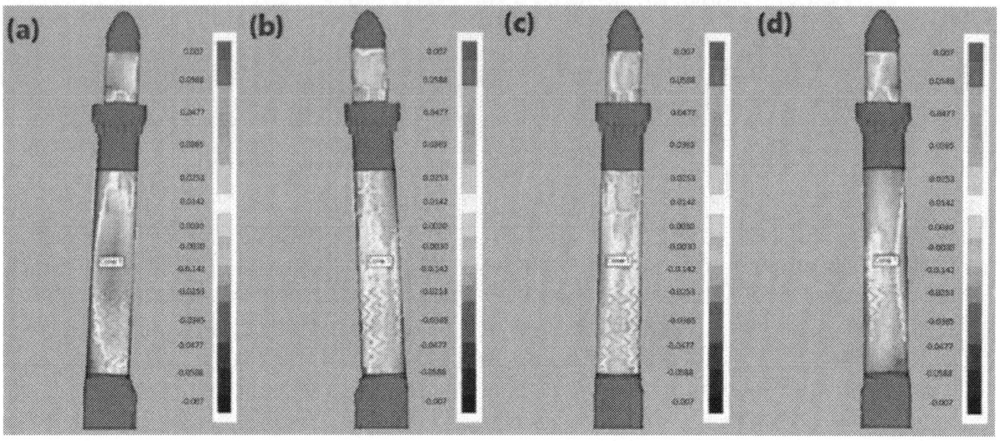

Figure 7. Surface maps of minaret trunk with best fit cone from four directions. (a) North-East (front, entrance) (b) South-West (back) (c) North-West (right) (d) South-East (Left).

7. Numerical analysis

The three-dimensional FE model of the minaret (Figure 8) was built based on the data discussed in previous section according to the macro-modelling technique. The ANSYS software was employed, and the FE model was aimed to investigate the effect of the actual tilt of the minaret on its seismic behaviour. Both the masonry trunk and the internal stairs were modelled by 8-nodes isoparametric solid elements (Solid65). In absence of a specific experimental campaign (tests, core drilling, mineralogical surveys, GPR survey, etc.) apart from the visual inspection, the mechanical properties of the walls (mainly specific weights and modulus of elasticity of masonry) were initially estimated based on literature data. The mean elastic modulus was estimated about 1500 N/mm^2, which is close to values already utilized in other works, and the specific weight of the material was assumed equal to 18 kN/m^3. To assess the effect of the tilt on the structural behaviour of the minaret, modal analyses and pushover analyses were performed.

Pushover analyses were performed considering time invariant load distributions for the lateral forces, namely: (i) a uniform distribution (U) and (ii) a triangular distribution (T). The uniform distribution is representative of the behaviour of the building near collapse, and the triangular distribution is representative of the behaviour of the building in its original undamaged configuration. U and T distributions were applied separately according to twelve equally spaced horizontal directions (with an angular scansion of 30°). The corresponding capacity diagrams were built considering the displacement of

IOP Conf. Series: Materials Science and Engineering **364** (2018) 012085 doi:10.1088/1757-899X/364/1/012085

the top level of the minaret and the base shear V_b. The analyses were performed employing a force controlled incremental Newton-Raphson algorithm that allows for the simulation of the ascending branches of the response up to the maximum base shear V_{bmax}. For the sake of brevity, in the following only the results obtained with the uniform distribution are reported and the capacity diagrams obtained with the twelve pushover analyses are reported in Figure 9a. As a consequence of the tilt of the minaret, it is possible to observe a significant variation of the maximum base shear that range between the 0.07 and the 0.18 of the total weight of the minaret depending on the assumed loading direction. In order to provide a synthetic representation of the performed pushover analyses, the results are also represented employing the three-dimensional domains called capacity baskets [18]. In such representation, each pushover curve is represented along the input direction in a three-dimensional plot where the radius of the capacity basket is equal to the base shear coefficient (or to the maximum displacement achieved among all the considered directions). Such simple representation allows an easy identification of the stronger and weaker directions of the minaret.

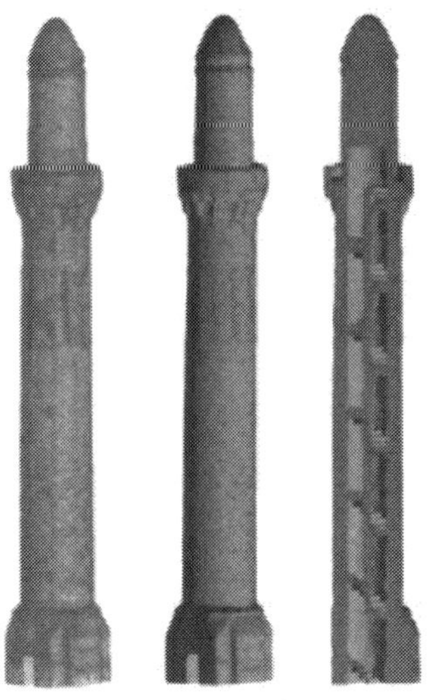

Figure 8. Views of the FE Model.

The capacity basket was built in order to synthetize three representative points of each pushover curve as illustrated in Figure 9b: i) the point correspondent to the end of the elastic branch ($[u_E, f_E]$ denoted with a red circle); ii) the intersection between the tangents of the elastic and plastic branches of the pushover curve ($[u_X, f_X]$ denoted with a magenta star) and iii) the point correspondent to the end of the pushover curve ($[u_L, f_L]$ denoted with a black x-mark). These points were subsequently represented in the capacity basket reported in Figure 10. The irregularity of the minaret both in terms of strength (Figure 10a; continuous lines starting from origin represent the base shear evolution during the analyses: lines are straight lines since a load control Newton-Raphson algorithm was employed) and displacements (Figure 10b; continuous curves represent the top displacement evolution during the analyses) for the uniform force distribution is clearly readable: significant variation of both the maximum base shear and the maximum displacement characterize the pushover curves along different directions. It is possible to observe the out-of-plane deviation of the control point with respect of the pushing direction. In addition, the capacity basket immediately shows the weakest direction of the minaret that, obviously, is the one corresponding to the tilting plane, which has been characterized with a good accuracy thanks to the availability of an exhaustive and precise TLS survey. In this respect, the TLS survey results a very important tool both to generate high resolution and precise 3D models of the structure and to provide the accurate information needed for the elaboration of 3D FE models allowing the investigation of the expected structural behaviour in extreme load conditions.

IOP Conf. Series: Materials Science and Engineering **364** (2018) 012085 doi:10.1088/1757-899X/364/1/012085

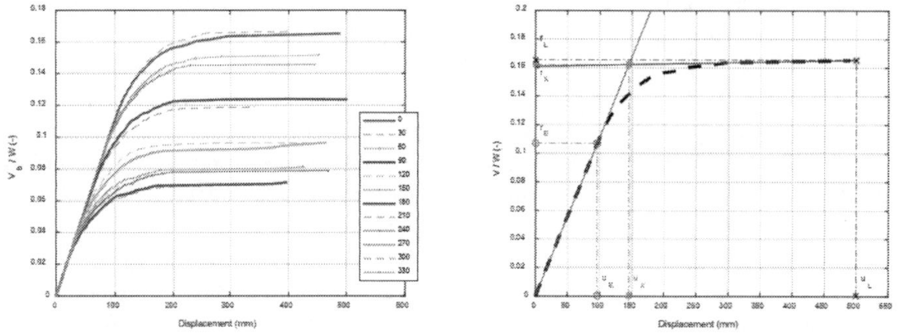

(a) (b)

Figure 9. (a) Capacity curves obtained according to the twelve 30° degrees spaced horizontal directions; (b) Selected representative points of the pushover curves.

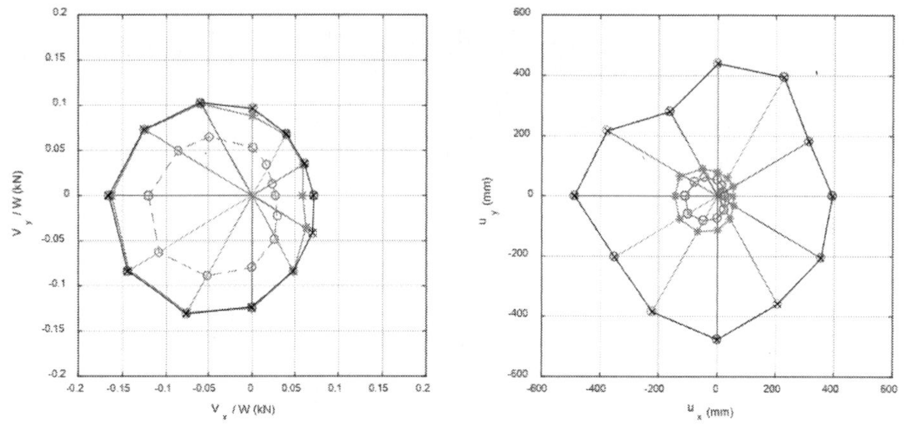

(a) (b)

Figure 10. Capacity basket: base shear (a); top displacement (b).

8. Conclusive remarks

In this paper, the health assessment of a brick minaret with both DA and FE model based on TLS was examined. TLS technologies can be considered as powerful tools for historical buildings analysis and monitoring projects. DA is a relatively fast and easy, innovative, contactless and non-destructive method for assessing building global and local deformations. With respect to the case study, DA provided the following result: i) an average deviation angle from vertical axis was measured and the leaning direction was determined; ii) partial anomalies, wall deformations and inclination maps were obtained easily and this data can support further analysis and decisions related to the structure; iii) partial sections and plans of structure were obtained directly from mesh models and they have been used for analysing anomalies of structures.

3D models of the minaret were obtained with tree different approaches and creating a mesh model and converting it to solid model, was recognized as a process fast, accurate and able to preserve the original geometry with a good accuracy. This approach resulted effective for FE modelling despite some difficulties related to transforming TLS data to mesh/solid models (and simplification of these models to computational models for FE analysis). The usefulness of the accuracy of the geometric survey by TLS has been discussed evaluating comparatively the seismic vulnerability of the tilted minaret through the pushover approach: it was observed a variation of the maximum base shear that range between the 0.07 and the 0.18 of the total weight of the minaret depending on the assumed loading direction, thus denoting the importance of the precise characterization of the minaret geometry.

References

[1] Bartoli G, Betti M, Facchini L, Orlando M 2012 Non-destructive characterization of stone columns by dynamic test: Application to the lower colonnade of the Dome of the Siena Cathedral *Engineering Structures* **45** 519-535.

[2] Lysandrou V, Agapiou A Comparison of documentation techniques for the restoration and re-habilitation of cultural heritage monuments: The example of Pyrgos 'Troulli' medieval tower in Cyprus. In: *Proceedings of EuroMed 2010*, Lemessos, Cyprus 2010.

[3] Arias P, Herraez J, Lorenzo H, Ordonez C 2005 Control of structural problems in cultural heritage monuments using close-range photogrammetry and computer methods *Computers and Structures* **83** 1754-1766.

[4] Pieraccini M, et al N 2014 Dynamic Identification of Historic Masonry Towers Through an Expeditious and No-Contact Approach: Application to the "Torre del Mangia in Siena (Italy) *Journal of Cultural Heritage* **15(3)** 275-282.

[5] Teza G, Pesci A 2013 Geometric characterization of cylinder-shaped structure from laser scanner data: development of an analysis tool and its use on a leaning bell tower *Journal of Cultural Heritage* **14(5)** 411–423.

[6] Castellazzi G, et al A 2015 From Laser Scanning to Finite Element Analysis of Complex Buildings by Using a Semi-Automatic Procedure *Sensors* **15** 18360-18380.

[7] Castellazzi G, D'Altri AM, de Miranda S, Ubertini F 2017 An innovative numerical modelling strategy for the structural analysis of historical monumental buildings *Engineering Structures* **132** 229-248.

[8] Ural A, Doğangün A, Meraki S 2013 Response evaluation of historical crooked minaret under wind and earthquake loadings *Wind and Structures* **17(3)** 345-359.

[9] Korumaz M, et al L 2017 An integrated Terrestrial Laser Scanner (TLS), Deviation Analysis (DA) and Finite Element (FE) approach for health assessment of historical structures A minaret case study *Engineering Structures* 153 224-238.

[10] Lemmens MI 2011 Geo-information. Technologies, Applications and the Environment *Springer, Netherlands.*

[11] Fregonese L, et al L 2013 Surveying and monitoring for vulnerability assessment of an ancient building *Sensors* **13(8)** 9747-9773.

[12] Pesci A, Teza G, Boschi E 2015 Laser scanning-based detection of morphological changes of a historical building occurred during a seismic sequence: Method and Case study *International Journal of Geomatics and Geosciences* **5(3)** 427-447.

[13] Riveiro B, et al Fully automatic approach for the diagnosis of masonry arches from laser scanning data and inverse finite element analysis. In: *Proceedings of the 10th SAHC*. Leuven, Belgium 2016.

[14] Tognaccini R. 2009 La Chiesa di Santa Maria del Mar a Barcellona - Dal Rilievo Tridimensio-nale all'Analisi Strutturale *Master's Thesis in Architectural Engineering*. (University of Pisa) (in Italian).

[15] Castagnetti C., Bertacchini E., Capra A., Dubbini A. Terrestrial Laser Scanning for Preserving Cultural Heritage: Analysis of Geometric Anomalies for Ancient Structures. In: *FIG Working Week 2012 - Knowing to Manage the Territory, Protect the Environment, Evaluate the Cultural Heritage*, Rome, Italy 2012.

[16] Remondino F 2011 Heritage recording and 3D modeling with photogrammetry and 3D scanning *Remote Sensing* **3(6)** 1104–1138.

[17] Pesci A, Casula G, Boschi E 2011 Laser scanning the Garisenda and Asinelli towers in Bologna (Italy): Detailed deformation patterns of two ancient leaning buildings *Journal of Cultural Heritage* **12(2)** 117–127.

[18] Caliò I, et al 2008 A new discrete-element approach for the assessment of the seismic resistance of composite reinforced concrete masonry buildings. In: *AIP Conference* **1020** 832-839.

IOP Conf. Series: Materials Science and Engineering **364** (2018) 012086 doi:10.1088/1757-899X/364/1/012086

Deformation measurement as a calibration tool for structural modelling of built heritage

E Coïsson[1], M Cotti and F Ottoni[1]

[1]Department of Engineering and Architecture, University of Parma, Parma, Italy

eva.coisson@unipr.it

Abstract. In spite of the important advances in structural modelling tools, the numerical analysis of masonry historical constructions is still a challenging activity, due to the significant difficulties in reliably describing their complex geometry, history, materials and damage evolution. A deep geometrical analysis, in combination with advanced numerical models, can represent a powerful tool for the analysis of masonry structures, able to take into account the outcome of a series of events suffered in time. Moreover, the precise measurement of deformation can give fundamental indication for masonry building structural interpretation, representing a crucial calibration tool for setting up a reliable structural model. In this paper two different case studies are presented, which represent as many possible ways to "use" the more and more precise survey methods for ancient structures in order to reliably interpret their structural models.

1. Introduction

The present geometry of historical masonry buildings is the outcome of a series of events suffered in time: construction phases, soil settlements, earthquakes, material deformations. These events alter not only the visible geometry of structures, but also the invisible internal stresses of the materials composing them, whose investigation is the main purpose of structural analyses. Therefore, it results clear that basing numerical models of complex masonry buildings only on their present geometrical survey - even extremely precise and complete - can lead to disregard the role of history in modifying their structural behavior; in many cases this misunderstanding causes negative results, as an overestimation of the real final safety levels.

Nevertheless, complicated and really detailed numerical models are commonly built just starting from the surveyed geometry in order to simulate the behavior of complex and deformed historic structures, simply ignoring the geometrical transformations occurred in the meantime [1].

In order to create more reliable models, able to interpret the real behavior of ancient monuments, a different approach can be adopted, based on the precise information we can have on the present geometry, making the most of the more and more evolving survey technologies and software. In this process, a fundamental role is played by the critical interpretation of the "distance" between the present and the original geometry: this becomes a crucial calibration tool for numerical model analysis. Through the analysis of two different case studies, the present paper aims to show two possible ways of carrying out this calibration, underlining the importance that a critical approach in interpreting the geometrical data as important evidences of damages and deformations is mandatory for a correct numerical analysis.

2. Calibrating structural models by deformation measurement: two possible approaches

It's a matter of fact that the numerical models used so far, frequently built starting directly from the present geometry, are not able to reliably simulate the complex behavior of ancient masonry buildings, whose resistant structure is substantially unknown. Due to the lack of resistance to tensile stress, in fact, there isn't an adequate diffusion of stress states and the resistance properties themselves change with the changing of the forces acting on masonry structures. Furthermore, it's really difficult to assign realistic values to the mechanical parameters of masonry, which are the results of craft and inconsistent procedures and which have been strongly and variously modified in time by damage and degradation phenomena [2]. Moreover, the large variability of masonry textures, depending on time of construction and on building elements (dome, pillars, walls), would require numerous tests spread on each zone characterized by different features, with an unsustainable impact on conservation.

In this paper, two are the possible paths proposed in order to take advantage of the precise survey of deformation in calibrating numerical models without disregarding the effects of history.

The first case study sees the correction of the results of even refined and accurate numerical analyses by examining the evidences of the real behavior of a building, starting from its deformation measurements; the second approach considers the stress produced on the structures by the movements occurred in time.

In both cases, the key point has been the translation of the complex relation between present and original geometry of ancient buildings into a crucial calibration tool for more realistic numerical model analyses, able to give results not so divorced from the reality.

2.1. Structural parameters from deformation survey, passing through metrology: the case of a 17th century dome.

It's well known that one of the main uncertainties in the numerical analysis of ancient buildings is which mechanical parameters have to be assigned to the materials in order to interpret at best their real behaviour. The issue is even more challenging in case of monuments, where the mechanical tests on materials have to be limited in order to preserve them. Hence, due to the lack of mechanical information - as the real stiffness of the various components, the internal masonry texture and, accordingly, their tensile and compressive strengths - it's necessary to operate parametrically by assigning a possible range of variation to the mechanical properties.

In this operation, an interesting method could be the "geometrical" one: indeed, as already stressed, the current shape of the building can be considered as the result of all the changes, damages and degradation phenomena (alterations in the soil, geometric variations, deformations and material degradation) occurred on it during times and the "distance" from its original geometry directly depends on the capacity, of the same building, to resist to all these actions: in a word, by its stiffness, and, in particular, by its modulus of elasticity (E). Therefore, the comparison between surveyed and original geometries, can help in defining the mechanical properties of the materials in reacting to all these changes through deformation. Hence, the best way to achieve the real building behavior and its possible deficiencies is to identify the original shape (by means of a metrological analysis), and studying the relationship between failures, consolidations and results.

An interesting application of this procedure has been the structural analysis of the 17th century church of Santa Maria del Quartiere in Parma, whose main structural element is represented by a masonry pavilion dome (9 meters in diameter), which stands up at the center of a hexagonal plan which recalls the "*tempio essagono*" prescribed by Sebastiano Serlio [3].

2.1.1. The metrology and the measured "distance" from the original geometry

In this case, a careful study on the geometry of the structure has represented the fundamental step for the understanding of its static behaviour and for the setting up of a reliable structural model. It's well known that digital survey techniques, and most notably terrestrial laser scanning, offer robust and realistic methods for adequately surveying heritage assets, and can provide very accurate base models for further investigations. Therefore, firstly, a precise architectural survey of the church has been carried out, in order to acquire the dimensional data necessary to describe the whole monument, by different

IOP Conf. Series: Materials Science and Engineering **364** (2018) 012086 doi:10.1088/1757-899X/364/1/012086

integrated methodologies. Outside, stereoscopic metric frames have been acquired through the materialization of support points and relative monographs for the photogrammetric survey of the main fronts of the church, while the continuous survey of the inside of the monument has been achieved by terrestrial laser scanner technology (Leica HDS 3000 with "flight time" technology, and a fixed net of 2 cm), with 5 different measurement stations - positioned along the main axes of the church - enabled to capture almost all the details of the structure. The overlapping of the five data sets ensured a complete 3D description of the dome and the underlying structures (software Cyclone 5.6). Moreover, the information on the curvature of the six wedges of the dome has been achieved by tracing lines along the intrados, using the point cloud model with only some horizontal cut planes plotted (Fig.1a), in order to understand better the deformation of the dome, referred to the hypothesized mechanism of dome thrust combined to a conceivable differential settlement towards the North direction.

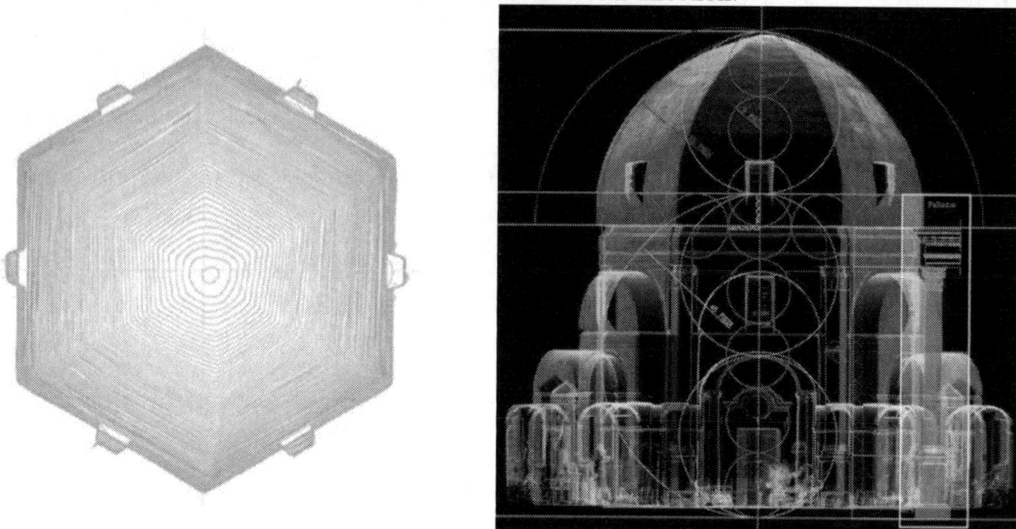

Figure 1. On the left, horizontal planes of the dome, highlighting its deformation (1a); On the right, the metrological analysis on the vertical section, which evidences the lowering of the dome and its deformation (1b).

Indeed, as stated before, the current configuration is the result of the deformations occurred to the monument during centuries and it's a matter of fact that the top of the dome has lowered sensibly since the time of construction. In order to identify, in absence of ancient project drawings, the original shape of the un-deformed dome, the geometrical investigation prosecuted by metrology.

Translating the quantitative results of the survey into the original measurement units ("*Braccio parmigiano*"/arm equal to 54,52 cm) has unravelled a rigorous scheme and a significant progression 18-6-6 in plan distribution which can be referred to a precise 17th century symbolism and which is also maintained vertically. However, it's very interesting to note that the current shape of the dome, even considering the change in level of the internal flooring (testified by a deep historical analysis), doesn't fit in this original re-constructed (and perfect) scheme (Fig.1b). In fact, the height of the dome was found to be 20 cm less than expected when compared to the original drawing and this discrepancy could further support the research of a structural parameter (E) which could reliably interpret the response of the monument to the actions and damages occurred on it in time, justifying this measured "distance" from its original shape.

2.1.2. The parametrical analysis, returning to metrology

Thanks to a fusion between historical analysis, precision geometrical survey and experimental investigations (like some dynamic tests carried out on the monument) the structural analysis of the dome has been then performed following a parametric procedure, in order to assign the more realistic mechanical values to the materials composing the monument, just starting from deformation evidences and working by difference between the original geometry and the current one.

Therefore, a simplified model has been built assuming the original geometry of the structure (as resulted by metrology) and taking advantage of the radial symmetry of the dome. However, even when the original geometry is not so different from the surveyed one, a strong critical reduction and simplification of the numerous (sometimes redundant) data obtainable by laser scanner survey is needed, in order to build a more serviceable numerical model of the ancient building. In the process of mechanical identification of the structure, a first virtual Finite Element Model of the dome alone has been built up with the commercial code ABAQUS, using solid tetrahedral iso-parametric finite elements in order to discretize a regular single shell solid on a symmetric hexagonal base, which grows in height following the original circular profiles of the six webs, with a variable thickness (from 30 cm at the abutment, to 20 cm at the top).
Then, different analyses have been performed, attributing to this simplified model the mechanical characteristics of an elastic, homogeneous, isotropic solid, assuming a linear behavior of masonry and operating parametrically by assigning different values for the modulus of elasticity (E) [4].

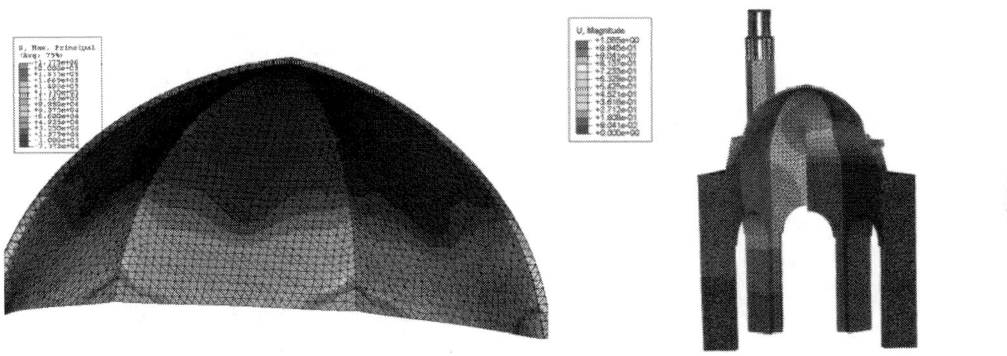

Figure 2. On the left, visualization of the FEM results on the un-cracked model of the dome, Smax principal stresses (2a); On the right, the modal analysis of the church (2b).

The damage and the crack pattern were analysed and used to calibrate the numerical model to fit the present shape of the dome. Starting from two different literature values for the elastic modulus (E) of the different parts of the structure, these were varied in order to reproduce reliable tensile stresses which could justify the presence of the cracks at the corners and in the middle of the webs (Fig.2a). Thereafter, the hypothesized mechanical parameters used for the numerical modeling have been further corroborated by a very simplified modal analysis model (Fig.2b) which has been related to the dynamic tests carried out on the monument.

Then, discrete cracks, matching with the crack pattern survey, have been inserted into the model, obtaining a final geometrical configuration coherent with the surveyed one. It is noteworthy to observe that the hypothesis on materials, boundary conditions and structural and non-structural masses of the building are so variable that very different results can be obtained by changing parameters and this procedure is only a first attempt to check the previous results. Anyhow, through a deep analysis of measurement variance, it has been possible to establish a connection between the actual configuration and the static movements in the monument, developed during the centuries, and to control and reduce, more reliably, the unavoidable uncertainties regarding the numerical modeling of a historic building.

2.2. The deformation data as information on the changing boundary conditions: the case study of a Romanesque parish church

Ancient masonry structures carry on their own shape the signs caused by historical movements, changes and damages. When numerical analyses are developed on this type of buildings, these "footsteps of history" [1] cannot be disregarded, or the results obtained even by the most up to date tools can turn out to be completely wrong. For this purpose, the first step represented by the direct contact with the

IOP Conf. Series: Materials Science and Engineering **364** (2018) 012086 doi:10.1088/1757-899X/364/1/012086

building, its precise geometrical survey, the interpretation of deformations and damages for the comprehension of these "footsteps" is fundamental.

In particular, the deformation measures can become an important input for the numerical model, in terms of changing boundary conditions, applied progressively to the un-deformed structure.

The case study of a thousand-years-old church, which only recently has shown severe damages, is here presented in order to point out the importance of this process and the influence that the deformation (in this case constituted by soil settlements) have both on the real building and on its virtual model.

2.2.1. Following the changes in deformations to understand the uneven evolution of cracks: the importance of historical monitoring

The parish church of San Pietro is located in Costa di Tizzano (Northern Italy), on the top of a hill, dominating the surrounding landscape. It dates back to the end of the 10th or beginning of the 11th century and, although it was modified in time, it is entirely built in local stone masonry (Fig. 3).

Figure 3. The parish church of San Pietro with its clocher-porche: outside historical view (left) and an inner damaged pillar recently subjected to a provisional intervention (right).

It is a typical Romanesque church – with three naves separated by rough stone masonry columns and covered by a timber roof – but its most peculiar feature is represented by the bell tower, which stands over the entrance, in the middle of the façade. This element, called clocher-porche, is quite widespread in France and in some parts of North-West Italy, but is very rare in this area. Usually the clocher-porches are either protruding outside of the church, with their own large sustaining walls and minor openings, or built over the first bay of the central nave, enlarging the first two pillars or adding specific supports. On the contrary, in the San Pietro church, the relevant weight of the clocher-porche is divided between the façade wall and the first two columns of the nave, which, however, have the same dimensions as the others, carrying instead a much larger load. This load is at the base of the present crack pattern, which has been increasing rather rapidly since the end of the summer of 2016, leading to the decision of inserting provisional steel hoopings around the columns (Fig. 3) and struts under the two longitudinal arches, carrying the clocher-porche. The two columns, in facts, showed vertical cracks all around their perimeter, rapidly increasing in number, length and width, clearly indicating that the compression stresses had reached the masonry strength and that a brittle, sudden collapse could be close to come. Once the safety conditions were re-established, the question about the origin of the damage arose: how could the columns be able to stand their load for one thousand years and then suddenly show such a severe crack pattern? Any standard calculation or numerical model would supply indications about the stress state which are time-independent, therefore they would not be able to explain the real present conditions of the church. The maximum compression was calculated manually in 1.35 MPa: a rather high value, indeed, but that was carried by the masonry for centuries with no signs of major damage until recent years. Two possible explanations could be hypothesized: the development of a creep phenomenon, decreasing the materials strength in time, or a change in the boundary conditions, increasing the actions on the columns.

The fact that the heavy worsening of the conditions followed a very dry summer, seems to validate the idea that the origin of the phenomenon should be searched for in the soil settlements, although the effects could have been possibly worsened by creep. Considering the second hypothesis, it should be also taken into account that about 20 m North of the church, a steep slope begins, characterised by a progressive detachment of materials, that is getting closer and closer to the church in the last decades. Indeed, in other parts of the church, cracks related to a differential soil settlement can be seen and, in the past, several interventions (the last ended in 2013) were made in order to reinforce the walls foundations with underpinning. Nevertheless, the information about the soil composition are limited and no investigations were made up to now about the slope stability. Under these circumstances, a possible contribution to explain the present situation can be obtained inserting survey data about the deformations of the structure inside the numerical model, thus taking into consideration not only the theoretical static compression on the columns, but also the increase in stress caused by the soil settlements and by the consequent deformations of the structure.

As already described in [8], the procedure in order to take into account the structural effects of deformations in the numerical model starts with a precision geometrical survey, which must be able to identify the changes in the geometry from an original, un-deformed condition. In the San Pietro church, the survey revealed a settlement of 3 cm between the capitals of the two cracked columns and the ones on the inner façade, which were clearly at the same level at the time of construction. Moreover, the difference between the levels measured on the flooring is of only 1 cm, as this was changed in 1992. This approach, defined "historical monitoring", allows to reconstruct the evolution of the deformation during centuries, supplying an even more refined information for the structural analysis and for the comprehension of the present stability conditions.

2.2.2. The settlement measures as changing boundary conditions in the model

Starting from these geometrical data, the most probable original shape of the arch was reconstructed with a parametric approach (Grasshopper software), based on the proportional analysis, on the comparison with other parts of the structure built in the same period and on considerations about the building features. In particular, the following assumptions have been made: original supports originally vertical and at the same level and round arches, originally undeformed.

An approach based on the Discrete Element Analysis of rigid blocks was chosen, with the aim of following the large deformations of the structures ([7,8]). The blocks interact with one another with unilateral, no-tension constraints, elastic in compression and sliding under a Coulomb friction model ([9,10]). In particular, two-dimensional plane elements in the UDEC environment were considered sufficient to describe the behaviour of the partial longitudinal section of the church. The real masonry texture of the stone arches was adopted as a base for the subdivision in discrete blocks (Fig. 4).

As far as the materials properties are concerned, only friction and density are relevant in this approach, as the blocks are considered rigid and the deformation is concentrated in the interfaces. For the stone masonry, the density was considered 21 kN/m^3 with a friction angle of 25°. The loads were then applied to this virtual original structure and the soil settlements were added, with the aim of understanding what these actions have induced in terms of stress state in the materials.

At first, the structure was analysed only under its own self load. The results in terms of stresses of this first analysis, carried out on the undeformed shape, show a good agreement with the first approximate calculations made, with a mean compression of 1.24 MPa, which do not explain the present crack pattern. Then, a first differential settlement of 2 cm was applied between the two springers of the arch, as an imposed vertical movement of the right support, corresponding to the façade. The results show a significant increase in the stresses inside the pillar (Fig.4), reaching 3.36 MPa, which is within the range of the expected mean strength of the material, as suggested by the Italian technical code for the roughly split stone masonry. At last, an increase in the differential settlement of further 1 cm was applied, reaching 3.8 MPa, which corresponds to the upper limit of the aforementioned range of strength.

IOP Conf. Series: Materials Science and Engineering **364** (2018) 012086 doi:10.1088/1757-899X/364/1/012086

This is perfectly consistent with the damage observed and also explains why the pillars have cracked only recently, rather than showing problems all along their history. The numerical analysis under these modified boundary conditions also show a crack pattern in the wall over-standing the arch which is coherent with the present conditions and a deformation of the arch which is the means of the increased stress state in the columns (Fig.4).

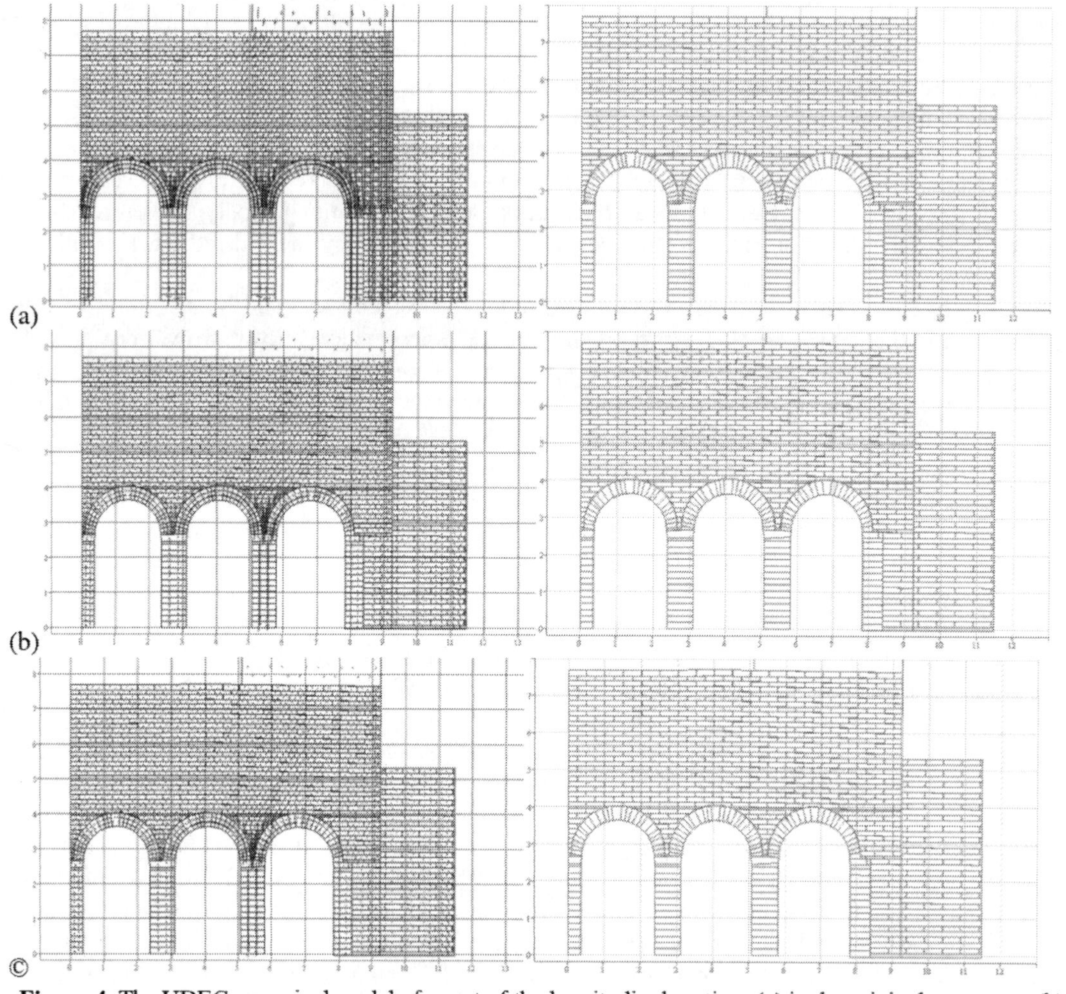

Figure 4. The UDEC numerical model of a part of the longitudinal section: (a) in the original geometry, (b) with 2 cm differential settlement, (c) with the final, 3 cm settlement. On the left, the stresses, on the right the crack pattern.

These results support the hypothesis that the damage is caused by the soil settlement and also explain the reasons for the sudden, recent onset of the damage. As a consequence, a deepening of the available knowledge in the geotechnical field has been proposed, together with a monitoring of the built structure, in order to reach a better understanding of the present evolution speed and to identify the most suitable solutions to hinder the soil movements and strengthen the church, thus limiting the interventions to the ones which are really needed for the preservation of the monument, unlike what has been done until now.

3. Conclusions

Masonry structures have been built for centuries with a design focused on the geometry and on the proportions of buildings, rather than on the strength of the materials. This strict relation between form and stability has been shading with the introduction of new materials, like steel and reinforced concrete, which allowed to build shapes more independent from the structural scheme. The development of new tools for the prevision of the behaviour of these new materials has progressively led the structural experts to disregard the aspects related to the geometry in favour of the materials properties. But when ancient masonry buildings are concerned, analysing their structures without taking into consideration their geometry and their changes in time risks to lead to dangerous conclusions.

The present paper focused on the important role that the deformations play in the definition of the present state of internal stresses and underlined, as a consequence, the necessity of adopting the deformation data as calibration tools for structural analysis, in order to represent at best not only what we see now, but also the historical path that lead to the present conditions. In particular, through the analysis of two case studies, two different, complementary approaches are presented: the use of the precise deformation survey data in order to calibrate the mechanical parameters of the masonry, limiting the need of destructive tests, and the adoption of the information on the displacement, possibly reconstructed in its evolution in time thanks to the historical monitoring method, as an input for the numerical analysis, in terms of changing boundary conditions.

The aim is reducing the distance, sometimes too large, between the real building and its virtual numerical model, recovering through the survey analysis the role of experience and of empiricism [10], which lead the actions of the ancient builders of our precious cultural architectural heritage.

References

[1] Roca P 2004 Considerations on the significance of history for the structural analysis of ancient constructions *Proc. of 4th Congress Structural analysis of historical constructions* 6373

[2] Roca P, Cervera M, Gariup G and Pelà L 2010 Structural analysis of masonry historical constructions. Classical and advanced approaches *Arch. Comput. Methods Eng.* **17** 299-325

[3] Ottoni F 2008 From geometrical and crack survey to static analysis method: the case study of Santa Maria del Quartiere dome in Parma (Italy) *Structural Analysis of Historical Construction* **1** 697-704

[4] Ottoni F, Blasi C, Betti M and Bartoli G 2016 The effectiveness of ancient "hidden" tie rods in masonry dome preservation: San Giovanni in Florence *Structural Analysis of Historical Construction* 1802-1809

[5] Coïsson E and Ottoni F 2015 Structural Monitoring of Historical Constructions: Increasing Knowledge to Minimize Interventions *Built Heritage: Monitoring Conservation Management* 83-92

[6] Bruno N, Coïsson E and Cotti M 2017 Laser-scanner survey of structural disorders: an instrument to inspect the history of Parma Cathedral's central nave *Int. Arch. Photogramm. Remote Sens. Spatial Inf. Sci.* **XLII-5/W1** 167-174

[7] Thavalingam A, Bicanic N, Robinson JI and Ponniah DA 2001 Computational frame work for discontinuous modeling of masonry arch bridges *Computers & Structures* **79** 19

[8] Giordano A, Mele E and De Luca A 2002 Modelling of historical masonry structures: comparison of different approaches through a case study *Engineering Structures*

[9] Lourenço PB, Milani G, Tralli A, Zucchini A, 2007 Analysis of masonry structures: review of and recent trends in homogenization techniques *Canadian Journal of Civil Engineering* **34**

[10] Ottoni F, Coisson E and Blasi C 2009 Empiricism reappraisal in ancient buildings structural analysis. The masonry domes, from historical debates to numerical model *Proc. REMO09* 197-206

IOP Conf. Series: Materials Science and Engineering **364** (2018) 012087 doi:10.1088/1757-899X/364/1/012087

Point clouds turned into finite elements: the umbrella vault of Castel Masegra

L Barazzetti, R Brumana, S Della Torre, G Gusmeroli, G Schiantarelli

Department of Architecture, Built environment and Construction engineering (ABC), Politecnico di Milano, via Ponzio 31, Milan, Italy

Email: luigi.barazzetti@polimi.it

Abstract. The paper presents a workflow for finite element model generation from point clouds acquired via photogrammetry and laser scanning. The case study is a room with an umbrella vault in Castel Masegra, located in Sondrio (Italy). The complexity of the room required an accurate geometric survey from a set of dense point clouds, from which a detailed BIM as well as 2D drawings were generated. Finally, a tetrahedral mesh was derived from the geometric model and was used to run a finite element analysis. The aim of this work is to describe the procedure developed and illustrate the main challenges found.

1. Introduction

This paper investigates the possibility to transform an accurate and dense point clouds into a 3D finite element model that preserves (as well as possible) the geometric irregularities captured by laser scanning or photogrammetry. The case study chosen for this purpose is Castel Masegra, the castle that dominates the city of Sondrio, originally built in the XI century and then modified several times.

The conversion of a point cloud into a finite element model may appear as a simple task. In fact, several automated procedures for mesh generation are available in commercial software. However, several aspects have to be taken into consideration to obtain a compatible and regular mesh that represents the constructive logic of the buildings (e.g. elements and joints), especially when geometric "anomalies" have to be preserved. Such approach is also underlined in [7], the possibility of enhancing working with more refined model allows to better understand the infinite anomalies of historic structures.

An additional issue taken into consideration during this project was the simultaneous creation of a BIM [1] for the whole castle. The idea behind our modeling work was to reuse the BIM for the generation of the FEM, instead of separating the production of the two models. On the other hand, the creation of BIM and FEM is much more than a "pure" geometric problem. The interconnection between the different constructive elements parts and their materials required particular attention. Modeling included geometrical aspects along with structural considerations, in order to understand and (numerically) replicate the constructive elements and their mutual interaction.

The information contained in the architectonical model were accurately investigated to obtain a model suitable for structural analysis. A large variety of issues had to be dealt with, such as the node-to-node correspondence (compatibility of the mesh), very distorted or small elements, and small local imprecisions. Such effects couldn't be neglected: any small error or local inconsistency effectively make it impossible either to generate a mesh or to obtain any sensible result from the analysis.

The paper will provide an overview of the steps of the work carried out on the south wing of the castle. Here, the tower has a significant out of plumb and other geometric anomalies, as shown in Fig. 1. Moreover, an umbrella vault is located at the first floor of the tower. It represents an interesting case study to test the proposed method, which aims at preserving the complexity of the shape. The next sections will briefly describe the workflow, from the digital surveying, to the modelling and analysis processes.

Figure 1. An image of the tower considered in this paper and the umbrella vault.

2. Geometric survey

The geometric survey was carried out with photogrammetric and laser scanning techniques. A geodetic network was also measured with a total station and adjusted via least squares. The network provided a stable reference system for registering images and laser scans.

A dense point cloud of the entire castle was acquired with a Faro Focus 3D. 176 scans were registered into a consolidated point cloud of about 7.5 billion points. Average registration precision was ±2-3 mm. The point cloud provided the data to generate a detailed BIM of the castle, which was realized in Autodesk Revit (see next sections).

Specific elements, such as the umbrella vault, were also surveyed from a set of images acquired with a calibrated camera. Fig. 2 shows the phases required to acquire the images and process the results for the umbrella vault. An image block of 35 images was acquired with a Nikon D700 and then processed with PhotoModeler. A dense point cloud was then extracted to obtain a 3D model of the vault with greater level of detail than that achievable from laser scanning. In addition, high resolution orthophotos were generated to document the conditions of the frescoes.

More details about the entire workflow for the survey are illustrated in [2].

Figure 2. The image acquisition and processing phases for the umbrella vault.

3. From point clouds to BIM

Starting from laser scanning and photogrammetric data, a geometric reconstruction with a high level of detail can be achieved. However, BIM is much more than a simple interpolation to represents the external surfaces. BIM requires the objects to be parametric and consistent in their relationship with other objects.

The methodology for parametric BIM generation used for the castle is based on a preliminary separation between simple and complex shapes. In the case of simple objects, the tools available in most commercial software (Revit in this case) can be sufficient. The case of irregular objects (e.g. vaults) is much more complicated. Indeed, the modelling tools of commercial BIM packages are not adequate to represent the geometric level of detail encapsulated into laser point clouds. For this reason, the procedure described in [3] was used.

This procedure was also used for the umbrella vault, for which the thickness was added as a parametric attribute. Pure modeling was carried out for the intrados captured by laser scanning, whereas the unknown thickness was parametrized to recreate the volumetric objects in a dynamic way. Unfortunately, the offset of a surface is not a trivial operation [4]. In fact, the normal direction has to be defined for each point of the surface. Changes in the curvature or geometric discontinuities can be a real challenge for an automatic offset, and lead to multiple object intrusions. This problem is still partially unsolved [5], so the previous pure modeling step was repeated, trying to smooth the surface and simplifying the generative lines used to create the original surface.

Fig. 3 shows the results for the umbrella vault modeled from point clouds (top) and then turned into a parametric object in Revit. It is rather clear that the model with surfaces cannot be the final BIM, but just an intermediate step in the process to create a real BIM object with relationships to other objects. Fig. 3 (bottom) shows the result obtained for the entire tower, which combined the proposed approach for complex shapes with the "simple objects" (in terms of geometry) modelled with the tools available in Revit.

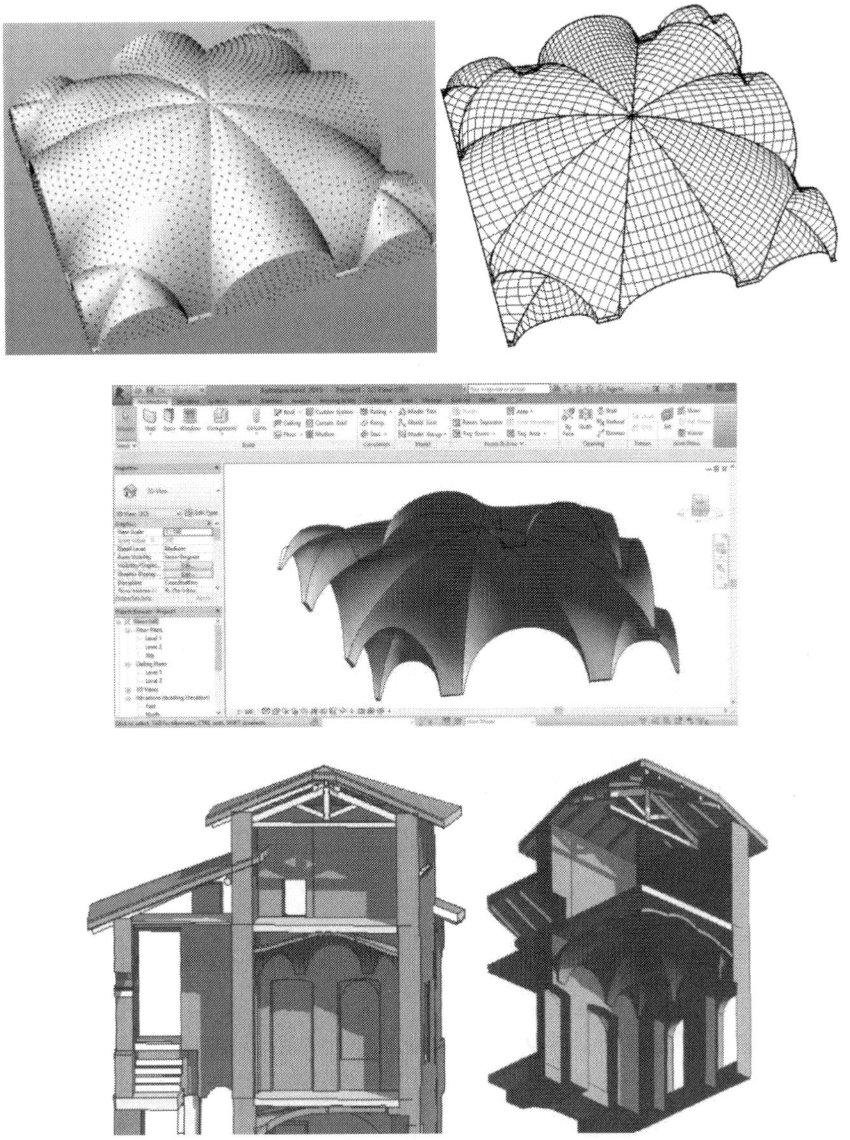

Figure 3. The point cloud used to generate the surface model of the umbrella vault (top). The creation of a parametric vault in Revit (middle) and the final model of the whole tower (bottom).

IOP Conf. Series: Materials Science and Engineering **364** (2018) 012087 doi:10.1088/1757-899X/364/1/012087

4. The finite element model

The objective of this section is to illustrate the procedure adopted to convert the complex shapes that characterize the building (mainly vaults and irregular walls) into a mesh for structural analysis. The software chosen was ©Midas FEA, because it deals with 3D elements. It also gives the possibility to import and mesh complex 3D geometry from existing models.

An accurate description of the distribution of masses and external forces is essential for both gravity and seismic analyses. Moreover, a good model of the structure should also provide a good description of the geometrical irregularities. Different works (e.g. [6]) have highlighted the importance of an accurate model based on photogrammetric or laser scanning technique as the first step for an accurate structural analysis of these types of structures. Given the complexity of the shape, it was decided to use tetrahedral elements. A mesh size of 0.2 m was set considering the thickness of the load-bearing walls, so that any wall would have at least 3-4 mesh elements along its thickness. Such choice was made to limit the number of elements (which otherwise could result to be unmanageable from a computational point of view) whilst still ensuring a good description of the stress distribution along the wall thickness.

The conversion of the BIM into a mesh for finite element analysis cannot be done directly in ©Midas. It was decided to use ©Rhinoceros and ©Inventor for specific operations, as illustrated in [2]. Then, a time-consuming adaptation of the model was carried out to obtain a mesh. Some operations were carried out for the openings, which were simplified. The imprecisions due to the ©Revit joining automatism were removed and corrected as well. The solids representing the different parts of the walls were joined together, toward the creation of a unique solid for the entire wall structure of the castle. Some of these operations were possible by using classical Boolean operations (union, subtraction and division), but a lot of time consuming work had to be done on the different surfaces forming the solids.

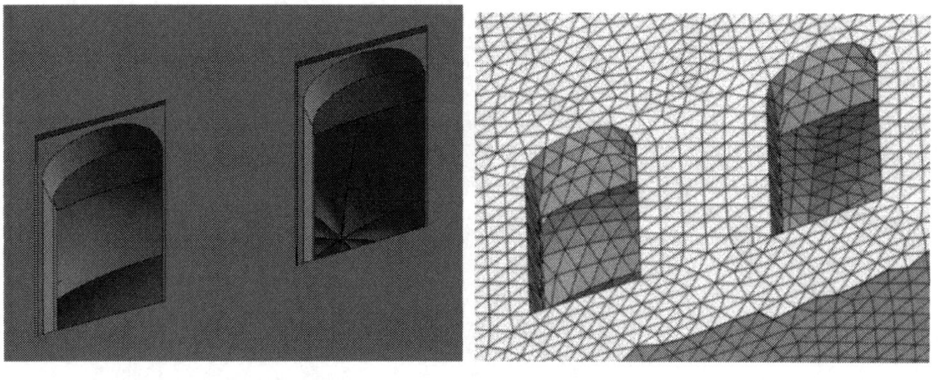

Figure 4. Simplification of the openings: model in ©Revit (a) and in ©Midas (b)

The vault required a significant amount of extra-work. In fact, the extrados had to be completely remodeled in ©Rhinoceros. The parametrization of the vault thickness obtained in ©Revit with the automatic offset had some errors (intersecting surfaces), that made it impossible to perform any Boolean operation to create a unique solid, so it was again necessary to adopt a pure modelling strategy. Some attempts were made using the automatic offset tool in ©Rhinoceros, that can deal with polysurfaces, but such approach works only with simple cases, while for the umbrella vault the offset surface generated had an inconsistent geometry with the wedge tips converging.

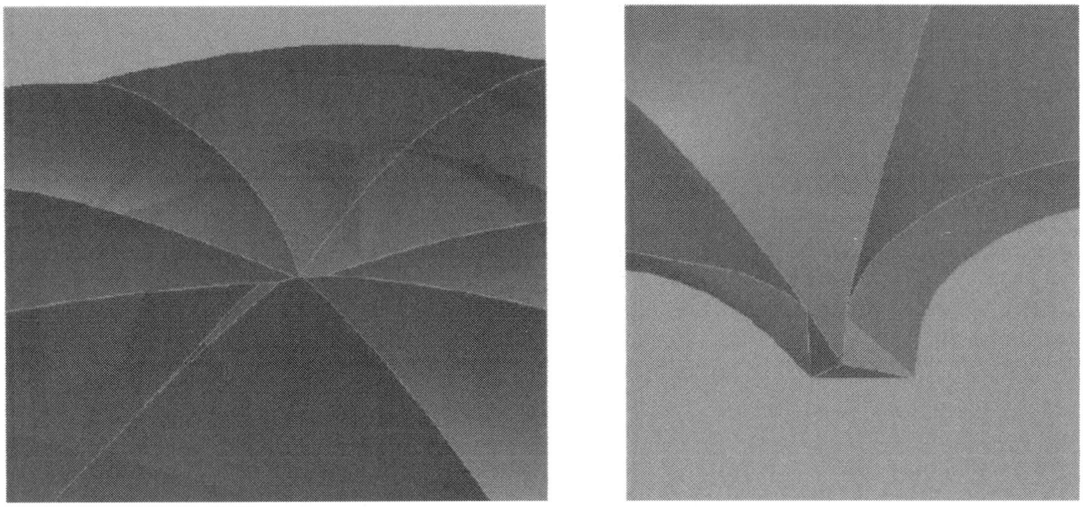

Figure 5. Problems related to the offset for the umbrella vault.

For this reason, it was necessary to extract the borders and isocurves of each wedge, to fix them, and to remodel each extrados wedge. At the beginning, the vault was kept separated from the rest of the model. After considering the difficulties encountered when trying to obtain a compatible mesh in these zones, it has been decided to include it in the whole solid.

Figure 6. Some images of the original model and the final mesh in Midas.

5. The Finite Element Analysis

Once the model was successfully meshed, the geometrical basis for a structural analysis was set. The next step included the setting of boundary conditions, material properties, and loads as a function of the type of analysis. In this work, it was decided to run a linear elastic analysis under self-weight. This is a fundamental step before any other more sophisticate analysis: it allows to adjust the boundary conditions and to check the model to detect any error, because strange stress or deformation patterns can easily be detected under self-weight only. It can be run many times (limited CPU cost), correcting errors or resetting boundary conditions. Moreover, the obtained stress values can be directly compared with single-flat jack measurements.

The reader is referred to [2] for more details about setting of material properties, loads and boundary conditions. This section will illustrate only some results obtained from the simulations run on the model described in the previous paragraph. Shown in Fig. 7 is the distribution of first principal stress. One interesting behavior can be found in correspondence of two openings (partially infilled) which weakens the wall itself right where the load from the vault is transferred to the walls. Here it is possible to notice a zone of tensile stress concentration exactly in correspondence of two repaired cracks.

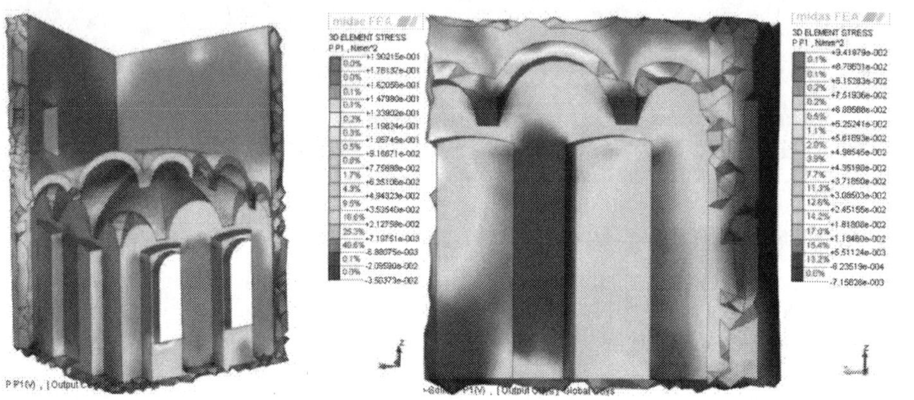

Figure 7. Simulation results (first principal stress) for room with the umbrella vault.

Some interesting results ware also observed by applying the live load provided by the Italian regulation (NTC, 2008) for the category C3 of spaces susceptible to crowding (q_k = 5 kN/m^2). Fig. 8 shows the difference with and without live load (the same scale for the color bar is set to highlight the differences). A significant increase of the tensile stresses is visible (almost doubled in the most critical zones), especially in the corner lunettes. A comparison of these results with photographs confirm that these areas have several cracks, which were repaired during previous restoration works.

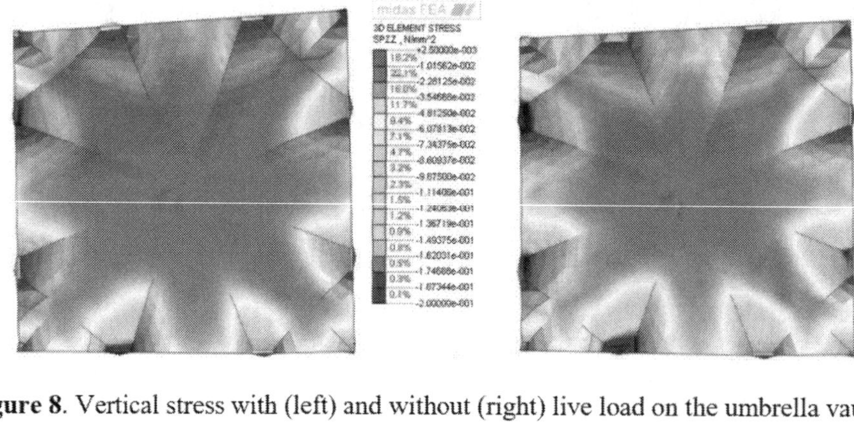

Figure 8. Vertical stress with (left) and without (right) live load on the umbrella vault.

6. Conclusions

This paper ha presented the workflow to generate a detailed BIM of an historical building from a laser scanning and photogrammetric survey. The BIM was then converted into an accurate 3D finite element model.

This approach can be useful for studying the structural behavior of historical buildings, especially when they present large geometric irregularities. In particular, an accurate geometric model can be a valid tool to understand the behavior of particular elements such as vaults or irregular walls. At the end we can state that the proposed approach can be an effective possibility, notwithstanding several difficulties and limitations found. Defining the entire workflow and the procedures during the different steps was quite problematic. However, after resolving the previous limitations, guidelines for operators can be prepared. We think that the effort required to transform a BIM into a compatible model for structural analysis can be acceptable.

The advantages in terms of quantity and level of detail of information that can be provided by such model advanced model is another aspect that requires further work. It would be interesting to compare the results given by this method with a more simplified model, to be able to quantify pros and cons for this approach.

Acknowledgments

This work was supported by the Interreg project "La Conservazione Programmata nello Spazio Comune Retico" (CPRE).

References

[1] Eastman, C., Teicholz, P., Sacks, R., Liston, K., 2008. BIM Handbook - A guide to Building Information Modeling for owners, managers, designers, engineers, and contractors. John Wiley & Sons, Inc.

[2] Gusmeroli, G., Schiantarelli, G., 2014. From Laser Clouds to BIM and Finite Element Analysis: the Case Study of Castel Masegra. Master Thesis in Civil Engineering for Risk Mitigation, Politecnico di Milano, 138 pages.

[3] Barazzetti, L., Banfi, F., Brumana, R., Previtali, M., 2015) Creation of Parametric BIM Objects from Point Clouds Using NURBS. The Photogrammetric Record, Vol. 30(152), pp. 339–362.

[4] Piegl, L. A., Tiller, W., 1999. Computing offsets of NURBS curves and surfaces. Computer-Aided Design, 31(2):147–156.

[5] Barazzetti, L., 2016. Parametric as-built model generation of complex shapes from point clouds. Advanced Engineering Informatics, 30, pp. 298-311.

[6] Riveiro, B., Caamano, J.C., Arias, P. and Sanz, E., 2011. Photogrammetric 3D modelling and mechanical analysis of masonry arches: An approach based on a discontinuous model of voussoirs. Automation in Construction, 20:380-388.

[7] Della Torre, S., 2017. Knowledge and modeling, that is the several ways to understand "constraints". in Construction art in Campania between restoration and structural safety, ISBN 978-88-8497-620-8, 5 pages.

IOP Conf. Series: Materials Science and Engineering **364** (2018) 012088 doi:10.1088/1757-899X/364/1/012088

The ARROWS Project: robotic technologies for underwater archaeology

B Allotta[1,2], R Costanzi[1,2], A Ridolfi[1,2], O Salvetti[3], M Reggiannini[3], M Kruusmaa[4], T Salumae[4], D M Lane[5], G Frost[5], N Tsiogkas[5], M Cocco[6], L Gualdesi[6], G Lacava[6], D Roig[7], H T Gundogdu[8], M I C Dede[13], S Baines[9], S Tusa[10], P Latti[11], D Scaradozzi[12,2]

[1] Department of Industrial Engineering Florence, University of Florence, Florence, Italy
[2] Interuniversity Center of Integrated Systems for the Marine Environment (ISME), Italy - www.isme.unige.it
[3] Institute of Information Science and Technologies (ISTI), National Research Council of Italy (CNR), Pisa, Italy
[4] Centre for Biorobotics, Tallinn University of Technology, Tallinn, Estonia
[5] Heriot-Watt University, Ocean Systems Laboratory, Edinburgh, Scotland
[6] Edgelab s.r.l., La Spezia, Italy
[7] Albatros Marine Technologies S.L., Palma de Mallorca, Baleares, Spain
[8] NESNE Electronic, Izmir, Turkey
[9] The Welding Institute, TWI Ltd, Cambridge, UK
[10] Regione Sicilia, Soprintendenza per i Beni culturali e Ambientali del Mare, Palermo, Italy
[11] Eesti Meremuuseum, Tallinn, Estonia
[12] Dipartimento di Ingegneria dell'Informazione, Universitá Politecnica delle Marche, Ancona, Italy
[13] Izmir Institute of Technology, Izmir, Turkey

E-mail: a.ridolfi@unifi.it

Abstract. The paper summarizes the main results achieved during the three-year European FP7 ARROWS project (ARchaeological RObot systems for the Worlds Seas). ARROWS concluded at the end of August 2015 and proposed to adapt and develop low-cost Autonomous Underwater Vehicle (AUV) technologies to reduce the operational cost of typical underwater archaeological campaigns. The methodology used by ARROWS researchers identified archaeologists requirements for all the phases of a campaign. These were based on guidelines issued by the project Archaeology Advisory Group (AAG), which comprised of many European archaeologists belonging to the consortium. One of the main goals of the ARROWS project was the development of a heterogeneous team of cooperating AUVs; these comprised of prototypes developed in the project and commercially available vehicles. Three different AUVs have been built and tested at sea: MARTA, characterized by flexible hardware modularity for easy adaption of payload and propulsion systems, U-CAT, a turtle inspired bio-mimetic robot devoted to shipwreck penetration and A-Size AUV, a small light weight vehicle which is easily deployable by a single person. The project also included the development of a cleaning tool for well-known artefacts and maintenance operations. Results from the official final demonstrations of the project, held in Sicily and in Estonia during Summer 2015, are presented in the paper as an experimental proof of the validity of the developed robotic tools.

IOP Conf. Series: Materials Science and Engineering **364** (2018) 012088 doi:10.1088/1757-899X/364/1/012088

1. Introduction

The interest in the underwater field development of specific tools for the sub-sea inspection and in particular the use of Autonomous Underwater Vehicles (AUVs) is increasing [1] [2] [3] [4] [5]. ARROWS project, acronym for ARchaeological RObot systems for the Worlds Seas (http://www.arrowsproject.eu), started in September 2012 and concluded in August 2015. Its consortium comprised expertise from underwater archaeology, underwater engineering, robotics, image processing and recognition from academia and industry. The challenge faced by ARROWS was to generate innovative solutions and to adapt existing technologies from the fields of military, security and offshore oil and gas applications, in order to develop several user friendly and low cost robotic technologies for underwater archaeological investigation. The underwater archaeology field as well as other research fields e.g. biology or geology are not able to exploit traditional AUV technologies because of prohibitive costs. The methodology that led the ARROWS researchers was to identify the archaeologists requirements in all the phases of a campaign, based on the guidelines issued by the project Archaeology Advisory Group (AAG), composed of many European archaeologists belonging to the consortium and not. The archaeological requirements and needs have been translated into technical specifications by the ARROWS engineers. The results related to the first half of the project are given in [6]. In the second half of ARROWS, the final demonstrations were performed both in the Mediterranean and Baltic Seas. Both acoustic sensors (e.g. forward-looking sonar, Side-Scan-Sonar) and optical cameras are integrated on the vehicles to deal with all the possible visibility conditions and to "see" even in the presence of high turbidity. As concerns the main requirements, the archaeologists expect the AUVs to return to identified targets with a high level of positioning accuracy, in order to produce precise archaeological underwater maps. Moreover, the option to produce smaller size AUVs which can be easily handled and guided by less specialized operators which, consequently, reduces operational and logistics expenses. The reduced size decreases costs, but also allows deployment and recovery from a small boat. In the design phase, to fulfill the requirements, two main decisions were taken: balancing different capabilities between distributed cooperating vehicles, constituting a heterogeneous team composed of pre-existing (commercial) vehicles and three new prototypes developed from scratch, in order to maximize the number of possible configurations; the other design phase considered the modularity of the hardware and software. This allowed for a rapid reconfiguration of the AUVs capabilities. Another of the required capabilities for the ARROWS system was for shipwreck penetration. The danger of being trapped inside the wreck is very serious and may cause the loss of expensive equipment, or even human life. To comply with the requirements, the development of a small biomimetic turtle-shape AUV was proposed.
ARROWS outcomes also fulfill the requirements of 2001 UNESCO Convention for the Protection of Underwater Cultural Heritage. According to such convention the archaeological objects found on the seabed will be left in their original context. Moreover, some classes of well-known sunken artefacts require periodic cleaning; in the framework of the project a cleaning device (Cleaning Tool), was developed and successfully tested as a standalone item [7]. The paper summarizes the main ARROWS results achieved at sea with the vehicles.

2. Developed Robotic Vehicles

In this paragraph a brief description of the vehicles developed within ARROWS to satisfy the identified archaeological requirements is given.

2.1. MARTA

MARTA AUV is one of the main outcomes of the ARROWS project. It is an AUV prototype developed and built from scratch. MARTA final version is shown in Figure 1 where the vehicle is navigating on surface of the Baltic Sea. It is composed of many modules, each one

IOP Conf. Series: Materials Science and Engineering **364** (2018) 012088 doi:10.1088/1757-899X/364/1/012088

Figure 1. MARTA AUV in autonomous navigation during ARROWS experiments (Estonia, July 2015)

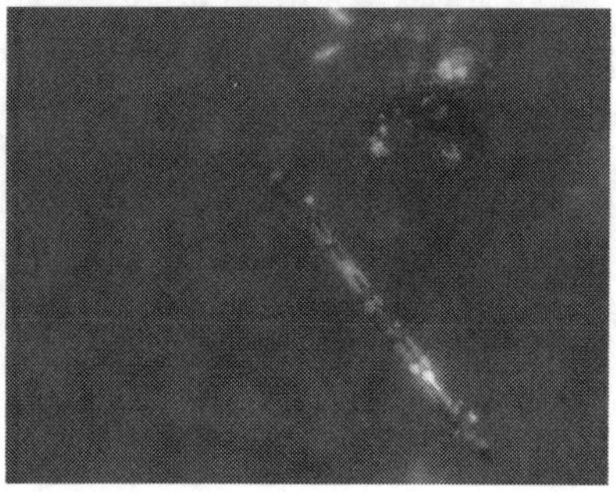

Figure 2. A-Size AUV in autonomous navigation during ARROWS experiments (Sicily, June 2015)

dedicated to a particular task (e.g. propulsion, sensor payloads, power supply, etc.). Due to its modularity and reconfiguration capability, MARTA AUV can be easily customized according to the archaeological mission to be performed. The vehicle is deployable from a small boat, has a total length of 3.7 m (in its longest and complete configuration), an external diameter of 180 mm and an in-air weight of 80 kg. Its maximum reachable depth is 120 m and the maximum speed 3 kn relates to an autonomy of about 4 hours. The typical cruise speed is 1 kn. MARTA, either on the surface or underwater, is able to perform hovering: it has 5 degrees of freedom (DOFs) fully controllable by means of 6 actuators (electric motors + propellers): 2 rear propellers, 2 lateral thrusters and 2 vertical thrusters. MARTA AUV becomes watertight only after the modules are connected together (the only wet sections are the final part of the bow and the final part of the stern). After the whole vehicle is assembled, a depressurization of 0.3 bar is created inside by means of a vacuum pump to reach a good alignment among the modules and to provide an adequate stiffness. MARTA can house both acoustic and optical payload.

2.2. A-Size

In the wide spectrum of underwater activities, suitable applications for A-Size vehicles (120 mm diameter) in the Underwater Cultural Heritage have been identified. Innovative aspects address underwater operations with a fully integrated and modular approach to optimize volume and weight distribution in the vehicles therefore making it easy to interface and host major AUV payloads. The 120 mm diameter solution called A-Size (Figure 2) corresponds to a vessel length of 1300 mm. A lightweight construction of the hull must be observed: carbon fibre was selected for its specific high Poisson modulus to withstand hydrostatic compression thus

Figure 3. U-CAT biomimetic 4-fin AUV

preserving lightness. The vehicle has a 300 Wh battery, with an endurance of 4 hours in transit navigation. To preserve maneuverability, directional propellers (bow/stern thrusters) were adopted as opposed to active vanes. Main propulsion is obtained through a brushless motor fitted with magnetic coupling. Propeller design variables were fixed after practical and experimental tests. The A-Size vehicle uses an ARM9 based micro-controller board, with Linux configured for embedded applications. While the vehicle is on the surface it may be connected through an optical fibre and communicate with a GPS receiver through an Ethernet connection or be in contact with the operator through a WiFi hotspot.

2.3. U-CAT

U-CAT is a highly experimental biomimetic AUV for shipwreck penetration (Figure 3). Shipwreck penetration is a difficult task that is not usually undertaken due to the high risk of losing equipment or human life. Divers can easily get lost or stuck in wrecks. Remotely Operated underwater Vehicles (ROVs) cannot enter wrecks because of tangling their tethers and current AUVs are mostly not designed for maneuvering and navigating in confined spaces. U-CAT is an AUV/ROV which attempts to fill the gap by providing a low cost tool for gathering video footage from inside shipwrecks. U-CAT is not meant to navigate in complex corridors of shipwreck, but it is designed to be used in relatively simple scenarios, which can still often offer large amounts of useful information. For example autonomous inspection of a single room in danger of collapse. Instead of propellers U-CAT uses 4 pitching fins for locomotion. Such biomimetic design has several advantages which help to meet the archaeologists requirements. One of the main advantages is the versatile maneuverability in tight spaces. The fins are configured such that the vehicle can move in all 6 DOFs using only 4 motors. Small actuators help to make the vehicle as small as possible allowing it to fit through narrow passages and to be easily portable. Another advantage of using fins is that they provide distributed propulsion, so reducing the amount of sediment brought up from the bottom. They also do not get tangled as easily as propellers and they are safe for the surrounding environment and for divers who may have to work together with the robot. To concurrently comply with the requirements of low final cost, small size and navigation in confined spaces, several sensors and vehicle modules have been custom developed during the project. U-CAT has 8 custom echo-sounders for obstacle detection and a custom hydrophone array for acoustic beacon localization. It also has custom buoyancy control pistons in each end allowing to actively trim the vehicle. From commercial sensors U-CAT contains a digital camera, several system health monitoring sensors, miniature Inertial Measurement Unit (IMU), depth sensor and an acoustic modem with range measurement capability. The battery allows at least 6 hours of autonomous operation. The control is implemented on a 1 GHz Quad-Core ARM computer using the Robot Operating System (ROS) middleware. The robot weighs around 19 kg, it is 0.6 m long and can dive to 100 meters.

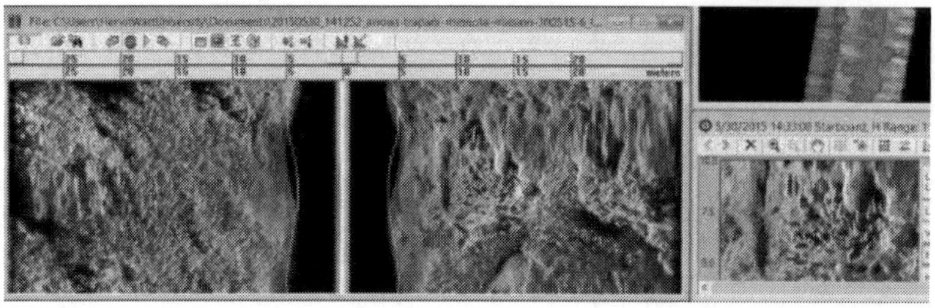

Figure 4. SSS image of the Amphoras at the Cala Minnola archaeological site

3. Final Demonstration in the Mediterranean Sea

During the trials throughout the ARROWS project, various missions were performed using the IVER-3 AUV. Its role as the search vehicle in the framework of ARROWS was to verify that given an approximate location of an archaeological site, we could accurately discover its location using an AUV and obtain good quality Side-Scan-Sonar (SSS) images; achieving these search goals allows closer inspection of the site using the ARROWS I-AUVs, MARTA or U-CAT. The first set of trials were performed in Sicily at a site called Cala Minnola. This site was known to contain archaeological artefacts of amphoras. In most locations, some prior information about the sub-surface topology is available. In some areas, this may be topographic maps or bathymetric data obtained by ships equipped with SSS. Even with such data known beforehand, caution should still be used when performing missions with AUVs. Using large-scale topographic maps, and therefore a low resolution, bathymetric map of the Cala Minnola site, we still assumed that the environment was initially unknown. Thus, the first mission was planned as reconnaissance, aiming to obtain detailed information of the natural features and sea floor. It was performed at a suitable altitude to ensure the safety of the vehicle and site, whilst able to record some sea floor data. As the uncertainty of the operating environment decreased, from the acquisition of more information, we could plan missions closer to the optimum SSS search parameters. Through experience using the IVER-3 AUV, we have found that an altitude of between 5-8 metres from the sea floor gives the best SSS images. The SSS images themselves can be used to determine the safety of the AUV by looking at the dark central region of the image. During one mission at Cala Minnola, shown in Fig. 4 which is very rocky, we noticed from the SSS image that the AUV only missed rocks by a approximately one metre. Learning of the steep rocks in the area, we determined safe mission parameters which allowed us to survey a large area and successfully find the Amphora site, shown in Fig. 4. The blurring in the image is hypothesized to be due to the high sea currents that were present in that area, coupled with the AUV's navigation being based only on dead reckoning (i.e. no acoustic positioning system was used). The AAG archaeologists at the sea trial were satisfied with the detail of the results.
After the acoustic search phase, MARTA was used as the inspection vehicle to acquire optical images of the site. Images of the field of amphorae in the Cala Minnola site were thus collected during autonomous missions and in Fig. 5 the related 3D reconstruction is reported (please see http://www.arrowsproject.eu/media-center/trials/3d-reconstructions-levanzo-sicily-and-rummu-quarry-estonia/) [8].

4. Final Demonstration in the Baltic Sea

The second set of tests was performed in Rummu, Estonia, at an artificial lake (flooded quarry).

Figure 5. 3D reconstruction of Cala Minnola site from images acquired by ARROWS AUVs

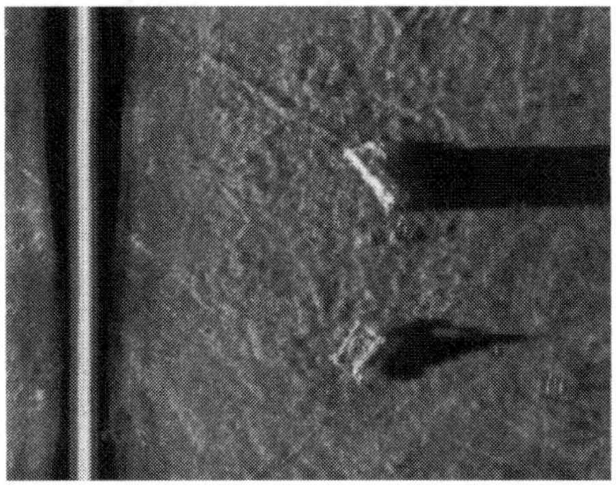

Figure 6. SSS image of the quarry machinery found in the Rummu quarry, Estonia

4.1. Acoustic and optical inspections in the Rummu quarry

The strategy employed during the Estonia trials was the same as used in the previous trials in Sicily. One of the main differences between the Cala Minnola site and Rummu quarry was their respective depth profiles. A substantial proportion of the Rummu quarry site is around ten metres depth or less. From the first SSS image, it was also apparent that there was a significantly larger number of features to be seen in the Rummu quarry than most sites we have visited at sea, consisting of roads, walls of buildings, and even traffic barriers were clearly visible in images taken of the quarry. During a mission which had been programmed to cover a large portion of the South-East bay of the quarry, several objects were particularly interesting. Through discussions with locals, we heard that there were several pieces of old quarry machinery that had not managed to be removed before the quarry was flooded but their locations were unknown. Having discovered likely candidates that could be the machinery, a mission was performed that took SSS images from all four sides of the objects. One of these images is shown in Fig. 6. From the objects' shape and size, including their height which from their shadows indicates that the top object is significantly larger than the bottom, it was determined that we had found the quarry machinery. Having obtained the geo-referenced data of the machinery, we could pass that data on to the IAUV operators who could inspect that area to validate that the coordinates we obtained using the IVER-3 AUV were correct, and to obtain further data of the machinery. After the acoustic search phase, the GPS points of the targets identified by the IVER3 were passed to MARTA. MARTA was thus used as the inspection vehicle (as in Sicily during the first Demo) to acquire optical images of the targets. An Inspection campaign was planned with MARTA AUV aimed at collecting optical images through its High Definition camera. The reference path and all the necessary mission parameters were defined through the graphical

IOP Conf. Series: Materials Science and Engineering **364** (2018) 012088 doi:10.1088/1757-899X/364/1/012088

Human Machine Interface (HMI) developed within the project with the aim of simplifying the interaction of a non-specialized end-user (archaeologists in ARROWS) with the developed tools [1]. The inspection mission was composed of two different lawn-mower submissions: the first one with path legs parallel to the line connecting the two target points and the second one with path legs orthogonal to it. Processing the dataset collected in the Rummu quarry by means of Structure from Motion resulted in the generation of a 3D reconstruction given in Fig. 7.

Figure 7. 3D reconstruction of the Rummu quarry machineries from images acquired by ARROWS AUVs

4.2. Penetrating Machinery and Buildings with U-CAT

One of the located and inspected machines was also further investigated using the U-CAT vehicle. The machine has a small engine room whose interior is fully visible from the hatches on the top and side. Therefore, there was no need to perform a fully autonomous penetration mission. The U-CAT vehicle was remotely controlled using a live feedback from the onboard camera. As the goal was to demonstrate the performance in confined areas, the vehicle was guided to different locations in and around the mining machine. For example, Figure 8a shows the U-CAT entering the machine room. During the mission no major sediment disturbance caused by the robot was detected. Also, the robot did not get tangled and the operator was able to guide it to desired points of interest. However, the manual control of the vehicle was relatively difficult due to

[1] A video tutorial describing the usage simplicity of the ARROWS HMI was produced in the framework of the project and is available at the link: https://www.youtube.com/watch?v=fHNRCCMtgFE

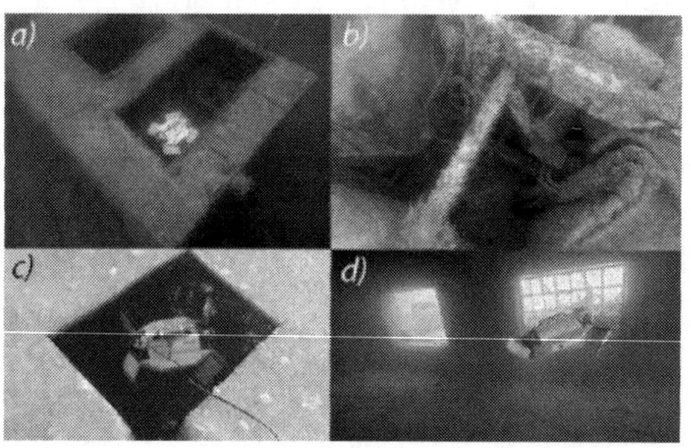

Figure 8. U-CAT on missions: a) Inspection of a machine room; b) Onboard camera image about the hydraulic mechanism of the machine; c) U-CAT launched through ice hole; d) autonomous inspection of a room

the high number of controllable DOFs. To simplify the inspection, the control was improved by implementing a stable autopilot for depth and yaw based on model-based controllers. To test the autopilot and to also test the vehicle in more difficult conditions, the mining machine inspection was repeated in March 2016. In March the lake was covered by 20 cm of ice. The robot had to be launched from a small hole cut into ice Figure 8c. Thanks to improved control and video quality, even more detailed visual information was gathered about the machine. Another mission with U-CAT was performed in Rummu quarry to demonstrate its usage in fully autonomous mode. The goal of the mission was to autonomously inspect a single room in one of the submerged prison buildings in the quarry (Figure 8b). The investigated room has dimensions of 10 m x 5.5 m. It has one large opening in one wall and several open doors and window holes on other walls. The U-CAT was launched by a diver from the large opening. In the first experiment the robot followed a predefined trajectory and then exited the building through the large opening. In the second experiment the vehicle also used its custom-made sonars to track and avoid the walls. The experiments demonstrate one possible scenario for using U-CAT in dangerous penetration missions.

5. Conclusions

The paper briefly summarized the main results achieved during the ARROWS (ARchaeological RObot systems for the Worlds Seas) European project, describing some of the tests at sea of the developed Autonomous Underwater Vehicle (AUV) technologies for archaeology. The tests started evaluating the performances of single AUVs before the final demos where the whole system, the heterogeneous fleet, was deployed at sea for a collaborative mission. Satisfying results at sea have been reached thanks to the collaboration among commercial vehicles and the ARROWS new prototypes. Good feedback were collected by the archaeologists during the tests, highlighting the potentiality of the proposed technologies.

Acknowledgments

This work has been supported by the European ARROWS project that has received funding from the European Unions Seventh Framework Programme for Research technological development and demonstration, under grant agreement no. 308724.

References

[1] Bingham B, Foley B, Singh H, Camilli R, Dellaporta K, Eustice R, Mallios A, Mindell D, Roman C and Sakellariou D. 2010 Robotic tools for deep water archaeology: Surveying an ancient shipwreck with an autonomous underwater vehicle *J. Field Robotics* **27** 702-17.

[2] Mahon I, Pizarro O, Johnson-Roberson M, Friedman A, Williams S and Henderson J 2011 Reconstructing Pavlopetri: mapping the world's oldest submerged town using stereo-vision *Proc. of ICRA 2011 (USA)*.

[3] Gracias N et al. 2013 Mapping the moon: Using a lightweight AUV to survey the site of the 17th century ship "La Lune" *Proc. of Oceans IEEE (Bergen, Norway)*.

[4] Roman C and Mather R 2010 Autonomous underwater vehicles as tools for deep-submergence archaeology *Proc. of the Institution of Mechanical Engineers, Part M: Journal of Engineering for the Maritime Environment* 224 pp 327-40.

[5] Wynn R B, Huvenne V A I, Le Bas T P, et. al 2014 Autonomous Underwater Vehicles (AUVs): Their past, present and future contributions to the advancement of marine geoscience, *Marine Geology* **352** 451-68.

[6] Allotta B et al. 2015 The ARROWS project: adapting and developing robotics technologies for underwater archaeology *IFAC-PapersOnline: 4th IFAC Workshop on Navigation, Guidance and Control of Underwater Vehicles (NGCUV 2015) (Girona, Spain)* 48(2) pp 194-99.

[7] Tolasa Gündoğdu H, Dede M I C, Taner B, Ridolfi A, Costanzi R and Allotta B 2015 An innovative Cleaning Tool for underwater soft cleaning operations *Proc. of MTS/IEEE OCEANS'15 (Genova, Italy)*.

[8] Sorbi L, Scaradozzi D, Conte G and Gambogi P 2012 Geoposition data aided 3D documentation of archaeological and biological sites *Proc. of the 11th European Conference on Underwater Acoustics Edinburgh (ECUA 2012) (Edinburgh, Scotland)*.

Enhancing Heritage fruition through 3D semantic modelling and digital tools: the INCEPTION project

Federica Maietti, Roberto Di Giulio, Emanuele Piaia, Marco Medici, Federico Ferrari

Department of Architecture, University of Ferrara, via Ghiara 36, 44121, Ferrara, Italy

federica.maietti@unife.it

Abstract. The INCEPTION project, "Inclusive Cultural Heritage in Europe through 3D Semantic Modelling", started in June 2015 and lasting four years, aims at developing advanced 3D modelling for accessing and understanding European cultural assets. One of the main challenges of the project is to close the gap between effective user experiences of Cultural Heritage via digital tools and representations, and the enrichment of the scientific knowledge. Within this framework, the INCEPTION project goals are consistently aligned while accomplishing the main objectives of accessing, understanding and strengthening European cultural heritage by means of enriched 3D models. At the end of the third year of activity, the project is now facing different challenging actions starting from already developed advancement in 3D data capturing and holistic digital documentation, under interdisciplinary and cross-cutting fields of knowledge. In this direction, the approach and the methodology for semantic organization and data management toward H-BIM modelling will be presented, as well as a preliminary nomenclature for semantic enrichment of heritage 3D models. According to the overall INCEPTION workflow, the H-BIM modelling procedure starts with documenting user needs, including experts and non-experts. The identification of the Cultural Heritage buildings semantic ontology and data structure for information catalogue will allow the integration of semantic attributes with hierarchically and mutually aggregated 3D digital geometric models for management of heritage information.

1. Introduction and overall framework

At the end of the third year of activity, the project "INCEPTION - Inclusive Cultural Heritage in Europe through 3D Semantic Modelling", funded by EC within the Programme Horizon 2020, is facing different challenging actions aimed at models deployment and widespread sharing, starting from a methodological and technical advancement in 3D data capturing and holistic digital documentation.

As part of 3D integrated survey applied to Cultural Heritage, digital documentation is gradually emerging as effective support of many different information in addition to the shape, morphology and dimensional data [1]. The implementation of data collection processes and the development of semantically enriched 3D models is an effective way to enhance the dialogue between ICT technologies, different Cultural Heritage experts, users and different disciplines, both social and technical. The possibility to achieve interoperable models able to enrich the interdisciplinary knowledge of European cultural identity is one of the main outcomes of INCEPTION.

The enhancement of digital cultural heritage accessibility is the ability to access cultural contents and resources for as many people as possible by using ICT functionalities and applications (web sites,

IOP Conf. Series: Materials Science and Engineering **364** (2018) 012089 doi:10.1088/1757-899X/364/1/012089

data-bases, digital libraries, virtual applications, etc.) overcoming cultural, environmental and management barriers for an easy and spread fruition. Nowadays it is possible to integrate different information in order to access cultural assets in many different ways and for many different purposes, thanks to new languages of interactive media aimed at innovative ways of communication of cultural heritage.

The overall project workflow is developed starting from requirements (what kind of data, information and visualization issues can be collected and managed by a 3D model according to specific users and needs), the integrated data capturing and holistic heritage documentation, the semantic enrichment via 3D modelling in H-BIM environment, and the models deployment and valorization through the INCEPTION platform.

This paper will shortly summarize the INCEPTION approach to the semantic modelling, focusing on heritage buildings and sites. Semantic modelling for Cultural Heritage buildings in H-BIM environment and the development of the INCEPTION platform for deployment and valorization of enriched 3D models will allow accomplishing the main objectives of accessing, understanding and strengthening European Cultural Heritage. In this direction, the approach and the methodology for semantic organization and data management toward H-BIM modelling will be presented, as well as a preliminary nomenclature for semantic enrichment of heritage 3D models. According to the overall INCEPTION workflow, the H-BIM modelling procedure starts with documenting user needs, including experts and non-experts. The identification of the Cultural Heritage buildings semantic ontology and data structure for information catalogue will allow the integration of semantic attributes with hierarchically and mutually aggregated 3D digital geometric models for management of heritage information.

2. Holistic digital documentation

Integrated digital documentation under INCEPTION is carried out through a holistic approach. According to "Principles for the recording of monuments, groups of buildings and sites" (ICOMOS 1996) holistic documentation needs to cover assets of critical information regarding a monument's main attributes and characteristics, able to define it as a whole and to identify its significance and main needs.

Beyond technical definitions and in-depth specifications regarding measuring instruments and data capturing devices, it has been essential to frame the workflow within the specific field of Cultural Heritage and within the overall concept of heritage documentation.

An interdisciplinary study of a monument leads to the creation of an enhanced methodology approach for the documentation. The overall approach tends to comprise of the documentation, analysis and management of all the different data, visualization, protection and preservation issues. An interdisciplinary process is required for the representation of the shape, the construction materials and the conservation condition. Complete recording addresses not only to the problem of 3D digitization of a monument but involves all the aspects of digital content management, representation and reproduction. Through the 3D digitization of a monument, various aspects of recording concerning diagnostics (building material and decay patterns), construction durability and architectural documentation, new digital content management may be achieved. In this context, interdisciplinary approach means that the advantages and the output of every method applied to a monument, can lead to the 3D modelling approach, improving an inclusive understanding of European cultural identity and diversity.

This is the main direction toward INCEPTION is moving: setting up a methodology able to break through the barriers caused by segmentation in collecting documentation data by establishing a common framework through 3D documentation of Cultural Heritage buildings and sites.

From one side, applications under development allow accessing digital information both from the heritage sites (via mobile devices) and off-site (offices, research labs, museums, etc.) in order to improve access and knowledge possibilities, thanks to interactive experiences and improved digital contents. On the other hand, enriched 3D digital models available in an open-standard format can be

IOP Conf. Series: Materials Science and Engineering **364** (2018) 012089 doi:10.1088/1757-899X/364/1/012089

easily accessible and reusable by researchers, scientists, experts, curators, and professionals in different cultural heritage fields, facilitating cross-disciplinary researches, dissemination, education and business opportunities.

2.1. Retrieving and aggregating data

Semantic 3D reconstruction of heritage buildings and sites supposes an interdisciplinary approach based on the collaboration of various experts from different disciplines towards the development of 3D models integrating semantic data. If 3D data can be captured following a more commonly shared procedure (set up, within INCEPTION, by the Data Acquisition Protocol), terminology and interpretation have a large diversity and variability when looking at semantics, due to different competencies, skills and languages.

Making data aggregable asks for a Semantic Web structure in order to give back to different users a new way to look at the information in a Cultural Heritage context. A semantic web structure means also that the relations, properties and composition of the "nodes" of the information give a new insight different from a list of the same concepts or a database structure. Since now, consistent structured data are missing in for Cultural Heritage architectures.

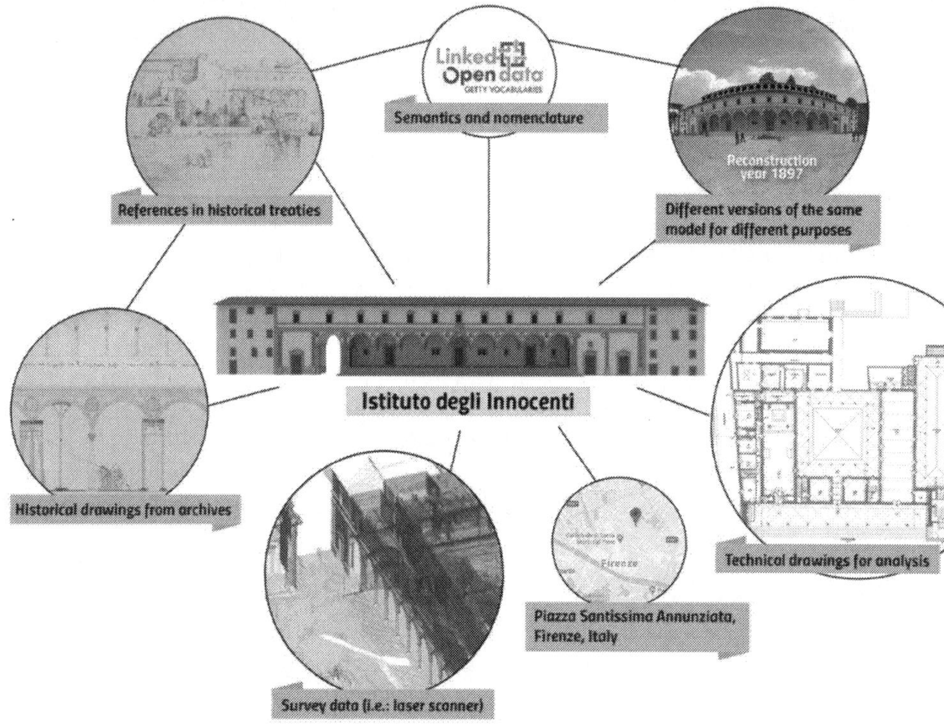

Fig. 1 Schema of the INCEPTION methodology, based on the example of Istituto degli Innocenti by Brunelleschi: from the holistic documentation and data capturing up to semantically enriched 3D models.

3. H-BIM modelling and semantic enrichment

One of the main challenges in 3D modelling is related to an effective BIM approach for cultural heritage knowledge, semantic enrichment and model management. Starting from an improved procedure of 3D data capturing [2] and documentation, INCEPTION proposes an improvement in this methodology by recognizing that buildings are a set of elements named by an architectural style nomenclature and organized by spatial relationships. At the moment, a shared library for historical elements does not exist. Starting from the Heritage Building Information Modelling (H-BIM)

IOP Conf. Series: Materials Science and Engineering **364** (2018) 012089 doi:10.1088/1757-899X/364/1/012089

approach the necessity of the libraries' implementation is under development within INCEPTION by avoiding the oversimplification of the shapes. When used in models of Cultural Heritage, semantic BIM [3] will be able to connect different users (e.g. scholars, technicians, citizens, governments), supporting the need for interpretation of the Cultural Heritage model, in addition to the common BIM features of 3D visualization, technical specification and dataset [4].

3.1. The INCEPTION approach

Starting from a definition of concepts and their relationship, the first step in creating semantic BIM for cultural heritage is defining the ontology: a formal representation of knowledge as a hierarchy of concepts within the cultural heritage domain. The linguistic definitions related to the ontology concepts can also be used as shared vocabulary to denote the types, properties and interrelationships of cultural heritage aspects [5].

Naming each architectural element that composes a building is not an easy task. In the field of classical architecture, it is possible to check several books that deeply analyse and summarize which "rules" were basically adopted in classical buildings. When mainly based on classical orders, ancient architecture is easier to understand and name, even if the building practice often differs from the theory. For instance, just to list some of the most known theoretical books from the past, we can mention *De architectura* by Vitruvio; *De re aedificatoria* by Leon Battista Alberti; and *I quattro libri dell'architettura* by Andrea Palladio, some of major treatises that influenced architectures all over the world for centuries. Nevertheless, local practices introduced several variations, as well there are a lot of different constructive techniques, shapes and decorations that were not standardized. Every building is the final result of different influences and combinations in order to solve practical problems, as well as further additions and changes over time.

For this reason, aiming at the standardization in heritage documentation, data handling and management, the INCEPTION project is developing common parameters, setting a nomenclature or "glossary of names" as a starting point to semantic enrichment and modelling in BIM environment [6]. The recognition of shapes, either manually or automatically performed, is possible only if single architectural elements (or their variations) are identified and univocally classified following a shared procedure.

Over the years, several architectural dictionaries have been produced. Thousands of architectural names have been collected and managed following a specific sorting. A traditional thematic dictionary, for example, collects names in alphabetical order, for a specific language, describing every single name and without the requirement of conceptually linking them together. On the other hand, an architectural treatise explained by graphics gives more consistency to the nomenclature, setting up specific relation between elements.

This defines two different kind of issues within INCEPTION: setting up of a common glossary of names for all Demonstration Cases under development by involving scholars and experts; organizing names following a structure that could ensure the linking between elements which could be re-used in the IT development phase.

3.2. The Nomenclature setting up

The need for the definition of an international "glossary'' for architectural heritage and by extension for Cultural Heritage has arose. This need has originated from the fact that there are various methodologies regarding heritage documentation. As previously explained, various vocabularies and thesauri are used in the field of conservation, while the variety of "uniqueness" of each cultural artefact turns its categorization into a difficult effort. In addition, not only spatial information needs to be standardized, but also the related metadata. In order to face this issue and start setting a methodology to define a shared vocabulary, an extensive analysis of the State of the Art was performed. Herein in the field of Architecture and Heritage, the UNESCO Thesaurus as a general Vocabulary, the UK Archival Thesaurus as a general archive, the Centre National d'Archaeologie Urbaine for Archaeology and Heritage sites, the Pactols Thesaurus by Frantiq in the Archaeology field

IOP Conf. Series: Materials Science and Engineering **364** (2018) 012089 doi:10.1088/1757-899X/364/1/012089

and the Getty Art & Architecture Thesaurus (AAT, but also TGN), are some of the Vocabularies checked.

The availability of so many different active sources providing valuable information that would be interesting to be reused, means both valuable but potentially competing information needs to be connected. Since this information is constantly in development, linking is preferable rather than copying a "static" version at a certain date. This is even truer for Heritage nomenclature.

Semantic Web technology and Linked Open Data principles make it possible to define an open H-BIM ontology without having the complexity of defining the complete schema as a copy of external sources. This State-of-the-Art technology is developed by members of the W3C organization and is at the moment a mature technology with mature supporting tools. INCEPTION makes use of the tools available supporting these standards.

The above mentioned vocabularies and thesauri define a nomenclature for the Cultural Heritage domain. Before this valuable knowledge can be used as linked open data the following steps need to be defined: (i) translate the vocabularies / thesauri in a technology that supports linked open data; (ii) interlink the different vocabularies / thesauri and link them together; (iii) filter what is relevant for the H-BIM ontology within INCEPTION.

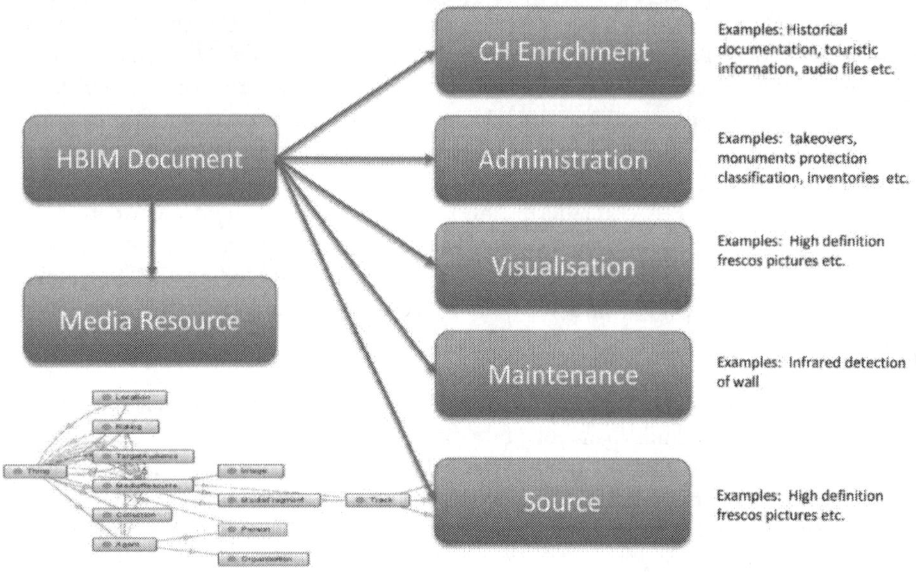

Fig. 2 H-BIM document ontology.

The starting point for semantic organization and glossary definitions for cultural heritage buildings is the integration of the glossary with ifcOWL (in order to start working on H-BIM modelling defining specific parts of the model), ifcOWL as Semantic Web serialization on schema level for IFC schemas will be followed where possible for BIM content stored/exchanged via IFC. Even if the analysis of available nomenclatures is continuously going on, by analysing national and international resources, for the Web service set up, the Getty Vocabularies were chosen as a starting point, including AAT - Art & Architecture Thesaurus, TGN - Getty Thesaurus of Geographic Names and ULAN - Union List of Artist Names.

The Getty vocabularies contain structured terminology for art, architecture, decorative arts, archival materials, visual surrogates, conservation, and bibliographic materials. Compliant with international IT standards, they provide authoritative information for cataloguers, researchers, and data providers. In the new linked, open environments, the vocabularies provide a powerful tool to be adopted within the INCEPTION project, even if a filtering activity is required in order to make them suitable.

Furthermore, where an existing vocabulary will not correctly cover a specific definition, it will be integrated. Current work is creating an ontology as a valuable base for possible upgrading about what is expected/required in H-BIM.

The AAT, TGN and ULAN contain structured terminology for art and other material culture, archival materials, visual surrogates, and bibliographic materials. Compliant with international standards, they provide authoritative information for cataloguers and researchers, and can be used to enhance access to databases and web sites. The Getty Vocabularies are produced by the Getty Vocabulary Program (GVP) and grow through contributions. The procedure under INCEPTION is aimed at extending the H-BIM ontology based on the classification available in Getty, integrating this with the 'BIM' classification generated from the IFC schemas.

4. Towards applications and digital tools for Heritage fruition

So far, the INCEPTION project has defined the approach and the methodology for semantic organization and data management toward H-BIM modelling, and the preliminary nomenclature for semantic enrichment of heritage 3D models. The organization of consolidated knowledge is performed following a specific workflow in order to get them suitable for their reuse into H-BIM semantic model, accordingly to digital documentation and capturing protocols that have been developed.

According to the INCEPTION workflow the H-BIM process starts documenting user needs, including and engaging not only experts but also non-experts. The demand has been leading us to "how" and "what" surveying information to be included in H-BIM. The surveying procedure produces a variety of different data, formats and outputs. It is essential to process that data without losing important information like metadata and paradata while editing and developing the digital elements of the H-BIM. A methodology of archiving digital data and linking them to the final product is one of the main outcomes. Before and during the creation of H-BIM, the nomenclature (vocabularies, thesaurus, etc.) is critical to maintain a common typology and to support interoperability.

Starting from the standardization for H-BIM modelling, the methodology for merging IFC models and semantic data has been defined. The development of a semantic 3D reconstructions, integrated with intangible information and social environment, structuring digital representation of buildings and sites will lead to the creation of models more accessible and implementable in a Heritage-BIM environment, based on Open BIM standard (IFC, IFD, etc.).

Fig. 3 Example of "on-site" application tested in the Italian Demonstration Case, the Istituto degli Innocenti in Florence

Thanks to the use of data collected by Demonstration Cases, the identified methodological workflow and the technical tools and procedures will be further tested and stressed by end of the project. This will allow, from one side, to check the real validity of what has been developed until now and, on the other side, to upload several contents to the platform, bringing it into life and make it a significant dissemination tool.

Now, each INCEPTION demonstration case will produce data that will be hosted by the brand-new platform, allowing to test the validity of the platform itself (3D models and versioning, animations, images, etc.) and further developments (immersive environments, Virtual Reality, Augmented Reality, etc.).

The INCEPTION platform will be populated by using validated Demonstration Cases' results and developing user-oriented applications to fulfil the user requirements, as well as to deliver exploitable innovation with a long-term impact. The deployment of 3D models with integral narratives will be tested on the heritage sites in on-site and off-site applications.

Fig. 4 Digital reconstruction of the façade of Istituto degli Innocenti in different eras through the use of a BIM model.

5. Conclusions

The current work is to create an ontology as a valuable base for possible up-grading about what is expected/required in H-BIM [7]. However, it is important to understand the difference between BIM standards and the Semantic Web technologies. Indeed, the ifcOWL is defined as a serialization of an IFC schema definition, in order to enable the use of contents semantically managed. For this reason, within INCEPTION, H-BIM is meant to be an ontology to support storage of semantic knowledge available for Cultural Heritage buildings and architectural complexes, as well as their related information [8]. The integration of Getty vocabularies and ifcOWL is a complex but feasible task, focusing on the actual queries, communications and current H-BIM work. Furthermore, within INCEPTION, we have the chance to test such integration, achieving positive results as well as beginning an implementation, thanks to the use of a "glossary of names" gathered by the Demonstration Cases analysis.

Once data are collected and aggregated thanks to the use of semantic technologies, the most valuable goal is strictly related to dissemination capabilities. Indeed, the richness of the INCEPTION project is represented by accessibility to a crowdsourced database of both scans and reconstructions, with different level of confidence by their source data.

Therefore, the INCEPTION platform will have to contain, visualize, manage, update, and exchange technical and divulgative information regarding historical heritage, through the use of 3D BIM models. Any form of digital or digitized content, stored or linked into a 3D H-BIM model, as well as semantically indexed by the use of the INCEPTION ontology, will allow the use of different navigation systems.

Acknowledgements
The project is under development by a consortium of fourteen partners from ten European countries led by the Department of Architecture of the University of Ferrara. Academic partners of the Consortium, in addition to the Department of Architecture of the University of Ferrara, include the University of Ljubljana (Slovenia), the National Technical University of Athens (Greece), the Cyprus University of Technology (Cyprus), the University of Zagreb (Croatia), the research centers Consorzio Futuro in Ricerca (Italy) and Cartif (Spain). The clustering of small medium enterprises includes: DEMO Consultants BV (The Netherlands), 3L Architects (Germany), Nemoris (Italy), RDF (Bulgaria), 13BIS Consulting (France), Z + F (Germany), Vision and Business Consultants (Greece).

The INCEPTION project has been applied under the Work Programme Europe in a changing world – inclusive, innovative and reflective Societies (Call - Reflective Societies: Cultural Heritage and European Identities, Reflective-7-2014, Advanced 3D modelling for accessing and understanding European cultural assets).

This research project has received funding from the European Union's H2020 Framework Programme for research and innovation under Grant agreement no 665220.

References
[1] Di Giulio R, Maietti F and Piaia E 2016 3D documentation and semantic aware representation of Cultural Heritage: the INCEPTION project. *EUROGRAPHICS Workshop on Graphics and Cultural Heritage* (The Eurographic Association) pp 195-198

[2] Di Giulio R, Maietti F, Piaia E, Medici M, Ferrari F and Turillazzi B 2017. Integrated data capturing requirements for 3D semantic modelling of Cultural Heritage: the INCEPTION Protocol *The International Archives of the Photogrammetry, Remote Sensing and Spatial Information Sciences*, Volume XLII-2/W3, pp 251-257

[3] Pauwels P et al. 2013 Integrating building information modelling and semantic web technologies for the management of built heritage information *Digital Heritage International Congress (DigitalHeritage)*, Vol. 1. IEEE

[4] Arayici Y, Counsell J and Mahdjoubi L 2017 *Heritage Building Information Modelling* (Oxford: Taylor & Francis)

[5] Cristiana A, Lombardini A and Tommasi C 2016 BIM and Cultural Heritage: Compatibility Tests in an Archaeological Site *International Journal of 3-D Information Modeling* (IJ3DIM) 5.1 pp 29-44

[6] Logothetis S, Delinasiou A and Stylianidis E 2015 Building information modelling for cultural heritage: a review *ISPRS Annals of the Photogrammetry, Remote Sensing and Spatial Information Sciences 2.5* (2015) p 177

[7] Bonsma P, Bonsma I, Maietti, F, Sebastian R, Ziri AE, Parenti S, Lerones PM, Llamas J, Turillazzi B and Iadanza E 2016 Roadmap for IT research on a Heritage-BIM Interoperable Platform within INCEPTION *Proceedings of the International Conference SBE Malta 2016. Europe and the Mediterranean: Towards a Sustainable Built Environment* ed RP Borg, P Gauci, C Spiteri Staines (Malta: Gutenberg Press) pp 283-290

[8] Hung-Ming C, Yang W and Yen Y 2015 BIM applied in historical building documentation and refurbishing *The International Archives of Photogrammetry, Remote Sensing and Spatial Information Sciences* 40.5 p 85

IOP Conf. Series: Materials Science and Engineering **364** (2018) 012090 doi:10.1088/1757-899X/364/1/012090

Comparing two 3D measurement techniques for documenting painted wooden panels surface deformations on a real test case: "Mystical Marriage of Saint Catherine" by Renaissance artist Piero di Cosimo

Anna Pelagotti[1], Francesca Uccheddu[2], Emanuela Massa[1], Monica Carfagni[2]

[1] Art-Test, via Santo Spirito 11 Firenze, Italy

[2] Department of Industrial Engineering, University of Florence, Via Santa Marta 3, Firenze, Italy

Abstract. Ancient wooden panel paintings are an essential part of our cultural heritage. Wood as artworks support has always been very popular and painted panels are largely diffuse in many churches and museums. However, depending on conservation conditions, the wooden panels shape may vary and, if not properly controlled, may lead to some sever damage to the artworks. This paper presents the results of a study on the measurement of paintings surface deformation carried out using two different three-dimensional acquisition devices both making use of structured light. The main goal was to highlight and measure the reliability of such 3D measuring techniques to evaluate deviations from planarity due to the curving and warping of the wood and to document spatial deformation suffered by the painting and monitoring its conservation status.

1. Introduction

The long history of panel paintings had a production peak in period from the Middle Ages to the Renaissance when a large number of them, sometimes of impressive dimensions, were produced. Many historically relevant and beautiful exemplars are still present in many European churches and museums. However, such environments are only rarely equipped with climate control systems and, often such artefacts suffer there from highly variable climatic conditions. Being wood a "living" material, it responds to different temperatures and humidity values and may significantly change its shape [1]. Such variations, which sometimes happen over a short period of time, induce stresses and strains that can, in the long term, irreversibly damage the painting [2][3]. In the past, it was believed that such movements had to be totally avoided and therefore a fixed frame was inserted into the back of the panel. However, time has proven this theory wrong, and sometimes, as in the case presented here, the wooden frame applied during a previous restoration intervention had to be removed, since it was causing unwanted cracks on the painted surface. To control the changes induced by this operation and to quantify the strain and stress applied to the wooden frame panel over time, shape measurements have been proposed to analyse how the panel modifies itself [4][5]. Shape analysis could allow researchers and restorers to accurately evaluate and prevent possible panel damages and help them in choosing the most suitable technique to control dangerous movements. Such new possibility is therefore envisaged to raise large interest, provided that it is possible to achieve it at reasonable operational costs, in terms of time and budget.

The use of three-dimensional scanning technology [6], so far applied to Cultural Heritage mainly for modelling of sculptural artworks [7][8], are reported to be able to provide a detailed map and a quantitative measurement of the painting deterioration; however the case studies presented in literature are mainly reporting on highly costly equipment results, and no before and after comparisons of induced deformation are described.

2. Experimental setup

The main purpose of the present study was to evaluate relative accuracy, and reliability, of two 3D measurement devices, (and of the commercial software provided to shape the model) for this sort of applications where the final accuracy has to be in the order of 1 centimetre and the acquisition time should not exceed half a day operation time. The two devices belong to very different price categories, although they are both based on structured light projection, one in the visible range and the other one in the infrared range.

2.1. Test object

The case study was a large wooden panel (210cm x 190cm) by Piero di Cosimo and painted in the fifteenth century. This large artwork, recently resurfed, had been restored several times and a large cradle had been applied to its back.

Figure 1. The "Mystical Marriage of Saint Catherine" by Piero di Cosimo, Fifteenth century

In particular, three horizontal wooden battens were inserted on the rear with the intent of flattening the artwork (see Figure 2). However, nowadays restores deem important that the wooden panel may move spontaneously, although in a controlled environmental condition to limit such movements to a minimum.

The wooden support was expected to deform after the battens removal, however none could predict how much. Such a parameter is instead crucial to assess the deformation and monitor its evolution and to timely allow intervention before irreversible damage takes place.

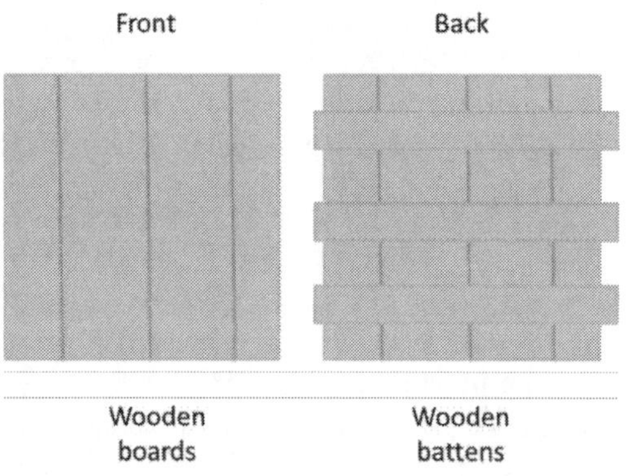

Figure 2. For large panel paintings, he artwork support is produced joining several wooden boards. In the illustration a schematic representation of the front and of the rear, where the three battens introduced during a restoration intervention are also depicted

2.2. Tested devices

The first device tested was a professional structured light-based scanner, the ScanProbe LT by Scan Systems (see Figure 3), while the second was a depth camera based on an infrared sensor system. The first scanner, working according to a method described in literature [9], makes use of a 3D triangulation system based on projecting onto the object under study eight black and white patterns in the visible spectrum at increasing frequency; the stripes projected on the surface warp depending on the surface surveyed and a camera records the images which are then processed. The price range of such sort of scanners is around the 20Keuros.

Figure 3. Data survey setup with the ScanProbe LT by Scan Systems

The second scanner (Kinect sensors family) consists of an infrared laser emitter, an infrared camera and an RGB camera [10]. The measurement of depth is a triangulation process in which the laser source emits a single beam which is split into multiple beams by a diffraction grating to create a constant pattern of speckles projected onto the scene. This pattern is captured by the infrared camera and is correlated to a reference pattern. When a speckle pattern is projected onto an object of which the distance from the sensor is smaller or larger than that of a known reference plane, the position of the speckle in the infrared image captured will be shifted in the direction of the baseline between the laser projector and the perspective centre of the infrared camera. These shifts are measured for all speckles by a simple image correlation procedure, which yields a disparity image. For each pixel the distance to the sensor

IOP Conf. Series: Materials Science and Engineering **364** (2018) 012090 doi:10.1088/1757-899X/364/1/012090

can then be retrieved from the corresponding disparity. Such system costs currently around 150 euros. For both devices, since a single range map taken at high resolution covers a limited surface, the number of images to be merged together in order to generate a complete three-dimensional model of the artwork has to be increased as the resolution grows. Another element to be taken into account is represented by measurement uncertainty, which depends on the squared distance between the camera and the object: a high precision is achieved by suitably limiting this distance, and thus the framed area.

2.3. Acquisition procedure

For the first acquisition campaign and the ScanProbe device, the scanner was approximately 90 cm far from the painting and 105 range maps were achieved to cover the entire area in 6 hours as shown in Figure 4. Much longer was the processing time. This is already an indication that at the current state of th art, this procedure has large drawback as routine monitoring. In order to refer all the acquisitions to the same reference system, adjacent range maps are aligned by means of specific software procedures that minimize the mean square deviation or find a best match between overlapping points. Such procedures, indicated in the literature as Iterative Closest Point (ICP) algorithms [11], use a cost function, for example consisting of the mean square deviation between groups of homologous points, that have to be minimized starting from a pre-alignment manually imposed by an operator.

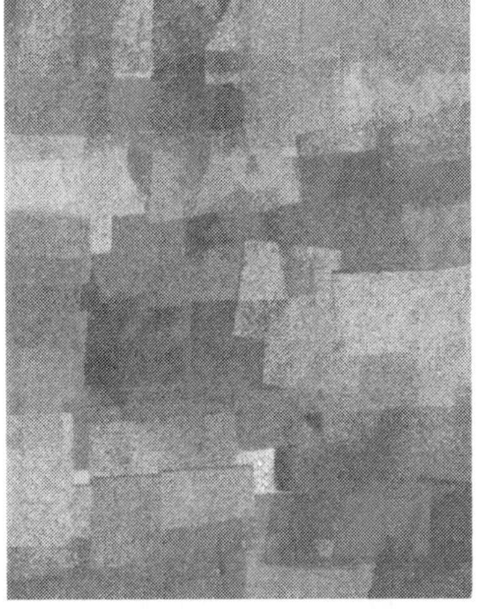

Figure 4. The wooden painting (210cmx190cm), surveyed with 105 scans. The final 3D model contains approximately 50 million faces.

The function value decreases steeply to a well recognizable absolute minimum in the case of surfaces with evident 3D features such as bumps or extrusions. It becomes less peaky, with many local minima, when the surfaces to be aligned have no evident 3D details, as on the flat surface of the front painting side. In the latter case the manual pre-alignment is a crucial step that influence the failure/success of the final convergence, the minimization becomes also much slower. Since in this case the aimed precision was not less than one centimetre, a decimated version of the 3D model was considered to perform such registration.

After the battens were removed, we then scanned the artwork both with the Scanprobe and with the Kinect infrared sensor system. In this second acquisition campaign the scanner was positioned 3 meters far from the painting. This allowed to obtain in a couple of seconds the model reported in Figure 5a. In fact, this depth sensor has the advantage of having a very big field of view thus allowing a very fast acquisition of the entire painting at once. In our experiment since the area to be covered was more than 2m x 2m, a distance of about 3 meters was needed to fully cover it.

a)

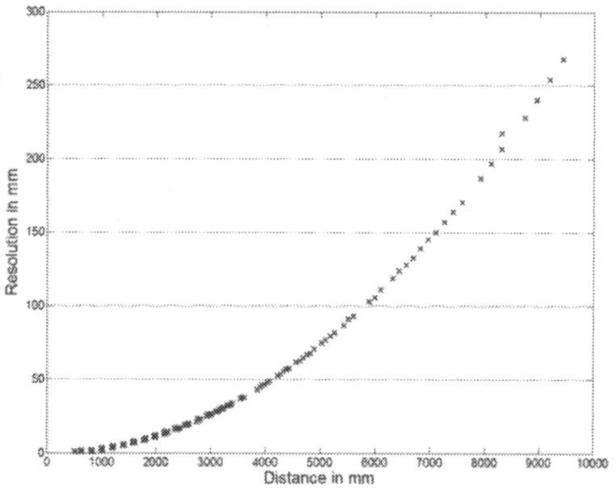

b)

Figure 5. a) The complete Kinect sensor acquisition of the painting b) The depth resolution is a function of distance and is getting coarser as the distance increases

3. Results

Using the Kinetic device would be the cheapest, easiest and fastest way to achieve a 3D model of such a panel painting, for a non-professionally trained operatos. However, the resulting accuracy cannot be the same as with the Scan Probe, not only because of the less accurate system but also because for every device, the accuracy of the measured data decreases with the squared distance (see Figure 5b). Arguably one could also stick together more acquisitions of the Kinetic camera, however the reverse would not be possible, ie. we could not acquire the whole panel surface with one acquisition covering the whole painting surface. Moreover we intended to challenge the low-cost system, in order to test its factual limits.

In order to perform a quantitative analysis, once the models were generated they were used to evaluate the painting deformations with respect to a planar reference. The reference was created by identifying the best-fitting plane of the cloud of points associated to the 3D model. The painting deformations, represented by the distances of the model points with respect to the reference plane, are here presented using color coding. The acquisition performed with the ScanProbe device before the battens removal, shows, as expected, a rather flat surface.

In Figure 6 is depicted the deviation from a local best-fit plane as acquired with the ScanProbe scanner; in the four corners of the painting (as we supposed, the corners are the more stressed areas). Only small variations (+/- 3mm almost everywhere) from local planarity are observed and the vertical wooden board are slightly visible. In Figure 7 is depicted the deviations from a local best-fit plane in the four corners of the painting after the battens have been removed, local deformation although not so pronounced (+/- 7mm) are to be observed, and it is clearly to be seen how the stress is focused mainly around the vertical wooden boards as expected).

The 3D model obtained by the less expensive system, after the battens removal, still retains the basic shape of the painting and its main structures (like the vertical wooden boards). However, due to intrinsic Kinetic system distortions (like the typical circular ripple [12] as visible in Figure 8), it is not possible to quantify the warping induced by removing the wooden battens; nonetheless this low-cost depth sensor allows a qualitative evaluation of the suffered warping as shown in Figure 9 (the four corners of the 3D model after removing the battens) in which the local Euclidean distance before and after the intervention is reported.

IOP Conf. Series: Materials Science and Engineering **364** (2018) 012090 doi:10.1088/1757-899X/364/1/012090

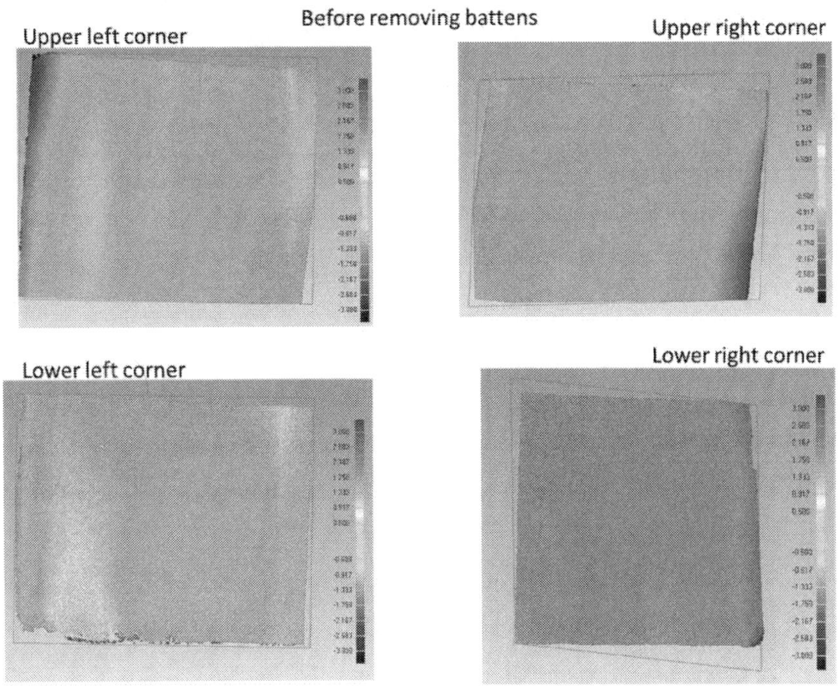

Figure 6. Image of deviations (mm) from the best-fit plane before removing the wooden battens

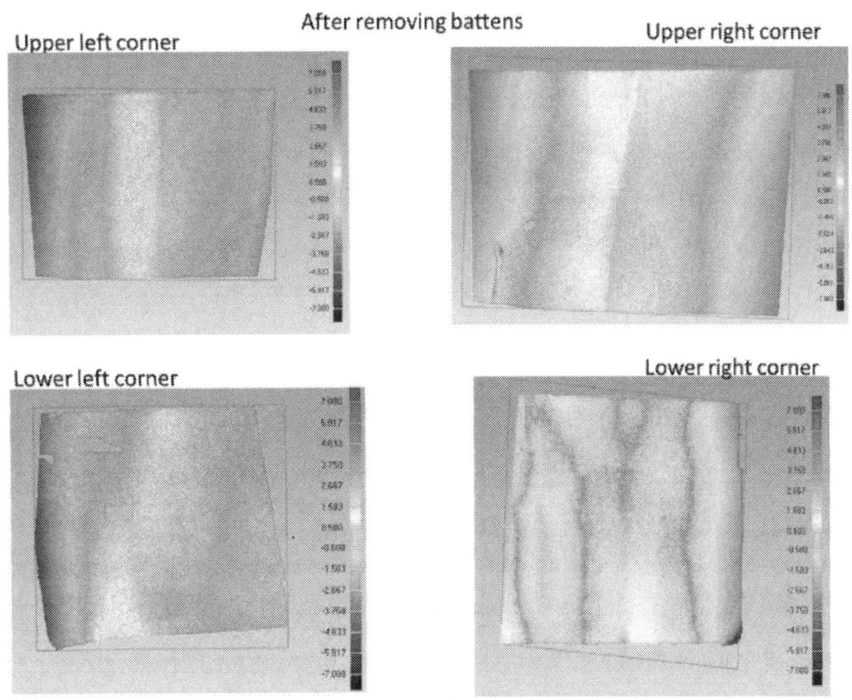

Figure 7: Image of deviations (mm) from the best-fit plane after removing the wooden battens

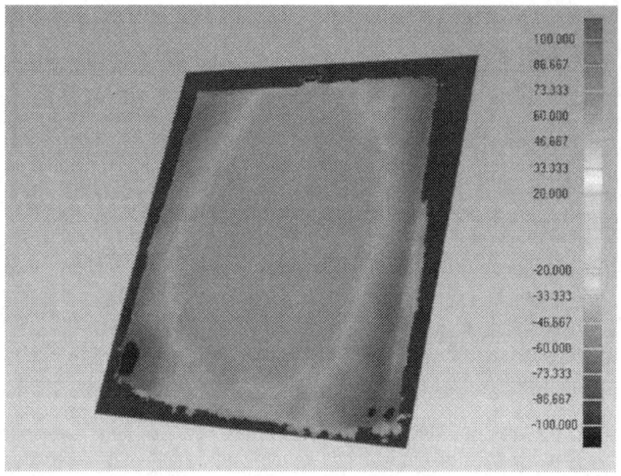

Figure 8. The low-cost Kinect sensor based acquisition compared to an ideal plane shows that depth estimate varies in the image as a "circular" ripple

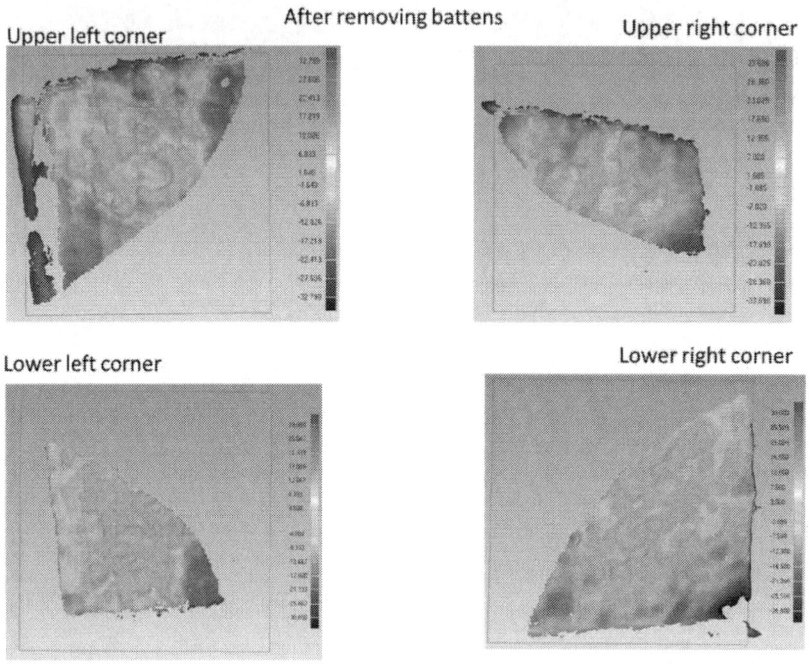

Figure 9*:* Kinect sensor deviations (mm) from planarity after removing the wooden battens at the four corners.

4. Conclusions

Typically, deformations of large wooden panels, which are a large part of our cultural heritage, are in the order of a few centimetres, thus not highly demanding in terms of measurements accuracy. The aim of the work presented in this paper was to assess the possibilities offered by the recently introduced depth camera system to be used as low-cost monitoring device for the deformation of such artworks. A more professional, high end system was also employed in order to have a reference measure. The wooden panel under test was a large XV century panel which was known to be likely to undergo some deformation after the battens introduced on the back during a previous restoration treatment, were to be removed. The test performed was carried out having in mind the simplest possibility which was a single shot acquisition form a fairly large distance. Such test showed that this low-cost device can be employed for qualitative monitoring, however, no quantitative data are to be inferred due to the intrinsic limits in measuring from far distances (as reported in Figure 5b). In order to have a higher accuracy more acquisitions from a shorter distance are possible, however this maybe not a suitable procedure for non-professionals. In order to have quantitative measures and assess the deformation properly the high-end scanner still proved to be the correct tool, although implying much higher costs and much longer processing time, together with the necessity of skilled and trained professionals.

References

[1] Gauvin, C., Jullien, D., Doumalin, P., Dupré, J. C., & Gril, J. (2014). Image correlation to evaluate the influence of hygrothermal loading on wood. Strain, 50(5), 428-435.

[2] Allegretti, O., Fioravanti, M., Dionisi-Vici, P., & Uzielli, L. (2014). The influence of dovetailed cross beams on the dimensional stability of a panel painting from the Middle Ages. Studies in Conservation, 59(4), 233-240.

[3] Cocchi, L., Marcon, B., Goli, G., Mazzanti, P., Castelli, C., Santacesaria, A., & Uzielli, L. (2017). Verifying the operation of an elastic crossbar system applied to a panel painting: the Deposition from the Cross by an anonymous artist from Abruzzo, sixteenth century. Studies in Conservation, 62(3), 150-161.

[4] Dureisseix, D., Colmars, J., Baldit, A., Morestin, F., & Maigre, H. (2011). Follow-up of a panel restoration procedure through image correlation and finite element modeling. International Journal of Solids and Structures, 48(6), 1024-1033.

[5] Spagnolo, G. S., Majo, R., Ambrosini, D., & Paoletti, D. (2003). Digital moiré by a diffractive optical element for deformation analysis of ancient paintings. Journal of Optics A: Pure and Applied Optics, 5(5), S146.

[6] Carfagni, M., Furferi, R., Governi, L., Servi, M., Uccheddu, F., & Volpe, Y. (2017). On the performance of the Intel SR300 depth camera: metrological and critical characterization. IEEE Sensors Journal, 17(14), 4508-4519.

[7] Remondino, F., Rizzi, A., Barazzetti, L., Scaioni, M., Fassi, F., Brumana, R. and Pelagotti, A. (2011), Review of Geometric and Radiometric Analyses of Paintings. *The Photogrammetric Record*, 26: 439–461

[8] Guidi, G., Atzeni, C., Seracini, M. and Lazzari, S., 2004. Painting survey by 3D optical scanning. The case of Adoration of the Magi by Leonardo da Vinci. *Studies in Conservation*, 49(1): 1–12

[9] Salvi J., Fernandez S., Pribanic T., Llado X. (2010), A state of the art in structured light patterns for surface profilometry, *Pattern Recognition*, Volume 43, Issue 8, August 2010, Pages 2666-2680

[10] Khoshelham K, Elberink SO. (2012), Accuracy and Resolution of Kinect Depth Data for Indoor Mapping Applications. *Sensors*. 2012; 12(2):1437-1454

[11] Segal A., Haehnel D., Thrun S., Generalized-ICP. In Proceedings of *Robotics: Science and Systems*, Seattle, USA, 2009

[12] Andersen M.R., Jensen T., Lisouski P., Mortensen A.K., Hansen M. K., Gregersen T., Ahrendt P., "Kinect Depth Sensor Evaluation for Computer Vision Applications", Electrical and Computer Engineering Technical Report ECE-TR-6, 2012

IOP Conf. Series: Materials Science and Engineering **364** (2018) 012091 doi:10.1088/1757-899X/364/1/012091

THE ACOUSTIC DESIGN AND REQUALIFICATION OF AN AUDITORIUM IN A CHURCH AND OF A HISTORICAL THEATER

Chiara Bartalucci, Francesco Borchi, Monica Carfagni
Department of Industrial Engineering, University of Florence Italy

Sergio Luzzi, Lucia Busa
Vie En.Ro.Se Ingegneria S.r.L, Italy

chiara.bartalucci@unifi.it

Abstract. Environmental sustainability is the ability to keep ecological processes within an ecosystem and its biodiversity in the future and, together with economic and social sustainability, contributes to the definition of well-being and progress. In this frame, circular economy and holistic approach in planning urban areas and designing public and private buildings play a decisive role. In the case of restoration, it can even be critical to achieve the best compromise between the desired acoustic quality and the architectural constraints imposed by the structure being restored. Software for simulating sound response in enclosed spaces can help to solve this issue.
The paper reports the numerical-experimental method used for the acoustic design and main results obtained in the acoustic requalification design of a church and a historic theater, both located in Tuscany (Italy). In the first case study, a requalification procedure was devoted to transform the church of Rispescia in a multifunctional auditorium by using sustainable materials. The study and design are referred to a building with a significant volume of approximately 1300 m^3, where the presence of finishing materials and reflecting furniture determined a high acoustic discomfort, making the hall difficult to use for events like lectures and concerts. After the implementation of the designed interventions, the church has been regularly used for different functions.
In the second case study, the proposed method has been applied for the acoustic restoration of the historic theater of Monsummano Terme by modifying its former acoustic design and thus extending its use to include concerts in addition to theatrical performances.

1. Introduction

Sensitive aspects of historical buildings concern restoration and renovation. In fact, in the case of the restoration of historic theaters and churches for example, it is not possible to create prototypes and negative results would be catastrophic [1-4]. Moreover, it can even be critical to obtain the desired acoustic quality when some architectural constraints imposed by the structure being restored are present. The use of the hybrid numerical-experimental acoustic design method proposed in this paper enables accurate calibration of the theater and church model so that any errors may be detected prior to restoration, thereby eliminating costly and most times, unsatisfactory retrofits, attempting to minimize the sound prediction uncertainties and the differences in the actual and simulated acoustic responses.
Referring to the design phase of possible solutions, sustainable growth to be referred to the three aspects of environment, economy and society, is the greatest challenge of the 21st Century. In this frame, circular economy and holistic approach in planning urban areas and designing public and private buildings play a decisive role. Circular economy is an economic system that can regenerate its own. In a circular economy,

material flows are of two types: biological ones, capable of being reintegrated into the biosphere, and technical ones, destined to be revalorized without entering the biosphere.
Holistic designer of urban areas and buildings try to apply to their designed works the idea that the whole is more than merely the sum of its parts, in theory, and above all, in practice. In the frame of circular economy, holistic design is also based on the attenuation of impacts by using materials characterized by reactive intelligence, as a coherent composition of natural and artificial materials which well adapts to the characterization of works in delicate contexts.
In the first example shown in this paper this specific matter of effectiveness has been applied by the designers in the acoustic requalification of the Rispescia Auditorium, located in a church. In fact, the proposed acoustic solutions have been designed keeping in mind the virtuous constraint of sustainability for all selected construction materials and artefact in order to create an eco-compatible and aesthetically attractive space. Moreover, only natural materials have been used, being them recycled and/or recyclable.

2. Methodology and parameters for the acoustic design

In this Paragraph the hybrid numerical-experimental acoustic design method useful to accurately calibrate the acoustic model adopted by both the case studies is presented. The advantage of using a model in this frame is to be able to detect errors and to minimize the sound prediction uncertainties and the differences in the actual and simulated acoustic responses prior to a potential restoration, thereby eliminating costly, and most times, unsatisfactory retrofits [5-9]. To method consists of the following steps:
Step 0. Definition of geometric and acoustic scenario and of design objectives, with reference to the various possible destinations of the building.
Step 1. Experimental measurements of sound pressure levels, reverberation time and impulse response with a previously characterized sound source.
Step 2. Construction of a three-dimensional geometric model simulating actual internal conditions.
Step 3. Calibration procedure which guides the user in modifying the absorption coefficient of each surface, starting from the most uncertain one. This step is especially critical, since, if the absorption coefficients are improperly assigned, the model could anyway be calibrated according to the reverberation time determined on the basis of previously acquired experimental data, but the error would render it worthless for representing the building's acoustic response and incapable of simulating structural modifications.
Step 4. Simulation and verification of the current acoustic conditions.
Step 5. Simulation of the design proposals and selection of the solution(s) to be implemented.
Step 6. Repetition of the experimental measurements carried out during Step 1 in the post-operam scenario and comparison with those collected during the ante-operam one.

3. The acoustic design of an auditorium into a church

3.1 Introduction

The study and design reported in this Paragraph are referred to the Legambiente Auditorium, located in the church of Rispescia, in the area of the National Centre for Sustainable Development "The Sunflower" [10] in western Tuscany. The church has a rectangular plant of about 8x19 m and variable height from a minimum of 8.86 m to a maximum of 9.92 m, in correspondence with the ridge line. The cover is made of two layers supported by wooden trusses. The room is equipped with an altar area currently used as a speaker zone when used for conferences and as stage for concerts and plays. The plastered masonry walls are provided of windows (1x2.82 m). The bottom wall has a large entrance with 1.90x 3.5 m wooden door of and two 0.78 x 1.5 m windows. The volume of the church is approximately 1300 m^3 and finishing materials and reflecting furniture (marble floor, plastered walls, stone in the altar area, cover bricks in the soffit tiles, wooden exterior carpentry and single glazing, wooden interior doors, metal benches drilled) are present. These aspects determined conditions of perceived acoustic discomfort that made difficult to use the hall for events including listening of speech and music and Festambiente, the Italian National festival of the environments. Moreover, the church, or parts of it, weren't easy to use neither for simple exhibitions, lectures and other cultural events.

3.2 The acoustic design

According to Step 0 of the method, the main objective of the acoustic project was to create a comfortable sound and a versatile environment for the different types of expected functions. This goal has been reached through the creation of variable solutions, able to guarantee for each intended use a proper acoustic performance, keeping in mind the virtuous constraint of sustainability for all selected construction materials and artefact in order to create a space that is eco-compatible and aesthetically attractive as well. Moreover, only natural, recycled and/or recyclable materials have been used (plants, wood, recycled rubber from discarded tires, polyester from recycled PET bottles).

3.2.1 Considered parameters

For the improvement of the acoustic performance of the space, with reference to its various possible destinations, the parameters reverberation time (TR), clarity (C50) and speech transmission (STI) have been selected. For the choice of target values for these parameters, the optimum ones defined in the Appendix C of the Italian Standard UNI 11367 [11] have been referred to. According to Step 1 of the method, the TR of the Auditorium has been measured according to the ISO3382 [12], while the optimum reverberation time in the frequency domain has been calculated according to the UNI 11367 [11]. However, since one of the project goals was the creation of optimal characteristics conditions for a multi-functional environment, a more general reference for TR has been established to a specific performance range (Table 1).

Table 1: Comparison between measured TR, optimal TR according to ISO 11367 and to a more general reference.

f [Hz]	125	250	500	1000	2000	4000
Measured TR [s]	3.54	3.53	3.47	3.52	2.75	1.89
Optimal TR [s] UNI 11367	1.23	1.23	1.03	1.03	1.23	1.23
Optimal TR [s] more general reference	1.3÷1.8		1.0÷1.2	1.0÷1.2	1.0÷1.2	

From Table 1 it can be noticed that the measured TR was far above the considered optimal values for the project. For the background noise, the ISO NR30 curve has been considered (Table 2).

Table 2: Background Noise according to NR30 curve.

f [Hz]	125	250	500	1000	2000	4000
NR30 [dB]	48	40	34	30	27	25

3.2.1.1 Validation of the acoustic model and designed solutions

The 3D model representing the current characteristics of the church has been created making use of Ramsete© 2.5 software. The model has been calibrated by comparing the TR measured in significant environmental points with the simulated TR in the same locations. With reference to Step 3, the sound absorption characteristics of materials were obtained via the iterative process of model calibration based on TR. The good validation of the model is proved by verifying that deviations in absolute value of the measurement and simulated average reverberation time data are lower than 0.1 s.

According to the main objectives of the acoustic design, the project of acoustic correction of the church involved design and construction of three integrated solutions summarized below, to be simulated during Step 5:

1. sound-absorbing evergreen plant coating for the improvement of absorption on the back wall;
2. sound absorbing curved acoustic panels suspended from the ceiling (36 baffles arranged in six parallel rows);
3. mobile network of double-face reflecting/absorbing panels on wheels for acoustic separation in different configurations of sub-areas.

Depending on their position and side, the reflecting/absorbing panels curved panels can perform a double function:

- sidewalls lining panels that are sound-absorbing or reflecting and diffusing sound, depending on the side facing the interior of the church;
- acoustic separation elements between different sub-areas of the hall.

Specifically, panels have radius of curvature of 3.5 m, height 2 m and a thickness of 106 mm, some with sound absorbing concave side and other with sound absorbing convex side. The system is supported by a wooden structure and floating through the use of wheels provided of brake system for a total weight of about 80 kg. In Figure 1 some pictures of the church are shown, according to the implemented solutions.

Figure 1: The church of Rispescia with the new designed implemented acoustic solutions.

3.3 Results

With respect to Step 6, in the post-operam scenario some sets of reverberation time measurements were carried out in several configurations, obtained changing the position and the orientation (side) of the double-face reflecting/absorbing panels on wheels. As first set of tests, the church has been considered as a whole auditorium and the effectiveness of the following three solutions has been considered:

a) only suspended sound-absorbing panels;
b) suspended sound-absorbing panels, evergreen coating and double-face panels in the minimum sound absorption configuration;
c) suspended sound-absorbing panels, evergreen coating and double-face panels in the maximum sound absorption configuration.

From the comparison of the results of measurements with the range of optimal values for speech derives that project objectives are fully respected.
Figure 2 shows how the configuration a) represents a huge improvement where compared with the ante-operam reverberation and how double-face movable panels and evergreen plants coating (b and c) have resulted in a further reduction of approximately 0.5 s at all frequencies for the reverberation time, compared to the configuration of only suspended panels.
Moreover, specific Sound Pressure Levels (SPL) measurements have been performed to assess the distribution of sound along the church, starting from an omnidirectional normalized source placed in a central position at the altar (stage), and in some places corresponding to listeners possible positions.
The results show levels varying from a maximum of 92.4 dB(A) at the receiver position closest to the source, to a minimum of 88.7 dB(A) at one of the more distant receiver.

The maximum difference between two measured level is therefore equal to 2.7 dB(A), in accordance to what calculated for this configuration.

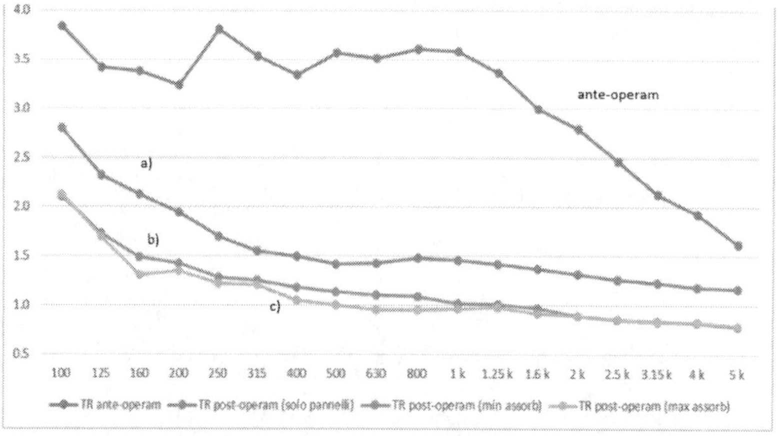

Figure 2: reverberation times in configurations a), b), c) and comparison with ante-operam.

The double-face panels can also be used as elements of separation and insulation between the different subareas of the room. This is particularly important when the room hosts several working groups operating simultaneously. The effectiveness of mobile acoustic panels as separation elements between adjacent sub-areas has been tested, carrying on SPL measurements aimed at identifying the acoustic attenuation determined by the presence of the panels in some of the catalogued configurations. A normalized sample source (dodecahedron) and receivers (sound meters) have been placed in different areas of the hall at ear height of a seated person. The SPL measurements have made possible to evaluate the differences present in the considered receivers points in both the basic configurations (panels at wall) and in workshop settings (panels as separating walls). The presence of the panels determines in the workshop schemes reductions in SPL from 5 dB to a maximum of almost 12 dB (A) in correspondence of all the receiver points in the sub-areas different from those in which the source is located. Conversely an increase of almost 1 dB (A) is measured in the receiver point located in the same sub-area where the source is.

4. The acoustic design of a historic theater in Monsummano Terme

4.1 Introduction and objectives

The historic Italian theater G. Giusti was built in the late nineteenth century in Monsummano Terme, approximately 25 miles northwest of Florence. Although it had been in disuse since the 1950s, it was considered an outstanding example of the horseshoe-shaped "teatro all'italiana" [13-15]. According to the Step 0 of the method described in Paragraph 2, the aims of the restorers were to return the building to its former glory while respecting current safety regulations and to improve on its original acoustics so that it could be used for concerts as well as theatrical performances. To accomplish this, it was necessary to improve its overall acoustic quality by distributing the sound uniformly throughout the building and optimizing the acoustic parameters in accordance with predetermined reference values [16-17].

4.2 Experimental measurements

In relation to the Step 1, for calibrating the simulation model, TR and SPL were experimentally measured. The TR was measured according to ISO3382 [12] prescriptions for the interrupted noise method and, although the standard prescribes a minimum of eighteen measurements to ensure proper coverage, twenty-one measures were carried out, three for each microphone position. Thanks to the symmetric configuration of the hall, it was necessary to carry out the measurements on only one side of the orchestra. Seven receiver points were selected in relation to the seating arrangement, the hall geometry and the ISO-imposed conditions of a minimum reciprocal distance of two meters and a minimum surface lateral distance of at least one meter. The microphone was positioned on a tripod 1.2 m above ground level. Three positions for the sound source enabled coverage of the whole theatrical area and verification of the symmetry of the configuration have been chosen:

- Source CS was positioned at centre stage at approximately two meters from the edge.
- Sources RS and LS were positioned four meters away from source CS, one on the right and one on the left side of the stage.

SPL at eleven positions in the orchestra and six positions in the balcony were also measured to provide the full coverage (Figure 3). The sound source used to perform reverberation time and sound pressure measurements in the theater was previously characterized. In particular, the sound source power spectrum and directivity were determined by performing SPL measurements in outdoor environment according to ISO 3744 [18].

4.3 Three-dimensional model designing and calibration

With reference to Step 2 of the procedure defined in Section 2, A three-dimensional geometric model was built aimed to duplicate the theater's actual acoustic conditions. The hall was reproduced with maximum definition levels in the balcony, orchestra, and stage sections. By accurately duplicating the volumes and solids, an object dimension accuracy of 30-40 cm was achieved (Figure 4). The surface absorption coefficients present in the hall were mostly set on the basis of indications available in the literature [19]. Nevertheless, some surface like the coverage was difficult to characterize, so, in this case, the sound absorption was firstly defined as a trial value to be optimized in the calibration phase (Step 3 of the procedure defined in Section 2) according to the measurements results.

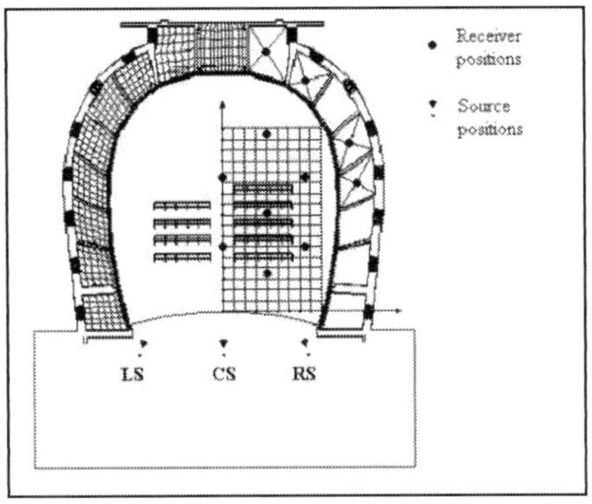

Figure 3: Positioning of the sources and the microphones.

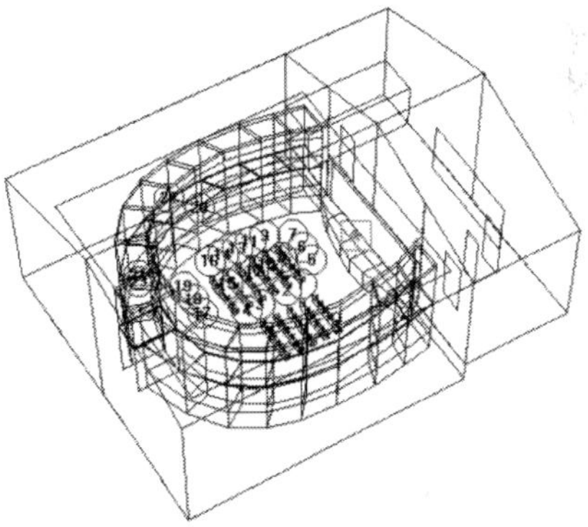

Figure 4: 3D model with sound source and microphones.

The source was set on the stage in the off-centre position previously selected for the SPL measurements. This setup was suitable for comparing the SPL and TR, since reverberation did not significantly vary in relation to the sound source position. Consequently, some receiver positions were defined in the 3D according to the receiving positions considered in the measurements of TR and SPL.

The model was calibrated so that its response would duplicate the actual acoustic conditions as closely as possible. It was decided to adopt the TR as calibration index, in particular it was felt advisable to refer to decay intervals such as T20 that did not involve a long period of sound response [20]. Finally, the model was checked in relation to the SPL. The calibration procedure [21] compared the simulated and measured T20 values by using the SD index:

$$SD = \sqrt{\frac{\sum_{1}^{n}\left[T_{20}(p,f)_{mis} - T_{20}(p,f)_{sim}\right]^2}{n}}$$

where:

n= number of measuring points per frequency band numbers

$T_{20}(p,f)_{mis}$ = reverberation time measured at position p and relative to frequency f

$T_{20}(p,f)_{sim}$ = reverberation time simulated at position p and relative to frequency f

IOP Conf. Series: Materials Science and Engineering **364** (2018) 012091 doi:10.1088/1757-899X/364/1/012091

To establish a priority of interventions, a weighted index was assigned to each surface that would allow accounting for:
- uncertainties in identifying the type of surface and in estimating its acoustic absorption;
- uncertainties in measuring the area of the surface;
- inaccuracies in geometric surface representation;
- effects of the acoustic absorption of the surface on the total environmental absorption.

Starting from the surface with the most critical index, the absorption coefficients was modified for each frequency band whenever the simulated reverberation time fitted the measured one. To evaluate the effects of corrections, a new simulation was performed for each correction step and the SD parameter was evaluated. This iterative procedure was replaced until SD parameter stopped decreasing and stabilized. When stabilization was reached, the absorption coefficients for the most critical element could be considered correct. We then examined the next element on the uncertainty list and proceeded in the same manner. To enhance the accuracy of the analysis, we observed the variations in the SPL at every step in the calibration procedure, comparing the values with the experimental data on hand.

4.4 Current acoustic condition and design proposals

According to the simulations of current scenario (Step 4 of the procedure defined in the Section 2), parameters such as TR, SPL, Sound clarity (C80) and Center time (Ts) have been evaluated [22-23] and it was found that it would be necessary to reduce its mean absorption and increase its reverberation time, according to the project design and the legislative references (Figure 5).
Regarding the design of an acoustic solution for the theater, the possibility of installing a specific acoustic ceiling was investigated and the acoustic effects of the proposed restoration solution were numerically simulated and evaluated using the simulation model (Step 5 of the procedure defined in the Paragraph 2). According to the simulation, the acoustic ceiling would retain the original reverberation time and reduce, albeit slightly, the sound definition. In keeping with the greater dispersion of the reflections, the total integration of the direct and indirect sounds would decrease to produce an improvement in music capability without a significant loss in clarity (Figure 6).

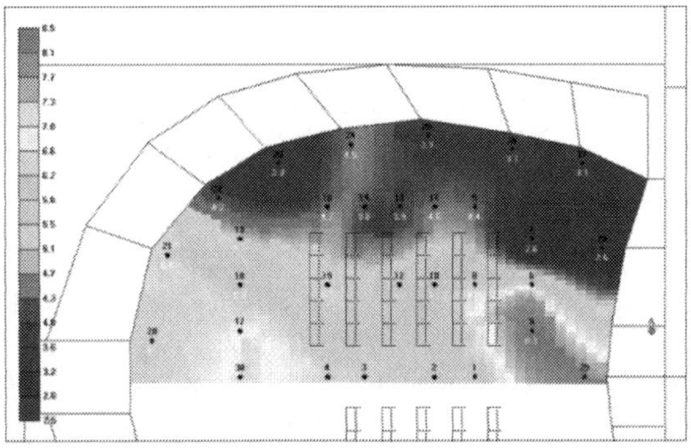

Figure 5: Map of sound clarity (C80).

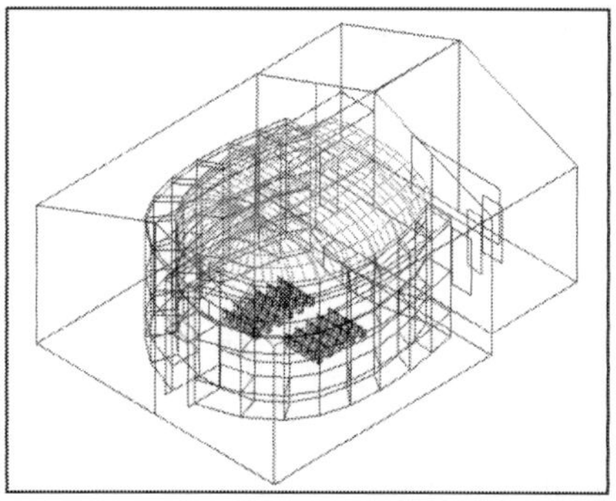

Figure 6: Monsummano Theater, designed solution.

Conclusions

A hybrid numerical-experimental acoustic design method can be considered as a precious instrument in order to obtain an accurate calibration of theaters and churches model so that errors may be detected and avoid prior to restoration. In this paper a numerical-experimental six-step method was described and applied to two pilot cases, respectively the church of Rispescia and the Monsummano Terme theater both located in the Tuscany Region, Italy. In the first case, after the application of the method, the church of Rispescia has been transformed in a real auditorium and in a multi-functional acoustic space as well. The reverberation time measured after design and construction shows full compliance with the project objectives, mainly aimed to create acoustic comfort conditions primarily in listening speech (conferences, lessons, workshops,

theatrical plays, cinema, expositions, etc.). The variable acoustics of the church, required to ensure optimal acoustic conditions in correspondence with the different uses is determined by the combination of three solutions and by the configuration of movable panels all using natural materials.
Concerning the theater of Monsummano Terme, the carried-out investigation showed that the theater was better suited to theatrical performances than concerts since, on the one hand, the acoustics characterized by deep lateral reflections combining with direct sound enhanced speech clarity and, on the other, the low reverberations rendered it unsuitable for music. The proposed acoustic ceiling produces a slight improvement in the overall musical quality through sound diffusion, without dramatically affecting the hall's acoustic properties.

References

1. C. Ianniello, An acoustic catalogue of historical Italian theaters for opera, Proceedings of Forum Acusticum, Budapest, 29 August–2 September (2005).
2. A. Magrini, R. Zecchin, A. Di Bella, A. Farina, A. Capra, L. Maffei, G. Iannace, C. Ianniello, R. Dragonetti, E. Cirillo, F. Martellotta, M. Masoero, A. Astolfi, R. Pompoli, N. Prodi, V. Tarabusi, L. Tronchin, Acoustical characteristics of Italian Historical Theaters: a research cooperation between the Universities of Bologna, Ferrara, Napoli Federico II, Napoli 2, Padua, Parma, Pavia, and Politecnico di Bari and Torino, Proceedings of the Conference: "Teatri d'Opera dell'Unità d'Italia", Associazione Italiana di Acustica, Venice, 23 November (2011).
3. F. Martellotta, Acoustics of Apulian Historical Theaters: further developments of the research, Proceedings of the Conference: "Teatri d'Opera dell'Unità d'Italia", Associazione Italiana di Acustica, Venice, 23 November (2011).
4. M. Garai, F. Morandi, D. D'Orazio, S. De Cesaris, L. Loreti, Acoustic measurements in eleven Italian opera houses: correlations between room criteria and considerations on the local evolution of a typology, Build. Environ. (2015), 10.1016/j.buildenv.2015.07.026.
5. M. Facondini, D. Ponteggia, Acoustics of the restored Petruzzelli Theater, Proceedings of 128th AES Convention, London (2010).
6. M. Facondini, Acoustic restoration of the Teatro Comunale Gioacchino Rossini in Pesaro, Proceedings of the 17th International Congress on Acoustics ICA, Rome (2001).
7. P. Fausti, N. Prodi, On the testing of renovations inside historical opera houses, J. Sound Vib., 258 (3) (2002), pp. 563-575.
8. A. Cocchi, M.C. Consumi, R. Shimokura, Considerations about the acoustical properties of "Teatro Nuovo in Spoleto after the restoration works, Proceedings of Acoustics08, Paris (2008), pp. 1391-1394.
9. K. Chourmouziadou, J.Kang, Acoustic evolution of ancient Greek and Roman theaters, Applied Acoustics, 69(6) (2002), pp. 514-529.
10. S. Luzzi, L. Busa, A. Gentili, G. Pisano, La correzione acustica dell'auditorium di Legambiente a Rispescia, Proceedings of 43° AIA National Congress, Alghero (2016).
11. Building acoustics, Acoustic classification of building units - Evaluation procedure and in situ measurements, UNI 11367, July 2010.
12. Acoustics. Measurement of the Reverberation Time of Rooms with Reference to Other Acoustical Parameters, ISO Standard 3382 (1996).
13. L. Tronchin, A. Farina, Acoustics of the former Teatro "La Fenice" in Venice, J. Audio Eng. Soc., 45 (12) (1997), pp. 1051-1062.
14. P. Fausti, R. Pompoli, N. Prodi, Acoustics of opera houses: a cultural heritage, J. Acoust. Soc. Am., 105 (1999), p. 929.
15. N.Prodi, R. Pompoli, Guidelines for acoustical measurements inside historical opera houses: procedures and validation, J. Sound Vib., 232 (2000), pp. 281-301.
16. G. Cammarata, A. Fichera, M.G. Rizzo, "Acoustical Prediction in Some Italian Theaters", Proceedings, 6th International Congress on Sound and Vibration, Copenhagen, Denmark, 173-180 (1999).
17. A. Magrini, P.Ricciardi, Churces as Auditoria: Analysis of acoustical parameters for a better understanding of sound quality, Building Acoustics, 10(2) (2003), pp. 135-158.
18. "Acoustics. Determination of Sound Power Levels of Noise Sources Using Sound Pressure. Engineering Method in an Essentially Free Field over a Reflecting Plane", ISO Standard 3744 (1994).
19. R. Lazzarin, M. Strada, Elementi di Acustica Tecnica. (Cleup, Padua, Italy, 1992).
20. P. Oliaro, M. Castellani "Studio numerico-sperimentale sulle potenzialità e sulle prestazioni di un codice di simulazione del campo sonoro in spazi confinanti", Proceedings, 27th Conference of the Associazione Italiana di Acustica, Perugia, Italy, 477-493 (1997).
21. A. Carbonari, G. Rossi, "Identificazione delle proprietà acustiche delle superfici in ambienti confinanti", Proceedings, 27th Conference of the Associazione Italiana di Acustica, Perugia, Italy, 64-67 (1999).
22. M. Garai, M.C. Rossi, L. Tronchin, "La percezione della qualità acustica dei teatri valutata mediante analisi statistica", Rivista Italiana di Acustica, 43, 3-18 (1997).
23. A. Farina, L. Tronchin, "Acoustics of the Former Teatro La Fenice in Venice", Journal of the Audio Engineering Society, 45, 1051-1062 (1997).

IOP Conf. Series: Materials Science and Engineering **364** (2018) 012092 doi:10.1088/1757-899X/364/1/012092

New insights on protective treatments for marble by FIB-SEM

Enrico Sassoni[1], Gabriela Graziani[1], Elisa Franzoni[1], George W. Scherer[2]

[1] Dept. Civil, Chemical, Environmental & Materials Engineering (DICAM), University of Bologna, Italy

[2] Dept. Civil and Environmental Engineering (CEE), Princeton University, USA

E-mail: enrico.sassoni2@unibo.it

Abstract. In this study, we investigated by FIB-SEM the microstructure of protective coatings formed by two inorganic treatments (namely, ammonium oxalate and ammonium phosphate solutions), with the aim of identifying the factors limiting the protective efficacy of the coatings and outlining possible improvement strategies. In the case of the oxalate treatment, the resulting layer of whewellite was found to contain vertical channels, which can allow water to reach the marble surface and trigger dissolution. Possible prevention of the formation of these channels by addition of a calcium source to the oxalate solution was attempted, but even micromolar additions led to rapid precipitation in the solution. In the case of the phosphate treatment, the resulting hydroxyapatite and octacalcium phosphate coating was found to be cracked and porous. To prevent cracks and pores, diminishing the ammonium phosphate concentration and adding ethanol to the solution were found to be effective strategies, as cracks were prevented and pores were reduced almost to zero. The resulting protective efficacy of the coating was found to be significantly improved, although still not perfect.

1. Introduction

Marble sculptures and architectural decorations exposed outdoors are subject to dissolution in rain, because of the slight solubility of calcite in water [1]. To prevent dissolution, the application of protectives is often necessary. Organic protectives have been widely investigated and applied in the past decades. In spite of their high efficacy in making marble hydrophobic, thus preventing dissolution in rain, organic protectives have exhibited severe compatibility and durability issues [2]. As a consequence, the use of inorganic treatments is often preferred [3].

Inorganic treatments are aimed at forming a superficial layer of a new mineral, able to prevent dissolution of the substrate. For this purpose, the newly formed layer needs to meet several requirements: (i) it must have reduced solubility, compared to calcite; (ii) it needs to be continuous, to prevent bare areas from triggering dissolution; (iii) it needs to be crack-free and pore-free, to prevent rain from reaching the substrate by permeation through the protective layer. Whereas fulfilment of the other requirements can be relatively easily evaluated by standard techniques (e.g., XRD and SEM), assessing the presence of pores inside the coating can be challenging. SEM observation of cross sections encapsulated in resin could be one route, but possible alteration of the cross section by resin penetration and polishing needs to be taken into account.

In this study, we present an example of the use of a focused ion beam (FIB) microscope to obtain cross sections of protective coatings with minimum alteration of the original microstructure. Two inorganic protectives were considered: (i) ammonium oxalate, which is currently the most widely used inorganic treatment; (ii) ammonium phosphate, which is receiving increasing attention because of its high potential. FIB cross sections revealed the factors limiting the efficacy of the two treatments (namely, the presence of channels in the resulting calcium oxalate layer and the presence of pores in the resulting calcium phosphate layer, respectively). This suggests possible strategies to overcome these limitations: in the case of the phosphate treatment, it was possible to significantly reduce pores and improve the protective efficacy by adding alcohol to the phosphate solution.

2. Materials and methods

2.1. Materials
Cubic samples (1 cm side) of Carrara marble (BasketweaveMosaics.com, USA) were used. Ammonium oxalate (AmOx, $(NH_4)_2C_2O_4{\cdot}H_2O$, assay > 99%, Sigma Aldrich), diammonium hydrogen phosphate (DAP, $(NH_4)_2HPO_4$, assay > 99%, Sigma Aldrich), calcium chloride ($CaCl_2{\cdot}2H_2O$, assay > 99%, Sigma Aldrich), ethanol (EtOH, Fisher-Scientific) and deionized water were used.

2.2. Treatments
All the treatments were applied by immersing each sample in 200 ml of solution and leaving it to react for 24 hours, the solution being continuously stirred by rotating the container with the solution and the sample. After 24 hours, the sample was rinsed with deionized water and dried at room temperature.

For the ammonium oxalate treatment (labelled "AmOx"), samples were treated with an aqueous solution of 5 wt.% AmOx, as proposed in [3,4]. The aim of the treatment is to transform calcite into calcium oxalate monohydrate (whewellite), according to the reaction [3]:

$$CaCO_3 + (NH_4)_2C_2O_4 \rightarrow CaC_2O_4{\cdot}H_2O + 2NH_3 + H_2O + CO_2.$$

For the first formulation of the ammonium phosphate treatment (labelled "DAP1"), samples were treated with an aqueous solution containing 1 M DAP and 1 mM $CaCl_2{\cdot}2H_2O$, as proposed in [5]. The aim is to transform calcite into hydroxyapatite (HAP, $Ca_{10}(PO_4)_6(OH)_2$), according to the reaction [6]:

$$10CaCO_3 + 5(NH_4)_2HPO_4 \rightarrow Ca_{10}(PO_4)_6(OH)_2 + 5(NH_4)_2CO_3 + 3CO_2 + 2H_2O.$$

$CaCl_2{\cdot}2H_2O$ was added to the DAP solution based on previous results, indicating that $CaCl_2$ helps achieve complete coverage of marble surface [5].

Based on FIB-SEM results, a possible improvement of the DAP1 formulation was then investigated, with the aim of reducing the porosity of the resulting coating. For the improved formulation (labelled "DAP2"), samples were treated with a water-ethanol (90-10 vol%) solution containing 0.1 M DAP and 0.1 mM $CaCl_2$. Ethanol was added based on previous results, indicating that HAP formation is favoured by addition of ethanol to the precipitation medium [7].

2.3. Characterization

2.3.1. Coating composition and microstructure. The coating composition was determined by grazing incidence X-ray diffraction (GID), using a Bruker D8 Discover X-Ray Diffractometer. By analysing the sample with an X-ray beam with very little inclination with respect to the analysed surface ($\theta = 0.5°$), the reduced depth of penetration of the X-ray beam into the sample yields information about the composition of the superficial coating with little interference from the substrate.

The coating morphology was analysed by an environmental scanning electron microscope (FEI Quanta 200 FEG ESEM), after making the samples conductive by carbon coating.

The coating microstructure was analysed by a focused ion beam SEM (FEI StrataTM DB 235 FIB), after carbon coating. The steps in cutting a cross section of a coating are illustrated in Figure 1 in the case of the DAP1 treatment: (1) the location for cutting the cross section is first identified; (2) a platinum layer is deposited to preserve the morphology of the coating; (3) the cross section is cut by milling with Ga^{3+} ions at high current (30 kV, 23 nA); (4) the cross section is finally polished by milling with Ga^{3+} ions at low current (30 kV, 40 pA)

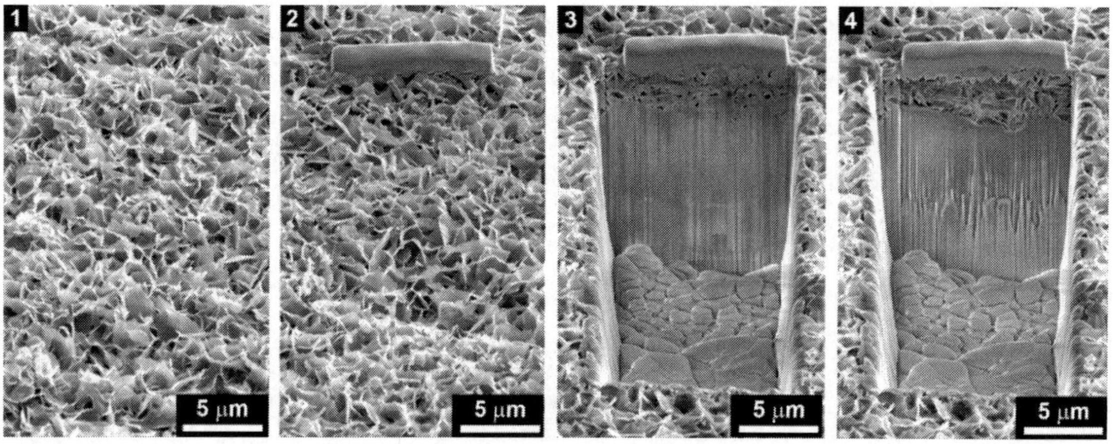

Figure 1. Stages of FIB-SEM analysis (here illustrated for the DAP1 coating): (1) surface of the sample to be analysed, tilted by 52° with respect to the horizontal; (2) platinum layer deposited over the surface, to protect the coating morphology; (3) rough cross section obtained by milling at high current; (4) polished cross section obtained by milling at low current.

2.3.2. Coating performance. The coating performance was assessed in terms of ability to delay dissolution of marble samples immersed in a HNO_3 aqueous solution at initial pH 4.8, simulating slightly acidic rain. Marble dissolution was monitored by measuring the increase in pH as a function of time, using an Oakton pH-meter (pH 1100 series). For better comparability, an area of 8×8 mm^2 was isolated in one face of the cubic samples, the other surfaces being coated by an impermeable layer of nail polish. For each condition, two samples were analysed and the results were averaged.

3. Results and discussion

3.1. Ammonium oxalate treatment

GID detected whewellite ($CaC_2O_4 \cdot H_2O$) as the only new phase formed after treatment, as expected (Figure 2). Whewellite has a solubility product ($K_{sp} \approx 10^{-9}$) lower than that of calcite ($K_{sp} \approx 5 \cdot 10^{-9}$) [1], so some protection of the underlying marble from dissolution in acid is expected.

From SEM observation of the sample surface, whewellite appeared to uniformly cover marble surface, with no visible bare areas or micro-cracks (Figure 3, "before" condition). Nonetheless, limited protective efficacy was achieved (Figure 4). In fact, initially the AmOx curve is below that of the untreated reference, which indicates that some delay in marble dissolution is achieved. However, after about 400 min the two curves overlap, which indicates that the protective action is lost. Consistently, when the AmOx sample was analysed by SEM at the end of the acid attack test, no trace of the whewellite coating was found and only bare marble was visible (Figure 3, "after" condition).

One reason for the limited protective efficacy could be the solubility of the new mineral. Although lower, the solubility product of whewellite is very similar to that of calcite, hence dissolution of the whewellite coating after prolonged exposure in the acid solution seems plausible (however, no data were found in the literature about the dissolution rate of whewellite in pure water [1]).

In addition, SEM observation of FIB cross sections (Figure 3, right) revealed a further possible reason for the measured performance. Whewellite crystals seem to be separated by small vertical channels, connecting the marble/whewellite interface with the upper whewellite surface. These channels can allow acid to reach the marble surface, thus triggering its dissolution and leading to pH increase over time. These vertical channels resemble those described in the case of mineral replacement [8], where the presence of channels is essential for the growth of the new mineral.

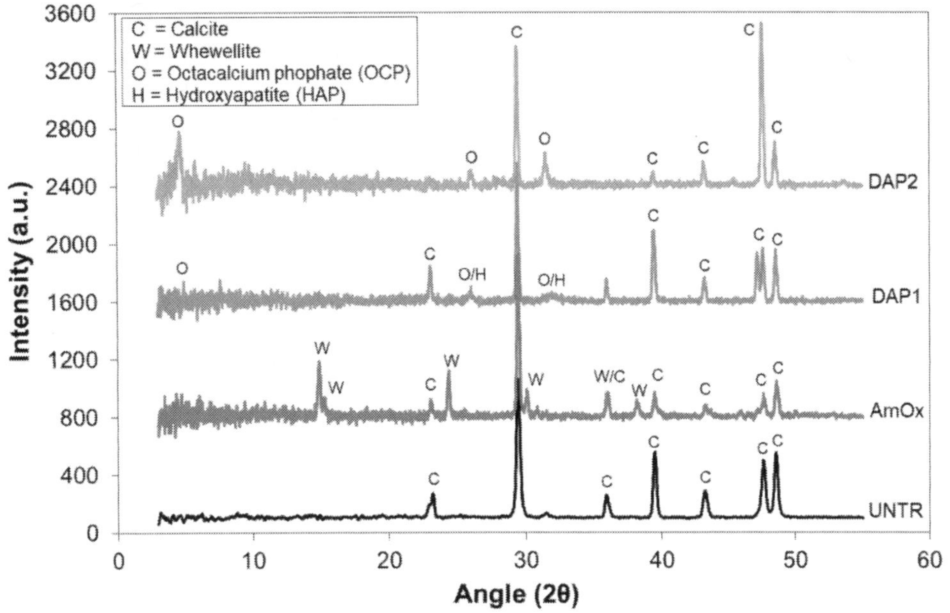

Figure 2. Mineralogical composition of untreated and treated samples.

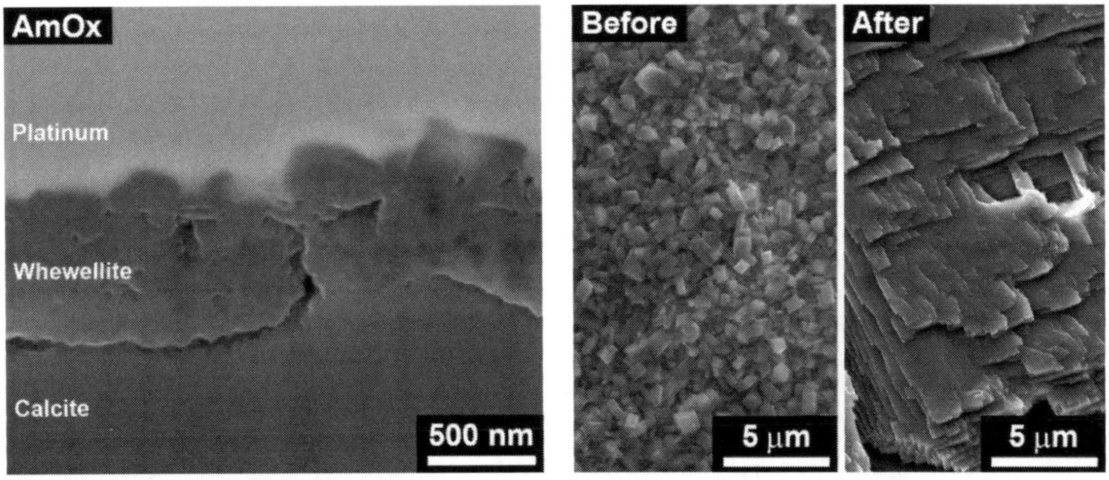

Figure 3. FIB cross section before acid attack (left) and surface (right), before and after acid attack, of AmOx sample.

Indeed, for the reaction to proceed and the new mineral to keep on forming, ions dissolved from the original mineral need to cross the new mineral and reach the interface between the new mineral and the solution [8]. In the present case, calcium ions necessary for whewellite formation only come from dissolution of the marble substrate, hence it is reasonable that the vertical channels are formed during whewellite formation to allow its growth.

To prevent formation of these vertical channels, an attempt was made to externally supply calcium ions (similarly to the case of the DAP1 treatment) [9]. However, because of the rapid precipitation of particles even for micromolar additions of $CaCl_2$ to the AmOx solution, the attempts to prevent formation of these vertical channels were unsuccessful [9].

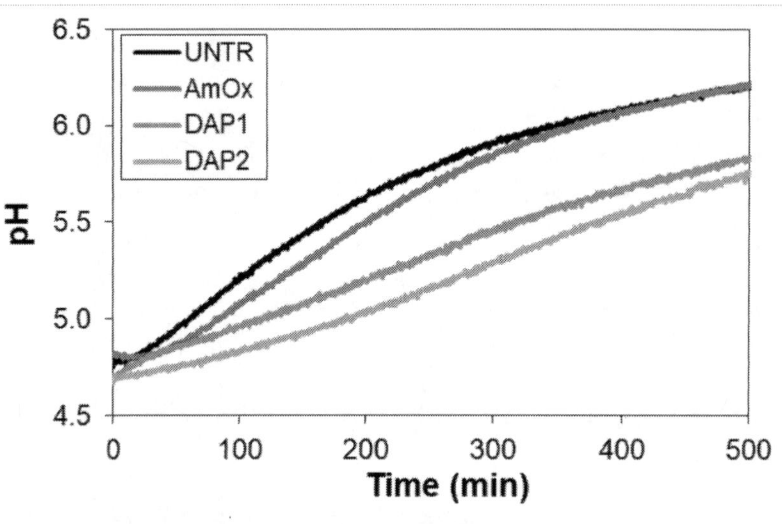

Figure 4. Increase of pH as a function of time for untreated and treated samples.

3.2. Ammonium phosphate treatment – First formulation

In the case of the DAP1 treatment, GID suggests formation of both HAP ($Ca_{10}(PO_4)_6(OH)_2$) and octacalcium phosphate (OCP, $Ca_8(PO_4)_6H_2{\cdot}5H_2O$) (Figure 2), consistently with previous results about the addition of $CaCl_2$ to the DAP solution [5,10]. The solubility product of HAP is many orders of magnitude lower than that of calcite ($K_{sp} \approx 10{\cdot}10^{-59}$ and $5{\cdot}10^{-9}$, respectively [1]), hence formation of HAP is expected to provide significant protection to marble surface. OCP has a solubility product ($K_{sp} \approx 10^{-48}$ [11]) higher than HAP, but still much lower than calcite, hence its formation is not expected to negatively affect the protectiveness of the coating, especially because it significantly helps achieve complete coverage of marble surface [5].

However, SEM analysis of the sample surface (Figure 5, "before" condition) revealed that the coating is cracked, which presumably occurred during drying, because its thickness exceeds the critical value below which cracking is thermodynamically inhibited [12]. This surely negatively affects the protective ability of the coating. Consistently, the acid attack test (Figure 4) indicates that the coating provides some protection to the marble surface, but still much less than expected based on the solubility products: the DAP1 curve is always well below that of the untreated reference, but a progressive increase of pH occurs with time. Accordingly, at the end of the test the coating exhibits visible degradation and bare areas are visible (Figure 5, "after" condition). To prevent cracking of the coating, decreasing the DAP concentration might be one route, with the aim of reducing the coating thickness. However, if the coating is also porous, permeation of acid through the coating might further reduce the protective ability.

FIB cross sections provided a fundamental piece of information about the presence of inherent porosity, as the DAP1 coating was found to be highly porous (Figure 5, left). The coating porosity surely reduces its protective efficacy, so it needs to be eliminated. This was attempted by developing the DAP2 treatment formulation.

3.3. Ammonium phosphate treatment – Improved formulation

Compared to DAP1, the DAP2 formulation has two differences: (i) the lower concentration of DAP (0.1 M instead of 1 M) and $CaCl_2$ (always in 1:1000 ratio to the DAP concentration), with the aim of forming a thinner coating, thus possibly preventing cracking; (ii) the addition of ethanol (10% in volume), with the aim of favoring formation of calcium phosphates and possibly preventing pore formation. Indeed, ethanol molecules are known to weaken the hydration sphere of phosphate ions in solution [13] and

IOP Conf. Series: Materials Science and Engineering **364** (2018) 012092 doi:10.1088/1757-899X/364/1/012092

previous tests have shown that coverage of calcite particles by HAP is improved when ethanol is added to the DAP solution [7].

Figure 5. FIB cross section before acid attack (left) and surface (right), before and after acid attack, of the DAP1 sample.

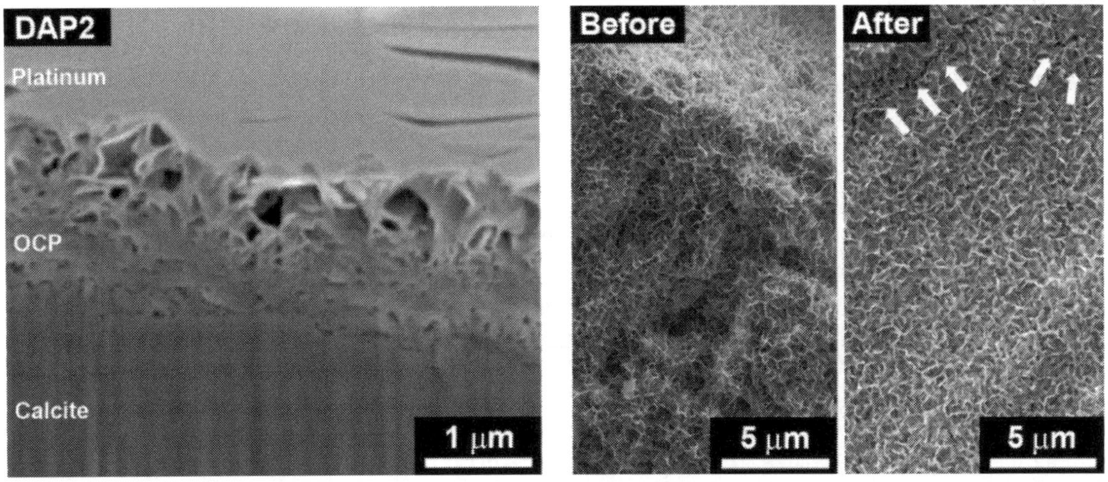

Figure 6. FIB cross section before acid attack (left) and surface (right), before and after acid attack, of the DAP2 sample.

In terms of composition of the coating, the DAP2 treatment led to formation of OCP, while no HAP peak was clearly evident (Figure 2). As discussed above, OCP is far less soluble than calcite, hence a good protecting ability can be expected, provided that the coating is crack-free and pore-free. SEM observation of the sample surface revealed no visible cracking (Figure 6, "before" condition), which is a further positive feature towards achieving a satisfactory protective action.

FIB-cross sections revealed a remarkable reduction in porosity of the DAP2 coating (Figure 6). The flower-like morphology typical of HAP and OCP is still visible on top of the coating, but the underlying part, in contact with marble surface, is significantly denser. Such a decrease in porosity is thought to derive from the beneficial effect of ethanol, which makes phosphate ions more reactive to form new calcium phosphate phases [13].

As a consequence of the absence of cracks and the densification of the microstructure, the acid resistance of the DAP2 coating results improved compared to DAP1 (Figure 4). With respect to the

IOP Conf. Series: Materials Science and Engineering **364** (2018) 012092 doi:10.1088/1757-899X/364/1/012092

untreated reference, the DAP2 sample reached the same value of pH after about twice as much time, which means that marble was provided with a significant improvement in durability.

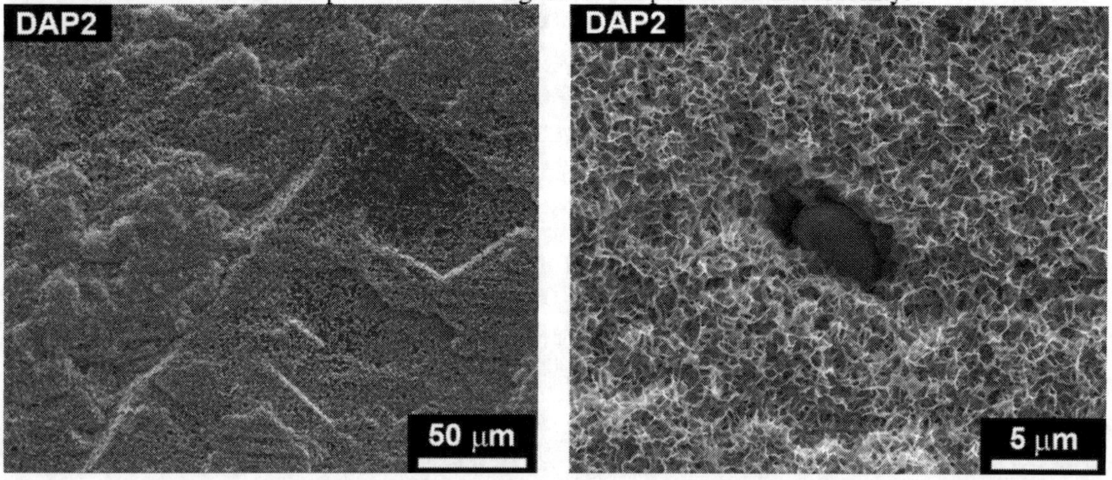

Figure 7. SEM images illustrating the presence of defects in the DAP2 coating in correspondence of calcite grains with unfavourable orientation (left) and impurities (right).

Nonetheless, the acid protection is not complete, as pH increases over time, although more slowly than in the case of untreated marble. Closer observation of the DAP2 sample by SEM revealed that, although the coating is crack-free and continuous, still some calcite grains exhibit poor coverage: an abrupt change is present between these isolated badly covered calcite grains and the surrounding ones, perfectly covered with the coating (Figure 7, left). This is thought to be a consequence of the different crystallographic orientation of calcite grains: whereas coverage of marble surface by the coating is generally good and even epitaxial growth of HAP (or OCP) might take place over calcite grains with favorable crystallographic orientation (e.g., when the distance between atoms in the calcite crystal is similar to the distance of atoms in the HAP or OCP crystals), the coating growth is more difficult over grains with unfavorable orientation. Similarly, the coating is interrupted where impurities are present (Figure 7, right). The presence of these interruptions in the coating continuity allows acid to trigger marble dissolution from these bare areas and then spread to the adjacent areas. This also leads to cracking of the coating, as revealed by SEM observation of DAP2 samples after acid attack (Figure 6, “after” condition – newly formed cracks are indicated by white arrows).

Possible strategies to promote the growth of the coating also over these calcite grains with unfavorable crystallographic orientation are currently being investigated.

4. Conclusions

In this paper, focused ion beam (FIB) microscopy was used to investigate the microstructure of two different inorganic coatings for marble protection, with the aim of identifying the factors limiting the performance of the treatments and outlining possible strategies for their improvement.

In the case of the ammonium oxalate treatment, calcium oxalate monohydrate (whewellite), slightly less soluble than calcite, was formed. Cross sections obtained by FIB-SEM revealed that vertical channels are present in the whewellite layer. These channels can allow acid to reach the marble/whewellite interface and hence trigger dissolution of the substrate. The channels are thought to originate from the mechanism of whewellite formation: calcium ions needed to form whewellite only come from dissolution of marble, hence the channels are needed to ensure that the calcium ions can cross the whewellite layer and reach the whewellite/solution interface, where the reaction is proceeding. To prevent formation of these vertical channels, an attempt was made to add calcium ions directly into

IOP Conf. Series: Materials Science and Engineering **364** (2018) 012092 doi:10.1088/1757-899X/364/1/012092

the ammonium oxalate solution. However, because of the rapid precipitation of particles even for micromolar additions of $CaCl_2$ to the ammonium oxalate solution, the attempts were unsuccessful.

In the case of the ammonium phosphate treatment, hydroxyapatite and octacalcium phosphate (both significantly less soluble than calcite) were formed when a 1 M DAP and 1 mM $CaCl_2$ solution was used. The coating exhibited cracks (presumably owing to the excessive thickness of the coating) and pores, whose detection was possible by FIB-SEM analysis. Even though the coating is able to delay dissolution of treated marble, still protection is not complete. The new insight made possible by FIB suggests a possible strategy to improve the coating performance: reducing the DAP concentration, to prevent excessive film growth and cracking, and adding ethanol to the DAP solution, to enhance the reactivity of phosphate ions in the solution. FIB cross sections confirmed that, by treatment with a 0.1 M DAP and 0.1 mM $CaCl_2$ solution also containing 10 vol% ethanol, cracks are prevented and pores are almost completely absent. Accordingly, the protective action was found to be improved compared to the first formulation of the phosphate treatment. Nonetheless, protection of marble is not complete yet, because a few calcite grains with unfavorable crystallographic orientation exhibit poor coverage by the coating. Further strategies to improve marble coverage by the coating and hence the protective performance of the coating are currently being investigated.

Acknowledgements
This project has received funding from the European Union's Horizon 2020 research and innovation programme under the Marie Sklodowska-Curie grant agreement No 655239 (HAP4MARBLE project, "*Multi-functionalization of hydroxyapatite for restoration and preventive conservation of marble artworks*"). The help of Dr. John J. Schreiber and Dr. Nan Yao (Princeton University's PRISM Imaging and Analysis Center, IAC) is gratefully acknowledged.

References

[1] Naidu S, Blair J and Scherer GW 2016 Acid-resistant coatings on marble *J. Am. Ceram. Soc.* **99** 3421-28
[2] Baglioni P, Carretti E and Chelazzi D 2015 Nanomaterials in art conservation *Nat. Nanotechnol.* **10** 287-90
[3] Matteini M 2008 Inorganic treatments for the consolidation and protection of stone artefacts *Conservation Science in Cultural Heritage* **8** 13-27
[4] Mudronja D, Vanmeert F, Hellemans H, Fazinic S, Janssens K, Tibljas D, Rogosic M, Jakovljevic S 2013 Efficiency of applying ammonium oxalate for protection of monumental limestone by poultice, immersion and brushing methods *Appl. Phys. A-Mater.* **111** 109-119
[5] Naidu S and Scherer GW 2014 Nucleation, growth and evolution of calcium phosphate films on calcite *J. Colloid. Interf. Sci.* **435** 128-37
[6] Sassoni E, Naidu S and Scherer GW 2011 The use of hydroxyapatite as a new inorganic consolidant for damaged carbonate stones *J. Cult. Herit.* **12** 346-55
[7] Graziani G, Sassoni E, Franzoni E and Scherer GW 2016 Hydroxyapatite coatings for marble protection: Optimization of calcite covering and acid resistance *Appl. Surf. Sci.* **368** 241-57
[8] Putnis A 2015 Transient porosity resulting from fluid-mineral interaction and its consequences *Rev. Mineral. Geochem.* **80** 1-23
[9] Wang S 2017 Formation of Nonporous Calcium Oxalate for Calcite Protection (Princeton University, unpublished work)
[10] Sassoni E, Graziani G and Franzoni E 2015 Repair of sugaring marble by ammonium phosphate: Comparison with ethyl silicate and ammonium oxalate and pilot application to historic artifact *Mater. Design* **88** 1145-57
[11] Tung MS, Eidelman N, Sieck B, Brown WE 1988 Octacalcium phosphate solubility product from 4 to 37 C *J. Res. Nat. Bur. Stand.* **93** 613-24
[12] Evans AG, Drory MD and Hu MS 1988 The cracking and decohesion of thin films *J. Mater. Res.* **3** 1043-49

IOP Conf. Series: Materials Science and Engineering **364** (2018) 012092 doi:10.1088/1757-899X/364/1/012092

[13] Lerner E, Azoury R, Sarig S 1989 Rapid precipitation of apatite from ethanol-water solution *J. Cryst. Growth* **97** 725-730

IOP Conf. Series: Materials Science and Engineering **364** (2018) 012093 doi:10.1088/1757-899X/364/1/012093

The capability of capacitive sensors in the monitoring relative humidity in hypogeum environments

F Frasca[1], A Caratelli[2*] and AM Siani[3]

[1] Sapienza Università di Roma, Department of Earth Sciences, P.le A. Moro 2, 00185 Rome, Italy.
[2] Tecno.EL S.r.l., Electronic Technologies company, via degli Olmetti 38, Formello (RM)
[3] Sapienza Università di Roma, Department of Physics, P.le A. Moro 2, 00185 Rome, Italy.

*Corresponding author: a.caratelli@tecno-el.it

Abstract. Hypogeum environments are characterized by high levels of relative humidity (RH). Most humidity sensors currently in use are based on the capacitive effect of the dielectric material to change according to water vapour uptake. In hypogeum environments the dielectric material can be saturated by water vapor, implying a significant error in the RH measurement. To improve the capacity of this type of humidity sensors, a modified hygrometer capacitive sensor, which uses a heating cycle to avoid the condensation, has been recently developed by Rotronic®. During four field campaigns in two different hypogea environments (the Monkey Tomb in Siena and the Mithreum of Caracalla Baths in Rome), RH was measured using the conventional capacitive sensor (CCS) and the heated capacitive sensor (HCS). The purpose of this study was to investigate the capability of HCS to detect RH variations when the environmental conditions were close to vapor saturation. Significant differences were found between the measurements of the two sensors: when RH was close to 100%, the CCS was not able to detect the RH decrease, giving only a measure of RH=100%, while HCS detected such a RH decrease. Therefore, these results encourage the use of HCS in the monitoring of RH levels in extreme humidity sites such as hypogea sites.

1. Introduction

Hypogea, usually located in subterranean structures, are characterized by specific thermo-hygrometric conditions [1-2]: the temperature (T) has low daily and seasonal fluctuations and rarely exceeds 20°C, whereas the relative humidity (RH) is always over 90%. This implies that such environment is often close to vapor saturation [3]. The presence of liquid water on the stone surface and inside the material can induce chemical reactions such as salt crystallization/dissolution or biological colonization. These dynamic processes are influenced by daily cycles of RH [4], with an increase or decrease of deterioration related to the RH range. Therefore, the microclimate monitoring within hypogea sites, plays a key role in the preventive conservation of the artifacts contained therein.

Most of humidity sensors currently in use [5] are based on the capacitive effect of the dielectric material to change according to the water vapor uptake. The main disadvantage of the capacitive sensor

IOP Conf. Series: Materials Science and Engineering **364** (2018) 012093 doi:10.1088/1757-899X/364/1/012093

is that it is strongly affected by the saturation of the dielectric material, introducing a significant error in the measurement. When the environment is close to saturation (RH=100%), the sensor hardly detects any further RH decrease and hence there is a lack of information on the RH behavior.

To improve the capacity of this type of humidity sensor, Rotronic® has integrated a heater in the conventional capacitive sensor (CCS) to prevent the saturation of the dielectric material. This modified capacitive sensor is generally defined as the heated capacitive sensor (hereafter called HCS). Some tests have demonstrated that, if the CCS sensor exposed to 100% RH, employs for example 4 hours to reach saturation, when exposed to a lower value of RH (95%) double of the time, i.e. 8 hours, is necessary to detect the RH decrease.

The aim of this study is to investigate the capability of HCS with respect to the conventional capacitive sensor, to detect RH variations when the environmental conditions are close to vapor saturation. Both sensors were installed in two hypogea heritage sites: The Monkey Tomb in Chiusi (Siena, Tuscany) and the Mithraeum within the Caracalla Baths in Rome. The hypogea sites considered in this study do not have the same architectural characteristics: The Monkey Tomb (Figure 1) is totally closed, no openings except for the main entrance; the Mithraeum of the Caracalla Baths (Figure 2) has the main entrance on the North side, and a big arch on the South side connected to the outdoors.

In this study, RH data measured using the HCS and the CCS were compared to investigate to which extent the HCS is more effective than the CCS.

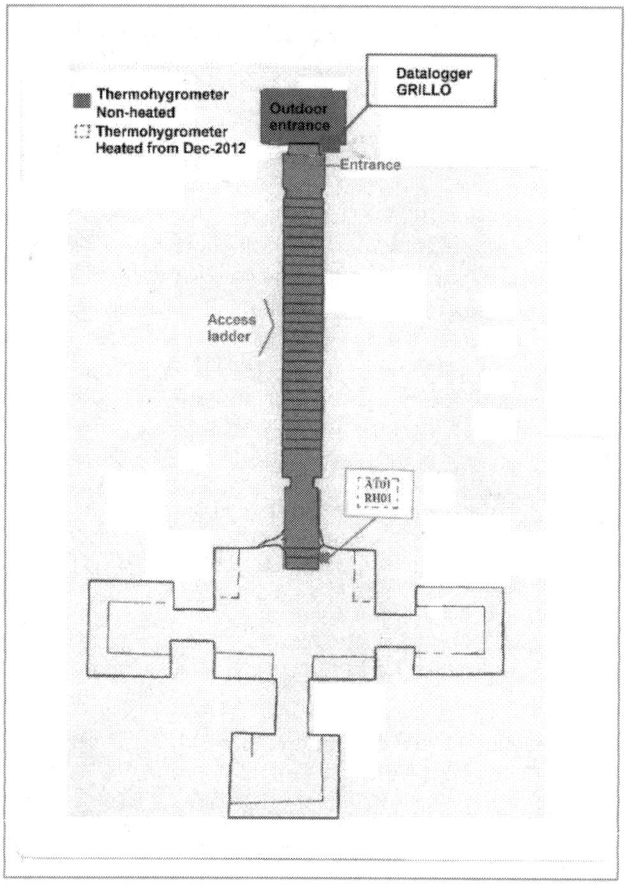

Figure 1. Plan of the Monkey Tomb in Chiusi (Siena, Tuscany), indicating the sensors installed. The thermo-hygrometer is indicated as AT01-RH01. Composed by the CCS in the first campaign and replaced by the HCS on 12 December 2012.

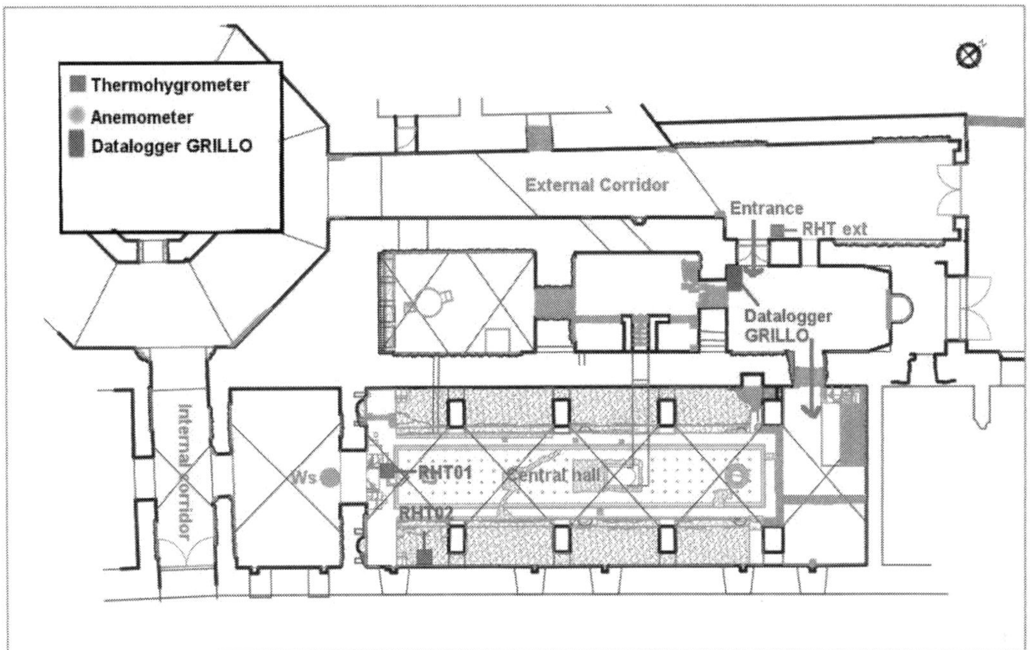

Figure 2. Plan of the Mithraeum of the Caracalla Baths in Rome, with the installed sensors. The indoor thermo-hygrometers (indicated as RHT01 and RHT02) used the CCS, replaced by the heated type on 11 November 2015. Ws indicates the biaxial sonic anemometer. RHT ext is the thermo-hygrometer installed outdoors.

2. Materials and methods

The microclimate monitoring on both sites was carried out using two electronic thermo-hygrometers developed and distributed by Rotronic®: the thermo-hygrometer model with the conventional capacitive sensor (CCS) and the heated capacitive sensor (HCS). The temperature is always measured by a platinum resistance sensor (Pt100), whereas for RH the same type of the capacitive sensor is used, but in the HCS a heater is installed under the capacitive sensor to prevent the dielectric saturation in environments with very high RH.

Two field campaigns were carried out for each heritage site: the first campaign using the CCS sensor, the second using the HCS sensor. Table 1 summarizes monitoring periods for each site. *In this study, the sensors have not been contemporaneously used in the identical period. In fact, we have assumed that in hypogeum environments the thermo-hygrometric conditions are very stable over a year, so that the comparison can be significantly carried out also taking into account the same time interval in different years. It should be noticed that the two sensors have been tested in a climatic chamber under the same thermo-hygrometric conditions to test their response time, before performing the field campaigns.*

Table 1. Duration of the two microclimate monitoring campaigns carried out with non-heated (CCS) and heated (HCS) sensors, for each site.

Hypogeum site	Period of monitoring	
	CCS	HCS
Monkey Tomb	22 September 2011 –11 December 201	12 December 2012 -27 February 2013 6 July 2016 – 17 June 2017
Mithraeum of Caracalla Baths	12 November 2013 – 27 July 2014	12 Nov 2015 – 27 April 2017

Table 2 summarizes the sensors' technical characteristics. The HCS internal working cycle consists of three phases: measuring - heating - waiting (before undertaking further measurements). The working cycle lasts 6 min: 1 min for measuring, 1 min for heating and 4 min for waiting, in order not to undertake the measuring just after heating. During the heating phase the sensor needs electrical supply. The output analog signal is updated during the measuring phase and remains constant for the other two phases.

All sensors were connected to the GRILLO MMTS datalogger, developed and distributed by Tecno.El s.r.l., with remote data transmission by GSM/GPRS technology (Global System for Mobile Communications/General Packet Radio System). Data are transmitted to the server allowing their visualization and download. For these studies, the acquisition time is set to 5 minutes and the processing time is set to 30 minutes, providing the minimum, maximum and the average of the recorded parameters.

Table 2. Technical characteristics of CCS and HCS Rotronic® sensors.

CCS	Physical variable	Accuracy
Thermo-hygrometer: Platinum resistance Sensor (PT 100)	Air temperature (T °C)	± 0.3 °C
Thermo-hygrometer: Capacitive hygrometer	Relative humidity (RH %)	± 1.5%
HCS	Physical variable	Accuracy
Thermo-hygrometer: Platinum resistance Sensor (PT 100)	Air temperature (T °C)	± 0.15 °C
Thermo-hygrometer: Capacitive hygrometer	Relative humidity (RH %)	± 1.3%

In the Monkey Tomb (Figure 1), the thermo-hygrometer CCS (indicated as AT01 - RH01), was installed in the main chamber and was later replaced by the HCS on 12 December 2012. In this study these data were considered in order to compare RH measurements carried out by both heated and non-heated sensors.

In the Mithraeum of Caracalla (Figure 2), two indoor thermo-hygrometers were installed: one at 1 cm from the wall (indicated as RHT02) and one in the central hall (indicated as RHT01). In addition, an outdoor thermo-hygrometer (indicated as RHT ext) was placed in the main entrance. A biaxial sonic anemometer (indicated as Ws), was installed in the adjacent room in communication with the central hall, in order to evaluate the influence of the air flows speed (accuracy of ±2% within the measurement range) and direction (accuracy of ±3°) on the RH behaviors.

3. Results

3.1. The Monkey Tomb

The RH data collected during the two monitoring campaigns show different trends.
In Figure 3(a), no fluctuations were observed and RH was always equal to 100% from September 2011 up to the beginning of December 2012. When the CCS was replaced by the HCS on 12 December 2012, RH values decreased to 96% and fluctuated until February 2013.
Figure 3(b) showed RH measurements by HCS. The variations were detected however the values were always higher than 94%.

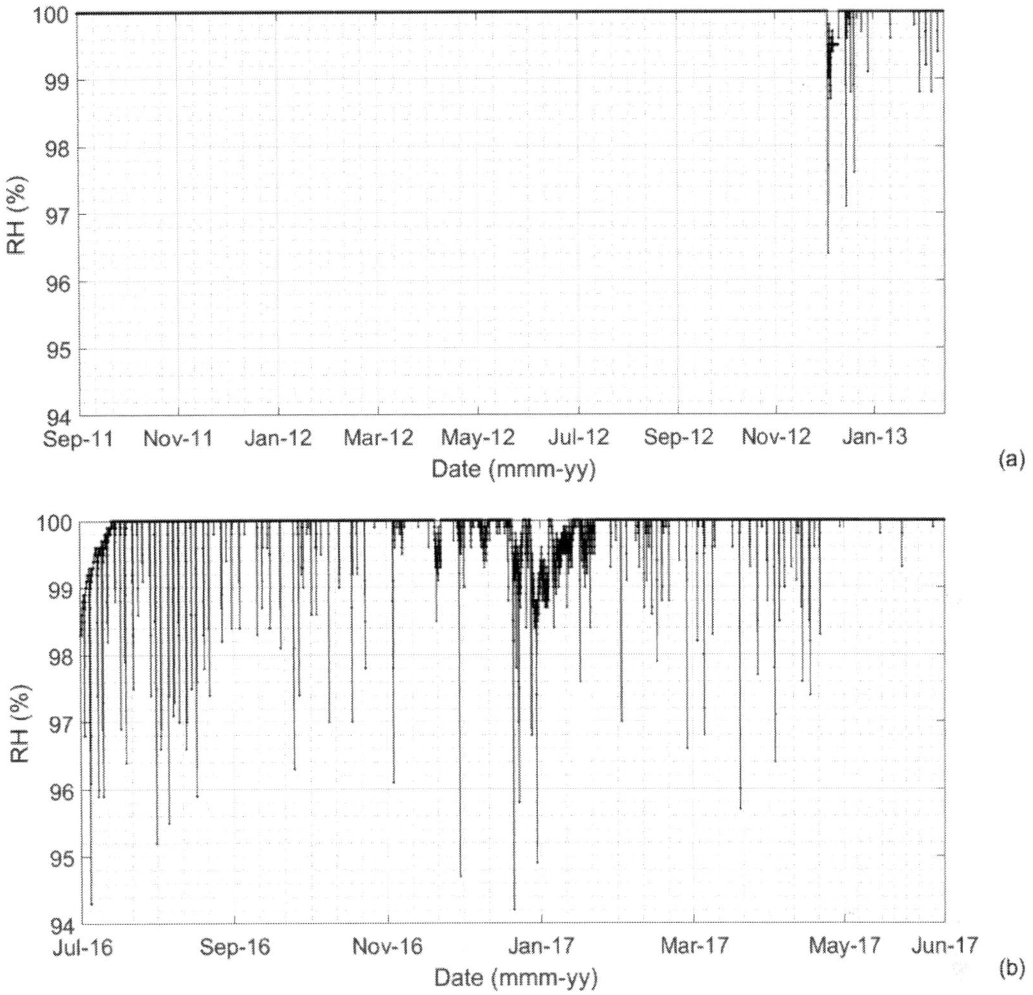

Figure 3. (a) Upper panel. RH trend in the Monkey Tomb during the period of 22 September 2011 - 27 February 2013. Measurements were carried out with CCS up to 11 December 2012, when CCS was replaced by HCS. (b) Lower panel. RH trend in the Monkey Tomb during the period of 16 July 2016 - 7 September 2017. Measurements were carried out only with the HCS.

3.2. The Mithraeum of the Caracalla Baths in Rome

The RH measurements, taken during the monitoring campaigns using CCS (Figure 4, upper panel) and HCS (Figure 4, lower panel) showed some differences.

In Figure 4 (upper panel) most values were over 90% and in few cases were below 70%, with a minimum value of 67%.

In Figure 4 (lower panel) most values were over 80% and in few cases were below 50%, with a minimum value of 31%.

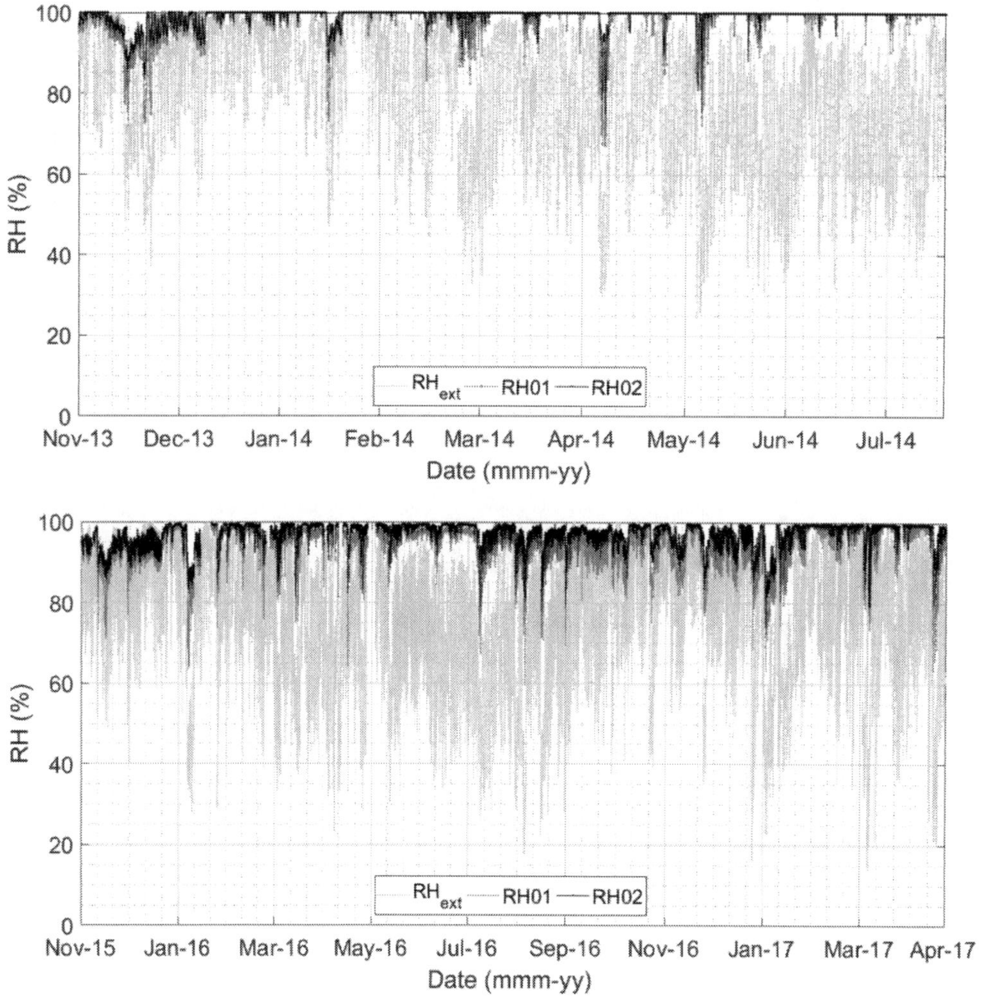

Figure 4. (a) Upper panel. Indoor and outdoor RH trends recorded inside the Mithraeum of the Caracalla Baths. Measurements were carried out with CCS. The RH was often over 90% and the minimum value reached was 67%. (b) Lower panel. Indoor and outdoor RH trends recorded inside the Mithraeum of the Caracalla Baths carried out with HCS. The RH is often over 80% and the minimum value is 31%.

To test whether there was a monotonic relationship between air flows (intensity and direction) and RH values (measured by both CCS and HCS), the Spearman's rank-order correlation [6] was used because any assumption on data frequency distribution is required.
A weak correlation was found with the air flow speed (Table 3), whereas a moderate correlation was found with the air flow direction. This may be attributed to the building structure which can influence the forces of the air flow directions, affecting RH fluctuations much more than the air flow intensity.

Table 3. Spearman's rank correlation coefficient values, between RH values and air speed, and direction values.

Spearman's rank correlation coefficient	CCS		HCS	
	Wd	Ws	Wd	Ws
RH01	0.37	-0.11	0.29	-0.19
RH02	0.37	-0.04	0.31	-0.15

4. Conclusions

This study has demonstrated that in the hypogeum environment, characterized by RH levels close to 100%, a thermo-hygrometric sensor with a heating cycle to prevent the saturation of the capacitive RH sensor, allows more accurate measurements. Differences in RH data collected by conventional RH capacitive sensors and heated capacitive sensors have been found in two different hypogea heritage sites, which suffer of degradation phenomena strictly related to high humidity.

In the Monkey Tomb, the RHs measured by the conventional capacitive sensor was always close to 100%, whereas those measured with the heated capacitive sensor show that RH can drop up to 94%. In the Mithraeum of the Caracalla Baths, the heated capacitive sensor measured decreases of RH up to 31%, while the conventional sensor measures a minimum value of 67%. Moreover, RH values seemed to be moderately affected by the air direction.

The heated capacitive sensor has resulted to be more effective in recording RH daily fluctuations. In fact, the heating cycle, allowing the desorption of water from the dielectric material, provides a RH measurement which is accurate and with a lower time response with respect to the capacitive sensor without heating cycle. This result should encourage the use of the heated capacitive sensor when indoor environmental conditions are characterized by such high RH values. A fast recording of RH drop is fundamental in hypogea environments, as it is able to detect short-term dry episodes which could affect their peculiar hygrometric conditions.

Acknowledgments

We are grateful to Dr. Giroldini of the SABAP-SI (Soprintendenza Archeologia, Belle Arti e Paesaggio per le province di Siena, Grosseto e Arezzo) for the access to microclimate data of the Monkey Tomb, and to Dr. Piranomonte of the Soprintendenza Speciale per i Beni Archeologici di Roma, for the access to microclimate data of the Mithraeum of the Caracalla Baths.

References

[1] Cardinale N and Ruggiero F 2002. A case study on the environmental measures techniques for the conservation in the vernacular settlements in Southern Italy. *Build Environ*, **37(4)**, 405-414.

[2] Cardinale N, Rospi G and Stazi A 2010. Energy and microclimatic performance of restored hypogeous buildings in south Italy: The “Sassi” district of Matera. *Build Environ*, **45(1)**, 94-106.

[3] Caratelli A, Siani A M, Casale G R, Paravicini A, Bertolin C and Camuffo D 2013. Indoor measurements of microclimate parameters in the Mithraeum in the Baths of Caracalla (Rome, Italy), *Proc. Conf. on BH2013 Monitoring conservation management (Milan).*

[4] Camuffo D 2014. *Microclimate for Cultural Heritage. Conservation, Restoration, and Maintenance of Indoor and Outdoor Monuments* (Elsevier), Second edition, p 560.

[5] Smit H, Kivi R, Vömel H and Paukkunen A 2013. Thin Film Capacitive Sensors in Monitoring Atmospheric Water Vapour In: Kämpfer N. (Ed.), *Monitoring atmospheric water vapour: ground-based remote sensing and in-situ methods*, Vol. 10 (pp. 11-38). Springer Science & Business Media.

[6] Spearman C 1904. The proof and measurement of association between two things. *Am J Psychol*, **15(1)**, 72-101.

An Augmented Reality Application for the Visualization and the Pattern Analysis of a Roman Mosaic

Mattia Santachiara[a], Francesco Gherardini[b]*, Francesco Leali[b]

[a]Santasco, Via Dardano Fenulli 19, 42123, Reggio Emilia, Italy
[b]Department of Engineering "Enzo Ferrari", University of Modena and Reggio Emilia Via Pietro Vivarelli 10, 41125, Modena, Italy

*corresponding author: francesco.gherardini@unimore.it

Abstract. The visualization and analysis of mosaics are often compromised by their large sizes, which do not enable the observer to perceive their whole arrangement or to focus on details placed in farthest areas from its boundaries. Moreover, the usual precarious state of conservation of these artefacts, often with damaged or missing areas, makes it difficult to perceive their original aesthetic value. To overcome these limitations, we propose an application of augmented reality able to support the observer in two ways: first, the application completes the missing surface of the mosaic by integrating the existent surface with a virtual reconstruction; second, it enables the analysis of the mosaic pattern by overlaying virtual lines and geometric shapes in order to explicit its geometric arrangements. The result is achieved via a custom Android application able to recognize and track the mosaic figures, thus obtaining a coordinate system used to perform a real-time rendering of the mosaic reconstruction. Such rendering is overlaid on the video stream of the real scene. The application runs on a standard smartphone embedded in a Google Cardboard-compatible Virtual Reality viewer and therefore is extremely affordable. As a case study, in order to reconstruct its aspects and to analyse its geometric pattern, we chose the roman mosaic re-found in Savignano sul Panaro (near Modena, Italy) in 2011, after 115 years from its first discovery, which is preserved less than half of its original 4.5 x 6.9 m surface.

Keywords. Augmented Reality, Photogrammetry, Real-time visualization, Roman mosaic, Geometric pattern, Cultural heritage.

1. Introduction

The overall perception of the whole composition of large artefacts such as floor mosaics and decorative pavements is generally limited by their size. It is not always possible to observe those artefacts from a point of view able to show them in their wholeness, nor it is possible to reach areas and details placed in farthest areas from their boundaries. Similarly, the state of conservation of artefacts, often damaged by time and by calamities, limits the observer, who cannot recognize the details, the geometric and aesthetic arrangements, the original colours, etc. These limitations are often increased by the collocation of the artefacts or by their lighting, which –far from being optimal- may limit their accessibility, visibility and therefore the possibility of perceiving their value.

The cultural heritage field recurs to technology not only in the field of materials and conservative techniques, but also as tools for visualization, analysis and study, in favour of scholars but also tourists

[1,2,3] and people with disabilities [4,5]. Among the most important applications, we can include the digitization of cultural heritage and the applications of Virtual Reality (VR) and Augmented Reality (AR), which allow a more immersive interaction between the observer and the artefact, which can go beyond its simple observation.

1.1. State of the art and research aim

The literature shows many applications of VR and AR in the field of cultural heritage, but only few of them deal with mosaics and pavement ornaments. The most of them recur to photogrammetry or image-based process in order to achieve the digital reconstruction of the artefact, which can be integrated within the VR or AR application. In [6], the close range photogrammetric survey of a roman mosaic in Palermo (Italy) is described, aiming at the production of a full-scale representation (scale 1:1) of the mosaic, useful for documentation and restoration processes. The research has allowed evaluating limit and potentiality of image-based approach using photogrammetric and computer vision (Structure for Motion) techniques in a context where the metric point of view is a very important factor. A similar photogrammetric reconstruction, but based on the use of tourist photographic images, is presented in [7], for the reconstruction of an ornamental portal covered of ceramic tiles, whose three-dimensionally arrangement in geometric patterns is then extracted and analysed. In [8], the authors propose an overview of image processing-based methods dedicated to the specificities of mosaic images for applications to restoration and cataloguing of ancient mosaics, focusing on the shape and the orientation of each "tessera", by means of a dedicated algorithm.

A survey of digital technologies in cultural heritage is presented in [9]: the authors investigate the potential of those technologies for providing instruments for sharing information about cultural heritage, which museums and institutions are called to preserve and promote. Their project aims to find the most suitable procedure to acquire archaeological artefacts, build their digital replica together with 3D printed prototypes, and derive simplified models to be visualized through stereoscopic devices, allowing the simultaneous viewing of real and digital 3D data through an AR environment, portable to mobile devices as well. With the aim of promoting the territory and its heritage, in [10] the authors recur to technologies (as 3D scanning, 3D printing and AR) for cultural tourism. Their aim is to document current and past historical ages and integrate 3D visualizations with renders capable of returning an immersive VR for a successful enhancement of the heritage. The case study is the archaeological complex of Massaciuccoli, one of the best preserved Roman site of the Versilia Area (Tuscany, Italy). As results, the project develop the 3D virtual reconstructions through the main historical periods on the basis of historical investigation and the analysis of the data acquired. Similarly, [11] presents a detailed account of the construction of a computerized model of the Roman theatre of Byblos, one of oldest continuously inhabited cities in the world. The starting point is a historical study, which yielded the formation of a hypothesis on the original shape and details of the theatre, which today retains very little of its original structure. Their final aim is the creation of both a VR and AR application of the Roman theatre, focusing also on the description of the floor ornaments and geometry.

The Literature review highlights that the development of novel integration between VR/AR technologies and cultural heritage are continuously developed, aiming at promoting the accessibility and the knowledge of artefacts. In this paper, our contribution will focus on 2D damaged artefacts, as mosaics and frescos, for completing their missing surface by a virtual reconstruction through AR techniques.

The paper is structured as follows. Section 2 describes the general method, Section 3 presents the case study and the method applied to a Roman mosaic, and finally Section 4 draws some final remarks.

2. Method

The proposed method is based on the application of AR to support the observer in two ways: first, the application completes the missing surface of ancient and damaged mosaics or, more generally, pavement or wall ornaments (e.g. frescos) by integrating the existent surface with a virtual reconstruction; second, it enables the analysis of the geometric pattern of the 2D ornaments by overlaying virtual lines and geometric shapes in order to explicit its geometric arrangements. Therefore, the method follows the steps listed below.

1) Digital reconstruction of the original artefact:

1.a) The first sub-step is the development of a digital model of the whole surface, which is –in the case of a floor mosaic– a 2D model. The 2D model can be achieved by means of an orthophoto, which is a parallel projection of the view of a photogrammetric textured model taken along a predetermined plan, or a photoplan, which is a photographic image corrected and rectified so that the image plane represents the plan of the real object in scale and metrically measurably [12]. A digital model of the artefact is required to implement the AR application.

1.b) In case of damaged or missing parts of the artefact, the missing areas may be reconstructed on the basis of the adjacent ones by directly modelling or reshaping them by an image editing software, according to archaeological studies. A high-resolution image of the ornament before damage (if available) may act in the same way in case of 2D artefacts.

2) Analysis of the artefact: the geometric pattern of the artefact is required to be analysed and highlighted by means of auxiliary geometric elements. This analysis is aimed at supporting the observer in understanding the whole pattern and its repetition of elements (e.g. if the missing element is identical with another one, etc.). Therefore, the auxiliary elements consist of virtual lines and geometric shapes that will be overlaid to the real artefact. By means of colours, each element may be identified and highlighted, and identical elements in the pattern may be represented with the same colour in order to help the observer in understanding the pattern.

3) Augmented Reality implementation: the AR application is implemented first by filling the missing parts of the real mosaic with the reconstructed parts and, once it is completed, by overlaying the virtual lines and the geometric shapes in order to explicit the geometric arrangements. Such rendering is overlaid to the video stream of the real scene. The application may run on a standard smartphone embedded in a VR viewer (e.g. a Google Cardboard-compatible viewer) and so is extremely affordable.

3. Case study: The Roman mosaic of Savignano sul Panaro

The ruins of a large late Roman structure were discovered for the first time in Savignano sul Panaro (near Modena, Italy) in the farm of Melda di Sotto, in compliance with via Magazzino (near the ancient via Claudia), in 1897. In that period, the director of the Civic Museum of Modena was Arsenio Crespellani, who was attracted by the exceptional floor mosaics. He promoted an archaeological excavation during which the mosaics have been documented with polychrome watercolours paintings and drawings by the young artist Giuseppe Graziosi [13]. At the end of the 19th-century the mosaics were covered in the same place of their discovery, up to a new excavation between 2010 and 2011, during the construction of a roundabout, which represented a new recovery opportunity. Thanks to the Graziosi's watercolour (see next paragraph) documenting the mosaic existence, the Archaeological Superintendence of Emilia-Romagna identified the ruins of the architectural complex, and the archaeological investigations was completed, allowing at the same time the separation of the mosaic of one of the rooms. The rooms currently investigated are four, but only the one described by Crespellani with the letter A (called room 1 in the last archaeological investigation), is completely discovered. All the found structures are very damaged by interventions of different ages: the scarce or absent indication of the walls makes it difficult to define exactly the size of the rooms. The room A is rectangular, approximately long m. 6.90 with a hypothetical width in m. 4.80 [14]. It seems to be the most important room, at least in this sector of the building.

The Roman mosaic originally measured about 6.90 x 4.50 meters, but it is preserved less than half of its original surface (Fig. 1), and is dated around the 5th century CE [14]. The combination of polychrome stone and terracotta tiles with emerald green and ruby red glass tiles shows the importance of the environment and the wealth of the commitment. It is decorated with intertwined elements, geometric and stylized plants that alternate with the knot of Solomon, with a central element framed by a laurel wreath that delimits a figurative decoration perhaps of a symbolic nature.

IOP Conf. Series: Materials Science and Engineering **364** (2018) 012094 doi:10.1088/1757-899X/364/1/012094

Figure 1. The Roman mosaic as permanently installed at the "Casa Natale Giuseppe Graziosi" (Savignano sul Panaro (Modena, Italy)). Photo credit: Marianna Grandi.

3.1. Digital reconstruction of the original artefact

This first step consists of two sub-steps:

a. Development of a digital model of the real artefacts: The difficulties in obtaining a photographic image of the whole mosaic, due to the small room in which the mosaic is preserved that causes a strong distortion in the photographs, and the non-optimal lighting condition, we decide to achieve the digital model from an orthophoto. A photogrammetric model of the Roman mosaic of Savignano sul Panaro is therefore achieved by means of 115 photos (Standard compact camera Nikon P310, 16.1MP CMOS Sensor, Sensor size: 1/2.3" (~ 6.16 x 4.62 mm), Max. image resolution: 4608 x 3456). The photogrammetric reconstruction is performed with the software PhotoScan (Agisoft) and consists of a textured detailed model of the mosaic. The final result, shown in Fig. 2 (Left), is an orthophoto of the mosaic (dimensions: 3982x3625 300 DPI).

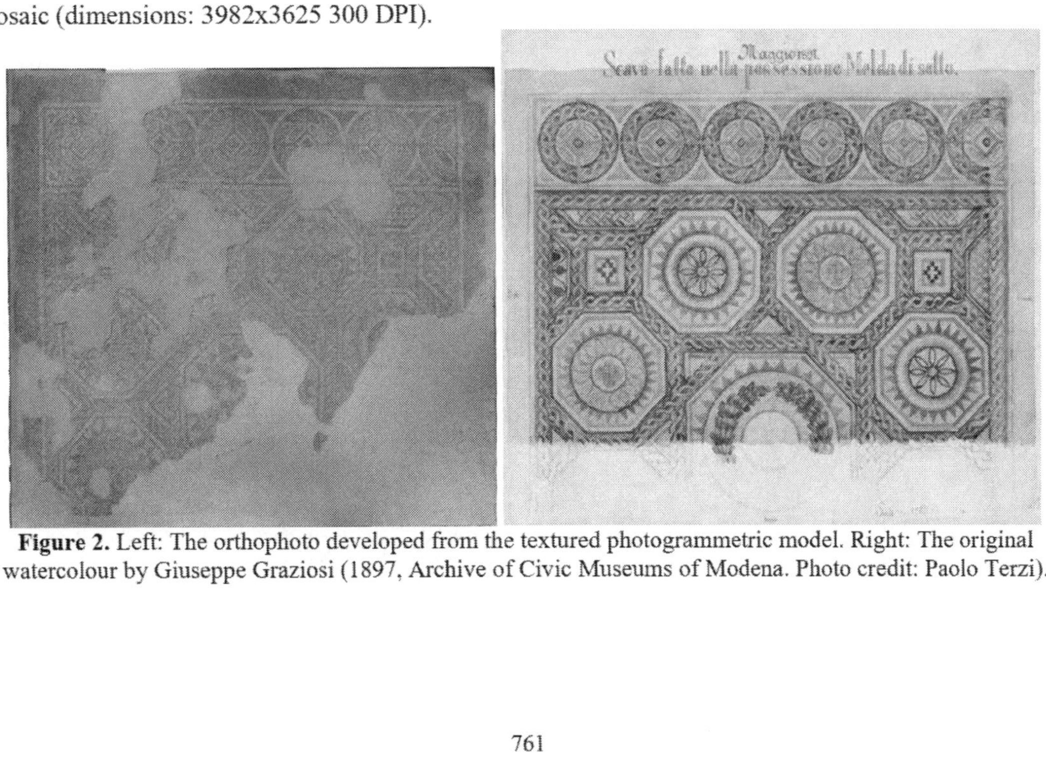

Figure 2. Left: The orthophoto developed from the textured photogrammetric model. Right: The original watercolour by Giuseppe Graziosi (1897, Archive of Civic Museums of Modena. Photo credit: Paolo Terzi).

IOP Conf. Series: Materials Science and Engineering **364** (2018) 012094 doi:10.1088/1757-899X/364/1/012094

b. Integration of the damaged or missing part: The Roman mosaic of Savignano sul Panaro is strongly damaged, because more than a half of its original surface is missing and, similarly, also the remaining one is incomplete. Even if the missing areas may be supposed by the symmetrical repetition of the decorative elements within the pattern, a further contribution to the whole reconstruction of the mosaic is determined by the original Graziosi's watercolour painting conserved in the archive of Civic Museums of Modena (Fig. 2, Right). This painting represents an original source in order to integrate the missing parts of the mosaic.

Starting from a high-resolution photographic image (dimensions: 4164x3688, 300 DPI) of the painting, first we mirrored the painting image, then we superimposed the orthophoto on it, and finally we extract the boundary lines of the real mosaic and we use them for cutting the painting image: in this way, we extract from the painting only the complementary part with respect to the real mosaic (Fig. 3). This "cut" image will be used in the AR application.

Figure 3. Mirrored and cut image of the Graziosi's painting.

3.2. Analysis of the artefact: pattern analysis

In order to create the auxiliary geometric elements for the AR application, we propose the analysis of the mosaic pattern according to [14,15], as shown in Fig. 4.

The mosaic shows a rotational symmetry with eight elements arranged around a central one of greater dimensions (Fig. 4a). The central element consists of an eight-pointed star, formed by two superimposed squares in order to form a central octagon with irregular sides (in purple, in Fig. 4b). The vertices of the star originate eight octagons, smaller in size, arranged in pairs of two on each side (in red, blue, and yellow, in Fig. 4b); the space between the octagons and the side walls is filled with different shapes, depending on the resulting space, which are trapezoidal and triangular (in red, in Fig. 4c).

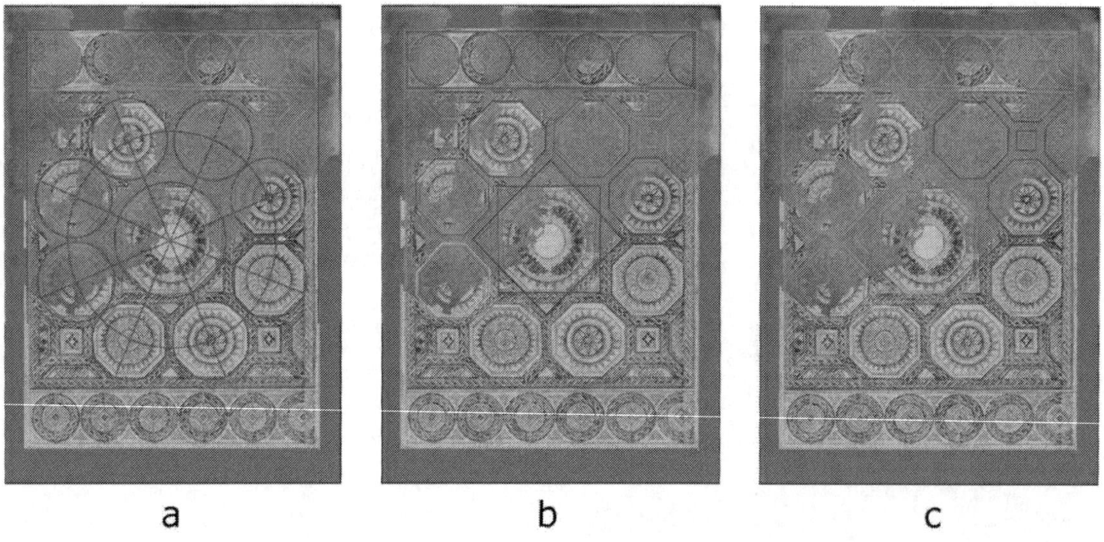

Figure 4. Auxiliary geometric elements built on the "reconstructed" Roman mosaic, in order to highlight and analyse its ornaments and their arrangement within the mosaic.

The division between the motifs is given by intertwined elements with shaded colours, one from white to black and the other from white to red, so that the diversity of tones makes the weave stand out better. The central octagon (in purple, in Fig. 4b) has a circular pattern with a white background containing a laurel wreath and, presumably, a figured centre; the line forming the outer edge is equipped with a series of triangular points of black colour and with different height so as to adapt to the available space. The inner, circular wreath seems to be made up of laurel leaves, mainly dark green, partially overlapping alternating bands of light and dark leaves. The importance of decoration is also confirmed by the presence of red and green glass tesserae, arranged randomly, but clearly visible to the light. The external octagons are only partially preserved, so that it is not possible to say whether they symmetrically repeated; all of them include internal circular motifs, with pointed triangles border in black on white.

3.3. Augmented Reality implementation

The AR application is implemented within the Unity engine platform (Unity Technologies) by means of EasyAR tracking technology library. The implementation follows the listed steps:

Step 1. Import the EasyAR plugin package into the Unity platform.

Step 2. Setup of the scene: The essential components for the scene setup are already available (as prefab) in the plugin package. In EasyAR, we can find the following tools: CameraDevice Behaviour, logical abstraction that allows interfacing with the hardware device camera; ARCamera Behavior, virtual camera of the scene that allows the overlap between video flow and realtime 3D rendering; Tracking ImageTracker, for target tracking; ImageTracker, component that detects and traces ImageTarget; ImageTarget, flat image that can be traced by an ImageTracker.

Step 3. Insert an image to be traced: Once a new ImageTarget is inserted in the scene, the image to be traced is set as follows: copy the image (in .jpg or .png format) in the subfolder of the "StreamingAsset" project, then set the X and Y proportions in the size of the ImageTarget in a manner consistent with those of the reference image; assign an ImageTracker present in the scene in the Loader field; then, the application may recognize and trace the image (Fig. 5 Left).

Step 4. Assign an object to render to the tracker of images: the object to visualize is set as a "child" of the ImageTracker. Specifically to the mosaic, the orthophoto of the mosaic is set as a tracking image which appears in the background and then a polygon with the texture of the mosaic painting is superimposed as a child. In this way, when the camera will frame the real mosaic, the mosaic painting will be rendered on the video stream (Fig. 5 Right) as well as the geometric patterns (Fig. 6). The possibility of having a preview of the tracking image allowed to place the child plan on the mosaic in order to obtain a perfect match with the missing parts of the real mosaic.

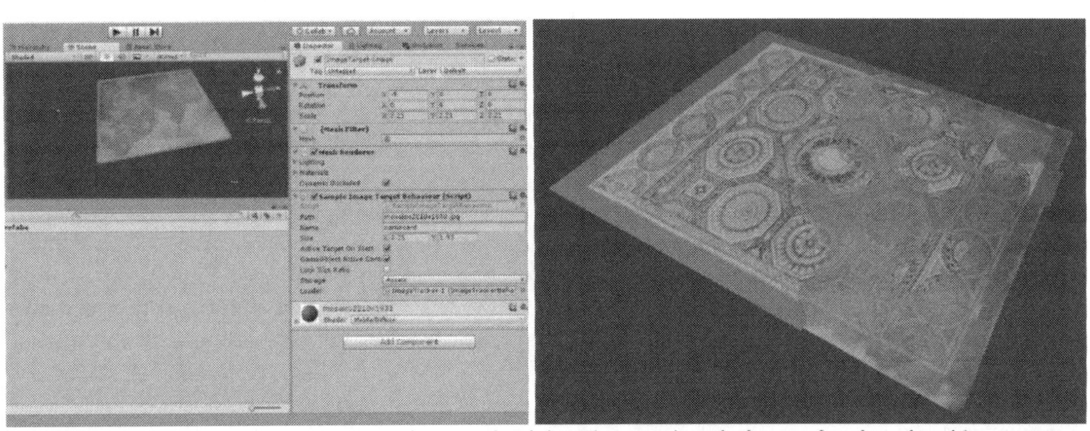

Figure 5. Left: Inserting an image to be traced. Right: The mosaic painting rendered on the video stream.

Step 5. Adaptation for stereographic rendering: In order to use the AR application with the VR viewer, we are required to generate a stereographic rendering, so that the display will be split in two parts: one will render for the left eye (left camera), the other one for right the eye (right camera). This is

IOP Conf. Series: Materials Science and Engineering **364** (2018) 012094 doi:10.1088/1757-899X/364/1/012094

achieved by inserting in the scene two virtual cameras spaced apart by 0.6 on the horizontal axis and assigning to each a specific area on which to render in the viewport respectively x = 0, y = 0, w = 0.5, h = 1 for the right camera and x = 0.5, y = 0, w = 0.5, h = 1 for the left camera. This requires a double rendering pipeline, so the scene is rendered twice.

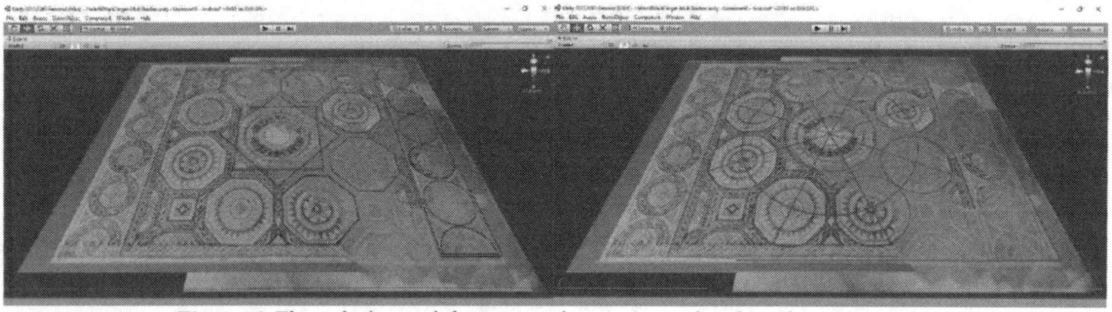

Figure 6. The painting and the geometric pattern rendered on the real mosaic.

4. Discussion and conclusion

This paper proposed a method aimed at reconstructing and integrating the missing areas of mosaics and similar artefacts by means of AR techniques. The real mosaic is "augmented" by means of images of the missing areas, and so virtually reconstructed, with the final aim of supporting the analysis of the mosaic pattern. The final result is a custom Android application able to recognize and track the mosaic figures, used to obtain a coordinate system for the real-time rendering of the mosaic reconstruction.

Some drawbacks of the method consist of: 1) The proposed integration of the areas of the damaged artefact with an image is limited to those cases in which an image is available, as in the case of recent damages or natural disasters or, as in the case of the Roman mosaic there is an original image (drawing, painting, old photo, etc.). 2) The image may differ from the real artefact due to colours (as in the case of the Roman mosaic, due to dust deposition or to lighting conditions), so to require to digitally modify the image. Moreover, the image may present differences in geometry, as shown in the comparison between the orthophoto and the painting, where the octagons are irregular (real) or regular (painted). Moreover, in the Roman mosaic, some elements are warped, probably due to damaging, therefore the watercolour image has required to be deformed and adapted to the real geometry. 3) In the specific case of the Roman mosaic, the small room in which it is installed causes great distortions in taking photos of the details and the overall geometry, so to require the use of photogrammetry for developing an orthophoto. 4) The use of a VR viewer requires a double rendering pipeline, so that the scene needs to be rendered twice: this could be a critical factor in the case of highly complex scenes, but negligible in the present case study.

On the other hand, the method is able to reconstruct damaged artefacts and their geometric patterns, supporting the visualization and analysis during their observation. The missing elements may be achieved from the existent ones (e.g. thanks to symmetry) or, in other cases, from different sources (e.g. images or photos) or be directly remodelled based on studies or hypothesis of scholars. Moreover, the use of AR supports the analysis of the artefact by means of interactions with observers, scholars or tourists, as an educational tool able to guide the understanding of what they observe. Finally, the method can be extended from floor mosaics and pavements to other similar artefacts, as wall mosaics and frescos, even if not planar: as further developments, we will investigate the application of the method in this last case.

Acknowledgments

The authors gratefully acknowledge the Superintendence of Archeology, Fine Arts and Landscape of the metropolitan city of Bologna and the provinces of Modena, Reggio Emilia e Ferrara (Italy); the Civic Museums of Modena (Italy), its Director Dr Francesca Piccinini and its curator Dr Cristiana Zanasi; the municipality of Savignano sul Panaro (Modena, Italy) and its instructor Elisabetta Zanettini; Prof.

Giorgio Pancaldi and the association "Ponte Alto Graziosi"; the professional photographer Marianna Grandi, for her contribution in the realization of the images.

Photo credits

All images are by the authors unless otherwise noted.

References

[1] Scopigno R, Corsini M, Callieri M, Dellepiane M 2011 Using Digital 3D Models for Study and Restoration of Cultural Heritage Artifacts *Digital Imaging for Cultural Heritage Preservation: Analysis, Restoration, and Reconstruction of Ancient Artworks* Ed F Stanco, S Battiato et al (Boca Raton, CRC Press) chapter 13 pp 353-384

[2] Noh Z, Shahrizal M, Pan Z 2009 A review on augmented reality for virtual heritage system *Learning by playing. game-based education system design and development* Ed M Chang, R Kuo et al (Heidelberg, Springer) pp 50-61

[3] Kečkeš A L, Tomičić I 2017 Augmented Reality in Tourism – Research and Applications Overview *Interdisciplinary Description of Complex Systems* 15(2) 157-167

[4] Drigas A, Koukianakis L, Glentzes J 2009 An e-culture – e-museums environment for common citizens and disabled individuals *Int. J. of Digital Culture and Electronic Tourism* 1(4) 267-279

[5] Volpe Y, Furferi R, Governi L, Tennirelli G 2014 Computer-based methodologies for semi-automatic 3D model generation from paintings *Int. J. Comput. Aided Eng. Technol.* 6(1) 88-112

[6] Lo Brutto M, Garraffa A, Pellegrino L, Di Natale B 2015 3D Mosaic documentation using close range photogrammetry *1st Int. Conference on Metrology for Archaeology* (Benevento, Italy) 82

[7] Gherardini F, Leali F 2016 A Framework for 3D Pattern Analysis and Reconstruction of Persian Architectural Elements *Nexus Network J.* 18(1): 133–167

[8] Benyoussef L, Derrode S 2011 Analysis of Ancient Mosaic Images for Dedicated Applications *Digital Imaging for Cultural Heritage Preservation: Analysis, Restoration, and Reconstruction of Ancient Artworks* Ed F Stanco, S Battiato et al (BocaRaton, CRC Press) chapt. 16 pp 453-478

[9] Garagnani S, Manferdini A M 2011 A reference Virtual and augmented reality applications for Cultural Heritage *SIGraDi Cultura aumentada 2011, XV Congreso de la Sociedad Iberoamericana de Gráfica Digital* (Santa Fe) Ed M Chiarella, M E Tosello pp 555–559

[10] Castagnetti C, Giannini M, Rivola R 2017 Image-based virtual tours and 3d modeling of past and current ages for the enhancement of archaeological parks: The visualversilia 3d project *Int. Arch. Photogramm. Remote Sens. Spatial Inf. Sci.* XLII- 5/W1 pp 639-645

[11] Younes G, Kahil R, Jallad M, Asmar D, Elhajj I, Turkiyyah G, Al-Harithy H 2017 Virtual and augmented reality for rich interaction with cultural heritage sites: A case study from the Roman Theater at Byblos *Digital Applications in Archaeology and Cultural Heritage* 5(June) 1-9

[12] Ippolito A, Cigola M 2017 Handbook of Research on Emerging Technologies for Digital Preservation and Information Modeling (Hersey PA, USA, Information Science Reference)

[13] Crespellani A 1899 Scavi nel Modenese (1896-1897) AMDST Province Modenesi Serie IV, IX, pp 269-288

[14] Maioli M G 2013 I mosaici di Savignano Il mosaico ritrovato Indagini archeologiche a Savignano sul Panaro (Firenze, Italy, All'insegna del Giglio) pp 55-65

[15] Corti C 2001 Atlante dei Beni Archeologici della Provincia di Modena. III, Collina e Alta Pianura. Book 1 (Firenze, Italy) pp 190-191

IOP Conf. Series: Materials Science and Engineering **364** (2018) 012095 doi:10.1088/1757-899X/364/1/012095

Study on the suitable lighting design of Beato Angelico's artworks displayed at the National Museum of San Matteo in Pisa (Italy)

F Leccese[1], G Salvadori[1], Feltrin[1], R Morozzi[2], P Nieri[2]
[1]Dept. of Energy Engineering, Systems, Territory and Constructions, University of Pisa, Italy
[2]National Museum of San Matteo, Pisa, Italy
email: f.leccese@ing.unipi.it

Abstract. The lighting design of exhibition space has a great impact on visual and colour perception and different lighting arrangements can create very different visual impression of artworks and, if not carefully designed, compromise the enjoyment of the viewers. This study involved the design of a new lighting solution for two of Beato Angelico's artworks displayed at the National Museum of San Matteo (Pisa, Italy). Multiple test lighting configurations were designed using different LED luminaires and different settings of the luminaires. The test lighting configurations were evaluated by a restricted group of observers through a survey in order to individuate the most suitable solution, able to enhance the two artworks simultaneously and to provide a good visual experience for museum visitors.

1. Introduction

Museum exhibitions are strongly affected by the features of the space where they are realized, the design of the exhibition path and, especially, the lighting design [1,2]. Architectural and lighting parameters have a huge impact on the visual perception of art exhibitions for museum visitors, and, if carefully designed, they can achieve the best enhancement of the displayed artworks and a great visual experience for the observers [3,4]. The beauty and uniqueness of each piece of art could not be properly perceived and appreciated without a good lighting project of the museum exhibition space, able to create the proper colour appearance of the artworks, the visibility of all details and the proper atmosphere in the exhibition space [5-8]. The lighting arrangement creates a determined relationship between the exhibits, the background and the surround of the exhibition space, depending on the purpose of the exhibition and the feelings that it should inspire to the viewers. In particular, the setting of the luminaires and the selection of the suitable type of luminaire (as well as a combination of different light sources) represent a crucial step when designing art exhibitions, because different layouts can create different lighting effects, firstly on the artworks, secondly on the overall exhibition space [9-11]. However light is one of the main causes of damage for the artworks, since it can cause fading and desegregation of sensitive materials, therefore the current legislation provides limitations of illuminance levels and annual luminous exposure, according to the level of sensitivity of the artworks' materials [12,13]. In this context, LED light sources represent a technological solution which allows a remarkable flexibility on the colorimetric choices, aimed to the improvement of the artworks perception [14], together with high energy performance, evermore required also in the lighting of cultural heritage [15,16].

In this paper a suitable lighting configuration aimed to guarantee the simultaneous properly perception of two artworks, displayed at the National Museum of San Matteo in Pisa (Italy), was investigated and discussed. The study is characterized as a scientific investigation, preparatory for an effective lighting design stage, from which guide information on the observer preference can be obtained. The artworks were painted by Beato Angelico during the 14th century (right on Figure 1): one gold-ground panel painting, the *Madonna dell'umiltà* (1423), and one old banner painted on fabric, the *Redentore benedicente* (1440-1445), that was recently restored. These artworks were classified by the Curator of the Museum of San Matteo as "low sensitivity" materials [13]. The two artworks, extremely different for style and dimensions, are displayed next to each other in a room, with other artworks, where the lighting system was not modified with respect to the previous mounting. The study tested multiple lighting configurations, obtained with different LED luminaires (with different optical systems) and different settings of the luminaires. The test lighting configurations were first measured and then evaluated by a group of observers, in order to point out the most suitable one for the enhancement of both artworks and a clear visibility of all their features and details.

2. Brief historical notes on the National Museum of San Matteo

The Museum of San Matteo in Pisa (Italy) is placed in the medieval monastery of San Matteo, which hosts the art collections since 1949. The exhibition halls are arranged on the first floor of the building, around the old courtyard of the monastery, the entrance of the museum overlooks the Arno River. The core of the art collection of the museum was owned by Sebastiano Zucchetti, who donated it to the Opera del Duomo of Pisa in 1796. Since the collection was enriched and extended through the years, it was moved to the former convent of San Francesco in 1893, which became the new Civic Museum of Pisa. During World War II, the Civic Museum was closed in order to protect the art collection form the potential bombing over the city, and next, between 1947 and 1949, the art collection was moved to the former monastery of San Matteo. The Museum consists of one of the most important collections of medieval painting and sculpture and it includes artworks of the most famous artists in Tuscany between the 11th century and the 13th century. This collection makes the Museum of San Matteo one of the main art galleries of medieval painting across Europe, especially for its collection of gold-ground panel paintings. During the years, the collection was enriched with more art pieces, such as the group of wooden sculptures, the sculptures from the Spina Church and the collection of ceramics.

Figure 1. Pictures of some exhibition rooms at the Museum of San Matteo: (top, left) Renaissance paintings room; (bottom, left) XIV-XV cen. paintings room; (right) the two Beato Angelico's artworks (panel painting *Madonna dell'umiltà* and old banner *Redentore benedicente* are easily recognizable).

3. Current lighting of the exhibition room

The current lighting of the exhibition rooms of the Museum of San Matteo is realized with halogen or LED spotlights set on tracks and fixed on the ceiling. Almost each artwork has its own direct lighting and there is a dim surround in most of the exhibition rooms, emphasised by the dark blue backgrounds of many panel paintings (see also Figure 1). It is important to mention that the Museum of San Matteo is a historic building, thus there are many restrictions related to the electric system and the overall management of the exhibitions. At the same time, the exhibition path should be as more flexible as possible, in order to allow to be easily adapted and modified when new artworks are displayed or some of them are loaned to other museums. Currently, the *Madonna dell'umiltà* is illuminated by one spotlight set frontally, whilst the *Redentore benedicente* is illuminated only indirectly by the light sources directed into the other displayed artworks and its details and the figure of Christ are not clearly visible. Therefore, the current lighting overshadows the *Redentore benedicente* and draws all the attention to the *Madonna dell'umiltà*, which appears more highlighted. In this context a new lighting design, able to reduce the unbalanced lighting conditions, is hence required to enhance visual experience for museum visitors.

4. Experimental activity: design and procedure

Four different lighting configurations were designed for Beato Angelico's artworks and were evaluated through a survey by a small group of observers, in order to individuate the most suitable and pleasant solution. For this study, the group of observers consisted of five people with expertise in lighting design or art. The aim of the survey was obtaining a preference pattern that could be used as a guideline for the design of a future permanent lighting system. Recent researches on museum lighting [5-6,9-10,14] point out that lighting measurements and the fulfilment of quantitative standards (i.e. the annual amount of light or the traditional setting of the luminaires) are not sufficient to provide a good visual experience of art for the viewers and qualitative standards (i.e. comfort and preference) must be taken into account.

4.1 Experimental setup

The test lighting configurations were designed using a support frame 3.20 m high, positioned in front of the two artworks at a distance of 1.50 m. The two artworks are displayed side by side, almost 1.50 m distant from each other. The support frame, 3.20 m large, allowed to create different settings of the luminaires, positioned laterally or frontally in relation to the artworks. However, the structure was not large enough to allow a frontal lighting for both the artworks, which was realized for most of the artworks displayed in the museum. For this reason, two different setups were designed (Figure 2): the first setup allowed to frontally illuminate the *Redentore benedicente* and laterally the *Madonna dell'umiltà*, whilst the second setup allowed to frontally illuminate the *Madonna dell'umiltà* and laterally the *Redentore benedicente*.

4.2 Luminaires and lighting configurations

All test lighting configurations for Beato Angelico's artworks were designed using Zumtobel products [17], all of them specifically designed and tailored for museum lighting. Since the current lighting design of the exhibition room is made of halogen lamps with low CCT, it was decided to use luminaires with a CCT of 3000 K and 4000 K. Two types of luminaires were used: the ARCOS LED expert spotlights, with different dimensions (ARCOS 2 and ARCOS 3), a Correlated Colour Temperature (CCT) of 3000 K and nominal power of 30-35 W; the SUPERSYSTEM II spotlights, with different dimensions (Mini, Midi and Maxi), different CCTs (3000-4000 K) and nominal power of 5-10 W. All the spotlights had a Colour Rendering Index higher than 90. The luminaires were fixed on tracks on the support structures and they were positioned and tilted in order to illuminate the two artworks. Four lighting configurations were designed, two for each setup (Table 1): the first two configurations (A and B Configurations) involved a frontal lighting for the *Redentore benedicente* and a lateral lighting for the *Madonna dell'umiltà*, whilst the other two configurations (C and D Configurations) involved a frontal lighting for the *Madonna dell'umiltà* and a lateral lighting for the *Redentore benedicente*. The description and the schemes of the four test lighting configurations are illustrated in Table 1.

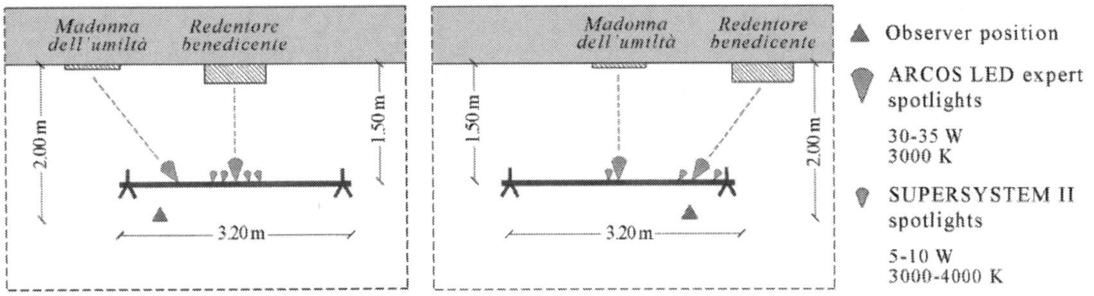

Figure 2. Plans of the luminaire setups designed for the test: (on the left) frontal lighting for the *Redentore benedicente*; (on the right) frontal lighting for the *Madonna dell'umiltà* (see also Table 1).

Table 1. Description and schemes of the four test lighting configurations.

Configuration A 	▪ The (lateral) lighting for the *Madonna dell'umiltà* was realized using an ARCOS 2 LED expert with a Flood optic and an opaque lens, able to illuminate the whole painting. ▪ The (frontal) lighting of the *Redentore benedicente* was realized with an ARCOS 3 LED expert with a Spot optic and a diffuse lens and a SUPERSYSTEM II (Midi) with a Spot optic, directed into the face of the Christ. ▪ All the luminaires had a CCT of 3000 K.
Configuration B 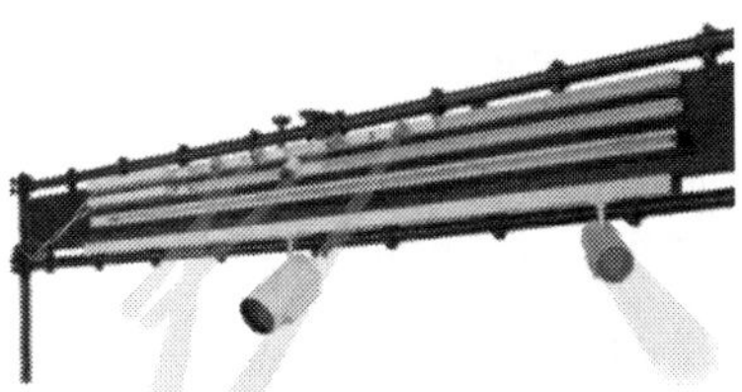	▪ The (lateral) lighting for the *Madonna dell'umiltà* was realized using an ARCOS 2 LED expert (3000 K) with a Flood optic and an opaque lens. ▪ The (frontal) lighting of the *Redentore benedicente* was realized with four SUPERSYSTEM II (Mini), two of them with a CCT of 3000 K and a Spot optic and the other two with a CCT of 4000 K and a Superspot optic. ▪ The warm and cool light were mixed and directed on the artwork to uniformly illuminate it: a 3000 K spotlight and a 4000 K were directed into the upper area of the artworks, whilst the other two 3000 K and 4000 K spotlights were directed into the lower area of the artwork.
Configuration C 	▪ The (frontal) lighting for the *Madonna dell'umiltà* was realized using an ARCOS 2 LED expert with a Flood optic and an opaque lens, plus a SUPERSYSTEM II (Mini) with a Spot optic directed into the upper golden decorations of the artwork, representing Christ. ▪ The (lateral) lighting of the *Redentore benedicente* was realized with two SUPERSYSTEM II (Maxi), both with a Spot optic. ▪ All the luminaires had a CCT of 3000 K.
Configuration D 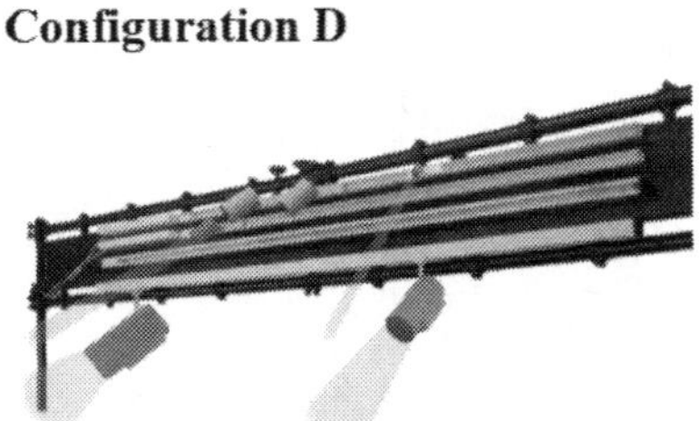	▪ The (frontal) lighting for the *Madonna dell'umiltà* was realized using an ARCOS 2 LED expert with a Flood optic and an opaque lens, plus a SUPERSYSTEM II (Mini) with a Spot optic directed into the upper golden decorations of the artwork, representing Christ. ▪ The (lateral) lighting of the *Redentore benedicente* was realized with a lateral ARCOS 3 LED expert with a Spot optic and a diffuse lens and a SUPERSYSTEM II (Midi) directed into the face of the Christ. ▪ All the luminaires had a CCT of 3000 K.

4.3 Experimental procedure

The experimental procedure for each configuration started with manually dimming the luminaires in order to uniform the new lighting solution to the current lighting of the whole exhibition room. Since it was not possible to turn off the single luminaires of the exhibition room, the test was conducted in a dark environment and there were no interferences with the lighting of the other exhibition rooms. Some monitoring measurements were done in significant selected points of the artworks (the most representative areas of the two artworks), indicated in Figure 3. The measurements were conducted with portable instruments made available by Lighting and Acoustics Laboratory of the University of Pisa: a luxmeter PRC Krochmann model Radiolux 111 and a luminance meter Hagner model L-202. These instruments have accuracy classes of "A" and "B" respectively, according to the Italian Technical Standard UNI 11142/2004. Moreover, the exposure of the artworks to UV light emitted by the lighting system was also assessed using a photoradiometer Delta Ohm model HD2102.1, equipped with a suitable irradiance probe. The measured illuminance and luminance values are summarized in Table 2, whilst the UV content was found to be negligible for all the lighting configurations.
The four lighting configurations were evaluated separately by a restricted group of five observers, with an average age 36 years. After turning on a configuration, it was first measured (measuring the illuminance, the luminance and the UV content on the surfaces of the two artworks) and then evaluated. The observers evaluated two parameters on a 6 points scale (without neutral response): the 'enhancement of the artworks' and their 'personal preference'. The observer position during the test was fixed between the two artworks, at a distance of 2.0 m from the wall. This position allowed to have a good visibility of both artworks at the same time and prevented any reflections or glare effect on the surfaces of the artworks that could have compromise the evaluations.

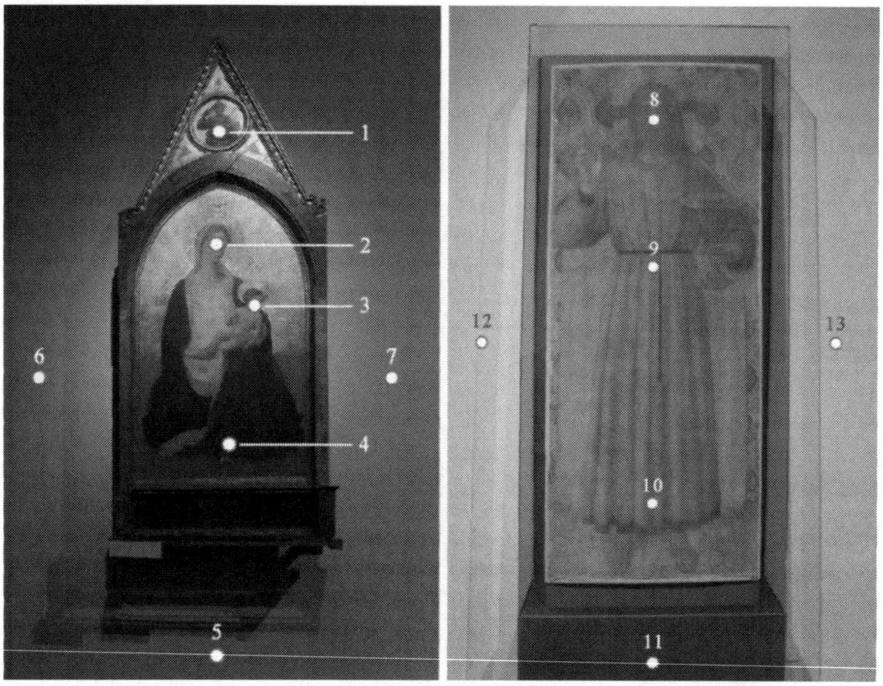

Figure 3. Measurement points selected on the *Madonna dell'umiltà* and the *Redentore benedicente*.

5. Results and discussion

The results were analysed by computing the means of the evaluations for each parameters, shown in Figure 4. From the graphs, it is clear that C and D Configurations were evaluated, on average, as more suitable for the enhancement of the artworks and more pleasant for the observers than A and B Configurations. C and D Configurations were appeared to be equally suitable for enhancing the artworks, whilst C Configuration was appeared to be the most appreciated lighting configuration (see also Figure 5). On the other hand, B Configuration was the least appreciated, probably because of the use of cooler light (4000 K).

Table 2. Illuminance and luminance measured values on the two artworks.

		Current Lighting		Test lighting configurations							
				A		**B**		**C**		**D**	
Artwork	Meas. Point	E (lx)	L (cd/m^2)	E (lx)	L (cd/m^2)	E (lx)	L (cd/m^2)	E (lx)	L (cd/m^2)	E (lx)	L (cd/m^2)
Madonna dell'umiltà	**1**	65	4.2	181	2.4	183	3.3	202	5.6	494	10.5
	2	140	9.0	296	14	272	14	213	14.3	355	31.5
	3	118	7.2	297	15.4	316	15.7	214	32.3	260	64.5
	4	35	1.3	187	5.1	186	4.9	168	4.6	171	4.8
	5	16	3.4	38	6.4	39	4.4	46	8.0	46	8.5
	6	25	4.1	121	18.5	137	16.5	72	11.2	89	15.3
	7	40	7.2	84	22.1	78	20.4	102	16.7	98	19.2
Redentore benedicente	**8**	80	4.1	396	12.7	923	26	814	24.4	725	23
	9	78	3.6	318	11.7	567	20	620	20.1	219	22.3
	10	67	2.8	122	3.2	151	4.2	312	7.1	229	6.0
	11	45	1.0	36	0.6	24	0.5	70	1.0	90	1.4
	12	55	8.4	58	10	33	9.8	62	9.8	95	13.5
	13	42	11	89	19.2	37	6.0	175	25.2	273	39.9

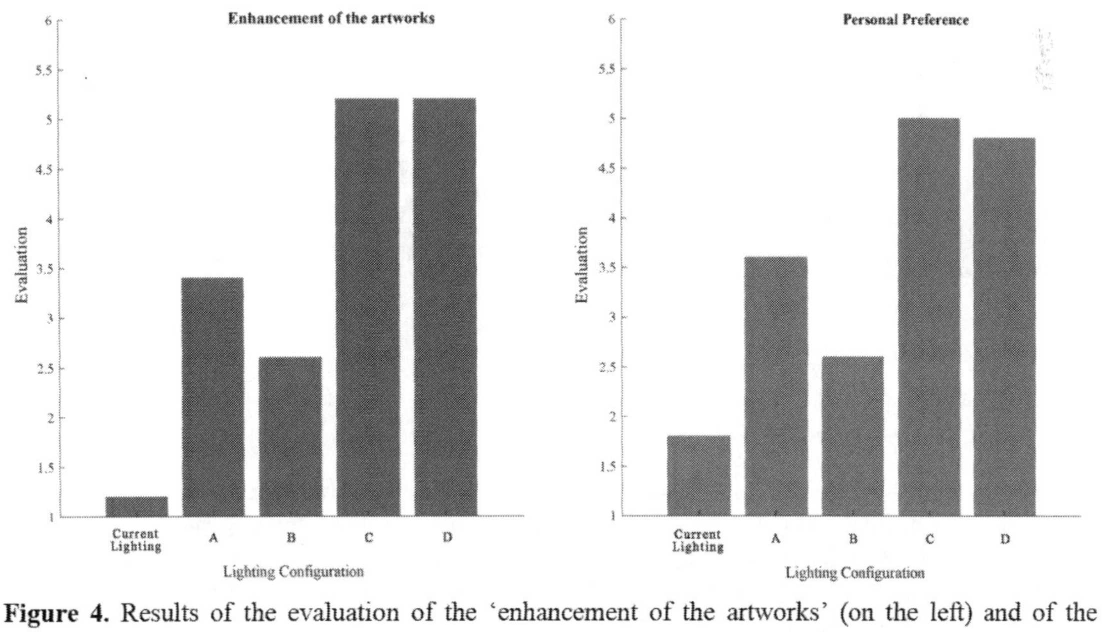

Figure 4. Results of the evaluation of the 'enhancement of the artworks' (on the left) and of the 'personal preference' (on the right) for the different test lighting configurations.

IOP Conf. Series: Materials Science and Engineering **364** (2018) 012095 doi:10.1088/1757-899X/364/1/012095

As regard the lighting measurements, it can be noticed from Table 2 that the illuminance levels on the artworks, for all the lighting configurations, were higher than those measured for the current lighting. In some cases the increases in the illuminance level were very significant (e.g. the average illuminance on the *Redentore benedicente* was increased from 75 lx of the current lighting to 582 lx of the C Configuration), but the higher illuminance level provided a better visual perception of the artworks ad a clear visibility of all their details. Although the illuminance levels are significantly increased, the annual luminous exposure limitation, fixed for this type of artworks [12], could be respected using control systems to limit the exposure time to the light, according to the annual opening hours. Since the lighting of each exhibition room is independent from the others, movement sensors could be installed in the exhibition room or in front of Beato Angelico's artworks, so that the lighting turns on only when visitors walk into the room. Among the test configurations, those preferred by the observers were characterized by the higher values of the luminance ratio between artworks and background, without overcoming the maximum limit value of 3. Usually, the range 1/3-3 can guarantee the adequate eye's adaptation [18]. It is possible to notice that A, C and D Configurations provided similar luminance ratios on the *Madonna del Latte* and the *Redentore Benedicente*, whilst B Configuration (the least appreciated) provided the higher luminance ratio on the *Redentore Benedicente* and a very low luminance ratio on the *Madonna del Latte,* creating a non-uniform lighting effect between the two artworks.

Figure 5. Test lighting configuration preferred (C Configuration).

6. Conclusions and future developments

This study analysed a preliminary lighting arrangement for two of Beato Angelico's artworks displayed at the National Museum of San Matteo. Multiple test lighting configurations were designed for the two artworks using Zumtobel LED luminaires with different characteristics. The four test lighting configurations were evaluated through a survey by five observers with expertise in lighting design or art. The observers had to assess the enhancement of the artworks and their personal preference for all test lighting configurations. The test results show that the C and D Configurations were the most appreciated by the observers, whilst B Configuration was the least appreciated and evaluated as the least suitable for the enhancement of the two artworks. Furthermore, C and D Configurations created more uniform solutions between the two artworks, providing similar luminance ratios between the artworks and their background.

The study described in this paper is part of a wider research activities about visual perception of artworks displayed in museum settings. The project was carried out at the Museum of San Matteo and consisted of the measurements of various lighting configurations designed for two artworks, the panel painting *Sacra Conversazione* by Ghirlandaio (1479) and the sculpture *Madonna del Latte* by Nino and Andrea Pisano (1343-1347), and the investigation of preference patterns for a group of observers, both expert and naïve. The study on Beato Angelico's artworks could be further developed by extending the number of observers of the experiment and involving naïve observers, whose evaluations would be compared to the ones made by expert observers presented in this paper.

Acknowledgments

The authors wish to thank the staff of the National Museum of San Matteo for their assistance with the setting of the experiment and for participating in the tests, and Zumtobel Lighting GmbH for providing the luminaires used for the study.

References

[1] Thompson G 1986 *The Museum Environment* Routledge London p 312.
[2] Tzorstzi K 2015 *Museum Space: Where Architecture Meets Museology* Routledge London p 320.
[3] Cuttle C 2007 *Light for art's sake. Lighting for artworks and museum displays* Butterworth p 288.
[4] Fördergemeinschaft Gutes Licht 2000 *Good Lighting for Museums, Galleries and Exhibitions* p 48.
[5] Feltrin F Hanselaer P Leccese F Smet K 2017 *Analysis of painted artworks' colour appearance under various lighting settings* 17th IEEE Int. Conf. on Environment and Electrical Engineering Milan (I) pp.1027-1032.
[6] Nascimento S Masuda O 2014 *Best lighting for visual appreciation of artistic paintings - Experiments with real paintings and real illumination* J Opt Soc Am Vol.31(4) pp.214-219.
[7] Zhai QY Luo MR Liu XY 2015 *The impact of LED parameters on viewing fine art paintings* Lighting Res. Technol. Vol.48 pp.711-725.
[8] Chen HS Chou CJ Luo HW Luo MR 2016 *Museum lighting environment: Designing a perception zone map and emotional response model* Lighting Res. Technol. Vol.48 pp.589–607.
[9] Leccese F Salvadori G Colli A 2011 *LED lighting in museum: the new diocesan museum in Piombino (Italy)* CISBAT 2011 Int. Conf. on Cleantech for Sustainable Buildings Lausanne (CH) pp.461-466.
[10] Druzik JR Michalski SW 2012 *Guidelines for selecting solid state lighting for museums* Canadian Conservation Institute and Getty Conservation Institute p 31.
[11] Magnusso Y 2014 *Museum lighting and LED technology* E-Conservation Journal pp.34-42.
[12] Commission International de l'Eclairage 2004 *Control of damage to museum objects by optical radiation* Wien (A) CIE 157 p 35.
[13] European Committee for Standardization 2014 *Conservation of Cultural Heritage – Guidelines and procedures for choosing appropriate lighting for indoor exhibitions* Brussels (B) CEN/TS 16163 p 32.
[14] Feltrin F Leccese F Hanselaer P Smet K 2017 *Take deLIGHT in Colours – Analysis of the factors affecting colour perception of paintings* 1st Int. Museum Lighting Symposium & Workshops London (UK) pp.88-95.
[15] Salvadori G Fantozzi F Rocca M Leccese F 2016 *The energy audit activity focused on the lighting systems in historical buildings* Energies Vol.9(12)-Article n.998 pp.1-13.
[16] Fantozzi F Leccese F Salvadori G Rocca M Capranelli I 2016 *Opportunities for Energy Savings with Interventions on the Lighting Systems of Historical Buildings: the Case of 'Palazzo Medici' in Pisa, Italy* IEEE 2016 – 16th IEEE Int. Conf. on Environment and Electrical Engineering Florence (I) pp.1076-1081.
[17] Zumtobel <http://www.zumtobel.com>, accessed 13 October 2017.
[18] Leccese F Salvadori G Rocca M 2016 *Visual discomfort among university students who use CAD workstations* Work Vol.55(1) pp.171-180.

IOP Conf. Series: Materials Science and Engineering **364** (2018) 012096 doi:10.1088/1757-899X/364/1/012096

THz imaging tool for characterizing both visible and concealed metals in paint layers

I. Cacciari[1], D. Ciofini[1], L. Gallo[2], S. Siano[1]

[1]Istituto di Fisica Applicata "Nello Carrara", Consiglio Nazionale delle Ricerche, v. M. del Piano 10, 50019 Sesto Fiorentino (FI), Italy

[2]Gallo Restauro, v. S. Zanobi 104/R, 50129, Florence, Italy

i.cacciari@ifac.cnr.it

Abstract. The Terahertz (THz) radiation can penetrate optically opaque materials, and in combination with low photon energy makes possible imaging applications in a variety of fields such as biomedicine, material science, security, etc.. This has motivated a growing interest in THz techniques during recent years and, in turn, has resulted in significant technological improvements and the introduction of several commercial-type systems. In particular, THz technology has been demonstrated to be particularly useful for the study of multi-layered non-metallic structures such as those encountered in easel and mural paintings, as well as of glued paper layers and of other similar ones. This work focuses on metallic layers concealed beneath paint layers in which THz imaging can be very effective thanks to the high reflection of metals. THz imaging measurements have been carried out using a time-domain THz spectrometer equipped with an XY translation stage. Both mock-ups and original multilayer samples were imaged. The former were prepared by gluing metal foil polygonal shapes onto a plastic substrate and then covering the whole of it with paint layers. The THz imaging revealed the metal shape beneath the paint layers, thus making possible some assessments on the layer depth. Original pictorial samples from a Sienese painting of the XIII-XIV century were also investigated. In this particular case, the THz imaging was capable of distinguishing between the original gold leaf and the paint layers (overpaint).

1. Introduction

The spectral region between 0.1 and 3 THz is commonly referred to as the Terahertz (THz) region. Its application for spectroscopy and imaging in the field of the cultural heritage has witnessed a growing interest that has generated research into a variety of new applications [1-9].

The THz wave has attracted particular interest within the field of the cultural heritage as it is completely safe being non-ionizing, and non-destructive since there is no contact with the artwork. It can also penetrate a wide variety of dielectric materials (*e.g.* stone, fabric, paper, plastic, leather, wood, and many others) and make it possible to "see" through opaque materials that are generally used in the fabrication of culturally valuable objects (such as stone, paper manuscript, wood, and many others).

In the case of a dielectric slab, due to the index change, the incident THz beam is partly reflected at the front air/surface interface, thus leading to a first reflected signal. The remaining part of the signal propagating through the sample thickness experiences a further partial reflection at the back of the sample, thus leading to a temporally-separated echo (second reflection). In the case of a multilayer sample, other reflections occur since there are other interfaces within the sample thickness.

In general, for each point of the sample under investigation, THz imaging records the amplitude of each reflected pulse and creates the corresponding false colour images. This approach is found to be extremely interesting, because the amplitude of the reflections is mostly determined by the material absorption and the surface smoothness, and hence, considerable information can be obtained. However, when the signal reflected is comparable to the system noise, i.e. when a high value of absorption characterizes the material, the imaging does not provide useful information. On the contrary, with metal surfaces, the THz pulse is totally reflected and there are no other subsequent reflections. We find this case extremely intriguing, due to the fact that the absorption does not have a role in providing information on layer thickness. In this regard, we focused our work on imaging multilayer samples made of visible and concealed metal. The THz setup described in section 2 have been used to image mock-ups in which the metal layers were concealed and original painting in which the metal layers were partially visible. The results of these measurements have been discussed in section 3 and, lastly, section 4 contains the conclusions drawn.

2. Materials and methods

2.1.1. THz imaging setup.

A THz time-domain spectrometer equipped with emitter and receiver in reflection geometry was used [10] for the refractive index and imaging measurements. This setup was chosen because it makes possible spectroscopic studies on multi-layered paint samples, i.e. samples which cannot be easily measured in transmission geometry. The incident and collection angles were placed at 30° to the normal sample surface. The THz pulse was emitted by illuminating a photoconductive antenna (PCA Emitter) with an ultrafast pulse (70 fs) from a fibre Erbium Laser (Toptica, Femto-FErb 1560) emitting at 1550 nm. The ultrafast pulse was then split into two guided beams: one was used for the generation and the other, for detection. A series of mirrors and lenses guided the ultrafast pulse to the emitter, which was coupled to a single mode fibre with a FC/PC connector for supplying the ultrafast pulse. The THz pulse emitted by the PCA was collimated and focused with plastic lenses on the sample, and the reflected signal from the sample was focused back by using a second pair of lenses on the detector (PCA Detector). The THz time domain spectrometer is schematically presented in figure 1.

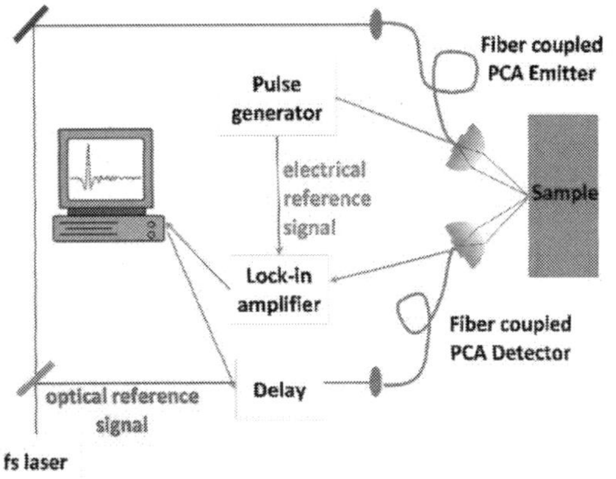

Figure 1. THz imaging setup.

In order to vary the arrival time of the signal with respect to the pulse used for detection, a delay stage was introduced into the generation arm. By scanning the delay line, both the electric field amplitude (A) and the phase of the THz waveform could be sampled as a function of time. A LabVIEW® software saved the electric field amplitude from a digital lock-in amplifier that measured the signal from the detector using a data acquisition card from National Instruments®. The bandwidth obtained with this setup ranged from 0.1 to 2.5 THz. In the present work, both the emitter and detector in reflection geometry were fastened on a two-axis translation stage that represents the only moving part of the imaging setup. The translation stage could move in a plane parallel to the sample surface. The THz image of a sample was taken by moving the stage in the x-y plane (the z axis is defined as the normal surface direction). The step sizes were 1 mm along both the x and y axes. Full THz temporal waveforms were recorded for each emitter-detector position, and the difference between the maximum and minimum amplitude for each reflected pulse was considered to form the corresponding THz image.

2.1.2. Samples

With the aim of exploring the penetrating properties of THz radiation in optically opaque artistic materials, purpose-designed laboratory samples reproducing a multi-layered stratigraphy were prepared and systematically investigated. A representative drawing illustrating the stratigraphic sequence of the two different types of prepared samples (samples A and B) is shown in figure 2. The first operative step, for both samples A and B, was that of creating rounded copper foils. For this purpose, 100 µm thick copper foils were first hole punched using a round metal tip of 8 mm diameter and were then glued onto the back side of Corning™ Falcon™ Petri Dishes. The latter consisted of crystal-grade polystyrene (PS), had a 35 mm diameter, an 11.8 cm^2 area and 500 µm of indentation which made it possible to exploit it as a mould for the final application of the paint coating. As shown in figure 1, three non-superimposed copper foils were applied for sample A, whereas a vertical stack of two and three copper foils were added to sample B. Paint coatings were then formulated using a mixture of pigment powder dispersed in a binding medium. The pigment selected was a commercial variety of synthetic cobalt blue (purchased from Zecchi, Florence), a cobalt aluminium oxide ($CoAl_2O_4$) manufactured in Europe in 1802 that was used from ancient times up to the nineteenth century as an artistic pigment for producing paints for decorating ceramic glazes and jewellery [11]. In particular, the choice was further motivated by the findings obtained in a previous work [12], in which cobalt blue, among the pigments studied, showed a relatively high refractive index and a significant THz reflection.

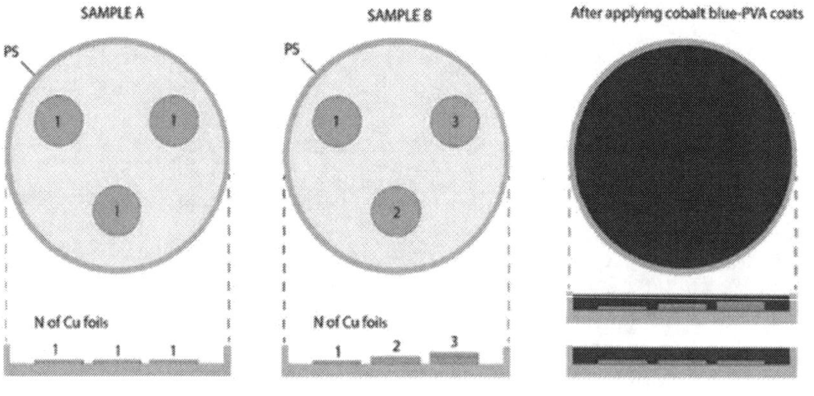

Figure 2. Multilayer Mockups preparation.

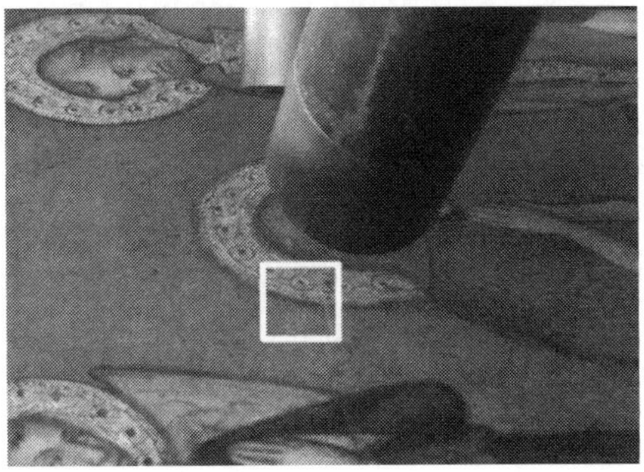

Figure 3. THz antennas analysing original pictorial sample. The box corresponds to the THz imaging area.

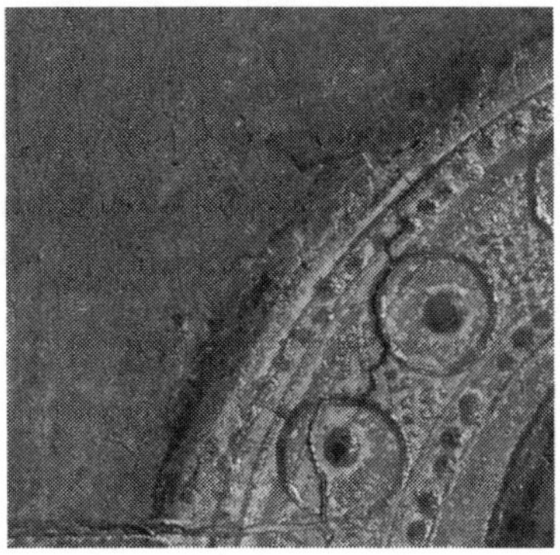

Figure 4. THz imaging area.

The pigment powder was finely ground, added in a 1:1 ratio to a 10 % (w/v) polyvinyl-alcohol (PVA)-water solution, and then thoroughly blended. PVA was selected because of its superior dispersing properties. Thus, after gluing the copper foils onto a PS substrate, the cobalt blue-PVA mixture was spread out so that it covered the PS mould with the three round copper foils. Usually, additional coats of paints were needed in order to fill the PS mould completely and to conceal the three copper sheets. After applying the cobalt blue-PVA coatings (third step of figure 2), samples A and B were left to be cured under laboratory conditions (23 ± 1 °C and 50% HR) for 7 days and were then analysed.

Moreover, an original pictorial sample from a Sienese painting of the XIII-XIV century was also investigated. In this sample, the gilding was performed by applying gold leaf to a red bole layer, in figure 3 is shown the THz antenna during the measurements and in figure the area considered for the imaging.

3. Results and discussion

Samples A and B were imaged using the THz setup shown in figure 1. For each pixel, two pulses could be observed: the first peak came from the air/painting interface; the second came from the painting/metal or from painting/polystyrene interfaces. The amplitude of the first peak (for all the pixels considered in samples A and B) is presented in figure 5 and 6.

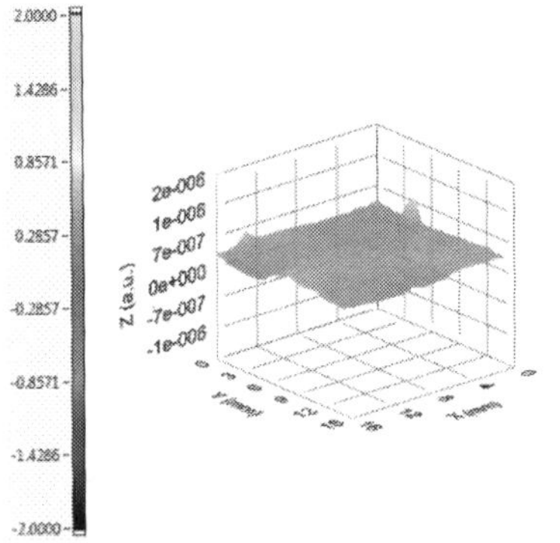

Figure 5. Amplitudes of the first peaks of sample A. The vertical scale is expressed in a.u., while the factor 10^{-6} is omitted for simplicity.

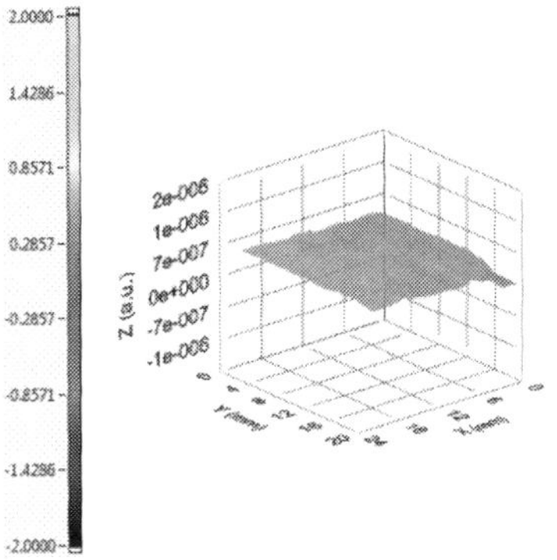

Figure 6. Amplitudes of the first peaks of sample B. The vertical scale is expressed in a.u., while the factor 10^{-6} is omitted simplicity.

We considered a 3D representation of these data in order to point out that the sample surface was smooth and the optical axis was normal for the surface. The first observation could be drawn by calculating the mean value of the amplitudes: we obtained $3.01 \times 10^{-7} \pm 10\%$ a.u. and $3.4 \times 10^{-7} \pm 10\%$ a.u. for samples A and B, respectively. The error associated was calculated as the standard deviation. On a rough surface the first peak amplitude is likely to be attenuated [10], and in general the corresponding standard deviation is more than 10%. Moreover, this observation is corroborated by the orthogonal position of the sample surface to the optical axis. In the case of a tilted sample surface, there are points closer and further away, and this determines an attenuation of the corresponding amplitudes that leads to a standard deviation of more than 10%.

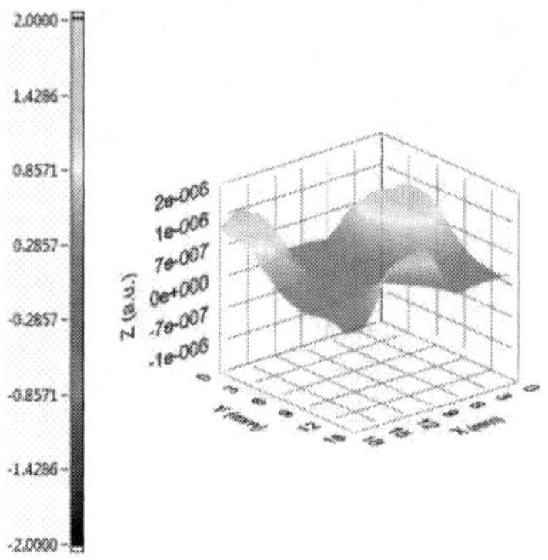

Figure 7. Amplitudes of the second peaks of sample A. The vertical scale is expressed in a.u., while the factor 10^{-6} is omitted for the sake of brevity.

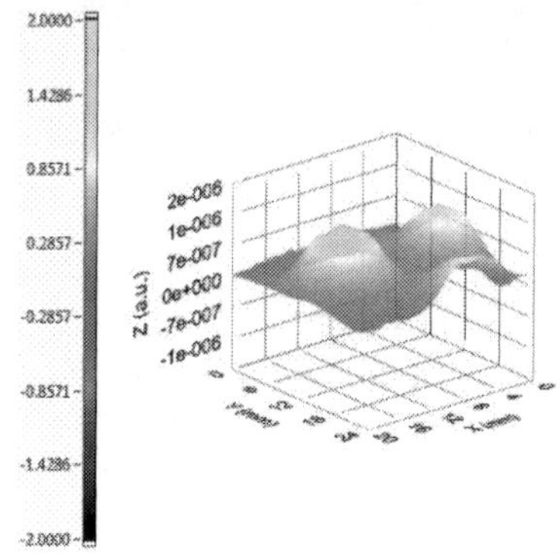

Figure 8. Amplitudes of the second peaks of sample B. The vertical scale is expressed in a.u., while the factor 10^{-6} is omitted for the sake of brevity.

The amplitudes of the second peak of samples A and B have been represented, respectively, in figures 7 and 8. The imaging clearly illustrates that, beneath the first absorbing layer, there are layers of highly reflective material (in this case, metal foil) that are round in shape. Where the amplitude is greater, there is a painting/metal interface and no more pulses can be evidenced because the metal foil blocks any further reflections. On the contrary, where the amplitude is less, the signal is absorbed by the paint layer and another reflection (the third) is due to the painting/polystyrene interface. It is interesting to note that, in the case of sample A (figure 7), the amplitudes of the THz signals reflected at the metal foils are almost the same: 1.31×10^{-6}, 1.30×10^{-6} and 1.29×10^{-6} a.u.. In this case, the THz signal travelled the same thickness of paint layer before reflecting the metal foil. It therefore underwent the same degree of attenuation. This was not true for sample B: the reflections at the metal foils (figure 8) were different. In correspondence with these points we obtained maximum amplitudes of 1.37×10^{-6}, 1.15×10^{-6} and 0.88×10^{-6} a.u.. This was compatible with the way in which the sample was prepared. In fact, the three metal foils were covered by different thicknesses of paint, the signal reflected by one metal foil was more attenuated because it travelled a greater thickness of paint. On the contrary, where it travelled a lesser thickness, the attenuation was lower. The corresponding paint thicknesses could be calculated by considering the dish indentation and the metal foil thickness: we obtained 200 μm, 300 and 400 μm. It was possible to establish a linear fit between the thickness of the paint layers and the maximum amplitudes in correspondence with the metal foils the goodness of this relationship was determined by a R^2 of 0.99. It should be noted that, the metal foil thickness prevents the THz pulse to penetrate into the metal, thus the glue used in the two mockups does not interfere with the measurements and in particular with the pulse attenuation.

IOP Conf. Series: Materials Science and Engineering **364** (2018) 012096 doi:10.1088/1757-899X/364/1/012096

In the case of original pictorial gilded samples, in which the gilding was performed by applying gold leaf to a red bole layer (figure 4), the potential of THz imaging appeared to be even more evident.

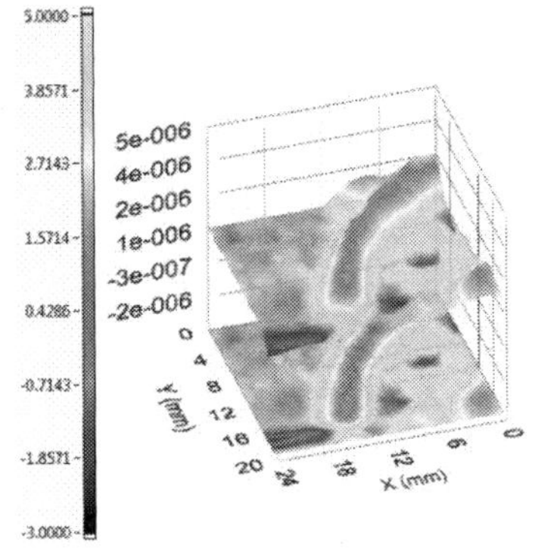

Figure 9. THz imaging of figure 4.

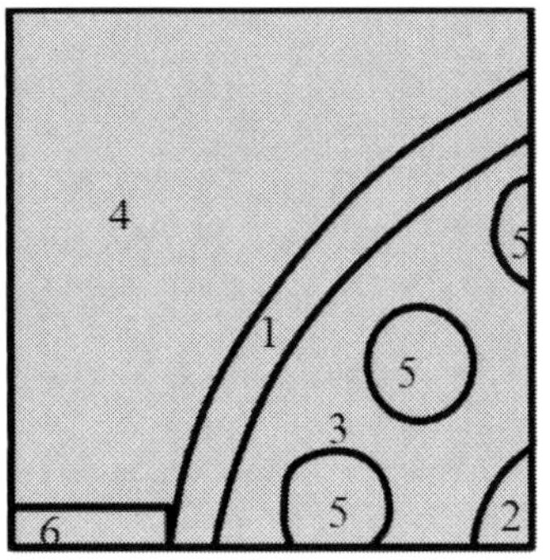

Figure 10. Different areas in terms of amplitude evidenced by THz imaging.

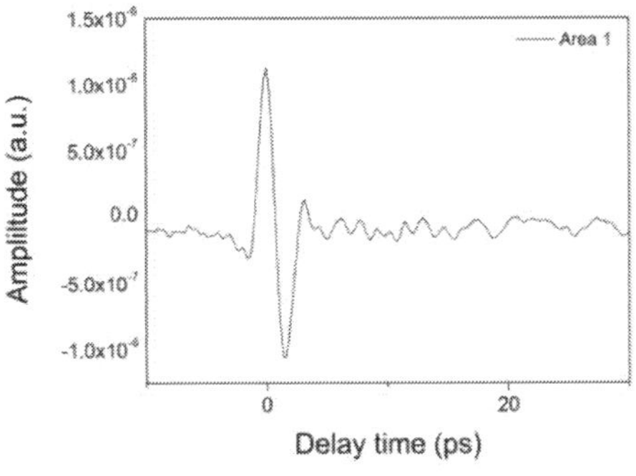

Figure 11. THz pulse reflected from area 1.

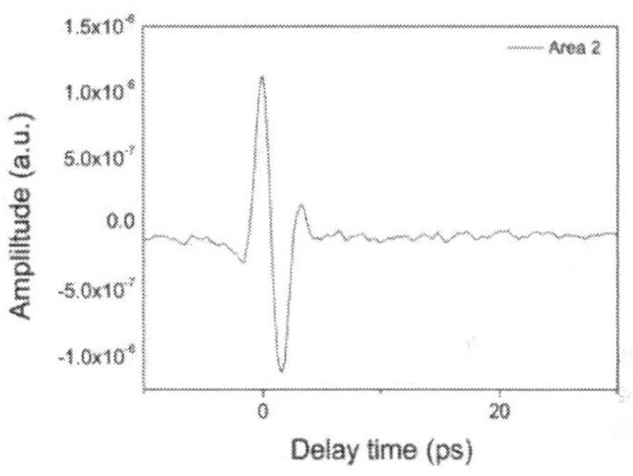

Figure 12. THz pulse reflected from area 2.

The THz imaging is shown in figure 9. Several areas, characterized by different amplitude values, have been highlighted and, for the sake of clarity, have been sketched in figure 10: in ascending order, these are areas 1 and 2 (~2×10^{-6} a.u.), areas 3 and 4 (~1×10^{-6} a.u.), and lastly areas 5 and 6 (~0.8×10^{-6} a.u.). The greater amplitudes of areas 1 and 2 of figure 9 could have been due to the reflection at a metal interface (gold leaf). This was confirmed by observing the corresponding THz signals (figures 11 and 12): only the first peak (due to the reflection at air/metal interface) existed. Moreover, it is worth noting that, in the original image (figure 4), the gold leaf is visible while, on the contrary, in areas 3 and 4 the gold leaf has been partially lost and the red bole is fairly observable. In terms of THz imaging, this led to an absorption of the THz pulse while traveling the red bole and, therefore, a lower amplitude than in areas 1 and 2 is appreciable in both areas 3 and 4.

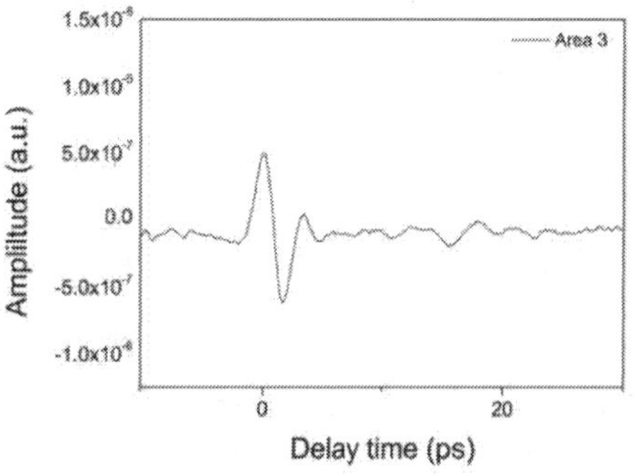

Figure 13. THz pulse reflected from area 3.

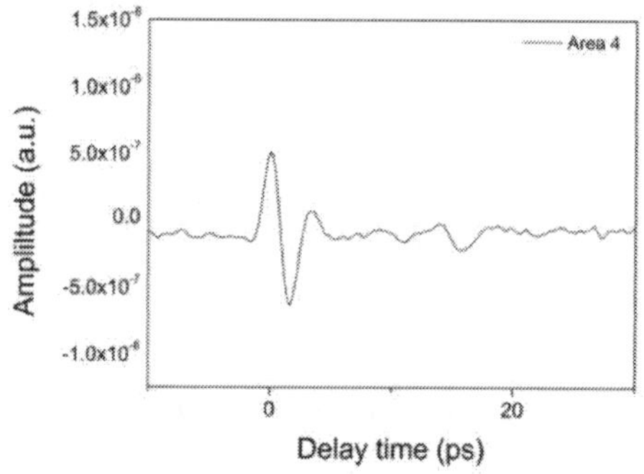

Figure 14. THz pulse reflected from area 4.

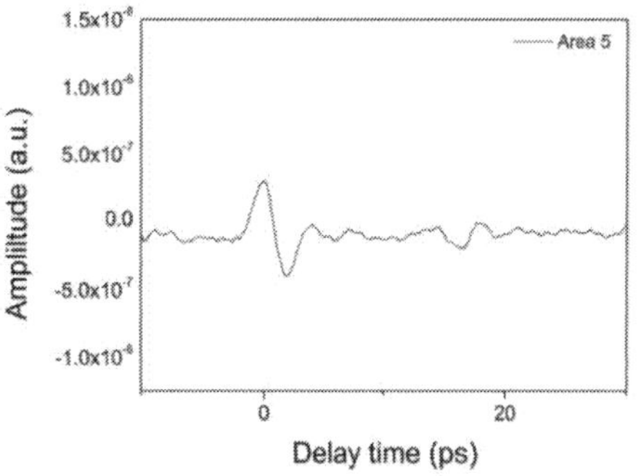

Figure 15. THz pulse reflected from area 5.

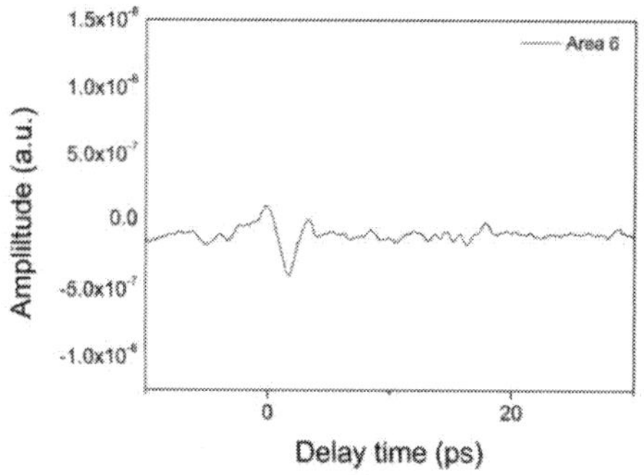

Figure 16. THz pulse reflected from area 6.

Moreover, the corresponding THz signals (figures 13 and 14) reveal the presence of a second pulse reflection, which is due to the red bole/preparation interface. Areas 5 and 6, which correspond to the punch work [13], are also extremely interesting. They are similar to areas 3 and 4, but show lesser amplitudes (figure 15). This may be due to the punch work itself: in that case, the surface is more distant than the previous areas from the emitter and the detector. This means that the incident THz pulse is not in the focal plane when it arrives on the surface, and that an absorption and possibly a small broadening could be affecting the propagation. This is accentuated in area 6, where the lesser amplitude (figure 16) is due to the horizontal craquelure.

4. Conclusions

This paper shows the feasibility of using THz imaging in the field of the characterization of artworks. A THz time-domain spectrometer was equipped to perform imaging measurements. This setup has been used to image samples that presented concealed or visible metal interfaces. In the first case, laboratory samples were prepared in order to study, starting from THz imaging, the layer stratigraphy. In particular, some considerations on the painting surface and layer thickness have been made by measuring the peak amplitudes of the THz pulse reflected at the air/pigment and pigment/metal interface. In the case of a visible metal interface, we imaged a portion of an original painting in which the metal is presented as a gold leaf on a red bole layer. The THz imaging distinguished the areas in which the metal is not lost from the areas in which the red bole or possibly an overpainting are present. This may be of great importance in the field of authentication.

5. Acknowledgements
The present work was carried out within the IPERION CH (H2020-INFRAIA-2014-2015, Grant No. 654028) project. The authors also acknowledge the financial support of the "Fondo per le infrastrutture produttive" of the Tuscan Region (D.G.R.T. n. 924 - 19/10/2009), for the purchase of the THz time-domain spectrometer used in the present study.

6. References
[1] Fukunaga K, Ogawa Y, Hayashi S and Hosako I 2007 *IEICE Electronics Express* **4** 258-263.
[2] Jackson J B, Mourou M, Whitaker J F, Duling I N, Williamson S L Menu M, Mourou G A 2008 *Opt.* Comm. **281** 527-532.
[3] Jackson J B, Bowen J, Walker G, Labaune J, Mourou G, Menu M and Fukunaga K 2011 *IEEE Trans. Terahertz Sci. Technol.* **1** 220-231.
[4] Jepsen P U, Cooke D G and Koch M 2011 *Laser & Photonics Reviews* **5** 124-166.
[5] Walker G C, Labaune J, Bowen J W, Jackson J, Hadjiloucas S, Mourou G, Menu M 2011 *36th International Conference on Infrared, Millimeter and Terahertz Waves (IRMMW-THz)* 1-2.
[6] Picollo M, Fukunaga K, Labaune J 2015 *J. Cult. Herit.* 2015 **16** 73-80.
[7] Dong J, Jackson J B, Melis M, Giovanacci D, Walker G C, Locquet A, Bowen J W and Citrin D S 2016 *Opt. Express* **24** 26972-26985.
[8] Cacciari I, Agresti J, Siano S 2016 *Microchem. J.* **126** 76-82.
[9] Zhang H, Sfarra S, Saluja K, Peeters J, Fleuret J, Duan Y, Fernandes H, Avdelidis N, Ibarra-Castanedo C, Maldague X 2017 *J Nondestruc Eval* **36** 1-12.
[10] Cacciari I, Siano S 2017 *J Infrared Milli Terahz Waves* **38** 503-517.
[11] Gettens R J, Stout G L, *Painting Materials: A Short Encyclopaedia* 1966 (Dover Publications).
[12] Cacciari I, Ciofini D, Siano S 2015 *IET Conf. Publ.* 1-4.
[13] Cacciari I, Nieri P, Siano S 2014 *J. Comput. Cult. Heritage* **7** 1-14

IOP Conf. Series: Materials Science and Engineering **364** (2018) 012097 doi:10.1088/1757-899X/364/1/012097

Artworks' Features Discovery Through Engaging Conversations for Children

Juan Carlos Jaramillo, Daniele Occhiuto, Franca Garzotto

i3lab – Politecnico di Milano – Via Golgi 39 (Milano)

juancarlos.jaramillo@mail.polimi.it, daniele.occhiuto@polimi.it, franca.garzotto@polimi.it

Abstract. Dialog-based interfaces have seen a renewed interest in the last years. Smart assistants like Siri, Google Assistant or Cortana are pervasive nowadays. Messaging platforms like Facebook Messenger, Telegram or Skype are integrating automatic responders commonly known as bots or chatbots. Industries started incorporating smart assistants as web-services to help, retain and profile their customers. We see a considerable potential in using conversational interfaces to enhance guided tours in cultural institutions. Museums user experiences may vary depending on the exhibition. Nonetheless, hearing is the common channel museums use to inform visitors, in particular by exploiting audio-guides. The existing guides are very well structured and provide indeed a complete guided experience but are generally intended for adults and neuro-typical users. As a consequence, museums may have difficulties in targeting all kinds of visitors, e.g. children or users with cognitive disability. They may require assistance to pass over their heritage to all audiences. Our research purpose is to design an infrastructure to help cultural institutions in broadcasting their heritage to all guests. We propose interactivity over information: the user takes an active part to the visit and is engaged though the whole experience. Using conversational interfaces, we switch from the perspective of a passive listener to the point of view of a dynamic "player": the user can ask for information. The engagement is a necessary aspect to maintain a high interest in children. We explore how bi-directional conversations can provide accessible pieces of information to users with special needs. We design a plug-and-play pervasive solution for a seamless adoption for the interested institutions. We allow holders of the system to fine-tune its characteristics based on the artwork, the exhibition and the target users.

1. Introduction

Conversational interfaces are a new interaction paradigm that is gaining popularity as the underneath technology becomes more and more robust. In the last decade, technological advancements have guaranteed new forms of interaction, some of them are re-shaping the user experience for a particular domain (e.g. smartphones impact in communications: SMS have almost completely been replaced by messaging apps). Still, we believe that shaping the user experience is not the outcome of technological breakthroughs alone, but it is strongly linked to the way tech advancements are put into place. In particular, our interests in human computer interaction (HCI) leads to the study of new interaction paradigms for empowering all users, with special attention in children and users with disability.

In this work we describe the design and development of MUSU a conversational application to trigger the user engagement in museums' exhibitions. MUSU is an app able to converse with the user and proactively ask questions about a recognized piece of artwork. The app purpose is not to replace classical audio-guides that excel in their task for adults, but to develop new forms of interest triggers in children or empower users with disability through conversations as in [1]. In fact, the role of museums is moving from the main conservation and exhibition responsibility to the enhancement of artworks in order to achieve engagement, entertainment and learning [2]. Some studies revealed that technology use during the visit may isolate the visitors from the environment [3]. Isolation during the visit is counterproductive in case of museums since a great effort is put in the environment itself. For example, the disposition of artworks is determined with care and the environmental lightning and sounds are the output of an attentive analysis. To avoid isolation, we decided to adopt a dominant interaction modality based on auditory cues, following the existing *de-facto* standard for museums tours that are audio-guides. However, one issue with audio-guides is the loss of interactivity. To cope with interactivity losses, we devised a system with customizable responses under the form of a conversational "agent". *Agents* are pieces of software integrated on different devices (smartphones, computers, boards) to assist users in general purpose tasks, e.g., event reminders, navigation, information retrieval. Of course, our guide cannot and does not aspire at replacing a human guide, while other works try to mimic a human interaction with robots [4][5], we try to trigger a comprehensive guided tour through conversations without displaying any representation of the agent. We decided to avoid humanoids to allow MUSU to be portable, we think of our system as a plug and play solution to be deployed on pervasive devices (smartphones or other museum proprietary solutions). Moreover, we bypass the agent graphical representation when possible to insist on the auditory channel only: visual stimuli shall come exclusively from the environment set on purpose for the visit.

Smartphones are by design interactive through the visual channel, since we target auditory interaction to let the user focus on the environment, we need to find a way to trigger the experience by limiting as much as possible the use of the graphical interface. The solution we came up with is the positioning of Bluetooth Low Energy (BLE) beacons in the environment. Bluetooth is a wireless technology that works in proximity, BLE radio signal range can arrive up to 100m but the signal is strongly reduced in presence of obstacles. Beacons are tags, under the form of cards, stickers, or badges, that act as continuous transmitters of a predefined message. In summary, beacons can be thought as lighthouses that broadcast messages to devices in range by means of radio signals, just as lighthouses broadcast light signals from a fixed position. In this way, when the visitor approaches an artwork, the device will automatically "wake-up", receive an identifier of the artwork in front of the user and start a relevant conversation with the visitor.

In the next sections we review exiting interactive solutions employed in museums nowadays, we explain the design and development of MUSU, we clarify its implementation process, we report the results of an exploratory study investigating what information users would like to know about artworks, and finally we discuss our findings and future works.

2. Related Work

Museums bear the responsibility to transfer their heritage to visitors, their task is complicated by a generation gap where every interaction is becoming digital while most artworks belong to an analogical world. Is there a way to "sample" artworks and make them accessible to users from the digital age? We would like to insist on the fact that technology is not a comprehensive answer, but we hope it may help in bridging the two worlds together. Should the digital and analogical world fail to communicate, there would be a considerable loss in heritage transfer.

IOP Conf. Series: Materials Science and Engineering **364** (2018) 012097 doi:10.1088/1757-899X/364/1/012097

Children born in the digital age are more engaged when dealing with interactive interfaces: in a research considering virtual assistants, children interacted much more than adults with the assistant (47% of gathered data from speakers belongs to children, compared to the 13% from adult males and 8% from adult females, while 31% did not respond to the virtual assistant) [6]. The same research showed an increase in children engagement and interests for the exhibit. Others demonstrated that experiences embedded in the context of the child's world, e.g. play and story, provide greater meaning to the exhibit [7]. Incorporating virtual guides in museums is not a new practice, in the past virtual guides have been extensively explored: Max is a screen-projected virtual guide enabling the interaction through a keyboard [8]. Tinker is another virtual guide situated in the Museum of Science in Boston, its particularity is the ability to recognize visitors over time, creating a bond with the visitors and engaging them using human relationship-building behaviors [9]. Anew, in Boston Museum of Science, Ada and Grace are twins virtual agents for guided tours with an educational goal for children [10]. MUSU exclusive channel for interaction will be auditory, hence, we shall put a great effort in making the experience pleasant, both by providing meaningful conversations and analyzing the user's sentiment in her/his responses.

Virtual agents are only one instance of museums guides; other studies leverage interactivity through physical objects instead. For example, sometimes physical robots are used as tour guides in presence of users, while in other examples tabletops and touchscreens are the means of interaction. Robots have the advantage of being mobile, they are a tangible product and do not need to be advertised to visitors. Indeed, as opposite to virtual agents that may go un-noticed if the users do not see the screen or do not start an eventual app, robots are self-explanatory in the sense they are explicitly apparent to visitors. Robots have been deployed in museums and fairs as guides in several studies [4][11]. One main motivation for robots' deployment is the possibility for multimodality: they capitalize on different interaction paradigms to simulate the natural multimodal interaction that takes place between humans. In fact, when people interact, they use gestures, body language and facial expressions alongside speech. Acquiring the features of natural human interaction is a challenging task for a machine; for now, humans can easily recognize robots, even if they are designed to look like and behave as real persons. As a consequence, the robot guide may not be seen as a peer, humans are not attracted to robots if they have a too much photo-realistic human-like appearance, this effect is known as "the uncanny valley" [12]. The author explains the uncanny valley as a function that is not strictly monotonic, providing the example of a climber arriving at the top of a mountain only to discover a valley on the other side with a bigger summit across it. Imagine the disappointment of the climber when finding that (s)he is not on the highest summit. Similarly, when humans interact with robots there is a point where they feel familiar enough with it to interact in a natural way. However, such seamless interaction is brittle, a simple mismatch in the expected communication would reveal that the interlocutor is not human, with the destructive consequence of breaking the illusion of interacting with a peer.

Strictly speaking, MUSU does not have any human-like aspects, in fact it does not even have a shape; it is pure software. Nonetheless, when speaking on the phone with somebody unknown, there are several cues that expose the interlocutor's human attitude. For example, the tone of voice, the duration of pauses, and the speed variation during speech, are all factors that influence how we picture and how we perceive the emotional state of the person talking with us. While there is no explicit definition of an uncanny valley for conversations we find affinities with the uncanny valley theory that are detectable in the conversational guidelines we follow. It is important to stress that in the uncanny valley theory, the recommendation when designing robots is to create artifacts that clearly do not appear as humans but share some features of human-like behaviors. For example, the user distinctly realizes that (s)he is interacting with a machine but is not bothered by it. Going back to the uncanny valley metaphor, it is like if the user aims at climbing the first summit and is perfectly satisfied with it because (s)he had no pretense to arrive at the highest top across the valley. To conclude this section,

IOP Conf. Series: Materials Science and Engineering **364** (2018) 012097 doi:10.1088/1757-899X/364/1/012097

we report an overview of other systems designed to augment the museum's interactivity that are further away from MUSU's approach but share the common goal of boosting the users' engagement. In particular, virtual reality (VR) applications are well suited for the creation of immersive environment. Immersion has been proven to augment the feeling of presence in the virtual environment, with the consequence of incrementing the user's participation to the experience [13]. Virtual reality (VR) and augmented reality (AR) effects are comprehensively investigated in recent studies [14][15]. One study focused on mixed realities experiences, combining both VR and AR, leveraging the experience economy (the focus is no longer in the product but is shifting to the visitor experience) [16]. Last but not least, touchscreen tables have been explored as well, and are reported to be effective for collaborative activities in a comparative study demonstrating that ICT plays a significant role in enhancing the user experience [17].

3. Design and development

When designing conversational interfaces, there are several requirements to keep in mind [18]. Agents need an appropriate task to be successful, "small-talk" capabilities are useful to humanize the agent (yet avoiding the uncanny valley), but in the end, agents shall prove their utility. In order to be successful, a conversational agent must answer the user's needs. Like in real-life conversations, there is a context of the dialog, and users expect the agent to remember the context. Shall the context fail, the agent can adopt social smarts, for example exhibiting its interest for the previous statement. The use of narration can be a trick to avoid context losses and enhancing the conversation: it is easier to talk about something the agent is prepared about. The conversational agent can keep up with the context if the focus of the discussion is known a priori e.g. the dialog is contextualized in a story. Finally, people tend to interact with peers, and don't have the same discussion at home or at the office. We can summarize five main requirements for natural language interaction with machines as follows:

1. The agent requires an appropriate task answering the users' needs
2. The agent awareness of the context is expected from users
3. The agent shall implement a reasonable number of social smarts (small-talk)
4. The agent shall support narrative conversations for bi-directional communication
5. The agent shall consider the dialog setting's influence on the conversation itself

The last point in particular was illuminating in the design of MUSU, conversations are not specifically suited to museums settings at first sight. Firstly, the user has no knowledge of the context and it is hard to participate to the narration: not all users are experts in the art domain, so what would they talk about with an agent? Secondly, people tend to respect silence in museums, thus conversations may not be the best form of interaction. The first objection where users have no knowledge on the conversation domain, is solved by curiosity: with real guides the interaction is mostly mono-directional: the guide explains and the visitors listen. Questions, if any, are raised at the end of the explanation. With real guides and group of visitors, the setting is collaborative, so even if a question may raise the awareness in the other visitors, asking it in the first place may be hard. In fact, the visitor can be shy or altruist, and avoid slowing down the tour for the others by asking questions. The same attitude is not present with audio-guides, the visitor can take all the time needed to appreciate the artwork and has no constraints with the group of visitors. The second objection, where users do not tend to talk in museums, limits the use of MUSU. Indeed, shall everyone talk with the guides, the visits would end up in a messy murmuring and visitors would no longer enjoy the experience. However, suppose you go visiting with a friend, you can share your opinion in a one-to-one talk without bothering the other visitors. Moreover, the number of users in front of an artwork is proportional to the artwork itself: as a consequence, there is a limited number of users asking information in front of each artwork, in such setting the noise would be strongly reduced, granting a pleasant tour and allowing an enjoyable use of MUSU.

We designed our system by prioritizing speech, nevertheless we kept in mind that in some situations speech is not the best channel. Since MUSU aims at being accessible, we diversified the interaction starting from the auditory channel, which is the default interaction modality of our system. MUSU received two subsequent enhancements, on the one hand to be adaptable to a museum setting, and on the other hand to be flexible for users who prefer chats over conversations or users who have speaking or hearing difficulties. MUSU interaction modalities are:

- *Sound vs Sound*: the interaction is on the auditory channel; the user speaks with MUSU through natural language like in a human dialog.
- *Text vs Sound*: the user inputs her/his request as text message, and the app answers with audio messages.
- *Text vs Text*: the interaction with MUSU takes place through textual messages like in chatting applications.

4. Prototype

The first version of MUSU was developed for iOS systems but we plan to develop a version for Android as well. The app presents an initial screen asking the user's permission to access the device location, this process is done only once after installation (Figure 1). Then it is possible to choose the preferred interaction modality to communicate with MUSU (Figure 2); the interaction modality can be changed at any time from the main menu (Figure 3).

Figure 1. MUSU home screen.

Figure 2. MUSU selection of interaction modality.

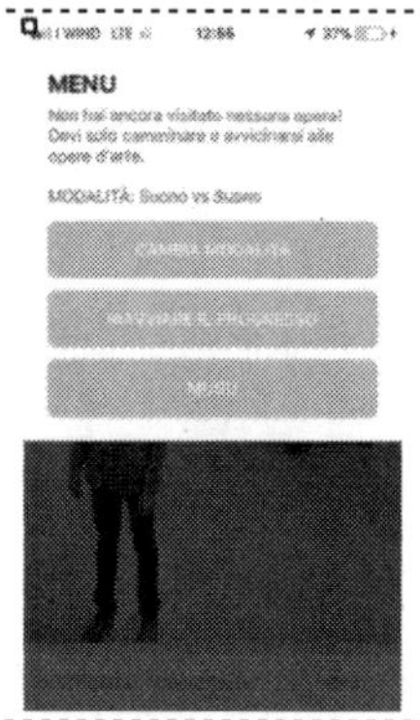

Figure 3. MUSU main menu.

At any time, the user can access MUSU's main functions from the bar menu at the bottom of the screen visible in Figure 3. The bottom bar holds a description of the museum, some suggestions about the use of MUSU and a button to open the main menu. The artwork dialogue is triggered by the user proximity to a beacon. To better understand the app's functions and its integration with the Bluetooth Low Energy (BLE) beacons, we report the general architecture of MUSU in Figure 4. The smartphone acts as end-point to communicate with the user, via chat messages, text-to-speech or speech-to-text, depending on MUSU's selected interaction modality. Through the smartphone's Bluetooth capabilities, MUSU receives the broadcasted identifier of a beacon, as shown by the line connecting MUSU app and the beacon tag. The received identifier corresponds to a precise target artwork, the identifier acts as key to retrieve the artwork information that will be provided during the conversation. We displayed dotted lines between the museum institution and the artwork since there is no digital communication among them. The same happens among MUSU app and the museum institution. The artwork information is stored in the museum catalogue which is matched in the cloud services used by MUSU, and is retrieved for the conversation when needed. In particular, we used IBM Watson conversations to implement the dialog with the users.

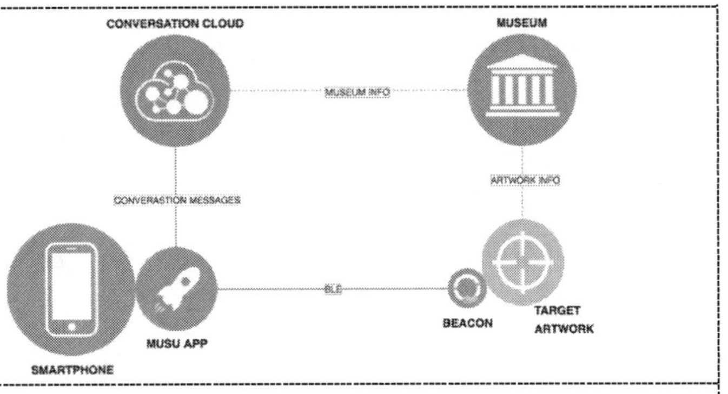

Figure 4. MUSU architecture overview.

5. Exploratory investigation

In this section we report an exploratory investigation about the contents that may be helpful in MUSU's conversations. Past research considered the information that users requested about a single artwork, in particular the painting of a military man, and classified the dialog statements based on who initiated the statement and on the type of the requested information (e.g. biographical questions, personal preferences, general purpose knowledge) [19]. One interesting finding was that 96,7% of user-initiated requests were about the portrayed individual biographical information.

Our interviews were divided in two questions with seven options each: the questions refer to the artwork in front of the user and aim at uncovering what users would like to learn about the artwork:

- *Question A*: What would you ask about this artwork?
- *Question B*: What is calling your attention?

Table 1. Participants' answers to Question A: What would you ask about this artwork?

Answer	*Question A*	*Children (%)*	*Adults (%)*
A1	What does it mean?	85%	67%
A2	Who is the portrayed person?	33%	55%
A3	Who painted it?	69%	45%
A4	Where is the painter from?	38%	42%
A5	What year is the painting from?	25%	27%
A6	How much does it cost?	4%	33%
A7	Other	46%	30%

Table 2. Participants' answers to Question B: What is calling your attention?

Answer	*Question B*	*Children (%)*	*Adults (%)*
A1	The colors in the painting	98%	67%
A2	The style of painting	44%	70%
A3	The shapes	60%	45%
A4	The realism	35%	58%
A5	A specific element (tree, eyes, dog)	25%	15%
A6	Nothing	4%	24%
A7	Other	33%	21%

Participants were assigned to two groups based on their age: children (age range 4-7 years) and adults (age range 24-55 years). The procedure followed a small interview were participants were asked to answer the two questions with a total of fourteen options. We recruited 33 adults (16 women and 17 men) from a university campus in Milan, and 48 children (21 girls and 27 boys) from a kindergarten in Milan (22 children) as well as in a day care center in Ecuador (26 children). As in [19] we chose one representative painting from a local museum (Museo del Novecento) portraying an old woman: "La signora Virginia" from Umberto Boccioni. We report in Table 1 the options for *Question A* and the percentage of participants who selected each possible answer (A1 to A7). We did the same for *Question B*, and report the results in Table 2. Figure 5 and Figure 6 present the answers (A1 to A7) for *Question A* and *Question B* in a visual form.

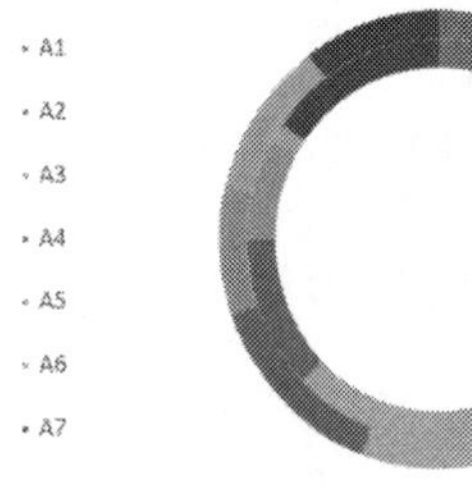

Figure 5. *Question A* – Children's percentages (inner circle) versus adults' percentages (outer circle) for each answer (A1 to A7).

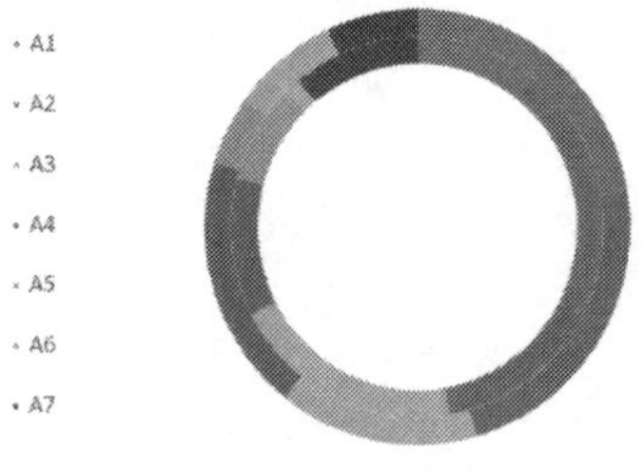

Figure 6. *Question B* – Children's percentages (inner circle) versus adults' percentages (outer circle) for each answer (A1 to A7).

6. Discussion and future work

We can see that children would ask more naturally an explanation about the meaning of the painting (answer A1 is selected by 85% of children against the 67% for adults), children were also interested in knowing who is the painter, on the contrary adults seemed to be less interested in the painter and stated they would ask biographical questions about the person being portrayed, in accordance with the findings in [19]. The biggest difference in the answers from children and adults for *Question A* can be seen for answer A6: asking about the price of the painting (33% of adults would ask the price - A6, against 4% for children). This could help MUSU in filtering what information it shall provide to its users, for adults some price information could be interesting, while it would be totally irrelevant for children. Concerning *Question B* instead, 98% of children (against 67% of adults) are attracted by the painting's colors, while adults seem to be predominantly attracted by the style of the painting (70% of adults against 44% of children). Adults accord more attention to the realism of the painting (58% pay attention to it, against 35% for children). As a result of *Question B*, we can design MUSU to talk more about colors and shapes when interacting with children, and explain the style of painting and the techniques used to create realistic portraits when conversing with adults.

In conclusion, we presented MUSU, a conversational application that acts as an audio guide designed to boost the visitors' engagement in interactive conversations. We leveraged BLE beacons as triggers to start the conversation seamlessly when the visitor is in front of the identified artwork. We did a small exploratory investigation to better understand what users would expect from a dialogue with MUSU and direct the context of conversation towards the visitors' needs. MUSU is at an early stage and it is a preliminary prototype investigating the role of the auditory channel in guided tours. We are aware of its limitations and hope we can overcome them in future versions of our app. Next steps include analyzing the user interaction more in depth, gathering further museums' needs and re-engineering the technology underneath the app. In the end, we hope MUSU could become an effective support tool to bridge the analogical and digital worlds and guarantee the heritage transfer to posterity.

References

[1] Buonamici, F., Carfagni, M., Furferi, R., Governi, L., & Volpe, Y. (2016). Are We Ready to Build a System for Assisting Blind People in Tactile Exploration of Bas-Reliefs?. *Sensors*, *16*(9), 1361.

[2] Hall, T., & Bannon, L. (2005, June). Designing ubiquitous computing to enhance children's interaction in museums. In Proceedings of the 2005 conference on Interaction design and children (pp. 62-69). ACM.

[3] van Dijk, E. M., Lingnau, A., & Kockelkorn, H. (2012, October). Measuring enjoyment of an interactive museum experience. In Proceedings of the 14th ACM international conference on Multimodal interaction (pp. 249-256). ACM

[4] Faber, F., Bennewitz, M., Eppner, C., Gorog, A., Gonsior, C., Joho, D., ... & Behnke, S. (2009, September). The humanoid museum tour guide Robotinho. In Robot and Human Interactive Communication, 2009. RO-MAN 2009. The 18th IEEE International Symposium on (pp. 891-896). IEEE.

[5] Boboc, R. G., Horaţiu, M., & Talabă, D. (2014). An Educational Humanoid Laboratory Tour Guide Robot. Procedia-Social and Behavioral Sciences, 141, 424-430.

[6] Traum, D., Aggarwal, P., Artstein, R., Foutz, S., Gerten, J., Katsamanis, A., ... & Swartout, W. (2012). Ada and Grace: Direct interaction with museum visitors. In Intelligent Virtual Agents (pp. 245-251). Springer Berlin/Heidelberg.

[7] Anderson, D., Piscitelli, B., Weier, K., Everett, M., & Tayler, C. (2002). Children's museum experiences: Identifying powerful mediators of learning. Curator: The Museum Journal, 45(3), 213-231.

[8] Kopp, S., Gesellensetter, L., Krämer, N. C., & Wachsmuth, I. (2005, September). A conversational agent as museum guide–design and evaluation of a real-world application. In International Workshop on Intelligent Virtual Agents (pp. 329-343). Springer, Berlin, Heidelberg.

[9] Bickmore, T. W., Vardoulakis, L. M. P., & Schulman, D. (2013). Tinker: a relational agent museum guide. Autonomous agents and multi-agent systems, 27(2), 254-276.

[10] Swartout, W., Traum, D., Artstein, R., Noren, D., Debevec, P., Bronnenkant, K., ... & Lane, C. (2010, September). Ada and Grace: Toward realistic and engaging virtual museum guides. In International Conference on Intelligent Virtual Agents (pp. 286-300). Springer, Berlin, Heidelberg.

[11] Burgard, W., Cremers, A. B., Fox, D., Hähnel, D., Lakemeyer, G., Schulz, D., ... & Thrun, S. (1999). Experiences with an interactive museum tour-guide robot. Artificial intelligence, 114(1-2), 3-55.

[12] Mori, M. (1970). Bukimi no tani [The uncanny valley], Energy,[Online] 7 (4), 33-35.

[13] Slater, M., Khanna, P., Mortensen, J., Yu, I.. (2009). Visual Realism Enhances Realistic Response in an Immersive Virtual Environment. IEEE Comput. Graph. Appl. 29, 76–84.

[14] Guttentag, D. A. (2010). Virtual reality: Applications and implications for tourism. Tourism Management, 31(5), 637-651.

[15] Jung, T., tom Dieck, M. C., Lee, H., & Chung, N. (2016). Effects of virtual reality and augmented reality on visitor experiences in museum. In Information and Communication Technologies in Tourism 2016 (pp. 621-635). Springer, Cham.

[16] Chung, N., Han, H., & Joun, Y. (2015). Tourists' intention to visit a destination: The role of augmented reality (AR) application for a heritage site. Computers in Human Behavior, 50, 588-599.

[17] Michael, D., Pelekanos, N., Chrysanthou, I., Zaharias, P., Hadjigavriel, L. L., & Chrysanthou, Y. (2010, November). Comparative study of interactive systems in a museum. In Euro-Mediterranean Conference (pp. 250-261). Springer, Berlin, Heidelberg.

[18] Luger, E., & Sellen, A. (2016, May). Like having a really bad PA: the gulf between user expectation and experience of conversational agents. In Proceedings of the 2016 CHI Conference on Human Factors in Computing Systems (pp. 5286-5297). ACM.

[19] Robinson, S., Traum, D. R., Ittycheriah, M., & Henderer, J. (2008, May). What would you Ask a conversational Agent? Observations of Human-Agent Dialogues in a Museum Setting.

IOP Conf. Series: Materials Science and Engineering **364** (2018) 012098 doi:10.1088/1757-899X/364/1/012098

Virtual prototyping of medieval weapons for historical reconstruction of siege scenarios starting from topography and archaeological investigations

G Annoscia[1], M Bici[2, 3], F Campana[2], L De Lellis[1]

[1] Dipartimento di Scienze dell'Antichità, Sapienza Università di Roma
[2] DIMA – Dip. Ingegneria Meccanica e Aerospaziale, Sapienza Università di Roma

Abstract. Chronicles of sieges to castles or fortresses, using "machinae", can often be found in historical sources. Moreover, archaeological excavations of castles or fortresses has brought to light rocks or projectiles whose carving suggests a military usage. Nevertheless, chronicles and discoveries alone, are seldom enough to propose a faithful reconstruction of these machines. Therefore, the aim of this research is the development of methodologies for reconstructing virtual scenarios of sieges, starting from the scarce information available. In order to achieve it, a procedure for the virtual reconstruction of the siege machine has been set up, focusing on typology and dimensions of the machines, also investigating possible fire positions according to topography. The entire procedure has been developed using the siege of Cervara di Roma's Rocca as a case study. Late medieval chronicles (end of 13th Century) report the siege brought by the papal army in order to restore the jurisdiction on the Cervara's stronghold, following the insurrection of a group of vassals headed by a monk named Pelagio. The discovery, in the area of the Rocca, of a stone that could have been used as a projectile confirms what reported. The proposed methodology is composed of two parts. The first one is connected to the study of the "internal ballistics", to understand the performances and to build virtual models of siege machines. The second part is the study of the "external ballistics", then to the positioning and shooting ability of possible machines, analysing the topography of the area. In this paper, we present the feasibility of this methodology through the preliminary results achieved correlating internal and external ballistics.

1. Introduction

This paper supports the multidisciplinary investigations about a better understanding of the dynamics and the possible development of sieges in the Middle Ages, as well as the type and the characteristics of the siege apparatus, also known in the written sources as "machina" or "engine". Engine is a term derived from Latin and the European vernaculars that stands for "ingenium", "an ingenious contrivance" and "ingeniators" are those who designed, made and used them [1].

From the methodological point of view, the approach that is here proposed starts with the analysis of the historical sources and the archaeological evidences, then proceeds with formulating siege scenario hypotheses. They are then evaluated through technical considerations related to territory topography and engineering feasibility. Virtual prototyping represents the engineering method to evaluate the apparatus performances according to the specific topography and siege scenarios discussed by the archaeologists. Unfortunately, the written sources give us very few technical details about the construction, structure

[3]Author to whom any correspondence should be addressed: michele.bici@uniroma1.it.

and effective capabilities of the ancient siege engines. For this reason, the virtual reconstruction is going to include also a preliminary design development of the possible apparatus according to requirements and technologies defined by the archaeologists.

The siege of Cervara di Roma is here reported as a case study for the proposed approach. In fact, its occurrence appears to have been a non-trivial example of military knowledge and practice, due to the strategical position of the Rocca of Cervara – built on a hilltop and characterised by the presence of a very rocky and harsh landscape – apparently very favourable to resist every attempt of conquering by the force.

The paper structure reflects the logical workflow of the proposed approach starting with the sources related to the siege of Cervara and the types of engine documented in that period (Section 2). Then in Section 3, a detailed description of the technical feasibility of the siege engine and its performances is given; and in Section 4 their application to the case of Cervara di Roma is treated, arriving to the presentation of results and their discussion, before conclusion.

2. Historical sources

2.1. The siege of Cervara di Roma

In 1273, Pelagio, a monk from the monastery of Santa Scolastica in Subiaco (Rome, Latium), rebelled against his abbot and took shelter with his companions inside of the Rocca of Cervara, a military stronghold erected on a mountaintop of the Simbruini mountains. To end the rebellion and re-establish his control over his lands, in 1276 the abbot Guglielmo – supported by the papacy – sent an army to besiege the Rocca. The army, leaded by Guglielmo di Borgogna, took the stronghold within two months, thanks also to the use of a siege engine, as reported by the written sources [2]. The archaeological excavations conducted inside of the remainings of the Rocca brought to light traces of the siege along with a stone projectile related with the presence of the siege engine at the siege itself [3]. Because of that, the role of the siege engine appears to have been determinant, as suggested also by the written source. The comprehension efforts have been focused on two main research questions related to the siege engine: the determination of the kind of machine used and of its presumed size and fire range capabilities, determined on the basis of the virtual prototyping of the machine, and the determination of the supposed deployment positions of the machine itself.

2.2. Siege engines

From an historical and archaeological point of view, the siege engine employed in the siege of the Rocca of Cervara was almost certainly a trebuchet. Sporadically attested starting from the Early Middle Ages, this kind of siege engine saw a widespread adoption during the Crusades period and by the 8th century had evolved from the traction trebuchet – driven exclusively by human power – into the counterweight trebuchet, where the propulsion force was supplied by a counterweight installed at the end of the swinging arm of the machine. Technically more sophisticated and more powerful than the traction trebuchet, it was probably a counterweight trebuchet the engine used to besiege Pelagio and his fellows. Despite its large diffusion on the medieval battlefields, today we possess very few details about the construction and the characteristics of that siege engine. Absent from the byzantine military treatises of the 6th and 10th centuries – more focused on the roman military tradition and in spite of its usage being attested in the byzantine world starting from the 6th century [4, 5] – the trebuchet makes its first appearance in the military literature with the famous treatise of Mardî al-Tarsûsî written for Salâh al-Dîn in the second half of the 12th century, where the author offers a thorough analysis of the traction trebuchet. The absence of military treatises in the medieval western world until the very late medieval or the Renaissance period, leaves us with very few details about the characteristics of the counterweight trebuchet. We have a certain number of mentions in written sources such as war chronicles along with a variety of drawing and illuminated codes illustrations.

Given the very few archaeological remains of these machines – due to the perishable nature of the materials they were made of – the reconstruction of the machine used in the virtual prototyping process has been based on the iconographical evidence available from the manuscripts and on the characteristics identified by the scientific literature on the matter [1, 4, 6, 7, 8]. Further elements were provided by the dimensions and weight of the stone projectile found during the archaeological excavation at the Rocca of Cervara, whose weight clearly suggested the need for a counterweight trebuchet for its throwing.

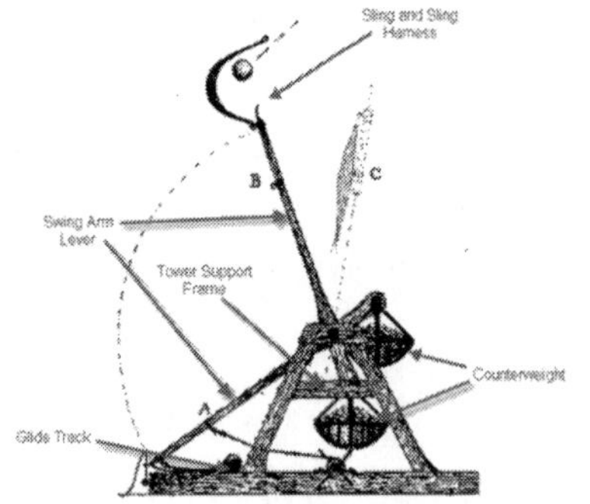

Figure 1. Example of counterweight trebuchet.

Figure 2. Example of traction trebuchet.

A preliminary analysis of the terrain characteristics of the area has been conducted to locate all the probable places for the deployment and operation of a machine satisfying our preliminary requirements. With the help of a GIS system, we analysed a digital terrain model to extract slope and altitude data and then we elaborated a model able to identify on the terrain all the areas having a suitable slope (< 20°) and altitude in relation to Rocca position. The preliminary results of the analysis confirmed the need for a machine able to reach a firing range greater than that of the traction trebuchet (> 100/125 m) [4], finally aiming our efforts toward the investigation of the capabilities of a counterweight trebuchet.

Trebuchets were often transported in pre-made pieces and the assembly was finalized on the siege location, therefore always requiring the presence of some specialized personnel to set up and operate the machine. As an example of this, see the presence of some German engineers in charge of the trebuchets in the army of the king Erik Emune of Denmark in the 12th century [9]. The Chronicle of Guglielmo Capisacchi da Narni [2, 3] reports the besieging army as being "of not small dimensions" and equipped with the siege engine, confirming us in the impression of the presence of a high degree of expertise and coordination in the execution of the siege.

The Chronicle doesn't give us any clue about the possible use of the trebuchet to open a breach in the walls of the fortification and from the actual condition of the masonry of the walls themselves (heavily restored in the last century) is impossible to establish if that was actually the case. From the weight and dimensions of the found projectile that could have been a real option, but certainly even the bombing of the battlements and the interior of the fortress must have had a devastating effect on the morale and the resistance capability of the besieged.

3. Technical feasibility

Obtaining a virtual model of a counterweight trebuchet suitable for a specific siege scenario means to provide type, structure, material and reasoning of its practical usage during the siege. In particular, it is necessary to evaluate solutions related to how they built and maintained it, considering constraints related to the topology of a suitable area where placing it and shooting safely a projectile on the target.

Due to the vastness of this kind of design problem, we decided to consider two main fields of analysis, conducted in parallel: the so-called internal and the external ballistics analysis. Although these analyses are typical of firearms, we decide to adopt them to distinguish the two basilar technical studies necessary to face this problem: the siege engine design and its selection according to the performances required by the target. Internal ballistics concerns with defining the way of functioning of the trebuchets and their dimensions. For the external ballistics analysis, instead, we are referring to all issues connected to the trajectory, the positioning and the topography of the area in which machine and target are placed.

Considering the timing of an attack, the internal ballistic is connected to all the issues starting from the activation of the machine to the moment when the projectile leaves. It defines the way the siege engine works. The external ballistics analysis is connected to the engine final performances, so that it is related to the choice of positioning area, and, mostly, to moments following the detachment of the projectile, during "flight", in terms of trajectory issues.

3.1. Internal ballistics: aim

Siege machines have had a significant development in their history, and obviously it conducted to a wide range of types and solutions. In this work, we limited the analysis onto trebuchets, due to the fact that, in the case of Cervara di Roma, this type of machine seems to be the used one, both for its diffusion at that age and for its launch capability.

In order to understand the capabilities of this typology of machines, three models have been studied, varying the level of technological progress. In fact, starting from a trebuchet with a fixed counterweight at the end of the shorter arm (Type A, in Figure 3), we added technical solutions oriented to the increase of performances, as reported in literature. One solution is a hinged counterweight (Type B in Figure 4), the other is a sling for positioning and launching the projectile (as added in Type C in Figure 5).

Figure 3. Type A model trebuchet with fixed counterweight.

Figure 4. Type B model trebuchet with hinged counterweight (circled in red).

Figure 5. Type C model trebuchet with added sling (circled in red).

From the structural point of view, considering loads as concentrated masses, the maximum bending moment is always placed in correspondence of the fulcrum of the launcher beam. This obviously defines the dimensioning of the structure in terms of resistant section.

Nevertheless, several input have to be defined for achieving the effective dimensions of the machines, and sensitivity analysis can be conducted varying these input into their ranges.

The main inputs are:

- Counterweight Mass (M_{cw}),
- Projectile Mass (M_p),
- Length of Launcher Beam (L),
- Ratio ($\frac{L_2}{L_1}$) between lengths of longer (L_2, the part that connect the projectile site and the fulcrum) and shorter arm (L_1, the connection between fulcrum and counterweight).

In addition, some assumption has to be done regarding:

- materials (generally wood, function of the geography of the site),
- initial and launching angle,
- height of the fulcrum,
- clearance between counterweight and the ground floor during motion (connected to the L_4, length of the arm that connects the hinge with the hinged counterweight, if present)
- length of the sling (L_3, only for Type C).

This type of sensitivity analysis can lead to the definition of the section lengths (assumed to be square), and of $\overrightarrow{V_0}$ values, the velocity of the projectile at the moment of launch that is directly connected to the maximum reachable range.

3.2. Internal ballistics: results

Sensitivity analysis has been preliminary approached with analytic and energetic formulation, inside Mathematica, spanning significant ranges for M_{CW}, M_P, and geometric configuration of the beam (L_1, L_2). It assumes poplar wood as material for all the pieces (with density of about $400\frac{kg}{m^3}$; shear yield stress of about 4.8MPa, static friction coefficient between 0.25 and 0.5 and kinetic friction coefficient of 0.19).

Virtual models of the three types of machine have been obtained into a parametric CAD, and they have been also imported in a multibody simulation software, Hyperworks Motionview, to evaluate, in addition, the influence of friction factor, which could be extremely complex with the analytic and energetic formulation. Results of this study are summarized in Figure 6. Assuming horizontal slope for the terrain, optimal launch conditions are related to 45°, so that the maximum range turns out to be: $R_{max} = \frac{V_0^2}{g}$.

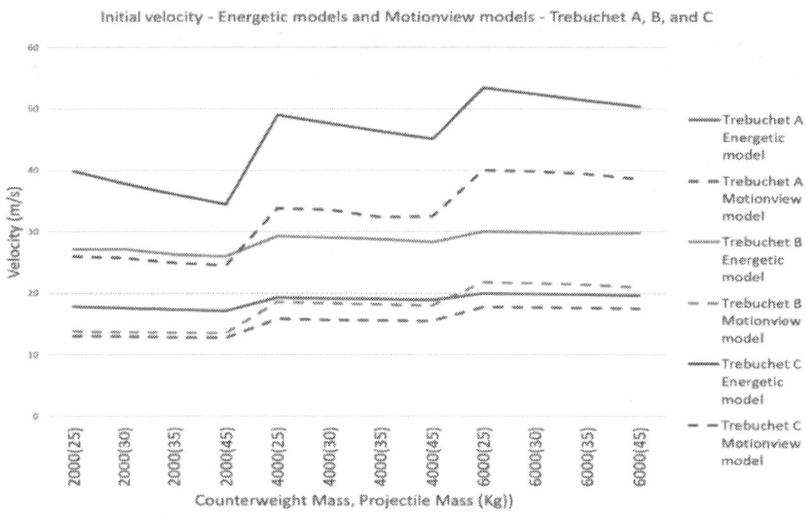

Figure 6. Sensitivity analysis according to M_{CW}, M_P, and type of machine, with and without friction.

As shown in Figure 6, changing the type of trebuchet, performances increase according to M_{CW} and decrease with the M_P (values fixed at 25, 30, 35 and 40 kg). Lower values of M_{CW} reduce the effect related to the type of machine, while higher values make more evident the improvements due to Type B and C. Obviously the presence of friction reduces the velocity (in the legend of Figure 6 it is related to Trebuchet with flag "Motionview"), nevertheless, additional studies must be done to evaluate the presence of metal parts in the hinged connections. In fact this solution, when suitably lubricated, would lead to lower friction coefficients than in the case of wood-wood, mitigating the reduction in the resultant performances.

3.3. External ballistics

The internal ballistic analysis have been conducted considering that the virtual attack could take place in a "flat" zone, admitting the target placed at the same height level of the launching point.

In real positioning, there are several differences with this ideal case, connected to issues not directly correlated to machine functioning. Generally, there is a difference in terms of height between the target and launching point, and three characteristic angles have to be defined (Figure 7):

- α, the launching angle that has to be optimised;
- β, the angle due to the inclination derived by planar and height differences between launching point and target (T),
- γ, the local slope of the area where the machine is placed.

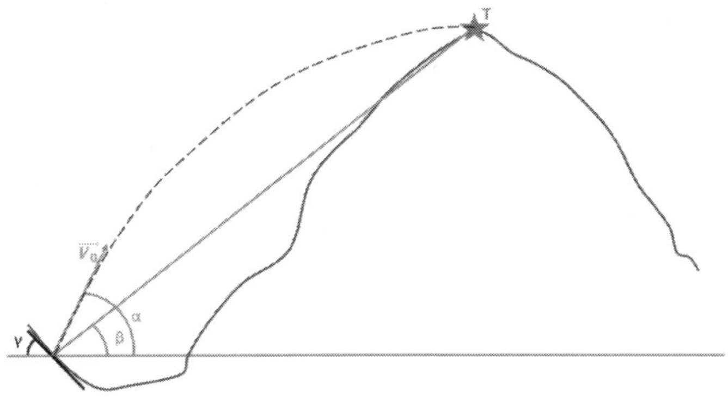

Figure 7. Characteristic angles in an external ballistic problem.

For what concerns γ, the local slope angle, there are several technical observations to do. Firstly, from the literature, it is known that often these types of machines were transported in parts and assembled *in situ*, so that, the local slope had to be limited to guarantee the possibility to move pieces of several meters of length and to assembly the machine safely. Analogous issues can be highlighted for the possibility of functioning, partially solvable through the usage of artificial basements, to avoid excessive slopes, or through operations of ground levelling. What has to be clear is the fact that the relative launching angle (in a reference system solid with the machine) is equal to the difference between α and γ, and it must be limited and close to 45° to guarantee the functioning. Due to this, only areas with $\gamma < 20°$ have been examined.

The optimal launching angle, α, can be optimised according to β, the inclination angle, to obtain the maximum range R_{MAX} (depending by the V_0). But, two different cases must be considered: one with the T at a higher level respect to the launching point, the other with the T at a lower level.

For the ascent case:

$$\tan\alpha_{opt} = \tan\beta + \sqrt{(\tan\beta)^2 + 1}; \quad \text{which gives } R_{MAX} = \frac{V_0^2}{g}\frac{(\tan\beta)^2+1}{(\tan\beta)^2+1+\tan\beta*\sqrt{(\tan\beta)^2+1}};$$

for the descent one:

$$\tan\alpha_{opt} = -\tan\beta + \sqrt{(\tan\beta)^2 + 1}; \quad \text{which gives } R_{MAX} = \frac{V_0^2}{g}\frac{(\tan\beta)^2+1}{(\tan\beta)^2+1-\tan\beta*\sqrt{(\tan\beta)^2+1}}.$$

It can be seen that, respect to the "flat" case, the one with the target in ascent has a R_{MAX} smaller, due to the multiplication of the $R_{MAX\text{-}FLAT}$ for a quantity lower than 1. Likewise, the second case has an R_{MAX} higher, due to a multiplication for a quantity higher than 1. So that, it is visible as the descent is favourable and the ascent has a negative effect on the launching capacity.

The developed procedure for external ballistics analysis starts with the GIS acquisition in an area close to the target. The first step is the analysis and the selection of areas with local slope $\gamma < 20°$, then, for all the selected zones, β and α_{opt} are calculated. Every distance with the target is also obtained. The substitution in the R_{MAX} formulation (for the respective case) furnishes the minimum value of V_0 to reach the target. These data can be crossed to what found in internal ballistics in order to select possible types of trebuchets and their dimensioning for each of the selected areas.

4. Application for the siege of Cervara di Roma

This section presents a resume of the developed procedures applied to Cervara di Roma [10]. The input data are:

- the archaeological finding, in the fortress area, of a limestone with an average diameter of about 0.3 m and a weight of about 30kg. The level of finishing and the quite symmetric shape of the stone have led to attribute it the role of projectile of a siege engine.
- the GIS acquisition of an area of about 250 m of radius centred nearby the fortress. This represents a credible range of action for the found projectile. We assume the area of the siege identical in topographical terms to what it was in the 13th century, nevertheless, no chronicles, documents or treatises have been found that report catastrophic events, as earthquakes, or landslides in the region.

In terms of internal ballistics, the parameters described in Section 3 can be set in the ranges reported in Table 1.

Table 1. Ranges of input parameters used for trebuchet models for Cervara di Roma analysis.

Input parameter	Range (Unit of Measurement)
Counterweight Mass (M_{CW})	2000 - 6000 (kg)
Projectile Mass (M_P)	25 - 40 (kg)
Length of Launcher Beam (L)	6 - 10 (m)
Ratio L_2/L_1	2 - 4

GIS acquisition of the area, with the fortress highlighted, is reported in Figure 8. In Figure 9, the local slope angle, γ, is also reported. Four areas (Z1÷4), with an average γ close to 20°, can be individuated. These are the suitable areas for positioning trebuchets. It has to be highlighted that data obtained by the GIS acquisition furnished local slope for sections of about 1 m^2 each. Resampling operations and the calculation of local slope average are useful to consider wide local areas that can allow the machine positioning.

In the Cervara di Roma case study, it has to be highlighted that no wide transformation and changes of the altimetry have been reported in written sources. Operations aimed at local levelling, in terms of slope, could be generally considered, but, in the case of Cervara di Roma, they are not applicable due to the imperviousness of the area, as reported in historical literature. So that, the current topography has been considered analogous to the ancient one.

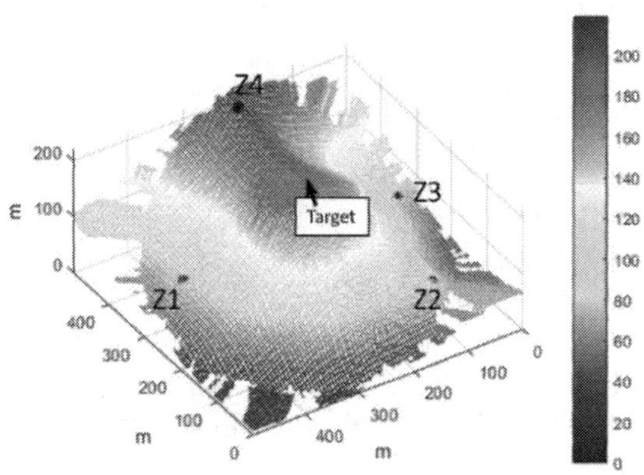

Figure 8 Altimetry of the zone..

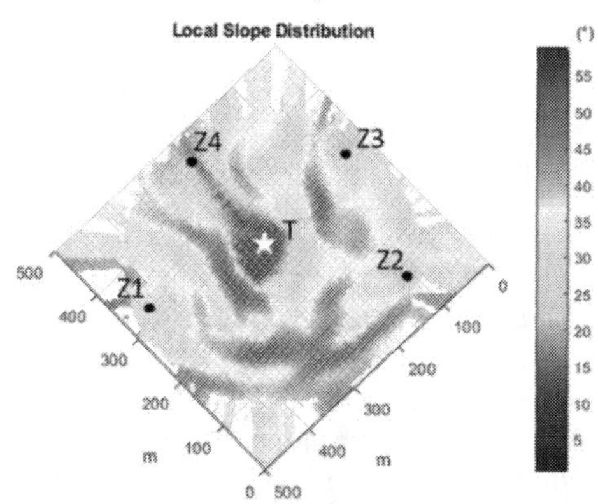

Figure 9. Local slope angle (°).

Extracting from the GIS data the central coordinates of the 4 areas of launch, average β and α_{OPT} can be evaluated. The distance between each zone and the target can be used to determine the minimum V_0 requested to reach it. Results are resumed in Table 2, using as central coordinates of the areas, the relative positions, *Rel*, from the point with lowest altitude.

Table 2. Data obtained for Z1, Z2, Z3, Z4 points of launch.

	X (Rel) [m]	*X (Rel) [m]*	*X (Rel) [m]*	*Local Slope [°]*	*Condition of Launch*	*Distance [m]*	*β [°]*	*α_{OPT} [°]*	*V_0 min [$m*s^{-1}$]*
Target	275 (0)	220 (0)	190 (0)						
Z1	296 (21)	431 (211)	77 (-113)	26.0	ascendent	212	28.1	59.0	55.3
Z2	84 (-191)	112 (-108)	78 (-112)	20.5	ascendent	219	27.0	58.5	56.0
Z3	296 (21)	39(-181)	80 (-110)	21.0	ascendent	182	31.1	60.1	52.1
Z4	453 (178)	230 (10)	220 (30)	11.0	descendent	178	9.6	40.2	38.2

Crossing these external ballistics data with the charts of trebuchet performances discussed in Section 3 (Figure 6), we obtain that:

- Type A trebuchet cannot be applied with the chosen range of parameters in the selected zones;
- Type B can be applied only in Z4 (descendent case) in few combinations of imposed parameters;
- Type C can be applied in every of the four areas of launch, with a high number of possible combinations of parameters, giving the possibility of a choice in function to easiness of functioning.

5. Conclusions

Through the developed procedures and the sensitivity analysis, the hypothesis of use of a trebuchet by the papal army in the siege of Cervara is reinforced. Trebuchet models confirmed the technical progresses, in terms of performances, due to the introduction of hinged counterweight and sling. The procedures developed, distinct in internal and external ballistics analysis, have shown 4 possible areas that may be successfully attacked by Type C, while Type B can be effective only in descendent terrain and Type A can work only assuming specific anti-friction solutions like metallic hinge.

This research work, in the hope of authors, will be the initial part of a wide study, future developments will be about the analysis of friction factors, connected to archaeological researches on metallic materials, trying to characterise them. Moreover, a better description and assembly analysis will be also made to focus the engine set-up and the men necessary to manoeuvre it.

In the next period, the study will focus on the air resistance analysis. In fact, it has not been considered because the obtained values of velocity are lower than the limit velocity calculated for an ideal sphere equivalent to the analysed projectiles. The consideration of air resistance could conduct to a reduction of about 10÷20% in terms of performances, thus maintaining the possibility of reaching the target from the selected areas. The goal is to consider specific wind maps of the siege area in order to develop an accurate analysis of the issue. This represents the one of the main further developments of the research, for improving the accuracy of the procedures.

The entire procedure, developed on the Cervara di Roma case, will be applied on other sites for other siege machine virtual reconstructions, focusing not only on trebuchet, but to the entire field of siege engines.

Notes and Acknowledgements

Introduction and Section 2.1 (*The siege of Cervara di Roma*) have been developed by Giorgia Annoscia.

Section 2.2 (*Siege engines*) has been developed by Lorenzo De Lellis.

Authors are pleased to thank Ahmad Anas Arifin, for the contribution to the development of models and sensitivity analysis, carried out during his thesis work.

References

[1] Chevedden P E, Eigenbrod L, Foley V and Soedel W 1995 The trebuchet *Scientific American* **273** 66–71

[2] Capisacchi G 2005 *Chronicon sacri monasterii Sublaci (anno 1573)* ed L Branciani (Subiaco)

[3] Annoscia G 2015 Guerra e territorio: "Cultura della difesa" nella Valle Sublacense *VII Congresso Nazionale di Archeologia Medievale* ed P Arthur and M L Imperiale (Firenze) 313–7

IOP Conf. Series: Materials Science and Engineering **364** (2018) 012098 doi:10.1088/1757-899X/364/1/012098

[4] Tarver W T S 1995 The traction trebuchet: a reconstruction of an early medieval siege engine *Technology and Culture* **36** 136–67
[5] Fulton M S 2017 A ridge too far: the siege of Saone/Sahyun in 1188 and contemporary trebuchet technology *Crusades* **16** 33–53
[6] Saimre T 2006 Trebuchet - A gravity operated siege engine *Estonian Journal of Archaeology* **10** 61–80
[7] Denny M 2005 Siege engine dynamics *European Journal of Physics* **26** 561–77
[8] Chevedden P E 2000 The invention of the counterweight trebuchet: a study in cultural diffusion *Dumbarton Oaks Papers* **54** 71–116
[9] Hansen P V 1992 Experimental reconstruction of a medieval trebuchet *Acta Archaeologica* **63** 189–268
[10] Arifin A A 2018 Virtual prototyping of medieval weapons for historical reconstruction based from topographic and archeological observation *Master Thesis in Mechanical Engineering at Sapienza Università di Roma*
[11] Pahl G, Beitz W, Feldhusen J and Grote K H 2006 Engineering Design: A Systematic Approach *Springer-Verlag*; 3 ed.

IOP Conf. Series: Materials Science and Engineering **364** (2018) 012099 doi:10.1088/1757-899X/364/1/012099

Application of binaural audio techniques for immersive fruition of cultural heritage

A Lapini[1], G Calamai[2], F Argenti[2] and M Carfagni[1]

[1] Department of Industrial Engineering, University of Florence, via S. Marta 3, Florence 50139, IT

[2] Department of Information Engineering, University of Florence, via S. Marta 3, Florence 50139, IT

E-mail: {alessandro.lapini@, giulio.calamai@stud., fabrizio.argenti@, monica.carfagni@}unifi.it

Abstract. In this paper we address the issue of enriching the fruition of museums and art shows for the visitors. We preliminary consider the design of an immersive audio environment by means of 3D audio rendering, aiming to provide each user a deeper connection with each exhibited artwork. By exploiting a real-time binaural audio system, artworks become virtual sources of the audio guide voice, making the user perceive they personally speak to her/him. As an advantage, the implementation of the proposed framework relies on basic audio and signal processing techniques, that is, no expensive or personalized equipment for the visitors is required. Furthermore, since the proposed method is independent of the recording stage, existing audio tracks or even real-time speech can be used.

1. Introduction

Immersive technologies are nowadays spreading across several fields and disciplines, improving the human perception in augmented reality or, more generally, in mixed reality environments. Remarkable engineering studies and applications in this broad context involve (but are not limited to) biomedicine, manufacturing processes, entertainment, gaming, military operations and also cultural heritage.

The efforts of technological progress in computer science have been historically focused firstly on improvements of the user' visual experience. Incidentally, this is nowadays witnessed by the wide spreading of stereoscopic displays [1, 2, 3, 4]. Nevertheless, human perception is particularly influenced by sound, too. Stereophonic audio (which has widespread much earlier than stereo vision) and the subsequent surround techniques are still far from providing a realistic experience. In the last years, the hardware and software improvements, especially in consumer equipment, have paved the way to introduce immersive audio technologies [5]. Spatial rendering techniques that allow the human auditory system to perceive the spatial localization of virtual sound sources are currently crossing the line between theory and practical implementations. Among them, binaural synthesis aims to control the sound directly in the human ears [6, 7].

The interest towards the application of binaural techniques to cultural heritage has grown both in institutions and private companies, as the EU financed project BINCI [8] demonstrates. The aim of this work is to describe a signal processing method for fast real-time binaural synthesis, whose main target application is, among others, the fruition of cultural heritage. The practical implementation of the system is beyond the scope of this work. The proposed method relies on the same logical approach proposed in [9], but, to the best of authors' knowledge, our framework presents some novelties.

2. Application scenario

A description of the application scenario based on binaural techniques is provided according to the following test case, whose sketch is reported in Figure 1.

We consider a visitor of a museum approaching to a specific artwork. We assume that he/she is provided with an audio–guide device and headphones or earphones; the audio–guide device must support the following capabilities:

(i) real–time processing of a monaural audio track that describes the artwork;
(ii) reproduction of a stereo signal through the headphones/earphones that are plugged into;
(iii) indoor positioning and head orienting system, briefly indicated as *positioning system*.

The visitor is required to select the audio track pertaining to the targeted artwork and the reproduction starts. A suitable binaural processing of the monaural audio track is then applied and the output signal is reproduced in the visitor's ears, in order to make him perceive the audio coming from the artwork itself. The visitor is allowed to freely walk around the artwork, getting closer or moving away, while the audio source accordingly moves with respect to him. This application can be straightforwardly generalized to two or more sources that can be located on the same artwork or even on different ones, obtaining, for instance, the effect of multiple voices incoming from different directions. Thanks to the earphones and to individual audio–guide, more visitors are allowed to target the artwork at the same time independently from each other.

A schematic view of the proposed system for the case of a single virtual source is depicted in Figure 2. For sake of generality, the visitor and the artwork are indicated as listener and the (virtual) source, respectively. At a given time instant n, the positioning system computes the source–listener reciprocal position and the head orientation of the listener with respect to the source, being the latter fixed or, in general, known a–priori. Research on indoor positioning systems is currently ongoing and the reader interested in indoor tracking might find useful to see [10, 11]. The information about position and orientation is then exploited to query and retrieve the corresponding couple of binaural filters, which are stored in a specific database. Such filters are applied to the monaural audio track in order to determine the left and right binaural audio signals until the time instant $n + M$, where M is the number of sampling instants between each position update. Finally, the signals are transmitted to the headphones/earphones of the listener. The algorithm then iterates to the following position update until the end of the audio track.

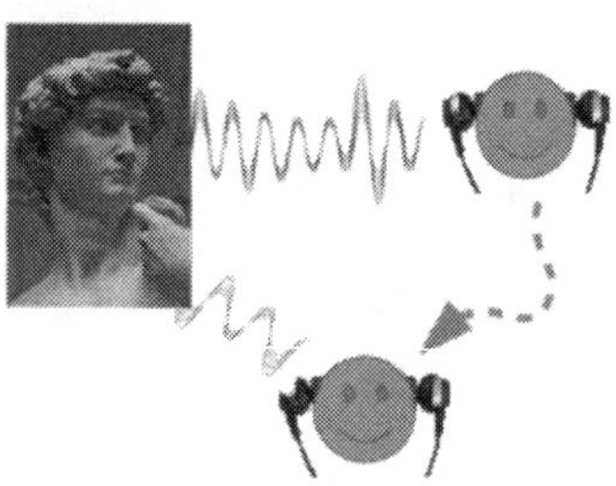

Figure 1. Sketch of the a possible application scenario for the proposed method.

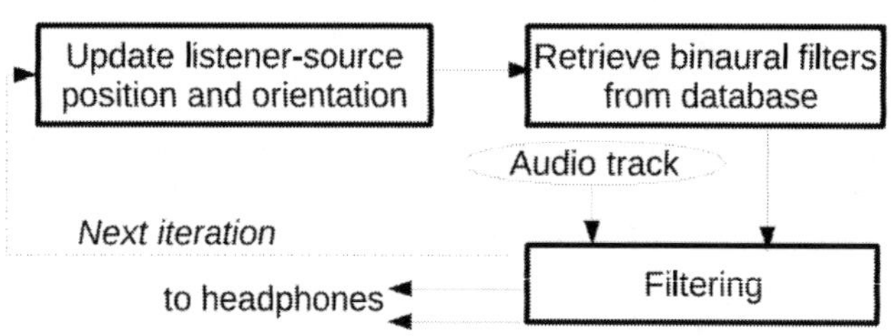

Figure 2. The proposed system for a real-time binaural processing considering a moving listener and a source located in a known position.

3. Time–varying binaural synthesis

We start this Section briefly introducing the modeling for human sound localization; the proposed method is subsequently described.

3.1. Characterization of auditory system

The modeling of human auditory system is a long term goal of several disciplines, being audio engineering among them. Nowadays it is known that human perception of sound spatiality is based on the decoding of the *binaural directional cues*, that is, the differences of sound between the ear channels. Time and level differences of the sound traveling into the listener's ears are due not only to the reciprocal position of the source with respect to the listener' head, but also to the interaction of the acoustic field with the listener's torso, head, and pinnae. These effects can be roughly thought as the incoming acoustic signal undergoing a *body filtering* stage before being captured by human auditory system.

The concept of *head–related impulse response* (HRIR) is commonly adopted to describe the relation between the source signal, namely x, and the signals in the listener's ears, namely y^L and y^R. We indicate the source–listener distance and the orientation of listener head with respect to the line passing trough the source and the listener with symbols r and ϕ, respectively. Thus, the output signals at the discrete time instants $n \geq 0$ are given by

$$y^{L,R}[n] = \sum_{m=0}^{\min(N-1,n)} h_{r,\phi}^{L,R}[m]x[n-m] = \sum_{m=\max(n-N+1,0)}^{n} h_{r,\phi}^{L,R}[n-m]x[m] \qquad (1)$$

where $h_{r,\phi}^L$ (respectively, $h_{r,\phi}^R$) is the left (right) ear HRIR, and N is its length in terms of number of samples. Equation (1) represents the input–output relation of a linear time-independent (LTI) system, where the HRIR is the impulse response of the associated LTI filter.

Given a known test signal x, HRIR can be estimated by recording the received signals y^L and y^R by means of ear microphones plugged in real human ears or in realistic mannequin heads. By moving the source and fixing the listener (or vice versa), HRIRs corresponding at different azimuths, elevations and distances are gathered. Incidentally, a number of research paper linked to free HRIR databases exists, e.g., [12, 13, 14].

Binaural synthesis techniques, instead, aim to control directly the sound in the ear canals to match a recorded real case or with a simulated virtual case, by using measured or modeled HRIRs [15]. Our scenario belongs to the latter case. Specifically, given a monaural signal x and a couple of filters ($h_{r,\phi}^L$,$h_{r,\phi}^R$), a virtual sound source, positioned at distance r and oriented at $-\phi$ with respect to the listener's head, is synthesized; the output signals y^L and y^R can be eventually reproduced by headphones/earphones.

3.2. Time–varying binaural modeling

Unfortunately, the HRIR defined in (1) is spatial dependent but not time dependent, that is, it is valid for constant r and ϕ across time n. Hence, it cannot be directly applied for a moving source or listener, as it would be required in our scenario. By means of a simple example, we initially show that extending (1) to a time–varying scenario is not trivial nor unambiguous; then we propose a solution in a constructive manner.

Let's assume that the source is fixed and the listener moves. Without loss of generality, we consider that the position updating occurs at each n (i.e., $M = 1$) and and we indicate the time–varying r and ϕ with r_n and ϕ_n, respectively. The output signal at $n = 0$ is then

$$y[0] = h_0[0]x[0],$$

where, for sake of brevity, the superscripts L, R have been omitted and the symbol h_0 is a short version of h_{r_0,ϕ_0}. At instant $n = 1$ one may think to extend (1) as

$$y[1] = h_1[0]x[1] + h_1[1]x[0]$$

or more generally

$$y[n] = \sum_{m=0}^{\min(N-1,n)} h_n[m]x[n-m] = \sum_{m=\max(n-N+1,0)}^{n} h_n[n-m]x[m]. \quad (2)$$

According to this model, at time n, there is no difference between the signal received by the moving listener and the signal that would be received by a fixed listener at (r_n, ϕ_n). Since $x[n_0]$ is somehow associated to an incident acoustic wavefront, this approach is equivalent to state that the acoustic scattering contribution of the human body at n_0 vanishes for $n > n_0$. We believe that the physical validity of such model is questionable and should be preliminarily validated.

In this paper we propose an alternative model to (2). The received signal is given instead by

$$y[n] = \sum_{m=0}^{\min(N-1,n)} h_{n-m}[m]x[n-m] = \sum_{m=\max(n-N+1,0)}^{n} h_m[n-m]x[m]. \quad (3)$$

Because $x[n]$ is weighted by h_n, in this case each incoming wavefront is affected by a time–invariant scattering contribution of the human body, that is, the listener is fixed with respect to each incoming wavefront. Accordingly, this model is equivalent to the case of a moving source and a fixed listener. It is our opinion that this assumption is more valid with respect to (2) and entails an approximation error that is acceptable for a listener moving much slower than the speed of sound in the air.

3.3. Linear real-time interpolation of HRIR

In a real scenario the updating rate of HRIRs is usually slower than to the sampling rate, that is, $M \gg 1$. This is mainly due by the time required by the positioning system update and the database query/retrieval process. By defining

$$p(n) \triangleq M \lfloor n/M \rfloor, \quad (4)$$

where $\lfloor \cdot \rfloor$ is the floor operator, (3) becomes

$$y[n] = \sum_{m=0}^{\min(N-1,n)} h_{p(n-m)}[m]x[n-m] = \sum_{m=\max(n-N+1,0)}^{n} h_{p(m)}[n-m]x[m]. \quad (5)$$

For $M = 1$, (5) suitably coincides with (3). Practical implementation of (5) is quite simple;however, it can introduce audio impairments due to abrupt head orientation changes whenever the listener sufficiently moves during the position updating period M.

We overcome this problem by considering the following modified version of (3):

$$y[n] = \sum_{m=0}^{\min(N-1,n)} \hat{h}_{n-m}[m]x[n-m] = \sum_{m=\max(n-N+1,0)}^{n} \hat{h}_m[n-m]x[m], \quad (6)$$

being $\hat{h}_m$ the *time–varying* HRIR defined as linear interpolation of the retrieved HRIRs, that is,

$$\hat{h}_m[n] \triangleq h_{p(m-M)}[n] + \frac{m - p(m)}{M}\left(h_{p(m)}[n] - h_{p(m-M)}[n]\right) \quad (7)$$

and assuming $h_{-M} = h_0$. According to (7), the computation of interpolated HRIRs conveniently demands to retrieve only one HRIR every M samples. It has to be noted that $\hat{h}_{p(m)} = h_{p(m-M)}$, i.e., the interpolated HRIR at index $m = p(m)$ is equal to the penultimate retrieved HRIR. This choice preserves causality at the expense of a time delay of M samples in the position of the virtual source. However, according to our tests, it has a lower perceptual impact than the reproduction of audio glitches.

3.4. Fast real-time block processing

Let's indicate $n = kM + l$, for $k \in \mathbb{N}_0$ and $0 \leq l < M$, and define $x_k[l] \triangleq x[kM + l]$ as the k-th input block. According to the linearity and the causality of (6), the contribution of $x_k[l]$ to the output signal y is given by

$$y_k[n] \triangleq \begin{cases} \sum_{m=\max(n-M+1,0)}^{\min(N-1,n)} \hat{h}_{n-m+kM}[m]x_k[n-m], & \text{for } 0 \leq n < M+N, \\ 0 & \text{otherwise,} \end{cases} \quad (8)$$

so that

$$y[n] = y[kM + l] = \sum_{p=0}^{k-1} y_p[l + (k-p)M] + y_k[l]. \quad (9)$$

The summation in (9) accounts for the contributions of input blocks prior to k, while the last term represents the update contribution of the k-th block. Substituting (7) in (8) and considering only the non–trivial case ($0 \leq n < M + N$) yields

$$y_k[n] = \sum_{m=\max(n-M+1,0)}^{\min(N-1,n)} \left[h_{(k-1)M}[m] + \frac{n-m}{M}\left(h_{kM}[m] - h_{(k-1)M}[m]\right)\right] x_k[n-m] \quad (10)$$

where, since $0 \leq n - m < M$, the identity $p(n - m + kM) = kM$ has been used. By defining the following quantities,

$$g_{kM}[n] \triangleq \frac{h_{kM}[n] - h_{(k-1)M}[n]}{M} \quad \text{and} \quad g'_{kM}[n] \triangleq n g_k[n], \quad (11)$$

and dropping the extrema of summations for sake of brevity, (10) is rewritten as

$$y_k[n] = \sum_m \left(h_{(k-1)M}[m] - g'_{kM}[m]\right) x_k[n-m] + n \sum_m g_{kM}[m]x_k[n-m]. \quad (12)$$

Both the summations in (12) are linear convolutions and can be potentially computed by means of Fast Fourier Transform (FFT) techniques to reduce the computational costs [16]. Since the monaural signal x_k is known a-priori[1] of and only the HRIRs at the time instants $(k-1)M$ and kM are required, all the quantities in (12) are available at time instant kM. Hence, the proposed method can take advantage by processing a block of M input samples every M time instants, without introducing further processing delay. The procedure is reported in Algorithm 1.

4. Simulation results

The proposed method has been tested by means of computer simulations that aim to reproduce the conditions of the real application scenario.

[1] In the case of real–time audio track, it sufficient to introduce a delay of M samples in x.

IOP Conf. Series: Materials Science and Engineering **364** (2018) 012099 doi:10.1088/1757-899X/364/1/012099

Algorithm 1 Fast real–time binaural synthesis. Steps (*) must be repeated for each source.

Require: x, $M \in \mathbb{N}$
  set $k = 0$ and $K = \lfloor (\text{length}(x) - 1) / M \rfloor$
  set $y^{L,R}[n] = 0$, for $0 \leq n \leq \text{length}(x) - 1$
  repeat
    extract x_k from x (*)
    acquire r_k and ϕ_k by positioning system
    for all channels L, R **do**
      retrieve h_k in the HRIRs database
      compute g_{kM} and g'_{kM} using eq.(11)
      compute $y_k[n]$, for $0 \leq n < M - N$, using eq.(12) (*)
      assign $y[n] \leftarrow y[n] + y_k[n]$, for $0 \leq n < M - N$ (*)
    end for
    assign $k \leftarrow k + 1$
  until $k > K$

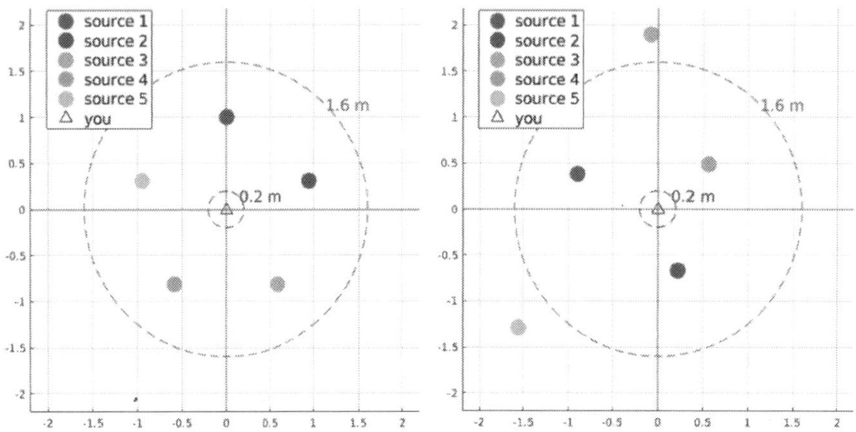

Figure 3. Two snapshots of the MATLAB GUI developed to test the binaural synthesis.

4.1. Setup

The real–time binaural synthesis has been implemented in MATLAB according to Algorithm 1 considering one or more source. After setting the sampling rate and position update rate, the audio track(s) is(are) preliminary loaded in the program, along with the HRIRs database. In our tests we have adopted the PKU-IOA HRTF database [13], which approximately covers elevation angles from -40 o to 90°, azimuth angles from 0 to 360°, and distances from 20 cm to 160 cm. For distances beyond 160 cm, HRIRs related to 160 cm are retrieved and a suitable extra delay is introduced. Since azimuth, elevation and distances are sampled in their respective ranges, the retrieval process requires an interpolation; incidentally, we have used a tetrahedral interpolation method described in [17] that provides an acceptable performances tradeoff in our case.

A 2D graphical user interface (GUI) has been developed to simulate the reciprocal movement on a plane between source(s) and listener, as shown in the snapshots reported in Figure 3. For sake of simplicity, the listener, which is the GUI user, is supposed to be fixed with respect to coordinate system. Source movements are implemented by means of mouse drag–and–drop and are real–time acquired by the GUI in order to simulate the positioning system.

4.2. Performance measurement

The proposed method has been tested on an system GNU/Linux 64–bit, kernel Linux4.4.0–112–generic SMP, mounting one Intel(R) Core(TM) i5–4200U CPU @ 1.60GHz. In the case of a single virtual source, the real–time system successfully works at a sampling rate up to $f_s = 65536$ Hz. For multiple virtual sources the system is capable to synthesize up to 5 source at $f_s = 32768$ Hz. As to the memory occupation, excluding the sources' audio tracks, it is mainly dictated by the HRIRs database, whose size scales approximately with f_s and it is about 26 MB at $f_s = 32768$ if stored in single precision (32–bit).

The audio quality has been objectively evaluated by considering the synthesis of a single tone at frequency f_0 emitted by a source moving with period $T = 8$ s on a circular trajectory around the listener. The performances have been evaluated in terms of Signal-to-noise and distortion ratio (SINAD) by means of the homologous function included in the MATLAB Signal Processing Toolbox. The SINAD is defined as

$$SINAD = (P_s + P_N + P_d)/(P_N + P_D),$$

being P_s, P_N and P_D the powers of the tone, of the background noise and of the distortion introduced by the the proposed method, in that order. The results obtained setting $f_s = 16384$ Hz for different f_0 (columns) and different position updating frequency f_{pos} (rows) are reported in Table 1. The *fixed* row refers to the SINAD for a fixed source, which is affected only by finite arithmetic noise. Interestingly, as a general rule, SINAD values increase from low to high frequency tones. Considering each single tone, we observe SINAD values for lower tones attain their best with lower values of f_{pos}, while for higher tones the best performance is for intermediate f_{pos}.

Table 1. Signal-to-noise and distortion ratio (dB) measured for the implemented method.

	$f_0 = 62.5$ Hz	125 Hz	250 Hz	500 Hz	1 kHz	2 kHz	4 kHz
fixed	137.2	137.1	136.5	136.1	136.2	136.5	136.4
$f_{pos} = 4$ Hz	15.4	17.1	18.1	21.9	31.8	28.0	24.5
8 Hz	10.1	12.0	14.4	17.1	26.2	36.1	34.9
16 Hz	8.8	11.4	13.8	16.8	25.4	34.8	34.8
32 Hz	8.2	10.6	13.2	16.1	24.6	33.6	33.5
64 Hz	8.0	10.4	13.0	16.1	24.4	33.3	33.0

5. Conclusions

In this paper, a real–time binaural method for an immersive fruition of the cultural heritage has been proposed. The realistic scenario of a visitor moving inside a museum hall has been considered as the case study. The proposed approach successfully exploits a database of statically recorder head–related impulse response (HRIR), as well as the position and orientation of the visitor, to synthesize one ore more virtual audio sources reciprocally moving with respect to a him/her. By using the real–time linear interpolation of consecutive HRIRs, a formulation of a fast block processing algorithm has been derived. The experimental results have shown that the system achieves a good audio quality at a computational cost that is affordable by consumer devices. Further optimization of the processing stage, possibly targeted to specific portable hardware, as well as the synthesis of environmental effects, e.g. room reverberations, would be worthwhile for future works. Furthermore, a psychoacoustic evaluation by means of a selected auditory panel should be carried out in order to detect and mitigate eventual undesirable sound artifacts.

6. References

[1] IJsselsteijn W A, de Ridder H and Vliegen J 2000 *IEEE Transactions on Circuits and Systems for Video Technology* **10** 225–233 ISSN 1051-8215
[2] Lin Y H and Wu J L 2014 *IEEE Transactions on Image Processing* **23** 1527–1542 ISSN 1057-7149
[3] Shao F, Lin W, Gu S, Jiang G and Srikanthan T 2013 *IEEE Transactions on Image Processing* **22** 1940–1953 ISSN 1057-7149
[4] Tam W J, Speranza F, Yano S, Shimono K and Ono H 2011 *IEEE Transactions on Broadcasting* **57** 335–346 ISSN 0018-9316
[5] Hacihabiboglu H, Sena E D, Cvetkovic Z, Johnston J and III J O S 2017 *IEEE Signal Processing Magazine* **34** 36–54 ISSN 1053-5888
[6] Baumgarte F and Faller C 2003 *IEEE Transactions on Speech and Audio Processing* **11** 509–519 ISSN 1063-6676
[7] Brown C P and Duda R O 1998 *IEEE Transactions on Speech and Audio Processing* **6** 476–488 ISSN 1063-6676
[8] URL `http://www.binci.eu`
[9] Ahnert W, Feistel S, Lentz T, Moldrzyk C and Weinzierl S 2004 *Audio Engineering Society Convention 117* URL `http://www.aes.org/e-lib/browse.cfm?elib=12932`
[10] Liu H, Darabi H, Banerjee P and Liu J 2007 *IEEE Transactions on Systems, Man, and Cybernetics, Part C (Applications and Reviews)* **37** 1067–1080 ISSN 1094-6977
[11] Martinelli A, Gao H, Groves P D and Morosi S 2018 *IEEE Sensors Journal* **18** 1600–1611 ISSN 1530-437X
[12] Algazi V R, Duda R O, Thompson D M and Avendano C 2001 *Proceedings of the 2001 IEEE Workshop on the Applications of Signal Processing to Audio and Acoustics (Cat. No.01TH8575)* pp 99–102
[13] Qu T, Xiao Z, Gong M, Huang Y, Li X and Wu X 2009 *IEEE Transactions on Audio, Speech, and Language Processing* **17** 1124–1132 ISSN 1558-7916
[14] BolaÃśos J G and Pulkki V 2012 *Audio Engineering Society Convention 133* URL `http://www.aes.org/e-lib/browse.cfm?elib=16501`
[15] Zölzer U, Amatriain X, Arfib D, Bonada J, De Poli G, Dutilleux P, Evangelista G, Keiler F, Loscos A, Rocchesso D *et al.* 2002 *DAFX - Digital Audio Effects* (John Wiley & Sons) ISBN 9780471490784 URL `https://books.google.it/books?id=h90HIV0uwVsC`
[16] Proakis J G and Manolakis D K 2006 *Digital Signal Processing (4th Edition)* (Upper Saddle River, NJ, USA: Prentice-Hall, Inc.) ISBN 0131873741
[17] Hugeng H, Anggara J and Gunawan D 2017 *2017 International Conference on Signals and Systems (ICSigSys)* pp 35–39

IOP Conf. Series: Materials Science and Engineering **364** (2018) 012100 doi:10.1088/1757-899X/364/1/012100

Linking HBIM graphical and semantic information through the Getty AAT: Practical application to the Castle of Torrelobatón

F J López[1, 4]**, P M Lerones**[2]**, J Llamas**[2]**, J Gómez-García-Bermejo**[3] **and E Zalama**[3]

[1] DISA, University of Valladolid, Valladolid 47011, Spain
[2] Cartif Foundation, Parque Tecnológico de Boecillo, Valladolid 47151, Spain
[3] ITAP-DISA, University of Valladolid, Valladolid 47011, Spain
[4] Corresponding author

Arq.lopezfacundo@gmail.com

Abstract. The current approaches for the built heritage documentation and modelling are to collecting, organizing and integrating immovable assets' data into a single graphic-semantic structure using BIM tools. This is the key issue to shifting from point clouds to HBIM. Although fully automation is not possible, practical approaches for the conversion of point clouds into HBIM elements grouped in libraries have recently appeared to meet the particular characteristics of historic buildings in an equivalent way to the libraries of elements already available for contemporary building and civil works. In spite of the graphic information on the built heritage elements is semantically rich in itself, linking the families of modelled elements with the Getty Art & Architecture Thesaurus (AAT) as worldwide well-known controlled vocabulary, allows not only the automation but the consistency in cataloguing of required elements, as well as more efficient retrieval of information in a standardized way. This so useful graphic-semantic linking is particularly applied to the Castle of Torrelobatón to make up the HBIM meaningful set of fundamental elements of the defensive architecture from the Middle Age to the Renaissance in Europe, of which this Castle is a representative example.

1. Introduction

Obtaining structured 3D digital models to document the built heritage is really an urgent necessity. Nowadays, a high level of automation can be achieved through the use of 3D surveying technologies, which allows accelerating the collection and subsequent processing of spatial data at adequate level of detail and accuracy. Point clouds can be incorporated into the Building Information Modelling (BIM) platforms that have the ability to import, manipulate and manage graphic and non-graphic information, as well as historical data [1]. In particular, the 3D virtual objects modelled using architectural historical documentation and 3D surveying data allows to generate a library of elements under the Heritage Building Information Modelling (HBIM) concept.

Today, the conversion of point clouds into HBIM elements grouped in libraries, adapted to the particular characteristics of historic buildings, is a relevant pending issue for the heritage sector [2, 3, 4, 5, 6]. Therefore, thanks to the qualities of HBIM for the management, documentation and reconstruction of the built heritage, a series of relevant publications have been generated during the last years.

A case study on the virtual modelling and reconstruction of a representative Romanesque church just following a straightforward procedure, so practical as easily reproducible to other kind of buildings of cultural interest is offered in [7, 8]. The different components are semi-automatically modelled within Autodesk REVIT commercial BIM software, and used to create a library of parametric elements under the HBIM concept. Another work [9] uses also the REVIT platform to develop a new approach that facilitates the temporal calculation, order and control of future architectural heritage restoration projects.

Authors such as [10, 11, 12, 13] focus their research on the development of a HBIM library that can help professionals to visualize a representative 3D model, learn the real information of each object and integrate and exchange information between stakeholders.

For their part [1, 14] emphasise the implementation of algorithms that work by selecting the point cloud surface and recognizing the shape, dimensions and characteristics of each element, reducing in this way the time and resources needed to model HBIM libraries of an already constructed building. These approaches would be affordable for any restoration project.

Other representative authors argue that BIM platforms are not currently able to solve all the issues that come from the modelling of architectural heritage, so they integrate other instruments or auxiliary tools. The works [15, 16, 17] propose a simple integration methodology between point clouds and historical data with the REVIT platform and the RHINOCEROS software to generate the HBIM library of the analysed building. These documents present case studies related to Italian Romanesque churches and certain objects were specifically developed for use in other projects with the same architectural style.

Authors such as [18, 19, 20] integrate BIM platforms and geographic information system (GIS) to perform a semantically rich 3D building model. In addition, they emphasize that the elements of the resulting HBIM libraries represent the reality of the monuments and assist other researchers to automatically perform the 3D virtual models of the historic buildings with the same architectural style.

Other works such as [21, 22, 23] are worth mentioning. These documents propose an innovative methodology for processing, cataloguing, analysing and identifying the building's historical and graphic information. In addition, they integrate this information with both Graphisoft ARCHICAD and REVIT software's to simplify the parametric modelling of the architectural heritage. The resulting models will serve as the basis for the creation of an HBIM library that guarantees the interoperability of the information with other interested professionals.

Additionally, an effective HBIM approach needs further semantic enrichment to embed cultural heritage knowledge. Defining an adequate ontology (a formal representation of knowledge as a hierarchy of concepts within the cultural heritage domain), using a shared vocabulary to denote the types, properties and interrelationships of cultural heritage aspects, is obliged.

The Getty AAT is the internationally recognized structured vocabulary responding to the gradual automation of records related to fine art, architecture, decorative arts, archival materials, and material culture [24]. It consists of around 44,000 concepts, including 131,000 generic terms, descriptions, bibliographic citations, and other information.

The Getty AAT is a thesaurus in compliance with ISO and NISO standards including ISO 2788:1986 (Guidelines for the establishment and development of monolingual thesauri); ISO 25964:2011 (Information and documentation-Thesauri and interoperability with other vocabularies) and ANSI/NISO Z39.19:2010 (Guidelines for the Construction, Format, and Management of Monolingual Controlled Vocabularies). It is made available via the Web browsers to support limited research and cataloguing efforts. Companies and institutions interested in more extensive use are advised to explore the Linked Open Data releases. APIs are available. Relational tables and XML releases are also available, but may be discontinued in the future. All releases are available under the Open Data Commons Attribution License (ODC-By) 1.0.

2. Methodological approach

The practical approach cited in [7 ,8] is applied right now to the Castle of Torrelobatón (Gothic castle of the XV century), linking the families of modelled elements with the Getty AAT, whenever possible. The global point cloud gathering internal and external surfaces is obtained and compared to the technical and historical construction details of the different architectural periods. This feedback between the spatial and semantic data allows the complexity of the modeling process required by the HBIM library to be reduced. Once the global point cloud is imported into REVIT, three main steps are applied.

The first step covers the collection of historical data (considered semantic information), as well as the technical documentation (considered graphic or spatial information). It was necessary to carry out a bibliographical study to take into consideration the rules, patterns and architectural proportions of the construction period the building belongs to. The works [25, 26, 27, 28, 29] help to understand and support the modeling of parametric objects that the virtual model will consist of. Another key aspect is the original graphic information collected from the surveying campaign (high-res pictures).

The second step is focused on the accurate and efficient modeling of the different architectural components that make up the HBIM library of the castle according to the Getty AAT, given that point clouds provide the actual deformations caused by the passage of time or structural problems, such as sloping walls, cracks or missing elements. To address this issue multiple cuts, sections, views, grids and reference lines were created on the point cloud. Cuts and sections are made in longitudinal and transverse directions, showing the composition and internal layout of the building, as well as to generate an estimation of the dimensions of the various elements and surfaces. The sections cross the center of the structural or supporting elements (curtain walls, keep, round towers, bartizan, barley, crenel, buttresses, vaults and other parts that represent the constructive logic). This strategy allows the deformations that affect the elements to be reflected more accurately and clearly. The finite horizontal planes constitute a reference for determining vertical distances. Subsequently, a set of grids are created to marking the outline of the cloud surfaces. When the geometry of a given surface is uniform and simple, the modeling is done using the parametric objects available in the REVIT internal library, such as walls, windows, doors or columns (Fig.1A). Moreover, the objects can be readily take part of the BIM project following the grid lines, levels and cloud sections (Fig.1B).

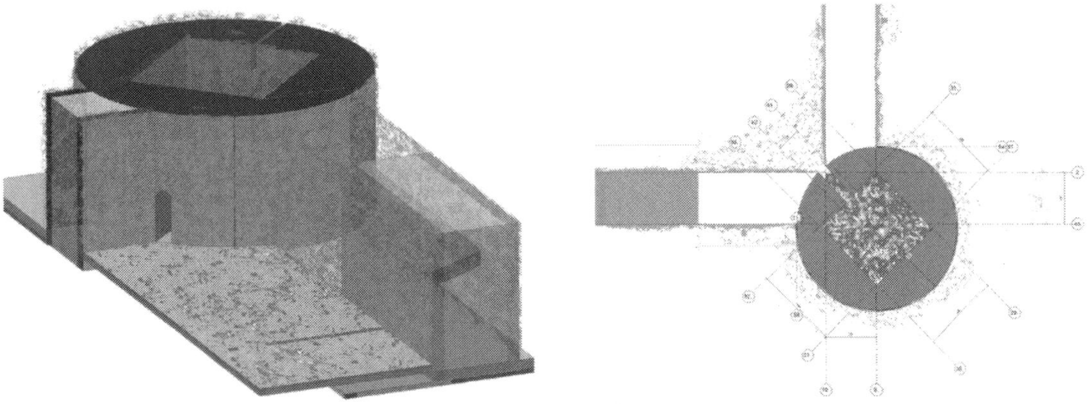

A) B)

Figure 1: A) Geometry modeled with the walls of the REVIT internal libraries. B) The wall is modeled on the grid that delimits the point cloud.
Source: Own figures.

In case of the geometry of a surface is irregular or complex, the modeling starts sketching of vertical and horizontal reference planes upon a 2D base plane. Subsequently, the use of reference lines is the key to drawing the profile of any element providing a solution for 3D modeling of a wide range of buildings in the same style. This is possible due to the flexibility of the modelled elements, which can change in form and proportions, thus adapting to new requirements and projects without having to start from scratch.

Finally, the third and last step is just to link the spatial data and the Getty AAT vocabulary to the designed parametric elements. The way to make this association depends on the BIM package in which it is worked on. In the case of REVIT, it is carried out through the option "Type of properties" tool from the menu bar. An example of the semantic organization and parameters of a given object can be seen in Fig. 2A. The properties for every element can be modified by changing the numeric values in the table. Conversely, the Getty AAT vocabulary can be linked to the elements modelled in two ways: (i) by associating the Getty AAT URL with the "Identity data" parameter located within the "Property type" tool (Fig. 2A); (ii) generating a link in the "Properties" palette through the use of "Instance Properties " (i.e. "Information" and "Family"), just associating the Getty AAT URL with the modelled element (Fig. 2B).

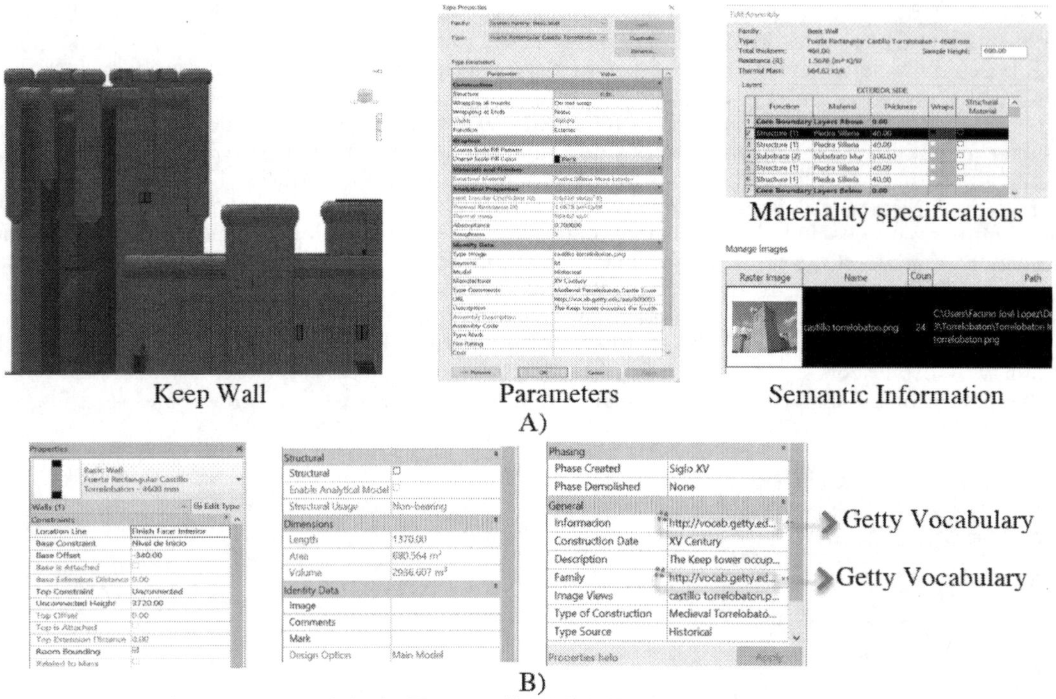

Figure 2: A) Parametric model of a Keep wall. B) Linking the Getty AAT vocabulary with the Keep wall in the properties palette.
Source: Own figures.

The components of the HBIM library are useful not only for cataloguing but for proper performance on the protection, conservation, restoration and dissemination actions for the built heritage. In addition, thanks to the parameters of each component, calculations and simulations on energy efficiency, decay assessment and structural analysis can be made when needed.

3. Results

Out of a total of 18 significant architectural elements (15 with correspondence in the Getty AAT), the most representative graphical-semantical linking is given in Table 1 according to the seven hierarchical architectural levels fixed by the AAT.

The corresponding results are given from Figure 3 to Figure 6, making up the set of fundamental elements of the defensive architecture from the Middle age to the Renaissance in Europe, of which the Castle of Torrelobatón is a representative example. These elements can be verified with the corresponding IFC and RVT files.

Element [English / Spanish]	AAT ID	Level	Result
Scarp / Alambor	http://vocab.getty.edu/aat/300003713	3	Figure 3
Battlements / Almenas	http://vocab.getty.edu/aat/300002595	2	Figure 3
Bartizan / Garitón	http://vocab.getty.edu/aat/300003636	4	Figure 4
Machicolation / Matacán Corbels / Ménsula	http://vocab.getty.edu/aat/300002695 http://vocab.getty.edu/aat/300003610	4	Figure 4
Curtain walls / Muro	http://vocab.getty.edu/aat/300002504	4	Figure 5
Round tower / Cubo	http://vocab.getty.edu/aat/300101468	3	Figure 5
Tower keep / Torre del homenaje	http://vocab.getty.edu/aat/300003694	2	Figure 6
Ribbed vault / Bóveda de crucería Barrel vault / Bóveda de medio punto	http://vocab.getty.edu/aat/300001393 http://vocab.getty.edu/aat/300001363	6	Figure 6

Table 1: Semantically enriched HBIM library for the Castle of Torrelobatón (Valladolid, Spain)

IOP Conf. Series: Materials Science and Engineering **364** (2018) 012100 doi:10.1088/1757-899X/364/1/012100

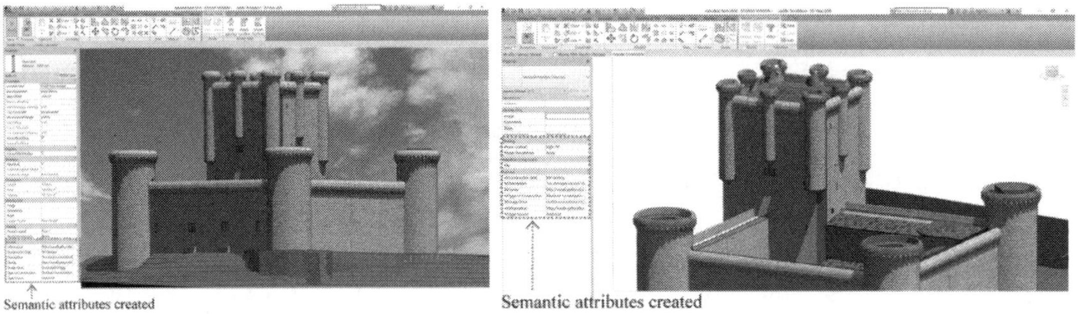

Figure 3: Highlighted in blue: Left: Scarp; Right: Battlements
Source: Own figures.

Figure 4: Highlighted in blue: Left: Bartizan; Right: Machicolation/Corbels
Source: Own figures.

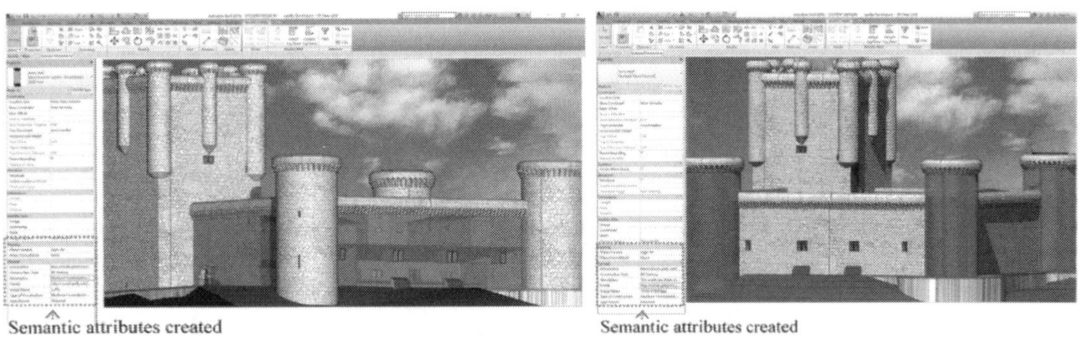

Figure 5: Highlighted in blue: Left: Curtain Walls; Right: Round towers
Source: Own figures.

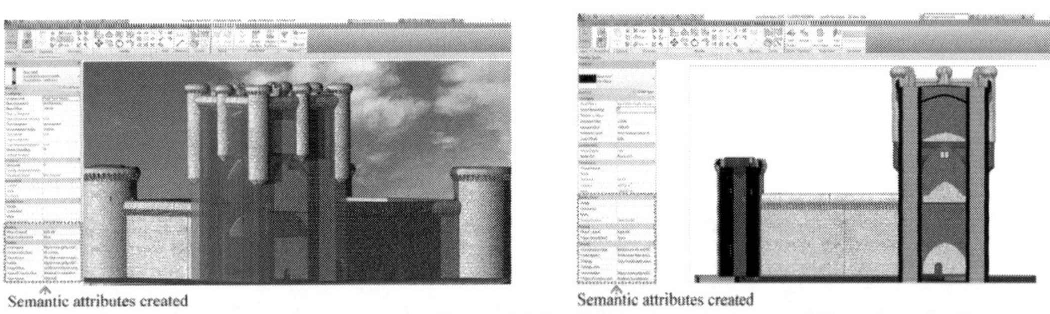

Figure 6: Highlighted in blue: Left: Keep; Right: Ribbed vault (up) and Barrel vault (down)
Source: Own figures.

IOP Conf. Series: Materials Science and Engineering **364** (2018) 012100 doi:10.1088/1757-899X/364/1/012100

4. Conclusions and Future Work

The multifactorial and multidisciplinary approach required for the research, protection, conservation, restoration and dissemination of the built heritage is increasingly demanding not only innovative BIM interoperable 3D parametric modelling methods keeping complexity and accuracy, but also semantically enriching this graphical information to facing useful HBIM projects. The Getty AAT is a taxonomy based on SKOS, that it is a specification based on RDF (base language of ontologies), formally recognized as standard framework to describe things that are not heritage objects in themselves, but a textual definition of these objects.

Although there is no complete listing for Europe, castles are one of the most widespread type of cultural asset. Thus, a realistic example on the combination of heritage meaningful 3D elements with the Getty AAT taxonomy has been shown for the particular case of the Castle of Torrelobatón (Valladolid, Spain), allowing not only the automation but the consistency in cataloguing of required elements, as well as more efficient retrieval of information in a standardized way as a basis for a trustworthy HBIM ontology.

The Getty vocabularies are compiled resources that grow through contributions from Getty projects and other institutions. Contributors generally represent institutions or consortia such as museums, archives, special collections, visual resource collections, large national or international translation or cataloguing projects, or others who catalogue art. In rare exceptions, a contributor may be an individual scholar.

INCEPTION, as a European representative RTD project, is willing to contribute by defining them to expand that AAT, but it must be done in XML format, according to the Schemas for AAT, TGN, ULAN and CONA. This work will be carried out as next step.

Acknowledge

The present applied research work is developed in the context of the INCEPTION project [www.inception-project.eu], funded from the EU's H2020 Reflective framework programme (grant agreement No. 665220).

References

[1] Cheng H M., Yang W B.and Yen Y N 2015 BIM applied in historical building documentation and refurbishing. *The International Archives of Photogrammetry, Remote Sensing and Spatial Information Sciences*. **40**(5), 85-90.

[2] European Commission 2015 Getting cultural heritage to work for Europe. *Report of the Horizon 2020* Expert Group on Cultural Heritage [accessed 10 April 2017]. Available at: < http://bookshop.europa.eu/es/getting-cultural-heritage-to-work-for-europe-pbKI0115128/

[3] Banfi F 2016 "Building Information Modelling. A Novel Parametric Modelling Approach Based on 3D Surveys of Historic Architecture". *Proc In Euro-Mediterranean Conference*. (Springer International Publishing) p. 116-127

[4] Barazzetti L 2016 Parametric as-built model generation of complex shapes from point clouds. *Advanced Engineering Informatics*. **30** 298–311.

[5] Tommasi C, Achille C, Fassi F 2016 From point cloud to BIM: a modelling challenge in the cultural heritage field. *The International Archives of the Photogrammetry, Remote Sensing and Spatial Information Sciences*. **41** B5.

[6] Arayici Y, Counsell J and Mahdjoubi L (Eds.) 2017 Heritage Building Information Modelling. *Taylor & Francis*. (London and New York) vol 1

[7] López F J, Martin Lerones P, Llamas J, Gómez-García-Bermejo J, Zalama E 2017a "A framework for using point cloud data of heritage buildings towards geometry modeling in a BIM context: a case study on Santa Maria la Real de Mave Church". *International Journal of Architectural Heritage*. (Accepted, press-pending http://www.tandfonline.com/doi/full/10.1080/15583058.2017.1325541)

[8] López F J, Martin Lerones P, Llamas J, Gómez-García-Bermejo J, Zalama E 2017b: "Semi-automatic generation of BIM models for cultural heritage". *Proc International Journal of Heritage Architecture*. (Accepted, press pending.)

[9] Biagini C, Capone P, Donato V and Facchini N 2016 Towards the BIM implementation for historical building restoration sites. *Automation in Construction*. **71** 74-86. https://doi.org/10.1016/j.autcon.2016.03.003

[10] Dore, C., Murphy, M., McCarthy, S., Brechin, F., Casidy, C., & Dirix, E. (2015). Structural simulations and conservation analysis-Historic building information model (HBIM). The

International Archives of Photogrammetry, Remote Sensing and Spatial Information Sciences, 40(5), 351. doi:10.5194/isprsarchives-XL-5-W4-351-2015

[11] Ma Y P, Hsu C C, Lin M C, Tsai Z W and Chen J Y 2015 Parametric Workflow (BIM) for the Repair Construction of Traditional Historic Architecture in Taiwan. *The International Archives of Photogrammetry, Remote Sensing and Spatial Information Sciences*. **40**(5), 315-322. doi:10.5194/isprsarchives-XL-5-W7-315-2015

[12] Lo Turco M, Santagati C, Parrinello S, Valenti G M and Inzerillo L 2016 BIM e beni architettonici: verso una metodologia operativa per la conoscenza e la gestione del patrimonio culturale-BIM and architectural heritage: towards an operational methodology for the knowledge and the management of Cultural Heritage. *DISEGNARECON* **9**(16), 161-169. : http://porto.polito.it/2659001/

[13] Themistocleousa K, Agapioua A and Hadjimitsisa D 2016 3D documentation and bim modeling of cultural heritage structures using uavs: the case of the foinikaria church. *International Archives of the Photogrammetry, Remote Sensing & Spatial Information Sciences*. **42**. doi:10.5194/isprs-archives-XLII-2-W2-45-2016

[14] Achille C, Lombardini N and Tommasi C 2015 BIM and cultural heritage: compatibility tests in an archaeological site. *WIT Transactions on The Built Environment*. **149**, 593-604. doi:10.2495/BIM150481

[15] Barazzetti L, Banfi F, Brumana R, Gusmeroli G, Oreni D, Previtali M and Schiantarelli G 2015 BIM from laser clouds and finite element analysis: combining structural analysis and geometric complexity. *The International Archives of Photogrammetry, Remote Sensing and Spatial Information Sciences*. **40**(5), 345. doi:10.5194/isprsarchives-XL-5-W4-345-2015

[16] Barazzetti L, Banfi F, Brumana R, Previtali M and Roncoroni F 2016 BIM from laser scans… not just for buildings: NURBS-based parametric modeling of a medieval bridge. *ISPRS Annals of the Photogrammetry and Remote Sensing*. **3**(5). doi:10.5194/isprsannals-III-5-51-2016

[17] Quattrini R, Malinverni E S, Clini P, Nespeca R and Orlietti E 2015 From TLS to HBIM. High quality semantically-aware 3D modeling of complex architecture. *The International Archives of Photogrammetry, Remote Sensing and Spatial Information Sciences*. **40**(5), 367. doi:10.5194/isprsarchives-XL-5-W4-367-2015

[18] Baik A, Yaagoubi R and Boehm J 2015 Integration of Jeddah Historical BIM and 3D GIS for Documentation and Restoration of Historical Monument. *The International Archives of Photogrammetry, Remote Sensing and Spatial Information Sciences*. ***40**(5) 29. doi:10.5194/isprsarchives-XL-5-W7-29-2015*

[19] Yang X, Koehl M, Grussenmeyer P and Macher H 2016 Complementarity of Historic Building Information Modelling and Geographic Information Systems. *ISPRS-International Archives of the Photogrammetry, Remote Sensing and Spatial Information Sciences*. **41** 437-443. doi:10.5194/isprsarchives-XLI-B5-437-2016

[20] Baik A H 2017 From point cloud to Jeddah Heritage BIM Nasif Historical House – case study. *Digital Applications in Archaeology and Cultural Heritage* http://dx.doi.org/10.1016/j.daach.2017.02.001

[21] Fregonese L, Achille C, Adami A, Fassi F, Spezzoni A and Taffurelli L 2015 BIM: An integrated model for planned and preventive maintenance of architectural heritage. *In Digital Heritage* vol **2** (Granada, Spain), p 77-80. doi: 10.1109/DigitalHeritage.2015.7419456

[22] Nieto J E, Moyano J J, Delgado F R and García D A 2016 Management of built heritage via HBIM Project: A case of study of flooring and tiling. *In Virtual Archaeology Review*. **7** (14), 1-12. DOI: https://doi.org/10.4995/var.2016.4349

[23] Rodríguez-Moreno C, Reinoso-Gordo J F, Rivas-López E , Gómez-Blanco A , Ariza-López F J and Ariza-López I 2016 From point cloud to BIM: an integrated workflow for documentation, research and modelling of architectural heritage. *Survey Review*. 1-20. Doi : http://dx.doi.org/10.1080/00396265.2016.1259719

[24] Contribute to the Getty Vocabularies. [accessed 10 September 2017]. Available at: http://www.getty.edu/research/tools/vocabularies/contribute.html

[25] Emery A 1975 The development of Raglan castle and keeps in late medieval England. *Archaeological Journal*. **132**(1), 151-186.

[26] León Muñoz A 2001 El castillo de Belalcázar: de Fortaleza señorial a residencia palaciega en el final del Medioevo.

[27] Chueca Goitia F and Navascues Palacios P 2001 *Historia de la arquitectura española: 1: Edad*

IOP Conf. Series: Materials Science and Engineering **364** (2018) 012100 doi:10.1088/1757-899X/364/1/012100

antigua, edad media: 2: Edad moderna, edad contemporanea. Fundacion Cultural Santa Teresa (Avila - España)

[28] Ruiz B A 2006 Villalonso, un castillo señorial en la campiña toresana, a la luz de la investigación arqueológica. I.E.Z Florián de Ocampo (Zaragoza – España) p 51-75

[29] Navarro Z E 2010 Un nuevo uso para el" Castillo Comunero" de Torrelobatón, Valladolid. *Estudios del Patrimonio Cultural*, (**5**), 31-53.

Tactile reproduction of paintings: the experience of the Department of Industrial Engineering of Florence

Francesco Buonamici[1], Rocco Furferi[1*], Lapo Governi[1], Luca Puggelli[1], Yary Volpe[1]

1. Dipartimento di Ingegneria Industriale, Università di Firenze, via di Santa Marta, 3 50134 Firenze (Italy)
*Corresponding author e-mail: rocco.furferi@unifi.it

Abstract. Within the T-VedO project, financed by Tuscany Region, the Reverse Engineering and Virtual Prototyping Lab team of the Department of Industrial Engineering of Florence (Italy) developed a number of methods for the semiautomatic generation of digital 2.5D models starting from paintings. Once such models are prototyped, they can be used to enhance visually impaired people tactile experience of artworks. Such methods, combined into a systematic procedure, allow to solve most of the typical problems arising when dealing with artistic representation of a painted scene. The present paper presents both an overview of the proposed procedure, including most recent updates, and the results obtained for a selected number of artworks of the Florentine Renaissance.

Key words: 3D Reconstruction; CAD; 2.5D models.

1. Introduction

Visually impaired and blind people are quite disadvantaged in enjoying artworks. These limitations can be easily overcome in the case of three-dimensional art pieces, by realizing tactile reproduction of statues or scale models of architectures, for instance by employing Reverse Engineering methods and 3D printing techniques. The translation of artworks into tactile models become impracticable when dealing with pictorial masterpieces, due to the lack of tools easily allowing the conversion of 2D images into 2.5D models.

On the basis of recent literature, a wide range of simplified tactile models can be built to provide visually impaired people with a faithful description of paintings [1, 2]. Among them, the use of 2.5D models has been demonstrated [3] as one of the most readable and meaningful for blinds, who perceive the bas-relief somehow faithful to their mental idea of the original artwork (described by an external guide).

The topic of 2.5D reconstruction starting from a single image is a long-standing issue in computer vision scientific literature and moves from the studies aiming at recovering 3D information from 2D pictures or photographs. Especially to represent complex scenes, in which more than one figure (i.e. human figures or objects) are displaced in a three-dimensional environment (e.g. architecture), the artist makes use two main stratagems to recreate the deepness-effect: perspective and shading. While a number of methods to obtain 2.5D models from perspective scenes are in literature [4, 5], an automatic method able to correctly convert shading information of a single image into 2.5D geometry doesn't exist, due to concave-convex ambiguity (e.g. a solid sphere and a spherical hollow produce the same image). The most promising method is Shape From Shading (SFS), for which recent studies have proved that moderate user-interaction highly increases the performances in terms of results quality [6-10].

Among the wide variety of methods dealing with bas-relief reconstruction starting from 2D images, a number of effective approach have been proposed by authors within the T-VedO Project, financed by Tuscany Region [11-14]. Such methods, combined into a systematic procedure, allow to solve most of the typical problems arising when dealing with artistic representation of a painted scene. The present paper presents both an overview of the proposed procedure, including most recent updates, and the results obtained for a selected number of artworks of the Florentine Renaissance.

2. Method

With the purpose of providing a robust 2.5D reconstruction of the scene and of the figures reproduced in the painting, the following 5 step computer-based modelling procedure has been developed:

1. Preliminary image processing: this operation is made on a digital image of the painting and it's devoted to a) image distortion correction, b) segmentation of the image by subjects and by colour, c) RGB to greyscale conversion of the image and d) albedo normalization of each colour segment.
2. Perspective scene reconstruction: this operation allows the reconstruction of the 2.5D virtual model of the scene, as a virtual *flat-layered bas-relief*. Mainly based on references [4, 11], but modelling also oblique planes, the procedure builds the framework on which subjects and objects are geometrically arranged by using perspective-related information (when available).
3. Volume reconstruction: this step allows the 2.5D reconstruction of the shape of each figure by exploiting shading related information of the image. In order to overcome incorrect shading issues, an approach based on SFS, image inflating and image embossing has been appositely devised [12].
4. Virtual bas-relief reconstruction: in this step, the definitive shape of the virtual bas-relief is retrieved by combining the results obtained in steps 2 and 3.
5. Rapid prototyping of the virtual bas-relief: in this step, the tactile model is 3D printed.

For sake of clarity, the description of each step will be described with reference to an exemplificative case study i.e. the reconstruction of "The Healing of the Cripple and the Raising of Tabitha" fresco by Masolino da Panicale (see Figure 1). This masterpiece is a typical example of Italian Renaissance paintings characterized by single-point perspective and is located in Brancacci Chapel (Church of Santa Maria del Carmine, Florence).

2.1. Preliminary image processing

The acquisition of the digital image I_0 of the painting to be reconstructed should be performed by means of proper photographic devices and illumination. Since the whole reconstruction process is based on this single input, the quality of the final result directly depends on the quality of the image. For this reason, the picture should be taken by using calibrated lenses and cameras, with a high resolution sensor. Referring to the case study used as example, a Canon EOS 6D camera (provided with a full-frame CMOS sensor with a resolution equal to 5472 x 3648 pixels) and a Canon EF 24-70mm f/2.8 L USM II calibrated lens have been used. The obtained image is then rectified in order to correct lens distortion [15].

Since both scene reconstruction and volumes definition are based on grey-scale information, the colour of such an image has to be discarded. This can be done by using any colour conversion tool available in any image processing software, such as Photoshop or Gimp. In this work, the grayscale image has been obtained by converting the image from sRGB to CIELAB colour space and by extracting L* channel. Let's call I_G the obtained image.

IOP Conf. Series: Materials Science and Engineering **364** (2018) 012101 doi:10.1088/1757-899X/364/1/012101

(a) (b)

Figure 1. a) Image of "The Healing of the Cripple and the Raising of Tabitha" fresco after lens correction; b) Segmented image I_S.

Successively, the different objects represented in the scene, such as human figures and architectural elements, are properly identified. This operation, called *image segmentation*, can be carried out by adopting one of the several methods available in literature (e.g. [16]). In the present work an interactive livewire boundary extraction algorithm has been used [17]. Let's call the segmented image I_S.

In a greyscale image, as I_G, the grey value of each pixel (i.e. the brightness) is given by two main factors: albedo ρ and shading S (eq. 1).

$$I_G(i,j) = \rho(i,j) * S(i,j) \tag{1}$$

The first one depends on the optical characteristics of the surface itself, like colour or material. The second one instead, which will be analysed in the successive SFS operations, directly depends on the orientation of the surface with respect to the light direction. Therefore, in order to isolate shading information, the albedo value of each segmented region needs to be normalized to 1. If the represented surface is somewhere perpendicular to light direction, this is obtained by dividing the grey value of each pixel by the maximum value of that segment. Otherwise, this can be done comparing the analysed segment with the surrounding ones: parallel or tangent areas must have equal grey value. Let's call the resulting image I_L.

Both I_L and I_S will be used for the 3D modelling of the virtual bas-relief.

2.2. *Perspective scene reconstruction*

Starting from the segmented image I_S, it is necessary to define the arrangement of each segment in a consistent 2.5D reconstruction of the scene, exploiting information derived from perspective analysis. It is a straight consequence that this kind of reconstruction can be obtained only for those paintings in which the artist used the perspective to simulate the deepness of the scene. The devised procedure starts with the definition of a reference coordinate system with the origin corresponding with the bottom left corner of the image, the *x-axis* parallel to the horizon line, from left to right and the *z-axis* outward pointing with respect to the painting. Each segment is then classified by means of a graphical user interface (GUI) in 4 types of planes: frontal, horizontal, vertical, oblique. Among the horizontal planes it is defined the *main plane*, which corresponds to the ground level. Successively, a hierarchy among the defined planes is created. Further details are available on [13].

At the end of the procedure, a depth-map Z_P of the flat layered bas-relief is obtained in the form of a depth map (i.e. a greyscale image), as depicted in Figure 2.

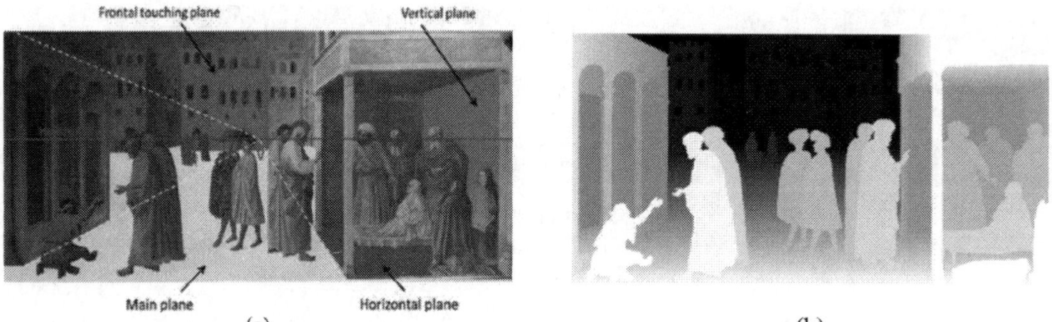

(a) (b)

Figure 2. a) Perspective scene reconstruction; b) Final greyscale depth map.

2.3. Volume reconstruction

Once the previous step is completed, the 2.5D shape of each depicted subject has to be retrieved. In the flat-layer representation, in fact, each figure is approximated with a plane, therefore no information about the actual shape of the figures (except the silhouette) is given.

First, all objects representing simple geometries, such as columns, arches or vaults, are directly retrieved by using specifically devised tools, implemented in the GUI. These elements can be easily reconstructed without a deep analysis of the shading, since a limited number of measurement on the picture itself and on Z_0 are required (i.e. diameter and length for a cylindrical column).

Referring to subject whose surface is not reproducible with simple geometric entities (e.g. human figures or clothes), the reconstruction is mainly obtained using SFS-based techniques. While this method is able to provide good results dealing with synthetic images, in which the surface is perfectly diffusive and the shading is free from errors, in the case of a handmade representation results can be far from satisfactory. For these reasons, the authors developed a simplified approach where the final solution i.e. the depth map Z_F of all the subjects in image is obtained as a linear combination of three different contributions (see eq. 2): main shape (Z_S), rough shape (Z_R) and fine detail shape (Z_D).

$$Z_M = \lambda_S Z_S + \lambda_R Z_R + \lambda_D Z_D \quad (2)$$

The main shape derives from shading analysis, by means of a specifically developed SFS procedure based on the minimization of a penalty function under wisely set boundary conditions [14]. The minimization approach was preferred to faster direct SFS-approaches (such as Fast Marching Method) due to a better handling of unprecise shading data. The main drawback of this approach is the over-flattening effect, i.e. the final result appears flattened with respect to the ground truth. In addition, this effect is emphasised in presence of ambient light added to a directional illumination, which is quite common in paintings. A secondary issue related to minimization approach is the loss of detail, due to an over-smoothing effect. Both the two auxiliary contribution, Z_R and Z_D are used to come up with these two issues [13] (see Figure 3).

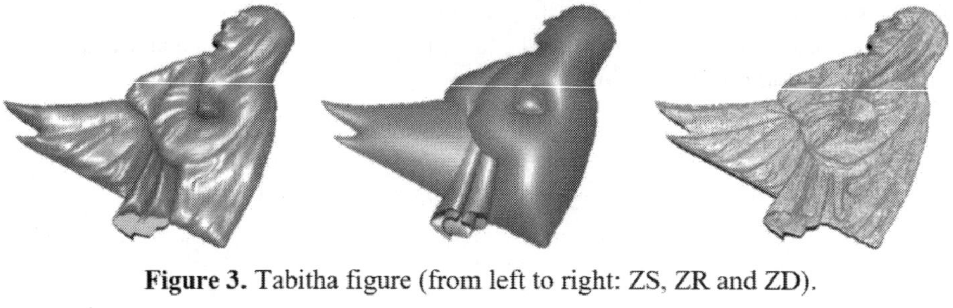

Figure 3. Tabitha figure (from left to right: ZS, ZR and ZD).

The rough shape definition aims at recovering the gross volume or the final shape. It derives from a two steps image inflating procedure developed by the authors, in which the initial rough inflated surface is then finely iteratively smoothed to avoid sharp edges that would be visible in the final surface.

The fine detail shape instead aims at recovering the loss of details due to over-smoothing effect. Inspired by some well-known techniques [18], a simple but effective strategy to recover the fine detail is to consider the brightness of the input image (I_G) as a depth map.

2.4. *Virtual bas-relief reconstruction*

Once the three contribution (Z_S, Z_R and Z_D) are obtained for each figure, a GUI based procedure helps users in defining the proper set of weights to achieve the better Z_M for each one of the *n* figures and, therefore, the better Z_F. An appositely devised GUI allows the user to regulate each weight value and to see in real time the effect on the 2.5D virtual bas-relief.

$$Z_F = \lambda_P Z_P + \sum_{i=1}^{n} \lambda_{S,i} Z_{S,i} + \lambda_{R,i} Z_{R,i} + \lambda_{D,i} Z_{D,i} \quad (3)$$

2.5. *Rapid prototyping of the bas-relief*

Once ZF has a satisfying quality, the procedure allows the user to export the STL file, ready to be manufactured by means of CNC milling machines or with rapid prototyping techniques, such as stereolithography (SLA), selective laser sintering (SLS), fused deposition modelling (FDM), Polyjet/Multi-jet technology. The main aspect that needs to be considered in order to select the most appropriate technology is the attainable level of details. However, other characteristics are relevant, such as surface finishing and the possibility of post-processing treatments (e.g. surface hardening, to prevent material consumption) [19, 20].

The bas-relief of "*The Healing of the Cripple and the raising of Tabitha*" was prototyped by milling and is sized about 900 mm x 400 mm x 80 mm (see Figure 4).

Figure 4. Tactile bas-relief realized for the artwork "The Healing of the Cripple and the raising of Tabitha".

3. Case studies

The described method was shared with both the Italian Union of Blind and Visually Impaired People in Florence (Italy) and experts in cultural heritage field (in particular from Uffizi Gallery, Musei Civici Fiorentini and Villa La Quiete). Under their expert suggestions, author realized several bas-reliefs of some of the best known artworks of the Italian Renaissance.

In chronological order, the first masterpiece to be translated into tactile bas-relief by the proposed procedure was "*The Annunciation*" fresco by Beato Angelico (see Figure 5). As many of the other ones that have been realised, the bas-relief is permanently displayed next to the original exhibit, in this case in the *Museo di San Marco*, in Florence.

IOP Conf. Series: Materials Science and Engineering **364** (2018) 012101 doi:10.1088/1757-899X/364/1/012101

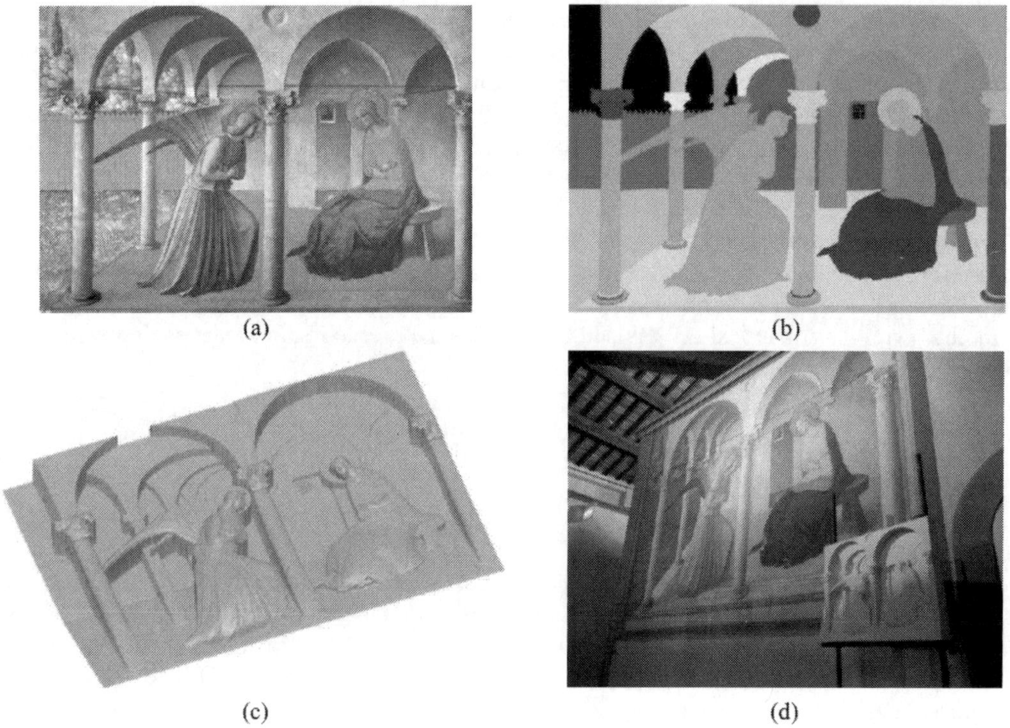

Figure 5. "The Annunciation" by Beato Angelico (Museo di San Marco, Florence): a) rectified grayscale image; b) clustered image; c) 2.5D model; d) final bas-relief exposed together with the original artwork.

Following this, two paintings from Villa La Quiete (Florence) collection have been converted into tactile models: *"Madonna with Child and angels"* by Niccolò Gerini (Figure 6) and some characters from "*Mystical marriage of Saint Catherine*" by Ridolfo del Ghirlandaio (Figure 7).

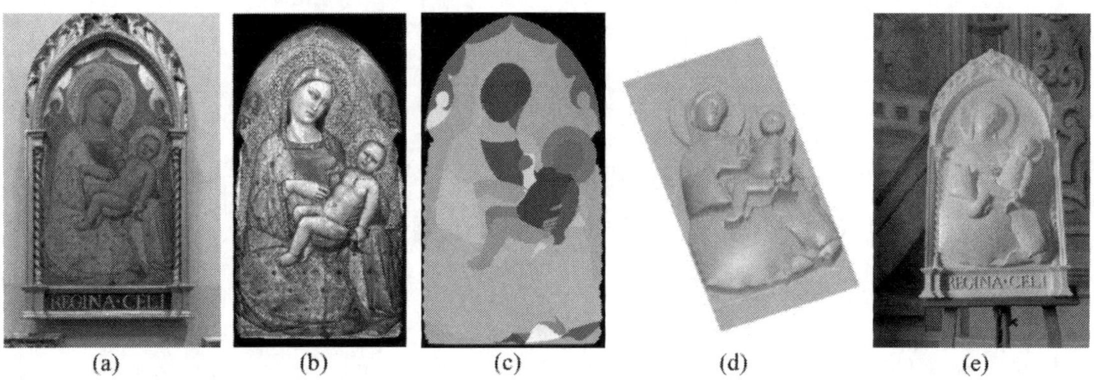

Figure 6. "Madonna with Child and Angels" by Niccolò Gerini, Villa La Quiete (Florence): a) original artwork; b) rectified grayscale image; c) clustered image; d)2.5D model; e) tactile bas-relief.

In the case of Niccolò Gerini artwork, also the frame of the painting has been included into the tactile representation, considered an integral part of the exhibit itself. Its 2.5D representation has been obtained by means of 3D laser scanner *Romer Absolute Arm 7520-SI.* Finally, it has been realised the tactile bas-relief of the painting *"Pala di Santa Lucia de'Magnoli"* by Domenico Veneziano, displayed in Uffizi Gallery in Florence (see Figure 8).

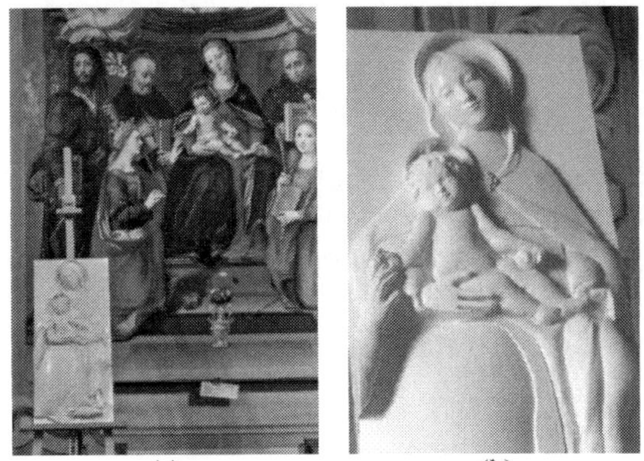

(a) (b)

Figure 7. "Mystical Marriage of Saint Catherine" by Ridolfo del Ghirlandaio (Villa La Quiete, Florence): a) original artwork; b) tactile bas-relief.

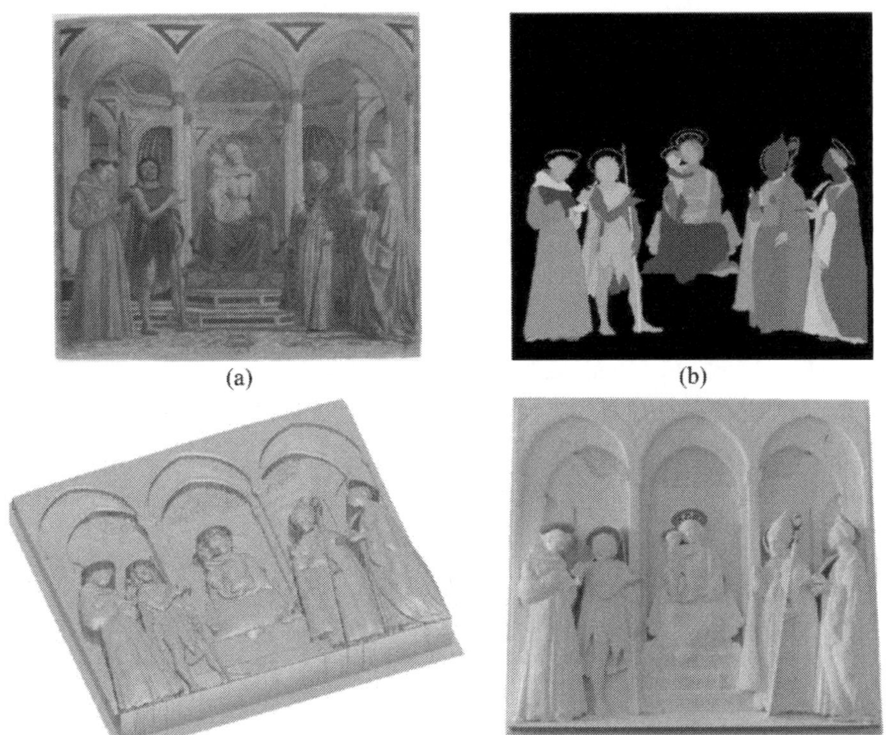

(a) (b) (c) (d)

Figure 8. "Pala di Santa Lucia de' Magnoli" by Domenico Veneziano (Uffizi Gallery, Florence): a) original artwork; b) clustered image; c) 2.5D model; d) tactile bas-relief.

Conclusions and discussion

Though technologies like 3D scanning and Rapid Prototyping are nowadays an important part of the digital age and have been successfully introduced to improve the 3D reproduction of sculptures, the research is not as mature as regards the realization of 3D models starting from paintings or photographs.

Taking into consideration most recent techniques dealing with this topic, the present paper provided an overview of the procedural CAD-based method developed by authors during the T-VedO project. Such a method was successfully applied to a number of important artworks of the Florentine

IOP Conf. Series: Materials Science and Engineering **364** (2018) 012101 doi:10.1088/1757-899X/364/1/012101

Renaissance. Manufactured bas-relief, located at museums that have collaborated in the experimentation, are now available to visually impaired people to enhance their tactile experience.

References

[1] A. Reichinger, M. Neumüller, F. Rist, S. Maierhofer, W. Purgathofer, Computer-aided design of tactile models – taxonomy and case studies 2013 *Lect Notes Comput. Sc.*, **7383**, 497–504.

[2] M. Carfagni, R. Furferi, L. Governi, Y. Volpe, G. Tennirelli, Tactile representation of paintings: an early assessment of possible computer based strategies 2012 *Lect Notes Comput. Sc.*, **7616** 261–270.

[3] Y. Volpe, R. Furferi, L. Governi, G. Tennirelli, Computer-based methodologies for semi-automatic 3D model generation from paintings 2014, *Int. J. Comput. Aided Eng. Technol.* **6** (1) 88–112.

[4] Y. Horry, K. Anjyo, K. Arai, Tour into the picture: using a spidery mesh interface to make animation from a single image 1997, *Proceedings of the 24th Annual Conference on Computer Graphics and Interactive Techniques*, ACM Press/Addison-Wesley.

[5] A. Reichinger, S. Maierhofer, W. Purgathofer, High-quality tactile paintings 2011, *J. Comput. Cult. Heritage* **4** (2), Art. No. 5.

[6] P. Daniel, J.-D. Durou, From deterministic to stochastic methods for shape from shading 2000, in: *Proc. 4th Asian Conf. on Comp. Vis*, 1–23.

[7] L. Di Angelo, P. Di Stefano, Bilateral symmetry estimation of human face 2013, *Int. J. Interactive Design Manuf.* **7**(4), 217-225.

[8] J.D. Durou, M. Falcone, M. Sagona, Numerical methods for shape-from-shading: a new survey with benchmarks 2008, *Comput. Vis. Image Underst.* **109**, 22–43.

[9] R.T. Frankot, R. Chellappa, A method for enforcing integrability in shape from shading algorithms 1988, *IEEE Trans. Pattern Anal. Mach. Intell.* **10** (4), 439–451.

[10] T.-P. Wu, J. Sun, C.-K. Tang, H.-Y. Shum, Interactive normal reconstruction from a single image 2008, *ACM Trans. Graphics (TOG)* **27** (5), 119.

[11] R. Furferi, L. Governi, N. Vanni, Y. Volpe, Tactile 3D bas-relief from single-point perspective paintings: a computer based method 2014, *J. Inform. Comput. Sci.* **11** (16), 1–14.

[12] L. Governi, M. Carfagni, R. Furferi, L. Puggelli, Y. Volpe, Digital bas-relief design: a novel shape from shading-based method, Comput. Aided Des. Appl. 11 (2) (2014) 153–164.

[13] R. Furferi, L. Governi, Y. Volpe, L. Puggelli, N. Vanni, M. Carfagni, From 2D to 2.5D i.e. from painting to tactile model 2014, *Graph. Models*, **76** (6), 706-723.

[14] L. Governi, R. Furferi, M. Carfagni, L. Puggelli, Y. Volpe, Digital bas-relief design: a novel shape from shading-based method 2014, *Comput. Aided Des. Appl.* **11** (2), 153–164.

[15] Y. Altunbasak, R. M. Mersereau and A. J. Patti, A fast parametric motion estimation algorithm with illumination and lens distortion correction 2013, IEEE Trans. Image Process., **12** (4), 395-408.

[16] E. Nadernejad, S. Sharifzadeh, H. Hassanpour, Edge detection techniques: evaluations and comparisons 2008, *Appl. Math. Sci.* **2** (31), 1507–1520.

[17] W.A. Barrett, E.N. Mortensen, Interactive live-wire boundary extraction 1997, *Med. Image Anal.* **1** (4), 331–341.

[18] K. Salisbury, D. Brock, T. Massie, N. Swarup, C. Zilles 1995, Haptic rendering: Programming touch interaction with virtual objects. *In Proc. of the 1995 symposium on Interactive 3D graphics*, 123-130.

[19] X. Yan, P. Gu, A review of rapid prototyping technologies and systems 1996, Comput. aided design, **28**(4), 307-318.

[20] M. Carfagni, L. Puggelli, Different strategies for rapid prototyping of digital bas-reliefs 2014, Appl. mech. mat., **510**, 163-167.